Reproductive Biology and Phylogeny of Fishes (Agnathans and Bony Fishes)

Phylogeny ▪ Reproductive System
Viviparity ▪ Spermatozoa

Reproductive Biology and Phylogeny Series

Series Editor: Barrie G. M. Jamieson

Vol. 1: Reproductive Biology and Phylogeny of Urodela
(Volume Editor: David M. Sever)

Vol. 2: Reproductive Biology and Phylogeny of Anura
(Volume Editor: Barrie G. M. Jamieson)

Vol. 3: Reproductive Biology and Phylogeny of Chondrichthyes
(Volume Editor: William C. Hamlett)

Vol. 4: Reproductive Biology and Phylogeny of Annelida
(Volume Editors: G. Rouse and F. Pleijel)

Vol. 5: Reproductive Biology and Phylogeny of Gymnophiona (Caecilians)
(Volume Editor: Jean-Marie Exbrayat)

Vol. 6A: Reproductive Biology and Phylogeny of Birds
(Volume Editor: Barrie G. M. Jamieson)

Vol. 6B: Reproductive Biology and Phylogeny of Birds
(Volume Editor: Barrie G. M. Jamieson)

Vol. 7: Reproductive Biology and Phylogeny of Cetacea
(Volume Editor: Debra L. Miller)

Vol. 8A: Reproductive Biology and Phylogeny of Fishes (Agnathans and Bony Fishes)
(Volume Editor: Barrie G. M. Jamieson)

Reproductive Biology and Phylogeny of Fishes (Agnathans and Bony Fishes)

Phylogeny ▪ Reproductive System
Viviparity ▪ Spermatozoa

Part A

Volume edited by
BARRIE G.M. JAMIESON
School of Integrative Biology
University of Queensland
St. Lucia, Queensland
Australia

Volume 8A of Series:
Reproductive Biology and Phylogeny

Series edited by
BARRIE G.M. JAMIESON
School of Integrative Biology
University of Queensland
St. Lucia, Queensland
Australia

CRC Press is an imprint of the
Taylor & Francis Group, an **informa** business

CRC Press
Taylor & Francis Group
6000 Broken Sound Parkway NW, Suite 300
Boca Raton, FL 33487-2742

First issued in paperback 2018

ISBN-13: 978-1-57808-580-4 (hbk)
ISBN-13: 978-1-138-11336-7 (pbk)

Library of Congress Cataloging-in-Publication Data

Reproductive biology and phylogeny of fishes (agnathans and bony fishes) / volume edited by Barrie G.M. Jamieson.
p. cm. -- (Reproductive biology and phylogeny series ; v. 8A)
Includes bibliographical references and index.
ISBN 978-1-57808-580-4 (hardcover)
1. Fishes--Reproduction. 2. Fishes--Phylogeny. I. Jamieson, Barrie G.M. (Barrie Gillean Molyneux)
QL639.2.R483 2009
597--dc22

2008050937

Visit the Taylor & Francis Web site at
http://www.taylorandfrancis.com

and the CRC Press Web site at
http://www.crcpress.com

Preface to the Series

This series was founded by the present series editor, Barrie Jamieson, in consultation with Science Publishers, Inc., in 2001. The series bears the title 'Reproductive Biology and Phylogeny' and this title is followed in each volume with the name of the taxonomic group which is the subject of the volume. Each publication has one or more invited volume editors (sometimes the series editor) and a large number of authors of international repute. The level of the taxonomic group which is the subject of each volume varies according to the amount of information available on the group, the advice of proposed volume editors, and the interest expressed by the zoological community in the proposed work. The order of publication of taxonomic groups reflects these concerns, and the availability of authors for the various chapters, and it does not proceed serially through the animal kingdom in a presumed "ladder of life" sequence. Nevertheless, a second aspect of the series is coverage of the phylogeny and classification of the group, as a necessary framework for an understanding of reproductive biology. Evidence for relationships from molecular studies is an important aspect of the chapter on phylogeny and classification. Other chapters may or may not have phylogenetic themes, according to the interests of the authors.

It is not claimed that a single volume can, in fact, cover the entire gamut of reproductive topics for a given group but it is believed that the series gives an unsurpassed coverage of reproduction and provides a general text rather than being a mere collection of research papers on the subject. Coverage in different volumes varies in terms of topics, though it is clear from the first volumes that the standard of the contributions by the authors will be uniformly high. The stress varies from group to group; for instance, modes of external fertilization or vocalization, important in one group, might be inapplicable in another.

This, in two parts, is the eighth volume in the series. Previous volumes in the series were devoted to 1. Urodela.; 2. Anura; 3. Chondrichthyes: Sharks, Batoids and Chimaeras. 4. Annelida.; 5. Gymnophiona (Caecilians); 6 A and B. Birds; 7. Cetacea (Whales, dophins and porpoises). Volume 8B will have 13 chapters covering testes, sperm, and sperm competition; endocrinology of reproduction; pheromones and reproduction; copulatory structures: taxonomic overview and the potential for sexual selection; sexual selection:

signaling and courtship; adaptation and evolution of reproductive mode in copulating cottoid species; fertilization; sex determination; parental care; reproduction in relation to conservation and exploitation of marine fishes; cryopreservation of gametes; embryogenesis and development; and molecular genetics of development.

My thanks are due to the School of Integrative Biology, University of Queensland, for facilities, and especially to the Executive Dean, Professor Mick McManus, for his encouragement. I thank Sheila Jamieson, who has supported me indirectly in so many ways in this work. I am grateful to the publishers for their friendly support and high standards in producing this series. Sincere thanks must be given to the volume editors and the authors, who have freely contributed their chapters, in very full schedules. This series would not have been possible without the excellent services of the University of Queensland Library. The editors and publishers are gratified that the enthusiasm and expertise of these contributors has been reflected by the reception of the series by our readers.

THE UNIVERSITY OF QUEENSLAND
AUSTRALIA

16 July 2008

Barrie Jamieson
The School of Integrative Biology
University of Queensland

Preface to this Volume

This volume is dedicated to the memory of Björn Afzelius 1925-2008

The animals loosely termed fish constitute more than half of all known vertebrate species. There are approximately 27,000 described living species of bony fishes (Euteleostomi =Osteichthyes) and, in the agnathans, about 70 species of hagfishes and some 34 species of lampreys. Approximately 970 species are chondrichthyans, the sharks and their relatives, which were the subject of Volume 3 in this series. It is perhaps because fishes live in a buoyant medium, whether it be fresh or sea water, that they show a diversity in body shapes that is unparalleled by other vertebrates. There is also a unique diversity in the modes of reproduction, whether by external or internal fertilization, and this, with the morphology and fine structure of the reproductive system and its components, is the subject of the present volume, Part A. We may here, albeit superficially, sample some of the many topics treated in the 17 chapters.

It is understandable that the classification of so large an assemblage of animals, though largely satisfactory, is under constant revision. The insights which molecular, chiefly DNA, analyses have given us into their relationships are the subject of Chapter 1. Evidence from molecular studies that support the cyclostome hypothesis that groups lampreys and hagfish is presented. Among basal actinopterygians, recent re-interpretation of morphology and new evidence from nuclear gene sequences support the Holostei (grouping amiids and lepisosteids). Relationships among the early-branching teleosts groups remain unresolved with elopomorphs, osteoglossomorphs, and clupeocephalans (ostarioclupeomorphs plus euteleosts) forming a polytomy at the base of the teleost tree. Our current understanding of supra-ordinal groups of euteleosts such as Protacanthopterygii, Paracanthopterygii, and Acanthopterygii is discussed. Molecular analysis of the percomorph crown group reveals profound polyphyly. Two major challenges for molecular systematics of fishes involve ancient gene and whole genome duplication events and systematic biases such as base compositional heterogeneity among DNA sequences sampled from distantly related taxa. Some possible solutions to these problems are presented.

In chapter 2 the ovary and the six major stages of oogenesis: oogonial proliferation, chromatin-nucleolus, primary growth, secondary growth (vitellogenesis), maturation, and ovulation are described. Hormone regulation of the functional and morphological events of maturation, both nuclear and ooplasmic changes are discussed.

The special morphological and physiological modifications associated with viviparity in teleosts are the subject of Chapter 3. Intrafollicular and intraluminal gestation, modifications of ovarian structure, and the development in some of an intromittent organ are discussed. It is shown that, lacking oviducts but having an ovarian lumen, intraovarian insemination, fertilization, and gestation occur in viviparous teleosts, a unique combination among vertebrates.

Chapter 4 deals with the testis, spermatogenesis and testicular cycles. Spermatogonia associate with Sertoli cells to form spermatocysts or cysts. It is shown that the testicular tissues form two compartments: the germinal compartment, formed by the germinal epithelium, which is composed by Sertoli cells and germ cells and, secondly, the interstitial compartment, integrated by the connective tissue; the two compartments separated by a basement membrane. The teleost testis is shown to be organized morphologically into three types: anastomosing tubular testis type, present in salmonids, cyprinids and lepisosteids; unrestricted spermatogonial testis type, found in neoteleosts except Atherinomorpha; and restricted spermatogonial testis type, characteristic of all Atherinomorpha. The morphology of the ovarian germinal epithelium and the testicular germinal epithelium during the annual reproductive cycle, reflecting reproductive seasonality, is described.

The efferent testicular duct system and accessory organs of the testis in bony fishes are the subject of Chapter 5. Adaptations in externally and internally fertilizing fishes are described. The anatomy, histology, fine structure and function of the efferent testicular duct system are discussed. Lastly, the anatomy, histology, fine structure, and function of the accessory glands of the efferent duct system, the testicular, glands, steroid glands, seminal vesicles, and testicular blind pouches is described.

This volume grew out of an intended revision of my book on *Fish Evolution and Systematics: Evidence from Spermatozoa* (Cambridge U. P. 1991) and therefore a large part of part A deals with the ultrastructure of spermatozoa, with some discussion of phylogenetic implications.

Despite molecular evidence in favor of recognition of the Cyclostomata, the sperm of Petromyzodontiformes (lampreys) and Myxiniformes (hagfishes) show no apomorphic similarities in sperm ultrastructure that would demand recognition of relationship between the two groups.

Spermatozoal ultrastructure yields little indication of the relationships of the enigmatic *Polypterus*. Major differences from the sperm of sarcopterygian fishes are described. The acrosome and perforatorium in the endonuclear canal, though plesiomorphic, do not detract from relationship with the sturgeons which is indicated from some molecular studies but there are no recognizable spermatozoal synapomorphies between *Polypterus* and acipenseriforms.

The sperm of the sturgeons (Acipenseridae) and Paddle fish (Polyodontidae) are closely similar. In Acipenseridae the acrosome bears a circlet of eight to ten posterolateral projections but these appear to be very short in *Polyodon*. The nucleus is penetrated for much of its anterior portion by one to three endonuclear canals that begin in the subacrosomal area, or (Pallid Sturgeon) in the anterior region of the acrosome. Neopterygians are spermatologically diagnosed from the agnathans through Dipnoi in having lost the acrosome. This correlates with the presence of an egg micropyle. The plesiomorphic sperm type for neopterygians (though it is not plesiomorphic for phylogenetically more basal fishes as exemplified by agnathans, Cladistia and Chondrostei) is a round-headed, acrosomeless spermatozoon, probably with lateral flagellar fins, which is spawned into the ambient water, the anacrosomal aquasperm. This first appears in the Holostei, in Amia and, with slight nuclear elongation, in *Lepisosteus*.

Osteoglossiformes exhibit a bewildering diversity of major modifications of the anacrosomal acrosome but this basic type is conserved in the notopteroid *Papyocranus afer*. The peculiar filiform sperm of the only other investigated osteoglossoid, *Pantodon bucholzi*, endorses the separate status of the Pantodontidae. The rare aflagellate condition of sperm seen in investigated Gymnarchidae and Mormyridae is in agreement with the association of these families in the Mormyroidei or as sister-groups in a wider Notopteroidei.

Two features constitute spermatozoal synapomorphies unequivocally linking the Elopiformes+Anguilliformes+Albuliformes and comprise autapomorphies of a monophyletic Elopomorpha. They are a 9+0 flagellum, a constant feature, and extension of the proximal centriole as two elongate bundles of 4 and 5 triplets. A third feature in most is the striated centriolar rootlet.

In the Clupeiformes, engraulid and clupeid spermatozoa differ from those of most members of the sister-group, the Ostariophysi, in the very deep nuclear fossa, the prevalence of a unilateral, often C-shaped mitochondrion and the presence of intratubular differentiations (ITDs) in the axoneme but share with most of them the absence of flagellar fins. However, some Ostariophysi also have a C-shaped mitochondrion, and/or flagellar fins, and *Chanos*, in the Anotophysi, has not only a C-shaped mitochondrion but also ITDs. Spermatozoa thus provide no clear evidence of a clupeiform-ostariophysan relationship.

Spermatozoa from all ostariophysan species analyzed again lack acrosomes. However, there is much diversity which is described in detail. Nine families of the Siluriformes produce bi-axonemal sperm. Evidence is given that sperm resembling Types I and II of percomorphs, and an additional Type III, are in reality novel forms that have arisen independently within the Ostariophysi. The four families that contain inseminating species, three in

Siluriformes and one in Characiformes, generally produce highly modified sperm.

Within the doubtfully valid grouping Protacanthopterygii, in the Argentiformes the mitochondrion is reported to be C-shaped as in Salmoniformes; ITDs are present but differ from those of *Salmo*. Esocid sperm differ from those of argentiforms and salmoniforms in having a number of separate mitochondria and approach cyprinid or cichlid sperm in morphology. They are unusual in having a single flagellar fin. In the sperm of Salmonidae, the single mitochondrion surrounds the cytoplasmic canal as a closed or incomplete ring, or a helix. The flagellum has two, or, in some sperm, three fins. The sperm of osmeroids differ notably from those of salmonids in having a single, unilateral mitochondrion.

In the Stenopterygii, stomiiform sperm are round-headed anacrosomal aquasperm. Within the great variety described, some myctophids have a highly distinctive biflagellate sperm type, with a 9+0 axonemal structure known elsewhere in fishes only in the Elopomorpha. No spermatozoal support for the questionable grouping Paracanthopterygii has been found. Batrachoidiformes also have biflagellate sperm. The Lophiiformes, considered from molecular evidence to be related to the percomorph Tetraodontiformes, have sperm varying from the simple anacrosomal aquasperm through sperm with two tiers of mitochondria, resembling those of some tetraodontids, to highly modified elongate sperm very different from known tetraodontiform sperm.

Mugilids, the putative plesiomorphic sister-group of the Atherinomorpha, are externally fertilizing and have sperm resembling Type II sperm of the Perciformes (including the presence of ITDs) or intermediate between Types I and II. It is suggested that the Type II sperm represent parallelism with perciforms by virtue of genetic relationship (paramorphy).

All atherinomorphs are united by the unique telogonic development of the ovary. Atheriniform families have modifications of the Type I sperm. Few are internally inseminating.

Reproductive modes in the Cyprinodontiformes are exceptionally varied, with oviparity, ovoviparity and the great diversity in sperm ultrastructure is described. Internally fertilizing families lacking true intromittent organs, have simple round-headed sperm but those of the intromittent organs have moderately to greatly modified sperm.

In the Beloniformes, where fertilization is external the spermatozoa are correspondingly round-headed aquasperm. The sperm of the internally fertilizing belonoids are profoundly modified. The fertilization biology and evolution of spermatozoa in the Atheriniformes, Beloniformes and Cyprinodontiformes spermatozoa is discussed.

The diversity of sperm types in the Percomorpha reflects the demonstrated polyphyly of the group and its diversity of fertilization modes and is detailed in Chapter 15. The vast majority of percomorphs are externally fertilizing but internal fertilization has arisen, in most cases independently, in several subgroups. Internal gametic association occurs in some Gasterosteiformes (Aulorhynchidae) and some Perciformes (Apogonidae). External fertilization occurs in almost all orders and suborders of the Percomorpha and involves one or other of two sperm types (Type I and Type II) or presumed modifications of these. The Type II spermatozoon, though round-headed, is considered an evolved type relative to Type I and is the commonest sperm type in the Perciformes. Sperm modifications include biflagellarity, and elongation of the nucleus and/or the midpiece. In terms of diversity of spermatozoal ultrastructure, the Gobioidei, despite being external fertilizing, are the most remarkable of the Perciformes and Percomorpha.

Within the sarcopterygian fishes, in the Ceratodontiformes, sperm ultrastructure reinforces separation of *Neoceratodus* (Ceratodontidae) from *Protopterus* and *Lepidosiren* while offering an autapomorphy (retronuclear body or its derivative) linking the three genera and, in modified form, the coelacanth. The exceptionally great length of the nucleus in Coelacanthimorpha and Ceratodontimorpha (70 μm long in *Neoceratodus*) is a synapomorphy of the Sarcopterygii. Internal relationships of sarcopterygian fish and their relationships with tetrapods are discussed.

Finally, in part A, Chapter 17 is devoted to sperm modifications related to insemination, with particular reference to the Ostariophysi. A distinction between insemination and internal fertilization is recognized because internal gametic association (insemination followed by external fertilization) is known for some species. The functional significance of evolutionary elongation of the nucleus is discussed It is suggested that longevity of the spermatozoon outside of the male body may be ensured by increase in the size of the sperm midpiece, glycogen storage within the spermatozoon, as well as secretions of both the male and female reproductive tracts. Morphological, chemical and behavioral modifications often associated with insemination are discussed. Many new illustrations are provided in this chapter and throughout the volume.

Summing up, the secondary simplification of fish sperm to the anacrosomal aquasperm from a complex sperm type retained, with modification, in chondrichthyans, Cladistia and Holostei, appears to have largely obscured phylogenetic relationships but many cases of synapomorphic resemblance within subgroups remain, of which the Elopomorpha offer the clearest example. This reduced phylogenetic signal in fish sperm contrasts with the striking phylogenetic picture seen in, for instance, avian sperm (see *Reproductive Biology and Phylogeny of Birds*).

The generosity of individuals and publishers who have allowed illustrations to be reproduced is most gratefully acknowledged. The kind collaboration of all of the contributing authors, who have borne uncomplainingly the requests of an editor attending the gestation and parturition of this work, has been greatly appreciated. The courteous and efficient participation of the publishers was indispensable to production of this volume.

It is hoped that this volume will be received with the interest accorded the 1991 book.

Barrie Jamieson
The School of Integrative Biology
University of Queensland

16 July 2000

Contents

CHAPTER 1

Phylogeny and Classification

Guillermo Ortí and Chenhong Li

1.1 INTRODUCTION

Our current knowledge about phylogeny and classification of "fishes" is in a state of flux. Most classification schemes which proposed to organize the vast fish biodiversity (Helfman *et al.* 1997; Nelson 2006) have been based on loosely formulated syntheses of many, largely disconnected phylogenetic studies among some of its components. An explicit cladistic analysis including representatives of all major taxonomic groups across the diversity of fishes has never been accomplished. As a consequence, phylogenetic relationships among the major groups of fishes are still controversial and unresolved, as are many of the proposed higher-level taxa (Greenwood *et al.* 1973; Lauder and Liem 1983; Jamieson 1991; Stiassny *et al.* 1996; Kocher and Stepien 1997; Chen *et al.* 2003; Meyer and Zardoya 2003; Miya *et al.* 2003; Chen *et al.* 2004; Cloutier and Arratia 2004; Stiassny *et al.* 2004). We expect this situation to change soon, as ongoing efforts by morphologists and molecular systematists are seeking to converge on a synthesis. While DNA sequence data are being collected rapidly and cost effectively, and provide a useful way to reconstruct phylogeny, the promise of a data-rich supermatrix approach to explicitly analyze phylogenetic relationships among representatives of *all* major groups of "fishes" still is unaccomplished. These are, therefore, exciting times for molecular systematics in general, and fish phylogenetics in particular. Molecular data sets are proliferating and rapidly transitioning towards phylogenomic proportions (Miya and Nishida 2000; Rokas *et al.* 2003; 2005; McMahon and Sanderson 2006; Comas *et al.* 2007) and a more thorough interpretation of morphological and paleontological material is also underway (Diogo 2007; Mabee *et al.* 2007). For example, a recent analysis of higher-level relationships among the major early-branching lineages of sarcopterygians and actinopterygians (Diogo 2007), based on osteological and

School of Biological Sciences, University of Nebraska, Lincoln, Nebraska 68588, USA.

myological characters, could pave the way for an expanded effort that may include the most diverse euteleostean taxa. We anticipate that in just a few years, efforts along these two fronts will converge to produce a well-supported phylogenetic classification based on genealogical analyses of large numbers of genes and a better understanding of morphological homologies based on detailed analysis of genetic and developmental pathways. In this chapter, we summarize some of the most recent results, with major emphasis on hypotheses for actinopterygian fishes derived from our own molecular studies.

1.2 PHYLOGENETIC ANALYSIS OF MOLECULAR DATA

Much of what we know of the relationships of fishes has been the result of a long history of morphological research (e.g., Rosen 1973; 1982; Stiassny 1986; Johnson 1992; Johnson and Patterson 1993). But at this time, however, there is no single resource that presents a comprehensive picture or synthesizes our current understanding of higher-level actinopterygian morphology, particularly within the species-rich percomorph crown group. Efforts underway (Diogo 2007; Ed Wiley, pers. com.), as stated above, are setting the stage to solving this shortcoming, especially with the insights that a combination of morphological and molecular data can make available. In this review, we compare hypotheses based largely on morphology with new proposals from molecular systematics. Although the relative merits of the different kinds of data commonly used for phylogenetic analysis remain in dispute—see Scotland *et al.* (2003) and subsequent reaction (Jenner 2004; Wiens 2004; Smith and Turner 2005) for a recent reincarnation of this debate—there is little doubt that molecular data are and will be most commonly used for phylogenetics. Part of the reason is the ease of collection and of establishing primary homology across vast taxonomic ranges (Li *et al.*, 2007). But molecular data, as any other kind of data, are not without problems.

Molecular phylogenies based on DNA sequences of a single locus or a few loci often suffer from low resolution and marginal statistical support due to limited character sampling. Individual gene genealogies also may differ from each other and from the organismal phylogeny under study. This discordance, know as the "gene-tree vs. species-tree" issue (Fitch 1970; Pamilo and Nei 1988) can be caused by several factors. In many cases, systematic biases leading to statistical inconsistency in phylogenetic reconstruction (i.e., base-compositional bias, long-branch attraction, heterotachy) may cause spurious results (Felsenstein 1978; Weisburg *et al.* 1989; Foster and Hickey 1999; Lopez *et al.* 2002). In other cases, discordance may be due to the actual history of gene duplication/extinction events leading to mistaken assumptions about orthology (Fitch 1970). Even though the correct gene tree may be obtained in the analysis, genealogical discordance between the history of the gene and the organsimal phylogeny may persist. Undetected paralogy (the relationship of homology among loci originating from gene duplication events) may result from sampling genomes of distantly related species using a direct-PCR approach. Ideally, to avoid this problem, only single-copy genes that did not undergo a complex history of duplication and extinction should be used for phylogenetic analysis. This condition may be hard to find among fishes in light of mounting evidence supporting a fish-specific whole-genome duplication event (Amores *et al.* 1998; Meyer and Van de Peer 2005) and the more general observation that gene duplications are a common mechanism of molecular evolution (Ohno 1970; Taylor and Raes 2004).

A phylogenomic approach—using genome-scale data sets to study evolutionary relationship—may provide the best solution to these problems (Eisen and Fraser 2003; Delsuc *et al.* 2005) but it requires compilation of large data sets that include many independent nuclear loci for many species (Bapteste *et al.* 2002; Rokas *et al.* 2003; Driskell *et al.* 2004; Philippe *et al.* 2004). Such data sets are less likely to succumb to sampling and systematic errors (Rokas *et al.* 2003) by offering the possibility to survey characters that are phylogenetically reliable and also to test phylogenetic results with alternative taxonomic samples. Some simple criteria can be used to assess the reliability of molecular markers (e.g. testing homogeneity of base composition, relative rates of evolution, saturation of base substitutions, etc). Taxonomic-rich data sets also allow the possibility of using different subsets of representative species for each group to test for consistency in the results. In spite of rapid success and initial optimism generated by phylogenomic approaches (Gee 2003; Rokas *et al.* 2003), large and complex data sets also exacerbate the unresolved methodological challenges (Li *et al.* 2008). Many long-standing challenges such as sparse taxon-sampling (Soltis *et al.* 2004), base compositional bias (Phillips *et al.* 2004), missing data (Wiens 2003; Waddell 2005) or incomplete lineage sorting (Kubatko and Degnan 2007) also increase in relevance as multi-locus data sets grow in size and complexity. We elaborate below on two major potential obstacles for recovering a comprehensive phylogenetic hypothesis for ray-finned fishes: base compositional bias and undetected paralogy due to gene duplications.

1.2.1 Base Compositional Bias

Stationarity (i.e. that evolutionary processes do not change significantly across or within lineages) and time-reversibility (i.e. that the rate of change from one nucleotide to another is the same in each direction) often are assumed in standard inference models used for phylogenetic analysis, in part to simplify computations and also due to the expectation that base frequencies in DNA sequences remain constant along the evolutionary path. However, highly variable base composition among orthologous DNA sequences sampled from different species is not uncommon (Jukes and Bhushan 1986; Bernardi 1993). This is especially true for the nuclear gene RAG-1 among fishes (Ortí *et al.* 2005), as shown in Figure 1.1. Some fish taxa show extremely high content of G and C at the third codon positions of this gene (e.g., elopomorphs, galaxiids, stomiiforms) while other taxa show extremely low frequencies (e.g., Ostariophysi). Significant variation in base composition also is evident within some groups (e.g., among acanthomorph fishes, within Clupeiformes and Elopiformes). Base compositional bias is a well-known source for systematic error in phylogenetic inference, usually resulting in groups with similar nucleotide frequencies that do not represent true evolutionary relationships. An example for fishes and several possible solutions that have been proposed (Steel *et al.* 1993; Lockhart *et al.* 1994; Gu and Li 1996; Foster and Hickey 1999; Foster 2004; Collins *et al.* 2005) are discussed below.

One simple way to address the potentially confounding effect of base compositional bias is to carefully choose genes that do not show this pattern (Collins *et al.* 2005). With phylogenomic-sized data sets containing large numbers of genes (>100), this may be a feasible approach, but usually most studies are confronted with limited data and some base compositional bias. The next simple solution in this case would be to recode the data as purines and pyrimidines (RY-coding, where R=G=A and Y=C=T). This approach homogenizes base composition among divergent sequences and removes the GC-bias (Woese *et al.* 1991; Phillips *et al.* 2004). However, the method also leads to loss of phylogenetic information. We applied this approach to assess the effect of base-composition in our study of clupeiform relationships based on DNA sequences of RAG1 and RAG2 genes (Li and Ortí 2007). Most clupeiform fishes have high (> 70%) GC content at the variable positions in these genes, except for *Denticeps clupeoides* (61%) and *Spratelloides delicatulus* (59%), that are closer to the average frequency observed among fishes (65%, Fig.1.1). In contrast, ostariophysans have a relatively low average GC content (55%). This pattern is repeated, albeit to a lesser degree for mitochondrial ribosomal genes (12S and 16S). Analyses of these sequences invariably grouped *Denticeps* with ostariophysans rather than with other clupeiforms (Li and Ortí 2007). Support for the *Denticeps*+ostriophysi clade should decrease significantly when using RY-coded data if this relationship is artificially obtained due to non-stationarity. Indeed, when branch weights (total number of characters supporting each alternative hypothesis) were calculated for RY-coded data, higher support was obtained for the *Denticeps*+Clupeiformes hypothesis, in contrast to the result obtained with non-coded data. Support for both hypothesis, however, was lower under the RY-coding strategy, consistent with the expected loss of phylogenetic information caused by this method (Li and Ortí 2007).

A more effective approach to avoid artifacts caused by base compositional biases involves accounting for the non-stationarity explicitly in the evolutionary model used for analysis. Several alternative models have been proposed, including the LogDet distance method (Lockhart *et al.* 1994), maximum-likelihood methods assigning local base frequencies to each branch (Yang and Roberts 1995; Galtier and Gouy 1998), and Bayesian methods assigning different base frequencies to predefined number of clades (Foster 2004). In many cases, the relatively simple LogDet distance approach has been

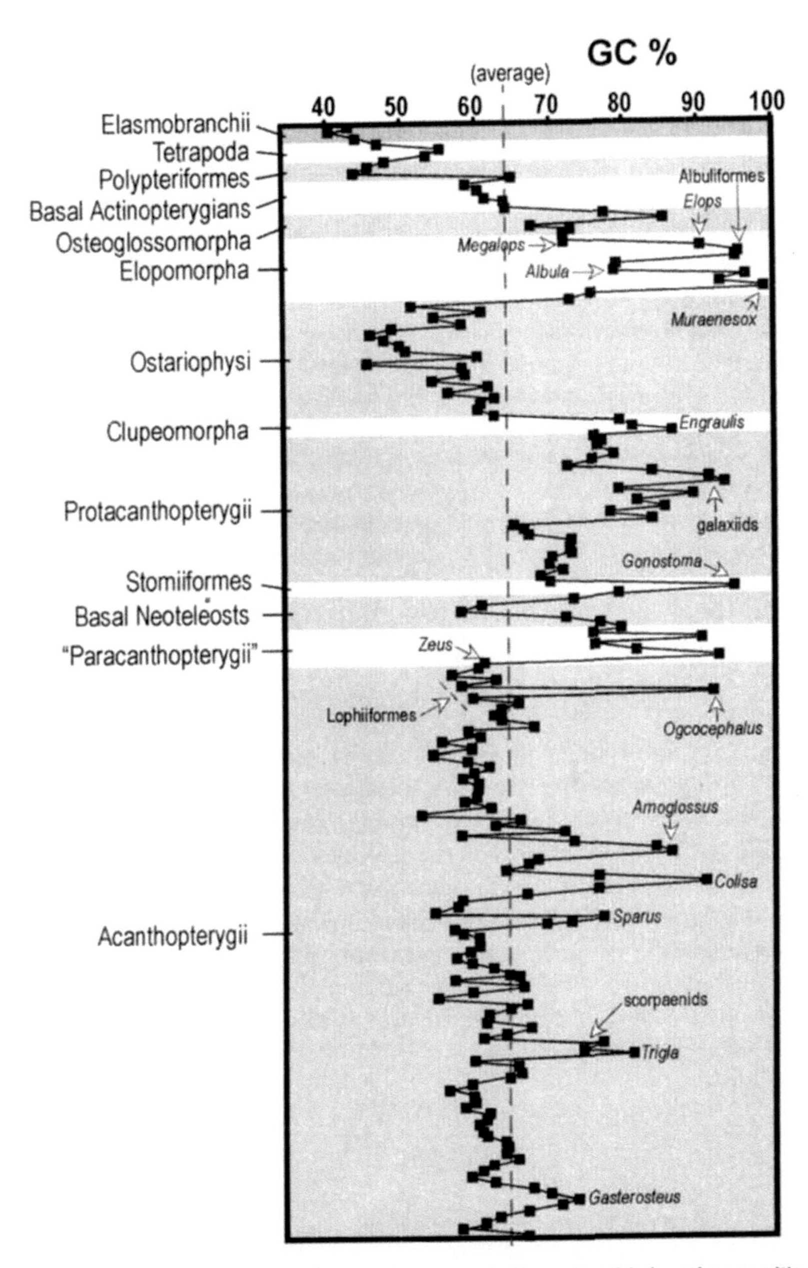

Fig. 1.1 Base composition (% content of G and C) at the third codon positions of the nuclear gene RAG-1 among fishes. Only sequences of exon 3 (*ca.* 1500 bp) of the RAG-1 gene were compared; most of the variation among these sequences is found at third codon positions. Taxonomic groups are indicated on the vertical axis following the sequential order presented in most current classifications; some representative genera or families are identified in the graph. For a complete list of taxa, please contact the authors. Original.

shown to fail and did not recover the expected tree topology (Foster and Hickey 1999). More realistic, parameter-rich, models that attempt to assign branch-specific base frequencies (Yang and Roberts 1995; Galtier and Gouy 1998) or clade-specific base frequencies (Foster 2004) may be too complex and over-parameterized to produce reliable results. Two recent developments to account for non-stationarity that were also aimed at reducing the high dimensionality of these models may provide promising options to address this problem. Blanquart and Latrillot (2006) proposed a new model that estimates variation among base frequencies across lineages by a stochastic process using a Poisson distribution. Their method is more realistic because it decouples the change of base frequencies from speciation events and also reduces the number of parameters to estimate. A second approach by Gowri-Shankar and Rattray (2007) extended Foster's (2004) methods by introducing a reversible-jump Monte Carlo Markov Chain (MCMC) sampler for efficient Bayesian inference of the model order along with other phylogenetic parameters of interest. The methods of Blanquart and Latrillot (2006) and Goweri-Shankar and Rattray (2007), implemented in the computer programs PhyloBayes and PHASE, respectively, should provide more robust phylogenetic results for large-scale analysis of nuclear genes when base compositional bias may be rampant.

1.2.2 Gene Duplication and Paralogy

Another important issue associated with the use of nuclear protein-coding genes for phylogeny inference is uncertainty about their orthology (Fitch 1970). This uncertainty may lead to the inference of erroneous relationships among species even when the true genealogical histories of specific loci are recovered in the analysis. As stated above, sophisticated phylogenetic methods exist and continue to be developed to identify and circumvent potential analytical artifacts, but confounding biological factors arising from the dynamic nature of the genome remain. Among these, the complex history of gene or genome duplication/extinction events that has been documented for ray-finned fishes (Van de Peer *et al.* 2003) is especially challenging for fish phylogenetics.

Most genes are represented in genomes by more than one copy, usually as members of a gene family. But some genes are unique ("single-copy"), meaning that no other region of the genome contains a sequence with high similarity to them. This definition of "single copy" is somewhat arbitrary and operational (it depends on definition of a threshold of similarity), since fragments of the genome with lower values of similarity to any gene may presumably be found. This suggests that no gene could be truly single copy unless duplicates have been lost from the genome or modified so drastically that they are no longer recognizable as such. Nonetheless, and to simplify interpretation of phylogenetic results, it is better to use single-copy nuclear genes to minimize the chance of sampling paralogous genes among taxa (Li *et al.* 2007). Even in the case that gene duplication events may have occurred during evolution of the taxa of interest (Van de Peer *et al.* 2003; Meyer and Van de Peer 2005), duplicated copies of a single-copy nuclear gene tend to be lost quickly, possibly due to dosage compensation, a mechanism that balances the phenotypic expression of genes with unequal copy number (Lynch and Conery 2000; Blomme *et al.* 2006). Almost 80% of the duplicated genes can be secondarily lost shortly after a genome-duplication event (Jaillon *et al.* 2004; Woods *et al.* 2005). Therefore, if duplicated copies are lost before the relevant speciation events occur (Fig. 1.2a, b), there will be no discrepancy among the inferred gene trees and the species tree. In contrast, if the unfortunate situation depicted in Fig. 1.2c occurs, paralogous comparisons will result in topological discordances among genes and among some of these and the species tree.

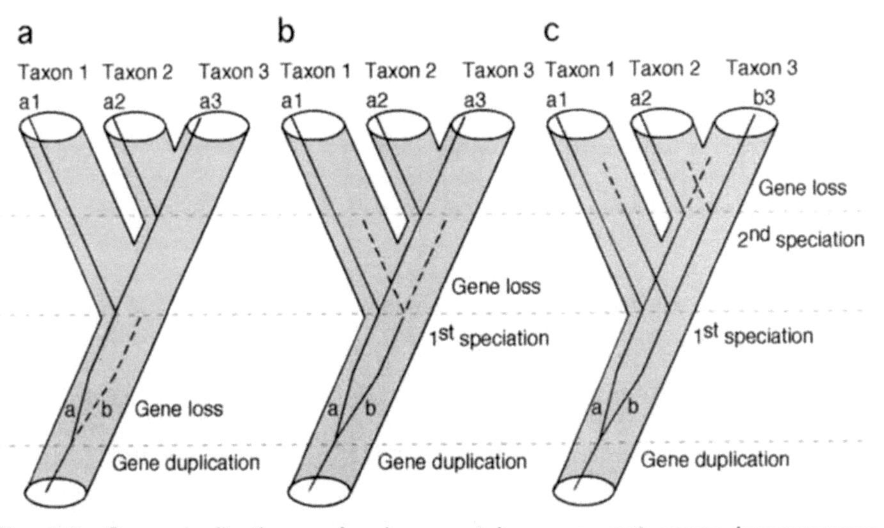

Fig. 1.2 Gene duplication and subsequent loss may not cause incongruence between gene tree and species tree if gene loss occurs before the first speciation event (a), or before the second speciation event (b). The only case that would cause incongruence is when the gene survived both speciation events and is asymmetrically lost in taxon 2 and taxon 3 (c). From Li, C., Ortí, G., Zhang, G., and Lu, G. 2007. A practical approach to phylogenomics: the phylogeny of ray-finned fish (Actinopterygii) as a case study. BMC Evolutionary Biology 7: 44, Fig. 1.

Phylogenomic approaches allow the comparison of such potential discordances for a large number of genes. Therefore, a gene-by-gene analysis of the topological distribution of the observed discordance may be used to reconstruct (reconcile) putative duplication/extinction events and avoid the pitfall of mistaken paralogy (Page and Cotton 2002). For example, gene duplication events that occurred before the inferred origin of the ingroup of interest that were followed by differential losses of the duplicates among ingroup taxa, may lead to the inclusion of *paralogous* sets of genes ("out-paralogs" sensu Remm *et al.*, (2001). In this case, an *a posteriori* examination of unrooted tree topologies and associated branch lengths may help detect the putative out-paralogs because they will form two highly divergent clades. If this pattern is detected, it would be safer to infer phylogenies only for the reduced taxonomic sets represented in each of the orthologous datasets. In contrast, if duplication/differential loss events occurred within the ingroup of interest ("in-paralogs") these will not be as easily detected by inspection of topology and branch lengths because these duplicates will be equidistant to other ingroup lineages that are not descendants of the ancestor in which the duplication took place. Genes that have this history must meet two conditions to remain undetected and have an effect on phylogenetic conclusions: (1) none of the taxa affected by the duplication maintain both copies of the gene or any existing duplicates remain undetected by PCR assays; and (2) the same taxonomic distribution of duplication and loss is repeated across multiple genes. Although possible, it seems unlikely that both of these conditions will be met.

In summary, although molecular characters are not free of many potential problems that usually confuse phylogenetic results, careful analysis of large numbers of single-copy nuclear genes (the phylogenomic approach) may provide a realistic means towards inferring the tree of life of "fishes" in the near future.

1.3 THE TREE OF LIFE OF "FISHES"

In this section, we outline some currently accepted hypotheses of relationships among the major groups of "fishes" relevant to subsequent chapters of this book. Although used freely in the literature, the term "fishes" does not refer to a natural group (a monophyletic lineage). The term is used to describe a heterogeneous collection of distantly related vertebrates such as hagfish, dogfish, knifefish, killifish, cowfish, and lungfish. The term could be restricted to a monophyletic group if it were applied only to the largest and most diverse clade of fishes (Actinopterygii). Because the tetrapods are always excluded, "fishes" form a paraphyletic group and classification schemes do not give this term taxonomic rank (Nelson 2006). Figure 1.3 is a summary of the most likely hypothesis upheld by phylogenetic analyses of morphology and molecular data. In the following subsections, we discuss evidence supporting or contradicting this hypothesis and provide more detailed phylogenetic relationships among some relevant groups of ray-finned fishes. For a discussion on sarcopterygian relationships, see Jamieson, **Chapter 16**.

1.3.1 Jawless Fishes (Agnathans)

The living jawless fishes (hagfishes and lampreys, Jamieson, **Chapter 6**) represent early-branching lineages at the base of the vertebrate tree of life. Their relationship to other long-extinct jawless fishes, to each other, and to the jawed vertebrates remains controversial. Most morphological and paleontological analyses (Hardisty 1982; Mallat 1984; Janvier 1996; Mallat 1997; Donoghue *et al.* 2000; Ota and Kuratani 2007) but see (Mallat 1997; Ota and Kuratani 2007) support the view that agnathans form a paraphyletic group, with lampreys more closely related to the gnathostomes than to hagfishes. Molecular evidence, in contrast, keeps mounting to overwhelmingly support a sister-group relationship between hagfishes and lampreys (Cyclostomata) as shown in Figure 1.3.

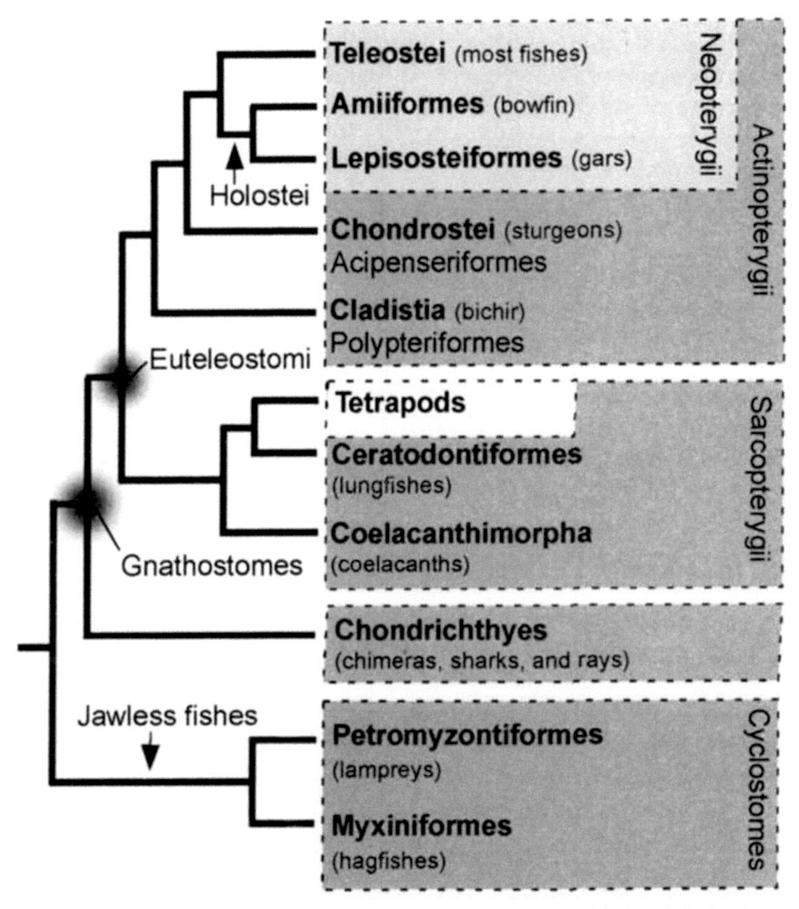

Fig. 1.3 Phylogenetic relationships among major groups of living "fishes" (in gray boxes), based on several morphological and molecular analyses discussed in the text where evidence for and against this hypothesis is discussed. Common names are given in parentheses. Original.

The first molecular study to address this issue was based on comparison of nucleotide sequences of 18S ribosomal RNA (Stock and Whitt 1992). Subsequently, Mallat and Sullivan (1998) added sequences of the 28S rRNA gene and also recovered Cyclostomata. These early studies, however, included only a single representative of hagfish (*Epatretus*) and lamprey (*Petromyzon*), risking potential analytical artifacts due to long-branch attraction. To address this issue, Mallat and Winchell (2007) added distantly related hagfish and lamprey taxa (*Myxine* and *Geotria*) and their analysis upheld the previous result with even higher support, rejecting the alternative "lampreys plus gnathostomes" hypothesis with confidence. Furlong and Holland (2002) reanalyzed 18S sequences using a Bayesian approach and also included in their analysis protein-coding mitochondrial DNA (mtDNA) and two protein-coding nuclear genes, triose phosphate isomerase (TPI) and superoxide dismutase (SOD). Their study provided strong support for Cyclostomata, again corroborating previous results with increasing confidence. To overcome potential biases in mtDNA sequences and alignment issues with rRNA genes, Takezaki *et al.* (2003) assembled an impressive data set with 35 loci to test cyclostome monophyly. The genes analyzed by Takezaki and collaborators represented a diverse group of nuclear protein-coding genes, including housekeeping and regulatory genes, with about half of them encoding ribosomal proteins which are known to evolve slowly. This study provided definitive evidence that molecular genetic data support the cyclostome hypothesis (Fig.1.3). Additional molecular studies in favor of this view include papers by Kuraru *et al.* (1999), Cotton and Page (2002), Delarbre *et al.* (2002), Blair and Hedges (2005), and Delsuc *et al.* (2006).

Agnathans and gnathostomes exhibit striking differences in their immune system. Neither hagfishes or lampreys possess the essential components that gnathostomes use for adaptive immunity, namely immunoglobulins (Ig), T cell receptors, recombination activating genes RAG1 and -2, and MHC class I and II molecules, but they share a fundamentally similar immune mechanism of generating variable lymphocyte receptors, VLRs (Pancer *et al.* 2004; Pancer *et al.* 2005). Although the VLR-based immune system could represent the plesiomorphic condition predating evolution of the vertebrate Ig-response, these two systems could have evolved simultaneously as early vertebrates experienced intense selective pressures to develop an anticipatory molecular recognition response. Whether the VLR system can be considered a synapomorphy supporting the monophyly of cyclostomes depends on the (largely unknown) condition observed in deuterostome outgroups (tunicates, hemichordates, echinoderms). A recent review of the immune system of the sea urchin based on comparative genomics shows that echinoderms exhibit immune signalling mediators and much of the gene regulatory toolkit for immunity known previously only for vertebrates, including a homologous Rag1/2 functional gene cluster (Rast *et al.* 2006). This finding suggests that the VLR system of hagfishes and lampreys could indeed be interpreted as a cyclostome synapomorphy.

1.3.2 Actinopterygii (Ray-finned Fishes)

This group contains nearly 27,000 described species, currently classified into three subclasses, 44 orders and 453 families (Nelson 2006). We review here some outstanding controversies regarding relationships among higher-level taxa that have become classic debates and discuss the incidence of new data. We begin by characterizing the early-branching lineages at the base of the tree of ray-finned fishes and progress towards the derived euteleostean crown groups, where the highest diversity among living actinopterygians can be found.

Despite previous hypotheses linking polypterids to sarcopterygians (dismissed by Jamieson and Mattei, **Chapter 7**), it is now quite well established that the extant sister group to all other ray-finned fishes is the lineage leading to the bichir (*Polypterus*) and its living relatives, some 11 species of African freshwater fishes (family Polypteridae). This view (Figs. 1.3 and 1.4) has been recently supported by molecular evidence (Venkatesh *et al.* 2001; Inoue *et al.* 2003; Kikugawa *et al.* 2004; Li *et al.* 2008) as well as morphological analysis (Diogo 2007). The classic concept of "Chondrostei" that grouped *Polypterus* and living sturgeons and paddlefishes (Acipenseriformes) and their fossil relatives (Schaeffer 1973), received some support in a recent analysis of 10 nuclear genes (Ortí and Li 2007; Li *et al.* 2008), albeit with low bootstrap (65%) and posterior probability (0.74) values (Fig. 1.4). Most evidence from both morphological (Grande and Bemis 1996; Gardiner *et al.* 2005; Grande 2007) and molecular data (see above) suggests that "Chondrostei" is actually a paraphyletic group. Therefore, the current consensus is that polypterids are the sister taxon to all other living actinoterygians, and Acipenseriformes (or Chondrostei *sensu stricto*; Jamieson, **Chapter 8**) are considered as the sister group to neopterygians (Nelson 2006; Jamieson and Mattei, **Chapter 7**; Figs. 1.3 and 1.4).

Relationships among the non-teleost actinopterygians have been somewhat controversial, but a consensus seems to be emerging (at least for the extant taxa). While most morphological (Regan 1923; Patterson 1973) and molecular evidence (Lê *et al.* 1993; Kikugawa *et al.* 2004; Hurley *et al.* 2007; Li *et al.* 2008) supports the monophyly of Neopterygii, a group represented by gars (Lepisosteiformes), *Amia*, and teleosts (Fig. 1.3), relationships among these three lineages are hotly debated; see Arratia (2001) for a review of alternative schemes of relationships. Historically, *Lepisosteus* and *Amia* were grouped in a single clade (Holostei), placed as the sister-group to Teleostei (Nelson 1969; Jessen 1972). Subsequently, Holostei was dissolved in favor of alternative hypotheses suggesting that either Amiiformes (Patterson 1973; Olsen 1984; Grande and Bemis 1996; Diogo 2007) or Lepisosteiformes (Olsen 1984) alone represent the sister-group of teleosts. Yet another hypothesis, derived from analysis of mitogenomic data or indel patterns in the nuclear gene RAG2, supports a monophyletic "ancient fish" group composed by Acipenseriformes, Lepisosteidae and *Amia* (Venkatesh *et al.* 2001; Inoue *et al.* 2003). This group was placed as the sister-group to Teleostei. Most recently, however, both molecular data (Kikugawa *et al.* 2004; Hurley *et al.* 2007; Ortí and Li 2007; Li *et al.* 2008) and a reassessment of morphology (Grande 2007) advocate the "resurrection" of Holostei as the sister group of Teleostei. This is our currently preferred hypothesis, presented in Figures 1.3 and 1.4 (see also Jamieson, **Chapter 9**).

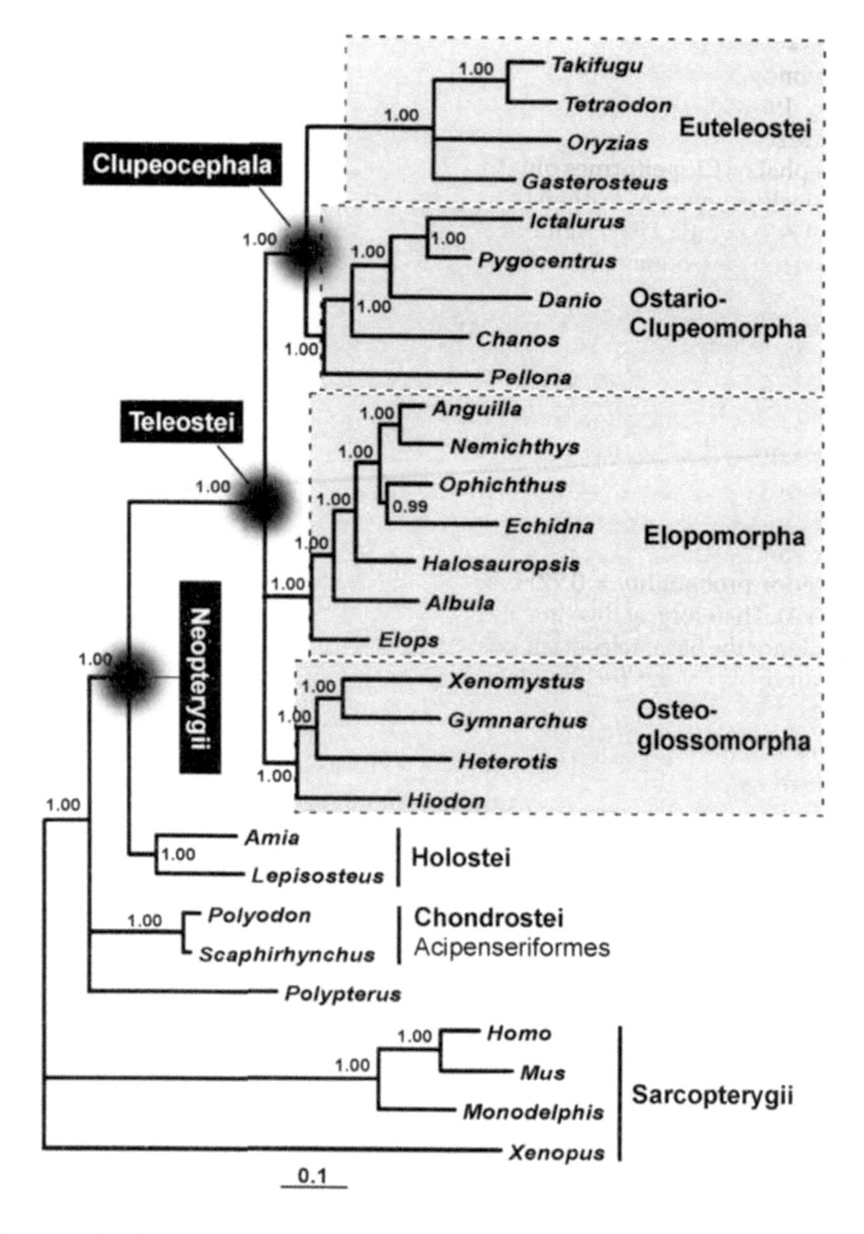

Fig. 1.4 Phylogeny of lower Actinopterygians and the early-branching teleost lineages. The tree is a consensus obtained by Bayesian analysis of nuclear gene DNA sequences (8 genes, 11,766 bp) for 29 representative taxa. Posterior probabilities (only values >0.95) are shown for well-supported groups and branches with low support were collapsed (e.g., a branch uniting *Polypterus* and acipenseriforms and a branch uniting elopomorphs and osteoglossomorphs). Original.

1.3.2.1 Teleostei

The monophyly of Teleostei is supported by many morphological characters (de Pinna 1996; Arratia 2000). Four major teleostean lineages are currently recognized: Elopomorpha, Osteoglossomorpha, Ostarioclupeomorpha (or Otocephala = Clupeiformes plus Ostariophysi), and Euteleostei (Nelson 2006). Ostarioclupeomorphs are generally placed as the sister-group of euteleosts (Lê *et al.* 1993; Arratia 1997; Inoue *et al.* 2001) a grouping named Clupeocephala, that excludes elopomorphs and osteoglossomorphs (Fig. 1.4). Interrelationships among elopomorphs, osteoglossomorphs, and clupeocephalans are still controversial. Both morphological (Patterson and Rosen 1977) and molecular (Inoue *et al.* 2001) studies support the position of osteoglossomorphs at the base of the teleosts, but this view was challenged by the alternative placing of elopomorphs as the living sister-group of all other teleosts (Arratia 1991; Shen 1996; Arratia 1997, 2000). A third alternative was suggested by Lê *et al.* (1993) based on relatively weak evidence from 28S ribosomal gene sequences, with osteoglossomorphs and elopomorphs forming a clade that is sister to clupeocephalans. We obtained the same result, albeit with low support (posterior probablility = 0.72) based on Bayesian analysis of 8 nuclear genes (Fig. 1.4). Therefore, at this time there is no unequivocal evidence to resolve with confidence the basal teleost trichotomy.

Support is strong for the Ostarioclupeomorpha hypothesis (Otocephala), placing the Clupeiformes as a sister group to Ostariophysi (Figs. 1.4 and 1.5). A recent result based on mitogenomic data (Saitoh *et al.* 2003), however, suggests that gonorynchiforms are more closely related to Clupeiformes, but this result could be due to poor taxonomic sampling or an analytical artifact. In most molecular studies (Dimmick and Larson 1996; Ortí and Meyer 1996; Saitoh *et al.* 2003), relationships within Ostariophysi are consistent with the traditional view (Fink and Fink 1981) placing Cypriniformes as a sister group to the rest, but relationships among Characiformes, Siluriformes, and Gymnotiformes cannot be resolved with confidence (see discussion in Saitoh *et al.* 2003). The close relationship shown in Fig. 1.5. between *Ictalurus* (Siluriformes) and *Pygocentrus* (Characiformes), to the exclusion of *Apteronotus* (Gymnotiformes) should, therefore, be taken with caution.

Our knowledge of the identity and relationships among major euteleostean lineages ranges from well corroborated to poorly understood. Johnson and Patterson (1996) and Lecointre and Nelson (1996) provide synapomorphies supporting euteleost monophyly, but mitogenomic data suggest an alternative definition (Ishiguro *et al.* 2003). Several early-branching euteleost lineages have been placed in the Protacanthopterygii (Greenwood *et al.* 1966), a supraordinal taxon that has undergone major re-definitions since its creation (Johnson and Patterson 1996). Recent molecular studies of basal euteleosts by Ishiguro *et al.* (2003) based on mitogenomic data and by Lopez *et al.* (2004) based on12S and 16S mitochondrial rRNA genes (815 bp) and exon 3 of the RAG-1 gene (1444 bp) examined protacanthopterygian taxa. Both corroborated the position of Esociformes (pikes, pickerels, and mudminnows) as the sister group of

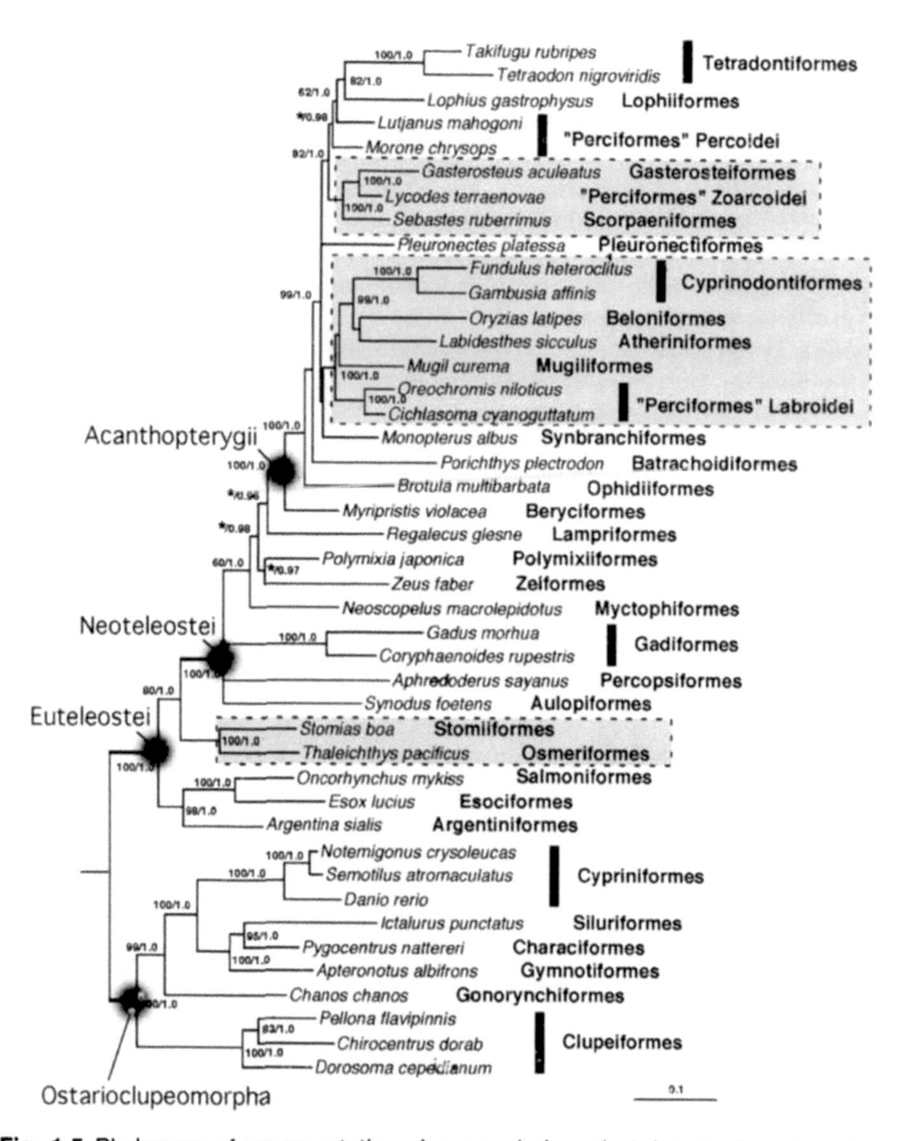

Fig. 1.5 Phylogeny of representative clupeocephalans (euteleosts plus otocephalans) based on analysis of 10 nuclear genes (7995 bp). The numbers on branches are maximum likelihood bootstrap values and Bayesian posterior probabilities. Asterisks indicate a bootstrap value < 50%. The names of representative species, orders, and supraordinal taxa are indicated. In grey boxes we highlight non-traditional groupings discussed in the text. Modified from Li, C., Lu, G., and Ortí, G. 2008. Systematic Biology 57(4): 519-539, Fig. 4.

Salmoniformes, a clade also supported by analysis of 10 nuclear genes (Li *et al.* 2008). This clade is either sister to or in a polytomy with the marine smelts (Argentiniformes), freshwater smelts (Osmeriformes), and Neoteleostei (Fig 1.5).

The identity, composition and relationships of Prothacanthopterygii still await analysis of taxon-rich basal euteleost data sets, but an interesting result of these studies is the suggestion that Osmeriformes and Stomiiformes are closely related, thus removing Stomiiformes from the Neoteleostei (Fig. 1.5, see also Jamieson, **Chapter 12**).

The monophyly and relationships of the basal neoteleost clades (Stomiiformes, Aulopiformes and Myctophiformes) to the crown group Acanthomorpha have been relatively well established based on morphology and mitogenomic data (Stiassny 1986; Johnson 1992; Patterson and Johnson 1995; Miya *et al.* 2003; Springer and Johnson 2004). Nuclear gene data necessary to fully assess these relationships are still unavailable, but our limited taxonomic sampling for 10 genes (Li *et al.* 2008) suggests that traditional hypotheses may not be supported. We discuss next the main implications of our study.

Rosen (1973) proposed Acanthomorpha (including more than 15,000 species) to comprise two subgroups, Acanthopterygii and Paracanthopterygii. Acanthomorph monophyly has since been corroborated by both morphology and DNA data (Stiassny 1986; Johnson and Patterson 1993; Smith and Wheeler 2006), but the monophyly of the two subgroups has been refuted. In addition to the placement of Stomiiformes outside of the Neoteleostei, our analysis rejects the notion of Paracanthopterygii, a classical grouping of neoteleosts that has been extensively debated in the literature (Greenwood *et al.* 1966; Patterson and Rosen 1989; Miya *et al.* 2003, 2005). Taxa traditionally included in this group are placed (Fig. 1.5) either among the early-branching lineages of Neoteleosts close to Aulopiformes and Myctophiformes (Gadiformes, Percopsiformes), closer to the base of Acanthopterygii (Ophidiiformes, and Batrachoidiformes), or in a derived position close to Tetraodontiformes (Lophiiformes). The paracanthopterygian hypothesis proposed by mitogenomic analyses (Miya *et al.* 2003, 2005) also was not supported in our study since *Polymixia* and *Zeus* did not form a monophyletic group with gadiforms and percopsiforms.

A problematic group missing in our analysis is the jellynoses (Ateleopodidae), which has been placed at the base of the Acanthomorpha (Miya *et al.* 2003). Lanternfishes (Lampriformes) are currently considered sister to the Acanthomorpha (Johnson 1992; Smith and Wheeler 2006), a position consistent with the placement obtained with the nuclear gene data (Fig. 1.5).

Acanthopterygii is strongly supported in our study as a monophyletic group only if ophidiiforms, batrachoidiforms and lophiiforms are included. Representative taxa for Beryciformes, Ophidiiformes and Batrachoidifromes branch off sequentially from the base of the acanthopterygians. The rest of the taxa included in our study formed a monophyletic group of crown acanthopterygians with a 99% bootstrap support (Fig. 1.5). Most notoriously without resolution is the crown of the teleost tree, represented by just a few species in our analysis. Relationships within the dominant acanthomorph group, Percomorpha, are essentially unknown. Percomorpha comprises more than 13,000 species of fishes (Nelson 2006). The majority of economically important fishes are percomorphs, yet their monophyly and relationships remain virtually unknown. In our study, taxa traditionally assigned to the order Perciformes (*Lutjanus, Morone, Lycodes, Oreochromis, and Cichlasoma*) do not form a monophyletic group, in agreement with several results suggesting the polyphyletic nature of this group (Lauder and Liem 1983; Johnson and Patterson 1993; Miya *et al.* 2003; Nelson 2006). Although the very limited sampling of perciform taxa in our study of 10 nuclear genes precludes general results, two noteworthy groupings emerge. First, some elements of the suborder Labroidei, such as cichlids, are placed in the same clade with atherinomorphs (Atheriniforms, Beloniformes, and Cyprinodontiformes) and Mugiliformes to the exclusion of other perciforms (Fig. 1.5). A close relationship among rice fish (*Oryzias*) and tilapia (*Oreochromis*) was first suggested by a phylogenomic study (Chen *et al.* 2004). Second, the three-spined stickleback (Gasterosteidae) is grouped with a zoarcid perciform (*Lycoides*), corroborating previous results suggesting this relationship by analysis of mitogenomic data (Miya *et al.* 2003). A recent, expanded study of 11 families of "Gasterosteiformes" based on the same type of data (Kawahara *et al.* 2008) clearly refutes the monophyly of this order, establishing that gasterosteiform fishes form indeed three separate lineages: Syngnathoidei, Gasterosteoidei (minus Indostomidae), and Indostomidae.

Much remains to be learned about the identity and relationships of many important groups of ray-finned fishes. An exhaustive review of the literature for all groups is beyond the scope of this chapter, so many taxa remain without mention. We are confident that within the next few years important discoveries will be made with insights from a concerted effort underway to combine molecular and morphological data. Resolution of the tree of life of fishes still is far away, but the stage is set for rapid progress to establish the branching pattern that explains their amazing diversity.

1.4 CHAPTER SUMMARY

Our knowledge of relationships among "fishes" ranges from well corroborated to very poorly understood. Fishes constitute a large, heterogeneous and paraphyletic assemblage of distantly related jawless and jawed vertebrates, with most of their living diversity found in the crown group of actinopterygians (the ray-finned fishes). As a group, they have fundamental relevance to understanding the evolution of vertebrate animals and their features, and they also carry great commercial importance. Yet, their phylogenetic relationships remain largely unknown. Currently, no explicit and comprehensive analyses featuring all groups are available to support a sound phylogenetic classification of fishes. This situation is likely to improve relatively soon in light of ongoing efforts to compile and analyze large phylogenomic data sets that span the diversity of fishes in concert with integrated studies of their morphology and development. In this chapter, we summarize some recent advances in these fields and discuss some of the challenges that lie ahead. We emphasize the molecular aspects and illustrate this with some recent and ongoing studies.

Two major challenges for molecular systematics of fishes involve ancient gene and whole genome duplication events and systematic biases such as base compostition heterogeneity among DNA sequences sampled from distantly related taxa. Whereas the latter may lead to spurious phylogenetic results, the former may impede proper interpretation of gene genealogies as indicators of species phylogeny. Some possible solutions to these problems are presented. We also summarize some results from recent studies concerning especially the relationships among jawless fishes, the early branching lineages of ray-finned fishes, and some unexpected relationships among the crown euteleost groups. We find compelling evidence from molecular studies that support the cyclostome hypothesis that groups lampreys and hagfish. Among basal actinopterygians, recent re-interpretation of morphology and new evidence from nuclear gene sequences support the Holostei (grouping amiids and lepisosteids). Relationships among the early-branching teleosts groups remain unresolved with elopomorphs, osteoglossomorphs, and clupeocephalans (ostarioclupeomorphs plus euteleosts) forming a polytomy at the base of the teleost tree. We discuss briefly our current understanding of supra-ordinal groups of euteleosts such as Protacanthopterygii, Paracanthopterygii, and Acanthopterygii. Relationships among the percomorph crown group are virtually unknown.

Much remains to be learned about the identity and relationships among the many groups of fishes, but we anticipate that the next few years will witness significant advances to establish the branching pattern of the tree of life of all fishes.

1.5 ACKNOWLEDGMENTS

We thank Wei-Jei Chen for help with the generation of sequence data and analyses of RAG-1, especially to compile Figure 1.1. This work was funded by grants DEB-9985045 and DEB-0732838 from the National Science Foundation (USA) to G.O.

1.6 LITERATURE CITED

Amores, A., Force, A., Yan, Y. L., Joly, L., Amemiya, C., Fritz, A., Ho, R. K., Langeland, J., Prince, V., Wang, Y. L., Westerfield, M., Ekker, M., and Postlethwait, J. H. 1998. Zebrafish hox clusters and vertebrate genome evolution. Science 282: 1711-1714.

Arratia, G. 1991. The caudal skeleton of Jurassic teleosts; a phylogenetic analysis. Pp. 249-340. In M.-M. Chang, H. Liu, and G.-R. Zhang (ed.), *Early vertebrates and related problems in evolutionary biology*. Science Press, Beijing.

Arratia, G. 1997. Basal teleosts and teleostean phylogeny. Palaeontologia Ichthyologica. 7: 5-168.

Arratia, G. 2000. Phylogenetic relationships of teleostei: past and present. Estudios Oceanológicos 19: 19-51.

Arratia, G. 2001. The sister group of Teleostei: consensus and disagreements. Journal of Vertebrate Paleontology 21: 767-773.

Bapteste, E., Brinkmann, H., Lee, J. A., Moore, D. V., Sensen, C. W., Gordon, P., Durufle, L., Gaasterland, T., Lopez, P., Muller, M., and Philippe, H. 2002. The analysis of 100 genes supports the grouping of three highly divergent amoebae: *Dictyostelium, Entamoeba,* and *Mastigamoeba*. Proceedings of the National Academy of Sciences USA 99: 1414-1419.

Bernardi, G. 1993. The vertebrate genome: isochores and evolution. Molecular Biology and Evolution 10: 186-204.

Blair, J. E., and Hedges, S. B. 2005. Molecular Phylogeny and Divergence Times of Deuterostome Animals. Molecular Biology and Evolution 22: 2275–2284.

Blanquart, S., and Lartillot, N. 2006. A Bayesian Compound Stochastic Process for Modeling Nonstationary and Nonhomogeneous Sequence Evolution. Molecular Biology and Evolution 23: 2058-2071.

Blomme, T., Vandepoele, K., De Bodt, S., Simillion, C., Maere, S., and Van de Peer, Y. 2006. The gain and loss of genes during 600 million years of vertebrate evolution. Genome Biology 7: R43.

Chen, W. J., Bonillo, C., and Lecointre, G. 2003. Repeatability of clades as a criterion of reliability: a case study for molecular phylogeny of Acanthomorpha (Teleostei) with larger number of taxa. Molecular Phylogenetics and Evolution 26: 262-288.

Chen, W. J., Ortí, G., and Meyer, A. 2004. Novel evolutionary relationship among four fish model systems. Trends in Genetics 20: 424-431.

Cloutier, R., and Arratia, G. 2004. Early diversification of actinopterygians. Pp. 217-270. In G. Arratia, M. V. H. Wilson, and R. Cloutier (eds), *Recent advances in the origin and early radiation of vertebrates*. Verlag Dr Friedrich Pfeil, Munich.

Collins, T. M., Fedrigo, O., and Naylor, G. J. P. 2005. Choosing the Best Genes for the Job: The Case for Stationary Genes in Genome-Scale Phylogenetics. Systematic Biology 54: 493-500.

Comas, I., Moya, A., and Gonzales-Candelas, F. 2007. From phylogenetics to phylogenomics: the evolutionary relationships of insect endosymbiotic gamma-Proteobacteria as a test case. Systematic Biology 56: 1-16.

Cotton, J. A., and Page, R. D. 2002. Going nuclear: gene family evolution and vertebrate phylogeny reconciled. Proceedings of the Royal Society of London B 269: 1555-1561.

de Pinna, M. C. C. 1996. Teleostean monophyly. Pp. 193-207. In M. L. J. Stiassny, L. R. Parenti, and G. D. Johnson (eds), *Interrelationships of fishes*. Academic Press, San Diego.

Delarbre, C., Gallut, C., Barriel, V., Janvier, P., and Gachelin, G. 2002. Complete mitochondrial DNA of the hagfish, *Eptatretus burgeri*: the comparative analysis of mitochondrial DNA sequences strongly supports the cyclostome monophyly. Molecular Phylogenetics and Evolution 22: 184–192.

Delsuc, F., Brinkmann, H., Chourrout, D., and Philippe, H. 2006. Tunicates and not cephalochordates are the closest living relatives of vertebrates. Nature 439: 965-968.

Delsuc, F., Brinkmann, H., and Philippe, H. 2005. Phylogenomics and the reconstruction of the tree of life. Nature Reviews Genetics 6: 361-375.

Dimmick, W. W., and Larson, A. 1996. A molecular and morphological perspective on the phylogenetic relationships of the otophysan fishes. Molecular Phylogenetics and Evolution 6: 120-133.

Diogo, R. 2007. The Origin of Higher Clades:Osteology, Myology, Phylogeny and Evolution of Bony Fishes and the Rise of Tetrapods. Science Publishers, Enfield, NH.

Donoghue, P. C., Forey, P. L., and Aldridge, R. J. 2000. Conodont affinity and chordate phylogeny. Biological Reviews (Cambridge) 75: 191-251.

Driskell, A. C., Ane, C., Burleigh, J. G., McMahon, M. M., O'Meara, B. C., and M.J., S. 2004. Prospects for building the tree of life from large sequence databases. Science 306: 1172-1174.

Eisen, J. A., and Fraser, C. M. 2003. Phylogenomics: intersection of evolution and genomics. Science 300: 1706-1707.

Felsenstein, J. 1978. Cases in which parsimony or capatibility methods will be positively misleading. Systematic Zoology 27: 401-410.

Fink, S. V., and Fink, W. L. 1981. Interrelationships of the ostariophysan fishes (Teleostei). Zoological Journal of the Linnean Society 72: 297-353.

Fitch, W. M. 1970. Distinguishing homologous from analogous proteins. Systematic Zoology 19: 99-113.

Foster, P. G. 2004. Modeling compositional heterogeneity. Systematic Biology 53: 485-495.

Foster, P. G., and Hickey, D. A. 1999. Compositional bias may affect both DNA-based and protein-based phylogenetic reconstructions. Journal of Molecular Evolution 48: 284-290.

Furlong, R. F., and Holland, P. W. 2002. Bayesian phylogenetic analysis supports monophyly of ambulacraria and of cyclostomes. Zoological Science 19: 593-599.

Galtier, N., and Gouy, M. 1998. Inferring pattern and process: maximum-likelihood implementation of a nonhomogeneous model of DNA sequence evolution for phylogenetic analysis. Molecular Biology and Evolution 15: 871-879.

Gardiner, B. G., Schaeffert, B., and Masserie, J. A. 2005. A review of the lower actinopterygian phylogeny. Zoological Journal of the Linnean Society 144: 511-525.

Gee, H. 2003. Evolution: ending incongruence. Nature 425:782.

Gowri-Shankar, V., and Rattray, M. 2007. A reversible jump method for Bayesian phylogenetic inference with a nonhomogeneous substitution model. Molecular Biology and Evolution 24: 1286-99.

Grande, L. 2007. Morphology Based Phylogenetic Study of Gars, Basal Neopterygian Interrelationships, and the Resurrection of Holostei. American Society of Ichthyologists and Herpetologists, Abstracts of Annual Meetings. ASIH, St. Louis, Missouri, USA.

Grande, L., and Bemis, W. E. 1996. Interrelationships of Acipenseriformes, with comments on "Chondrostei". Pp. 85-115. In M. L. J. Stiassny, L. R. Parenti, and G. D. Johnson (eds), *Interrelationships of fishes.* Academic Press, San Diego.

Greenwood, P. H., Miles, R. S., and Patterson, C. (eds) 1973. Interrelationships of fishes. Academic Press, London.

Greenwood, P. H., Rosen, D. E., Weitzman, S. H., and Meyers, G. S. 1966. Phyletic studies of teleostean fishes, with a provisional classification of living forms. Bulletin of the American Museum National History 131: 339-456.

Gu, X., and Li, W.-H. 1996. Bias-Corrected Paralinear and LogDet Distances and Tests of Molecular Clocks and Phylogenies Under Nonstationary Nucleotide Frequencies. Molecular Biology and Evolution 13: 1375-1383.

Hardisty, M. W. 1982. Lampreys and hagfishes: analysis of cyclosome relationships. Pp. 165–258. In M. W. Hardisy, and I. C. Potter (eds), *The biology of Lampreys*. Academic Press, London.

Helfman, G. S., Collette, B. B., and Facey, D. E. 1997. The Diversity of Fishes. Blackwell Science, Malden, MA.

Hurley, I. A., Mueller, R. L., Dunn, K. A., Schmidt, E. J., Friedman, M., Ho, R. K., Prince, V. E., Yang, Z., Thomas, M. G., and Coates, M. I. 2007. A new time-scale for ray-finned fish evolution. Proceedings of the Royal Society of London B 274: 489-498.

Inoue, J. G., Miya, M., Tsukamoto, K., and Nishida, M. 2001. A mitogenomic perspective on the basal teleostean phylogeny: resolving higher-level relationships with longer DNA sequences. Molecular Phylogenetics and Evolution 20: 275-285.

Inoue, J. G., Miya, M., Tsukamoto, K., and Nishida, M. 2003. Basal actinopterygian relationships: a mitogenomic perspective on the phylogeny of the "ancient fish". Molecular Phylogenetics and Evolution 26: 110-120.

Ishiguro, N. B., Miya, M., and Nishida, M. 2003. Basal euteleostean relationships: a mitogenomic perspective on the phylogenetic reality of the "Protacanthopterygii". Molecular Phylogenetics and Evolution 27: 476-488.

Jaillon, O., Aury, J. M., Brunet, F., Petit, J. L., Stange-Thomann, N. *et al.* 2004. Genome duplication in the teleost fish Tetraodon nigroviridis reveals the early vertebrate proto-karyotype. Nature 431: 946-957.

Jamieson, B. G. M. 1991. Fish Evolution and Systematics: Evidence from Spermatozoa. Cambridge University Press, Cambridge. 319 pp.

Janvier, P. 1996. Early vertebrates. Clarendon Press, Oxford.

Jenner, R. A. 2004. Accepting Partnership by Submission? Morphological Phylogenetics in a Molecular Millennium. Systematic Biology 53: 333 - 342.

Jessen, H. 1972. Schultergürtel und Pectoralflosse bei Actinopterygiern. Fossils and Strata 1: 1–101.

Johnson, G. D. 1992. Monophyly of the euteleostean clades: Neoteleostei, Eurypterygii, and Ctenosquamata. Copeia: 8-25.

Johnson, G. D., and Patterson, C. 1993. Percomorph phylogeny: a survey of acanthomorphs and a new proposal. Bulletin of Marine Sciences 52: 554-626.

Johnson, G. D., and Patterson, C. 1996. Relationships of lower euteleostean fishes. Pp. 251-332. In M. L. J. Stiassny, L. R. Parenti, and G. D. Johnson (ed.), *Interrelationships of fishes*. Academic Press, San Diego.

Jukes, T. H., and Bhushan, V. 1986. Silent nucleotide substitutions and G + C content of some mitochondrial and bacterial genes. Journal of Molecular Evolution 24: 39-44.

Kawahara, R., Miya, M., Mabuchi, K., Lavoué, S., Inoue, J. G., Satoh, T. P., Kawaguchi, A., and Nishida, M. 2008. Interrelationships of the 11 gasterosteiform families (sticklebacks, pipefishes, and their relatives): A new perspective based on whole mitogenome sequences from 75 higher teleosts. Molecular Phylogenetics and Evolution 46: 224-236.

Kikugawa, K., Katoh, K., Shigehiro, K., Sakurai, H., Ishida, O., Iwabe, N., and Miyata, T. 2004. Basal jawed vertebrate phylogeny inferred from multiple nuclear DNA-coded genes. BMC Biology 2: 1-11.

Kocher, T. D., and Stepien, C. A. 1997. Molecular systematics of fishes. Academic Press, San Diego.

Kubatko, L. S., and Degnan, J. H. 2007. Inconsistency of phylogenetic estimates from concatenated data under coalescence. Systematic Biology 56: 17-24.

Kuraku, S., Hoshiyama, D., Katoh, K., Suga, H., and Miyata, T. 1999. Monophyly of lampreys and hagfishes supported by nuclear DNA-coded genes. Journal of Molecular Evolution 49: 729-735.

Lauder, G. V., and Liem, K. F. 1983. The evolution and interrelationships of the Actinopterygian fishes. Bulletin of the Museum of Comparative Zoology 150: 95-197.

Lê, H. L., Lecointre, G., and Perasso, R. 1993. A 28S rRNA-based phylogeny of the gnathostomes: first steps in the analysis of conflict and congruence with morphologically based cladograms. Molecular Phylogenetics and Evolution 2: 31-51.

Lecointre, G., and Nelson, G. 1996. Clupeomorpha, sister group of Ostariophysi. Pp. 193-207. In M. L. J. Stiassny, L. R. Parenti, and G. D. Johnson (ed.), *Interrelationships of fishes*. Academic Press, San Diego.

Li, C., Lu, G., and Ortí, G. 2008. Optimal Data Partitioning and a Test Case for Ray-finned Fishes (Actinopterygii) Based on Ten Nuclear Loci. Systematic Biology 57(4): 519-539.

Li, C., and Ortí, G. 2007. Molecular phylogeny of Clupeiformes (Actinopterygii) inferred from nuclear and mitochondrial DNA sequences. Molecular Phylogenetics and Evolution 44: 386-398.

Li, C., Ortí, G., Zhang, G., and Lu, G. 2007. A practical approach to phylogenomics: the phylogeny of ray-finned fish (Actinopterygii) as a case study. BMC Evolutionary Biology 7: 44.

Lockhart, P. J., Steel, M. A., Hendy, M. D., and Penny, D. 1994. Recovering evolutionary trees under a more realistic model of sequence evolution. Molecular Biology and Evolution 11: 605-612.

Lopez, A. J., Chen, W. J., and Ortí, G. 2004. Esociform phylogeny. Copeia 2004: 449-464.

Lopez, P., Casane, D., and Philippe, H. 2002. Heterotachy, an important process of protein evolution. Molecular Biology and Evolution 19: 1-7.

Lynch, M., and Conery, J. S. 2000. The evolutionary fate and consequences of duplicate genes. Science 290: 1151-1155.

Mabee, P. M., Ashburner, M., Cronk, Q., Gkoutos, G. V., Haendel, M., Segerdell, E., Mungall, C., and Westerfield, M. 2007. Phenotype ontologies: the bridge between genomics and evolution. Trends in Ecology and Evolution 22: 345-350.

Mallat, J. 1984. Early vertebrate evolution: Pharyngeal structure and the origin of gnathostomes. Journal of Zoology 204: 169–183.

Mallat, J. 1997. Hagfish do not resemble ancestral vertebrates. J. Morphol. 232: 293.

Mallat, J., and Winchell, J. 2007. Ribosomal RNA genes and deuterostome phylogeny revisited: More cyclostomes, elasmobranchs, reptiles, and a brittle star. Molecular Phylogenetics and Evolution 43: 1005-1022.

Mallatt, J., and Sullivan, J. 1998. 28S and 18S rDNA sequences support the monophyly of lampreys and hagfishes. Molecular Biology and Evolution 15: 1706-1718.

McMahon, M. M., and Sanderson, M. J. 2006. Phylogenetic supermatrix analysis of GenBank sequences from 2228 papilionoid legumes. Systematic Biology 55: 818-836.

Meyer, A., and Van de Peer, Y. 2005. From 2R to 3R: evidence for a fish-specific genome duplication (FSGD). Bioessays 27: 937-945.

Meyer, A., and Zardoya, R. 2003. Recent advances in the (molecular) phylogeny of vertebrates. Annual Review of Ecology, Evolution, and Systematics 34: 311-318.

Miya, M., and Nishida, M. 2000. Use of mitogenomic information in teleostean molecular phylogenetics: a tree-based exploration under the maximum-parsimony optimality criterion. Molecular Phylogenetics and Evolution 17: 437-455.

Miya, M., Satoh, T. P., and Nishida, M. 2005. The phylogenetic position of toadfishes (order Batrachoidiformes) in the higher ray-finned fish as inferred from partitioned Bayesian analysis of 102 whole mitochondrial genome sequences. Biological Jourbnal of the Linnean Society of London 85: 289-306.

Miya, M., Takeshima, H., Endo, H., Ishiguro, N. B., Inoue, J. G., Mukai, T., Satoh, T. P., Yamaguchi, M., Kawaguchi, A., Mabuchi, K., Shirai, S. M., and Nishida, M. 2003. Major patterns of higher teleostean phylogenies: a new perspective based on 100 complete mitochondrial DNA sequences. Molecular Phylogenetics and Evolution 26: 121-138.

Nelson, G. J. 1969. Gill arches and the phylogeny of fishes, with notes on the classification of vertebrates. Bulletin of the American Museum National History 141: 475-552.

Nelson, J. S. 2006. *Fishes of the world*, 4th edition. John Wiley and Sons, Inc., New York. 601 pp.

Ohno, S. 1970. *Evolution by Gene Duplication*. Springer-Verlag. 160 pp.

Olsen, P. E. 1984. The skull and pectoral girdle of the parasemionotid fish *Watsonulus eugnathoides* from the Early Triassic Sakamena Group of Madagascar, with comments on the relationships of the holostean fishes. Journal of Vertebrate Paleontology 4: 481-499.

Ortí, G., Chen, W. J., Mahon, A., Holcroft Benson, N., Unmack, P., and others. 2005. Phylogeny of the ray-finned fish (Actinopterygii): Analysis of more than 500 RAG-1 exon 3 sequences. American Society of Ichthyologists and Herpetologists, Abstracts of Annual Meeting. ASIH, Tampa, Florida, USA.

Ortí, G., and Li, C. 2007. Chondrostei, Holostei, Teleostei? Nuclear DNA Sequence Data. American Society of Ichthyologists and Herpetologists, Abstracts of Annual Meeting. ASIH, St. Louis, Missouri, USA.

Ortí, G., and Meyer, A. 1996. Molecular Evolution of Ependymin and the Phylogenetic Resolution of Early Divergences Among Euteleost Fishes. Molecular Biology and Evolution 13: 556-573.

Ota, K. G., and Kuratani, S. 2007. Cyclostome embryology and early evolutionary history of vertebrates. Integrative and Comparative Biology 47: 329-337.

Page, R. D., and Cotton, J. A. 2002. Vertebrate phylogenomics: reconciled trees and gene duplications. Pacific Symposium of Biocomputing 2002: 536-547.

Pamilo, P., and Nei, M. 1988. Relationships Between Gene Trees and Species Trees. Molecular Biology and Evolution 5: 568-583.

Pancer, Z., Amemiya, C. T., Ehrhardt, G. R., Ceitlin, J., Gartland, G. L., and Cooper, M. D. 2004. Somatic diversification of variable lymphocyte receptors in the agnathan sea lamprey. Nature 430.

Pancer, Z., Saha, N. R., Kasamatsu, J., Suzuki, T., Amemiya, C. T., Kasahara, M., and Cooper, M. D. 2005. Variable lymphocyte receptors in hagfish. Proceedings of the National Academy of Sciences USA 102: 9224-9229.

Patterson, C. 1973. Interrelationships of holosteans. Pp. 207-226. In P. H. Greenwood, R. S. Miles, and C. Patterson (eds), *Interrelationships of fishes*. Academic Press, London.
Patterson, C., and Johnson, G. D. 1995. The intermuscular bones and ligaments of teleostean fishes. Smithsonian Contributions of Zoology 559: 1-85.
Patterson, C., and Rosen, D. E. 1977. Review of ichthyodectiform and other Mesozoic teleost fishes and the theory and practice of classifying fossils. Bulletin of the American Museum National History 158: 81-172.
Patterson, C., and Rosen, D. E. 1989. The Paracanthopterygii revisited: order and disorder. Pp. 5-36. In D. M. Cohen (ed.), *Papers on the systematics of gadiform fishes*. Natural History Museum of Los Angeles County, Los Angeles, California.
Philippe, H., Snell, E. A., Bapteste, E., Lopez, P., Holland, P. W., and Casane, D. 2004. Phylogenomics of eukaryotes: impact of missing data on large alignments. Molecular Biology and Evolution 21: 1740-1752.
Phillips, M. J., Delsuc, F., and Penny, D. 2004. Genome-scale phylogeny and the detection of systematic biases. Molecular Biology and Evolution 21: 1455-1458.
Rast, J. P., Smith, L. C., Loza-Coll, M., Hibino, T., and Litman, G. W. 2006. Genomic Insights into the Immune System of the Sea Urchin. Science 314: 952-956.
Regan, C. T. 1923. The skeleton of *Lepidosteus*, with remarks on the origin and evolution of the lower neopterygian fishes. Proceedings of the Zoological Society: 445-461.
Remm, M., Storm, C. E., and Sonnhammer, E. L. 2001. Automatic clustering of orthologs and in-paralogs from pairwise species comparisons. Journal of Molecular Biology 314: 1041-1052.
Rokas, A., Kruger, D., and Carroll, S. B. 2005. Animal evolution and the molecular signature of radiations compressed in time. Science 310: 1933-1938.
Rokas, A., Williams, B. L., King, N., and Carroll, S. B. 2003. Genome-scale approaches to resolving incongruence in molecular phylogenies. Nature 425: 798-804.
Rosen, D. E. 1973. Interrelationships of higher euteleostean fishes. . Pp. 397-513. In P. H. Greenwood, R. S. Miles, and C. Patterson (eds), *Interrelationships of Fishes*. Academic Press, London.
Rosen, D. E. 1982. Teleostean interrelationships, morphological function and evolutionary inference. American Zoologist 22: 261-273.
Saitoh, K., Miya, M., Inoue, J. G., Ishiguro, N. B., and Nishida, M. 2003. Mitochondrial genomics of ostariophysan fishes: perspectives on phylogeny and biogeography. Journal of Molecular Evolution 56: 464-472.
Schaeffer, B. 1973. Interrelationships of chondrosteans. Pp. 207-226. In P. H. Greenwood, R. S. Miles, and C. Patterson (eds), *Interrelationships of fishes*. Academic Press, London.
Scotland, R. W., Olmstead, R. G., and Bennett, J. R. 2003. Phylogeny Reconstruction: The Role of Morphology. Systematic Biology 52: 539-548.
Shen, M. 1996. Fossil "osteoglossomorphs" in East Asia and their implications in teleostean phylogeny. Pp. 261-272. In G. Arratia, and G. Viohl (eds), *Mesozoic fishes: systematics and plaleoecology*. Verlag Dr. F. Pfeil, München, Germany.
Smith, N. D., and Turner, A. H. 2005. Morphology's Role in Phylogeny Reconstruction: Perspectives from Paleontology. Systematic Biology 54: 166-173.
Smith, W. L., and Wheeler, W. C. 2006. Venom evolution widespread in fishes: a phylogenetic road map for the bioprospecting of piscine venoms. Journal of Heredity 97: 206-217.

Soltis, D. E., Albert, V. A., Savolainen, V., Hilu, K., Qiu, Y. L., Chase, M. W., Farris, J. S., Stefanovic, S., Rice, D. W., Palmer, J. D., and Soltis, P. S. 2004. Genome-scale data, angiosperm relationships, and "ending incongruence": a cautionary tale in phylogenetics. Trends in Plant Science 9: 477-483.

Springer, V. G., and Johnson, G. D. 2004. Study of the Dorsal Gill-arch Musculature of Teleostome Fishes, with Special Reference to the Actinopterygii. Bulletin of the Biological Society of Washington 11: 260.

Steel, M. A., Lockhart, P. J., and Penny, D. 1993. Confidence in evolutionary trees from biological sequence data. Nature 364: 440-442.

Stiassny, M. L. J. 1986. The limits and relationships of acanthomorph teleosts. Journal of Zoology B 1: 411-460.

Stiassny, M. L. J., Parenti, L. R., and Johnson, G. D. (eds) 1996. *Interrelationships of fishes*. Academic Press, San Diego. 500 pp.

Stiassny, M. L. J., Wiley, E. O., Johnson, G. D., and de Carvalho, M. R. 2004. Gnathostome fishes. Pp. 410-429. In J. Cracraft, and M. J. Donoghue (eds), *Assembling The Tree of Life*. Oxford University Press, New York.

Stock, D. W., and Whitt, G. S. 1992. Evidence from 18S ribosomal RNA sequences that lampreys and hagfishes form a natural group. Science 257: 787–789.

Takezaki, N., Figueroa, F., Zaleska-Rutczynska, Z., and Klein, J. 2003. Molecular phylogeny of early vertebrates: monophyly of the agnathans as revealed by sequences of 35 genes. Molecular Biology and Evolution 20: 287-292.

Taylor, J. S., and Raes, J. 2004. Duplication and Divergence: The Evolution of New Genes and Old Ideas. Annual Review of Genetics 9: 615-643.

Van de Peer, Y., Taylor, J. S., and Meyer., A. 2003. Are all fishes ancient polyploids? Journal of Structural and Functional Genomics 3: 65-73.

Venkatesh, B., Erdmann, M. V., and Brenner, S. 2001. Molecular synapomorphies resolve evolutionary relationships of extant jawed vertebrates. Proceedings of the National Academy of Sciences USA 98: 11382-11387.

Waddell, P. J. 2005. Measuring the fit of sequence data to phylogenetic model: allowing for missing data. Molecular Biology and Evolution 22: 395-401.

Weisburg, W. G., Giovannoni, S. J., and Woese, C. R. 1989. The Deinococcus-Thermus phylum and the effect of rRNA composition on phylogenetic tree construction. Systematic and Applied Microbiology 11: 128-134.

Wiens, J. J. 2003. Missing data, incomplete taxa, and phylogenetic accuracy. Systematic Biology 52: 528-538.

Wiens, J. J. 2004. The Role of Morphological Data in Phylogeny Reconstruction. Systematic Biology 53: 653-661.

Woese, C. R., Achenbach, L., Rouviere, P., and Mandelco, L. 1991. Archaeal phylogeny: reexamination of the phylogenetic position of Archaeoglobus fulgidus in light of certain composition-induced artifacts. Systematic and Applied Microbiology 14: 364-371.

Woods, I. G., Wilson, C., Friedlander, B., Chang, P., Reyes, D. K., Nix, R., Kelly, P. D., Chu, F., Postlethwait, J. H., and Talbot, W. S. 2005. The zebrafish gene map defines ancestral vertebrate chromosomes. Genome Res 15: 1307-1314.

Yang, Z., and Roberts, D. 1995. On the use of nucleic acid sequences to infer early branchings in the tree of life. Molecular Biology and Evolution 12: 451-458.

CHAPTER 2

The Ovary, Folliculogenesis, and Oogenesis in Teleosts

Harry J. Grier[1], *Mari Carmen Uribe Aranzábal*[2], *Reynaldo Patiño*[3]

2.1 INTRODUCTION

Oogenesis has been defined as the "processes involved in the growth and maturation of the ovum in preparation for fertilization."* Some biological definitions of oogenesis have diverged from the term's etymological origin, especially for teleosts where oogenesis has been sometimes narrowly defined as the transformation of oogonia into oocytes (deVlaming 1974; Selman and Wallace 1989; Wallace and Selman 1990; Tyler and Sumpter 1996; Khan and Thomas 1999). This chapter uses the term oogenesis to describe the morphological and functional processes by which oogonia transform into fertilizable eggs. This broader definition is consistent with the original meaning of the term and with its use in the current comparative reproduction literature (Lombardi 1998; Lessman 1999; Picton and Gosden 1999; Patiño and Sullivan 2002).

Dodd (1986) indicated that "ovarian terminology is confused and confusing," and Patiño and Sullivan (2002) pointed out that this "criticism is still relevant (in 2002)". Inconsistent or confusing terminology continues to be used today, not only in the basic fields of teleost ovarian physiology and morphology but also in the applied fisheries sciences (Blazer 2002). The "terminology may be confusing

[1]Division of Fishes, Department of Zoology, National Museum of Natural History, MRC 159, Smithsonian Institution, P.O. Box 3712, Washington D.C. 20012-7012, USA, and Florida Fish and Wildlife Research Institute, 100 8th Avenue, SE, St. Petersburg, FL 33701-5020, USA

[2]Laboratorio de Biología de la Reproducción, Departamento de Biología Comparada, Facultad de Ciencias, Universidad Nacional Autónoma de México, México, D.F. 04510 México

[3]U.S. Geological Survey Texas Cooperative Fish and Wildlife Research Unit and Departments of Natural Resources Management and of Biological Sciences, Texas Tech University, Lubbock, TX 79409-2120, USA

*McGraw-Hill Dictionary of Scientific and Technical Terms, McGraw-Hill Book Co., 1994.

due to our ignorance of comparative morphology, the use of synonyms, the still-emerging physiological and molecular events (regulating) oocyte growth, the now documented role of a germinal epithelium in follicle formation..." (Parenti and Grier 2004). Unfortunately, the lack of a generally accepted and consistent ovarian terminology limits communication, especially across fields. As noted by Patiño *et al.* (2001), confusion in the definition of terms can also restrict progress in our understanding of ovarian phenomena. Recent research has indicated that ovarian germinal epithelia and cellular and structural aspects of folliculogenesis are emerging constants throughout the vertebrates (Parenti and Grier 2004; Grier *et al.* 2007), thus indicating an important role for homology in the development of consistent nomenclature. Concepts and terminology should adapt and change as new knowledge is generated. But despite occasional calls for update and standardization (Dodd 1986; Patiño *et al.* 2001; Patiño and Sullivan 2002), there has been no deliberate assessment of teleost ovarian terminology in the context of ovarian functional morphology. In addition to describing current basic knowledge of the morphology and regulatory endocrinology of oogenesis, this chapter attempts to develop a standard terminology for oocyte and follicular development that can be used as a guide for studies of teleost oogenesis. The extent to which this terminology can be applied to fishes other than teleosts and also other vertebrates awaits future documentation.

2.2 OVARIAN STRUCTURE

In most teleosts, there are two ovaries, suspended dorsally within the coelom by the mesovarium. In some species there is only one ovary as the result of the fusion of both ovaries during embryological development, as in Medaka (*Oryzias latipes*) (Strüssmann and Nakamura 2002), most viviparous teleosts such as poeciliids and goodeids (Dodd 1977; Wourms 1981; Lombardi 1998; Uribe *et al.* 2005), and Swamp Eel (*Synbranchus marmoratus*) (Ravaglia and Maggese 2002). The somatic layer (epithelium) of the developing gonad is derived from germinal ridge epithelium, which in turn is derived from coelomic epithelium (Hoar 1969; Devlin and Nagahama 2002; Strüssmann and Nakamura 2002). The ovary of most teleosts is a saccular structure which contains a lumen or ovocoel. The developmental morphology of this saccular ovary in teleosts is unique among vertebrates. Although the initial development of the gonads occurs by extrusion of paired genital ridges as in other vertebrates, somatic outgrowths develop in the early ovary of these teleosts that then fuse to form the ovocoel. Fusion occurs between paired lateral outgrowths in each ovary, between a single lateral outgrowth from each ovary and the abdominal wall, or between lateral outgrowths in fused ovaries (one from each side of the fused ovary) and the abdominal wall (Strüssmann and Nakamura 2002). It is this developmental process that establishes the saccular structure of the ovary, the cystovarian condition (Hoar 1969; Dodd 1986). However, the ovaries of some groups of teleosts do not form an ovocoel, such as salmonids (Kendall 1921; Robertson 1953) and anguillids (Colombo *et al.* 1984). Ovaries without an ovocoel represent the gymnovarian condition. In cystovaries, lamellae form at the periphery of the ovocoel and protrude into it; in gymnovaries, lamellae form only on the germinal side of the developing ovary (Robertson 1953; Colombo *et al.* 1984). The lamellar epithelium becomes populated by germ cells and is known as the germinal epithelium.

In the adult ovary, the lamellae are irregular in shape. Ovarian lamellae are composed of the germinal epithelium and the stroma subjacent to the epithelium (Figs. 2.1A,B, 2.2A-D, 2.3A-D). A basement membrane supports the

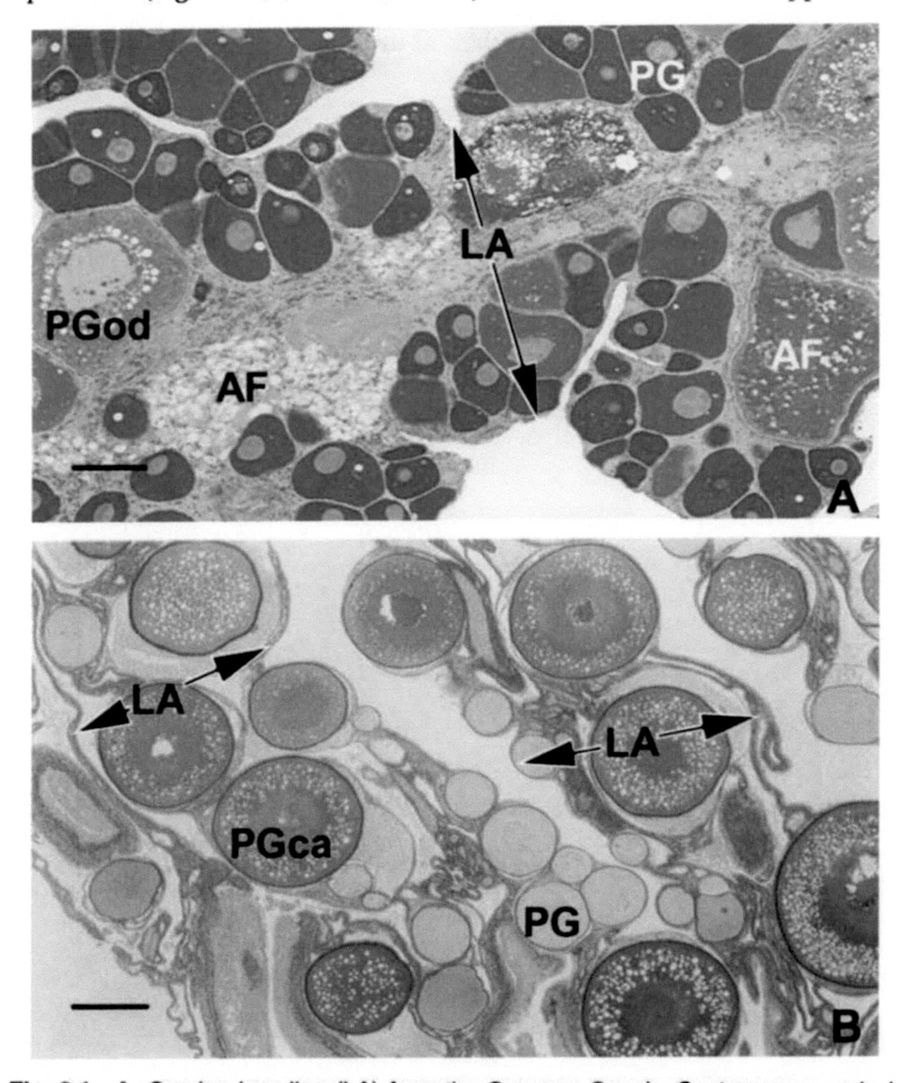

Fig. 2.1 A. Ovarian lamellae (LA) from the Common Snook, *Centropomus undecimalis*, from a regressing individual. Primary growth oocytes (PG) with basophilic ooplasm and some with oil droplets (PGod) are observed along with follicles whose oocytes are atretic (AF). Bar 100 µm. **B**. Lamella (LA) of developing ovary from Rainbow Trout, *Oncorhynchus mykiss*, with cortical alveolar primary growth oocytes (PGca) and less developed primary growth oocytes (PG). Bar 100 µm.

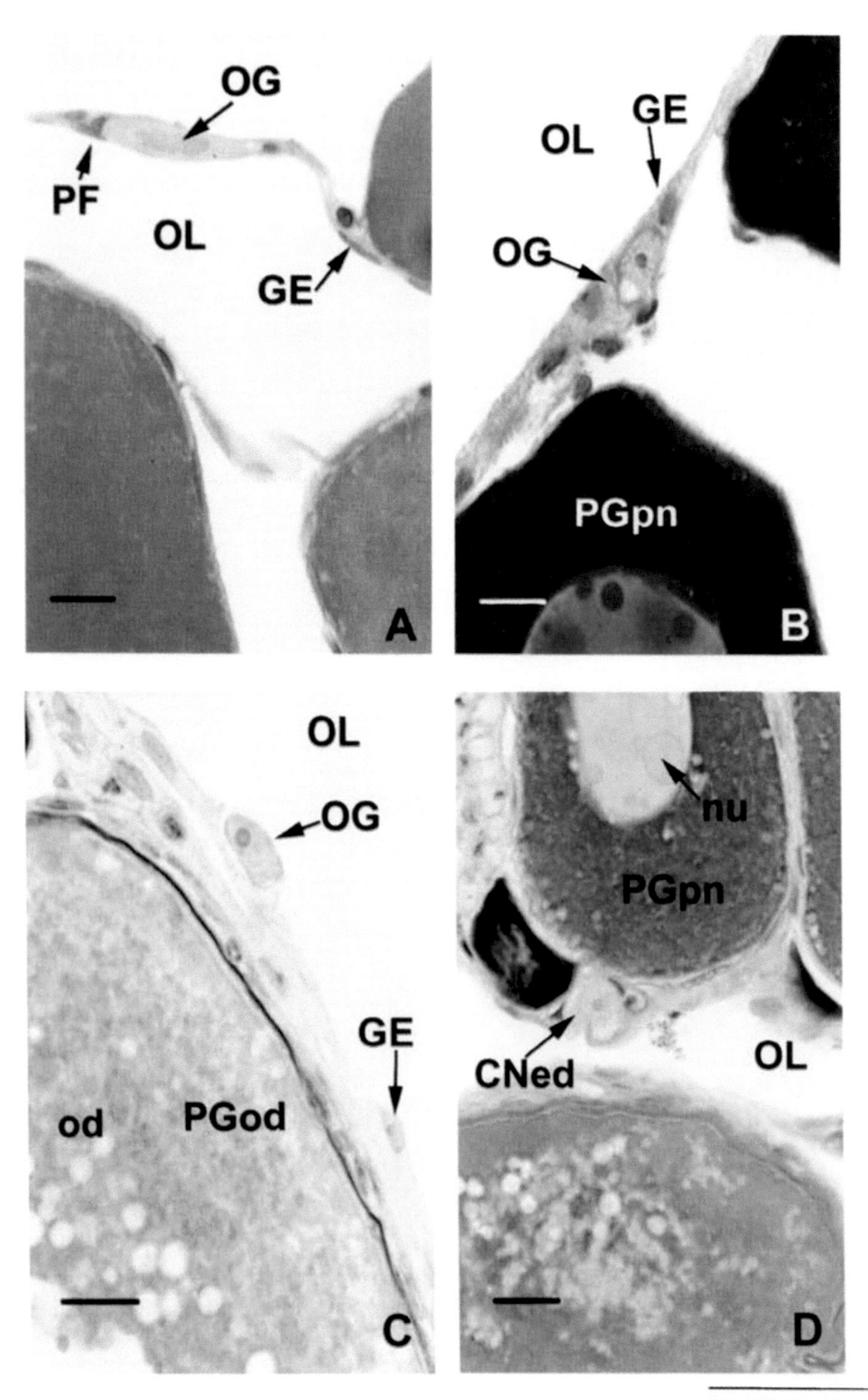

Fig. 2.2 Contd. ...

germinal epithelium (Fig. 2.3A-C) and also separates the two ovarian compartments. The germinal epithelium contains two cell types: somatic epithelial cells (Fig. 2.2A), originally derived from coelomic epithelium, and scattered germ cells (oogonia and early oocytes) (Figs. 2.2A-C, 2.3A,B). Within the stroma, the connective tissue contains diverse types of cells as undifferentiated, mesenchyme-like cells, fibroblasts and granulocytes, collagen fibers, and blood vessels.

Ovulation occurs into the ovocoel in adult cystovaries, and into the coelom in gymnovaries. Teleosts do not develop Müllerian ducts, unlike most other vertebrates. Consequently, in the cystovarian condition, the communication of the ovary to the exterior is through the caudal portion of the ovary, the gonoduct, which opens to the exterior at the genital pore (Hoar 1969; Dodd 1986; Wourms 1981; Nagahama 1983; Wake 1985; Patiño and Takashima 1995; Lombardi 1998; Uribe *et al.* 2005). In the gymnovarian condition, the ovulated eggs pass from the coelom to a discontinuous gonoduct that develops a caudal funnel to the genital pore (Kendall 1921; Lombardi 1998).

2.2.1 Germinal Epithelium

The germinal epithelium consists primarily of a monolayer of cells (Figs. 2.2A-D, 2.3A). In those regions where cell nests develop and folliculogenesis is an ongoing process, the germinal epithelium becomes multilayered (Fig. 2.3B). Cell nests consist of clusters of germ cells (oogonia, chromatin nucleolus oocytes [Fig. 2.3B,C], and oocytes beginning primary growth) and prefollicle cells (Fig. 2.3C). Prefollicle cells are epithelial cells that become associated with oogonia (Figs. 2.2A,B, 2.3A) or chromatin nucleolus oocytes (Fig. 2.3B,C). Considering the evolution and phylogeny of teleosts, Parenti and Grier (2004) described the ovarian germinal epithelium, at the level of light microscopy, in several species of atherinomorph teleosts including Gulf Killifish, *Fundulus grandis* (Cyprinodontidae), Ballyhoo, *Hemiramphus brasiliensis* (Hemiramphidae); and later Grier *et al.* (2005) described it in *Gnatholebias hoignei* (Cyprinodontidae), the viviparous atherinomorphs, Balsas Splitfin, *Ilyodon whitei* (Goodeidae), Sailfin Molly, *Poecilia latipinna*, Ten Spotted Live-bearer, *Cnesterodon decemmaculatus*

Fig. 2.2 Contd. ...

Fig. 2.2 Germinal epithelia (GE) border the ovarian lumen (OL). **A.** Germinal epithelium from the Goldfish, *Carassius auratus*, shows an oogonium (OG) flanked by prefollicle cells (PF). Bar 10 µm. **B.** Germinal epithelium from Common Snook, showing a single oogonium. A perinucleolar oocyte (PGpn) is seen. Bar 10 µm. **C.** Germinal epithelium from the Common Snook, with a single oogonium (OG). A follicle with a primary growth oocyte (PGod) having multiple oil droplets (od). Bar 10 µm. **D.** An early diplotene oocyte in the chromatin nucleolus stage (CNed) of growth is flanked by prefollicle cells whose nuclei are darkly-stained. A primary growth oocyte with nucleoli (nu) in the perinucleolar step of oocyte growth is portrayed (PGpn). Bar 10 µm. Original.

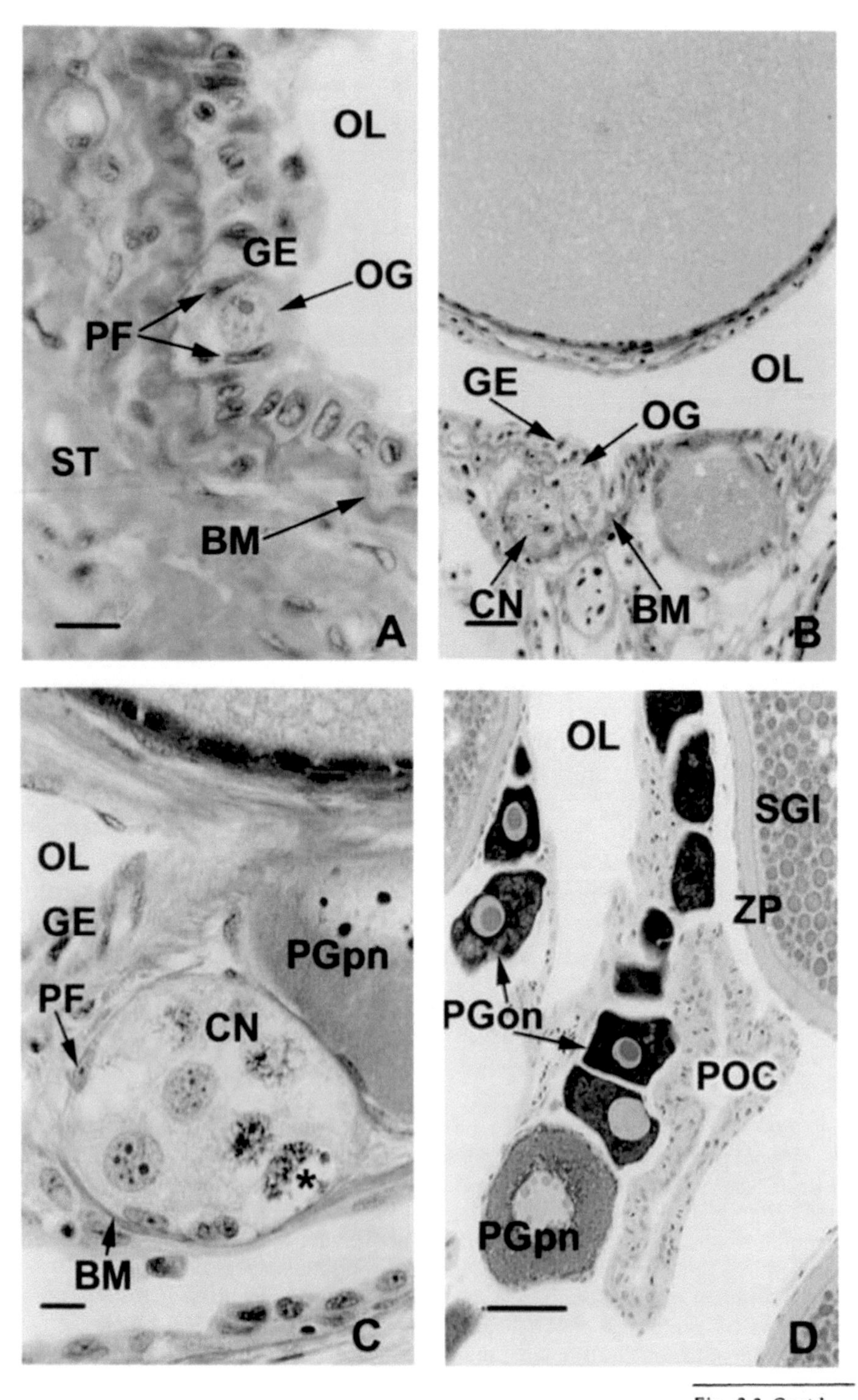
OL
GE
OG
PF
ST
BM
A
OL
GE
OG
CN
BM
B
OL
GE
PGpn
PF
CN
*
BM
C
OL
SGI
ZP
PGon
POC
PGpn
D

Fig. 2.3 Contd. ...

(Poeciliidae), Four-eyed Fish, *Anableps anableps* (Anablepidae), and the viviparous ophidiiform, *Dinematichthys* sp.

Ultrastructural analysis of the germinal epithelium in Common Snook (*Centropomus undecimalis*) (Grier 2000) has contributed to an understanding of the structural aspects of folliculogenesis, including differentiation of somatic and germ cells during folliculogenesis. As with all epithelia, the ovarian germinal epithelium "rests" upon a basement membrane which, under certain conditions, is periodic acid Schiff (PAS)-positive indicating the presence of polysaccharides. However, the PAS reaction may yield inconsistent results depending on the species or the fixative used for tissue processing. For example, in Rainbow Trout (*Oncorhynchus mykiss*), where egg diameters are between 4.5 mm and 5.5 mm, the basement membranes surrounding follicles are particularly well-stained after Bouin's or formalin fixation followed by the PAS procedure. The consistent staining of Rainbow Trout basement membranes was instrumental in revealing the clear separation between the germinal epithelium and its derivatives (cell nests, follicles throughout oocyte development, and postovulatory follicle) from the stroma, from which the theca is derived (Grier *et al.* 2007). However, the basement membrane is much more difficult to visualize in teleosts that produce smaller, pelagic eggs; in these follicles, basement membranes may or may not stain. There also seems to be a gradation in the PAS staining following the use of different fixatives with Bouin's producing the most intense staining, as in Lo Nostro *et al.* (2003), buffered formalin being intermediate, and Trump's fixative, a mixture of formaldehyde and glutaraldehyde, normally producing little if any demonstration of basement membranes but a superior fixation overall. Examination of basement membranes is essential to the understanding of folliculogenesis and the timing of follicle formation in the teleost ovary, although the process of folliculogenesis within a framework of a germinal epithelium appears to be the same among vertebrates (Parenti and Grier 2004).

Fig. 2.3 Contd. ...

Fig. 2.3 Germinal epithelium (GE) from Rainbow Trout. **A-C**. The germinal epithelium, supported by a basement membrane (BM) that separates if from the stroma (ST), borders the ovarian lumen (OL) which is in continuum with the coelom. Bar 10 µm. PAS, metanil yellow, hematoxylin. **B**. In the germinal epithelium, oogonia (OG) divide to produce cell nests (CN) with secondary oogonia. Ovarian lumen (OL). Bar 10 µm. PAS, metanil yellow, hematoxylin. **C**. A cell nest (CN) of chromatin nucleolus oocytes is surrounded by a basement membrane (BM) and has prefollicle cells (PF). An oocyte displaying a bouquet distribution of chromosomes (Asterisk) is observed in the nest. Perinucleolar step oocyte (PGpn) and ovarian lumen (OL). Bar 10 µm. PAS, metanil yellow, hematoxylin. **D**. Follicles from the Common Snook, with one-nucleolus oocytes (PGon) are beneath the germinal epithelium. A late secondary growth oocyte (SGl) with a well-developed zona pellucida (ZP) and a postovulatory follicle complex (POC) are observed. Bar 10 µm. PAS, metanil yellow, hematoxylin. Original.

2.3 OVARIAN FOLLICLE

Sensu stricto, the teleost ovarian follicle consists of an oocyte and encompassing follicle cells. This definition differs from other recent definitions for teleosts that included the basement membrane (Patiño and Sullivan 2002) and even the theca (Le Menn *et al.* 2007), but it is consistent with definitions used in the comparative reproduction literature (Lombardi 1998) and general histology texts (Grier and Lo Nostro 2000). In teleosts, follicle cells are arranged in a single layer enveloping the oocyte. Between the oocyte and the follicle cells is an acellular layer, appropriately termed "zona pellucida" in teleosts as previously noted by Guraya (1986) and Lo Nostro *et al.* (2003), but frequently referred to as the "chorion" (a term which this chapter does not favor). (In Murata, **Volume 8B, Chapter 7**, the term "egg envelope" has been substituted for zona pellucida or chorion). The nomenclature for proteins of the zona pellucida has been recently revised; the term now used to describe these proteins in teleost follicles is zona pellucida (ZP) proteins (Modig *et al.* 2007). The basement membrane separates the follicle from the stroma and stromal derivatives such as the theca (Grier 2000; Grier *et al.* 2007).

In this chapter, the term "folliculogenesis" is used to describe the formation of discrete follicle units, composed of the oocyte and encompassing layer of follicle cells, which are completely separated from germinal epithelium and stroma by a basement membrane. Folliculogenesis takes place early during oogenesis when the oocyte and encompassing prefollicle cells become fully enclosed within a basement membrane (Grier 2000). This definition of folliculogenesis is similar to that used for the formation of the mammalian primordial follicle (Guraya 1985). As the teleost follicle is surrounded by theca, the entire structure becomes the ovarian follicle complex composed of the oocyte, follicle cells, basement membrane and theca (Grier 2000; Grier and Lo Nostro 2000). The theca is wholly derived from the stroma in ovarian follicles of vertebrates (Tokarz 1978; Wallace and Selman 1990). Growth and differentiation of the follicle complex is a continuous process during oogenesis. After expulsion of the egg at ovulation, the remaining structure, the postovulatory follicle complex (POC), consists of postovulatory follicle cells, basement membrane and theca. The term postovulatory follicle (POF) has been used previously to define what this chapter calls POC, but as used in this chapter POF refers only to the postovulatory follicle cells present in the POC (without the oocyte, the ovarian follicle consists only of follicle cells). Thus, the terminology used here to describe the formation of the teleost ovum and ovarian follicle is such that, depending on the particular topic of research, the oocyte, the follicle, or the follicle complex can be referenced unambiguously. For example, in fish reproductive life history studies (fisheries), the staging of oocytes is a critical concern. In physiological studies, the ovarian follicle and sometimes the ovarian follicle complex often are the units of attention.

Accumulating evidence in teleosts indicates that throughout its growth and differentiation the follicle remains attached to the germinal epithelium by a region of shared basement membrane, as first observed in the goodeid Balsas Splitfin (Grier *et al.* 2005) and the Rainbow Trout (Grier *et al.* 2007). Thus, the ovarian follicle does not completely "pinch off" the basement membrane after folliculogenesis is completed, and consequently the theca never fully encompasses the follicle. Possible exceptions to this scenario may include the unique ovary of syngnathids, where the distribution of oogonia in the ovocoel epithelium seems to be restricted to a narrow ridge extending along the length of the ovary, which is the equivalent of the germinal epithelium in the typical teleost ovary. Ovarian follicles in these species grow as they migrate in clockwise fashion around the ovary, sandwiched between the ovocoel epithelium and the ovarian surface (Wallace and Selman 1990). However, the association of the developing follicle with the (germ cell-less) ovocoel epithelium has not been closely examined in syngnathids, and the possibility of shared basement membranes cannot be discounted for these species. In Rainbow Trout, the basement membrane always separates the germinal epithelium and its derivatives (cell nests, follicles and postovulatory follicles) from the stroma (Grier *et al.* 2007). Thus, germ cells, somatic epithelial cells, and follicle cells are never in direct contact with stroma in the adult ovary. Basement membranes should be recognized as structures that serve not only to support epithelia but also to separate them from other compartments, in this case the ovarian stroma. This is a new concept in the organization of the teleost ovary that has been documented in Rainbow Trout (Grier *et al.* 2007). Another new concept is the "shared" basement membrane between the teleost ovarian follicle and the germinal epithelium as in Balsas Splitfin (Grier *et al.* 2005) and Rainbow Trout (Grier *et al.* 2007). The description of folliculogenesis in this chapter assumes that these novel observations apply to all teleosts but studies of other species will be necessary to document this assumption.

When located beneath the basement membrane of the germinal epithelium, stromal cells have been called prethecal cells (Grier 2000). These cells are distinguished from other stromal cells only by location. They form a continuum of cells stretching from beneath the germinal epithelium and developing cell nests. Stromal cells form a reticular meshwork, being attached to each other by desmosomes (Grier unpublished). The theca is intimately involved with the development of the oocyte as capillaries develop within the theca interna (Grier 2000), and nutrients diffuse across the basement membrane as well as vitellogenin, a yolk precursor (Selman and Wallace 1989; Patiño and Sullivan 2002). Special thecal cells have been described in the theca of a number of teleosts (Guraya 1986), where they interact with follicle cells to produce steroid hormones. For example, special theca cells produce 17β-hydroxyprogesterone in the maturing ovarian follicle of salmonids, which is then transferred across the basement membrane to the follicle cells for conversion to a progestin with maturational activity, 17α, 20β-dihdroxy-4-pregnen-3-one (DHP) (Nagahama 1987). Thus, nutritionally and functionally, the follicular and stromal tissue compartments of the ovary are closely integrated despite their separation by a basement membrane. Considering the role of stromal cells in forming the theca, it is hypothesized here that special thecal cells are derived from undifferentiated stromal cells, as are the theca interna and the theca externa.

Juxtaposed follicles containing primary growth oocytes (see below) may share a single thecal cell between them. Initially the theca is composed of only one cell layer over the surface of the follicle, as in those follicles containing early perinucleolar oocytes. Multiple cell layers of theca exist between follicles during advanced secondary growth or in full-grown follicles, three being described in Common Snook (Grier 2000). Begovac and Wallace (1987) were the first to describe shared theca between follicles in the Gulf Pipefish (*Syngnathus scovelli*), indicating it to be unique, but this phenomenon is also observed in Red Drum (*Sciaenops ocellatus*) (Grier unpublished) and may be common among teleosts. Capillaries in the theca increase in number during oogenesis.

Intrafollicular communication also occurs within the growing ovarian follicle. Microvilli extend from the oocyte surface and, through pore canals of the zona pellucida, make contact with the overlaying follicle cells. Cytoplasmic continuity via heterologous gap junctions is often observed at the points of contact, thus allowing for close functional coordination between the oocyte and follicle cells (Patiño *et al.* 2001; Patiño and Sullivan 2002). Extracellular paracrine pathways of intrafollicular communication also exist and this communication is bidirectional, oocyte to follicle cells and vice versa (Wang and Ge 2004).

2.4 OVARIAN CYCLES AND OOGENESIS

A single wave of oogenic development is observed in the ovary of semelparous species such as Pacific salmonids, which reproduce only once in their lifetime; whereas oogenic "cycles" overlaid by daily, lunar, or seasonal environmental cycles are seen in most iteroparous species. Depending on the combination of oogenic and environmental cycles adopted by a particular species, three general patterns of ovarian development have been described in teleosts (de Vlaming 1974; Blazer 2002). For example, in semelparous species, the single wave of oogenic development yields a synchronous pattern of ovarian development, where all oocytes present in the ovary are at the same stage of oogenesis. Many temperate species with relatively short breeding periods and consisting of individuals that spawn only once per year exhibit a group-synchronous pattern of ovarian development with at least two germ cell stages (other than proliferating oogonia) present at any given time – primary and secondary growth as indicated in Table 2.1. Asynchronous ovarian development, where a continuum of stages of oogenesis are present simultaneously, is observed in species with protracted spawning periods consisting of individuals that spawn more than once in a breeding season, as has been reported for the Mummichog (*Fundulus heteroclitus*), where the ovary in reproductively active individuals appears to be a "random mixture of oocytes in every conceivable stage" (Wallace and Selman 1981).

Table 2.1 Oocyte development in fish is divided into three levels: periods, stages, and steps. Periods have five divisions that are common to all vertebrates. The six Stages are common to fish and other vertebrates that produce yolked eggs. Stages are subdivided into steps: those that are common pertain to the germinal vesicle (nucleus). Variable ooplasmic events are added in oocyte maturation, indicated with italic letters, which applies to those fish species whose pelagic eggs possess an oil globule. Upper case letters represent Stages while lower case letters represent Steps. The Codes, abbreviations for Stages and Steps, used in illustrations, are listed on the left of the table. Original.

CODES	STEPS	STAGES	PERIODS
OP	FREQUENTLY FORM CELL NESTS	OOGONIA PROLIFERATE	MITOSIS
CNl	LEPTOTENE	CHROMATIN NUCLEOLUS (CN)	ACTIVE MEIOSIS I
CNz	ZYGOTENE		
CNp	PACHYTENE		
CNed	EARLY DIPLOTENE		
PGon	ONE-NUCLEOLUS	PRIMARY GROWTH: (PG)	ARRESTED MEIOSIS I LATE DIPLOTENE OF THE PROPHASE I
PGmn	MULTIPLE NUCLEOLI		
PGpn	PERINUCLEOLAR		
PGod	CIRCUMNUCLEAR OIL DROPLETS		
PGca	CORTICAL ALVEOLAR		
SGe	EARLY SECONDARY GROWTH OR EARLY YOLKED OOCYTES	SECONDARY GROWTH: VITELLOGENESIS OR YOLKED OOCYTES (SG)	
SGl	LATE SECONDARY GROWTH OR LATE YOLKED OOCYTES		
SGfg	FULL-GROWN OOCYTE		
OMegv	ECCENTRIC GERMINAL VESICLE. *OIL DROPLETS COALESCE BECOMING ONE GLOBULE*	OOCYTE MATURATION (OM)	
OMgvm	GERMINAL VESICLE MIGRATTION TO THE ANIMAL POLE. *OOCYTE HYDRATES*		
OMgvb	GERMINAL VESICLE BREAKDOWN. *OOCYTE HYDRATION ALMOST COMPLETE*		
OMmr	MEIOSIS RESUMES. 2ND ARREST. *OOCYTE HYDRATION COMPLETE*		ACTIVE MEIOSIS II
OV	OOCYTE EMERGES FROM THE FOLLICLE, BECOMES AN EGG	OVULATION (OV)	ARRESTED MEIOSIS II

Oogenesis begins as oogonia proliferate and enter into meiosis becoming oocytes, and ends with the ovulation of eggs (ova) capable of fertilization. Following oogonial proliferation and entry into meiosis, progressive and specific morphological changes are observed in the oocyte's nucleus and cytoplasm that can be used to classify and study oogenesis in distinct stages. Not all morphological changes are related to meiotic progression; in fact, much of the growth of the teleost oocyte occurs during a relatively prolonged period of meiotic arrest. Thus, oogenesis occurs during five periods of germ cell development: Active Mitosis, Active Meiosis, Arrested Meiosis, a second period of Active Meiosis (also known as Oocyte Maturation), and a second period of Arrested Meiosis that generally precedes Ovulation (Table 2.1). Oocyte growth is primarily due to accumulation of yolk in the ooplasm. Yolk is defined as the "nutritive material stored in an ovum"**. After egg fertilization, it becomes the source of building materials and energy for the developing embryo. Yolk consists of derivatives of vitellogenin (**see Section 2.4.4**), which include protein and lipoprotein products stored in "yolk globules;" and, in many species, also oils rich in monounsaturated fatty acids and/or wax and steryl esters that are stored in "oil droplets" (Silversand and Haux 1995; Wiegand 1996; Patiño and Sullivan 2002).

The stages of oogenesis described in this chapter include oogonial proliferation, chromatin nucleolus, primary growth, secondary growth (vitellogenesis), maturation, and ovulation (Table 2.1). Most of these stages are subdivided into distinct steps, which are designed so that they can be applied to most teleosts while recognizing the variety of morphological patterns that exist, especially at the level of the ooplasm. Events occurring at the nuclear level (germinal vesicle) are quite constant among teleosts. Instead of having numbered stages, as is common in the literature, letters are used here to designate the stages and the steps within stages of oocyte development. The stages are represented by upper case letters while the steps are represented by lower case letters (Table 2.1). The letters are descriptive of either the stage or the step, as "PG" for "primary growth," etcetera. The present classification schema for oogenesis is similar to but in certain aspects of content, interpretation and terminology also different from other schemas previously used (e.g., Begovac and Wallace 1988; Iwamatsu *et al.* 1988; Wallace and Selman 1990; Patiño and Takashima 1995; Khan and Thomas 1999; Patiño and Sullivan 2002; Young *et al.* 2005; Grier *et al.* 2007; Le Menn *et al.* 2007). Current information is also provided concerning the basic cellular and endocrine regulation of oogenetic stages and, when available, their individual steps.

2.4.1 Oogenesis: Oogonial Proliferation (OP)

In most teleosts, oogonia continue to proliferate in the germinal epithelium into adult life. In species with recurring breeding cycles, oogonial numbers in

**McGraw-Hill Dictionary of Scientific and Technical Terms, McGraw-Hill Book Co., 1994.

the ovary typically peak soon after spawning in preparation for the next reproductive cycle (Tokarz 1978; Khan and Thomas 1999), but more protracted periods of oogonial mitoses have also been described (e.g., de Vlaming 1974). In Rainbow Trout (Grier *et al.* 2007), oogonia are very actively dividing during and after ovulation. These mitotic divisions produce germ cell nests (Fig. 2.3B,C) that appear as extensions of the germinal epithelium and are initially composed of prefollicle cells and oogonia (Grier 2000; Grier *et al.* 2007). In tilapias, oogonial proliferation also occurs soon after spawning (Smith and Haley 1987, 1988; Coward and Bromage 1998). In species such as the Mullet (*Liza aurata*), cytoplasmic bridges between oogonia have been reported following the mitotic cell divisions just preceding onset of meiosis (Le Menn *et al.* 2007). In Rainbow Trout, primary and secondary oogonia have been classified according to their presence as individual cells or as cells in nests within the germinal epithelium, respectively (Grier *et al.* 2007).

Regulation: Earlier studies based on hypophysectomy and hypophysation strategies suggested an involvement of the pituitary gland in the regulation of oogonial mitoses (Khan and Thomas 1999). There is also evidence indicating that the ovarian estrogen, estradiol-17β, induces oogonial proliferation in the cyprinids, Common European Minnow, *Phoxinus laevis* (Bullough 1942) and Common Carp, *Cyprinus carpio* (Miura *et al.* 2007); and the salmonid, Japanese Huchen (*Hucho perryi*) (Miura *et al.* 2007). A role for the postovulatory follicle complex during oogonial proliferation and other early stages of oogenesis is possible in some species. For example, in the Mozambique Tilapia (*Oreochromis mossambicus*), postovulatory follicle complexes not only remain in the ovary through the early stages of the next oogenic cycle (Smith and Haley 1987), but are also capable of producing estradiol-17β and testosterone (precursor for estradiol-17β) in response to gonadotropin (Smith and Haley 1988). Overall, the available information suggests that pituitary gonadotropins regulate oogonial proliferation via stimulation of ovarian estrogen, and in some species the postovulatory follicle complex may also contribute to the production of this steroid. For a discussion of the endocrinology of reproduction, see Schneider and Poehland, **Volume 8B, Chapter 2**.

2.4.2 Oogenesis: Chromatin Nucleolus Stage (CN)

When secondary oogonia are recruited into meiosis they become oocytes. This transformation occurs within the germinal epithelium. A review of the literature reveals a highly inconsistent terminology for the classification of early oocyte development, and particularly with the use of the term "chromatin nucleolus" stage. For example, the chromatin nucleolus "phase" has been considered to occur within the primary growth "stage" of oogenesis (Wallace and Selman 1981; Selman and Wallace 1986; Çek *et al.* 2001), when the oocyte has scant cytoplasm, a central nucleus, and single nucleolus. This definition also has been used in the fishery literature, as with the Arrowtooth Flounder (*Atheresthes stomias*) (Rickey 1995; Zimmerman 1997), the Sea Bass

(*Dicentrarchus labrax*) (Mayer *et al.* 1988), and the Anglerfish (*Lophius titulon*) (Yoneda *et al.* 2001). However, Matsuyama *et al.* (1988) considered the chromatin nucleolus stage to start in pachytene and extend until the "post-synaptic stage," which would be diplotene. In the review of Patiño and Takashima (1995), the chromatin nucleolus stage was defined as extending from leptotene to pachytene, thus including only early meiotic prophase. Selman and Wallace (1989) included preleptotene, leptotene, zygotene, pachytene and diplotene when "the chromosomes take on a 'lampbrush' configuration" in the chromatin nucleolus stage of the primary growth phase. Iwamatsu *et al.* (1988) regarded chromatin nucleolus stage as all oocytes preceding the perinucleolar step of development (see below). This chapter attempts to resolve the confusing terminology for early oogenesis by presenting recent information indicating that the early diplotene oocyte represents a distinct step preceding the onset of ooplasmic basophilia (Grier 2000; Grier *et al.* 2007). Thus, in the present schema, the chromatin nucleolus stage of oogenesis starts with the onset of meiosis, leptotene, and ends at early diplotene of Prophase I, prior to demonstrable ooplasmic basophilia (see below). The chromatin nucleolus stage is distinct from, and precedes, primary growth (Table 2.1).

Immediately following oogonial divisions and entry into meiosis, oocytes are in close physical contact with each other and their development is fairly synchronized (Fig. 2.3C). However, they are progressively isolated from one another as prefollicle cell processes begin to encompass each individual oocyte, marking the beginning of folliculogenesis (Grier 2000). Eventually, prefollicle cells completely surround individual oocytes within a cell nest as a basement membrane envelops this structural unit that will become the follicle. The basement membrane surrounding the cell nest is continuous with that of the germinal epithelium (Fig. 2.3B), as has been described for Rainbow Trout (Grier *et al.* 2007). Folliculogenesis is still incomplete by the end of the chromatin nucleolus stage.

The cytoplasm of chromatin nucleolus oocytes is fairly indistinct, but marked changes in nuclear morphology are observed during this oogenetic stage that can be used to describe its progression through the following steps (Table 2.1):

Leptotene Step (CNl): Chromosomes condense and become visible as thin threads. By the end of CNl, the telomeres begin to attach to the inner nuclear envelop in a polarized manner (on the same side of the nucleus) and the chromosomes loop outward, giving the appearance of a bouquet distribution of chromosomes in standard histological preparations (hematoxylin-eosin). The appearance of a chromosomal bouquet is a universal feature of meiosis during the leptotene-zygotene transition in all sexual organisms (Zickler and Kleckner 1998).

Zygotene Step (CNz): Chromosomes shorten and homologous chromosomes form pairs. Synaptonemal complexes (SCs) begin to form (visualized ultrastructurally). The bouquet distribution of chromosomes is most pronounced during this step of oogenesis (Patiño and Takashima 1995). In many teleost oocytes a single, prominent nucleolus is observed just over the chromosomal bouquet or at the opposite nuclear pole.

Pachytene Step (CNp): Chromosomes condense and lose their bouquet distribution. SCs complete their development, and crossing-over occurs.

Early Diplotene Step (CNed): SCs begin to breakdown and lampbrush chromosomes begin to form. The homologous chromosomes shorten and separate, each being formed by sister chromatids, but remain attached by chiasmata. There is some cellular growth as the oocyte advances through the chromatin nucleolus stage and by the time it reaches CNed, it is clearly larger in cell and nuclear size. The mechanism for this increase in size is unknown. The nucleus of a CNed oocyte is typically spherical and contains a single, prominent nucleolus and clear ooplasm (hyaline) when stained with PAS, metanil yellow, and hematoxylin (Fig. 2.2D), or hematoxylin and eosin. A basophilic ring of ribosomal RNA (rRNA) may be seen at the periphery of the nucleoplasm in CNed oocytes just prior to the commencement of primary growth. Early diplotene oocytes may have similar morphology to oogonia when observed with light microscopy and can be confused, except they are larger. The CNed oocyte has been recently described and was initially termed "preprimary growth oocyte" (Thacker and Grier 2005; Grier *et al.* 2007).

Regulation: Little information is available concerning the endocrine regulation of the chromatin nucleolus stage of oogenesis. However, a recent study with Japanese Huchen and Common Carp concluded that progestins stimulate oogonial recruitment into meiosis and their progression through the chromatin nucleolus stage (see also Schneider and Poehland, **Volume 8B, Chapter 2**). During the early reproductive development of Japanese Huchen, the first appearance of chromatin nucleolus oocytes coincided with increased blood levels of the progestin, DHP, which peaked at this time and declined thereafter (Miura *et al.* 2007). An examination of the photomicrographs provided by the study suggests that all steps of the chromatin nucleolus stage were present in the ovary at the time of increased DHP levels, from leptotene through early diplotene (see Fig. 1B in Miura *et al.* 2007). Also, ovarian fragments (containing oogonia) from Japanese Huchen and Common Carp showed an increased incidence of chromatin nucleolus oocytes following incubation with DHP in vitro (Miura *et al.* 2007). Although the incidence of SCs also increased in ovarian fragments from Japanese Huchen after DHP treatment in vitro, this response was not clear in Common Carp (Miura *et al.* 2007). The lack of a clear response of SCs to DHP in carp, or perhaps any other teleost, is not surprising because, following their assembly during the zygotene and pachytene steps, SCs begin to disassemble during early diplotene. Thus, their incidence during the chromatin nucleolus stage will vary depending on the specific step at which they are examined.

Overall, the evidence available suggests that progestins, perhaps of ovarian origin, induce the entry into meiosis and progression through the chromatin nucleolus stage. At the present time there is no information concerning higher level involvement, such as by pituitary gonadotropins. But if gonadotropins are involved in the regulation of oogonial proliferation (see **Section 2.4.1**), they would likely be also involved in the induction of meiosis. Because the so-called "maturational progestins" (or "maturation-inducing steroids"), such as DHP, may also be involved in non-maturational events of early oogenesis, this chapter simply refers to these steroids as "progestins." Curiously, progestins seem to regulate the meiotic process in both cases, entry into meiosis during the chromatin nucleolus stage and resumption of meiosis during maturation (see below). Whereas considerable information is available about mechanism of progestin-induced meiotic resumption during oocyte maturation, nothing is known at the present time about the cellular and molecular aspects of progestin-induced meiotic entry.

2.4.3 Oogenesis: Primary Growth (PG)

The morphological event normally taken as marker for the onset of primary oocyte growth is ooplasmic basophilia, which is the result of accumulation of rRNA and heterogeneous RNA in the ooplasm (Selman and Wallace 1989; Wallace and Selman 1990; Tyler and Sumpter 1996; Patiño and Sullivan 2002; Ravaglia and Maggese 2002). These RNAs play important roles in the regulation of oogenesis and embryogenesis (Patiño and Sullivan 2002; Lyman-Gingerich and Pelegri 2007).

Ooplasmic basophilia is first evident as oocytes reach late diplotene and still possess a single nucleolus (Figs. 2.3D, 2.4A). The initial appearance of these oocytes is similar to early diplotene oocytes except for the ooplasmic basophilia (Grier 2000; Grier *et al.* 2007). For this reason, the present schema classifies early and late diplotene oocytes within separate stages of oogenesis, the chromatin nucleolus and primary growth stages, respectively (Table 2.1). In primary growth oocytes of Common Snook, rRNA is seen extending from the single nucleolus to the perimeter of the germinal vesicle while the oocyte is still in the germinal epithelium; i.e., prior to the completion of folliculogenesis (see Fig. 7B in Grier 2000). However, it should be noted that a basophilic ring of rRNA develops around the inner membrane of the nucleus of early diplotene oocytes, when the ooplasm still has a hyaline appearance (see above, CNed), indicating that the mechanism responsible for ooplasmic basophilia is already active in early diplotene oocytes. This mechanism involves the production of rRNA by the single nucleolus of early diplotene oocytes and its accumulation within the germinal vesicle prior to moving into the ooplasm. Oocytes still contain a single nucleolus after the completion of folliculogenesis, where they are found lined up along the germinal epithelium (Fig. 2.3D). The nucleus of primary growth oocytes is known as the germinal vesicle (Patiño and Takashima 1995). A significant increase in oocyte size occurs during this stage of development, and meiosis enters its first of two meiotic arrests during oogenesis (Table 2.1).

The basement membrane continues to extend into cell nests thus further isolating prefollicle cells and their encompassed oocyte, as described for Common Snook (see Figs. 6, 7, 9A,B in Grier 2000) and the viviparous goodeid Balsas Splitfin (see Figs. 20c,e in Grier *et al.* 2005). Folliculogenesis ends soon after primary growth (ooplasmic basophilia) begins; namely, as meiosis arrests in the late diplotene oocyte, the oocyte and its monolayer of follicle cells are completely surrounded and separated from the germinal epithelium by the basement membrane. A similar temporal association between meiotic arrest and (primordial) follicle formation has been described in rodents (e.g., Paredes *et al.* 2005).

Upon completion of folliculogenesis there is usually a single Balbiani body observed in histological sections of oocytes, but soon numerous Balbiani bodies are observed in the ooplasm. In Rainbow Trout they are initially absent in perinucleolar oocytes (Fig. 2.4B), but later appear as polymorphic, basophilic bodies at the oocyte periphery (Fig. 2.4C,D). Their three-dimensional structure is unknown; i.e., does a single, irregular structure grow to be represented multiple times in sections? The Balbiani body or Balbiani bodies are temporary structures observed during early primary growth, and associated mitochondria are situated in the perinuclear region of the ooplasm. Balbiani bodies have been described as regions rich in nucleic acids (Hamaguchi 1993) and the less electron-dense endoplasmic reticulum, but their function is uncertain. Subsequent dispersal of the Balbiani bodies may help explain the loss of ooplasmic basophilia in advanced primary growth oocytes as observed with light microscopy. Gülsoy (2007) described the development of the Balbiani body in primary growth oocytes of Rainbow Trout, where a few dense granules first appear within a flocculent area of the ooplasm around the nucleus. Later, the dense granules increase in number and size and, at the end of primary growth, the Balbiani bodies disperse toward the peripheral ooplasm. The term "yolk nucleus" has been occasionally used to describe Balbiani bodies. This term is misleading because Balbiani bodies do not produce yolk, and it should not be used.

Cortical alveoli and oil droplets are the main inclusions (other than Balbiani bodies) that appear in the ooplasm during primary growth. In some species such as Mummichog (Selman and Wallace 1989), other Atherinomorpha (Grier unpublished), and Goldfish (*Carassius auratus*), cortical alveoli eventually fill most of the ooplasm (Fig. 2.4A, 2.6A-C), whereas in Rainbow Trout (Fig. 2.5A,B, 2.7A,B) as well as in teleosts that produce pelagic eggs that contain oil droplets (Figs. 2.5C, 2.7C,D, 2.8A,B), cortical alveoli seem to be always restricted to the oocyte periphery. The contents of cortical alveoli are synthesized within the oocyte (Wallace and Selman 1990). Nagahama (1983) considered the role of the cortical alveoli in the prevention of polyspermy after ovulation. In Rainbow Trout, Inoue and Inoue (1986)

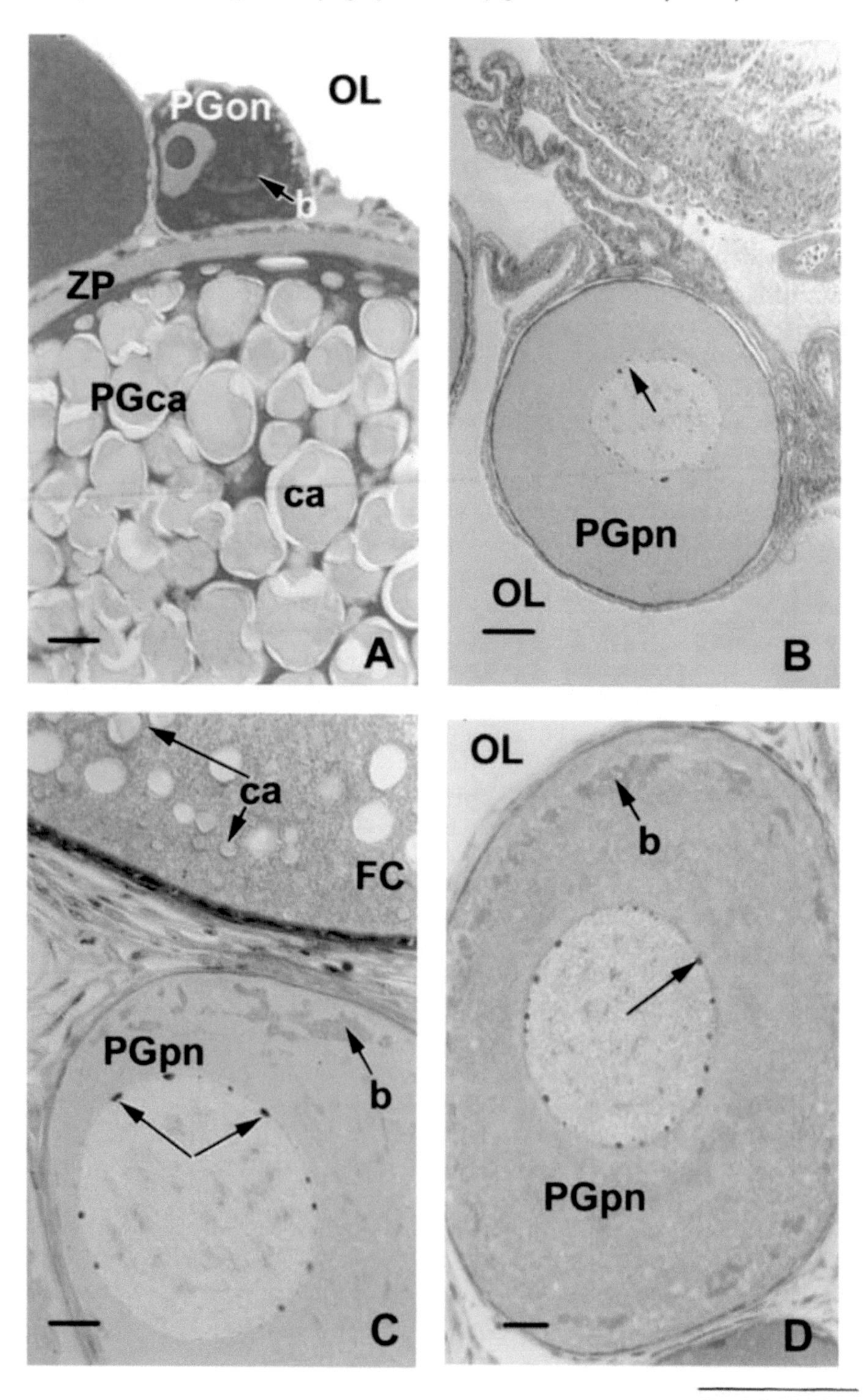

Fig. 2.4 Contd. ...

showed that the contents of cortical alveoli depolymerize into smaller fragments after fertilization and are released into the perivitelline space, and harden the zona pellucida. Shibata *et al.* (2000) later showed that the hardening-inducing constituent of cortical alveoli in Medaka is a metalloproteinase, which they termed alveolin. The hardened zona pellucida provides protection for the embryo against physical injury. Cortical alveoli are still sometimes referred in the literature as "yolk vesicles." This term is also a misnomer (Wallace and Selman 1990) and should not be used.

The origin of oil droplets is uncertain but they are believed to be derived from very low density lipoproteins taken up by the follicle from the blood stream (Le Menn *et al.* 2007). The important embryogenic role of the lipids associated with oil droplets and with lipoprotein (from yolk globules) has been recently reviewed (Wiegand 1996; Patiño and Sullivan 2002). The former seem to be primarily used in the development of cell structure while the latter are used to meet the energy requirements of the developing embryo. Oil droplets may also be the source of vitamin A for the onset of vision (Kunz 2004). Not all teleost eggs possess oil droplets. For example, they are lacking in oocytes and the demersal eggs of the freshwater cyprinids Zebrafish (*Danio rerio;* Selman *et al.* 1993) and Goldfish (Fig. 2.9A); or even in the pelagic eggs of some marine species such as the clupeids, Northern Anchovy (*Engraulis mordax;* Hunter and Macewicz 1985), Herring (*Clupea harengus;* reviewed by Kunz 2004), and Bay Anchovy (*Anchoa mitcheli;* Williams 1968), (Fig. 2.8C); and the labrids, Cunner (*Tautogolabrus adspersus*) and Tautog (*Tautoga onitis*) (Williams 1968). In species that have no oil droplets in their eggs, the source of all lipids necessary for embryo development is likely to be lipoprotein from yolk vesicles.

Close relationships are established between the oocyte and its follicle cells during primary growth. Initially, follicle cells are in close contact with the oocyte and may even be coupled cytoplasmically via gap junctions (Le Menn *et al.* 2007). The oocyte and follicle cells begin to separate in the early primary growth oocyte and the intervening space is filled with electron clear, flocculent material (Le Menn *et al.* 2007). Oocyte microvilli, and in some

---

Fig. 2.4 Contd. ...

Fig. 2.4 One-nucleolus and perinucleolar step oocytes. **A**. A one-nucleolus oocyte (PGon) from the Goldfish shows Balbiani body (b), which is dwarfed by a cortical alveolar oocyte (PGca) with well developed zona pellucida (ZP) and whose ooplasm is filled with cortical alveoli (ca). Ovarian lumen (OL). Bar 20 µm. Hematoxylin-Eosin. **B-D**. Perinucleolar oocytes (PGpn) in Rainbow Trout. In **B**, nucleoli (arrows) are around the inner membrane of the nucleus. The ooplasm is initially basophilic, but basophilia is masked by the PAS stain in **C**, about the time that cortical alveoli (ca) are seen. **D.** As growth proceeds, the Balbiani body (b) develops and disperses into the cortical ooplasm. Follicle cells (FC) encircle the follicle. B, PAS, metanil yellow, hematoxylin. Bar 50 µm. C, Bar 20 µm. PAS, metanil yellow, hematoxylin. D, Bar 10 µm. Thionin. Original.

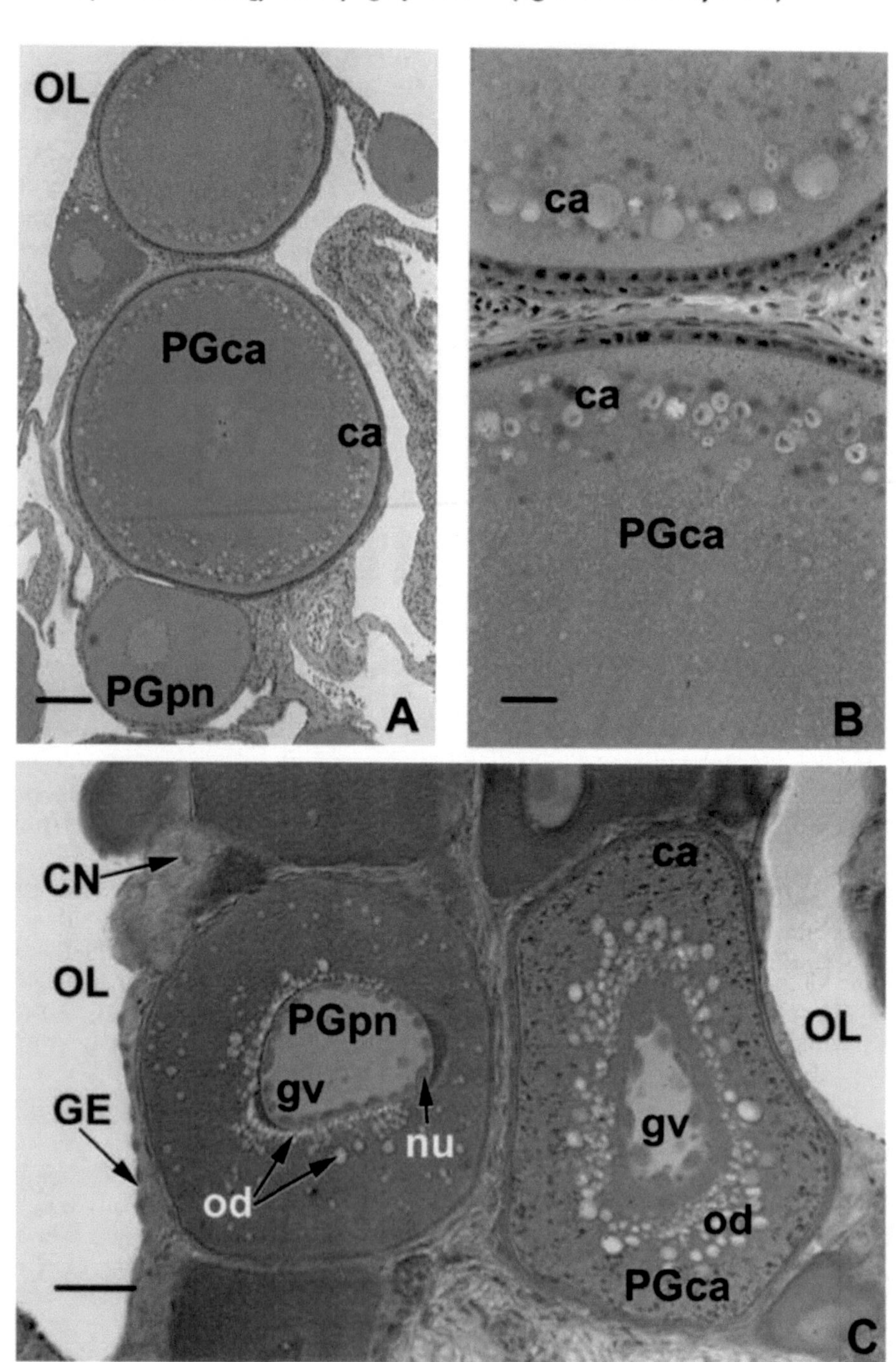

Fig. 2.5A-B Cortical alveolar oocytes in Rainbow Trout (PGca). Cortical alveoli (ca) are restricted to the peripheral ooplasm. Perinucleolar oocyte (PGpn) lacks cortical

Fig. 2.5A-B Contd. ...

species also follicle cell microvilli, can be seen extending into this space (Begovac and Wallace 1988; York *et al.* 1993; Le Menn *et al.* 2007). The zona pellucida is deposited around the microvilli. Depending on the species, ZP proteins are synthesized in the ovary, liver or both (Patiño and Sullivan 2002; Modig *et al.* 2007).

Primary growth has been divided into five steps according to the major nuclear and ooplasmic changes that are observed during this stage of oogenesis (Table 2.1). It is important to note that the relative timing of the onset of oil droplet and cortical alveoli formations seems to be reversed between species that produce pelagic versus demersal eggs (see below). Also, as noted already, oil droplets are absent in the oocytes and eggs of some species. Thus, the sequence of oogenic steps described below needs to be adjusted according to the species under study. The steps for the primary growth stage are:

One-Nucleolus Step (PGon): Meiosis arrests at late diplotene, SCs complete their disassembly, and lampbrush chromosomes fully develop. Like oocytes in early diplotene (CNed) (Fig. 2.2D), the nucleus (germinal vesicle) of PGon oocytes is spherical and has a single prominent nucleolus (Figs. 2.3D, 2.4A). However, unlike CNed oocytes which have a hyaline ooplasm, the ooplasm of PGon oocytes initially is finely basophilic when viewed with electron microscopy and homogeneously basophilic when viewed using light microscopy. PGon oocytes are first seen within the germinal epithelium prior to the completion of folliculogenesis (Grier 2000; Grier *et al.* 2007); they retain their appearance as folliculogenesis is completed when prefollicle cells, now known as follicle cells, show a squamous appearance around the periphery of the oocyte. Advanced PGon oocytes show an intensely basophilic ooplasm except in the less dense area of the Balbiani body (Fig. 2.4A), and their size has increased considerably. Initially the Balbiani body is associated with relatively few mitochondria, but it is subsequently surrounded by numerous mitochondria.

Multiple-Nucleoli Step (PGmn): PGmn oocytes have multiple nucleoli, two or more, irregularly situated within the germinal vesicle. The PGmn step is of short duration and is rarely seen. It has been reported that folliculogenesis is

---

Fig. 2.5A-B Contd. ...

alveoli. Ovarian lumen (OL). **A**. ar 100 µm. PAS, metanil yellow, hematoxylin. **B.** Bar 20 µm. PAS, metanil yellow, hematoxylin. **C**. Longitudinal section through the lamella of Common Snook showing two oocytes in primary growth. The left oocyte (PGpn) is in transition to PGod when oil droplets (od) will completely surround the germinal vesicle (gv). Nucleoli (nu) are arranged around the inner side of the germinal vesicle membrane. In the second oocyte (PGca), PAS+ cortical alveoli (ca) occupy the outer half of the ooplasm, and the germinal vesicle (gv) is surrounded by oil droplets (od). On either side, the ovarian lamella borders the ovarian lumen (OL). A cell nest (CN) is associated with the germinal epithelium (GE). Bar 20 µm. PAS, metanil yellow, hematoxylin. Original.

completed at a time equivalent to this step of oogenesis in Gulf Pipefish (Begovac and Wallace 1988), a syngnathid. However, in another syngnathid, the Lined Seahorse (*Hippocampus erectus*), folliculogenesis seems to be completed while oocytes still have a single *prominent* nucleolus and an intensely basophilic ooplasm (see Fig. 13 in Selman *et al.* 1991), just like the PGon oocyte of Common Snook (Grier 2000). Arguably, it may be difficult to distinguish an advanced PGon with a single nucleolus from a PGmn oocyte with few nucleoli when observing thin or ultrathin sections.

Perinucleolar Step (PGpn): Multiple nucleoli become oriented around the inner membrane of the germinal vesicle (Fig. 2.3C,D, 2.4B-D). This position of nucleoli facilitates the passage of ribosomal RNA to the ooplasm. The nucleoli are initially spherical (Fig. 2.3D), but some become flattened on the side along the nuclear membrane (Fig. 2.5C). The Balbiani body contains both mitochondria and smooth endoplasmic reticulum. Microvilli of oocyte origin are first observed in PGpn follicles (e.g., Begovac and Wallace 1988; York *et al.* 1993). The zona pellucida begins to form around the microvilli (Guraya 1986; Wallace and Selman 1990), although relatively late during this step in species with pelagic eggs (York *et al.* 1993). The theca is composed of a one-cell layer in the development of the follicle complex.

Oil Droplets Step (PGod): As already mentioned, the appearance of oil droplets (Figs. 2.2C; 2.5C) relative to the appearance of cortical alveoli varies among teleosts. Although the number of species examined in enough detail is insufficient for reliable conclusions, there seems to be a pattern emerging where oil droplets appear before cortical alveoli in marine species that produce pelagic eggs. Known examples include the sparids, Sea Bass (*Dicentrarchus labrax*; Mayer *et al.* 1988) and Red Seabream (*Pagrus major*; Matsuyama *et al.* 1991); and the centropomid, Common Snook (Neidig *et al.* 2000) (Fig. 2.5C). The ecophysiological significance of this observation is uncertain.

Regardless of the relative timing of their appearance, oil droplets have been reported in many other marine teleosts that produce pelagic eggs such as the sciaenids, Spotted Seatrout (*Cynoscion nebulosus*; Brown-Peterson *et al.* 1988) and Atlantic Croaker (*Micropogonias undulatus*; Patiño and Thomas 1990b); the sillaginid, Japanese Whiting (*Sillago japonica*; Matsuyama *et al.* 1990); the percichthyid, Striped Bass (*Morone saxatilis*; Berlinsky and Specker 1991); as well as other taxonomically diverse species. Oil droplets are also present in demersal eggs of freshwater and marine species including the Roach (*Rutilus rutilus*) (Cyprinidae) (Tyler and Sumpter 1996); Mummichog and Sheepshead Killifish (*Cyprinodon variegates*) (Cyprinodontidae: Atherinomorpha) (Wallace and Selman 1981, 1990); Swamp Eel (Synbranchidae) (Ravaglia and Maggese 2002); Largemouth Bass (*Micropterus salmoides*, Centrarchidae) (Chiu *et al.* 1991); as well as others.

Typically, oil droplets progressively increase in number and size around the germinal vesicle. However, in Striped Mullet (*Mugil cephalus*) the growing oocytes have oil droplets scattered among the yolk globules, not in a circumnuclear zone (Grier unpublished). In other teleosts, such as the poeciliids Guppy (*Poecilia reticulata*; reviewed by Kunz 2004) and Green Swordtail (*Xiphophorus helleri*; Bailey 1933) oil droplets form at the periphery of the ooplasm. Finally, as already mentioned, many teleosts lack oil droplets in their ooplasm regardless of the buoyancy of their ovulated eggs. Thus, there seems to be great variability in the patterns of oil droplet accumulation and distribution within the ooplasm of teleosts oocytes.

Cortical Alveoli Step (PGca): The available evidence seems to indicate that cortical alveoli appear before oil droplets in oocytes of freshwater and marine teleosts that produce demersal eggs. Species where this is known to occur include Medaka (Iwamatsu *et al.* 1988), Gulf Killifish (Fig. 2.6C), Mummichog (Selman and Wallace 1986, 1989), and Gulf Pipefish (Begovac and Wallace 1988). In Coho Salmon (*Oncorhynchus kisutch*) (freshwater, demersal eggs), cortical alveoli not only seem to appear first but the timing of oil droplet formation also is negatively associated with the growth rate of individual fish; namely, the appearance of oil droplets, but not of cortical alveoli, is delayed in oocytes of slow-growing individuals (Campbell *et al.* 2006). Thus, at least in Coho Salmon, the timing of oil droplet formation seems to be specifically influenced by nutritional or metabolic factors. Curiously, in Sea Bass oocytes – a species where oil droplets appear first (see above) – cortical alveoli become evident only after secondary growth (vitellogenesis) has begun (Mayer *et al.* 1988). Interestingly, an initial distribution of cortical alveoli at the periphery of the ooplasm and of oil droplets in the vicinity of the germinal vesicle is observed in Rainbow Trout (Fig. 2.7A,B) as well as Common Snook (Fig. 2.7C,D). As mentioned already, in species such as Goldfish (Figs. 2.4A, 2.6A,B) and others cortical alveoli are the only major ooplasmic inclusion observed in primary growth oocytes.

In this chapter, the order of appearance of oil droplets (PGod) and cortical alveoli (PGca) is meant to be reversible, and in the case of oil droplets the entire step is also dispensable, depending on the species of interest. This schema retains wide applicability while recognizing that differences exist among species.

Regulation: Meiotic arrest at late diplotene and completion of folliculogenesis are temporally associated in teleosts (Begovac and Wallace 1988; Grier 2000; Grier *et al.* 2007) as well as rodents (Paredes *et al.* 2005). Synaptonemal complex protein-1 (SCP1) is part of the SC that holds chromosomes closely apposed during crossing-over in pachytene oocytes. Ovaries from 21-day-old Rat (*Rattus norvegicus*) fetuses, which contain no follicles, when cultured in the presence of antisense oligonucleotide against SCP1 mRNA not only undergo premature SC disassembly and increased expression of a diplotene oocyte-specific marker, but also show an increased number of newly formed follicles relative to ovaries treated with scrambled oligonucleotides (Paredes *et al.* 2005). These observations indicate that it is the late diplotene oocyte which

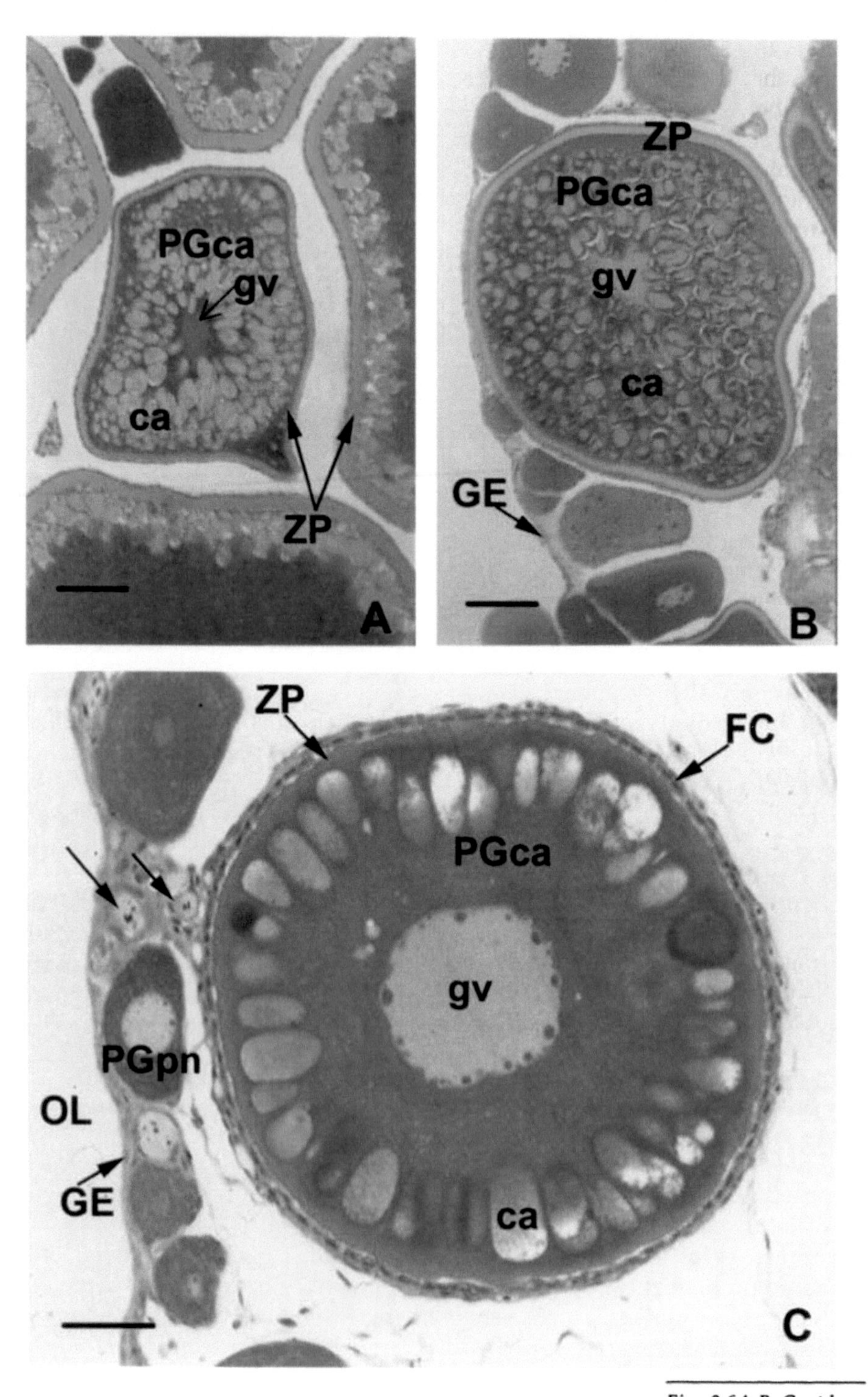

Fig. 2.6A-B Contd. ...

directs the formation of the ovarian follicle in rodents (Paredes *et al.* 2005). Given the similarities in the basic structure of the follicle and in the relative timing between meiotic arrest and completion of folliculogenesis in both taxa, it seems reasonable to suggest that similar cellular and molecular mechanisms regulate folliculogenesis in teleosts and rodents. However, the notion that meiotic arrest is necessary for completion of folliculogenesis was recently questioned for the bovine ovary (Yang and Fortune 2008). Additional research will be necessary to determine the general relationship between meiotic arrest and completion folliculogenesis in the vertebrate ovary.

Little information is available concerning the endocrine regulation of primary growth. It is possible that the transition from CNed (last step of chromatin nucleolus) to PGon (first step of primary growth) is preprogrammed at the onset of the progestin-dependent chromatin nucleolus stage (see above). A general role for ovarian-derived insulin-like growth factors has been suggested during primary growth of teleosts oocytes based on the finding that these factors and their receptors are produced by the follicle at this stage of oogenesis (e.g., Kagawa *et al.* 1995; Maestro *et al.* 1997). In Short-Finned Eel (*Anguilla australis*), in vitro incubation of ovarian fragments with insulin-like growth factor-I or androgen (11-ketotestosterone), but not estrogen (estradiol-17β), stimulated the primary growth of ovarian follicles (Lokman *et al.* 2003). In addition, in vivo implants of 11-ketotestosterone stimulated the formation cortical alveoli in primary growth oocytes of Short-Finned Eel (Rohr *et al.* 2001). In Coho Salmon, increased blood levels of the gonadotropin, FSH, and of estradiol-17β correlated with the formation of cortical alveoli (Campbell *et al.* 2006). In the same study with Coho Salmon, correlative evidence was presented suggesting a role for FSH, estrogen, and insulin-like growth factors (systemic and ovarian) in the regulation of oocyte lipid deposition and of fish body growth. These findings with Coho Salmon differ from those with Short-Finned eel in that estrogens were implicated during cortical alveoli formation in the former (Campbell *et al.* 2006) and androgens in the latter (Rohr *et al.* 2001; Lokman *et al.* 2003). Unfortunately, androgens were not measured in the study with Coho Salmon (Campbell *et al.* 2006) and it is therefore difficult to compare results. Oocyte-derived epidermal growth factor (Wang and Ge 2004) and

---

Fig. 2.6A-B Contd. ...

Fig. 2.6A-B Cortical alveolar oocytes in the Goldfish may double in size, but always possess an ooplasm replete with cortical alveoli (ca) extending from the germinal vesicle (gv) to the periphery of the oocyte, where a zona pellucida (ZP) is observed encircling the oocyte. Germinal epithelium (GE). **A**. Bar 100 μm. **B**. Bar 100 μm. **C**. Primary growth cortical alveolar oocyte (PGca) from *Fundulus grandis* shows cortical alveoli (ca) at the periphery. The germinal vesicle (gv) is seen. The zona pellucida (ZP) and the follicle cells (FC) surround the oocyte. Perinucleolar oocyte (PGpn), two oocytes in the germinal epithelium (GE) are seen bordered by the ovarian lumen (OL). Bar 50 μm. Original.

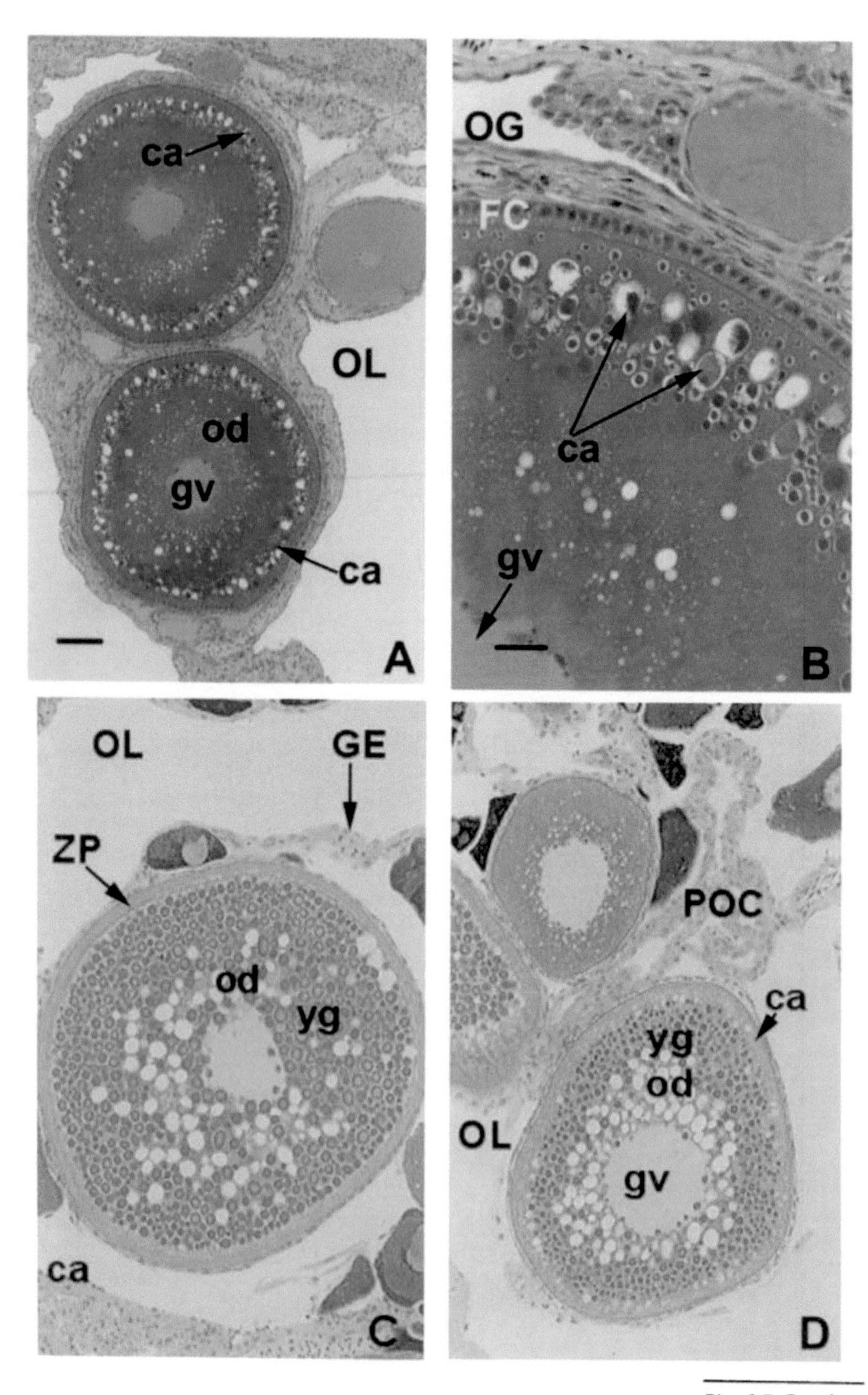

Fig. 2.7 Contd. ...

growth differentiation factor 9 (Liu and Ge 2007) have been also associated with the regulation of primary oocyte growth in Zebrafish. Finally, in those species where there is hepatic synthesis of ZP proteins during primary growth, they seem to be regulated by ovarian estrogen (Patiño and Sullivan 2002; Modig *et al.* 2007).

Overall, while the information currently available for teleosts is insufficient to determine what humoral factors control oogenesis through the early steps of primary growth (from PGon to PGpn), current data for several species consistently suggest that insulin-like growth factors play an important role during the later steps (advancement to PGod and PGca). However, it is likely that oil droplets and cortical alveoli depositions are independently regulated either by common or different factors, especially because they rely on different sources for their synthesis (cortical alveoli, endogenous; oil droplets, blood lipids) and their relative timing of appearance differs among species. As more knowledge becomes available concerning the regulation of these ooplasmic inclusions, and the ecophysiological significance and mechanisms for the difference in their relative timing of appearance among species is better understood, it may become necessary to revise the classification schema for primary growth presented in this chapter.

2.4.4 Oogenesis: Secondary Growth (SG)

Vitellogenin is the term used to describe a complex glycophospholipoprotein that is produced by the liver (Silversand and Haux 1995; Wiegand 1996; Patiño and Sullivan 2002). Vitellogenesis refers to the uptake of vitellogenin and its transformation into lipoprotein and protein yolk by the oocyte, where is stored in yolk globules. *Sensu stricto*, vitellogenesis should also be concerned with the process of oil droplets (lipid yolk) formation in those species where they are present. But until the origin and mechanisms of oil droplets formation are clarified, it seems prudent to continue to apply the standard definition of vitellogenesis (formation and accumulation of vitellogenin-derived yolk). Vitellogenesis is sometimes confused with the hepatic synthesis of vitellogenin (e.g., Le Menn *et al.* 2007). However, the etymological roots of the term are the Latin words "vitellus" and "genesis," which stand for "yolk" and "formation," respectively.

Fig. 2.7 Contd. ...

Fig. 2.7 **A**. Oocytes of Rainbow Trout. Oil droplets (od) begin to accumulate around the germinal vesicle while cortical alveoli (ca) are peripheral in the ooplasm. Bar 100 μm. **B**. Enlargement to show cortical alveoli (ca) at the periphery. The follicle cells (FC) are columnar in shape. An oogonium (OG) is observed in the germinal epithelium. Bar 10 μm. **C-D**. In the Common Snook, oil droplets (od) may be scattered in the ooplasm, but typically encircle the germinal vesicle (gv), peripheral to which are yolk globules (yg) and then cortical alveoli (ca) beneath the zona pellucida (ZP). The germinal epithelium (GE) borders the ovarian lumen (OL). A postovulatory follicle complex (POF) is seen. Bar 50 μm in both.

Secondary growth is marked by the onset of vitellogenesis, when vitellogenin-derived yolk globules begin to accumulate in the ooplasm (Fig. 2.7C,D, 2.8A). In a number of species, yolk globule accumulation is the primary reason for the growth (increase in size) of the oocyte (Figs. 2.7C,D, 2.8A-C, 2.9A-D). However, deposition of other yolk materials, especially oil droplets in those species in which they occur, continues during this stage and in many species is as significant as the deposition of yolk globules (Patiño and Sullivan 2002). Thus, the present use of the term "secondary growth" as synonym for "vitellogenesis" does not imply that materials other than vitellogenin-derived products do not contribute to the increasing mass of the oocyte during this stage of oogenesis.

It has been reported that yolk globule deposition may occur simultaneously with, and as early as, the appearance of both cortical alveoli and oil droplets in species such as Rainbow Trout (Perazzolo *et al.* 1999) and Black Goby (*Gobius niger*) (Le Menn *et al.* 2007). However, observations using light and electron microscopy suggest that this is not the case in many other teleosts where the order of appearance of these cytoplasmic inclusions seems to relate to reproductive life history and may even be affected by nutritional or metabolic status (see above). Also, in Rainbow Trout, significant deposition of yolk globules does not occur for up-to-6- months after the appearance of cortical alveoli in first-time spawners, and up-to-1-month in repeat-spawners (Scott 1990); in fact, cortical alveoli stage oocytes for the next batch of eggs are already present in the ovary at the time of ovulation (Grier *et al.* 2007). These observations suggest that cortical alveoli and yolk globule formation in the trout oocyte are distinctly regulated both temporally and functionally. Also, it has been shown that oil droplets and yolk globule depositions can be photothermally uncoupled in Striped Bass, indicating that they are normally separately regulated (Clark *et al.* 2005). Thus, the information currently available seems to justify the classification of vitellogenesis (secondary growth) as a distinct stage of oogenesis. Justification for secondary growth as a separate stage of oogenesis is also based on its distinct regulatory mechanisms concerned with the uptake of vitellogenin into the oocyte and storage as yolk globules.

Some earlier interpretations of oogenesis also separated cortical alveoli deposition and vitellogenic growth as distinct but provided both events equal

Fig. 2.8 **A**. Cross section of an ovarian lamella in Common Snook. There is an oocyte in early secondary growth (SGe) and another in late secondary growth (SGl). These steps in development are discerned by the size of yolk globules. The germinal vesicles (gv) are central. The oocytes are surrounded by an extensive extravascular space (EVS) which stains positively for periodic acid Schiff. The germinal epithelium (GE) covers the lamella, forming the border to the ovarian lumen (OL). Bar 50 µm. PAS, metanil yellow, hematoxylin. **B**. Cortical alveoli (ca)

Fig. 2.8 Contd. ...

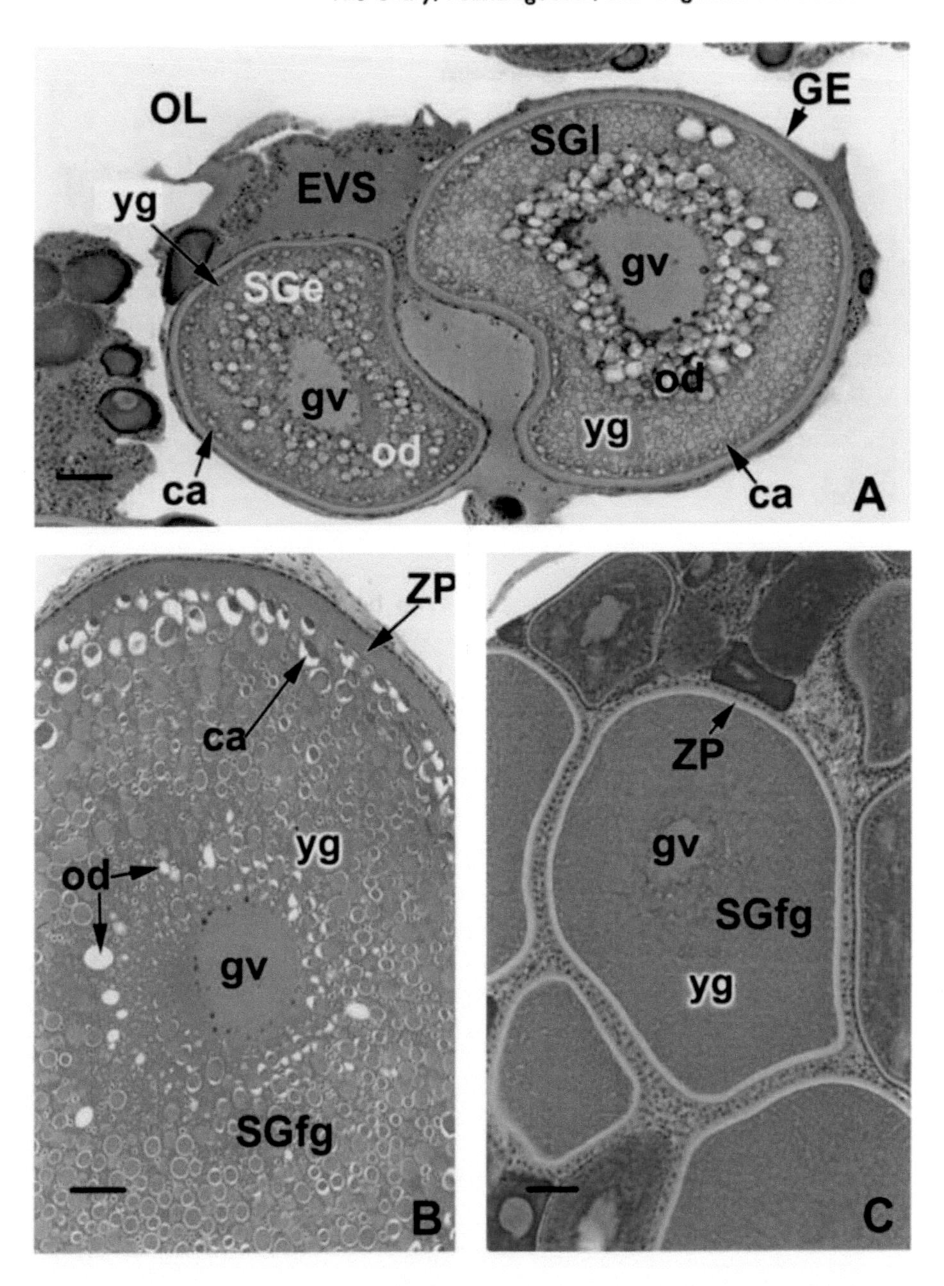

and yolk globules (yg), and oil droplets (od) are the major ooplasmic inclusions observed in the Rainbow Trout full-grown oocyte (SGfg). The germinal vesicle (gv) is seen. The oocyte has a well-developed zona pellucida (ZP). Bar 50 μm. PAS, metanil yellow, hematoxylin. **C**. Full-grown oocytes (SGfg) from Bay Anchovy. A well-developed zona pellucida (ZP), the germinal vesicle (gv) and yolk globules (yg) are seen. Bar 100 μm. PAS, metanil yellow, hematoxylin. Original.

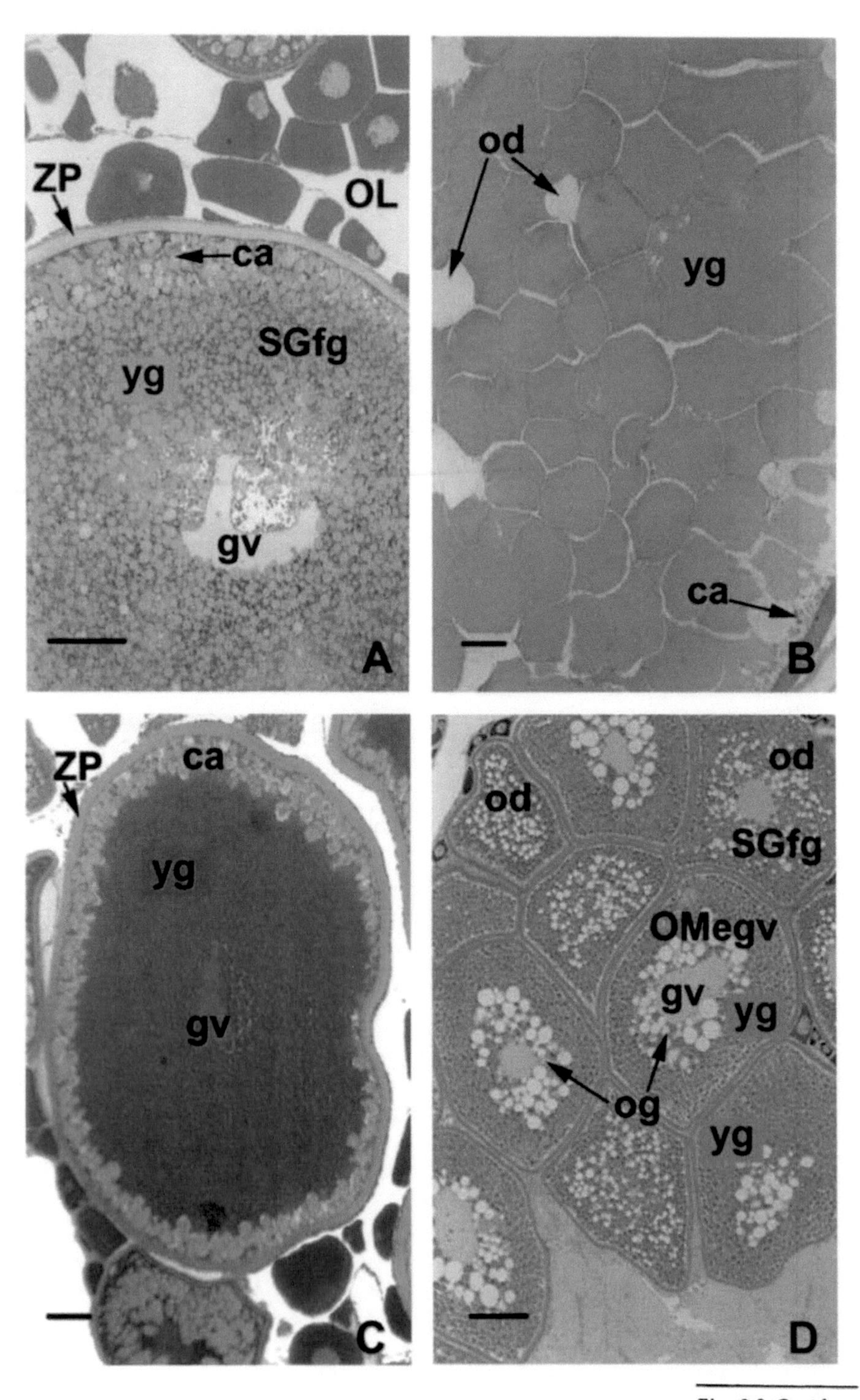
ZP
OL
ca
SGfg
yg
gv
A
od
yg
ca
B
ZP
ca
yg
gv
C
od
od
SGfg
OMegv
gv
yg
og
yg
D

Fig. 2.9 Contd. ...

rank in classification schemas (Sumpter *et al.* 1984; Begovac and Wallace 1988; Selman and Wallace 1989; Tyler and Sumpter 1996). In the present schema, cortical alveoli oocytes are classified as a step under primary growth just prior to the commencement of secondary growth. This classification is consistent with the conclusions of Patiño and Sullivan (2002), who noted that ovarian follicle growth in teleosts can be broadly classified into previtellogenic (generally coinciding with primary growth) and vitellogenic (secondary growth) stages.

During secondary growth, vitellogenin is taken into the oocyte by specific receptors located in clathrin-coated pits. The pits then move into the ooplasm by endocytosis, forming vesicles. These vesicles fuse with multivesicular bodies (lysosomes) where the vitellogenin is broken down by proteolytic cleavage into yolk components (Wallace and Selman 1990; Patiño and Sullivan 2002; Jalabert 2005; Le Menn *et al.* 2007). The multivesicular bodies increase in size and transform into small yolk globules and subsequently into large yolk globules (Le Menn *et al.* 2007). In some species of teleosts yolk is stored, in a relatively insoluble crystalline form, in what has been termed "platelets," as in Striped Mullet (Fig. 16d in Wallace and Selman 1990). However, in most teleosts, vitellogenin-derived yolk is stored as fluid-filled yolk globules (Wallace and Selman 1981, Selman and Wallace 1989).

In the cyprinid Zebrafish (Kessel *et al.* 1985, 1988), the atherinomorph Mummichog (Cerdá *et al.* 1993), and the perciforms Atlantic Croaker (York *et al.* 1993; Yoshizaki *et al.* 2001) and Red Seabream (*Sparus aurata*) (Patiño and Kagawa 1999), heterologous gap junctions develop between the microvilli of the oocyte and those of the follicle cells, being documented with both ultrastructure and fluorescence microscopy (Cerdá *et al.* 1993; Yoshizaki *et al.* 2001). Homologous gap junctions also form between follicle cells (York *et al.* 1993). Gap junctions, forming cytoplasmic continuity between cells, are

---

Fig. 2.9 Contd. ...

Fig. 2.9 **A**. Full-grown oocyte (SGfg) from the Goldfish, showing yolk globules (yg) in the ooplasm and peripheral cortical alveoli (ca). The germinal vesicle (gv) is irregular. Zona pellucida (ZP) and ovarian lumen (OL) are seen. Bar 50 µm. PAS, metanil yellow, hematoxylin. **B**. Yolk globules (yg) in an oocyte of Rainbow Trout. Cortical alveoli (ca) at the oocyte periphery and oil droplets (od) distal to it are observed. Bar 100 µm. PAS, metanil yellow, no hematoxylin. **C**. Oocyte from the Goldfish, showing abundant yolk globules (yg) in the ooplasm and peripheral cortical alveoli (ca). The germinal vesicle (gv) is irregular. Zona pellucida (ZP). Bar 100 µm. Hematoxylin-Eosin. **D**. Ovarian lamella from the Common Snook. Some oocytes are full-grown (SGfg) while others are in early oocyte maturation (OMegv) where oil droplets are fusing to become oil globules (og). Oil droplets and globules surround the germinal vesicle (gv) and are mixed with yolk globules. Distally, there is a homogeneous population of yolk globules (yg). At the magnification of the micrograph, cortical alveoli (ca) only appear as clear areas at the periphery of an oocyte. Bar 100 µm. PAS, metanil yellow, hematoxylin. Original.

important to an understanding of intrafollicular physiology during oocyte maturation as they may be involved in the transfer of factors that regulate vitellogenesis and maturation (Patiño *et al.* 2001; Patiño and Sullivan 2002).

The zona pellucida continues to develop during secondary growth (Figs. 2.10A,B). At the animal pole of the oocyte, a special structure can be observed, the micropyle. The micropyle has been described in several species such as Goldfish (Yamamoto and Yamazaki 1961), Gulf Killifish (Figs. 2.10C,D) (Brummett and Dumont 1981), Zebrafish (Hart and Donovan 1983), Swamp Eel (Ravaglia and Maggese 2002) and others. The opening diameter of the micropyle differs between species, commonly being 2-3 µm. Frequently, the micropyle is an invagination of the zona pellucida and the follicle cells, developing a cone shaped canal, terminating at the oolema. Follicle cells located around the opening of the micropyle, the micropylar cells, are modified in size and shape being described as larger and thinner cells or pyramidal-shaped. The micropyle is the site where the spermatozoon enters the egg during fertilization (Riehl 1980; Nagahama 1983; Guraya 1986; Patiño and Takashima 1995). The opening diameter of the micropyle coincides with the size of the spermatozoon, being as large as sufficient to permit the entrance of a single spermatozoon.

The sub-classification of secondary growth into early and late steps (see below) is somewhat arbitrary. However, this separation is made to extend the applicability of the present functional anatomy schema into the fisheries field, where terms such as "early-yolked" versus "advanced-yolked" (Abookire 2005) or "partially yolked" versus "yolked" (Hunter and Macewicz 1985) are used to assist in estimates of fecundity in fisheries management (Hunter *et al.*

Fig. 2.10 **A**. Oocyte surface from the Goldfish appears to be quite typical. The germinal epithelium, in which only epithelial cells (E) are observed, borders upon the ovarian lumen (OL). There is one resolvable basement membrane (bold arrow) between the germinal epithelium and follicle cells (F) of the ovarian follicle. The Z3 layer of the zona pellucida (ZP) has a striated appearance and a PAS-positive outer layer, there being only two layers of the zona pellucida resolved. The ooplasm is intensely PAS-positive and replete with less intensely PAS-positive cortical alveoli (ca). Bar 10 µm. PAS, metanil yellow, hematoxylin. **B**. Surface of full-grown oocytes from Common Snook. An oogonium (OG) is observed in the germinal epithelium. Beneath the zona pellucida (ZP), cortical alveoli (ca) and yolk globules (yg) are observed. Bar 10 µm. Thionin. **C**. Maturing oocytes from Gulf Killifish. The germinal vesicle (gv) has migrated to the animal pole of the oocyte and is still surrounded by oil droplets (od). Cortical alveoli (ca) are located at the oocyte periphery. The micropyle (bold arrow) is seen. Bar 100 µm. Hematoxylin-Eosin. **D**. Oocyte in maturation from Gulf Killifish showing the micropyle (bold arrow) in the zona pellucida (ZP). The germinal vesicle (gv) has migrated to just beneath the micropyle at the animal pole of the future egg. Oil droplets (od) and fluid yolk globules (yg) are observed. Bar 10 µm. Hematoxylin-Eosin. Original.

Fig. 2.10 Contd. ...

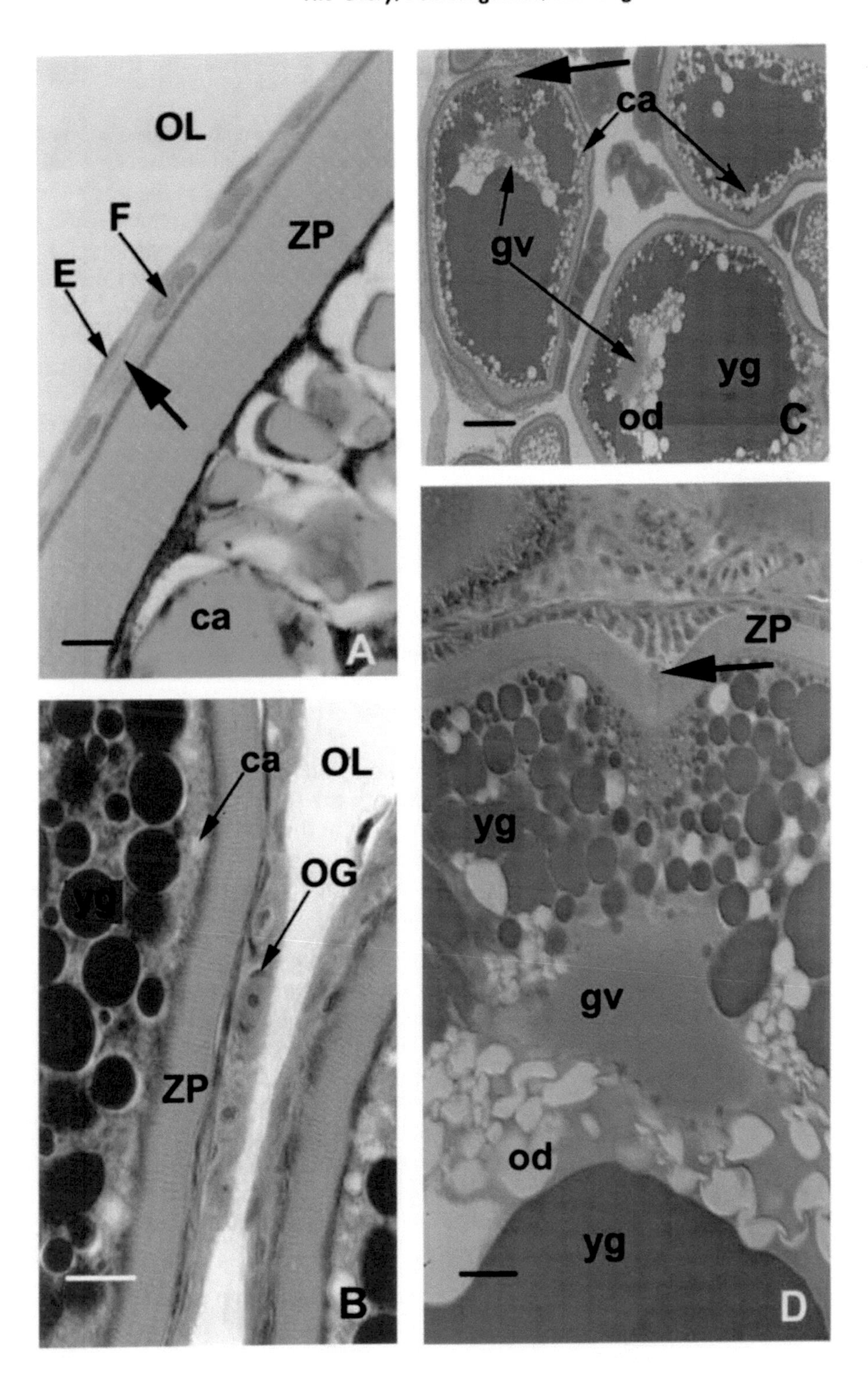
OL
F
ZP
E
ca
A
ca
gv
yg
od
C
ca
OL
OG
yg
ZP
B
ZP
yg
gv
od
yg
D

1992). Thus, secondary growth (vitellogenesis) is divided into three steps based mainly on the size of the follicle:

Early Secondary Growth Step (SGe): The oocyte begins to accumulate yolk globules, which appear initially at the periphery of the ooplasm but later accumulate around the germinal vesicle. The yolk globules at the periphery are typically small and intermingle with the cortical alveoli; whereas those closer to the germinal vesicle are larger and, in species with oil droplets, are found within and around the oil droplets, as in Rainbow Trout and Common Snook (Fig. 2.7A-D). The increasing size of the yolk globules may be a useful reference to distinguish a SGe oocyte that has reached the mid-point of secondary growth, as in Common Snook (Fig. 2.8A). In species such as Common Carp, as the oocyte reaches its midpoint of vitellogenesis, the accumulation of large yolk globules around the nucleus displaces the cortical alveoli to the periphery of the ooplasm causing a clear concentric separation into inner yolk globules and outer cortical alveoli layers (Blazer 2002).

Yolk globules in oocytes of many marine teleosts that produce pelagic eggs normally do not fuse until maturation (see below). However, in some species with demersal eggs such as Threespine Stickleback (*Gasterosteus aculeatus*) and Fourspine Stickleback (*Apeltes quadracus*), and several syngnathids, fusion of yolk globules begins during early vitellogenesis (Wallace and Selman 1981; Selman and Wallace 1989). In fact, fusion of yolk globules occurs in all the Atherinomorpha throughout vitellogenesis and is *prima facie* support for the monophyly of the group along with a number of other characters (Rosen and Parenti 1981; Parenti and Grier, 2004).

The zona pellucida continues to thicken and microvilli of the oocyte continue to elongate to maintain contact with the follicle cells. Homologous and heterologous gap junctions are often seen in these follicles (Iwamatsu *et al.* 1988; York *et al.* 1993). Homologous gap junctions are those that exist between follicle cells, and heterologous gap junctions are those between an oocyte and the surrounding follicle cells.

Late Secondary Growth Step (SGl): During late secondary growth, the diameter of the oocyte increases significantly due to increased deposition of yolk globules, and oil droplets in some species (Fig. 2.8A-C). The cortical alveoli are displaced to a relatively thin layer at the periphery of the ooplasm by the yolk globules as oocyte growth continues. In those species with extensive accumulation of oil droplets (Fig. 2.8A), these droplets are seen primarily encircling the germinal vesicle, but yolk globules and oil droplets co-mingle in Striped Mullet (Grier unpublished). In Goldfish, fusion of yolk globules begins during late vitellogenesis, prior to maturation (Yamamoto and Yamazaki 1961); however, oil droplets are absent in secondary growth oocytes of this species (Fig. 2.9A).

Full-Grown Step (SGfg): The follicle reaches its maximum pre-maturational diameter at this step. The term "full-grown" is commonly used in studies of fish reproductive physiology to refer to follicles just prior to the initiation of maturation (Nagahama 1987; York *et al.* 1993; Thomas *et al.* 2002; Carnevali *et al.* 2006). Morphologically, full-grown follicles (Figs. 2.8B,C, 2.9A,D) are similar to late vitellogenic oocytes, differing only in diameter. The oocytes contain abundant yolk globules distributed in most of the ooplasm, except the peripheral ooplasm which contains the cortical alveoli mixed with generally smaller yolk globules. In species with extensive accumulation of oil droplets, these will still be seen clustered around the germinal vesicle. One known exception is the Striped Mullet, a species where yolk globules and oil droplets intermix in the ooplasm and there is no oil droplets "zone" surrounding the germinal vesicle (Grier unpublished). In the full-grown follicle, the germinal vesicle is typically situated at the center of the oocyte. As in earlier stages of oogenesis, in most species examined only oocyte microvilli make it fully across the zona pellucida. As the follicle reaches its maximum diameter, the interactions between oocyte and granulosa cell microvilli, as well as the incidence of heterologous and homologous gap junctions, seem to decrease in species such as Atlantic Croaker (York *et al.* 1993). Physiologically, the full-grown follicle is typically insensitive to stimulation by progestins; namely, it will not mature if directly stimulated with these steroids (Patiño and Thomas 1990ab; Patiño *et al.* 2001; Patiño and Sullivan 2002).

Regulation: The endocrine and molecular regulation of secondary growth has been well studied in teleosts and extensively reviewed elsewhere (Patiño and Sullivan 2002; Jalabert 2005; Babin *et al.* 2007). Ovarian estrogens induce the production of hepatic vitellogenin as well as other materials required for secondary growth of the follicle, such as hepatic ZP proteins. The primary estrogen produced is estradiol-17β but other estrogens, such as estrone, have been also implicated in the regulation of vitellogenesis (van Bohemen *et al.* 1982). In many teleosts, ovarian estrogen production during secondary growth is regulated by the gonadotropin FSH, but there may exceptions such as Red Seabream where LH may be the primary gonadotropin at any time in development (Patiño and Sullivan 2002). The uptake of vitellogenin by its receptor on the oocyte surface may also be hormonally regulated, but further study seems to be necessary before generalizations can be made about uptake regulation (Patiño and Sullivan 2002).

2.4.5 Oogenesis: Oocyte Maturation (OM)

Oocyte maturation involves a sequence of morphological and physiological events in the ovarian follicle complex that precede ovulation. These events include distinct changes in the germinal vesicle and ooplasm (Patiño *et al.* 2001; Patiño and Sullivan 2002). Among the nuclear changes, the germinal vesicle first takes an eccentric position, then migrates to the periphery of the ooplasm at the animal pole, and finally breaks down just prior to the resumption and completion of the first meiotic division. These nuclear morphological changes are common to teleosts and other vertebrates. Therefore, they are used in the present schema to define the various steps of the Oocyte Maturation Stage. Morphological changes of the ooplasm that accompany maturation vary among species, but this variation seems to be explained, at least partly, by reproductive life history. Namely, ooplasmic changes are generally pronounced during oocyte maturation in species that produce pelagic eggs, whereas they are fairly indistinct in most species that produce demersal eggs.

Notable ooplasmic changes that occur during oocyte maturation in marine pelagic spawners include a significant increase in follicular diameter due to hydration (Figs. 2.11A-D, 2.12A-C), and an attending increase in egg buoyancy as it is released into seawater during spawning (water being less dense than seawater). It has been reported that the degree of ooplasmic hydration among various species of teleosts correlates with the degree of protein hydrolysis in yolk globules (Greeley *et al.* 1986; Carnevali *et al.* 2006; Cerdá *et al.* 2007). Thus, in most marine teleosts that produce buoyant eggs in saltwater, hydration of the ooplasm during maturation is associated with extensive hydrolysis of yolk-protein (Greeley *et al.* 1986; Carnevali *et al.* 2006; Cerdá *et al.* 2007); whereas in species such as Goldfish and others which are substrate or bottom spawners and produce demersal eggs, little or no detectable ooplasmic hydration is associated with minimal yolk-protein hydrolysis during maturation (Greeley *et al.* 1986). Hydration of marine pelagic eggs is driven, at least in part, by the osmotic gradient created by the free amino acids released during the proteolytic reaction (Patiño and Sullivan 2002; Cerdá *et al.* 2007). In fact, it has been proposed that yolk-protein hydrolysis in marine pelagic eggs is a *prima facie* factor that allowed fish to invade the saltwater seas about 100 million years ago (Fyhn *et al.* 1999; Fyhn and Kristoffersen 2007). There have been no observed substantial differences in oocyte development, including oocyte maturation and egg morphology, revealed histologically in marine

Fig. 2.11 Oocyte maturation in the Common Snook. **A**. Oocyte during OMegv. Oil droplets begin to fuse becoming oil globules (og) surrounding the germinal vesicle (gv). The ooplasm contains yolk globules (yg). Bar 100 µm. PAS, metanil yellow, hematoxylin. **B**. In OMegv, oil globules fuse until there is but a single oil globule (og) and the germinal vesicle (gv) is displaced from the center of the oocyte. Oil globules are surrounded by yolk globules (yg). Bar 50 µm. Thionin. **C**. During OMgvm, yolk globules (yg*) begin to fuse. The oil globule (og) is generally central, and the germinal vesicle (gv) has migrated to the animal pole of the oocyte. Bar 100 µm. Hematoxylin-Eosin. **D**. In OMgvb, the oocyte is preovulatory. Typical of Common Snook preovulatory oocytes and ovulated eggs, the ooplasm contains a web-like array of basophilic ooplasm. The oil globule (og) is central. Bar 100 µm. Hematoxylin-Eosin. Original.

Fig. 2.11 Contd. ...

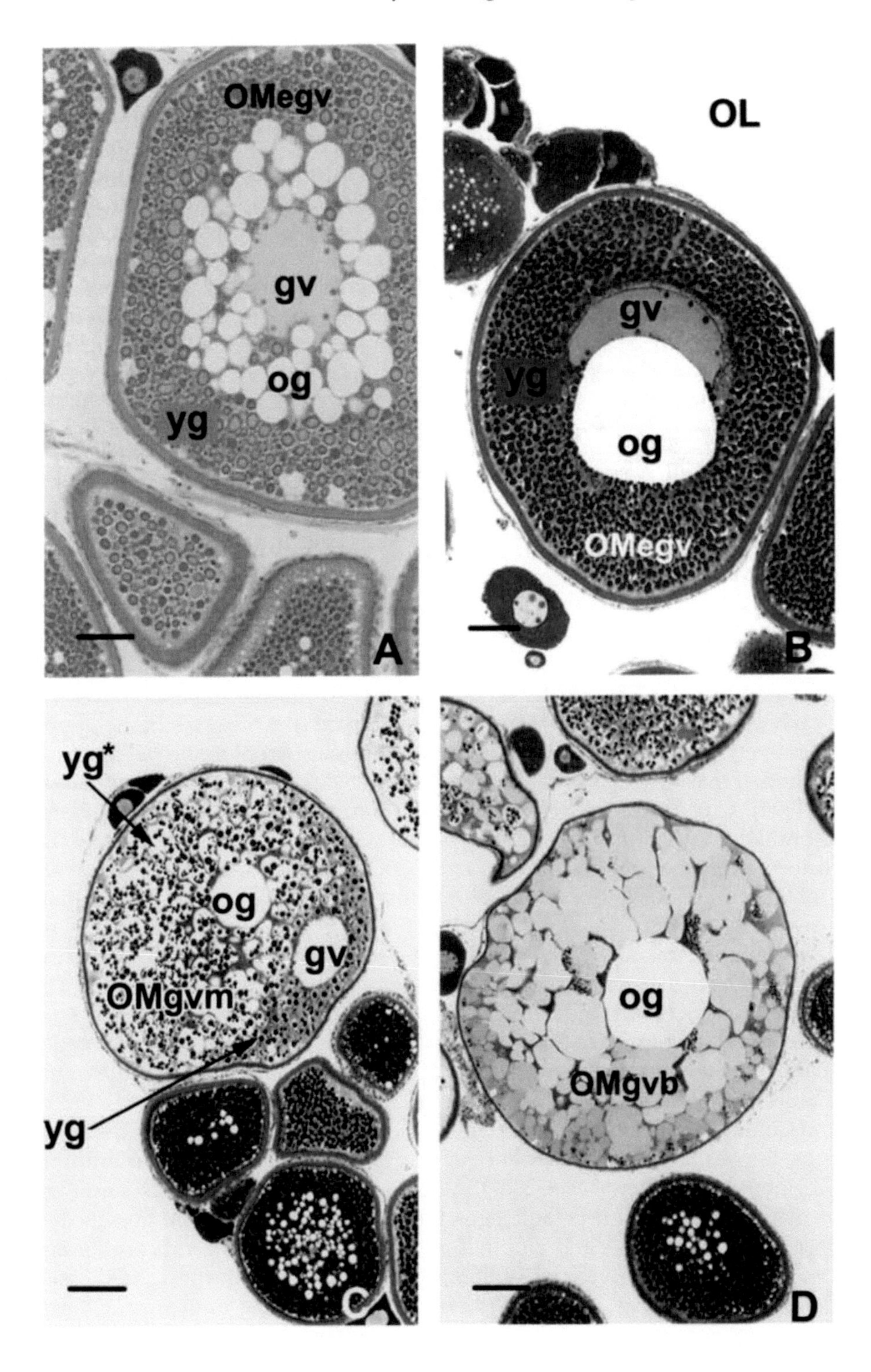
OMegv
gv
og
yg
A
OL
gv
yg
og
OMegv
B
yg*
og
gv
OMgvm
yg
C
og
OMgvb
D

teleosts of the families Centropomidae, Mugilidae, Pomacentridae, Sciaenidae, Serranidae, Tetradontidae (Grier unpublished); thus supporting the conclusion that significant uptake of water into oocytes during maturation in marine species is highly conserved (Fyhn *et al.* 1999). However, both in pelagic and in some demersal eggs, mechanisms other than yolk-protein hydrolysis may also contribute to the creation of osmotic gradients and hydration, such as passive and active ion intake by the oocyte (Cerdá *et al.* 2007). In fact, a small but significant degree of hydration (17 percent increase in water content) has been reported during maturation of Rainbow Trout oocytes, which produce demersal eggs in freshwater, in the absence of any detectable hydrolysis of yolk-protein (Milla *et al.* 2006). Although the entry of water into the maturing oocyte was initially believed to occur by simple passive diffusion, recent research indicates that water channels (aquaporins) on the oolema of the oocyte microvilli may actively participate in the water uptake driven by osmotic gradients (Cerdá *et al.* 2007).

Another dramatic ooplasmic change associated with maturation of marine pelagic eggs is the fusion of yolk globules (Fig. 2.11C). This fusion is normally accompanied by the development of ooplasmic transparency. Yolk globule fusion has been described as occurring in a centripetal direction, eventually forming a yolk mass at the center of the oocyte that displaces the ooplasm towards the periphery (Cerdá *et al.* 2007). However, the path of ooplasmic clearing in maturing oocytes of Atlantic Croaker (Yoshizaki *et al.* 2001) and Common Snook (Fig. 2.11A-D), which are also marine pelagic spawners, clearly occurs along the polar axis starting at the vegetal hemisphere. Moreover, a polarized fusion of yolk globules is apparent in the pelagic eggs of another marine teleost, the Japanese Whiting (see Fig. 2E,F in Matsuyama *et al.* 1990). Curiously, the yolk globules in demersal eggs of the freshwater Zebrafish also clear during maturation but without the globules undergoing fusion (Selman *et al.* 1993; Cerdá *et al.* 2007). As mentioned already, yolk globules in the oocytes of some species fuse much earlier, during secondary growth (see above).

Yet another major ooplasmic change seen during maturation of marine pelagic eggs is the coalescence of oil droplets to form larger oil globules. The designation of coalescing oil droplets as "globules" serves to separate oocytes in maturation from those that are growing or are full-grown. As the globules coalesce they become larger globules until there is a single oil globule in the mature oocyte, as described for Atlantic Croaker (Patiño and Thomas 1990c) and Common Snook (Grier 2000). However, in a marine species with demersal eggs, the Mummichog, coalescence of oil droplets yields not one but multiple oil globules (e.g., Cerdá *et al.* 2007). Also, as already noted there are a number of marine species with pelagic eggs that lack oil globules. The presence or absence of oil droplets and globules generally is not a good indicator of egg buoyancy, as noted earlier by Williams (1968). However, in specialized cases, the presence of large oil globules can contribute to egg buoyancy, even in freshwater species (reviewed by Cerdá *et al.* 2007).

The wide range of patterns of ooplasmic morphologic change, alluded to thus far, precludes the formulation of a classification schema, based on ooplasmic morphology, which would be applicable to describe maturation outside of specific groups of teleosts with shared traits of life history. It is perhaps due partly to this variability in ooplasmic morphology that the terminology for oocyte maturation is, and has been for many years, "confused and confusing" (see **Section 2.1**). There is one term in particular that is still widely used in the literature despite being obsolete and highly misleading: Final Oocyte Maturation (FOM). As currently used, FOM would include most if not all the steps of maturation described in this chapter (see below). But if one accepts the validity of FOM, then one must also define the corollary term, "Preliminary" Oocyte Maturation. This preliminary-final dichotomy is inconsistent with current knowledge of oocyte maturation, both at morphological and physiological levels. Thus, as indicated earlier by Patiño *et al.* (2001) and Patiño and Sullivan (2002), FOM is a misnomer and should no longer be used (the origin of the term was described by Patiño *et al.* 2001).

In contrast with ooplasmic morphology, nuclear morphological changes associated with oocyte maturation are constant across species and can be used as general landmarks (Steps) to describe and study maturation. Observations of nuclear morphology or position within the oocyte can be easily made under the dissecting microscope in chemically cleared follicles (e.g., with Serra solution; ethanol/60:formalin/30:acetic acid/10) or in follicles sandwiched between glass slides, as described by Neidig *et al.* (2000). In large, relatively opaque oocytes that do not clear during maturation and cannot be made to clear chemically, alternatives include hardening of the oocyte by boiling or by fixation in an appropriate solution followed by slicing (cutting with scalpel or sharp blade) under a dissecting microscope to determine the position or presence-absence of the germinal vesicle. Preparation of histological sections is also an option. Steps of teleost oocyte maturation (OM) can be classified as follows:

Eccentric Germinal Vesicle Step (OMegv): The first nuclear change associated with maturation is the appearance of an eccentric germinal vesicle in the direction of the animal pole. In species with little or no oil droplets in their ooplasm, this is also the only overall morphological change that characterizes OMegv oocytes. However, in marine pelagic eggs with extensive oil droplets around their germinal vesicle, such as Atlantic Croaker (Patiño and Thomas 1990b) and Common Snook (Neidig *et al.* 2000), the droplets also begin to coalesce, becoming oil globules (Figs. 2.9D, 2.11A), around the central germinal vesicle. As the coalescing oil globules become larger and fewer, they seemingly displace the germinal vesicle to an eccentric position (Fig. 2.11B). In Common Snook, the end of OMegv is marked by the coalescence of oil droplets into a single globule (Figs. 2.11A,B); Neidig *et al.* (2000). OMegv step combines the "stages" of Final Maturation-1 (FM-1) and Final Maturation-2 (FM-2) formerly used by Neidig *et al.* (2000). Ooplasmic clearing is not yet evident at this step of oocyte maturation (Fig. 2.11B) but slightly increased follicular diameters, indicating the onset of hydration, are evident in Atlantic Croaker (Yoshizaki *et al.* 2001). Dissociation between ooplasmic hydration and clearing has been also reported during oocyte maturation in Black Sea Bass (*Centropristes striata*) (Selman *et al.* 2001).

At the ultrastructural level, increased interactions between oocyte microvilli and follicle cells as well as gap junctional contacts (both homologous and heterologous) are observed during OMegv relative to full-grown follicles (York *et al.* 1993; Patiño and Kagawa 1999). Functionally, whereas the full-grown follicle is insensitive to progestins (see above), the OMegv oocyte is able to resume meiosis if stimulated with the appropriate progestin; namely, it has developed maturational competence (Patiño *et al.* 2001; Patiño and Sullivan 2002). Thus, the eccentricity of the germinal vesicle and, in oocytes of marine pelagic spawners, oil globule coalescence, can serve as simple morphological indices of an important functional change in the oocyte and follicle. The OMegv oocyte is not yet committed to resume meiosis unless stimulated with the appropriate progestin, as has been shown for Atlantic Croaker (Patiño and Thomas 1990ab) and a number of other species from diverse taxa including freshwater species that produce demersal eggs (Patiño *et al.* 2001).

Germinal Vesicle Migration Step (OMgvm): Upon stimulation with progestin, the germinal vesicle moves towards the animal pole and reaches the periphery of the ooplasm (Fig. 2.11C). At the same time, the follicular machinery responsible for meiotic resumption typically becomes irreversibly activated. Thus, functionally, the OMgvm follicle becomes committed to complete maturation and resume meiosis. In marine species that produce pelagic eggs, such as Common Snook (Neidig *et al.* 2000) and others, ooplasmic changes include the onset of yolk globule coalescence, a phenomenon which is quite pronounced in wet mounts (Fig. 2.12 A,B) (Neidig *et al.* 2000). But as noted earlier, in demersal eggs of a number of freshwater and marine species the fusion of yolk globules occurs or begins well before maturation, during secondary growth (see above). Thus, yolk globule fusion is not a universal feature of ooplasmic maturation among teleosts. In oocytes of some marine species that produce pelagic eggs, as Common Snook, a single oil globule (Fig. 2.11B) is present at the beginning of germinal vesicle migration, but sometimes oil globules may continue to fuse during this step in hormone-treated (time-release gonadotropin-releasing hormone) individuals (Grier unpublished). After one single oil globule has formed (Fig. 2.11B), yolk globules begin to coalesce (Fig. 2.11C). However, in other species such as Atlantic Croaker, ooplasmic clearing may begin before oil globules have fully coalesced (Yoshizaki *et al.* 2001). The OMgvm step ends when the germinal vesicle is at the animal pole. Ooplasmic hydration results in increased oocyte diameters during this step. As noted earlier, ooplasmic clearing occurs in a polarized pattern in species such as Atlantic Croaker (Yoshizaki *et al.* 2001) and Common Snook (Fig. 2.11).

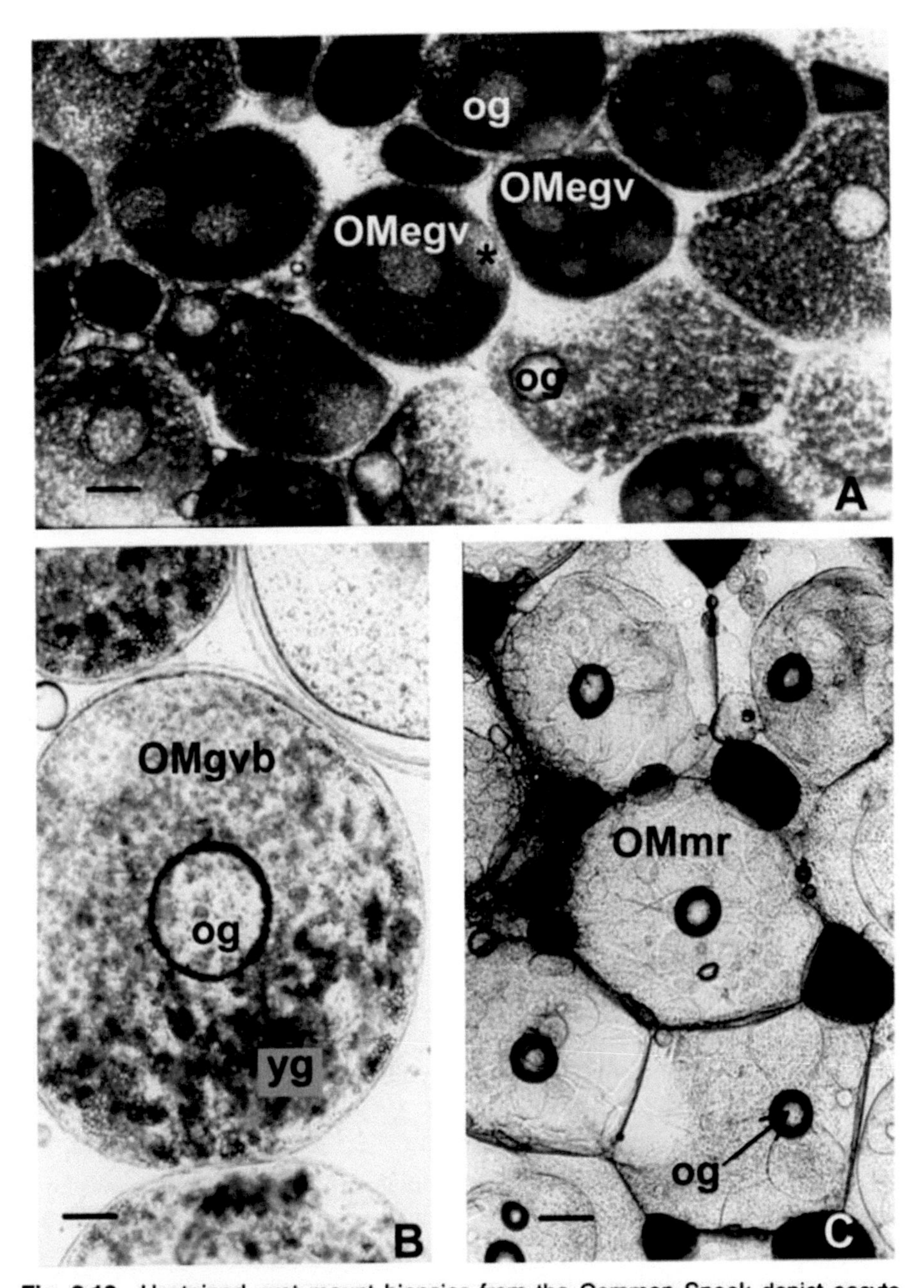

Fig. 2.12 Unstained, wet mount biopsies from the Common Snook depict oocyte maturational stages. **A**. Coalescing oil globules (OMegv), a single oil globule (og) have a dark border. Bar 20 µm. **B**. As yolk globules (yg) clear in OMgvb, the germinal vesicle is no longer observed. The oil globule (og) is central in location. Bar 50 µm. **C**. Preovulatory oocytes in OMmr have a single oil globule (og). "Lines" observed in the fluid yolk are ooplasm. Bar 100 µm. Original.

Germinal Vesicle Breakdown Step (OMgvb): The germinal vesicle "breaks down" shortly after reaching the animal pole. It is no longer visible using any of the standard visualization techniques (see above). This step is commonly referred to as germinal vesicle breakdown (GVBD) in the literature. Dissolution of the nuclear membrane is necessary for meiotic cell division, and this landmark occurrence is widely used as index of oocyte maturation in studies of teleosts and other vertebrates. In oocytes of marine pelagic spawners, the fusion of yolk globules is almost complete by the time of GVBD, and their contents are dissolved into a clear transparent yolk. In Common Snook, the preovulatory oocyte (Fig. 2.11D, 2.12C) has streams of basophilic ooplasm throughout, an uncommon character among teleosts that produce pelagic eggs. Ooplasmic clearing is at an advanced state and the oocyte has increased further in size (Yoshizaki *et al.* 2001). Microvilli extending from the oocyte to the follicle cells have began to retract, as described in Atlantic Croaker (York *et al.* 1993) and Red Seabream (Matsuyama *et al.* 1991). However, gap junctional coupling between oocyte microvilli and follicle cells is still observed although at reduced levels compared to follicles at earlier steps of maturation (Yoshizaki *et al.* 2001; Bolamba *et al.* 2003).

Meiotic Resumption Step (OMmr): The first meiotic division is completed producing the first polar body (which subsequently degenerates), and the oocyte advances to second metaphase, when the oocyte becomes arrested for the second time during oogenesis. In most marine species that produce pelagic eggs, the yolk globules have formed a continuous yolk mass and cytoplasmic clearing is complete. One exception is the Northern Anchovy (Hunter and Macewicz 1985), where yolk globules become large "platelets" in the matured oocyte perhaps similar to those in the matured, demersal eggs of Zebrafish (Selman *et al.* 1993). In some species, the ooplasm is restricted to a thin, basophilic band at the animal pole, immediately beneath the zona pellucida. Oocyte microvilli are completely retracted and heterologous gap junctional communication has ended. The oocyte has reached its maximum preovulatory size. As noted already, a hydration-dependent increase in the size of matured oocytes is observed even in some freshwater species which produce demersal eggs and which do not undergo proteolytic hydrolysis of yolk (Milla *et al.* 2006).

Regulation: The endocrine regulation of oocyte maturation has been extensively studied (Patiño *et al.* 2001; Patiño and Sullivan 2002; Thomas *et al.* 2007). Briefly, it appears that all functional and morphological changes associated with maturation, both nuclear and ooplasmic, are ultimately regulated by pituitary LH. Further, with the notable exception of the first step of maturation (OMegv), the action of LH is mediated by follicular progestins, either DHP or 17α, 20β, 21-trihydroxy-4-pregnen-3-one, depending on the species.

The LH-dependent ability of the oocyte to respond to progestin is acquired during OMegv. This ability is termed "oocyte maturational competence" and it develops independently of steroid production. In fact, in species such as Atlantic Croaker (Patiño and Thomas 1990a), OMegv follicles are unable to produce progestins. Maturational and progestin-production competencies are both acquired during OMegv and they are among the principal physiological transformations of the ovarian follicle during the first step of oocyte maturation. All subsequent steps (OMgvm through OMmr) require follicular production of progestin, which acts by binding to a membrane receptor on the surface of the oocyte (Patiño and Thomas 1990c; Thomas *et al.* 2007). The irreversible commitment to resume meiosis in OMgvm oocytes is likely due to the onset of endogenous production of progestin and consequent irreversible stimulation of the oocyte's machinery (Patiño and Thomas 1990b). Although LH and LH-dependent progestin are the main hormones regulating oocyte maturation, other factors have also been implicated (Patiño *et al.* 2001; Pang and Ge 2002; Patiño and Sullivan 2002; Ge 2005; see also Schneider and Poehland, **Volume 8B, Chapter 2).**

2.4.6 Oogenesis: Ovulation (OV)

Prior to ovulation, the shared basement membrane that separates the follicle cells from the germinal epithelium also separates the preovulatory oocyte from the ovarian lumen. At ovulation, the shared basement membrane, follicle cells and overlying germinal epithelium break, creating an opening through which the oocyte moves from the follicle into the ovarian lumen, becoming an egg (Fig. 2.13A-C). During ovulation, only the oocyte is voided into the lumen, leaving behind the postovulatory follicle complex (Fig. 2.14A,B). The size (diameter) of the egg increases further during ovulation due to continued hydration, but this increase is relatively small compared to maturation (Cerdá *et al.* 2007). The follicle cells, now part of the postovulatory follicle complex, become a continuum with the epithelial cells of the germinal epithelium, from which they were originally derived during folliculogenesis (Fig. 2.14C). The shape of postovulatory follicle complexes is irregular, as a result of the collapse of the follicle at ovulation, but their lumen is continuous with that of the ovarian lumen. The preceding description of the morphological interactions among ovarian follicle components during ovulation is based on observations with Rainbow Trout (Grier *et al.* 2007) and Common Snook (Grier 2000). However, it seems probable that the same process occurs in other teleosts. The ovulated egg remains in meiotic arrest until fertilization, at which time the second meiotic division is completed with the expulsion of the second polar body and the haploid pronuclei come together to form a diploid zygote.

Because postovulatory follicle complexes are a source of steroid hormones, including estrogens (Nagahama *et al.* 1982; Smith and Haley 1987, 1988), the possibility has been raised that they are involved in the induction of the next oogenic cycle (Smith and Haley 1987, 1988; see above). In Mozambique Tilapia, postovulatory follicle complexes can persist for more than 3 weeks after spawning (Smith and Haley 1987); in Rainbow Trout, for at least 2 weeks

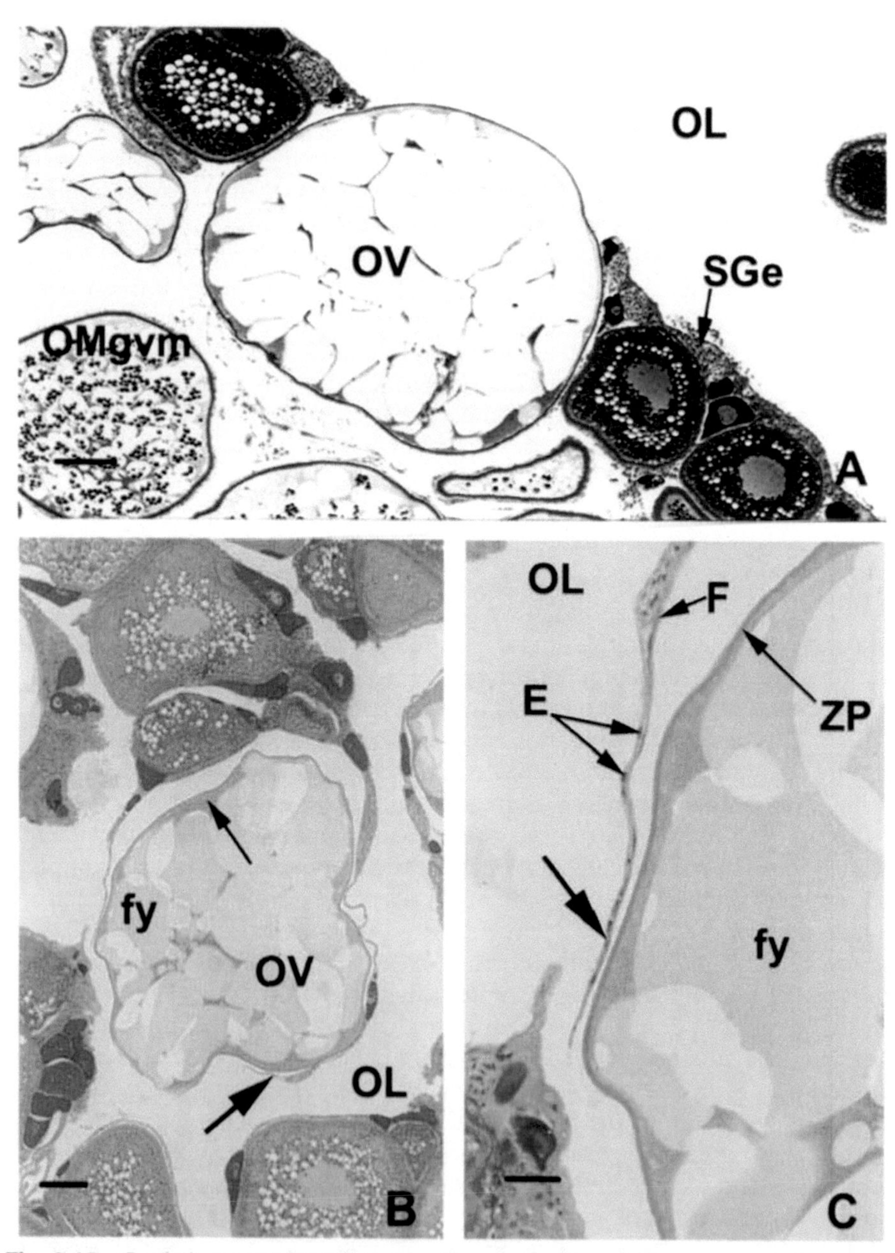

Fig. 2.13 Ovulating eggs from Common Snook. **A**. At ovulation an oocyte becomes an egg (OV) which is emerging from an ovarian follicle and entering the ovarian lumen (OL). An oocyte with clearing yolk globules (OMgvm) is observed as are follicles in early secondary growth (SGe). Bar 50 μm. Thionin. **B**. An ovulating egg (OV) entering the ovarian lumen (OL). Ooplasm is primarily at the oocyte periphery (arrow), but thin bands of ooplasm are seen in the fluid yolk (fy) in this species.

Fig. 2.13 Contd. ...

(Grier *et al.* 2007); and in Northern Anchovy (Hunter and Macewicz 1985) and Japanese Whiting (Matsuyama *et al.* 1991), for only 24-48 hours. The mechanism of postovulatory follicle degeneration seems to be apoptosis (Drummond *et al.* 2000). The presence and morphology of postovulatory follicle complexes can be used to estimate spawning frequency (Hunter and Macewicz 1985), an important parameter of the reproductive biology of fishes with application in fisheries management.

An interesting behavioral correlate of postovulatory follicle complex function has been described for Mozambique Tilapia. In this species, a mouthbrooder, the postovulatory follicle complex will survive and possibly synthesize steroids for at least 25 days after ovulation in females that continue to hold their developing young in their mouths (Smith and Haley 1987, 1988). However, in females that ate the fertilized eggs and therefore showed no parental behavior, postovulatory follicle complexes disappeared within 10 days (Smith and Haley 1987).

Regulation: Ovulation is regulated by the same primary endocrine factors that control oocyte maturation: pituitary LH and LH-dependent follicular progestin (Patiño and Sullivan 2002; Patiño *et al.* 2003ab). However, the action of progestin on ovulation is mediated not by a membrane receptor on the oocyte but by a classical intracellular steroid receptor in somatic cells, possibly the follicle cells. Like oocyte maturational competence, the ability of the follicle to ovulate in response to progestin (ovulatory competence) is also induced by LH early during the process of oocyte maturation, just after the acquisition of maturational competence (Patiño *et al.* 2003b).

The mechanisms by which progestin induces ovulation are uncertain. However, it is clear that rupture of the follicle, at the point of shared basement membrane with the germinal epithelium, is a necessary outcome of these mechanisms. In the ovarian follicle of Medaka, ovulation is accompanied by increased activities of gelatinase A, which degrades collagen type IV present in basement membranes, and membrane-type 2 metalloproteinase, which degrades type I collagen thought to be expressed in the theca (Ogiwara *et al.* 2005). These observations were interpreted as an indication that rupture of the basement membrane and the theca are both necessary for ovulation (Ogiwara *et al.* 2005). The increased activity of gelatinase A is consistent with current

---

Fig. 2.13 Contd. ...

Bold arrow points to both epithelial and follicle cells that separated so that the egg could enter the ovarian lumen. Bar 100 μm. PAS, metanil yellow, hematoxylin. **C.** Magnification from previous micrograph (**B**) showing epithelial (E) cells of the germinal epithelium and follicle (F) cells (bold arrow) separated from the surface of the emerging egg. The zona pellucida (ZP), ooplasm (arrow), and fluid yolk (fy) are visible. Note the follicle cells (F) are no longer on the egg surface. Their nuclei are small, squamous, and their cytoplasm is thinly-stretched. Bar 10 μm. PAS, metanil yellow, hematoxylin. Original.

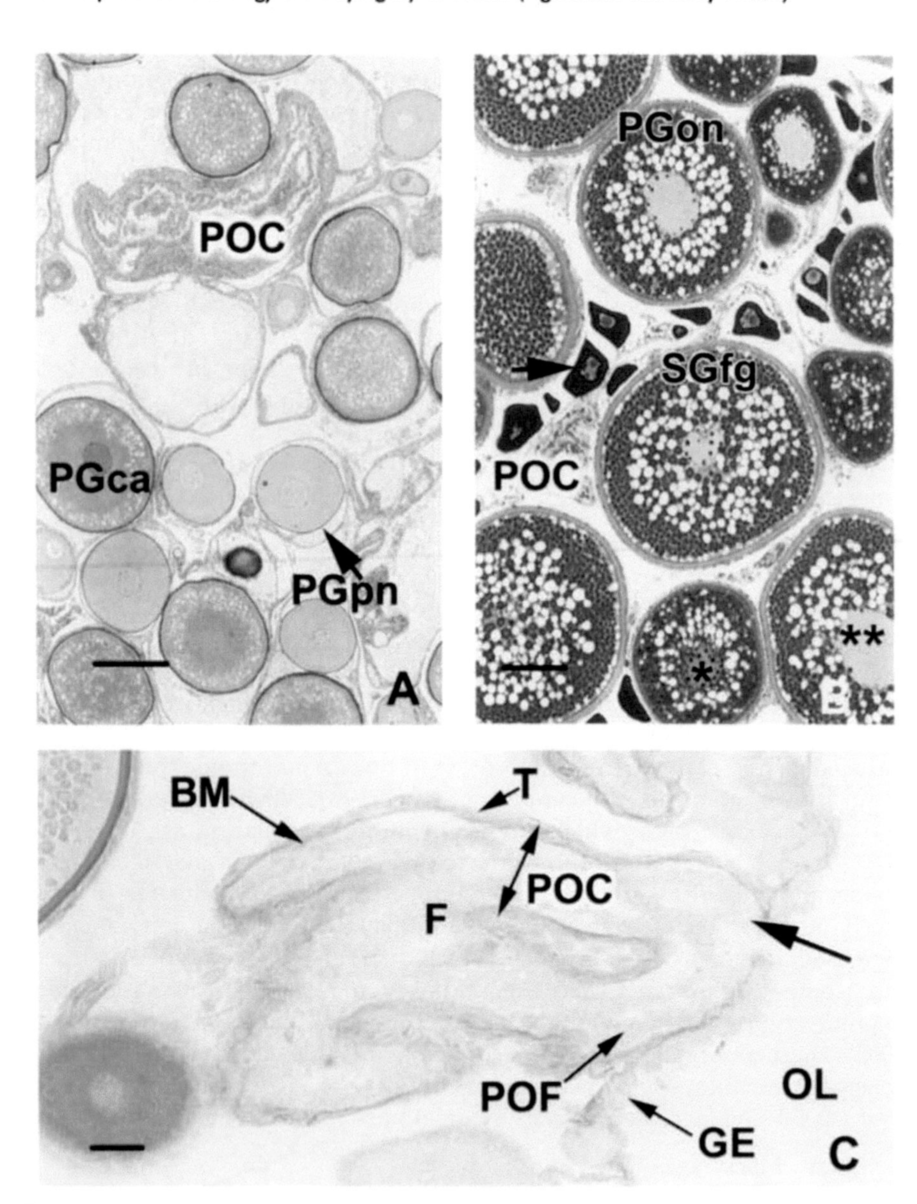

Fig. 2.14 After ovulation, postovulatory follicle complexes (POC) are common in lamellae. **A**. A postovulatory follicle complex from the Rainbow Trout among primary growth follicles, some in the perinucleolar step (PGpn) and others in the cortical alveolar step (PGca) of primary growth. Bar 100 μm. PAS, metanil yellow, hematoxylin. **B**. A postovulatory follicle complex (POC) is observed among full-grown oocytes (SGfg) and oocytes with multiple nucleoli (bold arrow), early (*) and late (**) secondary growth oocytes in Common Snook. Bar 100 μm. Thionin. **C**. A postovulatory follicle complex (POC), designated by a double arrow, from Common

Fig. 2.14 Contd. ...

understanding of the morphology of ovulation; namely, the shared basement membrane that separates the follicle from the germinal epithelium must break for the oocyte to be released into the ovocoel. However, the relevance of increased activity membrane-type 2 metalloproteinase, if in fact is found in the theca, is uncertain because rupture of the theca is not physically necessary for ovulation; namely, the theca does not encompass the ovarian follicle at the region of shared basement membrane with the germinal epithelium (where ovulation occurs). In addition to proteases, a number of other factors are suspected to play a role in ovulation, such as increased follicular production of prostaglandins as in Rainbow Trout (Jalabert and Szöllösi 1975), Goldfish (Stacey and Pandey 1975), and Atlantic Croaker (Patiño *et al.* 2003a). The specific mechanisms activated by prostaglandins to induce ovulation are uncertain.

2.5 ATRESIA

Atresia is the process of degeneration and removal of ovarian follicles from the ovary. Although not a "stage" of oogenesis, it is an important aspect of ovarian cycles and physiology. It plays an essential role in regulating fecundity, by reducing the number of oocytes that attain ovulation compared to those which initiate oogenesis during the reproductive cycle (Lambert 1970; Saidapur 1978; Van den Hurk and Peute 1979; Guraya 1986; Miranda *et al.* 1999). The absorption of oocytes during atresia permits the recovery and recycling of their components.

Follicles may become atretic at any stage of oogenesis, but it is more frequent in secondary growth follicles (Blazer 2002). Increased atresia, particularly in vitellogenic follicles, can indicate a pathological condition (Johnson *et al.* 1988). After spawning, oocytes that do not ovulate also become atretic. The follicle cells play a very active role during atresia in digesting and removing the oocyte. Diverse environmental or physiological conditions can increase the incidence of atretic follicles such as insufficient nutrition (e.g., Rainbow Trout, Bromage *et al.* 1992) or exposure to contaminants (e.g., English Sole, *Parophrys vetulus*, Johnson *et al.* 1988; Zebrafish, Van den Belt *et al.* 2002). Van den Belt *et al.* (2002) described a reduction in the number of vitellogenic follicles and increase in atresia in the ovary of Zebrafish in response to exposure to the potent estrogenic compound, ethynylestradiol-17α, a synthetic analogue of estradiol-17β.

Fig. 2.14 Contd. ...

Snook that was fixed within a half hour after ovulation. There is a well-defined basement membrane (BM) separating the follicle cells (F) from the thecal cells (T). A postovulatory follicle (POF) refers only to the follicle cell layer of the POC. Germinal epithelium (GE) cells and those of the postovulatory follicle complex (POC) are continuous (bold arrow). Bar 10 μm. PAS, metanil yellow, no hematoxylin to emphasize basement membranes. Original.

Several morphological changes are observed during atresia (Fig. 2.15A-D). These include alterations in the appearance of the ooplasm, disintegration of the germinal vesicle and fragmentation and folding of the zona pellucida (Fig. 2.15C). Atretic follicles during secondary growth initially possesses yolk globules (Fig. 2.15A), but these soon disappear (Fig. 2.15B-D). Atresia is associated with two successive events: a) development of the atretic follicle, which involves degeneration and resorption of the oocyte, and proliferation and hypertrophy of the follicle cells which phagocytize the degenerating oocyte (follicle cells move from the follicle layer to the ooplasm; later, cells of the theca layer may also penetrate into the atretic follicle); and b) phagocytosis of follicle cells by stromal cells (after resorption of the oocyte is completed). Because follicles are attached to the germinal epithelium [e.g., Rainbow Trout (Grier *et al.* 2007), so too are atretic follicles (Fig. 2.15A-C). This observation has not been reported previously. Atretic follicles possess numerous oil globules (Fig. 2.15 A,B) that increase in number and diameter as atresia progresses (compare Figs. 2.15A,C with 2.15B,D).

Atretic follicles have been identified in numerous species of teleosts such as Goldfish (Yamamoto and Yamazaki 1961; Nagahama 1983); *Gobius giuris* (Rajalakshmi 1966); *Eucalia inconstans* (Braekevelt and McMillan 1968); the Cuchia, *Amphipnous cuchia* (Rastogi 1969); Guppy (Lambert 1970); *Acanthobrama terreasanctae*; Nile Tilapia, *Tilapia nilotica* (Yaron 1971); Fourspine Stickleback; Threespine Stickleback (Wallace and Selman 1979); English Sole (*Parophrys vetulus*) (Johnson *et al.* 1988); King Weakfish (*Macrodon ancylodon*) (Vizziano and Berois 1990); Rainbow Trout (Van den Hurk and Peute 1979; Tyler *et al.* 1990; Bromage *et al.* 1992); Atlantic Cod (*Gadus morhua*) (Kjesbu *et al.* 1991); Two-spot Astyanax (*Astyanax bimaculatus lacustris*); the characiform (*Leporinus reinhardtii*)

Fig. 2.15 Follicular atresia. **A.** In Common Snook, early atresia is indicated by the disorganization of the follicle cell layer, yolk globules (yg), and oil droplets (od) in the oocyte of the atretic follicle (AF), which is attached (bold arrow) to the germinal epithelium (GE). Ovarian lumen (OL). Bar 50 μm. Thionin. **B**. Ovary from the Red Grouper *Epinephelus morio* shows the ooplasm of the atretic oocyte invaded by phagocytic follicle cells (arrows). The basement membrane (BM) that surrounded the follicle appears to have remained intact. Ovarian lumen (OL) and the germinal epithelium (GE) are seen. Bar 10 μm. PAS, metanil yellow, hematoxylin. **C**-**D**. Follicle atresia in Rainbow Trout. **C**. The zona pellucida (ZP) of the atretic follicle (AF) breaks up and becomes "internalized" within the atretic oocyte. The atretic follicle remains attached to the germinal epithelium (bold arrow) by a column of cells in which blood vessels (arrows) are observed. Bar 10 μm. PAS, metanil yellow, hematoxylin. **D**. The breakdown of ooplasmic components causes lipids (li) to accumulate, becoming the major component in the atretic follicle (AF). Follicles with perinucleolar oocytes (PGpn), cortical alveolar oocytes (PGca), and early vitellogenic oocytes (SGe) are observed surrounded by the stroma (ST). Bar 100 μm. PAS, metanil yellow, no hematoxylin. Original.

Fig. 2.15 Contd. ...

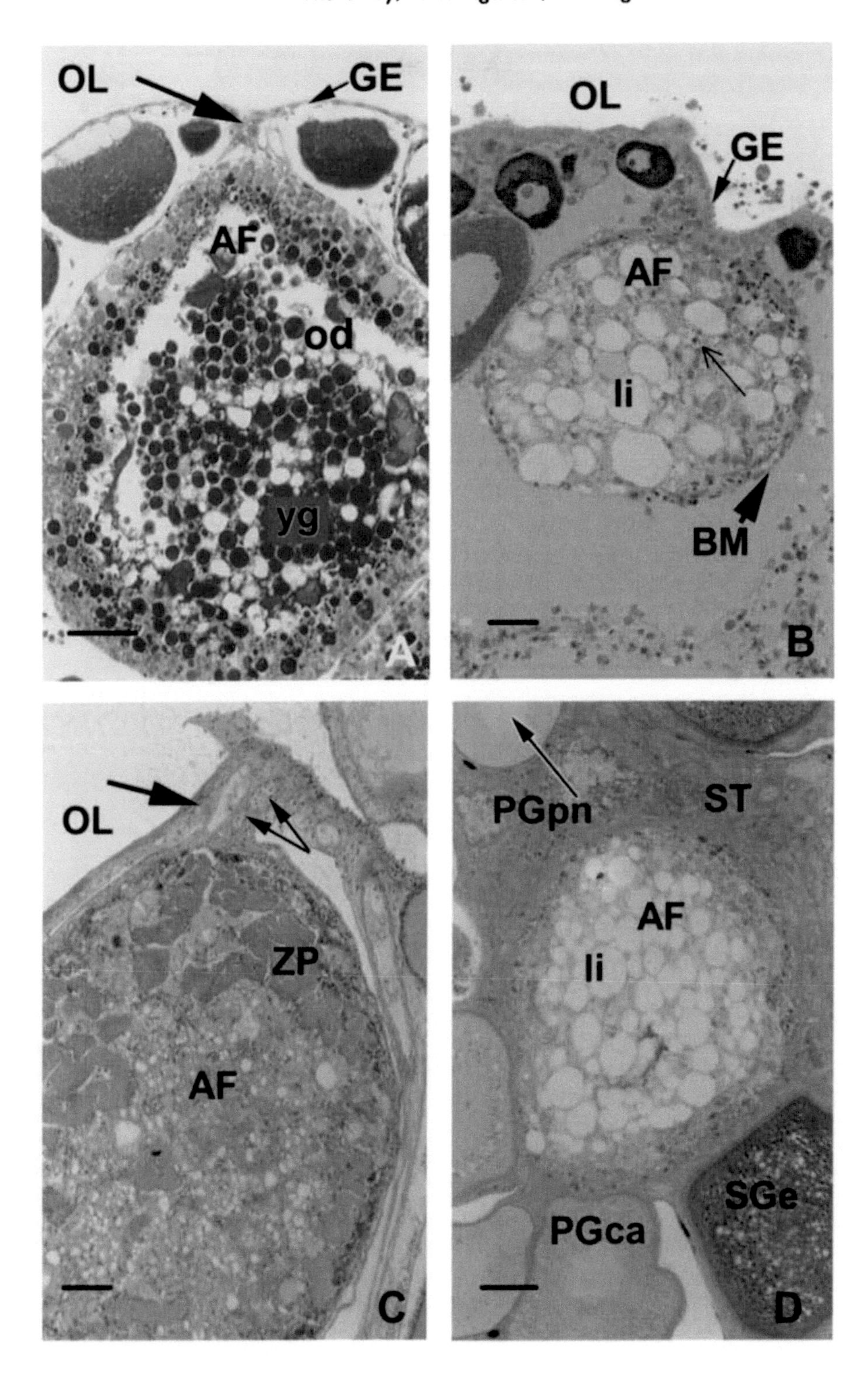
OL
GE
AF
od
yg
A
OL
GE
AF
li
BM
B
OL
ZP
AF
C
PGpn
ST
AF
li
SGe
PGca
D

(Miranda *et al.* 1999); Common Carp; Largemouth Bass (Blazer 2002); Balsas Splitfin (Uribe *et al.* 2006); and others.

Regulation: There is little information concerning the endocrine and other physiological mechanisms controlling atresia in the teleost ovarian follicle. In mammals, it is generally believed that atresia is associated with apoptosis (Habibi and Andreu-Vieyra 2007).

2.6 CHAPTER SUMMARY

In most teleosts, the ovaries are paired, elongated, saccular structures with a central lumen or ovocoel. From the ovarian wall, highly branched, irregular lamellae project into the ovocoel. These lamellae consist of a stroma and numerous ovarian follicles and are covered by an epithelium, the germinal epithelium. The germinal epithelium is composed of oogonia and oocytes scattered individually or in cell nests among somatic epithelial cells. Oogonia divide mitotically forming clusters of oogonia and prefollicle cells, the latter being derived from epithelial cells. Meiosis begins in oocytes and folliculogenesis is initiated as prefollicle cells begin to surround individual oocytes. Folliculogenesis is completed when the oocyte and its encompassing prefollicle cells become completely surrounded by a basement membrane. The prefollicle cells then become follicle cells. During folliculogenesis and thereafter, the stroma forms the theca around the basement membrane. The theca is first a single layer of thecal cells but eventually becomes multilayered and also contains capillaries and collagen fibers. Collectively, the ovarian follicle (oocyte and follicle cells), the basement membrane, and the theca form the follicle complex. There are six major stages of oogenesis: oogonial proliferation, chromatin nucleolus, primary growth, secondary growth (vitellogenesis), maturation, and ovulation. Oogonial proliferation typically occurs at, or soon after, the end of the preceding reproductive cycle and may be under the control of ovarian estrogens. Onset of meiosis marks the beginning of the chromatin nucleolus stage, which then ends at early diplotene of Prophase I. Early diplotene oocytes begin to develop lampbrush chromosomes and have a single prominent nucleolus, and their cytoplasm still has a hyaline appearance. The chromatin nucleolus stage of development may be regulated by progestins. Primary growth, marked histologically by the onset of ooplasmic basophilia, commences in late diplotene as meiosis becomes arrested. Initially, these oocytes possess a single nucleolus as do early diplotene oocytes but are distinguished by ooplasmic basophilia and the fact that they continue to grow in size. Folliculogenesis is completed soon after meiotic arrest. Cortical alveoli and, in many species also oil droplets (lipid yolk), appear in the ooplasm during primary growth, and there is also a significant increase in oocyte diameter. There is evidence suggesting that pituitary gonadotropin (FSH), insulin-like growth factors, and perhaps sex steroids and other hormones regulate follicular development during primary growth. Secondary growth, or vitellogenesis, is characterized by the onset of deposition of yolk globules in the ooplasm. During vitellogenesis, FSH-dependent ovarian estradiol-17β induces the production of hepatic vitellogenin, which is taken up by the oocyte, processed, and deposited as yolk globules. The follicle reaches its maximum (prematurational) size at the end of secondary growth as the oocyte becomes full-grown. Oocyte maturation, culminating with the resumption and completion of the first meiotic division, involves a series of nuclear and ooplasmic events. The germinal vesicle (nucleus) first becomes eccentric, and later migrates to the periphery of the ooplasm at the animal pole. Germinal vesicle breakdown is followed by the resumption of meiosis. The oocyte acquires sensitivity to progestin (maturational competence) coincidentally with the eccentric position of the germinal vesicle and, in some species, also with the ability of the ovarian follicle complex to produce progestin. Progestin then induces germinal vesicle migration and the resumption of meiosis. In oocytes of marine species that produce pelagic eggs, remarkable ooplasmic changes are also evident during maturation, including the coalescence of oil droplets, becoming oil globules and eventually a single oil globule; the clearing of ooplasm due to proteolysis and fusion of yolk globules; and a significant increase in oocyte size due to ooplasmic hydration. When the ooplasm has become completely cleared, the first meiotic division has been completed and the oocyte has arrested again, this time at metaphase of the second meiotic division. The primary hormone regulating all of the functional and morphological events of maturation, both nuclear and ooplasmic changes, is pituitary LH. During ovulation, the basement membrane separating the follicle from the germinal epithelium breaks and the follicle cells come in contact with the epithelial cells of the germinal epithelium; the oocyte is then released into the ovocoel. Ovulation is also under the control of LH-dependent ovarian progestin, but the cellular and molecular mechanisms of ovulation are different from those of maturation. Postovulatory follicle complexes consist of postovulatory follicle cells, basement membrane and theca. In some species they continue to produce steroid and may be involved in the induction of the next oogenic cycle. Atresia, a degenerative and removal process, is often observed in vitellogenic follicles and full-grown follicles especially at the end of a breeding season.

2.7 ACKNOWLEDGMENTS

The authors are indebted to and thank the histology staff of the Florida Fish and Wildlife Research Institute, Catalina Brown, Noretta Perry, Illiana Quintero, Ruth Reese, and Yvonne Waters for their excellence in histological processing. We also thank Drs. Craig Sullivan and Vicki Blazer for their critical review of the chapter. We thank Gabino De la Rosa-Cruz and Adriana García-Alarcón, Facultad de Ciencias, Universidad Nacional Autónoma de México, for their contribution in the preparation of the plates.. The Texas

Cooperative Fish and Wildlife Research Unit is jointly supported by the U.S. Geological Survey, Texas Tech University, Texas Parks and Wildlife, The Wildlife Management Institute, and the Fish and Wildlife Service. We are grateful to Professor Barrie Jamieson for his advice throughout preparation of this chapter and his substantial editorial aid.

2.8 LITERATURE CITED

Abookire, A. A. 2005. Reproductive biology, spawning season, and growth of female rex sole (*Glyptocephalus zachirus*) in the Gulf of Alaska. Fish Bulletin 104: 350-359.

Babin, P. J., Carnevali, O., Lubzens, E. and Schneider, W. J. 2007. Molecular aspects of oocyte vitellogenesis in fish. Pp 39-76. In P. J. Babin, J. Cerdá and E. Lubzens (eds), *The Fish Oocyte: From Basic Studies to Biotechnological Applications*. Springer. Dordrecht, The Netherlands.

Bailey, R. J. 1933. The ovarian cycle in the oviparous teleost *Xiphophorus helleri*. Biological Bulletin 64: 206-225.

Begovac, P. C. and Wallace, R. A. 1987. Ovary of the pipefish, *Syngnathus scovelli*. Journal of Morphology 193: 117-133.

Begovac, P. C. and Wallace, R. A. 1988. Stages of oocyte development in the pipefish, *Syngnathus scovelli*. Journal of Morphology 197: 353-369.

Berlinsky, D. L. and Specker, J. L. 1991. Changes in gonadal hormones during oocyte development in the striped bass, *Morone saxatilis*. Fish Physiology and Biochemistry 9: 51-62.

Blazer, V. S. 2002. Histopathological assessment of gonadal tissue in wild fishes. Fish Physiology and Biochemistry 26: 85-101.

Bolamba, D., Patiño, R., Yoshizaki, G. and Thomas, P. 2003. Changes in homologous and heterologous gap junction contacts during maturation-inducing hormone-dependent meiotic resumption in ovarian follicles of Atlantic croaker. General and Comparative Endocrinology 131: 291-295.

Braekevelt, C. R. and McMillan, D. B. 1968. Cyclic changes in the ovary of the brook stickleback *Eucalia inconstans* (Kirtland). Journal of Morphology 123: 373-396.

Bromage, N. R., Jones, J., Randall, C., Thrush, M., Davis, B., Springate, J., Duston, J and Barker, G. 1992. Broodstock management, fecundity, egg quality and the timing of egg production in the Rainbow trout (*Oncorhynchus mykiss*). Aquaculture 100: 141-166.

Brown-Peterson, N. J., Thomas, P. and Arnold, C. R. 1988. Reproductive biology of the spotted seatrout, *Cynoscion nebulosus*, in sout Texas. Fishery Bulletin 86: 373-388.

Brummett, A. R. and Dumont, J. N. 1981. A comparison of. chorions from eggs of northern and southern populations of *Fundulus heteroclitus*. Copeia 1981:607-614.

Bullough, W. S. 1942. Gametogenesis and some endocrine factors affecting it in the adult minnow (*Phoxinus laevis*). Journal of Endocrinology 3: 211-219.

Campbell, B., Dickey, J., Beckman, B., Young, G., Pierce, A., Fukada, H. and Swanson, P. 2006. Previtellogenic oocyte growth in salmon: relationships among body growth, plasma insulin-like growth factor-1, Estradiol-17beta, follicle-stimulating hormone and expression of ovarian genes for insulin-like growth factors, steroidogenic-acute regulatory protein and receptors for gonadotropins, growth hormone and somatolactin. Biology of Reproduction 75: 34-44.

Carnevali, O., Cionna, C., Totsi, L., Lubzens, E. and Maradonna, F. 2006. Role of cathepsins in ovarian follicle growth and maturation. General and Comparative Endocrinology 146: 195-203.

Çek, S., Bromage, N., Randall, C. and Rana, K. 2001. Oogenesis, hepatosomatic and gonadosomatic indexes, and sex ratio in rosy barb (*Puntius conchonius*). Turkish Journal of Fish Aquaculture Science. 1: 33-41.

Cerdá, J. L., Cetrino, T. R. and Wallace, R. A. 1993. Functional heterologous gap junctions in *Fundulus* ovarian follicles maintain meiotic arrest and permit hydration during oocyte maturation. Developmental Biology 160: 228-235.

Cerdá, J. L., Fabra, M. and Raldúa, D. 2007. Physiological and molecular basis of fish oocyte hydration. Pp 349-396. In P. J. Babin, J. Cerdá and E. Lubzens (eds), *The Fish Oocyte: From Basic Studies to Biotechnological Application*. Springer. Dordrecht: The Netherlands.

Chiu, C. R., Sakai, K. and Takashima, F. 1991. Gonadal maturation and its induction in largemouth bass, *Micropterus salmoides*. Suisanzoshoku 39: 343-351.

Clark, R. W., Henderson-Arzapalo, A. and Sullivan, C. A. 2005 Disparate effects of annually-cycling daylength and water temperature on reproductive maturation of striped bass (*Morone saxatilis*). Aquaculture 249: 497-513.

Coward, K. and Bromage, N. R. 1998. Histological classification of oocyte growth and the dynamics of ovarian recrudescence in *Tilapia zillii*. Journal of Fish Biology 53: 285-302.

Colombo, G., Grandi, G. and Rossi, R. 1984. Gonad differentiation and body growth in *Anguilla anguilla* L. Journal of Fish Biology 24: 215-228.

de Vlaming, V. L. 1974. Environmental and endocrine control of teleost reproduction. Pp. 13-83. In C. B. Schreck (ed.), *Control of Sex in Fishes*. Sea Grant, Extension Division, Virginia Polytechnic Institute and State University, Blacksburg, VA, USA.

Devlin, R. H. and Nagahama, Y. 2002. Sex determination and sex differentiation in fish: an overview of genetic, physiological, and environmental influences. Aquaculture 208: 191-364.

Dodd, J. M. 1977. The structure of the ovary of nonmammalian vertebrates. Pp. 219-263. In S. Zuckerman and B. J. Weir (eds), *The Ovary*. Vol. 1, Academic Press, New York, USA.

Dodd, J. M. 1986. The ovary. Pp. 351-397. In P. K. T. Pang and M. P. Schreibman (eds), *Vertebrate endocrinology: fundamentals and biochemical implications*. Academic Press, San Diego, USA.

Drummond, C. D., Bazzoli, N., Rizzo, E. and Sato, Y. 2000. Postovulatory follicle: model for experimental studies of programmed cell death or apoptosis in teleosts. Journal of Experimental Zoology 287: 176-182.

Fyhn, H. J., Finn, R. N., Reith, M. and Norberg, B. 1999. Yolk protein hydrolysis and oocyte free amino acids as key features in the adaptive evolution of teleost fishes to seawater. Sarsia 84: 451-456.

Fyhn, H. J. and Kristoffersen, B.A. 2007. Vertebrate vitellogenin gene duplication in relation to the "3R hypothesis": correlation to the pelagic egg and the oceanic radiation of teleosts. PLoS ONE 2: e169. doi:10.1371/journal.pone.0000169.

Ge, W. 2005. Activin and its receptors in fish reproduction. Pp 128-154. In P. Melamed and N. Sherwood (eds), *Molecular Aspects of Fish and Marine Biology*, Vol. 4, World Scientific Publishing Company. Singapore.

Greeley, M. S., Calder, D. R. and Wallace, R. A. 1986. Changes in teleost yolk protein during oocyte maturation: correlation of yolk proteolysis with oocyte hydration. Comparative Biochemistry and Physiology 84B: 1-9.

Grier, H. J. 2000. Ovarian germinal epithelium and folliculogenesis in the common snook, *Centropomus undecimalis* (Teleostei: Centropomidae). Journal of Morphology 243: 265-281.

Grier, H. J. and Lo Nostro, F. 2000. The germinal epithelium in fish gonads, the unifying concept. Pp 233-236. In B. Norberg, O. S. Kjesbu, G. L. Taranger, E. Andersson and S. O. Stefansson (eds), *Proceedings of the 6th International Symposium on the Reproductive Biology of Fish.* University of Bergen, Norway.

Grier, H. J., Uribe, M. C., Parenti, L. R. and De la Rosa-Cruz, G. 2005. Fecundity, the germinal epithelium, and folliculogenesis in viviparous fishes. Pp. 191-126. In M. C. Uribe and H. J. Grier (eds), *Viviparous Fishes.* New Life Publications Homestead, FL, USA.

Grier, H. J., Uribe, M. C. and Parenti, L. R. 2007. Germinal epithelium, folliculogenesis, and postovulatory follicles in ovaries of rainbow trout, *Oncorhynchus mykiss* (Walbaum, 1792) (Teleostei, Protacanthopterygii, Salmoniformes). Journal of Morphology 268: 293-310.

Gülsoy, N. 2007. Development of the yolk nucleus of previtellogenic oocytes in rainbow trout, *Oncorhynchus mykiss,* studied by light microscopy. Journal of Applied Biological Sciences 1: 33-35.

Guraya, S. S. 1985. Primordial follicle. Pp. 3-14. In S. S. Guraya (ed.), *Biology of Ovarian Follicles in Mammals.* Berlin, Springer-Verlag.

Guraya, S. S. 1986. The cell and molecular biology of fish oogenesis. Pp 223. Karger. Basel, Switzerland.

Habibi, H. R. and Andreu-Vieyra, C. V. 2007. Hormonal regulation of follicular atresia in teleost fish. Pp 235-253. In P. J. Babin, J. Cerdá and E. Lubzens (eds), *The Fish Oocyte: From Basic Studies to Biotechnological Applications.* Springer. Dordrecht: The Netherlands.

Hamaguchi, S. 1993. Alterations in the morphology of nuages in spermatogonia of the fish, *Oryzias latipes,* treated with puromycin and actinomycin D. Reproduction, Nutrition and Development 33: 137-141.

Hart, N. H. and Donovan, M. 1983. Fine structure of the chorion and site of sperm entry in the egg cortex of *Brachydanio rerio.* Cell and Tissue Research 265:317-328.

Hoar, W. S. 1969. Reproduction. Pp 1-72. In W. S. Hoar and D. J. Randall (eds), *Fish Physiology,* Vol. 3. Academic Press. New York and London.

Hunter, J. R. and Macewicz, B. J. 1985. Measurement of spawning frequency in multiple spawning fishes. Pp. 79-94. In R. L. Lasker (ed), *An egg production method for estimating spawning biomass of pelagic fish: application to the northern anchovy.* 36 Department of Commerce. National Oceanic Atmospheric Administration. Technical Report. National Marine Fisheries Service, USA.

Hunter, J. R., Macewicz, B. J., Lo, N. C. and Kimbrell, C. A. 1992. Fecundity, spawning, and maturity of female Dover sole, *Microstomus pacificus,* with an evaluation of assumptions and precision. Fishery Bulletin 90: 101-128.

Inoue, S. and Inoue, Y. 1986. Fertilization (activation)-induced 200- 9kDa depolymerization of polysialoglycoprotein, a distinct component of cortical alveoli of rainbow trout eggs. Journal of Biological Chemistry 261: 5256-5261.

Iwamatsu, T., Ohta, T., Oshima, E. and Sakai, N. 1988. Oogenesis in the medaka *Oryzias latipes* – Stages of oocyte development. Zoological Science 5: 353-373.

Jalabert, B. 2005. Particularities of reproduction and oogenesis in teleost fish compared to mammals. Reproduction, Nutrition and Development 45: 261-279.

Jalabert, B. and Szöllösi, D. 1975. *In vitro* ovulation of trout oocytes: effects of prostaglandins on smooth muscle-like cells of the theca. Prostaglandins 9: 765-778.

Johnson, L. L., Casillas, E., Collier, T. K., McCain, B. B. and Varanasi, U. 1988. Contaminant effects on ovarian development in English Sole (*Parophrys vetulus*) from Puget Sound, Washington. Canadian Journal of Fisheries and Aquatic Sciences 45: 2133-2146.

Kagawa, H., Moriyama, S. and Kawauchi, H. 1995. Immunocytochemical localization of IGF-1 in the ovary of d seabream. *Pagrus major*. General and Comparative Endocrinology 99: 307-315.

Kendall, W. C. 1921. Peritoneal membranes, ovaries and oviducts of salmonids fishes and their significance in fish-cultural practices. Bulletin of the Bureau of Fisheries 37: 184-208.

Kessel, R. G., Tung, H. N., Roberts, R. and Beams, H. W. 1985. The presence and distribution of gap junctions in the oocyte-follicle cell complex of the zebrafish, *Brachydanio rerio*. Journal of Submicroscopic Cytology 17: 239-253.

Kessel, R. G., Roberts, R. L. and Tung, H. N. 1988. Intercellular junctions in the follicular envelope of the teleost, *Brachydanio rerio*. Journal of Submicroscopic Cytology and Pathology 20: 415-424.

Khan, I. A. and Thomas, P. 1999. Ovarian cycle, teleost fish. Pp. 552-564. In E. Knobil and J. D. Neill (eds), *Encyclopedia of Reproduction*, Vol 3. Academic Press, San Diego, CA, USA.

Kjesbu, O. S., Klungsoir, J., Kryvi, H., Witthames, P. R. and Greer Walker, M. 1991. Fecundity, atresia, and egg size of captive Atlantic cod (*Gadus morhua*) in relation to proximate body composition. Canadian Journal of Fisheries and Aquatic Sciences 48: 2333-2343.

Kunz, Y. W. 2004. Developmental Biology of Fishes. p 636. Springer. Dordrecht, The Netherlands.

Lambert, J. G. D. 1970. The ovary of the guppy *Poecilia reticulata*. The atretic follicle, a corpus atreticum or a corpus luteum praeovulationis. Zeitschrift für Zellforschung und mikroskopische Anatomie 107: 54-67.

Le Menn, F., Cerdá, J. and Babin, P. J. 2007. Ultrastructural aspects of the ontogeny and differentiation of ray-finned fish ovarian follicles. Pp 1-37. In P. J. Babin, J. Cerdá and E. Lubzens (eds), *The Fish Oocyte: From Basic Studies to Biotechnological Applications*. Springer. Dordrecht, The Netherlands.

Lessman, C. A. 1999. Oogenesis, in nonmammalian vertebrates. Pp 498-508. In E. Knobil and J. D. Neill (eds), *Encyclopedia of Reproduction*, Vol 3, Academic Press. San Diego, CA. USA.

Liu, L. and Ge, W. 2007. Growth differentiation factor 9 and its spatiotemporal expression and regulation in the zebrafish ovary. Biology of Reproduction 76: 294-302.

Lokman, P. M., George, K. A. N. and Young, G. 2003. Effects of steroids and peptide hormones on in vitro growth of previtellogenic oocytes from eel, *Anguilla australis*. Fish Physiology and Biochemistry 28: 283-285.

Lo Nostro, F., Grier, H., Andreone, L. and Guerrero, G. A. 2003. Involvement of the gonadal germinal epithelium during sex reversal and seasonal testicular cycling in the protogynous swamp eel, *Synbranchus marmoratus* Bloch 1795 (Teleostei, Synbranchidae). Journal of Morphology 257: 107-126.

Lombardi, J. 1998. *Comparative Vertebrate Reproduction.* p 1-469. Kluwer Academic Publications. Boston, Dordrecht, London.

Lyman-Gingerich, J. and Pelegri, F. 2007. Maternal factors in fish oogenesis and embryonic development. Pp. 141-174. In P. J. Babin, J. Cerdá and E. Lubzens (eds), *The Fish Oocyte: From Basic Studies to Biotechnological Applications.* Springer. Dordrecht: The Netherlands.

Maestro, M. A., Méndez, E., Párrizas, M. and Gutiérrez, J. 1997. Characterization of insulin and insulin-like growth factor-1 during the reproductive cycle of carp. Biology of Reproduction 56: 1126-1132.

Matsuyama, M., Adachi, S., Nagahama, Y. and Matsuura, S. 1988. Diurnal rhythm of oocyte development and plasma steroid hormone levels in the female red sea bream, *Pagrus major,* during the spawning season. Aquaculture 73: 357-372.

Matsuyama, M., Adachi, S., Nagahama, Y., Maruyama, K. and Mansura, S. 1990. Diurnal rhythm of steroid hormone levels in the Japanese whiting, *Sillago japonica,* a daily-spawning teleost. Fish Physiology and Biochemistry 8: 329-338.

Matsuyama, M., Nagahama, Y. and Matsuura, S. 1991. Observations on ovarian follicle ultrastructure in the marine teleost, *Pagrus major,* during vitellogenesis and oocyte maturation. Aquaculture 92: 67-82.

Mayer, I., Shackley, S. E. and Ryland, J. S. 1988. Aspects of the reproductive biology of the bass, *Dicentrarchus labrax* L. I. An histological and histochemical study of oocyte development. Journal of Fish Biology 33: 609-622.

Milla, S., Jalabert, B., Rime, H., Prunet, P. and Bobe, J. 2006. Hydration of rainbow trout oocyte during meiotic maturation and *in vitro* regulation by 17α,20β-dihydroxy-4-pregnen-e-one. Journal of Experimental Biology 209:1147-1156.

Miranda, A. C. L., Bazzoli, N., Rizzo, E. and Sato, Y. 1999. Ovarian follicular atresia in two teleost species: a histological and ultrastructural study. Tissue and Cell 31: 480-488.

Miura, C., Higashino, T. and Miura, T. 2007. A progestin and a steroid regulate early stages of oogenesis in fish. Biology of Reproduction 77: 822-828.

Modig, C., Westerlund, L. and Olsson, P. E. 2007. Oocyte zona pellucida proteins. Pp. 113-139. In P. J. Babin, J. Cerdá and E. Lubzens (eds), *The Fish Oocyte: From Basic Studies to Biotechnological Applications.* Springer. Dordrecht: The Netherlands.

Nagahama, Y. 1983. The functional morphology of teleost gonads. Pp 223-275. In W. S. Hoar, D. J. Randall and E. M. Donaldson (eds), *Fish Physiology,* Vol. IX. Academic Press, New York, USA.

Nagahama, Y. 1987. 17α, 20β-dihydroxy-4-pregene-3-one: a teleost maturation-inducing hormone. Developmental, Growth and Differentiation. 29: 1-12.

Nagahama, Y., Kagawa, H. and Young, G. 1982. Cellular sources of sex steroids in teleost gonads. Canadian Journal of Fisheries and Aquatic Sciences 39: 56-64.

Neidig, C. L., Skapura, D. P., Grier, H. J. and Dennis, C. W. 2000. Techniques for spawning common snook: broodstock handling, oocyte staging, and egg quality. North American Journal of Aquaculture 62: 103-113.

Ogiwara, K., Takano, N., Shinohara, M., Murakami, M. and Takahashi, T. 2005. Gelatinase A and membrane-type matrix metalloproteinases 1 and 2 are responsible for follicle rupture during ovulation in the medaka. Proceedings of the National Academy of Sciences. USA. 102: 8442-8447.

Pang, Y. and Ge, W. 2002. Epidermal growth factor and TGFα promote zebrafish oocyte maturation in vitro: potential role of the ovarian activin regulatory system. Endocrinology 143: 47-54.

Paredes, A., Garcia-Rudaz, C., Kerr, B., Tapia, V., Dissen, G. A., Costa, M. E., Cornea, A. and Ojeda, S. R. 2005. Loss of synaptonemal complex protein-1, a synaptonemal complex protein, contributes to the initiation of follicular assembly in the developing rat ovary. Endocrinology 146: 5267-5277.

Parenti, L. R. and Grier, H. J. 2004. Evolution and phylogeny of gonad morphology in bony fishes. Integrative and Comparative Biology 44: 333-348.

Patiño, R. and Kagawa, H. 1999. Regulation of gap junctions and oocyte maturational competence by gonadotropin and insulin-like growth factor-I in ovarian follicles of red seabream. General and Comparative Endocrinology 115: 454-462.

Patiño, R. and Sullivan, C. V. 2002. Ovarian follicle growth, maturation, and ovulation in teleost fish. Fish Physiology and Biochemistry 26: 57-70.

Patiño, R. and Takashima, F. 1995. Gonads. Pp. 128-153. In F. Takashima and T. Hibiya (eds), *An Atlas of Fish Histology, Normal and Pathological Features*. Kodanska Ltd/Gustav Fisher Verlag. Tokyo/Stuttgart/New York.

Patiño, R. and Thomas, P. 1990a. Effects of gonadotropin on ovarian intrafollicular processes during the development of oocyte maturational competence in a teleost, the Atlantic croaker: evidence for two distinct stages of gonadotropic control of final oocyte maturation. Biology of Reproduction 43: 818-827.

Patiño, R. and Thomas, P. 1990b. Induction of maturation of Atlantic croaker oocytes by 17α,20β,21-trihydroxy-4-pregnen-3-one in vitro: Consideration of some biological and experimental variables. Journal of Experimental Zoology 255: 97-109.

Patiño, R. and Thomas, P. 1990c. Characterization of membrane receptor activity for 17α,20β,21-trihydroxy-4-pregnen-3-one in ovaries of spotted seatrout (*Cynoscion nebulosus*). General and Comparative Endocrinology 78: 204-217.

Patiño, R., Yoshizaki, G., Bolamba, D. and Thomas, P. 2003a. Role of arachidonic acid and protein kinase c during maturation-inducing hormone-dependent meiotic resumption and ovulation in ovarian follicles of Atlantic croaker. Biology of Reproduction 68: 516-523.

Patiño, R., Yoshizaki, G. and Thomas, P. 2003b. Ovarian follicle maturation and ovulation: An integrated perspective. Fish Physiology and Biochemistry 28: 305-308.

Patiño, R., Yoshizaki, G., Thomas, P. and Kagawa, H. 2001. Gonadotropic control of ovarian follicle maturation: the two-stage concept and its mechanisms. Comparative Physiology and Biochemistry B 129: 427-439.

Perazzolo, L. M., Coward, K., Davail, B., Normand, E., Tyler, C. R., Pakdel, F., Schneider, W. J. and Le Menn, F. 1999. Expression and localization of messenger ribonucleic acid for the vitellogenin receptor in ovarian follicles throughout oogenesis in the rainbow trout, *Oncorhynchus mykiss*. Biology of Reproduction 60: 1057-1068.

Picton, H.M. and Gosden, R. G. 1999. Oogenesis, in mammals. Pp 488-497. In E. Knobil and J. D. Neill (eds), *Encyclopedia of Reproduction,* Vol 3. Academic Press. San Diego, CA, USA.

Rajalakshmi, M. 1966. Atresia of oocytes and ruptured follicles in *Gobius giuris* (Hamilton-Buchanan). General and Comparative Endocrinology 6: 378-385.

Rastogi, R. K. 1969. The occurrence and significance of ovular atresia in the fresh water mud-eel, *Amphipnous cuchia* (Ham.). Acta Anatomica 73: 148-169.

Ravaglia, M. A. and Maggese, M. C. 2002. Oogenesis in the swamp eel *Synbranchus marmoratus* (Bloch, 1795), (Teleostei, Synbranchidae). Ovarian anatomy, stages of oocyte development and micropyle structure. Biocell (Mendoza) 26: 325-337.

Rickey, M. H. 1995. Maturity, spawning, and seasonal movement of arrowtooth flounder, *Atheresthes stomias*, of Washington. Fishery Bulletin 93: 127-138.

Riehl, R. 1980. Micropyle of some salmonids and coregonids. Env Biol Fishes 5: 59-66.

Robertson, JG. 1953. Sex differentiation in the Pacific salmon *Oncorhynchus keta* (Walbaum). Canadian Journal of Zoology 31: 73-79.

Rohr, D. H., Lokman, P. M., Davie, P. S. and Young, G. 2001. 11-ketotestosterone induces silvering-related changes in immature female short-finned eels, *Anguilla australis*. Comparative Biochemistry and Physiology 130A: 701-714.

Rosen, D. E. and Parenti, L. R. 1981. Relationships of *Oryzias*, and the groups of Atherinomorph fishes. American Museum Novitates. 2719: 1-25.

Saidapur, S. K. 1978. Follicular atresia in the ovaries of nonmammalian vertebrates. International Review of Cytology 54: 225-244.

Scott, A. P. 1990. Salmonids. Pp. 33-51. In A. D. Munro, A. P. Scott and T. J. Lam (eds), *Reproductive Seasonality in Teleosts: Environmental Influences*. CRC Press, Inc., Boca Raton, FL, USA.

Selman, K. and Wallace, R. A. 1986. Gametogenesis in *Fundulus heteroclitus*. American Zoologist 26: 173-192.

Selman, K. and Wallace, R. A. 1989. Cellular aspects of oocyte growth in teleosts. Zoological Science 6: 211-231.

Selman, K., Wallace, R. A. and Player, D. 1991. Ovary of the seahorse, *Hippocampus erectus*. Journal of Morphology 209: 285-304.

Selman, K., Wallace, R. A. and Cerdá, J. 2001. Bafilomycin A1 inhibits proteolytic cleavage and hydration but not yolk crystal disassembly or meiosis during maturation of sea bass oocytes. Journal of Experimental Zoology 290: 265-278.

Selman, K., Wallace, R. A., Sarka, A. and Xiaoping. Q. 1993. Stages of oocyte development in the zebrafish, *Brachydanio rerio*. Journal of Morphology 218: 203-224.

Shibata, Y., Iwamatsu, T., Oba, Y., Kobayashi, D., Tanaka, M., Nagahama, Y., Suzuki, N. and Yoshikuni, M. 2000. Identification and cDNA cloning of alveolin, an extracellular metalloproteinase, which induces chorion hardening of medaka (*Oryzias latipes*) eggs upon fertilization. Journal of Biological Chemistry 275: 8349-8354.

Silversand, C. and Haux, C. 1995. Fatty acid composition of vitellogenin from four teleost species. Journal of Comparative Physiology 164 B: 593-599.

Smith, C.J. and Haley, S. R. 1987. Evidence of steroidogenesis in postovulatory follicles of the tilapia, *Oreochromis mossambicus*. Cell and Tissue Research 247: 675-687.

Smith, C. J. and Haley, S. R. 1988. In vitro stimulation and inhibition of steroid hormone release from postovulatory follicles of the tilapia, *Oreochromis mossambicus*. Cell and Tissue Research 254: 439-447.

Stacey, N. E. and Pandey, S. 1975. Effects of indomethacin and prostaglandins on ovulation of goldfish. Prostaglandins 9: 597-608.

Strüssmann, C. A. and Nakamura, M. 2002. Morphology, endocrinology, and environmental modulation of gonadal sex differentiation in teleost fishes. Fish Physiology and Biochemistry 26: 13-29.

Sumpter, J. P., Scott, A. P., Baynes, S. M. and Witthames, P. R. 1984. Early stages of the reproductive cycle in virgin female rainbow trout (*Salmo gairdneri* Richardson). Aquaculture 43: 235-242.

Thacker, C. and Grier, H. J. 2005. Unusual gonad structure in the paedomorphic teleost, *Schindleria praematura* (Teleostei: Gobioidei): a comparison with other gobioid fishes. Journal of Fish Biology 66: 378-391.

Thomas, P., Tubbs, C., Berg, H. and Dressing, G. 2007. Sex steroid hormone receptors in fish ovaries. Pp 203-233. In P. J. Babin, J. Cerdá and E. Lubzens (eds), *The Fish Oocyte: From Basic Studies to Biotechnological Applications.* Springer. Dordrecht, The Netherlands.

Thomas, P., Zhu, Y. and Pace, M. 2002. Progestin membrane receptors involved in the meiotic maturation of teleost oocytes: A review with some new findings. Steroids 67: 511-517.

Tokarz, R. R. 1978. Oogonial proliferation, oogenesis, and folliculogenesis in nonmammalian vertebrates. Pp 145-179. In R. Jones (ed.), *The Vertebrate Ovary.* Plenum Press, New York/London,

Tyler, C. R. and Sumpter, J. P. 1996. Oocyte growth and development in teleosts. Reviews in Fish Biology and Fisheries 6: 287-318.

Tyler, C. R., Sumpter, J. P. and Witthames, P. R. 1990. The dynamics of oocyte growth during vitellogenesis in the Rainbow trout (*Oncorhynchus mykiss*). Biology of Reproduction 43: 202-209.

Uribe, M. C., De la Rosa-Cruz, G. and Garcia-Alarcon, A. 2005. The ovary of viviparous teleosts. Morphological differences between the ovaries of *Goodea atripinnis* and *Ilyodon whitei.* Pp. 217-235. In M. C. Uribe and H. J. Grier (eds), *Viviparous Fishes.* New Life Publications, Homestead, FL, USA.

Uribe, M. C., De la Rosa-Cruz, G., Garcia-Alarcon, A., Guerreo-Estevez, S. M. and Aguilar-Morales, M. 2006. Características histológicas de los estados de atresia de folículos ováricos en dos especies de teleósteos vivíparos *Ilyodon whitei* y *Goodea atripinnis* (Goodeidae). Hidrobiológica 16: 67-73.

Van Bohemen, C. G., Lambert J. G. D. and Van Oordt, P. G. W. J. 1982. Vitellogenin induction by estradiol in estrone-primed rainbow trout, *Salmo gairdneri.* General and Comparative Endocrinology 46: 136-139.

Van den Belt, K., Wester, P. W., Van der Ven, L. T. M., Verbeyen, R. and Witters, H. 2002. Effects of ethynylestradiol on the reproductive physiology in zebrafish (*Danio rerio*) time dependency and reversibility. Environmental Toxicology and Chemistry 21: 767-775.

Van den Hurk, R. and Peute, J. 1979. Cyclic changes of the ovary of the rainbow trout *Salmo gairdneri,* with special reference to sites of steroidogenesis. Cell and Tissue Research 199: 289-306.

Vizziano, D. and Berois, N. 1990. Histología del ovario de *Macrodon ancylodon* (Bloch y Schneider, 1801) (Teleostei: Scianediae). Ovogénesis. Folículos post-ovulatorios. Atresia. Revista Brasileña de Biología 50: 523-536.

Wake, W. H. 1985. Oviduct structure and function in non-mammalian vertebrates. Fortschritte der Zoologie 30: 427-435.

Wallace, R. A. and Selman, K. 1979. Physiological aspects of oogenesis in two species of sticklebacks, *Gasterosteus aculeatus* (L.) and *Apeltes quadracua* (Mitchill). Journal of Fish Biology 14: 551-554.

Wallace, R. A. and Selman, K. 1981. Cellular and dynamic aspects of oocyte growth in teleosts. American Zoologist 21: 325-343.

Wallace, R. A, Selman, K. 1990. Ultrastructural aspects of oogenesis and oocyte growth in fish and amphibians. Journal of Electron Microscopy Techniques 16: 175-201.

Wang, Y. and Ge, W. 2004. Cloning of epidermal growth factor (EGF) and EGF receptor from the zebrafish ovary: evidence for EGF as a potential paracrine factor from the oocyte to regulate activin/follistatin system in the follicle cells. Biology of Reproduction 71: 749-760.

Wiegand, M. D. 1996. Composition, accumulation and utilization of yolk lipids in teleost fish. Review of Fish Biology Fish 6: 259-286.

Williams, G. C. 1968. Distribution of planktonic fish eggs in Long Island Sound. Limnol Oceanography 13: 382-385.

Wourms, J. P. 1981. Viviparity: The maternal-fetal relationship in fishes. American Zoologist 21: 473-515.

Yamamoto, K. and Yamazaki, F. 1961. Rhythm of development in the oocyte of the goldfish *Carassius auratus*. Bulletin of the Faculty of Fisheries Hokkaido University 12: 93-100.

Yang, M. Y. and Fortune, J. E. The capacity of primordial follicles in fetal bovine ovaries to initiate growth *in vitro* develops during mid-gestation and is associated with meiotic arrest of oocytes. Biology of Reproduction 78: 1153-1161.

Yaron, Z. 1971. Observations on the granulosa cells of *Acanthobrama terrea-sanctae* and *Tilapia nilotica* (Teleostei). General and Comparative Endocrinology 17: 247-252.

Yoneda, M., Tokimura, M., Fujita, H., Takeshita, N., Takeshita, K., Matsuyama, N. and Matsuura, S. 2001. Reproductive cycle, fecundity, and seasonal distribution of the anglerfish *Lophius litulon* in the east China and Yellow seas. Fishery Bulletin 99: 356-370.

York, W. S., Patiño, R. and Thomas, P. 1993. Ultrastructural changes in follicle cell-oocyte associations during development and maturation of the ovarian follicle in Atlantic croaker. General and Comparative Endocrinology 92: 402-418.

Yoshizaki, G., Patiño, R., Thomas, P., Bolamba, D. and Chang, X. 2001. Effects of maturation-inducing hormone on heterologous gap junctional coupling in ovarian follicles of Atlantic croaker. General and Comparative Endocrinology 124: 359-366.

Young, G., Kusabe, M., Nakamura, I., Lokman, P. M. and Goetz, F. W. 2005. Gonadal steroidogenesis in teleost fish. Pp 155-223. In P. Melamed and N. Sherwood (eds), *Molecular Aspects of Fish and Marine Biology*, Vol. 4. World Scientific Publishing Co. Singapore.

Zickler, D. and Kleckner, N. 1998. The leptotene-zygotene transition of meiosis. Annual Review of Genetics 32: 619-697.

Zimmermann, M. 1997. Maturity and fecundity of arrowtooth flounder, *Atheresthes stomias*, from the Gulf of Alaska. Fishery Bulletin 95: 598-611.

CHAPTER 3

Modifications in Ovarian and Testicular Morphology Associated with Viviparity in Teleosts

Mari Carmen Uribe Aranzábal[1], *Harry J. Grier*[2], *Gabino De la Rosa Cruz*[1], *Adriana García Alarcón*[1]

3.1 INTRODUCTION

3.1.1 Occurrence of Viviparity

The viviparity in teleosts is essential in the study of the evolution of vertebrate viviparity (Amoroso 1981; Wourms *et al.* 1988; Guillette 1987). Wourms and Lombardi (1992) consider that fishes are very important in the understanding of viviparity in vertebrates, because they are the first viviparous vertebrates, and they manifest the maximum diversity in the degree of maternal-embryonic relationships. In the evolution of viviparity from oviparity there is a change in the site of embryonic development, from the external environment to the internal female reproductive system. Therefore, viviparity is the reproductive mode in which eggs are fertilized internally and are retained to complete their embryonic development within the maternal reproductive system. This transition involves new maternal-embryonic relationships that include modification of trophic, osmoregulatory, excretory, respiratory, endocrinological and immunological processes (Wourms 2005).

[1]Lab. Biología de la Reproducción, Departamento de Biología Comparada. Facultad de Ciencias, Universidad Nacional Autónoma de México, Ciudad Universitaria, 04510 México, D.F. México.

[2]Division of Fishes, Department of Zoology, National Museum of Natural History, MRC 159, Smithsonian Institution, P.O. Box 3712, Washington, D.C. 20012-7012, USA. Florida Fish and Wildlife Research Institute, 100 8th Avenue, SE, St. Petersburg, FL 33701-5020, USA.

Viviparity in teleosts is restricted to an estimated 510 species belonging to 13 families: Poeciliidae, Clinidae, Labrisomidae, Anablepidae, Zoarcidae, Parabrotulidae, Bythitidae, Aphyonidae, Zenarchopteridae, Goodeidae, Scorpaenidae, Comephoridae and Embiotocidae. Given current estimates of the total species of teleosts, over 27,000, viviparous species comprise fewer than 2% of all bony fishes (Wourms 1988, 2005; Wake 1989).

Four families, Poeciliidae, Anablepidae, Goodeidae and Zenarchop-teridae, are members of the taxon Atherinomorpha (first recognized by Rosen 1964 as the Atheriniformes) diagnosed as monophyletic by an array of largely reproductive and osteological characters. Among the reproductive characters, atherinomorphs (both oviparous and viviparous) share a unique type of testis and egg. Parenti (2005) correlated the unique restricted lobular testis and fluid yolk in the eggs with other reproductive modifications such as coupling during mating, sperm-bundle formation, internal fertilization, viviparity, and superfetation.

This chapter aims to illustrate the gonadal structures, and diverse characteristics of gestation of viviparous atherinomorphs, such as intrafollicular gestation in poeciliid species, and intraluminal gestation in goodeid species.

3.1.2 General Aspects of Testicular Structure

The testes of viviparous teleosts corresponds to the restricted spermatogonial lobular type (Figs. 3.1A-F, 3.2A-F), present in atherinomorphs fishes (Billard 1969; van den Hurk and Barends 1975; Grier 1975a,b, 1981, Grier *et al.* 1978; Parenti and Grier 2004, Parenti 2005). This type of testis, described in poeciliids and goodeids by Grier (1975a,b, 1981) and Grier *et al.* (1978), is characterized by spermatogonia, interspersed between Sertoli cells, being located exclusively at the testis periphery in clusters at the distal end of the lobules (Figs. 3.1B,D, 3.2A,B). Spermatogonia are located immediately beneath the tunica albuginea rather than being distributed along the length of the lobules as in the unrestricted spermatogonial lobular type of testis, present in other teleosts, as in all Perciformes. Exceptions to the rule occur as in the goby, *Schlinderia* (Thacker and Grier 2005) which has a restricted spermatogonial testis type, as occurs in

Fig. 3.1 Restricted lobular spermatogonial testis type. Testis of the goodeid *Xenotoca eiseni*. **A**. Structure of the testis with active spermatogenesis (Spg) at the distal part of the testis, and deferent ducts with spermatozoa (Sz) at the proximal part of the testis. **B-F**. In the lobules, the germinal cells in spermatogenesis are within cysts, and the later stages are located progressively closer to the deferent ducts. Stroma (ST) is located between the lobules. The spermatogonia (Sg) are restricted exclusively at the peripheral part of the testis. A spermatogonium in mitotic metaphase is observed (arrow). Cysts containing spermatocytes (Sc), spermatids (St) and spermatozoa (Sz) are seen. The spermatozoa are situated in groups forming spermatozeugmata. The heads of the spermatozoa are oriented to the center of the spermatozeugmata. The deferent ducts (DD) contain spermatozeugmata. **A**. Bar 0.5 mm. **B**. Bar 20 µm. **C**. Bar 20 µm. **D**. Bar 10 µm. E. Bar 20 µm. F. Bar 50 µm. Original.

Fig. 3.1 Contd. ...

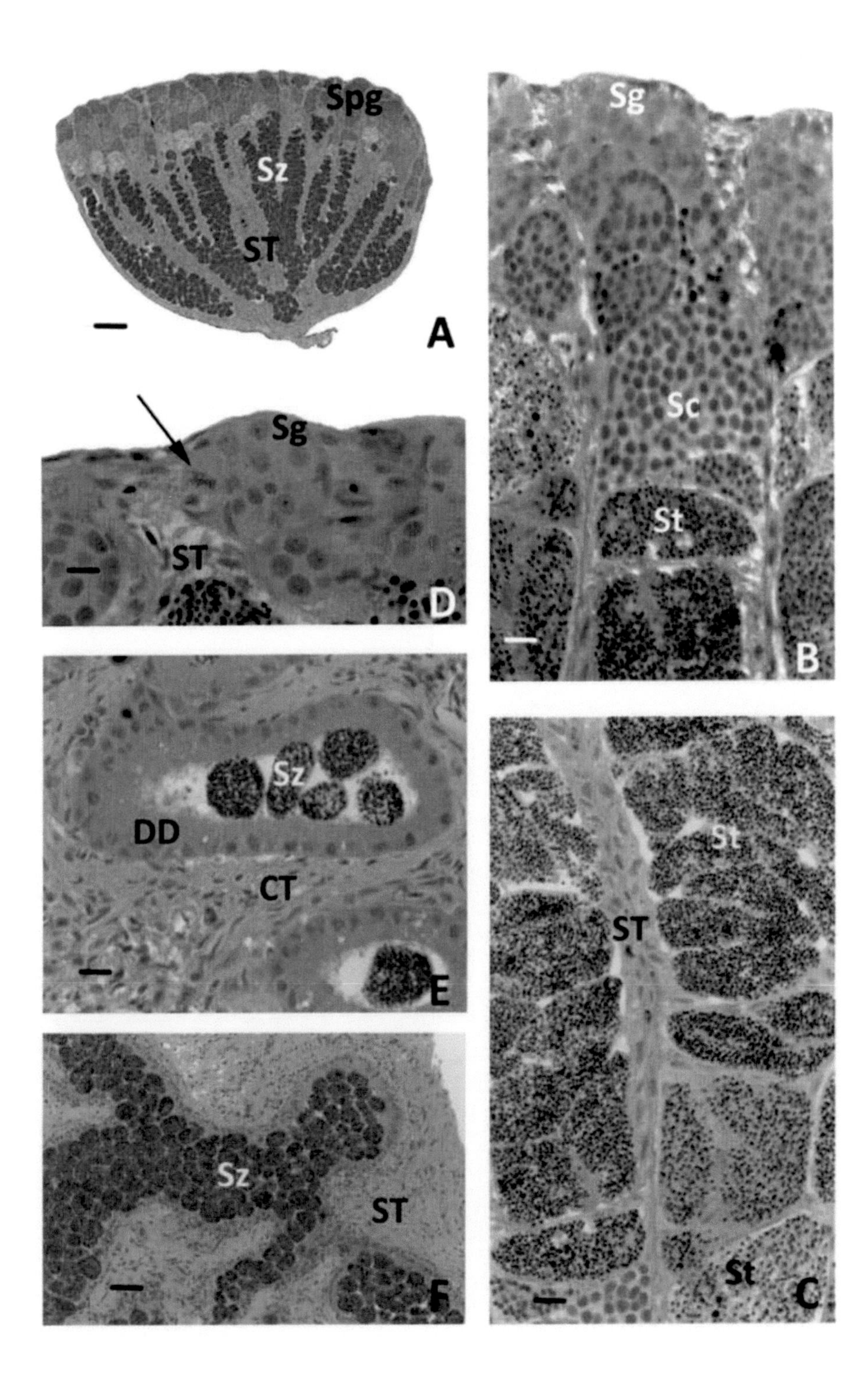
Spg
Sz
ST
A
Sg
ST
D
Sz
DD
CT
E
Sz
ST
F
Sg
Sc
St
B
St
ST
St
C

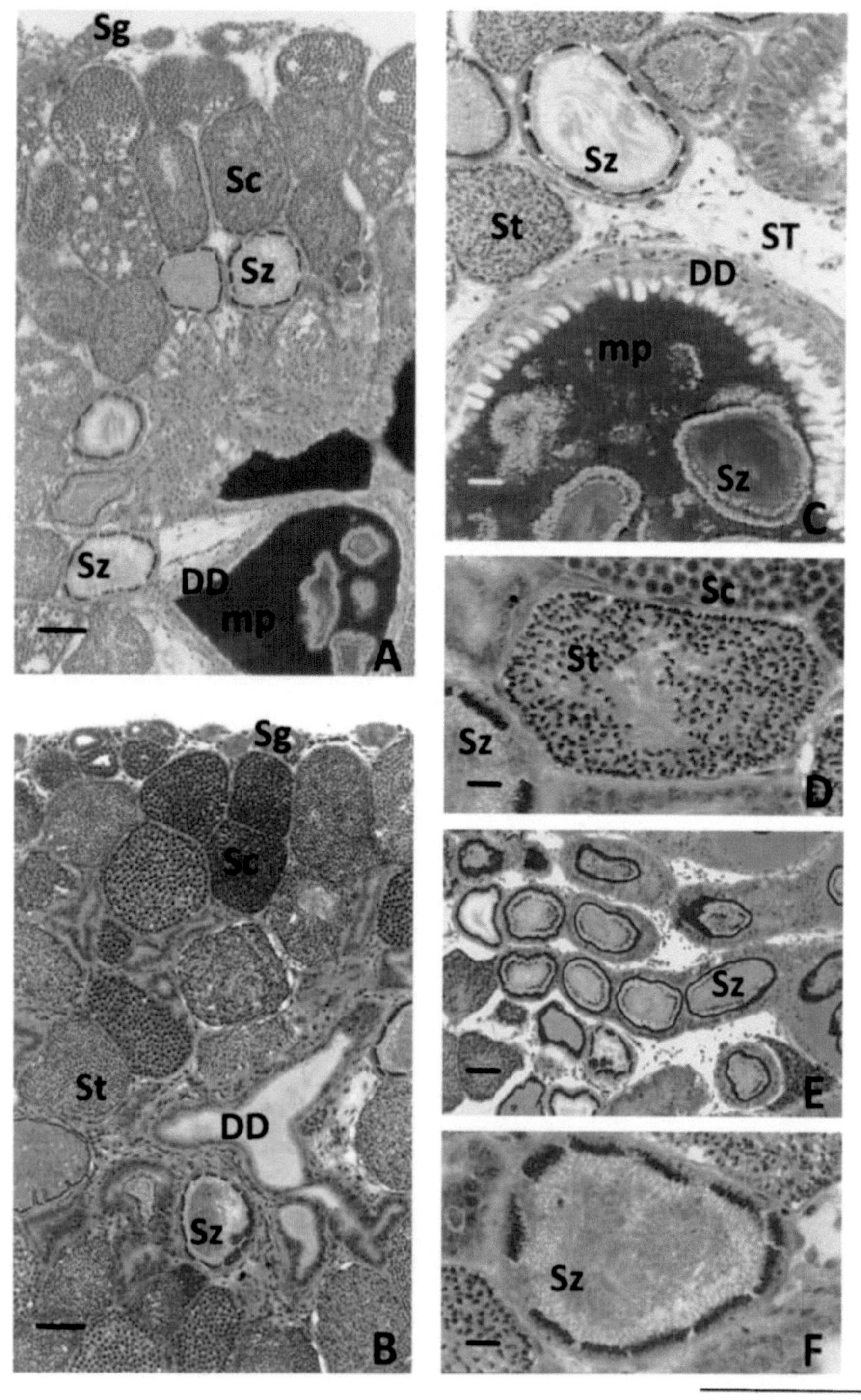
Sg
Sc
Sz
Sz
DD
mp
A
Sg
Sc
St
DD
Sz
B
Sz
St
ST
DD
mp
Sz
C
Sc
St
Sz
D
Sz
E
Sz
F

Fig. 3.2 Contd. ...

atherinomorphs and unlike other gobies. Spermatogenesis in fish is cystic, each spermatocyst contains synchronously-developing spermatogenic cells (Figs. 3.1A-C, 3.2A-D). The cysts develop when germ cells become surrounded by a surface layer of Sertoli cells. As spermatogenesis proceeds, the spermatocysts move in the testis toward the efferent ducts (Figs. 3.1B,C, 3.2A-C). In poeciliids, the arrangement of the late spermatids and spermatozoa within the spermatocyst is quite regular, as it is seen in *Poecilia latipinna* (Grier 1975a; van den Hurk and Barends 1975), and in *Xiphophorus helleri* (Jamieson 1991). Sperm heads are in contact with the apical end of the Sertoli cells and the flagella are oriented to the center of the spermatocyst (Figs. 3.2C,E,F). In contrast, in goodeid species, the disposition of the late spermatids and spermatozoa within the cyst is irregular, the flagella are associated with Sertoli cells and sperm heads are in the center of the cyst (Figs. 3.1E,F). In both poeciliids and goodeids, the spermatocysts open at spermiation, and the bundles of spermatozoa are shed into the efferent ducts, as spermatozeugmata, which are unencapsulated or naked sperm bundles (Figs. 3.1E,F, 3.2E,F). In spermatozeugmata each sperm mass is surrounded by a gelatinous secretion of the efferent duct cells. According to van den Hurk and Barends (1975), the secretion in poeciliids is a mucoprotein complex, positive to the periodic acid-Schiff reaction (Fig. 3.2A,C). Each spermatozeugma contains, approximately, 5000 spermatozoa, and measure about 0.2 mm in diameter (Constanz 1989; Potter and Kramer 2000). Grier *et al.* (1978) considered that spermatozeugmata evolved to resolve some requisites necessary for efficient transfer of sperm to the female reproductive tract to accomplish internal fertilization (see **Section 3.2.2**)

3.1.3 General Aspects of Ovarian Structure

The ovary of viviparous teleosts is similar to those of oviparous species. It is suspended to the dorsal wall by the mesovarium, between the swim bladder and the intestine. The ovary is of the cystovarian type, consisting of a saccular structure, with a central lumen (Figs. 3.3A-C, 3.4A,B). The communication of

---

Fig. 3.2 Contd. ...

Fig. 3.2 Restricted lobular spermatogonial testis type. Testis of the poeciliid *Poecilia latipinna*. **A-C**. The germinal cells in spermatogenesis are in cysts, situated progressively into the lobules to the deferent ducts. Stroma is located between the cysts (ST). The spermatogonia (Sg) are restricted exclusively at the peripheral part of the testis. Cysts containing spermatocytes (Sc), spermatids (St) and spermatozoa (Sz) are seen. The spermatozoa are situated in groups forming spermatozeugmata. The heads of the spermatozoa are oriented to the periphery of the spermatozeugmata. The deferent ducts (DD) contain spermatozeugmata. In the deferent ducts, the spermatozeugmata are surrounded by a PAS+ mucoprotein complex (mp). **D**. A cyst with spermatids (St). Parts of cysts are observed containing spermatocytes (Sc) and spermatozoa (Sz). **E,F**. Cysts with spermatozoa whose nuclei are oriented towards the apical end of Sertoli cells. **A**. Bar 50 μm. **B**. Bar 50 μm. **C**. Bar 20 μm. **D**. Bar 10 μm. **E** Bar 50 μm. **F** Bar 10 μm. Original.

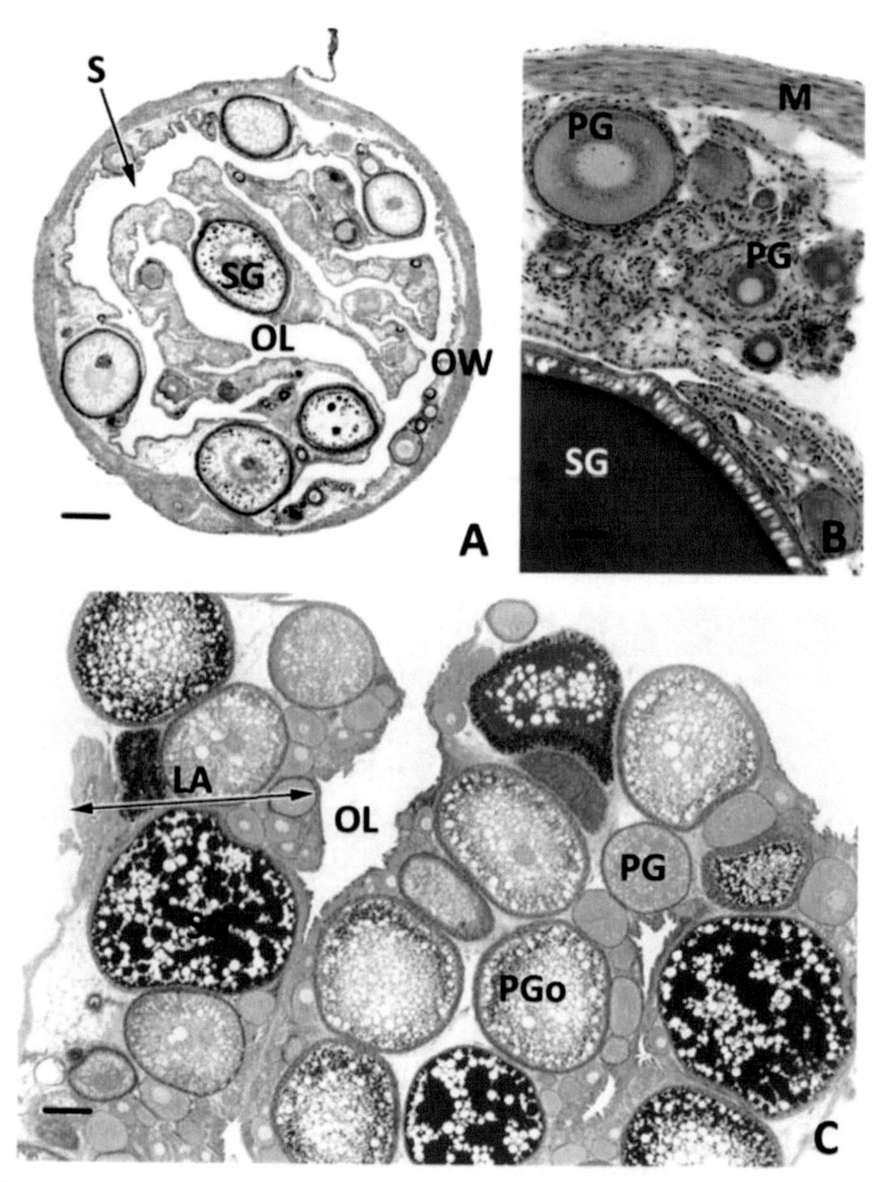

Fig. 3.3 Ovaries during non-gestation. **A**. Transverse section of the ovary of the goodeid *Goodea atripinnis*. The ovarian wall (OW) forms small folds with ovigerous tissue. The septum (S) is highly folded to the ovarian lumen (OL). Bar 50 µm. **B.** Ovary of the goodeid *Xenotoca eiseni*, containing peripheral circular smooth muscle (M), and oocytes in different stages of development as primary growth or previtellogenesis (PG) and secondary growth or vitellogenesis (SG). Bar 20 µm. **C.** Transverse section of the ovary of the poeciliid *Poecilia latipinna* showing several lamella (LA), folds of the wall in the ovarian lumen (OL) with oocytes in different stages of development as primary growth or previtellogenesis (PG), primary growth with oil droplets (PGo), and secondary growth or vitellogenesis (SG). Bar 50 µm. Original.

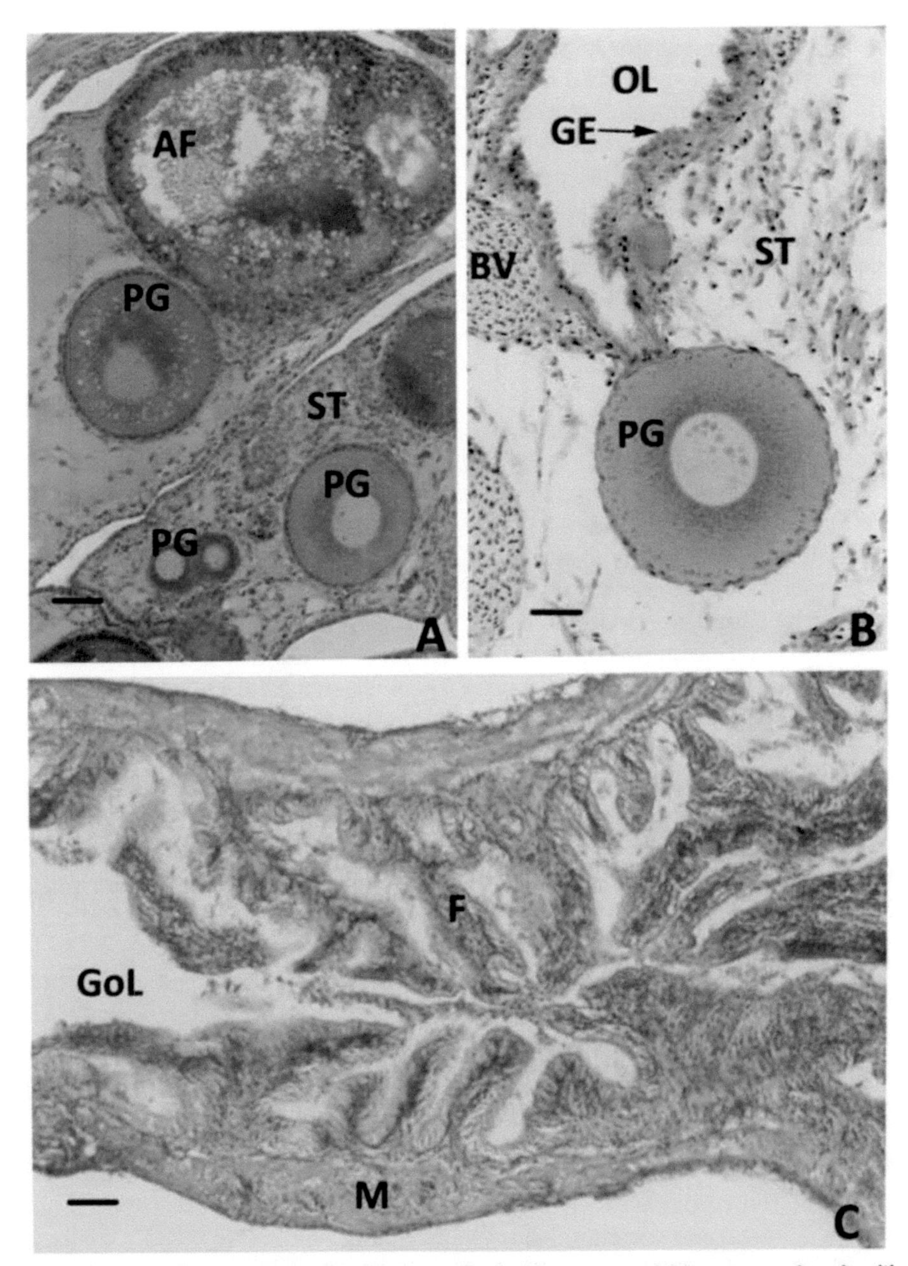

Fig. 3.4 Ovarian components. **A**. Lamella in the ovary of *Xenotoca eiseni*, with oocytes in developmental steps of primary growth (PG) and an atretic follicle (AF) surrounded by stroma (ST). Bar 50 μm. **B**. The ovarian lumen (OL) is lined by germinal epithelium (GE). An oocyte in primary growth (PG) is surrounded by stroma (ST), which contains large blood vessels (BV). Bar 20 μm. **C**. Gonoduct of the poeciliid *Heterandria formosa*, with thick layer of smooth muscle (M) and numerous folds (F) of the mucosa to the lumen of the gonoduct (GoL). Bar 50 μm. Original.

the ovary to the exterior is by the gonoduct (Fig. 3.4C). The ovarian mucosa forms folds that extend into the ovarian lumen, called lamellae. These folds contain stroma with follicles in different stages of development (Figs. 3.3A-C, 3.4A,B). They have a surface germinal epithelium (Figs. 3.4B, 3.5A,B, 3.6A,B). As a result of the internal position of the germinal epithelium, lining the ovarian lumen, ovulation occurs into the central lumen, instead of into the coelom as it is in all other vertebrates (Turner 1933; Mendoza 1940, 1943; Wallace and Selman 1981; Wourms 1981; Dodd and Sumpter 1984; Guraya 1986; Grier *et al.* 2005; Uribe *et al.* 2005).

As occurs in oviparous species (see Grier *et al.*, **Chapter 2**), the germinal epithelium contains oogonia (Fig. 3.5A) and primary oocytes between somatic epithelial cells (Figs. 3.5A,B, 3.6A,B). The somatic epithelial cells become prefollicle cells when associated with oogonia. The oogonia form cell nests, and initiate meiosis thus becoming primary oocytes (Figs. 3.5B, 3.6A). The meiotic process advances until early diplotene when the meiotic arrest occurs (Selman and Wallace 1989). The oocyte remains in this nuclear stage throughout previtellogenic and vitellogenic growth until oocyte maturation, when meiosis resumes. In the goodeid *Ilyodon whitei*, proliferation of oogonia and primary oocytes is very active around the time of birth, when there are numerous cell nests in the ovary (Uribe *et al.* unpublished). In *Heterandria formosa*, a poeciliid with superfetation (see **Section 3.2.10**), the proliferation of the oogonia was observed throughout the reproductive cycle. In this species, oocytes in the cell nests are associated with prefollicle cells which become follicle cells when the follicle is surrounded by the basement membrane. Along with the theca, these form the follicle complex.

Viviparous teleosts, as is characteristic of atherinomorphs fishes, have homogeneous, fluid yolk, additionally, in poeciliids the oocyte has peripheral oil globules (Mendoza 1940, 1943; Dodd Sumpter 1984; Parenti and Grier 2004), e.g. in *Poecilia reticulata* (Droller and Roth 1966).

Oogenesis is the process initiated by the oogonia, and ending in maturation. Several authors describe oogenesis in viviparous teleosts (Bailey 1933; Mendoza 1940; Droller and Roth 1966; Lambert 1970a; Gardiner 1978; Thibault and Schultz 1978; Schindler *et al.* 1988; Wourms *et al.* 1988; Pavlov 2005; Uribe *et al.* 2005). In viviparous teleosts, oogenesis is similar to that described for oviparous species (Wallace and Selman 1981; Guraya 1986; Selman and Wallace, 1989; Grier *et al.*, **Chapter 2**) which consists of sequence of stages including folliculogenesis, primary growth or previtellogenesis (Figs. 3.5A,B, 3.6A,B, 3.7A-D) and secondary growth or vitellogenesis (Figs. 3.8A-C, 3.9A-D) followed by oocyte maturation when meiosis resumes. During folliculogenesis, the ooplasm is hyaline and the nucleus contains one nucleolus. During primary growth the ooplasm increase of the significantly, becoming basophilic, especially, because of the increase in ribosomes (Fig. 3.5A,B). The follicle cells form a squamous layer around the oocyte (Figs. 3.5A,B). The nucleus contains several nucleoli, progressively situated at the periphery of the nucleus (Figs.

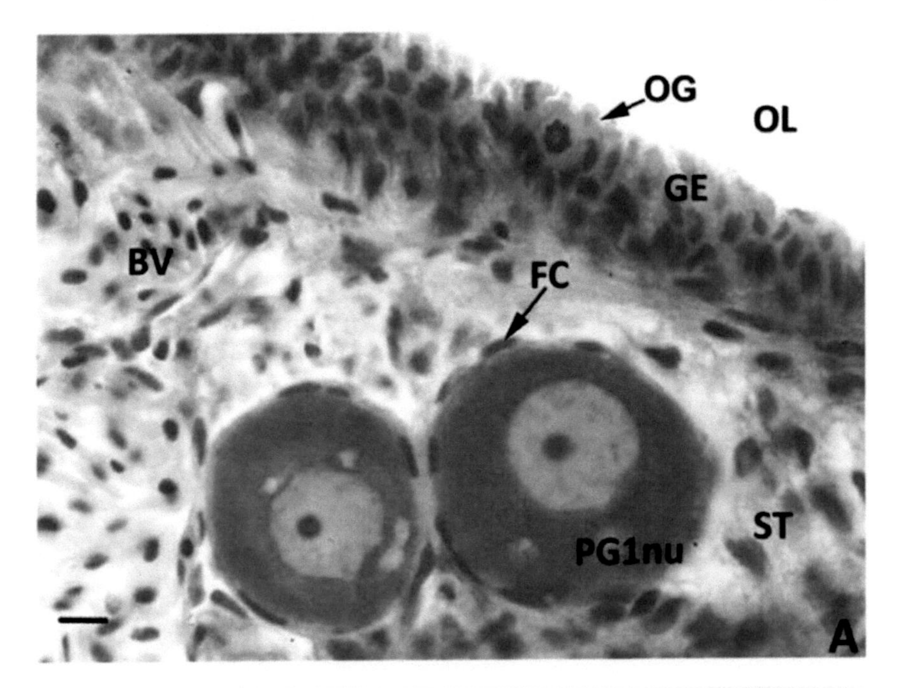

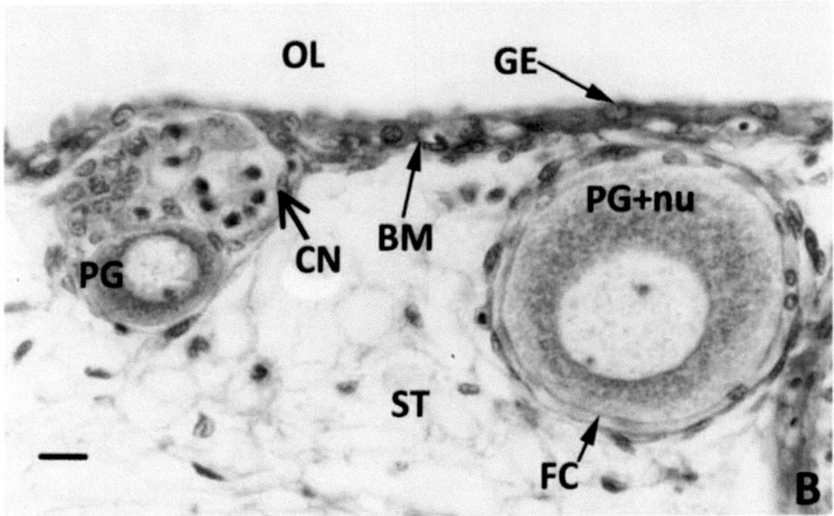

Fig. 3.5 Germinal epithelium. **A**. Germinal epithelium (GE) of the ovary of *Xenotoca eiseni* borders the ovarian lumen (OL), an oogonium (OG) in mitotic division is seen. Oocytes in primary growth with one nucleolus (PG1nu) are surrounded by follicle cells (FC) and stroma (ST) with blood vessels (BV). Bar 10 μm. **B**. Germinal epithelium (GE) of the ovary of the goodeid *Ilyodon whitei*, exhibits a basement membrane (BM) subjacent to the epithelium. A cell nest (CN) containing one oocyte in early primary growth (PG). A second oocyte in primary growth with multiple nucleoli (PG+nu) is seen. Follicle cells (FC), stroma (ST) and ovarian lumen (OL). Bar 10 μm. Original.

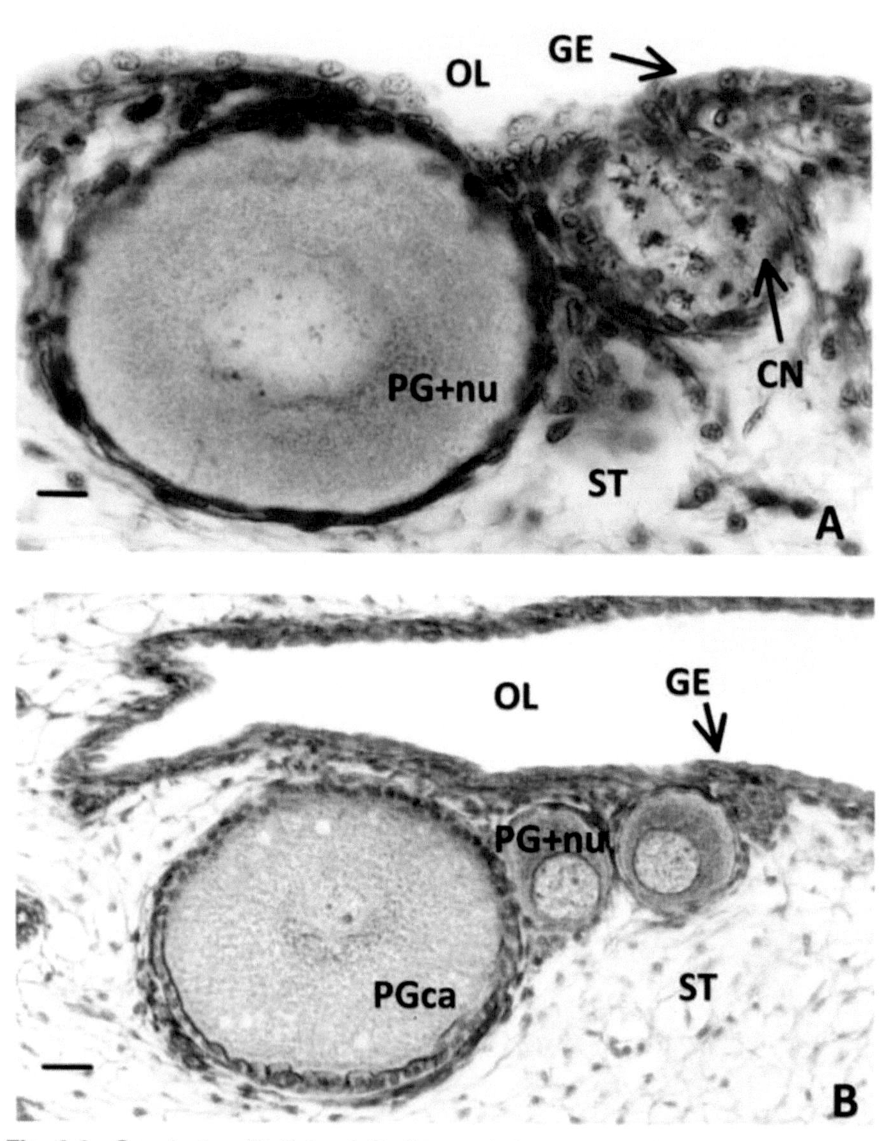

Fig. 3.6 Germinal epithelium. **A**,**B**. Ovary of *Ilyodon whitei*. Germinal epithelium (GE), a cell nest (CN) and oocytes in primary growth with multiple nucleoli (PG+nu) and cortical alveoli (PGca) are seen. Stroma (ST) and ovarian lumen (OL). **A**. Bar 10 μm. **B**. Bar 20 μm. Original.

3.5B, 3.6A, 3.7A-C). The follicle cells gradually become cuboidal (Fig. 3.7C) and columnar (Fig. 3.7D). During secondary growth abundant yolk accumulates in the ooplasm, which later provides nutrients for the developing embryo (Figs. 3.8A-C, 3.9A-D) and the yolk becomes fluid (Figs. 3.8B, 3.9D).

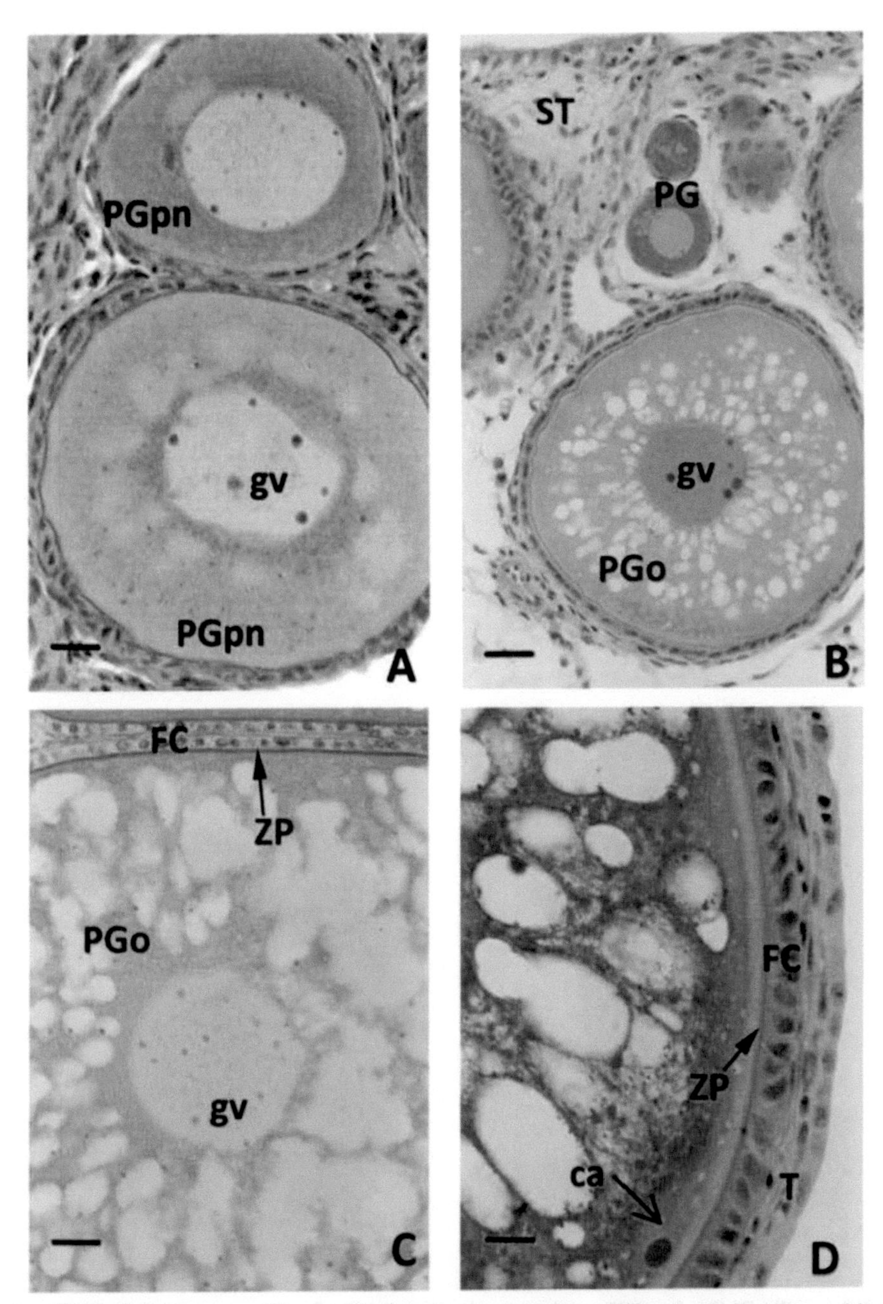

Fig. 3.7 Primary growth. **A.** Perinucleolar oocytes (PGpn) of *Ilyodon whitei*, nucleoli are peripheral in the germinal vesicle (gv). Bar 10 µm. **B.** Ovary of *Heterandria formosa*, oocytes in early primary growth (PG), and when oil droplets (PGo) begin to accumulate around the germinal vesicle (gv). Bar 20 µm. **C,D.** Ovary of *Xenotoca eiseni* with cortical alveoli (ca), the zona pellucida (ZP), the follicle cells (FC) and the theca (T) are seen. **C.** Bar 20 µm. **D.** Bar 10 µm. Original.

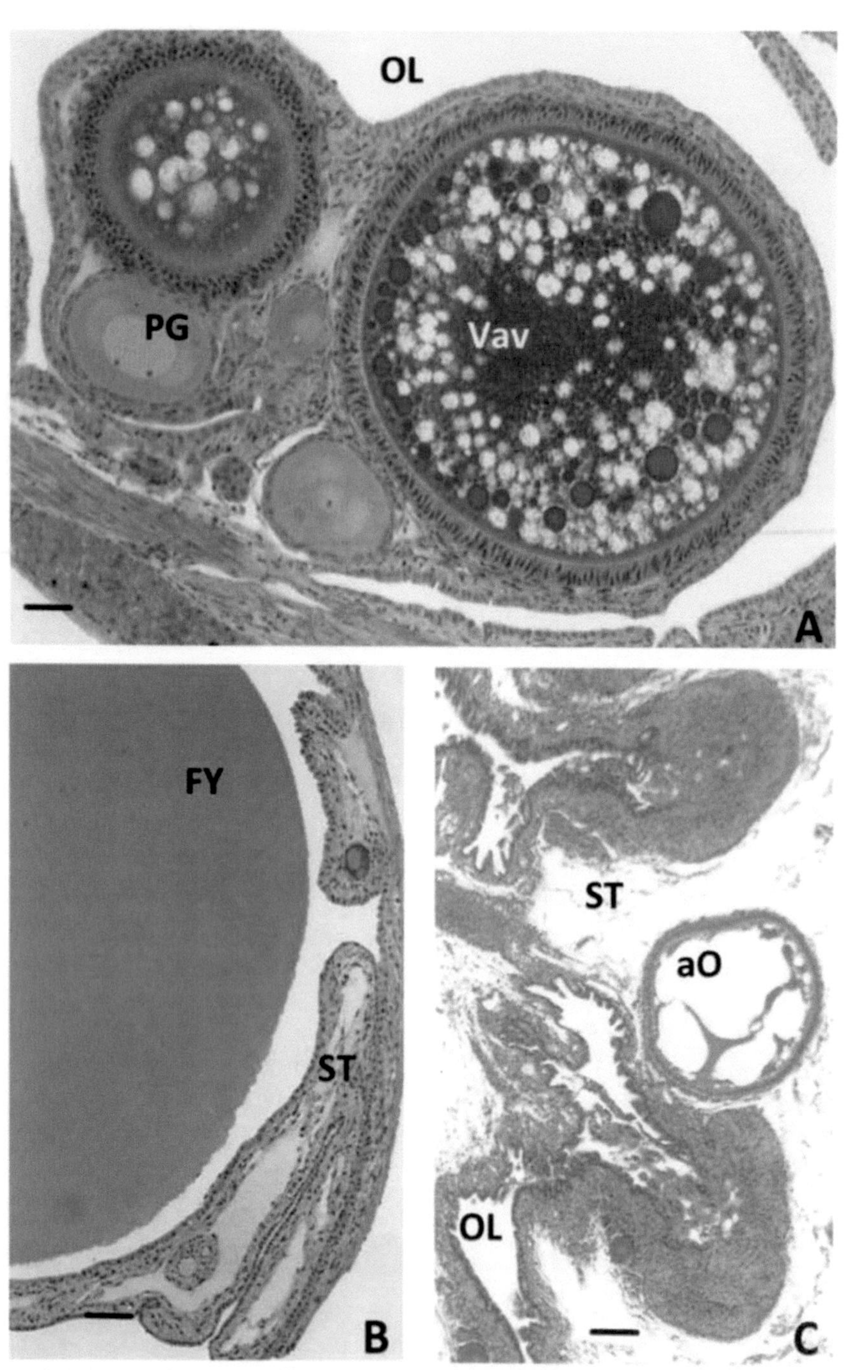

Fig. 3.8 Secondary growth. **A.** Oocytes of *Xenotoca eiseni* in advanced vitellogenesis (Vav). The lipoprotein yolk globules begin to fuse. Oocytes in primary

Fig. 3.8 Contd. ...

Atresia, the process of degeneration and removal of ovarian follicles, occurs during any stage of follicle development (Fig. 3.10A-D). The morphological characteristics of atretic follicles were summarized by Lambert (1970b), who described atresia in *Poecilia reticulata*. He divided atresia into two parts: 1) resorption of the oocyte characterized by fragmentation of the zona pellucida (Fig. 3.10D) and hypertrophy of the follicle cells which become phagocytes (Fig. 3.10A-D), engulfing the oocyte and, 2) the gradual regression and decrease of the follicle cells. During gestation in goodeids *Ilyodon whitei* and *Goodea atripinnis*, atretic follicles are frequently observed (Fig. 3.10C,D). Specially, during early gestation, the vitellogenic follicles which were not fertilized become atretic (Uribe *et al.* 2006). Tyler and Sumpter (1996), considered that atresia plays an important role in the control of the number of oocytes which attain maturity, therefore, defining the number of embryos.

3.2 MODIFICATIONS OF REPRODUCTIVE SYSTEMS IN VIVIPAROUS TELEOSTS

3.2.1 Ovary and Gonoduct

Most viviparous teleosts have a single, ovoid, fused ovary (Fig. 3.3A). The origin of the single ovary has been mentioned by several authors: Turner (1947), Mendoza (1940, 1965), Hoar (1969), Amoroso (1960), Wourms (1981), Nagahama (1983), Wourms *et al.* (1988). They suggested that the two ovaries fuse during early embryonic development, except the species of the genera *Sebastodes* (Scorpaeniformes), *Dermogenys* (Atheriniformes) and *Stygicola* (Ophidiiformes), which have two ovaries, or the ovaries are partially fused (Wourms 1981).

The lack of Müllerian ducts in teleosts results in the absence of oviducts (Turner 1947; Amoroso 1960, 1981; Wake 1985; Wourms *et al.* 1988, Constanz 1989). The communication of the ovary to the exterior occurs by the gonoduct (Fig. 3.4C) which opens to the exterior by the genital pore. The gonoduct is not an oviduct because this term is restricted to the structures derived from Müllerian ducts (Turner 1947; Wake 1985; Wourms *et al.* 1988, Uribe *et al.* 2005). The gonoduct is a single structure, even in species that have double or partially fused ovaries. The lack of oviducts in teleosts implies that, in viviparous species, gestation occurs inside the ovary. Therefore, in viviparous teleosts the ovary performs a gestational role (intraovarian gestation) unique among vertebrates (Mendoza 1940; Turner 1947; Amoroso 1981; Wourms 1981, 2005; Wourms *et al.* 1988). Consequently, the ovary of viviparous teleosts

---

Fig. 3.8 Contd. ...

growth (PG) are also seen. Bar 50 µm. **B**. Egg of *Ilyodon whitei* where the yolk globules are completely fused, forming fluid yolk (FY). Bar 50 µm. **C**. Advanced oocyte (aO) of *Heterandria formosa*, with large lipid globules and scarce lipoprotein yolk. Ovarian lumen (OL) and stroma (ST). Bar 100 µm. Original.

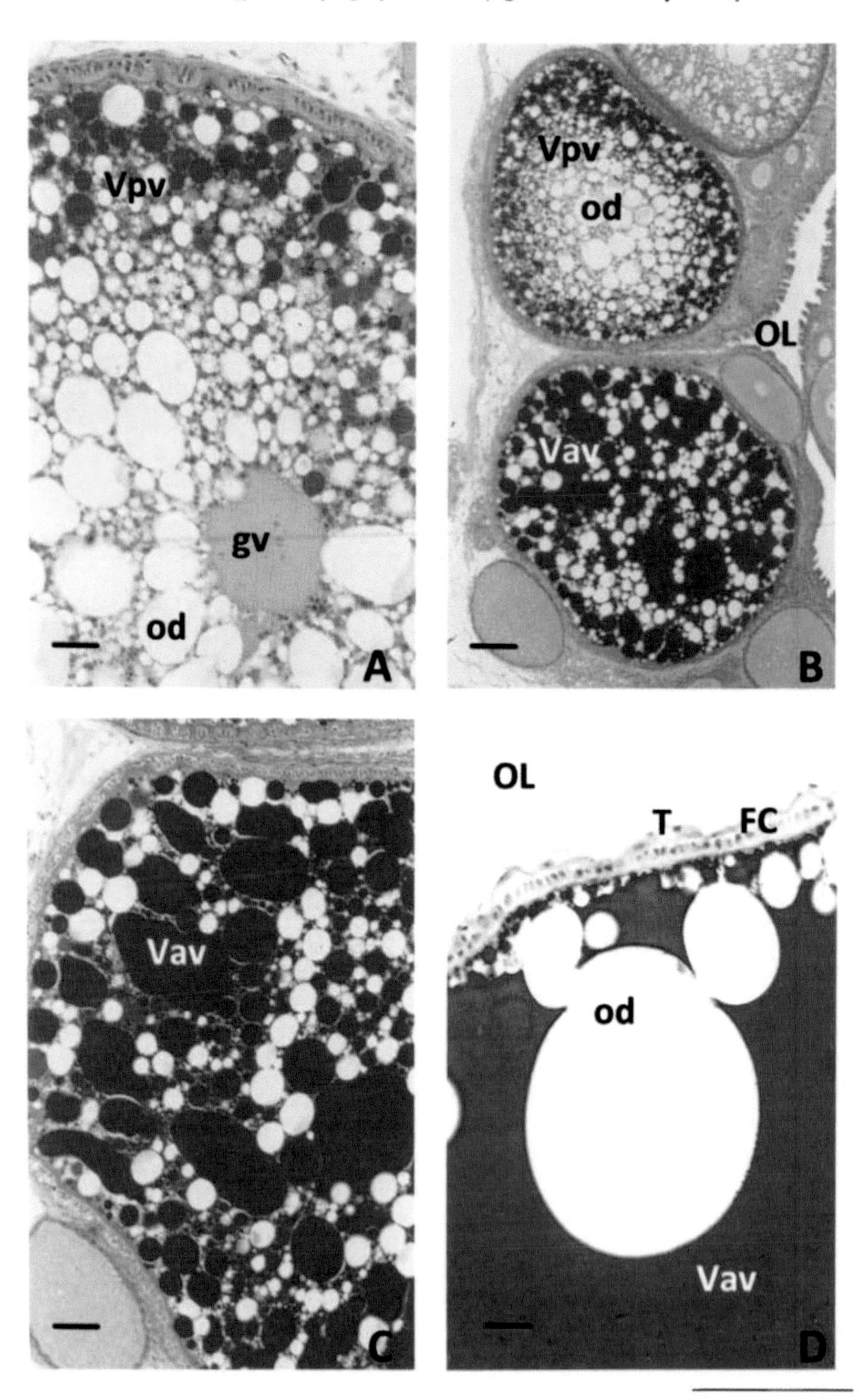

Fig. 3.9 Contd. ...

differs from those of all other vertebrates because it is the site, not only for production of eggs, but also for internal fertilization, gestation and parturition (Mendoza 1940, 1965; Turner 1947; Wourms 1981, 2005; Wourms *et al.* 1988; Constanz 1989; Wourms and Lombardi 1992; Schindler and Hamlett 1993). Intraovarian gestation in teleost involves specific morphological and physiological characteristics of the ovary that change according to the sequential stages during the reproductive cycle.

3.2.2 Insemination and Fertilization

Viviparity would never have been possible without the co-evolution of reproductive behavior, insemination and internal fertilization (see also Burns and Javonillo, **Chapter 17**; Abe and Munehara, **Volume 8B, Chapter 6**; Murata, **Volume 8B, Chapter 7**). These are essential requirements for viviparity. Insemination is the transfer of sperm from the male to the female gonoduct, complemented by internal fertilization, the fusion of sperm and egg within the ovary (Burns and Weitzman 2005; Meisner 2005; Greven 2005). Burns and Weitzman (2005) extensively reviewed the characteristics of spermatozoa and testes that are associated with insemination in ostariophysan fishes. They described the majority of inseminating species as having specializations such as elongation of the sperm nuclei and midpieces, distinct sperm packets or spermatozeugmata, and testes with large posterior sperm storage areas.

After insemination, the spermatozeugmata disassociate and the spermatozoa move from the gonoduct to the interior of the ovary where conditions must be adequate to maintain the sperm alive, at least for a short period of time, in the fluids of the ovarian lumen (Turner 1947). Fertilization then occurs. For a discussion of insemination and internal fertilization in oviparous species, see Burns and Javonillo, **Chapter 17**, Abe and Munehara, **Volume 8B, Chapter 6** and Murata, **Volume 8B, Chapter 7**).

Besides insemination, internal fertilization is an obligatory aspect of viviparity (Turner 1947; Amoroso 1960; Turner *et al.* 1962; Thibault and Schultz 1978; Wourms 1981; Wourms *et al.* 1988; Constanz 1989; Wake 1989; Wourms and Lombardi 1992; Greven 2005).

In most viviparous teleosts, fertilization occurs, inside the follicle when sperm penetrate the follicular wall. Fertilization is intrafollicular and precedes ovulation (Turner 1947; Wourms 1981, 1988, 2005; Constanz 1989). Amoroso (1960) and Dodd and Sumpter (1984) mention the exceptions of Scorpaenidae

---

Fig. 3.9 Contd. ...

Fig. 3.9 Secondary growth. **A-D**. Oocytes of *Poecilia latipinna* in partial vitellogenesis (Vpv) and advanced vitellogenesis (Vav). The oocytes contain abundant oil droplets (od) and yolk globules that fuse progressively as oocyte growth advances. The germinal vesicle (gv), the ovarian lumen (OL), follicle cells (FC) and theca (T) are seen. **A**. Bar 20 μm. **B**. Bar 50 μm. C. Bar 20 μm. D. Bar 10 μm. Original.

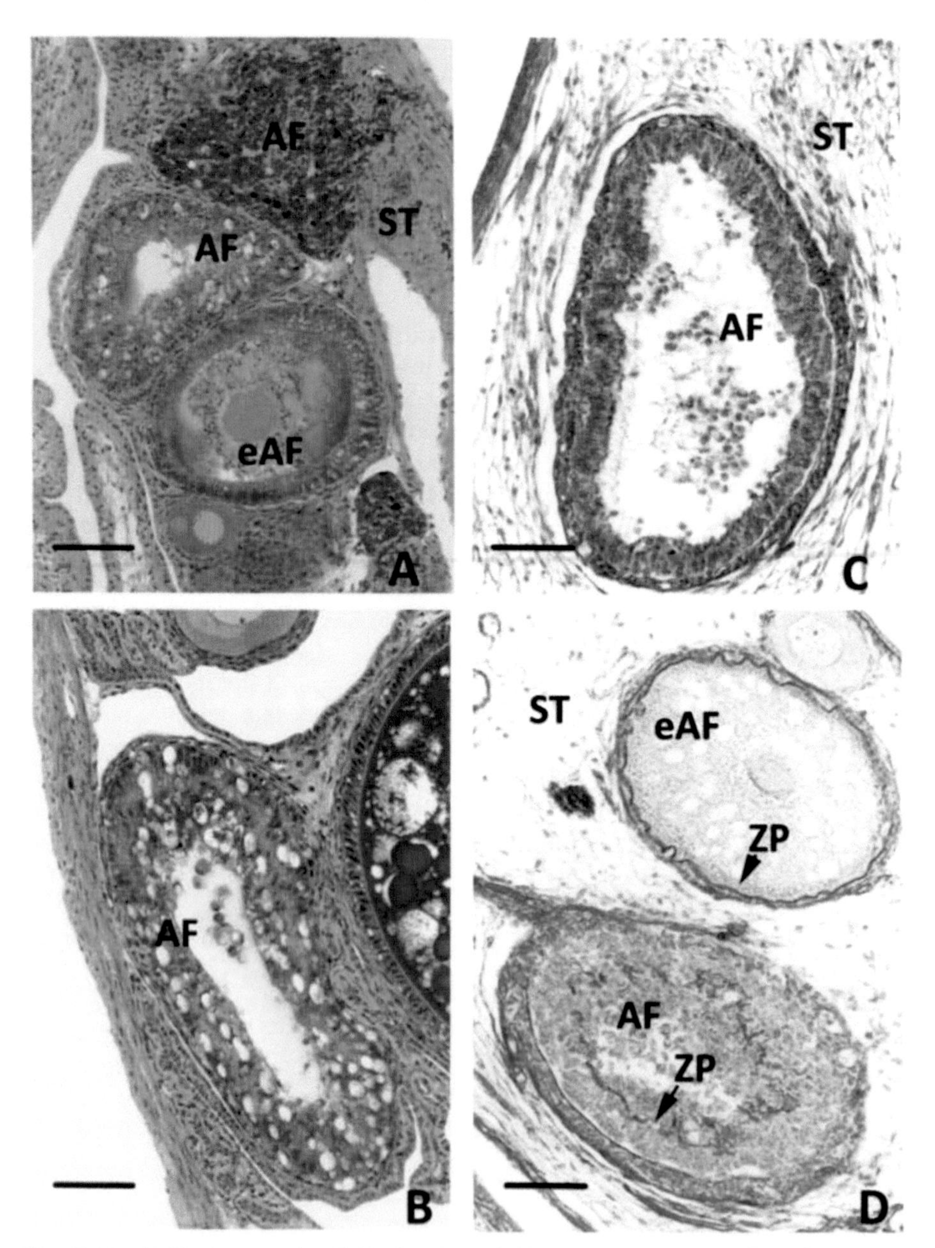

Fig. 3.10 Follicular atresia. **A**,**B**. Ovaries of *Xenotoca eiseni* and **C**,**D**. Ovaries of *Ilyodon whitei*. Follicles during several stages of atresia. Early atresia (eAF) is indicated by disorganization of the ooplasm, and of the zona pellucida (ZP). Advanced atresia (AF) shows the ooplasm being invaded by phagocytic follicle cells, and the zona pellucida has been broken. Stroma (ST) surrounds the follicles. **A**. Bar 100 µm. **B**. Bar 50 µm. **C**. Bar 50 µm. **D**. Bar 50 µm. Original.

and *Zoarces viviparus*. After ovulation into the ovarian lumen, the eggs are fertilized.

Pavlov (2005) observed insemination in zoarcids in which there was a close apposition of the male and female genital openings. However, males of several species of viviparous teleosts, such as poeciliids (Fig. 3.12), goodeids and anablepids, have developed the gonopodium for deposition and transfer the sperm to female ovarian gonoduct. For a discussion of copulatory structures, see Evans and Meisner, **Volume 8B, Chapter 4**).

3.2.3 Gonopodium

The gonopodium (Fig. 3.12A-B) is formed by the modification of the unpaired anal fin, where some anal fin rays fuse during maturity (Turner *et al.* 1962; Wourms *et al.* 1988; Constanz 1989; Burns 1991; Greven 2005; Rosa-Molinar 2005; Evans and Meisner, **Volume 8B, Chapter 4**).

The gonopodium is a diverse structure in size and shape, according to the species. In goodeids, Hubbs and Turner (1939), Turner *et al.* (1962), and in poeciliids, Rosen and Gordon 1953; Greven (2005) described the formation of the gonopodium by the fusion of the 3, 4, and 5 anal fin rays. The gonopodium also develops specific structures such as hooks, spines or serrations. This diversity of gonopodia corresponds to the structure of the female gonoduct, which also presents morphological diversity (Greven 2005). These species-specific morphological features of the gonopodia, have been used for taxonomic purposes (Hubbs and Turner 1939; Turner *et al.* 1962; Meisner 2005).

3.2.4 Sperm Storage

For insemination and fertilization to occur, it is necessary that spermatozoa be able to live in the ovary and maintain the capacity for fertilizing the oocytes. Sperm storage has been described in ovaries of poeciliids, in particular, species of the genera *Xiphophorus, Girardinus* and the species *Heterandria formosa*. The females spermatozoa store in special invaginations, or folds of the ovarian wall (Fig. 3.11A-D) and gonoduct (Turner 1933, 1947; Constanz 1989; Potter and Kramer 2000). During sperm storage there is an immunological barrier whereby the spermatozoa are isolated from the female's immune system (Potter and Kramer 2000; Koya *et al.* 1995, 1997). The possibility for spermatozoa to remain alive within the ovary is, mainly, a function of the ovary. The secretory activity of the ovarian epithelium, having a trophic function for the spermatozoa, contributes essentially in their survival for varying periods of time (Turner 1933; Mendoza 1943; Amoroso 1960; Hoar 1969; Koya *et al.* 1995; Potter and Kramer 2000). The activity of the ovarian epithelium in the maintenance of spermatozoa has been considered by Potter and Kramer (2000), Koya *et al.* (1995, 1997), and Koya (2008). Potter and Kramer (2000) observed the morphological relationship of sperm heads

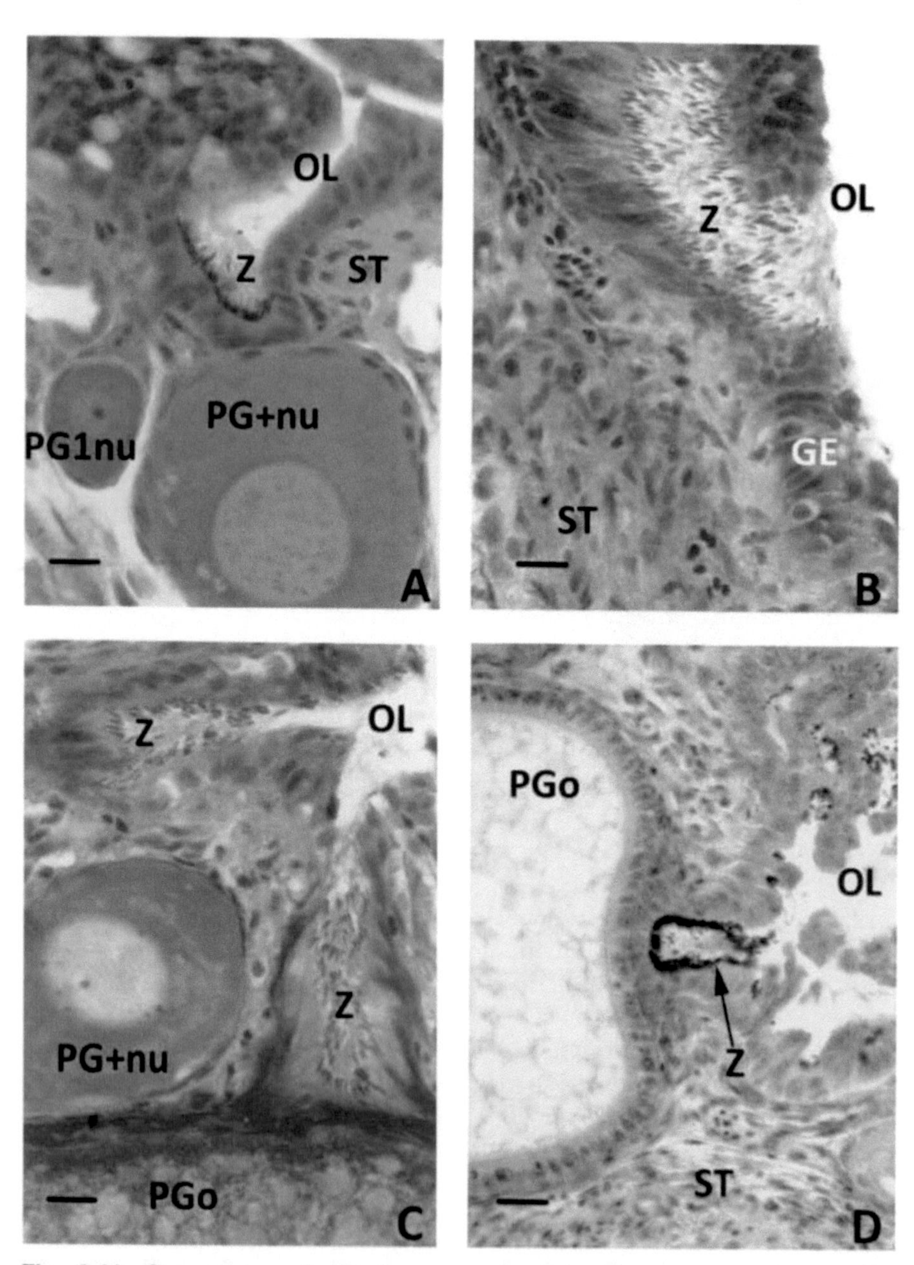

Fig. 3.11 Spermatozoa inside the ovary in **A,B.** *Xenotoca eiseni,* and **C,D.** *Heterandria formosa.* The spermatozoa (Z) are in the ovarian lumen (OL), close to the apical end of the epithelial cells of the germinal epithelium (GE). Some of them are inserted in infolds that communicate with the ovarian lumen and the follicles. Oocytes in primary growth with one nucleolus (PG1nu), with multiple nucleoli (PG+nu), and with oil droplets (PGo) are seen. **A**. Bar 10 µm. **B**. Bar 10 µm. **C**. Bar 10 µm. **D**. Bar 20 µm. Original.

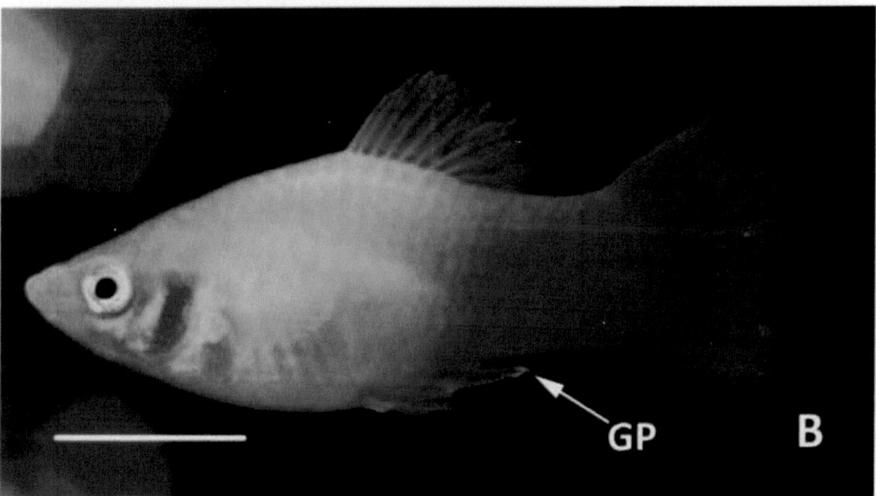

Fig. 3.12 Gonopodium (GP) of adult males of poeciliids. **A**. *Xiphophorus helleri*. Bar 1 cm. **B**. *Xiphophorus maculatus*. Bar 1 cm. Original.

incorporated in the cytoplasm at the apical end of ovarian epithelial cells in the ovary of *Xiphophorus maculatus* with stored sperm, and described the structure of special epithelial cells which were associated with spermatozoa. These special cells show depths in their apical end and intracellular pockets. Sperm heads may be found close to the basal end within these cells.

It has been suggested that sperm storage may contribute to regulate sperm competition, to separate the number of inseminations and of broods. In *Xiphophorus maculatus*, isolated females may produce five broods (Potter and

Kramer 2000). In the ovary of *Heterandria formosa*, live sperm have been found for ten months after insemination (Turner 1947; Amoroso 1960).

3.2.5 Number and Size of Oocytes

The number of matured oocytes produced by the ovary of viviparous fishes during reproductive cycles is, in general, lower compared to oviparous species. The viviparous fishes ovulate thousands of eggs, as in *Sebastodes*, a few hundred eggs, as in zoarcids, or dozens or less eggs, as in poeciliids and goodeids (Amoroso 1960; Thibault and Schultz 1978; Dodd and Sumpter 1984). The restricted number of eggs is characteristic of intraovarian gestation. The number of embryos that are developed in viviparous fish is evidently lower than that of oviparous species.

Oocyte diameter of viviparous species of teleosts varies between taxa as commented upon by Thibault and Schultz (1978). They compared genera of poeciliids and goodeids, such as *Poecilia, Gambusia* or *Xiphophorus* with eggs of about 2 mm in diameter, eggs of *Poeciliopsis gracilis*, or the goodeids *Ilyodon whitei, Goodea atripinnis* or *Xenotoca eiseni*, with eggs about 0.8 mm in diameter, or *Heterandria formosa* with eggs of about 0.5 mm in diameter. The reduction of yolk parallels the development of new structures that provide various exchanges between the embryo and the mother. According to Turner (1947) the most advanced type of viviparity in fishes occurs in species in which the yolk is greatly reduced. Therefore, complex transfers of nutrients from the mother to the embryo develop.

In matrotrophic species, the relationships between yolk and embryo dry-weight indicates that most nutrients derive from the mother rather than from yolk (Thibault and Schultz 1978; Wourms 1981; Wourms *et al.* 1988). Thibault and Schultz (1978) examined the eggs of *Poecilia reticulata*, moderately large (1.7 mm), and the total length of the young at birth, which is 6.5 mm, indicating a strong dependency on yolk rather than maternal nutrients. Wourms (2005) compared matrotrophy in poeciliids with low level of matrotrophic relationships such as *Xiphophorus* where the maternal contribution to embryonic weight attains 40% to term embryonic weight with a poeciliid with high level of matrotrophic relationships as in *Heterandria formosa* in which the maternal contribution constitutes 3,900% of the term embryonic weight (Grove and Wourms 1991, 1994). Scrimshaw (1944) described in embryos of *Heterandria formosa* the increase in weight from 0.017mg at the time of fertilization to 6.8mg at the time of parturition. Therefore, *Heterandria formosa*, because the small egg (Fig. 3.8C) presents a high degree of matrotrophic gestation, the follicular layer is very thin, amply vascularized, in close apposition to the external embryonic surface and there is a net of capillaries around the embryo; the capillaries are particularly localized around the head. These morphological characteristics are evidence of a high level of matrotrophy in this species (Fraser and Renton 1940; Turner 1947; Amoroso 1960; Wourms 1981; Wourms *et al.* 1988; Grove and Wourms 1994).

3.2.6 Intraovarian Gestation

After intrafollicular fertilization occurs within the ovary, intraovarian gestation occurs (Figs. 3.13A-D, 3.14A-E, 3.15A,B, 3.16A). The embryonic development is initiated within the ovarian follicle and may follow two different pathways of gestation: 1) intrafollicular gestation, in which the embryos remain in the follicle during their development (Fig. 3.16A) and move into the ovarian lumen immediately before birth, as in poeciliids (Amoroso 1960; Hoar 1969; Wourms 1981, 2005; Wourms and Lombardi 1992; Wourms *et al.* 1988; Constanz 1989); or, 2) intraluminal gestation, in which the embryos move from the follicle into the ovarian lumen during early development (cleavage, neurula or somites), where they continue to develop (Fig. 3.13A-C, 14B-E, 15A,B) until birth (Turner 1947; Amoroso 1981; Wourms *et al.* 1988; Wourms 2005). The embryos in the ovarian lumen are surrounded by the fluid secreted by the ovarian epithelium.

Wourms *et al.* (1988) classified the families of viviparous teleosts according to the type of gestation exhibited. Intrafollicular gestation occurs in the families Poeciliidae, Clinidae and Labrisomidae (some species). Intraluminal gestation occurs in the majority of the families, these are Zoarcidae, Parabrotulidae, Bythitidae, Aphyonidae, Zenarchopteridae, Goodeidae, Anablepidae, Scorpaenidae, Comephoridae and Embiotocidae. Wourms (2005) considered maternal-embryonic relationships, either lecithotrophy or matrotrophy, of viviparous teleosts developed in both types of gestation, intrafollicular and intraluminal are the most diverse relationships described for metabolic interchanges described in vertebrates.

3.2.7 Lecithotrophy and Matrotrophy

In lecithotrophic species, such as poeciliids, the yolk is gradually spent but is present throughout embryonic development. In matrotrophic species, such as goodeids, the yolk is consumed, approximately, during the first third of gestation; thereafter, embryonic growth is progressively dependent on maternal nutrients. In matrotrophic gestation, several forms of transfer of maternal nutrients to the embryos have been evolved (Schindler and de Vries 1987; Wourms *et al.* 1988; Wourms and Lombardi 1992; Schindler and Hamlett 1993). Reznick *et al.* (1996) discussed the difference in time, during the reproductive process, when maternal provisioning is determined between lecithotrophic and matrotrophic gestations. In lecithotrophic gestation, the provisions are determined during the formation of the oocyte in the oogenesis, thus, prior to fertilization. In matrotrophic gestation, the provisions are complemented after fertilization. For this type of gestation, with little reserve of yolk storage in the egg, maternal sources of nutrients are required together with stable environmental conditions for a constant transfer of nutrients from the mother to the embryo.

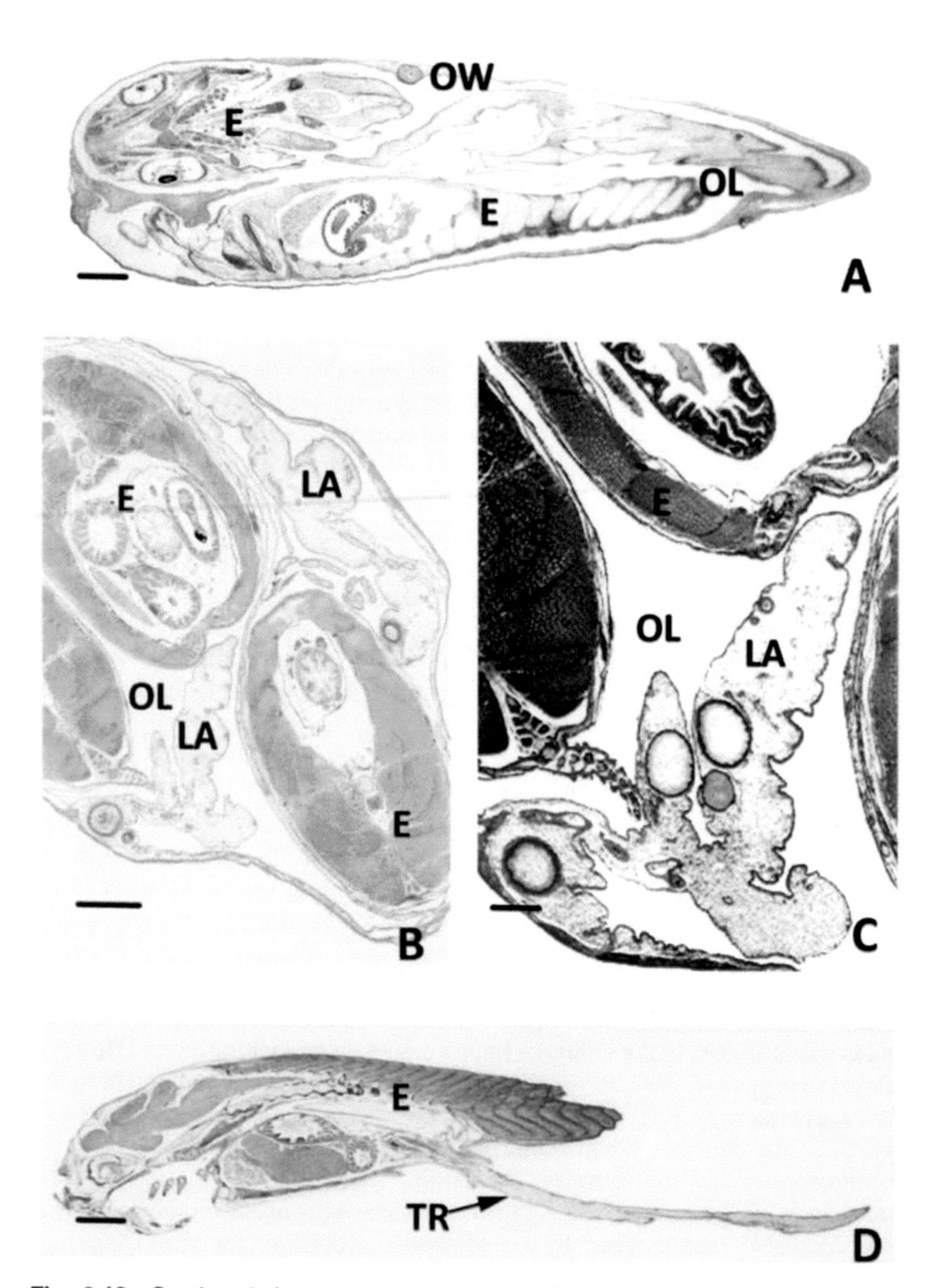

Fig. 3.13 Ovaries during gestation and illustrating trophotaenia development in goodeids. **A**. Longitudinal section of the ovary and embryos of *Ilyodon whitei* where the embryos (E) are developing in the ovarian lumen (OL). The ovarian wall (OW) is well defined all around the periphery. **B,C**. Transverse section of the ovary and

Fig. 3.13 Contd. ...

3.2.8 Intrafollicular and Intraluminal Gestation

Intrafollicular gestation in species of poeciliids has been studied by several authors: Turner (1940, 1947), Jollie and Jollie (1964), Lambert (1970a), Thibault and Schultz (1978), Wourms *et al.* (1988), Grove and Wourms (1991, 1994), Hollenberg and Wourms (1994), Reznick *et al.* (1996). Intraluminal gestation in species of goodeids has been studied by authors, including Turner (1933, 1937, 1947), Mendoza (1940, 1943, 1965), Schindler and De Vries (1987), Schindler *et al.* (1988), Lombardi and Wourms (1988), Wourms *et al.* (1988), Greven *et al.* (1993). Schindler and Hamlett (1993), Hollenberg and Wourms (1994).

In the process of intrafollicular gestation, after the fertilization of the oocyte, its surrounding layers, squamous follicle cells and theca, develop into the follicular placenta. During this process, follicle cells persist during gestation, becoming columnar, and the theca increases greatly in the number and size of blood vessels. In this type of gestation, the yolk provides the main nourishment for the embryo, but respiration and excretion functions are provided by exchange through the follicular placenta (Turner 1947; Jollie and Jollie 1964; Constanz 1989; Wourms 1981; Wourms *et al.* 1988; Grove and Wourms 1991, 1994).

As mentioned previously, in intraluminal gestation, the yolk is completely consumed during the first third of gestation. Subsequently, in goodeids, one parabrotulid *Parabrotula dientiens*, one embiotocid genus, *Rhacochilus*, and two genera of the family Aphyonidae, *Grammonus (=Oligopus)* and *Microbrotula* (Wourms 2005), the embryos develop trophotaenia, flattened folds growing from the caudal portion of the gut into the ovarian lumen (Figs. 3.13D, 3.14A-D) where they are surround by ovarian fluids. There are two types of trophotaenia in goodeids: 1) the rosette type of trophotaenium consists of short, lobulated, folded processes (Fig. 3.14A), as in genera *Allotoca, Goodea, Neoophorus, Xenoophorus,* and 2) the ribbon type consists of long, thin, flattened processes (Figs. 3.13D, 3.14B,C), as in genera *Allophorus, Ameca, Chapalichthys, Characodon, Girardinichthys, Ilyodon, Skiffia, Xenotoca, Zoogoneticus* (see reviews by Wourms *et al.* 1988; Greven *et al.* 1993; Wourms 2005). In terms of nutrient uptake, the ribbon type trophotaenia, as in *Ilyodon,* have an apical endocytotic complex and function in the uptake of macromolecules, as proteins, and small molecules, e.g. sugars and fatty acids. However, rosette type trophotaeniae, as in *Goodea,* lack an apical endocytotic complex; they only take up small molecules (Wourms and Lombardi 1992; Hollenberg and Wourms 1994; Schindler and deVries 1987; Greven *et al.*

Fig. 2.13 Contd. ...

embryos of *Ilyodon whitei*, two lamella (LA), one at each side of the ovary are seen. **D**. Embryo (E) of *Ilyodon whitei* with the ribbon type trophotaenia (TR), as a large extension of the gut. From Uribe *et al.* 2005. *Viviparous Fishes.* New Life Publications, Homestead, FL, USA. Pp: 217-235. Fig. 10a. **A**. Bar 1 mm. **B**. Bar 100 µm. **C**. Bar 50 µm. **D**. Bar 1 mm. Original.

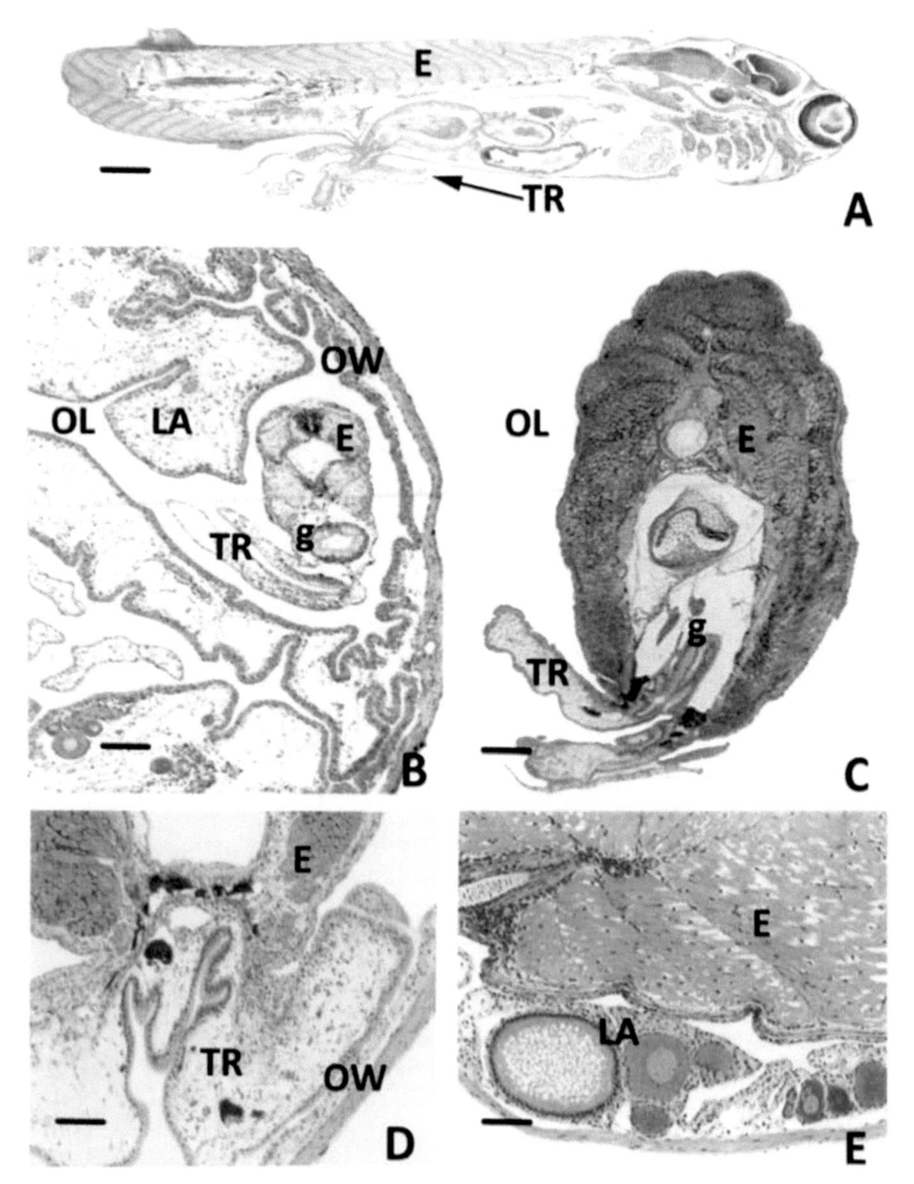

Fig. 3.14 Trophotaenia of goodeids. **A**. Embryo of *Goodea atripinnis* with the rosette type trophotaenia (TR), as a folded extension of the gut. Bar 1 mm. **B**. Ovary of *Skiffia multipunctata* during gestation. The ovarian wall (OW) and lamella (LA) are seen. The embryo (E), in transverse section, shows the trophotaenia (TR) emerging from the gut (g), as two folds that extend into the ovarian lumen (OL). Bar 50 µm. **C**. Ovary of *Xenotoca eiseni*, with an embryo (E) showing trophotenia (TR) extending from the gut (g) into the ovarian lumen (OL). Bar 50 µm. **D**. Trophotaenia (TR) of an embryo (E) of *Xenotoca eiseni* is in apposition to the ovarian wall (OW). Bar 20 µm. **E**. Lateral wall of the embryo (E) of *Xenotoca eiseni* is in apposition to the ovarian wall (OW). Bar 20 µm. Original.

1993). Ribbon trophotaenia are considered to be the evolutionarily most advanced type (Wourms 2005).

The trophotaenia are lined by an absorptive columnar epithelium that is enhanced by abundant microvilli at the apical ends of the epithelial cells. At the same time, the ovarian epithelium becomes higher, and the subjacent connective tissue swollen. Ovarian fluids contain a complex material formed by secretions, desquamated epithelial cells, cells of the immunological system, and diverse substances in solution originating from diffusion from maternal blood vessels. Histological analysis in goodeid embryos revealed that large portions of the trophotaenia are in apposition to the ovarian epithelium, facilitating the absorption of nutrients by the trophotaenial epithelium. The maternal tissue may be folded at these regions, forming invaginations where the epithelium is hypertrophied and the subjacent capillaries increase in number. In addition to the trophotaenia, the wall of the embryo may also be in apposition to the ovarian epithelium. These characteristics suggest secretory activity by the maternal epithelium and absorptive activity by the embryonic tissues.

Wourms (2005) described the process of development of trophotaenia as the culmination of a sequence of changes of embryonic gut adaptations, preceded by a) a simple tubular gut, b) enlargement of the gut, c) hypertrophy of intestinal villi, d) exteriorization of the gut by formation of trophotaenia in a rosette, and e) increase in number and length of trophotaenia, developing the ribbon type.

Several authors considered that the development of trophotaenia for metabolic interchange between maternal and embryonic tissues (Fig. 3.14E) in intraluminal gestation, described in goodeids, is the most complex system of relationships between mother and embryos in all viviparous teleosts (Turner 1947; Mendoza 1940, 1965, 1972; Schindler and de Vries 1987; Schindler *et al.* 1988; Wourms *et al.* 1988; Wourms and Lombardi 1992; Wourms 2005).

3.2.9 Oophagy and Adelphophagy

Complementary sources of nourishment described in goodeids are provided by oophagy (Fig. 3.15A,B), ingestion of eggs by the embryos, and adelphophagy (Fig. 3.15B), ingestion of other embryos, or fragments of other embryos (Turner 1947; Wourms 1988; Greven and Groâherr 1992). Turner (1947) mentioned that during gestation, some embryos die and are ingested by the surviving embryos, providing supplementary nutrients to healthy embryos. Both eggs and embryos ingested by late embryos are observed in its gut, during evident process of digestion. Additionally, adelphophagy may avoid the adverse consequences of degeneration of dead embryos among healthy embryos in the ovarian lumen.

3.2.10 Superfetation

Superfetation, observed in poeciliids species, is the development of two or more broods of embryos at different ages at the same time in which maturation

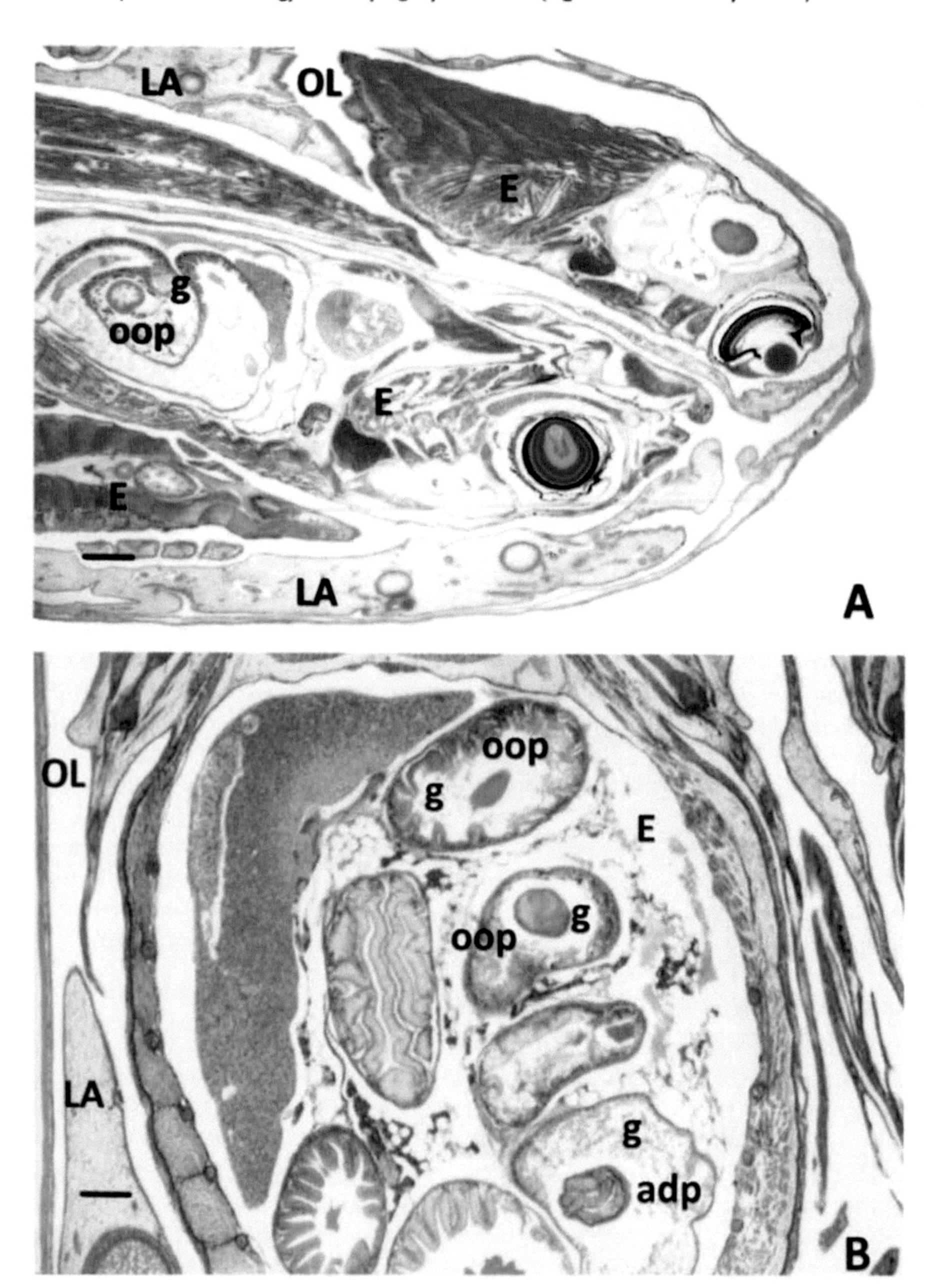

Fig. 3.15 Oophagy and adelphophagy. **A,B**. Ovaries of *Ilyodon whitei,* containing embryos (E) in the ovarian lumen (OL), during advanced stage of gestation. Lamella (LA) are at the periphery of the ovary. Oophagy (oop, ingestion of eggs), and adelphophagy (adp, ingestion of other embryos), occur as is evidenced by examination of the gut (g) contents in developing embryos. **A**. Bar 100 μm. **B**. Bar 100 μm. Original.

and fertilization of new ova overlaps the previous gestational period. Superfetation may occur several times, forming several groups of embryos, increasing the number of broods of different ages, at the same time (Turner 1947).

The poeciliid, *Heterandria formosa*, presents the maximum level of superfetation (Fig. 3.16A,B). Up to eight broods of embryos in different stages of development were described from the same ovary (Turner 1937, 1940; Scrimshaw 1944; Dodd and Sumpter 1984; Cheong *et al.* 1984; Grove and Wourms 1991, 1994). A factor that favors superfetation is the capacity of the ovary for sperm storage, maintaining them in viable condition (Turner 1940). Amoroso (1981) suggested that superfetation is possible because each follicle has some autonomy in regulating the development of oocytes and each embryo within its follicle cells is an independent physiological unit. Cheong *et al.* (1984) suggest that superfetation is a particularly interesting phenomenon because it represents a change to a near-continuous reproduction during the reproductive season.

The reproductive cycle in viviparous species, is controlled by endocrinological factors (for review see Koya 2008), similar to that described in other vertebrate groups. In viviparity, the cycle is regulated seasonally, coinciding with the time of birth to be most convenient for survival of the new born (Turner 1947; Wourms 1981; Dodd and Sumpter 1984, Contreras-McBeath and Ramírez-Espinoza 1996).

3.3 CHAPTER SUMMARY

During the evolution of viviparity in teleosts, several unique morphological and physiological characteristics have developed that are associated with intraovarian gestation. These include intrafollicular gestation in poeciliid species and intraluminal gestation in goodeid species. The cystovarian type of ovary is common to both oviparous and viviparous teleosts; it is a saccular organ with a central lumen. Most viviparous teleosts have a single ovary whereas ovaries are normally paired. The ovary contains germinal cells (oogonia and oocytes during different stages of development) and somatic tissues (epithelial cells, vascularized stroma, smooth muscle and serosa). As common to all fish, oogenesis has previtellogenic and vitellogenic stages. Follicular atresia may occur in any stage of oogenesis. All teleosts lack Müllerian ducts from which oviducts develop in other vertebrates. Lacking oviducts, but having an ovarian lumen, intraovarian insemination, fertilization, and gestation occur in viviparous teleosts which collectively are unique among vertebrates. Males of viviparous species have testicular characteristics that co-evolved along with changes in the ovary that are related to internal fertilization and gestation. In most viviparous teleosts, the type of testis is spermatogonial restricted lobular. Spermatozoa occur within spermato-zeugmata. An intromittent organ, the gonopodium, develops as a modification of the anal fin for transport of spermatozoa to the female

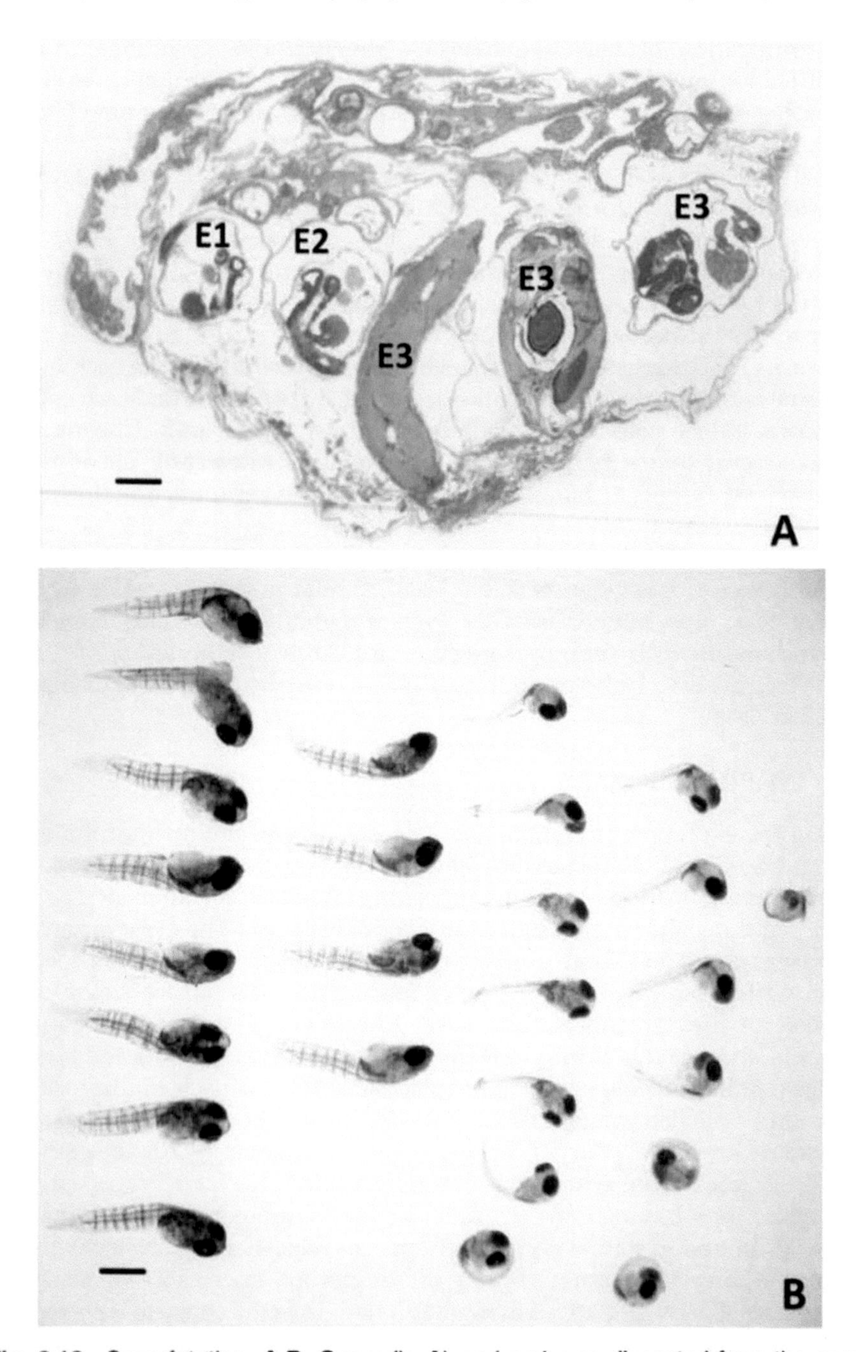

Fig. 3.16 Superfetation. **A,B**. Ovary (in **A**) and embryos dissected from the same ovary (in **B**) of *Heterandria formosa.* After the extraction of the embryos, the remaining ovarian tissue contains three different stage of development of embryos (E1, E2, E3), additionally, there are 4 groups of embryos according to the stage of development. **A**. Bar 100 μm. **B**. Bar 2 mm. Original.

gonoduct. Fertilization in the majority of viviparous teleosts is intrafollicular. Sites for sperm storage occur in poeciliids. In goodeids, spermatozoa are only detected into the ovarian lumen around the time of fertilization, these from recent mating. The yolk is gradually spent during the early development of the embryos. Intraovarian gestation defines diverse types of maternal-embryonic interchanges for the development of the embryos. Embryonic development may be within the follicle (intrafollicular gestation), in which the embryo develops inside the follicle until birth. Embryonic development may also occur within the ovarian lumen (intraluminal gestation); during an early stage of gestation, the embryos are evacuated into the ovarian lumen where their development continues. The evolution of intraovarian gestation ranges between lecithotrophy (nourishment by yolk resources of the oocyte) to matrotrophy (nourishment by supply of maternal nutrients). Superfetation, observed in poeciliids, is the development of two or more broods of embryos at different ages at the same time.

3.4 ACKNOWLEDGMENTS

We are grateful to Professor Barrie Jamieson for his advice throughout preparation of this chapter and his substantial editorial aid. We thank Marcela Esperanza Aguilar Morales who assisted with tissue processing of *Heterandria formosa* specimens with excellent results.

3.5 LITERATURE CITED

Amoroso, E. C. 1960. Viviparity in fishes. Symposium. Zoological Society (London) 1: 153-181.

Amoroso, E. C. 1981. Viviparity. Pp. 3-24. In S. R. Glasser and D. W. Bullock (eds), *Cellular and molecular aspects of implantation*, Plenum Press, New York, USA.

Billard, R. 1969. La spermatogenese de *Poecilia reticulata* II. La production spermatogenètique. Annals of Biological Animal Biochemical and Biophysics 9: 307-313.

Burns, J. R. 1991. Testis and gonopodium development in *Anableps dowi* (Pisces: Anablepidae) correlated with pituitary gonadotropic zone area. Journal of Morphology 210: 45-53.

Burns, J. R. and Weitzman, S. H. 2005. Insemination in Ostariophysan fishes. Pp. 105-132. In M. C. Uribe and H. J. Grier (eds), *Viviparous Fishes*. New Life Publications, Homestead, Fl., USA.

Cheong, R. T., Henrich, S., Farr, J. A. and Travis, J. 1984. Variation in fecundity and its relationship to body size in a population of the least killifish, *Heterandria formosa* (Pisces: Poeciliidae). Copeia (3): 720-726.

Constanz, J. 1989. Reproductive biology of the poeciliid fishes. Pp. 33-50. In G. K. Meffe and F. F. Snelson (eds), *Ecology and evolution of live bearing fishes (Poeciliidae)*. Prentice Hall. Englewood Cliffs, New Jersey, USA.

Contreras-MacBeath, T. and Ramírez-Espinoza, H. 1996. Some aspects of the reproductive strategy of *Poeciliopsis gracilis* (Osteichtyes: Poeciliidae) in the Cuautla River, Morelos, México. Journal of Freshwater Ecology 11(3): 327-338.

Dodd, J. M. and Sumpter, J. P. 1984. Fishes. Pp. 1-126. In G. E. Lamming (ed.), *Marshall's Physiology of Reproduction*, Vol 1. London: Churchill Livingstone.

Droller, M. J. and Roth, T. F. 1966. An electron microscope study of yolk formation during oogenesis in *Lebistes reticulates* Guppy. Journal of Cell Biology 28: 209-232.

Fraser, E. A. and Renton, R. M. 1940. Observation on the breeding and development of the viviparous fish, *Heterandria formosa*. Quarterly Journal of Microscopical Science 81: 479-520.

Gardiner, D. M. 1978. Cyclic changes in fine structure of the epithelium lining the ovary of the viviparous teleost, *Cymatogaster aggregata* (Perciformes: Embiotocidae). Journal of Morphology 210156: 367-380.

Greven, H. 2005. Structural and behavioral traits associated with sperm transfer in Poeciliinae. Pp. 145-163. In M. C. Uribe and H. J. Grier (eds), *Viviparous Fishes*. New Life Publications, Homestead, Fl., USA.

Greven, H. and Groâherr, M. 1992. Adelphophagy and oophagy in *Ameca splendens* Miller and Fitzimons, 1971 (Goodeidae: Teleostei). Zeitschrift für Fischkunde 1: 193 197.

Greven, H., Passia, D. and Marani, E. 1993. CD15 in trophotaeniae of the viviparous goodeid fish *Xenotoca eiseni* (Cyprinodontiformes, Teleostei). European Journal of Morphology 31(4): 267-273.

Grier, H. J. 1975a. Aspects of germinal cysts and sperm development in *Poecilia latipinna* (Teleostei: Poeciliidae). Journal of Morphology 146(2): 229-249.

Grier, H. J. 1975b. Spermiogenesis in the teleost *Gambusia affinis* with particular reference to the role placed by microtubules. Cell and Tissue Research 165: 89-102.

Grier, H. J. 1981. Cellular organization of the testis and spermatogenesis in fishes. American Zoologist 21: 345-357.

Grier, H. J., Fitzsimons, J. M. and Linton, J. R. 1978. Structure and ultrastructure of the testis and sperm formation in goodeid teleosts. Journal of Morphology 156(3): 419-437.

Grier, H. J., Uribe, M. C., Parenti, L.R. and De la Rosa-Cruz, G. 2005. Pp: 191-126. Fecundity, the germinal epithelium, and folliculogenesis in viviparous fishes. Pp: 145-163. In M. C. Uribe and H. J. Grier (eds), *Viviparous Fishes*. New Life Publications, Homestead, Fl., USA.

Grove, B. D. and Wourms, J. P. 1991. The follicular placenta of the viviparous fish, *Heterandria formosa*: ultrastructure and development of the embryonic absorptive surface. Journal of Morphology 209: 265-284.

Grove, B. D. and Wourms, J. P. 1994. Follicular placenta of the viviparous fish, *Heterandria formosa*: ultrastructure and development of the follicular epithelium. Journal of Morphology 220: 167-184.

Guillette, L. J. 1987. The evolution of viviparity in fishes, amphibians and reptiles: an endocrine aproach. Pp 523-562. In D. O. Norris and R. E. Jones (eds), *Hormones and Reproduction in Fishes, Amphibians and Reptiles*. Plenum Press, New York-London.

Guraya, S. S. 1986. *The Cell and Molecular Biology of Fish Oogenesis*. Pp 5-223. Karger. Basel, München, Germany.

Hoar, W. S. 1969. Reproduction. Pp 1-72. In W. S. Hoar and D. J. Randall (eds), *Fish physiology*, Vol. 3, Academic Press, New York, USA.

Hollenberg, F. and Wourms, J. P. 1994. Ultrastructure and protein uptake of the embryonic trophotaeniae of four species of goodeid fishes (Teleostei: Atheriniformes). Journal of Morphology 219: 105-119.

Hubbs, C. L. and Turner, C. L. 1939. Studies of the fishes of the order Cyprinodontes. XVI. A revision of the Goodeidae. Miscelaneous Publications of the Museum of Zoology University of Michigan 42: 1-80.

Hurk, R. van den and Barends, P. 1975. Cytochemistry of the intratesticular efferent duct system of the Black Molly (*Mollienisia latipinna*). Koninkl. Nederland Akademie van Wetenschappen. Amsterdam C, 77 (3): 205-214.

Jamieson, B. G. M. 1991. *Fish Evolution and Systematics: Evidence from Spermatozoa.* Cambridge University Press, Cambridge. 319 pp.

Jollie, W. P. and Jollie, L. G. 1964. The fine structure of the ovarian follicle of the ovoviviparous poeciliid fish, *Lebistes reticulates*. II. Formation of follicular pseudoplacenta. Journal of Morphology 114: 503-526.

Koya, Y. 2008. Reproductive physiology in viviparous teleosts. Pp. 245-275. In M. J. Rocha, A. Arukwe and B. G. Kapoor (eds), *Fish Reproduction.* Science Publishers. Enfield, NH, Jersey, Plymouth.

Koya, Y., Takano, K. and Takahashi, H. 1995. Annual changes in fine structure of inner epithelial living of the ovary of a marine sculpin, *Alcichthys ancicornis* (Teleostei: Scorpaeniformes), with internal gametic association. Journal of Morphology 223: 85-97.

Koya, Y., Munehara, H. and Takano, K. 1997. Sperm storage and degradation in the ovary of a marine copulating sculpin, *Alcichthys ancicornis* (Teleostei: Scorpaeniformes): role of intercellular junctions between inner ovarian epithelial cells. Journal of Morphology 233: 153-163.

Lambert, J. G. D. 1970a. The ovary of the guppy *Poecilia reticulata.* The granulosa cells as sites of steroid biosynthesis. General and Comparative Endocrinology 21: 287-304.

Lambert, J. G. D. 1970b. The ovary of the guppy *Poecilia reticulata.* The atretic follicle, a *corpus atreticum* or a *corpus luteum praeovulationis*. Zeitschrift für Zellforschung und mikroskopische Anatomie. 107: 54-67.

Lombardi, J. and Wourms, J. P. 1988. Embryonic growth and trophotaenial development in goodeid fishes (Teleostei: Atheriniformes). Journal of Morphology 197: 193-208.

Meisner, A. D. 2005. Male modifications associated with insemination in teleosts. Pp. 165-190. In M. C. Uribe and H. J. Grier (eds), *Viviparous Fishes.* New Life Publications, Homestead, Fl., USA.

Mendoza, G. 1939. The reproductive cycle of the viviparous teleost *Neotoca bilineata,* a member of the family Goodeidae. I. The breeding cyclic. Biological Bulletin 76: 359-370.

Mendoza, G. 1940. The reproductive cycle of the viviparous teleost *Neotoca bilineata,* a member of the family Goodeidae. II. The cyclic changes in the ovarian soma during gestation. Biological Bulletin 78: 349-365.

Mendoza, G. 1943. The reproductive cycle of the viviparous teleost *Neotoca bilineata,* a member of the family Goodeidae. IV. The germinal tissue. Biological Bulletin 84: 87-97.

Mendoza, G. 1965. The ovary and anal processes of *Characodon eiseni,* a viviparous cyprynodont teleost from Mexico. Biological Bulletin 129: 303-315.

Mendoza, G. 1972. The fine structure of an absorptive epithelium in a viviparous teleost. Journal of Morphology 136: 109-130.

Nagahama, Y. 1983. The functional morphology of teleost gonads. Pp. 223-275. In W. S. Hoar, D. J. Randall and E. M. Donaldson (eds), *Fish Physiology*, Vol. IX, Academic Press, New York, USA.

Parenti, L. R. and Grier, H. J. 2004. Evolution and phylogeny of gonad morphology in bony fishes. Integrative and Comparative Biology 44: 333-348.

Parenti, L. R. 2005. The phylogeny of atherinomorphs: evolution of a novel fish reproductive system. Pp. 13-30. In M. C. Uribe and H. J. Grier (eds), *Viviparous Fishes.* New Life Publications, Homestead, Fl., USA.

Pavlov, D. A. 2005. Reproductive biology of wolffishes (Anarhichadidae) and the transition from oviparity to viviparity in the Suborder Zoarcoidei. pp: 13-30.

Potter, H. and Kramer, C. R. 2000. Ultrastructural observations on sperm storage in the ovary of the platyfish, *Xiphophorus maculatus* (Teleostei: Poeciliidae): the role of the duct epithelium. Journal of Morphology 245: 110-129.

Reznick, D., Callahan, H. and Llauredo, R. 1996. Maternal effects on offspring quality in poeciliid fishes. American Zoologist 36: 147-156.

Rosa-Molinar, E. 2005. Goodrich's theory of transportation revisited: the shift to a sexually dimorphic axial formulae and nervous system in a poeciliine fish. Pp. 59-70. In M. C. Uribe and H. J. Grier (eds), *Viviparous Fishes*, New Life Publications, Homestead, Fl., USA.

Rosen, D. E. 1964. The relationships and taxonomic position of the halfbeaks, killifishes, silversides and their relatives. Bulletin of American Museum of Natural History 127: 217-268.

Rosen, D. E. and Gordon, M. 1953. Functional anatomy and evolution of male genitalia in poeciliid fishes. Zoologica NY 38: 1-47.

Schindler, J. F. and De Vries, U. 1987. Maternal-embryonic relationships in the goodeid teleost *Xenoophorus captivus.* Embryonic structural adaptations to viviparity. Cell and Tissue Research 247: 325-338.

Schindler, J. F., Kujat, R. and De Vries, U. 1988. Maternal-embryonic relationships in the goodeid teleost *Xenoophorus captivus.* The internal ovarian epithelium and the embryotrophic liquid. Cell and Tissue Research 254: 177-182.

Schindler, J. F. and Hamlett, W. C. 1993. Maternal-embryonic relations in viviparous teleosts. Journal of Experimental Zoology 266: 378-393.

Scrimshaw, N. S. 1944. Embryonic growth in the viviparous poeciliid, *Heterandria formosa.* Biological Bulletin 87: 37-51.

Selman, K. and Wallace, R. A. 1989. Cellular aspects of oocyte growth in teleosts. Zoological Science 6: 211-231.

Thacker, C. and Grier, H. J. 2005. Unusual gonad structure in the paedomorphic teleost, *Schindleria praematura* (Teleostei: Gobioidei): a comparison with other gobioid fishes. Journal of Fish Biology 66: 378-391.

Thibault, R. E. and Schultz, R. J. 1978. Reproductive adaptations among viviparous fishes (Cyprinodontiformes: Poeciliidae). Evolution 32: 320-333.

Turner, C. L. 1933. Viviparity superimposed upon ovoviparity in the Goodeidae, a family of cyprinodont teleosts fishes of the Mexican Plateau. Journal of Morphology 55: 207-251.

Turner, C. L. 1937. The trophotaeniae of the Goodeidae, a family of viviparous cyprinodont fishes. Journal of Morphology 61: 495-523.

Turner, C. L. 1940. Pseudoamnion, pseudochorion and follicular pseudoplacenta in poeciliid fishes. Journal of Morphology 67: 59-89.

Turner, C. L. 1947. Viviparity in teleost fishes. Science Monthly 65: 508-518.

Turner, C. L., Mendoza, G. and Reiter, R. 1962. Development and comparative morphology of the gonopodium of goodeid fishes. Iowa Academic Science 1962: 571-586.

Tyler, C. R. and Sumpter, J. P. 1996. Oocyte growth and development in teleosts. Reviews in Fish Biology and Fisheries 6: 287-318.

Uribe, M. C., De la Rosa-Cruz, G. and Garcia-Alarcon, A. 2005. The ovary of viviparous teleosts. Morphological differences between the ovaries of *Goodea atripinnis* and *Ilyodon whitei*. Pp. 217-235. In M. C. Uribe and H. J. Grier (eds), *Viviparous Fishes.* New Life Publications, Homestead, Fl., USA.

Uribe, M. C., De la Rosa-Cruz, G., Garcia-Alarcon, A., Guerrero-Estevez, S. M. and Aguilar Morales, M. 2006. Características histológicas de los estados de atresia de folículos ováricos en dos especies de teleósteos vivíparos *Ilyodon whitei* y *Goodea atripinnis* (Goodeidae). Hidrobiológica 16(1): 67-73.

Wake, M. H. 1985. Oviduct structure and function in non-mammalian vertebrates. Fortschritte der Zoologie 30: 427-435.

Wake, M. H. 1989. Phylogenesis of direct development and viviparity in vertebrates. Pp. 235-250. In D. B. Wake and G. Roth (eds), *Complex Organismal Functions: Integration and Evolution in Vertebrates.* John Wiley and Sons Ltd. Chichester, England.

Wallace, R. A. and Selman, K. 1981. Cellular and dynamic aspects of oocyte growth in teleosts. American Zoologist 21: 325-343.

Wourms, J. P. 1981. Viviparity: The maternal-fetal relationship in fishes. American Zoologist 21: 473-515.

Wourms, J. P. and Lombardi, J. 1992. Reflections on the evolution of piscine viviparity. American Zoologist 32: 276-293.

Wourms, J. P., Grove, B. D. and Lombardi, J. 1988. The maternal-embryonic relationships in viviparous fishes. Pp. 1-134. In W. S. Hoar and D. J. Randall (eds), *Fish Physiology,* Vol. XI, Academic Press, New York, USA.

Wourms, J. P. 2005. Functional morphology, development, and evolution of trophotaeniae. Pp. 217-242. In M. C. Uribe and H. J. Grier (eds), *Viviparous Fishes.* Homestead, FL: New Life Publications.

CHAPTER 4

The Testis and Spermatogenesis in Teleosts

Harry J. Grier[1] *and Mari Carmen Uribe Aranzábal*[2]

4.1 INTRODUCTION

In most teleosts, testes are paired elongated organs, attached to the dorsal wall of the body by a mesorchium. Both testes join caudally, converging in a central efferent duct system which is open to the exterior through the urogenital pore. The testes change morphologically during the annual reproductive cycle. Functions of teleost testes are basically conserved, as in other vertebrates, regarding three main aspects: 1) the formation of spermatozoa during spermatogenesis, 2) delivering them to the efferent ducts during spermiation, and 3) the secretion of male sex hormones (Pudney 1995).

Histological examination of teleost testes reveals that they possess an interstitial tissue and a germinal component, the latter being a germinal epithelium composed of germ cells that are supported by somatic Sertoli cells. Germ cells include all of the stages of differentiation of the spermatogenic cells: spermatogonia, spermatocytes, spermatids, and sperm. The somatic cells are the Sertoli cells that are homologous among vertebrates based on a common origin from the germinal ridge (Grier *et al*. 2005) and the phagocytosis of residual bodies cast off by developing spermatids (Grier 1993). Within the interstitium, the steroidogenic Leydig cells produce testosterone (Billard 1986; Grier 1993; Patiño and Takashima 1995).

[1]Division of Fishes, Department of Zoology, National Museum of Natural History, MRC 159, Smithsonian Institution, P.O. Box 3712, Washington, D.C. 20012-7012, USA. Florida Fish and Wildlife Research Institute, 100 8th Avenue, SE, St. Petersburg, FL 33701-5020, USA.

[2]Laboratorio de Biología de la Reproducción, Departamento de Biología Comparada, Facultad de Ciencias, Universidad Nacional Autónoma de México, México, D.F. 04510 México.

Spermatogenesis includes a sequence of morphological and physiological changes in the germ cells as they differentiate from spermatogonia into spermatozoa through the formation of intermediate cell types: primary spermatocytes, secondary spermatocytes, and spermatids. Spermatogonia are diploid stem cells that proliferate by mitotic divisions increasing the number of germ cells within the testis germinal compartment. Once spermatogonia replicate their chromosomes and initiate meiosis, they are primary spermatocytes. Primary spermatocytes are involved in the first meiotic division, during which the homologous chromosomes pair and separate, after which they are secondary spermatocytes that become haploid spermatids after the second division of meiosis. The spermatids undergo morphological transformation during spermiogenesis becoming spermatozoa (for spermatozoal ultrastructure, see **Chapters 6-17**).

During spermatogenesis in all teleosts, spermatogonia are associated with the somatic cell in the germinal compartment, the Sertoli cells. Initially, Sertoli cell processes surround individual spermatogonia, the primary or Type A spermatogonia. After mitotic divisions of the spermatogonia, Sertoli cell processes surround an isogenic clone of secondary or Type B spermatogonia, enclosing them within a cyst, or spermatocyst, in which meiosis and spermiogenesis occur. Cystic spermatogenesis is characteristic of teleosts; Sertoli cells provide a permeability barrier during spermatogenesis (Grier 1993). This barrier results from the formation of tight junctions between adjacent Sertoli cells (Patiño and Takashima 1995). The barrier is also known as a male germ cell protective barrier or blood testis barrier (Abraham 1991). Sertoli cells of the Northern Pike, *Esox lucius*, phagocytize residual bodies cast off by developing spermatids (Grier 1977), as do those of the Redfin Pickerel, *E. americanus* (Grier *et al.* 1989) (Esociformes). The phagocytosis of residual bodies is a common Sertoli cell function proposed to designate homology between Sertoli cells of different vertebrate taxa (see above). In similar fashion to the somatic cells in the male, the Sertoli cells, the somatic follicle cells in the teleost ovary also become phagocytes during follicular atresia (Hunter and Macewicz 1985; Miranda *et al.* 1999, Uribe *et al.* 2006). Both gonadal somatic cells (the Sertoli cells and the follicle cells) of common embryonic origin, possess similar functions as both serve gonad clean-up functions in the testis or ovary, respectively, by becoming phagocytes.

4.2 MORPHOLOGY OF GERM CELLS DURING SPERMATOGENESIS

Spermatogenesis in teleosts was studied by several authors who described all stages of germ cell maturation in different species (see review by Pudney 1995).

Primary spermatogonia: Primary spermatogonia are the largest germ cells in the germ line. They are localized in the germinal epithelium and have light granular cytoplasm and a central, spherical nucleus. Their nuclei possess delicate granular diploid chromatin and one or two nucleoli. Primary spermatogonia proliferate mitotically renewing the pool of germ cells and giving rise to secondary spermatogonia. Primary spermatogonia are single cells or are arranged in small groups.

Secondary spermatogonia: Secondary spermatogonia are spherical cells, smaller that the primary spermatogonia. Secondary spermatogonia are surrounded by Sertoli cell processes, thus being enclosed within a spermatocyst. Secondary spermatogonia divide mitotically several times before the initiation of meiosis. As with primary spermatogonia, secondary spermatogonia have a spherical nucleus with diploid chromosomes and one or two nucleoli. They replicate their chromosomes and enter into meiosis, becoming primary spermatocytes.

Primary and secondary spermatocytes: Primary spermatocytes are spherical cells, similar in size to secondary spermatogonia. Whereas spermatogonia divide by mitosis, spermatocytes have initiated meiosis, two consecutive cell divisions in which the first is the reduction division where the chromosome number is halved. Meiosis in males is a continuous process, and four haploid sperm are produced rather than one egg and the polar bodies in females. In the nucleus of primary spermatocytes, the homologous chromosomes have replicated and are in different phases of meiotic prophase I: leptotene (fine reticular chromatin), zygotene (fine fibrillar pattern of replicated, homologous chromosomes in synapsis forming the synaptonemal complex), pachytene (paired homologous chromosomes condensed, and synaptonemal complexes readily defined ultrastructurally, and crossing-over occurs), and diplotene (separation of homologous, replicated chromosomes that remain attached at the chiasmata). Within any spermatocyst, meiosis is fairly synchronous. When viewed histologically, cysts with primary spermatocytes at the pachytene of meiotic prophase I are relatively abundant, whereas those in leptotene, zygotene, and especially in diplotene are rarely seen, quite unlike the female pattern of meiosis (See Grier *et al.*, **Chapter 2**, for staging and definition of "Step" of meiotic prophase). This reflects the duration of meiotic Steps, among which pachytene is the longest Step, leptotene and zygotene are shorter, and diplotene is the shortest Step. The primary spermatocytes enter metaphase I, anaphase I, and telophase I, when the first division of meiosis is completed, resulting in two secondary spermatocytes. Secondary spermatocytes are spherical cells, the nucleus contains the haploid number of chromosomes, but duplicated. They are seen less frequently than primary spermatocytes as they divide rapidly, after a short interphase between the two meiotic divisions, each giving rise to two spermatids.

Spermatids and spermatozoa: Early spermatids are haploid cells that are spherical in shape. They do not divide but transform into spermatozoa through the process called spermiogenesis. As spermiogenesis proceeds, the nuclei of spermatids become smaller while the chromatin shows increasing degrees of condensation. The cells become motile by development of a flagellum, retain mitochondria that become localized within the sperm midpiece, and void excess cytoplasm as a residual body that is phagocytozed by Sertoli cells.

Spermatogenesis, as in other vertebrates, is a complex process that occurs in the special biochemical environment defined by the Sertoli cells associated with the germ cells. The Sertoli cells surround the germ cells, communicate with them, and separate them from the rest of the somatic tissues by a permeability barrier. These interactions between Sertoli and germ cells are controlled by the endocrine axis: hypothalamus – hypophysis – testis. Fish spermatogenesis, from spermatogonial stem-cell renewal to spermatozoa is controlled by the sex steroid hormones. Schulz and Miura (2002); Miura and Miura (2003); Miura *et al.* (2006) and Ozaki *et al.* (2006) described hormonal aspects of fish spermatogenesis (see also Schneider and Poehland, **Volume 8B, Chapter 2**). Sertoli cells express receptors for sex steroids and follicle-stimulated hormone (FSH), mediating the hormone's effect to the germ cells. Testosterone acts via feedback mechanisms on FSH-dependent steroidogenesis. LH is essential for the regulation of androgen production (Schulz and Miura 2002). Miura (2003) and Miura *et al.* (2006) categorized the mitotic division of spermatogonia by (a) spermatogonial stem-cell renewal, the mitosis of primary spermatogonia, and (b) spermatogonial proliferation, the mitosis of secondary spermatogonia inside cyst prior to the initiation of meiosis. Both processes are regulated by steroid hormones in different ways. The first one is regulated by estradiol-17β, and the second one is regulated by an androgen, 11-ketotestosterone that is induced by FSH stimulation. 11-ketotestosterone also induces meiosis and spermiogenesis, but the mechanisms of these processes are not clear (Miura and Miura 2003; Ozaki *et al.* 2006). Progestins that are produced under the influence of LH appear to play an important role in the control of the contents of the seminal plasma (Schulz and Miura 2002), and the induction of spermiation and sperm maturation (Miura *et al.* 2006).

In several species of teleosts which are internal fertilizers, the spermatozoa are arranged into packets which are formed in the spermatocysts and released during spermiation into the efferent duct system. These sperm packets are termed spermatophores when they are covered by a capsule. When these packets are unencapsulated or naked they are known as spermatozeugmata (plural) or spermatozeugma (singular) (Grier *et al.* 1978, 1981; Gardiner 1978; Grier 1993; Downing and Burns 1995; Pecio *et al.* 2001). The formation of spermatozeugmata always involves the association of Sertoli cells with the spermatozoa. In the Poeciliidae, the sperm nuclei associate with Sertoli cells, and spermatozeugmata have outwardly-directed sperm nuclei. In the Goodeidae, the spermatid flagella associate with the Sertoli cells, and sperm flagella cover the surface of the spermatozeugma (Grier *et al.* 2005). The presence of secretions in the efferent ducts as PAS-positive material has been described in testes of teleosts which produce sperm packets (see Uribe *et al.*, **Chapter 3**). These secretions have been found accompanying spermatozeugmata in species of Poeciliidae (Grier 1981), Hemirhamphidae (Downing and Burns 1995), and the Glandulocaudinae (Pecio *et al.* 2001).

The testicular tissues form two compartments, 1) the germinal compartment, which is composed by Sertoli cells and germ cells and 2) the interstitial compartment, integrated by the connective tissue. These two compartments are always separated by a basement membrane. Therefore, the cells in one compartment do not intermix with the cells of the other compartment.

4.3 GERMINAL COMPARTMENT

Germinal epithelium: The germinal epithelium forms the germinal compartment. This epithelium fulfills the characteristics of any epithelium except that it is specialized in having two cell types: somatic cells and germ cells (Grier 2000; Grier and Lo Nostro 2000). As in all epithelia, the germinal epithelium rests upon a supporting layer, the basement membrane that separates the epithelium from underlying connective tissue, the interstitium (Fawcett and Jensh 2002). The analysis of the germinal epithelium in testes and ovaries manifests the development of an emerging constant in vertebrate reproduction (Parenti and Grier 2004). This concept contributes to an understanding of the morphology and physiology of male and female gametogenesis, both spermatogenesis and oogenesis (see Grier *et al.*, **Chapter 2**).

Embryonic origin of the germinal epithelium: The origin of the germinal epithelium is similar in males and females. Primordial germ cells migrate to the germinal ridge during embryonic development where, along with coelomic somatic cells, a germinal epithelium develops (Strüssmann and Nakamura 2002; Yoshizaki *et al.* 2002). The primordial germ cells and the somatic cells possess the capacity to develop into male or female cell lineages and produce either ovaries or testes. Given a common origin, germ cells in males and females are homologous as are the somatic cells of testes and ovaries, the Sertoli cells and the follicle cells, respectively. Follicles in the teleost ovary are wholly derived from the germinal epithelium (Grier 2000; Parenti and Grier 2004; Grier *et al.* 2005), whereas the theca surrounding the follicles is derived from the stroma (Wallace and Selman 1990), a fact true in the ovaries of all vertebrate groups (Tokarz 1978). The basement membrane is more than a support structure as it separates tissue compartments of different embryonic origins. In males, a basement membrane separates the germinal compartment from the interstitium. In females, a basement membrane separates the germinal epithelium from the underlying stroma and also the derivatives of the female germinal epithelium, the follicles and follicle cells. Rainbow Trout was the first teleost species in which the permanent separation between the follicle and stroma was demonstrated throughout the whole of oocyte growth, maturation, and extending to the postovulatory follicle (Grier *et al.* 2007) (see Grier *et al.*, **Chapter 2**).

4.4 INTERSTITIAL COMPARTMENT

The interstitial compartment is formed by connective tissue containing Leydig cells, nerve fibers, fibroblasts, collagen fibers, smooth muscle cells and blood vessel. It was interpreted that the fishes *Esox lucius*, *Salmo salar*, and *Labeo* sp. lacked typical interstitial Leydig cells (Marshall and Lofts 1956). Instead, the testicular endocrine cells were distributed as Leydig cell homologs, as lobule boundary cells. This interpretation was made using low-resolution, frozen sections followed by the Sudan stain which appeared to outline the germinal compartment, particularly in *E. lucius*, with sudanophilic lipids. However, a later ultrastructural study on testis structure in *E. lucius* and *E. americana* (Grier *et al.* 1989) revealed Leydig cells within the interstitium and accumulations of testicular lipids within the germinal compartment, in the Sertoli cells of *E. lucius*. Lipids were not detected in the testis of *E. americana*, apparently a real inter-specific difference. In both species, the typical arrangement of a germinal compartment in the testes being separated from the interstitial compartment by a basement membrane was demonstrated along with interstitial Leydig cells after the common vertebrate pattern. In some species, as poeciliids and goodeids, the Leydig cells are abundant around the efferent ducts, as it is described in *Poecilia latipinna* (Hurk van den 1973; Grier 1981).

4.5 TESTIS TYPES OF TELEOSTS

The fish testis evolved from being polyspermatocystic in agnathans and elasmobranchs, to an anastomosing tubular type in some phylogenetically lower teleosts and to a lobular type in advanced neoteleosts. Two different lobular types have been described: in the unrestricted lobule type, the spermatogonia occur along the walls of the lobules. In the restricted lobular type (the telogonic testis, *sensu* Jamieson 1991), spermatogonia only occur at the distal termini of the lobules. Given a great deal of comparative morphology, a "tubular" designation was applied to testis structure in order to distinguish between anastomosing tubules that do not terminate at the testis periphery. The designation "tubular" was retained from the mammalian condition where tubules do not terminate at the testis periphery and form continuous loops beginning at and terminating at the efferent ducts, but without branching or forming anastomoses. The distinction between this condition in mammals and the anastomosing condition in fish is remarkable (Grier 1993). A "lobular" designation was applied to describe testis structure where the germinal compartments do terminate at the testis periphery (Grier 1993).

Therefore, the teleost testis is organized morphologically as three types of testis (Grier 1993; Parenti and Grier 2004): 1) anastomosing tubular testis type (Fig. 4.1); 2) unrestricted spermatogonial testis type (Figs. 4.2-4.5); and 3) restricted spermatogonial testis type (Figs. 4.6-4.9).

Anastomosing tubular testis: The anastomosing tubular testis type is characterized by the germinal compartments forming loops at the testis periphery and double back to the efferent ducts, forming a highly branched and anastomosing network. This testis type is present in lower fishes, as salmonids, cyprinids and lepisosteids. As examples, anastomosing tubular testes were described in the gar, *Lepisosteus platyrhinchus* (Lepisosteidae), *Ictalurus natalis* (Siluridae), *Opisthonema oglinum* and *Dorosoma petense* (Clupeidae) (Fig. 4.1A) (Grier 1993); the Rainbow Trout, *Oncorhynchus mykiss* (Salmonidae) (Fig. 4.1B), *Esox niger* (Esocidae) (Grier 1993) and the Tarpon, *Megalops atlanticus* (Elopidae) (Fig. 4.1C).

Unrestricted spermatogonial testis: The unrestricted spermatogonial testis type is found in neoteleosts, except for the atherinomorphs. In this testis type the germinal compartment extends to the periphery of the testis terminating blindly forming a lobule. It is characterized by the occurrence of spermatogonia along the lengths of the lobules. During spermiation the spermatozoa are released into the lobular lumen continuing to the efferent ducts. As examples, this type of testis was described in higher teleosts, such as *Oreochromis mossambicus* (Fig. 4.2A-D), the Common Snook, *Centropomus undecimalis* (Centropomidae) (Fig. 4.3A-D) and the Cobia, *Rachycentron canadum* (Fig. 4.5A-C) (Brown-Peterson *et al.* 2002).

Restricted spermatogonial testis: The restricted spermatogonial testis type is found in all Atherinomorpha and *prima facie* evidence supporting the monophyly of that group (Parenti and Grier 2004). In this testis type, similar to that mentioned for the unrestricted type, the germinal compartments extend to the periphery of the testis and terminate blindly. However, spermatogonia are never distributed along the lengths of the lobules, but are restricted to the distal termini of the lobules, at the testis periphery. During spermatogenesis, the cysts migrate toward the efferent duct system, as germ cells mature inside. As examples, the restricted distribution of spermatogonia occurs in the goodeids, *Xenotoca eiseni* (Fig. 4.A-C) and *Ilyodon whitei* (Fig. 4.7A), in the belonid, *Hemiramphus brasiliensis* (Fig. 4.7B), in the cyprinodontid, *Fundulus grandis* (Fig. 4.7C-D), and in the poeciliid, *Poecilia latipinna* (Fig. 4.9A-C).

The differences between anastomosing tubular and lobular testis types is significant regarding their systematic implications, distinguishing between gonad types in lower and higher fishes (Parenti and Grier 2004), but essentially these differences are a matter of degree as in all fish the unit of function is the spermatocyst, or cyst (Grier 1993). Spermatocysts line the walls of anastomosing tubules (Fig. 4.1A,B) and lobules (Fig. 4.2A-D; 4.3B-D; 4.4A-C; 4.5B-D), and sperm collect in the lumina of either tubules or lobules except in one large group of teleosts, the Atherinomorpha. These fishes have lobular testes (Figs. 4.6, 4.7, 4.8, 4.9) and the lobules lack lumina as Sertoli cell processes bridge the widths of the lobules.

Instead of spermatogonia being distributed along the walls of the lobules, as in most teleosts (Figs. 4.2D; 4.4A; 4.5D), and also at the distal termini of lobules depending on time during the annual reproductive cycle (Figs. 4.3C; 4.4A,B) spermatogonia are entirely restricted to the distal termini of lobules in

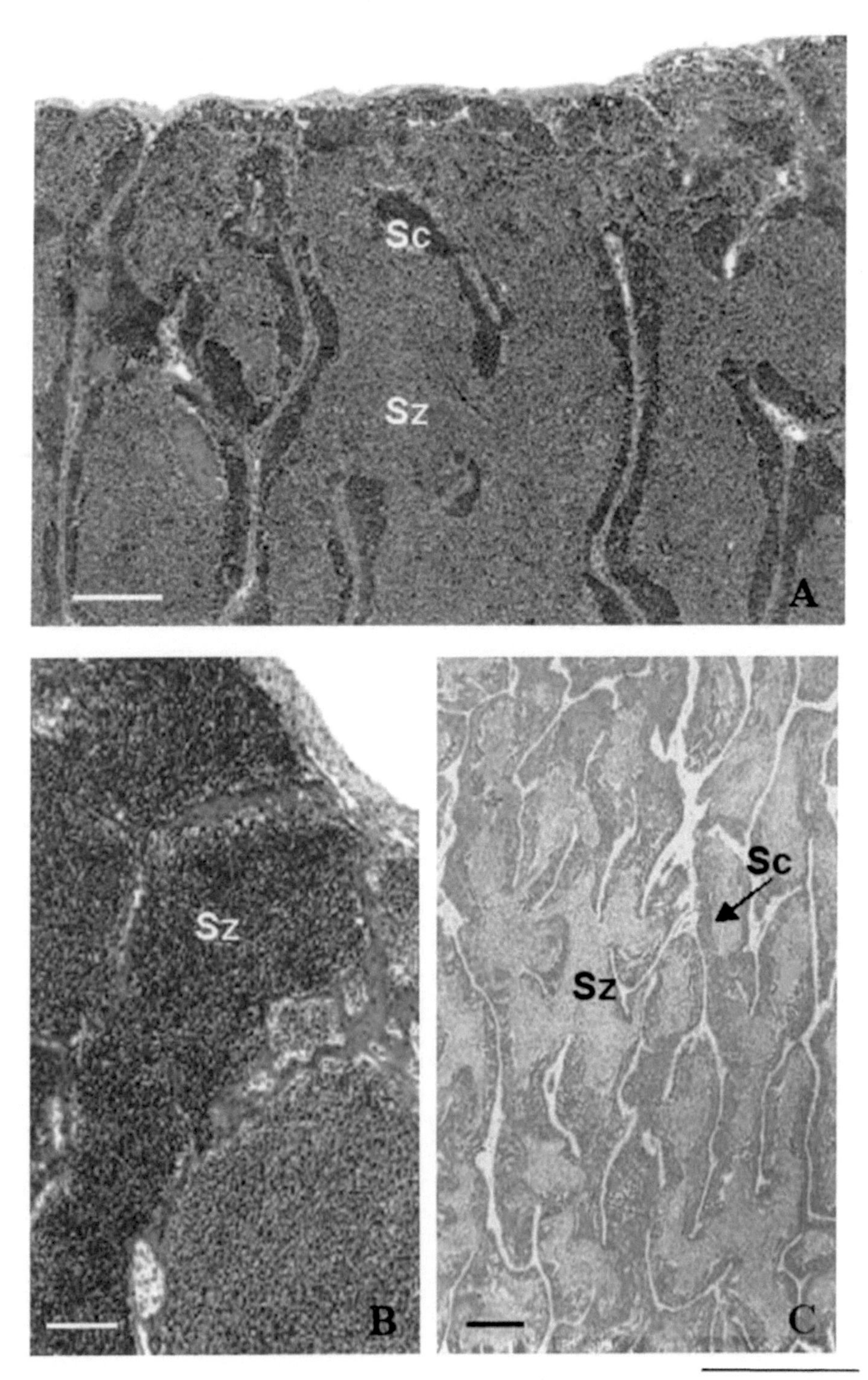

Fig. 4.1 Contd. ...

the atherinomorphs. Therefore, the mitotic divisions of spermatogonia are entirely restricted to the testis periphery (Fig. 4.8A) while meiotic divisions (Fig. 4.8B), and the later stages of sperm development, are closer to the testis ducts than earlier stages (Figs. 4.6A-C; 4.7B-D; 4.8A-D; 4.9A,B). In the viviparous species of atherinomorphs, *Xenotoca eiseni* (Goodeidae) (Fig. 4.8C,D) and *Poecilia latipinna* (Poeciliidae) (Figs. 4.9A-C), sperm are bundled as spermatozeugmata, seen in the testis ducts. Interestingly, sperm flagella course over the surface of spermatozeugmata in the species of Goodeidae (Fig. 4.8C,D), while sperm nuclei form the surface of spermatozeugmata in species of Poeciliidae (Grier *et al.* 1978, 1981; Grier *et al.* 2005) (Figs. 4.9A-C). The difference in morphology is best explained by the fact that sperm nuclei associate with Sertoli cells in poeciliids, while sperm flagella associate with Sertoli cells in goodeids.

4.6 ANNUAL REPRODUCTIVE CYCLES

As most teleosts have well-defined annual reproductive cycles. The morphology of both the ovarian germinal epithelium and the testicular germinal epithelium changes during the annual reproductive cycle, reflecting reproductive seasonality. It has long been assumed that when fish were not spawning, the gonads were in a "resting stage," but, rather than resting, Murphy and Taylor (1989) described the gonads of non-spawning Red Drum, *Sciaenops ocellatus*, as "rejuvenating," but did not present histological detail. More than a decade later, dividing spermatogonia were depicted in the testes of the swamp eel, *Synbranchus marmoratus*, during the so-called resting period when the fish was not reproducing (Lo Nostro *et al.* 2003). These authors also used an immunocytochemical technique to detect proliferating cell nuclear antigen (PCNA) which labeled both spermatogonia and Sertoli cells during the "resting stage," indicating that these cells would divide.

The annual reproductive cycle in male fish has been divided into "mitosis dominated" and "meiosis dominated" periods, these being the non-reproductive phase and the reproductive phase of the annual reproductive

---

Fig. 4.1 Contd. ...

Fig. 4.1 Anastomosing tubular testis of **A.** *Dorosoma petenense* (Clupeidae). Basophilia of germ cell nuclei, the spermatocytes (Sc) and the Spermaotzoa (Sz) stains the germinal compartment blue while the interstitium is acidophilic, pink. The germinal compartment is observed to branch and to also rejoin forming both a branching and anastomosing network. Bar 100μm. **B.** When the Rainbow Trout, *Oncorhynchus mykiss* (Salmonidae), is ready to spawn, the anastomosing tubules are filled with sperm, and early stages of spermatogenesis are not observed. Bar 50μm. **C.** In the Tarpon, *Megalops atlanticus* (Elopidae), the germinal compartments both branch and form an anastomosing network that stains more intensely than the interstitum. The lumen of the tubules is filled with stored spermatozoa (Sz). At the periphery of the lumen there are spermatocysts with developing germ cells (Sc). Bar 100μm. Original.

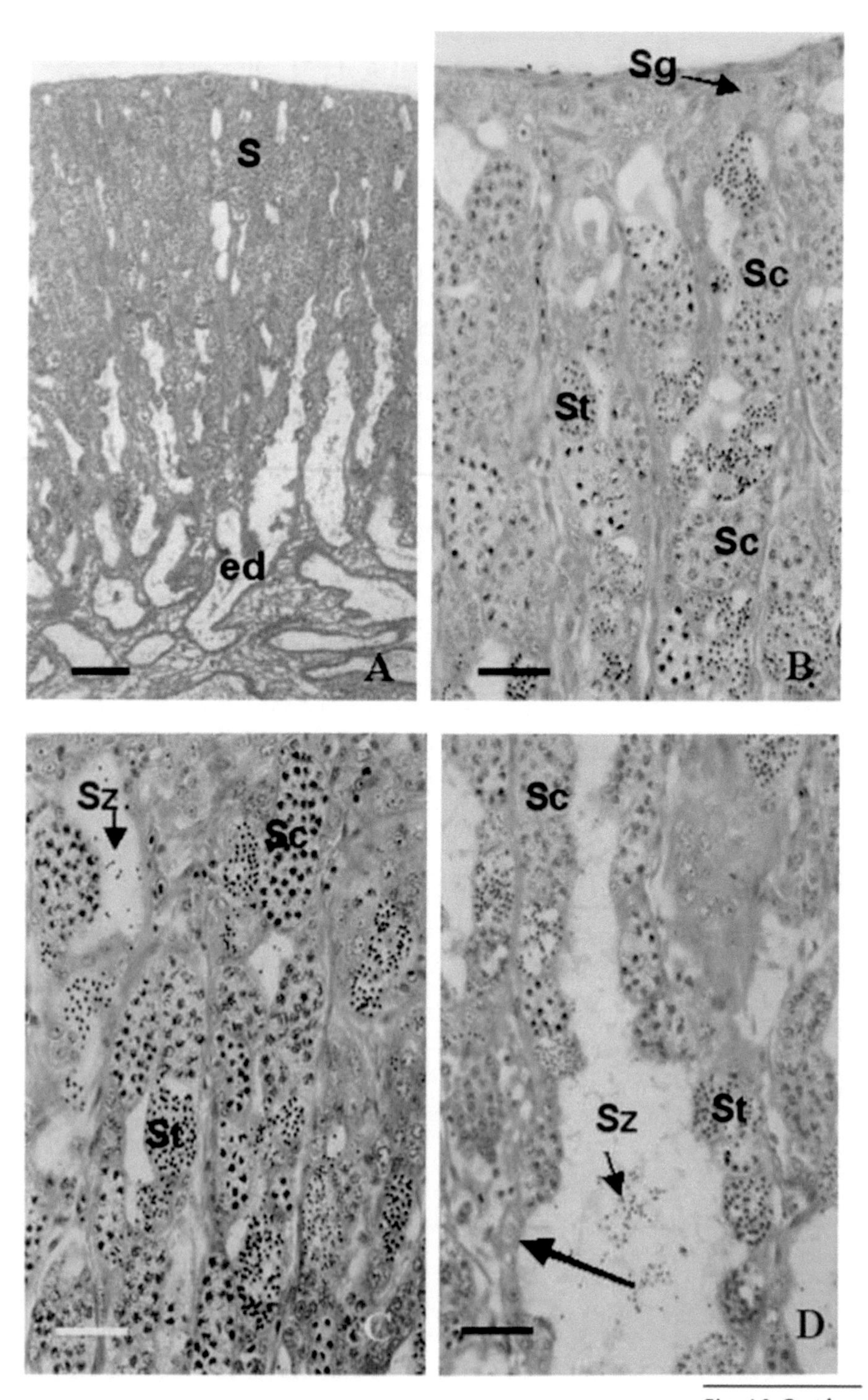

Fig. 4.2 Contd. ...

season (Grier 2002), respecitvely. However, this simple categorization does not apply to females. The only study conducted on the germinal epithelium of a non-reproductive female fish, Common Snook (Grier 2000) revealed synaptonemal complexes in meiotic oocytes and active folliculogenesis, thus revealing a basic difference in the germinal epithelia between males and females. In non-reproductive males, diploid spermatogonia are the only germ cells in the testes, and these cells may divide mitotically when the gonad is "resting." In females, meiosis and folliculogenesis occur during this phase.

While folliculogenesis is an ongoing process during the nadir of the reproductive cycle in female fish, this does not preclude testicular growth in males at the same phase in the reproductive cycle as lobules in sexually mature, regressed male fish may develop by a process of elongation. This process has hardly been examined, but in Common Snook it has been proposed that mitotic divisions of both spermatogonia and Sertoli cells at the distal termini of lobules results in lobular elongation (Grier and Taylor 1998). The fundamental process requires that both spermatogonia and Sertoli cells divide, a different condition from that in the mammalian testis where Sertoli cell divisions cease entirely after spermatogenesis commences (Gondos and Berndstom 1993). However, in fish with recycling gonadal development, it is possible to assume that both spermatogonia and Sertoli cells divide in a coordinated fashion to affect testis development and sperm production during annual reproductive cycles. An illustration of a metaphase Sertoli cell in the poeciliid, *Poecilia latipinna* (see Fig. 25 in Grier 1993) and Sertoli cell nuclei labeled with tritiated thymidine in the protogynous Bluehead Wrasse, *Thalassoma bifasciatum* (Koulish *et al.* 2002), an indication of pending cell division, are evidences of different life cycles between fish and mammalian Sertoli cells. The gonads of teleosts never "rest," but their cellular activities largely elude classical histological procedures. Thus, the frequently used term "resting stage" does not apply to fish gonads and should no longer be used,

---

Fig. 4.2 Contd. ...

Fig. 4.2 Unrestricted spermatogonial lobular testis of *Oreochromis mossambicus*. **A.** The panoramic view shows lobules where spermatocysts with developing germ cells in spermatogenesis (S) are distal to the efferent ducts (ed). Bar 100μm. **B**,**C**,**D**. Bar 50μm. **B**. Longitudinal sections of testicular lobules where spermatogonia (Sg) are along the walls of the lobules and spermatocysts with developing germ cells have different stages of spermatogenesis: spermatogonia (Sg), primary spermatocytes (Sc), spermatids (St). **C**. Near the periphery of the lobule, there is a continuous germinal epithelium represented by a continuum of spermatocysts containing different stages of synchronously-developing germ cells. The lumina with some scattered sperm (Sz) are not continuous. **D**. The junction between that portion of a lobule with a distal continuous germinal epithelium and a proximal discontinuous germinal epithelium (large arrow) where there are no germ cells. Original.

as is indicated by newly developed information on cell cycles and accumulating evidence that fish Sertoli cells do divide.

4.7 NOMENCLATURE

As new information on physiology and morphology reveals the mechanisms by which the teleost testis works, so does this information need to be used to upgrade and redefine the nomenclature applied to the description of testis structure. Histological examination of testis structure in the Common Snook, *Centropomus undecimalis* (Taylor *et al*. 1998; Grier and Taylor 1998) led to the definition of continuous and discontinuous germinal epithelia, as in this species (Fig. 4.3D), *Oreochromis mossambicus* sp. (Fig. 4.2D), and Cobia (Fig. 4.5C,D). The changes in the male germinal epithelium were used to define annual reproductive classes. As the period of reproductive abatement ends, the fish testis begins to develop and lobules elongate. There is a continuous population of Sertoli cells in association with germ cells along the lengths of lobules. That is, the germinal epithelium is continuous between the testis ducts and the testis periphery. As development continues however, the germinal epithelium remains continuous distally, near the edge of the testis (Fig. 4.2A-C), but becomes discontinuous proximally (Fig. 4.2D) where germ cells mature, becoming sperm which are released into the lumina of the lobules. However, there are no divisions of spermatogonia to act as a renewal of germ cells within the proximal germinal epithelium. Therefore, as sperm mature and are released, newly formed spermatocysts do not replace those from which sperm were released. Those spermatocysts that do exist become separated by a Sertoli cell cytoplasm, and a discontinuous germinal epithelium results from this process. This change defined the mid maturation class of testis development. Later, a discontinuous germinal epithelium was observed to extend to the testis periphery, the character defining late maturation (Grier and Taylor 1998).

As defined, the reproductive "classes" early maturation, mid maturation, and late maturation (Taylor *et al*. 1998; Grier and Taylor 1998) describe

Fig. 4.3 Seasonal changes in the unrestricted spermatogonial lobular testis of *Centropomus undecimalis*. **A.** In early testis development, lobule lumina are discontinuous and spermatozoa are lacking. Spermatocysts with primary spermatocytes (Sc) are seen. Bar 50μm. **B.** As the testis develops, the lobules become filled with sperm (Sz) and spermatocysts, some with spermatocytes (Sc) are located near the testis periphery. Bar 100μm. **C.** Near the end of the breeding season, spermatogonia become established at the distal ends of lobules and along the lengths of lobules. A discontinuous germinal epithelium has spermatocysts with spermatocytes (Sc), spermatids (St) and spermatozoa (Sz). Bar 20μm. **D.** The germinal epithelium is discontinuous with scattered spermatocytes along its walls, some with spermatocytes (Sc). Sperm (Sz) occupy the center of lobules. Bar 50μm. Original.

Fig. 4.3 Contd. ...

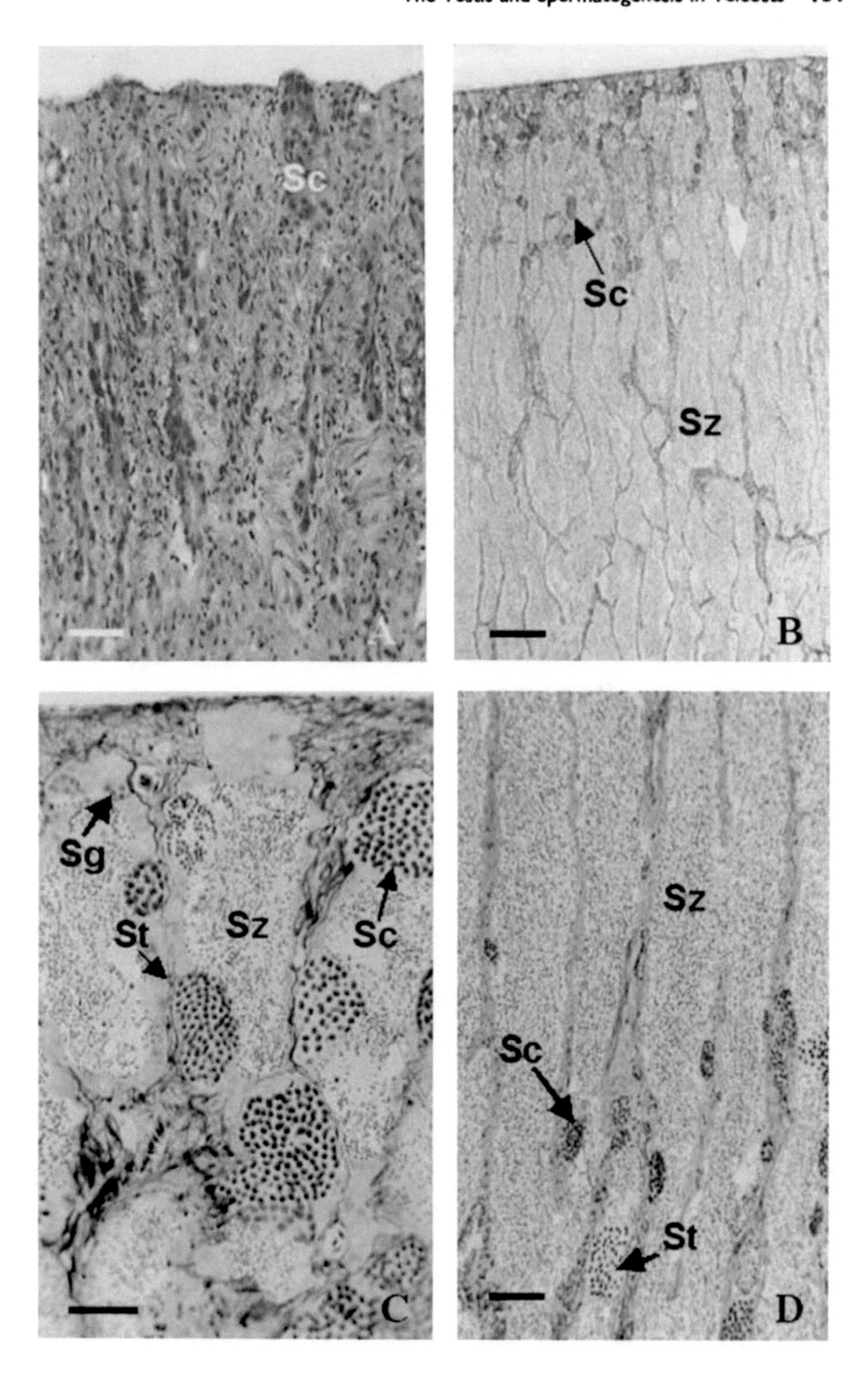
Sc
A
Sc
Sz
B
Sg
St
Sz
Sc
C
Sz
Sc
St
D

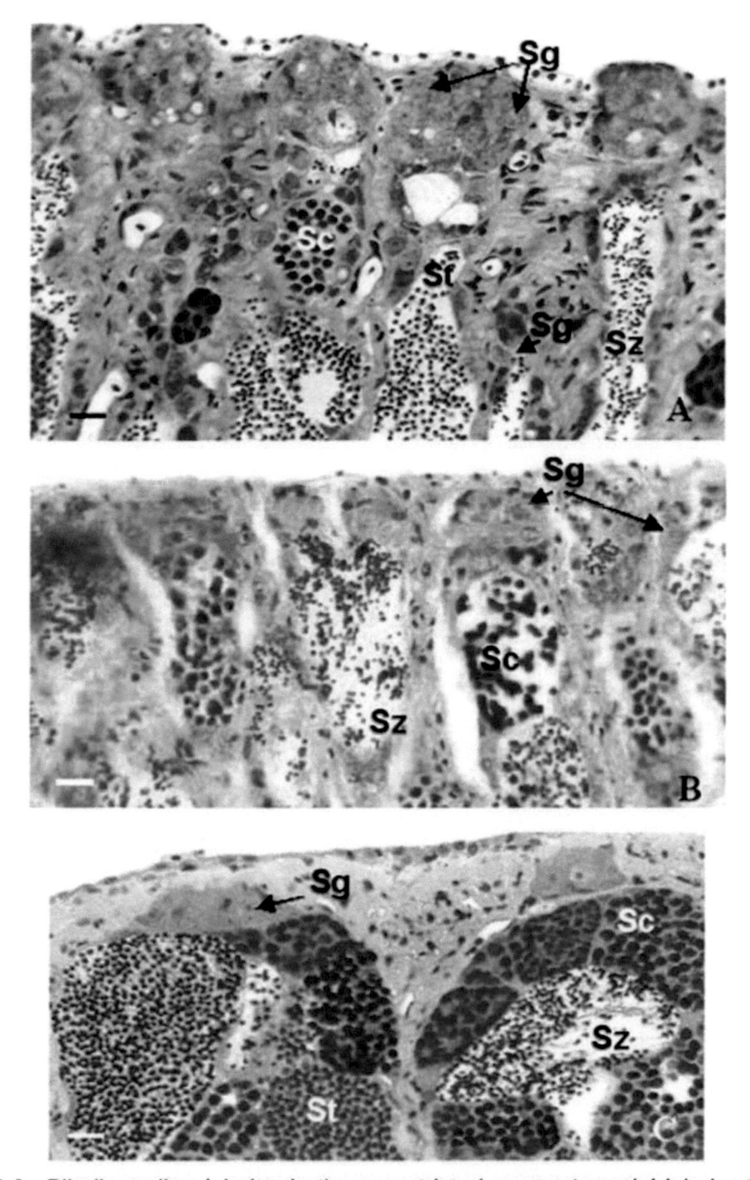

Fig. 4.4 Blindly-ending lobules in the unrestricted spermatogonial lobular testis of *Centropomus undecimalis*. **A,B,C**. Bar 10μm. **A.** Sections of testicular lobules containing cysts of germ cells with several stages of spermatogenesis. Spermatogonia are at the distal termini of lobules and along the lobule walls. Spermatogonia (Sg), primary spermatocytes (Sc), spermatids (St), and spermatozoa (Sz). **B**. Occasionally, a branched lobule is observed at the testis periphery indicating they may elongate by a branching process. Other lobules end blindly without branching. **C**. As the spawning season advances, spermatogonia (Sg) remain at the distal end of lobules, and a continuous germinal epithelium with spermatocytes (Sc), spermatids (St), largely devoid of spermatogonia, is observed at the distal end of the lobules. In the lumen, spermatozoa are seen (Sz). Original.

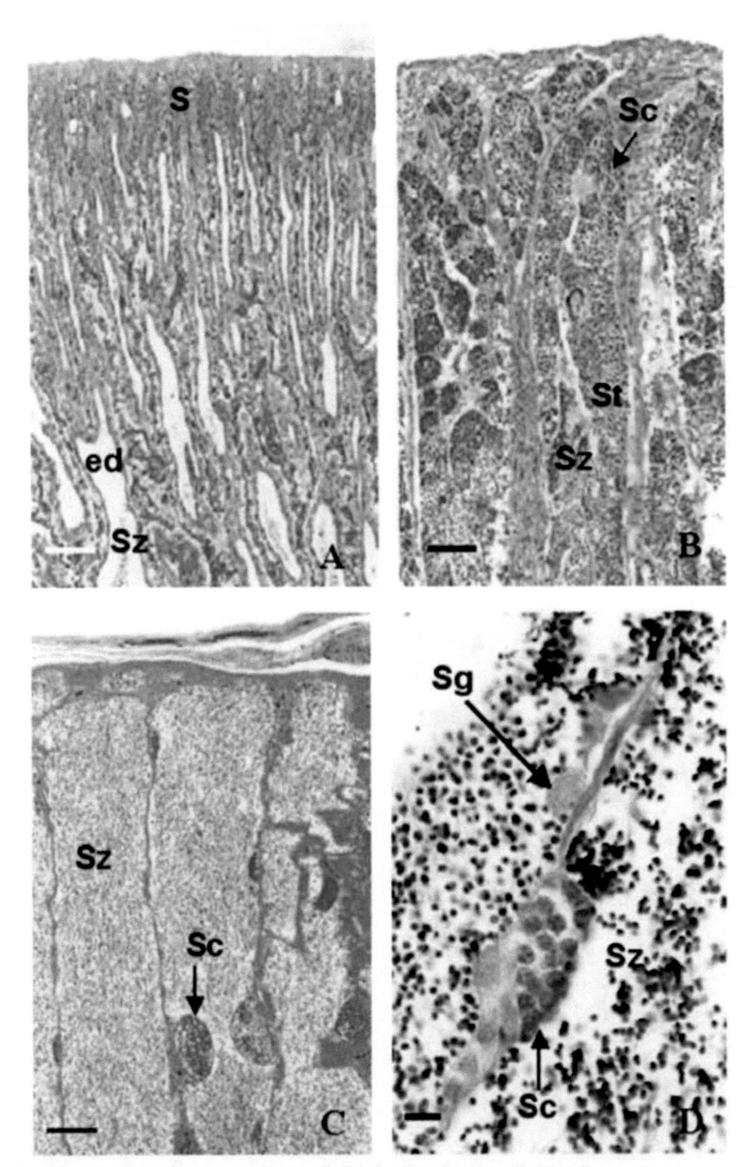

Fig. 4.5 Unrestricted spermatogonial lobular testis of *Rachycentron canadum*. **A**. The panoramic view shows blind-ending lobules with germ cells in spermatogenesis (S) near the testis periphery while sperm (Sz) are in the efferent ducts (ed). Bar 100μm. **B**. The continuous germinal epithelium at the testis periphery with a continuum of spermatocysts having spermatocytes (Sc), spermatids (St), and spermatozoa (Sz). Bar 50μm. **C**. Towards the end of the breeding season, spermatocysts are rare along the walls of lobules, some of them contain (Sc), while sperm (Sz) fill the lumina. Bar 50μm. **D**. Magnified view of a discontinuous germinal epithelium. Spermatogonia (Sg) are scattered along the lobule wall and there is not a continuum of spermatocysts, some with spermatocytes (Sc). Spermatozoa (Sz) fill the lobule lumina and extend to the lobule wall. Bar 10μm. Original.

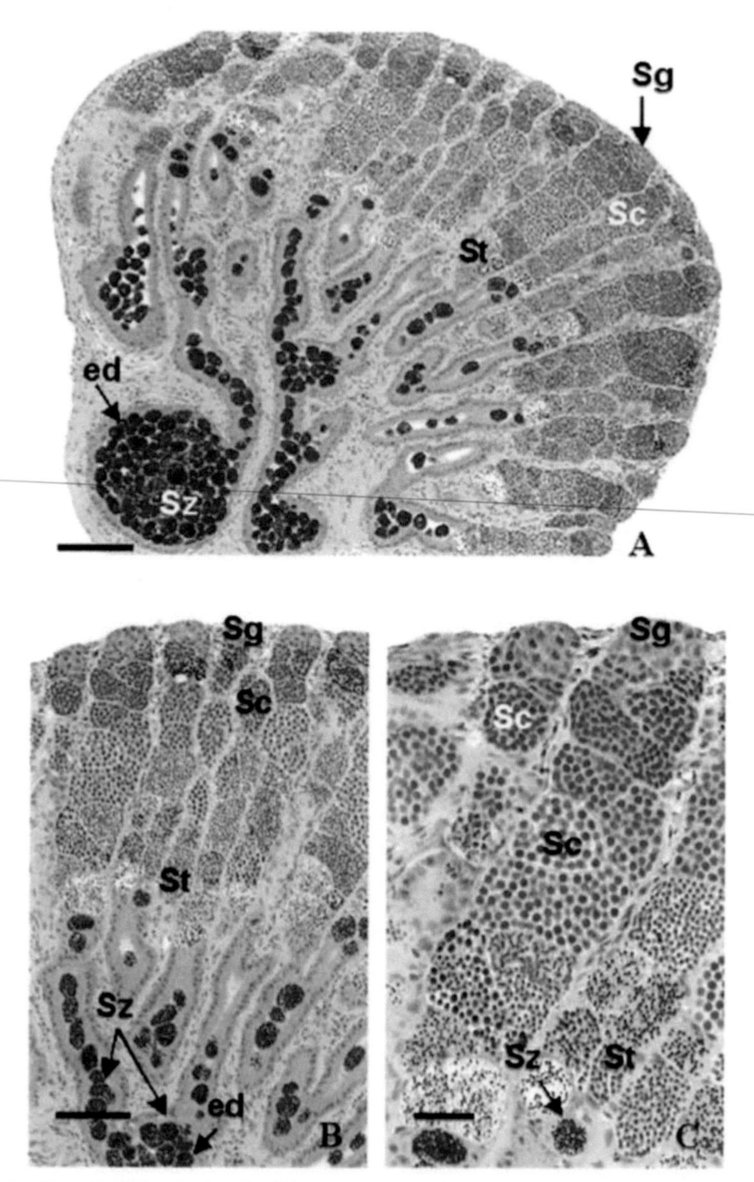

Fig. 4.6 Restricted spermatogonial lobular testis of *Xenotoca eiseni* (Goodeidae). **A**. Spermatogonia (Sg) are restricted to the distal termini of lobules, and there is a progression of spermatocysts with spermatocytes (Sc), then spermatids (St) ever closer to the efferent ducts (ed) Sperm (Sz) are packaged as spermatozeugmata. Bar 100µm. **B**. Higher magnification showing peripheral Spermatogonia (Sg), then spermatocysts with spermatocytes (Sc) and spermatids (St) closest to the ducts (ed) in which spermatozeugmata (Sz) are observed. Bar 50µm. **C**. At the testis periphery are spermatogonia (Sg). The lobules lack a lumen, and spermatocysts extend across their widths containing spermatocytes (Sc), spermatids (St). Sperm (Sz) are packaged into spermatozeugmata (Sz) in the testis ducts. Bar 20µm. Original.

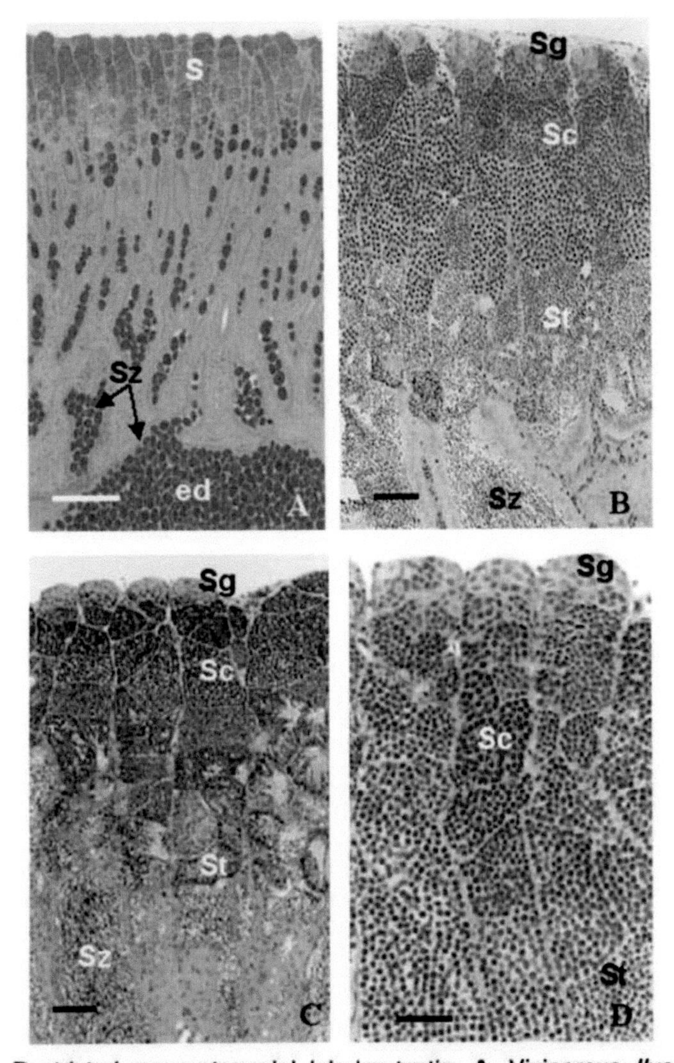

Fig. 4.7 Restricted spermatogonial lobular testis. **A**. Viviparous *Ilyodon whitei* (Goodeidae).The panoramic view shows lobules with germinal cells in early spermatogenesis (S), at the opposite side the efferent ducts (ed) containing abundant spermatozoa in groups forming spermatozeugmata, bundles of spermatozoa (Sz) where the nuclei are seen irregularly situated to the center. Bar 100µm. **B**. Oviparous *Hemiramphus brasiliensis* (Belonidae), illustrating spermatogonia (Sg) at the testis periphery, spermatocytes (Sc) and spermatids (St) more medial. Mature sperm (Sz) are voided into the testis ducts. Bar 50µm. **C**. Oviparous *Fundulus grandis* showing spermatogonia (Sg) at the lobule termini and spermatocytes (Sc) and spermatids (St) closer to the testis ducts in which free sperm (Sz) are located. Bar 50µm. **D**. *Fundulus grandis* (Cyprinodontidae) lobules, illustrating the lack of a lobule lumen, spermatogonia (Sg) restricted to the distal termini of the lobules while progressively later stages of spermatogenesis (spermatocytes, Sc and spermatids, St) are more distal. Bar 50µm. Original.

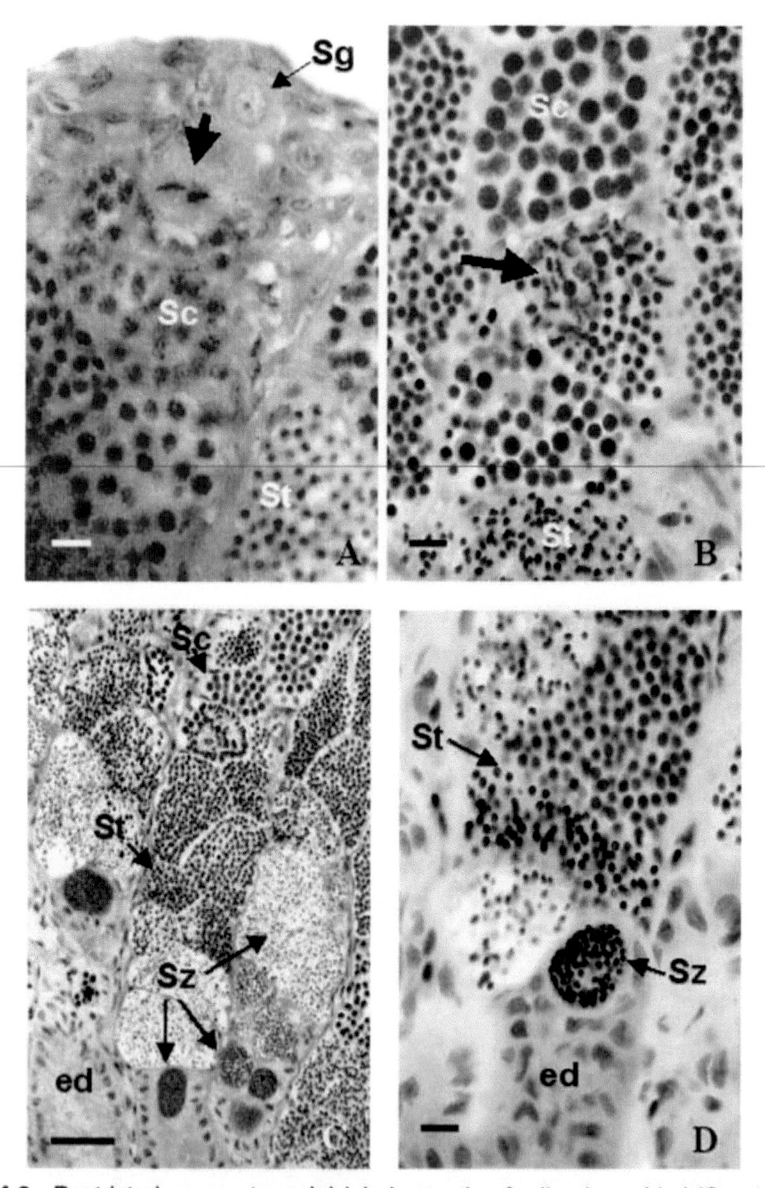

Fig. 4.8 Restricted spermatogonial lobular testis. **A**. *Ilyodon whitei* (Goodeidae). Spermatogonia (Sg) at the periphery of the testis. Two spermatogonia during mitotic metaphase (thick arrow). Spermatocytes (Sc) and Spermatids (St). Bar 50μm. **B,C,D**. *Xenotoca eiseni* (Goodeidae). Spermatocysts with spermatocytes (Sc) and spermatids (St) are observed. Metaphase spermatocytes (thick arrow) are observed. Bar 10μm. Lobules with spermatocysts having free spermatozoa (Sz) that are "packaged" into spermatozeugmata in the efferent ducts (ed). Bar 50μm. A spermatocyst with spermatids (St) is observed. A spermatozeugma with mature spermatozoa (Sz) is moving into the efferent ducts (ed). Bar 10μm. Original.

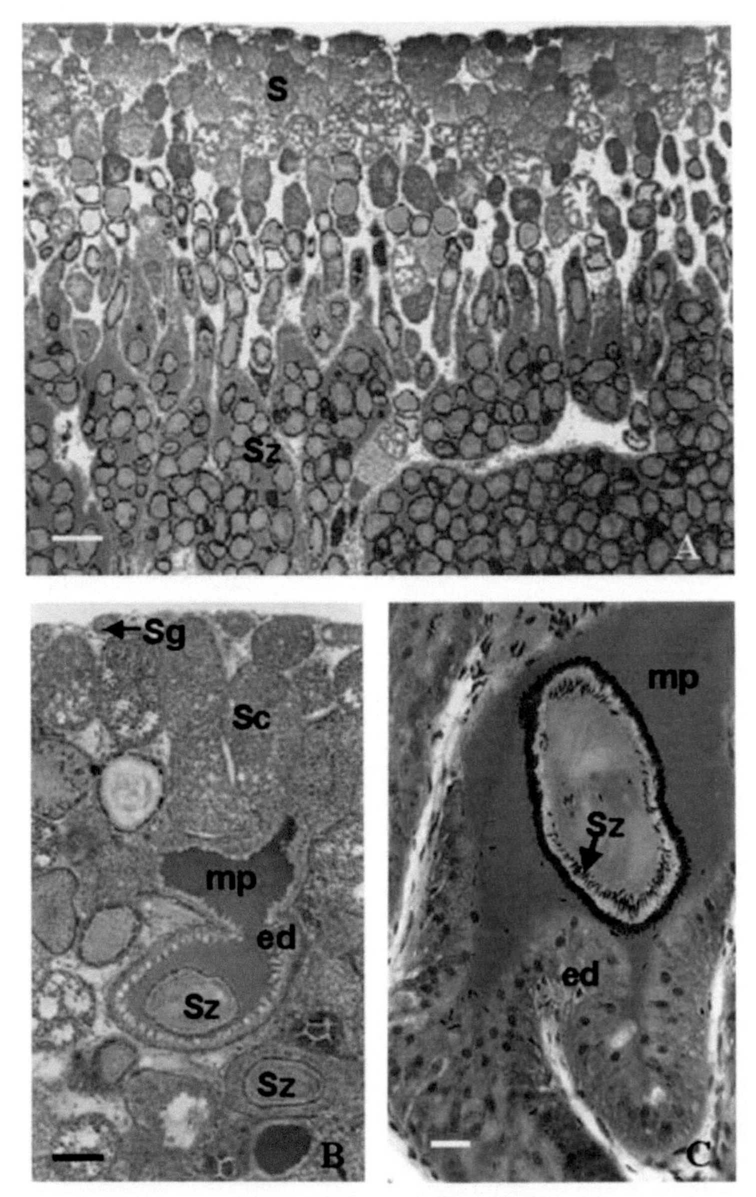

Fig. 4.9 Restricted spermatogonial lobular testis of *Poecilia latipinna* (Poeciliidae). **A**,**B**. The panoramic views show lobules with germ cells in early spermatogenesis (S), containing cysts of germ cells, spermatogonia (Sg), primary spermatocytes (Sc) and spermatozoa (Sz) in spermatozeugmata. In the efferent ducts (ed), the spermatozeugmata are surrounded by a PAS+ mucoprotein complex (mp). **A**. Bar 100μm. **B**. Bar 50μm. **C**. Cyst with spermatozoa (Sz) in a spermatozeugma, which are different to those seen in goodeids (see Figs. 4.8C,D). In the efferent ducts (ed), the spermatozeugmata are surrounded by a PAS+ mucoprotein complex (mp). Bar 10μm. Original.

naturally occurring phenomena and present a new approach to the description of testis analysis during annual reproductive cycles. However, it is now realized that these designations also present a nomenclatural conflict: there is an extensive literature surrounding the term "maturation" as it applies to the resumption of meiosis in oocytes (see Grier *et al.*, **Chapter 2**). "Maturation" should not be used to define an annual reproductive "class" in male and female fish (Taylor *et al.* 1998) while, at the same time, being used to define a "stage" of oocyte development that is completely different, namely the resumption of meiosis known as oocyte maturation. Therefore, it is suggested that "maturation" as it refers to male reproductive classes in Common Snook (Taylor *et al.* 1998; Grier and Taylor 1998), in Cobia (Brown-Peterson *et al.* 2002), and in Seatrout (Brown-Peterson 2003) should be replaced with the terms early GE development, mid GE development, and late GE development, where "GE" means germinal epithelium. This change is proposed in recognition of the extensive literature surrounding the definition of "oocyte maturation" and to prevent conflict with newly introduced terms based on updating terminology due to the accumulation of new knowledge. It is to be recognized that changes in the male germinal epithelium during annual reproductive cycling in numerous species of perciform teleosts are valid. Changes in the male germinal epithelium, in view of new evidence that teleost Sertoli cells do divide (Koulish *et al.* 2002; Schulz *et al.* 2005; Chavez-Pozo *et al.* 2005) in concert with germ cells during annual reproductive cycles, are valid criteria upon which to base reproductive classification and to revise nomenclature. Nomenclature needs to be changed to reflect new information and to avoid conflict in meanings.

4.8 CHAPTER SUMMARY

In most teleosts, testes are paired elongated organs that are attached to the dorsal wall of the body by a mesorchium. Histological examination of teleost testes, as in other vertebrates, shows the testes formed by germ cells and somatic cells. The germ cells are integrated during all of the stages of differentiation of the spermatogenic cells that produce sperm: spermatogonia, spermatocytes, spermatids, and spermatozoa. The process of spermatogenesis includes a sequence of morphological and physiological changes of germ cells that differentiate the spermatogonia into spermatozoa. Spermatogonia associate with Sertoli cells to form spermatocysts or cysts. The cyst is the unit of spermatogenic function, composed of a cohort of isogenic germ cells surrounded by encompassing Sertoli cell. The testicular tissues form two compartments, 1) the germinal compartment, formed by the germinal epithelium, which is composed by Sertoli cells and germ cells and 2) the interstitial compartment, integrated by the connective tissue. These two compartments are always separated by a basement membrane. The teleost testis is organized morphologically in three types of testis: 1) anastomosing tubular testis type, present in lower teleosts as salmonids, cyprinids and lepisosteids; 2) unrestricted spermatogonial testis type, found in neoteleosts except Atherinomorpha; and 3) restricted spermatogonial testis type, characteristic of all Atherinomorpha. The morphology of both the ovarian germinal epithelium and the testicular germinal epithelium changes during the annual reproductive cycle, reflecting reproductive seasonality.

4.9 ACKNOWLEDGMENTS

The senior author dedicates this work to Karen A. Steidinger. Thank you Karen for your leadership, inquisitiveness, and encouraging my pursuit of comparative gonad morphology. The authors thank the histology staff of the Florida Fish and Wildlife Research Institute, Catalina Brown, Noretta Perry, Illiana Quintero, Ruth Reese, and Yvonne Waters for their excellence in histological processing. We also thank Gabino De la Rosa-Cruz and Adriana García-Alarcón, Facultad de Ciencias, Universidad Nacional Autónoma de México, for their contribution to the preparation of the plates. We are grateful to Professor Barrie Jamieson for his advice throughout preparation of the chapter and his substantial editorial aid.

4.10 LITERATURE CITED

Abraham, M. 1991. The male germ cell protective barrier along phylogenesis. International Review of Cytology 130: 111-190.

Billard, R. 1986. Spermatogenesis and spermatology of some teleost fish species. Reproduction, Nutrition and Development 26: 877-920.

Brown-Peterson, N. 2003. The reproductive biology of spotted seatrout. Pp. 99-133. In S. A. Bortone (ed.), *Biology of the Spotted Seatrout*. CRC Press, Boca Raton, FL.

Brown-Peterson, N., Grier, H. J. and Overstreet, R. 2002. Annual changes in the germinal epithelium determine reproductive classes in male cobia, *Rachycentron canadum*. Journal of Fish Biology 60: 178-202.

Chavez-Pozo, E., Mulero, V., Meseguer, J. and Ayala, A. G. 2005. An overview of cell renewal in the testis throughout the reproductive cycle of a seasonal breeding teleost, the Gilthead Seabream (*Sparus aurata* L.). Biology of Reproduction 72: 593-601.

Downing, A. L. and Burns, J. R. 1995. Testis morphology and spermatozeugma formation in three genera of viviparous halfbeaks: *Nomorhamphus*, *Dermogenys*, and *Hemirhamphodon* (Teleostei: Hemirhamphidae). Journal of Morphology 225: 329-343.

Fawcett, D. W. and Jensh, R. P. 2002. Epithelium. *Concise Histology*. 2nd Edition. Pp. 29-41. Arnold Publishers, London.

Gardiner, D. M. 1978. The origin and fate of spermatophores in the viviparous teleost *Cymatogaster aggregata* (Perciformes: Embiotocidae). Journal of Morphology 155: 157-172.

Gondos, B. and Berndston, W. E. 1993. Postnatal and pubertal development. Pp115-154. In L.D. Russell and M.D. Griswald (eds), *The Sertoli Cell*. Cache River Press, Clearwater, FL.

Grier, H. J. 1977. Ultrastructural indentification of the Sertoli cell in the testis of the Northern Pike, *Esox lucius*. American Journal of Anatomy 149: 282-288.

Grier, H. J. 1981. Cellular organization of the testis and spermatogenesis in fishes. American Zoologists 21: 345-357.

Grier, H. J. 1993. Comparative organization of Sertoli cells including the Sertoli cell Barrier. Pp. 704-739. In L. D. Russell and M. D. Griswald (eds), *The Sertoli Cell*. Cache River Press, Clearwater, FL.

Grier, H. J. 2000. Ovarian germinal epithelium and folliculogenesis in the common snook, *Centropomus undecimalis* (Teleostei: Centropomidae). Journal of Morphology 243: 265-281.

Grier, H. J. 2002. The germinal epithelium: Its dual role in establishing male reproductive classes and understanding the basis for indeterminate egg production in female fishes. Pp. 537-552. In Creswell, R. L. (ed.), *Proceedings of the Fifty-third Annual Gulf and Caribbean Fisheries Institute, November 2000*. Mississippi/ Alabama Sea Grant Consortium, Fort Pierce, Florida.

Crier, H. J. and Lo Nostro F. 2000. The germinal epithelium in fish gonads: the unifying concept. Pp. 233-236. In Norberg, B., Kjesbu, O. S. Taranger, G. L., Andersson E., and Stefansson S. O. (eds), *Proceedinjgs of the 6th International Symposium on the Reproductive Biology of Fish*. University of Bergen, Norway.

Grier, H. J. and Taylor, R. G. 1998. Testicular maturation and regression in the common snook. Journal of Fish Biology 53: 521-542.

Grier, H. J., Hurk van den, R. and Billard, R. 1989. Cytological identification of cell types in the testis of *Esox lucius* and *E. niger*. Cell and Tissue Research 257: 491-496.

Grier, H. J., Fitzsimmons, J. M. and Linton, J. R. 1978. Structure and ultrastructure of the testis and sperm formation in goodeid teleosts. Journal of Morphology 156: 419-438.

Grier, H. J., Uribe, M. C., Parenti, L. R. and De la Rosa-Cruz, G. 2005. Fecundity, the germinal epithelium, and folliculogenesis in viviparous fishes. Pp 191-126. In M. C. Uribe and H. J. Grier (eds), *Viviparous Fishes*. New Life Publications Homestead, FL, USA.

Grier, H. J., Uribe, M. C. and Parenti, L. R. 2007. Germinal epithelium, folliculogenesis, and postovulatory follicles in ovaries of Rainbow Trout, *Oncorhynchus mykiss* (Walbaum, 1792) (Teleostei, Protacanthopterygii, Salmoniformes). Journal of Morphology 268: 293-310.

Hunter, J. R. and Macewicz, B. J. 1985. Measurement of spawning frequency in multiple spawning fishes. Pp 79-94. In R. L. Lasker (ed.). *An egg production method for estimating spawning biomass of pelagic fish: application to the northern anchovy*. 36 Department of Commerce. National Oceanic Atmospheric Administration. Technical Report. National Marine Fisheries Service, USA.

Hurk van den, R. 1973. The localization of steroidogenesis in the testis of oviparous and viviparous teleosts. Proceedings Koninklijke Nederlandse Akademie Wetensch Series C 76: 270-279.

Jamieson, B. G. M. 1991. *Fish Evolution and Systematics: Evidence from Spermatozoa*. Cambridge University Press, Cambridge. 319 pp.

Koulish, S., Kramer, C. R. and Grier, H. J. 2002. Organization of the male gonad in a protogynous fish, *Thalassoma bifasciatum* (Teleostei: Labridae). Journal of Morphology 254: 292-311.

Lo Nostro, F., Grier, H. J., Andreone, L. and Guerrero, G. A. 2003. Involvement of the gonadal germinal epithelium during sex reversal and seasonal testicular cycling in the protogynous swamp eel, *Synbranchus marmoratus* Bloch, 1795. (Teleostei, Synbranchidae). Journal of Morphology 258: 107-126.
Marshall, A. J. and Lofts, B. 1956. The Leydig cell homologue in certain teleost fishes. Nature 177: 704-705.
Miranda, A. J. L., Bazzoli, N., Rizzo, E. and Sato, Y. 1999. Ovarian follicular atresia in two teleost species: a histologial and ultrastructural study. Tissue and Cell 31: 480-488.
Miura, T. and Miura, C. I. 2003. Molecular control mechanisms of fish spermatogenesis. Fish Physiology and Biochemistry 28: 181-186.
Miura, T., Higuchi, M., Ozaki, Y., Ohta, T. and Miura, C. 2006. Progestin is an essential factor for the initiation of the meiosis in spermatogenetic cells of the eel. Developmental Biology 103 (19): 7333-7338.
Murphy, M. D. and Taylor, R. G. 1989. Reproduction, growth, and mortality of red drum, *Sciaenops ocellatus*, in northeast Florida waters. Fish Bulletin 88: 531-542.
Ozaki, Y., Higuchi, M., Miura, C., Yamaguchi, S., Tozawa, Y. and Miura, T. 2006. Roles of 11β-hydroxysteroid dehydrogenase in fish spermatogenesis. Endocrinology 147 (11): 5139-5146.
Parenti, L. R. and Grier, H. J. 2004. Evolution and phylogeny of gonad morphology in bony fishes. Integrative and Comparative Biology 44: 333-348.
Patiño, R. and Takashima, F. 1995. Gonads. Pp 128-153. In F. Takashima and T. Hibiya (eds). *An Atlas of Fish Histology, Normal and Pathological Features.* Kodanska Ltd/ Gustav Fisher Verlag. Tokyo/Stuttgart/New York.
Pecio, A., Lahnsteiner, F. and Rafinski, J. 2001. Ultrastructure of the epithelial cells in the aspermatogenic part of the testis in *Mimagoniates barberi* (Teleostei: Characidae: Glandulocaudinae) and the role of their secretions in spermatozeugmata formation. Annals of Anatomy 183: 427-435.
Pudney, J. 1995. Spermatogenesis in nonmammalian vertebrates. Microscopy Research and Techniques 32: 459-497.
Schulz, R. W. and Miura, T. 2002. Spermatogenesis and its endocrine regulation. Fish Physiology and Biochemistry 26: 43-56.
Schulz, R. W., Menting, S., Bogerd, J., França, L. R., Vilela, D. A. R. and Godinho, H. P. 2005. Sertoli cell proliferation in the adult testis—evidence from two fish species belonging to different orders. Biology of Reproduction 73: 891-898.
Strüssmann, C. A. and Nakamura, M. 2002. Morphology, endocrinology, and environmental modulation of gonadal sex differentiation in teleost fishes. Fish Physiology and Biochemistry 26: 13-29.
Taylor, R. G., Grier, H. J. and Whittington J. A. 1998. Spawning rhythms of common snook in Florida. Journal of Fish Biology 53: 502-520.
Tokarz, R. R. 1978. Oogonial proliferation, oogenesis, and folliculogenesis in nonmammalian vertebrates. Pp 145-179. In R. Jones (ed.). *The Vertebrate Ovary.* Plenum Press: New York/London,
Uribe, M. C., De la Rosa-Cruz, G., García-Alarcón, A., Guerreo-Estévez, S. M. and Aguilar-Morales, M. 2006. Características histológicas de los estados de atresia de folículos ováricos en dos especies de teleósteos vivíparos *Ilyodon whitei* y *Goodea atripinnis* (Goodeidae). Hidrobiológica 16: 67-73.

Yoshizake, G., Takeuchi, Y., Kobayashi, T., Ihara, S. and Takeuchi, T. 2002. Primordial germ cells: the bludprint for a piscine life. Fish Physiology and Biochemistry 26: 3-12.

Wallace, R. A. and Selman, K. 1990. Ultrastructural aspects of oogenesis and oocyte growth in fish and amphibians. Journal of Electron Microscopy Technique 16: 175-201.

Male Reproductive System: Spermatic Duct and Accessory Organs of the Testis

Franz Lahnsteiner and Robert A. Patzner

5.1 INTRODUCTION

The efferent duct system of the male gonad is defined as the duct system transporting the spermatozoa from the testis to the external environment. It consists of the efferent ducts and of accessory glands which have specific functions in the formation of the seminal fluid and in maintaining the viability of spermatozoa. In fishes (Pisces) the efferent duct system of the male gonads is not homogenous but differs in relation to the systematic position and to the mode of reproduction.

Under phylogenetic considerations the efferent duct system of the Cyclostomata can be considered as the most basic one (Fig. 5.1): The initially paired testes fuse during sexual maturation to form an unpaired organ which extends throughout the coelomic cavity. In Petromyzontia only the right gonad is fully developed while the left one is rudimentary. Spermatozoa are released from the testes into the coelomic cavity from where they are transported into the environment via genital funnels. These funnels open via pores into the cloaca, a chamber receiving from the intestines, from the ducts of the kidneys and from the gonads (Patzner 1997).

In the phylogenetic development of vertebrates there exist two tendencies, either to use the primary ureters as efferent ducts for the semen or to develop secondary spermatic ducts (Blüm 1985). Within the Eutelestomi (=Osteichthyes) which consist of two classes, the Actinopterygii with the subclasses Cladistia, Chondrostei and Neopterygii (Holostei and Teleostei) and the Sarcopterygii with the subclasses Coelacanthimorpha and Dipnotetrapodomorpha, the

Department of Organismic Biology, University of Austria, Hellbrunnerstr. 34, A-5020, Austria

Chondrostei, Ceratodontimorpha (=Dipnoi) and Holostei have connections between the renal and genital system. In the Chondrostei (Blüm 1985) the seminiferous tubules of the testis empty into a longitudinal testicular channel (Fig. 5.1). From this channel ductuli efferentes arise which are connected to a

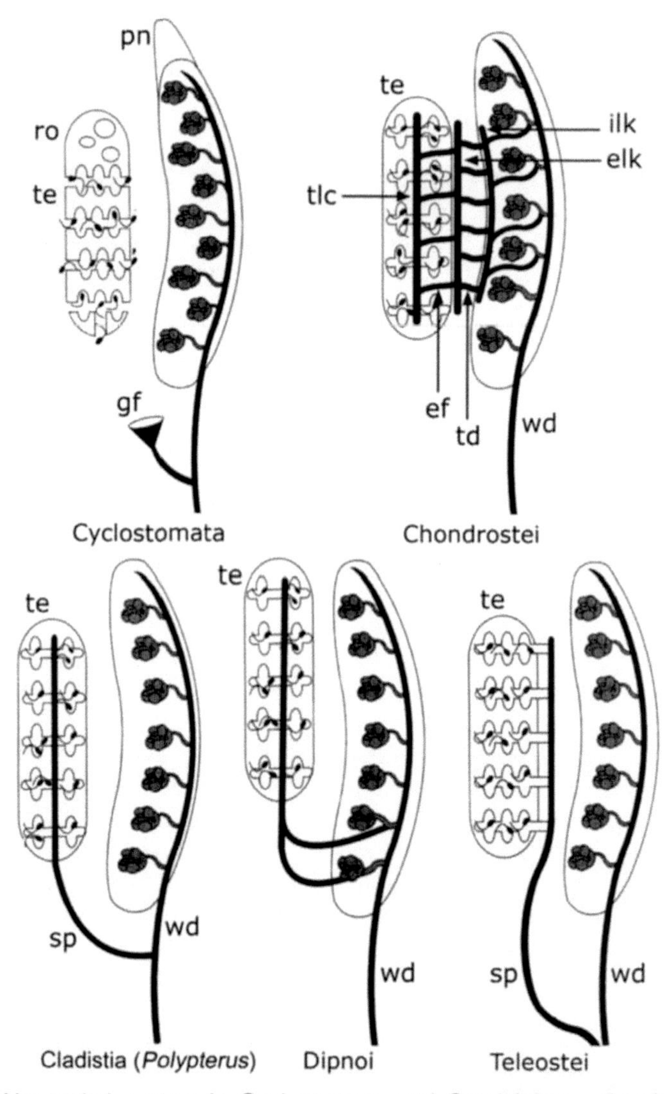

Fig. 5.1 Urogenital system in Cyclostomata and Osteichthyes showing different phylogentic steps in development of a secondary spermatic duct. ef, efferent transverse duct; elk, exterior longitudinal marginal cannel of the kidney; ilk, Interior longitudinal marginal cannel of the kidney; gf, genital funnel; k, kidney; pn, pronephros; ro, rudimentary ovary; sp, secondary spermatic duct; td, mesorchial transverse duct; te, testis; tlc, testicular longitudinal channel; wd, Wolffian duct.

mesonephric duct running longitudinally on the mesonephros. Also the renal ducts empty into this longitudinal channel. The Wolffian ducts therefore are urogenital ducts in their whole extent.

In dipnoans a testicular main duct arises from the testis (Blüm 1985). However, this main duct ends blind and is connected with the mesonephros via one (e.g., *Protopterus annectens annectens*) or several ductuli efferentes (e.g., *Lepidosiren paradoxa*) (Fig. 5.1).

In the Holostei and Cladistia the testes form a secondary spermatic duct which empties into the terminal portion of the Wolffian duct indicating that the genital ducts and the nephric ducts are almost separated (Blüm 1985) (Fig. 5.1). In the Teleostei the renal system is completely separated from the genital system (Fig. 5.1). The seminiferous tubules of the testis empty into the testicular main ducts. These ducts are longitudinally running channels located adjacent to the testis which continue into the spermatic ducts (Grier 1981). The sperm ducts empty separately from the renal duct in the urogenital papilla (Blüm 1985).

In the Chondrichthyes the situation is principally similar to the Chondostei. Mature sperm are released from the seminiferous follicles of the testis through the efferent ducts (Wourms 1977). The efferent ducts join the epididymis, which expands to form a long tube with complex convolutions. In comparison to the Chondrostei the mesonephros is highly developed and differentiated into a pars sexualis and a pars renalis. The pars sexualis is exclusively involved in the storage and transport of spermatozoa and is termed the epididymis. It is connected to the Wolffian duct (sperm duct). These ducts are continuous with the seminal vesicle (ampulla ductus deferens). The sperm duct and seminal vesicle are storage organs for seminal products and in some species the spermatozoa are packaged into either spermatozeugmata or spermatophores here (Wourms 1977). The terminal portion of the ureter empties into the seminal vesicle and both empty into the urogenital sinus. The urogenital sinus vents into a common cloaca, where two accessory glands are located, the Leydig glands and the alkaline glands. Leydig glands are a series of branched tubular glands that secrete seminal fluids into the epididymis and ductus deferens. The function of the alkaline gland of batoids is uncertain; it may be involved in sperm protection (Hamlett 1999).

While in the past the efferent duct systems as well as the accessory glands were mainly characterized on a morphological or histological base, they were investigated in recent decades in greater detail using also electron microscopical, enzyme histochemical and biochemical techniques. Most studies were performed on the efferent duct system of the Teleostei. Although its morphology is relatively homogenous within the investigated species there exist specific adaptations in size and function in relation to the mode of reproduction. Furthermore, several species have developed accessory reproductive organs, as seminal vesicles, testicular glands, or testicular blind pouches which resulted in modifications of the efferent duct system. Knowledge about the efferent duct system of the testis is important for a better understanding of gonad function, the processes of sperm maturation, the formation of the seminal fluid, and the fertilization processes. These data are important under basic as well as under applied aspects. Therefore, the present book chapter summarizes the recent data on the efferent duct system of the male gonads of the Euteleostomi (=Osteichthyes).

5.2 CHONDROSTEI

The male genital systems of *Acipenser ruthenus* and *Acipenser baerii baerii* were studied with histological and scanning electron microscopical methods by Wrobel and Jouma (2004). Further, in the sterlet, *Acipenser ruthenus*, early gonadal development was studied histologically and by scanning electron microscopy (Wrobel *et al.* 2002) to elucidate the relations between gonadal and renal tissue. The development of the efferent testicular duct system begins in 8-month-old animals in the pregonadal area of the gonadal fold. In the period from 8 to 18 months, the testicular excurent duct system reaches the adult state. The following description of the male genital system of sturgeons is based on the study of Wrobel and Jouma (2004).

5.2.1 The Testicular Excurrent Duct System of the Adult Sturgeon

In Chondrostei (sturgeons and paddlefishes) paired testes extend from the cranial end of the swim bladder to the level of the spiral intestine (Wrobel and Jouma 2004). In the region where the swim bladder is located, both testes are widely separated from each other. Caudal to the swim bladder, the testes are located closer to each other. The coiled seminiferous tubules converge towards the interior face of the testis. There they continue in short terminal segments which form a network of anastomosing channels at the inner side of the testis (the side located towards the mesorchium) (Fig. 5.1). Mesorchial transverse ducts originate from the terminal segment ducts which run through the mesorchium to the ventral aspect of the pars sexualis of the kidney (Fig. 5.1). The transversal ducts are not present along the entire length of the testis but are restricted to its middle segment. They occasionally anastomose with each other. Distally from the testis they increase in diameter and thickness. Contacts between nephrostomial tubules and mesorchial transversal ducts do not exist (Wrobel and Jouma 2004). The terminal segments are lined by a simple flat epithelium, the transversal ducts by a cubical or columnar epithelium.

Along the ventral side of the pars sexualis of the opisthonephros the mesorchial transversal ducts form a network of canals and solid strands, i. e. the longitudinal marginal network of the kidney (Fig. 5.1). The ducts enter the kidney forming a multipartite system, consisting of (a) centropapillary seminal ducts, (b) lacunary basal sinuses, and (c) intracolumnar ducts running inside the renal columns, the latter representing typical functional units of the adult sturgeon kidney. The centropapillary seminal ducts are situated in the center of a group of urinary collecting ducts. Each centropapillary seminal duct ends in a lacunary basal sinus located at the base of an opisthonephric column. The cuboidal cells which line the basal sinus are in direct contact with the opisthonephric blastema tissue. Renal corpuscles are connected to the testicular efferent duct system (Fig. 5.1). These connections transform into intracolumnar seminal ducts when the corresponding renal corpuscles move towards the apex of a renal column due to the development of new generations of nephrons at the base of the column. The contacts between the vascular poles of the renal corpuscles and the intracolumnar seminal ducts represent the urogenital junction. The nephrons of the pars sexualis which are involved in sperm transport have not lost their urinary function and are histologically identical to those of the pars excretoria which are solely urinary. In the pars sexualis and in the pars excretoria of the opisthonephros, the urinary collecting ducts of neighboring renal columns unite to form larger ducts that are surrounded by connective tissue to papilla-like formations. In the center of such a papilla, a centropapillary duct of the testicular efferent duct system is located. On their course to the exterior, the spermatozoa have to pass through Bowman's capsule, the different parts of the opisthonephric tubule, the urinary collecting duct system and finally the Wolffian duct (Wrobel and Jouma 2004) (Fig. 5.1).

5.2.2 The Wolffian Duct

The Wolffian duct of the Chondrostei is exclusively an opisthonephric component. The pronephric and cranial opisthonephric portions are not developed in adult males, parallel to the transformation of pronephros and anterior opisthonephros into hemopoietic organs. The Wolffian duct begins at the level of the caudal end of the swim bladder by the union of the most anterior urinary collecting tubules. Initially, the duct has a small diameter and is situated on the lateral edge of the kidney complex. Further caudally, the Wolffian duct is attached to the ventral aspect of the kidney and its lumen enlarges continuously. Dorsal to the rectum, both Wolffian ducts join with each other to form the sinus urogenitalis which has its own opening to the exterior immediately caudal to the anus. Histologically, the epithelium of the Wolffian duct varies along its course. In the cranial portion of the pars sexualis, it is simple columnar and similar as in the renal collecting ducts. Further caudally, the epithelium becomes stratified and in the pars excretoria, it is a transitional epithelium. Within the stratified and the transitional epithelium, large spherical secretory cells are regularly observed, whereby their number increases towards the urogenital sinus. These cells represent intraepithelial glands, staining positively with periodic acid Schiff and alcianblue. In the pars sexualis, the Wolffian duct possesses a thin muscular layer, the thickness of it increases steadily towards the urogenital sinus. In both sexes, the Müllerian duct runs parallel to the ventral side of the Wolffian duct. The caudal ending of the Müllerian duct is regularly divided into two or three smaller terminal ends which penetrate the ventral wall of the Wolffian duct. In males the distal ends of the Müllerian terminals are closed and covered by a thin layer of Wolffian epithelium (Wrobel and Jouma 2004).

5.3 CLADISTIA

In the Cladistia the testes form a secondary spermatic duct which empties into the terminal portion of the Wolffian duct, indicating that the genital ducts and the nephric ducts are almost separate (Blüm 1985) (Fig. 5.1). No additional information is available on the testicular efferent duct system in Cladistia. Bichir (*Polypterus senegalus senegalus*) and Reedfish (*Erpetoichthys calabaricus*) have sexually dimorphic anal fins, as males have a pronounced bulge at the anal fin origin (anal fin is broader and more muscular) (Britz and Bartsch 1998). This modification develops gradually with sexual maturity. The anal fin is important during spawning, as the male uses the anal and caudal fins to envelop the genital opening of the female, thereby forming a receptacle in which they fertilize the eggs. Eggs are then released by the male, through vigorous shaking of the anal fin, and quickly adhere to vegetation (Britz and Bartsch 1998).

5.4 CERATODONTIMORPHA (=DIPNOI)

The urogenital system of dipnoi has been studied in the genera *Lepidosiren*, *Protopterus*, and *Neoceratodus* on a morphological and histological base (Wake 1986) (Fig. 5.1). Lungfish kidneys are paired, somewhat lobed, elongate, highly pigmented retroperitoneal structures. Kerr (1901) distinguished the genital or vesicular posterior portion of the kidney from the anterior urinary part, the latter recognizable in part by reduced pigment, especially in *Protopterus*. The posterior kidney is involved in sperm transport, and the nephrons are essentially unmodified from typical urinary structures. The nephron structure in both the anterior and vesicular parts of the kidney consists of a nephron, a typical corpuscle and glomerulus, a ciliated neck segment, a proximal tubule of two components, a ciliated intermediate segment, a distal segment, and a collecting duct system (Wake 1986). In the vesicular portion of the kidney the nephrons have greater tubular diameters. The tubule dilatation may be due to the presence of sperm, for capsule and tubule may be packed with sperm in the breeding season. The morphology suggests that the posterior nephrons are capable of a urinary function. It is not known specifically whether they are functional at any time, nor is it known whether lungfish urine would affect sperm viability (Wake 1986).

Dipnoan testes are elongate structures, bound to the kidney and dorsal body wall by mesenteries, and overlain by fat, especially before the dry season (Kerr 1901). The testes are stout medially, tapering laterally (Wake 1986). In *Protopterus* and *Lepidosiren* the posterior part of the testis is not spermatogenic, but vesicular (Kerr 1901). The vesicular testis invades kidney tissue. *Neoceratodus* lacks a vesicular posterior region (Jespersen 1969). The vesicular region is spongy, trabecular, and lacks spermatogenic tubules. During the breeding season the spermatogenic part of the testis is filled with tubules containing mature sperm. The tubules open into a longitudinal testis duct that extends posteriorly to a series of lateral ducts, the vasa efferentia, which carry sperm to the nephrons.

In all three genera the Wolffian (archinephric, mesonephric) duct is separate and transports urine from the adult mesonephric kidney to the cloaca (Wake 1986) (Fig. 5.1). In *Neoceratodus* where the testis does not have discrete sperm-producing and vesicular components "vasa efferentia" (lateral ducts) connect the longitudinal testis duct to nephrons over the entire length of the kidney. Thus the Wolffian duct evacuates urine, and also sperm during the breeding season, from the entire extent of the kidney. In *Lepidosiren* sperm are transported from a tubular longitudinal testis duct extending through the vesicular part of the testis, where 5 or 6 lateral ducts lead to nephrons of the posterior part of the kidney (Wake 1986). *Protopterus* has a reduced testicular net in which a single lateral duct extends from the longitudinal testis duct to several posterior nephrons (Wake 1986).

Males of all three genera of lungfish retain Müllerian ducts, at least to some extent (Wake 1986). Those of *Neoceratodus* extend from the anterior end of the testis to the cloaca, where they fuse. They do not open to the cloaca but are thin, non-functional, connective tissue strips. In *Protopterus,* adult males retain vestiges of the infundibulum at the anterior end of the kidney, and a few millimeters of each duct posteriorly. The ducts unite in the wall of the cloaca. The ducts are even more reduced in males of *Lepidosiren* as only the anterior infundibula are present (Wake 1986).

All three genera of dipnoans retain a simple cloaca with the urinary ducts opening by a single aperture. Kerr (1901) reported that each aperture is marked by a prominent papilla during the breeding season in *Protopterus annectens annectens* and *Lepidosiren.* Lagios and McCosker (1977) report the presence of a cloacal gland in *Protopterus aethiopicus aethiopicus* and *Protopterus dolloi,* and its absence in *Neoceratodus* and *Lepidosiren.* The gland is dorsal and posterior to the urinary caecum and opens into the cloaca at the juncture of the caecum with the cloaca. Lagios and McCosker (1977) suggest that the gland may be cation-secreting, functionally similar to the rectal glands of elasmobranchs, holocephalans, and *Latimeria,* though the anatomical relationship differs from that of the latter groups.

5.5 TELEOSTEI

5.5.1 Testicular Efferent Duct System of Species with External Fertilization and Separate Sex

5.5.1.1 Morphology of the efferent duct system

In the Teleostei the renal system is completely separated from the genital system and a secondary spermatic duct has developed (Fig. 5.2A). In most species the testes are located cranially from the genital papilla. The testicular main ducts run craniocaudally and continue into the spermatic ducts at the caudal end of the testes (Lahnsteiner *et al.* 1993a, 1993b; Lahnsteiner *et al.* 1994; Vicentini *et al.* 2001; Lahnsteiner 2003). The spermatic ducts are paired and run towards the genital papilla (Fig. 5.2A). Just before emptying out at the genital papilla the spermatic ducts join to an unpaired part which discharges via the urogenital papilla. In some species, as e.g. *Uranoscopus scaber* and *Trachinus draco,* the paired testes are located caudally from the genital papilla in the coelomic cavity (Lahnsteiner 2003) (Fig. 5.2B). In the cranial third of the testis the testicular main ducts join to form an unpaired spermatic duct which runs cranially towards the genital papilla (Fig. 5.2A).

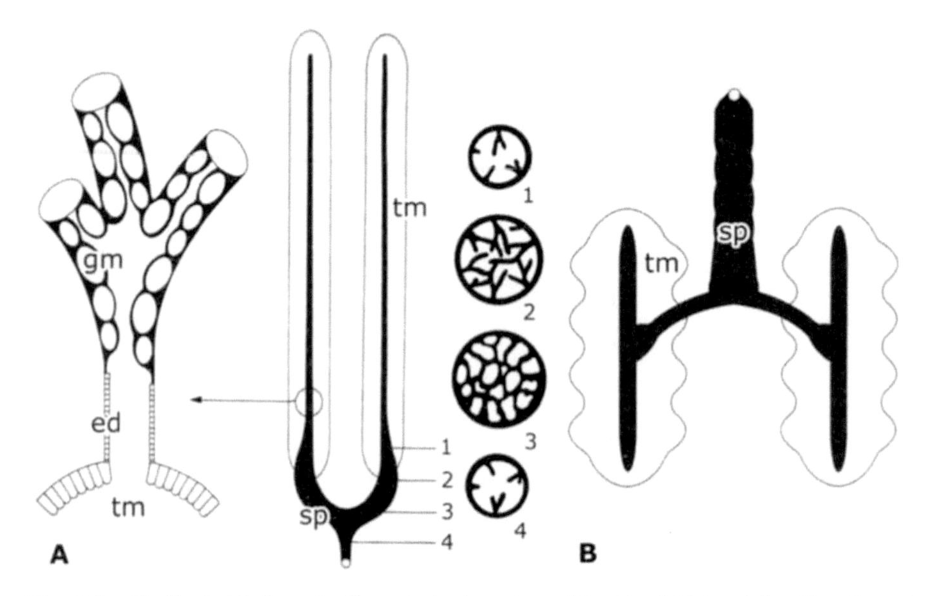

Fig. 5.2 Testis (white) and efferent duct system (black) of Teleostei with external fertilization and without accessory glands. tm, testicular main duct; ed, efferent duct of the testis (note that these ducts occur only in Esocidae and Salmonidae); gm, germinal cyst; sp, spermatic duct; gp, genital papilla. **A**. Testis located cranially to the urogenital pailla as usual in Teleostei (e.g. in Salmonidae, Cyprinidae, Esocidae, Sparidae, Mullidae, Labridae, Synodontidae). Circles represent schematic cross sections made in different portions of the spermatic duct showing variations in the epithelial folds. **B**. Testis located caudally to the genital papilla as found in Trachinidae.

The efferent duct system consists of testicular efferent ducts, testicular main ducts and spermatic ducts (Fig. 5.2A). Testicular efferent duct have been described only in the Salmonidae (Lahnsteiner *et al.* 1993a) and Esocidae (Lahnsteiner *et al.* 1993b). They represent a transitional stage between the seminiferous ducts of the testes and the testicular main ducts. The epithelial cells of the testicular efferent duct are regarded as modified Sertoli cells which differentiate only during spawning (Grier *et al.* 1980). In the other investigated teleost species the seminiferous tubules empty directly into the testicular main ducts at the ventral side of the testis. The latter situation is found in the investigated Cyprinidae (*Alburnus alburnus, Leuciscus cephalus, Vimba vimba,* Lahnsteiner *et al.* 1994) and in *Diplodus sargus, Mullus barbatus, Sparisoma cretense, Synodon saurus, Trachinus draco, Uranuscopus scaber, Thalassoma pavo* (Lahnsteiner 2003), and *Prochilodus lineata* (= *scrofa*) (Vicentini *et al.* 2001). The testicular main ducts extend longitudinally ventral of the testis. This is similar for all species investigated. At the caudal end of the testis the testicular main ducts increase in diameter and continue into the spermatic ducts. The spermatic ducts are persisting organs of extratesticular origin (Grier *et al.* 1980; Billard 1986).

5.5.1.2 Histology, fine structure, histochemistry and function of the testicular efferent duct

The testicular efferent ducts are located within the testis and represent a transition zone between seminiferous tubules and testicular main ducts (Grier *et al.* 1980; Lahnsteiner *et al.* 1993b) (Fig. 5.2A). Whereas the seminiferous tubules are formed by the typical Sertoli cells and their processes, the testicular efferent ducts are formed by a monolayered, cuboidal epithelium which has a height of circa 5 μm (Fig. 5.2A). The testicular efferent ducts have been studied in spawning Salmonidae: *Oncorhynchus mykiss, Salvelinus alpinus alpinus, Thymallus thymallus, Coregonus* sp. (Lahnsteiner *et al.* 1993a) and in spawning Northern pike, *Esox lucius* (Lahnsteiner *et al.* 1993b). In the histological and fine structural characteristics of the testicular efferent ducts of the studied species no differences were found. The epithelium has an intensely folded basal lamina and is sometimes covered with microvilli. The nuclei are irregularly shaped and have an eccentric nucleolus. The cytoplasm contains an abundance of smooth and rough endoplasmatic reticulum, mitochondria with tubular cristae, and Golgi apparatuses which produce secretory vesicles. Lipid vacuoles, lamellar bodies and heterophagic vacuoles are also found in the cytoplasm. The secretion mode of the epithelial cells is mainly eccrine. The testicular efferent ducts differ from the testicular main ducts and from the spermatic ducts in the epithelial height and the mode of secretion (Lahnsteiner *et al.* 1993b); all other fine structural and histochemical parameters and the functions are similar to the testicular main ducts and spermatic ducts.

5.5.1.3 Testicular main ducts and spermatic ducts

Morphology and histology: The testicular main ducts and spermatic ducts consist (from outside to inside) of three layers, a peritoneum, a transversal orientated connective tissue layer containing smooth muscle cells and an epithelium. The epithelium is smooth in the testicular main ducts (Fig. 5.3A). At the origin from the testicular main ducts the spermatic duct epithelium is also smooth or reveals only some longitudinal folds (Fig. 5.2A). In caudal direction the up-folding of the spermatic duct epithelium becomes more intensive. Often several folds fuse with each other to form a complicated,

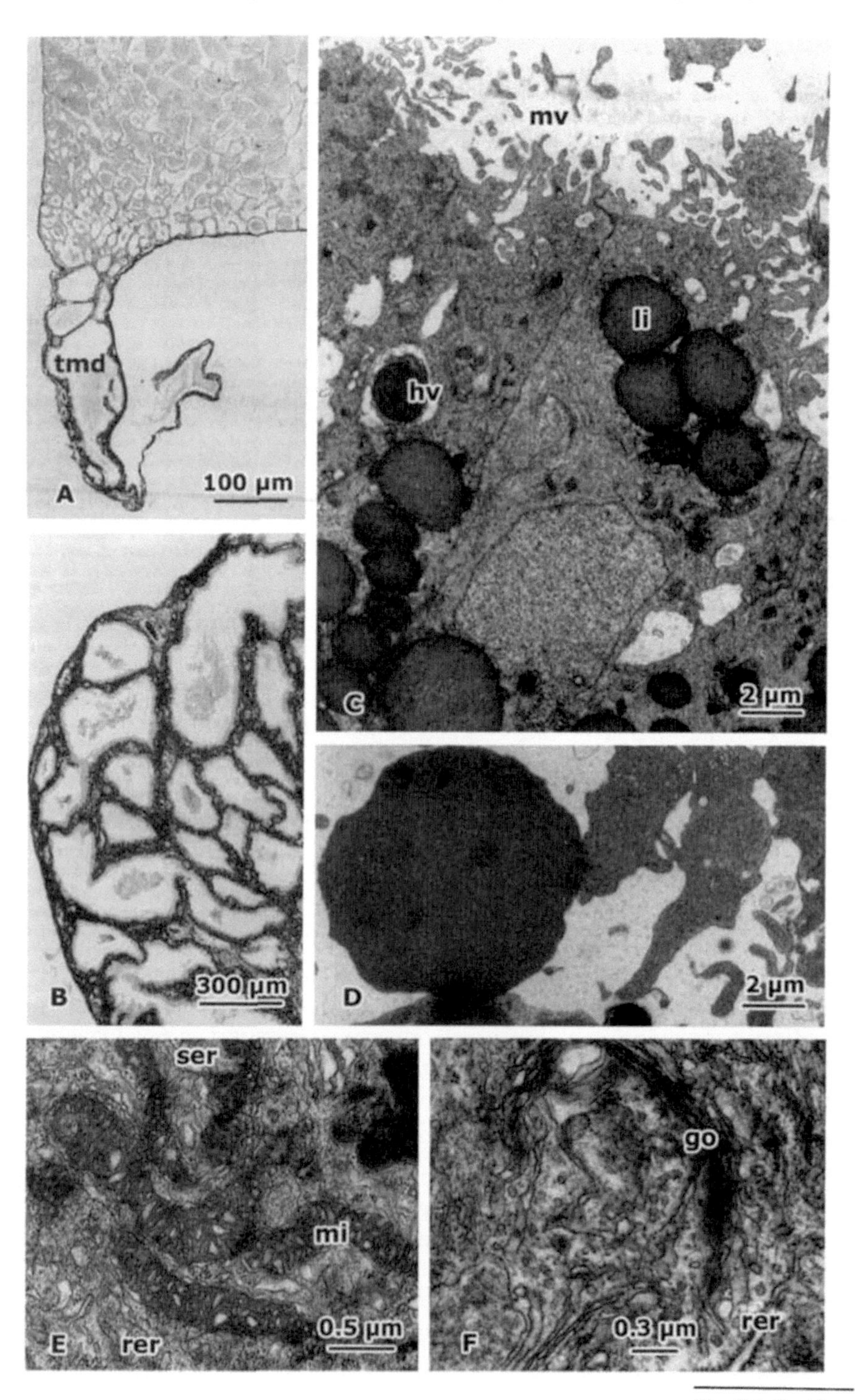
tmd
100 µm
A
B
300 µm
mv
li
hv
C
2 µm
D
2 µm
ser
mi
E
rer
0.5 µm
go
F
0.3 µm
rer

Fig. 5.3 Contd. ...

mostly longitudinally orientated tubules system (Figs. 5.2A, 5.3B). This is similar for the investigated species of the Cyprinidae (Lahnsteiner *et al.* 1994), Salmonidae (Lahnsteiner *et al.* 1993a), Esocidae (Lahnsteiner *et al.* 1993b), Sparidae (Lahnsteiner 2003), Mullidae (Lahnsteiner 2003), Scaridae (Lahnsteiner 2003), Trachinidae (Lahnsteiner 2003), Labridae (Rasotto and Shapiro 1998; Lahnsteiner 2003), and Synodontidae (Lahnsteiner 2003). This epithelial up-folding enlarges the internal surface of the epithelium of the spermatic ducts and could reflect the high activity of these organs. Distally, just before empting on the genital papilla the unpaired portion of the spermatic duct changes into a simple tube epithelium which finally continues into the cutaneous epidermis. The length of the spermatic ducts in relation to the testis size and body length differs species specifically. Thus, in the Northern pike the spermatic ducts are very short organs (Lahnsteiner *et al.* 1993b) whereas in other species, such as the Salmonidae and Cyprinidae, they are tubes of considerable length (Lahnsteiner *et al.* 1993a, 1994). During spawning the testicular main ducts and spermatic ducts are filled with spermatozoa. Therefore, the ducts can be considered to be the main storage organs for spermatozoa. This hypothesis has also been confirmed by Dulka and Demski (2005). In the latter study the contractile activity of the spermatic ducts and testes was investigated in goldfish (*Carassius auratus*) by electrical stimulation of the brain to determine the mechanism of sperm release. The results indicate that the spermatic duct and not the testes primarily mediate sperm release in goldfish. Testicular contractions serve to load the spermatic duct with semen (Dulka and Demski 2005). Males vary the volume of semen they release in successive spawnings depending on females and clutch size. Sperm release is probably controlled by musculature in the urogenital papilla. It has been studied in the coral reef fish, *Thalassoma bifasciatum* (Rasotto and Shapiro 1998). Urinary and spermatic ducts pass separately through a common urogenital papilla and are lined by a striated sphincter muscle and a pair of thin, smooth ligament muscles arising from the first proximal anal fin radial and passing laterally around the sperm duct.

Fine structure, histochemistry and functions: The testicular main ducts and spermatic ducts of several teleost species were investigated by fine structural

---

Fig. 5.3 Contd. ...

Fig. 5.3 Efferent duct system of *Esox lucius* (Esocidae) during spawning. **A**. Cross section of the testicular main duct (tmd). **B**. Cross section of the spermatic duct showing the extensive up-folding of the epithelium forming a chamber-like structure. Transmission electron micrograph. **C**. Epithelial cell with large lipid vesicles (li); microvilli processes (mv); and a heterophagic vacuole (hv). Transmission electron micrograph. **D**. Secretion (se). Mv microvilli processes. Transmission electron micrograph. **E**. Mitochondria (mi) and smooth (ser) and rough endoplasmatic reticulum (rer). Transmission electron micrograph. **F**. Golgi field (go) and rough endoplasmatic reticulum (rer). Transmission electron micrographs.

and (enzyme-) histochemical methods. Most studies were conducted on spawning fish where the described organs have the highest activity. Changes in the spermatic ducts occurring during the reproductive cycle have only been investigated in *Salaria* (= *Blennius*) *pavo* (Blenniidae) in detail (Lahnsteiner and Patzner 1990). In this species four activity phases could be distinguished based on histological, fine structural and histochemical studies: (a) An inactive phase during interspawning period, (b) a proliferation phase during prespawning period which was characterized by the onset of synthesis of secretion products, (c) a phase of high secretory activity during the spawning period and (d) a regression phase of the epithelium during the post-spawning period which was associated with extensive resorption processes of the remaining secretory products.

The fine structure and histochemistry of testicular main ducts and spermatic ducts during the spawning period has been investigated in *Diplodus sargus, Mullus barbatus* and *Sparisoma cretense* (Lahnsteiner 2003), *Alburnus alburnus, Leuciscus cephalus* and *Vimba vimba* (Lahnsteiner *et al.* 1994), *Esox lucius* (Lahnsteiner *et al.* 1993b), *Oncorhynchus mykiss, Salvelinus alpinus alpinus, Thymallus thymallus* and *Coregonus* sp. (Lahnsteiner *et al.* 1993a). Within a species the histological and fine structural features and the functions of the epithelial cells of the testicular main ducts and spermatic ducts were similar and also on the interspecies level only few differences were found. The epithelial cells of the testicular main ducts and spermatic ducts have an approximately cylindrical shape and adhere to an intensely folded basal lamina. The height of the epithelium is 10–15 μm, depending on the species. In all investigated species the epithelium is covered with microvilli (Fig. 5.3C). The nuclei are irregularly shaped and consist of heterogenous chromatin material. They have eccentrically located irregularly shaped nucleoli (Fig. 5.3C). The nuclear envelope has nuclear pores and numerous ribosomes are closely adjacent to the cytoplasmic side of the exterior nuclear membrane. The cytoplasm contains smooth and rough endplasmatic reticulum (Fig. 5.3E), mitochondria (Fig. 5.3E), Golgi apparatuses (Fig. 5.3F) and different types of vacuoles and secretory products. During spawning the epithelium has a high secretory activity. Three types of secretions are observed in the epithelia of the testicular main ducts and spermatic ducts whereby the occurrence and frequency of the different types depends on the species: (a) An eccrine secretion of secretory vesicles. (b) An apocrine secretion of apical cell portions: Cell organelles and microfilaments disappear from apical cell portions and the apical cell portions are released into the lumina of the spermatic ducts and testicular main ducts (Fig. 5.3D). (c) Complete cells can separate from the epithelium in the form of a holocrine secretion. These cells have structural features similar to those located in the epithelium and are lysed in the lumina of the testicular main ducts and spermatic ducts. Based on the fine-structural and enzyme-histochemical studies functions of the testicular main duct and spermatic duct epithelium in protein and steroid synthesis, in nutrition of spermatozoa, in formation of an ionic gradient in the seminal fluid, and in phagocytosis have been described (Lahnsteiner *et al.* 1993a, 1993b, 1994; Lahnsteiner 2003).

Protein synthesis: Within the cytoplasm of the epithelial cells of the testicular main ducts and spermatic ducts of all investigated species rough endoplasmatic reticulum which is densely coated with ribosomes is frequently observed (Fig. 5.3E). Free ribosomes and dense cored vesicles are abundant. Golgi apparatuses are also frequent (Fig. 5.3F). Within the Golgi cisternae and closely adjacent to them electron-dense material is often present. The Golgi cisternae release the primary secretory vesicles. In the surroundings of the Golgi apparatuses these secretory vesicles join to form larger secretory vacuoles which are mostly found in apical cell regions. The described fine structural features are clear indications for the synthesis of proteins. These results were ascertained by histochemical data which demonstrated that the epithelium of the testicular main ducts and spermatic ducts of all species investigated stained positively with different protein staining methods (mercury bromophenol blue, acrolein/Schiff) in which very intensive reactions were observed in apical parts and in apocrine secreted cell portions. SDS page electrophoresis and biochemical enzymatic assays proofed that the seminal fluid contains a specific mixture of different proteins and enzymes which have distinct functions in controlling and regulating sperm viability (Lahnsteiner 2007).

Steroid synthesis: Beside rough endoplasmatic reticulum the cytoplasm of the epithelial cells of the testicular main ducts and spermatic ducts contains also smooth endoplasmatic reticulum (SER) (Fig. 5.3E). Mitochondria are also frequent. They have an elongated shape and are found throughout the cytoplasm. Their matrix is electron-dense, the cristae are tubular and paralelly or irregularly arranged. The cytoplasm contains lipid vesicles (Fig. 5.3C) the abundance of which depends on the species (see below). The occurrence of SER, of mitochondria with tubular cristae and of lipid vacuoles is an indication of steroidogenetic activity in the epithelial cells. Occurrence of steroid synthesis was also demonstrated enzyme-histochemically (Lahnsteiner *et al.* 1993a, 1993b, 1994) and biochemically (Lahnsteiner 2003) by detection of activities of 3ß-steroid dehydrogenase and glucose-6-phosphate dehydrogenase. As also uridine-diphosphoglucose dehydrogenase activity was detected, steroids occur as steroid glucuronides. Through glucuronidation the generally apolar steroids become more water soluble (Van den Hurk and Resink 1992) and therefore might act as pheromones. Generally, steroid glucuronide synthesis is known from specialized accessory glands such as the testicular glands of Gobiidae (Colombo *et al.* 1977) and the seminal vesicles of the catfish (Van den Hurk and Resink 1992) where they have an attractive effect on the females. Glucuronated steroids also occur in the urine of spawning fish, and have meanings in the attraction of sex partners and synchronization of reproduction (Scott and Vermeirssen 1994). Similar functions could be supposed for the steroid glucuronides produced in the spermatic duct.

Auto-and heterophagocytotic activity: In addition to secretory vesicles the Golgi apparatuses produce also lysosomes and the cytoplasm contains autophagocytotic and heterophagocytotic vacuoles. Autophagocytotic processes are characterized by crescent shaped double membranes which enclose zones of the cytoplasm including cell organelles and secretory vesicles. Gradually, their contents increase in electron density and lamellar bodies become visible. During heterophagocytotic processes spermatozoa are encircled by microvilli processes (Fig. 5.3C) and transported into the interior of the cells in heterophagic vacuoles. There they are lysed and lamellar bodies remain as residuals in the heterophagic vacuoles. The auto-and heterophagocytotic cell portions contain phospholipids and glycolipids as demonstrated by the histochemical stainings for Nile blue and periodic acid Schiff. The lytic potency of the testicular main duct and spermatic duct epithelium is also ascertained enzyme-histochemically by the occurrence of acid phosphatase, ß-glucuronidase and aminopeptidase activity. A similar pattern of enzyme activity has been detected in the seminal fluid by biochemical enzymatic assays (Lahnsteiner *et al.* 1999, 2000). Autophagocytotic processes may play a role in regulation of the secretory activity (Lahnsteiner *et al.* 1994). Heterophagocytotic processes are from importance when spermatozoa undergo natural cell death and are eliminated. During postspawning spermatozoa remaining in the sperm duct are phagocytozed by the epithelial cells (Billard and Takashima 1983; Besseau and Faliex 1994). Also immune cells such as eosinophilic granulocytes and macrophages are involved in the elimination of residual spermatozoa (Besseau and Faliex 1994).

Secretion of lipids and monosaccharides for nutrition of spermatozoa: The cytoplasm of the epithelial cells of the testicular main ducts and spermatic ducts contains lipid vacuoles the size and abundance of which is species-specific. Histochemically they stain positively with Sudan black B. In the Salmonidae and in the Northern pike lipid vacuoles are numerous (Fig. 5.3C) and they are also actively secreted into the lumina of the testicular main ducts and spermatic ducts. Contrary, in the Cyprinidae and in *Diplodus sargus*, *Mullus barbatus* and *Sparisoma cretense* the epithelial cells of the testicular main ducts and of the spermatic ducts contain only few and small lipid vacuoles. Enzyme-histochemically, the epithelium of the testicular main ducts and spermatic ducts exhibits glucose-6-phosphatase and glucose-6-phosphate dehydrogenase activity indicating the occurrence of gluconeogenesis and the synthesis of free glucose. Lipids, glucose and other monosaccharides are actively secreted into the seminal fluid (Lahnsteiner *et al.* 1993a, 1993b, 1994) as they have been detected in the seminal fluid of the different species in varying concentrations (Lahnsteiner *et al.* 1999a, 2000). Lipids and monosacharides are energy substrates for teleost spermatozoa (Lahnsteiner *et al.* 1992a, 1993c, 1999b). Spermatozoa of the Salmonidae and of the Northern pike use triglicerides and glucose as energy substrates (Lahnsteiner *et al.* 1993c, 1999b) and therefore in the epithelium of their efferent duct system a high abundance of lipid vacuoles as well as a high activity of glucose-6-

phosphatase activity occurs. In contrast, spermatozoa of the Cyprinidae and of the investigated marine species utilize mainly glycolysis for energy production (Lahnsteiner *et al.* 1992a, 1999, unpublished data for marine species) and therefore lipid vacuoles in the epithelium of the testicular main ducts and spermatic ducts and lipid secretion are very rare (Lahnsteiner *et al.* 1992a).

Ion transport and formation of an ionic gradient: A high activity of Na^+/K^+ dependent ATPase and of Ca^{++} dependent ATPase was demonstrated enzyme-histochemically in the epithelium of the testicular efferent duct and of the spermatic ducts of the Cyprinidae, Salmonidae, and Esocidae and biochemically in the epithelium of the testicular efferent duct and of the spermatic ducts of the Sparidae, Mullidae, and Scaridae. In all investigated species the epithelial cells contain an abundance of mitochondria which are necessary for energy supply in these transport systems. These results indicate that the described tissues are involved in the formation of a specific ionic gradient for the seminal fluid which in teleosts differs significantly from that of the blood plasma (Morisawa 1983; Morisawa *et al.* 1983). The ionic composition of the seminal plasma plays an important role to maintain the viability of sperm cells, to control and regulate fine maturation processes and to inhibit sperm motility (Ohta *et al.* 1997, 2001). In the Salmonidae and in the pike the high potassium content of the seminal fluid and in the Cyprinidae the high osmolarity is responsible for the inhibition of the motility of spermatozoa (Morisawa 1983). In marine fish sperm motility is inhibited in all species with exception of *Mullus barbatus* by an osmolality lower than that of sea water (*Diplodus sargus* - Lahnsteiner *et al.* 1998; other species, Lahnsteiner unpublished data). In *D. sargus* it is inhibited by very high osmolality extending that of sea water (Lahnsteiner *et al.* 1998).

5.5.2 Testicular Efferent Duct System of Simultaneous Hermaphrodite Species

In simultaneous hermaphrodite species ovary and testis are present and mature at the same time (Patzner 2007). The morphology of the gonads has been studied in different species of the Serranidae (*Serranus atricauda* - García-Díaz *et al.* 2002; *Serranus scriba* - Zorica *et al.* 2005; *Diplectrum bivittatum* - Touart and Bortone 1980; *Diplectrum pacificum* - Bortone 1977). The ovotestis consists of two lobes of similar size with an approximately cylindrical shape, which are fused posteriorly. A capsule of smooth muscle and connective tissue, the tunica albuginea, externally covers each lobe. Most of the gonad is occupied by ovarian tissue (García-Díaz *et al.* 2002; Zorica *et al.* 2005). Both gonadal lobes vent to a common duct that leads to a genital opening located posterior to the anus. Moreover, in *S. atricauda* an accessory structure, the post-ovarian sinus, is located in the dorso-posterior region of the gonad. This structure is lobe-like, surrounded by smooth-walled epithelium, and is most prominently developed in spawning individuals. In some individuals, a few mature oocytes were observed within this structure and therefore it may be a storage organ for ovulated eggs prior to their release (García-Díaz *et al.* 2002). The duct of the postovarian sinus and the oviduct join prior to the urogenital papilla. The accessory structure found in *S. atricauda* has not been described in other species of the genus *Serranus*. However, this sinus is also found in *Diplectrum bivittatum* and *D. pacificum* (Bortone 1977; Touart and Bortone 1980) in which numerous well developed villi project into the sinus from the sinus wall. A more simple, vestigial structure was described in *Hemanthias vivanus* and *Serranus subligarius* (Hastings and Bortone 1980; Hastings 1981). Testicular tissue is located ventrolaterally. The seminiferous tubules of the testis are irregularly arranged and form a network that converges into a deferent duct (= testicular main duct) at the level of the gonad wall (García-Díaz *et al.* 2002). Sperm duct and oviduct are completely separated and there is no possibility of internal self-fertilization (García-Díaz *et al.* 2002). The oviduct, sperm duct and mesonephric duct open into the urogenital papilla. Two smooth muscle bands associated with the infracarinalis medius muscle bands, the collagenous tissue surrounding the gonoduct, and the intestinal lining are implicated in the extrusion of gametes and urine (Hastings 1981). Serranidae have complicated mating systems (serial monogamy or harems) depending on the species (e.g., *Hypoplectrus nigricans*, Fischer 1980; *Serranus tortugarum*, Fischer 1984). The possibility of self-fertilization in serranid species was experimentally demonstrated (e.g. *Serranus scriba*, Abd-el-Aziz and Ramadan 1990). Furthermore, Clark (1959) obtained embryos and larvae from artificial fertilization as well as from isolated fish kept in aquaria, and no developmental differences were observed between self-fertilized and cross-fertilized eggs.

5.5.3 Testicular Efferent Duct System of Species with Internal Fertilization

In viviparous fish the ripe eggs are fertilized and develop within the oviducts. The fertilizing sperm must penetrate the follicular envelope of each egg to reach the ovum. In viviparous species not only the female genital tract but also the male gonad system reveals modifications in comparison with species with external fertilization, i.e. (a) the package of spermatozoa to sperm bundles, (b) the development of sperm storage organs and (c) the development of an organ to transmit the spermatozoa to the female.

5.5.3.1 Package of spermatozoa in sperm bundles

Spermatozoa of internally fertilizing species are packed into encapsulated spermatophores or unencapsulated spermatozeugmata (Burns 1991; Burns *et al.* 1995; Pecio *et al.* 2005). During formation of spermatophores or spermatozeugmata spermatozoa are embedded in proteins and mucoproteins. Therefore, the cells and organs involved in sperm bundling (Sertoli cells, spermatic duct epithelium or seminal vesicles) reveal modification in developing a high secretory activity of the described substances. In some species of the Characidae (Burns *et al.* 1995), Hemiramphidae (Grier and Collette 1987), and Ageneiosidae (Loir *et al.* 1989) the bundles are formed in the sperm ducts whereas in the Clinidae and the Adrianichthyidae (*Horaichthys setnai*, Grier 1984) the spermatozeugmata and spermatophors, respectively, are formed in the germinal cysts. Also in *Dermogenys* sp. spermatozeugmata are formed in the testes and the fully developed spermatozeugmata are released into the testicular ducts (Downing and Burns 1995). Some species (e.g. Auchenipteridae) have seminal vesicles which produce mucous secretion used for the formation of spermatozeugmata (Loir *et al.* 1989). Sperm bundles are considered to minimize sperm loss during copulation (Fishelson *et al.* 2007). They might also protect the delicate needle-shaped elongate sperm heads from damage during passage during the male genital duct.

5.5.3.2 Morphology of the male gonad and of the efferent duct system

The efferent duct system of viviparous fish has been studied in 16 species of the Clinidae by Fishelson *et al.* (2007). Clinidae belong to the Blennioidei. The testes of viviparous blennies lack testicular glands which are typical for oviparous blennies (see below). The testes of the Clinidae are paired and situated dorsally in the coelomic cavity (Fig. 5.4 C,D,E,F). Their shape is elongate or triangular. The testicular seminiferous tubules form a delicate branching network similar to that in internally inseminating viviparous Poeciliidae (Grier *et al.* 1981; Grier and Collette 1987). During the process of final maturation spermatozoa are packed in an enveloping layer of mucoid substances secreted by the Sertoli cells (Fishelson *et al.* 2007). Two or three spermatozeugmata are formed in one cyst. The testicular main ducts extend along the inner margin of each testis and continue into the spermatic ducts at the caudal end of the testis (Fishelson *et al.* 2007) (Fig. 5.4 C,D). Distally, the two ducts merge to form an unpaired sperm duct. The walls of the ducts are composed of circular layers of muscles and surrounded by a ring of connective tissue, their luminal epithelium is columnar, and the epithelial cells are covered with microvilli and located on a basal lamina. The epithelium reveals some folds and protrusions. Groups of secretory cells are located within the epithelium. Final structural characteristics of the spermatic duct epithelium have not previously been investigated. Distally the spermatic ducts continuously increase in diameter and the walls become stronger with thicker rings of muscles and connective tissue. The unpaired part of the spermatic duct reveals the most intensive foldings and the largest protrusions and vents into the ampulla (Fig. 5.4 C,D,E,F). The ampulla has a so-called epididymal gland (Fishelson *et al.* 2007). Its epithelium is coiled and arranged in branches (Fig. 5.4 H). Mucoid substances staining positively with periodic acid Schiff and alcian blue are produced by the epithelium of the epididymal gland as well as by the epithelium of the sperm ducts. The epididymis is a storage organ for the spermatozeugmata where they are stored between the epithelial folds and embedded in mucotic substances. The testicular blind

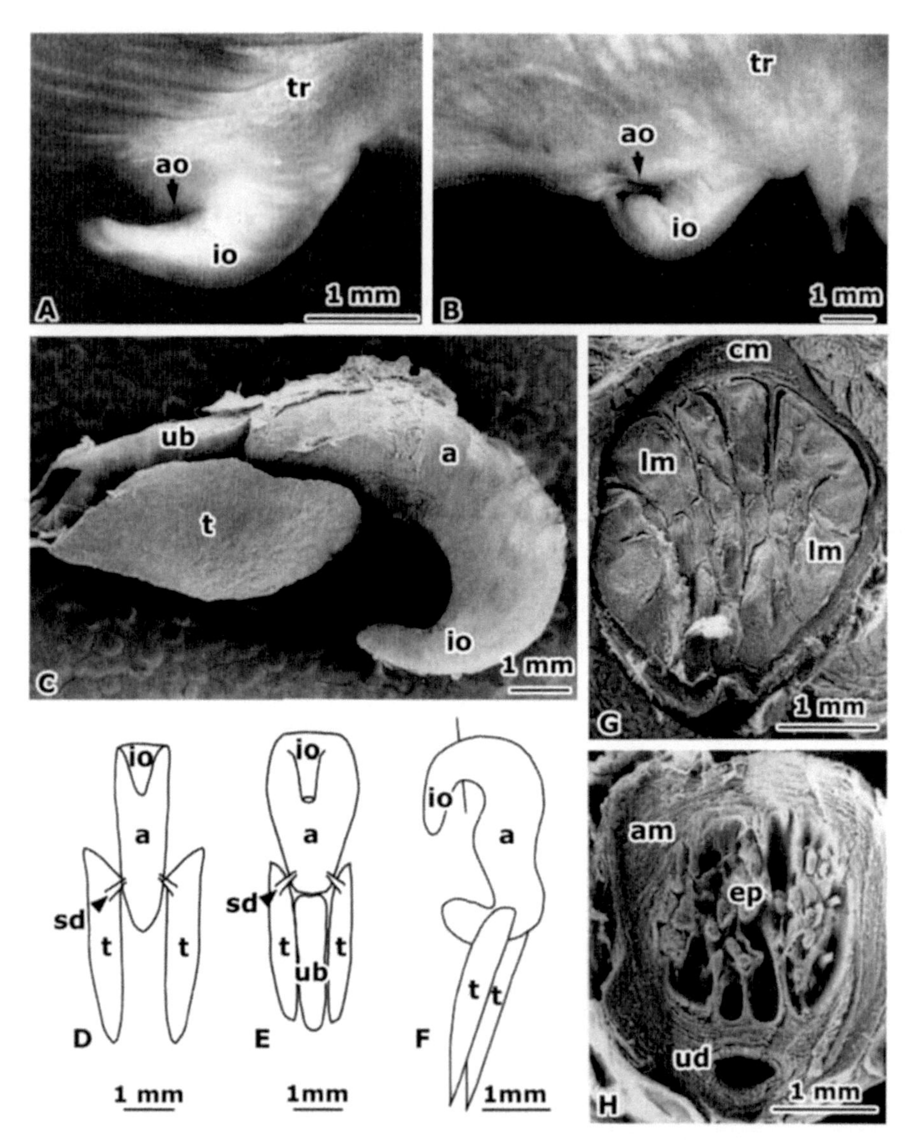

Fig. 5.4 *Reproductive organs and* intromittent organs of clinid fishes. a, ampulla; am, ampulla muscles; ao, anal opening; cm, circular muscles; ep, coils of epididymis; io, intromittent organ; lm, longitudinal bundles of muscles; sd, sperm duct; t,testes; tr, fish trunk; ub, urinary bladder ud, urinary duct. **A**. Intromittent organ of *Clinus superciliosus* (SEM). **B**. Intromittent organ of *Clinus cottoides*. **C**. The reproductive organs of *Muraenoclinus dorsalis* (SEM). **D**. Schema of reproductive organ of *C. supercillosus*. **E**. Schema of reproductive organs of *Clinus taurus*. **F**. Schema of reproductive organs of *Clinus arborescens*. **G**. Cross section of basal part of the ampulla in *Clinus acuminatus* (SEM). **H**. Cross section of ampulla in *C. acuminatus* (SEM). Modified after Fishelson, L., Gon, O., Holdengreber, V. and Delarea, Y. B. 2007. Journal of Morphology 290: 311-323, Figs. 2, 3.

pouches of the blenny *Salaria pavo* resemble the epithelium of the epididymis of clinids (Fishelson *et al.* 2007). Spermatozeugmata are transported via the testicular main ducts and the spermatic ducts into the epididymal gland where they are stored until copulation. In non-reproducing males remnants of sperm are found in the spermatic duct (Fishelson *et al.* 2007).

The male genital system was also studied in the internally fertilizing auchenipterid catfishes, *Trachelyopterus lucenai* and *Trachelyopterus galeatus* (Downing Meisner *et al.* 2000) and in *Auchenipterus nuchalis* (Loir *et al.* 1989; Mazzoldi *et al.* 2007). According to Downing Meisner *et al.* (2000) the male genital system of the Auchenipteridae has four main regions (Fig. 5.5): (a) In the anterior part of the gonads spermatogenic lobes involved in spermatogenesis and divided into numerous finger-like lobes are located. (b) Posterior to the spermatogenic area is the sperm storage region of the seminal vesicle, a large. median structure with a honeycomb-like appearance. (c) Attached to the storage region of the seminal vesicle, secretory lobes are found which are composed of tubules lined by a secretory epithelium and which produce mucins or mucoproteins. Here, the spermatozoa are organized into spermatozeugmata. (d) Finally, storage regions for the spermatozeugmata are located ventral to the storage region of the seminal vesicle. They contain the highly compact mature spermatozeugmata. The anal fin is modified as an intromittent organ (Downing Meisner *et al.* 2000; Mazzoldi *et al.* 2007).

In *Auchenipterus nuchalis* the seminal vesicles are short and broad and packed as a mass around the ampulla, which has a reduced diameter and volume but a thick conjunctive wall (Loir *et al.* 1989). They consist only of some very wide tubules lined by a thick epithelium. Usually the tubules are filled with spermatozoa embedded in a compact or sometimes bubbly material also containing cellular remnants of holocrine or merocrine secretion.

In the viviparous characid fish, *Brittanichthys axelrodi,* the testes have three distinct regions (Javonillo *et al.* 2006): an anterior spermatogenic region, an aspermatogenic middle region lined with a simple squamous epithelium and used for storage of mature spermatozoa, and a posterior region of coiled chambers lined with a high simple cuboidal epithelium. The most posterior region appears to be involved in the formation and storage of spermatozeugmata. The gonopore opens posterior to the anus, with the urinary pore having a separate opening posterior to the gonopore. Bands of skeletal muscle are found in the area of the male gonopore.

Only limited information is available on the efferent duct system of other viviparous species. In the viviparous halfbeaks of the genera *Nomorhamphus, Dermogenys,* and *Hemirhamphodon* the testes are paired organs running laterally along the body wall on either side of the gut (Downing and Burns 2005). The short efferent ducts empty into a single longitudinal testicular main duct in each testis. In viviparous glandulocaudine species mature testes have a large posterior testis portion, which is devoid of developing germ cells and spermatocysts and is therefore aspermatogenic (Burns *et al.* 1995). It is devoted to sperm storage (Burns *et al.* 1995).

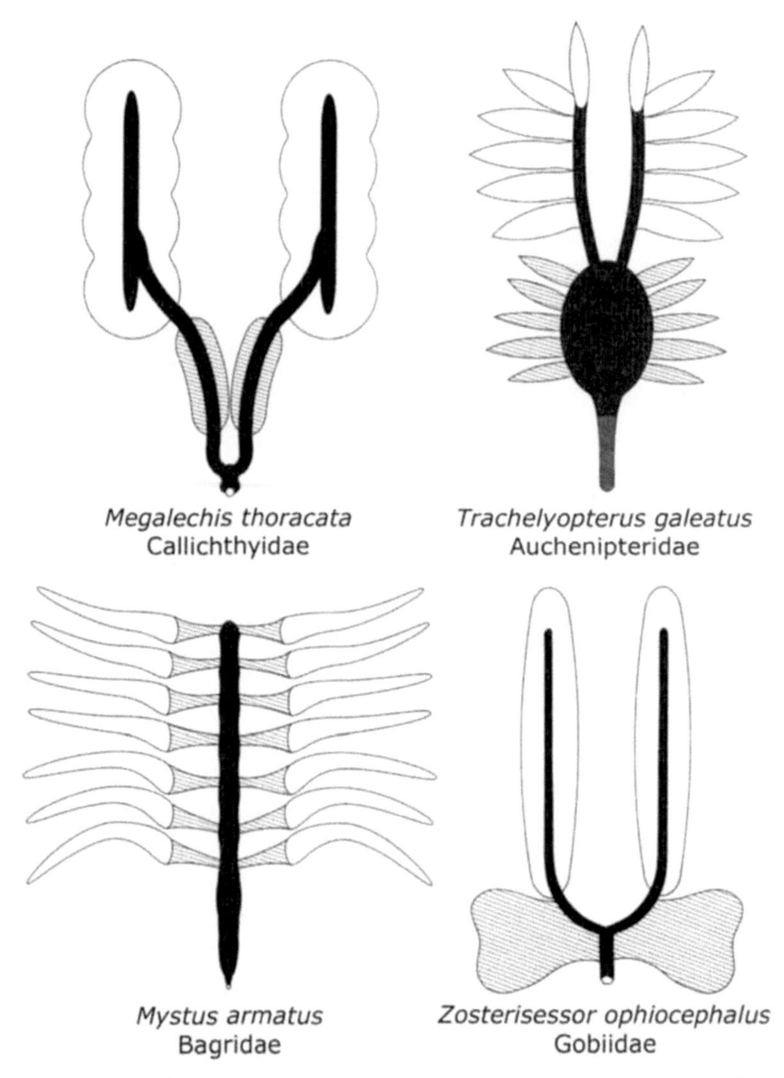

Fig. 5.5 Male genital system of different species with seminal vesicle. White, testis; black, efferent duct system; shaded, seminal vesicle; gp, gonopodium. The vesicle like enlarged portion of the spermatic duct of Auchenipteridae is called sperm storage region (see text).

5.5.3.3 Intromittent organs

In the Clinidae spermatozoa are introduced into the female receptive organ by means of a special intromittent muscular papilla (Fig. 5.4 A,B) that culminates in a genital pore as described by Penrith (1970) and Shen (1971). In the intromittent papillae the cohesive material forming the spermatozeugmata dissolves and the single spermatozoa are released into the female receptive tube. The strong muscular wall of the ampulla appears to generate the pressure and propulsion to inject the sperm across the intromittent organ into the female's oviduct. In viviparous cyprinodont fishes gonopodia are found for sperm transfer into the female oviduct (Grier *et al.* 1980; Constanz 1989). The viviparous halfbeak, *Dermogenys* sp., has an andropodium for transferring sperm into the female reproductive tract (Downing and Burns 1995). The andropodium is a modification of the anal fin and it possesses a pair of well-developed spines and a single, segmented, sickle-shaped spiculus on the second ray. In the viviparous fish, *Anableps dowei* the anal fin is transformed into the intromittent gonopodium, spermatozeugmata are formed in the testes and found in the main testicular ducts (Burns 1991). In the characid fish, *Brittanichthys axelrodi* bony hooks can be found on the fourth unbranched ray and on the branched rays 1-4 of the anal fin in mature males (Javonillo *et al.* 2006). For further discussion of intromittent organs see Evans and Meisner, **Volume 8B, Chapter 4**).

5.5.4 Testicular Efferent duct System of Species with Accessory Glands

Several teleost species with external fertilization have accessory glands which are a part of the efferent duct system. Testicular glands can be considered to be modified testicular efferent ducts and are located at the ventral side of the testis, seminal vesicles are lobe-like glands emptying into the spermatic ducts, the testicular blind pouches are tube-like evaginations of the spermatic ducts. The male gonads of Blenniidae have testicular glands and testicular blind pouches (Eggert 1931) (Fig. 5.6). Testicular glands but no blind pouches are found in the Tripterygiidae (Eggert 1931; Rasotto 1995), Labrisomidae (Rasotto 1995; Patzner and Hastings unpublished data), externally fertilizing Clinidae (Guitel 1893; Eggert 1931; Patzner unpublished data), and Dactyloscopidae (Rasotto 1995). Chaenopsidae have testicular glands and seminal vesicles (Patzner 1991; Rasotto 1995). Also some Gobiidae have seminal vesicles and glands attached to the testes (see below) (Figs. 5.5, 5.6), however, these glands are not homologous with those of the Blenniidae (Seiwald and Patzner 1989). The glands of the Gobiidae are compact steroid producing tissues located on the ventral side of the testis (Fig. 5.8A). The testicular main duct is located between testis and steroid producing gland (Colombo *et al.* 1977, Seiwald and Patzner 1989) (Fig. 5.8B). Seminal vesicles (= spermatic duct glands) are paired accessory genital glands, which empty into the spermatic duct (Fig. 5.5). They are found in several families of teleost fish, in the Clariidae (Van den Hurk *et al.* 1987), Gobiidae (Fishelson 1991; Lahnsteiner *et al.* 1992b; Miller 1992), Bagridae (Loir *et al.* 1989; Mansour and Lahnsteiner 2003), Pimelodidae (Loir *et al.* 1989), Heteropneustidae (Chowdhury and Joy 2001), Ictaluridae (Sneed and Clemens 1963), Siluridae (Van Tienhoven 1983, Mansour and Lahnsteiner 2003), Callichthyidae, (Loir *et al.* 1989, Mansour and Lahnsteiner 2003; Mazzoldi *et al.* 2007),

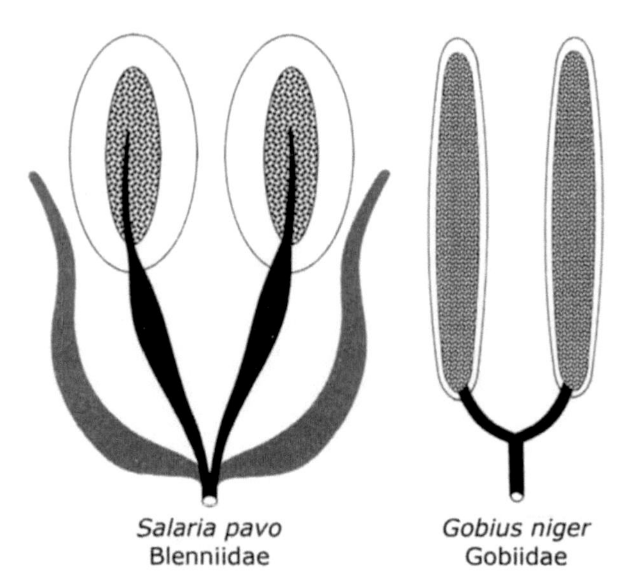

Fig. 5.6 Male genital systems of species with glands attached to the testis. White, testis; black, efferent duct system; coarse dotted, testicular gland (Blenniidae); grey, testicular blind pouches; fine dotted, steroid gland (Gobiidae).

Auchenipteridae (Loir *et al.* 1989; Mazzoldi *et al.* 2007) and Blenniidae (Patzner 1989). They exist in externally as well as in some internally fertilizing teleosts (see below) and reveal interspecific and interfamily variations in size and morphology (Fig. 5.5). Seminal vesicles can be separated from the testes and connected to the spermatic duct (e.g., Clariidae - Van den Hurk *et al.* 1987, Gobiidae - Lahnsteiner *et al.* 1992b) or are in close association with testicular tissue (e.g., Bagridae and Pimelodidae, Loir *et al.* 1989) (Fig. 5.5): They function as secretory glands or serve as a sperm storage organ (Chowdhury and Joy 2007).

5.5.4.1 Efferent duct system of the Blenniidae as an example for species with testicular glands and testicular blind pouches

The efferent duct system including the accessory organs was best studied in the Blenniidae (Lahnsteiner and Patzner 1990; Lahnsteiner *et al.* 1990, 1993d). The seminiferous tubules of the testis continue into the ducts of the testicular gland in these species (Fig. 5.7). Therefore, the ducts of the testicular gland can be considered as modified testicular efferent ducts. The glandular ducts are tortuous and irregularly arranged and empty into the glandular main duct on the ventral side of the testicular gland which is homologous with the testicular main ducts in species without testicular glands (Figs. 5.7, 5.9A, 5.9C). In the testes the spermatids do not differentiate to mature spermatozoa, but the spermatids are released from the germinal cysts via the seminiferous tubules into the ducts of the testicular gland (semicystic type of spermiogenesis) (Figs. 5.7, 5.9B). In the ducts of the testicular gland and in the spermatic ducts further differentiation processes to spermatozoa occur (Fig. 5.7). The main ducts of the testicular gland continue into the spermatic ducts at the caudal end of the testicular gland. The spermatic ducts are paired and run towards the genital papilla. Just before emptying out at the genital papilla the spermatic ducts join to form an unpaired duct. From this unpaired portion the paired tube or sac-like testicular blind pouches arise.

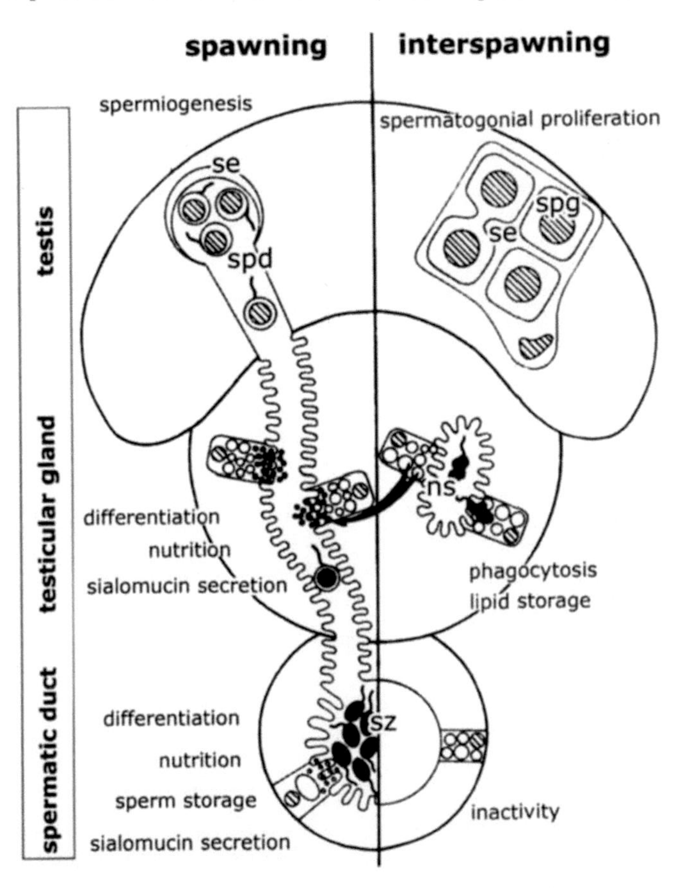

Fig. 5.7 The male gonads of Blenniidae and their functions in reproduction during spawning (left) and interspawning period (right); cross section. ns, necrotic spermatid; se, Sertoli cell; spd, spermatid; spg, spermatogonium; sz, spermatozoon.

Testicular gland: The testicular gland consists of (a) the glandular tubules which are formed by a monolayered epithelium and (b) of the glandular interstitium, which contains beside fibrocytes, blood capillaries, and steroid producing cells homologous to the Leydig cells of the testes. The size of the gland differs species specifically (Figs. 5.8A, 5.9C). In some species (e.g. in

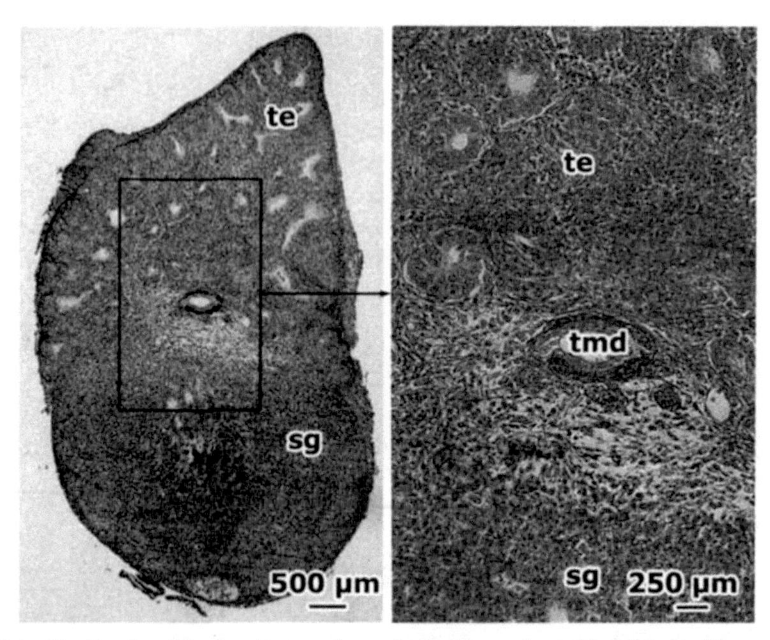

Fig 5.8 Testis-steroid gland complex of *Gobius niger* (Gobiidae). The steriod gland (sg) is a compact organ which is pervaded by the testicular main duct (tmd). te, testis. Paraffin section.

Salaria pavo) the testicular gland is small (volume only 15 – 30 % of the testicular volume) while in *Microlipophrys* (= *Lipophrys*) species the testis is remarkably reduced and the testicular gland is voluminous (220 – 250 % of the testicular volume) (Lahnsteiner *et al.* 1990; Richtarski and Patzner 2000).

The epithelial cells of the testicular gland have a prominent layer of microvilli (Lahnsteiner *et al.* 1990) (Figs. 5.9B,C). Their nuclei are spherical, consist of heterogenous chromatin material and have an electron-dense nucleolus. Mitochondria are abundant (Fig. 5.9B). Endoplasmic reticulum is found in a smooth tubulovesicular form as well as covered with ribosomes. Rough endoplasmatic reticulum (RER) is located around heterophagic vacuoles originating from reabsorbtion of necrotic spermatids. They are sparse during the spawning period, located only in the basal portions of the gland cells, and contain mainly phospholipids. Golgi apparatuses are most preferably situated in the supranuclear region. They produce secretory vesicles with osmiophilic contents. When the secretory vesicles reach the apex of the gland cells, their contents transform into osmiophilic granulae. Granular glycogen is dispersed in the cytoplasm, often in close vicinity to the SER. Sometimes, and especially in the *Microlipophrys* species, apical cell portions contain larger vacuoles filled with glycogen particles. Secretory vesicles and glycogen are released by apocrine secretion. During spawning, apical cell portions as well as the contents of the ducts react strongly to periodic acid Schiff indicative for polysaccharids (Lahnsteiner *et al.* 1990).

These polysaccharids are present in the form of glycogen and of non-sulfated mucosubstances, as they stain positively for Best's glycogen detection and for alcian blue at pH 2.5 and negatively after neuraminidase treatment. At the end of the spawning period, in the cells of the testicular gland, the synthesis of glycogen and secretory vesicles stops and the amount of RER and of lysosomes in the cytoplasm increases. Spermatids and secretion remaining in the ducts of the testicular gland from spawning time are reabsorbed into heterophagic vacuoles. The testicular gland cells become filled with these vacuoles (Fig. 5.9D) and the whole testicular gland has a strong reaction for acid phosphatase, the histochemical proof for lysomatic processes, and for Sudan black B, indicative of lipids (Lahnsteiner *et al.* 1990).

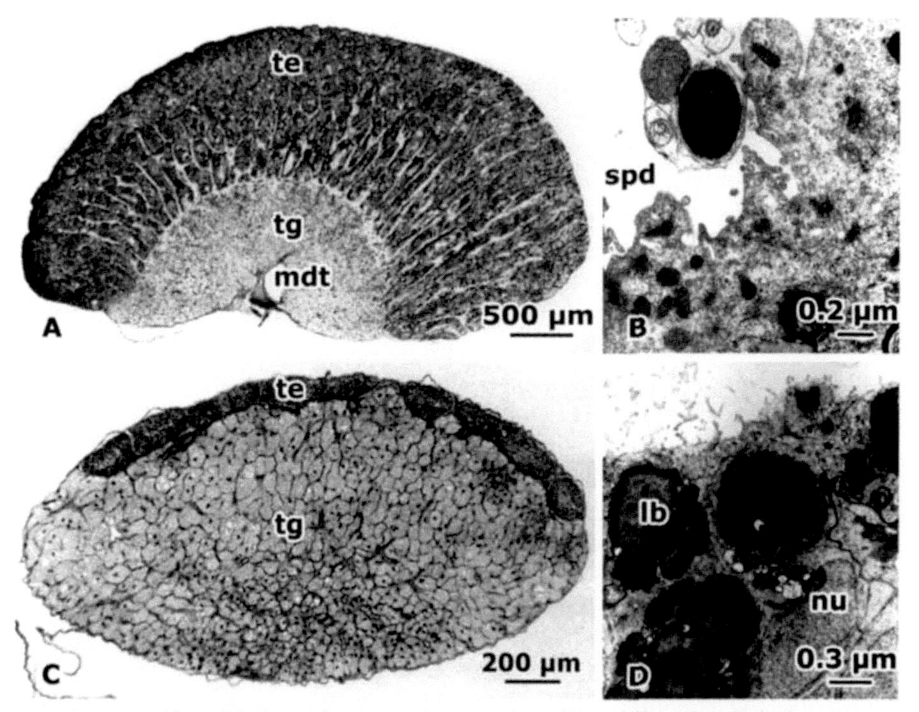

Fig. 5.9 Testis-testicular gland complex of the Blennidae. **A**. *Salaria pavo*, a species with a small testicular gland (tg) and a well developed testis (te). Paraffin sections. mdt, main duct of the testicular gland. **B**. *Microlipophrys dalamtinus*, a species with a voluminous testicular gland and a reduced testis. Paraffin section. **C**. Testicular gland cell of *Salaria pavo* during spawning. In the duct a spermatid is visible. Transmission electron micrograph. **D**. Testicular gland cell of *Salaria pavo* during post-spawning. The testicular gland cells are occupied by lamellar bodies (lb); nu, nucleus. Transmission electron micrographs.

Therefore, the testicular gland has functions (Fig. 5.7) in (a) nutrition of germ cells with glycogen, (b) secretion of sialomucins to the seminal fluid, (c) regulation of differentiation processes of spermatids, (d) phagocytosis of remaining spermatids and lipid storage during postspawning. The secreted sialomucins could increase the viscosity of the seminal fluid. As eggs of blenniid fish adhere to the substrate with the micropylar region (Patzner 1984), the mucosubstances of the testicular gland and of the spermatic ducts are regarded to function in the formation of a layer of sperm on which the eggs are deposited. This hypothesis was established in recent studies in Gobiidae (Ota *et al.* 1996; Marconato *et al.* 1996). Gobiidae have seminal vesicles which produce mucus of very similar composition as the testicular glands and spermatic ducts of the Blenniidae (Lahnsteiner *et al.* 1992b). The fact that in blenniid fishes the sexually mature testes contain no spermatozoa but differentiation processes of spermatids occur in the testicular gland and spermatic ducts is a very special feature and is a further case of semicystic spermiogenesis. Also Callichthyidae (Mattei *et al.* 1993), some Gobiidae (Mazzoldi *et al.* 2005) and Scorpaenidae (Muñoz *et al.* 2002) are characterized by a semicystic mode of spermiogenesis. This type of spermatogenesis leads to a reduced number of simultaneously mature gametes compared to cystic spermatogenesis and it has been up to now observed only in external fertilizers performing male parental care. In some of these species, low female fecundity, the absence of sperm competition and the occurrence of fertilization in a relatively enclosed environment indicate that males need only few sperm to successfully fertilize the limited number of eggs laid by the females (Marconato and Rasotto 1993; Mazzoldi *et al.* 2005). Therefore, males show a very low investment in gonads (Stockley *et al.* 1997; Kohda *et al.* 2002). In Blenniidae this is seen in the *Microlipophrys* species where the testicular gland is voluminous and the testis is much reduced in comparison to other blenniid species. The spermatids which are released into the testicular gland have an earlier stage of maturation than spermatids of species with more voluminous testes, which is indicated by incomplete nuclear condensation, an undifferentiated flagellar complex and the presence of SER.

The testicular glands of Tripterygiidae have been studied in *Tripterygion tripteronotus* (= *nasus*) by light microscopy (Eggert 1931) and electron microscopy (de Jonge *et al.* 1989; Lahnsteiner *et al.* 1990). De Jonge *et al.* (1989) investigated the testicular gland of *Tripterygion delaisi* histochemically. Rasotto (1995) gave brief descriptions of the testicular organs of the *Enneanectes boelkei* and *Enneanectes pectoralis*. The structure of the testicular gland is slightly different from that of the Blenniidae. Radiating seminiferous tubules drain into an efferent duct system, situated against the testicular gland and penetrating it at several sites. The glandular mass consists of radially oriented strands of polyhedral cells surrounded by capillaries or sinusoids. The glandular cells contain smooth endoplasmic reticulum, numerous lipid droplets and mitochondria with large cristae. No steroidogenic activity was found within the testicular gland (Rasotto 1995).

In the Labrisomidae the testicular glands were investigated in *Labrisomus bucciferus, Labrisomus haitensis, Labrisomus nuchipinnis, Malacoctenus macropus,*

Malacoctenus triangulatus and *Paraclinus nigripinnis* by Rasotto (1995) and in *Labrisomus xanti* and *Paraclinus sini* by Patzner and Hastings (unpublished data). The testis-testicular gland complex is similar to that of the Blenniidae. The gland is very large in these species, its volume ranging between less or up to about 100 % (in *Labrisomus bucciferus, L. xanti, P. sini*) and more than 200 % (in *L. haitensis, Malacoctenus versicolor*) that of the testis.

Concerning the Clinidae, the testis-testicular gland complex has been studied in *Clinitrachus* (=*Cristiceps*, =*Clinus*) *argentatus* (Guitel 1893; Eggert 1931; Patzner unpublished data). In gross morphology the testicular gland of *Clinitrachus* resembles that of Blenniidae.

Spermatic ducts: The spermatic ducts of Blenniidae were investigated in *Aidablennius sphynx, Microlipophrys* (= *Lipophrys*) *adriaticus, Microlipophrys dalmatinus, Parablennius incognitus* and *Salaria pavo* (Lahnsteiner and Patzner 1990). They have modifications in comparison to other species with external fertilization. As the testicular glands the spermatic ducts differ considerabely in size. They are smallest in species where also the testicular gland is small (e.g. *Salaria pavo*: spermatic duct volume is about 0.02 % of the body volume) and largest and vesicular-like in species with big testicular glands (e.g. *Microlipophrys adriaticus* and *Microlipophrys dalmatinus*: 0.3 % of the body volume) (Lahnsteiner and Patzner 1990; Richtarski and Patzner 2000). Species with the largest testicular glands and spermatic ducts have the smallest testes and spermatids are released from the germinal cysts in an early differentiation stage.

The spermatic duct consists of (a) a peritoneum, (b) a transversal orientated collagen fiber layer intermingled with smooth muscle cells and (c) a one layered cylindrical epithelium (Lahnsteiner and Patzner 1990) (Fig. 5.10A). The epithelial cells of the spermatic duct have an intensively folded basal lamina and are connected by complex membrane specializations. They have a prominent microvilli layer. Their nuclei are located in the middle of the cells (Fig. 5.10C).

During the prespawning period the height of the epithelium of the spermatic duct of *Salaria pavo* increases gradually (Lahnsteiner and Patzner 1990). The nuclei have a spherical or cubical shape and an eccentric, electron-dense nucleolus. Large amounts of tubular SER are found throughout the cytoplasm. RER is less abundant and restricted to the apical cell portions. Golgi fields are frequent; they are closely associated with electron-dense material and with secretory vesicles. Microfilaments are observed preferably in the apical cell portions. During prespawning cytoplasmic areas differentiate which are electron-lucent or slightly flocculent in structure (Fig. 5.10B). They are surrounded by SER and mitochondria. Lamellar bodies consisting of concentric layers of membranes become visible in the cytoplasm, preferably in close vicinity to the described electron-lucent portions of the cytoplasm (Fig. 5.10B). Later during prespawning, the described cytoplasmic areas become filled with glycogen particles (Lahnsteiner and Patzner 1990).

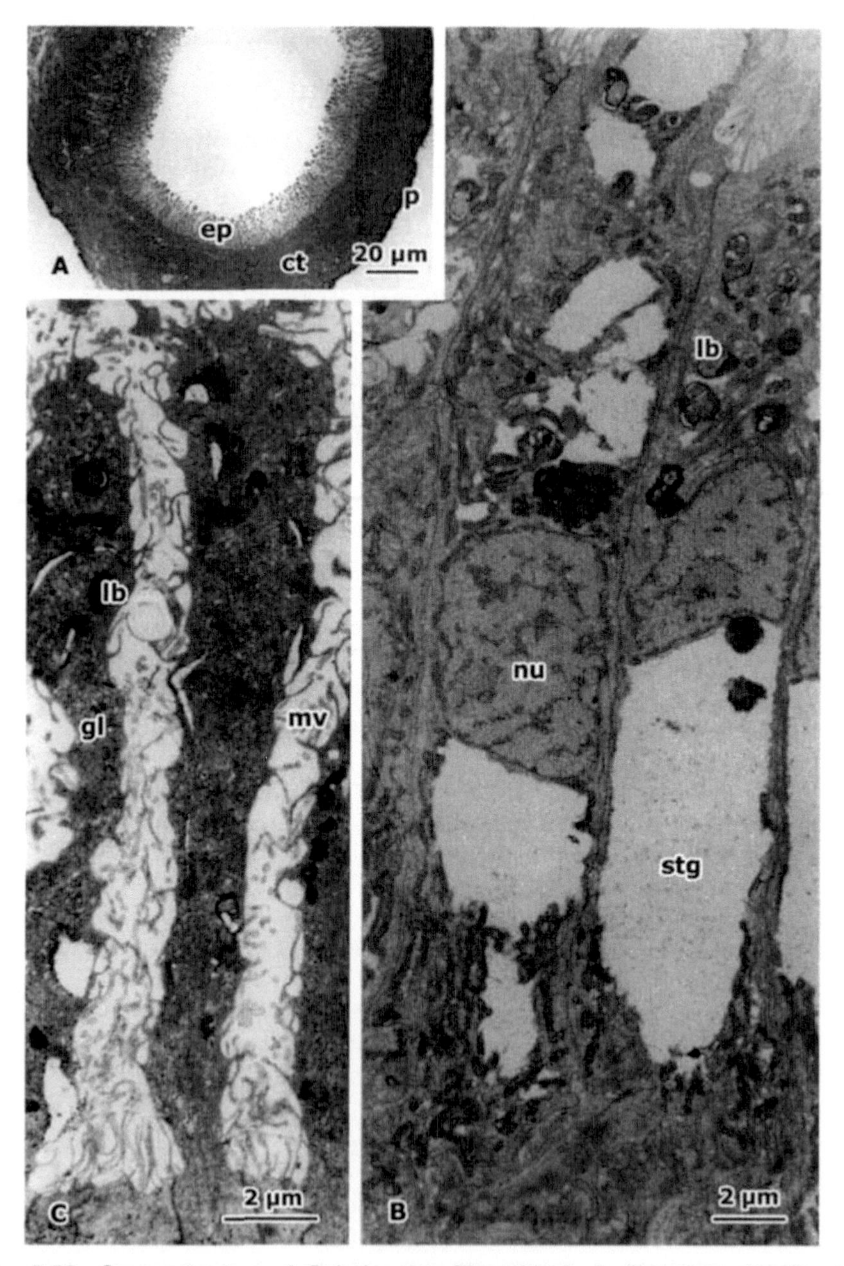

Fig. 5.10 Spermatic duct of *Salaria pavo* (Blennidae). **A**. Overview showing the peritoneum (p), the conncective tissue layer (ct) and the unfolded epithelium (ep) with large apical processes. Paraffin section. **B**. Epithelial cell of the spermatic duct during prespawning. mi mitochondria; stg sites of glycogen synthesis; nu nucleus; nul nucleolus; lb lamellar body. Transmission electron micrograph. **C**. Apical processes of an epithelial cell with lamellar bodies (lb) and glycogen granula (gl) during spawning. mv, microvilli. Transmission electron micrographs.

During the spawning period the spermatic ducts contain differentiating spermatids and spermatozoa (Lahnsteiner and Patzner 1990). The epithelial cells have a high secretory activity and the secretion mode is apocrine. The cells show a differentiation into four distinct zones by now (Lahnsteiner and Patzner 1990). (a) The basal cell portion contains an extensive amount of tubular ser, an abundance of mitochondria and large accumulations of glycogen. (b) In the nuclear region a spherical or cuboidal nucleus is located. (c) The supranuclear region is characterized by smooth and RER, Golgi fields, and accumulations of glycogen particles, preferably in close vicinity to lamellar bodies. (d) The apical portions of the epithelial cells have processes which are covered with microvilli (Fig. 5.10C). These regions contain an abundance of glycogen and some lamellar bodies but only few cell organelles (Fig. 5.10C). During the spawning period the epithelium stains positively with the histochemical staining for periodic acid Schiff and the method of Best's glycogen detection indicative of glycogen and with the staining for alcianblue at pH 1.0 and at 0.4 to 0.8 M electrolyte concentrations indicative of sulfated mucins. The enzyme histochemical lead salt method for ATPase activity is also positive in the epithelial cells,. The enzyme-histochemical reactions for 3ß-steroid dehydrogenase, glucose-6-phosphate dehydrogenase and uridine-diphosphoglucose dehydrogenase are negative (Lahnsteiner and Patzner 1990).

During postspawning extensive regressive processes occur in the epithelium of the spermatic duct whereby remaining secretion, spermatids and spermatozoa are reabsorbed similar as described for the testicular gland (Lahnsteiner and Patzner 1990).

These morphological and histochemical results indicate that the spermatic ducts of the Blenniidae have very similar functions as the testicular glands, namely in the nutrition of germ cells with glycogen, in secretion of sulfomucins to the seminal fluid, in regulation of differentiation processes of spermatids, and during postspawning in phagocytosis of remaining spermatids, spermatozoa and secretion (Lahnsteiner and Patzner 1990) (Fig. 5.7). Additionally, the spermatic ducts are also storage depots for spermatozoa. The occurrence of ATPase activity indicates that the spermatic duct epithelium is involved in the regulation of the ionic composition of the seminal fluid (Lahnsteiner and Patzner 1990). This may be also true for the testicular gland, however, ATPase activity has not been investigated for this organ. The function of the spermatic ducts differs from other teleost species with external fertilization (Cyprinidae, Lahnsteiner *et al.* 1994; Salmonidae, Lahnsteiner *et al.* 1993a; Esocidae, Lahnsteiner *et al.* 1993b; Sparidae, Lahnsteiner 2003; Mullidae, Lahnsteiner 2003; Scaridae, Lahnsteiner 2003; Trachinidae, Lahnsteiner 2003; Labridae, Rasotto and Shapiro 1998; Lahnsteiner 2003; and Synodontidae, Lahnsteiner 2003) as the spermatic duct epithelium does not produce steroid glucoronides and lytic enzymes (acid phosphatase, aminopeptidase) durig spawning.

For the Tripterygiidae (de Jonge *et al.* 1989; Rasotto 1995), Labrisomidae (Rasotto 1995; Patzner and Hastings unpublished data), and Clinidae (Eggert 1931) no data on the fine structure and histochemistry and function of the spermatic duct are available. In the Tripterygiidae the efferent ducts passing the testicular gland form a longitudinal running glandular main duct, located ventrally and centrally in the testicular gland. At the posterior end of the gland the sperm duct is dilated and leaves the gland caudally. The spermatic ducts are slightly enlarged and fuse together before entering the genital papilla (Rasotto 1995). In the labrisomid *Paraclinus sini* the spermatic duct is enlarged but rather short (Patzner 2007). In *Labrisomus* spp. and *Malacoctenus* spp. the ducts are longer and convoluted. In the Clinidae the spermatic ducts are not enlarged (Patzner 2007).

Testicular blind pouches: The testicular blind pouches are paired evaginations of the distal portion of the spermatic ducts which extend rostrally to about half the length of the testis (Eggert 1931) (Fig. 5.7). Their wall consists of three layers, an inner monolayered epithelium, a transversally orientated connective tissue layer with smooth muscle cells and a peritoneum (Fig. 5.11A). With the exception of *Salaria pavo*, the epithelium forms a smooth luminal surface in all investigated species of Blenniidae (Eggert 1931; Blüm 1972; Richtarski and Patzner 2000). In *S. pavo* the epithelium exhibits prominent longitudinal and transverse folds which reach a height of up to 400 μm (Fig. 5.11A,D). Transverse folds always fuse with the next longitudinal fold and therefore develop a complicated system of protrusions (Lahnsteiner *et al.* 1993d). The monolayered epithelium of the testicular blind pouches has a height of circa 10 μm and is formed by cylindrical cells covered with microvilli (Lahnsteiner *et al.* 1993d) (Fig. 5.11B). At the core of each microvillus, bundles of microfilaments are located which extend towards the apical regions of the cytoplasm. The epithelial cells are linked by membrane specializations. The irregularly shaped nuclei are situated in the basal portions of the cells , consist of heterogenous chromatin material and have an eccentric electron-dense nucleolus (Fig. 5.11B). Smooth endoplasmatic reticulum is found only rarely and preferably in association to lipid vacuoles. Elongated mitochondria with an electron-dense matrix and tubular, parallel cristae are frequent (Fig. 5.11C). Golgi apparatuses are preferably situated in the apical cell portions (Lahnsteiner *et al.* 1993d) (Fig. 5.11C).

During spawning, the testicular blind pouches are not storage organs for spermatozoa as spermatozoa were never observed within the pouches (Lahnsteiner *et al.* 1993d). The occurrence of an abundance of tubular RER, of mitochondria with tubular, parallelly arranged cristae and of lipid vacuoles as well as the enzyme histochemical detection of activities of 3β-steroid dehydrogenase, glucose-6-phosphate dehydrogenase, and uridine-diphosphoglucose dehydrogenase indicate a function in steroid glucuronides synthesis (Lahnsteiner *et al.* 1993d). The blind pouches therefore could have functions in pheromone synthesis which is also ascertained by observations of Laumen *et*

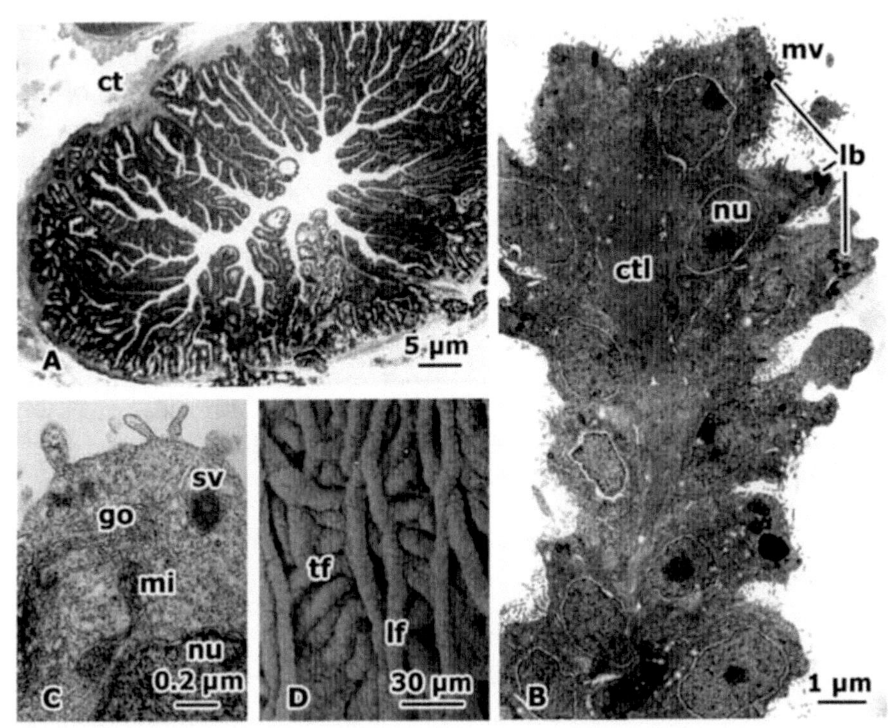

Fig. 5.11 Testicular blind pouches of *Salaria pavo* (Blenniidae). **A**. Cross section showing the intensely folded epithelium. ct, connective tissue layer. Paraffin section. **B**. Overview of the epithelium. ctl, connective tissue layer; lb, lamellar body; nu, nucleus. Transmission electron micrograph. **C**. Apical portion of an epithelial cell with Golgi apparatus (go) and mitochondria (mi). nu, nucleus; sv, secretory vesicle. Transmission electron micrograph. **D**. Arrangement of epithelial fold. Lf, longitudinal fold; tf, transversal fold. Scanning electron micrograph.

al. (1974) that mature male *Salaria pavo* are able to attract females by releasing pheromones into the water. Pheromones are also produced by other accessory reproductive organs like the steroid glands of *Gobius niger* (= *jozo*) (Colombo *et al*. 1977) and the seminal vesicles of *Clarias gariepinus* (Resink *et al*. 1989).

During spawning the epithelium of the testicular blind pouches has a secretory activity (Lahnsteiner *et al*. 1993d): The Golgi apparatuses release secretory vesicles with electron-dense contents which join to form larger secretory vacuoles (Fig. 5.11C). The secretory vacuoles are transferred out of the cells by exocytosis. In some areas apocrine secretion is also observed. Apical portions of the epithelium of the testicular blind pouches containing the secretory products locally react positively with periodic acid Schiff and the alcian blue staining at pH 2.5 and at critical electrolyte concentrations from 0.1 to 0.4 mM. This suggests the synthesis of sialomucin in the blind pouches of the Blenniidae (Lahnsteiner *et al*. 1993d). However, the secretory vesicles often are not secreted but eliminated by autophagocytosis: The Golgi apparatuses also produce primary lysosomes containing fine osmiophilic granula. They fuse with secretory vacuoles and form secondary lysosomes. The contents of these secondary lysosomes change into lamellar bodies consisting of concentric layers of membranes. By this process large lamellar bodies up to 3 µm originate in the epithelial cells. Therefore, mucin secretion of the blind pouches is very low in comparison to the testicular glands and spermatic ducts and is of minor importance (Lahnsteiner *et al.* 1993d).

The epithelial cells of the testicular blind pouches contain different lytic enzymes as indicated by enzyme-histochemical stainings, i.e. acid phosphatase, ß-glucuronidase and aminopeptidase (Lahnsteiner *et al.* 1993d). It might be possible that the lytic enzymes are added to the seminal fluid and that they have a role in the elimination of necrotic sperm cells from the spermatic ducts. They may also play a role in liberation of spermatozoa from the mucus-semen layer on which the eggs are deposited. On the other hand there was no evidence that lysosomes are secreted and activities of acid phosphatase, aminopeptidase and ß-glucuronidase could not be detected in the lumen of the blind pouches of blennies by histochemical methods (Lahnsteiner *et al.* 1993d).

Synthesis of steroid glucuronides and production of lytic enzymes are functions which are typical for the spermatic duct epithelium of many teleost fish species with external fertilization (Lahnsteiner *et al.* 1993a, 1993b, 1994; Lahnsteiner 2003). Therefore, it might be possible that in the Blenniidae the testicular blind pouch overtakes functions of the spermatic duct epithelium and the spermatic duct specializes on the functions described above.

During the postspawning period necrotic processes occur in the epithelium (Lahnsteiner *et al.* 1993d). In the necrotic epithelial cells the organelles degenerate and the cell size decreases. Then the necrotic cell material becomes detached from the epithelium and accumulates either at the basal lamina or in the lumina of the testicular blind pouches. It is eliminated by macrophages which invade the testicular blind pouches via the circulatory system (Lahnsteiner *et al.* 1993d). Acid phosphatase, β-glucuronidase and aminopeptidase activity are strongly positive during this phase. Also the staining of vacuoles for lipids (Sudan black B, Nile blue) is most intensive in regions where necrotic processes occur. Production of steroid glucuronides is no longer detectable and staining for mucins is negative, too (Lahnsteiner *et al.* 1993d).

5.5.4.2 Efferent duct system of the Gobiidae as an example for species with seminal vesicles

The efferent duct system has been studied in the grass goby, *Zosterisessor ophiocephalus* (Lahnsteiner *et al.* 1992b; Loidl 1992) and is described in the present chapter to exemplify the structural organization and function of a seminal vesicle and of the relate efferent duct system. Fine structural studies on the seminal vesicle have also been conducted on other (non-gobiid) species as e.g. *Clarias gariepinus* (Van den Hurk *et al.* 1987; Fishelson *et al.* 1994). The

Grass goby has paired, compact testes (Fig. 5.5). The seminiferous tubules of the testes vent into the testicular main ducts at the ventral side of the testis which continue into the spermatic ducts. Distally both spermatic ducts coelascence to form an unpaired portion which empties on the genital papilla (Fig. 5.5).

Steroid gland: Some gobiid species (not *Zosterisessor ophiocephalus*) have a gland situated adjacent to the testis (Figs. 5.6, 5.8A). However, this gland is not homologous to the testicular gland described in the Blenniidae (Seiwald and Patzner 1989), therefore we term it the 'steroid gland'. In *Gobius niger* the testicular main duct is situated between the testis and the steroid gland (Figs. 5.5, 5.8B). The function of the steroid gland is exclusively endocrine (Colombo *et al.* 1977; Seiwald and Patzner 1989). Characteristics of the gland cells of this species are well developed smooth endoplasmatic reticulum and tubulovesicular or paracrystalline mitochondria. The stainings for 3ß-steroid dehydrogenase, glucose-6-phosphate dehydrogenase, and uridine-diphosphoglucose dehydrogenase give strong positive results in the whole steroid gland, indicating the presence of steroids and steroid glucuronides (Colombo *et al.* 1977; Seiwald and Patzner 1989).

Seminal vesicles: In the Gobiidae the seminal vesicles (= sperm duct glands) are paired lobular organs, which empty into the spermatic duct (Fig. 5.5) without sphincter muscles (Patzner *et al.* 1991; Lahnsteiner *et al.* 1992b). They are separated from the testicular tissue. Seminal vesicles show cyclic changes during the reproductive phase with a high activity during prespawning and spawning, a regression phase during post spawning and a non-active phase during interspawning (Cinquetti *et al.* 1990; Patzner *et al.* 1991). They consist of multi-chambered lobes (Fig. 5.12A) which empty into the spermatic ducts and are involved in secretion. The seminal vesicles of the Grass goby have been investigated by histological, fine structural and histochemical methods (Patzner *et al.* 1991; Lahnsteiner *et al.* 1992b). They are formed of chambers which are interconnected with each other by wide openings (Fig. 5.12A) and consist of a monolayered epithelium underlined by a basal lamina (Fig. 5.12A, inset). The interstitium of the seminal vesicles consists of a thin layer of connective tissue fibers (Fig. 5.12A, inset) which contains blood vessels and nervous fibers.

The seminal vesicle epithelium is composed of cuboidal cells which sometimes are covered with microvilli. Adjacent epithelial cells overlap in their apical portions by membrane specializations, the zonula occludens, the macula occludens and the zonula adherens. The nuclei are situated in the basal or middle portions of the cells (Fig. 5.12B). They consist of heterogenous chromatin material and have an eccentric electron-dense nucleolus. The cytoplasm contains an abundance of tubular, RER (Lahnsteiner *et al.* 1992b) (Fig. 5.12C). Its tubules are arranged parallel to each other, and are densely coated with ribosomes. The mitochondria are elongated, their matrix is electron-dense and the tubular cristae are irregularly arranged. As most

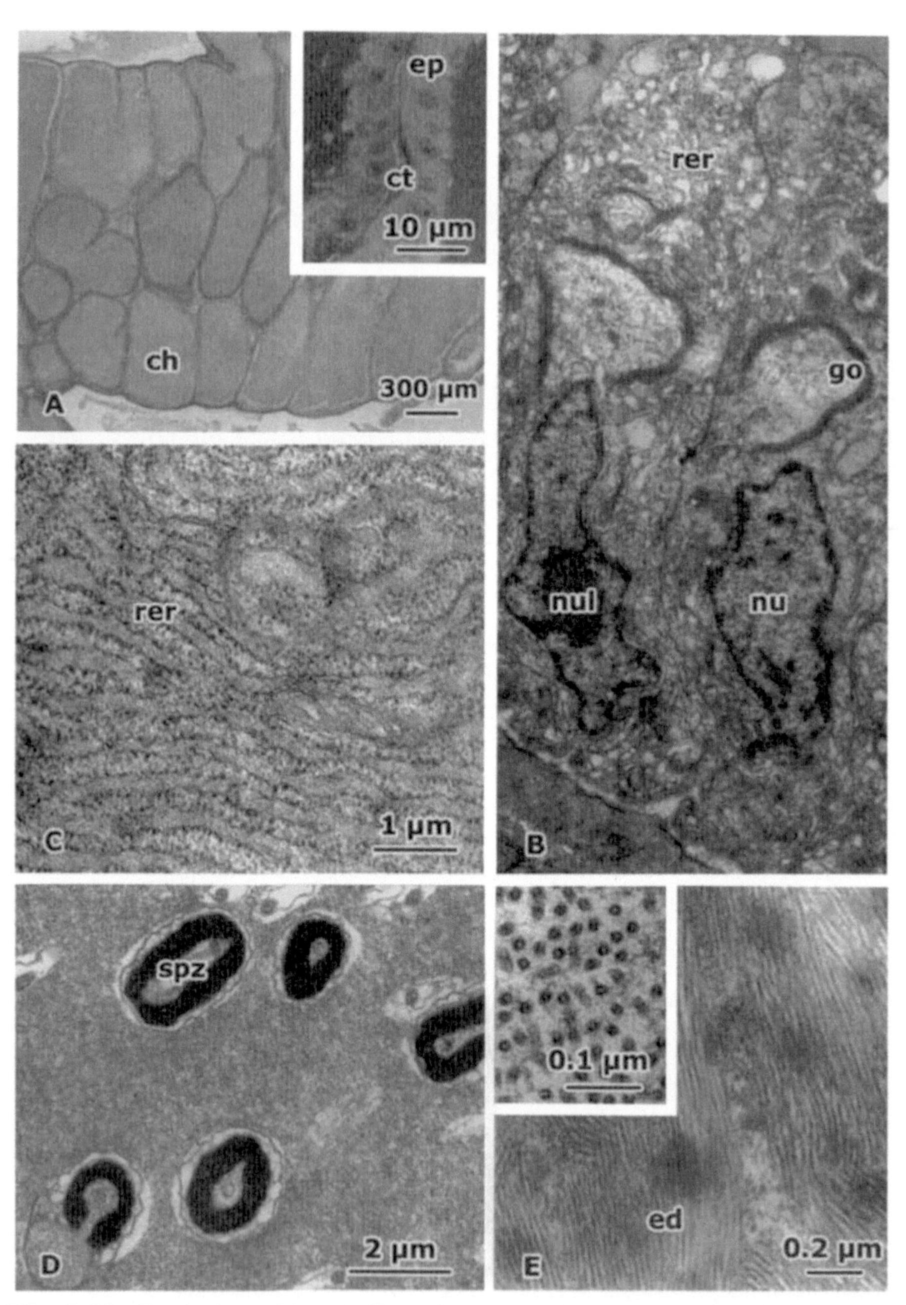

Fig. 5.12 Seminal vesicle of *Zosterisessor ophiocephalus* (Gobiidae) during spawning. **A**. Overview showing the seminal vesicle chambers (ch) filled with secretion. Inset: Detail of the epithelium (ep). ct, connective tissue layer. Paraffin section. **B**. Epithelial cell showing the nucleus (nu) with the nucleolus (nul) and the

Fig. 5.12 Contd. ...

prominent feature a seminal vesicle cell contains one or two very large-sized Golgi complexes (Fig. 5.12B). In addition, the cells contain numerous smaller Golgi fields (Lahnsteiner *et al.* 1992b).

During spawning the seminal vesicles have their maximal extensions and their chambers are completely filled with secretion. The Golgi complexes release vesicles containing fibrous structures. These vesicles join to form larger vacuoles which are released into the lumina of the seminal vesicle chambers by exocytosis (Lahnsteiner *et al.* 1992b). Also electron-dense granula or electron-dense material which is not enclosed in vesicles but forms irregularly arranged spots is produced by the Golgi cisternae. Apically this electron-dense material is transferred out of the cells through the plasmalemma. In some areas also an apocrine secretion is observed (Lahnsteiner *et al.* 1992b).

The lumina of the seminal vesicles contain a secretion consisting of a fine granular matrix in which bundles of fibrils and electron-dense material are intermingled (Fig. 5.12E). The secretion and the apical portions of the epithelium show an intensive staining with periodic acid Schiff, alcian blue at pH 2.5 and at electrolyte concentrations of 0.1 to 0.2 M. The whole epithelium as well as the secretion reacts positively to mercury bromophenol blue (Lahnsteiner *et al.* 1992b). The staining with Sudan black B and Nile blue for lipid and phospholipid are both negative. These results demonstrate that the secretion consists of sialomucins and of proteins. As in the seminal vesicles the histochemical stainings for mucin and protein are found in precisely the same areas, it seems obvious that sialomucin is conjugated with the protein to form a sialoglycoprotein (Lahnsteiner *et al.* 1992b). Secretion of mucins or mucoproteins is typical for the seminial vesicles of all investigated species (Chowdhury and Joy 2007). Biochemical analysis of the seminal vesicle secretion of *Heteropneustes fossilis* (Heteropneustidae) (Nayyar and Sundararaj 1970; Chowdhury and Joy 2001), and *Clarias batrachus* (Clariidae) (Mansour *et al.* 2004) established the histochemical results obtained for *Zosterisessor ophiocephalus*. In *Z. ophiocephalus* the secretion of the seminal vesicles is released together with spermatozoa (Fig. 5.12D). The sialoglycoproteins of the seminal vesicles probably cause the very viscous and sticky consistency of the semen (Giulianini *et al.* 2001). When the semen of *Z. ophiocephalus* is suspended in seawater the spermatozoa do not disperse but remain enclosed

Fig. 5.12 Contd. ...

prominent Golgi field (go). The apical cell portion contains extensive amounts or rough endoplasmatic reticulum (rer) and mitochondria (mi). Transmission electron micrograph. **C**. Apical cell portion with rough endoplasmatic reticulum (rer). Transmission electron micrograph. **D**. Spermatozoa (spz) embedded in seminal vesicle secretion in the basal portions of the seminal vesicle. Transmission electron micrograph. **E**. Detail of the seminal vesicle secretion showing its fibrilous composition (ed). Inset: Cross section of the fibres. ed, electron-dense material. Transmission electron micrographs.

in clumps together with the secretion of the seminal vesicles. As male Z. *ophiocephalus* rub their urogenital papilla over a sea grass root before the females deposit there the eggs (Miller 1984), it is suggested that sperm layers are glued with the sticky seminal vesicle secretion on the plants. Then the eggs are deposited with the micropyle towards the sperm layers on the roots of the sea grass. Several gobies lay sperm in the form of sperm trails that slowly release sperm into the water for several hours (Ota *et al.* 1996; Marconato *et al.* 1996) and whose longevity depends on the trail's mucin content (Mazzoldi *et al.* 2005). Females can take a long time to lay eggs (Scaggiante *et al.* 1999) and males release sperm trails on the nest surface immediately before and during egg deposition (Marconato *et al.* 1996; Scaggiante *et al.* 1999). The intermittent gamete release enables parental males to spend part of the mating time patrolling nest entrances against incursions by predators or opportunistic males trying to sneak fertilizations (Mazzoldi *et al.* 2007).

Seminal vesicles of *Zosterisessor ophiocephalus* (a) are not storage reservoir for spermatozoa as only some sperm cells are found in their basal portions (Lahnsteiner *et al.* 1992b). However, in other species seminal vesicles have also functions in sperm storage as the whole seminal vesicle contains spermatozoa embedded in the seminal vesicle secretion (Loir *et al.* 1989). (b) They produce no steroids as their fine structural characteristics and enzyme histochemical features are atypical of steroid producing cells (Lahnsteiner *et al.* 1992b). In contrast, in *Clarias gariepinus* and *Heteropneustis fossilis*, the seminal vesicles secrete glucuronated steroids, which act as olfactory stimuli in females (Resink *et al.* 1989). (c) The seminal vesicles do not contribute to the regulation of the ionic composition of the seminal fluid, as they have no ATPase activity (Lahnsteiner *et al.* 1992b). (d) They do not play a role as reservoir of enzymes commonly produced by the spermatic duct epithelium of many teleost fish such as alkaline and acid phosphatase, aminopeptidase, 5-nuclueosidase and ß-glucuronidase (Lahnsteiner *et al.* 1992b). (e) The absence of lipids, monosaccharides, and of a glucose-6-phosphatase activity in the seminal vesicles negates a function in nutrition of spermatozoa (Lahnsteiner *et al.* 1992b).

During the postspawning period the secretory activity of the seminal vesicle epithelium stops and the size of the seminal vesicles decreases significantly owing to degenerative processes whereby either apical cell portions containing the secretion or complete cells and groups of cells become necrotic cells and are detached from the epithelium. The necrotic cell material is eliminated by macrophages, which invade the seminal vesicles during postspawning via the circulatory system. The lumina of the seminal vesicles decrease drastically or collapse completely. During postspawning the epithelium has an intensive acid phosphatase and ß-glucuronidase activity demonstrating the occurrence of lysomatic processes. The lumina mainly contain lipids and phospholipids as they stain positively with Sudan black B and Nile blue (Lahnsteiner *et al.* 1992b).

In comparison with the seminal vesicle of Grass goby the seminal vesicle of the African catfish, *Clarias gariepinus*, consists of 36 – 44 fingerlike lobes built up of tubules in which a fluid is secreted containing acid polysaccharides, acid-, neutral- and basic proteins, and phospholipids (Van den Hurk *et al.* 1987). In this fluid sperm cells are stored. The seminal vesicle fluid is secreted by the epithelium lining the tubules. The tubules in the proximal part of the lobes are predominantly lined by a simple cylindrical and those of the distal part by a simple squamous epithelium. Interstitial cells between the tubules have enzyme-histochemical and ultrastructural features indicative of steroid biosynthesis. The most rostral seminal vesicle lobes and the most caudal testicular efferent tubules form a network of tubules that opens at the point where the paired parts of the sperm ducts fuse with each other. The tubules of most seminal vesicle lobes, however, form a complex system that fuses with the unpaired part of the sperm duct.

In many silurid species the relation between testis and seminal vesicle is very complex. For instance, for *Megalechis* (= *Hoplosternum*) sp. (Callichthyidae), the male genital tract consists of two elongated slightly lobated and flattened testes which join posteriorly in a segment which goes towards the genital papilla. The posterior part of this segment is surrounded by a mass composed of several packed seminal vesicles. The tract ends in a retractile external appendix. The anterior testicular lobes consist entirely of germinal tissue. Further posteriorly, the tubules become progressively devoid of germ cells and may be considered as equivalent to efferent ducts (Loir *et al.* 1989). At the end of the anterior part there is no more testicular tissue but only tubules, the basal part of which is similar to efferent ducts and the apical part of which secretes homogeneous material of a consistency similar to that of the seminal vesicle tubules. In the posterior part of the tract, seminal vesicles are tightly packed around a single spermatic duct. In this part, tubules are wide and filled with a secretory material. Spermatozoa located in the basal part of the tubules and in the spermatic duct are embedded in the secretory substance (Loir *et al.* 1989).

Spermatic duct: According to the description of Loidl (1992) the testicular efferent duct system of *Zosterisessor. ophiocephalus* is similar as in other teleosts. The seminiferous tubules vent into the testicular main ducts which continue into the spermatic ducts. Distally, the spermatic duct forms an unpaired portion which empties into the urogenital papilla. The epithelium is one layered and cylindrical. It is not folded. The epithelium of the testicular main ducts and of the spermatic ducts is similar with the exception that the cells of the testicular main duct are not covered with microvilli. The epithelial cells contain irregularly shaped nuclei with a nucleolus. RER and mitochondria with tubularvesicular cristae are observed in the cytoplasma. Golgi apparatuses, which produce electron-dense secretory vesicles, are also frequent. However, the epithelial cells do not contain SER or lipid vacuoles. The secretion mode is apocrine. During spawning the spermatic duct is filled with spermatozoa which are embedded in the secretion of the seminal vesicles (Loidl 1992).

Owing to the absence of SER and lipid vacuoles and the lack of of 3ß-steroid dehydrogenase and glucose-6-phosphate dehydrogenase activity, the spermatic duct epithelium does not produce steroids. The secretion of the spermatic ducts stains positively for polysaccharides and proteins. However, there exists no information if the polysaccharides are mucins as the alcian blue staining was not performed. Enzyme histochemically a positive reaction for Na^+/K^+ ATPase was detected indicating functions in the ionic regulation of the seminal fluid. Storage of spermatozoa together with seminal vesicle secretion is considered to be the main function of the spermatic ducts of *Zosterisessor ophiocephalus*, polysacharides and proteins which are secreted by the epithelium are considered to have functions in maintaining the viability of spermatozoa (Loidl 1992).

5.6 CHAPTER SUMMARY

This chapter has summarized the recent data on the efferent duct system of the male gonads of the Euteleostomi (=Osteichthyes). It describes the anatomy, histology, and fine structure of the efferent testicular duct system (testicular main ducts, spermatic ducts) and informs about the functions of these organs. The adaptations of the efferent duct system developed in external and internal fertilizing fish and in simultaneous hermaphrodites are described. Finally, the anatomy, histology, fine structure, and function of the accessory glands of the efferent duct system, the testicular glands, steroid glands, seminal vesicles, and testicular blind pouches is described.

5.7 ACKNOWLEDGEMENTS

Parts of our cited studies were funded by the Austrian Fonds zur Förderung der wissenschaftlichen Forschung (FWF) and by the Austrian Bundesministerium für Land-und Forstwirtschaft, Umwelt und Wasserwirtschaft.

5.8 LITERATURE CITED

Abd-el-Aziz, S. H. and Ramadan, A. A. 1990. Sexuality and hermaphroditism in fishes. I. Synchronous functional hermaphroditism in the serranid fish *Serranus scriba* L. Folia Morphologica 38: 86-103.

Besseau, L. and Faliex, E. 1994. Resorption of unemitted gametes in *Lithognathus mormyrus* (Sparidae, Teleostei): a possible synergic action of somatic and immune cells. Cell and Tissue Research 276: 123-132.

Billard, R. 1986. Spermatogenesis and spermatology of some teleost fish species. Reproduction, Nutrition, Developement 26: 877-920.

Billard, R. and Takashima, F. 1983. Resorbtion of spermatozoa in the spermatic duct of the rainbow trout during the postspawning period. Bulletin of the Japanese Society of Sciences of Fish 49: 387-392.

Blüm, V. 1972. The influence of ovine follicle stimulating hormone (FSH) and luteinizing hormon (LH) on the male reproductive system of the Mediterranean blenniid fish *Blennius sphynx* (Valenciennes). Journal of Experimental Zoology 181: 203-216.

Blüm V. 1985. *Vergleichende Reproduktionsbiologie der Wirbeltiere.* Springer Verlag, Berlin, Heidelberg, 387 pp.

Bortone, S. A. 1977. Gonad morphology of the hermaphroditic fish *Diplectrum pacificum* (Serranidae). Copeia 1977: 448-453.

Britz, R. and Bartsch, P. 1998. On the reproduction and early development of *Erpetoichthys calabaricus, Polypterus senegalus,* and *P. ornatipinnis* (Actinopterygii: Polypteridae). Ichthyological Exploration of Freshwaters 9: 325-334.

Burns, J. R. 1991. Testis and gonopodium development in *Anableps Dowi* (Pisces: Anablepidae) correlated with pituitary gonadotropic zone area. Journal of Morphology 210: 45-53.

Burns, J. R., Weitzman, S. H., Grier, H. J. and Menezes, N. A. 1995. Internal fertilization, testis and sperm morphology in glandulocaudinae fishes (Teleostei: Characidae: Glandulocaudinae). Journal of Morphology 224: 131-145.

Cinquetti, R., Rinaldi, L., Azzoni, C. and de Ascentiis, M. 1990. Cycling changes in the morphology of the sperm duct glands of *Padogobius martensi* (Pisces, Gobiidae). Bollettino di Zoologica 57: 295-302.

Chowdhury, I. and Joy, K. P. 2001. Seminal vesicle and testis secretions in *Heteropneustes fossilis* (Bloch): composition and effects on sperm motility and fertilization. Aquaculture 193: 355-371.

Chowdhury, I. and Joy, K. P. 2007. Seminal vesicle and its role in the reproduction of teleosts. Fish Physiology and Biochemistry 33: 383-398.

Clark, E. 1959. Functional hermaphroditism and self-fertilization in a serranid fish. Science 129: 215-216.

Colombo, L., Belvedere, P. C. and Pilati, A. 1977. Biosynthesis of free and conjugated 5ß-reduced androgens by the testis of the black goby, *Gobius jozo* L. Bollettino di Zoologica 44: 131-134.

Constanz, G. D. 1989. Reproductive biology of poeciliid fishes. Pp. 33–68. In G. K. Meffe and F. F. Snelson (eds), *Ecology and evolution of life bearing fishes (Poecillidae).* Prentic Hall, New Jersey.

de Jonge, J., de Ruiter, A. J. H. and van den Hurk R. 1989. Testis-testicular gland complex of two *Tripterygion* species (Blennioidei, Teleostei): differences between territorial and non-territorial males. Journal of Fish Biology 35: 497-508.

Downing, A. L. and Burns, J. R. 1995. Testis morphology and spermatozeugma formation in three genera of viviparous halfbeaks: *Normorhamphus, Dermogenys,* and *Hemirhamphodon* (Teleostei: Hemiramphidae). Journal of Morphology 225: 329-343.

Downing, A. and Burns, J. R. 2005. Testis morphology and spermatozeugma formation in three genera of viviparous halfbeaks: *Nomorhamphus, Dermogenys,* and *Hemirhamphodon* (Teleostei: Hemiramphidae). Journal of Morphology 225: 329-343.

Downing Meisner, A., Burns, J. R., Weitzman, S. H. and Malabarba, L. R. 2000. Morphology and histology of the male reproductive system in two species of internally inseminating South American catfishes, *Trachelyopterus lucenai* and *T. galeatus* (Teleostei: Auchenipteridae). Journal of Morphology 246: 131-141.

Dulka, J. G. and Demski, L. S. 2005. Sperm duct contractions mediate centrally evoked sperm release in goldfish. Journal of Experimantal Zoology 237: 271-279.

Eggert, B. 1931. Die Geschlechtsorgane der Gobiiformes und Blenniiformes. Zeitschrift für wissenschaftliche Zoologie 139: 249-558.

Fischer, E. A. 1980. The relationship between mating system and simultaneous hermaphroditism in the coral reef fish, *Hypoplectrus nigricans* (Serranidae). Animal Behaviour 28: 620-633.

Fischer, E. A. 1984. Egg trading in the chalk bass, *Serranus tortugarum*, a simultaneous hermaphrodite. Zeitschrift für Tierpsychologie 66: 143-151.

Fishelson, L. 1991. Comparative cytology and morphology of seminal vesicles in male gobiid fishes. Japanese Journal of Ichthyology 38: 17-30.

Fishelson, L., Van Wren, J. H. J. and Tyran, A. 1994. Ontogenesis and ultrastructure of seminal vesicles of the catfish, *Clarias gariepinus*. Journal of Morphology 219: 59-71.

Fishelson, L., Gon, O., Holdengreber, V. and Delarea, Y. B. 2007. Comparative morphology and cytology of the male sperm-transmission organs in viviparous species of clinid fishes (Clinidae: Teleostei, Perciformes). Journal of Morphology 290: 311-323.

García-Díaz, M. M., Lorente, M. J., González, J. A. and Tuset, V. M. 2002. Morphology of the ovotestis of *Serranus atricauda* (Teleostei, Serranidae). Aquatic Sciences - Research Across Boundaries 64: 87-96.

Grier, H. J. 1981. Cellular organization of the testis and spermatogenesis in fishes. American Zoologist 21: 345-357.

Grier, H. J. 1984. Testis structure and formation of spermatophores in the atherinimorph teleost *Heraichthys setnai*. Copeia 1984: 833-839.

Grier, H. J. and Collette, B. B. 1987. Unique spermatozeugmata in testes of halfbeaks of the genus *Zenarchopterus* (Teleostei: Hemiramphidae). Copeia 1987: 300-311.

Grier, H. J., Linton, J. R., Leatherland, J. F. and de Flaming, V. L. 1980. Structural evidence for two different testis types in teleost fishes. American Journal of Anatomy 159: 331-345.

Grier, H. J., Burn, J. R. and Flores, J.A. 1981. Testis structure in three species of teleosts with tubular gonopodia. Copeia 1981: 797-801.

Giulianini, P. G., Ota, D., Marchesan, M. and Ferrero, E. A. 2001. Can goby spermatozoa pass through the filament adhesion apparatus of laid eggs? Journal of Fish Biology 58: 1750-1752.

Guitel, F. 1893. Description des orifices génito-urinaires de quelques *Blennius*. Archives de Zoologie expérimentale et générale, Série 3.1: 611-658.

Hamlett, W. C. 1999. Male reproductive system. Pp 444-470. In W. C. Hamlett (ed.), *Sharks, skates and rays the biology of elasmobranch fishes*. The Johns Hopkins University Press: Baltimore.

Hastings, P. A. 1981. Gonad morphology and sex succession in the protogynous hermaphrodite *Hemanthias vivanus* (Jordan and Swain). Journal of Fish Biology 18: 443-454.

Hastings, P. A. and Bortone, S. A. 1980. Observations on the life history of the belted sandfish, *Serranus subligarus* (Serranidae). Environmental Biology of Fishes 5: 365-374.

Javonillo, R., Burns, J. and Weitzman, S. 2006. Reproductive morphology of *Brittanichthys axelrodi* (Teleostei: Characidae), a miniature inseminating fish from South America. Journal of Morphology 268: 23-32.

Jespersen, A. 1969. On the male urogenital organs of *Neoceratodus forsteri*. Biologiske Meddelelser: Kongelige Danske Videnskabernes Selskab = Royal Danish Academy of Sciences and Letters 16: 1-11.

Kerr, J. G. 1901. On the male genito-urinary organs of the *Lepidosiren* and *Protopterus*. Proceedings of the Zoological Society of London 1901: 484-498.

Kohda, M., Yonebayashi, K., Nakamura, M., Ohnishi, N., Seki, S., Takahashi, D. and Takeyama, T. 2002. Male reproductive success in a promiscuous catfish *Corydoras aeneus* (Callichthyidae). Environmental Biology of Fishes 63: 281-287.

Lagios, M. D. and McCosker, J. E. 1977. A cloacal excretory gland in the lundish *Protopterus*. Copeia 1977: 176-178.

Lahnsteiner, F. 2003. Morphology, fine structure, biochemistry, and function of the spermatic ducts in marine fish. Tissue and Cell 35: 363-373.

Lahnsteiner, F. 2007. Characterization of seminal plasma proteins stabilizing the sperm viability in rainbow trout (*Oncorhynchus mykiss*). Journal of Animal Reproduction, 97: 151-164.

Lahnsteiner, F. and Patzner, R. A. 1990. The spermatic duct of blenniid fishes. Fine structure, histochemistry and function. Zoomorphology 110: 63-73.

Lahnsteiner, F., Richtarski, U. and Patzner, R. A. 1990. Function of the testicular gland in two blenniid fishes, *Salaria* (=*Blennius*) *pavo* and *Lipophrys* (=*Blennius*) *dalmatinus* (Blenniidae, Teleostei) as revealed by electron microscopy and enzyme histochemistry. Journal of Fish Biology 37: 85-97.

Lahnsteiner, F., Patzner, R. A. and Weismann, T. 1992a. Monosaccharids as energy resources during motility of spermatozoa in *Leuciscus cephalus* (Cyprinidae, Teleostei). Fish Physiology and Biochemistry 10: 283-289.

Lahnsteiner, F., Seiwald, M., Patzner, R. A. and Ferrero E. A. 1992b. The seminal vesicles of the male grass goby, *Zosterisessor ophiocephalus*. Fine structure and histochemistry. Zoomorphology 111: 239-248.

Lahnsteiner, F., Patzner, R. A. and Weismann, T. 1993a. The spermatic ducts of salmonid fishes (Salmonidae, Teleostei). Morphology, histochemistry and composition of the secretion. Journal of Fish Biology 42: 79-93.

Lahnsteiner, F., Patzner, R. A. and Weismann, T. 1993b. The efferent duct system of the male gonads of the European pike (*Esox lucius*): Testicular efferent ductTesticular efferent ducts, testicular main ducts and spermatic ducts. Journal of Submicroscopic Cytology and Pathology 25: 487-498.

Lahnsteiner, F., Patzner, R. A. and Weismann T. 1993c. Energy resources of spermatozoa of the rainbow trout (*Oncorhynchus mykiss*) (Pisces, Teleostei). Reproduction, Nutrition, Development 33: 349-360.

Lahnsteiner, F., Nussbaumer, B. and Patzner, R. A. 1993d. Unusual testicular accessory organs, the testicular blind pouches of blennies (Teleostei, Blenniidae). Fine structure, (enzyme-) histochemistry and possible functions. Journal of Fish Biology 42: 227-241.

Lahnsteiner, F., Patzner, R. A. and Weismann T. 1994. The testicular main ducts and spermatic ducts in cyprinid fishes. Morphology, fine structure and histochemistry. Journal of Fish Biology 44: 937-951.

Lahnsteiner, F., Weismann, T. and Patzner, R. A. 1998. Sperm motility of the marine teleosts *Boops boops*, *Diplodus sargus*, *Mullus barbatus*, and *Trachurus mediterraneus*. Journal of Fish Biology 52: 726-742.

Lahnsteiner, F., Weismann, T. and Patzner, R. A. 1999a. Physiological and biochemical parameters for egg quality determination in lake trout, *Salmo trutta lacustris*. Fish Physiology and Biochemistry 20: 375-388.

Lahnsteiner, F., Berger, B. and Weismann, T. 1999b. Sperm metabolism of the teleost fishes *Oncorhynchus mykiss* and *Chalcalburnus chalcoides* and its relation to motility and viability. Journal of Experimental Zoology 284: 454-465.

Lahnsteiner, F., Urbanyi, B., Horvath, A. and Weismann, T. 2000. Bio-markers for egg quality determination in cyprinid fishes. Aquaculture 195: 331-352.

Laumen, J., Pern, U. and Blüm, V. 1974. Investigations on the functions and hormonal regulation of the anal appendices in *Blennius pavo* (Risso). Journal of Experimental Zoology 190: 47-56.

Loidl, B. 1992. Histologische und histochemische Untersuchungen am Samenleiter der Grasgrundel *Zosterisessor ophiocephalus*, Pallas 1811 (Teleostei, Gobiidae) im Jahreszyklus. Msc. Thesis, University of Salzburg, Austria.

Loir, M., Cauty, C, Planquett, P. and Le Bai, P. Y. 1989. Comparative study of the male reproductive tract in seven families of South-American catfishes. Aquatic Living Resources 2: 45-56.

Mansour, N. and Lahnsteiner, F. 2003. Morphology of male genitalia and sperm fine structure in siluroid fish. Journal of Submicroscopic Cytology and Pathology 35: 277-285.

Mansour, N., Lahnsteiner, F. and Patzner, R. A. 2004. Seminal vesicle secretion of African catfish, its composition, its behaviour in water and saline solutions and its influence on gamete fertilizability. Journal of Experimental Zoology 301A: 745-755.

Marconato, A. and Rasotto, M. B. 1993. The reproductive biology of *Opistognathus whitehurstii* (Pisces, Opistognathidae). Biologia Marina Mediterranea 1: 345-348.

Marconato, A., Rasotto, M. B. and Mazzoldi, C. 1996. On the mechanism of sperm release in three gobiid fishes (Teleostei: Gobiidae). Environmental Biology of Fishes 46: 321-327.

Mattei, X., Siau, Y., Thiaw, O. T. and Thiam, D. 1993. Peculiarities in the organization of testis of *Ophidion* sp. (Pisces Teleostei). Evidence for two types of spermatogenesis in teleost fish. Journal of Fish Biology 43: 931-937.

Mazzoldi, C., Petersen, C. W. and Rasotto, M. B. 2005: The influence of mating system on seminal vesicle variability among gobies (Teleostei, Gobiidae). Journal of Zoological Systematics and Evolutionary Research 43: 307-314.

Mazzoldi, C., Lorenzi, V. and Rasotto, M. B. 2007. Variation of male reproductive apparatus in relation to fertilization modalities in the catfish families Auchenipteridae and Callichthyidae (Teleostei: Siluriformes). Journal of Fish Biology 70: 243-256.

Miller, P. J. 1984. The tokology of gobioid fishes. Pp 119-154. In G. W. Potts and R. J. Wootton (eds), *Fish reproduction: strategies and tactics*. Academic Press: London, New York.

Miller, P. J. 1992. The sperm duct gland: a visceral synapomorphy for gobioid fishes. Copeia 1992: 253-256.

Morisawa, M. 1983. Effects of osmolality and potassium on motility of spermatozoa from fresh water cyprinid fishes. Journal of Experimental Zoology 95-103.

Morisawa, M., Suzuki, K., Shimizu, H., Morisawa, S. and Yasuda, K. 1983. Effects of osmolality and potassium on spermatozoan motility of salmonid fishes. Journal of Experimental Zoology 107: 105-113.

Muñoz, M., Casadevall, M. and Bonet, S. 2002. Testicular structure and semicystic spermatogenesis in a specialized ovuliparous species: *Scorpaena notata* (Pisces, Scorpaenidae). Acta Zoologica 83: 213-219.

Nayyar, S. K. and Sundararaj, B. I. 1970. Seasonal reproductive activity in the testes and seminal vesicles of the catfish *Heteropneustes fossilis* (Bloch). Journal of Morphology 130: 207-226.

Ohta, H., Ikeda, K. and Izawa, T. 1997. Increase in concentrations of potassium and bicarbonate ions promote acquisition of motility in vitro by Japanese eel spermatozoa. Journal of Experimental Zoology 277: 171-180.

Ohta, H., Unuma, T., Tsuji, M., Yoshioka, M. and Kashiwagi, M. 2001. Effects of bicarbonate ions and pH on acquisition and maintenance of potential for motility in ayu, *Plecoglossus altivelis* Temminck et Schlegel (Osmeridae), spermatozoa. Aquaculture Research 32: 385-392.

Ota, D., Marchesan, M. and Ferrero, E. A. 1996. Sperm release behaviour and fertilization in the grass goby. Journal of Fish Biology 49: 246-256.

Patzner, R. A. 1984. The reproduction of *Blennius pavo* (Teleostei, Blenniidae) II. Surface structures of the ripe egg. Zoologischer Anzeiger, Jena 213: 44-50.

Patzner, R. A. 1989. Morphology of the male reproductive system of two Indo-Pacific blenniid fishes, *Salarias fasciatus* and *Ecsenius bicolor* (Blenniidae, Teleostei). Zeitschrift für Zoolgische Systematik und Evolutionsforschung 27: 135-141.

Patzner, R. A. 1991. Morphology of the male reproductive system of *Coralliozetus angelicus* (Pisces, Blennioidei, Chaenopsidae). Journal of Fish Biology 39: 867-872.

Patzner, R. A. 1997. Gonads and reproduction in hagfishes. Pp 378-395. In J. M. Jorgensen, J. P. Lomholt, R. E. Weber and H. Malte (eds), *Biology of hagfishes*. Chapman and Hall: New York.

Patzner, R. A. 2007. Reproductive strategies of fish. Pp 311-350. In M. J. Rocha, A. Arukwe and B. G. Kapoor (eds), *Fish reproduction: cytology, biology and ecology*. Science Publishers, Enfield, New Hampshire, USA.

Patzner, R. A., Seiwald, M., Angerer, S., Ferrero, E. and Giulianini P. 1991. Genital system and reproductive cycle of the male grass goby, *Zosterisessor ophiocephalus* (Teleostei, Gobiidae), in the northern Adriatic Sea. Zoologischer Anzeiger, Jena 226: 205-219.

Pecio, A., Burns, J. R. and Weitzmann, S. H. 2005. Sperm and spermatozeugma ultrastructure in the inseminating species *Tyttocharax cochui, T. tambopatensis*, and *Scopaeocharax rhinodus* (Pisces: Teleostei: Characidae: Glandulocaudinae: Xenurobryconini). Journal of Morphology 263: 216-26.

Penrith, M. L. 1970. The systematic of the fishes of the family Clinidae in South Africa. Annals of the South African Museum 55: 1-121.

Rasotto, M. B. 1995. Male reproductive apparatus of some Blennioidei (Pisces: Teleostei). *Copeia* 1995: 907-914.

Rasotto, M. B. and Shapiro, D. Y. 1998. Morphology of gonoducts and male genital papilla, in the bluehead wrasse: implications and correlates on the control of gamete release. Journal of Fish Biology 52: 716-725.

Resink, J. W., Voorthuis, P. K., van den Hurk, R., Peters, R. C. and van Oordt, P. G. W. J. 1989. Steroid glucuronides of the seminal vesicle as olfactory stimuli in African catfish, *Clarias gariepinus*. Aquaculture 83: 153-166.

Richtarski, U. and Patzner, R. A. 2000. Comparative morphology of male reproductive systems in Mediterranean blennies (Blenniidae). Journal of Fish Biology 56: 22-36.

Scaggiante, M., Mazzoldi, C., Petersen, C. W. and Rasotto, M. B. 1999. Sperm competition and mode of fertilization in the grass goby *Zosterisessor ophiocephalus* (Teleostei: Gobiidae). Journal of Experimental Zoology 283: 81-90.

Scott, A. P. and Vermeirssen, E. M. 1994. Production of conjugated steroids by teleost gonads and their role as pherormones. Pp 645-654. In K. G. Davey, R. E. Peter and S. S. Tobe (eds), *Perspectives in comparative endocrinology*. National Research Council: Ottawa.

Seiwald, M. and Patzner, R. A. 1989. Histological, fine structural and histochemical differences in the testicular glands of gobioid and blennioid fishes. Journal of Fish Biology 34. 631-640.

Shen, S. C. 1971. A new genus of clinid fishes from the Indo-Pacific, with a rediscription of *Clinus nematopterus*. Copeia 1971: 697-707.

Sneed, K. E. and Clemens, H. P. 1963. The morphology of the testes and accessory reproductive glands of the catfishes (Ictaluridae). Copeia 1963: 606-611.

Stockley, P., Gage, M. J. G., Parker, G. A. and Møller, A. P. 1997. Sperm competition in fishes: the evolution of testis size and ejaculate characteristics. American Naturalist 149: 933-954.

Touart, L. W. and Bortone, S. A. 1980. The accessory reproductive structure in the simultaneous hermaphrodite *Diplectrum bivittatum* (Pisces: Serranidae). Journal of Fish Biology 16: 397-403.

Van den Hurk, R. and Resink, J.W. 1992. Male reproductive system as sex pheromone producer in teleost fish. Journal of Experimental Zoology 261: 204-213

Van den Hurk, R., Resink, J. W. and Peute, J. 1987. The seminal vesicle of the African catfish, *Clarias gariepinus*. A histological, histochemical, enzyme-histochemical, ultrastructural and physiological study. Cell and Tissue Research 247: 573-582.

van Tienhoven, A. 1983. Reproductive physiology of vertebrates, second edition. Cornell University Press, Ithaca, New York, pp. 491.

Vicentini, C. A., Franceschini-Vicentini, I. B., Benetti, E. J. and Orsi, A. M. 2001. Testicular ultrastructure and morphology of the seminal pathway in *Prochilodus scrofa*. Journal of Submicroscopic Cytology and Pathology 33: 357-362.

Wake, M. H. 1986. Urogenital morphology of dipnoans, with comparisons to other fishes and to amphibians. Journal of Morphology Suppl. 1: 199-216.

Wourms, J. P. 1977. Reproduction and development in chondrichthyan fishes. American Zoologist 17: 379-410.

Wrobel, K. H., Hees, I., Schimmel, M. and Stauber, E. 2002. The genus *Acipenser* as a model system for vertebrate urogenital development: nephrostomial tubules and their significance for the origin of the gonad. Anatomy and Embryology 205: 67-80.

Wrobel, K. H. and Jouma, S. 2004. Morphology, development and comparative anatomical evaluation of the testicular excretory pathway in *Acipenser*. Annals of Anatomy 186: 99-113.

Zorica, B., Sinovčić, G. and Čikes Keč, V. 2005. Reproductive period and histological analysis of the painted comber, *Serranus scriba* (Linnaeus, 1758), in the Trogir Bay area (eastern mid-Adriatic). Acta Adriatica 46: 77-82.

CHAPTER 6

Ultrastructure of Spermatozoa in Agnathans

B. G. M. Jamieson

6.1 JAWLESS FISHES (AGNATHANS)

Hagfishes and lampreys (adults, Fig. 6.1) have traditionally been placed together in the Agnatha, the jawless craniates. A biting apparatus differing from that of gnathostomes in not being derived from gill arches is present in some fossil forms (Nelson 2006). The terms Agnatha or Cyclostomata are not formally recognized here in view of albeit controversial evidence that it is a paraphyletic group (see below). Figure 6.2 illustrates two alternative views of their relationships with each other and with gnathostomes.

The 'cyclostome hypothesis' assumes that lampreys and hagfishes are a monophyletic clade, the cyclostomes, the sister-group of which is the jawed vertebrates (gnathostomes). In contrast, the 'vertebrate hypothesis' assumes that lampreys and gnathostomes are a clade (the vertebrates), the sister-group of which is the hagfishes.

The cyclostome hypothesis has been supported for protein-coding genes of mtDNA (Rasmussen *et al.* 1998); for four nuclear DNA-coded single-copy genes and Mn superoxide dismutase sequences (Kuraku *et al.* 1999), for 18S and 28S rRNA genes (Delarbre *et al.* 2000), for 35 nuclear protein-encoding genes (Takezaki *et al.* 2003), from nuclear and mitochondrial protein sequences (Blair and Hedges 2005) and from nearly complete 28S and 18S rRNA genes (Mallatt and Winchell 2007). Other works supporting the cyclostome hypothesis are Stock and Whitt (1992), Mallat and Sullivan (1998), Kuraru *et al.* (1999), Cotton and Page (2002) and Delsuc *et al.* (2006) (see Ortí and Li, **Chapter 1**). Ortí and Li consider that there is definitive molecular genetic evidence in support of the cyclostome hypothesis. They also review evidence that the immune system shared by hagfishes and lampreys may be a

School of Integrative Biology, University of Queensland, Brisbane, Queensland 4072, Australia

cyclostome synapomorphy. In contrast, molecular support for the vertebrate hypothesis appears sparse. It was said to be supported by Delarbre *et al.* (2000) when, alternatively, sequencing mtDNA.

Nelson (2006) gives a valuable discussion of these conflicting issues and, while recognizing that the matter is not yet settled, opts for paraphyly of the Agnatha. Most morphological and paleontological analyses also support paraphyly and the vertebrate hypothesis.The 'vertebrate hypothesis' favors the view that the morphological and physiological similarities shared between lampreys and gnathostomes, but not hagfishes, are due to common ancestry and not convergent evolution. These similarities are: gill-arch jaws absent, no pelvic fins, highly differentiated kidney tubules, absence of a persistent pronephros, one or two vertical semicircular canals, large exocrine pancreas, photosensory pineal organ, notochord but no vertebral centra present, gills covered with endoderm, gill-arch skeleton joined to neurocranium and external to gills, gills with pores but not slits; histological structure of the adenohypophysis, and composition of the body fluid (Nelson 1984, 2006).

Lampreys (Fig. 6.1B) and, it is believed, hagfishes (Fig. 6.1A) have external fertilization yet their sperm show features often associated with internal fertilization: long endonuclear perforatorium in *Lampetra* (but also in ect-aquasperm in some polychaetes); elongate nucleus; in hagfishes especially, an

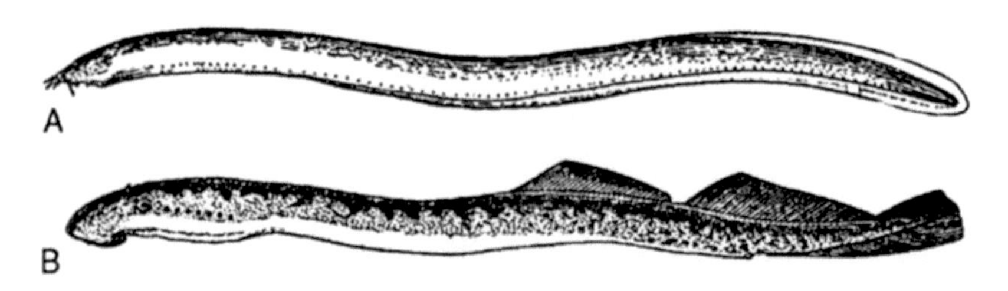

Fig. 6.1 **A**. The hagfish, *Myxine* (Myxinidae). **B**. the Brook Lamprey, *Lampetra planeri* (Petromyzontidae). From Romer, A.S. and Parsons, T. 1977. *The Vertebrate Body*. W. B. Saunders Company, Philadelphia, Fig. 16B,C. After Dean.

Fig. 6.2 Phylogeny of the Craniata. Above, according to the cyclostome hypothesis. Below, according to the vertebrate hypothesis. Major spermatozoal apomorphies of agnathans and chondrichthyans, only, are indicated. Some features which may be plesiomorphic retentions are indicated for their descriptive value. For spermatozoa, the cyclostome hypothesis is less parsimonious as it requires endonuclear canals (and preformed perforatorial rods) to have appeared independently (homoplasically) in petromyzontiforms and gnathostomes, or that myxiniforms have lost them. Numbering of characters refers to the analysis in Jamieson, B. G. M. 1991. *Fish Evolution and Systematics: Evidence from Spermatozoa*. Cambridge University Press, Fig. 6.2.

Fig. 6.2 Contd. ...

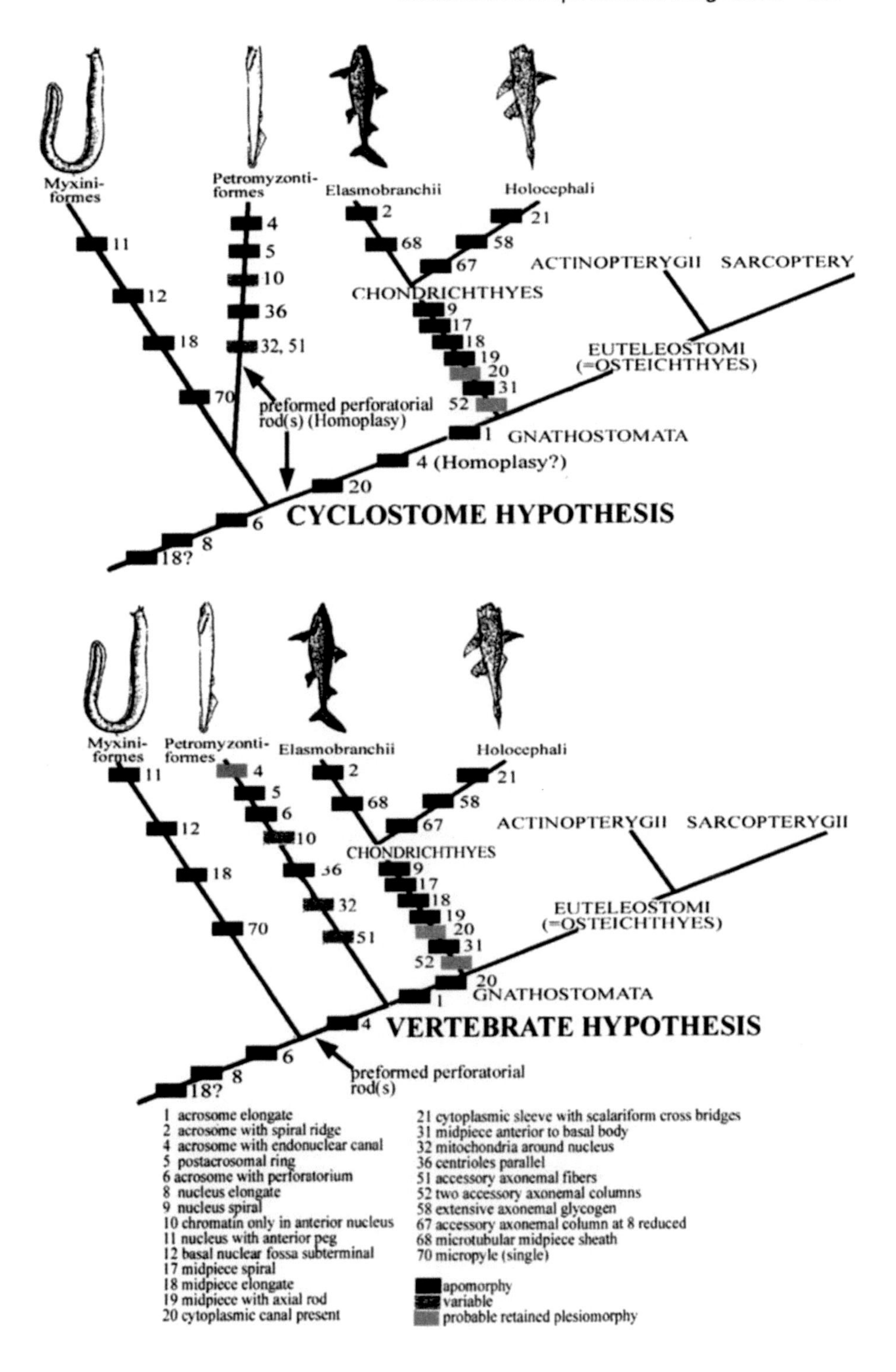

Myxini-formes
Petromyzonti-formes
Elasmobranchii
Holocephali
ACTINOPTERYGII
SARCOPTERY
CHONDRICHTHYES
EUTELEOSTOMI (=OSTEICHTHYES)
GNATHOSTOMATA
4 (Homoplasy?)
preformed perforatorial rod(s) (Homoplasy)
CYCLOSTOME HYPOTHESIS
Myxini-formes
Petromyzonti-formes
Elasmobranchii
Holocephali
ACTINOPTERYGII
SARCOPTERYGII
CHONDRICHTHYES
EUTELEOSTOMI (=OSTEICHTHYES)
GNATHOSTOMATA
VERTEBRATE HYPOTHESIS
preformed perforatorial rod(s)
1 acrosome elongate
2 acrosome with spiral ridge
4 acrosome with endonuclear canal
5 postacrosomal ring
6 acrosome with perforatorium
8 nucleus elongate
9 nucleus spiral
10 chromatin only in anterior nucleus
11 nucleus with anterior peg
12 basal nuclear fossa subterminal
17 midpiece spiral
18 midpiece elongate
19 midpiece with axial rod
20 cytoplasmic canal present
21 cytoplasmic sleeve with scalariform cross bridges
31 midpiece anterior to basal body
32 mitochondria around nucleus
36 centrioles parallel
51 accessory axonemal fibers
52 two accessory axonemal columns
58 extensive axonemal glycogen
67 accessory axonemal column at 8 reduced
68 microtubular midpiece sheath
70 micropyle (single)
apomorphy
variable
probable retained plesiomorphy

elongate mitochondrial sheath around the axoneme; and, in lampreys, nine accessory fibers around the axoneme. Afzelius (in Nicander 1970) suggested that these features have been retained from "cyclostome" ancestors which were internally fertilizing. In possible support of this view, lampreys have a penial tube in the male (Fig. 6.9A) and copulation (Breder and Rosen 1966; Hardisty and Potter 1971) although fertilization is not internal. It thus appears plausible that the ancestors of 'agnathans', or even of gnathostomes as a whole, had internal fertilization.

The sperm of lampreys and hagfishes show no similarities that would demand recognition of relationship between the two groups. For spermatozoa, the cyclostome hypothesis is less parsimonious than the vertebrate hypothesis as it requires endonuclear canals (and preformed perforatorial rods) to have appeared independently (homoplasically) in petromyzontiforms and gnathostomes, or that myxiniforms have lost them. This gives support to the view that the Agnatha is an artificial group unified by the symplesiomorphy jawlessness. We have seen, nevertheless, that molecular analyses strongly favor the vertebrate hypothesis (see Ortí and Li, **Chapter 1**).

Remarkable homogeneity of sperm structure is seen within *Lampetra* (but see *Mordacia*, below) and, of a very different nature, within the hagfish *Eptatretus*.

6.2 SUPERCLASS MIXINIMORPHI, CLASS MYXINI

Diagnosis as for the order Myxiniformes.

6.3 ORDER MYXINIFORMES

Unlike lampreys and their fossil relatives, myxiniforms have only one semicircular canal on each side. Like the lampreys they lack bone, though bone is present in cephalaspidiform relatives of lampreys. A remarkable resemblance to petromyzontiforms is the single dorsal nostril, giving additional support to the cyclostome hypothesis. The single family, Myxinidae, consists of the hagfishes, marine in the temperate zones and the Gulfs of Mexico and Panama (Nelson 2006).

6.3.1 Family Myxinidae

Sperm literature: Spermatogenesis to tne spermatid stage has been described ultrastructurally for the Atlantic, or European, Hagfish *Myxine glutinosa* (Alvestad-Graebner and Adam 1977) and some data on the mature sperm of this species have been given by Nicander (1968; 1970). Spermiogenesis to the mature spermatozoon has been described for eastern Pacific hagfish, *Eptatretus stoutii, E. deani* and *E.* sp. (Jespersen 1975) and for *E. burgeri* (Morisawa 2005). The elongating spermatid of *Eptatretus stoutii* is illustrated by Pfeiffer and Vogl (2002) in a study of adhesion junctions in Sertoli cells. The mature spermatozoon of *E burgeri* is described by Morisawa (1995) and its acrosome reaction by Morisawa (1999a) and the acrosome reaction of both species by Morisawa and Cherr (2002).

Eptatretus stoutii: The mature sperm of *Eptatretus stoutii*, the Pacific Hagfish (Fig. 6.3), is an elongate structure, about 40-60 μm long. It consists of a head (acrosome and nucleus), a midpiece and an endpiece. The acrosome (Fig. 6.5A), apical on the nucleus, has the form of a conical cap about 1 μm long enclosing subacrosomal material into which projects a central cylinder which is a narrow peg-like extension of the nucleus. This anterior process is also

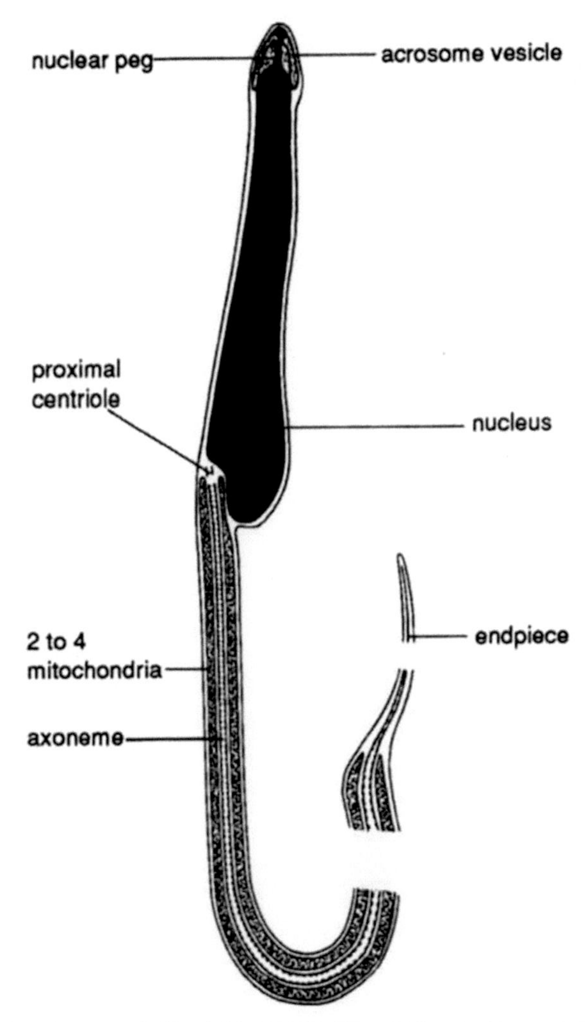

Fig. 6.3 *Eptatretus stoutii* (Myxinidae). Diagram of the mature spermatozoon. From Jamieson 1991, after Jespersen, Å. 1975. Acta Zoologica (Stockholm) 56: 189-198, Fig. 19.

seen in *E. deani* and *E.* sp. (Fig. 6.5B, C). The nucleus is lanceolate, 12 µm long and maximally, near it base, about 1.5 µm wide, tapering anteriorly and posteriorly and with highly condensed chromatin. On one side, for about 1 µm from the posterior end, the nucleus is indented to receive the centriole and the base of the flagellum. The midpiece, which is about 1 µm wide and at least 20 µm long, consists of a 9+2 flagellum, lacking accessory fibers outside the doublets, surrounded by 2, 3 or 4 long and slightly twisted irregular mitochondria. Behind this is the free flagellum (narrowed to form the endpiece) of undetermined length (Jespersen 1975).

***Eptatretus burgeri*:** The entire spermatozoon of *Eptatretus burgeri* is shown by scanning electron microscopy in Figure 6.4 (Morisawa 2005; see also Morisawa 1995). The spermatozoon closely resembles that of *E. stoutii*. The acrosome contains electron-dense and less dense materials in two different compartments. Amorphous subacrosomal material lies between the acrosome

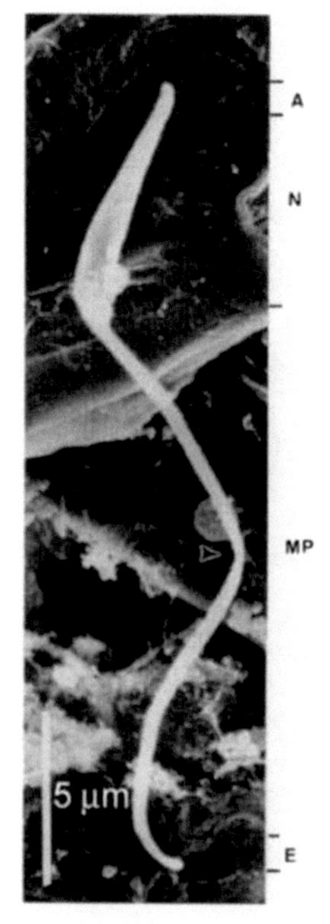

Fig. 6.4 *Eptatretus burgeri* (Myxinidae). A scanning electron micrograph of a spermatozoon. A, acrosome; arrowhead, a twisting portion; E, endpiece; MP, midpiece; N, nucleus. From Morisawa, S. 2005. Biological Bulletin (Woods Hole) 209: 204-214, Fig. 1. Reprinted with permission from the Marine Biological Laboratory, Woods Hole, MA.

and the nucleus. No distinct perforatorial rod or filamentous structure is observable within the subacrosomal material but an acrosomal process is formed on reaction (see below). The midpiece is extremely long; two of the four midpiece mitochondria extend through almost the entire length of the tail. Two centrioles lie almost end to end in the nuclear fossa near, but not at, the posterior end of the nucleus (Morisawa 1995; Morisawa 1999a).

Other Eptatretus species: Sperm structure in *Eptatretus deani* and *E.* sp. (Fig. 6.5B,C) is said to be similar to that of *E. stouti* (Fig. 6.5A) but the distribution of electron-dense material in the acrosome vesicle of *E.* sp. differs in that instead of being uniform it is chiefly apical, with pale flocculent material filling the remainder (Jespersen 1975).

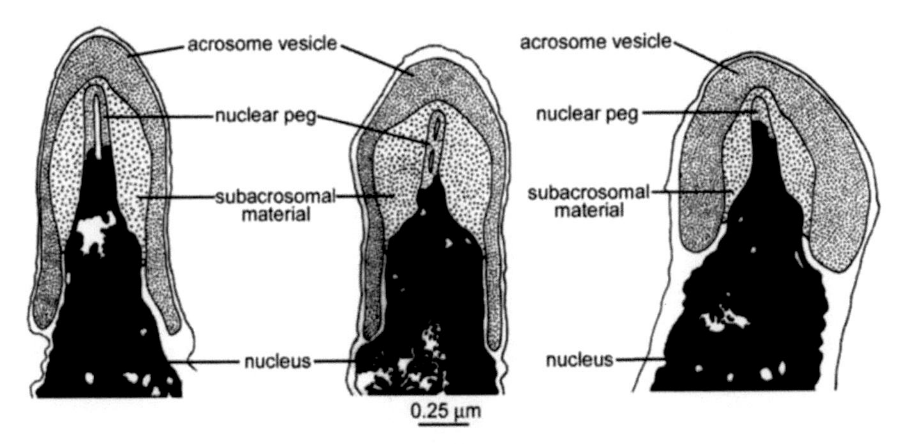

Fig. 6.5 *Eptatretus* (Myxinidae). Longitudinal sections of the acrosome and nuclear peg. **A.** *E. stoutii.* **B.** *E. deani.* **C.** *E.* sp. From Jamieson 1991, after micrographs by Jespersen, A. 1975. Acta Zoologica (Stockholm) 56: 189-198, Figs. 14-16.

Myxine: The account of spermatogenesis for *Myxine glutinosa* by Alvestad-Graebner and Adam (1977) contains no data on the mature sperm but an acrosome illustrated for the late spermatid and by Nicander (1970) for the mature sperm (Fig. 6.6) resembles that of the *Eptatretus* sperm, including the presence of a central nuclear peg embedded in subacrosomal material.

Alvestad-Graebner and Adam (1977) note an association of the chromatoid body with the inception of the acrosome. Jespersen (1975) refers to unpublished work indicating similarity of *Myxine* and *Eptatretus* sperm. Nicander (1968) refers to a true mitochondrial sheath in the sperm of *M. glutinosa* and gives a micrograph of a transverse section showing five round mitochondrial profiles symmetrically distributed around the axoneme (Fig. 6.7); groups of microtubules are present peripheral to the mitochondria in what is regarded as a mature spermatozoon. External fertilization is deduced from the absence of copulatory organs (Walvig 1963).

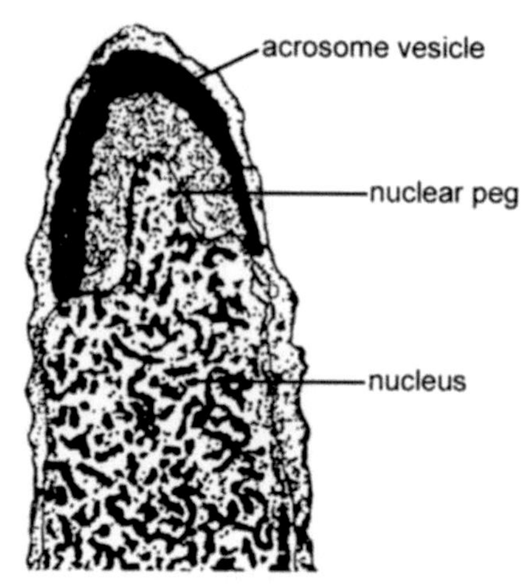

Fig. 6.6 *Myxine glutinosa* (Myxinidae). Acrosome. From Jamieson 1991, after a micrograph by Nicander, L. 1970. Pp. 47-55. In B. Baccetti (ed.), *Comparative Spermatology*, Academic Press, New York, Fig. 2.

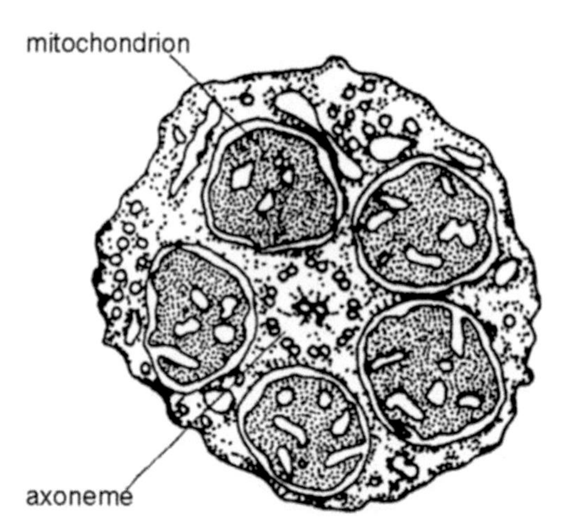

Fig. 6.7 *Myxine glutinosa* (Myxinidae). Cross section of sperm showing midpiece with mitochondrial sheath and microtubules. From Jamieson 1991, after a micrograph by Nicander, L. 1968. *Proceedings of the 6th International Congress on Artificial Animal Reproduction Paris* 1: 89-107, Fig. 13.

Eptatretus micropyle: The egg of the hagfish *Eptatretus burgeri*, unlike that of the lamprey, has a micropylar funnel. This is said by Fernholm (1975), using scanning electron microscopy, to be surrounded by eight radiating capitate filaments. At the bottom of the funnel is a honeycomb pattern of about 3,500 'cells' (chambers) of which only one is open at the bottom. The successful sperm presumably is that which enters this cell which, at 3.5 µm, is wide enough for only a single sperm cell (Fernholm 1975). Using TEM, Morisawa

(1999b) gave a detailed account of the micropylar region during oogenesis in this species but concluded that 'the ultrastructure of the mature micropyle of hagfishes is unknown and that 'the pit region appeared likely to have a role in fertilization'.

It is contended (Jamieson 1991) that there is a high positive correlation between absence of a preformed perforatorium (or of an acrosome, Longo 1987) and presence of egg micropyles in fishes, of which hagfishes are the first example in the present study.

Eptatretus acrosome reaction: Induction of an acrosome reaction for spermatozoa of the hagfish, *Eptatretus burgeri* (Fig. 6.8) and *E. stouti* occurs on treatment of mature spermatozoa with the ionophore ionomycin and excess Ca^{2+}. The spermatozoon produces an acrosomal process, with a filamentous core that elongates from the apex of the long sperm head (Morisawa 1999a; Morisawa and Cherr 2002; for details see these papers).

The acrosomal region and midpiece in *E. stouti* exhibit immunofluorescent labeling using an actin antibody. The acrosomal region showed a similar labeling pattern when sperm were probed with tetramethylrhodamine isothyocyanate (TRITC)-phalloidin; the midpiece did not label. Following induction of the acrosome reaction with ionomycin, TRITC-phalloidin labeling was more intense in the acrosomal region, suggesting that the polymerization of actin occurs during formation of the acrosomal process, as seen in many invertebrates. During acrosomal exocytosis, the outer acrosomal membrane and the overlying plasma membrane disappear and are replaced by an array of vesicles; these were considered to resemble an early stage of the acrosome reaction in spermatozoa of higher vertebrates in which no formation of an acrosomal process occurs, suggesting that spermatozoa of hagfishes, is intermediate in this respect between invertebrates and higher vertebrates (Morisawa and Cherr 2002).

6.4 SUPERCLASS PETROMYZONTOMORPHI, CLASS PETROMYZONTIDA

Diagnosis as for the Petromyzontiformes

6.5 ORDER PETROMYZONTIFORMES

Petromyzontiformes (lampreys) resemble myxiniforms in lacking bone. Whereas myxiniforms have a single semicircular canal on each side, lampreys have two semicircular canals thus differing from gnathostome fishes, which have three semicircular canals and have paired fins. Lampreys are anadromous and freshwater in cool regions (Nelson 2006). They include the families Petromyzontidae (northern lampreys), Geotriidae (southern lampreys) and Mordaciidae (southern top-eyed lampreys).

Lampreys are unique among vertebrates in having no male ducts; they discharge the sperm, like the ova, into the coelom for egress through pores to

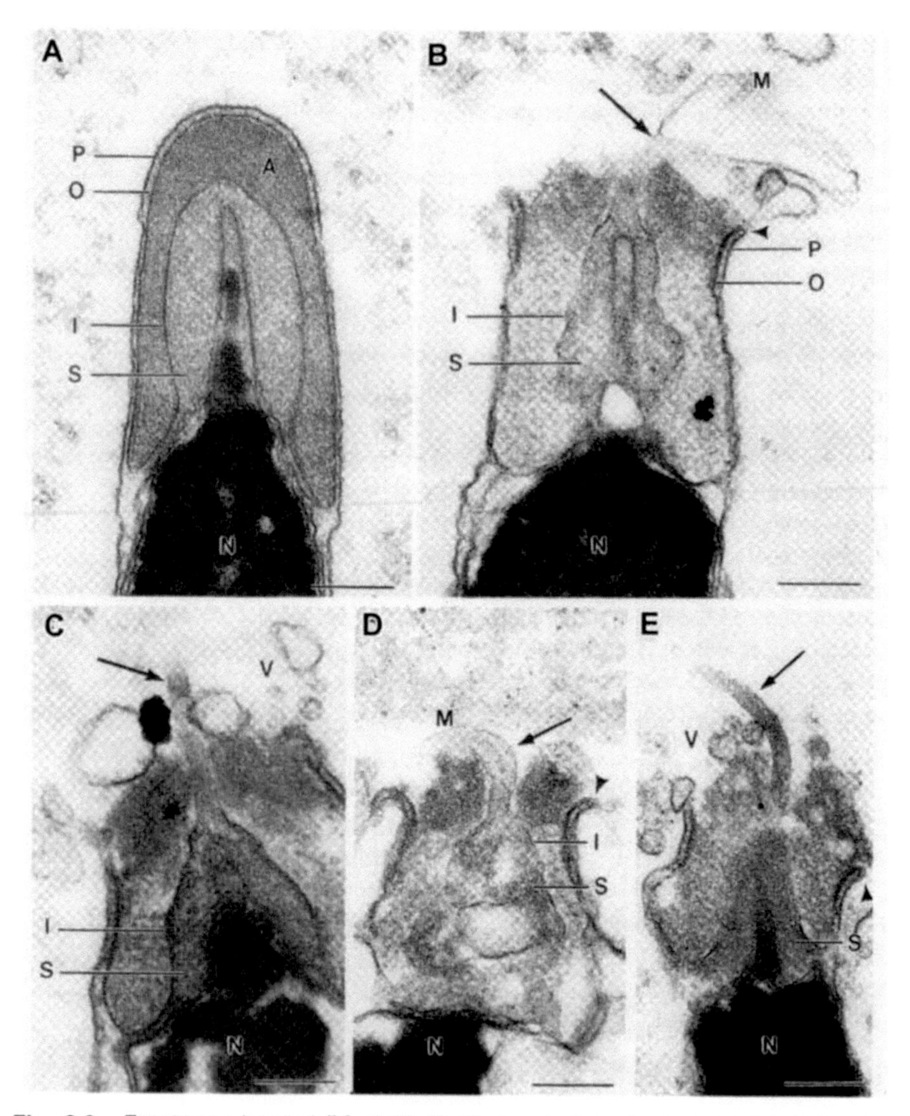

Fig. 6.8 *Eptatretus burgeri* (Myxinidae). Transmission electron micrograph of an acrosome reaction in spermatozoa from the hagfish. **A**. Intact acrosomal region of a spermatozoon. No distinct filamentous structures are detected within the subacrosomal material (S) between the bell-shaped acrosomal vesicle (A) and the anterior portion of the nucleus (N). (P) plasma membrane; (O) outer and (I) inner acrosomal membranes. **B-E**. Reacted acrosomal region. Acrosomal process (arrow) extends from the subacrosomal material (S), covered with membrane (M) continuous with the inner acrosomal membrane (I). The outer acrosomal membrane (O) and the plasma membrane (P) have fused at the rim of the opening of the acrosomal vesicle (arrowhead), leaving some vesicles (V). N, nucleus. Bar, 200 nm. After Morisawa, S. 1999. Development Growth and Differentiation 41: 109-112, Fig. 2. With permission of Wiley-Blackwell.

the exterior. One might speculate that the penis-like structure in the male (Fig. 6.9A) is a substitute for a ductus ejaculatorius or other muscular component of the normal vertebrate male duct and is not evidence of former internal fertilization. From comparison with other Metazoa with accessory axonemal fibers which are generally internally fertilizing, the presence of accessory fibers in the lamprey sperm axoneme remains inconsistent with external fertilization, however.

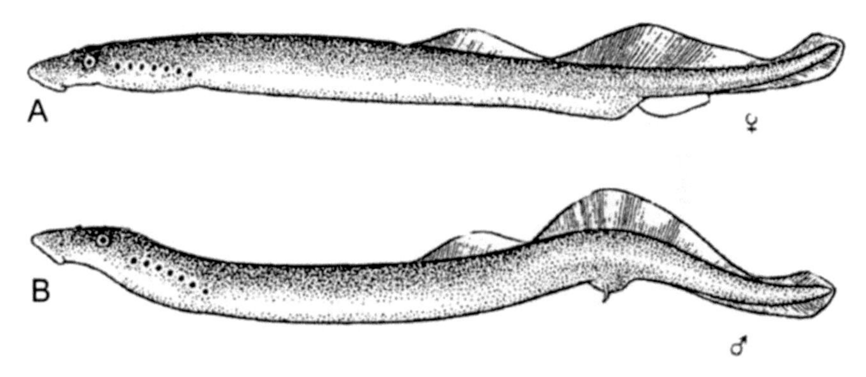

Fig. 6.9 *Lampetra planeri*, the Brook Lamprey (Petromyzontidae). **A**. Ripe female with anal fold. **B**. Ripe male, showing copulatory papilla with penis-like structure. From Jamieson 1991, after Young, J. Z. 1981. *The Life of Vertebrates*. Third edition, Clarendon Press, Oxford, Fig. 4.1.

6.5.1 Family Petromyzontidae

Sperm literature: The spermatozoon of the non-parasitic *Lampetra planeri* (adults, Fig. 6.9) has been described ultrastructurally by Follenius (1965) and Stanley (1967) (Fig. 6.10) and that of the parasitic *L. (=Petromyzon) fluviatilis,* which possibly represents its ancestral stock, by Nicander (1968, 1970) and Nicander and Sjöden (1968, 1971). That of *Lampetra japonica* has been described by Jaana and Yamamoto (1981).

Lampetra: The sperm of *Lampetra* has a length of 130 µm (*L. japonica*) or about 140 µm (*L. planeri*). The length of the rod-shaped head is about 8 µm (*L. japonica*) or 14-16 µm (*L. planeri*), with a maximum diameter, posteriorly, of 1.0 µm.

Acrosome: The acrosome vesicle is subovoidal with its greatest width transverse (Fig. 6.10C). Apposed to the posterior face of the acrosome vesicle there is a small ring of very dense material (Fig. 6.10C,D) (apical corpuscle of Follenius 1965) which, in *L. planeri* at least, is attached by short finger-like extensions to the nuclear membrane. From the center of this postacrosomal ring (subacrosomal ring, Nicander 1970; Nicander and Sjödén 1968; Jaana and Yamamoto 1981) a long 'central fiber' (subacrosomal fiber of Nicander 1970), extends posteriorly throughout the length of an endonuclear canal to

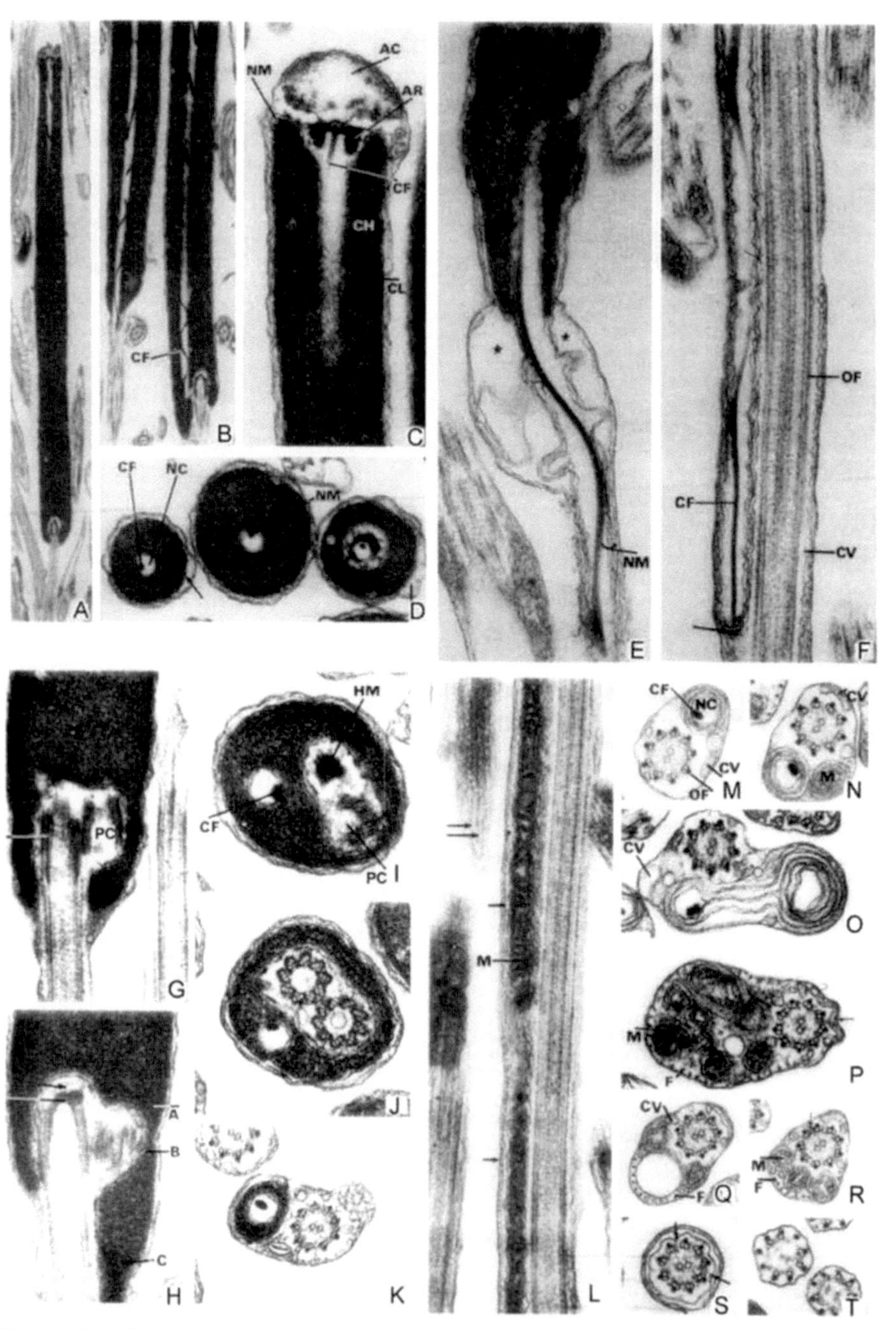

Fig. 6.10 *Lampetra planeri* spermatozoa (Petromyzontidae). **A**. Longitudinal section (LS) through the sperm head, x 4190. **B**. LS through posterior portions of two sperm heads. The head canal (endonuclear canal) passes through the entire

Fig. 6.10 Contd. ...

where this is terminated posteriorly by the indented nuclear membranes, well into the tail (*L. planeri*) (Fig. 6.10B,E,F)and *L. japonica*). Numerous vesicles reported in the immature acrosome of *L. planeri* were not observed by Stanley (1967).

Fig. 6.10 Contd. ...

chromatin mass. The central fiber (perforatorium) follows an undulating course through the canal, x 5265. **C**. LS at the anterior tip of the sperm head. The postacrosomal ring (AR) lies between the acrosome vesicle (AC) and the nucleus. The chromatin cylinder (CH) of the nucleus possesses a dense layer (CL) at its periphery, x 2270. **D**. Transverse sections of three sperm heads. At left and right the section passes through the smaller anterior end (with postacrosomal ring at right; at the center a spermatozoon is sectioned through the larger, posterior end. Endonuclear canal (NC); perforatorium (CF); nuclear membranes (NM). The arrows indicate the space between the leaflets of the nuclear membrane covering the outer side of the nucleus. **E**. Section at posterior end of chromatin cylinder showing the continuation of the endonuclear canal and perforatorium posterior to the chromatin. The bracket spans the four leaflets of the nuclear membrane (NM), two external and two internal, bounding the posterior portion of the endonuclear canal. Asterisks mark a space between the inner and outer leaflets of the external nuclear envelope, x 16880. **F**. Section of the tail showing the posterior, closed end of the endonuclear canal (arrow). OF, outer dense fiber of axoneme, x 17960. **G**. LS near posterior end of nucleus, showing the implantation fossa containing the basal body (BB) and the accessory (proximal) centriole, x 18330. **H**. A section similar to that in **G** but more clearly showing the hemisphere of dense material (cap) (HM) over the proximal end of the basal body and the array of fine filaments (arrow) extending between the hemisphere and the nuclear membrane. Letters A, B, and C mark levels corresponding to the transverse sections illustrated in **I**, **J** and **K**, respectively, x 19500. **I**. TS showing the cap (HM) of the basal body, the accessory (proximal) centriole (PC), and the perforatorium (CF) within the endonuclear canal. The section corresponds to level A of **H**, x 22425. **J**. A section just posterior to that shown in **I** passing through both centriolar structures. Note the dense material apposed to the outer surfaces of each triplet. The section corresponds to level B of **H**, x 19110. **K**. A section posterior to the centriole region showing the single axoneme at right and at the left the posterior projection of the chromatin mass as a tube around the endonuclear canal and perforatorium. The section corresponds to level C in **H**, x 22425. **L**. LS through the mitochondrial areas of several spermatozoa showing the beaded cytoplasmic filaments (arrows) just under the plasma membrane. M, Mitochondrion, x 17340. M-T. A series of TS along the tail placed in approximate order from proximal to distal. CF, perforatorium; NC, endonuclear canal; M, mitochondria; OF, outer coarse fibers of the axoneme; CV, cytoplasmic vesicle; F, cytoplasmic filament. The arrows in R and S indicate possible radial attachments of the axoneme to the plasma membrane or to the cytoplasmic vesicle membrane encircling the axoneme, respectively, x 29200. Adapted from Stanley, H. P. 1967. Journal of Ultrastructure Research 19: 84-99, Figs. 1-20. With permission of Elsevier.

The central fiber is clearly a perforatorium and has been shown to be capable of extrusion as a 50 μm long "head filament" in ultrastructural investigation of the river lamprey, *Lampetra fluviatilis* (Afzelius and Murray 1957) and in an optical study of this species and *L. planeri* (Kille 1960). Retzius (1921) had earlier noted protrusion of a head filament in *L. fluviatilis.* For *L. planeri,* it is termed an "acrosomal filament" by Jaana and Yamamoto (1981). In sperm apparently reacted by fixation, the nuclear membranes of the canal, and the acrosome vesicle, disappear (Stanley 1967), the posterior acrosomal membrane being pushed into contact with the anterior membrane which in turn fuses with the plasma membrane (Jaana and Yamamoto 1981) whereas the plasma membrane is drawn out into a slender sheath containing the central fiber and projecting anteriorly through the postacrosomal ring (Stanley 1967). The acrosomal filament undergoes no observable change on extrusion (Follenius 1965). It is possibly extruded by a spring-like action as it has an undulating course before reaction (Stanley 1967) (Fig. 6.10B,E,F). In *L. fluviatilis,* spermatozoal ultrastructure of which agrees (Nicander 1968, 1970) with that of *L. planeri,* sperm in the egg coatings show a true acrosome reaction: the acrosome vesicle bursts and the central fiber is extended, surrounded by a membrane continuous with the plasma membrane. Thus an acrosomal tubule is formed and penetrates the egg envelopes to reach the egg surface; the membrane covering the fiber is 6-7 nm thick compared with 9-10 nm for the plasma membrane in testicular spermatozoa (Nicander 1968; Nicander and Sjödén 1968). In lampreys, unlike hagfishes, there is no micropyle in the egg envelopes (Okkelberg 1914; Kille 1960; Afzelius and Sjöden 1968; Jaana and Yamomoto 1981), though presence was claimed by Rothschild (1958). As noted by Follenius (1965), the apical corpuscle and long perforatorium have remarkable parallels in the sperm (ect-aquasperm) of *Limulus.*

Nucleus: The chromatin-containing part of the rod-like nucleus is 7 μm (*Lampetra japonica*) or 14 μm long (*L. planeri*) (Fig. 6.10A) but the nuclear membranes, bounding the nucleus externally and lining the canal, extend at least 6 μm or 34 μm, respectively, behind this, as does the central fiber (Stanley 1967, Fig. 6.10F,J,K,M,N,O; Jaana and Yamamoto 1981). Within the nucleus the canal is loosely helical with, in *L. planeri,* about 15 to 17 turns (Follenius 1965). Behind the chromatin the central fiber is thus surrounded by four concentric nuclear membranes (Follenius 1965; Stanley 1967, Fig. 6.10O; Jaana and Yamomoto 1981). There is a wide deep basal fossa (Fig. 6.10G,H, I, J; see below).

Mitochondria: In *Lampetra planeri* the elongate mitochondria, cristate with dense matrices, are arranged longitudinally along the proximal portion of the tail (Fig. 6.10K) although few lie adjacent to the central fiber-endonuclear canal complex. More distally they are numerous (Fig. 6.10P), seven or eight sometimes being observed in a single cross section (Stanley 1967). In *L.*

japonica the mitochondria are either at the beginning of the flagellum or further posterior (Jaana and Yamomoto 1981).

Cytoplasmic vesicles: In *Lampetra planeri* posteriorly to the central fiber complex one or more large, often elongate, cytoplasmic vesicles are present. Longitudinally beaded filaments lie between the mitochondria or vesicles and the plasma membrane (Fig. 6.10L). In some posterior regions of the tail flattened vesicles partly or completely surround the axoneme (Stanley 1967).

Centrioles and flagellum: Independently of the endonuclear canal, the chromatin cylinder in the sperm of *Lampetra planeri, L. fluviatilis and L. japonica* has a deep eccentric posterior invagination, the implantation fossa (Fig. 6.10G,H,I,J), containing two parallel centrioles of triplet construction (Fig. 6.10J). One of these, the basal body, gives origin to the typical 9+2 flagellum which has 9 accessory fibers (Follenius 1965; Nicander 1970; Jaana and Yamamoto 1981). As shown for *L. planeri* by Stanley (1967), each axonemal doublet has hollow subtubules and two [dynein] arms. The accessory fibers, small dense rods (outer dense fibers) adherent to the outer surface of each doublet (Fig. 6.10K,M-T) appear to be narrow continuations of dense material which surrounds each triplet of the basal body and is in turn continuous with an "epicentriolar body" which caps the basal body (6.10H,I). Fine filaments extend from the basal body to the nuclear membrane which lines the implantation fossa (Fig. 6.10I). These, with the epicentriolar body, are probably an anchoring apparatus, as suggested by their absence from the proximal (accessory) centriole. The dense fibers, in their small size and close proximity to the doublets, resemble those of reptilian rather than mature avian or mammalian sperm although similar to those of spermatids of these latter two groups (Stanley 1967). The occurrence of accessory fibers, known elsewhere only in internally fertilizing sperm, affords some support for the view that lampreys were formerly internally fertilizing. This would not necessitate regarding internal fertilization as basic to fishes as whole, though this is not improbable.

At the posterior tip of the flagellum, in *Lampetra planeri,* the axonemal elements terminate in a consistent proximal to distal order (Fig. 6.10S,T): the central two filaments, subtubule B, subtubule A of the doublets, the outer dense fibers and finally granular material of medium density central to and binding together and preventing disarray of the doublets (Stanley 1967). Whether triads of microtubules paralleling the axoneme in *L. fluviatilis* (Nicander 1968) indicate immaturity or are present in the definitive sperm is uncertain.

6.5.2 Family Mordaciidae

Mordacia: Some data on spermatogenesis of the parasitic mordaciid lamprey *Mordacia mordax* and the sympatric, and possibly descendant, non-parasitic M. *praecox* are given in an interesting paper comparing gametogenesis in these 'paired' species by Hughes and Potter (1969). The mature sperm of *Mordacia* *praecox* were not observed but the structure of the advanced spermatid was described and is presumably definitive.

It is not known whether a postacrosomal ring is present. The nucleus of the advanced *Mordacia* spermatid superficially bears at least three deep longitudinal folds or ridges which are somewhat spirally twisted. In the grooves between these are located mitochondria and extensive vacuoles, beneath the plasma membrane, a notable difference from the periaxonemal location of the mitochondria in *Lampetra* (though this may have some mitochondria around the base of the nucleus) and other vertebrates. The small acrosomal cap is separated from the nucleus by a complex of vacuoles. As in *Lampetra,* there is a deep implantation fossa containing two parallel triplet centrioles (a notable synapomorphy of *Mordacia* and *Lampetra*) and an endonuclear ('intranuclear') canal containing a central fiber which penetrates the tail. No mention is made of extension of the nuclear membranes around the fiber posterior to the chromatin. Reported absence of outer dense fibers accompanying the doublets, if confirmed, would be an interesting but presumably symplesiomorphic agreement with the hagfishes *Myxine* and *Eptatretus* (see **Section 6.3**, above).

Sperm phylogeny: Lampreys represent the first occurrence of an endonuclear canal and perforatorium in vertebrates. As these structures are present in most major groups up to and including Lissamphibia and Chelonia, it seems reasonable to conclude that they were acquired in the common ancestor of lampreys and gnathostomes and therefore that they are plesiomorphic for gnathostomes, as indicated in Fig. 6.11.

6.6 CHAPTER SUMMARY

The sperm of Petromyzodontiformes (lampreys) and Myxiniformes (hagfishes) show no similarities that would demand recognition of relationship between the two groups. The shared presence of an acrosome is considered a symplesiomorphy, being present in protochordates and cephalochordates. The absence of synapomorphies endorses other morphological and paleontological evidence that the Agnatha is an artificial group unified by the symplesiomorphy jawlessness. However, molecular analyses overwhelmingly support monophyly of the Agnatha. Remarkable homogeneity of sperm structure is seen within *Lampetra* and, of a very different nature, within the hagfish *Eptatretus.* In *Mordacia,* as in *Lampetra,* there is a deep implantation fossa containing two parallel triplet centrioles (a notable synapomorphy of *Mordacia* and *Lampetra*) and an endonuclear ('intranuclear') canal containing a central fiber which penetrates the tail. The sperm of *Mordacia* appears to differ from that of *Lampetra* in lacking outer dense fibers in the flagellum. Myxinforms lack an endonuclear canal but, like lampreys, produce an acrosomal process in the acrosome reaction. Unlike lampreys, a nuclear peg extends into the acrosome and two to four mitochondria extend almost the

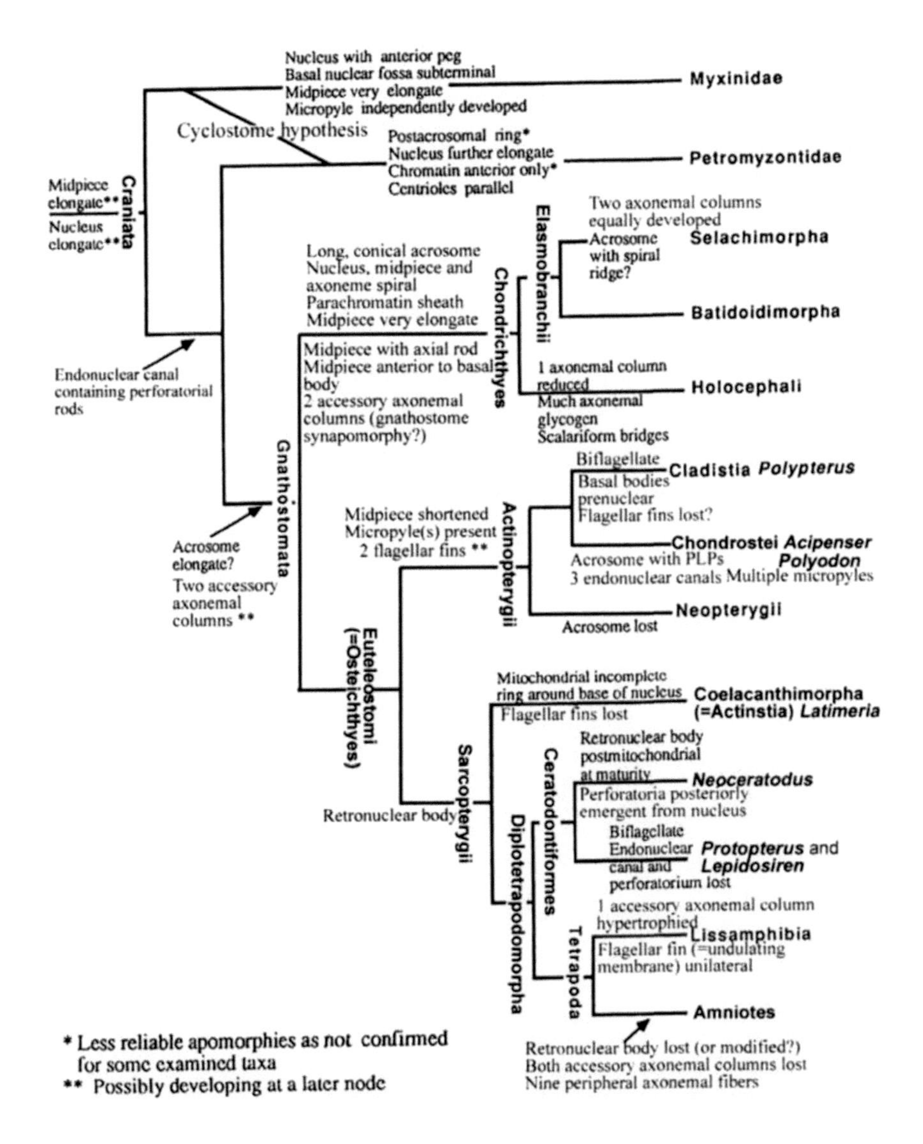

Fig. 6.11 Chief phylogenetic trends in vertebrate spermatozoa. Slightly modified after Jamieson, B. G. M. 1991. *Fish Evolution and Systematics: Evidence from Spermatozoa*. Cambridge University Press, Fig. 18.7.

entire length of the flagellum. In contrast, in *Lampetra*, although the mitochondria are elongate, they are restricted to the proximal part of the tail and some may surround the base of the nucleus. Whereas *Lampetra* and *Mordacia* have parallel centrioles they lie end to end in the myxinid *Eptatretus*.

6.7 LITERATURE CITED

Afzelius, B. A. 1970. Discussion in Nicander, L. 1970. Comparative studies on the fine structure of vertebrate spermatozoa. Pp. 47-55. In B. Baccetti (ed.), *Comparative Spermatology*. Academic Press, New York.

Afzelius, B. A. and Murray, A. 1957. The acrosomal reaction of spermatozoa during fertilization or treatment with egg water. Experimental Cell Research 12: 325-337.

Afzelius, B. A. and Sjöden, I. 1968. Fine structure of egg envelopes and the activation changes of cortical alveoli in the river lamprey, *Lampetra fluviatilis*. Journal of Embryology and Experimental Morphology 19: 311-318.

Alvestad-Graebner, I. and Adam, H. 1977. Zur Feinstruktur der spermatogenetischen Stadien von *Myxine glutinosa* L (Cyclostomata). Zoologica Scripta 6: 113-126.

Blair, J. E. and Hedges, S. B. 2005. Molecular phylogeny and divergence times of deuterostome animals. Molecular Biology and Evolution 22: 2275-2284.

Breder, C. M. and Rosen, D. E. 1966. *Modes of reproduction in fishes*. The American Museum of Natural History, The Natural History Press, New York. 941 pp.

Delarbre, C., Escriva, H., Gallut, C., Barriel, V., Kourilsky, P., Janvier, P., Laudet, V. and Gachelin, G. 2000. The complete nucleotide sequence of the mitochondrial DNA of the agnathan Lampetra fluviatilis: Bearings on the phylogeny of cyclostomes. Molecular Biology and Evolution 17: 519-529.

Fernholm, B. 1975. Ovulation and eggs of the Hagfish *Eptatretus burgeri*. Acta Zoologica (Stockholm) 56: 199-204.

Follenius, E. 1965. Particularités de structure des spermatozoïdes de *Lampetra planeri*: etude au microscope électronique. Journal of Ultrastructure Research 13: 459-468.

Hardisty, M. W. and Potter, I. C. 1971 (eds). The Biology of Lampreys. Academic Press, New York. 466 pp.

Hughes, R. L. and Potter, I. C. 1969. Studies on gametogenesis and fecundity in the lampreys *Mordacia praecox* and *M. mordax* (Petromyzonidae). Australian Journal of Zoology 17: 447-464.

Jaana, H. and Yamamoto, T. S. 1981. The ultrastructure of spermatozoa with a note on the formation of the acrosomal filament in the Lamprey *Lampetra japonica*. Japanese Journal of Ichthyology 28: 135-147.

Jamieson, B. G. M. 1991. *Fish Evolution and Systematics: Evidence from Spermatozoa*. Cambridge University Press, Cambridge. 319 pp.

Jespersen, Å. 1975. Fine structure of spermiogenesis in eastern Pacific species of Hagfish (Myxinidae). Acta Zoologica(Stockholm) 56: 189-198.

Kille, R. A. 1960. Fertilization of the lamprey egg. Experimental Cell Research 20: 12-27.

Kuraku, S., Hoshiyama, D., Katoh, K., Suga, H. and Miyata, T. 1999. Monophyly of lampreys and hagfishes supported by nuclear DNA-coded genes. Journal of Molecular Evolution 49: 729-735.

Longo, F. J. 1987. *Fertilization*. Chapman and Hall, New York. 183 pp.

Mallatt, J. and Winchell, C. J. 2007. Ribosomal RNA genes and deuterostome phylogeny revisited: More cyclostomes, elasmobranchs, reptiles, and a brittle star. Molecular Phylogenetics and Evolution 43: 1005-1022.

Morisawa, S. 1995. Fine structure of spermatozoa of the Hagfish *Eptatretus burgeri* (Agnatha). Biological Bulletin (Woods Hole) 189: 6-12.

Morisawa, S. 1999a. Acrosome reaction in spermatozoa of the hagfish *Eptatretus burgeri* (Agnatha). Development Growth and Differentiation 41: 109-112.

Morisawa, S. 1999b. Fine structure of micropylar region during late oogenesis in eggs of the hagfish Eptatretus burgeri (Agnatha). Development Growth and Differentiation 41: 611-618.

Morisawa, S. 2005. Spermiogenesis in the Hagfish *Eptatretus burgeri* (Agnatha). Biological Bulletin (Woods Hole) 209: 204-214.

Morisawa, S. and Cherr, G. N. 2002. Acrosome reaction in spermatozoa from hagfish (Agnatha) *Eptatretus burgeri* and *Eptatretus stouti*: Acrosomal exocytosis and identification of filamentous actin. Development Growth and Differentiation 44: 337-344.

Nelson, J. S. 1984. *Fishes of the World*, 2nd edition. John Wiley and Sons, New York. 523 pp.

Nelson, J. S. 2006. *Fishes of the World*, 4th edition. John Wiley and Sons, Inc., Hoboken, New Jersey. 601 pp.

Nicander, L. 1968. Gametogenesis and the ultrastructure of germ cells in vertebrates. Proceedings of the 6th International Congress on Artificial Reproduction in Animals, Paris 1: 89-107.

Nicander, L. 1970. Comparative studies on the fine structure of vertebrate spermatozoa. Pp. 47-55. In B. Baccetti (ed.), *Comparative Spermatology*. Academic Press, New York.

Nicander, L. and Sjödén, L. 1968. The acrosomal complex and the acrosomal reaction in spermatozoa of the river lamprey. Scandinavian Society for Electron Microscopy, Journal of Ultrastructure Research 25: 167-168.

Nicander, L. and Sjödén, L. 1971. An electron microscopical study of the acrosomal complex and its role in fertilization in the river lamprey, Lampetra fluviatilis. Journal of Sumbmicroscopic Cytology 3: 309-317.

Okkelberg, P. 1914. Volumetric changes in the egg of the brook lamprey, *Entosphenus (Lampetra) wilderi* (Gage), after fertilization. Biological Bulletin (Woods Hole) 26: 92-99.

Pfeiffer, D. C. and Vogl, A. W. 2002. Actin-related intercellular adhesion junctions in the germinal compartment of the testis in the hagfish (*Eptatretus stouti*) and lamprey (*Lampetra tridentatus*). Tissue and Cell 34: 450-459.

Rasmussen, A.-S., Janke, A. and Arnason, U. 1998. The mitochondrial DNA molecule of the hagfish (*Myxine glutinosa*) and vertebrate phylogeny. Journal of Molecular Evolution 46: 382-388.

Retzius, G. 1921. Die Spermien der Cyklostomen. Biologische Untersuchungen von G Retzius Nf 19: 43-56.

Rothschild, Lord 1958. Fertilization in fish and lampreys. Biological Reviews of the Cambridge Philosophical Society 33: 372-389.

Romer, A. S. and Parsons, T. S. 1977. *The Vertebrate Body*. W.B. Saunders Company, Philadelphia. 624 pp.

Stanley, H. P. 1967. The fine structure of spermatozoa in the lamprey *Lampetra planeri*. Journal of Ultrastructure Research 19: 84-99.

Takezaki, N., Figueroa, F., Zaleska-Rutczynska, Z. and Klein, J. 2003. Molecular phylogeny of early vertebrates: Monophyly of the Agnathans as revealed by sequences of 35 genes. Molecular Biology and Evolution 20: 287-292.

Walvig, F. 1963. The gonads and the formation of the sexual cells. Pp. 530-580. In A. Brodal and R. Fänge (ed.), *The Biology of Myxine*. Universitetsforlaget, Oslo.

CHAPTER 7

Ultrastructure of Spermatozoa: Euteleostomi (=Osteichthyes): Cladistia

B. G. M. Jamieson[1] *and X. Mattei*[2]

7.1 EUTELEOSTOMI

All the remaining, non-agnathan, non-chondrichthyan, extant fishes are members of the Subgrade (here termed Superclass) Euteleostomi (=Osteichthyes, the bony fishes). In strict cladistic terms this group also contains the tetrapods (the amphibian-reptilian-avian-mammalian assemblage), as rightly recognized by Lauder and Liem (1983). If the tetrapods are arbitrarily excluded a paraphyletic group is obtained.

Nelson (2006) recognizes division of the Euteleostomi into two classes, the Actinopterygii and the Sarcopterygii. The Actinopterygii are the ray-finned fishes and contain the subclasses Cladistia, Chondrostei and Neopterygii. The Sarcopterygii are divided into the subclass Coelacanthimorpha (Actinistia) and the subclass Dipnotetrapodomorpha (named for the contained lungfishes and tetrapods) (Fig. 7.1).

7.2 CLASS ACTINOPTERYGII

The Actinopterygii (ray-finned fishes) are so named because of the presence of dermal, segmented, ray-like supports within the fins. Among living forms, they are considered to be the sister-group of the Sarcopterygii. Inclusion of the Cladistia (e.g. *Polypterus)* in the Actinopterygii, as advocated by Wiley (1979); Patterson (1982); Jamieson (1991) and Nelson (2006) and their position within the ray-finned fishes have, however, been the subject of debate.

[1]School of Integrative Biology, University of Queensland, Brisbane, Queensland 4072, Australia

[2]Parasites et écosystèmes méditerranéens, Université de Corse, F-20250 Corte, France

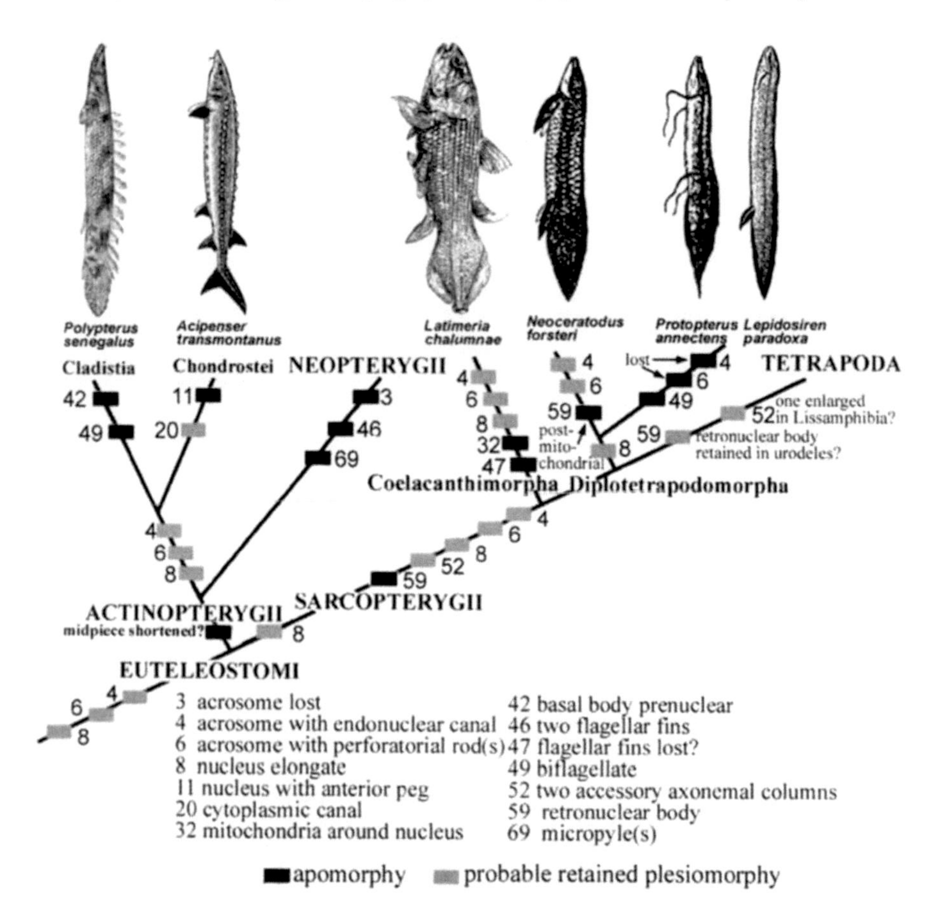

Fig. 7.1 Phylogeny of the Euteleostomi (=Osteichthyes) based on somatic analyses referred to in this chapter and, for Cladistia and Acipenseriformes, the molecular analysis of Li, Lu, and Ortí, G. (2008) based on ten nuclear loci. Spermatozoal apomorphies are here added. Some spermatozoal plesiomorphies are repeated on the tree for their descriptive value. Note that Cladistia and Chondrostei are depicted as sister-groups but that Ortí and Li, in Chapter 1, represent Chondrostei (Acipenseriformes) as the sister group of the Neopterygii and the Cladistia as the sister group of all other Actinopterygii (Chondrostei+Neopterygii). Character numbers refer to a previous analysis in Jamieson, B. G. M. 1991. Fish Evolution and Systematics: Evidence from Spermatozoa. Cambridge University Press, Cambridge, Fig. 9.5. Original. Adult fishes after Norman and Greenwood.

Phylogenetic analysis of the structure and organization of the mitochondrial genome of *Polypterus* supports its placement as the most basal living member of the ray-finned fishes and to rule out its placement with the lobe-finned fishes [Sarcopterygii] (Noack *et al.* 1996; Inoue *et al.* 2003) (Fig. 7.2) as supported by Ortí and Li (**Chapter 1** and references therein) (Fig. 1.3).

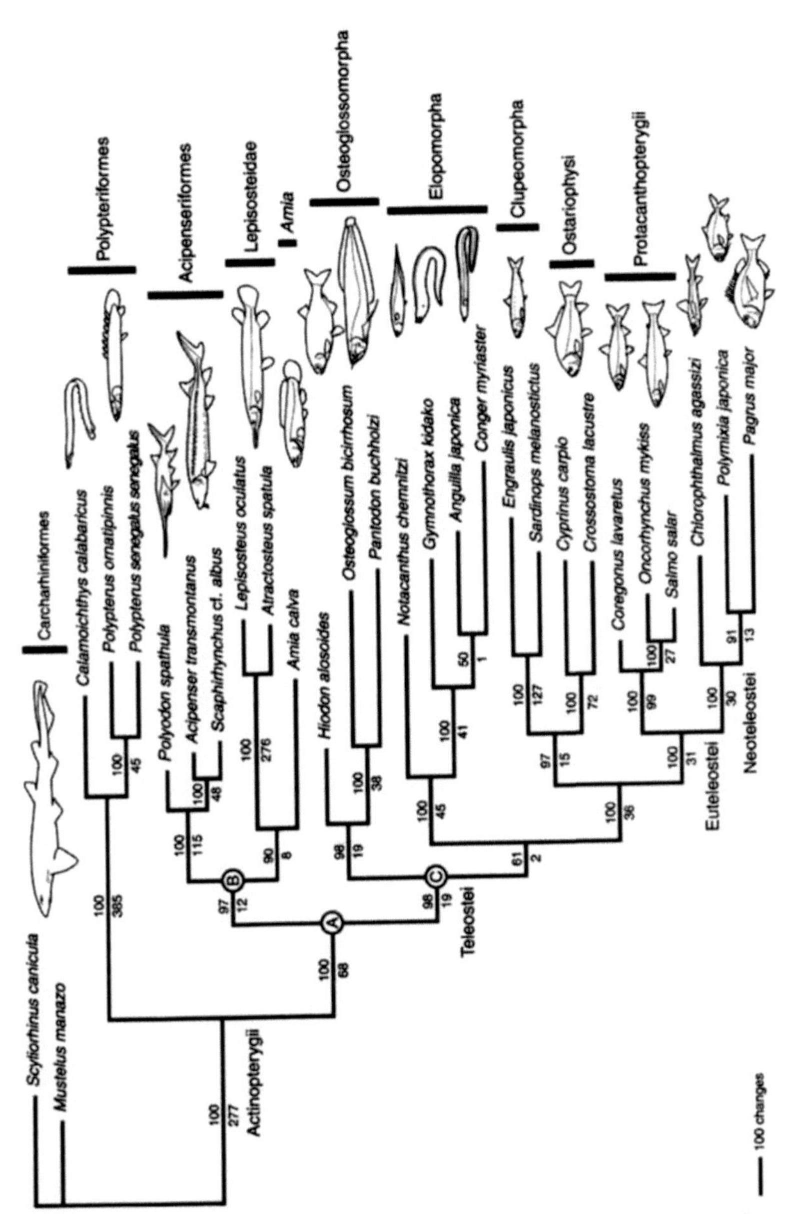

Fig. 7.2 A single most-parsimonious (MP)tree derived from unweighted analysis of mitogenomic data comprising concatenated nucleotide sequences from 12 protein-coding (excluding the ND6 gene and third codon positions)and 22 transfer RNA (tRNA)genes (stem regions only) from 28 fish species. Numbers above and below internal branches indicate jackknife values and decay indices,respectively. Here, as in Chapter 1, the Cladistia (*Calamoicththys+Polypterus*) are shown as the sister group of all other Actinopterygii. From Inoue, J. G., Miya, M., Tsukamoto, K. and Nishida, M. 2003. Molecular Phylogenetics and Evolution 26: 110-120, Fig. 2. With permission of Elsevier.

However, an analysis by Li, Lu and Ortí (2008) while supporting the actinopterygian status of *Polypterus*, places it as the sister-taxon of the Acipenseriformes (represented, however, only by *Polyodon*). In figure 7.1, representing spermatozoal evolution, a sister relationship of *Polypterus* and Acipenseriformes is also shown but the evidence for the more conventional view that the Cladistia, including *Polypterus*, are sister group of all other ray-finned fishes is strong. Nevertheless, the phylogenetic position of *Polypterus*, as is so often the case for enigmatic groups, awaits definitive resolution.

Analysis of nearly complete 28S and I SS rRNA genes also placed *Polypterus* in a monophyletic Actinopterygii but appeared equivocal as to relationships of these to the Sarcopterygii (Mallatt and Winchell 2007).

It is of interest that Bachmann *et al.* (1972) showed that *Polypterus palmas* has the highest level of DNA per nucleus that has been reported for any species of fishes except the Dipnoi. His conclusion that this lends support to the view that the Polypteridae are better grouped with the Sarcopterygii than with the Actinopterygii may be questioned, however. If polypterids were at the base of the Actinopterygii, near the point of departure of the Sarcopterygii from a common ancestor, it might be expected that the basal members of each group would share a similar (high) DNA content.

7.3 SUBCLASS CLADISTIA: ORDER POLYPTERIFORMES (BRACHYOPTERYGII)

Polypteriforms are the bichirs, in the single family Polypteridae with two genera, *Polypterus* (Fig. 7.3) and *Erpetoichthys*, restricted to tropical African fresh waters.

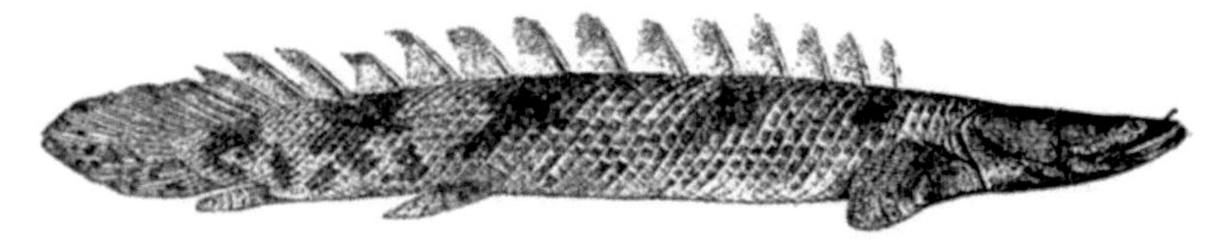

Fig. 7.3 *Polypterus*, the Bichir (Polypteridae). From Norman, J. R. 1937. *Illustrated Guide to the Fish Gallery*. Trustees of the British Museum, London, Fig. 18.

Sperm literature: The ultrastructure of the sperm of *Polypterus senegalus*, the Bichir, has been briefly described by Mattei (1969, 1970) in accounts chiefly referring to the spermatid and has been summarized in an overview of fish sperm by Mattei (1991).

Polypterus senegalus: The spermatozoon of *P. senegalus* (Fig. 7.5) is biflagellate, a curious resemblance, but the only notable spermatological one, to the dipnoan *Protopterus* and *Lepidosiren*. This resemblance appears to be convergent (homoplasic) and not a synapomorphy (see DNA phylogeny above).

Biflagellarity is also seen in 16 teleost families: Amblycipitidae, Apogonidae, Ariidae, Aspredinidae, Batrachoididae, Cetopsidae, Cichlidae, Doradidae, Gobiesocidae, Gobiidae, Heptapteridae Ictaluridae, Malapteruridae, Myctophidae, Nematogenyiidae and Zoarchidae, as in some Amphibia.

General sperm ultrastructure: The *Polypterus* sperm nucleus (Figs. 7.4, 7.5B) is elongate tear-shaped, with a small, apically rounded acrosome on the narrow anterior end. In the acrosome a structure identified by Mattei (1969; 1970) as the acrosome granule is always present (Fig. 7.5B,C); this is perhaps homologous with subacrosomal material in some sarcopterygian sperm. The two centrioles are situated at the same level, slightly below the acrosome (Fig. 7.5B). Migration of the mitochondria towards the centrioles which generally

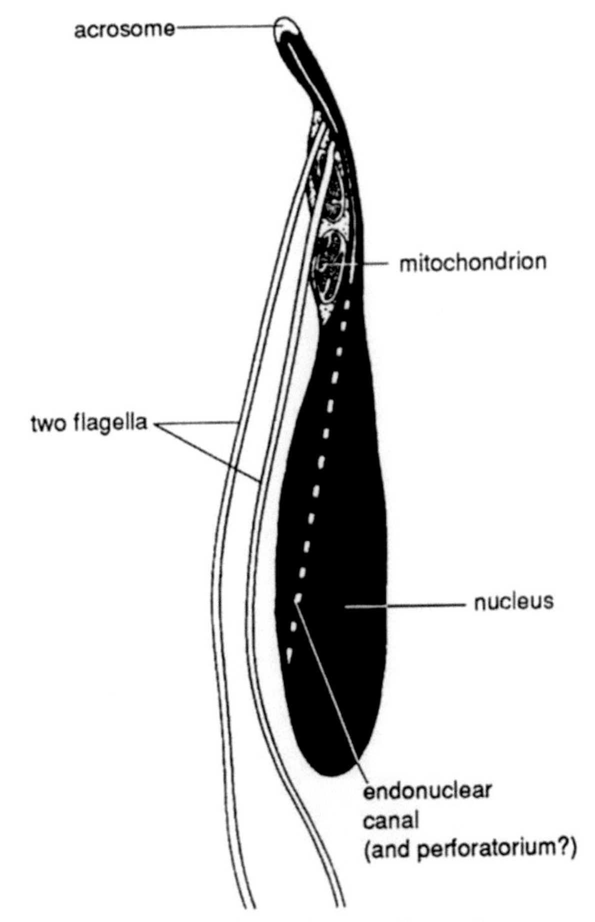

Fig. 7.4 *Polypterus senegalus*. Spermatozoon. *After a drawing by Mattei, X. 1969.* Contribution à l'étude de la spermiogenèse et des spermatozoïdes de poissons par les méthodes de la microscopie électronique. *D.Sc. thesis, Université de Montpellier, Fig. 7c.*

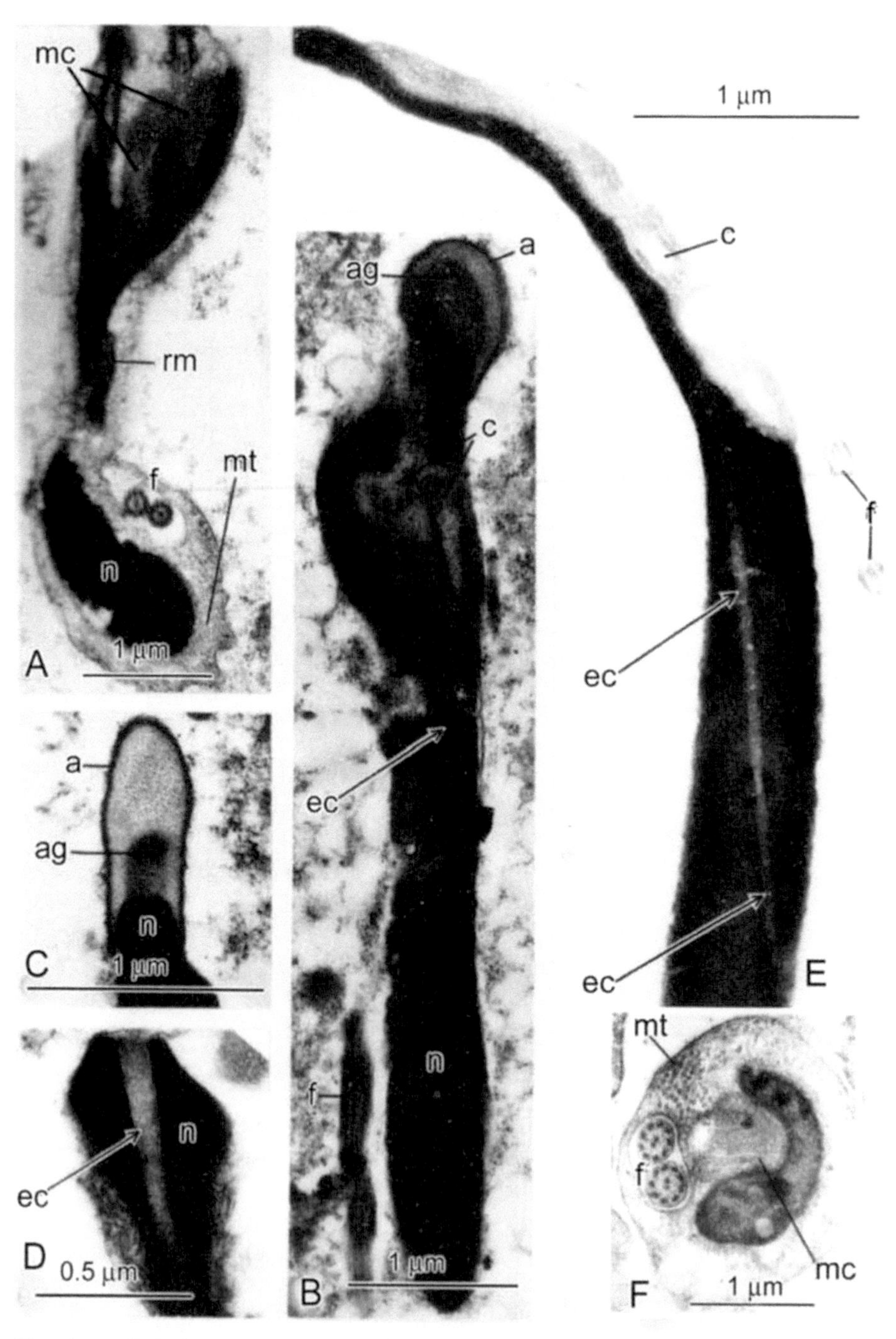

Fig. 7.5 *Polypterus senegalus* (Polypteridae). Sagittal sections: **A.** Of a late spermatid showing the disposition of the mitochondria constituting the chondriome.

Fig. 7.5 Contd. ...

occurs late in spermiogenesis occurs earlier in *Polypterus*, before the centrioles have elaborated their flagella. The flagella become entirely free. The chondriome resolves into two or three large mitochondria occupying the concavity which the nucleus presents behind the centrioles. An endonuclear canal [and perforatorium?] penetrates the nucleus (Fig. 4.7B,D,E); the canal is axial in the anterior region of the nucleus; posteriorly it crosses the nucleus obliquely and ends against the nuclear membrane 2 or 3 µm from the distal end of the nucleus (Fig. 4.7E). (A course, curiously, seen also in *Neoceratodus*). The two flagella and the chondriome are illustrated for a late spermatid in Figure 7.5F. Flagellar fins have not been observed.

Egg micropyle: Egg micropyles were thought to be absent in Cladistia but a single micropyle has been demonstrated in *Erpetoichthys calabaricus*, *Polypterus senegalus*, and *P. ornatipinnis* (Bartsch and Britz 1997; Britsch and Bartsch 1998). As micropyles are now known for the Cladistia, Acipenseriformes and Neopterygii but are apparently absent from the Petromyzontiformes, it is here concluded that their presence is a synapomorphy for the Euteleostomi, assuming that development in myxiniforms was convergent. Evolution of the micropyle must therefore have long predated loss of the acrosome which is diagnostic of the Neopterygii.

7.4 CHAPTER SUMMARY

Spermatozoal ultrastructure yields little indication of the relationships of the enigmatic *Polypterus*. The absence of a retronuclear body and of flagellar fins are notable differences from sarcopterygian fishes but with these and the sturgeons it retains the acrosome and the endonuclear canal (with contained perforatorium) here considered to be basic plesiomorphies of gnathostomes. The acrosome and perforatorium in the endonuclear canal, though plesiomorphic, do not detract from relationship with the sturgeons which is indicated from molecular studies (see Ortí and Li, **Chapter 1**) but there are no recognizable spermatozoal synapomorphies with acipenseriforms. The biflagellate condition is clearly an independent development in *Polypterus*

Fig. 7.5 Contd. ...

A residual mitochondrion seems to persist against the nucleus; it would probably be eliminated with the cytoplasm and the manchette. **B.** Of the anterior region of a spermatozoon. **C.** At the level of the acrosome. **D.** Of the anterior region, showing the endonuclear canal. **E.** Of the nucleus, showing the endonuclear canal. **F.** Transverse section of a spermatid at the level of the chondriome, showing the two flagella. a, acrosome; ag, acrosome granule; c, centrioles; ec, endonuclear canal; f,flagella; mc, mitochondria associated with the centrioles; mt, microtubules of the manchette; n, nucleus; rm, residual mitochondrion. Adapted from Mattei, X. 1969. Contribution à l'étude de la spermiogenèse et des spermatozoïdes de poissons par les méthodes de la microscopie électronique. D.Sc. thesis, Université de Montpellier, Plates XIIe and XIIIa-e.

relative to the at least 15 neoteleost families in which it is known as is origin of the flagella laterally near the anterior end of the nucleus, seen, for instance, in some Characiformes and Scorpaeniformes.

7.5 LITERATURE CITED

Bachmann, K., Goin, O. B. and Goin, C. J. 1972. The Nuclear DNA of *Polypterus palmas*. Copeia 1972: 363-365.

Bartsch, P. and Britz, R. 1997. A single micropyle in the eggs of the most primitive living actinopterygian fish *Polypterus*. Journal of Zoology (London) 241: 589-592.

Britz, R. and Bartsch, P. 1998. On the reproduction and early development of *Erpetoichthys calabaricus, Polypterus senegalus*, and *P. ornatipinnis* (Actinopterygii: Polypteridae). Ichthyological Explorations of Freshwaters 9: 325-334.

Jamieson, B. G. M. 1991. *Fish Evolution and Systematics: Evidence from Spermatozoa*. Cambridge University Press, Cambridge. 319pp.

Lauder, G. V. and Liem, K. F. 1983. The evolution and interrelationships of the actinopterygian fishes. Bulletin of the Museum of Comparative Zoology 150: 95-197.

Mallatt, J. and Winchell, C. J. 2007. Ribosomal RNA genes and deuterostome phylogeny revisited: More cyclostomes, elasmobranchs, reptiles, and a brittle star. Molecular Phylogenetics and Evolution 43: 1005-1022.

Mattei, X. 1969. *Contribution à l'étude de la spermiogenèse et des spermatozoïdes de poissons par les méthodes de la microscopie électronique*. D.Sc. thesis, Université de Montpellier.

Mattei, X. 1970. Spermiogenèse comparée des poissons. Pp. 57-69. In B. Baccetti (ed.), *Comparative Spermatology*. Academic Press, New York.

Mattei, X. 1991. Spermatozoon ultrastructure and its systematic implications in fishes. Canadian Journal of Zoology 69: 3038-3055.

Nelson, J. S. 2006. *Fishes of the World*, 4th edition. John Wiley and Sons, Inc., Hoboken, New Jersey.

Noack, K., Zardoya, R. and Meyer, A. 1996. The complete mitochondrial DNA sequence of the bichir (*Polypterus ornatipinnis*), a basal ray-finned fish: Ancient establishment of the consensus vertebrate gene order. Genetics 144: 1165-1180.

Patterson, C. 1982. Morphology and interrelationships of primitive actinopterygian fishes. American Zoologist 22: 241-259.

Wiley, E. O. 1979. Ventral gill arch muscles and the interrelationships of gnathostomes, with a new classification of the Vertebrata. Zoological Journal of the Linnean Society 67: 149-179.

CHAPTER 8

Ultrastructure of Spermatozoa: Chondrostei

Barrie G. M. Jamieson

8.1 SUBCLASS CHONDROSTEI

Living members of the Chondrostei are the northern hemisphere anadromous and freshwater sturgeons (Fig. 8.1) and the paddlefish of freshwaters of N. America and China (Nelson 2006) (Fig. 8.7). Despite the fact that several synapomorphies are reductive, the Chondrostei appear to comprise a monophyletic group. Among living forms they are usually considered to be the sister-group of the Neopterygii (Lauder and Liem 1983; Nelson 2006) and this is supported by Ortí and Li (**Chapter 1**) from an analysis of nuclear DNA sequences for 8 genes and 11,766 base pairs in 29 taxa. However, Li, Lu and Ortí (2008), in an analysis of 10 nuclear genes and 7995 base pairs, find the Chondrostei (represented in their analysis only by *Polyodon*) to be the sister group of *Polypterus*. If so, placement of *Polypterus* in the Chondrostei would be permissible as in the spermatozoal phylogeny in **Chapter 7** (Fig. 7.1). However, the evidence for a sister relationship of Cladistia and all other ray-finned fishes has strong support (see **Chapters 1** and **7**).

The only spermatozoa apomorphy shared between Chondrostei and Neopterygii appears to be the presence of flagellar fins. These are, absent presumed lost, in Cladistia and are absent from many Neopterygii and the possibility of homoplasy between Chondrostei and Neopterygii cannot be ruled out.

8.2 ORDER ACIPENSERIFORMES

There are two extant families: the Acipenseridae and the Polyodontidae.

8.2.1 Family Acipenseridae

These are the sturgeons, anadromous and freshwater in the Northern Hemisphere (Nelson 2006).

School of Integrative Biology, University of Queensland, Brisbane, Queensland 4072, Australia

Sperm literature: Sturgeon sperm for which ultrastructural data are available are: Russian Sturgeon, *Acipenser guldenstadtii* (Ginsburg 1968; Mitina *et al.* 1992); White Sturgeon, *A. transmontanus* (Cherr and Clark 1984a,b; Cherr and Clark 1985) (adult, Fig. 8.1A); Stellate Sturgeon, *A. stellatus* (Ginsburg 1968; Mitina *et al.* 1992); Atlantic Sturgeon, *A. oxyrinchus* [correction for *A. oxyrhynchus*](DiLauro *et al.* 1998) (adult, Fig. 8.1B); Shortnose Sturgeon, *A. brevirostrum* (DiLauro *et al.* 1999); Lake Sturgeon, *A. fulvescens* (DiLauro *et al.* 2000) (adult, Fig. 8.1C); Pallid Sturgeon, *Scaphirhynchus albus* (DiLauro *et al.* 2001); Siberian Sturgeon, *Acipenser baerii* (Psenicka *et al.* 2006a; Psenicka 2007); Chinese Sturgeon, *Acipenser sinensis* (Wei *et al.* 2007); and the Huso, *Huso huso* (Ginsburg 1968). The fine structure of spermiogenesis in a hybrid sturgeon, the so-called bester (*Huso huso* female x *Acipenser ruthenus* male) has also been investigated (Amiri and Takahashi 2006).

Fig. 8.1 *Acipenser* (Acipenseridae). **A**. *A. transmontanus*, White sturgeon. **B**. *A. oxyrinchus*, Atlantic sturgeon. **C**. *A. fulvescens*, Lake sturgeon. By Norman Weaver. With kind permission of Sarah Starsmore.

Comparative account of sturgeon sperm: This account is based on the ultrastructural works listed above.

General morphological characteristics: Sturgeon sperm are differentiated into a head with an acrosome, a midpiece and a flagellar region with 9+2 pattern of microtubules and a narrow endpiece. Comparisons of morphological parameters in different species of sturgeon (Table 8.1) reveal interspecies variation (Psenicka 2007) (Fig. 8.3). Scanning electron micrographs are given for *Scaphirhynchus albus*, Pallid Sturgeon (Fig. 8.2A) and *Acipenser oxyrinchus*, Atlantic Sturgeon (Fig. 8.4). The general ultrastructural features are shown in Fig. 8.5, drawn from (*Acipenser transmontanus*, the White Sturgeon).

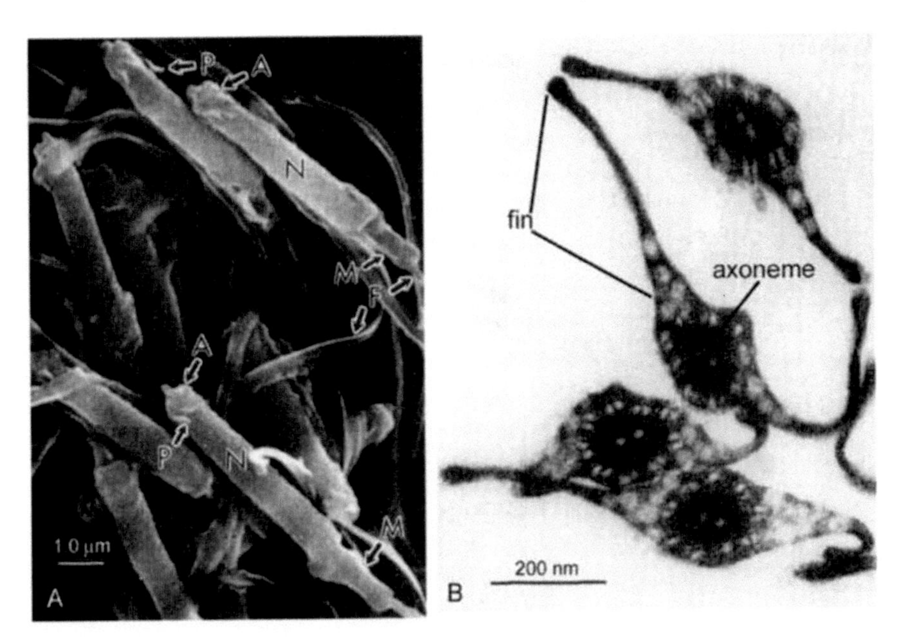

Fig. 8.2 *Scaphirhynchus albus*, Pallid Sturgeon (Acipenseridae). **A**. Scanning electron micrograph of spermatozoa. A, acrosome; F, flagellum; M, midpiece; N, nucleus; P, posterolateral projections. **B**. TEM transverse sections of flagella. Adapted from Dilauro, M. N., Walsh, R. A., Peiffer, M. and Bennett, R. M. 2001. Canadian Journal of Zoology 79: 802-808, Figs. 2 and 7. With permission of NRC Research Press.

Table 8.1 Morphological comparisons of spermatozoa in sturgeon species. Slightly modified after Psenicka, M. 2007. Biology of the Cell 99: 103-115, Table 4, augmented from Wei, Q., Li, P., Psenicka, M., Alavi, S., Shen, L., Liu, J., Peknicova, J. and Linhart, O. 2007. Theriogenology 67: 1269-1278.

Sturgeon species	*Acrosome length µm*	*Acrosome width µm*	*Nuclear length µm*	*Anterior nucleus width µm*	*Posterior nucleus width µm*	*Midpiece length µm*
Stellate	0.97	1.22	6.66	0.98	1.49	3.43
White	1.31	1.34	9.21	1.25	1.44	1-2.13*
Atlantic	0.83	1.00	3.15	0.92	0.55	1.37
	±0.11	±0.07	±0.36	±0.06	±0.08	±0.16
Shortnose	0.78	0.91	6.99	0.75	1.21	1.91
	±0.08	±0.06	±0.83	±0.11	±0.12	±0.35
Lake	0.73	0.81	5.69	0.86	1.04	2.88
	±0.14	±0.07	±0.43	±0.07	±0.08	±0.43
Pallid	1.07	0.82	3.78	0.68	0.89	1.23
	±0.10	±0.06	±0.33	±0.04	±0.08	±0.18
Siberian	0.95	0.93	4.98	0.87	0.87	1.09
	±0.17	±0.12	±0.83	±0.13	±0.18	±0.43

Table 8.1 Contd. ...

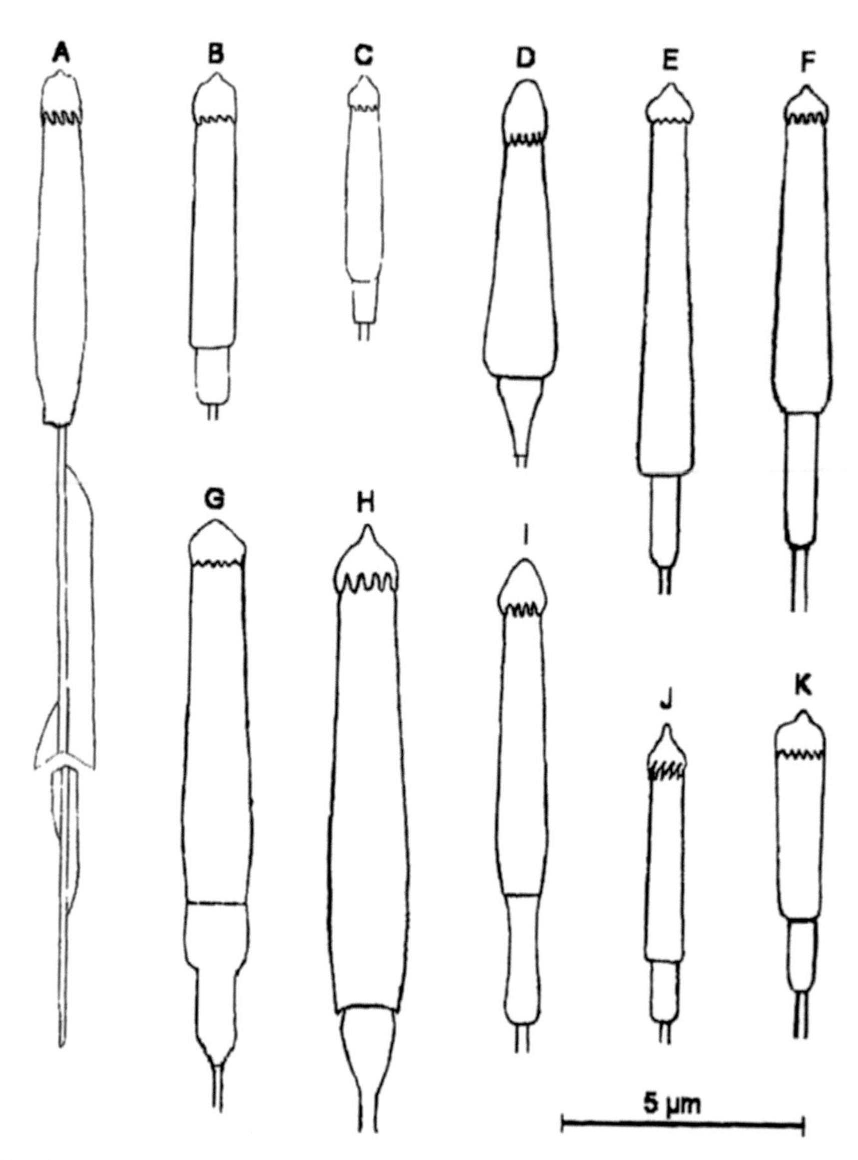

Fig. 8.3 Drawings, to scale, of the various described sperm cells of several different sturgeon species **A**. *Acipenser baerii* (Psenicka 2007). **B**. *Polyodon spathula* (Psenicka 2007). **C**. *A. ruthenus* (Psenicka 2007). **D**. *A. sinensis* (Xu and Xiong, 1988). **E**. *A. brevirostrum* (DiLauro *et al.*, 1999). **F**. *A. fulvescens* (DiLauro et al., 2000). **G**. *A. stellatus* (Ginsburg, 1977). **H**. *A. transmontanus* (Cherr and Clark, 1984; 1985). **I**. *A. gueldenstaedti colchicus* (Ginsburg, 1968). **J**. *Scaphirhynchus albus* (DiLauro et al., 2001). **K**. *A. oxyrinchus* (DiLauro *et al.*, 1998). Slightly modified from Psenicka, M. 2007. Biology of the Cell 99: 103-115, Fig. 7. Reproduced with permission of Portland Press.

Table 8.1 Contd. ...

Sturgeon species	Acrosome length μm	Acrosome width μm	Nuclear length μm	Anterior nucleus width μm	Posterior nucleus width μm	Midpiece length μm
Chinese	0.54 ±0.15	0.68 ±0.06	2.73	0.59 ±0.05	1.84 ±0.45	2.17 ±0.36

Sturgeon species	Midpiece width μm	Acrosome head & midpiece length μm	Flagellum length μm	Total length μm	n	Reference
Stellate	1.38	11.05	40.70	51.75	1	(Ginsburg 1977)
White	1.08	11.82	30-40	41.82-51.82	1	(Cherr and Clark 1984a; Cherr and Clark 1985)
Atlantic	0.51 ±0.07	5.66 ±0.37	37.08	42.74	12	(DiLauro *et al.* 1998)
Shortnose	0.81 ±0.09	9.71 ±0.73	36.70	48.41	15	(DiLauro *et al.* 1999)
Lake	0.70 ±0.08	9.10 ±0.53	47.53	56.63	14	(DiLauro *et al.* 2000)
Pallid	0.67 ±0.08	8.07 ±0.48	37.16	43.23	16	(DiLauro *et al.* 2001)
Siberian	0.81 (A), 0.57 (P)	7.01 ±0.83	44.75 ±4.93	51.76	8	(Psenicka 2007)
Chinese	1.57 ±0.27	5.44	33.26 ±2.74	38.7 ±0.37	–	(Wei *et al.* 2007)

* Mitochondria extend into an 8 μm long collar (Cherr and Clark 1984a)

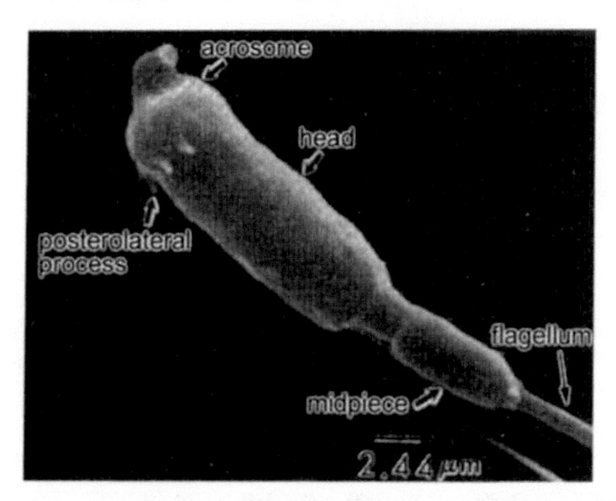

Fig. 8.4 *Acipenser oxyrinchus*, Atlantic Sturgeon (Acipenseridae). Scanning electron micrograph of a spermatozoon. Adapted from Dilauro, M. N., Kaboord, W., Walsh, R. A., Krise, W. F. and Hendrix, M. A. 1998. Canadian Journal of Zoology 76: 1822-1836, Fig. 3B. With permission of NRC Research Press.

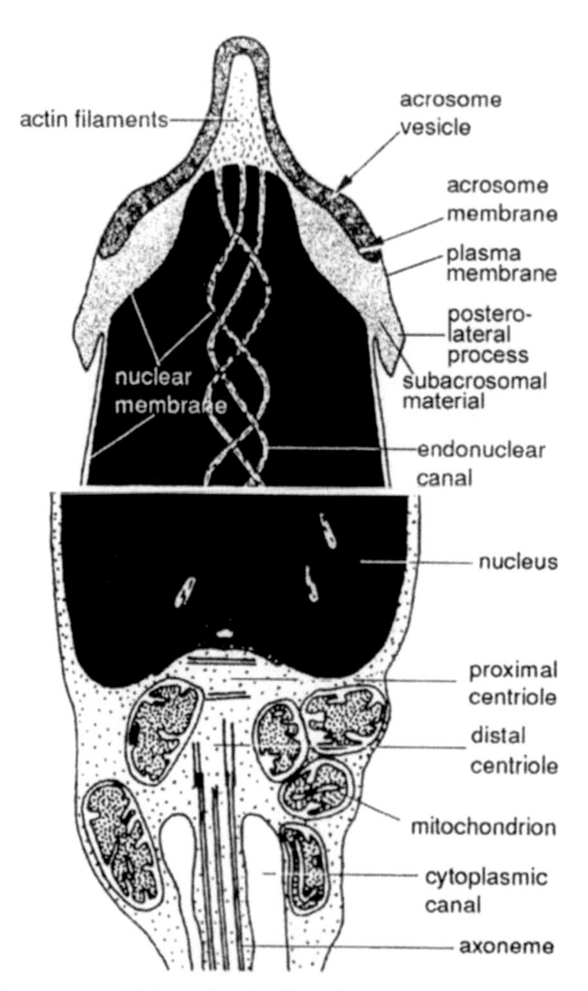

Fig. 8.5 *Acipenser transmontanus*, White Sturgeon (Acipenseridae). Diagram of a longitudinal section of the spermatozoon. The long middle region of the nucleus is omitted. From Jamieson 1991, adapted from Cherr, G. N. and Clark, W. H. 1984. The Journal of Experimental Zoology 232: 129-139, Figs. 3,7 and 10.

Acrosome: The sturgeons have not only egg micropyles but also a sperm acrosome. The egg has a thick impenetrable envelope perforated by numerous micropyles, the number of which varies within and among individual fish, and between species. This may allow polyspermic fertilization which, if it occurs, is pathological (Ginsburg 1968, p. 61; p. 183).

As previously noted (**Chapter 6**), there is a high positive correlation between absence of a preformed perforatorium or of an acrosome (Longo 1987; Jamieson 1991) and presence of egg micropyles in fishes. The micropyle provides a passage for the spermatozoon through the egg membranes (Yamamoto 1961; Ginsburg 1968; Jamieson 1991; Kudo *et al.* 1994; Linhart *et* *al.* 1995; Riehl and Patzner 1998; Lahnsteiner and Patzner 1999). An acrosomal process occurs in lampreys, sharks, and lungfish, none of which possesses an egg micropyle. The neopterygian *Lepidogalaxias,* unique in teleosts in possessing a true acrosome (Leung 1988), also lacks a micropyle. Exceptions in which an acrosome and micropyle coexist are myxinids, *Polypterus,* sturgeons and paddlefishes. In myxinids an acrosomal process is produced but there are no endonuclear canals. *Polypterus* has an endonuclear canal and putative perforatorium and a single micropyle (Bartsch and Britz 1997) (see **Chapter 7**). Multiple micropyles in the egg of sturgeons (Markov 1975; Poduschka 1993a,b; Debus *et al.* 2002) and paddlefish (Linhart 1997) coexist with one (?) to three endonuclear canals and perforatoria. These data suggest that in ancestral Actinopterygii, as represented by the phylogenetically basal Cladistia and Acipenseriformes, an acrosome with perforatoria in endonuclear canals coexisted with one or more micropyles. More derived actinopterygians dispensed with the acrosome while keeping a single (rarely multiple) micropyle and shortened the nucleus which is still slightly elongated in *Lepisosteus* but is round-headed, as the anacrosomal aquasperm, in *Amia* and in externally fertilizing teleosts (see **Chapters 9-15**). It is not known whether the block to polyspermy which exists in teleosts, in contrast to *Acipenser,* developed only in teleosts. It should be noted that the type of filamentous or rod-like perforatorium in lampreys, *Polypterus,* acipenseriforms, coelacanth, *Neoceratodus* and some amphibians may be distinguished as the 'preformed acrosome filament' *sensu* Afzelius (2006) and that the term perforatorium may be limited to subacrosomal material, also capable of undergoing the acrosome reaction, which is not organized as a filament.

The acrosome of the Pallid Sturgeon supposedly differs from that of the Atlantic and Shortnose Sturgeons in forming a hollow cone whereas the acrosomes of the other species resemble the cap of an acorn (DiLauro et al. 1998, 1999, 2000), but this difference is not apparent in Fig. 8.3.

Acrosomal dimensions vary moderately in length and width (Table 8.1). In all described sturgeon sperm cells, the acrosome is fringed posteriorly by radial lobe-like posterolateral projections (PLPs) (DiLauro *et al.* 1998, 1999, 2000, 2001; Psenicka *et al.* 2006a). Numbers and sizes of PLPs vary between species. Ten PLPs are present in Sibèrian Sturgeon (Psenicka *et al.* 2006a; Psenicka 2007) (Fig. 8.6) which differs from the Pallid Sturgeon which has eight PLPs (DiLauro *et al.* 2001). The ten PLPs in *A. baeri* have mean lengths of 0.94 ± 0.15 μm and widths of 0.93 ± 0.11 μm (Psenicka *et al.* 2006b; Psenicka 2007).

There is high inter-specific variation in the sizes of PLPs: the lengths of the PLPs in the Lake *(Acipenser fulvescens*), Atlantic *(A. oxyrinchus*), Shortnose *(A. brevirostrum*), Siberian (*A. baerii*) and Pallid Sturgeon (*Scaphirhynchus albus*) have been reported as 324, 233, 246, 600 and 760 nm respectively. Their function remains to be demonstrated. DiLauro *et al.* (1998, 1999, 2000) ascribed a barb or holdfast function to these PLPs during the fertilization

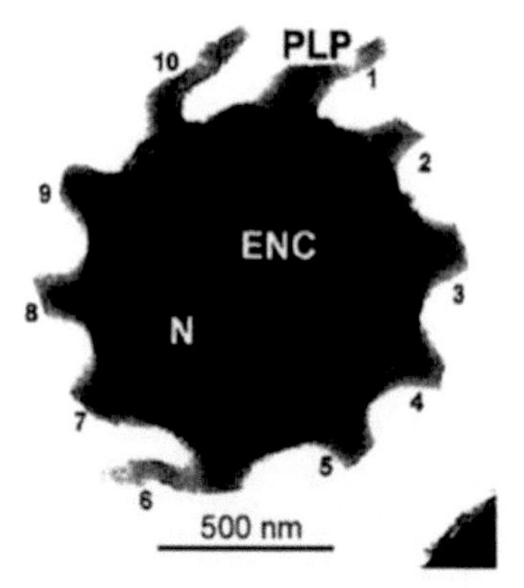

Fig. 8.6 *Acipenser baerii*, Siberian Sturgeon (Acipenseridae). TEM transverse section at region between the acrosome and the nucleus, showing the posterolateral projections (PLP) and the endonuclear canals (ENC) penetrating the nucleus (N). After Psenicka, M. 2007. Biology of the Cell 99: 103-115, Fig. 4b. With permission of Portland Press Ltd.

process. They further suggested (DiLauro *et al.* 2001) that the spiral form of the PLPs in the Pallid Sturgeon (*Scaphirhynchus albus*) might represent yet another adaptation for attachment to the egg cell during fertilization and that its unusual PLP arrangement may represent an adaptation to some structural factor, perhaps within the egg cell. The length of the PLPs of Pallid Sturgeon, at 0.76 μm, is also longer than that of any of the other sturgeons studied and fully twice those of Atlantic Sturgeon and 58% longer than the mean length of the Lake Sturgeon PLP. The hollow conical form of the acrosome, the triangular form and spiral configuration of the PLPs, and their greater length, are differences in Pallid Sturgeon sperm consistent with its placement in a separate genus, *Scaphirhynchus*, from *Acipenser*.

The acrosome of the hybrid, Bester, resembles that of other sturgeons. Other features of the spermatozoon, for instance the perpendicular centrioles, are also similar though the number of mitochondria appears to be increased and was related to a longer viability of the spermatozoa than in other species (Amiri and Takahashi 2006).

Acrosome reaction: The acrosome reaction in sturgeon sperm and its natural inducer ionic controls have been described (Detlaf and Ginzburg 1963; Cherr and Clark 1984a,b; Cherr and Clark 1985) but are imperfectly understood. Although *Acipenser* sperm are spawned in freshwater, the ionic controls of the acrosome reaction appear to be similar to those reported in the sperm of marine invertebrates (Cherr and Clark 1984a).

The egg envelope consists of four layers, the outermost layer being an adhesive jelly coat (Cherr and Clark 1984b). This jelly does not contain acrosome reaction inducing activity. However, an acrosome reaction is experimentally induced by ionophore A23187 with obligate presence of Ca^{2+} and Mg^{2+}; high Ca^{2+}; high pH, optimally 9.0; or egg water, of which the active component is a 66 KD glycoprotein (Cherr and Clark 1984a). Detlaf and Ginzburg (1963) also reported that Ca^{2+} is required for the acrosome reaction in sperm of sturgeon and other species; their finding that that Mg2+ could be substituted for Ca^{2+} was not confirmed for *A. transmontanus*. The acrosome undergoes exocytosis and a process approximately 10 µm long and 0.1 µm wide is produced from the subacrosomal material. Some of the filamentous material, presumably actin, extends through the subacrosomal region and into the acrosomal process. No observable change occurs in the nuclear canals as a result of the acrosome reaction (though the acrosomal process is said to originate from the nuclear canals in addition to the subacrosomal region). However, the lateral margins of the subacrosomal region exhibit a decrease in the thickness of the granular region which lies lateral to the subacrosome and below the acrosome vesicle. It is suggested that H+ influx (internal alkalinization) occurs during this acrosome reaction (Cherr and Clark 1984a). Wei *et al.* (2007) have indicated an "actin filament", projecting from the tip of the acrosome in Chinese Sturgeon.

In a valuable study on semen biology and sperm cryopreservation in the sterlet, *Acipenser ruthenus*, Lahnsteiner *et al.* (2004) found that the acrosome reaction could not be artifcially induced, in contast with *A. transmontanus*.

Presence of actin: The major protein isolated from the sperm head in *A. transmontanus* comigrates on gels with rabbit muscle actin and has a molecular weight of 43 KD. In conjunction with the occurrence of 6 nm microfilaments this indicates that actin present in sperm heads is the protein responsible for formation of the acrosomal process. Following the acrosome reaction, the material in the endonuclear canals remains closely associated with the midpiece and the centriolar extensions. Cherr and Clark (1984b) and Dilauro *et al.* (1999) consider that this may ensure that a centriole is incorporated into the egg after sperm-egg fusion and it is therefore uncertain that they regard the material in the nuclear canals as perforatorial.

Nucleus: The sturgeon sperm nucleus is an elongated trapezoid, but appreciable differences in the shapes and sizes of the nuclei occur (Table 8.1, Fig. 8.3). Differences between the posterior and anterior nucleus widths vary among species, possibly reflecting differences in their components (Psenicka 2007). It narrows, sometimes only slightly, in an anterior direction (Fig. 8.4), except in the Atlantic Sturgeon, *A. oxyrinchus* (Fig. 8.3, 8.4), in which it narrows in a posterior direction. There is also variation in size, thus the nucleus of the Pallid Sturgeon sperm is only slightly larger than that of the Atlantic Sturgeon and smaller than those of all other sturgeons (Table 8.1). The smaller nuclei evidently have a tighter packaging of the chromatin and nucleic acids (DiLauro *et al.* 2001).

Three endonuclear canals occur in Stellate, *Acipenser stellatus* (Ginsburg 1977); White, *A. transmontanus* Cherr and Clark (1984); Lake, *A. fulvescens* (DiLauro *et al.* 2000), Pallid, *Scaphirhynchus albus* (DiLauro *et al.* 2001) Siberian, *A. baerii* (Psenicka 2007), and Chinese Sturgeon, *A. sinensis* (Wei *et al.* 2007) as in the Paddlefish, *Polyodon spathula* (Zarnescu 2005), but only two occur in Atlantic Sturgeon, *A. oxyrinchus* (DiLauro *et al.* 1998). A single canal is mentioned for one unspecified sturgeon species (DiLauro *et al.* 2001). The membrane-lined endonuclear canals traverse the nucleus from the acrosomal end to the basal nuclear fossa region. However, in Pallid Sturgeon, the endonuclear canals extend into the anteriormost reaches of the acrosome, much farther anterior than in the other species. Differences have been reported in the diameters of canals between sturgeon species, with values of 35 (Atlantic), 97 (Shortnose), 40-49 (Lake), 57.41 (Pallid), 44.59 nm (Siberian) and 80 nm (Chinese) Sturgeon. As noted above, it has been suggested that the endonuclear canals play a role in transferring the centriole into the egg after sperm-egg interaction during fertilization (DiLauro *et al.* 1999. Cherr and Clark 1984a) and Jamieson (1991) suggested that the material in the nuclear canals plays a perforatorial role during penetration of the sperm into the egg.

There is an implantation fossa at the posterior end of the nucleus in Siberian Sturgeon, as observed in other sturgeon species (Ginsburg 1977; Cherr and Clark 1984a; DiLauro *et al.* 1998, 1999, 2000, 2001). It has somewhat questionably attributed with the function of connecting the endonuclear canals with the centrioles of the midpiece (Psenicka *et al.* 2006a; Psenicka 2007).

Midpiece: The midpiece of sturgeon sperm contains variable numbers of mitochondria, the proximal and distal centrioles or only part of the proximal centriole in *A. transmontanus* (Cherr and Clark 1984a) (Fig. 8.5), and certain ovoidal vacuoles (Ginsburg 1968; Cherr and Clark 1984a; DiLauro *et al.* 1998, 1999, 2000, 2001; Psenicka *et al.* 2006a; Psenicka 2007).

There are between three and six mitochondria and two vacuoles composed of lipid droplets in the midpiece of *Acipenser baerii* (Psenicka 2007) and three to eight mitochondria in *A. sinensis* (Wei *et al.* 2007).

The posterior region of the midpiece extends as a cytoplasmic collar around the base of the flagellum, being separated from it by the cytoplasmic canal (Fig. 8.5). The proximal and distal centrioles differ in size within and between species. The structural arrangement of the distal centriole is identical with that of the proximal centriole: nine sets of microtubular triplets form the periphery of the centriole (Pallid Sturgeon) (DiLauro *et al.* 2001). In Lake, Atlantic, Shortnose, Siberian, and probably all sturgeon species, the proximal centriole is attached to the nucleus by several thin strands in the center of the nuclear fossa, termed the fibrous body (e.g. DiLauro *et al.* 1998; Psenicka 2007) which appears to connect the nuclear fossa with the proximal centriole. Microfilaments connect the proximal and distal centriole with each other or with the nucleus. The two centrioles are mutually at right angles (e.g. *A. transmontanus*, Cherr and Clark 1984a) (Fig. 8.5). Vacuoles, composed of lipid droplets, are often observed in the midpiece region of the sperm cells in various fish species (e.g. Garfish sperm, (Afzelius 1978), and have been detected in Atlantic Sturgeon (DiLauro *et al.* 1998) and, as noted above, Siberian and White Sturgeon but are not reported for the Lake Sturgeon (DiLauro *et al.* 2000).

With lengths ranging from 1.09 ± 0.43 in the Siberian to 3.43 in the Stellate Sturgeon (Table 8.1), the midpiece is scarcely elongated compared with an ect-aquasperm.

Centriolar proteins: Mitina *et al.* (1992) compared the structure and protein composition of centrioles from spermatozoa of sturgeon (*Acipenser stellatus, A. guldenstadti*) and salmon (*Oncorhynchus gorbuscha*). The total protein content of the extracted fractions was studied by Na-SDS electrophoresis. Proteins with molecular weights from 15 to 170 kD were detected. In both cases the major protein of centrioles has a molecular weight equal to that of tubulin. A protein with the molecular weight corresponding to actin was also detected. In both cases ATPase activity stimulated by Ca^{2+} and Mg^{2+} ions was revealed. Electron microscopic studies showed differences in the ultrastructure of centrioles from sturgeon and salmon spermatozoa.

Flagellum: The flagellum, arising from the distal centriole, has the 9+2 pattern of axonemal microtubules in all sturgeons. Its anterior portion is separated from the midpiece by the cytoplasmic canal, as in many other fish spermatozoa. The flagella of Siberian (Psenicka *et al.* 2006a; Psenicka 2007), Short-nose (DiLauro *et al.* 1999), Stellate (Ginsburg 1977), White (Cherr and Clark 1984a), Atlantic (DiLauro *et al.* 1998) and Pallid (DiLauro *et al.* 2001) sturgeons have lateral fins (Fig. 8.2B). They extend distally up to 0.53 µm from the central longitudinal axis of the flagellum in the Pallid Sturgeon (DiLauro *et al.* 2001). These have been considered an adaptation for locomotion during external fertilization (Psenicka *et al.* 2006a; Psenicka 2007) but their homologue, the undulating membrane, persists in internally fertilizing Amphibia, the urodeles (Scheltinga and Jamieson 2003). It is estimated that the surface area of the flagellum, with the shape of a simple cylinder (34×10^6 nm^2), is approximately one quarter of that of the same cylinder with fins (160×10^6 nm^2) and this has large implications for water exchange/osmotic regulation at activation (Psenicka 2007). However, high permeability of the sperm membranes cannot be assumed. Thus, the sperm of *Acipenser transmontanus* remain motile in freshwater for 5 minutes, compared with less than 30 seconds for sperm of other freshwater fish and this has been attributed to a very impermeable plasma membrane as inferred from difficulty in solubilizing the sperm for electrophoresis and poor penetration of fixatives (Cherr and Clark 1984b). The present author considers that the possibility that the fins impede motility and therefore reduce dispersion of the sperm compared with those lacking fins deserves examination.

8.2.2 Family Polyodontidae

These are the paddlefishes, in fresh, rarely brackish, water in China and the United States (Nelson 2006).

Polyodon spathula: Sperm of the Paddlefish, *Polyodon spathula* (adult, Fig. 8.7), have the same characteristic architectural features as sturgeon spermatozoa.

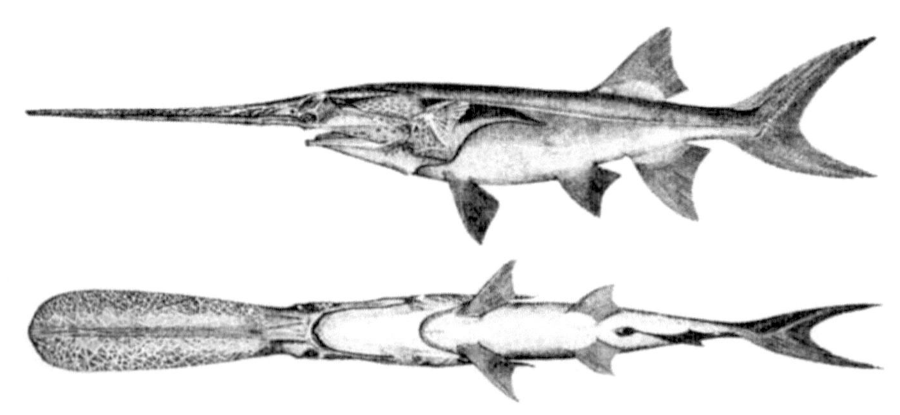

Fig. 8.7 *Polyodon spathula* (Polyodontidae). Adapted from NOAA archive.

The external characters have been illustrated diagrammatically by Psenicka (2007) (Fig. 8.3B). They consist of a rod-shaped head capped by the acrosome, a midpiece and a long flagellum. The PLPs are very short. The head is about 5.15 μm long and contains the nucleus and an apical acrosomal complex. There are three endonuclear canals that begin in the subacrosomal area [though separated from the apical plasma membrane by only a thin layer of the acrosome vesicle] and have a triple helical arrangement (Zarnescu 2005), each apparently with a putative perforatorium (Zarnescu 2007) (Fig. 8.8). A nuclear fossa is present centrally, at the posterior end of the strongly electron-

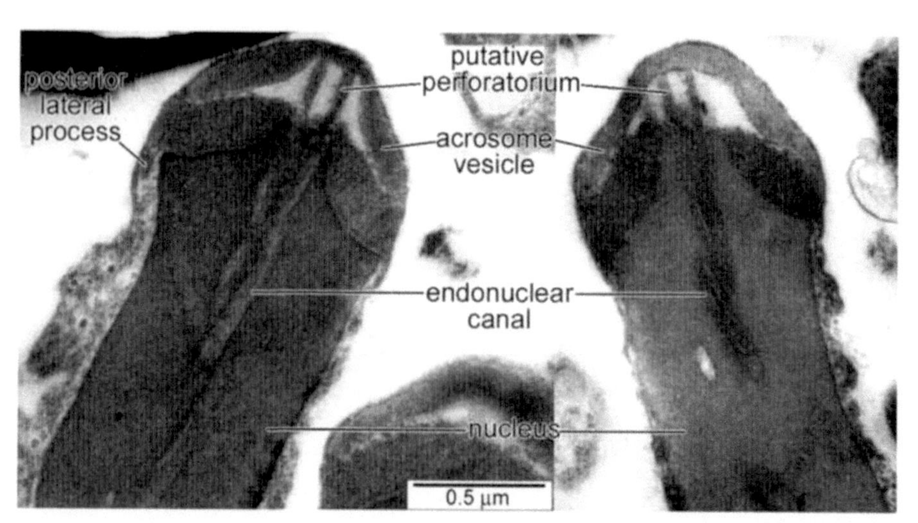

Fig. 8.8 *Polyodon spathula*. Acrosomal and anterior nuclear region of two spermatozoa. The spermatozoon contains completely condensed, electron-opaque chromatin. Three endonuclear canals are present and each appears to contain a putative perforatorium. Relabeled from micrographs courtesy of Professor Otilia Zarnescu.

dense nucleus. The midpiece contains a pair of centrioles in a perpendicular arrangement, mitochondria and a narrow cytoplasmic sleeve. The flagellum has a central axoneme with a 9+2 pattern and two lateral projections or fins (Zarnescu 2005).

8.3 CHAPTER SUMMARY

The sperm of the sturgeons (Acipenseridae) and Paddlefish (Polyodontidae) are closely similar. They share with hagfishes and cladistians possession of an acrosome despite the presence of egg micropyles which in acipenseriforms are multiple. There is a cap-like or conical acrosomal vesicle; a long, cylindrical to elongate trapezoidal, electron-dense nucleus; a midpiece of moderate length; and a long flagellum. In Acipenseridae the acrosome bears a circlet of eight to ten posterolateral projections but these appear to be very short in *Polyodon*. The nucleus is penetrated longitudinally by one to three endonuclear canals that begin in the subacrosomal area, or (Pallid Sturgeon) in the anterior region of the acrosome, and have, if more than one, a helical arrangement, each containing a putative perforatorium. A nuclear fossa is present centrally, at the posterior end of the nucleus. The midpiece contains a pair of triplet centrioles in a perpendicular arrangement, several separate mitochondria and a narrow cytoplasmic sleeve enclosing a cytoplasmic canal. The flagellum has a central axoneme with a 9+2 configuration of microtubules and two long lateral fins.

8.4 LITERATURE CITED

Afzelius, B. A. 1978. Fine structure of the garfish spermatozoon. Journal of Ultrastructure Research 64: 309-314.

Afzelius, B. A. 2006. Preformed acrosome filament. A chronicle. Brazilian Journal of Morphological Science 23: 279-285.

Amiri, B. M. and Takahashi, H. 2006. Fine structure of spermatogenesis in a hybrid sturgeon, bester (*Huso huso* female x *Acipenser ruthenus* male). Journal of Applied Ichthyology 22: 406-410.

Bartsch, P. and Britz, R. 1997. A single micropyle in the eggs of the most primitive living actinopterygian fish *Polypterus*. Journal of Zoology (London) 241: 589-592.

Cherr, G. N. and Clark, W. H. 1984a. An acrosome reaction in sperm from the White Sturgeon, *Acipenser transmontanus*. The Journal of Experimental Zoology 232: 129-139.

Cherr, G. N. and Clark, W. H. 1984b. Jelly release in the eggs of the white sturgeon, *Acipenser transmontanus*: An enzymatically mediated event. The Journal of Experimental Zoology 230: 145-149.

Cherr, G. N. and Clark, W. N. 1985. Gamete interaction in the white sturgeon *Acipenser transmontanus*: a morphological and physiological review. Environmental Biology of Fishes 14: 11-22.

Debus, L., Winlker, M. and Billard, R. 2002. Structure of micropyle surface on oocytes and caviar grains in sturgeons. International Review of Hydrobiology 87: 585-603.

Detlaf, T. A. and Ginzburg, A. S. 1963. Acrosome reaction in sturgeons and the role of calcium ions in the union of gametes. Doklady Akademii Nauk Sssr 153: 1461-1464.

Dilauro, M. N., Kaboord, W., Walsh, R. A., Krise, W. F. and Hendrix, M. A. 1998. Sperm-cell ultrastructure of North American sturgeons. I. The Atlantic sturgeon (*Acipenser oxyrhynchus*). Canadian Journal of Zoology 76: 1822-1836.

Dilauro, M. N., Kaboord, W. S. and Walsh, R. A. 1999. Sperm-cell ultrastructure of North American sturgeons. II. The shortnose sturgeon (*Acipenser brevirostrum,* Lesueur, 1818). Canadian Journal of Zoology 77: 321-330.

Dilauro, M. N., Kaboord, W. S. and Walsh, R. A. 2000. Sperm-cell ultrastructure of North American sturgeons. III. The lake sturgeon (*Acipenser fulvescens* Rafinesque, 1817). Canadian Journal of Zoology 78: 438-447.

Dilauro, M. N., Walsh, R. A., Peiffer, M. and Bennett, R. M. 2001. Sperm-cell ultrastructure of North American sturgeons. IV. The pallid sturgeon (*Scaphirhynchus albus* Forbes and Richardson, 1905). Canadian Journal of Zoology 79: 802-808.

Ginsburg, A. S. 1968. Fertilization in fishes and the problem of polyspermy. Akademiya Nauk SSSR, Institut Biologii Razvitiya. Moscow. 354 pp. Translated from Russian, by Israel Program for Scientific Translations, Jerusalem 1972.

Ginsburg, A. S. 1977. Fine structure of the spermatozoan and acrosome reaction in *Acipenser stellatus*. Pp. 246-256. In D. K. Beljaev (ed.), *Problemy eksperimental noj biiologii*. Nauk, Moscow.

Jamieson, B. G. M. 1991. *Fish Evolution and Systematics: Evidence from Spermatozoa*. Cambridge University Press, Cambridge. 319 pp.

Kudo, S., Linhart, O. and Billard, R. 1994. Ultrastructural studies of sperm penetration in the egg of the European catfish, *Silurus glanis*. Aquatic Living Resources 7: 93-98.

Lahnsteiner, F. and Patzner, R. A. 1999. Characterization of spermatozoa and eggs of the rabbitfish. Journal of Fish Biology 55: 820-835.

Lahnsteiner, F., Berger, B., A Horvath, A. and B Urbanyi, B. 2004. Studies on the semen biology and sperm cryopreservation in the sterlet, *Acipenser ruthenus* L. Aquaculture Research 35: 519-528.

Lauder, G. V. and Liem, K. F. 1983. The evolution and interrelationships of the actinopterygian fishes. Bulletin of the Museum of Comparative Zoology 150: 95-197.

Leung, L. P. 1988. The ultrastructure of the spermatozoon of *Lepidogalaxias salamandroides* and its phylogenetic significance. Gamete Research 19: 41-49.

Li, C., Lu, G. and Ortí, G. 2008. Optimal data partitioning and a test case for ray-finned fishes (Actinopterygii) based on ten nuclear loci. Systematic Biology: (In press).

Linhart, O., Kudo, S., Billard, R., Slechta, V. and Mikodina, Y. V. 1995. Morphology composition and fertilization of carp eggs: a review. Aquaculture 129: 75-93.

Linhart, O. and Kudo, S. 1997. Surface ultrastructure of paddlefish eggs before and after fertilization. Journal of Fish Biology 51: 573-582.

Longo, F. J. 1987. *Fertilization*. Chapman and Hall, New York. 191 pp.

Markov, K. P. 1975. Scanning electron microscope study of the microstructure of the egg membrane in the Russian sturgeon (*Acipenser gueldenstaedti*). Journal of Icthyology 15: 759-749.

Mitina, N. A., Vostrovskaya, M. and Shtein-Margolina, V. A. 1992. Comparative study of the structure and protein composition of centrioles from spermatozoa of true sturgeon and salmon fishes. Prikladnaya Biokhimiya i Mikrobiologiya 28: 462-467.

Nelson, J. S. 2006. *Fishes of the World*, 4th edition. John Wiley and Sons, Inc., Hoboken, New Jersey. 601 pp.

Norman, J. R. 1937. *Illustrated Guide to the Fish Gallery*. British Museum (Natural History) Trustees of the British Museum, London.

Poduschka, S. B. 1993a. Variability in the number of micropyles in the eggs of *Acipenser persicus* in the Volga River. Journal of Icthyology 33: 145-146.

Poduschka, S. B. 1993b. The variability of the number of micropyles in the eggs of Volga Stellate sturgeon *Acipenser stellatus*. Fish Physiology and Biochemistry 33: 152-155.

Psenicka, M. 2007. Morphology and ultrastructure of Siberian sturgeon (*Acipenser baerii*) spermatozoa using scanning and transmission electron microscopy. Biology of the Cell 99: 103-115.

Psenicka, M., Alavis, M. H., Rodina, M., Nebesarova, J. and Linhart, O. 2006a. Morphology and ultrastructure of sperm in fish: a comparative study between teleost (tench *Tinca tinca*) and chondrost (Siberian sturgeon *Acipenser baerii*). In (ed.), *IX Czech Ichthyological Conference. Collection of contributions from the specialist international conference held in Vodnany on 4.-5.5.2006 within the framework of workshops on fisheries and on the occasion of the 85th anniversary of the founding of VURH*. VURH JU.

Psenicka, M., Rodina, M., Nebesarova, J. and Linhart, O. 2006b. Ultrastructure of spermatozoa of tench *Tinca tinca* observed by means of scanning and transmission electron microscopy. Theriogenology 66: 1355-1363.

Riehl, R. and Patzner, R. A. 1998. Minireview: The modes of attachment in eggs of teleost fishes. Italian Journal of Zoology 65: 415-420.

Romer, A. S. 1950. *The Vertebrate Body*. W.B. Saunders, Philadelphia and London. 643 pp.

Scheltinga, D. M. and Jamieson, B. G. M. 2003. The Mature Spermatozoon. Pp. 203-274. In D. M. Sever (ed.), *Reproductive Biology and Phylogeny of Urodela*, vol. 1. Science Publishers, Enfield, New Hampshire, U.S.A.

Wei, Q., Li, P., Psenicka, M., Alavi, S., Shen, L., Liu, J., Peknicova, J. and Linhart, O. 2007. Ultrastructure and morphology of spermatozoa in Chinese sturgeon (*Acipenser sinensis* Gray 1835) using scanning and transmission electron microscopy. Theriogenology 67: 1269-1278.

Yamamoto, T. 1961. Physiology of fertilization in fish eggs. International Reviews in Cytology 12: 361-405.

Zarnescu, O. 2005. Ultrastructural study of spermatozoa of the paddlefish, *Polyodon spathula*. Zygote 13: 241-247.

Zarnescu, O. 2007. Immunohistochemical distribution of hyperacetylated histone H4 in testis of paddlefish *Polyodon spathula*: ultrastructural correlation with chromatin condensation. Cell and Tissue Research 328: 401-410.

CHAPTER 9

Ultrastructure of Spermatozoa: Neopterygii: Holostei through Osteoglossomorpha

Barrie G. M. Jamieson

9.1 SUBCLASS NEOPTERYGII

The Neopterygii contain all living actinoperygians except the Polypteriformes and Acipenseriformes. They thus contain the Holostei (as defined below) and the Teleostei as sister-groups (Fig. 9.1).

It is widely agreed that the Neopterygii constitute a monophyletic group (Patterson 1973; Wiley 1976; Bartram 1977; Patterson and Rosen 1977; Lauder and Liem 1983; Nelson 2006). This is supported, from molecular analysis, by Ortí and Li (**Chapter 1**, see Fig. 1.3). Additional inclusion of the Acipenseriformes, in the mitogenomic analysis of Inoue *et al.* (2003), (see **Chapter 7**, Fig. 7.2), is debatable.

Irrespective of whether internal fertilization is primitive or derived in fishes as a whole, neopterygians are spermatologically diagnosed from the agnathans through Dipnoi in having lost the acrosome, giving an anacrosomal aquasperm. The secondary simplification of fish sperm to the anacrosomal aquasperm, first seen in Holostei, from a complex sperm type retained, with modification, in chondrichthyans, Cladistia and Holostei, appears to have largely obscured phylogenetic relationships. Nevertheless, many cases of synapomorphic resemblance within subgroups remain, of which the Elopomorpha (**Chapter 10**) offer the clearest example.

With rare exceptions, only one micropyle is present in the neopterygian egg, presumably as an adaptation preventing polyspermy (Ginzburg 1968, p. 62; p. 200). Divergence from the simple anacrosomal sperm type, though only in *Lepidogalaxias* with redevelopment of the acrosome, is found in the relatively few

---

School of Integrative Biology, University of Queensland, Brisbane, Queensland 4072, Australia

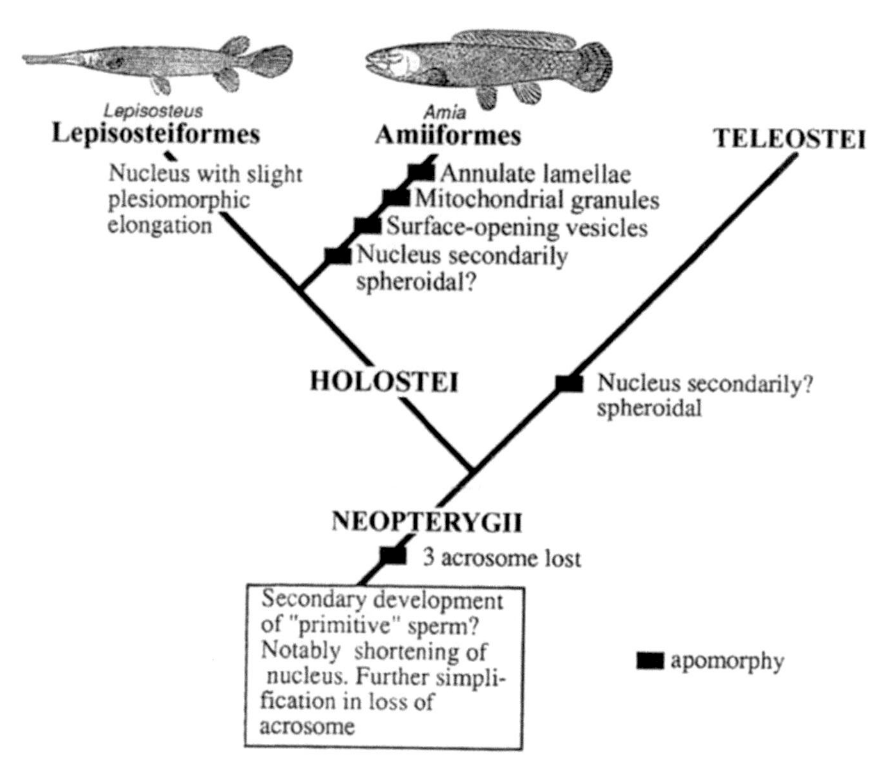

Fig. 9.1 Phylogeny of the main groups of the Neopterygii with spermatozoal apomorphies added. Evidence is accepted that the Amiiformes is the sister-group of the Lepisosteiformes. Original.

neopterygians which have redeveloped internal insemination or have external fertilization of a presumably specialized type (Jamieson 1991). Internal insemination is known for one species of the Salmoniformes (*Lepidogalaxias*), some Characiformes (all Glanduolocaudinae), Siluriformes (Scoloplacidae, Astroblepidae and Auchenipteridae); Ophidiiformes (Bythitidae), Atheriniformes (many species), Gasterosteiformes (one species of Aulorhynchidae); some Scorpaeniformes (in the Sebastinae, Cottidae, Hemitripteridae), and a few Perciformes (Apogonidae, *Santoperca* in the Cichlidae), Embiotocidae, Zoarchidae, Anarhichadidae, Clinidae and Labrisomidae) (for details see these groups) and viviparity is reported for the Poeciliidae, Clinidae, Labrisomidae, Anablepidae, Zoarcidae, Parabrotulidae, Bythitidae, Aphyonidae, Zenarchopteridae, Goodeidae, Scorpaenidae, Comephoridae and Embiotocidae (see Uribe Aranzábal *et al.*, Chapter 3).

An environmental change from seawater to freshwater has been suggested as a possible cause in the loss of the acrosome in neopterygian spermatozoa (Baccetti 1979; 1985). This hypothesis is not supported by the fact that the Petromyzontidae, Cladistia, Chondrostei and Dipnoi spawn in freshwater and yet possess sperm acrosomes (Leung, unpublished).

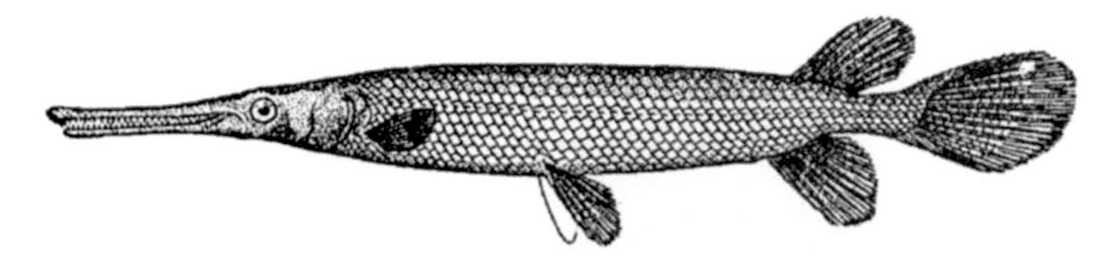

Fig. 9.2 *Lepisosteus osseus* (Lepisosteidae). The Garpike. After Norman, J.R. 1937. *Illustrated Guide to the Fish Gallery.* British Museum (Natural History. Trustees of the British Museum, London, Fig. 19.

9.2 HOLOSTEI

Traditionally, *Lepisosteus* (Garpike) and *Amia* (Bowfin) were placed together in the Holostei (Goodrich 1909; Romer 1966). Monophyly of the Holostei was supported from somatic morphology by G. J. Nelson (1969) and Jessen (1973). However, most workers, including Lauder and Liem (1983), followed Patterson (1973) and considered Holostei to be paraphyletic and *Amia* to be the sister taxon to the Teleostei. From a parsimony analysis of mitochondrial DNA Normark *et al.* (1991) supported the traditional view of a monophyletic Holostei (*Amia* + *Lepisosteus*) and this relationship has been confirmed, from molecular analysis, by Ortí and Li (**Chapter 1**, see Fig. 1.3) and by Hoegg and Meyer (2007) from analysis of KCNA gene clusters. The category Holostei, in this sense, is therefore recognized in the present work. The terms Halecomorphi, Halecostomi and Ginglymodi, which are not formally recognized by Nelson (2006), are not used here and will not therefore be defined.

The sperm of *Lepisosteus*, with that of *Amia*, is the first example of an externally fertilizing, anacrosomal aquasperm in extant fishes.

9.3 ORDER LEPISOSTEIFORMES

One extant family (Lepisosteidae) with two extant genera (*Lepisosteus* and *Atracosteus*), comprising the redundant Ginglymodi, the gars. Freshwater, occasionally brackish, very rarely in marine water; N. and Central America and Cuba (Nelson 2006).

***Lepisosteus osseus*:** The spermatozoon of the Garpike *Lepisosteus osseus* (adult, Fig. 9.2; sperm, Figs. 9.3, 9.4) is described by Afzelius (1978) as being of the "primitive" type (here considered probably derived) like that of most teleosts, that is an anacrosomal aquasperm.

The bullet-shape of the nucleus, 2.5 μm long and 1.1 μm wide, is, however, a slight departure from the more rounded form of the nucleus in most teleostean aquasperm. It has a posterior indentation which houses the proximal centriole.

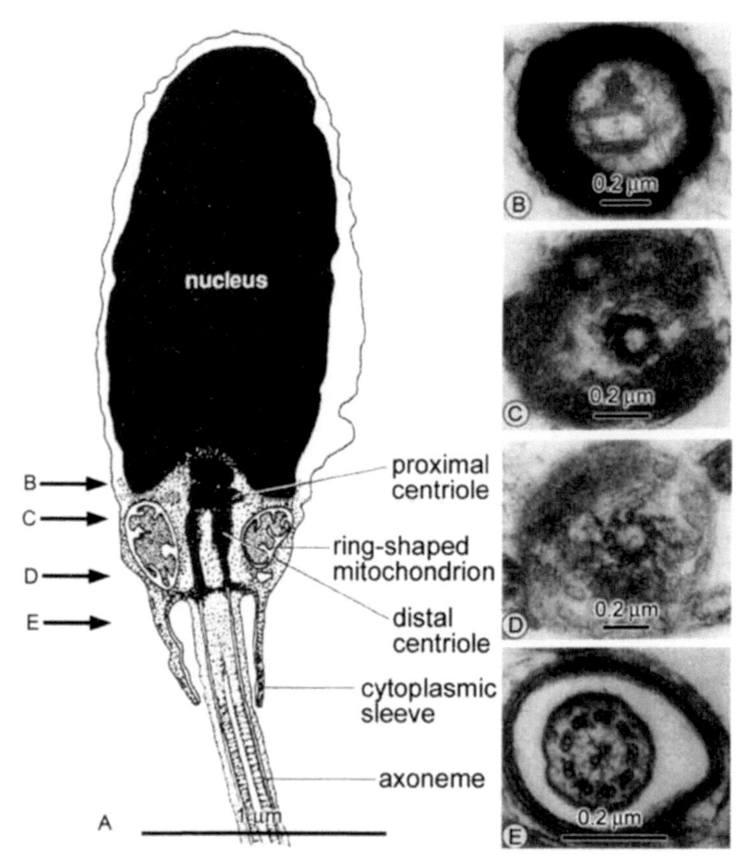

Fig. 9.3 *Lepisosteus osseus* (Lepisosteidae) **A**. Drawn from a micrograph. Longitudinal section through head and midpiece of a spermatozoon. An acrosome is absent. Arrows indicate the levels sectioned in **B-E**. **B**. Transverse section (TS) at the level of the proximal centriole. Along the centriole there is a fibrous body. **C**. TS at the level of the distal centriole and the midpiece. The mitochondrion is ring-shaped. **D**. TS at the posterior border of the midpiece. At this level the proximal centriole has nine doublets, each with a short hook-like extension and with a longer projection toward the cell membrane. **E**. TS through the proximal portion of the sperm tail and the surrounding cytoplasmic sleeve. Adapted from micrographs by Afzelius B.A. 1978. Journal of Ultrastructure Research 64: 309-314, Figs.1-5. With permission of Elsevier.

The ring- or C-shaped mitochondrion (Afzelius 1978), described as a single annular mitochondrion by Mattei (1988), is not usual for neopterygians; it is known in only five teleost families (see Clupeomorpha, below); in *Branchiostoma* and, *inter alia*, in echinoderms, cnidarians and some polychaetes (see Jamieson 1984) and has doubtful cladistic significance. In some *L. osseus* sperm the midpiece also contains a lipid droplet. A thin sleeve projects posteriorly for 0.6 µm from the midpiece around the anterior end of the flagellum (Afzelius 1978).

The two triplet centrioles have the mutually perpendicular arrangement considered plesiomorphic for the animal kingdom (Afzelius 1979). Along one side of the proximal centriole there is a cross-striated fibrous body (Fig. 9.4) that appears to be attached to the center of the basal nuclear fossa by an extension consisting of several thin strands. Afzelius equates this with a body [the "button"] seen in the sperm of *Neoceratodus* but it has alternatively been regarded as a short flagellar root (Mattei, C. *et al.* 1981; Mattei 1988). The two views may be compatible. Each centriole has a doublet-containing distal extension, each doublet having a hook-like projection. In addition, the distal centriole has a satellite apparatus. The doublet portion of the proximal centriole has been interpreted as an abortive flagellum (Mattei, C. *et al.* 1981; Mattei 1988) but there seems no reason to accept the implication that biflagellarity is plesiomorphic for actinopterygians.

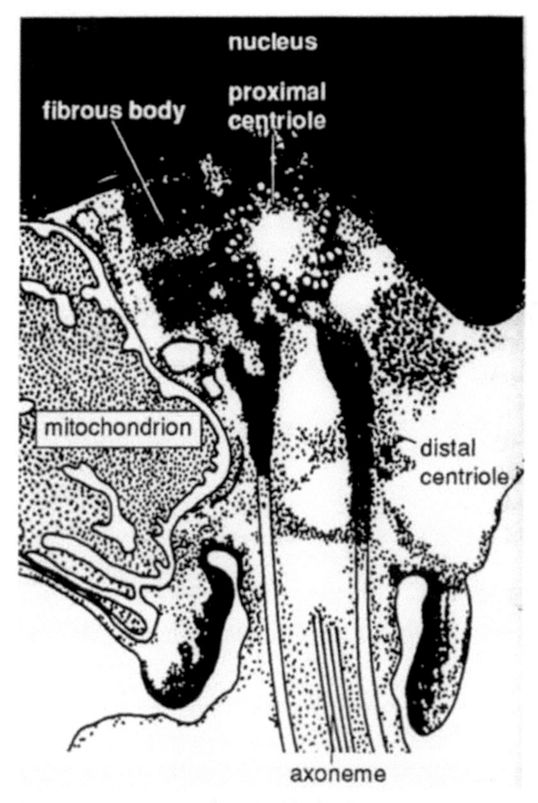

Fig. 9.4 *Lepisosteus osseus* (Lepisosteidae). Longitudinal section through the sperm midpiece, transversely cutting the proximal centriole and showing the fibrous body and its extension to the nucleus. From Jamieson 1991, after a micrograph by Afzelius, B.A. 1978. Journal of Ultrastructure Research 64: 309-314, Fig. 6.

The 9+2 flagellum, 45 μm long, has a pair of lateral fins (Fig. 9.5). Each doublet appears to have two dynein arms (Afzelius, 1981). The doublets are not all identical. The A tubule of doublets 1,2,6 and 7, contains a dense substance whereas the other A tubules and the B tubules appear empty. This densification is due to the presence of an intratubular differentiation (ITD) which has the form of a partition, inside the tubule, in line with the outer arms of the doublet (Mattei 1988).

Fig. 9.5 *Lepisosteus* osseus (Lepisosteidae). Cross section through the axoneme, showing the pair of lateral fins. From Jamieson 1991, after a micrograph by Afzelius, B.A. 1978. Journal of Ultrastructure Research 64: 309-314, Fig. 9.

9.4 ORDER AMIIFORMES

The Amiiformes contains fossil forms and, in the suborder Amioidei, family Amiidae, the single extant species *Amia calva*, the Bowfin (adult, Fig. 9.6), in freshwater in eastern N. America (Nelson 1984, 2006).

Fig. 9.6 *Amia calva* (Amiidae). Photo Stan Shebs, Steinhart Aquarium, San Franscico.

Sperm structure: The anacrosomal aquasperm of *Amia calva* (Amiidae) was described optically by Retzius (1905). It is seen to differ from that of *Lepisosteus* in the spheroidal nucleus. It was examined ultrastructurally by Afzelius and Mims (1995).

The sperm head is spherical with a diameter of 2 μm (Fig. 9.7A). There is no acrosome but a nuclear envelope covers all of the nucleus and consists of two membranes separated by a distinct perinuclear cisterna. In most spermatozoa a lateral region of the cross-sectioned nuclear membrane can be seen to have nuclear pore complexes. Outside this region there are some layers of annulate lamellae (Figs. 9.7 A-C). The chromatin of the nucleus has an even electron-density.

The midpiece is about 2 μm wide at its base and about 1 μm long. It contains 12-16 mitochondria arranged in two rings around the two centrioles. The mitochondria have the same appearance as those found in the somatic cells, even to the extent that some mitochondrial matrix granules are seen. Another peculiarity of the midpiece is the presence of several small vesicles. Some of them have apparently fused with the cell membrane, which at these sites shows inpocketings, similar to large caveolae (Figs. 9.7 A-D). At the cytoplasmic side the inpocketings seem to have a thin lining of some electron-dense material. The cytoplasm between the organelles in the midpiece contains what appear to be ribosomes.

In the midpiece there are two mutually perpendicular centrioles that are joined by a cross-striated ribbon (Fig. 9.7E). A mushroom-like projection extends from each of the centrioles (Figs. 9.7 B,E). Both centrioles lie at a distance from the nucleus. No connection can be found between any of the centrioles and the nucleus. Nine satellite fibers attach the distal centriole to the cell membrane (Fig. 9.7 F).

The 9+2 flagellum emerges from the distal centriole. The A- and B-subtubules all have electron-lucid lumina. There are no fins on the flagellum (Afzelius and Mims 1995).

Remarks: Some structures of the bowfin spermatozoa have no equivalents in any other sperm type as far as is known: annulate lamellae, mitochondrial matrix granules, and vesicles that appear to open at the cell surface. The presence of what might be ribosomes is also an unusual finding. It is possible these structures indicate that bowfin spermatozoa retain features from an earlier developmental stage, somewhat analogous to neoteny in the soma of some animals. Annulate lamellae are commonly found in spermatocytes of various animal groups, but not in spermatozoa (Afzelius and Mims 1995).

The structure of the bowfin spermatozoon differs from that of the Garfish in several important connections: a round rather than oval head shape, centrioles distant from the nucleus, absence of flagellar fins, absence of ITDs. The absence in both of an acrosome is a synapomorphy of the Neopterygii. *Amia* and *Lepisosteus* are represented in Fig. 9.1 as the sister-group of the Teleostei. If, however, a sister relationship to the Acipenseriformes were accepted (Inoue *et al.* 2003), it would be necessary to accept that the acrosome was lost independently in the Holostei or, less likely, that acipenseriforms have secondarily developed an acrosome.

Electron microscopy shows that, contrary to an observation of Retzius (1905), the bowfin spermatozoon is not similar to those of the primitive teleosts, such as the herring (Clupeidae) and the pike (Esocidae) except for the rounded shape of the head (Afzelius and Mims 1995).

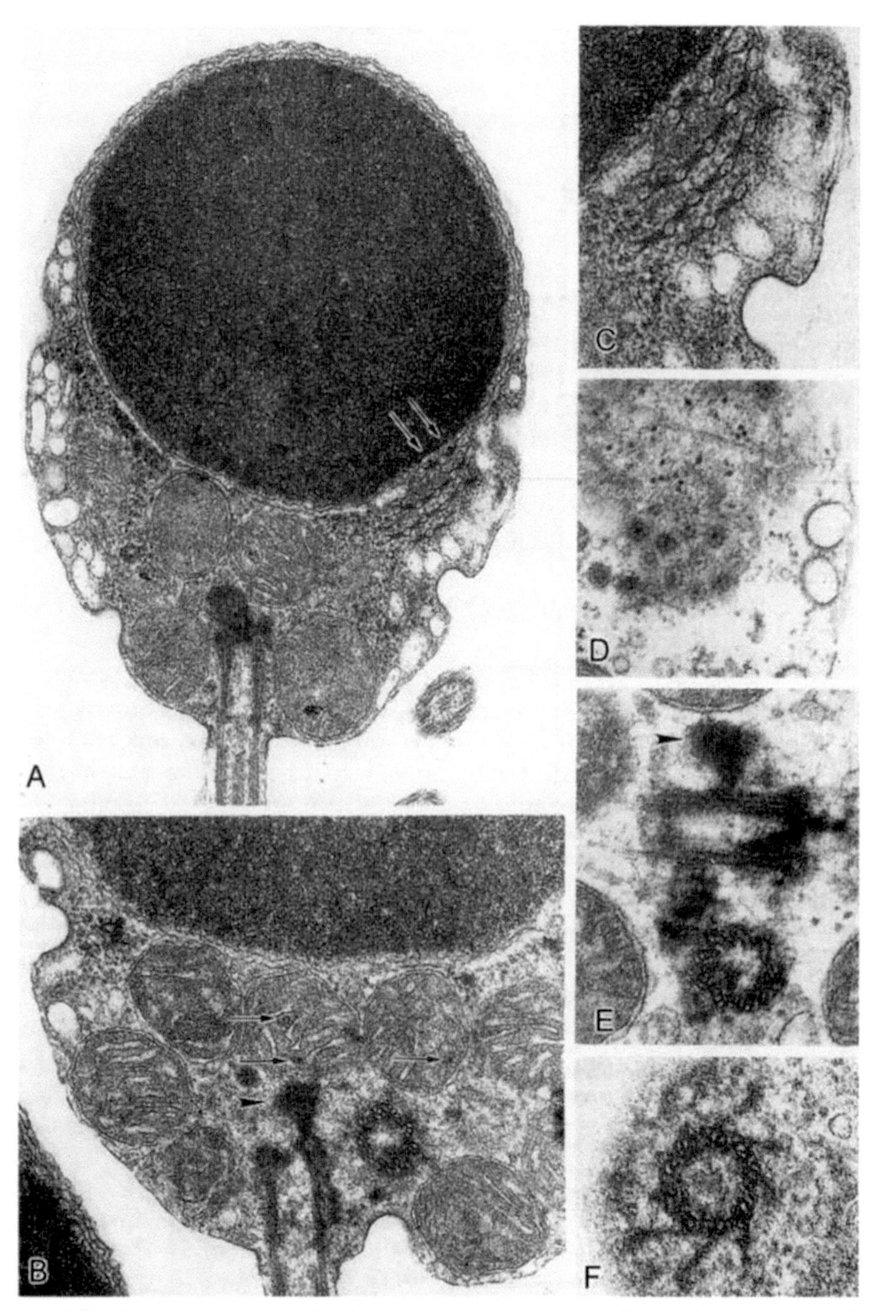

Fig. 9.7 A. *Amia* calva (Amiidae). Longitudinal section through a bowfin spermatozoon. Note the nuclear pore complexes (arrows) in the nuclear envelope and the annulate lamellae outside this region. Along the periphery of the midpiece

Fig. 9.7 Contd. ...

9.5 DIVISION TELEOSTEI

The Teleostei, with an estimated 26,840 species, is by far the most diverse group of the Actinopterygii. Teleosts are distinguished by elongation of the ural neural arches into uroneurals which function to stiffen the upper tail lobe and to support a series of dorsal fin rays. These and other features indicate the monophyly of teleosts (Lauder and Liem 1983; Nelson 1984, 2006).

9.6 SUBDIVISION OSTEOGLOSSOMORPHA

There are two orders and five families. Most osteoglossomorphs exhibit some kind of parental care; with some mouthbrooders (Nelson 2006).

Living or fossil osteoglossomorphs are known from every continent except Europe. The osteoglossomorph fishes take their name from the "tongue bite" in which the basihyal, covered by a massive toothplate, "bites" against the roof of the mouth which bears large teeth. Most osteoglossomorphs have parental care sometimes (*Osteoglossum* and *Scleropages*) with mouth brooding. Those with parental care possess only the left ovary. *Pantodon* and *Hiodon* lack parental care and retain both ovaries (Britz 2004; Nelson 2006).

Lauder and Liem (1983) recognize four major groups of living teleosteans: the Osteoglossomorpha, Elopomorpha, Clupeomorpha and Euteleostei and regard the osteoglossomorphs as the most primitive group of living teleosts; but contrast Patterson (1973), who regards *Elops* as the most primitive living teleost. The other three groups, sometimes termed the elopocephalans, are united by the presence of only two uroneurals. Whole mitogenomic analysis appears to have established the basal position of the Osteoglossomorpha as the sister group of all other teleosts (Lavoué *et al.* 2005).

Fig. 9.7 Contd. ...

there are several vesicles, some of which apparently fused with the cell membrane. x 29,500. **B**. Longitudinal section through another bowfin spermatozoon. The proximal centriole is transversely sectioned and is seen to the right of the distal centriole from which the tail flagellum emerges. Note also the mushroom-like projection (arrowhead) from the distal centriole and the matrix granules (small arrows) in the mitochondria. x 40,500. **C**. Enlarged detail from Fig. 1 showing the annulate lamellae at a higher magnification. x 55,000. **D**. Oblique section through the sperm midpiece including a tangentially cut annular lamella. Two cell membrane inpocketings are seen to the right. x 33,000. **E**. Transverse section through the midpiece. The proximal centriole (above) and the cross-cut distal centriole (below) are joined by a cross-striated connection. Note also that a mushroom-like connection (arrowhead) is also seen at the proximal centriole. x 51,500. **F**. Transverse section through the distal centriole close to the cell membrane. Obliquely oriented projections, so called satellites, extend from the centriole to the cell membrane. x 59,000. After Afzelius, B. A. and Mims, S. D. 1995. Journal of Submicroscopic Cytology and Pathology 27: 291-294, Figs. 1-6.

Taxonomic interrelationships within Osteoglossomorpha are controversial (see Greenwood 1973; Lauder and Liem 1983; Jamieson 1991; Li and Wilson 1996; Nelson 2006; and references therein). Recent molecular analyses are those of Kumazawa and Nishida (2000) and Lavoué and Sullivan 2004) (Fig. 9.8).

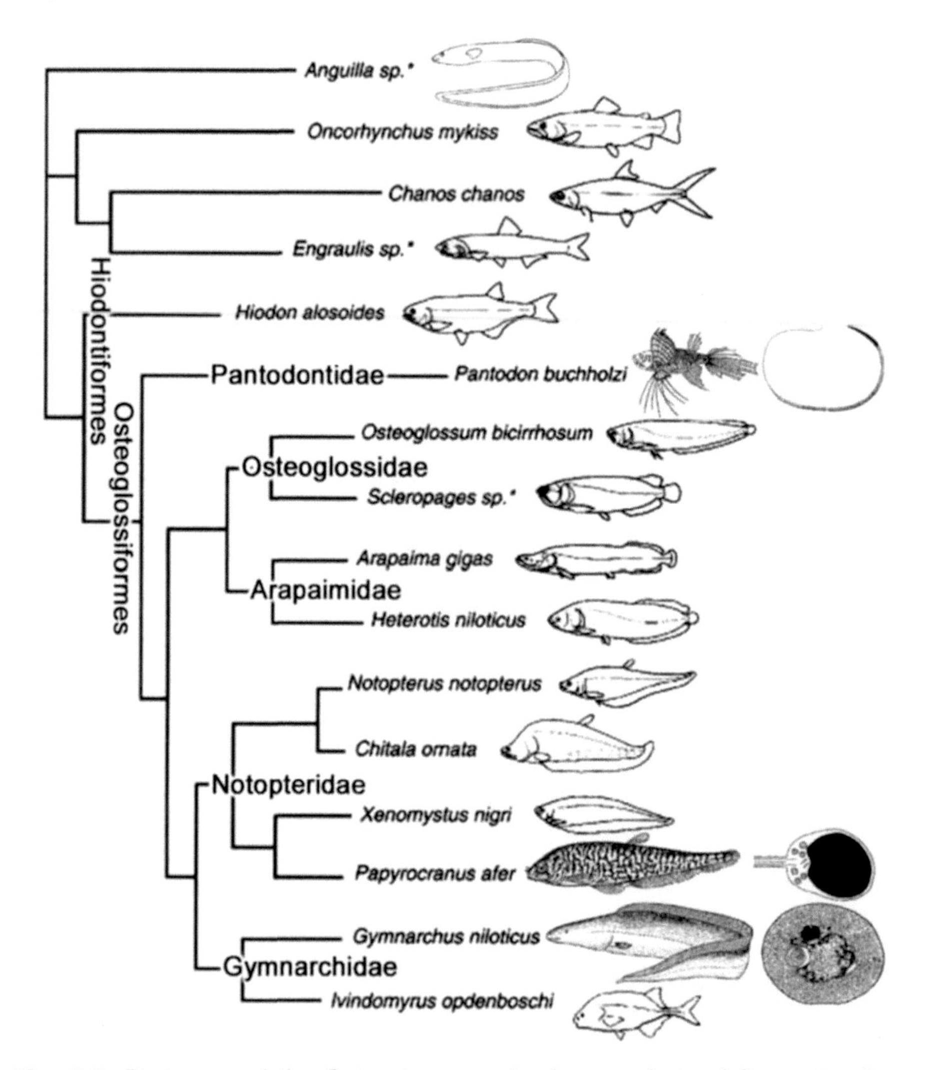

Fig. 9.8 Phylogeny of the Osteoglossomorpha from analysis of five molecular markers (the mitochondrial cytochrome b, 12S and 16S rRNA genes, and the nuclear genes RAG2 and MLL). 50% majority rule consensus of 920 trees. Asterisks indicate chimeric taxa produced from combination of sequence data from congeneric species. The few sperm types known for the represented genera of the Osteoglossomorpha are here appended. Adapted, with addition of known sperm types, from Lavoué, S. and Sullivan 2004. Molecular Phylogenetics and Evolution 33: 171-185, Fig. 3. With permission of Elsevier.

9.7 ORDER HIODONTIFORMES

9.7.1 Family Hiodontidae

The Hiodontidae, Mooneyes, were included in the Notopteroidei *sensu lato* of Lauder and Liem (1983) and *sensu stricto* of Nelson (1984), the latter author excluding the mormyroids. However, Nelson (2006) recognizes their placement in a distinct order. As in the Notopteridae they have an elongated anal fin but this is only moderately long and is not confluent with the caudal fin. They are the only osteoglossomorphs from North American waters (Lauder and Liem 1983; Nelson, 1984, 2006). Hiodontids have not been investigated for sperm ultrastructure.

9.8 ORDER OSTEOGLOSSIFORMES

These are the bonytongues.

Sperm relationships: Spermatozoal ultrastructure of the few osteoglossomorphs in which it has been investigated promised to throw light on the phylogeny of these fishes but there have been no new descriptions of their sperm since those noted in Jamieson (1991). The rare aflagellate condition of sperm seen in Gymnarchidae (*Gymnarchus niloticus*) and in Mormyridae (*Gnathonemus niger, G. senegalensis, Hyperopisus bebe, Mormyrus rume, Petrocephalus bovei*) is in agreement with the association of these families in the Mormyroidei (Nelson 1984) or as sister-groups in a wider Notopteroidei (Lauder and Liem 1983) but Nelson (2006) refrains from subordinal classification (For spermatozoal phylogeny see Fig. 9.9). The simple anacrosomal aquasperm of the notopteroid *Papyocranus afer*, the African Knife Fish, appears to be a plesiomorph condition, contrasting with the apomorphic, aflagellate sperm of mormyroids, but does not in itself necessitate exclusion of the mormyroids from the Notopteroidea in which Greenwood (1973) placed them. The peculiar filiform sperm of the only other investigated osteoglossoid, *Pantodon bucholzi*, endorses the separate status of the Pantodontidae recognized by (Nelson 1984) though it was later (Nelson 2006) placed in the Osteoglossidae. The Pantodontidae is recognized here in view of its separate status in the phylogram of Lavoué and Sullivan (2004) (Fig. 9.8). It should be noted that *Pantodon* was grouped in the Notopteridae in the analysis of Kumazawa and Nishida (2000) which used the more appropriate outgroup of *Amia*, together with chondrichthyans and the coelacanth. It is therefore noteworthy that no members of the Osteoglossidae have been investigated for sperm ultrastructure.

Spermatozoal apomorphies are superimposed on a slight modification of the phylogeny of Lauder and Liem (1983) in Fig. 9.9. The phylogeny is modified to separate the Gymnarchidae from the Mormyridae but branching relationships are unchanged.

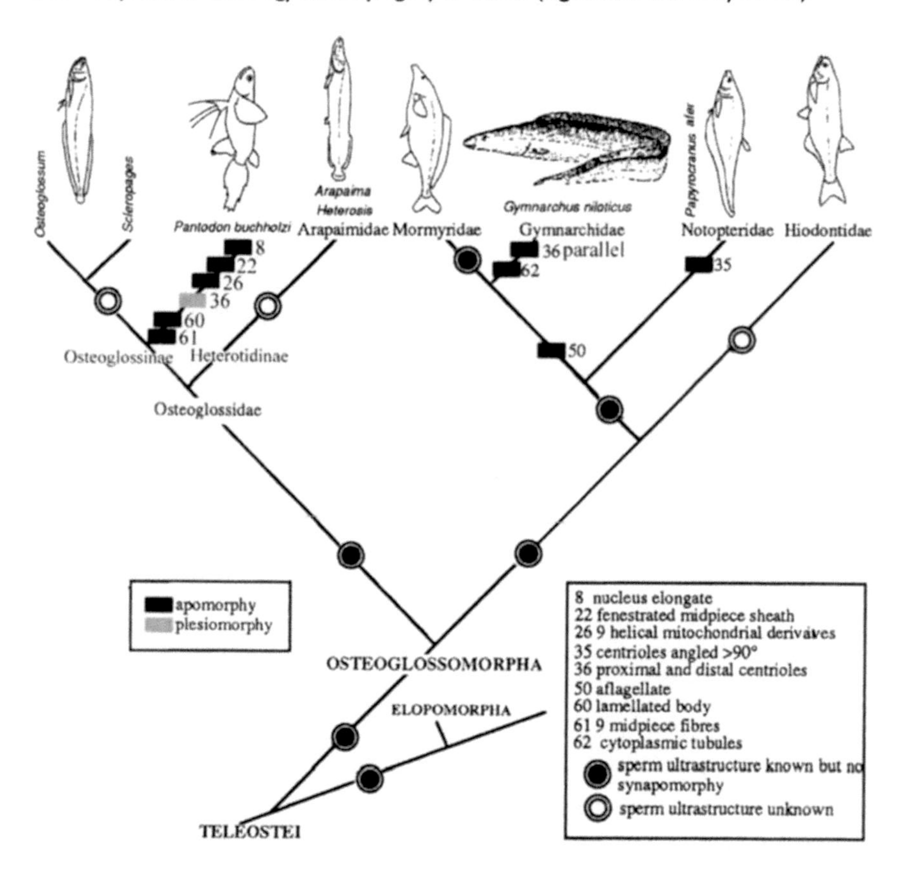

Fig. 9.9 Phylogenetic relationships of the Osteoglossomorpha. Chiefly after Lauder and Liem (1983), with classification modified according to Nelson (2006) and with spermatozoal apomorphies superimposed. Numerals before characters refer to a previous cladistic analysis of Jamieson (1991). Original.

9.8.1 Family Pantodontidae

Pantodon buchholzi: The Butterfly fish (adult, Fig. 9.10A), is the only species of the Pantodontidae though subsumed in the Osteoglossidae (Nelson 2006). Its distinctive nature is nevertheless indicated, *inter alia*, by its being the only osteoglossiform with one caecum (rather than two caeca) and the unique and highly apomorphic structure of the spermatozoon and is underlined by molecular analysis (Fig. 9.8).

Its anacrosomal spermatozoon is 80-85 μm long (Deurs 1973; Deurs and Lastein 1973; Deurs 1974) (Fig. 9.11).

The nucleus (Fig. 9.11A) is elongated, comprising a cylindrical, electron-dense chromatin rod, about 7 μm long and 0.6 μm wide, tapering anteriorly. At maturity the nuclear envelope is not visible. Between the nuclear rod and the cell membrane some granular material is normally seen.

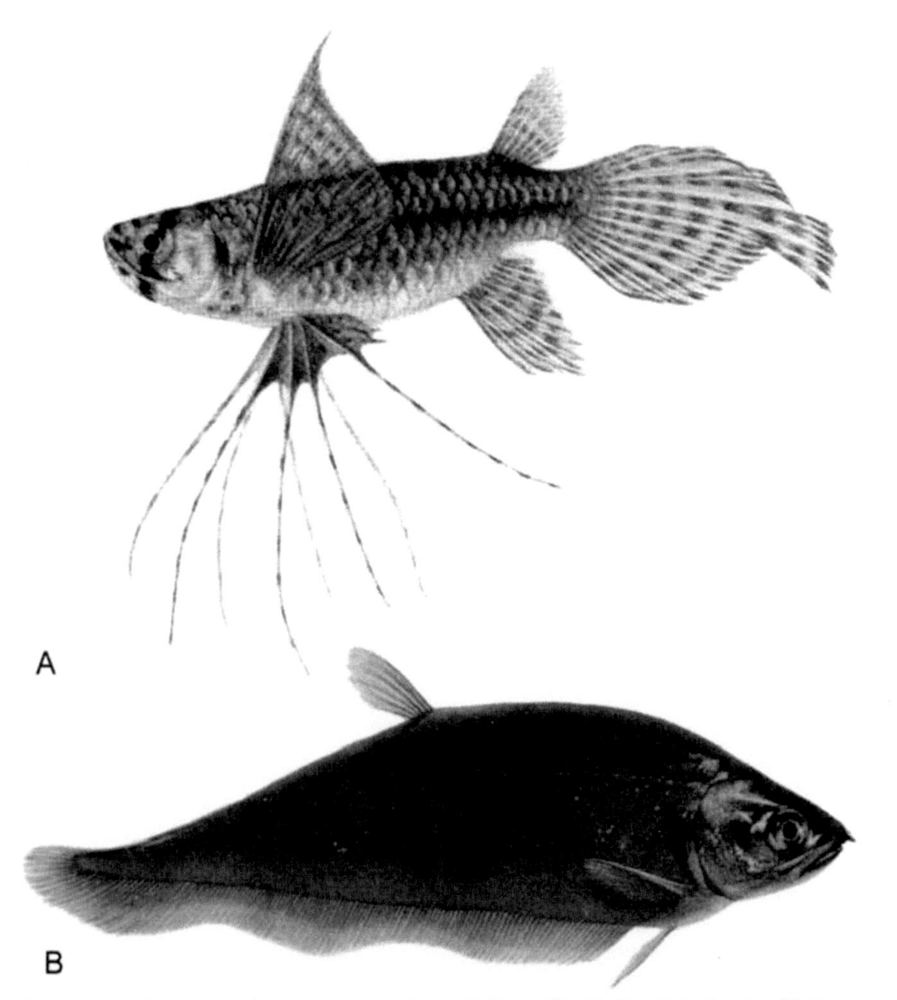

Fig. 9.10 **A.** *Pantodon bucholzi*, Butterflyfish (Pantodontidae). **B**. *Notopterus chitala*, Asian featherback (Notopteridae). By Norman Weaver.With kind permission of Sarah Starsmore.

The midpiece (Fig. 9.11C, G) is very elongate, length about 45 μm, and consists of 9 helical mitochondrial derivatives, each describing about 20-25 turns with an inclination of 60-70° and each about 100 nm thick. The cristae are modified as columnar derivatives. End to end fusion of the original mitochondria to form these derivatives is also seen in snakes, passeridan birds and mammals.

Nine helical dense fibers (Fig. 9.11C, G) are also present in the midpiece, alternating with the mitochondria in the spermatid but between the mitochondria and the axoneme in the sperm; they are not part of the axonemal complex; each is cross striated, with a major period of 75-90 nm. They are equated with flagellar rootlets (van Deurs, 1973).

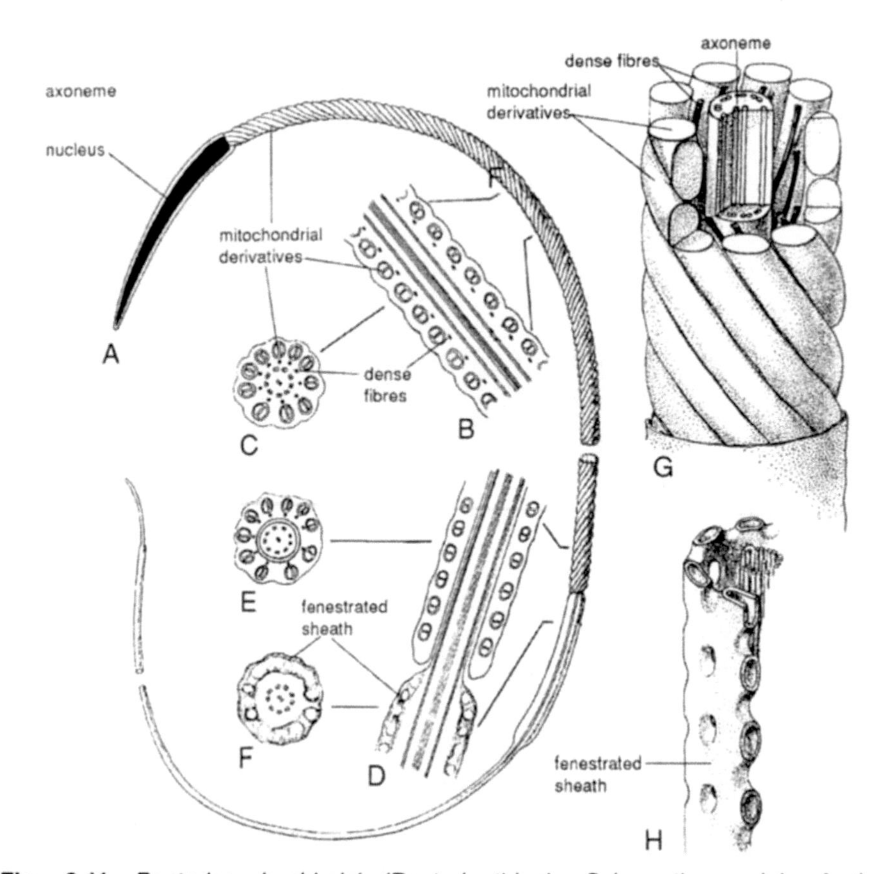

Fig. 9.11 *Pantodon buchholzi* (Pantodontidae). Schematic model of the spermatozoon. **A**. The nucleus is elongated. The centriolar complex is covered by the anterior part of a sheath formed by the 9 helical mitochondrial derivatives ("threads"). Behind the midpiece follows the fenestrated sheath region and the free flagellum. **B**. Longitudinal section of the midpiece anterior to the flagellar canal in which the flagellum runs. Axoneme, dense fibers and mitochondria are seen. **C**. Cross-section corresponding to **B**. **D**. Longitudinal section through the junction between the midpiece and the anterior part of the of the fenestrated sheath region. Through the posterior part of the midpiece, the flagellum runs in a flagellar canal. Leaving the canal, the flagellum swells and a membrane structure is found below the flagellar membrane. **E**. Cross section of the posterior part of the midpiece, showing axoneme, flagellar canal, dense fibers, and mitochondria. **F**. Cross section of the fenestrated sheath region. **G**. Schematic three dimensional reconstruction of part of the midpiece, as seen in the mature spermatozoon, showing the helical arrangement of the 9 mitochondrial threads and the 9 dense fibers surrounding the flagellum (cristae omitted). **H**. Schematic three-dimensional reconstruction of part of the fenestrated sheath, showing that the sheath is composed of an outer and an inner membrane penetrated by pores. The pore walls connect the outer and inner membranes (plasma membrane omitted). After Deurs, B. V. and Lastein, U. 1973. Journal of Ultrastructure Research 42: 517-533, Figs. l, 28 and 29. With permission of Elsevier.

Behind, but not as an extension of, the midpiece there is a fenestrated sheath, 6 µm long (Fig. 9.11D,H). This is a 0.5 µm wide swelling of the flagellum containing a membranous structure around the axoneme. Sections tangential to the surface show a fenestrated plate, with pores 35 nm in diameter and 85 nm apart, center-to-center; the sheath is a double membrane penetrated by the pores. Both the midpiece and the sheath are separated from the axoneme by a space (cytoplasmic canal).

The centriolar complex consists of a single centriole (basal body), although an additional centriole, presumably the proximal centriole, is sometimes observed parallel to it. A lamellate body between the nucleus and the basal body is reminiscent of the fibrous body of the sperm of *Lepisosteus* and possibly of that of *Neoceratodus* and (Deurs 1973) is particularly similar to the "intracentriolar lamellated body" in the spermatozoon of *Poecilia.* The lamellated body is a curved cap, about 0.5 µm wide and 130-150 nm thick. It is composed of four layers: (1) an electron-dense lamella, 20-30 nm thick; (2) a more electron-lucent layer, 45-52 nm thick, striated perpendicularly to the electron-dense lamella; (3) a further electron-dense lamella, 20-30 nm thick; and (4) a more electron-lucent striated layer, 30-40 nm thick. From this layer some dense material appears to connect with the centriole (Deurs and Lastein 1973). About 25 µm of the 9+2 flagellum is free; no fins are described.

***Pantodon fertilization biology*:** From sperm structure the mode of fertilization in *Pantodon* is deduced by van Deurs and Lastein (1973) to be either internal fertilization or deposition of bundles of spermatozoa near the female genital opening or on the eggs. It has been reported that the male places its large anal fin under the female so that the genital duct is in contact with that of the female and that the eggs are fertilized on extrusion (See Breder and Rosen 1966: 144). The similarity of this form of fertilization to that of the lamprey is noteworthy and the possession in the sperm of both of 9 accessory fibers around the axoneme, although these are of a different nature, is a striking convergence, conceivably for a similar, if unknown, function. Whether the osteoglossomorphs, an ancient group, originally had internal fertilization is a matter of conjecture.

9.8.2 Family Notopteridae

Notopterids are elongate and laterally flattened nocturnal fishes that propel themselves by undulations of the long anal fin and are capable of breathing air. *Notopterus* (adult, Fig. 9.10B) occurs in India and Southeast Asia; the closely related *Papyocranus* (adult, Fig.9.12) and *Xenomystus* are tropical African (Lauder and Liem 1983; Nelson 1984). Only the monotypic *Papyocranus* has been examined for sperm ultrastructure.

***Papyocranus afer*:** The freshwater African Knife fish, *Papyocranus* (=*Notopterus*) *afer*, has an anacrosomal aquasperm in present terminology (Mattei 1970) (Fig. 9.13).

Fig. 9.12 *Papyocranus afer* (Notopteridae). After Sterba, G. 1962. *Freshwater Fishes of the World.* Vista Books, London, Fig. 53.

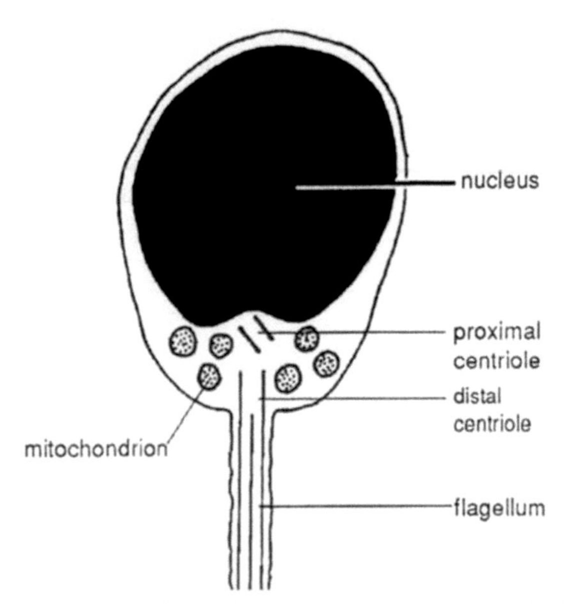

Fig. 9.13 *Papyocranus afer* (Notopteridae). Diagrammatic longitudinal section of the spermatozoon. After Mattei, X. 1970. Pp. 59-69. In B. Baccetti (ed.), *Comparative Spermatology,* Academic Press, New York, Fig. 4: 21.

The proximal centriole is partly contained in the small nuclear fossa and is in line with but tilted relative to the distal centriole and flagellum. There are several irregularly arranged small mitochondria. No cytoplasmic canal is indicated.

9.8.3 Family Mormyridae

The most outstanding features of the Mormyridae and Gymnarchidae (adults, Fig. 9.14) are the electrogenic organs, derived from the caudal muscles, and the greatly enlarged cerebellum. Both families inhabit fresh waters in tropical Africa and the Nile. Mormyrids, unlike the other osteoglossiform families, which have very few species, have undergone an evolutionary radiation which has produced over 200 extant species (Lauder and Liem 1983; Nelson 2006).

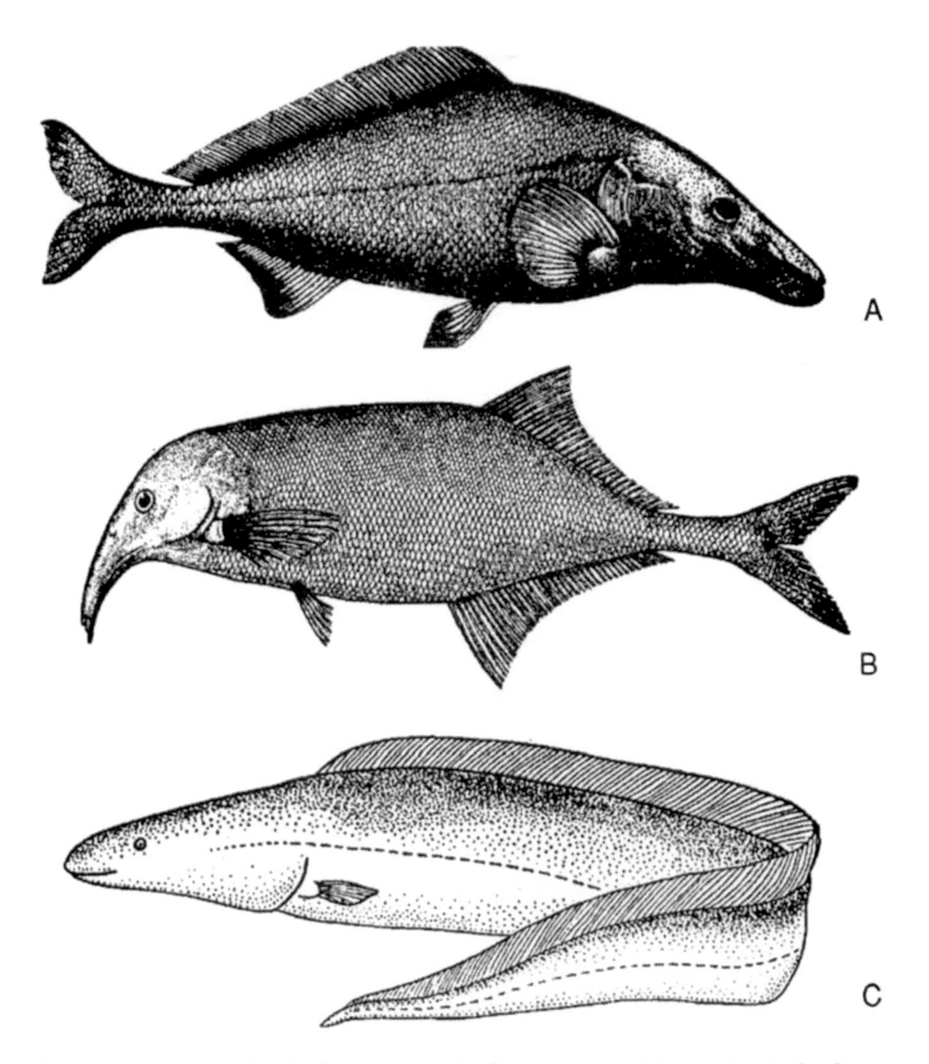

Fig. 9.14 Mormyroids. **A**. *Mormyrops*. **B**. *Gnathonemus* (Mormyridae). **C**. *Gymnarchus* (Gymnarchidae). After Grassé, P. 1958. Traité de Zoologie. In *Anatomie, Systématique, Biologie. XIII Agnathes et Poissons. Masson*, Paris, Fig. 1576.

Sperm literature: The five examined species of Mormyridae, *Brienomyrus niger*, *Marcusenius (=Gnathonemus) senegalensis*, *Hyperopisus bebe*, *Mormyrus rume* and *Petrocephalus bovei* (Fig. 9.15) have relatively uniform aflagellate sperm (Mattei *et al.* 1967, 1972; Mattei 1991).

Sperm ultrastructure: The five species from different genera of Mormyridae all possess aflagellate spermatozoa. The sperm cell is rounded and at its base are the two centrioles, the mitochondria and an abundant vesicular cytoplasm. The gametes has not only lost its flagellum but no longer has the appearance

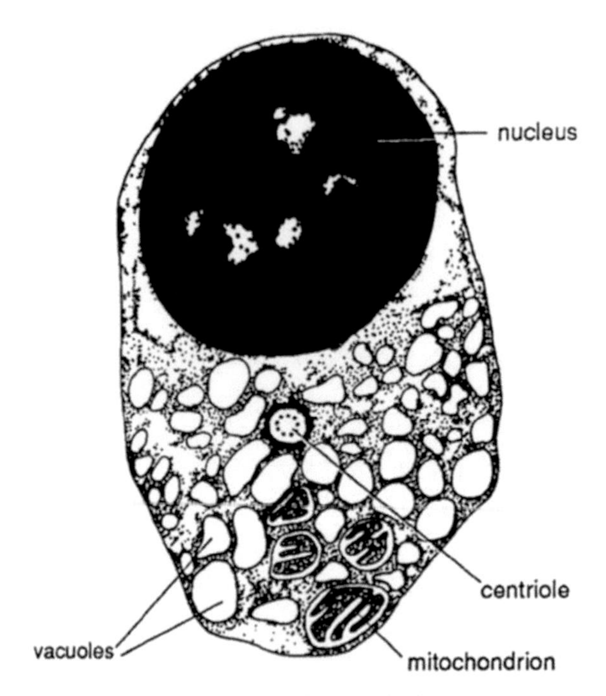

Fig. 9.15 *Petrocephalus bovei* (Mormyridae). Longitudinal section of a spermatozoon. From Jamieson 1991, after a micrograph of Mattei *et al.* 1972. Journal de Microscopie (Paris) 15: 67-78.

of a spermatic cell; the nucleus is uncondensed, the cytoplasm is abundant and the plasma membrane is lined internally by a network of microtubules. It maintains is mobility by means of the microtubular network (Mattei *et al.* 1972; Mattei 1991).

In *Petrocephalus*, the nucleus occupies one pole of the cell; the centrioles are situated between it and the mitochondria which are grouped at the opposite pole, as in flagellate teleost sperm. These mormyrid sperm are intermediate between the typical teleost anacrosomal aquasperm and the more modified sperm of *Gymnarchus* (below).

9.8.4 Family Gymnarchidae

Gymnarchus niloticus: The sperm of G. *niloticus*, the only species of its genus, has a very polymorphic nucleus with a double envelope with some nuclear pores; the internal membrane is adherent to the chromatin (Figs. 9.16, 9.17). The cell membrane is thick and osmiophilic, lined internally by a regular arrangement of "subcuticular" 25-30 nm thick tubular fibrils. There is abundant cytoplasm consisting mostly of sinuous tubules which in some sections have the appearance of small vesicles. In the vicinity of the nucleus there are some globular mitochondria, 1 µm wide, lipid masses and myelin

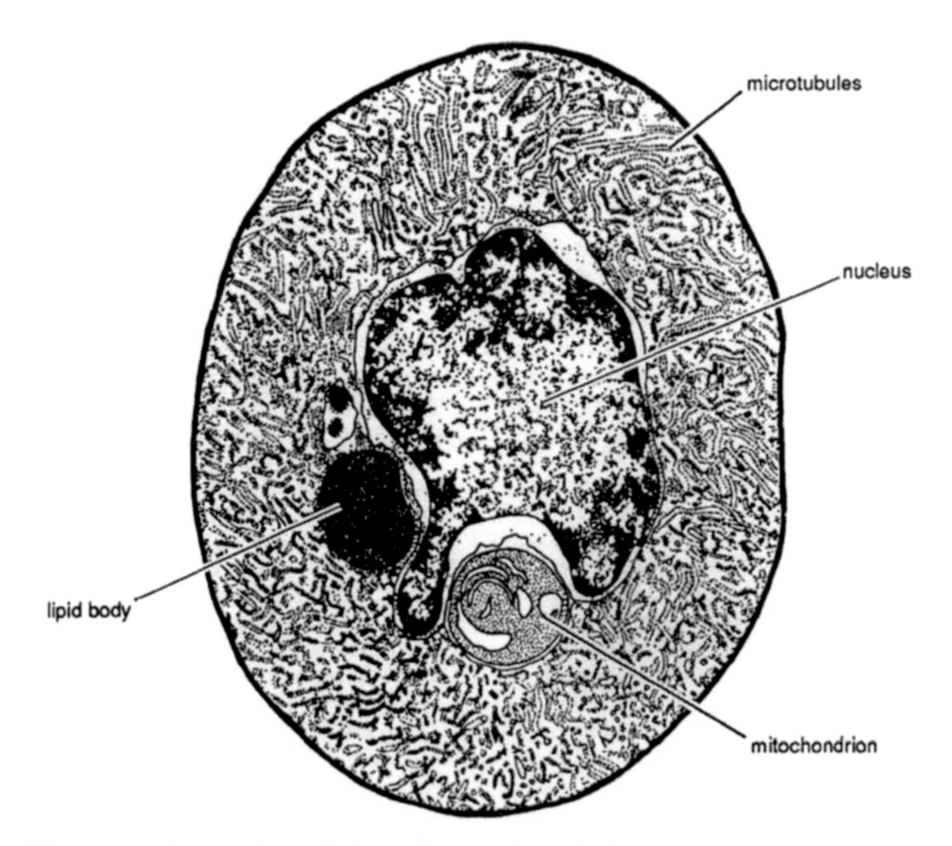

Fig. 9.16 *Gymnarchus niloticus* (Gymnarchidae). Section of a spermatozoon showing the cytoplasm containing abundant microtubules and the quiescent, polymorphic nucleus which is pierced in places by nuclear pores. From Jamieson 1991, drawn from a micrograph by Mattei, X., Boisson, C., Mattei, C. and Reizer, C. 1967. Comptes Rendus Hebdomadaires des Séances de l'Académie des Sciences D 265: 2010-2012, Pl. II, Fig. 4.

vesicles. The two centrioles lie parallel at an equal distance from the nucleus (Mattei *et al.* 1967).

Mormyroid fertilization biology: The mode of fertilization in mormyroid fishes has received little attention. Their lack of copulatory organs strongly suggested that fertilization is external, as it was known to be in *Notopterus* and *Papyocranus* (Jamieson 1991). Free spawning of apparently unfertilized eggs has been confirmed for six mormyrid species, including *Petrocephalus soudanensis* and *Mormyrus rume* (Kirschbaum and Schugardt 2002). The correlation of aflagellarity and external fertilization in mormyroids is intriguing. Amoeboid movement has been suggested as a means of motility for these aflagellate sperm during fertilization (Mattei 1970, 1988, 1991)

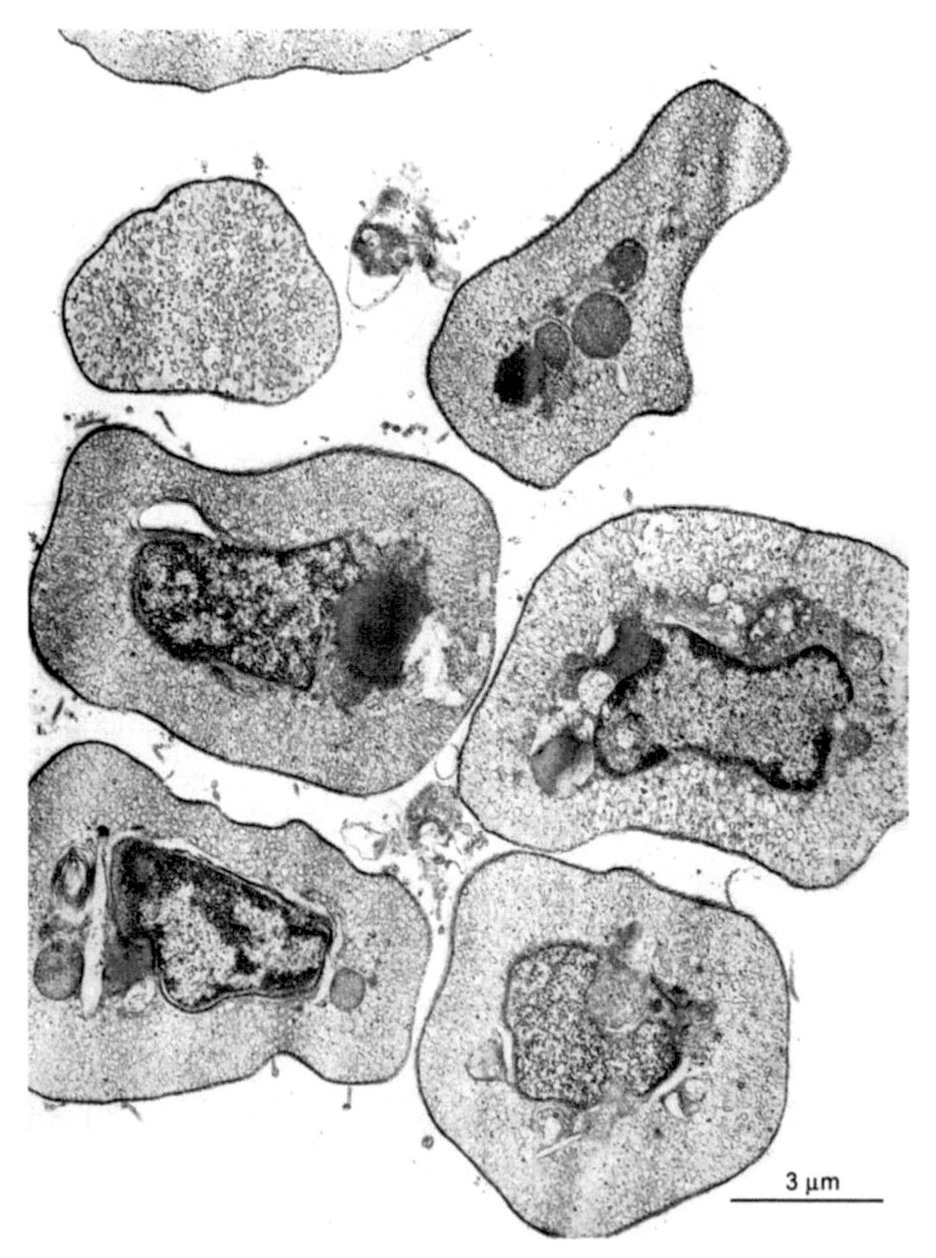

Fig. 9.17 *Gymnarchus niloticus* (Gymnarchidae). TEM section of several spermatozoa. Courtesy of Professor Xavier Mattei.

9.9 CHAPTER SUMMARY

Neopterygii: Neopterygians are spermatologically diagnosed from the agnathans through Dipnoi in having lost the acrosome. This correlates with the presence of an egg micropyle. The plesiomorphic sperm type for neopterygians (though it is probably not plesiomorphic for phylogenetically more basal fishes as exemplified by agnathans, Cladistia and Chondrostei) is a round-headed, acrosomeless spermatozoon, with short midpiece, two triplet centrioles, and long 9+2 flagellum, probably with lateral flagellar fins, which is spawned into the ambient water. This is termed the anacrosomal aquasperm.

Holostei: Molecular analyses indicate that the Holostei is a monophyletic group. The Garpike, *Lepisosteus osseus* (Lepisosteiformes) has an anacrosomal aquasperm but departs from the basic morphology in slight elongation of the nucleus and in the single, ring- or C-shaped mitochondrion. Occurrence of intratubular differentiations (densities in some A subtubules of the axoneme) may also be apomorphic. The Amiiformes, represented by the Bowfin, *Amia calva*, appear to be the sister-taxon of the Lepisosteiformes. The sperm of *Amia* has a spherical nucleus and in this and other respects is an anacrosomal aquasperm. Some cytological peculiarities (mitochondrial granules, annulate lamellae, exocytotic vesicles) are noted, however.

Teleostei. Osteoglossomorpha: Osteoglossiformes exhibit a bewildering diversity of major modifications of the anacrosomal acrosome but this basic type is conserved in the notopteroid *Papyocranus afer*. The peculiar filiform sperm of the only other investigated osteoglossoid, *Pantodon bucholzi*, endorses the separate status of the Pantodontidae also supported by a molecular analysis. The rare aflagellate condition of sperm constant in investigated Gymnarchidae and Mormyridae is in agreement with the association of these families in the Mormyroidei or as sister-groups in a wider Notopteroidei. It is unfortunate that no members of the Osteoglossidae or Hiodontidae have been investigated for sperm ultrastructure.

9.10 LITERATURE CITED

Afzelius, B. A. 1978. Fine structure of the garfish spermatozoon. Journal of Ultrastructure Research 64: 309-314.

Afzelius, B. A. 1979. Sperm structure in relation to phylogeny in the lower Metazoa. Pp. 243-251. In D. W. Fawcett and J. M. Bedford (eds), *The Spermatozoon: Maturation, Motility, Surface Properties and Comparative Aspects; 3rd International Symposium, Woods Hole, Mass., USA*. Urban and Schwarzenberg, Baltimore, Md., USA; Munich, West Germany.

Afzelius, B. A. and Mims, S. D. 1995. Sperm structure of the bowfin, *Amia calva* L. Journal of Submicroscopic Cytology and Pathology 27: 291-294.

Baccetti, B. 1979. The evolution of the acrosome complex. Pp. 305-329. In D. W. Fawcett and J. M. Bedford (eds), *The Spermatozoon*. Urban and Schwarzenberg, Baltimore-Munich.

Baccetti, B. 1985. Evolution of the sperm cell. Pp. 3-58. In C. B. Metz and A. Monroy (eds), *Biology of fertilization*. Academic Press, New York.

Bartram, A. W. H. 1977. The Macrosemiidae, a Mesozoic family of holostean fishes. Bulletin of the British Museum of Natural History (Geology) 29: 137-234.

Breder, C. M. and Rosen, D. E. 1966. *Modes of reproduction in fishes*. The American Museum of Natural History, The Natural History Press, New York. 941 pp.
Britz, R. 2004. Egg structure and larval development of *Pantodon buchholzi* (Teleostei: Osteoglossomorpha), with a review of data on reproduction and early life history in other osteoglossomorphs. Icthyological Exploration of Freshwaters 15: 209-224.
Deurs, B. V. 1973. Helical, striated rootlets in the midpiece of a teleost fish spermatozoon. Zeitschrift für Anatomie Entwicklungsgeschichte 140: 11-17.
Deurs, B. V. 1974. The sperm cells of *Pantodon* (Teleostei) with a note on residual body formation. Pp. 311-318. In B. A. Afzelius (ed.), *The functional anatomy of the spermatozoon*. Pergamon Press, Oxford.
Deurs, B. V. and Lastein, U. 1973. Ultrastructure of the spermatozoa of the teleost *Pantodon buchholzi* Peters, with particular reference to the midpiece. Journal of Ultrastructure Research 42: 517-533.
Goodrich, E. S. 1909. Vertebrata Craniata. Part 9. In E. R. Lankester (ed.), *A Treatise on Zoology*. Adam and Charles Black, London.
Greenwood, P. H. 1973. Interrelationships of osteoglossomorphs. Pp. 307-332. In P. H. Greenwood, R. S. Miles and C. Patterson (ed.), *Interrelationships of Fishes*. Journal of the Linnean Society of London Zoology 53 (suppl. 1) Academic Press, New York.
Jamieson, B. G. M. 1984. Spermatozoal ultrastructure in *Branchiostoma moretonensis* Kelly, a comparison with *B. lanceolatum* (Cephalochordata) and with other deuterostomes. Zoologica Scripta 13: 223-229.
Jamieson, B. G. M. 1991. *Fish Evolution and Systematics: Evidence from Spermatozoa*. Cambridge University Press, Cambridge. 319 pp.
Jessen, H. L. 1973. Interrelationships of actinopterygians and brachiopterygians: evidence from pectoral anatomy. Pp. 227-232. In P. H. Greenwood, R. S. Miles and C. Patterson (ed.), *Interrelationships of fishes*. Academic Press, New York.
Kirschbaum, F. and Schugardt, C. 2002. Reproductive strategies and developmental aspects in mormyrid and gymnotiform fishes. Journal of Physiology Paris 96: 557-566.
Kumazawa, Y. and Nishida, M. 2000. Molecular biology of osteoglossoids: a new model for Gondwanan origin and plate tectonic transportation of the Asian Arowana. Molecular Biology and Evolution 17: 1869–1878.
Lauder, G. V. and Liem, K. F. 1983. The evolution and interrelationships of the actinopterygian fishes. Bulletin of the Museum of Comparative Zoology 150: 95-197.
Lavoué, S., Miya, M., Inoue, J. G., Saitoh, K., Ishiguro, N. B. and Nishida, M. 2005. Molecular systematics of the gonorynchiform fishes (Teleostei) based on whole mitogenome sequences: Implications for higher-level relationships within the Otocephala. Molecular Phylogenetics and Evolution 37: 165-177.
Lavoué, S. and Sullivan 2004. Simultaneous analysis of five molecular markers provides a well-supported phylogenetic hypothesis for the living bony-tongue fishes (Osteoglossomorpha: Teleostei). Molecular Phylogenetics and Evolution 33: 171-185.
Li, G.-Q. and Wilson, M. V. H. 1996. Phylogeny of Osteoglossomorpha. Pp. 163-174. In M. L. J. Stiassny, L. R. Parenti and G. D. Johnson (eds), *Interrelationships of fishes*. Academic Press, San Diego.
Mattei, C., Mattei, X., Marchand, B. and Billard, R. 1981. Réinvestigation de la structure des flagelles spermatiques: cas particulier des spermatozoïdes à mitochondrie annulaire. Journal of Ultrastructure Research 74: 307-312.

Mattei, X. 1970. Spermiogenèse comparée des poissons. Pp. 57-69. In B. Baccetti (ed.), *Comparative Spermatology*. Academic Press, New York.

Mattei, X. 1988. The flagellar apparatus of spermatozoa in fish Ultrastructure and evolution. Biology of the Cell 63: 151-158.

Mattei, X. 1991. Spermatozoon ultrastructure and its systematic implications in fishes. Canadian Journal of Zoology 69: 3038-3055.

Mattei, X., Boisson, C., Mattei, C. and Reizer, C. 1967. Spermatozoïdes aflagellés chez un poisson: *Gymnarchus niloticus* (Téleostéen, Gymnarchidae). Comptes Rendus Hebdomadaires des Séances de l'Académie des Sciences D 265: 2010-2012.

Mattei, X., Mattei, C., Reizer, C. and Chevalier, J. L. 1972. Ultrastructure of Nonflagellated Spermatozoa of Teleost Family Mormyridae. Journal de Microscopie (Paris) 15: 67-78.

Nelson, G. J. 1969. Gill arches and the phylogeny of fishes with notes on the classification of vertebrates. Bulletin of the American Museum of Natural History 141: 475-552.

Nelson, J. S. 1984. *Fishes of the World*, 2nd edition. John Wiley and Sons, New York.

Nelson, J. S. 2006. *Fishes of the World*, 4th edition. John Wiley and Sons, Inc., Hoboken, New Jersey. 601 pp.

Normark, B. B., Mccune, A. R. and Harrison, R. G. 1991. Phylogenetic relationships of Neopterygian fishes, inferred from mitochondrial DNA sequences. Molecular Biology and Evolution 8: 819-834.

Patterson, C. 1973. Interrelationships of holosteans. Pp. 233-306. In P. H. Greenwood, R. S. Miles and C. Patterson (eds), *Interrelationships of Fishes*. Journal of the Linnean Society of London Zoology 53 (suppl 1). Academic Press, New York.

Patterson, C. and Rosen, D. E. 1977. Review of the ichthyodectiform and other Mesozoic teleost fishes and the theory and practice of classifying fossils. Bulletin of the American Museum of Natural History 158: 83-172.

Retzius, G. 1905. Die Spermien der Leptokardier, Teleostier und Ganoiden. Biologische Untersuchungen von G Retzius Nf 12: 103-115.

Romer, A. S. 1966. *Vertebrate paleontology*, 3rd edition. University of Chicago Press, Chicago. 468 pp.

Wiley, E. O. 1976. Phylogeny and biogeography of fossil and recent Gars (Actinopterygii: Lepisosteidae). University of Kansas Museum of Natural History Miscellaneous Publications 64: 1-111.

CHAPTER 10

Ultrastructure of Spermatozoa: Elopomorpha and Clupeomorpha

Barrie G. M. Jamieson[1] *and X. Mattei*[2]

10.1 SUBDIVISION ELOPOMORPHA

For a discussion of the composition and relationships of the Elopomorpha (see Lauder and Liem 1983; Smith 1984; Jamieson 1991; Nelson 2006). They contain the Elopiformes, Albuliformes (in which the Notacanthiformes are subsumed), and Anguilliformes (Nelson 2006). They all have a leptocephalus larva. That this is a synapomorphy of a monophyletic group (Greenwood *et al.* 1966) has been questioned but sperm structure unequivocally unifies the group. The great value of spermatology for taxonomy and phylogeny is thus strikingly demonstrated by this assemblage. Molecular analysis of mitochondrial 12S ribosomal RNA sequences (Wang *et al.* 2003) (Fig. 10.18) and of 10 nuclear genes (Orti and Li, **Chapter 1**) and a combined molecular and morphological analysis (Inoue *et al.* 2004) have confirmed elopomorph monophyly. Monophyly is further supported by whole mitogenomic analysis which indicates that elopomorphs are the plesiomorph sister-group of the remaining teleosts (Clupeocephala), excepting the more basal Osteoglossomorpha (Lavoué *et al.* 2005) (Fig. 10.1).

Sperm literature: Some 22 named species of the Elopiformes, Albuliformes and Anguilliformes have now been investigated for sperm ultrastructure (see details under groups), viz. Elopiformes, Elopidae, 1 species; Albuliformes Albuloidei, Albulidae, 2 species; Notacanthoidei, 1 species; Anguilliformes, Anguilloidei Anguillidae, 5 species; Muraenoidei, Muraenidae, 3 species;

[1]School of Integrative Biology, University of Queensland, Brisbane, Queensland 4072, Australia
[2]Parasites et écosystèmes méditerranéens, Université de Corse, F-20250 Corte, France

Congroidei, Congridae, 3 species and unnamed species; Ophichthyidae, 4 species Heterenchelydae, 1 species; Muraenesocidae, 2 species.

Elopomorph sperm autapomorphies: Sperm ultrastructure in Elopomorpha shows unique features and, with some exceptions (Muraenidae and, less notably, the Nettastomatidae and Halosauridae) is remarkably uniform.

Two features constitute spermatozoal synapomorphies unequivocally linking the Elopiformes, Anguilliformes and Albuliformes and comprise autapomorphies of the Elopomorpha. They are a 9+0 flagellum, a constant feature, and extension of the proximal centriole as two elongate bundles of 4 and 5 triplets, running from the centriolar region towards the tip of the elongate nucleus, which may extend as a free pseudoflagellum. A third feature is the striated centriolar rootlet, seen in all but the Muraenidae, Nettastomatidae and Halosauridae. In the last three families it is not reported whether the extension of the proximal centriole is divided. These apomorphies are superimposed on a phylogram in Fig. 10.1.

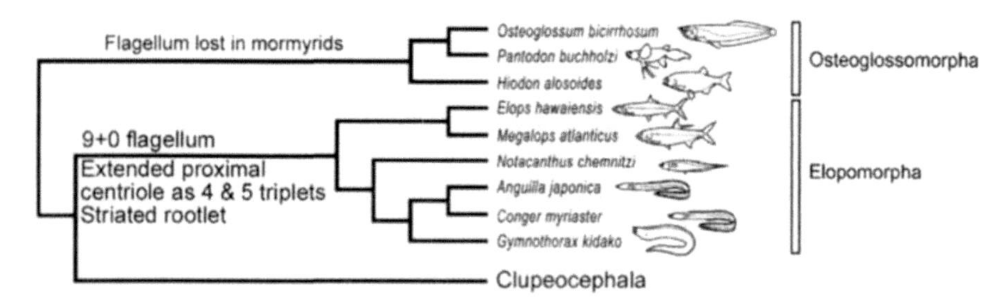

Fig. 10.1 Spermatozoal apomorphies superimposed on a phylogeny of the Elopomorpha. Phylogeny adapted from Lavoué, S., Miya, M., Inoue, J. G, Saitoh, K., Ishiguro, N. B. and Nishida, M. 2005. Molecular Phylogenetics and Evolution 37: 165-177, Fig. 3 (part). With permission of Elsevier.

General elopomorph sperm ultrastructure: Because of the relative uniformity of the elopomorph spermatozoon, a general description, embracing all investigated species but indicating the chief variations will be given here. It is derived from the accounts of Mattei and Mattei (1974) and Mattei (1991) augmented from other references. The sperm of *Albula vulpes* well exemplifies general elopomorph features (Fig. 10.2) but variation is shown diagrammatically in Figures 10.7, 10.9 and 10.18.

Nucleus: The Muraenidae, e.g. *Lycodontis afer* (Fig. 10.9E) and Nettastomatidae (e.g. *Facciolella physonema*) (Fig. 10.9I) are exceptional in having a rounded nucleus, as in most teleosts, with, in muraenids, the distal area slightly depressed. The other elopomorphs investigated have an elongate nucleus of variable shape and size (4-20 µm long). As the "lateral" limit of the nucleus is furthest from the basal body (distal centriole) it is reasonably termed the anterior end by Todd (1976), although the head is often carried at right angles or a more acute angle to the flagellum. The shape is gently curved with a curved or hook-shaped "lateral"(here termed anterior) end which is directed either dorsally (here taking the striated rootlet to be dorsal relative to the axoneme), as in *Congermuraena* (Fig. 10.9H), or ventrally, as in *Albula vulpes* (Fig. 10.2) and other species (Mattei and Mattei (1974).

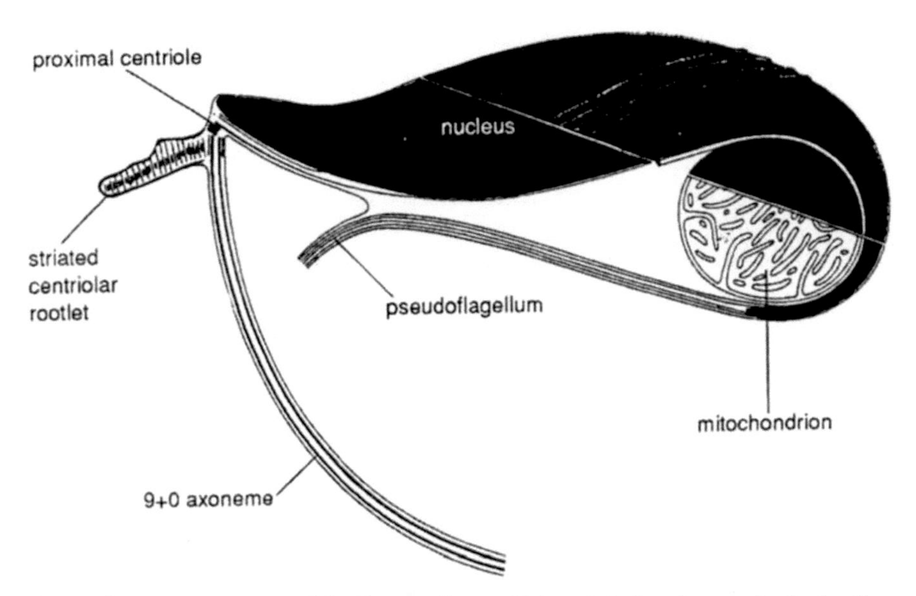

Fig. 10.2 *Albula vulpes* (Albulidae). Exemplifying full development of elopiform spermatozoal characteristics. After Mattei, C. and Mattei, X. 1973. Zeitschrift für Zellforschung und Mikroskopische Anatomie 142: 171-192. Fig. 11. With kind permission of Springer Science and Business Media.

Except for *Elops* (Fig. 10.7A) and the albulid *Pterothrissus* (Fig. 10.7C), the nucleus shows a perceptibly rounded mid-region and two edges of smaller section. It has a concavity in which the base of the proximal centriole is located. At its anterior ("lateral") end the nucleus terminates with a slender or depressed area. In *Elops* and *Pterothrissus* the centriolar and mid-region of the nucleus are attenuated, whereas the anterior end is rounded. In all species the nuclear material is not compact but has areas of clear nucleoplasm.

Okamura *et al.* (2000) have questioned why the sperm of elopomorph fish [with the exception of muraenids], though externally fertilizing, have elongate nuclei. They ascribe this to genetic drift and neutral evolution. However, it is not here considered that elongation of the nucleus, reduction to a single mitochondrion, migration of this mitochondrion away from the centrioles, development of two bundles of microtubules from the proximal centriole and modification of the flagellum to the 9+0 condition, with its enhanced beat frequency, could be ascribed to neutral evolution. If such profound changes commenced owing to genetic drift (or some cytogenetic 'accident') it is likely that they would have been opposed by selective pressures unless, as was presumably the case, they conferred some advantages which led to their subsequent selection with further modification of the sperm.

Mitochondria: In *Muraena* the mitochondria (unusual for elopomorphs in being multiple) remain in the centriolar area. In the other investigated elopomorphs (*Anguilla anguilla* and *A. japonica* are now known not to be exceptions despite earlier reports), the mitochondrion is single (rarely accompanied by a further small mitochondrion)and is located at a well-defined site along the nucleus (Mattei and Mattei 1974) (Fig. 10.9A-D, F-H). The observation that *Anguilla anguilla* (adult, Fig. 10.5A; sperm, Fig. 10.5B) has several mitochondria throughout the length of the nucleus (Gibbons *et al.* 1983) has not been confirmed by other workers who note a single mitochondrion in this species (Billard and Ginsburg 1973a; Okamura *et al.* 2000). It seems unlikely that the species identified by Gibbons *et al.* (1983) was a different species and it seems possible that the multiple mitochondria represented an immature condition. However, presence of a protruding pseudoflagellum in this entity is a further difference. The accounts of Çolak and Yamamoto (1974a,b) which claim that in the Japanese Eel, *Anguilla japonicaa,* the posterior sleeve of cytoplasm surrounding the tail contains the mitochondrion (Çolak and Yamomoto 1974a) or mitochondria (Çolak and Yamomoto 1974b) were later refuted (Gwo *et al.* 1994; Okamura *et al.* 2000).

In *Paraconger, Mystriophis, Pythonichthys* (Fig. 10.9B) and *Echelus* the mitochondrion reaches the anterior extremity or tip ("lateral edge") of the nucleus. In *Albula* (Fig. 10.2, 10.7B), it is located in the curve of the tip of the nucleus, much as in *Anguilla.* In *Myroconger* (formerly the undetermined congrid) (Fig. 10.9D) the nucleus presents a tubercle close to the tip on which the mitochondrion is anchored (Mattei and Mattei 1974; Mattei 1991) and Gwo *et al.* (1992) have demonstrated similar intrusion of nuclear material into the mitochondrion in *Anguilla japonica,* confirmed in a TEM micrograph. In other cases the mitochondrion has not migrated to the end of the nucleus but has stopped at its mid-region. It may rest against the nucleus, as in *Pterothrissus* (Fig. 10.7C), or penetrate into a recess in the nucleus as in *Congermuraena* (Fig. 10.9H). In *Pisodonophis* (Fig. 10.15A), two mitochondria are present, one at the anterior curve of the nucleus [as in *Anguilla*], the other in a recess of the mid-region of the nucleus at the opposite ("anterior") face (Mattei and Mattei 1974).

Proximal centriole and pseudoflagellum: The well-developed proximal centriole persists. A diverticulum of the nuclear membrane penetrates into the centriolar cylinder during late spermiogenesis and provokes its dissociation into two bundles, one of five triplets, the other of four (Fig. 10.3) (Mattei and Mattei 1973, 1974; Todd 1976). This dissociation denied by Çolak and Yamamoto (1974a,b) for *Anguilla japonica* is now known to occur in this species as in *A. anguilla, A. australis* and *A. dieffenbachii.* It is also seen in the Notacanthoidei as described below (**Section 10.3.2**) under that suborder

(Mattei and Mattei 1973, 1974; Todd 1976; Mattei 1991) (Fig. 10.7D). The fate of the five triplet bundle varies with the species and even the individual.

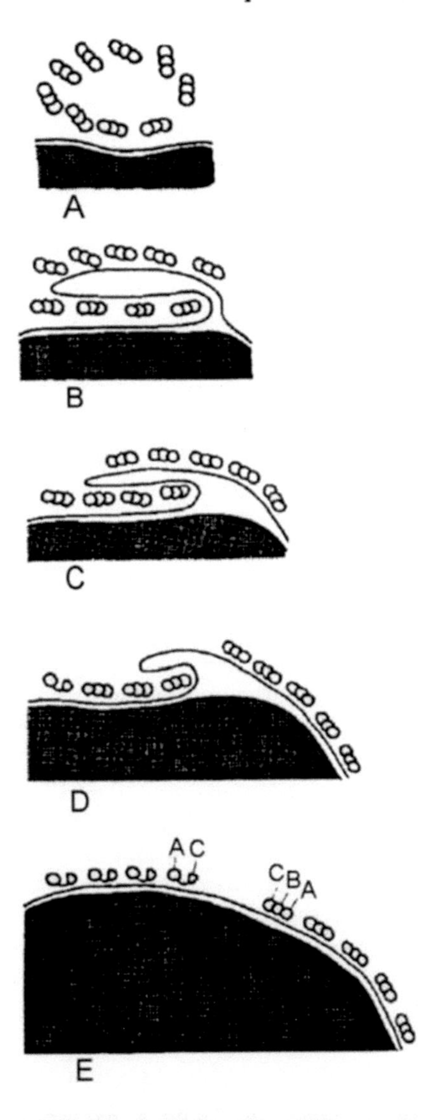

Fig. 10.3 *Albula vulpes* (Albulidae). Dislocation of the proximal sperm centriole by penetration of a diverticulum of the nuclear membrane. A,B and C are subtubules of the centriolar triplets. From Mattei, C. and Mattei, X. 1973. Zeitschrift für Zellforschung und Mikroskopische Anatomie 142: 171-192, Fig. 19. With kind permission of Springer Science and Business Media.

In *Albula vulpes* (Mattei and Mattei 1973, 1974) (Figs. 10.3, 10.4) and supposedly in *Anguilla anguilla* (Ginzburg and Billard 1972) the five triplets retain their structure as far as the tip of the nucleus; beyond it the components may appear as doublets A-C. (However, (Gibbons *et al.* 1983) indicate that in *Anguilla anguilla* A-C doublets coexist with triplets on the nucleus). In the other species, the transformation of the five triplet bundle is achieved in the mid or distal part of the nucleus. In *Myroconger*, the five triplets yield five doublets A-C; in *Paraconger, Mystriophis, Pisodonophis, Pythonichthys, Ophichthus* and *Elops* four triplets are altered into doublets A-C, whereas the fifth one, through loss of the C subtubule, transforms into doublet A-B, the latter being the one nearest the four triplet bundle. In *Pythonichthys* the transformation of the triplet into a doublet A-B is achieved where the proximal centriole dissociates into two bundles.

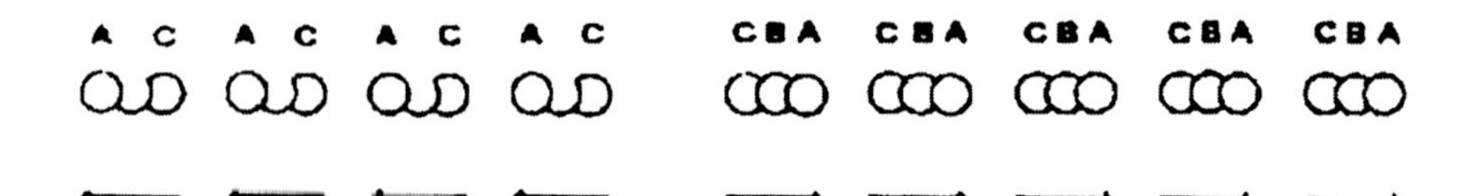

Fig. 10.4 *Albula vulpes* (Albulidae). Constitution of the extension of the proximal centriole in the mid-region of the nucleus. Subtubules A, B and C are indicated on each of the elements (basically triplets) of the centriole. The arrows indicate the orientation of each of these. From Mattei, C. and Mattei, X. 1973. Zeitschrift für Zellforschung und Mikroskopische Anatomie 142: 171-192, Fig. 25. With kind permission of Springer Science and Business Media.

At the tip ("lateral edge") of the nucleus the two bundles tend to join. In *Albula vulpes*, the centriolar bundles may extend as a free pseudoflagellum beyond the nucleus, ensheathed by the nuclear membrane (Fig. 10.2). The free pseudoflagellum may include doublets A-C or doublets A-C and triplets (Mattei and Mattei 1974). In the putative *Anguilla anguilla* sperm described by Gibbons *et al.* (1983) the split bundle also extends, albeit for a short distance, as a free pseudoflagellum; it contains only 9 doublets and dynein arms are absent (Gibbons *et al.* 1983). In *Elops Cynoponticus* and *Albula* the free pseudoflagellum is conspicuous. In *Pisodonophis* some elements of the proximal centriole form an intracellular cylinder surrounding what seems to be an extension of the outer membrane of the nuclear envelope (Mattei and Mattei 1974). The arrangement of the bundles in *A. japonica* and *A. anguilla* is described by Gwo *et al.* (1992) and (Okamura *et al.* 2000), see legend to Fig. 10.12.

It has been suggested that the function of the non-emergent pseudoflagellum may be to stiffen and stabilize the orientation of the large head as the sperm are propelled by the true flagellum. The function of the striated rootlet is uncertain (Gibbons *et al.* 1983). Alternatively, Billard and Ginsburg (1973) suggest that the striated centriolar rootlet (described below), and the extension of the proximal centriole, exist as temporary structures playing a consolidating and stabilizing role in joining the centriolar-flagellar complex to the sperm head during elongation. However, it has been noted for *Anguilla japonica* that the tubules of the pseudoflagellum end on the mitochondrion (Gwo *et al.* 1992; Okamura *et al.* 2000) and this has led to the interesting suggestion that the nine sets of tubules, which originate from the proximal centriole, may act as ducts connecting the mitochondrion with the flagellum and that, if so, ATP transport occurring through such a pipeline should be more efficient than via the cytoplasm (Okamura and Motonobu 1999; Okamura *et al.* 2000). This hypothesis would explain why it is possible for the mitochondrion, the source of oxidative phosphorylation, to be so far removed from the flagellum. These alternative hypotheses require evaluation from additional studies.

Distal centriole: This has a classical triplet structure and emits a motile flagellum though of the 9+0 type. Cytoplasmic microtubules, abundant in spermatids, persist in *Congermuraena* and in the case of *Pterothrissus* make up a prominent bundle along the nucleus.

Centriolar rootlet: A rootlet exists in all examined elopomorph sperm except the Muraenidae, Nettastomatidae and Halosauridae. It is shorter than 1 μm in *Elops* and *Pterothrissus*, approximately 1 μm in *Albula*, 1-1.8 μm in *Anguilla australis*, 1.6-2.0 μm in *Anguilla dieffenbachii*, 1-2 μm in *Congermuraena*, *Paraconger* and *Ophichthus*, 2-4 μm in *Mytriophis*, *Pisodonophis* and *Pythonichthys* and longer than 4 μm in *Myroconger* (Mattei and Mattei 1974; Todd 1976). It shows conspicuous periodicity (160-200 Å in *Anguilla australis* and *A. dieffenbachii*, Todd 1976) and is supported by a central axis (Mattei and Mattei 1974).

Gibbons *et al.* (1983) suggest, at least for *A. anguilla*,that projection of the rootlet beyond the cell body may be a shrinkage artifact during fixation as it is almost never seen by light microscopy of unfixed sperm. Its strong projection in other anguilliforms (Fig. 10.9) is clearly not artifactual. Whether exposed or not, the rootlet remains a striking feature of all non-muraenid elopomorph sperm.

Axoneme: The doublets of the 9+0 true axoneme run close to the plasma membrane (Mattei and Mattei 1983; Gibbons *et al.* 1983), except proximally where Y links clearly span the gap between tubules and membrane. The length of the flagellum varies from 24-66 μm in *Anguilla* species, and is 25 μm in *Muraenesox* (See Table 10.1). The axoneme has a flexible endpiece (Gibbons *et al.* 1983).

The elopomorph sperm axoneme, and specifically that of *Anguilla anguilla*, lacks the central-sheath and -tubules (central singlets), the radial spokes and the outer dynein arms (Baccetti *et al.* 1981; Gibbons *et al.* 1983), though a short outer arm is said to be present on doublets in the extreme proximal part of the axoneme (Baccetti *et al.* 1981). Mattei and Mattei (1973) found two arms to be present on a doublet in certain images but both were in the internal position; outer arms were never seen. The presence of inner arms only, at least for the main part of the axoneme, was clearly illustrated by Gwo *et al.* (1992) for *A. japonica*. Electrophoretic deficiencies in dynein bands have been demonstrated in various animal groups for such axonemes with doublets lacking an arm (Baccetti *et al.* 1979a).

Axonemal deficiencies and motility: The structural and molecular deficiencies of the elopomorph centriolar and axonemal apparatus have not impaired the frequency of beat of the eel flagellum, at 95Hz, is the highest known in eukaryotic flagella. This rapidity is tentatively ascribed to absence of a regulatory response to Ca^{2+}, such as occurs in most cilia and flagella with a complete 9+2 structure (Gibbons *et al.* 1983).

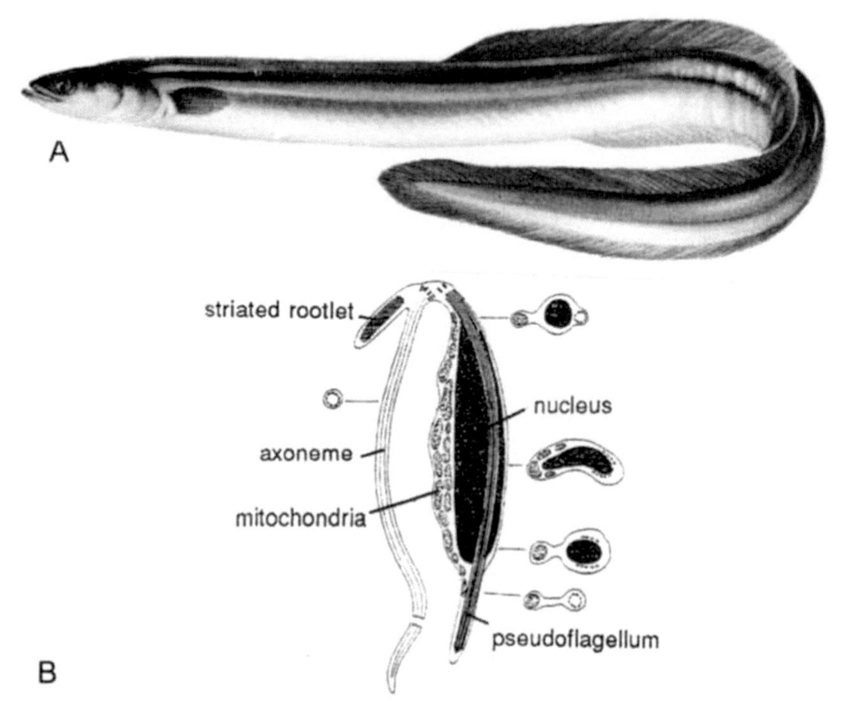

Fig. 10.5 *Anguilla anguilla* (Anguillidae). **A.** Adult. By Norman Weaver. With kind permission of Sarah Starsmore. **B.** Structure of a spermatozoon shown diagrammatically. Note the free pseudoflagellum and multiple mitochondria, features not described for this species by other authors. After Gibbons, B. H., Gibbons, I. R. and Baccetti, B. 1983. Journal of Submicroscopic Cytology 15: 15-20, Fig. 10.

The sperm progress forwards at 140 µm per sec, at about 21° C, as a result of a left-handed helical wave propagated distally along the flagellum at a true beat frequency of about 95Hz. Simultaneously the sperm rolls at a frequency of approximately 18 Hz. Thus these three dimensional waves, which have a propulsive efficiency of about 1.5 µm/beat are significantly less effective than planar waves, which typically have efficiencies of 4.5 µm/beat (Gibbons *et al.* 1983).

Genetic control: In contrast with the sperm tails, ependymal cilia of *Anguilla* have a normal 9+2 axoneme with two dynein arms on each doublet. This indicates that the structure of ciliary and flagellar axonemes in *Anguilla* is controlled by at least partly different groups of genes (Baccetti *et al*. 1981) both in terms of numbers of central singlets and representation of dynein on the doublets. In man 9+0 sperm flagella may, similarly, coexist with 9+2 cilia in the same individual (Baccetti *et al*. 1979b), indicating control of microtubule numbers by at least partly different genomes, but, in contrast, the "armless" syndrome affects both cilia and flagella and is thought to indicate a single genetic control of human dynein (Afzelius *et al*. 1975).

10.2 ORDER ELOPIFORMES

The elopiforms here include the Elopidae (ten pounders) and the Megalopidae (tarpons) (Nelson 2006), the latter family placed in a separate order by Smith (1984). Pelvic fins abdominal; body slender, usually compressed; gill openings wide; caudal fin deeply forked. With a small leptocephalus larva. Marine, tropical and subtropical, with some extension into brackish and fresh waters (Nelson 2006).

Family Elopidae

Elops: The sperm of *Elops lacerta*, a species regarded by Patterson (1973, p. 235), as the most archaic of the teleosts (adult, Fig. 10.6), has been briefly referred to by Mattei and Mattei (1974) (see general account, above).

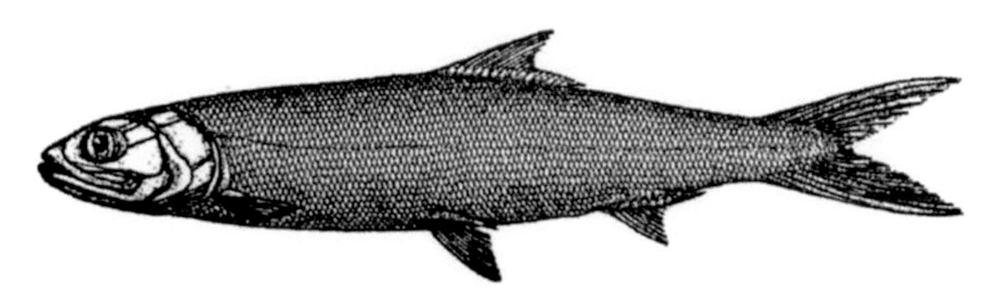

Fig. 10.6 *Elops lacerta* (Elopidae). After Grassé, P .-P. (ed.) 1958. Traité de Zoologie. XIII Agnathes et Poissons. Masson et Cie, Paris, Fig. 1565.

In *Elops* (Fig. 10.7A), as in the albulid *Pterothrissus* (Fig. 10.7C) the centriolar and mid-region of the sperm nucleus are attenuated whereas the anterior ("lateral") end is rounded. The two genera agree also in having the shortest striated rootlet at less than 1 μm.

10.3 ORDER ALBULIFORMES

With two suborders: the Albuloidei, containing the Bone Fish or Queen Fish and relatives from tropical seas and off Japan, and the Notacanthoidei, including the Notacanthidae or Spiny eels and relatives, deep-sea and worldwide (Nelson 2006).

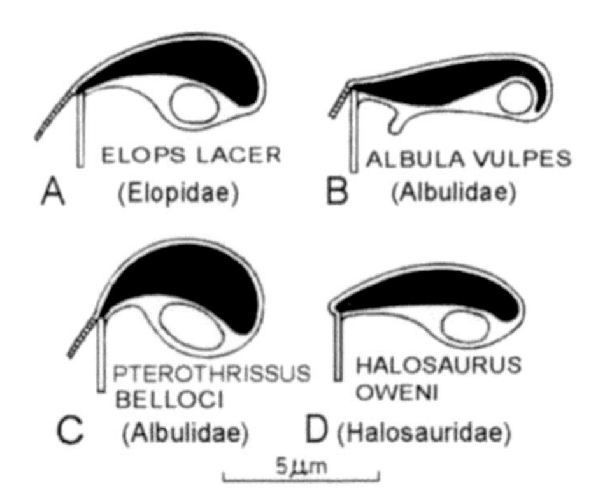

Fig. 10.7 Diagrams of spermatozoa of **A**. Elopiformes. **B-D**. Albuliformes. Adapted from Mattei, X. 1991. Canadian Journal of Zoology 69: 3038-3055, Fig. 6 (part).

10.3.1 Suborder Albuloidei

Albulidae: Two albulid species have been investigated (*Albula vulpes*, Mattei and Mattei 1972, 1973, 1974) (adult, Fig. 10.8) and *Pterothrissus belloci*, Mattei and Mattei 1974). With small variations, their sperm (Fig. 10.7B,C) conform to the general elopomorph pattern described above.

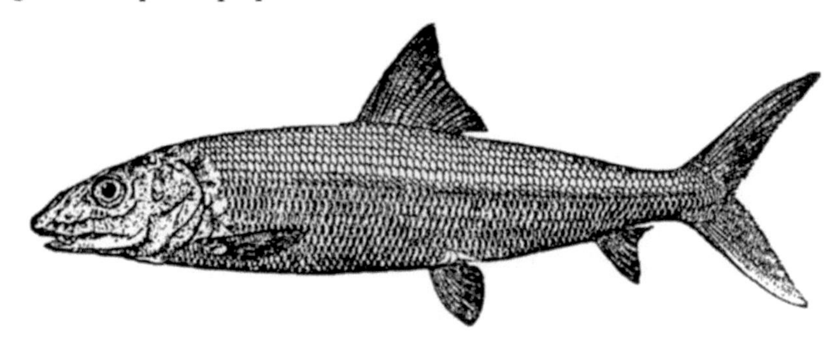

Fig. 10.8 *Albula vulpes* (Albulidae). After Jordan, D.S. (1907). Fishes. Henry Holt, New York, Fig. 203.

***Albula*:** This pattern finds full expression in *Albula* in which a striated rootlet is developed (albeit modestly) and (as in *Elops* and *Cynoponticus*) the pseudoflagellum is well developed and migration of the mitochondrion towards the tip (lateral edge, *sensu* Mattei and Mattei 1974) of the nucleus is pronounced. The tip of the nucleus is unusual in *Albula* in its strong posterior hook. *Albula* may be unique in the investigated elopomorphs in persistence to the end of the nucleus of five intact triplets in one of the extensions of the proximal centriole (Mattei and Mattei 1972, 1973, 1974).

Pterothrissus: (Fig. 10.7C) agrees with *Albula* in the modest development of the striated rootlet. In *Pterothrissus*, as in *Elops*, the centriolar and mid-region of the nucleus are attenuated, whereas the lateral end is rounded. Persistence

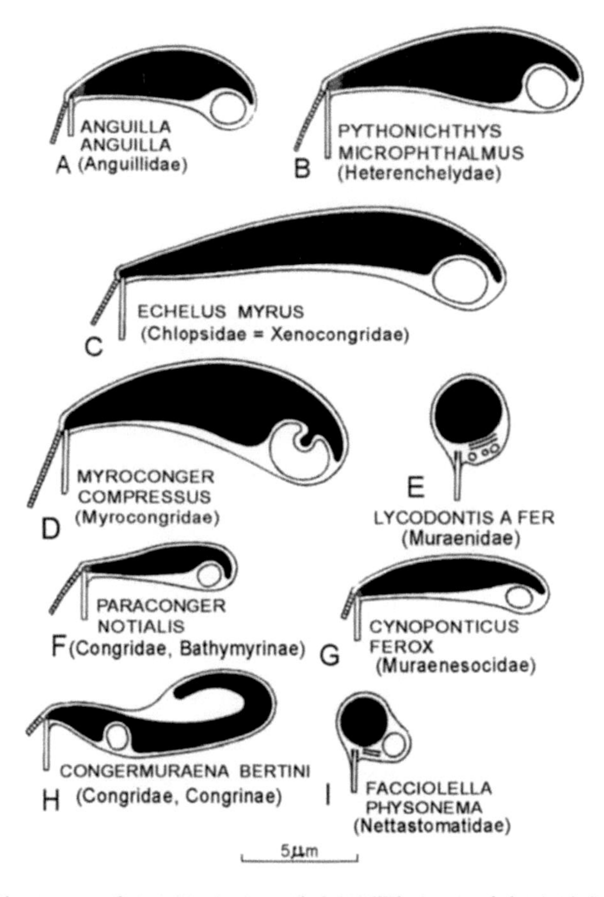

Fig. 10.9 Diagrams of spermatozoa of Anguilliformes. Adapted from Mattei, X. 1991. Canadian Journal of Zoology 69: 3038-3055, Fig. 6 (part).

of microtubules from the spermatid as a prominent bundle along the nucleus is peculiar to *Pterothrissus*. It retains one to several mitochondria (Mattei and Mattei 1974).

10.3.2 Suborder Notacanthoidei

Notocanthids are regarded as members of the Albuloidei within the Anguilliformes in phylogenies by Lauder and Liem (1983) and Smith (1984) but are here excluded from the Anguilliformes in accordance with Nelson (2006).

Halosauridae: These are deep-sea eel-like fish. The spermatozoon of *Halosaurus oweni*, the only notocanthoid studied, has most of the basic elopomorph apomorphies (Fig. 10.7D): the proximal centriole extends along the nucleus and divides into two bundles of tubules arranged in five triplets and four doublets, the latter arising from tubules A and C of the centriole; the doublets of the 9+0 axoneme have a single arm in an internal position and are in contact with the flagellar membrane. However, the 'striated rod' (rootlet) is absent, presumably by loss which is considered to have occurred independently of loss in the muraenid-nettastomatid clade (Mattei 1988, 1991).

10.4 ORDER ANGUILLIFORMES

Anguilliforms are the marine and freshwater eels, with the suborders Anguilloidei, Muraenoidei and Congroidei; finless, with a leptocephalus larva (Nelson 2006).

10.4.1 Suborder Anguilloidei

10.4.1.1 Family Anguillidae

Anguilla: Sperm ultrastructure has been described in four species of the freshwater eel genus *Anguilla*, viz. *A. anguilla* (Baccetti 1979; Baccetti *et al.* 1981; Billard and Ginsburg 1973b; Ginzburg and Billard 1972 (Fig. 10.10);

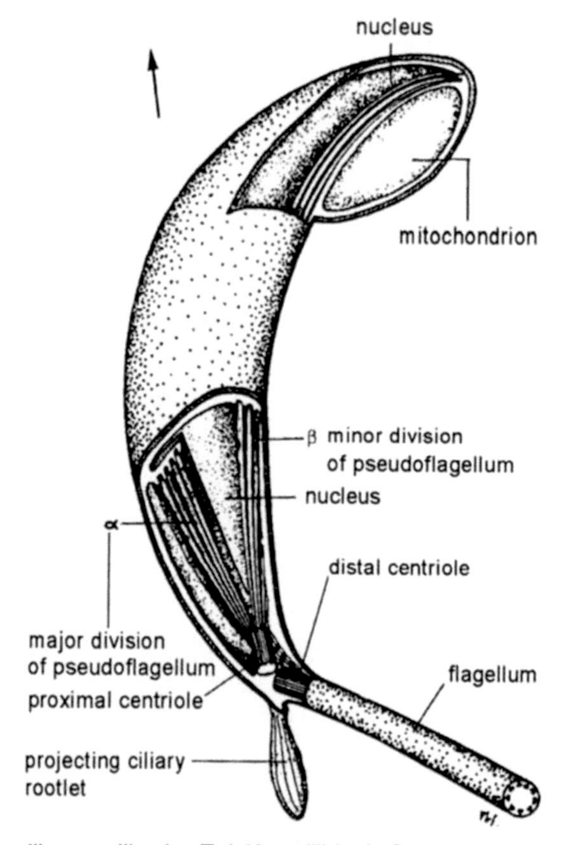

Fig. 10.10 *Anguilla anguilla*, the Eel (Anguillidae). Schematic representation of the spermatozoon. The arrow indicates the direction of movement. From Jamieson 1991, after Billard, R. and Ginsburg, A. S. 1973. Annales de Biologie Animale, Biochemie, Biophysique 13: 523-534, Fig. 1.

Gibbons *et al.* 1983 (Fig. 10.5); Mattei 1991 (Fig. 10.9A); Okamura *et al.* 2000). *A. australis schmidtii* (Fig. 10.11) and *A. dieffenbachia* (Todd 1976); and *A. japonica* (Çolak and Yamamoto 1974a; b; Gwo *et al.* 1992; Gwo *et al.* 1994; Hara and Okiyama 1998; Okamura *et al.* 2000) (Fig. 10.12). Dimensions for the four species of *Anguilla* and for *Muraenesox cinereus* are given in Table 10.1.

The sperm of *Anguilla anguilla* is illustrated in Fig. 10.10, that of *A. australis* in Fig. 10.11 and that of *A. japonica* in Fig. 10.12.

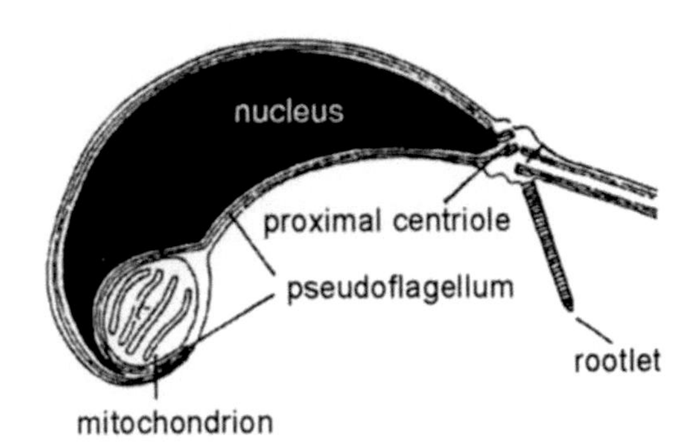

Fig. 10.11 *Anguilla australis schmidtii* (Anguillidae). Schematic reconstruction of a spermatozoon. From Jamieson 1991, after Todd, P. R. 1976. Cell and Tissue Research 171: 221-232, Fig. 20.

The sperm of *Anguilla japonica* as described by Çolak and Yamamoto (1974a,b), appeared to be the simplest in the Elopomorpha excepting the Muraenidae. However, it has since been shown to be a typical *Anguilla* sperm (Gwo *et al.* 1992; Hara and Okiyama 1998; Okamura *et al.* 2000) (Fig. 10.12).

Baccetti *et al.* (1979a) describe the axoneme of *Anguilla anguilla*. It is composed only of 9 doublets and lacks central singlets as well as any system of radial connections. The sperm nevertheless move actively and the tail beats with large frequent waves propagating backwards. Each doublet possesses only the inner, straight arm which is about 20 nm long. A very short segment of outer arm in present in the proximal segment of the axoneme. Short Y-links bind doublets to the plasma membrane. The inner arms and Y-links are strongly ATPase positive with the Wachstein and Meisel test. Electrophoretically *Anguilla* axonemes have a predominant band corresponding with the B band of sea urchin sperm and clearly relating to the presence in the eel sperm of only the inner dynein arm.

The sperm of the following families are described chiefly in Mattei and Mattei (1974), summarized in the general description, above.

10.4.1.2 Heterenchelydae

The sperm of *Pythonichthys microphthalmus* (Mattei 1970; Mattei and Mattei 1974; Mattei 1991) (Fig. 10.9B) is briefly discussed in the general account, above.

arison of morphometric measurements of the spermatozoa of *Anguilla* species and *Muraenesox cinereus*. Adapt
al of Fish Biology 57: 161-169, Table 1.

Head length (μm)	*Head width (μm)*	*Flagellum Length (μm)*	*Rootlet Length (μm)*	*Mitochondrion diameter (μm)*	*Method*
8-11		24-36			TEM
6-9		35			TEM & SEM
5-4 (± 0.4)	1 (± 0.2)	25 (± 5.5)	1.2 (± 0.2)	0.8 (± 0.3)	SEM
6-3	1	30.5			TEM & LM§
4-5(± 0.5)#	1-2(± 0.1)	29 (±6.9)*	1.3(±0.1)	1.0 1.1 (± 0.3)#	TEM SEM
6	2	26-30	1-1.8		TEM & LM
8	3	28-44	1.6-2.0		TEM & LM
5-1(± 0.5)	1 (± 0.2)	25 (± 4.4)	1.1 (± 0.2)	0-7 (± 0-2)	SEM

esents the average and standard deviation (± S.D.) of 20 to 30 spermatozoa.
ficantly different from the data of Okamura *et al.*(2000) for *A. anguilla* and *M. cinereus* at: #P<0.01, *P<0.05.

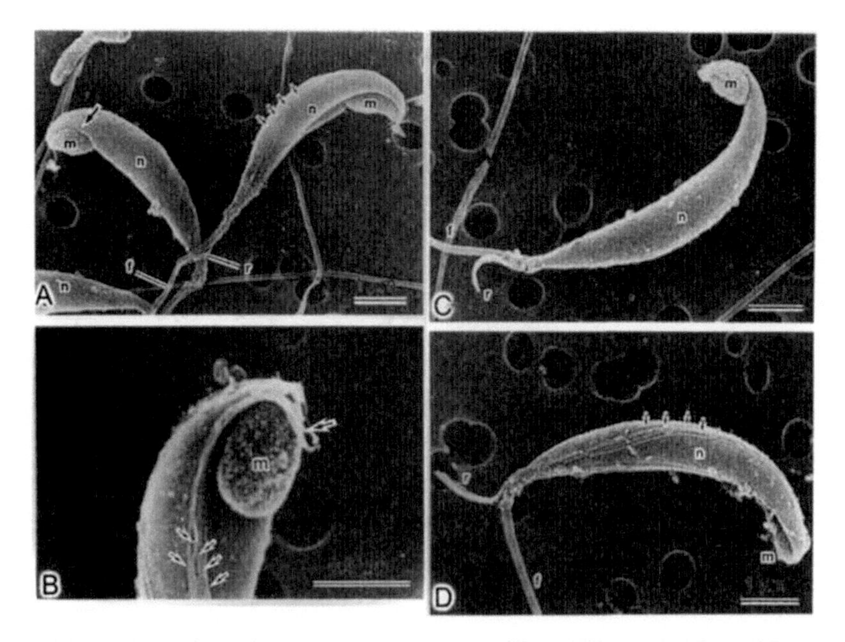

Fig. 10.12 A, B *Anguilla japonica* spermatozoa (Anguillidae). **A**. Viewed from both right and left sides. Four filaments (small arrows) forming the outer bundle are seen only on the right wide surface. A constriction (large arrow) is seen on the anterior portion of the left side of the nucleus (n). m. mitochondrion; r, rootlet; f, flagellum. **B**. Viewed from the concave aspect. A large spherical mitochondrion (m) on the superoanterior aspect of the nucleus is seen. The inner bundle consists of five filaments (small arrows) and their ends are located on the mitochondrion (large arrow). **C**, **D**. *Anguilla anguilla* spermatozoa. **C**. Viewed from the left side. The nuclear surface (n) is smooth. **D.** Viewed from the right side. The outer bundle also consists of four filaments (arrows). f, flagellum; m, mitochondrion; n, nucleus; r, rootlet. After Okamura, A., Zhang, H., Yamada, Y., Tanaka, S., Horie, N., Mikawa, N., Utoh, T. and Oka, H. P. 2000. Journal of Fish Biology 57: 161-169, Figs. 1-4. With permission of Blackwell Publishing.

10.4.2 Suborder Muraenoidei

10.4.2.1 Family Chlopsidae (= Xenocongridae)

The typical elopomorph spermatozoon of *Echelus myrus* is illustrated diagrammatically in Fig. 10.9C (Mattei 1991).

10.4.2.2 Family Muraenidae

Examined species are *Lycodontis afer* (Mattei 1970; Mattei and Mattei 1974; Mattei 1991) (Figs. 10.9E, 10.14), *Muraena helena* (adult, Fig. 10.13) and *Muraena robusta* (see Mattei and Mattei 1974).

The isolated position of the Muraenidae (e.g. *Lycodontis afer*, Figs. 10.9E, 10.14) is confirmed by their lacking a fundamental elopomorph spermatozoal apomorphy: development of a striated rootlet. A flagellar rootlet is probably plesiomorphic for the teleost spermatid but its retention as a projection from

Fig. 10.13 *Muraena helena* (Muraenidae). From Norman, I. R. (1937). Illustrated Guide to the Fish Gallery. British Museum (Natural History). Trustees of the British Museum, London. Fig. 26.

the sperm body in the elopomorph spermatozoon must be considered apomorphic. Mattei (1991) may be correct in holding that absence of the striated rootlet, together with failure of the nucleus to elongate are apomorphic reversals in muraenids as in nettastomatids and that multiple mitochondria in muraenids also represent a reversal from the single elopomorph mitochondrion. He sees the first two conditions as muarenid-nettastomatid synapomorphies.

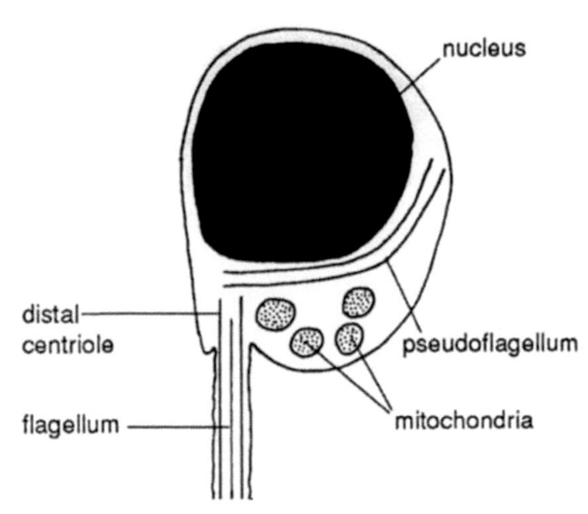

Fig. 10.14 *Lycodontis afer* (Muraenidae). Diagrammatic longitudinal section of the spermatozoon. After Mattei, X. (1970). Pp. 567-569. In B. Baccetti (ed.), *Comparative Spermatology*. Academic Press, New York, Fig. 4:28.

10.4.3 Suborder Congroidei

10.4.3.1 Family Ophichthyidae

Examined ophichthyid (snake eel) species are *Ophichthus ophis, Pisodonophis semicinctus, Mystriophis rostellatus* and *Ophysurus serpens* (Figs. 10.15A-D) (Mattei 1970; Mattei and Mattei 1974; Mattei 1991). The sperm, typical of anguilliforms, are discussed in the general account, above.

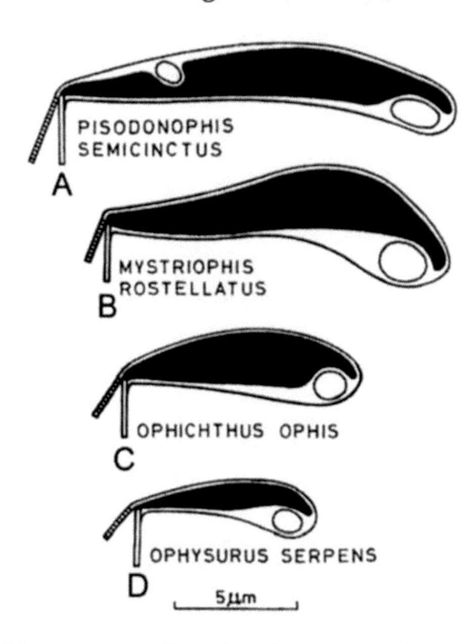

Fig. 10.15 Anguilliformes continued. Diagrams of spermatozoa of the Ophichthidae. Adapted from Mattei, X. 1991. Canadian Journal of Zoology 69: 3038-3055, Fig. 6 (part).

10.4.3.2 Family Muraenesocidae

The sperm of the muraenesocids (pike eels) *Cynoponticus ferox*, placed in the Congridae by Mattei (1991) (Fig. 10.9G); and *Muraenesox cinereus* (Okamura *et al.* 2000), resemble the sperm of *Anguilla* (for details see Table 10.1).

10.4.3.3 Family Congridae

Examined congrid species are *Myroconger compressus* (Fig. 10.9D, *Paraconger notialis* (Fig. 10.9F), *Congermuraena bertini* (Fig. 10.9H) (Mattei and Mattei 1974; Mattei 1991; see general account, above) and *Conger myriaster* (Figs. 10.15, 10.16) (Okamura and Motonobu 1999). *Paraconger notialis* (Fig. 10.9F) and *Conger myriaster* (Fig. 10.16) have an *Anguilla*-type sperm. *Congermuraena bertini* (Fig. 10.9H) is unusual in 'dorsal' flexure of the nucleus (arbitrarily here designating the striated rootlet as dorsal to the flagellum) and location of the mitochondrion in a dorsal pocket of the nucleus (Mattei and Mattei 1974).

Conger myriaster sperm (Figs. 10.16, 10.17) investigated by scanning electron microscopy (SEM) and transmission electron microscopy (TEM) have a total length of ca 40 µm, with the head 3 µm, neck 0.5 µm and flagellum ca 37 µm long. The head is crescent-shaped, with a mitochondrion on the concave surface. The neck, with an attached rootlet, consists of two constrictions close to the base of the flagellum. As in other elopomorphs, two bundles of microtubules extending from the proximal centrioles occur on both the convex and concave surfaces of the sperm head and the flagellum has a 9+0 axonemal pattern. Some flagella showed a coiled form, the developmental mechanism of which is unclear (Okamura and Motonobu 1999).

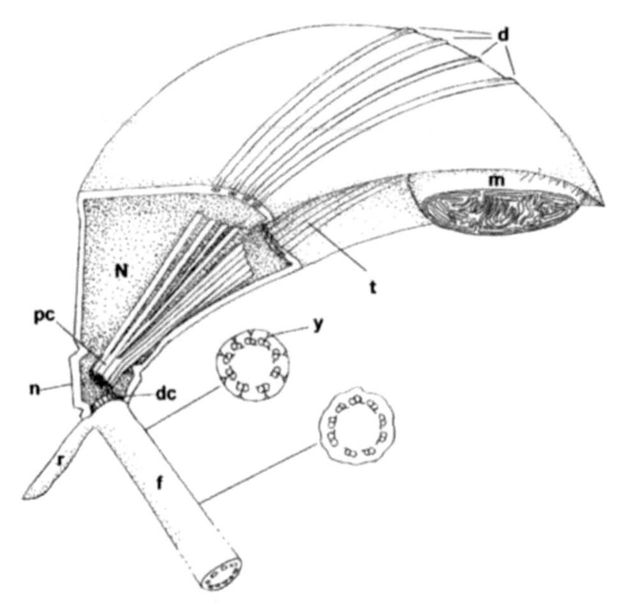

Fig. 10.16 *Conger myriaster* (Congridae). Schematic drawing of a spermatozoon. D, four doublet bundle; dc, distal centriole, f, flagellum; pc, proximal centriole; r, rootlet; M, mitochondrion; N, nucleus; n, constricted neck; t, five triplet bundle; Y, Y-links. From Okamura, A. and Motonobu, T. 1999. Zoological Science (Tokyo) 16: 927-933, Fig. 7. With permission of the Zoological Society of Japan.

10.4.3.4 Family Nettastomatidae

These are the duckbill eels; Marine; Atlantic, Indian and Pacific (Nelson 2006). The spermatozoon of *Facciolella physonema* illustrated diagrammatically by Mattei (1991) (Fig. 10.9I) is atypical for the Elopomorpha in having a round nucleus and in lacking the striated rootlet, as in muraenids.

10.5 ELOPIFORM-ANGUILLIFORM SPERM RELATIONSHIPS

Greenwood *et al.* (1966) advocated uniting Elopiformes and the Anguilliformes in the Elopomorpha with the presence of a leptocephalus larva as a shared

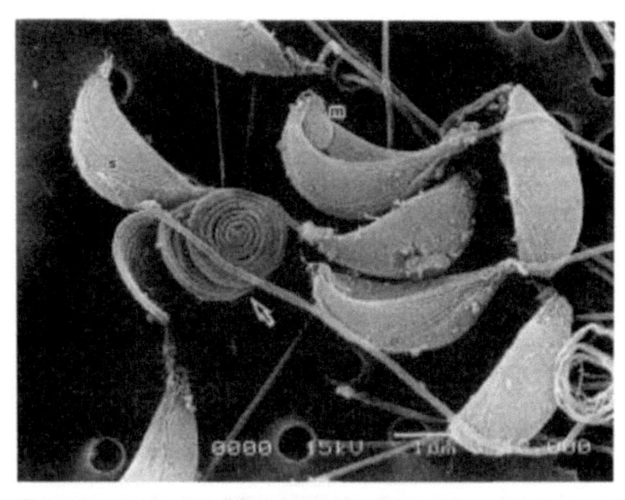

Fig. 10.17 *Conger myriaster* (Congridae). Scanning electron micrograph of spermatozoa. Four striae (s), a filament bundle, are seen on the convex surface of the head and a spherical mitochondrion (m) is seen on the convex surface. Coiled flagella of uncertain significance (arrow) are also seen. After Okamura, A. and Motonobu, T. 1999. Zoological Science (Tokyo) 16: 927-933, Fig. 2. With permission of the Zoological Society of Japan.

character. The distinctive spermatozoal synapomorphies unequivocally support this union. The close congruence of spermatozoal ultrastructure with a molecular phylogeny of the Elopomorpha is demonstrated in Figure 10.18.

Sperm morphology thus unifies the Elopomorpha and has resolved some taxonomic and phylogenetic problems. The interrelationship of elopiforms, anguilliforms and "notacanthiforms" and inclusion of the albuloids among these is confirmed. Nybelin considered the mutual possession of a leptocephalus as a poor, possibly plesiomorphic, character for relationship of *Elops* and *Albula* which were considered unrelated on the basis of features of their somatic anatomy, the albulids being supposedly closer to clupeoids (Nybelin 1973 and references therein). More recently, Filleul and Lavoué (2001) rejected monophyly of the Elopomorpha, considering elopiforms, anguilliforms, albuliforms and notacanthiforms to be four separate monophyletic groups (see, however, monophyly confirmed by Inoue *et al.* 2004). We have seen that sperm structure clearly unites the groups to which *Albula* and *Elops* belong, as upheld from somatic anatomy by Forey (1973) and others, while indicating no close affinity with clupeoids (*q.v.*). The leptocephalus is now confirmed as an elopomorph synapomorphy.

The elopomorph spermatozoon is a complex anacrosomal aquasperm probably derived from an aquasperm with the "primitive" morphology. It is nevertheless possible that the relatively simple sperm of muraenids and nettastomatids are derived by secondary simplification from the typical elopomorph spermatozoon.

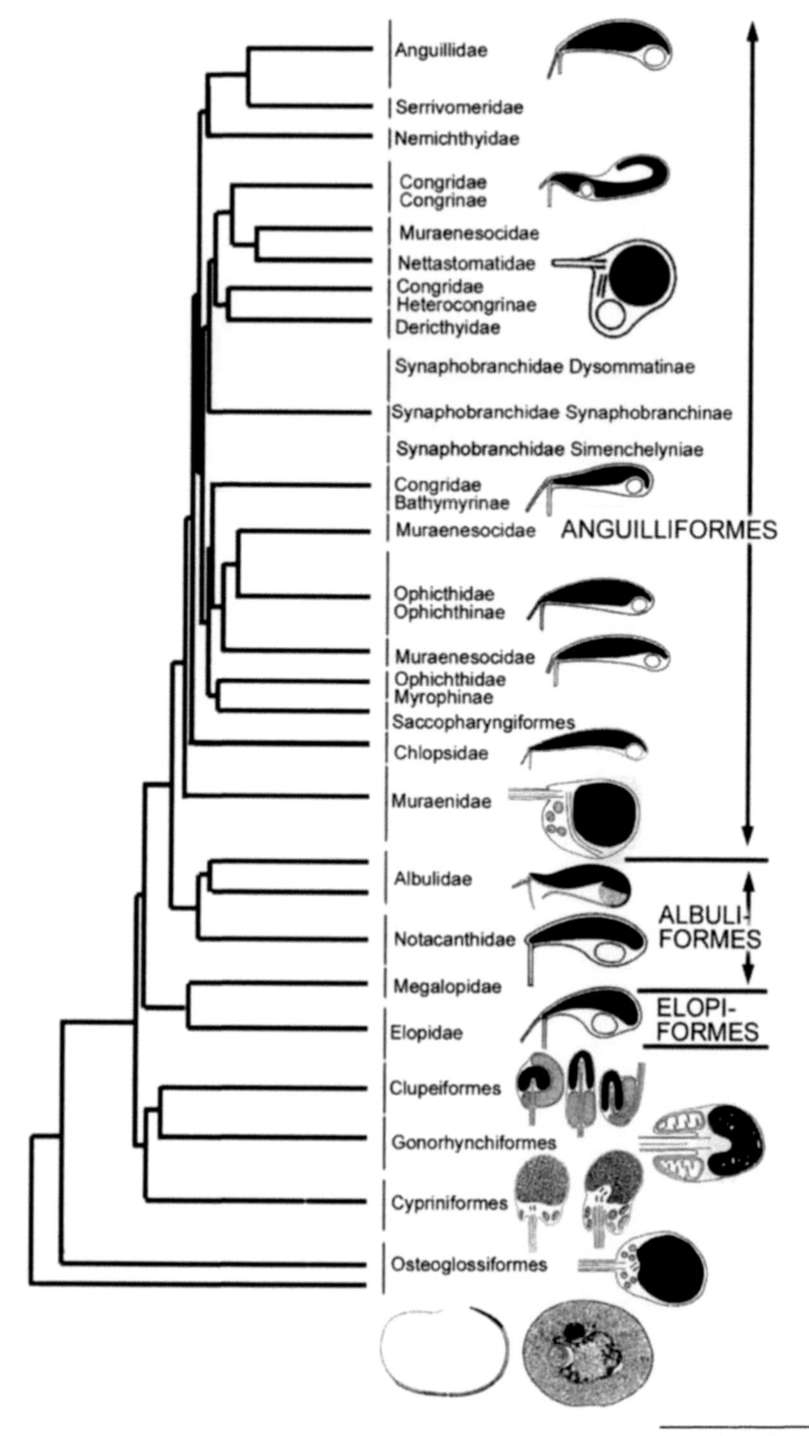

Fig. 10.18 Contd. ...

10.6 SUBDIVISION OSTARIOCLUPEOMORPHA

The Clupeomorpha and the entire Euteleostei (Lauder and Liem 1983) (see also Fig. 10.1) have traditionally been grouped in the Clupeocephala but this has been dropped as a formal rank by Nelson (2006) who recognizes the subdivision Ostarioclupeomorpha on the basis of strong phylogenetic evidence for a sister-group relationship Clupeomorpha and Ostariophysi on molecular and morphological grounds. Clupeomorpha and Ostariophysi are seen as the sister-group to the Euteleostei. Orti and Li (**Chapter 1**), confirm clupeiform+ostariophysan monophyly.

10.7 SUPERORDER CLUPEOMORPHA

Clupeomorphs are unique in having the otophysic connection involving diverticula of the swim bladder which penetrate the exoccipital and extend into the prootic within the lateral wall of the brain case. They possess components of the Weberian apparatus which typifies Ostariophysi (Greenwood *et al.* 1966; Nelson 2006). A generalized spermatozoon is shown in Figure 10.22.

10.8 ORDER CLUPEIFORMES

Clupeiforms are herrings (adult, Fig. 10.19) and their relatives, mostly plankton feeders, with numerous gill rakers which serve as efficient straining

Fig. 10.19 *Clupea harengus* (Clupeidae). The Herring. By Norman Weaver. With kind permission of Sarah Starsmore.

Fig. 10.18 Contd. ...

Fig. 10.18 A phylogeny (Neighbor-joining) of elopomorph fishes after Wang, C. H., Kuo, C. H., Mok, H. K. and Lee, S. C. 2003. Zoologica Scripta 32: 231-241, Fig. 3, with spermatozoal ultrastructure from this chapter superimposed. Note the high consistency of the unique spermatozoal ultrastructure with the monophyletic Elopomorpha, with the exception of the Muraenidae and Nettastomatidae which, although possessing the peculiar elopomorph pseudoflagellum, differ from typical elopomorphs in lacking the striated rootlet, failure of the nucleus to elongate, and, in muraenids, presence of multiple mitochondria. Sperm of the arbitrarily selected outgroups: clupeiforms, gonorynchiforms, cypriniforms and osteoglossiforms are also illustrated. Original.

devices. Primarily marine, many move easily into brackish and freshwater. There are about 364 Recent species in 84 genera but they make up a large proportion of the world's total commercial fishing catch (Nelson 1984; Grande 1985; Nelson 2006).

Li and Orti (2007) have recently tested relationships among clupeiforms using mitochondrial rRNA genes (12S and 16S) and nuclear RAG1 and RAG2 sequences for 37 clupeiform taxa representing all five extant families and all subfamilies of Clupeiformes, excepting Pristigasterinae. The results confirmed monophyly of the families Engraulidae (consisting of two monophyletic subfamilies Engraulinae and Coiliinae) and Pristigasteridae. It is of great interest that the sister group of the Clupeiformes was a clade consisting of Ostariophysi and *Denticeps*, the latter genus traditionally considered basal to the clupeiforms. This gave support to recognition of the Ostarioclupeomorpha. However, monophyly of Clupeidae was not supported. Some clupeids were more closely related to taxa assigned to Pristigasteridae and Chirocentridae.

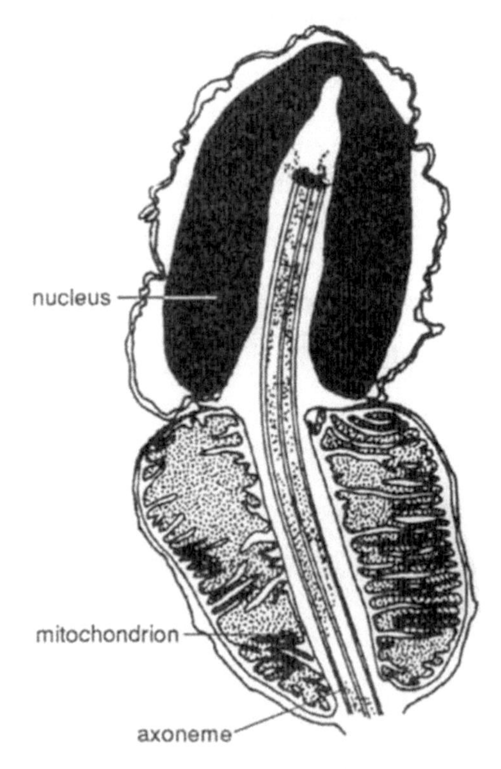

Fig. 10.20 *Anchoa guineensis* (Engraulidae). Diagrammatic longitudinal section of the spermatozoon. From Jamieson (1991) after Mattei, C., Mattei, X., Marchand, B. and Billard, R. 1981. Journal of Ultrastructure Research 74: 307-312. Fig. 1.

Sperm literature: The order Clupeiformes contains five extant families (comprising the Clupeoidei): Denticipitidae, Pristigasteridae, Engraulidae, Chirocentridae and Clupeidae (Nelson 2006) but spermatozoal ultrastructure is known in only three of these families: the Clupeidae, Pristigasteridae and Engraulidae. Sperm ultrastructure has been illustrated diagrammatically by Mattei (1970), for two clupeids, *Ethmalosa fimbriata* (Fig. 10.21A) and the sardine, *Sardinella aurita* (Fig. 10.21C) and an engraulid, the anchovy, *Anchoa* (=*Engraulis*) *guineensis* (Fig. 10.21B); that of *A. guineensis* particularly the axoneme and mitochondrion has been described by Mattei *et al.* (1981) (Fig. 10.20); the sperm of the clupeids *Clupea harengus*, the Atlantic herring, has been described by Rajasilta *et al.* (1997); *Sardinops melanostictus* and the engraulid *Engraulis japonicus* by Hara *et al.* (1994) and Hara and Okiyama (1998); the clupeid *Spratelloides gracilis* by Gwo *et al.* (2006); and the pristigasterid *Ilisha africana*, Long-fin Herring, by Mattei (1991). The sperm of *C. harengus*and *C. pallasi*, the Pacific Herring, were described by light microscopy by Retzius (1905) and Griffin *et al.* (1996).

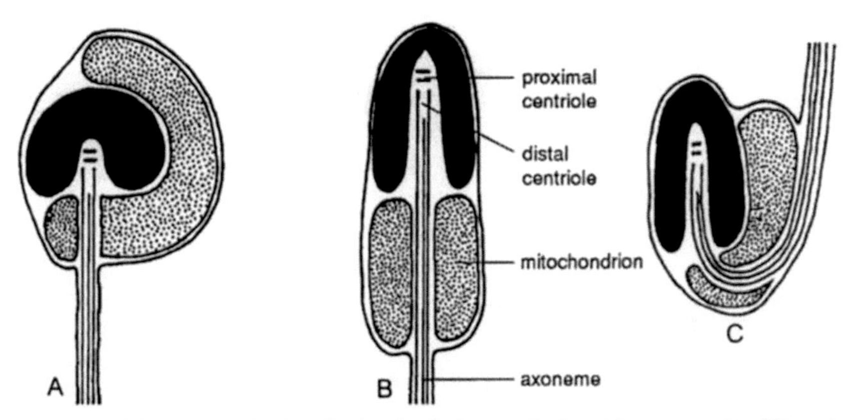

Fig. 10.21 Diagrammatic longitudinal sections of clupeid sperm. **A**. *Ethmalosa fimbriata* (Clupeidae). **B**. *Anchoa (=Engraulis?) guineensis* (Engraulidae). **C**. *Sardinella aurita* (Clupeidae). From diagrammatic longitudinal section of the spermatozoon. After Mattei, X. 1970. Pp 59-69. In B. Baccetti (ed.), *Comparative Spermatology*. Academic Press, New York, Fig. 4: 25, 27 and 22.

General sperm ultrastructure: The engraulids (Figs. 10.20, 10.21B) and clupeids (Fig. 10.21C), at least, show a departure from the commoner teleostean aquasperm ultrastructure in penetration of the nucleus almost to its tip by the basal fossa and contained axoneme, a tendency present but less developed in the clupeid *Ethmalosa* (Fig. 10.21A). Deep penetration is also seen in mullids, citharids, soleids, monacanthids, balistids and one tetraodontid. The single mitochondrion is ring- or C-shaped, a feature known in the holostean *Lepisosteus* and in teleosts in four families in addition to the Clupeidae (but not *Spratelloides*) and Engraulidae namely the salmoniform families Alepocephalidae (*Xenodermichthys*sp.), Searsidae(*Searsia*sp.), Salmonidae and Galaxiidae. Each of these has some axonemal doublets (varying with the family) with an A subtubule which appears solid (an intratubular differentiation, ITD) owing to intrusion of a septum. Among these teleost

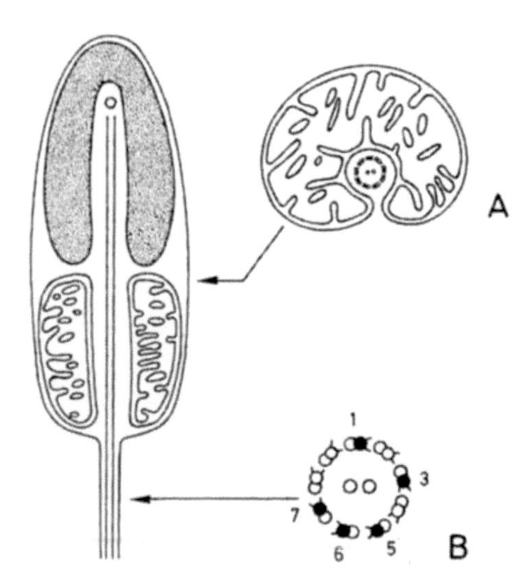

Fig. 10.22 A generalized spermatozoon of the Clupeomorpha. From Mattei, X. 1991. Canadian Journal of Zoology 69: 3038-3055, Fig. 7.

families (*q.v.*), the affected doublets are 1, 2, 5, 6, 7, or 1, 3, 5, 6, 7, as in the clupeomorphs (though ITDs are absent in *Spratelloides*), or 1, 2, 3, 5, 6, 7. Association of a ring-shaped mitochondrion and septate A subtubules is seen also in *Lepisosteus* (1, 2, 6 and 7 septate) and in the holothurian echinoderm *Cucumaria* (1, 2, 5, 6, 7 septate) but is attributed to the general asymmetry of the spermatozoon rather than to a direct correlation with the axonemal condition (Mattei *et al.* 1981). Occluded A subtubules also occur in families in which the mitochondria are not annular, including the Carangidae and Hemiramphidae. From micrographs (Mattei *et al.* 1981) no definite flagellar fins appear to be present in the clupeomorphs.

10.8.1 Family Engraulidae

***Engraulis japonicus*:** The spermatozoon of *Engraulis japonicus* (Fig. 10.23) has an ovoid nucleus with a projection at the tip and a deep basal fossa. The distal centriole lies in the fossa but a proximal centriole has not been seen. A single, large, thickly C-shaped mitochondrion surrounds the base of the flagellum. The flagellum shows the 9+2 pattern of microtubules but the A subtubules of doublets 1, 3, 5, 6 and 7 are septate as ITDs. No flagellar fins are present (Hara and Okiyama 1998).

10.8.2 Family Clupeidae

***Clupea harengus*:** The ovoid, acrosomeless head of the spermatozoon of *Clupea harengus*, the Herring (adult, Fig. 10.19), described and illustrated by Rajasilta *et al.* (1997) is ca 1.5 μm long and 2.0 μm wide, with heterogeneous condensation of chromatin. The centriolar complex lies in a 0.6 μm deep

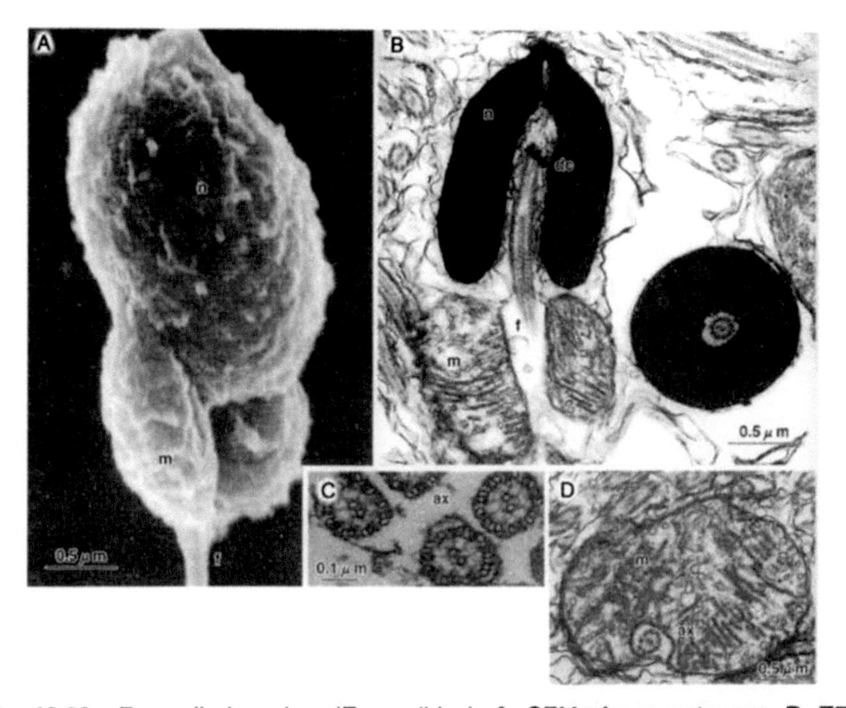

Fig. 10.23 *Engraulis japonicus* (Engraulidae). **A**. SEM of spermatozoon. **B**. TEM longitudinal and transverse sections of a spermatozoon. **C**. Transverse section of flagella. **D**. Transverse section of the mitochondrion almost completely surrounding the axoneme. ax, axonemes; dc, distal centriole; f, flagellum; m, mitochondrion; n, nucleus. Adapted from Hara, M. and Okiyama, M. 1998. Bulletin of The Ocean Research Institute, University of Tokyo 33: 1-138, Fig. 12B-E. With permission.

medial indentation of the head, the basal nuclear fossa. The base of the flagellum is almost completely surrounded by a single large thickly C-shaped mitochondrion up to 1.6 μm in width.

***Sardinops melanostictus*:** The spermatozoon of the Japanese sardine, *Sardinops melanostictus* (Fig. 10.24), has a total length of about 50 μm.

The head is ovoid and slightly depressed with a deep basal fossa containing the proximal centriole slanted at about 45° to the distal centriole (Hara *et al.* 1994); the two centrioles were later said to be mutually at right angles (Hara and Okiyama 1998) as appears to be approximately the case in Figure 10.24B. A single, extremely large mitochondrion in close contact with the posterior border of the head is C-shaped in cross-section, with its inner membranes modified into variously-shaped cristae, including characteristic walls along the longitudinal axis of the mitochondrion, as well as tubular and vesicular structures. Microtubules of axonemal doublets 1, 3, 5, 6 and 7 are characterized by intratubular differentiation as in most other clupeids. Flagellar fins are absent (Hara *et al.* 1994; Hara and Okiyama 1998).

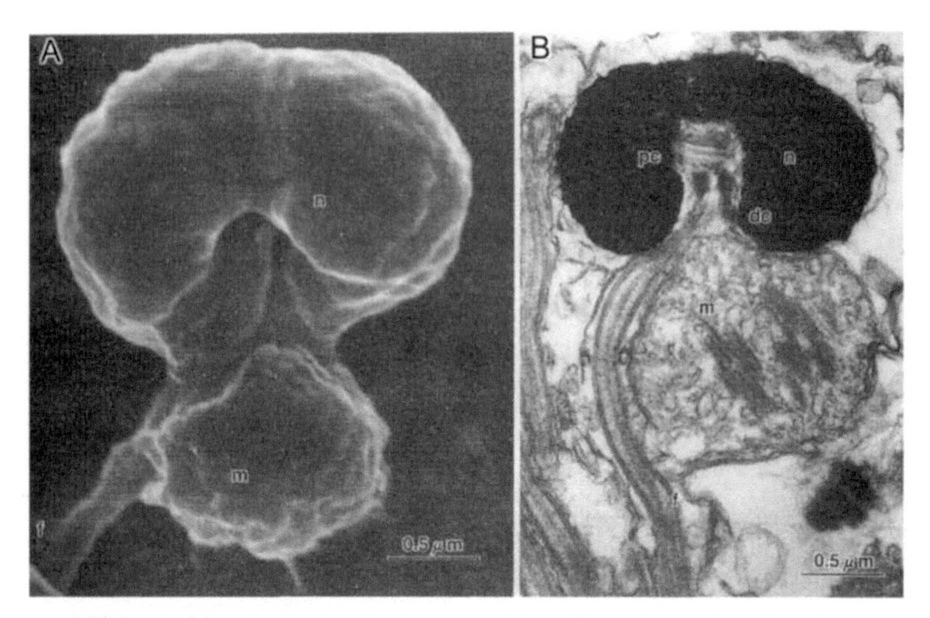

Fig. 10.24 *Sardinops melanostictus* spermatozoa (Clupeidae). A. SEM of nuclear region. B. TEM longitudinal section of same. dc, distal centriole; f, flagellum; m, mitochondrion; n, nucleus; pc, proximal centriole. Adapted from Hara, M. and Okiyama, M. 1998. Bulletin of The Ocean Research Institute, University of Tokyo 33: 1-138, Fig. 11B and C. With permission.

***Spratelloides gracilis*:** The spermatozoon of the clupeid *Spratelloides gracilis* has an oliviform nucleus (Fig. 10.25A-E). The head, lacking an acrosome, is 1.66 ± 0.16 μm long and 1.41 ± 0.15 μm wide. The tubiform nuclear fossa (Fig. 10.25B,E) penetrates about seven tenths of the nuclear axis. The chromatin is electron-dense and granular with irregular lacunae (Fig. 10.25F-N).

No proximal centriole has been identified. The distal centriole has a classic 9+0 microtubular triplet construction (Fig. 10.25F,G,J,K), lies within the anterior of the deep nuclear fossa and is connected to the nuclear membrane

---

Fig. 10.25 *Spratelloides gracilis* spermatozoa. **A–E** and **P**. Scanning electron micrographs of spermatozoa showing the head (h), midpiece (mp) and flagellum (f). **E**. Arrow shows the nuclear fossa. A single mitochondrion (m) is situated laterally in relation to the flagellum (f). **F** and **G**. Sagittal longitudinal section showing the oliviform and acrosomeless nucleus (n) with the axial nuclear fossa containing the distal centriole (dc). **G**. The distal centriole (dc) gives rise to the flagellum (f). **H–L**. The distal centriole (dc) has a classic 9+0 microtubular triplet construction, lies anteriorly in the deep nuclear fossa and is connected to nuclear membrane by electron-dense filaments (arrow). **H**. The distal centriole within the nuclear fossa. **J–N**. A series of circular transverse sections of the nuclear fossa. **J**. An electron-dense ring (arrow) surrounds the anterior end of the distal centriole (dc). **O**. A

Fig. 10.25 Contd. ...

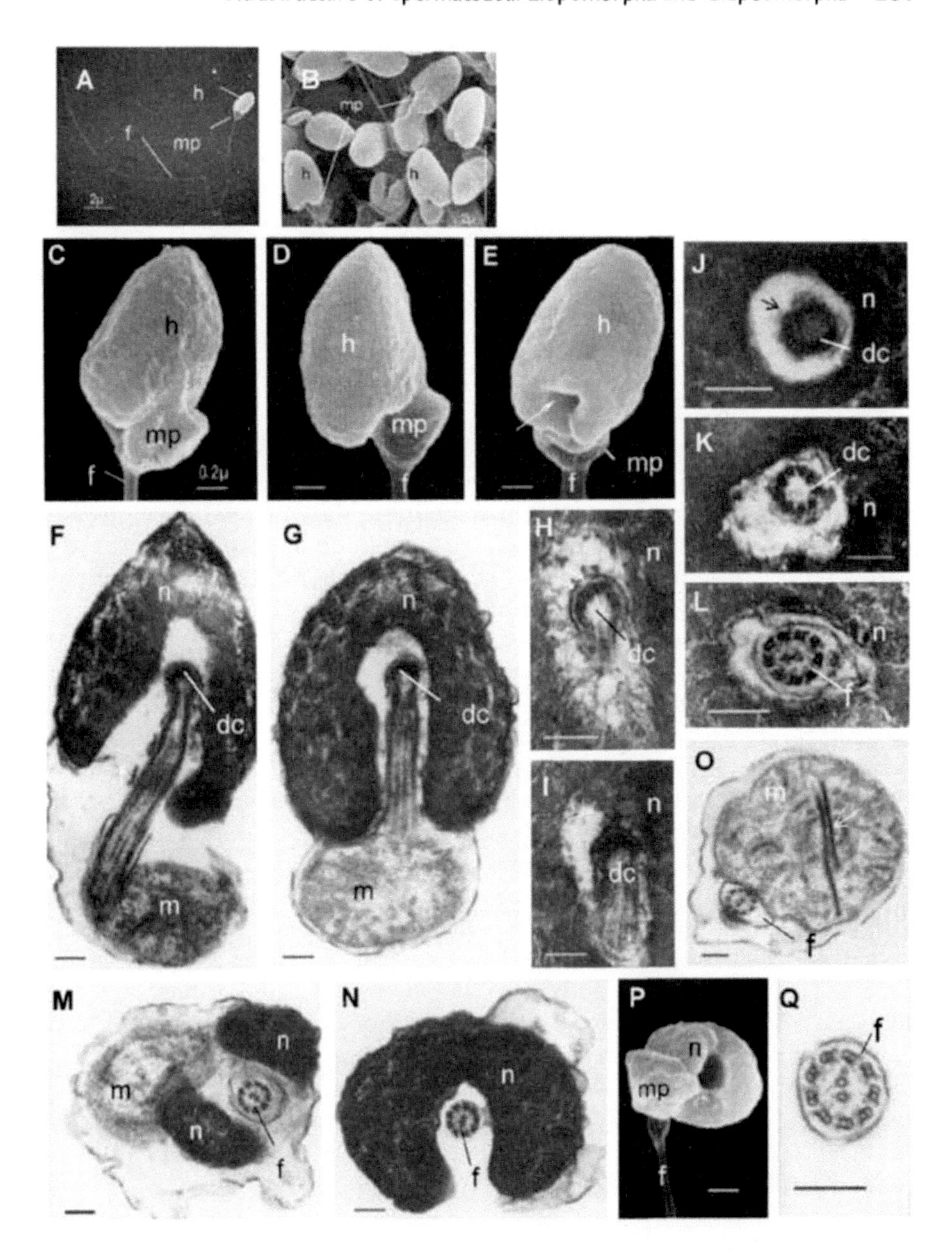

single mitochondrion (m), round in cross section. No cytoplasmic canal separates the flagellum (f) from the midpiece. **Q**. The flagellum with the 9+2 axonemal doublet configuration. In the transition region, neither central tubules nor ITDs are present in the flagellum. dc, distal centriole; f, flagellum; m, mitochondrion; mp, midpiece; n, nucleus. Scales = 0.2 µm (A–P). From Gwo, J. C., Lin, C. Y., Yang, W. L. and Chou, Y. C. 2006. Tissue and Cell 38: 285-291, Fig. 1. With permission of Elsevier.

by electron-dense filaments (Fig. 10.25H,I). The outline of the nuclear fossa is scalloped (Fig. 10.25H). An electron-dense ring surrounds the anterior end of the distal centriole and covers the outer surface of each triplet in the transverse section (Fig. 10.25J).

A single mitochondrion is situated laterally to the flagellum (Fig. 10.25A,B,D,E,O,P) and its matrix shows well-developed cristae and a smooth matrix (Fig. 10.25O). The midpiece (but not the mitochondrion itself) encircles the flagellum and no cytoplasmic canal is present (Fig. 10.25G,O). The tail is 34 ± 0.6 μm long. The 9+2 axoneme has inner and outer dynein arms (Fig. 10.25Q). In the transition region, no central tubules are present (Fig. 10.25K). ITDs (intratubular differentiations), flagellar fins and vesicles are absent (Fig. 10.25A, P and Q) (Gwo *et al.* 2006).

Conclusion for clupeoids: As noted by Gwo *et al.* (2006), clupeoid sperm possess an oliviform head with a distinct deep nuclear fossa, a midpiece with one mitochondrion and a posterior flagellum. Mattei (1991) concluded the spermatozoa of the families Engraulidae and Clupeidae are characterized by two apomorphies: an annular or C-shaped mitochondrion and intratubular differentiations (ITDs) in the A subtubules of the axonemal doublets. 1, 3, 5-7. However, their presence does not appear to be an essential character of the Clupeomorpha-type spermatozoon. Thus the spermatozoon of *Spratelloides gracilis* differs from those of other investigated Clupeiformes in lacking ITDs. As a further difference, the single mitochondrion of *S. gracilis* is approximately spherical, not C-shaped in cross section as in other clupeiforms, being situated laterally to the flagellum. Furthermore, there is no known proximal centriole in *S. gracilis.* Hara and Okiyama (1998) also found the proximal centriole of *Engraulis japonicus* (Engraulidae) to be indistinct (Gwo *et al.* 2006).

Engraulid and clupeid spermatozoa differ from those of most members of the sister-group, the Ostariophysi, in the very deep nuclear fossa, the prevalence of the unilateral, often C-shaped mitochondrion and the presence usually of ITDs but share with them the absence of flagellar fins. However, some Ostariophysi also have a C-shaped mitochondrion, some siluriforms have flagellar fins, and *Chanos*, in the Anotophysi has not only a C-shaped mitochondrion but also ITDs. The spermatozoon of the clupeiform *Denticeps*, embedded phylogenetically in the Ostariophysi, is unknown.

10.9 CHAPTER SUMMARY

Elopomorpha: Two features constitute spermatozoal synapomorphies unequivocally linking the Elopiformes+Anguilliformes+Albuliformes and comprise autapomorphies of a monophyletic Elopomorpha. They are a 9+0 flagellum, a constant feature, and extension of the proximal centriole as two elongate bundles of 4 and 5 triplets, running from the centriolar region towards the tip of the elongate nucleus, which may extend as a free pseudoflagellum. A third feature is the striated centriolar rootlet, seen in all but the Muraenidae, Nettastomatidae and Halosauridae. In the last three families it is not reported whether the extension of the proximal centriole is divided.

Ostarioclupeomorpha, Clupeomorpha, Clupeiformes: Spermatozoal ultrastructure is known in only three extant families (Clupeoidei): the Clupeidae, Pristigasteridae and Engraulidae. Clupeoid sperm, possess an oliviform head with a distinct deep nuclear fossa, a midpiece with one mitochondrion and a posterior flagellum.

Engraulid and clupeid spermatozoa differ from those of most members of the sister-group, the Ostariophysi, in the very deep nuclear fossa, the prevalence of a unilateral, often C-shaped mitochondrion and the presence of ITDs but share with them the absence of flagellar fins. However, some Ostariophysi also have a C-shaped mitochondrion, some siluriforms have flagellar fins, and *Chanos*, in the Anotophysi has not only a C-shaped mitochondrion but also ITDs. Spermatozoa thus provide no clear evidence of a clupeiform–ostariophysan relationship.

10.10 LITERATURE CITED

Baccetti, B. 1979. The evolution of the acrosome complex. Pp. 305-329. In D. W. Fawcett and J. M. Bedford (eds), *The Spermatozoon*. Urban and Schwarzenberg, Baltimore-Munich.

Baccetti, B., Burrini, A. G. and Pallini, V. 1981. Different axoneme patterns in cilia and flagella of the same animal. Journal of Submicroscopic Cytology 13: 479-481.

Billard, R. and Ginsburg, A. S. 1973a. La spermiogenèse et le spermatozoïde d' *Anguilla anguilla* L. Étude ultrastructurale. Annales de Biologie Animale, Biochemie, Biophysique 13: 523-534.

Billard, R. and Ginsburg, A. S. 1973b. Spermiogenesis and Spermatozoon of *Anguilla anguilla* L. Study of ultrastructure. Annales de Biologie Animale Biochimie Biophysique 13: 523-534. (In French).

Çolak, A. and Yamamoto, K. 1974a. An electron microscope study of spermiogenesis in the Japanese Eeel, *Anguilla japonica*. Bulletin of the Faculty of Fisheries, Hokkaido University 25: 1-5.

Çolak, A. and Yamamoto, K. 1974b. Ultrastructure of the Japanese eel spermatozoon. Annotationes Zoologicae Japonenses 47: 48-54.

Filleul, A. and Lavoué, S. 2001. Basal teleosts and the question of elopomorph monophyly. Morphological and molecular approaches. Comptes Rendus Hebdomadaires des Séances de l'Académie des Sciences III 32: 393-399.

Forey, P. L. 1973. Relationships of elopomorphs. Pp. 351-368. In P. H. Greenwood, R. S. Miles and C. Patterson (eds), *Interrelationships of Fishes*. Journal of the Linnean Society of London Zoology 53 (suppl. 1) Academic Press, New York.

Gibbons, B. H., Gibbons, I. R. and Baccetti, B. 1983. Structure and motility of the 9+0 flagellum of eel spermatozoa. Journal of Submicroscopic Cytology 15: 15-20.

Ginzburg, A. S. and Billard, R. 1972. Ultrastructure du spermatozoide d'Anguille. Journal de Microscopie et de Biologie Cellulaire (Paris) 14: 50a-51a.

Grande, L. 1985. Recent and fossil clupeomorph fishes with materials for revision of the subgroups of clupeoids. Bulletin of the American Museum of Natural History 181: 231-372.

Grassé, P. (ed.) 1958. Traité de Zoologie. Anatomie, Systématique, Biologie. XIII Agnathes et Poissons. Masson, Paris.

Greenwood, P. H., Rosen, D. E., Weitzman, S. H. and Myers, G. S. 1966. Phyletic studies of teleostean fishes, with a provisional classification of living forms. Bulletin of the American Museum of Natural History 131: 339-456.

Griffin, F. J., Vines, C. A., Piilai, M. C., Yanagimachi, R. and Cherr, G. N. 1996. Sperm motility initiation factor is a minor component of the Pacific herring egg chorion. Development Growth and Differentiation 38: 193-202.

Gwo, J. C., Gwo, H. H. and Chang, S. L. 1992. The spermatozoon of the Japanese eel, *Anguilla japonica* (Teleostei, Anguilliformes, Anguillidae). Journal of Submicroscopic Cytology and Pathology 24: 571-574.

Gwo, J. C., Gwo, H. H. and Kao, Y. S. 1994. Spermatozoon ultrastructure of *Anguilla japonica* (Teleostei, Anguilliformes, Anguillidae). Pp. 797-801. In L. M. Chou, A. D. Munro, T. J. Lam, T. W. Chen, I. K. K. Cheong, J. K. Ding, K. K. Hooi, H. W. Khoo, V. P. E. Phang, K. F. Shim and C. H. Tan (eds), *The third Asian fisheries forum: proceedings of the Third Asian Fisheries Forum 26-30 October 1992, Singapore.* Asian Fisheries Society, Manila, Philippines.

Gwo, J. C., Lin, C. Y., Yang, W. L. and Chou, Y. C. 2006. Ultrastructure of the sperm of blue sprat, *Spratelloides gracilis*; Teleostei, Clupeiformes, Clupeidae. Tissue and Cell 38: 285-291.

Hara, M., Ishijima, S. and Okiyama, M. 1994. Ultrastructure and motility of spermatozoa of the Japanese sardine, *Sardinops melanostictus*. Japanese Journal of Ichthyology 41: 322-325 (In Japanese with English abstract).

Hara, M. and Okiyama, M. 1998. An ultrastructural review on the spermatozoa of Japanese fishes. Bulletin of The Ocean Research Institute, University of Tokyo 33: 1-138.

Inoue, J. G., Miya, M., Tsukamoto, K. and Nishida, M. 2004. Mitogenomic evidence for the monophyly of elopomorph fishes (Teleostei) and the evolutionary origin of the leptocephalus larva. Molecular Phylogenetics and Evolution 32: 274-286.

Jamieson, B. G. M. 1991. *Fish Evolution and Systematics: Evidence from Spermatozoa.* Cambridge University Press, Cambridge. 319 pp.

Lauder, G. V. and Liem, K. F. 1983. The evolution and interrelationships of the actinopterygian fishes. Bulletin of the Museum of Comparative Zoology 150: 95-197.

Lavoué, S., Miya, M., Inoue, J. G., Saitoh, K., Ishiguro, N. B. and Nishida, M. 2005. Molecular systematics of the gonorynchiform fishes (Teleostei) based on whole mitogenome sequences: Implications for higher-level relationships within the Otocephala. Molecular Phylogenetics and Evolution 37: 165-177.

Li, C. and Orti, G. 2007. Molecular phylogeny of Clupeiformes (Actinopterygii) inferred from nuclear and mitochondrial DNA sequences. Molecular Phylogenetics and Evolution 44: 386-398.

Mattei, C. and Mattei, X. 1973. La spermiogenèse d'*Albula vulpes* (L. 1758) (Poissin Albulidae). Zeitschrift für Zellforschung und Mikroskopische Anatomie 142: 171-192.

Mattei, C. and Mattei, X. 1974. Spermatogenesis and spermatozoa of the elopomorpha (teleost fish). Pp. 211-221. In B. A. Afzelius (ed.), *The functional anatomy of the spermatozoon.* Pergamon Press, Oxford.

Mattei, C., Mattei, X., Marchand, B. and Billard, R. 1981. Réinvestigation de la structure des flagelles spermatiques: cas particulier des spermatozoïdes à mitochondrie annulaire. Journal of Ultrastructure Research 74: 307-312.

Mattei, X. 1970. Spermiogenèse comparée des poissons. Pp. 57-69. In B. Baccetti (ed.), *Comparative Spermatology*. Academic Press, New York.
Mattei, X. 1988. The flagellar apparatus of spermatozoa in fish Ultrastructure and evolution. Biology of the Cell 63: 151-158.
Mattei, X. 1991. Spermatozoon ultrastructure and its systematic implications in fishes. Canadian Journal of Zoology 69: 3038-3055.
Mattei, X. and Mattei, C. 1972. L'appareil centriolaire et flagellaire du spermatozoïde d'*Albula vulpes* (Poissin, Albulidae). Journal de Microscopie 14: 67a-68a.
Nelson, J. S. 1984. *Fishes of the World*, 2nd edition. John Wiley and Sons, New York. 523 pp.
Nelson, J. S. 2006. *Fishes of the World*, 4th edition. John Wiley and Sons, Inc., Hoboken, New Jersey. 601 pp.
Nybelin, O. 1973. Comments on the caudal skeleton of actinopterygians. Pp. 369-372. In P. H. Greenwood, R. S. Miles and C. Patterson (eds), *Interrelationships of Fishes*. Journal of the Linnean Society of London Zoology.
Okamura, A. and Motonobu, T. 1999. Spermatozoa of *Conger myriaster* observed by electron microscopy. Zoological Science (Tokyo) 16: 927-933.
Okamura, A., Zhang, H., Yamada, Y., Tanaka, S., Horie, N., Mikawa, N., Utoh, T. and Oka, H. P. 2000. Re-examination of the spermatozoal ultrastructure of eels: Observations of the external morphology of spermatozoa in three species. Journal of Fish Biology 57: 161-169.
Patterson, C. 1973. Interrelationships of holosteans. Pp. 233-306. In P. H. Greenwood, R. S. Miles and C. Patterson (eds), *Interrelationships of Fishes*. Journal of the Linnean Society of London Zoology 53 (suppl 1). Academic Press, New York.
Rajasilta, M., Paranko, J. and Laine, P. T. 1997. Reproductive characteristics of the male herring in the northern Baltic Sea. Journal of Fish Biology 51: 978-988.
Retzius, G. 1905. Die Spermien der Leptokardier, Teleostier und Ganoiden. Biologische Untersuchungen von G Retzius Nf 12: 103-115.
Smith, D. G. 1984. Elopiformes, Notacanthiformes and Anguilliformes: Relationships. Pp. 94-102. In H. G. Moser, W. J. Richards, D. M. Cohen, M. P. Fahay, A. W. Kendall, Jr., and S. L. Richardson (eds), Ontogeny and Systematics of Fishes. Special Publication Number 1. American Society of Ichthyologists and Herpetologists.
Todd, P. R. 1976. Ultrastructure of the spermatozoa and spermiogenesis in New-Zealand Fresh Water eels (Anguillidae). Cell and Tissue Research 171: 221-232.
Wang, C. H., Kuo, C. H., Mok, H. K. and Lee, S. C. 2003. Molecular phylogeny of elopomorph fishes inferred from mitochondrial 12S ribosomal RNA sequences. Zoologica Scripta 32: 231-241.

CHAPTER 11

Ultrastructure of Spermatozoa: Ostariophysi

John R. Burns[1], *Irani Quagio-Grassiotto*[2], *Barrie G. M. Jamieson*[3]

11.1 SUPERORDER OSTARIOPHYSI

The Ostariophysi, marine and freshwater worldwide, with nearly 8,000 species, contain 68 families of which the largest are the Cyprinidae, Characidae, Loricariidae and Balitoidae (Nelson 2006). They contain nearly three quarters of the freshwater fishes of the world, from carps (Cypriniformes) and the Neotropical tetras (Characiformes) to freshwater catfishes (Siluriformes) and the weakly electric gymnotids (Gymnotiformes). A small proportion, about 123 species, including the chanids and gonorynchids, are marine (Lauder and Liem 1983; Fuiman 1984; Nelson 2006). The Ostariophysi appears to be monophyletic on the basis of at least seven synapomorphies of which one is the presence in the epidermis of cells which exude an alarm substance when wounded; this causes a fright reaction which is evoked even in non-ostariophysans (references in Lauder and Liem 1983; Stacey, **Volume 8B, Chapter 3**). Monophyly of Ostariophysi has been endorsed by mitogenomic analysis (Lavoue *et al.* 2005) and by analysis of 8 nuclear genes, including 11,766 base pairs (see Orti and Li, **Chapter 1**, Fig. 1.4, and Fig. 11.1 below).

Spermatozoal overview: Spermatozoa of the Ostariophysi display tremendous variability in their anatomy. The final form of these cells depends on the process of differentiation (spermiogenesis), which appears to be correlated with the reproductive habits of the species. In addition, the form and location of the cell organelles within the spermatozoa are also highly

[1]Department of Biological Sciences, George Washington University, Washington DC 20052
[2]Departamento de Morfologia, Instituto de Biociências, Universidade Estadual Paulista, 18618-000 Botucatu, SP, Brazil.
[3]School of Integrative Biology, University of Queensland, Brisbane, Queensland 4072, Australia

variable. This diversity generates a series of characters that are likely related to the evolutionary history of these species.

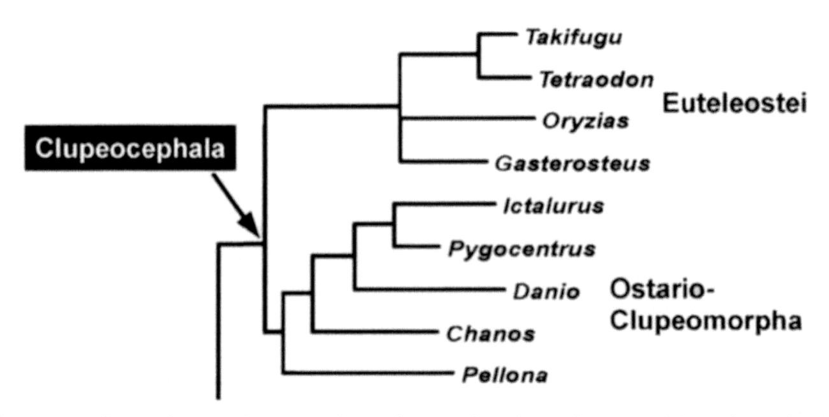

Fig. 11.1 Part of a phylogeny of ray-finned fish based on partitioned analyses of eight nuclear genes, here showing monophyly of the Ostarioclupeomorpha and Ostariophysi and their joint inclusion with the Euteleostei in a monophyletic Clupeocephala. Adapted from Ortí and Li, Chapter 1, Fig. 1.4.

The majority of ostariophysan species that are externally fertilizing produce Type I aquasperm, although currently there are several known that produce cells resembling Type II aquasperm (Mattei 1970; Mattei *et al.* 1993). In addition, among a number of externally fertilizing Characiformes, during spermiogenesis nuclear rotation, a characteristic of Type I spermiogenesis, is often incomplete resulting in different degrees of eccentricity of the flagellum with regard to the nucleus, so that intermediate types of spermatozoa are produced. Another variation in spermiogenesis, type III, was first observed in the siluriform family Pimelodidae, in the catfish *Sorubim lima* (Quagio-Grassiotto and Carvalho 2000). However, it was not designated as type III until that same pattern of spermiogenesis was seen in some species of the siluriform family Heptapteridae by Quagio-Grassiotto *et al.* (2005). In that report, the differences between type III and the previously described Types I and II of Mattei (1970) are discussed. In a more recent publication (Quagio-Grassiotto and Oliveira 2008), type III spermiogenesis is again reported in other members of the Pimelodidae, and a review of the spermatozoa of catfishes is presented that includes a schematic drawing of this process. Type III spermiogenesis, with some variations, appears to be common among siluriforms, and it is now known to also occur in some characiforms (Veríssimo-Silveira 2007).

Finally, an increasing number of inseminating species are being identified, especially in the characiform family Characidae. The possible selective advantages of characters seen in these modified introsperm are discussed in Javonillo *et al.* (**Chapter 17**).

The main difference in type III sperm development relates to the early spermatids. In most teleosts, including the majority of ostariophysan species thus far analyzed that undergo either Type I or II spermiogenesis, the flagellar axis is parallel to that of the nucleus in the early spermatids. In early spermatids of type III species, on the other hand, the flagellar axis is perpendicular to that of the nucleus. Nuclear rotation does not occur, with the result that the flagellar axis remains perpendicular to that of the nucleus in the spermatozoon, similar to that seen in type I sperm. If a cytoplasmic canal forms in a type III cell it is due to displacement of cytoplasm, rather than centriolar movement as is the case in types I and II sperm (Quagio-Grassiotto and Oliveira 2008). Finally, although intracystic spermatogenesis is characteristic of the great majority of ostariophysans, as is the case with most teleosts, semicystic development, where the terminal stages of spermatid development occur within the luminal compartment of the testis, has now been reported in a number of siluriform species.

Among the Gymnotiformes thus far analyzed, sperm cells resemble those seen among the Characiformes. In both taxa, Type I aquasperm with a more or less eccentric position of the flagellum predominate, although some species do produce cells resembling Type II aquasperm *sensu* Mattei (1970). Insemination, which is known for a number of taxa and species *incertae sedis* in the Characiformes, has never been reported in the Gymnotiformes.

Among the Siluriformes, the most common form of aquasperm is also Type I, where complete nuclear rotation takes place resulting in a more central position of the flagellum, or flagella in biflagellate species. However, in some species the central position of the flagellum may also be due to the absence of nuclear rotation characteristic of type III spermiogenesis. Currently, only two siluriform species have been shown to produce cells resembling Type II aquasperm *sensu* Mattei (1970). Highly modified introsperm are characteristic of the families Scoloplacidae, Astroblepidae and Auchenipteridae.

Two main patterns of chromatin condensation can be recognized in the Ostariophysi. In one, condensation of chromatin during spermiogenesis leads to the formation of electron-dense, flocculent aggregates surrounded by electron-lucent areas. In the other, condensation results in a homogeneous, electron-dense granular chromatin. In some groups with this latter type, electron-lucent areas are still visible among the condensed granules (e.g., the characiform family Curimatidae), whereas in others condensation is so extensive that no electron-lucent areas are visible (e.g., the siluriform family Pimelodidae).

Among the Characiformes, the flocculent pattern is found in the Citharinoidei, a superfamily considered to be sister to the Characoidei which contains all the other families of Characiformes. The flocculent pattern is also found in the Characoidei, mainly in some subfamilies of the Characidae. In Siluriformes, this pattern is seen in the Diplomystidae, the most basal family, as well as in some other families such as Clariidae. Finally, the flocculent pattern appears to be the dominant form in Gymnotiformes.

The spermatozoa of Characiformes and Gymnotiformes that exhibit the flocculent pattern of chromatin condensation often have unusually long midpieces that have posterior cytoplasmic sleeves but few vesicles. Some also have flagella with fins. Mitochondria may be unusually large, long, branched and C-shaped. Occasionally a single, large, C-shaped mitochondrion is present. In Siluriformes, species with this chromatin pattern also tend to have very long mitochondria and sometimes a single, C-shaped organelle.

Most Characiformes, Gymnotiformes and Siluriformes that exhibit either variant of the homogeneous pattern of chromatin condensation tend to have abundant mitochondria that are small and spherical, ovoid or slightly elongate. Midpieces in these species are highly variable in shape, with some having well developed tubular-vesicular systems.

Caution is required when analyzing and interpreting these sperm characters, especially when attempting to determine homologies. For example, the presence or absence of a cytoplasmic canal may be due to two different processes. In one case the canal forms as a result of the migration of the centriolar complex toward the nucleus during Types I and II spermiogenesis. Therefore, absence of a cytoplasmic canal in these spermatozoa may be interpreted as a lack of formation during spermiogenesis. A cytoplasmic canal forming in the spermatids at the beginning of spermiogenesis may eventually be absent in the spermatozoon. Examples of this loss are found in Characiformes (Erythrinidae and the *incertae sedis* genus *Triportheus*) and Gymnotiformes (*Brachyhypopomus cf. pinnicaudatus*). On the other hand, in type III spermiogenesis centriolar migration does not occur and a cytoplasmic canal may be absent in the spermatozoon (the siluriform *Pimelodus maculatus*). When a canal does form, it is due to the displacement of cytoplasm or vesicles of the midpiece in the direction of the flagellum (the siluriforms *Cetopsis coecutiens* and *Pimelodella gracilis*).

Among the Siluriformes, it should be noted that the high incidence of biflagellate sperm tends to be associated with semicystic spermatogenesis. Spermiogenesis in these groups also presents unique characters that may be considered to be a variation of type III spermiogenesis (Quagio-Grassiotto *et al.* 2005; Shahin 2006b; Spadella *et al.* 2006a; Quagio-Grassiotto and Oliveira 2008).

In spite of the diverse forms of spermatozoa seen among the Ostariophysi, for those taxa whose monophyly has been established through rigorous phylogenetic analyses, spermatozoa often display remarkably consistent morphological patterns, depending on the taxonomic level. This is the case with the characiform families Anostomidae, Curimatidae and Erythrinidae whose member species have spermatozoa with many common characters. A progressive variability in the anatomy of these cells is evident when comparing species within the same genus, or genera within the same family. However, at higher taxonomic levels, such as for the apparently monophyletic superorder Anostomoidea, comprising the families Curimatidae, Prochilodontidae, Anostomidae and Chilodontidae, not all morphological characters are shared among all taxa. For example, the flocculent pattern of chromatin condensation is found only in the Prochilodontidae and Anostomidae and flagellar fins are seen only in the Anostomidae and Chilodontidae. Again within Characiformes, particularly in taxa not yet shown to be monophyletic such as many groups within the family Characidae, consistent patterns of sperm cell morphology may be seen among the species in some subfamilies but not in others.

Within Siluriformes, in a large number of taxa that are considered to be monophyletic, such as the family Loricariidae comprised of various subfamilies, morphological variability of sperm cells is less when comparing species within a given subfamily, indicating a possible common evolutionary history of these cells. Similarly for the family Trichomycteridae, in spite of there being variation in the overall form of the spermatozoa, a consistent pattern is seen for certain cell organelles.

Unfortunately, data on sperm ultrastructure in most species of Gymnotiformes are still lacking. Nonetheless, the variations seen in the sperm cells of the species thus far analyzed appear to be more similar to those observed within the Characiformes, rather than the Siluriformes. Hypotheses on the phylogenetic relationships among these three orders based on molecular data also support these observations (Dimmick and Larson 1996; Orti 1997; Lavoué *et al.* 2005).

11.2 SERIES ANOTOPHYSI

Anotophysi have a primitive Weberian apparatus (Nelson 2006). Only gonorynchiforms have been examined for sperm ultrastructure. The Gonorynchiformes and the Otophysi have been found to be sister groups (Lavoue *et al.* 2005).

11.3 ORDER GONORYNCHIFORMES

Gonorynchiformes, milkfishes, constitute a small, predominantly freshwater order containing four families of which only one species, in the Chanidae, has been investigated for sperm ultrastructure.

Chanos chanos: The anacrosomal aquasperm of the Milkfish, *Chanos chanos* (adult, Fig. 11.2A), is a relatively simple, elongated cell composed of a head, a short midpiece, and a tail (Fig. 11.2B). The nucleus is ovoid, measuring ca 1.2 μm in its longest axis and 1.5 μm in diameter. The chromatin is highly electron-dense, granular, and heterogeneous in texture. A single mitochondrion, forming an incompletely closed ring, is located posterior to the nucleus and surrounds the cytoplasmic canal. The two conventional 9 triplet centrioles are mutually perpendicular and lie in a deep basal nuclear fossa which extends almost to the equator of the nucleus. The axoneme has the classic 9+2 microtubular doublet construction; inner and outer dynein arms are present. The A tubule of doublets 1, 5 and 8 contains a dense substance that appears darker than in other doublets.

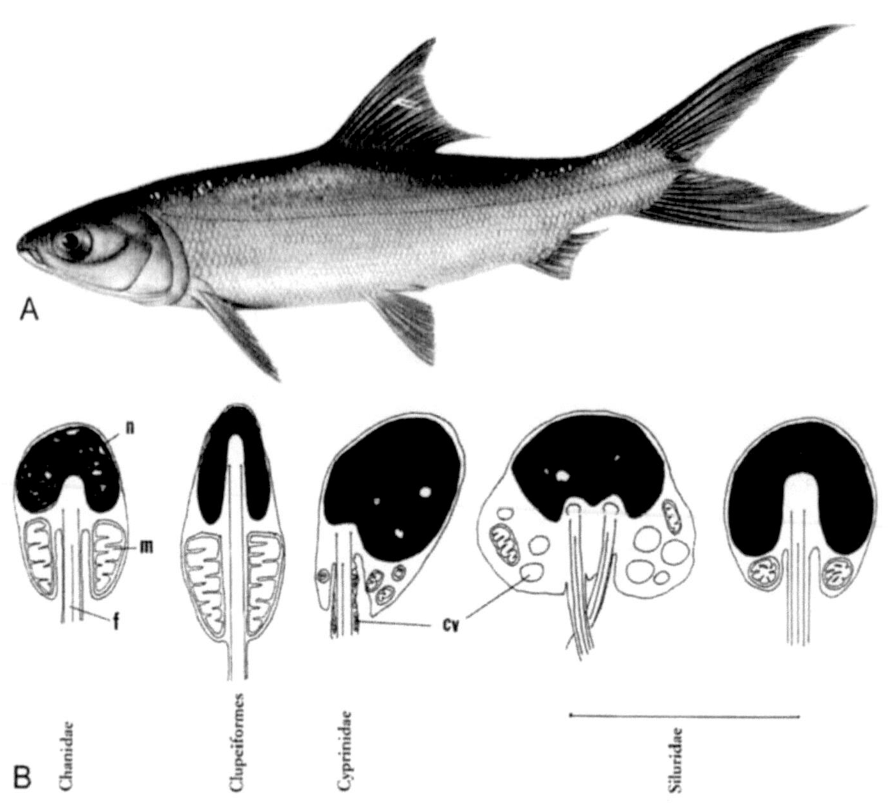

Fig. 11.2 *Chanos chanos*, Milkfish (Gonorynchiformes). By Norman Weaver. With kind permission of Sarah Starsmore. **B**. Diagrams of longitudinal sections through the sperm of *Chanos chanos* and some other orders. cv, cytoplasmic vesicle; f, flagellum; m, mitochondrion; n, nucleus. From Gwo, J. C., Kao, Y. S., Lin, X. W., Chang, S. L. and Su, M. S. 1995. Journal of Submicroscopic Cytology and Pathology 27: 99-104, Fig. 11.

Remarks: The single, ring-shaped mitochondrion and these ITD (intratubular differentiations of the doublets; occluded A subtubules in the flagellum) in spermatozoa of milkfish and of *Anchoa guineesis, Ethmalosa fimbriata,* and *Sardinella aurita* (Clupeiformes) are considered to support placement of Chanoidei closer to Clupeiformes than to the Ostariophysi (Gwo *et al.* 1995) although the ITDs in the clupeiforms are 1, 2, 5, 6, 7, or 1, 3, 5, 6, 7. These authors cite evidence that gonorynchiforms and clupeiforms are also similar in egg and larval morphology. This relationship is endorsed from mitogenomic analysis by Ishiguro *et al.* (2003; but see Lavoue *et al.* 2005).

11.4 SERIES OTOPHYSI

Otophysi are diagnosed by a Weberian apparatus in which movable bony ossicles connect the swim bladder to the inner ear. They consist of four orders,

Cypriniformes Siluriformes, Characiformes and Gymnotiformes (Nelson 2006), all of which have been examined for sperm ultrastructure.

11.5 ORDER CYPRINIFORMES

Cypriniforms are mainly freshwater fishes that have their greatest diversity in Southeast Asia but are native to all continents excepting Australia. The Weberian apparatus typical of the Ostariophysi (see above) is present. On adult somatic characters, the group appears monophyletic (Fink and Fink 1981 and references in Lauder and Liem 1983; Nelson 2006). Of the cypriniform families only the Cyprinidae and Cobitidae have been examined for sperm ultrastructure.

11.5.1 Family Cyprinidae

Cyprinids are the minnows or carps. Freshwater, rarely brackish water, in North America (to southern Mexico), Africa and Eurasia (Nelson 2006).

Sperm literature: Most subfamilies of the Cyprinidaae have been examined for sperm ultrastructure:

Acheilognathinae: *Acheilognathus rhombeus* (Ohta *et al.* 1993); *Acheilognathus (=Pseudoperilampus) typus* (Ohta *et al.* 1994) (Fig. 11.8H,I); *Rhodeus ocellatus* (Ohta and Iwamatsu 1983); *R. ocellatus ocellatus* (Ohta 1991; Ohta *et al.* 1998) (Fig. 11.6), *R. sericeus sinensis* (Guan and Afzelius 1991).

Cyprininae: *Carassius auratus* (Fribourgh *et al.* 1970; Munoz-Guerra *et al.* 1982; Baccetti *et al.* 1984; Ohta *et al.* 1993; Hara and Okiyama 1998) (Figs. 11.4, 11.7,11.10); *Cyprinus carpio* (Fujimura *et al.* 1957; Billard 1970; Kudo 1980; Stein 1981; Zhang, X. *et al.* 1991; Gwo *et al.* 1993) (adult, Fig. 11.3A).

Barbinae: *Barbus barbus plebejus* (Baccetti *et al.* 1984) (Fig. 11.4); *B. longiceps* and (not certainly assigned to this subfamily) *Capoeta damascina* (Stoumboudi and Abraham 1996, immature, spermatogenic stages only); *Spinibarbus caldwelli* (Lin, Dan-Jun *et al.* 2003).

Labeoninae: *Labeo victorianus* (Rutaisire *et al.* 2006); *L. rohita* (Gopalakrishnan *et al.* 2000).

Squalobarbinae: *Ctenopharyngodon idella* (Zhang, X. *et al.* 1991).

Tincinae: *Tinca tinca* (Psenicka *et al.* 2006) (adult, Fig. 11.3B).

Xenocyprinae: **Hypophthalmichthys molitrix* (Zhang, X. *et al.* 1991; see also Fürböck *et al.*, pers.comm. who place it in the Leuciscinae) (Fig. 11.5F).

Gobioninae: *Squalidus chankaensis* (Kim, K. H. *et al.* 1998).

Rasborinae: *Danio* rerio (Patil and Khoo 1995); *Danio (=Brachydanio) rerio* (Wolenski and Hart 1987; Patil and Khoo 1995); *Hemigrammocypris rasborella* (Ohta *et al.* 1994) (Fig. 11.8C,D); *Zacco platypus* (Fig. 11.8A,B) and *Z. temminckii* (Fig. 11.8E,G) (Ohta *et al.* 1994).

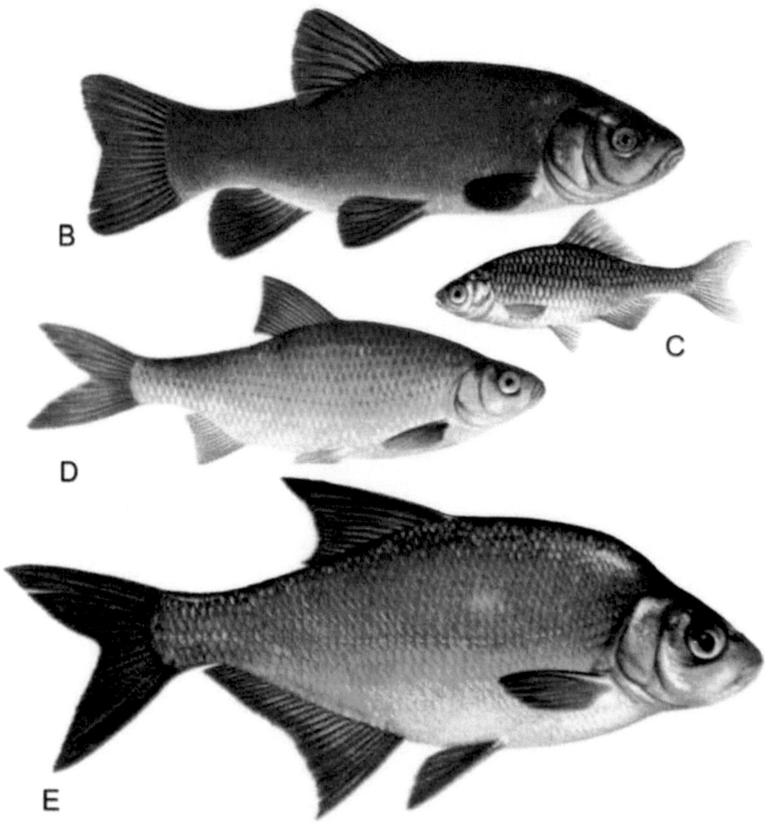

Fig. 11.3 Cyprinidae. **A**. *Cyprinus carpio*, Carp. Photo Bruce Cowell. Courtesy of the Queensland Museum. **B**. *Tinca tinca*, Tench. **C**. *Rhodeus sericeus*, Bitterling. **D**. *Rutillus rutilus*, Roach. **E**. *Abramis brama*, Bream. **B-E**. By Norman Weaver. With kind permission of Sarah Starsmore.

Leuciscinae: **Abramis brama* (Stein 1981; Fürböck *et al.*, pers.comm.) (Fig. 11.5A; adult Fig. 11.3E); *Alburnus alburnus* (Lahnsteiner and Patzner 1991; Fürböck *et al.*, pers.comm.).(Fig. 11.5C); **Alburnoides bipunctatus* Fürböck *et al.*, pers.comm.) (Fig. 11.5B); **Alburnus alburnus alborella* (Fig. 11.4); *Blicca bjorkna* (Stein 1981); **Chalcaburnus chalcoides* Fürböck *et al.*, pers.comm.) (Fig. 11.5D); **Chondrostoma nasus* Fürböck *et al.*, pers.comm.) (Fig. 11.5E); *Chondrostoma toxostoma* (Baccetti *et al.* 1984) (Fig. 11.4); *Leuciscus cephalus* (Stein 1981; Baccetti

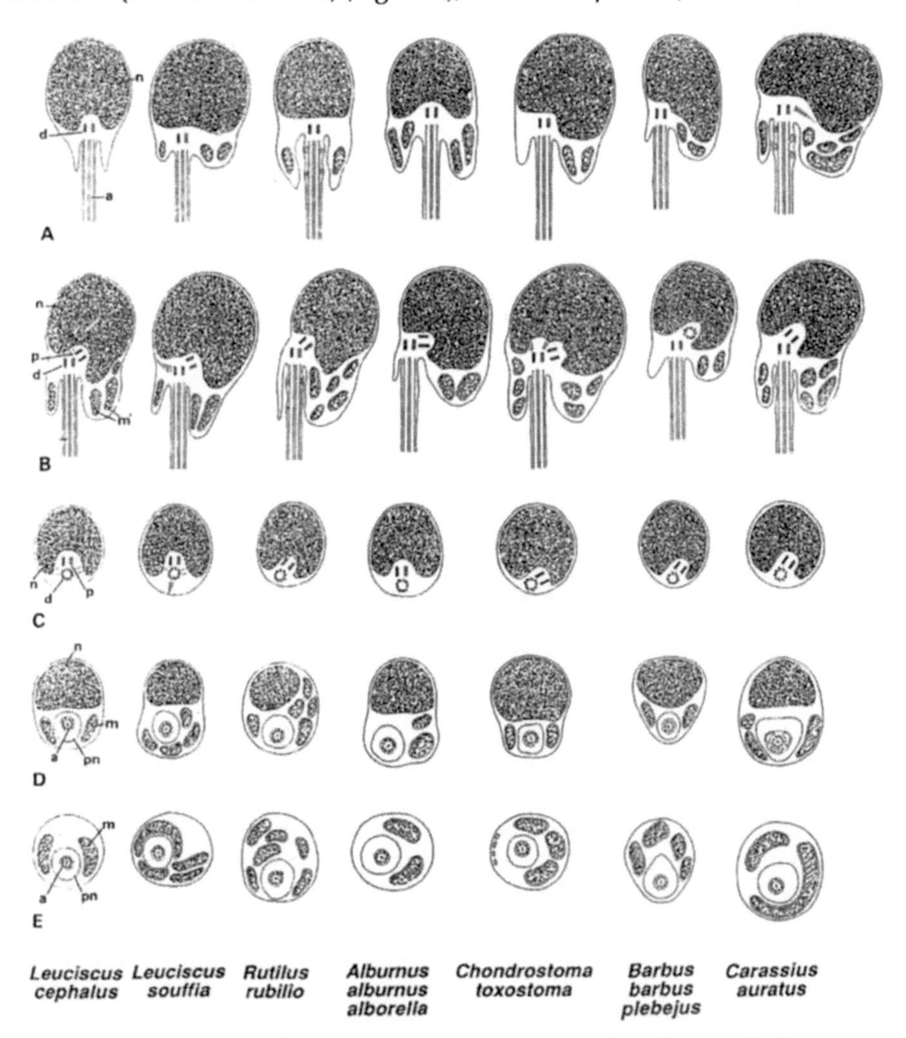

Fig. 11.4 Variations in sperm of seven cyprinid species. A. Longitudinal frontal section. B. Longitudinal sagittal section. C. Cross section at centriolar region. D. Postnuclear cross section. E. Cross section of the midpiece region. a, axoneme; de, distal centriole; m, mitochondrion; n, nucleus; pc, proximal centriole; pn, postnuclear (cytoplasmic) canal. From Jamieson 1991 after Baccetti *et al.* 1984. Gamete Research 10: 373-396, Figs. 2-8.

et al. 1984; Lahnsteiner *et al.* 1992) (Fig. 11.4, 11.9); **Leuciscus cephalus* Fürböck *et al.*, pers.comm.) (Fig. 11.5G); *Leuciscus souffia* (Baccetti *et al.* 1984) (Fig. 11.4); *Leuciscus leuciscus* (Stein 1981; Lahnsteiner *et al.* 1992); **Phoxinus phoxinus* (Stein 1981; Fürböck *et al.*, pers.comm.) (Fig. 11.5H); **Rutilus meidingerii* (Fürböck *et al.*, pers.comm.) (Fig. 11.5J); *R. rubilio* Baccetti *et al.* 1984) (Fig. 11.4); **R. rutilus* (Fig. 11.5I; adult Fig. 11.3D) and **Scardinius erythrophthalmus* (Stein 1981; Fürböck *et al.*, pers.comm.) (Fig. 11.5K); **Vimba vimba* (Lahnsteiner and Patzner 1991; Fürböck *et al.*, pers.comm.) (Fig. 11.5L).

Recently, a valuable study of the spermatozoa of 12 species (asterisked above) of the Leuciscinae, some of which had previously been studied, has been conducted by Fuerbock, Lahnsteiner and Patzner (pers. comm.) using SEM and TEM.

Spermatozoal ultrastructure: Stein (1981) recognized a "type 2 sperm" (not to be confused with the Type II sperm of Mattei 1970) for cyprinids alone, characterized by an almost spherical head, on the basis of ultrastructural investigation of the sperm of the cyprinids *Abramis brama, Blicca bjorkna, Cyprinus carpio, Leuciscus cephalus, L. leuciscus, Phoxinus phoxinus, Rutilus rutilus* and *Scardinius erythrophthalmus*. However, although the nucleus was spheroidal in most species examined by Baccetti *et al.* (1984), it was slightly ellipsoidal in *Leuciscus souffia* and *Barbus barbus*. (For type 1 sperm of Stein see Salmoniformes, Jamieson, **Chapter 12**).

The studies by Baccetti *et al.* (1984), on seven cyprinid species (Figs. 11.4A-E), by Fürböck *et al.* (pers. comm.) on 12 species attributed to the Leuciscinae (Fig. 11.5A-L), and others reported here exemplify variation in sperm ultrastructure within a family.

Cyprinids have simple anacrosomal aquasperm which show significant differences, even intragenerically. Variation occurs with regard to the shape and dimensions of the nucleus, the position of the centrioles and of the proximal centriole relative to the distal centriole and to the nucleus, the dimensions of the midpiece and the number and arrangement of the mitochondria, a number related to the depth of the cytoplasmic canal, the inclusions of the midpiece (occurrence of glycogen and vesicles) and the length of the tail. The sperm have a spheroidal or slightly elliptical nucleus, always eccentrically placed on the tail; two variously orientated centrioles, and a postnuclear cytoplasmic region of varying size which contains the mitochondria (2 to 10) and surrounds the periaxonemal postnuclear canal (cytoplasmic canal). The spermatozoa are relatively short. They were shown to have a length of about 42 µm in *Carassius auratus*, 35 ± 2.8 µm in *Zacco platypus*, 42 ± 6.2 µm in *Z. temminckii*, 33.2 ± 3.7 in *Hemigrammocypris rasborella* and 37.0 ± 0.5 µm in *Acheilognathus (=Pseudoperilampus) typus* by Ohta *et al.* (1994) and in a detailed table of dimensions in the Leuciscinae, which greatly adds to the data of Baccetti *et al.* (1984) and Emelyanova and Makeeva (1985), Fürböck *et al.* (pers. comm.) list a shortest sperm length of 26.6 µm (for *Abramis bramis*) and a greatest length of 57.3 µm (for *Leuciscus souffia*). That of *Tinca*

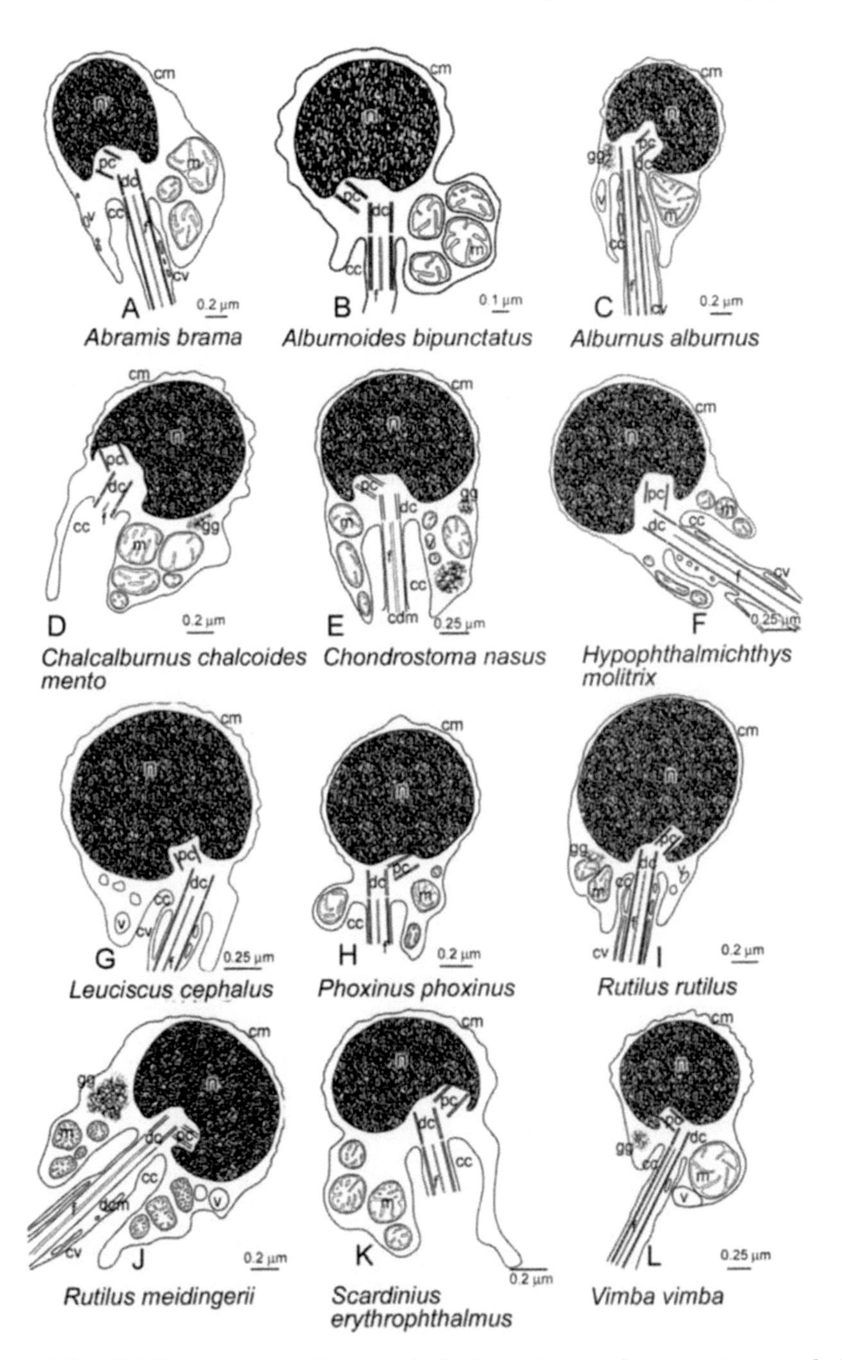

Fig. 11.5 Slightly diagrammatic longitudinal sections of spermatozoa of 12 species of Leuciscinae. cc, cytoplasmic canal; cm, cytoplasmic membrane; cv, cytoplasmic vesicles; dc, distal centriole; f, flagellum; gg, glycogen granules; v, vesicles; m, mitochondria; n, nucleus; Adapted from figures courtesy of Fürböck, Lahnsteiner and Patzner.

tinca appears to be one of the shortest cyprinid sperm at 26.1 ± 3.8 µm total length (Psenicka *et al.* 2006).

Head: The head diameter is said (Baccetti *et al.* 1984) to be uniformly 2 µm, in approximate agreement with *Rhodeus sericeus sinensis* (Guan and Afzelius 1991), 1.9 µm for *Rhodeus ocellatus* (Ohta and Iwamatsu 1983) and *Carassius auratus* (Ohta *et al.* 1993) (though dubiously given as 3.2 µm by Fribourgh *et al.* 1970), and a length of 1.6 µm for *Acheilognathus rhombeus*. The diameter is similar at 1.7. 1.6, 1.5, and 1.6 µm, respectively, in *Zacco platypus*, *Z. temminckii*, *Hemigrammocypris rasborella* and *Acheilognathus (=Pseudoperilampus) typus* (Ohta *et al.* 1994) and 1.64 ± 0.11 µm in *Labeo victorianus* (Rutaisire *et al.* 2006). In their detailed table Fürböck *et al.* (pers. comm.) list a greatest nuclear length of 1.8 µm (for *Chondrostoma toxostoma*) and a largest diameter of around 1.79 µm (for *Leuciscus souffia*). The shortest nuclear length was 1.35 µm (for *L. souffia*) and the smallest diameter around 1.49 µm (for *Alburnus alburnus*). It was considered that this variation could have phylogenetic value.

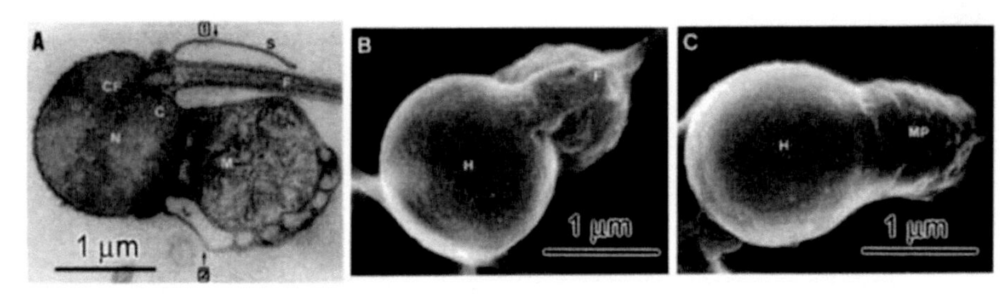

Fig. 11.6 Spermatozoa of *Rhodeus ocellatus ocellatus* (Cyprinidae). **A**. TEM of a spermatozoon. C, centriole (basal body); CF, centriolar fossa; F, flagellum; M, mitochondrion; N, nucleus; *S*, sleeve. x 27,000. **B**. SEM photo of a spermatozoon viewed from the flagellar side. F, flagellum; H, head. x 30,000. **C**. A SEM photo of a spermatozoon viewed from the mitochondrial side. H, head; MP, midpiece. x 30,000. Ohta, T. 1991. Anatomical Record 229: 195-202, Fig. 1.

In leuciscine sperm the nucleus is spherical in *Abramis brama*, *Chalcalburnus chalcoides mento*, *Leuciscus cephalus*, *Phoxinus phoxinus* and *Scardinius erythrophthalmus* but is ovoid in *Alburnoides bipuctatus*, *Alburnus alburnus*, *Chondrostoma nasus*, *Hypophthalmichthys mobilix*, *Rutilus rutilis*, *R. meidingerii* and *Vimba vimba* (Fürböck *et al.* pers. comm.).

The structure of the chromatin varies; it forms dense masses with scattered pale lacunae in *Carassius auratus* (Hara and Okiyama 1998) and is homogeneously granular and strongly electron-dense in *Labeo victorianus* (Rutaisire *et al.* 2006). A small basal nuclear fossa is always present. It is fairly homogeneous except for occasional vacuoles in *Tinca tinca* (Psenicka *et al.* 2006).

Intramembrane particles (IMPs) have been described for *Carassius auratus* (Goldfish), *Misgurnus anguillicaudatus* (Loach) and *Archaeilognatus rhombeus* (Flat bitterling) by Ohta *et al.* (1993) (Fig. 11.7) and for *Zacco platypus* (Common minnow), *Hemigrammocypris rasborella* (Golden venus chub), *Zacco temminckii* (River chub) and *Pseudoperilampus typus* (Netted bitterling) by Ohta *et al.* (1994) but the details are beyond the scope of this chapter. Guan and Afzelius (1991) give a detailed, illustrated account of intramembrane particles and their distribution in *Rhodeus sericeus sinensis*. They cover the anterior portion of the sperm head and the same side as the flagellum. Whereas there are several patches of hexagonal gratings in the sperm of the zebrafish (Kessel *et al.* 1983) or of the Goldfish or other examined cyprinids (Guan 1990), there is only one grating in the Chinese Bitterling sperm albeit a large one (see details in (Guan and Afzelius 1991).

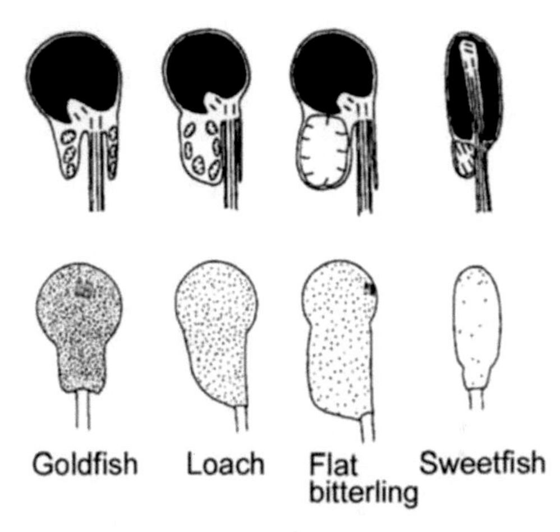

Fig. 11.7 Diagrammatic representation of the spermatozoa of four species of fish. Goldfish (*Carassius auratus*); Loach (*Misgurnus anguillicaudatus*); Flat Bitterling, *Archaeilognatus rhombeus*) and Sweetfish (*Plecoglossus altivelis*). The upper half shows fine structure. The lower half shows the distribution of intramembranous particles. After Ohta, T., Kato, K. H., Abe, T. and Takeuchi, T. 1993. Tissue and Cell 25: 725-735, Fig. 27. Wth permission of Elsevier.

Mitochondria: Asymmetry in addition to that of the centrioles is seen in the distribution of the mitochondria. Most of them are located in the area adjacent to the nucleus, in the opposite area only one (in *Leuciscus cephalus*, *L. souffia*, and *Carassius auratus*) or none (as in all other species, including *L cephalus* in Fig. 1 of Lahnsteiner *et al.* 1992) is present (Baccetti *et al.* 1984). A single large unilaterally located mitochondrion occurs in the sperm of the Rose bitterling, *Rhodeus ocellatus* (Ohta and Iwamatsu 1983; Ohta 1991) (Fig. 11.6), the Chinese bitterling (Guan and Afzelius 1991), the Flat bitterling, *Acheilognatus rhombeus* (Ohta *et al.* 1993) and *Pseudoperilampus typus* (Fig. 11.8I) Ohta *et al.* (1994). The midpiece was also unilateral in *Zacco platypus* (Fig. 11.8A,B), *Z. temmincki* (Fig. 11.8E,F) and *Hemigrammocypris rasborella* (Fig. 11.8C,D) (Ohta *et al.* 1994) though containing several mitochondria. In *C. auratus* although many mitochondria were observed by sections, they were not necessarily separate

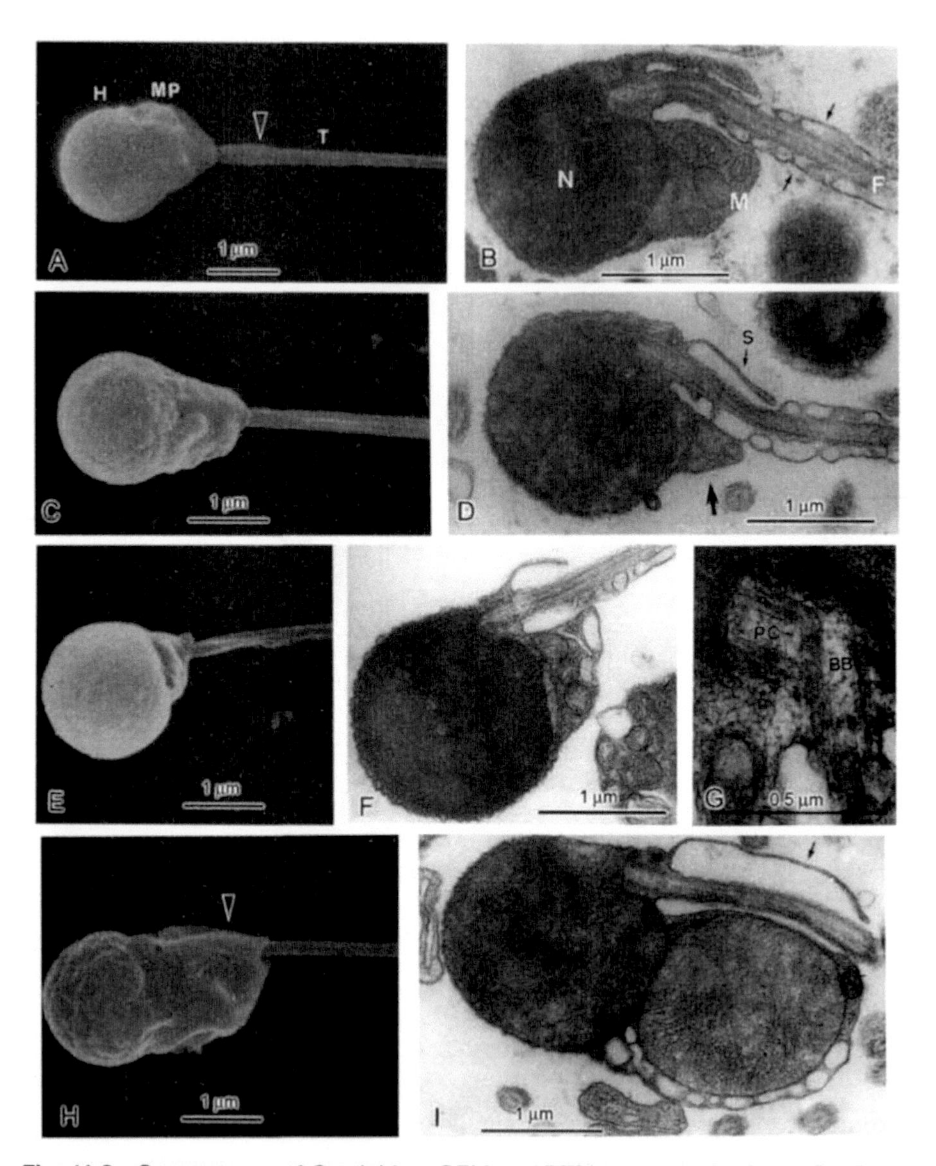

Fig. 11.8 Spermatozoa of Cyprinidae. SEM and TEM respectively. **A** and **B**. *Zacco platypus*, Common minnow. Arrows, portion of the flagellum which is thicker owing to presence of tubular smooth endoplasmic reticulum. **C** and **D**. *Hemigrammocypris rasborella*, Golden venus chub. Arrow, showing mitochondria of unilateral midpiece. **E-G**. *Zacco temminckii*, River chub. **H** and **I**. *Pseudoperilampus typus*, Netted bitterling. Arrows, portion of midpiece lacking cytoplasmic components, i.e. sleeve. F, flagellum; H, head; M, mitochondrion; MP, midpiece; N, nucleus; T, tail. Adapted from Ohta, T., Mizuno, T., Mizutani, M. and Matsuda, M. 1994. Journal of Submicroscopic Cytology and Pathology 26: 181-189, Figs. 1,2,7,8,13,14,15,21,22.

from each other; nine mitchondria were seen in section but the number of mitochondria was estimated to be four (Ohta *et al.* 1993). Hara and Okiyama (1998) observed spherical and cylindrical mitochondria in a sheath surrounding a deep cytoplasmic canal. At least six small spherical mitochondria are present in *Labeo victorianus* (Rutaisire *et al.* 2006). Different arrangements of mitochondria are illustrated by Fürböck *et al.* (pers. comm.) (Fig. 11.6), all 12 of which are asymmetrical. The number of mitochondria varied considerably in the examined leuciscines. The number was low for *Vimba vimba* and *Rutilus rutilus* with 1 or 2; medium for *Abramis brama, Alburnoides bipunctatus, Alburnus alburnus, Chalcalburnus chalcoides mento, Leuciscus cephalus* and *Phoxinus phoxinus* with 3 or 4; and high for *Chondrostoma nasus, Hypophthalmichthys mobilix* and *Scardinius erythrophthalmus* with more than five. The highest number of mitochondria, 7, was seen in *Rutilus meidingerii*. These authors note fair agreement with other accounts: Bacetti *et al.* (1984) counted 3 or 4 mitochondria in *Alburnus alburnus alborella*, 3 or 4 in *Chondrostoma toxostoma*, 2 or 3 in *Leuciscus cephalus*, 4 in *Leuciscus souffia* and 5 or 6 in *Rutilius rubilio.* EmelÝanova and Makeeva (1985) reported 4 or 5 mitochondria in *Hypophthalmichthys molitrix.* Small discrepancies in numbers may be due to duplicated counts where a mitochondrion is tortuous so as to appear more than once in a section. The midpiece is cylindric/cone-shaped, 0.86 ± 0.27 µm in length and 1.17 ± 0.24 µm in width proximally in *Tinca tinca.*

A relationship between the number of mitochondria and axonemal length is not demonstrable but a clear correlation exists between the number of mitochondria and the length of the cytoplasmic canal, which exceeds 1.5 µm in *Rutilus* and *Carassius*, and is only 0.1 µm long in *Barbus* (Baccetti *et al.* 1984).

The portion of the midpiece lacking mitochondria may extend as a sleeve around the proximal part of the flagellum as in *Acheilognatus rhombeus* (also seen in the cobitids) (Ohta *et al.* 1993). In *Rhodeus sericeus sinensis* the sleeve is 2 µm long and it contains an inner scaffolding consisting of a cytoskeletal lattice of helically wound filaments with a diameter of about 10 nm (a less likely interpretation is that this part of the cytoskeleton consists of a stack of separate rings (Guan and Afzelius 1991). This lattice is presumably the lining to the inner membrane of the cytoplasmic canal reported in some other teleosts.

Proximal centriole: The proximal centriole lies in the sagittal plane of the sperm in only one case (*Leuciscus cephalus*) and is inclined at 40° to this plane in *Leuciscus souffia* and *Rutilus*, 50° in *Carassius auratus* (a proximal centriole was not observed by Hara and Okiyama (1998), 60° in *Alburnus*, 80° in *Barbus* and 90° in *Chondrostoma toxostoma*. It is rarely perpendicular to the distal centriole as in *Alburnus alburnus* (Lahnsteiner and Patzner 1991), *Barbus* (Baccetti *et al.* 1984); and *Labeo victorianus* (Rutaisire *et al.* 2006) but, to exemplify the variation in selected species, is reported to be inclined at an angle of 60° in *Zacco platypus* and *Z. temminckii* (Ohta *et al.* 1994); 105° in **Vimba vimba*; 110° in **Alburnus alburnus, Leuciscus souffia, Misgurnus*

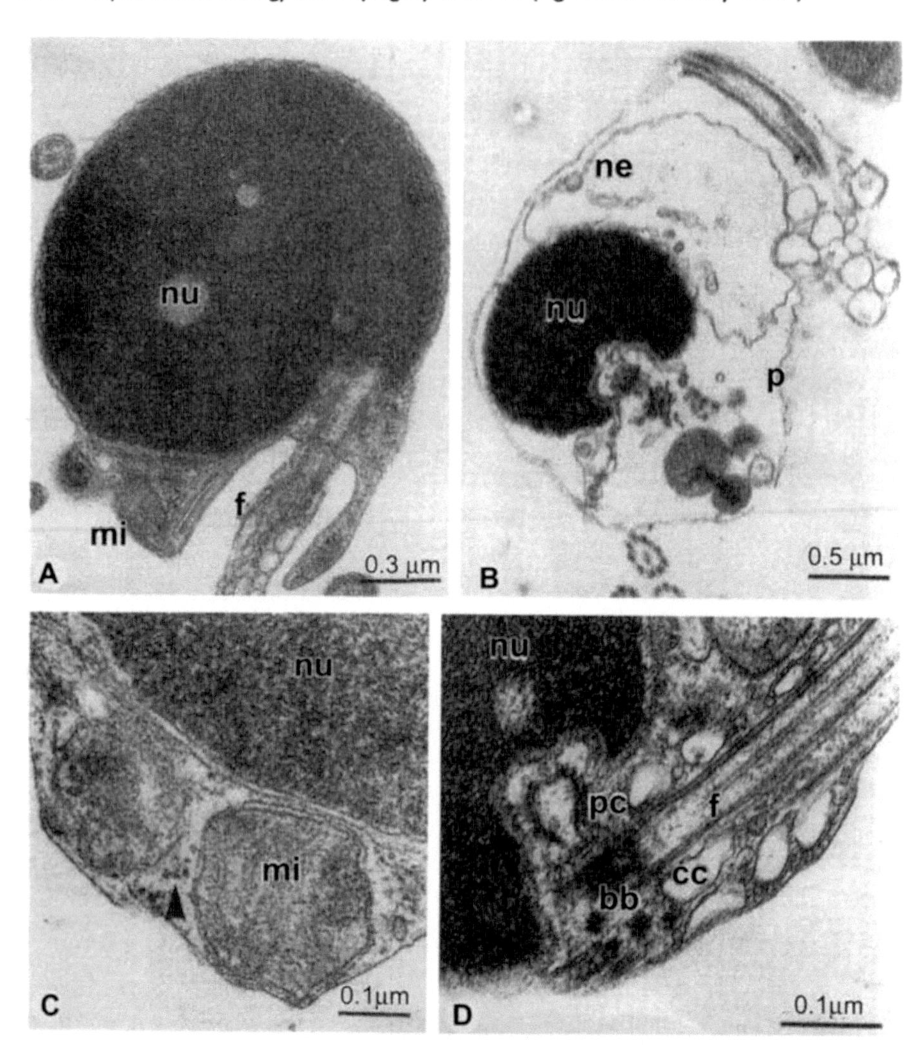

Fig. 11.9 *Leuciscus cephalus* (Cyprinidae) spermatozoa. **A**. Longitudinal section, after 10 min in physiological saline. **B**. After motility for 10 min in saline. **C**. After motility for 60 seconds in deionized water. **D**. Flagellar complex, after 60 seconds in saline. Bb, basal body (distal centriole); cc, cytoplasmic canal; f, flagellum; mi, mitochondrion; ne, nuclear envelope; nu, nucleus; p, plasmalemma; pc, proximal centriole; arrow, skeletal microfilaments. Adapted from Lahnsteiner, F., Weismann, T. and Patzner, R. A. 1992. Fish Physiology and Biochemistry 10: 283-289, Figs. 1-4.

anguilicaudatus and **Rutilus meidingerii;* 110° (or 70°) or 125° in *Carassius auratus;* 120° in **Alburnoides puctatus*, **Leuciscus cephalus*,**Chondrostoma toxostoma*, **Phoxinus phoxinus* and **Hypophthalmichthys molitrix* (but 110° in this species according to Emelynova and Makeeva 1985); 125° in **Abramis brama*, **Chalcalburnus chalcoides mento* and **Chondrostoma nasus;* 130° in *Acheilognathus (=Pseudolampanus) typus*, *Hemigrammocypris rasborella* (Ohta *et* *al*. 1994), **Rutilus rutilus* and **Scardinius erythrophthalmus*; and 140° in *Rutilus rubilio* (Baccetti *et al*. 1984; Ohta *et al*. 1993, and, asterisked, Fürböck *et al*., pers. comm.). These variations in centriolar geometry are clearly correlated with the position of the nucleus with respect to the axis of the tail (Baccetti *et al*. 1984).

Distal centriole and rootlet: The distal centriole is linked to the surface of the nuclear fossa by fibers in *Leuciscus cephalus, L. souffia, Chondrostoma toxostoma* and *Carassius auratus*. Fibers may also connect the two centrioles, as in *L. victorianus*. *Leuciscus souffia* is unique in at least the seven species in having a large striated rootlet which links the distal centriole to the adjacent plasma membrane. Such rootlets are widely reported for spermatids, for instance those of the salmonid *Oncorhynchus mykiss (=Salmo gairdneri)* (Billard 1983) and the gonorynchiform *Chanos chanos*. Membranous vesicles intervene between the plasma membrane and the axoneme in the anterior region of the flagellum in *L. cephalus* and *Barbus barbus plebejus*, for almost its whole length in *Rutilus rubilio*, or are numerous over an unspecified length in *Carassius auratus* (Baccetti *et al*. 1984). In *C. auratus*, at least, they extend from the cytoplasmic canal for the proximal region of the flagellum, as illustrated by Ohta *et al*. (1993) who describe them as a "tubular fin". In *Tinca tinca* the proximal centriole is located in the implantation fossa. The distal centriole (basal body) is almost tangential to the nucleus, at an orientation of 140° to the distal centriole. A vesicle was observed, attached to the most basal region of the flagellum just under its plasma membrane (Psenicka *et al*. 2006).

Glycogen: Glycogen granules are reported for the cytoplasm of several species and are probably general for cypriniform sperm.

Flagellum: The tail is of moderate length with a 9+2 axoneme; both dynein arms are present. The length of the tail varies from 25.45 ± 2.47 µm in *Tinca tinca* (Psenicka *et al*. 2006) to 60 µm in *Barbus barbus* (Baccetti *et al*. 1984). Because the proximal centriole is always eccentric, the axonemal axis is tangential to the nucleus in *L. cephalus, Rutilus* and *Alburnus*, lateral to it in *Chondrostoma, Barbus* and *Carassius* (as also in *Rhodeus ocellatus*, Ohta and Iwamatsu 1983) and almost central with respect to the nucleus in *L. souffia* (Baccetti *et al*. 1984) and a single mitochondrion in *Rhodeus* (see above).

All cyprinid sperm examined to date lack the one or more fins (Fig. 11.10) seen on the flagellum of most teleost sperm. However, in *Carassius auratus* Hara and Okiyama (1998) interpreted an arrangement of vesicles under the flagellar membrane proximally as a 'tubular fin' and in *Labeo victorianus* the flagellum is flanked by lateral vesicles (Rutaisire *et al*. 2006).

In view of the widespread occurrence of flagellar fins throughout the Euteleostomi (=Osteichthyes), absence in cyprinids and other investigated ostariophysans was interpreted as a loss and was considered an ostariophysian synapomorphy (Jamieson 1991) but later demonstration of flagellar fins in at least one species in each of the other three ostariophysan orders may indicate that presence is plesiomorphic for the group with apomorphic loss in most members. Nevertheless, it could represent an apomorphic reversal.

Fig. 11.10 *Carassius auratus* (Cyprinidae). Transverse section of axoneme, showing absence of flagellar fins in Cypriniformes. Drawn from a micrograph of Fribourgh, J. H. (1970). Copeia 2: 274-279, Fig. 6. From Jamieson, B. G. M. 1991. *Fish Evolution and Systematics: Evidence from Spermatozoa*. Cambridge University Press, Cambridge, Fig. 12.8.

11.5.2 Family Cobitidae

These are the loaches, in Eurasian and Moroccan freshwaters (Nelson 2006).

Sperm literature: The sperm of two cobitids have been examined ultrastructurally: *Pangio semicincta* (=*Acanthophthalmus semicinctus*), the Coolie Loach (adult, Fig. 11.11), Jamieson (1991), and the Loach, *Misgurnus anguillicaudatus* (Fig. 11.8) (Ohta *et al.* 1993).

The spermatozoon of *Misgurnus anguillicaudatus* is 28 µm long.

Fig. 11.11 *Pangio semicincta (=Acanthophthalmus semicinctus)* (Cobitidae). After Sterba, G. 1962. Freshwater *Fishes of the World.* Vista Books, London. Fig. 484. From Jamieson, B. G. M. 1991. *Fish Evolution and Systematics: Evidence from Spermatozoa*. Cambridge University Press, Cambridge, Fig. 12.9.

In *Pangio semicincta* the nucleus is approximately spherical, 1.7 µm wide, but has an eccentric fossa sufficiently large to house the entire basal body (Fig. 11.12) and, anteriolateral to this, the proximal centriole. The regularity of the spherical nucleus is enhanced by the homogeneous nature of the chromatin despite some small pale lacunae. In *M. anguillicaudatus* the nucleus, 1.7 µm long, is also smoothly rounded; the fossa is again highly eccentric but is smaller and does not fully contain the distal centriole.

In *Pangio semicincta* the cytoplasmic collar, 2.7 µm long, is very slender on the side of the sperm further from the proximal centriole but on the side occupied by this is widened proximally to contain a group of mitochondria at

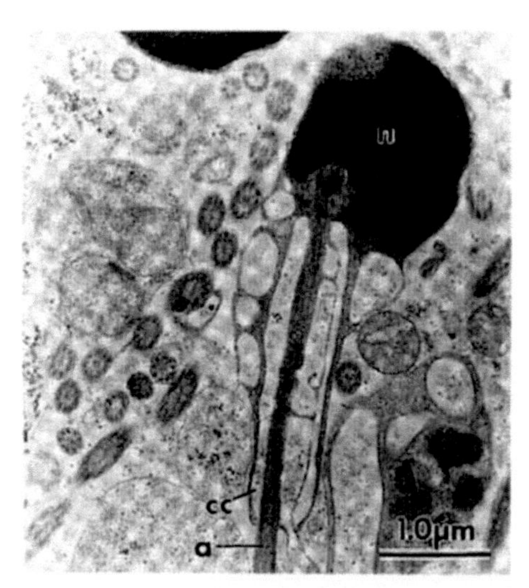

Fig. 11.12 *Pangio semicincta (=Acanthophthalmus semicinctus)* (Cobitidae). Longitudinal section of a spermatozoon. From Jamieson, B. G. M. 1991. *Fish Evolution and Systematics: Evidence from Spermatozoa*. Cambridge University Press, Cambridge, Fig. 12.10.

least three deep anteroposteriorly. These are mutually adpressed and vary from spherical to angular. In *Misgurnus anguillicaudatus* many mitochondria were continuous, as in the goldfish. A maximum of 13 or so mitochondria are observable in one section. The two centrioles are situated at about 110° angles to each other as in the goldfish (Fig. 11.4). As in *Pangio semicincta*, cytoplasmic components such as mitochondria are one-sided in the midpiece. Therefore, the part of the midpiece lacking them is attenuated and forms a sleeve-like structure.

In both species because of the eccentric location of the implantation fossa and its contained centrioles, the long axis of the flagellum is well to one side of the long axis of the nucleus but does not have the truly tangential disposition of the Type II spermatozoon *sensu* Mattei (1970). Furthermore, the centrioles lie in the fossa, a major departure from the Type II condition.

Remarks: The spherical head of the two cobitid species conforms with the type 2 sperm recognized by Stein (1981) for the related cyprinids. In this and other respects, including the thin, trailing midpiece collar, and usual absence of flagellar fins, the cypriniform sperm resembles the characiform sperm described below.

11.6 ORDER CHARACIFORMES

Characiformes contains freshwater fishes found exclusively on the American and African continents, and comprises one of the dominant groups in the Neotropical region. In spite of the large number of species and their wide distribution, the relationships among many groups of characiforms are not satisfactorily resolved (Buckup 1998). This is particularly true for the family Characidae, the group with the greatest taxonomic diversity, which contains many fishes less than 15 cm in standard length. Characidae contains 12 subfamilies, in addition to at least 88 genera not assigned to a particular subfamily and therefore considered *incertae sedis* in the family (Reis *et al.* 2003). Buckup (1998) recognizes two sub-orders in Characiformes: Citharinoidei, composed of two families, and Characoidei, with eight super-families and 18 families. Currently, information on sperm ultrastructure is available for at least one species in most families that comprise the order. The only families lacking such information are Distichondontidae of the suborder Citharinoidei (African), and the families Hepsetidae (African) and Ctenoluciidae of the superfamily Erythrinoidea, suborder Characoidei. Most externally fertilizing characiforms produce Type I aquasperm, although several have been shown to produce sperm resembling Type II aquasperm *sensu* Mattei (1970). Type III aquasperm (*sensu* Quagio-Grassiotto and Oliveira 2008) have recently been described in the family Lebiasinidae (Veríssimo-Silveira 2007). Inseminating species that often produce highly modified introsperm are restricted to the family Characidae, subfamilies Cheirodontinae, Glandulocaudinae, and Stervardiinae, as well as a number of species *incertae sedis* (Burns and Weitzman 2005).

11.6.1 Suborder Citharinoidei

11.6.1.1 Family Citharinidae

***Citharinus* sp.** The spermatozoon of *Citharinus* sp., described by Mattei *et al.* (1995), is a Type I aquasperm with a slightly elongate, conical nucleus measuring 2 μm in length and 1.5 μm in width (Fig. 11.13A) and containing highly condensed flocculent chromatin and a shallow nuclear fossa (Fig. 11.13C).

The proximal centriole, which is nearly parallel to the distal, is located at the periphery of the cell and associated with the nuclear envelope. The anterior portion of the centrally located distal centriole is contained within the shallow nuclear fossa (Fig. 11.13C). The anterior portion of the midpiece contains several small mitochondria (Fig. 11.13A,B), whereas the posterior region is occupied by a large, well-organized "lattice tubule" or "tubular-vesicular system" (Fig. 11.13B,D) formed from vesicles during spermiogenesis (Fig. 11.13E). The initial segment of the flagellum is contained within a cytoplasmic canal. The single flagellum contains a classic 9+2 axoneme with all microtubules electron-lucent, and no fins (Fig. 11.13B).

11.6.2 Suborder Characoidei Parodontoidea

11.6.2.1 Family Parodontidae

The spermatozoon of *Apareiodon affinis* has been described by Verissimo-Silveira (2007). It is a Type I aquasperm with an ovoid nucleus (maximum

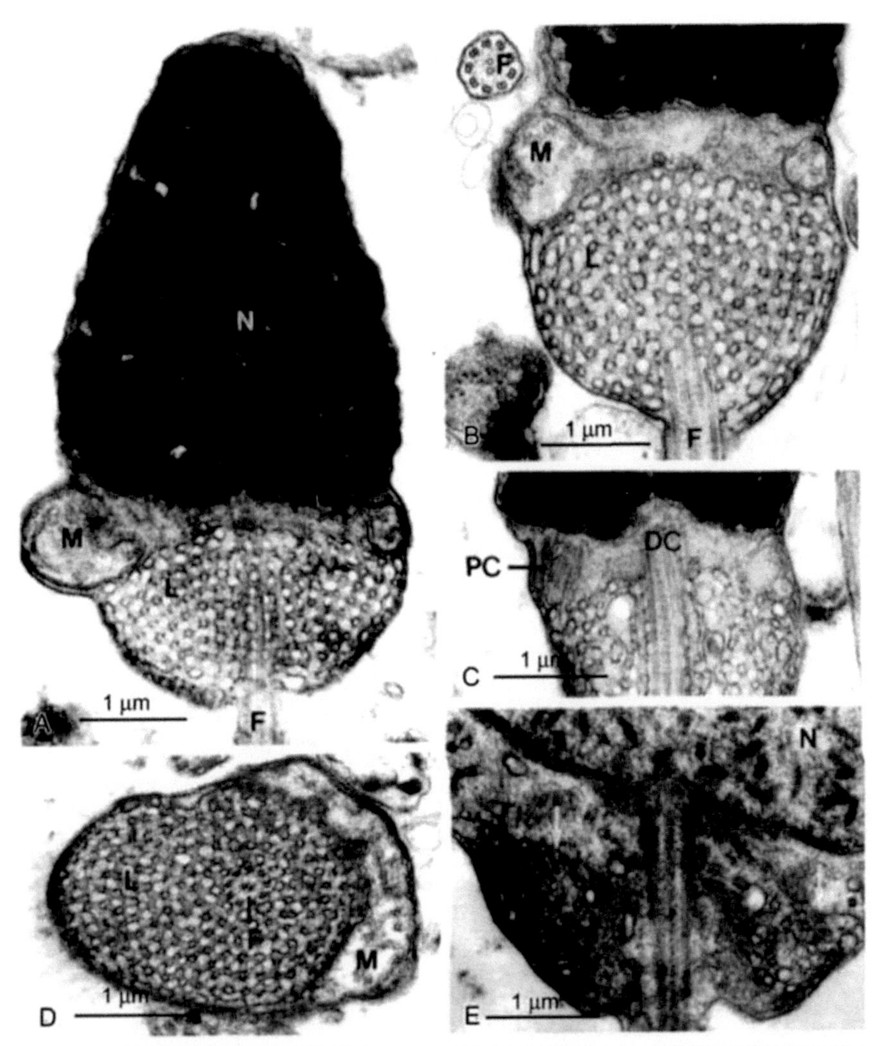

Fig. 11.13 *Citharinus* sp. spermatozoa (Citharinidae). **A**. Longitudinal section (LS). **B**. LS showing the concentric orientation of the lattice tubules. **C**. LS showing the disposition of the centrioles at the base of the nucleus. **D**. Transverse section at level of midpiece. **E**. LS spermatid; numerous vesicles (arrows) are present around the base of the flagellum. DC, distal centriole; F, flagellum; L, lattice tubule; M, mitochondrion; N, nucleus; PC, proximal centriole. After Mattei, X., Marchand, B. and Thiaw, O. T. 1995. Journal of Submicroscopic Cytology and Pathology 27: 189-191, Figs. 1-5.

dimension 1.7 µm) containing condensed granular chromatin and a deep, eccentrically located nuclear fossa. The proximal centriole is anterior, perpendicular and slightly lateral to the distal, and both are contained within the nuclear fossa. The long, asymmetric midpiece contains several elongate mitochondria with a dense matrix and a few vesicles randomly distributed.

The initial segment of the flagellum in contained within a cytoplasmic canal that is surrounded by several concentric membranous rings. The posterior portion of the midpiece tapers to a cytoplasmic sleeve. The single, eccentrically located flagellum contains a classic 9+2 axoneme with all microtubules electron lucent, and no fins.

ANOSTOMOIDEA

11.6.2.2 Family Curimatidae

Species analyzed ultrastructurally include *Curimata inornata* (Matos *et al.* 1998); *Cyphocharax gillii, Cyphocharax modestus, Cyphocharax spilotus, Potamorhina altamazonica* and *Steindachnerina insculpta* (Quagio-Grassiotto *et al.* 2003); and *Psectrogaster essequibensis* (Veríssimo-Silveira 2007). An original description is provided for *Psectrogaster* sp.

***Psectrogaster* sp.** The spermatozoa of all curimatids analyzed have the same basic morphological pattern as *Psectrogaster* sp. They are Type I aquasperm with spherical nuclei, averaging 1.8 μm in diameter, containing highly condensed granular chromatin and a moderately deep nuclear fossa that houses the proximal centriole and most of the distal centriole (Fig. 11.14A-D). The proximal centriole is anterior, medial to slightly lateral, and perpendicular to the distal centriole (Fig. 11.14C). The short midpiece contains several elongate mitochondria and a short cytoplasmic collar containing the initial segment of the flagellum (Fig. 11.14B,E). The eccentrically located flagellum has the classic 9+2 axoneme with all tubules electron-lucent, and no fins (Fig. 11.14H). A well-developed system of vesicles can be found within the midpiece, and these often continue into the anterior region of the flagellum (Fig. 11.14B,E-G).

11.6.2.3 Family Prochilodontidae

Prochilodus scrofa (=*P. lineatus*) (Vicentini *et al.* 2001; Vicentini 2002); *P. lineatus* (Veríssimo-Silveira 2007). The spermatozoon of *Prochilodus* is a Type I aquasperm with a spherical nucleus, 1.7 μm in diameter, containing condensed, flocculent chromatin and a deep nuclear fossa housing both centrioles. The proximal centriole is anterior, slightly lateral, and perpendicular to the distal. The long, thin midpiece contains several elongate C-shaped mitochondria that have a dense matrix, and some vesicles. The initial segment of the flagellum is contained within a cytoplasmic canal that is surrounded by concentric membranous rings. The flagellum has a classic 9+2 axoneme and no fins.

11.6.2.4 Family Anostomidae

Anostomids analyzed to date include *Leporinus friderici* (Matos *et al.* 1999a); *Leporinus macrocephalus* (Amaral 2003); *Leporinus lacustris* and *Abramites* sp. (Veríssimo-Silveira 2007); *Schizodon isognatus, S. nasutus* and *Schizodon* sp. (original descriptions).

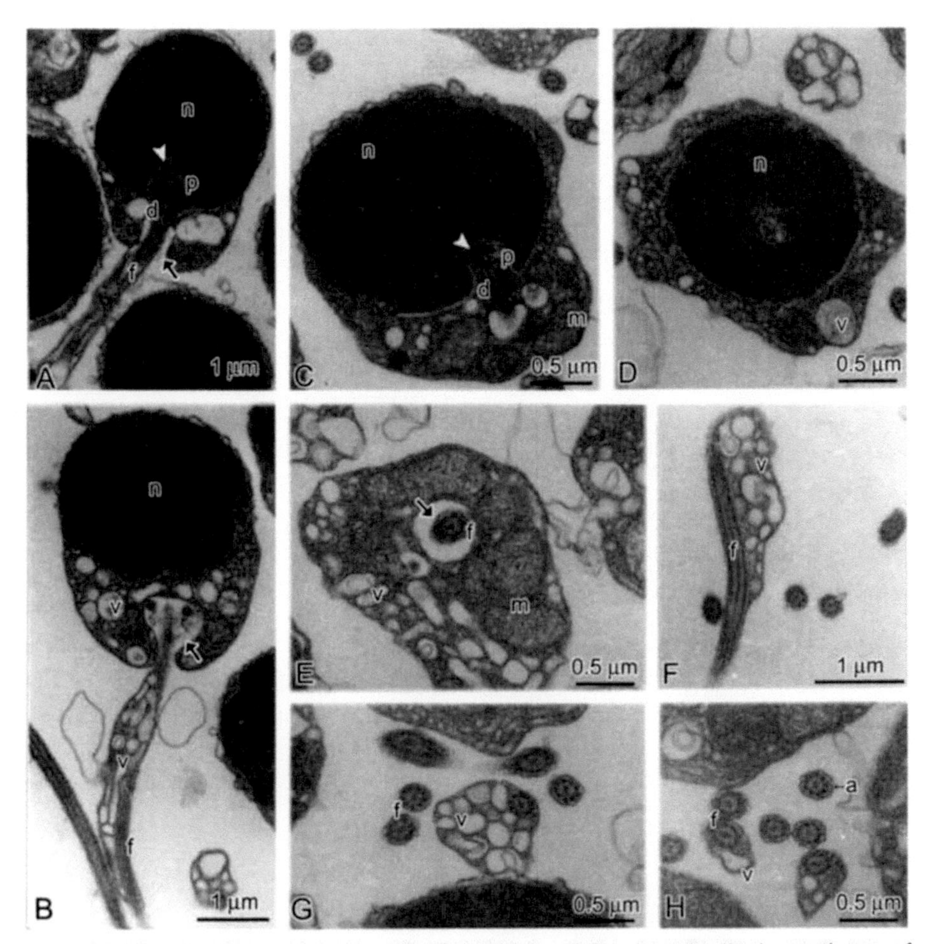

Fig. 11.14 *Psectrogaster* sp. (Curimatidae). **A-B**. Longitudinal sections of spermatozoa. Note the well- developed vesicular system within the midpiece and anterior region of the flagellum. **C**. Longitudinal section of nucleus and midpiece; elongate mitochondria are located close to nucleus. **D**. Transverse section of base of the nucleus. **E**. Transverse section of the midpiece with the mitochondria surrounding the cytoplasmic canal. **F-G**. Longitudinal and transverse sections of the anterior region of the flagellum. **H**. Transverse section of the region where the flagellum loses the vesicles and becomes comprised of only the axoneme surrounded by the flagellar membrane. a, axoneme; d, distal centriole; f, flagellum; m, mitochondrion; n, nucleus; p, proximal centriole; v, vesicle; arrowhead, nuclear fossa; arrow, cytoplasmic canal. Original.

***Schizodon* spp.** Sperm ultrastructure of anostomids resembles that described for the three species of *Schizodon*. During spermiogenesis, nuclear rotation occurs such that the centriolar complex and flagellum assume a final position posterior to the nucleus, while the chromatin condenses into large flocculent masses (Figs. 11.15A-D, 11.16A,B). The spermatozoon is a Type I aquasperm

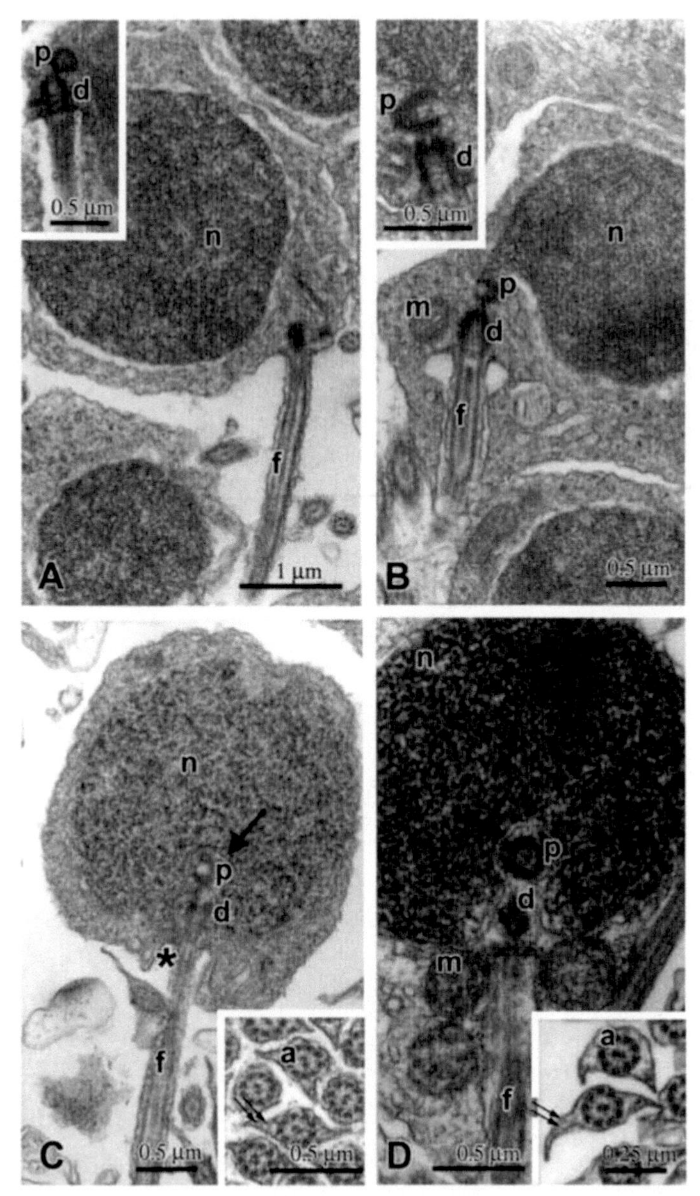

Fig.11.15 *Schizodon* spp. (Anostomidae). Type I spermiogenesis. **A-D**. Longitudinal sections of the earliest, early, mid and late spermatids, respectively. Note the lateral position of the flagellum in **A,** nuclear rotation in **B,** and the central location of the flagellum relative to the nucleus in **C** and **D**. **A, B insets**. Centriolar complex. **C, D insets**. Transverse sections through the flagella showing the fins. **A, B**. *S. nasutus*. **C**. *S.* sp. **D**. *S. isognatus*. a, axoneme; d, distal centriole; f, flagellum; m, mitochondrion; n, nucleus; p, proximal centriole; v, vesicle; arrow, nuclear fossa; double arrow, fins; asterisk, cytoplasmic canal. Original.

with a spherical nucleus and a relatively deep nuclear fossa housing the entire centriolar complex (Fig. 11.16A,B). Nuclear diameters are 1.7 µm in *L. lacustris*, 1.9 µm in *Abramites* sp., and 1.5-1.9 µm in *Schizodon*. The proximal centriole is anterior and perpendicular to the distal (Fig. 11.16B). The midpiece contains numerous mitochondria, some being elongate and C-shaped that

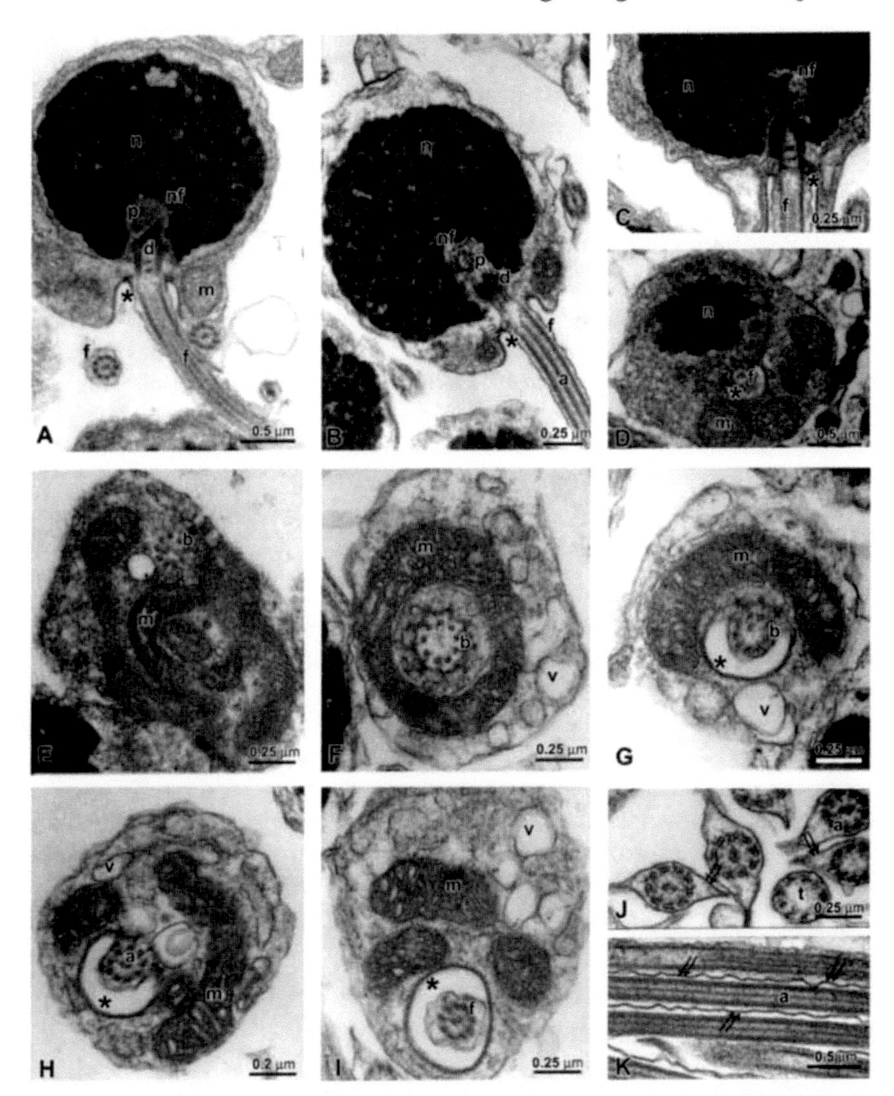

Fig. 11.16 *Schizodon* spp. (Anostomidae). Type I aquasperm. **A-C**. Longitudinal sections of spermatozoa. **D-I**. Transverse sections at different levels of the midpiece. Note the long branched, C-shaped mitochondria. **J-K**. Transverse and longitudinal sections through flagella, respectively. Note the fins. **A, C, D, E, K**. *S. nasutus*. **B, F, G, H, J**. *S. isognatus*. a, axoneme; b, basal body; d, distal centriole; f, flagellum; m, mitochondrion; n, nucleus; p, proximal centriole; v, vesicle; nf, nuclear fossa; double arrows, fins; asterisk, cytoplasmic canal. Original.

encircle the cytoplasmic canal containing the anterior segment of the flagellum (Fig. 11.16D-I). Some vesicles are also present in the midpiece (Fig. 11.16F-I). The flagellum has a typical 9+2 axoneme with all tubules electron-lucent and two fins (Fig. 11.16J,K).

11.6.2.5 Family Chilodontidae

The spermatozoa of two species of chilodontids have been examined ultrastructurally, *Chilodus punctatus* (Pecio 2003) and *Caenotropus labyrinthicus* (Verissimo-Silveira 2007). They are Type I aquasperm with spherical nuclei, measuring 2.4 μm in diameter in *C. punctatus* and 1.6 μm in *C. labyrinthicus*, and containing deep nuclear fossae. The nuclei contain homogeneous, highly condensed granular chromatin. The centrioles are perpendicular or slightly oblique to one another, with the proximal anterior and sometimes slightly lateral to the distal. The proximal and all or part of the distal are contained within the nuclear fossa. Electron-dense anchoring fibers or lateral spurs are associated with the distal centriole. The midpiece in *C. punctatus* contains several mitochondria and vesicles, whereas that of *C. labyrinthicus* contains a single, elongate mitochondrion encircling the cytoplasmic canal. The initial segment of the flagellum is contained within a cytoplasmic canal. The single flagellum contains a classic 9+2 axoneme with all microtubules electron-lucent. Two flagellar fins are present.

CRENUCHOIDEA

11.6.2.6 Family Crenuchidae

The spermatozoon of *Characidium gomesi*, described by Verissimo-Silveira (2007), is a Type I aquasperm with a spherical nucleus (1.9 μm in diameter) containing highly condensed, homogeneous granular chromatin and a deep nuclear fossa. The proximal centriole is anterior and perpendicular to the distal and both are located within the nuclear fossa. The short midpiece appears to have a single, large spherical mitochondrion and abundant, relatively large vesicles. The initial segment of the flagellum is contained within a cytoplasmic canal. The single flagellum contains a classic 9+2 axoneme with all microtubules electron lucent, a membranous compartment in the anterior segment, and no fins.

HEMIODONTOIDEA

11.6.2.7 Family Hemiodontidae

The spermatozoon of *Anodus elongatus*, described by Verissimo-Silveira (2007), is a Type I aquasperm with a spherical nucleus (1.9 μm in diameter) containing highly condensed, homogeneous granular chromatin and a moderately deep nuclear fossa. The proximal centriole is anterior and oblique to the distal. The proximal and part of the distal are located within the nuclear fossa. The relatively large midpiece contains several spherical to elongate mitochondria and a system of elongate, interconnected vesicles, some of which fuse with the plasma membrane. The initial segment of the flagellum is contained within a cytoplasmic canal. The single flagellum contains a classic 9+2 axoneme with all microtubules electron lucent, and no fins.

ALESTOIDEA

11.6.2.8 Family Alestidae

Alestids examined for sperm ultrastructure are *Alestes dentex* (Shahin 2006b); *Micralestes* sp. and *Phenacogrammus interruptus* (Veríssimo-Silveira 2007). The spermatozoon of alestids is a Type I aquasperm with a spherical nucleus containing a moderately deep nuclear fossa. Nuclear diameters are 1.7 μm in *A. dentex*, 1.9 μm in *Micralestes* sp., and 2.0 μm in *P. interruptus*. Chromatin is highly condensed and granular in *Micralestes* sp. and *P. interruptus*, whereas in *A. dentex* it appears to be flocculent. The proximal centriole is anterior and perpendicular to the distal. The proximal centriole and all or most of the distal are contained within the nuclear fossa. The midpiece contains several spherical to elongate mitochondria, with the matrix markedly electron-dense in *P. interruptus*. Abundant elongate vesicles are also present. The single flagellum contains a classic 9+2 axoneme, with all microtubules electron-lucent, and no fins. Posterior to the cytoplasmic collar the flagellum also contains elongate vesicles.

CYNODONTOIDEA

11.6.2.9 Family Acestrorhynchidae

Acestrorhynchids examined for sperm ultrastructure are *Acestrorhynchus falcatus* (Matos *et al.* 2000); *A. pantaneiro* (Veríssimo-Silveira 2007). The spermatozoon resembles a Type II aquasperm *sensu* Mattei (1970), with a spherical nucleus containing highly condensed, homogeneous granular chromatin with several electron lucent areas, a lateral centriolar complex, and no nuclear fossa. Nuclear diameter is 2.0 μm in *A. pantaneiro*; no measurement is available for *A. falcatus*. The proximal centriole is lateral and perpendicular to the distal, from which thin striated rootlets radiate. Several slightly elongate mitochondria with an electron-dense matrix, as well as many small vesicles, are located in the abundant cytoplasm that surrounds the nucleus. Thus, no distinct posterior "midpiece" is present. A series of parallel cisternae encircles the entire nucleus, resembling a cluster of vesicles in transverse section. No cytoplasmic canal is evident. The single, laterally located flagellum contains a classic 9+2 axoneme along with nine peripheral dense bodies, each being immediately distal to a pair of axonemal microtubules. All microtubules are electron-lucent and no fins are present.

11.6.2.10 Family Cynodontidae

The spermatozoon of *Rhaphiodon vulpinus*, described by Verissimo-Silveira (2007), is a Type I aquasperm with a spherical nucleus (1.4 μm in diameter) containing condensed flocculent chromatin and a deep nuclear fossa. The proximal centriole is anterior and apparently perpendicular to the distal and both are located inside the nuclear fossa. The midpiece, which contains several elongate mitochondria and vesicles, tapers posteriorly to a cytoplasmic sleeve. The initial segment of the flagellum is contained within a cytoplasmic canal. The single flagellum contains a classic 9+2 axoneme with all microtubules electron-lucent, and two fins.

ERYTHRINOIDEA

11.6.2.11 Family Erythrinidae

Erythrinids examined for sperm ultrastructure are *Hoplerythrinus unitaeniatus* (Matos *et al.* 1999b; Veríssimo-Silveira 2007); *Hoplias malabaricus* (Quagio-Grassiotto *et al.* 2001a); and *Erythrinus erythrinus* (original description).

Erythrinus erythrinus: All erythrinid spermatozoa resemble that of *Erythrinus erythrinus* (Figs. 11.17, 11.18), being Type I aquasperm with spherical nuclei containing highly condensed, homogeneous granular chromatin with occasional electron lucent areas and a shallow nuclear fossa. Nuclear diameters are 1.7 µm in *E. erythrinus*, 1.4 µm in *H. unitaeniatus*, and 1.8 µm in *H. malabaricus*. In some, the centriolar complex remains slightly lateral to the nucleus indicating that nuclear rotation may not be complete. The proximal centriole is anterior, slightly lateral and oblique to the distal (Figs. 11.17A,C, 11.18A,C,D). Only the proximal lies within the nuclear fossa. The relatively long midpiece contains several elongate mitochondria, often with a dense matrix, concentrated in the anterior region (Fig. 11.17B,D). A cytoplasmic canal is absent, having closed during late spermiogenesis (Fig. 11.17A-D). The posterior midpiece is occupied by dilated, elongate vesicles (Fig. 11.18A-D). The lateral flagellum has a classic 9+2 axoneme with all microtubules electron lucent and no fins. The elongate vesicles seen in the midpiece continue into the anterior segment of the flagellum (Fig. 11.18E).

11.6.2.12 Family Lebiasinidae

This family contains the subfamilies Lebiasininae and Pyrrulininae. Species in Lebiasininae examined for sperm ultrastructure are *Nannostomus digrammus*, *N. marginatus* and *N. unifasciatus* (Verissimo-Silveira 2007). The spermatozoa of *Nannostomus digrammus* and *N. marginatus* are type III aquasperm with slightly ovoid nuclei, while those of *N. unifasciatus* resemble the Type II aquasperm of Mattei (1970). All three have highly condensed, homogeneous granular chromatin with occasional electron lucent areas. A shallow nuclear fossa is present in *N. digrammus* but absent in the other two species. In *N. digrammus* the proximal centriole is anterior and nearly perpendicular to the distal, with only the proximal contained within the nuclear fossa. In the other two species, the proximal centriole is anterior, lateral and oblique to the distal and no nuclear fossa is present. In all three species, anchoring fibers attach the distal centriole to the plasma membrane. The midpiece contains several to more numerous elongate mitochondria.

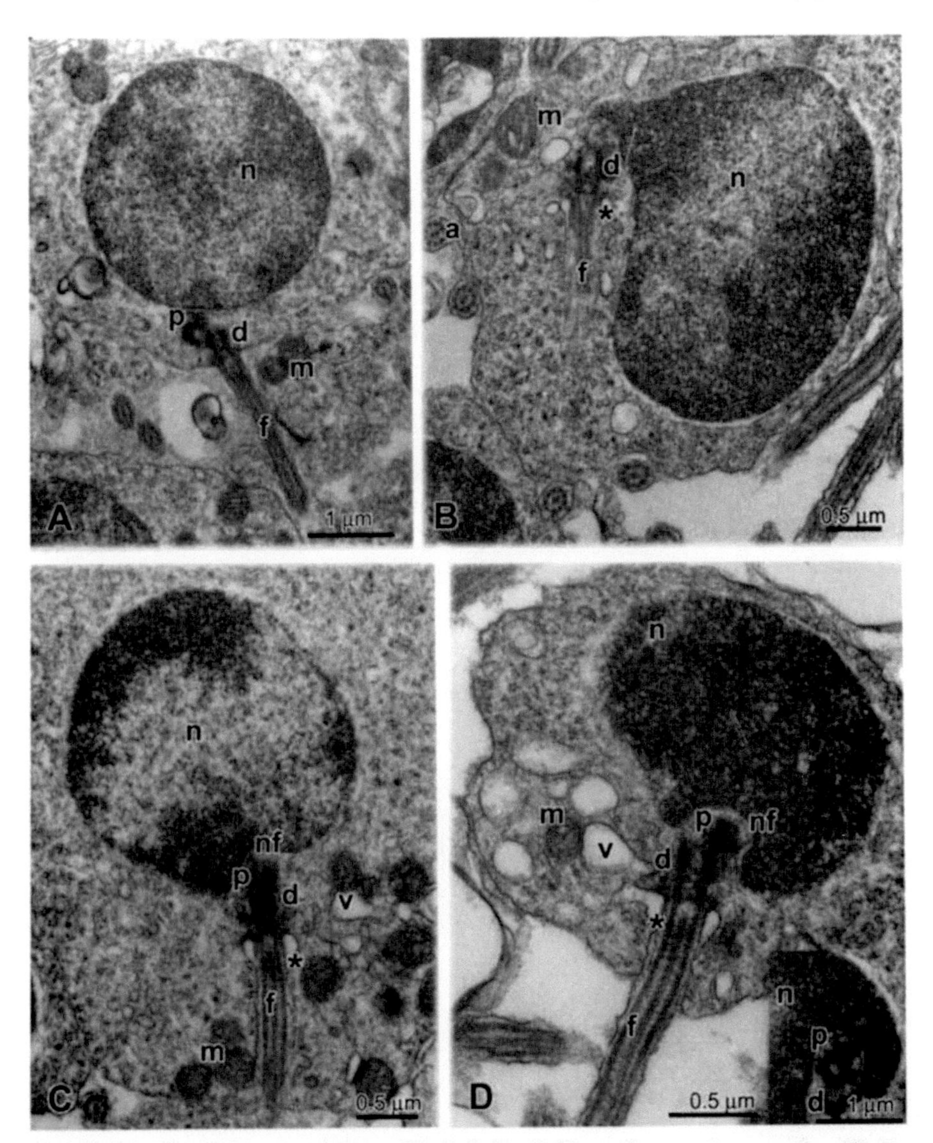

Fig. 11.17 *Erythrinus erythrinus* (Erythrinidae). Type I spermiogenesis with the cytoplasmic canal closing prior to spermiation. **A-D**. Longitudinal sections of the earliest, early, mid and late spermatids, respectively. Note the rotation of the nucleus between **B** and **C**. Closure of the cytoplasmic canal is evident in **C** and **D**. **D inset**. Centriolar complex. a, axoneme; d, distal centriole; f, flagellum; m, mitochondrion; n, nucleus; nf, nuclear fossa; p, proximal centriole; v, vesicle; asterisk, cytoplasmic canal. Original.

Abundant elongate vesicles are present in *N. marginatus* and *N. unifasciatus*, but absent in *N. digrammus*. A very shallow posterior depression in *N. digrammus* may be a very short cytoplasmic canal. No cytoplasmic canal is

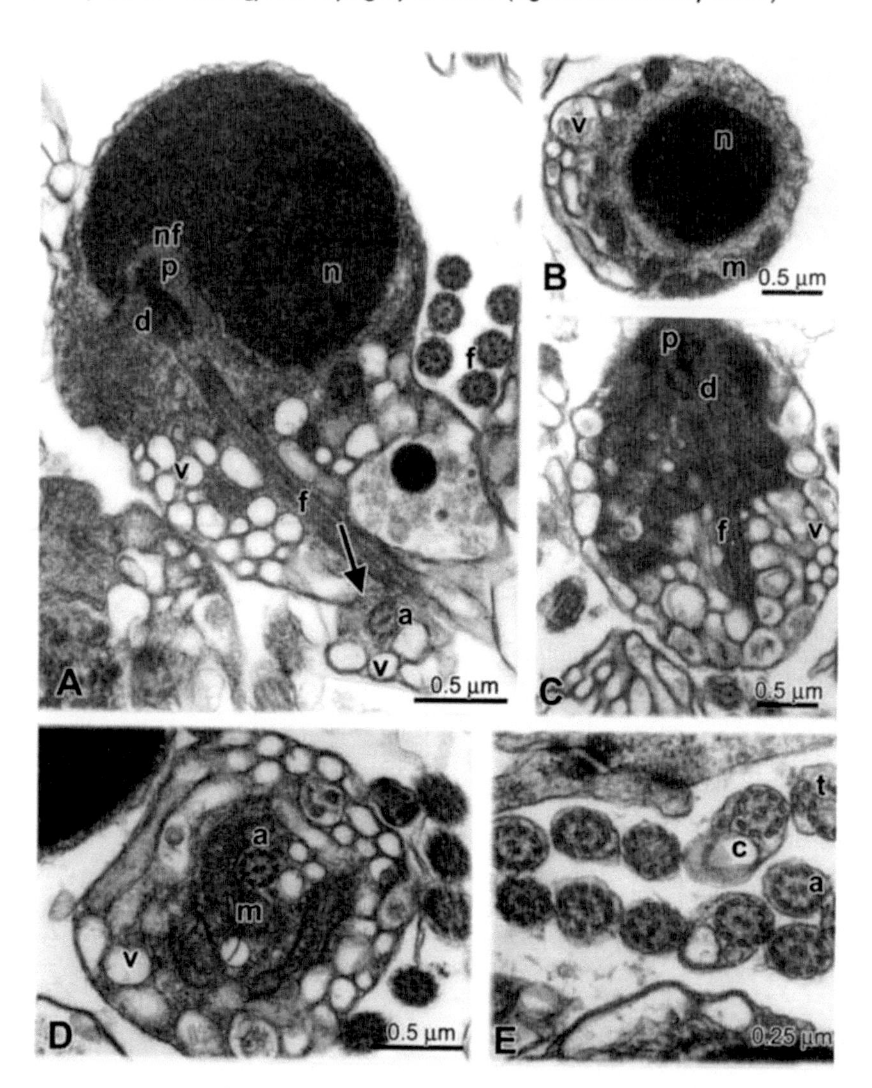

Fig. 11.18 *Erythrinus erythrinus* (Erythrinidae). Type I aquasperm. **A**. Longitudinal section of a spermatozoon. Note the eccentric position of the flagellum relative of the nucleus, the absence of a cytoplasmic canal, and the abundant vesicles. **B**. Transverse section of the posterior end of the nucleus. Note the distribution of the mitochondria. **C**. Longitudinal section along the side of the spermatozoon. **D**. Transverse section through the anterior third of the midpiece. Note the long branched mitochondria. **E**. Transverse sections of flagella. Note the vesicular compartment in this segment of the flagellum just posterior to the midpiece and the absence of the central microtubules in terminal end of the axoneme. a, axoneme; c, vesicular compartment; d, distal centriole; f, flagellum; m, mitochondrion; n, nucleus; nf, nuclear fossa; p, proximal centriole; v, vesicle; t, terminal end of axoneme; arrow, axoneme. Original.

present in *N. marginatus*, but a short canal alongside the nucleus is found in *N. unifasciatus*. The single flagellum contains a classic 9+2 axoneme and no fins. All microtubules are electron lucent in *N. digrammus* and *N. unifasciatus*, whereas in *N. marginatus* some sections through the flagellum show intratubular differentiation with the A-tubules of the doublets being electron-dense. An elaborate system of elongate vesicles is present in the flagellum *N. digrammus*. Several elongate vesicles are seen in *N. marginatus* but none is observed in *N. unifasciatus*. It is unusual to see such a high degree of variability in sperm ultrastructure within a single genus.

For the subfamily Pyrrulininae, an original description is presented below for *Pyrrulina australis*.

***Pyrrulina australis*:** Fig. 11.19 demonstrates a process similar to Type II spermiogenesis *sensu* Mattei (1970) in this species. The centriolar complex in the early spermatid in located lateral to the nucleus (Fig. 11.19A) and remains in this position through sperm development, a nuclear fossa being absent (Fig. 11.19H). The spermatozoon of *Pyrrulina australis* resembles a Type II aquasperm with a spherical nucleus (1.6 μm in diameter) containing highly condensed, homogeneous granular chromatin and no nuclear fossa (Fig. 11.20A). An electron-lucent region is also evident in the center of the nucleus (Fig. 11.20A). The centriolar complex lies to one side of the nucleus. The proximal centriole is lateral and oblique to the distal (Figs. 11.19A,E,F, 11.20B). The highly asymmetric midpiece is not well defined, with several spherical to ovoid mitochondria distributed throughout. The initial segment of the flagellum is contained within a cytoplasmic canal. Unique to this species is the presence of abundant spherical, strongly electron-dense bodies and several small interconnected vesicles arranged in a reticular pattern in the posterior half of the midpiece (Fig. 11.19B-H,I). The single, lateral flagellum contains a classic 9+2 axoneme and no fins (Fig. 11.20H-I).

CHARACOIDEA

11.6.2.13 Family Gasteropelecidae and *Triportheus* (genus *incertae sedis* in Characidae)

As the spermatozoa of *Carnegiella strigata*, described by Verissimo-Silveira (2007), and *Triportheus paranensis*, described by Gusmão-Pompiani (2003), share a number of characters, they will be described together. Although gasteropelecids and Triportheus both have expanded coracoid bones (Toledo-Piza 2000), Weitzman (1954) hypothesizes that these two taxa acquired these characters independently. The relationships between these two groups are currently unresolved. Spermatozoa in both species are Type I aquasperm with antero-posteriorly compressed ovoid nuclei containing highly condensed, homogeneous granular chromatin and deep nuclear fossae, which branch in *T. paranensis*. The maximum nuclear dimension is 1.7 μm in *C. strigata* and 2.1

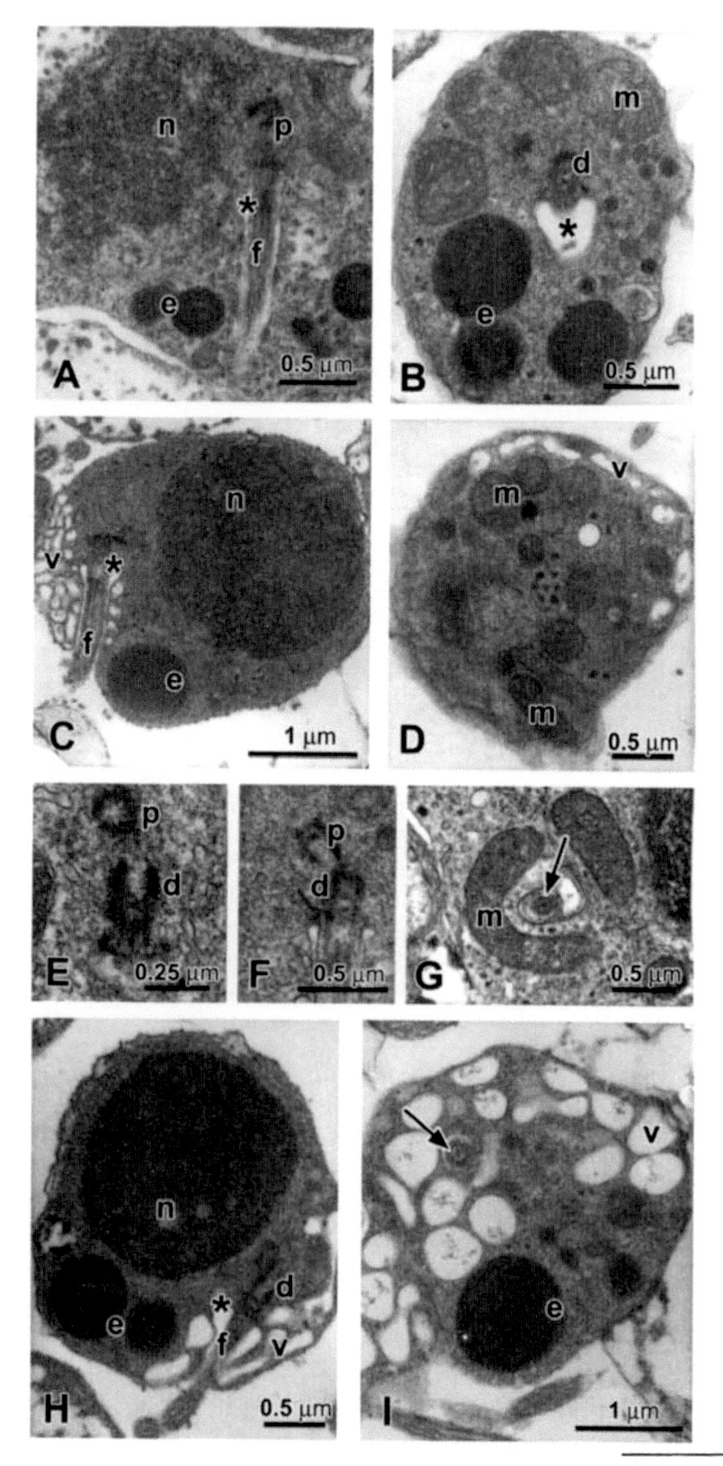

Fig. 11.19 Contd. ...

μm in *T. paranensis*. The proximal centriole is anterior and perpendicular to the distal and both are located within the nuclear fossa. Electron-dense material is associated with both centrioles. The midpiece contains several spherical to elongate mitochondria. In *T. paranensis,* the long cytoplasmic canal that formed during the early stages of spermiogenesis closes in the late spermatids, whereas it remains open in the spermatozoon of *C. strigata*. Abundant vesicles containing some stained material are present in the midpiece. The *T. paranensis* midpiece also contains a well developed tubular-vesicular system that is lacking in *C. strigata*. The cytoplasmic canal in *C. strigata* is surrounded by concentric membranous rings. The single flagellum contains a classical 9+2 axoneme with all microtubules electron-lucent, no flagellar fins. In *C. strigata,* abundant extracellular stained material surrounds the spermatozoa as well as being present within the cytoplasmic canals.

11.6.2.14 Family Characidae

The spermatozoa of over 25 externally fertilizing species and more than 20 inseminating species of the family Characidae have been analyzed to date. The nine subfamilies that constitute Characidae appear as three polytomies in the classification proposed by Malabarba and Weitzman (2003). Information on sperm ultrastructure is still lacking for the subfamilies Iguanodectinae and Rhoadsiinae. The phylogenetic relationships of over 629 species in 88 genera are currently unresolved and these are reported as *incertae sedis* in the family (Lima *et al.* 2003). An original description is here provided for *Astyanax fasciatus*. Characidae is the only family in Characiformes that contains species that engage in the unusual reproductive habit of insemination, where the male is able to transfer sperm to the female reproductive tract (Burns and Weitzman 2005). The majority of inseminating species thus far analyzed produce highly modified introsperm. Original descriptions will be provided for three inseminating species, one in the genus *Chrysobrycon* and two in *Gephyrocharax* (subfamily Stevardiinae). The ultrastructure of the spermatozoa of the externally fertilizing and inseminating species will be treated separately below.

Fig. 11.19 Contd. ...

Fig. 11.19 *Pyrrulina australis* (Lebiasinidae, Pyrrulininae). Spermiogenesis resembling that of Type II aquasperm *sensu* Mattei (1970). **A, C, H**. Longitudinal sections of early, mid and late spermatids, respectively. Note the lateral position of the flagellum. **B, D, I**. Transverse sections through the midpiece of early, mid and late spermatids, respectively. **E, F**. Centriolar complex. **G**. Mitochondria. a, axoneme; d, distal centriole; e, spherical electron dense material; f, flagellum; m, mitochondrion; n, nucleus; p, proximal centriole; v, vesicle; arrow, axoneme; asterisk, cytoplasmic canal. Original.

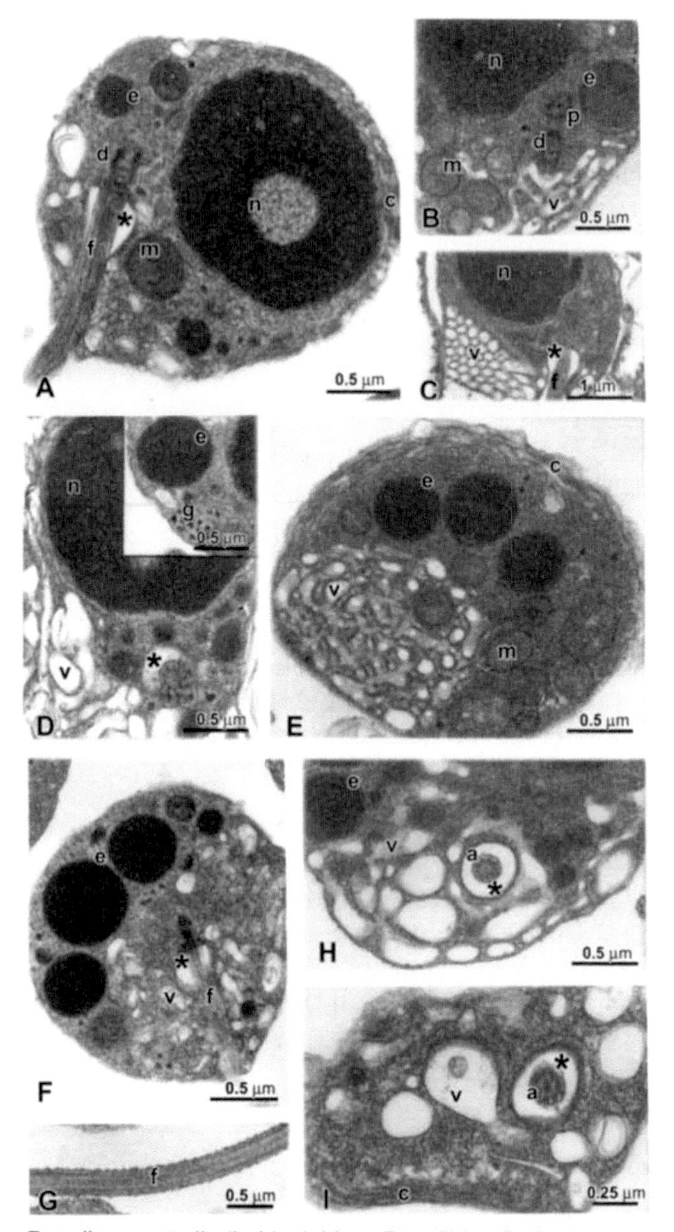

Fig. 11.20 *Pyrrulina australis* (Lebiasinidae, Pyrrulininae). Aquasperm resembling Mattei's (1970) Type II. **A, C**. Longitudinal sections of spermatozoa. Note the lateral position of the flagellum and the tubule vesicular compartment. **B**. Centriolar complex. **D, E, F, H, I**. Transverse sections through the midpiece. **G**. Longitudinal section of flagellum. a, axoneme; c, concentric cisternae; d, distal centriole; e, spherical electron dense material; f, flagellum; g, granules; m, mitochondria; n, nucleus; p, proximal centriole; v, vesicle; asterisk, cytoplasmic canal. Original.

A. Externally fertilizing Characidae

The spermatozoa of many species of externally fertilizing Characidae have been examined ultrastructurally. The species investigated are too numerous to be individually described but are listed here under their subfamilies. A brief overview of their ultrastructure is provided below and some species are exemplified by illustrated descriptions.

Serrasalminae: *Metynnis maculatus* (Matos *et al.* 1993a); *Metynnis mola, Myleus duriventri(for M. levi)* and *Serrasalmus maculatus (= spilopleura)* (Gusmão-Pompiani 2003); *Piaractus mesopotamicus* (Cruz-Landim *et al.* 2003; Gusmão-Pompiani 2003).

Bryconinae: *Brycon cephalus* (Romagosa *et al.* 1999); *B. hilarii* (Zaiden 2000); *B. orbignyanus* (Aires *et al.* 2000; Gusmão-Pompiani 2003;Veríssimo-Silveira *et al.* 2006); *B. microlepis* (Veríssimo-Silveira *et al.* 2006).

Characinae: *Galeocharax knerii* (Magalhães *et al.* 2004; Gusmão-Pompiani 2003); *G. humeralis* and *Roeboides bonariensis* (Gusmão-Pompiani 2003).

Stethaprioninae: *Poptella paraguayensis* (Gusmão-Pompiani 2003).

Aphyocharacinae: *Aphyocharax anisitsi* (Burns *et al.* 1998; Gusmão-Pompiani 2003).

Cheirodontinae: *Serrapinnus kriegi* (Burns *et al.* 1998; Burns and Weitzman 2005); *Cheirodon interruptus, Serrapinnus calliurus* and *S. heterodon* (Oliveira 2007).

Tetragonopterinae: *Tetragonopterus argenteus* (Gusmão-Pompiani 2003).
Several species in genera which are *incertae sedis* in the Characidae have also been examined for sperm ultrastructure: *Astyanax scabripinnis* (Vicentini 2002); *Astyanax fasciatus* (original description); *Bryconamericus stramineus* (Gusmão-Pompiani 2003); *Bryconops affinis* (Andrade *et al.* 2001; Gusmão-Pompiani 2003); *Hemigrammus erythrozonus* (Pecio et al. 2007); *Hyphessobrycon eques* (Gusmão-Pompiani 2003); *Moenkhausia sanctafilomenae* (Gusmão-Pompiani 2003); *Odontostilbe pequira* (Oliveira 2007); *Paracheirodon innesi* (Jamieson 1991); *Salminus maxillosus* (Veríssimo-Silveira *et al.* 2006).

Overview: With the exception of *Bryconamericus stramineus* that has an aquasperm that resembles a form intermediate between types I and II, the externally fertilizing characids studied to date produce Type I aquasperm. Nuclei vary from spherical to slightly elongate, with maximum dimensions ranging from 1.5 to 2.1 μm. Chromatin is condensed into dense flocculent masses in the Serrasalminae, Bryconinae and *Salminus maxillosus*, whereas in most other characid taxa it is highly condensed, homogeneous and granular. Most species have a single, posterior nuclear fossa that is deep in Serrasalminae, Bryconinae and Characinae, relatively shallow in Stethaprioninae, Tetragonopterinae and *Hyphessobrycon eques*, and very shallow or lacking in Aphyocharacinae and Cheirodontinae. Two posterior shallow fossae are present in *Bryconops affinis* and *Moenkausia sanctafilomenae*, whereas in *Bryconamericus straminues* two shallow fossae are located laterally on the nucleus. In all species studied, the proximal centriole is anterior to the distal. It can be directly anterior or slightly lateral, and vary from perpendicular to oblique in orientation. However, in *Serrapinnus kriegi* (Fig. 11.22B) the centrioles are parallel to one another. In species with deep nuclear fossae, the proximal centriole and all or part of the distal centriole are located within the fossa, those with relatively shallow fossae have only the proximal located within it and those with very shallow fossae encompass only part of the proximal centriole. Some species have electron-dense material associated with the centrioles. The main part of the midpiece tends to be relatively short. In all cases the proximal portion of the flagellum is located within a cytoplasmic canal that in several species may continue posteriorly as a thin sleeve. The cytoplasmic canal in Bryconinae is surrounded by several concentric membranous rings. Mitochondria tend to be few in number and vary in shape from spherical to elongate. Two of the most variable characters are the number and morphology of vesicles. For example, within the Serrasalminae, some species nearly lack vesicles, whereas in others they are abundant, often appearing as large vacuoles. Species in Characinae and Cheirodontinae and *Bryconamericus straminues* have well developed tubular-vesicular systems similar to that initially described in *Citharinus* (Mattei *et al.* 1995). Similar systems may also be present in Aphyocharacinae, Tetragonopterinae, *Bryconops affinis* and *Moenkausia sanctafilomenae* but more data are needed to confirm this. All species have a single flagellum containing a classic 9+2 axoneme with all microtubules electron-lucent. Two fins are present on the flagellum of the serrasalmin genera *Serrasalmus* and *Piaractus*, with two to three on *Mylosoma*. All other species lack flagellar fins. In some species, vesicles may continue into the proximal portion of the flagellum.

Examples from three subfamilies and *insertae sedis* follow.

Serrasalminae. *Piaractus mesopotamicus*: The spermatozoon of *P. mesopotamicus* (Cruz-Landim *et al.* 2003; Gusmão-Pompiani 2003) is a Type I aquasperm with a conical nucleus (1.8 µm in length) containing condensed flocculent chromatin and a deep nuclear fossa (Fig. 11.21A,B). The proximal centriole is anterior, perpendicular and slightly lateral to the distal and both are located inside the nuclear fossa (Fig. 11.21B). The midpiece, which contains several elongate branched, C-shaped mitochondria (Fig. 11.21D), tapers posteriorly to a cytoplasmic sleeve (Fig. 11.21A). The initial segment of the flagellum is contained within a cytoplasmic canal (Fig. 11.21A,D,E). The single flagellum contains a classic 9+2 axoneme with all microtubules electron-lucent, and two fins (Fig. 11.21C).

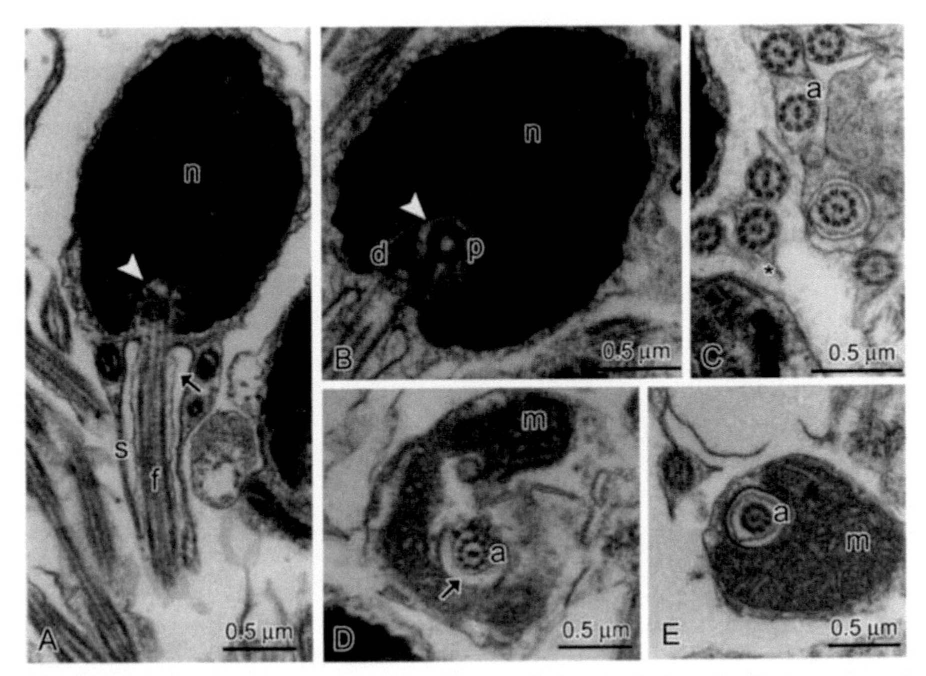

Fig. 11.21 *Piaractus mesopotamicus* (Serrasalminae). **A.** Longitudinal section of spermatozoon. Note the conical nucleus with condensed flocculent chromatin and the cytoplasmic sleeve at the end of the midpiece. **B.** Longitudinal section of the nucleus and midpiece with both centrioles located within the nuclear fossa. **C.** Transverse sections of flagella with short fins. **D-E.** Transverse sections at different levels of the midpiece. Note the elongate, branched, C-shaped mitochondrion surrounding the cytoplasmic canal. a, axoneme; d, distal centriole; f, flagellum; m, mitochondrion; n, nucleus; p, proximal centriole; s, cytoplasmic sleeve; arrowhead, nuclear fossa; arrow, cytoplasmic canal. asterisk, flagellar fins. Original, courtesy of Priscila Gusmão-Pompiani).

Aphyocharacinae. ***Aphyocharax anisitsi*:** The spermatozoon of *A. anisitsi* is a Type I aquasperm, having a spherical nucleus, approximately 1.6 µm in diameter, and short cytoplasmic collar with the flagellum initially contained within a cytoplasmic canal (Fig. 11.22A). A portion of the proximal centriole is located within a shallow nuclear fossa. The nucleus is electron-dense indicating marked condensation of the chromatin. Mitochondria are located posterior to the nucleus. The axoneme of the flagellum has the classic 9+2 arrangement (Burns *et al.* 1998). Within the posterior midpiece, abundant vesicles are present (Fig. 11.22A). Whether or not these vesicles are equivalent to the tubular-vesicular system first described in *Citharinus* (Mattei *et al.* 1995) remains to be determined.

Cheirodontinae. ***Serrapinnus kriegi*:** The spermatozoon of *Serrapinnus kriegi* is a Type I aquasperm with a spherical nucleus, approximately 1.5 µm in diameter, containing highly condensed granular chromatin (Fig. 11.22B). The centrioles are parallel to one another and both are located posterior to the nucleus; no fossa is present. The anterior midpiece contains several spherical mitochondria, whereas the posterior midpiece is comprised of a well-developed tubular-vesicular system (Fig. 11.22B) that is essentially identical to that described in *Citharinus* (Mattei *et al.* 1995). The flagellum lacks fins.

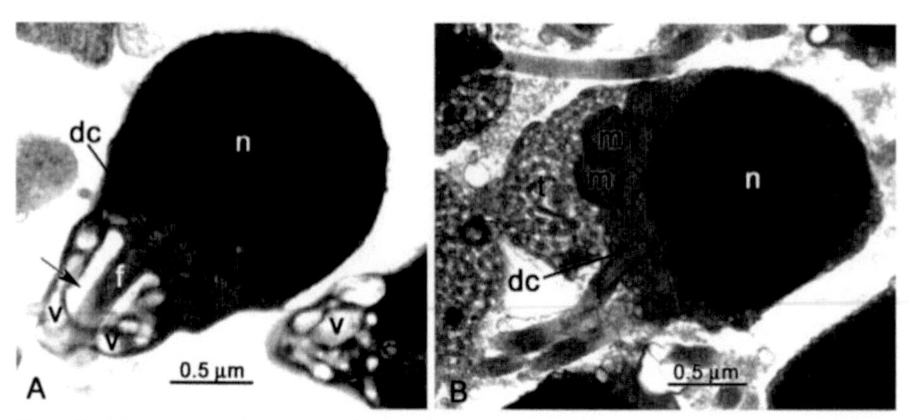

Fig. 11.22 Longitudinal sections of spermatozoa. **A**. *Aphyocharax anisitsi* (Characidae, Aphyocharacinae). SL 33.0 mm, Brazil, MCP 18583. **B**. *Serrapinnus kriegi* (Characidae, Cheirodontinae, Cheirodontini). SL 24.2 mm, pet shop specimen. dc, distal centriole, f, flagellum, m, mitochondrion, n, nucleus, t, tubular-vesicular system, v, vesicle, arrow, cytoplasmic canal. Original.

Insertae sedis

The spermatozoa of *Paracheirodon innesi* and *Astyanax fasciatus* are described here.

***Paracheirodon innesi*:** There is no acrosome but some sperm (Fig. 11.23A) have small vesicles anterior to the nucleus. The nucleus is an electron-dense sphere, 2 µm wide, which may show some small pale lacunae, and has two contiguous small embayments, one for the proximal, the other for the distal centriole. Only the tips of the centrioles lie within these fossae (Fig. 11.23A). The cytoplasmic canal and corresponding collar are long, commencing shortly behind the nucleus and extending for about 2.8 µm. Mitochondria are confined to the anterior half of the collar and the posterior half consists of a thin sleeve consisting of little more than two apposed plasma membranes (Fig. 11.23A, B). The two membranes are occasionally separated by large vesicles. The thick part of the collar consists of a wide zone of cytoplasm around the cytoplasmic canal and axoneme (Fig. 11.23C) in which are embedded, in transverse section, several simple cristate mitochondria in an irregular layer. Various inclusions are scattered amongst and posterior to them. The basal body (distal centriole) and flagellum are at a slight angle to the apparent anteroposterior axis of the nucleus. The proximal centriole is anteriolateral and almost parallel to the basal body, the long axes of the two being slightly convergent anteriad (Fig. 11.23A). The lumina of the microtubules of the 9+2 axoneme are electron-lucent (Fig. 11.23B).

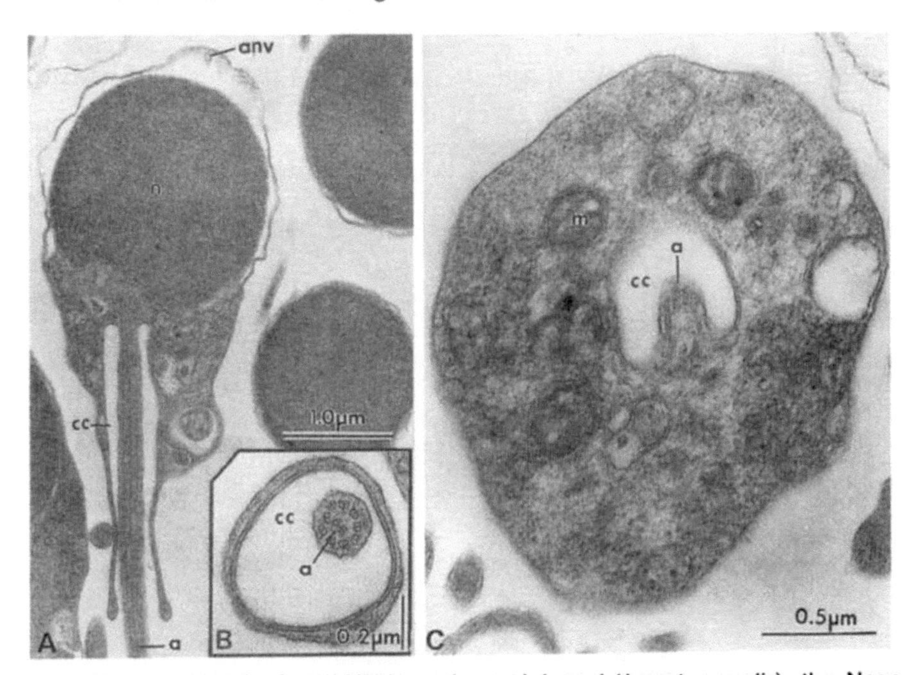

Fig. 11.23 *Paracheirodon (=Hyphessobrycon) innesi* (*Incertae sedis*), the Neon Tetra. **A**. Longitudinal section of the spermatozoon. **B**. Transverse section of the cytoplasmic canal and axoneme. **C**. Transverse section of the midpiece at beginning of the cytoplasmic canal. a. axoneme. anv. anterior vesicles. cc. cytoplasmic collar. m. mitochondrion. n. nucleus. From Jamieson, B. G. M. 1991. *Fish Evolution and Systematics: Evidence from Spermatozoa*. Cambridge University Press, Cambridge, Fig. 12.11.

Astyanax fasciatus (Clarianna Martins Baicere-Silva, personal information). *Astyanax*, one of the most speciose genera in the Characidae, is reported as *incertae sedis* in the family (Lima *et al.* 2003). The spermatozoon of *A. fasciatus* is a Type I aquasperm with a spherical nucleus containing highly condensed, homogeneous granular chromatin and a very shallow and double nuclear fossa (Fig. 11.24A-C). The proximal centriole is slightly lateral of the distal and only part of the proximal is within the fossae. The relatively large midpiece contains several spherical to elongate mitochondria and a system of elongate, interconnected vesicles, some of which fuse with the plasma membrane. The single flagellum contains a classical 9+2 axoneme with all microtubules electron lucent, and no fins (Fig. 11.24B, G-I). In *A. fasciatus*, abundant extracellular stained material surrounds the spermatozoon as well as being present within the cytoplasmic canal and the midpiece vesicles.

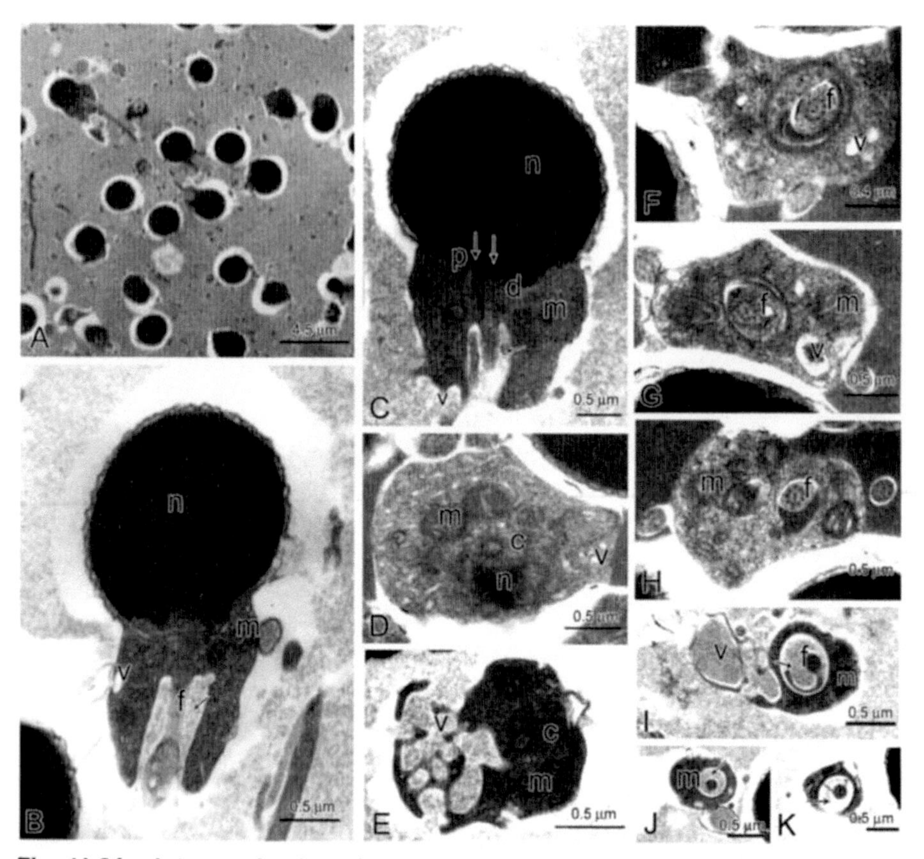

Fig. 11.24 *Astyanax fasciatus* (genus *incertae sedis* in the family Characidae). **A.** Note the abundant extracellular stained material that surrounds the spermatozoa in testicular lumen. **B-C.** Longitudinal sections of spermatozoa. Note the extracellular stained material within the cytoplasmic canal. **D-K.** Transverse sections at different levels of the midpiece. Note the spherical to elongate mitochondria concentrated in the two anterior thirds of the midpiece, whereas the vesicles which contain stained material are located in its posterior third. c. centrioles. d, distal centriole; f, flagellum; m, mitochondrion; n, nucleus; p, proximal centriole; v, vesicle; double arrows, nuclear fossa; arrow, cytoplasmic canal. Original, courtesy of Clarianna Martins Baicere-Silva.

B. Inseminating Characidae

Glandulocaudinae

Examined species include *Mimagoniates barberi* (Pecio and Rafiñski 1994, 1999; Burns *et al.* 1998); *Mimagoniates microlepis* (Burns *et al.* 1998).

***Mimagoniates microlepis* and *M. barberi*:** The ultrastructure of the spermatozoa of both species of *Mimagoniates* is essentially identical. Using light microscopy, mean length estimates of the greatly elongate nuclei were

13.8 µm for *M. microlepis* and 17.1 µm for *M. barberi* (Burns *et al.* 1995). The nucleus is electron-dense, and the elongate mitochondrial region (midpiece) extends along one side of the nucleus (Fig. 11.25B-D). In transverse section, the nucleus is greatly flattened at the anterior end of the cell (Fig. 11.23B). More posteriorly the nucleus initially becomes oval in section (Fig. 11.23C) and eventually flattened (Fig. 11.25D). The centrioles, which are perpendicular to one another and not contained within a nuclear fossa, are located at the anterior (advancing) end of the cell and the lumina of all microtubule triplets are electron-lucent and clearly defined (Fig. 11.25A). Most of the elongate cytoplasmic collar with its contained flagellum, present in the developing spermatid, degenerates prior to spermiation leaving only a short cytoplasmic collar in the spermatozoon (Fig. 11.25B) which terminates near the anterior end of the cell, thus the flagellum is free along most of its length. Cristate mitochondria are initially located to one side of the nucleus (Fig. 11.25D) and extend posterior to the nuclear region (Fig. 11.25). The axoneme of the flagellum has the typical 9+2 arrangement of microtubules. In the region of the cytoplasmic collar the lumina of all tubules are electron-lucent, whereas in most of the free flagella the lumina of the A tubules of the peripheral doublets are electron-dense (Fig. 11.25B). Throughout the entire centriolar, nuclear and mitochondrial regions of the cell, numerous microtubules, orientated parallel to the long axis of the cell, are present at one side of the cell (Fig. 11.25A-E).

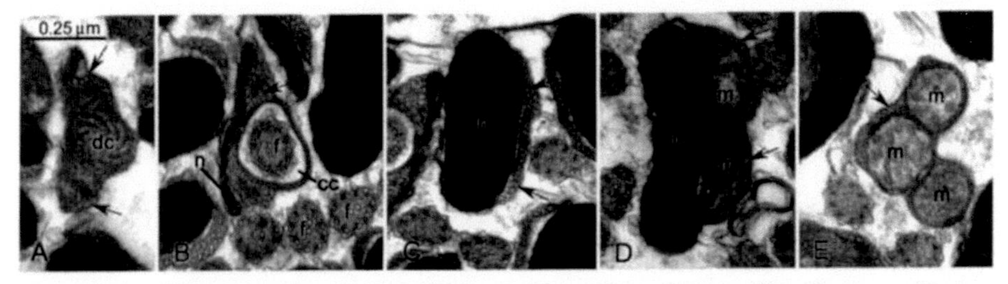

Fig. 11.25 *Mimagoniates barberi* (Characidae, Glandulocaudinae) spermatozoa. SL 33.7 mm, aquarium specimen from stock originally collected in Paraguay. **A-E**. Successively more posterior transverse sections. cc, cytoplasmic canal; dc, distal centriole; f, flagellum; m, mitochondrion; n, nucleus; arrow, accessory microtubule. Original.

Tribe Diapomini

Examined diapomins are *Diapoma speculiferum* and *Diapoma* sp. (Burns *et al.* 1998). Both species have essentially identical sperm ultrastructure.

Diapoma speculiferum: A longitudinal section clearly shows the elongation of the nucleus of the spermatozoon of *D. speculiferum* (Fig. 11.26A), which reaches approximately 4.3 µm in length. During spermiogenesis the nucleus may begin to rotate slightly but further rotation appears to be inhibited. Thus, spermiogenesis may be a highly modified Type I. Since both centrioles are localized to one side of the nucleus near its anterior end, nuclear elongation appears to be in the posterior direction. The chromatin is highly condensed (electron-dense).The microtubules of the centrioles are poorly resolved in the micrographs (Fig. 11.26A), but electron-dense material can be found between the centrioles. Multiple cristate mitochondria that are initially located along one side of the elongate nucleus extend posterior to the nucleus (Fig. 11.26B). In both the nuclear and mitochondrial regions the cytoplasmic collar, with the contained flagellum, remains attached to the cell and continues posterior to the mitochondrial area for some distance before terminating, beyond which point the flagellum is free. An electron-dense ring encircles the entrance to the cytoplasmic collar. The axoneme has the typical 9+2 arrangement of microtubules, and the lumen of the A-tubule of each peripheral doublet is electron-dense (Burns *et al.* 1998).

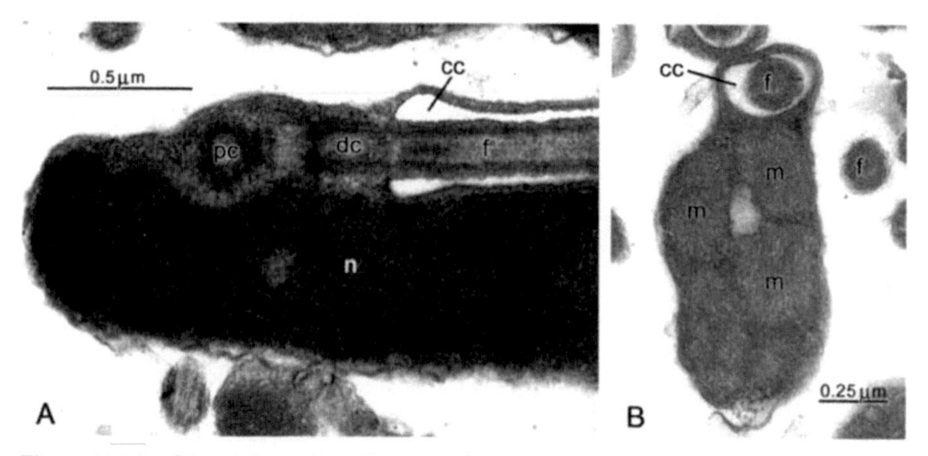

Fig. 11.26 *Diapoma speculiferum* (Characidae, Stevardiinae, Diapomini) spermatozoa. SL 39.0 mm, MCP 18498, Brazil. **A**. Longitudinal section through anterior of cell. **B**. Transverse section through mitochondrial region. cc, cytoplasmic canal, dc, distal centriole, f, flagellum, m, mitochondrion, n, nucleus, pc, proximal centriole. Original.

Tribe Stevardiini

Examined stervardiins are *Corynopoma riisei* (Burns *et al.* 1998; Pecio *et al.* 2007); *Gephyrocharax atricaudatus* and *G. intermedius* (original descriptions).

Corynopoma riisei: The nucleus of the spermatozoon of *C. riisei*, approximately 8.4 µm in length and 1.5 µm in maximal width, appears to elongate posteriorly in a manner similar to that seen in the Diapomini. This cell may also be a highly modified Type I spermatozoon that undergoes only partial nuclear rotation during spermiogenesis. Although the chromatin in the study of Burns *et al.* (1998) appears uniformly electron-dense, the more recent analysis of Pecio *et al.* (2007) demonstrated that the chromatin retains a flocculent pattern in the spermatozoon. In frontal section the nucleus appears spatulate (Fig. 11.27A), whereas in sagittal section it appears club-shaped, being narrow at the anterior (advancing) end and widening at the middle and posterior regions (Fig. 11.27B). In transverse sections the nucleus, which is flattened at the anterior end, becomes increasingly more rounded toward the posterior end (Figs. 11.28A-D). Part of the proximal centriole, the entire distal centriole and the proximal part of the flagellum are housed within a lateral nuclear fossa (Figs. 11.27A, 11.28B). The centrioles are approximately perpendicular to one another and their microtubules are poorly resolved in the micrographs (Fig. 11.27A). The multiple, elongate cristate mitochondria are initially localized to one side of the nucleus (Figs. 11.28C-D) and then extend posterior to the nucleus (Fig. 11.26E). Beyond this point the flagellum continues to be located within the cytoplasmic canal of a thin cytoplasmic sleeve (Fig. 11.26F) before exiting the canal (Figs. 11.27E, 11.28G). Some glycogen rosettes were observed among the mitochondria. The flagellum, which lacks fins, has the typical 9+2 axoneme with all tubules electron-lucent.

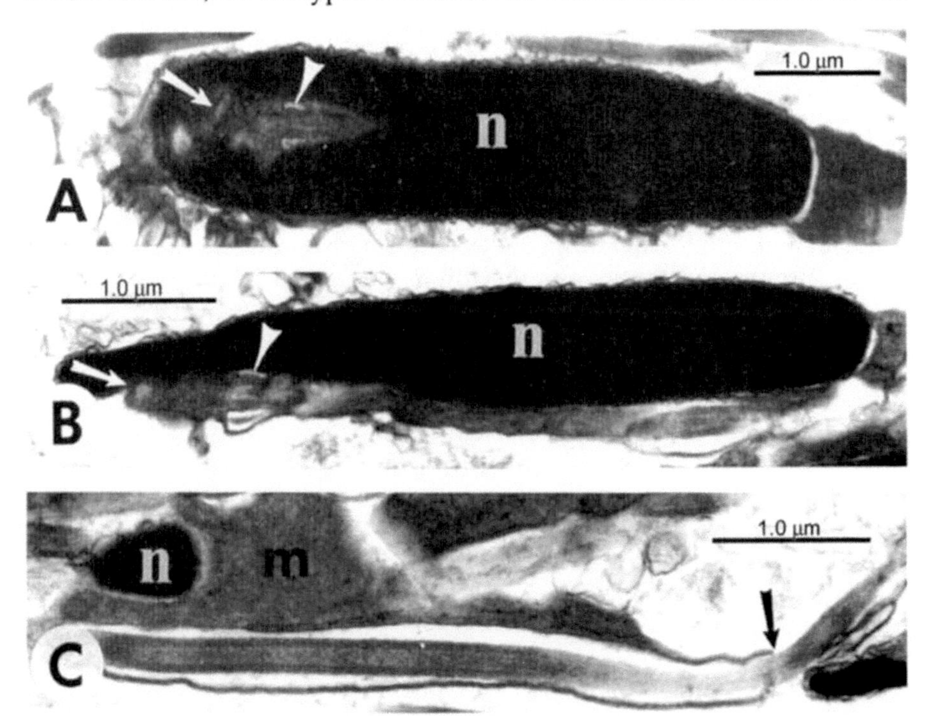

Fig. 11.27 *Corynopoma riisei* (Characidae, Stevardiinae, Stevardiini) spermatozoa, longitudinal sections. **A**. Frontal section through nucleus. **B**. Sagittal section through nucleus. **C**. Section immediately posterior to nucleus. m, mitochondrion; n, nucleus; white arrow, proximal centriole; black arrow, point of exit of flagellum from cytoplasmic canal; arrowhead, cytoplasmic canal. Modified from Burns, J. R., Weitzman, S. H., Lange, K. R. and Malabarba, L. R. 1998. Sperm ultrastructure in characid fishes (Teleostei, Ostariophysi). Pp. 235-244. In L. R. Malabarba, R. E. Reis, R. P. Vari, Z. M. S. Lucena and C. A. S. Lucena (eds), *Phylogeny and Classification of Neotropical Fishes*. Edipucrs, Porto Alegre, Fig. 5.

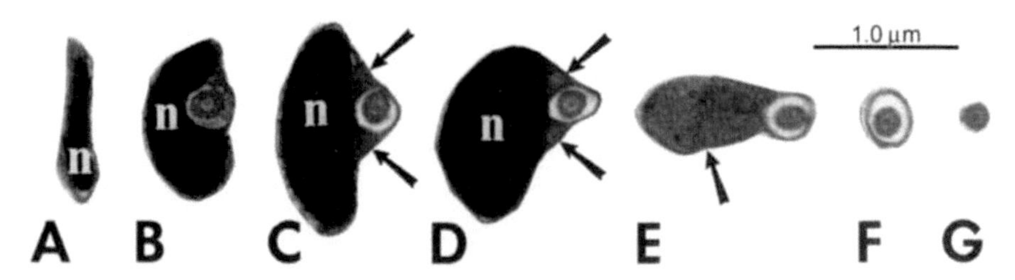

Fig. 11.28 *Corynopoma riisei* (Characidae, Stevardiinae, Stevardiini) spermatozoa, successively more posterior transverse sections (left to right). **A**. Anteriormost section showing flattened nucleus. **B**. Section showing distal centriole within nuclear fossa. **C** and **D**. Sections through nuclear area showing flagellum within cytoplasmic canal. **E**. Mitochondrial region posterior to nucleus. **F**. Posterior section lacking mitochondria but flagellum still within cytoplasmic canal. **G**. Free flagellum. n, nucleus; arrow, mitochondrion. Modified from Burns, J. R., Weitzman, S. H., Lange, K. R. and Malabarba, L. R. 1998. Pp. 235-244. In L. R. Malabarba, R. E. Reis, R. P. Vari, Z. M. S. Lucena and C. A. S. Lucena (eds), *Phylogeny and Classification of Neotropical Fishes*. Edipucrs, Porto Alegre, Fig. 6.

***Gephyrocharax atricaudatus* and *G. intermedius*:** Sperm ultrastructure of these two species is essentially identical. Due to the poor fixation of the *G. intermedius* testis, only micrographs of *G. atricaudatus* are presented here. The nuclei in both species measure 4.5-4.6 µm in length and 1.3-1.4 µm in maximal width and the chromatin, although highly condensed into large masses, remains flocculent. Along most of its length the nucleus is oval in transverse section (Fig. 11.29B). A lateral nuclear fossa houses the proximal centriole (Fig. 11.29A). The proximal centriole is approximately perpendicular to the distal. Elongate mitochondria that have an electron-dense matrix are located to one side of the nucleus beginning in the region of the nuclear fossa and extending posterior to the nucleus (Fig. 11.29A-B). The flagellum, which is initially contained within an elongate cytoplasmic canal that remains attached to the cell along the nuclear and mitochondrial regions and more posteriorly, exits the canal at its posterior terminus (Fig. 11.29A-B). The flagellum has a typical 9+2 axoneme with all tubules electron-lucent and lacks fins. The abundant stained secretion seen among the spermatozoa in the *G. atricaudatus* testis (Fig. 11.29) was absent in the testis of *G. intermedius* perhaps reflecting different stages of a reproductive cyclicity.

Tribe Hysteronotini

The only hysteronotin examined to date is *Pseudocorynopoma doriae* (Burns *et al.* 1998).

[1] Institutional abbreviations used in several figure legends in this section are: MCP, Museu de Ciências y Tecnologia, Pontifícia Universidade Católica do Rio Grande do Sul, Porto Alegre, Brazil, and USNM, National Museum of Natural History (formerly United States National Museum), Smithsonian Institution, Washington, DC, USA.

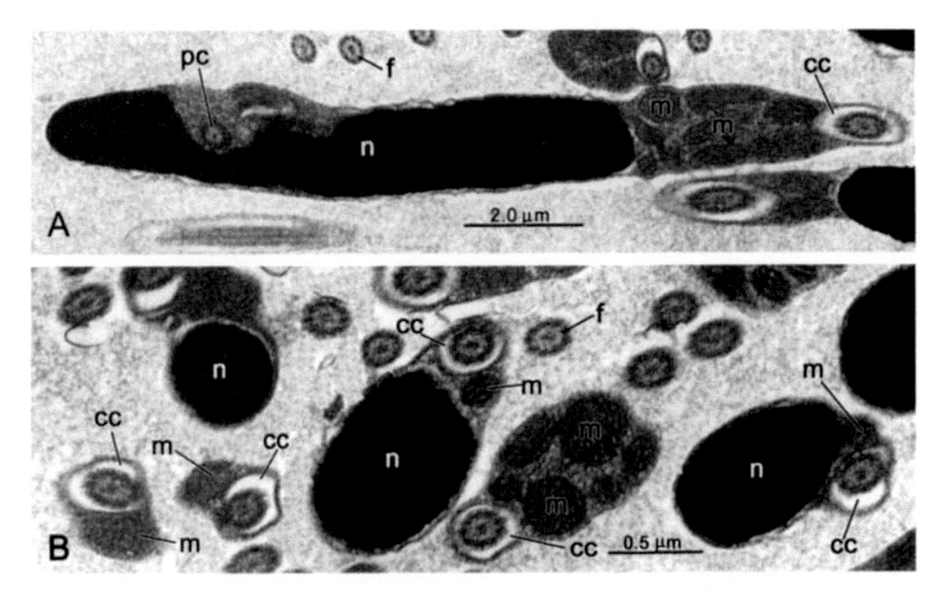

Fig. 11.29 *Gephyrocharax atricaudatus* (Characidae, Stevardiinae, Stevardiini) spermatozoa. SL 36.5 mm, USNM 348759[1], Panama. **A**. Longitudinal section. **B**. Transverse sections. Note the abundant secretion among the cells. cc, cytoplasmic canal; f, flagellum; m, mitochondrion; n, nucleus; pc, proximal centriole. Original.

Pseudocorynopoma doriae: The overall morphology of the spermatozoon of *P. doriae* is similar to that seen in diapomins and stevardiins. The nuclei, which may represent the longest for any teleost species thus far investigated, are so elongate that length measurements were not possible with TEM. Based on measurements from light microscopy, nuclear length averaged 31.6 µm (Burns *et al.* 1995), while from TEM width was approximately 0.9 µm. Nuclear chromatin is flocculent, remaining less condensed that that seen in most characid species, and no nuclear fossa is present. The centrioles, whose microtubules are poorly resolved, are perpendicular to one another (Fig. 11.30A). The flagellum, which arises near the anterior end of the cell (Fig. 11.30A), is initially contained within the cytoplasmic canal of the elongate cytoplasmic collar which is attached to the cell along the nuclear and mitochondrial regions. The collar, with the contained flagellum, extends posterior to the mitochondrial region before terminating, at which point the flagellum is free. A distinct electron-dense ring comprises the inner wall of the collar (Fig. 11.30B). The axoneme of the flagellum, which lacks fins, has the typical 9+2 structure, with the lumina of the A-tubule of each peripheral doublet being electron-dense along the entire length. The mitochondria, which are also extremely elongate, have a electron-dense matrix and each appears to contain a single, tube-like crista running along its entire length (Fig. 11.30B-C). In some cells peripheral accessory microtubules which run parallel to the long axis of the cell are present (Fig. 11.30B).

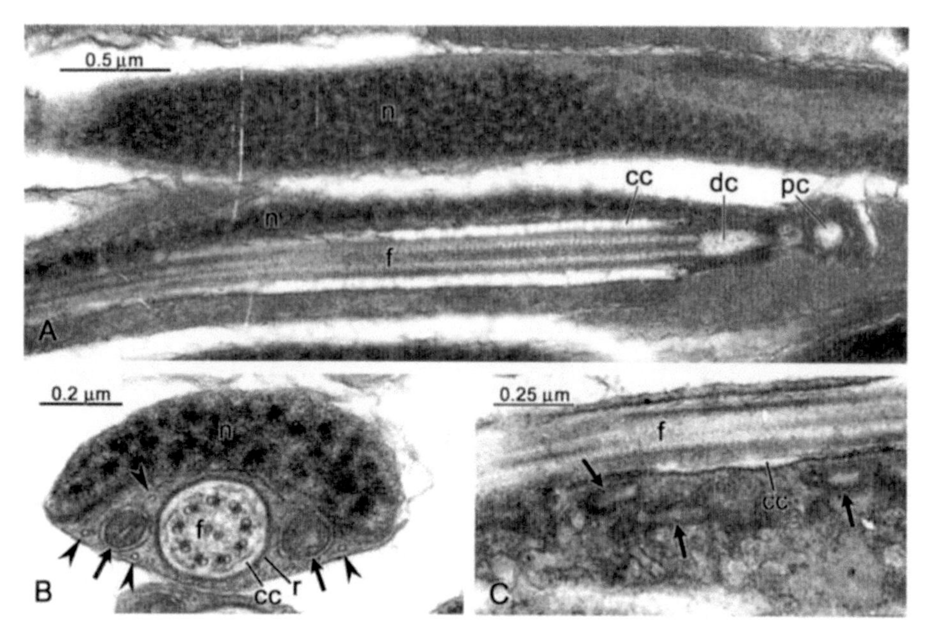

Fig. 11.30 *Pseudocorynopoma doriae* (Characidae, Stevardiinae, Hysteronotini) spermatozoa. SL 60.5 mm, MCP 18470, Brazil. **A**. Longitudinal section through anterior portion of cell. **B**. Transverse section through mitochondrial region. **C**. Longitudinal section of cell showing longitudinal sections through mitochondria. cc, cytoplasmic canal; dc, distal centriole; f, flagellum; n, nucleus; pc, proximal centriole; r, dense ring; arrow, mitochondrion; arrowhead, accessory microtubule. Original.

Tribe Xenurobryconini

Xenurobryconins examined include *Scopaeocharax rhinodus*, *Tyttocharax cochui* and *T. tambopatensis* (Pecio *et al.* 2005); *Chrysobrycon* sp. (original description).

***Tyttocharax tambopatensis* and *Scopaeocharax rhinodus*:** The sperm of these species, described by (Pecio *et al.* 2005), are packaged into spermatozeugmata (Fig. 11.31A-C). Neither genus appears to undergo nuclear rotation. As the nucleus elongates during spermiogenesis, the attached cytoplasmic collar with its contained flagellum also elongates. However, prior to spermatozeugma formation, which occurs within the spermatocysts prior to spermiation, the posterior portion of the collar degenerates, leaving a short collar at the anterior end of the spermatozoon (Fig. 11.32). The anterior terminus of the cells lacks any distinguishable organelles. The centrioles, which are located near the anterior end of the cell, are nearly parallel to one another, with their long axes parallel to the long axis of the cell. The distal centriole gives rise to the single flagellum which lacks fins. The greatly elongate nuclei in both genera, which are flattened along most of their lengths (Fig. 11.31A-C), taper to a point anteriorly (Fig. 11.32). The chromatin is

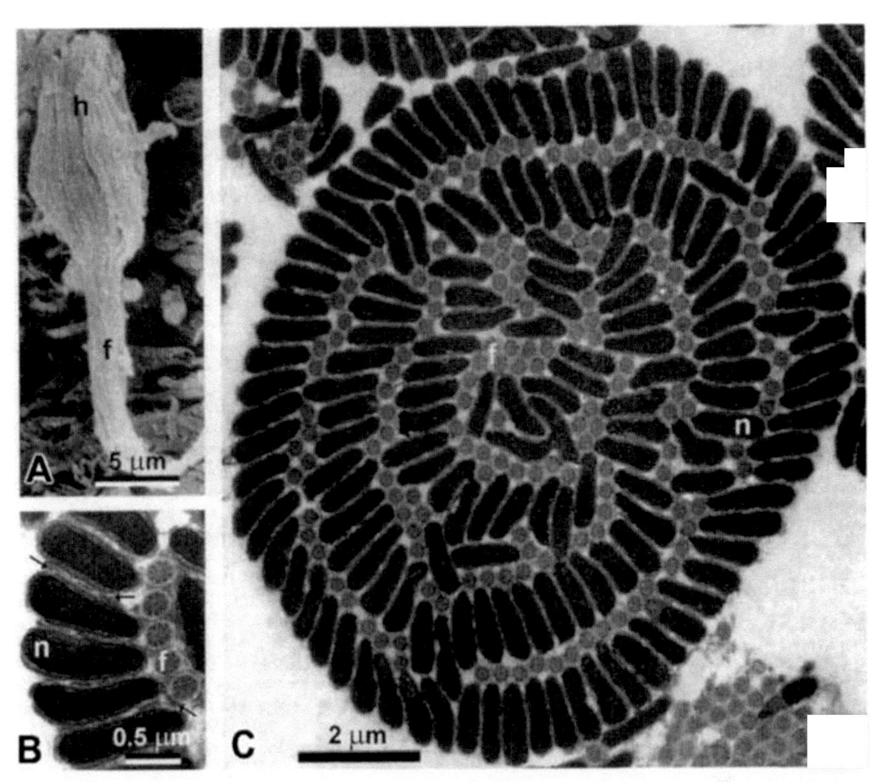

Fig. 11.31 Testis and spermatozeugmata of *Scopaeocharax rhinodus* **A**. A single spermatozeugma showing the structured arrangement with sperm heads (h) located in the anterior portion and flagella (f) trailing behind. SEM. **B**. Peripheral region of a transverse section through a spermatozeugma showing nucleus (n), flagellum (f) and electron-dense material (arrows) among the nuclear and flagellar regions of the various spermatozoa constituting the packet. **C**. Transverse section through the anterior portion of a single spermatozeugma within the sperm duct lumen (lu). TEM. Sperm nuclei are located around the periphery of the anterior packet, while the core of the anterior packet consists of concentric rings of nuclei (n) with flagella (f) in between these rings. TEM. Modified from Pecio, A., Burns, J. R. and Weitzman, S. H. 2005. Journal of Morphology 263: 216-226, Fig. 2B,D and C.

relatively condensed. Mean nuclear lengths were 16.0 µm (n = 8) for *Tyttocharax tambopatensis* , 16.9 µm (n = 8) for *T. cochui* and 11.2 µm (n = 10) for *Scopaeocharax rhinodus*. Numerous elongate mitochondria (length range = 4.0-4.6 µm) with lamellar cristae are found alongside the nuclei in both genera (Fig. 11.32). However, in *T. tambopatensis* the mitochondria are located on either side of the flattened nucleus, as well as posterior to the nucleus. On the other hand, the mitochondria in *S. rhinodus* are only located at one end of the flattened nucleus and they do not appear to be found posterior to the nucleus. In *T. tambopatensis*, in the proximal part of the flagellum the A-tubules of the peripheral doublets are electron-dense, whereas in the posterior part of the flagellum they are electron-lucent. It was not possible to discern the electron-density of the flagellar microtubules in the other two species. Flagellar length was approximately 22 μm in the two species of *Tyttocharax* and 25 μm in *S. rhinodus*. Accessory microtubules are found in the spermatozoa of all three species studied (Fig. 11.32). In both *Tyttocharax* and *Scopaeocharax* they are located on either side of the flattened nuclei just beneath the plasma membrane and orientated parallel to the long axis of the cell (Pecio *et al*. 2005).

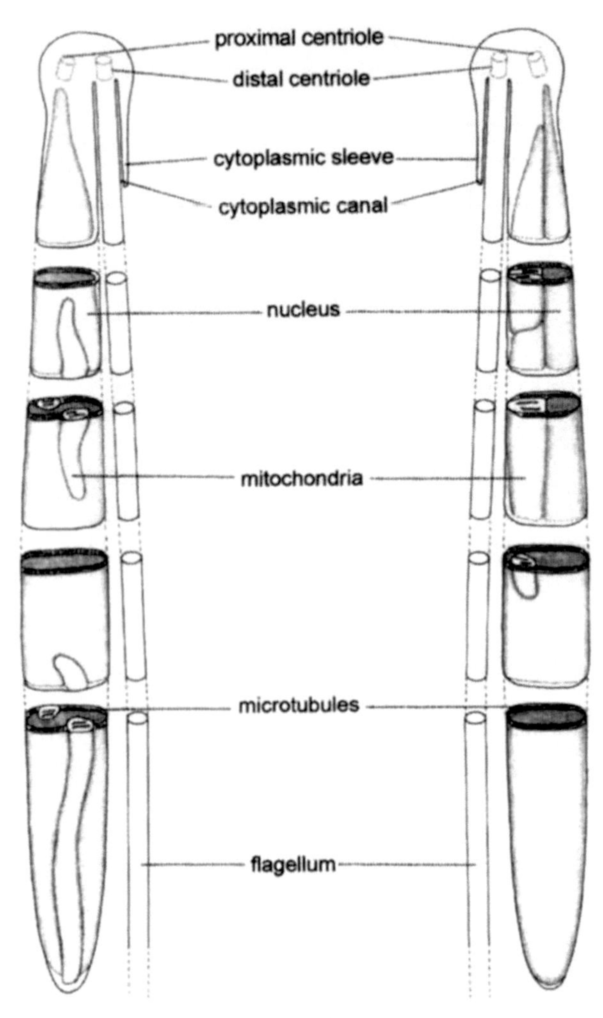

Fig. 11.32 Drawings based on electron micrographs, showing 3D views through spermatozoa of *Tyttocharax tambopatensis* (left) and *Scopaeocharax rhinodus* (right). From Pecio, A., Burns, J. R. and Weitzman, S. H. 2005. Journal of Morphology 263: 216-226, Fig. 4.

***Chrysobrycon* sp.** The spermatozoon of this species resembles those of the stevardiins, *Gephyrocharax atricaudatus* and *G. intermedius*, with several differences. This cell may also represent a highly modified Type I spermatozoon with most of the nucleus elongating posterior to the centrioles (Fig. 11.33A). The nuclear chromatin is highly electron-dense. The maximal nuclear length measured is 7.6 µm and maximal width 1.9 µm. In transverse section the nucleus tends to be oval (Fig. 11.33C,D). The proximal centriole and part of the distal are located within a nuclear fossa which appears as an eccentric notch in Fig. 11.33B). The proximal centriole is anterior and perpendicular to the distal. Elongate mitochondria, which have an electron-dense matrix, are located along the nucleus and more posteriorly (Fig. 11.33A, C-D). The cytoplasmic collar containing the flagellum is attached along the nuclear and mitochondrial regions and continues beyond that point as a thin sleeve before terminating, at which point the flagellum is free (Fig. 11.33A, C-D). A dense ring surrounding the collar, similar to that seen in the

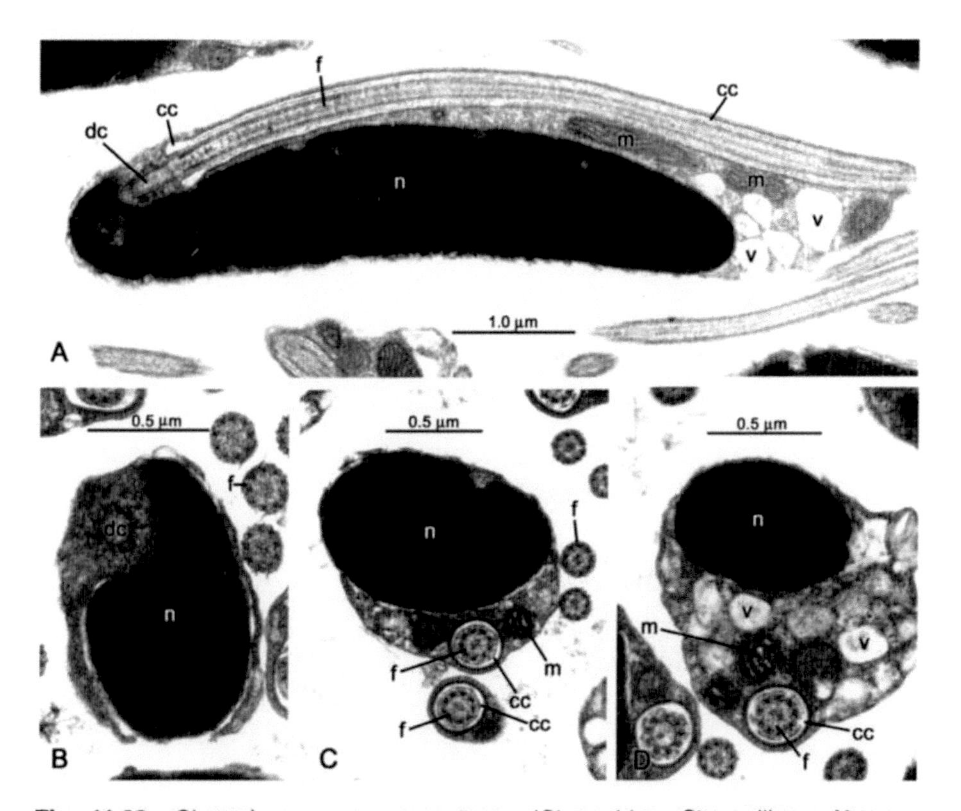

Fig. 11.33 *Chrysobrycon* sp. spermatozoa (Characidae, Stevardiinae, Xenurobryconini). Specimens collected in Peru 1996 by J. R. Burns and H. Ortega. **A**. SL 49.5 mm. **B**. SL 45.9 mm. C and D. SL 47.5 mm. **A**. Longitudinal section. **B-D**. Transverse sections. cc, cytoplasmic canal; dc, distal centriole; f, flagellum; m, mitochondrion; n, nucleus; v, vesicle. Original.

hysteronotin, *Pseudocorynopoma doriae*, is also present in *Chrysobrycon*. The flagellum has a typical 9+2 axoneme but no fins. However, the A-tubules of each doublet are electron-dense throughout the entire flagellum. In addition, in the region of the posterior nucleus and mitochondria abundant vesicles containing stained material are seen (Fig. 11.33A, D).

CHEIRODONTINAE

Tribe Compsurini

Species examined to date include *Acinocheirodon melanogramma, Compsura heterura, Kolpotocheirodon theloura, "Odontostilbe"dialeptura, "Odontostilbe" mitoptera*, and *Saccoderma hastatus* (Oliveira 2007); *Macropsobrycon uruguayanae* (Burns *et al.* 1998; Burns and Weitzman 2005; Oliveira 2007; Oliveira *et al.* 2008). Data for all species except *"Odontostilbe"dialeptura* (original figure, this chapter) and *Macropsobrycon uruguayanae* (Burns *et al.* 1998; Burns and Weitzman 2005; Oliveira 2007; Oliveira *et al.* 2008) remain unpublished at this time.

"Odontostilbe"dialeptura: Although fixation of the specimen in Fig. 11.34 is poor, it is the only figure available at this time but fortunately most of the major characters of the spermatozoon are visible. This species has an elongate, bullet-shaped nucleus, with highly condensed chromatin, that measures

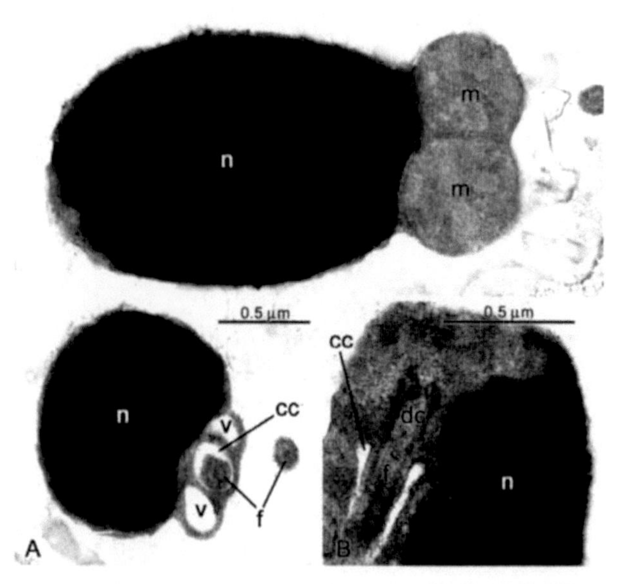

Fig. 11.34 *"Odontostilbe" dialeptura* (Characidae, Cheirodontinae) spermatozoa. SL 27.0 mm, aquarium specimen collected in Costa Rica courtesy of D. W. Fromm. **A**. Longitudinal section (upper cell) and transverse section (lower cell). **B**. Longitudinal section through anterior tip of cell. cc, cytoplasmic canal; dc, distal centriole; f, flagellum; m, mitochondrion; n, nucleus; v, vesicle. Original.

approximately 3.9 μm in length and 1.6 μm in width (Burns *et al.* 1997). During spermiogenesis, a slight degree of nuclear rotation make occur, but most of this process appears to be absent given the location of the centrioles lateral to the nucleus, as is the case in many glandulocaudines and stevardiines. The proximal centriole is anterior and oblique to the distal and both are located outside of a nuclear fossa (Oliveira 2007). The flagellum is initially contained within a cytoplasmic canal that is attached along the nuclear and mitochondrial regions of the cell (Fig. 11.34A-B). Several large, spherical mitochondria are located immediately posterior to the nucleus (Fig. 11.34A) where portions of a tubular-vesicular system are present (Oliveira 2007). The flagellum lacks fins and all tubules of the typical 9+2 axoneme are electron-lucent.

***Macropsobrycon uruguayanae*:** The bullet-shaped, electron-dense nucleus of the spermatozoon of *M. uruguayanae* is approximately 3.0 μm in length and 1.3 μm in width (Oliveira *et al.* 2008). Since the centrioles, mitochondria and cytoplasmic collar are all posterior to the nucleus (Fig. 11.35, n1), it appears that complete nuclear rotation takes place as the nucleus elongates anteriorly. The cytoplasmic collar, which may be slightly elongated, contains multiple elongate mitochondria with lamellar cristae, as well as a tubular-vesicular system. The two centrioles are parallel to one another with the proximal centriole slightly anterior to the longer distal (Oliveira *et al.* 2008). Striated rootlets radiate anteriorly and posteriorly from the distal centriole (Fig. 11.34). The axoneme of the flagellum, which lacks fins, has the typical 9+2 microtubular arrangement, and only in the proximal portion of the flagellum the A-tubule of each peripheral doublet is electron-dense (Burns *et al.* 1998). Nine peripheral accessory microtubules surround the axoneme in the proximal region of the flagellum (Burns and Weitzman 2005; Oliveira *et al.* 2008). Atypical spermatozoa are also released into the sperm ducts at spermiation (Fig. 11.35, n2). These cells do not appear to be simply spermatids but instead unique cells that are also transferred to the female (personal observation). The nuclei of these cells have an irregular shape and the granular chromatin is much less condensed than that of the spermatozoa (Fig. 11.35). In some cases the spacing between the mitochondrial cristae is also greater in these cells (Fig. 11.35). Unfortunately, neither the source of these cells nor their function is known at this time (Oliveira *et al.* 2008).

Incertae sedis

Inseminating species currently *incertae sedis* within Characidae for which some ultrastructural information is available are *Brittanichthys axelrodi* (Javonillo *et al.* 2007); *Bryconadenos tanaothoros* (Weitzman *et al.* 2005); *Hollandichthys affinis* (TEM) and *H. perstriatus* and four undescribed *Hollandichthys* species (all SEM) (Azevedo 2004). Data for the species of *Hollandichthys* have not yet been published.

Brittanichthys axelrodi: The elongate, spatulate nucleus, approximately 5 µm in length, is curved at its anterior end (Javonillo *et al.* 2007) and has highly condensed chromatin (Fig. 11.36A-B). The centrioles, which are perpendicular to one another, are partially located within a lateral nuclear fossa. A large striated rootlet (Fig. 11.36A), that originates at the anterior end of the distal centriole, wraps around the entire anterior ventral area of the nucleus (Javonillo *et al.* 2007). The flagellum is contained within a cytoplasmic collar that is attached along the entire nuclear (Fig. 11.36B) and mitochondrial regions, and continues posteriorly as a thin sleeve before the flagellum exits. Several large, spherical mitochondria, approximately 0.6 µm in diameter, with lamellar cristae overlap the posterior end of the nucleus and continue posteriorly. The flagellar axoneme has the typical 9+2 arrangement, with the A-tubules of the peripheral doublets being electron-lucent in the anterior

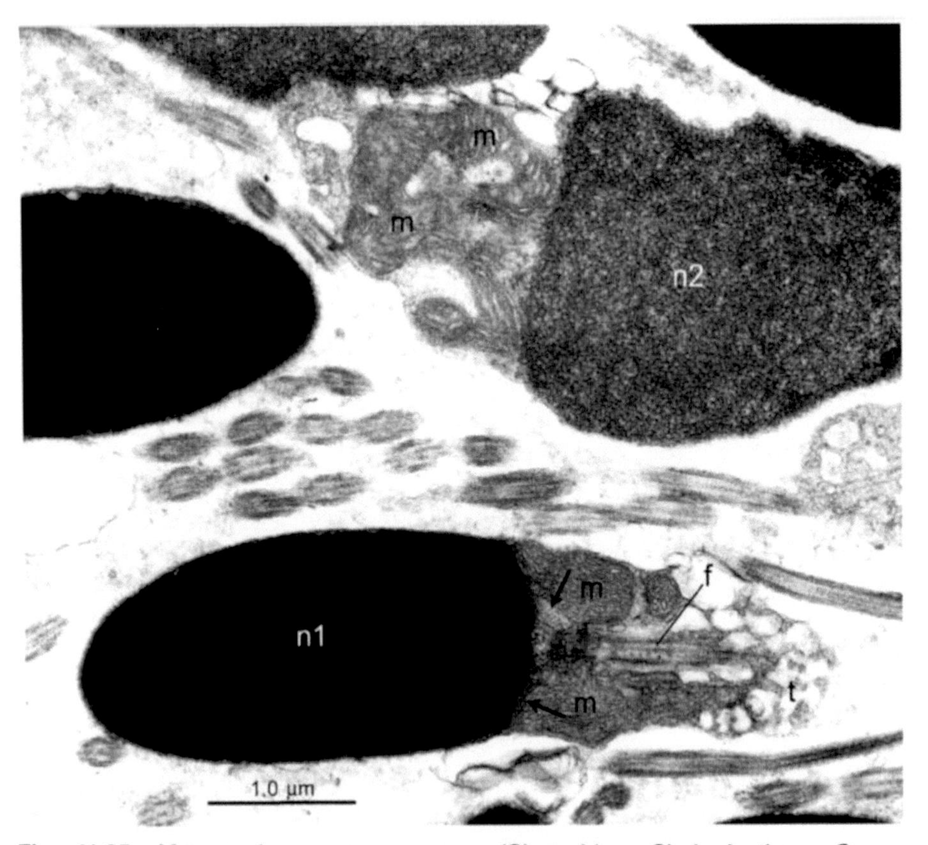

Fig. 11.35 *Macropsobrycon uruguayanae* (Characidae, Cheirodontinae, Compsurini) spermatozoa. SL 39.0 mm, MCP 18588, Brazil. Note the marked differences in the shape and degree of chromatin condensation between nuclei n1 and n2. f, flagellum; m, mitochondrion; n1, nucleus of spermatozoon; n2, nucleus of atypical spermatozoon; t, tubular-vesicular system; arrow, striated rootlet. Original.

segment but electron-dense in the posterior segment (Fig. 11.36B). In transverse section, the anterior nucleus has an unusual triangular shape (Fig. 11.36B, left cell), whereas posteriorly its shape is more oval (Fig. 11.36B, right cell). In addition to having several unique sperm ultrastructural characters, the testis itself is tripartite, being divided into an anterior spermatogenic region, an intermediate sperm storage region and a coiled posterior region that appears be instrumental in sperm packaging (Javonillo *et al.* 2007). *B. axelrodi* produces feather-like spermatozeugmata that bear a striking resemblance to those of species in the xenurobryconin (Stevardiinae) genus *Xenurobrycon* (Burns *et al.* 2008).

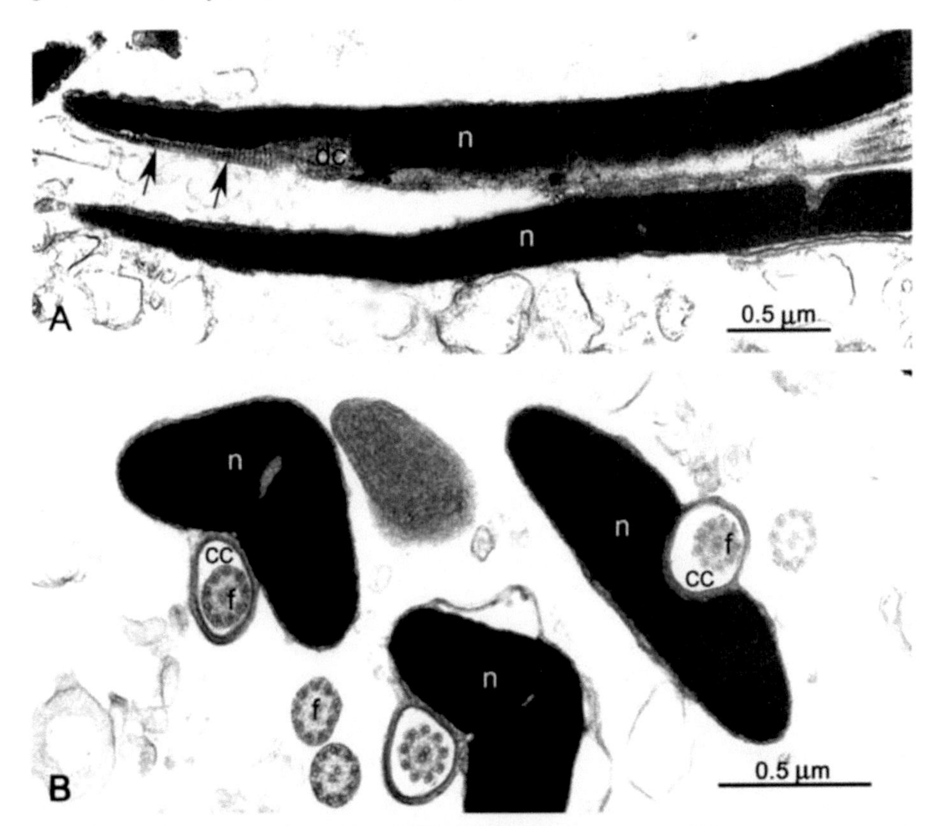

Fig. 11.36 *Brittanichthys axelrodi* (Characidae, *Incertae Sedis*) spermatozoa. SL 27.8 mm, aquarium specimen collected in the Rio Negro, Brazil, courtesy of L. Chao. **A**. Longitudinal sections. **B**. Transverse sections. cc, cytoplasmic canal; dc, distal centriole; f, flagellum; n, nucleus; arrow, striated rootlet. Original.

Bryconadenos tanaothoros: The elongate, spatulate sperm cell of *B. tanaothoros* tapers at either end. An attached cytoplasmic collar, with its contained flagellum, terminates approximately midway along the length of the cell, at which point the flagellum is free (Weitzman *et al.* 2005). The elongate nucleus with its moderately condensed, granular chromatin measures 6.7 μm in length. The centrioles are parallel to one another and portions of each are located within separate lateral nuclear fossae (Fig. 11.37A). The flagellum is initially contained within a cytoplasmic collar along the anterior nucleus (Fig. 11.37F) but is free more posteriorly (Fig. 11.37B-C). Multiple mitochondria are located alongside the nucleus (Fig. 11.37B-C, E-F) and posteriorly (Fig. 11.37D). Peripheral accessory microtubules that run parallel to the long axis of the cell are found through most of the nuclear and mitochondrial regions (Fig. 11.37B-D). The sperm are packaged into compact spermatozeugmata (Weitzman *et al.* 2005).

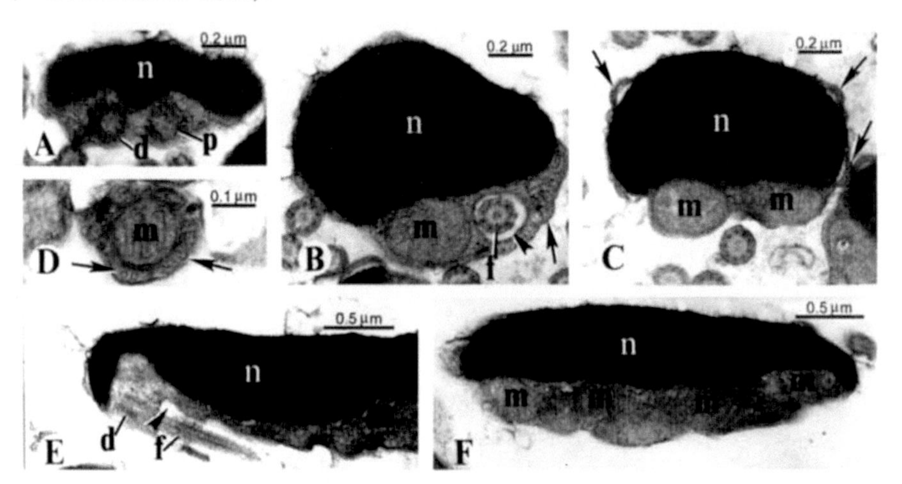

Fig. 11.37 *Bryconadenos tanaothoros* (Characidae, *Incertae Sedis*). SL 50.0 mm, USNM 352061, Brazil. **A-D**. Successively more posterior transverse sections. **E** and **F**. Longitudinal sections. d, distal centriole; f, flagellum; m, mitochondrion; n, nucleus; p, proximal centriole; arrow, accessory microtubule; arrowhead, cytoplasmic canal. Modified from Weitzman, S. H., Menezes, N. A., Evers, H.-G. and Burns, J. R. 2005. Neotropical Ichthyology 3: 329-360, Fig. 18.

Fertilization biology in Characidae: The great majority of inseminating (including internally fertilizing) fishes exhibit sperm and testis modifications that appear to be adaptations related to this mode of reproduction (Jamieson 1991; Burns *et al.* 1998). Discussions of such adaptations are available in Burns and Weitzman (2005) for ostariophysans and Pecio *et al.* (2005) for characiforms, as well as in **Chapter 17**. Testis modifications include development of special areas devoted to sperm storage and sperm packet formation. The most common sperm adaptation appears to be elongation of the nucleus. In addition, the midpiece or mitochondrial capacity may be enlarged through increasing the number, size or length of mitochondria. Basal body (distal centriole) reinforcement may take the form of electron-dense material, spurs, rings and striated rootlets. Basic cell reinforcement and stiffening often results from longitudinal arrays of accessory microtubules. Finally, some species store glycogen in the spermatozoon. The possible selective advantages of these specializations will be discussed in more detail in **Chapter 17**. Characters obtained from testis structure and sperm ultrastructure have been found to be increasingly useful in hypothesizing phylogenetic relationships among characiform fishes, particularly those that are inseminating (Malabarba 1998; Weitzman and Menezes 1998; Weitzman *et al.* 2005). The types of characters that may prove useful in such studies include: nuclear size, shape and degree of chromatin condensation; centriolar orientation; location, shape and size of mitochondria; presence and location of accessory microtubules; intratubular differentiation of the flagellar axoneme; presence, length and location of cytoplasmic collars; tubular-vesicular system and size and distribution of other vesicles. The addition of reproductive characters to existing morphological data sets, together with results from ongoing molecular studies, will hopefully shed light on the evolutionary history of these diverse fishes. The question of how many times insemination has evolved has yet to be answered. Results to date suggest that within the family Characidae, insemination has arisen independently from externally fertilizing ancestors at least three times: in the subfamily Glandulocaudinae, in the subfamily Stevardiinae, and in the tribe Compsurini of the subfamily Cheirodontinae (Malabarba 1998; Weitzman *et al.* 2005). However, preliminary results from ongoing molecular analyses indicate that this number may increase (personal observations).

11.7 ORDER SILURIFORMES

Siluriformes (catfishes) is the most diverse and widely distributed ostariophysan order. Although the relationships among catfish families have been studied extensively, the higher-level taxonomy of Siluriformes is still controversial (Diogo 2003, 2004a,b). However, a general consensus holds that the family Diplomystidae is sister to all other groups within Siluriformes. According to Nelson (2006), the order Siluriformes comprises 12 superfamilies, 37 families, and one *incertae sedis* species. Both Siluriformes and the related Gymnotiformes are morphologically the most peculiar and highly modified ostariophysans and show a substantial number of synapomorphies. Both have electroreceptive capability (references in Lauder and Liem 1983). They exhibit great diversity in trophic mechanisms and are far more diverse than characiforms and cypriniforms in the structure of the Weberian complex, swim bladder and caudal fin skeleton (Roberts 1973; Fuiman 1984; Nelson 2006).

Sperm ultrastructure is known for at least one species in 26 of the 37 families that comprise the order, as well as in *Conorhynchos conirostris*. Unfortunately some of these reports do not provide sufficiently detailed descriptions. External fertilization is the dominant reproductive mode in the majority of families within this order. In external fertilizing species, most spermatozoa are Type I aquasperm (*sensu* Mattei 1991), these are exemplified, *inter alia*, (Mattei 1991) by the Bagridae, Clariidae, Mochokidae, and Schilbidae, in contrast with the Ariidae, Ictaluridae, and Malapteruridae in which the spermatozoon is biflagellate.

A new type of aquasperm, type III (Quagio-Grassiotto and Carvalho 2000; Quagio-Grassiotto *et al.* 2005; Spadella *et al.* 2006a; Quagio-Grassiotto and Oliveira 2008), has been described in some Neotropical species along with several intermediate forms. Highly modified introsperm are produced by the inseminating families Auchenipteridae (Loir *et al.* 1989; Meisner *et al.* 2000; Burns *et al.* 2002) Scoloplacidae (Burns and Weitzman 2005; Spadella *et al.* 2006b), and Astroblepidae (Burgess 1989; Spadella 2007). Semicystic spermatogenesis, mainly associated with the production of biflagellate sperm, has been reported in some families (Spadella *et al.* 2006a; Shahin 2006a) and in the Doradidae and Ariidae (**Sections 11.7.9.3 and 11.7.10.1**).

11.7.1 DIPLOMYSTOIDEA

11.7.1.1 Family Diplomystidae

Diplomystes mesembrinus: The spermatozoon of *D. mesembrinus*, described by Quagio-Grassiotto *et al.* (2001b), is a Type I aquasperm with a spherical nucleus (diameter 1.5 μm). Chromatin consists of highly condensed, flocculent masses surrounded by a less dense matrix. In mid-longitudinal section, a long and narrow, centrally located nuclear fossa extends almost to the anterior margin of the nucleus. The proximal centriole is anterior and perpendicular to the distal centriole and both are contained within the nuclear fossa. The short midpiece contains a single, large, C-shaped mitochondrion that encircles the anterior portion of the axoneme. Because the cytoplasmic canal present in the early spermatid closes near the end of spermiogenesis, it is lacking in the spermatozoon. No vesicles are evident in the midpiece. The single flagellum, which contains a classic 9+2 axoneme, has two short fins.

11.7.2 CETOPSOIDEA

11.7.2.1 Family Cetopsidae

Cetopsis coecutiens: The spermatozoon of *C. coecutiens*, described by Spadella *et al.* (2006a) is an aquasperm with a slightly ovoid head (length 1.3 μm, width 1.4 μm). The bell-shaped nucleus contains highly condensed, granular chromatin and a wide, deep, centrally located nuclear fossa. The centrioles are lateral and parallel to one another, and both are located within the nuclear fossa. Each centriole gives rise to an axoneme. The midpiece is short and slightly wider than the base of the nucleus. A short cytoplasmic canal is present. Several spherical mitochondria, located at the base of the nucleus, form a peripheral ring around the initial region of the flagellum. A well organized system of parallel, elongate vesicles is found throughout the midpiece. The two centrally located flagella have a classic 9+2 axoneme and lack fins.

11.7.3 LORICARIOIDEA

11.7.3.1 Family Trichomycteridae

Although all species analyzed in the Trichomycteridae appear to produce Type I aquasperm, the degree of nuclear rotation that takes place during spermiogenesis varies, being incomplete in *Ituglanis amazonicus*, *Trichomycterus* aff *iheringi*, *Trichomycterus* sp. 1 and *Trichomycterus reinhardti*, and nearly complete in *Copionodon orthiocarinatus*, *Trichomycterus areolatus* and *Trichomycterus* sp. 2 (Spadella 2007).

***Copionodon orthiocarinatus*,*Trichomycterus areolatus* and *Trichomycterus* sp. 2:** The spermatozoa of these three species are Type I aquasperm, with ovoid nuclei (*C. orthiocarinatus*, length 1.9 μm, width 1.8 μm; *T. areolatus* and *Trichomycterus* sp. 2, length 1.7 μm, width 1.6 μm) that contain highly condensed, homogeneous chromatin, and in mid-longitudinal section, a central, shallow to moderately deep nuclear fossa. The proximal centriole is slightly anterior, lateral and oblique to the distal one. The entire proximal centriole and anterior the tip of the distal are contained within the nuclear fossa. The midpiece contains several spherical to elongate mitochondria. The anterior part of the flagellum is located within a cytoplasmic canal. In *T. areolatus* some small vesicles can be found in the basal region of the midpiece. In *C. orthiocarinatus* and *Trichomycterus* sp.2, abundant large vesicles are distributed around the peripheral and basal regions of the midpiece; vesicles also surround parts of the nucleus in the latter species. The single flagellum contains a classic 9+2 axoneme. The flagellum in *C. orthiocarinatus* and *T. areolatus* has two fins. In *Trichomycterus* sp.2, two to three irregular fins are present and a membranous compartment can be seen in some regions of the flagellum.

***Ituglanis amazonicus* and *Trichomycterus* aff. *iheringi*:** The spermatozoa of *Ituglanis amazonicus* and *Trichomycterus* aff *Iheringi*, are aquasperm with ovoid nuclei (*I. amazonicus*, length 1.6 μm, width 1.8 μm; *T.* aff *iheringi*, length 1.5 μm, width 1.3 μm) located lateral to the centriolar complex. The nucleus contains highly condensed, homogeneous chromatin and a shallow nuclear fossa. In *I. amazonicus* the proximal centriole is lateral and perpendicular to the distal, whereas in *T.* aff *iheringi* the proximal centriole is lateral and oblique to the distal. In *I. amazonicus* the proximal centriole and most of the distal are inside the nuclear fossa, whereas in *T.* aff *iheringi* only the proximal centriole is located within the nuclear fossa. Midpieces are strongly asymmetric, with several spherical to elongate mitochondria distributed throughout. The anterior part of the flagellum is located within a cytoplasmic canal. In *I. amazonicus* large vesicles are seen within the peripheral and basal regions of the midpiece. In *T.* aff *iheringi* small vesicles form an organized system located at the peripheral and basal regions, with occasional vesicles present around the nucleus. The single flagellum, which has a strongly eccentric location with respect to the nucleus, contains a classic 9+2 axoneme and two to three irregular fins.

Trichomycterus reinhardti **and** ***Trichomycterus*** **sp. 1:** Spermatozoa of *Trichomycterus reinhardti* and *T.* sp.1 approach the morphology of Type II aquasperm but differ from this in having the proximal centriole located in the nuclear fossa. The ovoid nuclei (*T. reinhardti*, length 1.7 μm, width 1.6 μm; *Trichomycterus* sp.1, length 1.7 μm, width 1.5 μm) are located lateral to the flagella. The nucleus contains granular, homogeneous and highly condensed chromatin, and a lateral shallow nuclear fossa. The proximal centriole is lateral and perpendicular to the distal in *T. reinhardti*, whereas in *Trichomycterus* sp.1 the proximal centriole is lateral and oblique to the distal. Only the proximal centriole is located within the nuclear fossa. The short midpiece, located at the base of the nucleus, is strongly asymmetric. The anterior part of the flagellum is located within a short cytoplasmic canal, and a short narrow cytoplasmic sleeve is present on the side distal to the nucleus. In both species several spherical to elongate mitochondria are concentrated near the base of the nucleus, while in *T. reinhardti* mitochondria can also be found lateral to the nucleus near the centrioles. Some small vesicles are present throughout the midpiece of *T. reinhardti*, whereas in the midpiece of *Trichomycterus* sp.1 several large vesicles are distributed peripherally and a well organized system of elongate vesicles or tubules is located in the basal region. The single, lateral flagellum has a classic 9+2 axoneme and two to three irregular fins. A membranous compartment is found in some regions of the flagellum of *Trichomycterus* sp.1, but not *T. reinhardti*.

11.7.3.2 Family Nematogenyiidae

Nematogenys inermis**:** The spermatozoon of *N. inermis*, described by Spadella *et al.* (2006a) is an aquasperm that may be formed by a process similar to type III spermiogenesis (Fig. 11.54).

The cell has an ovoid nucleus containing highly condensed, granular chromatin. The centrally located centrioles are lateral and parallel to one another and anchored to the plasma membrane at the periphery of the midpiece. Each gives rise to one axoneme. The midpiece has a very short cytoplasmic canal. Numerous long and sometimes branching mitochondria are mainly concentrated in the basal region of the midpiece, but few vesicles are seen. The two flagella, each with a classic 9+2 axoneme, are initially separate from one another, but more posteriorly become enveloped by a single plasma membrane. Flagella lack fins.

11.7.3.3 Family Callichthyidae

Corydoradinae: ***Aspidoras poecilus, Corydoras aeneus, C. flaveolus*** **and** ***Scleromystax lacerda*****.** Spermiogenesis in these Corydoradinae, described by Spadella (2007), has characteristics that as appear to be intermediate between Types II and III. Consequently the spermatozoa in this group of fishes are also intermediate types.

Spermatozoa are aquasperm with ovoid nuclei (*A. poecilus*, length 1.6 µm, width 1.5 µm; *C. aeneus*, length 1.6 µm, width 1.7 µm; *C. flaveolus*, length 1.9 µm, width 2.0 µm; *S. lacerdai*, length 1.5 µm, width 1.6 µm). Nuclei contain highly condensed, homogeneous chromatin and shallow nuclear fossae. The proximal centriole is directly anterior to the distal one, and separated from it by a horizontal, electron-dense, laminated body. Both centrioles are outside the nuclear fossa. The asymmetric midpiece has a short cytoplasmic canal. Numerous elongate mitochondria are located near the nucleus and distributed throughout most of the midpiece with the exception of the basal region, where abundant small vesicles are observed. The single, eccentrically located flagellum, which contains a classic 9+2 axoneme, lacks fins but has a membranous compartment in some regions.

Callichthyinae: Species which have been examined for sperm ultrastructure are *Callichthys callichthys* and *Megalechis thoracata* (Spadella *et al.* 2007); *C. thoracatus* (Mansour and Lahnsteiner 2003); and *Hoplosternum littorale* (Matos *et al.* 1993b; Spadella *et al.* 2007).

The spermatozoa of the Callichthyinae are Type I aquasperm with ovoid nuclei in *C. callichthys* and *H. littorale* (length 1.7-1.8 µm, width 1.6-2.0 µm) and spherical nuclei in *M. thoracata* (diameter 1.8 µm). The nucleus contains highly condensed, homogeneous chromatin and a shallow nuclear fossa that is eccentric in *C. callichthys* and *H. littorale*, and central in *M. thoracata*. The proximal centriole is lateral and oblique (variable angle) to the distal. In *H. littorale* and *M. thoracata* both centrioles are inside the nuclear fossa, whereas in *C. callichthys* only the proximal centriole lies within the nuclear fossa. The midpiece is asymmetric in *C. callichthys* and *H. littorale* and symmetric in *M. thoracata*. A cytoplasmic canal is present. Several elongate mitochondria are located near the nucleus and mainly concentrated in the anterior third of the midpiece. Large vesicles are found in the posterior two-thirds of the midpiece. The posterior midpiece of all species analyzed contains a structure very similar to the tubular-vesicular system initially described in the characiform, *Citharinus* sp. (Mattei *et al.* 1995). The single, eccentrically located flagellum contains a classic 9+2 axoneme but lacks fins. A membranous compartment is present in some segments along the flagellum in *C. callichthys*, but none is observed in *H. littorale* or *M. thoracata*.

11.7.3.4 Family Scoloplacidae

***Scoloplax distolothrix*:** The spermatozoon of *S. distolothrix*, described by Spadella *et al.* (2006b) is an introsperm with a greatly elongate, conical and arched nucleus (length 6.0 µm, width 0.4 µm).

A well developed membranous compartment is present along the concave face of the nucleus. The nucleus contains homogeneous, highly condensed chromatin and a central, deep nuclear fossa. The proximal centriole is anterior and perpendicular to the distal, and both are located within the nuclear fossa. The long, thin, midpiece has a central cytoplasmic canal containing the flagellum. The very abundant, ring-shaped mitochondria, which are organized in rows along the midpiece, surround the cytoplasmic canal. Several elongate vesicles occupy the peripheral and terminal regions of the midpiece. The single, centrally located flagellum has a classic 9+2 axoneme and two short fins. The A-tubules of the axonemal doublets appear to be electron-dense.

11.7.3.5 Family Astroblepidae

***Astroblepus cf. mancoi*:** Given that insemination is now reported in females of two species of astroblepids by Javonillo *et al.* (**Chapter 17**), the spermatozoon of *A.* cf. *mancoi* is also likely to be an introsperm (Spadella 2007). It has an elongate conical nucleus (length 6.0 µm, width 0.6 µm) whose tip is slightly curved. The nucleus contains homogeneous, highly condensed chromatin and a deep nuclear fossa. The proximal centriole is anterior and perpendicular to the distal centriole and both are located within the nuclear fossa. The symmetric midpiece is long and thin, with a cytoplasmic sleeve at the posterior end. Numerous elongate and apparently fused mitochondria are distributed along the midpiece, with the exception of the cytoplasmic sleeve. The mitochondria form a ring around the cytoplasmic canal which contains the flagellum. No vesicles are present. The single, centrally located flagellum has a classic 9+2 axoneme and two medium sized fins whose terminal ends are dilated and filled with electron-dense material. The A-tubules of the axonemal doublets appear to be electron-dense in the posterior flagellum.

11.7.3.6 Family Loricariidae

Hypoptomatinae: *Corumbataia cuestae, Hisonotus* sp., *Hypoptopoma guentheri* and *Schizolecis guntheri*. The spermatozoa of these four Hypoptomatinae (Spadella 2007) are Type I aquasperm with spherical nuclei in *Hisonotus* sp. (diameter 1.5 µm) and *H. guentheri* (diameter 1.6 µm) and ovoid nuclei in *C. cuestae* and *H. guentheri* (length 1.6-1.8 µm , width 1.4-1.5 µm). The nucleus contains homogeneous, highly condensed chromatin and a deep nuclear fossa. The proximal centriole is anterior and perpendicular to the distal, and both are located within the nuclear fossa. The anterior region of the midpiece contains several spherical to elongate mitochondria, while the posterior region is filled with several vesicles, some of which are interconnected. Vesicles are absent in *Hisonotus* sp. A cytoplasmic canal contains the flagellum. The single flagellum has a classic 9+2 axoneme and two fins. The electron density of the axonemal doublets could not be determined.

Hypostominae: *Ancistrus triradiatus* and *Hypostomus ancistroides* and *Hypostomus* sp. The spermatozoa of the hypostomines *Ancistrus triradiatus* (Mansour and Lahnsteiner 2003), *Hypostomus ancistroides* (Spadella 2007) and *Hypostomus* sp. (Fig. 11.38) are type I aquasperm with ovoid nuclei in *H. ancistroides* (length 2.0 µm, width 2.3 µm) and spherical nuclei in *Hypostomus* sp. (diameter 2.4 µm) that contain homogeneous, highly condensed chromatin (Fig. 11.38A-C) and shallow to moderately deep nuclear fossae. The proximal centriole is anterior, slightly lateral and perpendicular to the distal (Fig.

11.38D). Either the proximal centriole is located within the nuclear fossa, or both centrioles lie outside it. The midpiece contains several spherical to elongate mitochondria that surround the centrioles in the anterior half of the midpiece (Fig. 11.38A,B). Numerous vesicles are located at the periphery and in the posterior half of the midpiece (Fig. 11.38C,E). The single flagellum has a classic 9+2 axoneme and two fins (Fig. 11.38F).

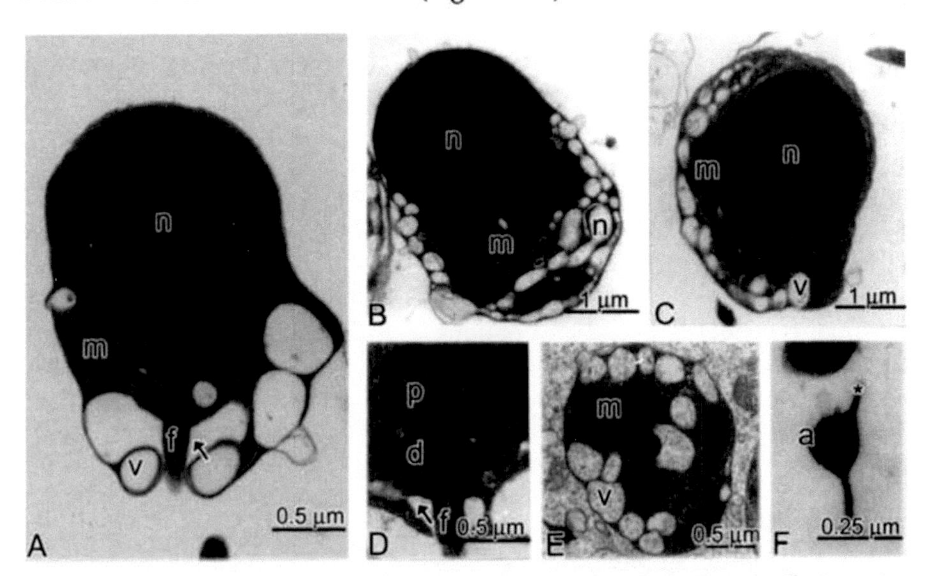

Fig. 11.38 *Hypostomus* sp. (Loricariidae, Hypostominae). **A**. Longitudinal section of spermatozoon. Note the numerous large vesicles in the midpiece and short cytoplasmic canal. **B-C**. Longitudinal and transverse sections of the nucleus and midpiece with the elongate mitochondria close to the nucleus and the numerous vesicles distributed at the periphery and in the posterior half of the midpiece. **D**. Centriolar arrangement with the proximal centriole anterior, slightly lateral and perpendicular to the distal. **E**. Transverse section of the midpiece where the cytoplasmic canal begins. Mitochondria surround the initial portion of the axoneme and are bordered by vesicles. **F**. Transverse section of flagellum with classic 9+2 axoneme surrounded by the flagellar membrane that has two fins. a, axoneme; d, distal centriole; f, flagellum; m, mitochondrion; n, nucleus; p, proximal centriole; v, vesicles; arrow, cytoplasmic canal; asterisk, flagellar fin. Original, courtesy of Gisleine Fernanda França.

Loricariinae: Two different types of spermiogenesis occur among the Loricariinae analyzed, giving rise to two different types of spermatozoa.

***Loricariichthys platymetopon*:** The spermatozoon of *L. platymetopon* (Spadella 2007) is a type III aquasperm with an ovoid nucleus (length 1.9 μm, width 2.0 μm) that has homogeneous, highly condensed chromatin and no nuclear fossa. The proximal centriole is anterior and perpendicular to the distal. The midpiece has a short cytoplasmic canal containing the flagellum and several spherical to elongate mitochondria peripherally distributed. Several vesicles are seen in the posterior region of the midpiece. The single flagellum has a classic 9+2 axoneme and no fins.

***Loricaria* sp.** The spermatozoon of *Loricaria* sp. (Spadella 2007) is a Type I aquasperm with a spherical nucleus (diameter 2.0 μm) that contains homogeneous, highly condensed chromatin and a shallow nuclear fossa. The proximal centriole is anterior and perpendicular to the distal. Only the proximal centriole lies within the nuclear fossa. The midpiece contains several elongate mitochondria and vesicles randomly distributed. A short cytoplasmic canal contains the single flagellum that has a classic 9+2 axoneme but no fins.

Neoplecostominae. *Kronichthys heylandi* and *Neoplecostomus paranensis*: The spermatozoa of *K. heylandi* and *N. paranensis* described by Spadella (2007) are Type I aquasperm with spherical nuclei in *K. heylandi* (diameter 1.5 μm) and ovoid nuclei in *N. paranensis* (length 1.6 μm, width 1.7 μm) that contain homogeneous, highly condensed chromatin and deep nuclear fossae. The proximal centriole is lateral and oblique to the distal and both are located within the nuclear fossa. The anterior region of the short midpiece contains several spherical to elongate mitochondria and some vesicles. A short cytoplasmic canal contains the flagellum. The single flagellum has a classic 9+2 axoneme and two fins.

11.7.4 SISOROIDEA

11.7.4.1 Family Amblycipitidae

***Liobagrus mediadiposalis*:** The spermatozoon of *L. mediadiposalis* (Lee and Kim 1999) is a biflagellate aquasperm with an ovoid nucleus that contains homogeneous, highly condensed fibrous chromatin and two moderately deep nuclear fossae. The centrioles are lateral and parallel one another. The anterior part of each centriole is located within one of the nuclear fossae. Each centriole gives rise to one axoneme. The short midpiece contains several spherical mitochondria located at the base of the nucleus and in 2-3 layers around the initial segment of the flagellum. No vesicles were observed. The two flagella, each within a separate cytoplasmic canal, have classic 9+2 axonemes and two medium sized fins.

11.7.4.2 Family Aspredinidae

***Bunocephalus coracoideus* and *B. amazonicus*.** The spermatozoa of *Bunocephalus(=Dysichthys) coracoideus* (Mansour and Lahnsteiner 2003) and *B. amazonicus* (Spadella 2006a) are Type I aquasperm with a conical nucleus (length 1.9 μm, width 1.6 μm in *B. amazonicus*) that contains condensed granules of flocculent chromatin surrounded by a lighter matrix, and a deep nuclear fossa.

The centrioles are lateral and parallel to one another, and both are located within the nuclear fossa. Each gives rise to one axoneme. The midpiece is long and thin with a short posterior cytoplasmic sleeve. Two separate cytoplasmic canals, each containing one flagellum, pass through the center of the midpiece. Several elongate mitochondria are distributed throughout most of the midpiece with the exception of the cytoplasmic sleeve. An organized system of vesicles is found throughout the midpiece. The two flagella have classic 9+2 axonemes but no fins.

11.7.5 PSEUDOPIMELODOIDEA

11.7.5.1 Family Pseudopimelodidae

***Microglanis* aff. *parahybae*:** The spermatozoon of *M.* aff. *parahybae* (Quagio-Grassiotto *et al.* 2005) is a Type I aquasperm with a spherical nucleus that contains homogeneous, highly condensed chromatin and a deep nuclear fossa (Fig. 11.39A). The proximal centriole is directly in front of and aligned with the distal centriole, and the proximal centriole and anterior part of the distal are contained within the nuclear fossa (Fig. 11.38A). The long, slightly asymmetric midpiece is highly folded due to the presence of numerous large vesicles or tubules that connect with the plasma membrane. Several elongate mitochondria are distributed throughout the midpiece (Fig. 11.38B,E). A relatively long cytoplasmic canal contains the single flagellum that has a classic 9+2 axoneme but no fins (Fig. 11.38A-D).

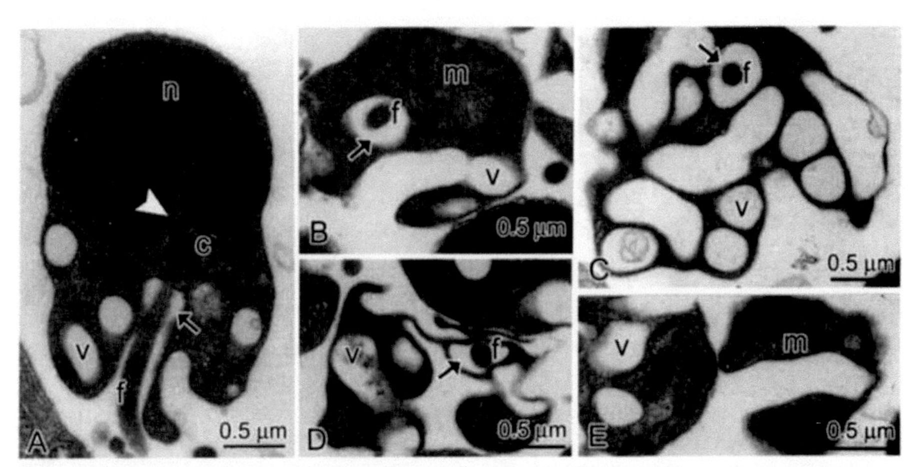

Fig. 11.39 *Microglanis* aff. *parahybae*. **A.** Longitudinal section of spermatozoon. **B-E.** Transverse sections at different levels of the midpiece. Note the numerous large vesicles or tubules that connect with the plasma membrane and the distribution of the elongate mitochondria. c, centrioles; f, flagellum; m, mitochondrion; n, nucleus; v, vesicles; arrowhead, nuclear fossa; arrow, cytoplasmic canal. Original.

11.7.6 HEPTAPTEROIDEA

11.7.6.1 Family Heptapteridae

Sperm ultrastructure has been described in three species of heptapterids. The fine structure of the spermatozoon of *Rhamdia sapo* is known only from SEM

(Maggese *et al.* 1984). Because of their distinct morphologies, the ultrastructure of the spermatozoa of *Pimelodella gracilis* and *Rhamdia quelen* (Quagio-Grassiotto *et al.* 2005) will both be presented.

Pimelodella gracilis: The spermatozoon of *P. gracilis* (Quagio-Grassiotto *et al.* 2005) is a type III aquasperm with a spherical nucleus containing homogeneous, highly condensed chromatin and no nuclear fossa (Fig. 11.40A). The proximal centriole is lateral and parallel to the distal and both lie close to the central axis of the nucleus. The midpiece is short and thin (Fig. 11.40A). The cytoplasmic canal observed in the spermatid closes prior to the completion of spermiogenesis, thus no canal is present in the spermatozoon (Fig. 11.40A,F). Several elongate, fused mitochondria encircle the centrioles and initial segment of the axoneme (Fig. 11.40C-E). Long tubules or vesicles form membranous loops at the peripheral and terminal regions of the midpiece (Fig. 11.40A-F). The single flagellum has a classic 9+2 axoneme but no fins (Fig. 11.40G).

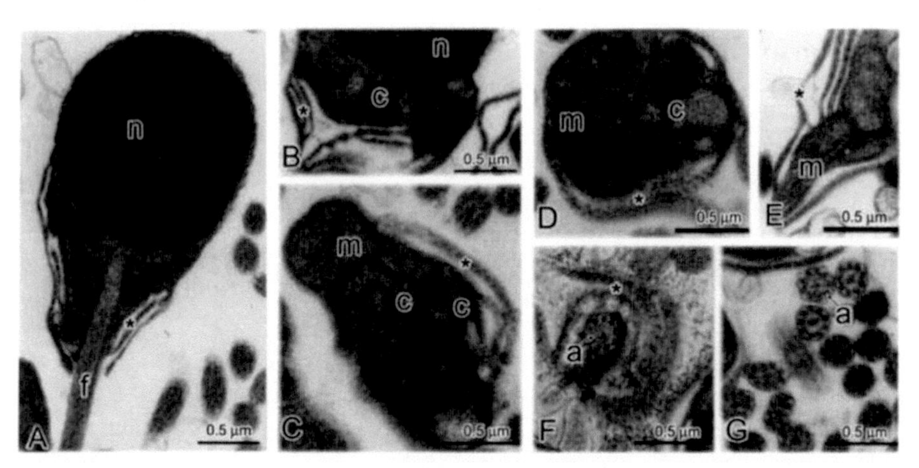

Fig. 11.40 *Pimelodella gracilis.* **A**. Longitudinal section of spermatozoon. Note the membranous loops at the peripheral and terminal regions of the midpiece. **B**. Situated close to the nucleus the proximal and distal centriole are lateral and parallel to one another. Note the absence of nuclear fossa. **C**. Transverse section of the midpiece at the level of the centrioles which are encircled by the mitochondria. **D**. Transverse section of the midpiece at the level of the initial segment of the axoneme which is also encircled by mitochondria. Note the absence of a cytoplasmic canal. **E**. Transverse section at the terminal end of the midpiece. **F**. Detail of the elongate fused mitochondria and the membranous loops. **G**. Transverse sections through flagella. a, axoneme; c, centrioles; f, flagellum; m, mitochondrion; n, nucleus; asterisk, membranous loops. Original.

Rhamdia quelen: The spermatozoon of *R. quelen* (Quagio-Grassiotto *et al.* 2005) is a type III aquasperm with an elongate, bullet-shaped nucleus that contains homogeneous, highly condensed chromatin but no nuclear fossa (Fig. 11.41A). The proximal centriole is lateral and perpendicular to the distal (Fig. 11.41B)

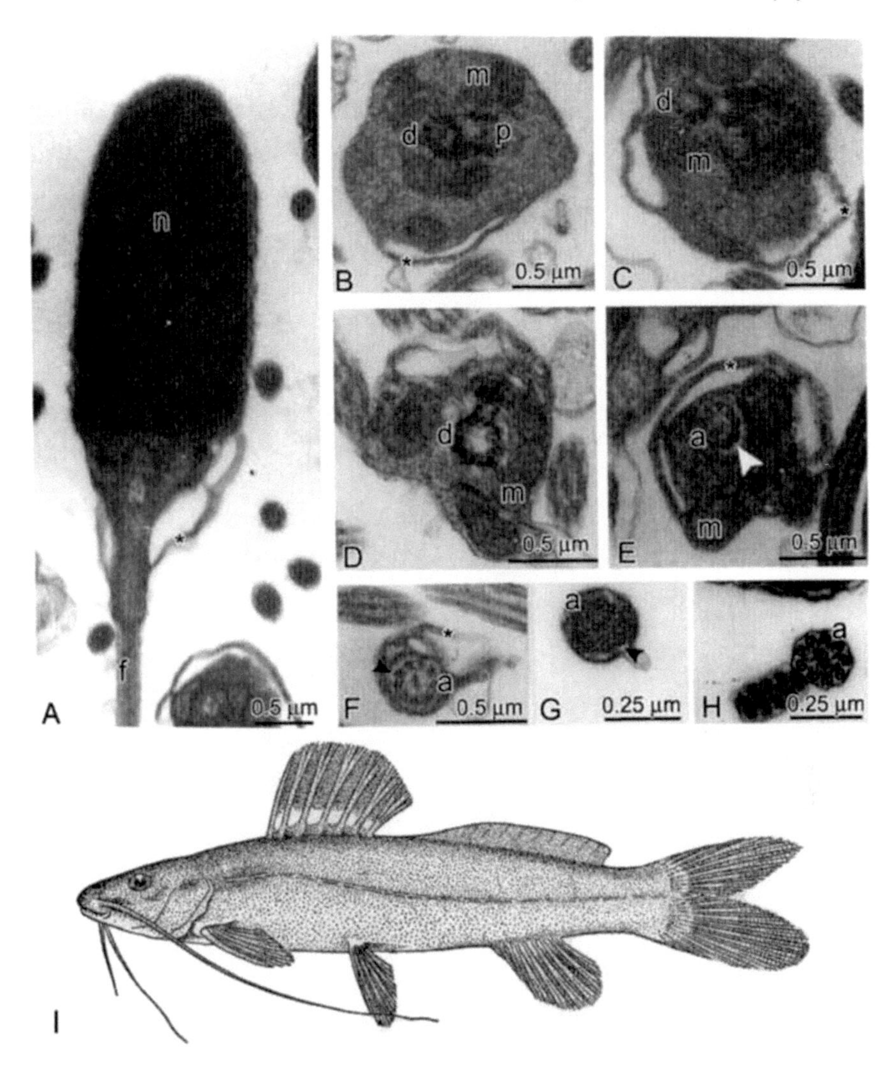

Fig. 11.41 *Rhamdia quelen*. **A**. Longitudinal section of spermatozoon. Note the bullet-shaped nucleus and absence of a nuclear fossa. **B**. Transverse section of midpiece showing the proximal centriole located lateral and perpendicular to the distal. **C-E**. Transverse sections at different levels of midpiece. Elongate mitochondria encircle the initial segment of the axoneme which is surrounded by narrow cisternae. Note the membranous loops. **F-G**. Transverse sections through the terminal end of the midpiece; the axoneme is surrounded by narrow cisternae and membranous loops. **H**. Transverse sections of flagellum. Note the classic 9+2 axoneme and absence of fins. At the terminal end the flagellum (lower left) becomes laterally compressed and the central pair of microtubules is lost. a, axoneme; d, distal centriole; f, flagellum; m, mitochondrion; n, nucleus; p, proximal centriole; arrowhead, cisternae surrounding the axoneme; asterisk, membranous loops. Original. I. *Rhambdia quelen*. From Norman, J. R. 1937. *Illustrated Guide to the Fish Gallery*. Trustees of the British Museum, London, Fig. 85A.

and both lie close to the central axis of the nucleus. The midpiece is short and thin. No cytoplasmic canal is present (Fig. 11.41A). Several elongate, fused mitochondria encircle the centrioles and initial segment of the axoneme, and narrow cisternae surround the axoneme separating it from the mitochondria (Fig. 11.41B-E). Long tubules or vesicles form membranous loops at the peripheral and terminal regions of the midpiece (Fig. 11.41B-G). The single flagellum has a classic 9+2 axoneme but no fins (Fig. 11.41H).

11.7.7 ICTALUROIDEA

11.7.7.1 Family Ictaluridae

Two ictalurids have been examined for sperm ultrastructure: *Amiurus nebulosus* (Emel'yanova and Makeyeva 1991a) And *Ictalurus punctatus* (Jaspers *et al.* 1976; Poirier and Nicholson 1982; Emel´yanova and Makeyeva 1991a) (adult, Fig. 11.42; sperm, Fig. 11.43).

Fig. 11.42 *Ictalurus punctatus.* By Norman Weaver. Courtesy of Sarah Starsmore.

Jaspers *et al.* (1976) give the following dimensions for 525 sperm of *Ictalurus punctatus* from mixed wild and domestic stocks of this commercially important species: total length of head and midpiece 3.9 μm; head length 2.3 μm; midpiece length 1.6; head width 2.4 μm; midpiece width 3.1 μm; flagellum length 94.9 μm. The standard error was very small. The head and contained nucleus are rounded and each of the two basal bodies lies in a moderate basal nuclear fossa (Fig. 11.43). The condensed chromatin is granular in appearance. The double nuclear envelope and the plasma membrane are tightly apposed over the anterior one half to two thirds of the sperm head. In this region the surface is slightly undulated and the individual membranes may be difficult to distinguish. Slightly more posteriorly in the head, the nuclear membrane appears wavy and the space between the two membranes varies. The nuclear membrane lining the nuclear fossa is distinct and constitutes a basal plate. The midpiece extends anteriorly, enveloping up to half of the nucleus in a collar-like arrangement. The mitochondria, located at the periphery of the midpiece at some distance from the centrioles, are not

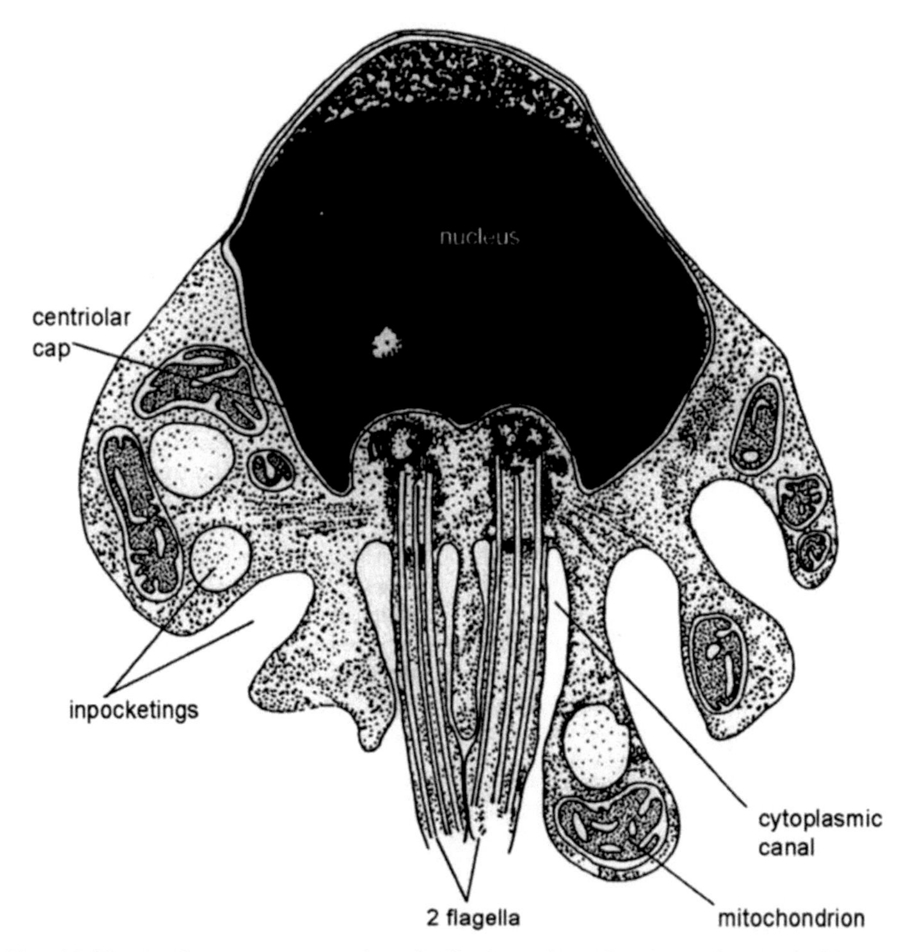

Fig. 11.43 *Ictalurus punctatus*. Longitudinal section of a spermatozoon. After a micrograph by Poirier, G. R. and Nicholson, N. (1982). Journal of Ultrastructure Research 80: 104-110, Fig. 2. From Jamieson, B. G. M. 1991. *Fish Evolution and Systematics: Evidence from Spermatozoa*. Cambridge University Press, Cambridge, Fig. 12.13.

fused and are circular to oblong in section. Inpocketings occur at irregular intervals along the surface of the midpiece and in cross section have the appearance of double-walled vacuoles. The cytoplasmic canal surrounding each flagellum also appears double-walled; its inner wall is thinner than the outer wall lining the canal. Putative beta glycogen granules, 150-300 Å in diameter, are present throughout the midpiece. A centriolar cap, circular in section, 0.35 µm wide with a 62 Å wall, extends from the proximal end of each of the two centrioles, while satellite-like structures surround the distal portion. Thin filaments connect each cap with the outer nuclear membrane in the fossa. Microtubules which extend from satellites in bundles of up to 20 radiate throughout the length and width of the midpiece. The axonemes are of the 9+2 type but the outer dynein arms are absent; inner arms are tentatively recognized. Structures similar in location to Y links, interdoublet links, and radial spokes [Afzelius rays] are present in the axoneme within the midpiece. Projections from the A subtubule curve below the adjacent doublet to connect with one of the central singlets (Poirier and Nicholson 1982). A peripheral sheet or ridge [fin] is absent (Jaspers *et al.* 1976). The statement (Poirier and Nicholson 1982) that the centriolar caps in *Ictalurus* are unique structures has been questioned as they strongly resemble those of *Lepisosteus* and, apparently, of *Neoceratodus* (*q.v.*) though no phylogenetic link is apparent from this (Jamieson 1991).

11.7.8 DORADOIDEA

11.7.8.1 Family Mochokidae

These are the Upside down catfish or squeaker in African freshwaters.

***Synodontis membranaceus* and *S. schall*:** These two species have been examined by Mattei (1970) and *S. schall* by Shahin (2007b). The spermatozoa of *Synodontis* are Type I aquasperm (Mattei 1970), though considered close to Type III postulated for *Chrysichthys* (Shahin 2007a). The spherical nucleus, 1.5 ± 0.2 μm in diameter, has a moderately deep basal fossa about two-fifths the length of the nucleus. The chromatin is highly electron-dense and consists of a compact homogeneous matrix that is devoid of fibers and contains electron-lucent lacunae. The proximal centriole is anterior and oblique (Mattei 1970) or perpendicular (Shahin 2007b) to the distal centriole, the former being continued in the fossa, the latter with its anterior end in the fossa; the two are interconnected by osmiophilic filaments and both have the 9 + 0 arrangement of triplets. In addition, the proximal centriole and the proximal part of the distal centriole are anchored to the nucleus by numerous fibrils while the basal body is attached laterally to the nucleus on both sides by a basal foot and is traversed by a thick basal plate. The midpiece measure 1.2 ± 0.22 μm in diameter and 0.6 ± 0.01 μm in length and contains numerous (up to ten) spherical to elongate mitochondria distributed throughout and separated from the single 9+2 flagellum by the cytoplasmic canal; there are no vesicles. The flagellum is 43.8 ± 1.23 um long; flagellar fins are absent (Shahin 2007b).

11.7.8.2 Family Auchenipteridae

Sperm ultrastructure has been examined in three species of driftwood catfish in the genus *Trachelyopterus*: *T. lucenai* (Burns *et al.* 2002), *T. galeatus* (Parreira and Godinho 2004), and *T. striatus* (Rinaldo José Ortiz unpublished).

***Trachelyopterus lucenai*:** The spermatozoon of *T. lucenai* (Burns *et al.* 2002) is an introsperm with a very elongate conical nucleus measuring 10.1 μm in length and 0.34 μm in width (Fig. 11.44A). The nucleus of the spermatozoon of *T. galeatus*, on the other hand, measures 12.9 μm in length (Parreira and Godinho 2004). In all three species analyzed, the nucleus does not reach the

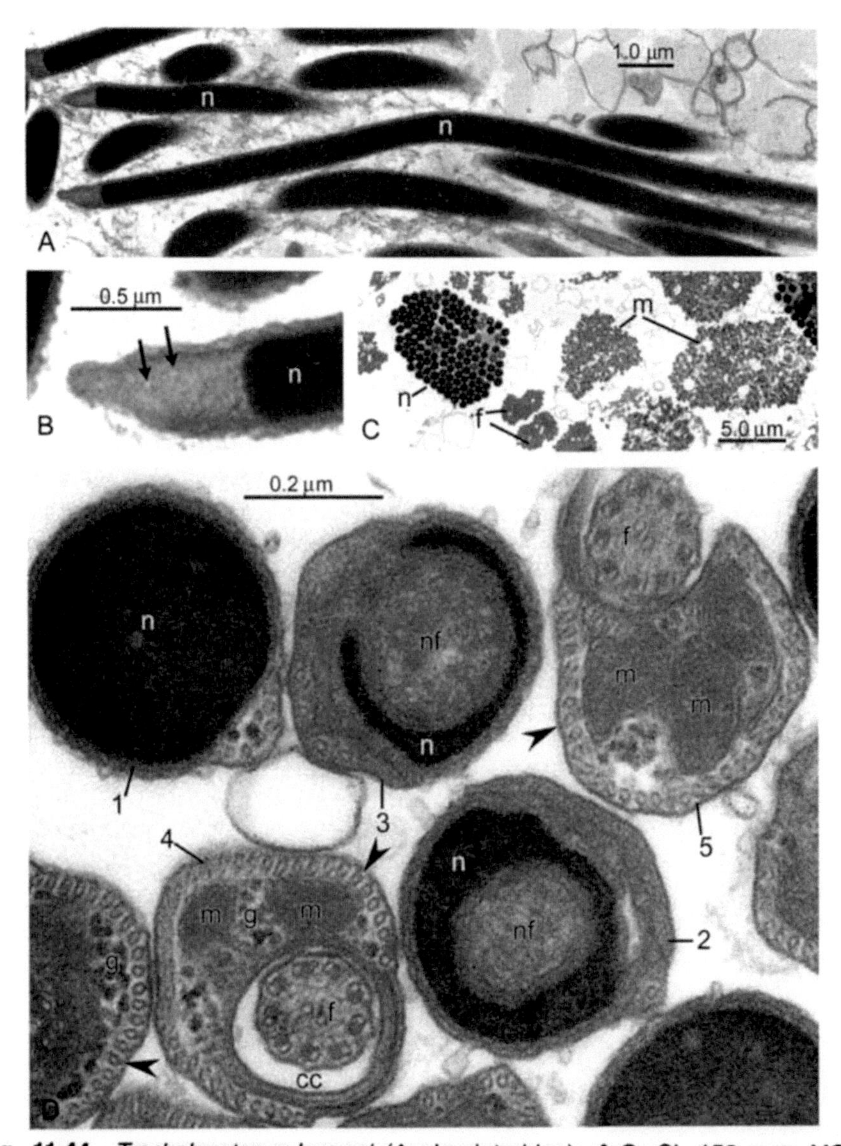

Fig. 11.44 *Trachelyopterus lucenai* (Auchenipteridae). **A-C**. SL 153 mm, MCP 18410, Brazil. D. SL 147 mm, MCP 18469, Brazil. **A**. Longitudinal sections through spermatozoa showing elongate nuclei and anterior tips of cells at left. **B**. Higher magnification of cell tip from A showing small vesicles (arrows) in cytoplasm. **C**. Transverse sections through the tightly packed spermatozeugmata at the level of the sperm nuclei (n), midpieces (m) and flagella (f). **D**. Successively more posterior transverse sections (1-5) through spermatozoa. 1, mid-nucleus; 2, anterior part of nuclear fossa; 3, posterior part of nuclear fossa; 4, midpiece with cytoplasmic canal; 5, midpiece at opening of cytoplasmic canal; cc, cytoplasmic canal; f, flagellum; g, glycogen rosettes; m, mitochondrion; n, nucleus; nf, nuclear fossa; arrow, vesicle; arrowhead, accessory microtubule. Original.

anterior tip of the cell where the cytoplasm contains small vesicles approximately 0.02 µm in diameter (Fig. 11.44B). Parreira and Godinho (2004) reported that these vesicles also contained "small inner vesicular membranes" and that their shape was round in spermatozoa present within the sperm ducts, but elongate in spermatozoa within the ovary, and suggested that these vesicles may be equivalent to an acrosome. A recent study on *T. striatus* (Rinaldo José Ortiz unpublished) revealed that these vesicles probably develop during final chromatin condensation at the end of spermiogenesis and represent expansions of the nuclear envelope. Thus, no acrosome is present. In addition, Rinaldo José Ortiz (unpublished) showed that *T. striatus* undergoes Type I spermiogenesis. The nucleus of *T. lucenai* contains highly condensed, homogeneous chromatin, and an eccentric, moderately deep nuclear fossa at the posterior end (Fig. 11.44D). The proximal centriole is perpendicular to the distal. The proximal centriole and part of the distal are contained within the nuclear fossa. The long midpiece, which begins immediately posterior to the nucleus, initially has a short (1.4 µm) cytoplasmic collar at one side which contains the flagellum. Numerous elongate mitochondria with lamellar cristae are found throughout the midpiece. A single row of accessory microtubules, which run antero-posteriorly around most of the periphery of the cell beginning just posterior to the nucleus, eventually encircle the entire midpiece outside the cytoplasmic canal (Fig. 11.44D). No vesicles are seen, but abundant glycogen rosettes are present throughout the midpiece. The carbohydrate nature of the latter was confirmed by an intense positive reaction of the midpieces for the periodic acid-Schiff technique using light microscopy (Meisner *et al.* 2000; Burns and Weitzman 2005). The single flagellum lacks fins. The axoneme has the classic 9+2 arrangement with the A-tubules of each peripheral doublet being electron-dense. The elongate spermatozoa of *T. lucenai* form highly compact spermatozeugmata consisting of upwards of 130 tightly packed cells aligned in parallel (Fig. 11.44C).

11.7.9 SILUROIDEA

11.7.9.1 Family Siluridae

Silurids, sheatfishes, investigated for sperm ultrastructure are *Silurus glanis* (Emel'yanova and Makeyeva 1991b; Kudo *et al.* 1994); *S. asotus* (Kwon et al. 1998; Kwon and Kim 2004) (Fig. 11.45); *S. microdorsali* (Lee and Kim 2001; Kwon and Kim 2004) and *S. soldatovi* (Yin *et al.* 2000). The spermatozoon of *Silurus* is an aquasperm with a spherical to ovoid nucleus containing highly condensed, flocculent chromatin and a moderately deep nuclear fossa. The proximal centriole is anterior and oblique to or directly in front of the distal. Both centrioles are contained within the nuclear fossa. The midpiece is short with several spherical mitochondria distributed throughout. Regularly arranged tubules or vesicles can be found in the posterior third of the midpiece. A short cytoplasmic canal contains the flagellum. The single flagellum has a classic 9+2 axoneme but no fins.

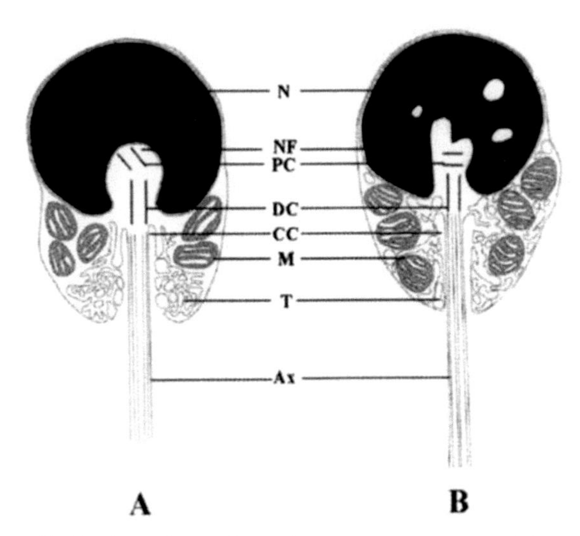

Fig. 11.45 Longitudinal sections through spermatozoa. **A**. *Silurus asotus*. **B**. *S. microdorsalis*. Ax, axoneme; CC, cytoplasmic canal; DC, distal centriole; M, mitochondria; N, nucleus; NF, nuclear fossa; PC, proximal centriole; T, tubules. From Kwon, A. S. and Kim, K. H. 2004. Korean Journal of Limnology 16: 128-134, Fig. 3.

11.7.9.2 Family Malapteruridae

The spermatozoon of *Malapterurus electricus*, an electric catfish (adult, Fig. 11.46A), examined by Mattei (1991) and Shahin (2006a), is an aquasperm with an ovoid nucleus containing highly condensed fibrous chromatin with some pale lacunae and two shallow nuclear fossae. The present study reveals a broad, less electron-dense apical zone in the nucleus (Fig. 11.46).

The centrioles are lateral, and parallel to one another. The anterior part of each centriole is located within one of the nuclear fossae. Each centriole gives rise to one axoneme. The long midpiece has numerous spherical to slightly elongate mitochondria mainly concentrated near the base of the nucleus. Each flagellum is contained within a separate cytoplasmic canal. Small vesicles occur at the posterior end and in a narrow layer at the periphery of the midpiece. Each of the two flagella has a classic 9+2 axoneme; there are no fins but (present study) a broad band of cytoplasm surrounds the axoneme.

11.7.9.3 Family Doradidae

In the Doradidae, Thorny catfishes, *Anadoras wedellii*, as in the ariid *Genidens genidens*, spermatogenesis is semicystic and spermiogenesis occurs in the luminal compartment of the testis. There are two flagella which, unlike those of *G. genidens*, lie in independent cytoplasmic canals.

Remarks: The type of spermatogenesis of Ariidae and Doradidae, and of the others families of catfish with the same type of spermatogenesis, indicates that in Siluriformes the occurrence of biflagellate sperm, or one flagellum with two axonemes, and semicystic spermatogenesis are associated characters.

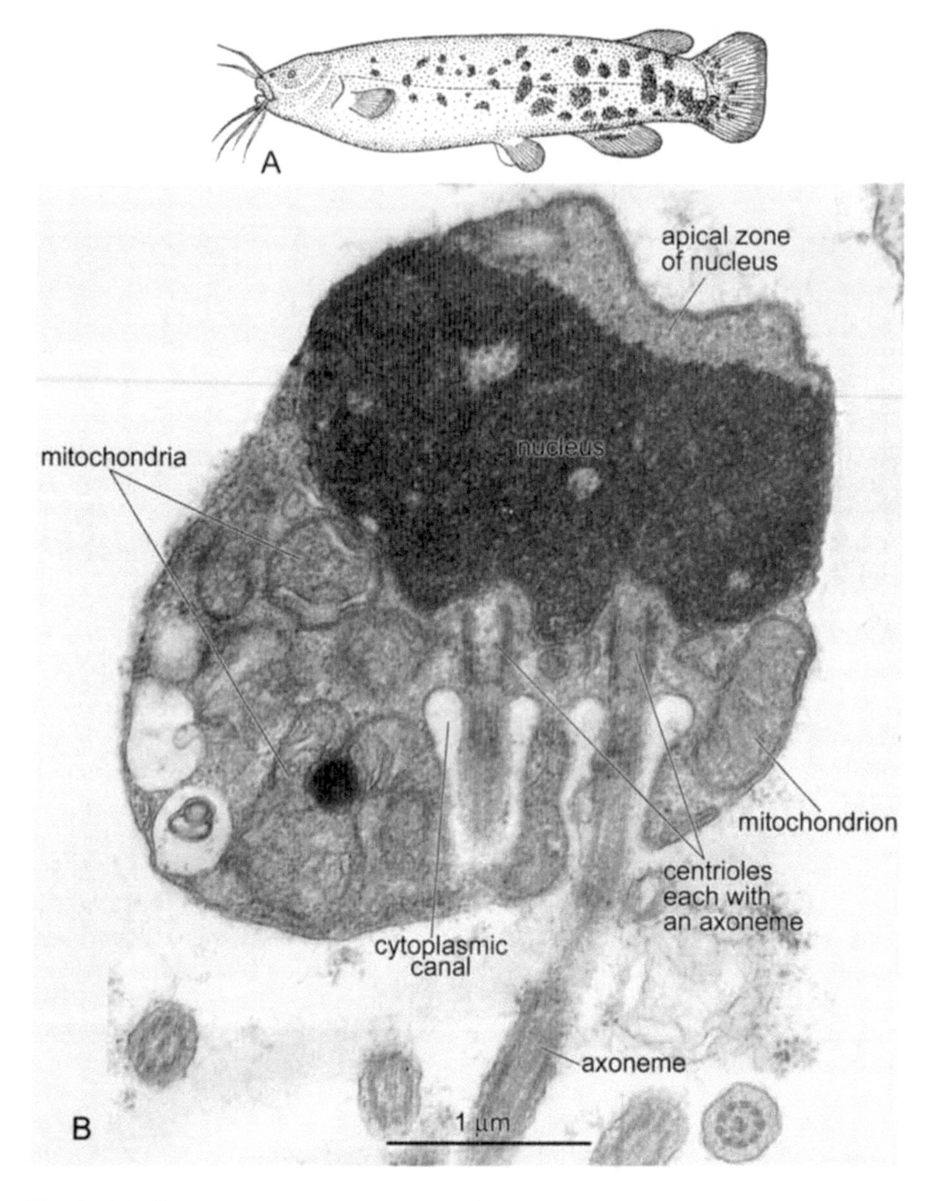

Fig. 11.46 *Malapterurus electricus* (Malapteruridae). **A**. Adult fish. From Norman, J. R. *Illustrated Guide to the Fish Gallery*. Trustees of the British Museum, London, Fig. 95B. **B**. Longitudinal section of a spermatozoon, showing the biflagellate condition. Original, from a micrograph courtesy of Professor Xavier Mattei.

11.7.9.4 Family Plotosidae

***Plotosus lineatus*:** In the anacrosomal aquasperm of *Plotosus lineatus* (= *P. anguillaris*), an eeltail catfish, the nucleus is bell-shaped with a deep, stepped nuclear fossa. The cross-shaped transverse sectional profile of the fossa appears to be an apomorphy of this species. The chromatin is described as forming dense masses with scattered pale lacunae but in micrographs (Fig. 11.47) appears strongly and homogeneously condensed. The mitochondria are small and spherical and an extension of the mitochondrial collar is said to surround the base of the nucleus. The two mutually perpendicular centrioles are located in the fossa approximately at the center of the nucleus. A lateral centriolar connection develops around the distal centriole (Hara and Okiyama 1998). There are no distinct flagellar fins but a wide band of cytoplasm invests the axoneme (Fig. 11.47B).

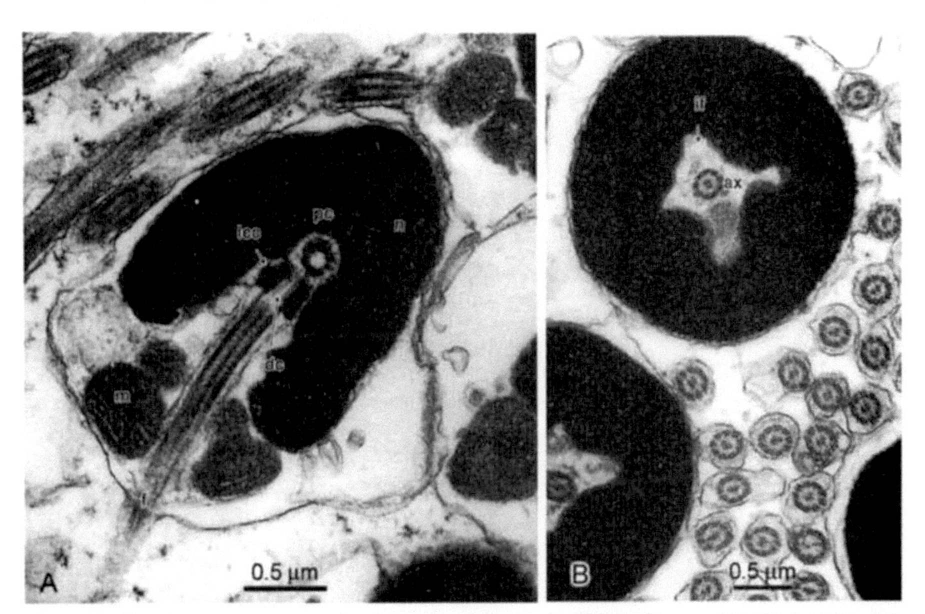

Fig. 11.47 *Plotosus anguillaris* spermatozoa (Plotosidae, Siluriformes). **A.** Longitudinal section. **B.** Transverse sections of nuclei and flagella. ax, axoneme; dc, distal centriole; if, implantation fossa; m, mitochondrion; lcc, lateral centriolar connection; n, nucleus; pc, proximal centriole. Adapted from Hara, M. and Okiyama, M. 1998. Bulletin of The Ocean Research Institute, University of Tokyo 33: 1-138: Fig. 27.

11.7.9.5 Family Clariidae

***Clarias anguillaris, C. gariepinus* and *C. lazera*.** Clariids, air breathing catfishes, examined for sperm ultrastructure are *Clarias anguillaris* (=*C. senegalensis*) (Mattei 1970; 1991) (adult, Fig. 11.48; sperm, Fig. 11.50); *C. gariepinus* (Mansour *et al.* 2002; Mekkawy and Osman 2006; Lahnsteiner and Patzner 2007) (Fig. 11.49), *C. lazera* (Ismial and Khalifa 1995; Lin 1995).

Fig. 11.48 *Clarias anguillaris (= C. senegalensis).* After Grassé, P.P. (1958). Traité de Zoologie. XIII Agnathes et poissons. Masson et Cie, Paris, Fig. 1651.

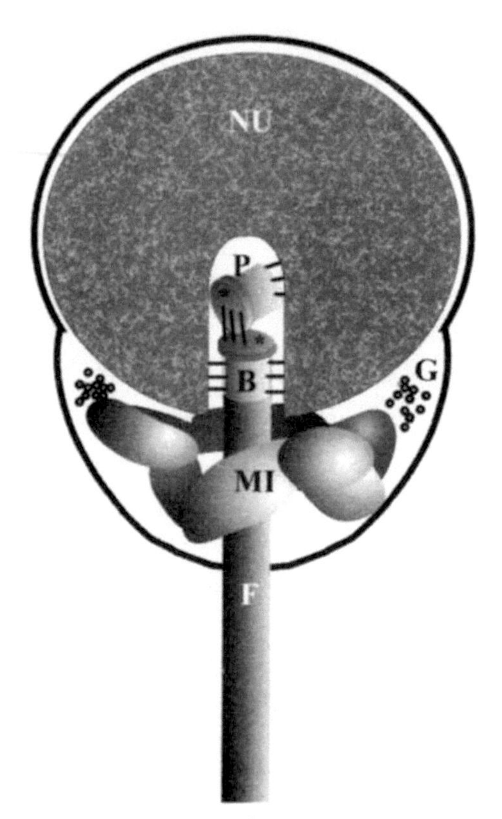

Fig. 11.49 *Clarias gariepinus.* Schematic reconstruction of the spermatozoon. B, Basal body; F, flagellum; G, glycogen; MI, mitochondria; NU, nucleus; P, Proximal centriole; *, Electron-dense disk. From Mansour, N., Lahnsteiner, F. and Patzner, R. A. 2002. Journal of Fish Biology 60: 545-560, Fig. 1. With permission of Blackwell.

The spermatozoon of *Clarias* is an aquasperm with a spherical to ovoid nucleus containing highly condensed, flocculent chromatin and a moderately deep nuclear fossa. The proximal centriole is anterior and oblique to the distal, and both are contained within the nuclear fossa. The midpiece is short and lacks a cytoplasmic canal or, in *C. anguillaris* (Fig. 11.50), this is present.

Several elongate to irregularly shaped mitochondria fuse to form a ring around the initial segment of the axoneme. A well-organized vesicular system is located at the peripheral and basal regions of the midpiece. Contrary to Mansour *et al.* (2002), Mekkaway and Osman (2006) state that mitochondria in *C. gariepinus* are distributed separately within this reticular structure but this requires confirmation. The single flagellum has a classic 9+2 axoneme but no fins. The topography of the spermatozoon of *C. gariepinus* seems to conform to the diameter of the inner aperture of the micropylar canal and to provide binding facets for the attachment of the spermatozoon on the chorion surface by a ligand-receptor mechanism (Mekkawy and Osman 2006).

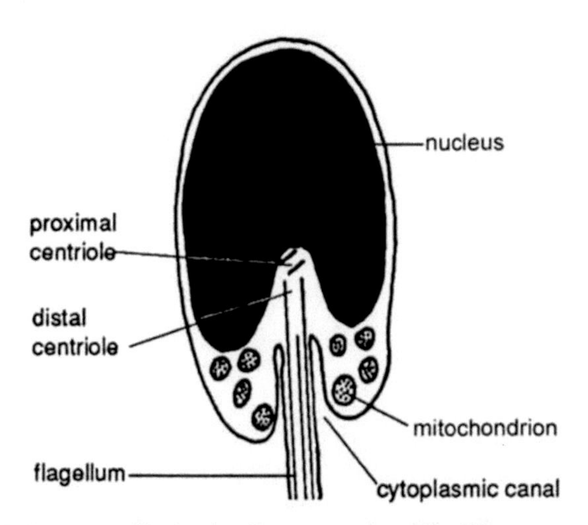

Fig. 11.50 *Clarias anguillaris (= C. senegalensis).* Diagrammatic longitudinal section of the spermatozoon. After Mattei, X. (1970). Pp 59-69. In B. Baccetti (ed.) *Comparative Spermatology.* Academic Press: New York. Fig. 4: 10.

11.7.9.6 Family Heteropneustidae

***Heteropneustes fossilis*:** The spermatozoon of *H. fossilis* (Nath and Chand 1998), an air sac catfish, is an aquasperm with an ovoid nucleus containing highly condensed, flocculent chromatin and a deep nuclear fossa. Little information is available regarding the centrioles. However, both appear to be contained within the nuclear fossa. The midpiece has a cytoplasmic canal. No information is available on either mitochondria or vesicles. The single flagellum has a classic 9+2 axoneme but no fins.

11.7.10 BAGROIDEA

11.7.10.1 Family Ariidae

Examined ariids, sea catfishes, are *Arius heudeloti* (Mattei 1991) and *Genidens genidens* (Rinaldo José Ortiz, personal communication). Sperm ultrastructure is provided for *G. genidens* below.

***Genidens genidens*:** Spermatogenesis in *G. genidens* has been shown to be semicystic with spermiogenesis occurring in the luminal compartment of the testis (Fig. 11.51A). The spermatozoon is an aquasperm with an ovoid nucleus that contains homogeneous, highly condensed chromatin and a deep nuclear fossa (Fig. 11.51B). A cytoplasmic canal is absent. The centrioles are lateral

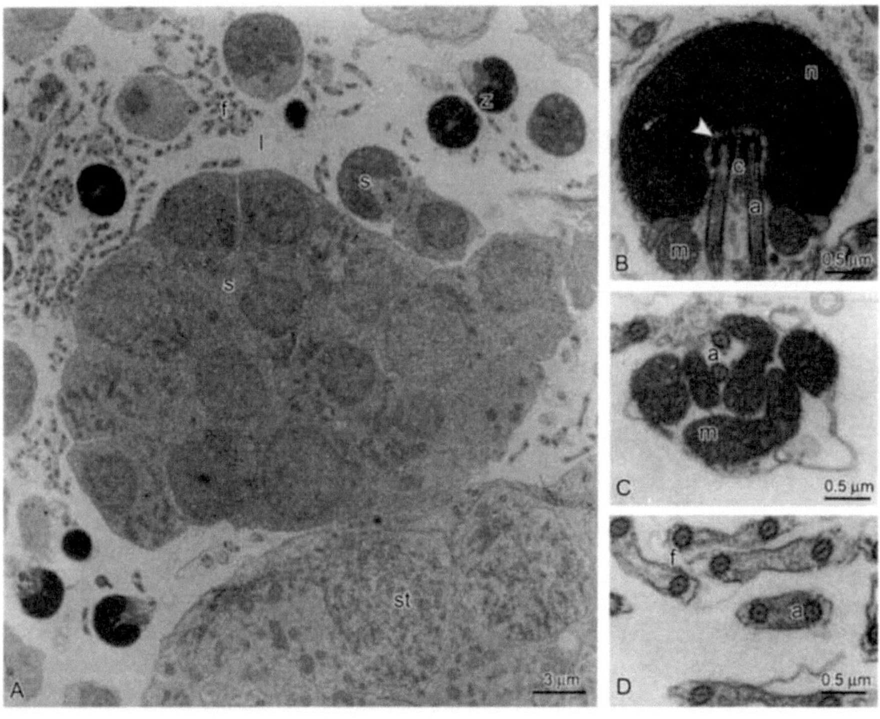

Fig. 11.51 *Genidens genidens* (Ariidae). **A.** Section through the luminal compartment of the testis. In semicystic spermatogenesis, first described by Mattei (1993), although spermatogonial proliferation and the meiotic divisions occur inside the spermatocysts, spermatid differentiation is extracystic and occurs outside the cysts in the luminal compartment of the testis. In the luminal compartment, clusters of spermatids recently released from the cysts probably remain connected to one another by cytoplasmic bridges. Spermatids gradually lose these connections and differentiate. Spermatid differentiation is not synchronous and cells in distinct phases of development can be seen together in the luminal compartment. Spermatozoa are also present. **B.** Longitudinal section of the nucleus and midpiece with the centrioles and initial segments of both axonemes inside the deep nuclear fossa. **C.** Transverse sections of the midpiece showing large elongate mitochondria surrounding the axonemes. Note the absence of a cytoplasmic canal. **D.** Transverse sections at different levels of the flagellum. Note that both axonemes are contained within a common, flattened flagellum. a, axoneme; c, centrioles; f, flagellum; l, luminal compartment; m, mitochondrion; n, nucleus; s, spermatids; st, primary spermatocytes; z, spermatozoa, arrowhead, nuclear fossa. Original, courtesy of Rinaldo José Ortiz.

and parallel to one another, and both centrioles and the initial segment of the axoneme are contained within the nuclear fossa, with each centriole giving rise to one axoneme (Fig. 11.51B). The midpiece has no cytoplasmic canal or vesicles. The several large, elongate mitochondria are concentrated around the initial segments of the axonemes (Fig. 11.51C). There is only a single, centrally located, flatened flagellum which contains both axonemes (Fig. 11.51D). Each axoneme has a classic 9+2 arrangement. No flagellar fins are present.

11.7.10.2 Family Schilbeidae

Physailia pellucida: The spermatozoon of *P. pellucida* (Mattei 1991), a schilbeid catfish, is an aquasperm with a spherical nucleus which has a moderately deep nuclear fossa. There is no information on chromatin condensation. The proximal centriole is anterior and perpendicular to the distal, and both are contained within the nuclear fossa. The symmetric midpiece contains numerous spherical to ovoid mitochondria distributed throughout. The single flagellum has a classic 9+2 axoneme and no fins.

11.7.10.3 Family Bagridae

Several bagrids have been examined for sperm ultrastructure: *Bagrus* sp. (Mattei 1991 and this account) (Fig. 11.52); *Auchenoglanis* sp., and *Chrysichthys* sp. (Mattei 1991); *Leiocassis longirostris* (Zhang *et al.* 1992, 1993; Wei *et al.* 1997);

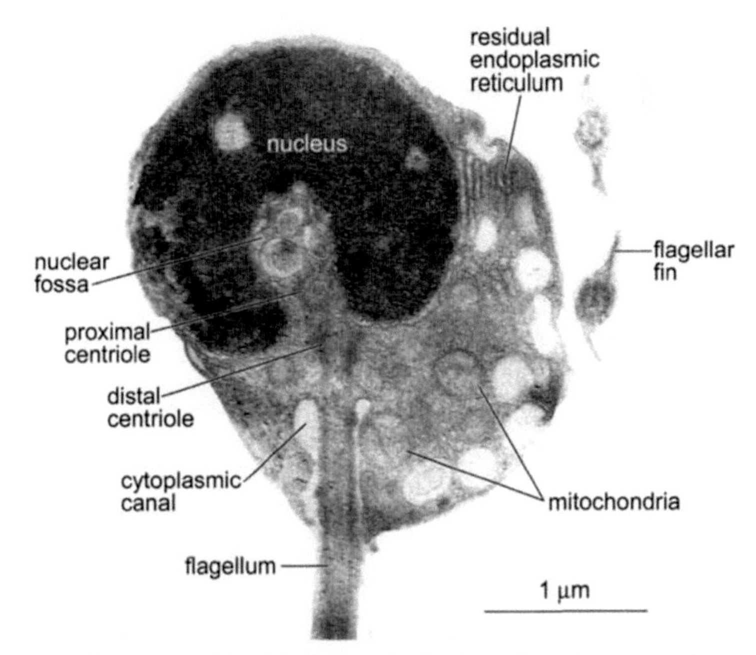

Fig. 11.52 *Bagrus* sp. (Bagridae). Longitudinal section of a spermatozoon. This example has residual endoplasmic reticulum and may not be fully developed. It is also unusual in having mitochondrion-like structures within the nuclear fossa in addition to the mitochondria in the broad midpiece. Original, from an unpublished negative courtesy of Professor Xavier Mattei.

L. ussuriensis (Emel'yanova and Makeyeva 1991; Kim, K.H. and Lee 2000); *Pseudobagrus fulvidraco* (Emel'yanova and Makeyeva 1991b; Lee 1998), *Mystus armatus* (Mansour and Lahnsteiner 2003); *Mystus armatus* (Mansour and Lahnsteiner 2003); *Pseudobagrus fulvidraco* (Lee 1998); *Pseudobagrus brevicorpus* (Kim and Lee 2003) (Fig. 11.53).

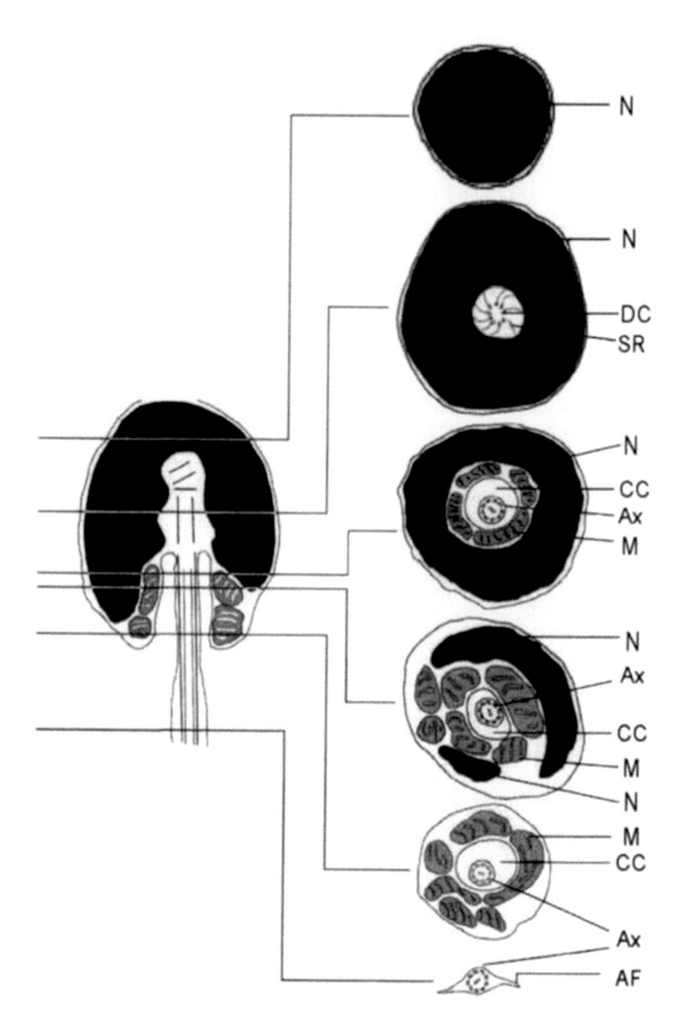

Fig. 11.53 *Pseudobagrus brevicorpus.* Longitudinal and transverse sections through the spermatozoon. AF, flagellar fins; Ax, axoneme; CC, cytoplasmic canal; DC, distal centriole; M, mitochondria; N, nucleus; NF, nuclear fossa; SR, satellite rays. From Kim, K. H. and Lee, J. I. 2003. Journal of the Korean Fisheries Society 36, Fig. 3.

Like the Clariidae, Mochokidae, and Schilbidae, spermatozoa of these Bagridae are aquasperm with ovoid nuclei that contain highly condensed, fibrous chromatin and a deep nuclear fossae. The proximal centriole is anterior and oblique to (at variable angles) or directly in front of the distal. Both centrioles and the initial segment of the flagellum are contained within the nuclear fossa or, in *Bagrus* sp. (Fig. 11.52) the distal centriole lies shortly outside the fossa. The short midpiece contains several spherical to elongate mitochondria and some vesicles. The initial segment of the flagellum is contained within a cytoplasmic canal. The single flagellum has a classic 9+2 axoneme and two long fins.

11.7.10.4 Family Pimelodidae

Several pimelodids have been examined for sperm ultrastructure: *Pimelodus maculatus* (Quagio-Grassiotto and Oliveira 2008); *Pimelodus ornatus* (original description); *Pseudoplatystoma fasciatum* (Batlouni *et al.* 2006; Quagio-Grassiotto and Oliveira 2008); *Sorubim lima* (Quagio-Grassiotto and Carvalho 2000) and *Iheringichthys labrosus* (Santos *et al.* 2001). Sperm ultrastructure is described below for three species.

***Sorubim lima, Pimelodus ornatus* and *Pseudoplatystoma fasciatum*:** The spermatozoa of all pimelodids analyzed undergo type III spermiogenesis (Fig. 11.54A-C). Spermatozoa also have the same basic morphology, being type III aquasperm with spherical nuclei that contain homogeneous, highly condensed chromatin (Fig. 11.54D-F). Only *I. labrosus* has a very shallow nuclear fossa (Santos *et al.* 2001); fossae are absent in the other species (Fig. 11.54D-F). The proximal centriole is anterior and perpendicular to the distal and both lie close to the central axis of the nucleus. The short midpiece contains several elongate, apparently fused mitochondria that form a ring around the centrioles (Fig. 11.54G,H). A short cytoplasmic canal is present only in *I. labrosus* (Santos *et al.* 2001). Numerous large vesicles or tubules are found throughout the midpiece (Fig. 11.54D-I). The single flagellum has a classic 9+2 axoneme and no fins.

***Conorhynchos conirostris*:** *C. conirostris* has long been regarded as *incertae sedis* but from *rag1* and *rag2* nuclear gene sequences Sullivan *et al.* (2006) have shown it to lie within a Pimelodoidea clade which has two subclades Pimelodidae + Pseudopimelodidae, and Heptapteridae + *Conorhynchos*. The spermatozoon of *C. conirostris* (Lopes *et al.* 2004) is an aquasperm with a slightly elongate conical nucleus that appears to contain homogeneous, highly condensed fibrous chromatin and a very deep nuclear fossa. Both centrioles, and probably the initial segment of the flagellum, are contained within the nuclear fossa. No other information is available on the centrioles. The short midpiece appears to contain a single large mitochondrion that encircles the axoneme. The cytoplasmic canal formed at the beginning of spermiogenesis appears to close prior to spermiation. Some small vesicles can be seen in the peripheral region of the midpiece. The single flagellum has a classic 9+2 axoneme and no fins. The ring-shaped mitochondrion supports affinity with the Pimelodidae.

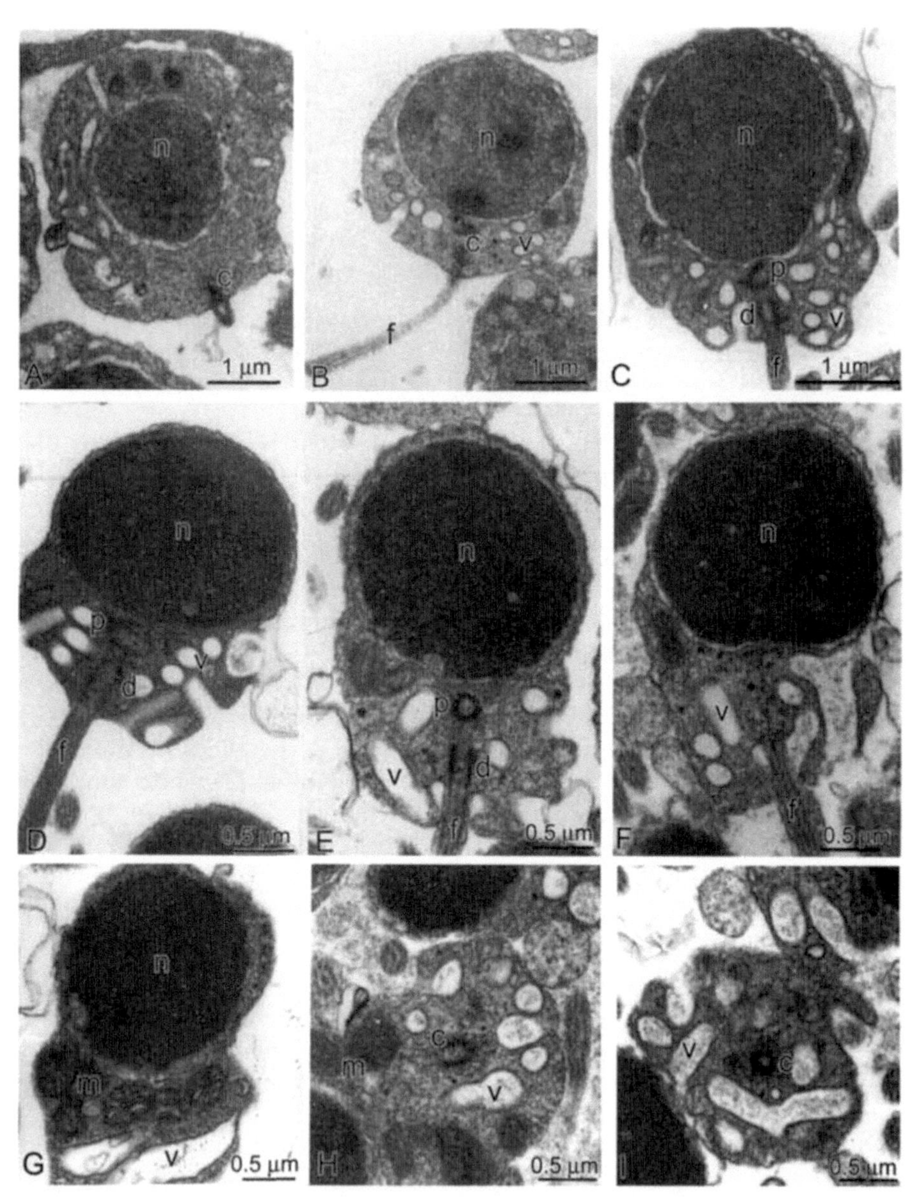

Fig. 11.54 Type III spermiogenesis and sperm ultrastructure in Pimelodidae. **A-C**. Longitudinal sections through spermatids of *Sorubim lima*. In type III spermiogenesis, first described in the Pimelodidae (Quagio-Grassiotto and Carvalho, 2000; Quagio-Grassiotto and Oliveira 2008),and in the Heptapteridae (Quagio-Grassiotto *et al.*, 2005) the centrioles are originally perpendicular to the central axis of the nucleus, centriolar migration and nuclear rotation do not occur. **D-F**. Longitudinal sections through spermatozoa of *Sorubim lima, Pimelodus ornatus* and *Pseudoplatystoma fasciatum*, respectively. Note the similar

Fig. 11.54 Contd. ...

11.7.10.5 Family Claroteidae

Claroteids are worldwide, tropical to warm temperate.

***Chrysichthys* sp. and *C. auratus*:** The spermatozoon of *Chrysichthys* sp. examined by Mattei (1991) was included in bagrids with a simple spermatozoon. It has a classical Type I spermatozoon (This account, Fig. 11.55). The spheroidal nucleus has strongly condensed chromatin and a

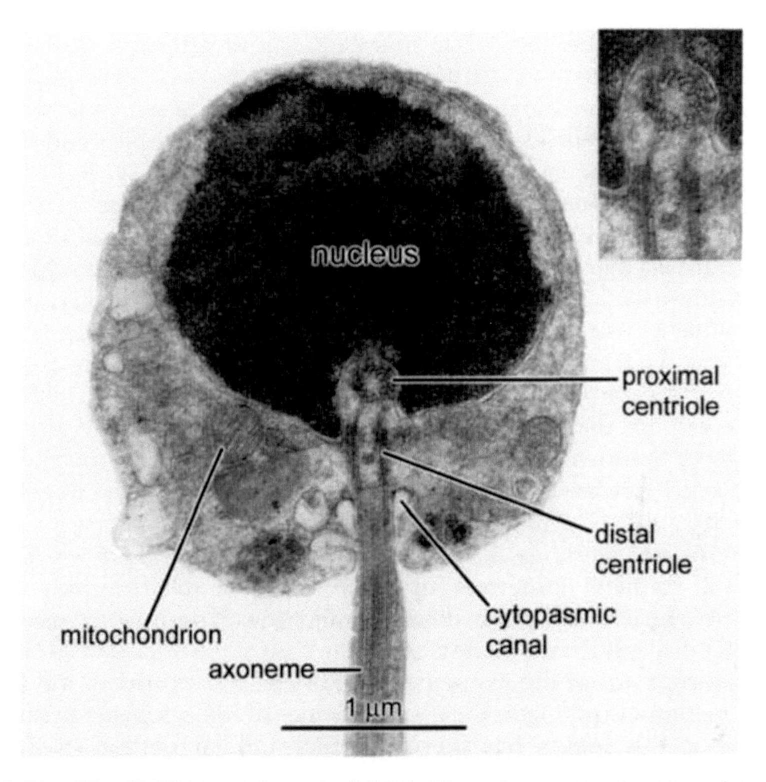

Fig. 11.55 *Chrysichthys* sp. Longitudinal section of a spermatozoon, showing a Type I morphology. Inset: detail of the proximal centriole showing 9 triplets of microtubules. Original, from a TEM negative courtesy of Professor Xavier Mattei.

Fig. 11.54 Contd. ...

morphologies of these three spermatozoa and the absence of nuclear fossae. Situated close to the nucleus, the proximal centriole is anterior and perpendicular to the distal. **G**. Longitudinal section through the nucleus and midpiece of *Pimelodus ornatus*. **H-I.** Transverse sections at different levels through the midpiece in *Pseudoplatystoma fasciatum*. Note the numerous large vesicles or tubules that connect with the plasma membrane, the peripheral distribution of the mitochondria and the absence of a cytoplasmic canal. c, centriole; d, distal centriole; f, flagellum; m, mitochondrion; n, nucleus; p, proximal centriole; v, vesicles. Original.

well defined though small basal fossa which contains the proximal centriole and the anterior tip of the distal centriole (basal body). The two centrioles are mutually perpendicular and are interconnected by stout fibers. There is a small but anteriorly wide cytoplasmic canal within a wide midpiece which contains several small cristate mitochondria.

The spermatozoon of *Chrysichthys auratus* described by Shahin (2007a) has an elongated conical-shaped head, a long midpiece and a long tail or flagellum. The nucleus, measuring 1.8 ± 0.01 μm in diameter and 2.1 ± 0.01 μm in length, has the form of an inverted U in longitudinal section and is surrounded by a narrow layer of cytoplasm with no organelles (Fig. 11.56A, B). There is a shallow basal nuclear fossa, the length of which is about one-third of that of the nucleus. It contains the centriolar complex and the basal body of the axoneme, is filled by an electron-dense material (Fig. 11.56A, B) and appears bell-shaped in longitudinal section and circular in transverse section. The chromatin is highly electron-dense and consists of compact coarse granular patches, with irregularly-shaped electron-dense thick fibers and electron-lucent lacunae. The proximal and distal centrioles are connected to each other by osmiophilic filaments. The proximal centriole is anchored to the nucleus by numerous anchoring fibrils and the basal body is attached laterally to the nucleus on both sides by the basal foot (Fig. 11.56B). Each centriole exhibits the classical (9 + 0) nine-microtubular triplet pattern. The basal part of the basal body is not traversed by a basal plate and the axes of the two centrioles are perpendicular to each other and to the flagellar axis (11.56A, B).

The midpiece measures 1.9 ± 0.02 μm in diameter and 1.8 ± 0.01 μm in length and contains numerous (up to six) unequal spherical mitochondria (about 0.1-0.6 ± 0.05-0.03 μm in diameter) and several peripheral vesicles (Fig. 11.56A-C). The mitochondria lie close to the base of the nucleus and surround the initial segment of the axoneme and are closely applied to the flagellar plasma membrane owing to the disappearance of the cytoplasmic canal. The axoneme in this region has the same microtubular pattern (9+2) as the flagellum (Fig. 11.56C,D) which is 30.1 ± 0.03 μm in length and 0.3 ± 0.01 μm in diameter; inner and outer dynein arms are present. The flagellum has neither lateral fins nor a membranous compartment but its base is surrounded by a backward extension of the peripheral vesicles located in the midpiece (Fig. 11.56BC).

Remarks: The spermatozoa of the two species of *Chrysichthys* differ more than is usual in a genus.

11.7.11 Occurrence of Siluriform Biflagellarity

Nine families of the Siluriformes, in five clades, have biflagellate sperm, so far as is known from sampling to date. At least 14 families, in disparate siluriform groups, have only uniflagellate sperm whereas nine have biflagellate sperm and one (Heptapteridae) has both conditions. The

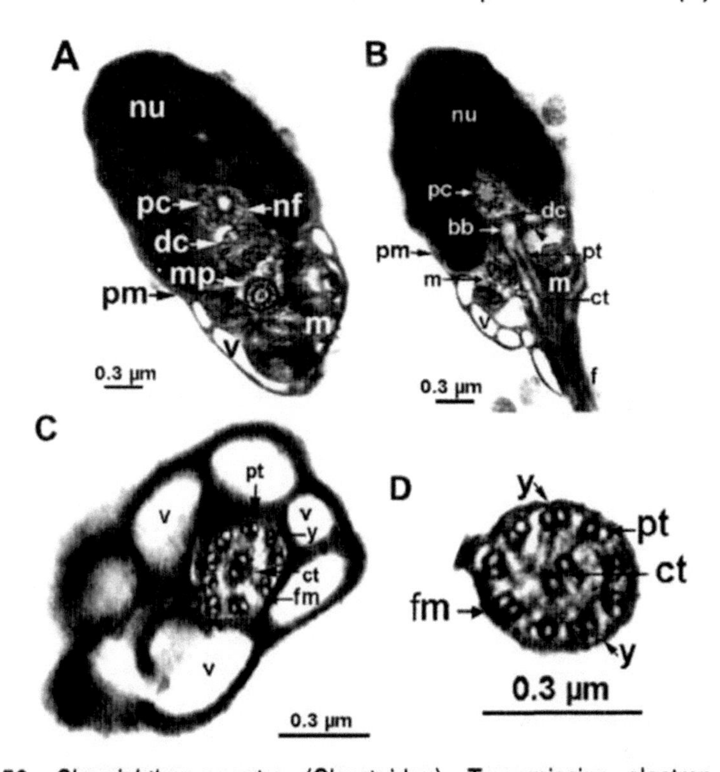

Fig. 11.56 *Chrysichthys auratus* (Claroteidae). Transmission electron micrographs of the spermatozoon. **A**. Oblique longitudinal section in the head and midpiece regions showing the inverted U-shaped nucleus (nu) surrounded by the plasma membrane (pm) that is applied very tightly to the nuclear envelope, the proximal (pc) and distal (dc) centrioles, which are located in the nuclear fossa (nf), and the midpiece region (mp) where there are three mitochondria (m) and numerous vesicles (v). Note the anchoring fibrils connecting the proximal centriole to the nucleus and the microtubular structure of the axoneme in the midpiece region. **B**. Longitudinal section showing the inverted, conical U-shaped nucleus (nu) surrounded by the plasma membrane (pm), which is closely applied to the nuclear envelope, the medial shallow nuclear fossa (nf) that contains both the proximal (pc) and distal (dc) centrioles and part of the basal body (bb), which is attached to the nucleus on both sides by the basal foot (arrowhead), a short midpiece (mp) with no cytoplasmic canal, the mitochondria, which are separated from the flagellum by the flagellar membrane, and a long tail, of which its base is enclosed by the vesicles extending backwards from the midpiece. **C**. Transverse section in the initial region of the flagellum showing the numerous vesicles (v), which are in close contact with the flagellar plasma membrane (fm), and the central (ct) and peripheral microtubular doublets of the flagellum. Note the Y-shaped link (y) that connects the peripheral doublets to the flagellar membrane. **D**. Transverse section in the flagellum showing its basic structure of nine peripheral (pt) and two central (ct) microtubular doublets. Note that each of the nine outer doublets is connected to the flagellar plasma membrane (fm) by a Y-shaped link (y) and consists of subfibers A and B. From Shahin, A. A. B. 2007a. Zoological Research 28: 193-206, Fig. 5 (scales modified).

distribution of the two conditions is shown in Figure 11.57 superimposed on a cladistic analysis of somatic characters presented by Diogo (2004b). The topology depicted requires only one more step if biflagellarity is presumed to be the plesiomorphic condition for catfish sperm but the flagellarity has not been reported for several families. Nevertheless, the slightly more parsimonious result where uniflagellarity in considered plesiomorphic (assuming biflagellarity to be irreversible) is supported by the clear plesiomorphy of uniflagellarity in fishes as a whole.

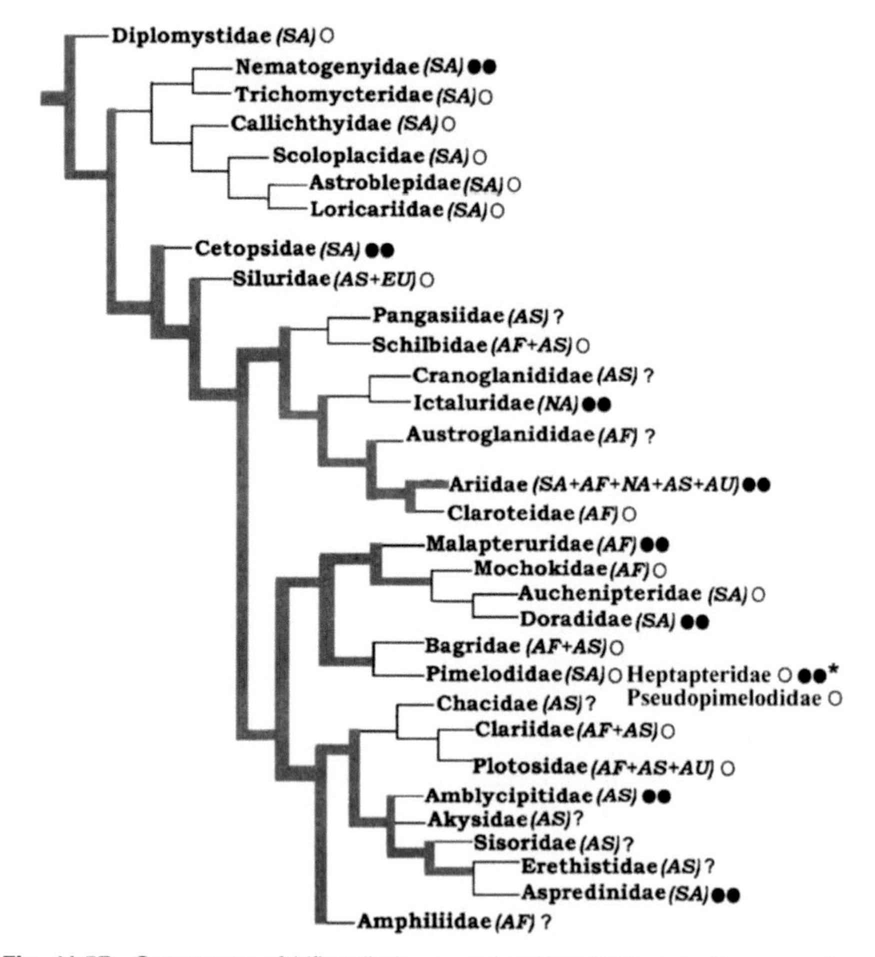

Fig. 11.57 Occurrence of biflagellarity •• and uniflagellarity o in the spermatozoa of catfishes. * Biflagellarity in the Heptapteridae is known only for *Rhamdia sapho*. ?, spermatozoa unknown. SA = South America; AF = Africa; NA = North America; EU = Europe; AS = Asia; AU = Australia. Original, superimposed on a cladistic analysis of somatic characters by Diogo, R. 2004b. Animal Biology 54: 331-351. Fig. 1. The Heptapteridae, Pseudopimelodidae and Heteropnenustidae have been added here in the approximate phylogenetic positions they occupy according to Sullivan, J. P., Lundberg, J. G. and Hardman, M. 2006.

The somatic phylogeny of Diogo (2004b) is presented here in order to demonstrate the distribution of biflagellarity. It differs in major respects from the molecular phylogenies of Sullivan *et al.* (2006) based on rag1 and rag2 nuclear gene sequences and resolution of these differences is awaited. Analysis of a wider range of genes would be desirable.

11.8 ORDER GYMNOTIFORMES

The order Gymnotiformes constitutes an important component of the Neotropical freshwater fish fauna of South and Central America. These fishes have a knife-like appearance and an elongate anal fin. Gymnotiformes has been hypothesized to be monophyletic based on a number of synapomorphies that includes the ability to produce and detect electrical fields which are used in navigation and intraspecific communication and in *Electrophorus* for stunning or killing the prey. The relationships among the families of this order are still being debated. According to Nelson (2006), the order Gymnotiformes is made up of two suborders, two super families and five families. Sperm ultrastructure has been reported for at least one species in each of the families that comprise the order. A study of seven gymnotiform species has shown that they spawn the apparently unfertilized eggs; their genera include *Eigenmannia, Rhamphychthys, Apteronotus* and *Gymnotus* mentioned in this chapter. *G. carapo* is shown to be a male mouth-brooder (Kirschbaum and Schugardt 2003).

11.8.1 GYMNOTOIDEI

11.8.1.1 Family Gymnotidae

Gymnotus* cf. *anguillaris: The spermatozoon of *G. cf. anguillaris*, a knifefish, described by França *et al.* (2007), resembles a type II aquasperm with a spherical nucleus (approximately 1.5 µm in diameter) located lateral and parallel to the flagellum (Fig. 11.58A). Chromatin is homogeneous and highly condensed (Fig. 11.58A-C,D,F). The proximal centriole is anterior, slightly lateral and perpendicular to the distal (Fig. 11.58C). A nuclear fossa in the form of two arcs is present; however, both centrioles are located outside this structure. The short midpiece contains several elongate mitochondria and well organized arrays of small vesicles located at the periphery of the cell (Fig. 11.58B,E,F). These vesicular arrays lack the basket-like appearance of the "tubular-vesicular system" first described in the characiform *Citharinus* sp. (Mattei *et al.* 1995) and thus may not be equivalent structures. Electron-dense spurs and other material are associated with both centrioles. In addition, microtubular arrays, which may comprise upwards of 100 closely packed parallel accessory microtubules in a single 800 nm section, radiate from the centriolar complex. One such array ran in a straight path from the centriolar complex, under the nucleus, to become associated with the opposite plasma membrane. The single flagellum, initially located within a short cytoplasmic canal (Fig. 11.58A,C), has a classic 9+2 axoneme, with all doublets electron-lucent, and no fins (Fig. 11.58D). Intertubular differentiations, typical of the A subtubules of Type II axonemes, have not been reported.

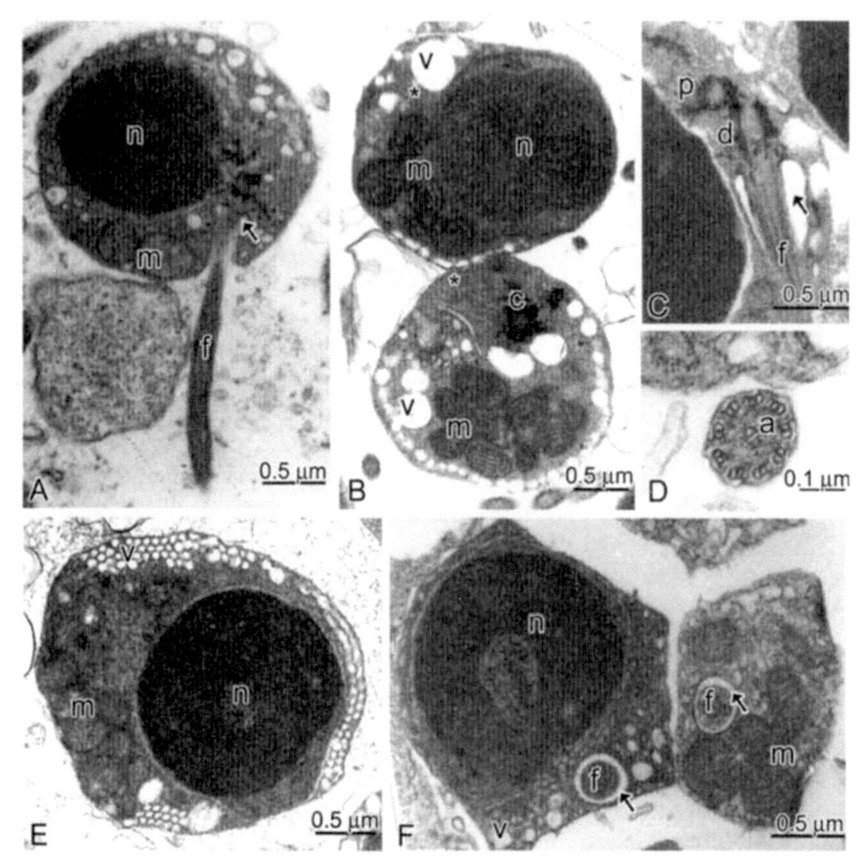

Fig. 11.58 *Gymnotus* cf. *anguillaris* (Gymnotidae). **A**. Longitudinal section of spermatozoon. Note the location of the nucleus lateral and parallel to the flagellum, and the short midpiece. **B**. Longitudinal section of nucleus and midpiece, and transverse section of the midpiece showing elongate mitochondria close to the nucleus. **C**. The centriolar arrangement showing the proximal centriole anterior, slightly lateral and perpendicular to the distal. **D**. Transverse section of the flagellum with the classic 9+2 axoneme. **E-F**. Transverse sections at different levels of the nucleus and midpiece. Note the well-organized arrays of small vesicles located at the periphery of the cell. c, centrioles; d, distal centriole; f, flagellum; m, mitochondrion; n, nucleus; p, proximal centriole; v, vesicles; arrow, cytoplasmic canal; asterisk, microtubules. Original, courtesy of Gisleine Fernanda França.

11.8.2 STERNOPYGOIDEI-Rhamphichthyoidea

11.8.2.1 Family Rhamphichthyidae

Rhamphichthys cf. hahni: The spermatozoon of *R. cf. hahni* (França 2006), a sand knifefish, is a Type I aquasperm with a semi-spherical nucleus (length

1.2 μm, width 1.3 μm) containing condensed flocculent chromatin and a deep nuclear fossa containing both centrioles. The proximal centriole is anterior, slightly lateral and perpendicular to the distal and both are connected to the nuclear envelope by means of anchoring fibers. Extensive electron-dense material is also associated with the distal centriole. The midpiece, which is long, thin and asymmetric, has a short cytoplasmic sleeve at the posterior end. Several elongate and branched mitochondria, some with a dense matrix, are found throughout the midpiece. Abundant elongate vesicles are confined to the posterior portion of the midpiece. The cytoplasmic canal, which contains the flagellum, is short and restricted to the cytoplasmic sleeve. The single flagellum contains a classic 9+2 axoneme with all doublets electron-lucent and two fins of moderate length.

11.8.2.2 Family Hypopomidae

Brachyhypopomus cf. pinnicaudatus. The spermatozoon of *B. cf. pinnicaudatus* (França *et al.* 2007), Bluntnose knifefish, is a Type I aquasperm with a slightly elongate, ovoid nucleus (approximate length 2 μm and width 1 μm). Nuclear rotation takes place during spermiogenesis (Fig. 11.59A), and just after its completion (Fig. 11.59B), the nucleus elongates lateral and posterior to the centriolar complex such that the final shape of the nucleus is highly asymmetric (Fig. 11.59C,D). Nuclear chromatin tends to be highly condensed and granular. A deep, radially branching nuclear fossa contains both centrioles (Fig. 11.59C,D). The proximal centriole is lateral and perpendicular to the distal. The peripheral microtubular triplets of both centrioles are embedded in electron-dense material that also connects both centrioles together. In longitudinal sections through both centrioles, electron-dense, ladder-like bodies can be seen in the core regions. The elongate midpiece contains several elongate mitochondria with a dense matrix, as well as abundant vesicles containing some stained material (Fig. 11.59E,F). Clusters of irregularly arranged accessory microtubules run adjacent and parallel to the axoneme. The posterior portion of the midpiece becomes narrow but still contains abundant vesicles (Fig 11.59G). The single flagellum has a classic 9+2 axoneme with all doublets electron-lucent and no fins. Most of the flagellum retains a rim of cytoplasm that is clearly thicker than that seen in most ostariophysan species analyzed (Fig. 11.59H).

11.8.3 Apteronotoidea (Sinusoidea)

11.8.3.1 Family Apteronotidae

Apteronotus albifrons: The sperm of *Apteronotus (=Sternarchus) albifrons*, a ghost knifefish, examined by Jamieson (1991) (adult, Fig. 11.60), and, as *A. cf. albifrons*, by França (2006), is a simple Type I anacrosomal aquasperm. The flocculent chromatin of the subspheroidal nucleus (length 1.4 μm, width 2.0 μm) (Fig. 11.61A,B,C,F) consists of numerous dense masses in a paler granular

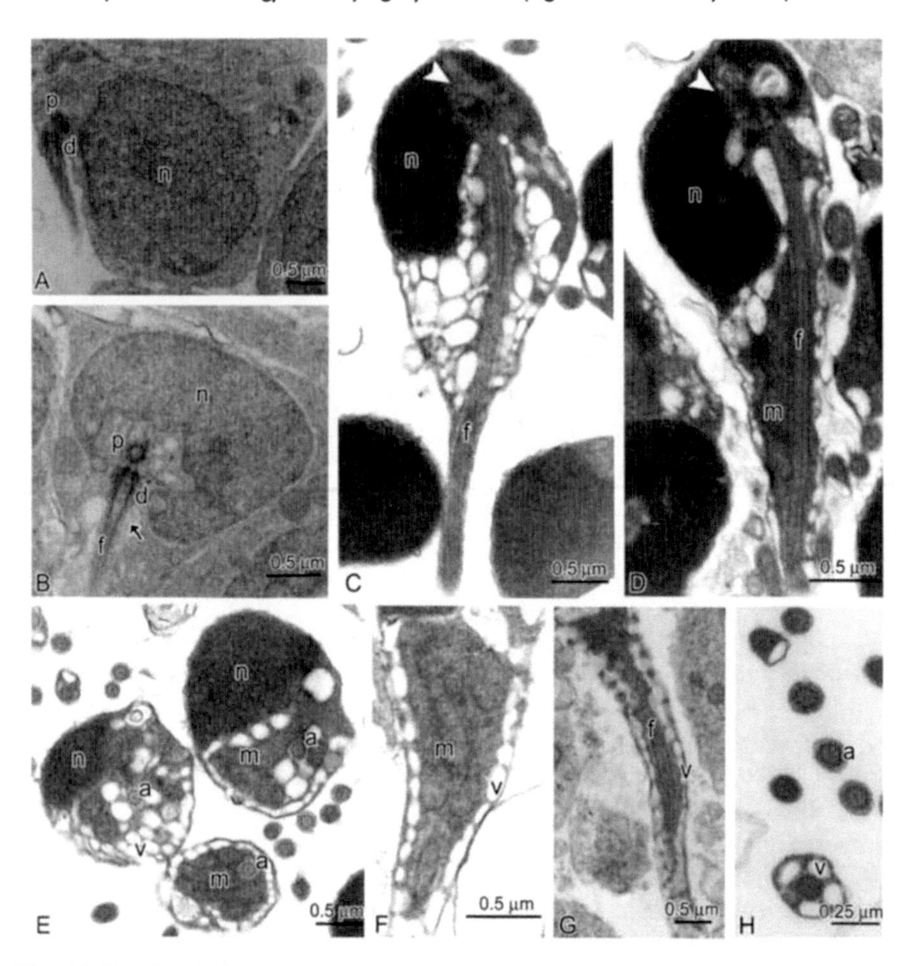

Fig. 11.59 *Brachyhypopomus cf. pinnicaudatus* (Hypopomidae). **A-B**. Longitudinal sections of early spermatids at the time of the centriolar migration (**A**) and when nuclear rotation is complete (**B**). **C-D**. Longitudinal sections of spermatozoa. Note the highly asymmetric shape of the nucleus that is elongated lateral and posterior to the centriolar complex, the deep, radially branching nuclear fossa, and the elongate midpiece with several elongate mitochondria, as well as abundant vesicles. **E**. Transverse sections at different levels of the midpiece. Note the absence of a cytoplasmic canal, mitochondria surrounding the initial portion of the axoneme, and the distribution of vesicles. **F**. Lateral and longitudinal view of the midpiece with elongate mitochondria in a cluster bordered by the peripheral vesicles. **G**. Longitudinal section where the posterior portion of the midpiece becomes narrow but still contains abundant vesicles. **H**. Transverse sections through the region where the flagellum loses vesicles but retains a thick rim of cytoplasm. a, axoneme; d, distal centriole; f, flagellum; m, mitochondrion; n, nucleus; p, proximal centriole; v, vesicles; arrow, cytoplasmic canal; arrow head, nuclear fossa. Original, courtesy of Gisleine Fernanda França.

Fig. 11.60 *Apteronotus (=Sternarchus) albifrons*. After Sterba, G. 1962. *Freshwater Fishes of the World*. Vista Books, London, Fig. 514.

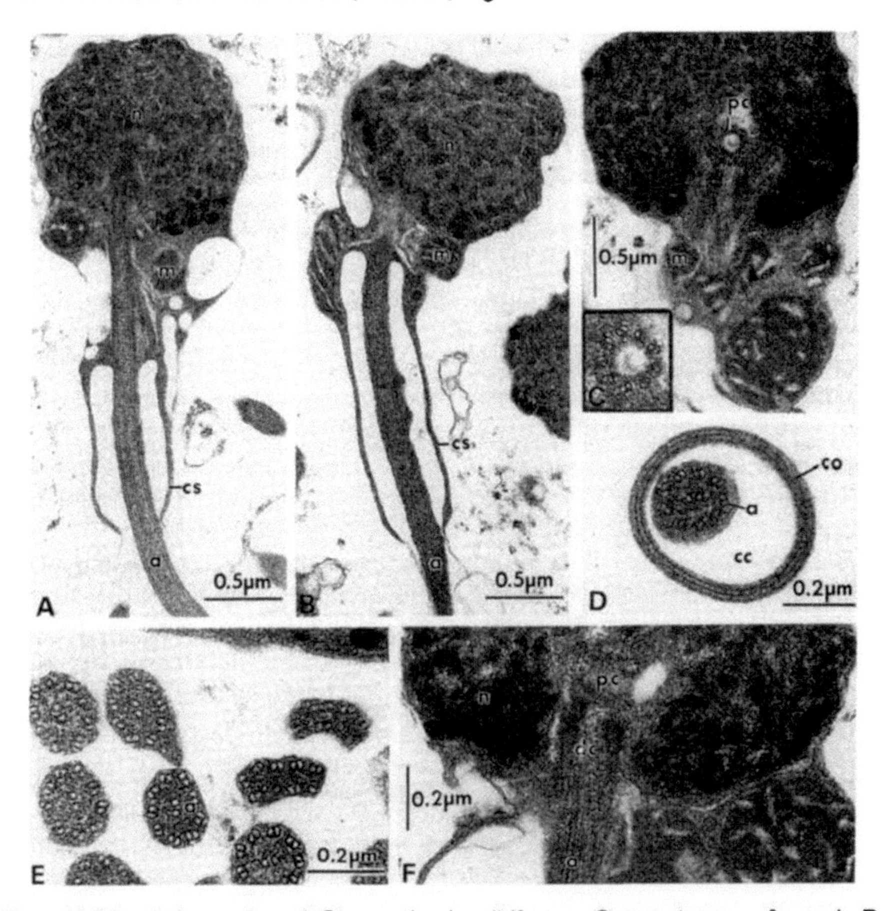

Fig. 11.61 *Apteronotus (=Sternarchus) albifrons*. Spermatozoa. **A** and **B**. Longitudinal sections (LS) of two spermatozoa. **C**. LS of the head and midpiece passing transversely through the proximal centriole which is shown in the inset at higher magnification. **D**. Transverse section of the posterior, amitochondrial extension of the collar, enclosing the cytoplasmic canal and axoneme. **E**. Transverse sections through the endpiece of the axoneme. **F**. LS through the nucleus and midpiece at right angles to C. a, axoneme; cc, cytoplasmic canal; m, mitochondrion; n, nucleus. From Jamieson, B. G. M. 1991. *Fish Evolution and Systematics: Evidence from Spermatozoa*. Cambridge University Press, Cambridge, Fig. 12.17.

matrix. The mitochondria (Fig. 11.61A,B,C,F) are grouped irregularly around the long cytoplasmic canal for a much as half of its length. They are in one or two tiers longitudinally and vary from subspherical to elongate. The intra-mitochondrial matrix is strongly electron-dense. This is reflected by unusual density of the general cytoplasm and of the axoneme. The cytoplasmic collar is approximately one and a half times the length of the nucleus. Its posterior half, behind the mitochondria, forms a thin trailing sheath, seen in longitudinal section in Figs. 11.61A and B and in transverse section in Figure. 11.61D. The proximal centriole is approximately at right angles to, and in the same longitudinal axis as, the basal body (Fig. 11.61C and inset, 11.61F) and both are contained with a nuclear fossa. The flagellum contains a classic 9 + 2 axoneme with all doublets electron-lucent, and no fins. At the endpiece (Fig. 11.61E) it becomes roughly oblong in transverse section where the two central singlets are lost.

11.8.3.2 Family Sternopygidae

***Eigenmannia* cf. *virescens*:** In the glass knifefishes, two entities designated *Eigenmannia cf. virescens* 1 and 2 have been examined by França (2006)). They have Type I aquasperm with semi-spherical nuclei (length 1.2 μm, width 1.5-1.6 μm) containing condensed flocculent chromatin and a deep nuclear fossa containing both centrioles. The proximal centriole is anterior, slightly lateral and perpendicular to the distal centriole and both are connected to the nuclear envelope via stabilizing fibers. In addition, extensive electron-dense material is associated with the microtubules of the distal centriole. The midpiece is long, thin and asymmetric with a short cytoplasmic sleeve at the posterior end and few vesicles. Several elongate and branched mitochondria, most of which appear to have a dense matrix, are distributed throughout the midpiece. The cytoplasmic canal, containing the flagellum, is short and restricted to the cytoplasmic sleeve. The single flagellum contains a classic 9+2 axoneme with the doublets appearing to be electron-lucent, and two relatively long fins.

11.9 CHAPTER SUMMARY

Spermatozoa from all ostariophysan species analyzed lack acrosomes. Flagella have typical 9+2 axonemes and, in the majority of cases, the tubules of all doublets are electron-lucent. Sperm ultrastructure has been studied in only one species of Gonorhynchiformes, *Chanos chanos*. The cell resembles a simple Type I aquasperm *sensu* Mattei (1970).

For the order Cypriniformes, upwards of 36 species have been studied in the family Cyprinidae and two in the Cobitidae. In all cases, spermatozoa appear to be variations of Mattei's (1970) Type I aquasperm. Nuclei vary from spherical to elliptical and in all cases at least part of the centriolar complex lies within a nuclear fossa. Chromatin is flocculent to strongly electron-dense and granular. Mitochondria vary in number from one to about six. With respect to the nucleus, the flagellar axis may be central, lateral or tangential. Several species have been shown to have glycogen granules. Distinct flagellar fins are lacking.

Sperm ultrastructure has been analyzed in fifteen families comprising the order Characiformes. Regarding externally fertilizing species, nearly all produce Type I aquasperm with nuclei ranging from spherical to conical and slightly elongate. However, spermatozoa resembling Mattei's (1970) Type II aquasperm are reported for some species in the families Acestrorhynchidae and Lebiasinidae, and possibly *Bryconamericus stramineus* (*incertae sedis* in Characidae). The novel type III aquasperm is found in the Lebiasinidae. Among the families, chromatin varies from flocculent to highly condensed and granular. In most cases, centrioles are perpendicular or oblique to one another, the exception being the subfamily Cheirodontinae (Characidae) where they are parallel in some species. Mitochondria tend to be few in number. The abundance and characteristics of vesicles vary greatly, with a distinct tubular-vesicular system, first described by Mattei *et al.* (1995), present in the Citharinidae, in subfamilies Characinae and Cheirodontinae of the Characidae, and in *Bryconamericus stramineus* (*incertae sedis* in Characidae). Although the A-tubules of the peripheral axonemal doublets tend to be electron-dense in some species of Lebiasinidae, in all other externally fertilizing species they are electron-lucent. Two to three flagellar fins are present in the Anostomidae, Chilodontidae, Cynodontidae, and some species in the subfamily Serrasalminae (Characidae).

The family Characidae of the Characiformes also contains numerous inseminating species. With few exceptions where classic type I aquasperm are produced, all other inseminating species have what appear to be highly modified Type I sperm with nuclei varying from slightly to extremely elongate. Some species have several large spherical to extremely elongate mitochondria, whereas in others multiple mitochondria are often located alongside an elongate nucleus or more posteriorly. Chromatin may be flocculent or highly condensed and granular. Centriolar orientiation varies from perpendicular to parallel. Distinct tubular-vesicular systems are seen in the subfamily Cheirodontinae. Many of these inseminating species have electron-dense A-tubules in at least some portion of the axoneme, but flagellar fins are lacking in all. Cytoplasmic collars vary from short to extremely long. Other characters unique to individual species are described.

Sperm ultrastructure has been analyzed in 26 of the 37 families reported to comprise the order Siluriformes. Most families are externally fertilizing, with the majority producing Type I aquasperm. However, cells similar to Mattei's (1970) Type II aquasperm have been reported in some species in the Callichthyidae, and the recently described type III aquasperm are produced by other species in the Callichthyidae, as well as the families Loricariidae, Heptapteridae and Pimelodidae. Sperm nuclei vary from spherical to conical.

Chromatin is highly condensed and granular in most taxa, but flocculent in five families. Centriolar orientation generally varies from perpendicular to oblique, but is parallel in all biflagellate groups. Some species in the Callichthyidae, Corydoradinae, have the unusual arrangement where the proximal centriole is directly anterior to, and in the same longitudinal axis as, the distal centriole. The abundance and characteristics of vesicles vary greatly. Most species have several to numerous mitochondria, with two groups having cells with a single fused mitochondrion. Five families have members with flagellar fins. Nine families produce biflagellate sperm, including the Ariidae where two axonemes are contained within a single flattened flagellum. A-tubules of axonemal doublets appear to be electron-lucent.

The three inseminating siluriform families, Scoloplacidae, Astroblepidae and Auchenipteridae, produce highly modified type I sperm with elongate nuclei containing condensed granular chromatin. Centrioles are mutually perpendicular and mitochondria numerous. Scoloplacid spermatozoa contain numerous vesicles, while those of auchenipterids have a peripheral row of accessory microtubules. Only scoloplacids and astroblepids have flagellar fins, but all three families have the A-tubules of the axonemal doublets electron-dense.

Sperm ultrastructure has been analyzed in five families of the order Gymnotiformes. With the exception of the gymnotids that produce aquasperm resembling Mattei's (1970) Type II, all other have Type I aquasperm. Nuclei vary from spherical to slightly elongate, with chromatin being flocculent to highly condensed and granular. Centrioles are mutually perpendicular, and usually several mitochondria are present. Gymnotids and hypopomids have numerous accessory microtubules in the cells. Flagellar fins are present in the Rhamphichthyide and Sternopygidae.

Based on recent phylograms (e.g. Fig. 11.1), all taxa among the Characiformes, Siluriformes and Gymnotiformes reported to produce spermatozoa similar to the Type II aquasperm of Mattei (1970) are nested within groupings whose more basal species form Type I aquasperm. This strongly suggests that these sperm types are in reality novel forms that have arisen independently within the Ostariophysi. If this is the case, the similarities with the "true" Type II aquasperm of perciforms is the result of convergence.

In conclusion, the sperm cells of the Ostariophysi display a remarkable degree of variability. The great majority of species are externally fertilizing, producing aquasperm of several types. The four families that contain inseminating species, three in Siluriformes and one in Characiformes, generally produce highly modified sperm that appear to be adapted to this unique reproductive habit. Since the earlier work of Jamieson (1991) on fish sperm ultrastructure, a remarkable number of new species have been analyzed contributing to the ever-growing database of sperm ultrastructural characters. This information will be invaluable for developing new hypotheses on the phylogenetic relationships among these fishes, as well as studies on the basic reproductive biology of these groups.

11.10 LITERATURE CITED

Aires, E. D., Stefanini, M. A. and Orsi, A. M. 2000. Ultrastructural features and differentiation of the spermatids in *Brycon orbgyanus* during the spermiogenesis. Brazilian Journal of Veterinary Research and Animal Science 37.

Amaral, A. A. 2003. Aspectos ultra-estruturais da espermatogênese do piauçu Leporinus macrocephalus Garavello and Bristski, 1988 (Teleostei, Characiformes, Anostomidae). Doctoral thesis. Universidade Estadual Paulista, Jaboticabal.

Andrade, R. F., Bazzoli, N., Rizzo, E. and Sato, Y. 2001. Continuous gametogenesis in the neotropical freshwater teleost, *Bryconops affinis* (Pisces:Characidae). Tissue and Cell 33: 524-532.

Azevedo, M. A. 2004. Análise comparada de caracteres reprodutivos em três linhagens de Characidae (Teleostei: Ostariophysi) com inseminação. Doctoral thesis. Universidade Federal do Rio Grande do Sul, Porto Alegre, Brazil.

Baccetti, B., Burrini, A. G., Callaini, G., Gibertini, G., Mazzini, M. and Zerunian, S. 1984. Fish Germinal Cells 1. Comparative Spermatology of 7 Cyprinid Species. Gamete Research 10: 373-396.

Batlouni, S. R., Romagosa, E. and Borella, M. I. 2006. The reproductive cycle of male catfish *Pseudoplastystoma fasciatum* (Teleostei: Pimelodidae) revealed by changes of the germinal epithelium. An approach addressed to aquaculture. Animal Reproduction Science 96:116-132.

Billard, R. 1970. Ultrastructure comparée de spermatozoïdes de quelques poissons Téléostéens. Pp. 71-79. In B. Baccetti (ed.), *Comparative Spermatology*. Academic Press, New York.

Billard, R. 1983. Spermiogenesis in rainbow trout (*Salmo gairdneri*). Cell and Tissue Research 233: 265-284.

Buckup, P. A. 1998. Relationships of the Characidiinae and the phylogeny of characiform fishes (Teleostei: Ostariophysi). In R. Malabarba, R. E. Reis, R. P. Vari and C. A. S. Lucena (eds), *Phylogeny and Classification of Neotropical Fishes*. Edipucrs, Porto Alegre, Brazil.

Burgess, W.E., 1989. An atlas of freshwater and marine catfishes. A preliminary survey of the Siluriformes. T.F.H. Publications, Inc., Neptune City, New Jersey, USA. 784 pp.

Burns, J. R. and Weitzman, S. H. 2005. Insemination in ostariophysian fishes. Pp.107-134. In H. J. Grier and M. C. Uribe (eds). *Viviparous Fishes*, New Life Publications, Homestead, FL, USA.

Burns, J. R., Weitzman, S. H., Grier, H. J. and Menezes, N. A. 1995. Internal fertilization, testis and sperm morphology in glandulocauline fishes (Teleostei: Characidae: Glandulocaulinae). Journal of Morphology 224: 131-143.

Burns, J. R., Weitzman, S. H. and Malabarba, L. R. 1997. Insemination in eight species of cheirodontine fishes (Teleostei: Characidae: Cheirodontiae). Copeia 1997:433-438.

Burns, J. R., Weitzman, S. H., Lange, K. R. and Malabarba, L. R. 1998. Sperm ultrastructure in characid fishes (Teleostei, Ostariophysi). Pp. 235-244. In L. R.

Malabarba, R. E. Reis, R. P. Vari, Z. M. S. Lucena and C. A. S. Lucena (ed.), *Phylogeny and Classification of Neotropical Fishes*. Edipucrs, Porto Alegre, Brazil.

Burns, J. R., Meisner, A. D., Weitzman, S. H. and Malabarba, L. R. 2002. Sperm and spermatozeugma ultrastructure in the inseminating catfish, *Trachelyopterus lucenai* (Ostariophysi: Siluriformes: Auchenipteridae). Copeia 2002: 173-179.

Burns, J. R., Pecio, A. and Weitzman, S. H. 2008. Sperm and spermatozeugma structure in *Xenurobrycon* (Teleostei: Characidae: Stevardiinae: Xenurobryconini). Copeia 2008: 657-661.

Cruz-Landim, C., Abdalla, F. C. and Cruz-Höfling, M. A. 2003. Morphological study of the spermatogenesis in the teleost *Piaractus mesopotamicus*. BioCell 27: 319-328.

Dimmick, W. W. and Larson, A. 1996. A molecular and morphological perspective on the phylogenetic relationships of the otophysan fishes. Molecular Phylogenetics and Evolution 6: 120-133.

Diogo, R. 2003. Higher-level phylogeny of Siluriformes: an overview. Pp. 353-384. In B. G. Kapoor, G. Arratia, M. Chardon and R. Diogo (ed.), *Catfishes*. Science Publishers, Enfield, NH, USA.

Diogo, R. 2004a. Adaptations, homoplasies, constraints, and evolutionary trends: catfish morphology, phylogeny and evolution, a case study on theoretical phylogeny and macroevolution. Science Publishers, Enfield, NH, USA.

Diogo, R. 2004b. Phylogeny, origin and biogeography of catfishes: support for a Pangean origin of 'modern teleosts' and reexamination of some Mesozoic Pangean connections between the Gondwanan and Laurasian supercontinents. Animal Biology 54: 331-351.

Emel'yanova, N. G. and Makeeva, A. P. 1985. The Ultrastructure of spermatozoa in some Cyprinidae. Voprosy Ikhtiologii 25: 459-468.

Emel´yanova, N. G. and Makeyeva, A. P. 1991a. Morphology of gametes in the channel catfish, *Ictalurus punctatus*. Voprosy Ikhtiologii 31: 143-148 (In Russian, see English version below).

Emel'yanova, N. G. and Makeyeva, A. P. 1991b. Ultrastructure of spermatozoa in some Siluriformes species. Voprosy Ikhtiologii 31: 1014-1019 (In Russian, see English version below).

Emel'yanova, N.G. and Makeyeva, A.P. 1991. Morphology of the gametes of the channel catfish *Ictalurus punctatus*. Journal of Ichthiology 31: 143-148.

Fink, S. V. and Fink, W. L. 1981. Interrelationships of the ostariophysan fishes (Teleostei). Zoological Journal of the Linnean Society of London 72: 297-353.

França, G. F. 2006. Ultraestrutura da espermiogênese e dos espermatozóides de representantes da ordem Gymnotiformes (Teleostei, Ostariophysi) com considerações filogenéticas. Masters thesis. Instituto de Biologia. Universidade Estadual de Campinas.

França, G. F., Oliveira, C. and Quagio-Grassiotto, I. 2007. Ultrastructure of spermiogenesis and spermatozoa of *Gymnotus* cf. *anguillaris* and *Brachyhypopomus* cf. *pinnicaudatus* (Teleostei : Gymnotiformes). Tissue and Cell 39: 131-139.

Fribourgh, J. H., Mcclendon, D. E. and Soloff, B. L. 1970. Ultrastructure of the goldfish, *Carassius auratus* (Cyprinidae):spermatozoon. Copeia 1970: 274-279.

Fuiman, L. A. 1984. Ostariophysi: development and relationships. Pp. 126-137. In (ed.), *Ontogeny and Systematics of Fishes*. Special Publication Number 1 American Society of Ichthyologists and Herpetologists.

Fujimura, W., Harutsugu, M., Nishiki, T. and Ito, K. 1957. Electron microscope study of sections of carp sperms. Journal of Nara Medical Association 7: 122-124.

Gopalakrishnan, A., Ponniah, A. G. and Lal, K. K. 2000. Fine structural changes of Rohu (*Labeo rohita*) sperm after dilution wiith cryoprotectants. Indian Journal of Fisheries 47: 21-27.

Guan, T. L. 1990. Regional specificity within plasma membrane and nuclear membrane of goldfish sperm. Acta biologica experimentalis Sinica 23: 17-27.

Guan, T. L. and Afzelius, B. A. 1991. The spermatozoon of the chinese bitterling *Rhodeus sericeus sinensis* (Cyprinidae, Teleostei). Journal of Submicroscopic Cytology and Pathology 23: 351-356.

Gusmão-Pompiani, P. 2003. Ultraestrutura da espermiogênese e dos espermatozóides de peixes da ordem Characiformes, família Characidae (Teleostei, Ostariophysi): uma abordagem filogenética.Doctoral thesis, Universidade Estadual Paulista, Botucatu.

Gwo, J. C., Kao, Y. S., Lin, X. W., Chang, S. L. and Su, M. S. 1995. The ultrastructure of milkfish, *Chanos chanos* (Forsskal), spermatozoon (Teleostei, Gonorynchiformes, Chanidae). Journal of Submicroscopic Cytology and Pathology 27: 99-104.

Gwo, J. C., Kurokura, H. and Hirano, R. 1993. Cryopreservation of spermatozoa from rainbow trout, common carp, and marine puffer. Nippon Suisan Gakkaishi 59: 777-782.

Hara, M. and Okiyama, M. 1998. An ultrastructural review on the spermatozoa of Japanese fishes. Bulletin of the Ocean Research Institute, University of Tokyo 33: 1-138.

Ishiguro, N. B., Miya, M. and Nishida, M. 2003. Basal euteleostean relationships: A mitogenomic perspective on the phylogenetic reality of the "Protacanthopterygii". Molecular Phylogenetics and Evolution 27: 476-488.

Ismial, M. E. and Khalifa, S. A. 1995. Ultrastructure of the spermatozoa in the catfish *Clarius lazera* (Claridae - Teleostei). Proceedings of the Zoological Society of Egypt 26: 144-150.

Jamieson, B. G. M. 1991. *Fish Evolution and Systematics: Evidence from Spermatozoa.* Cambridge University Press, Cambridge.

Jaspers, E. J., Avault, J. W. J. and Roussel, J. D. 1976. Spermatozoal morphology and ultrastructure of Channel catfish *Ictalurus punctatus*. Transactions of the American Fisheries Society 105: 475-480.

Javonillo, R., Burns, J. R. and Weitzman, S. H. 2007. Reproductive morphology of *Brittanichthys axelrodi* (Teleostei: Characidae), a miniature inseminating fish from South America. Journal of Morphology 268:23-32.

Kessel, R. G., Beams, H. W., Tung, H. N. and Roberts, R. 1983. Unusual particle arrays in the plasma membrane of zebrafish spermatozoa. Journal of Ultrastructure Research 84: 268-274.

Kim, K. H., Ae, S. and Lee, Y. H. 1998. Spermatozoal ultrastructure and phylogenetic relationships of the subfamily Gobioninae (Cyprinidae). 2. Ultrastructure of spermatozoa in the Korean gudgean [sic]. Korean Journal of Limnology 31: 159-164.

Kim, K. H. and Lee, J. I. 2003. Ultrastructure of spermatozoa of the slender catfish, *Pseudobagrus brevicorpus* (Teleostei, Bagridae) with phylogenetic considerations. Journal of the Korean Fisheries Society 36.

Kim, K. H. and Lee, Y. H. 2000. The ultrastructure of spermatozoa of the Ussurian bullhead, *Leiocassis assuriensis* (Teleostei, Siluriformes, Bagridae) with phylogenetic considerations. Korean Journal of Limnology 33: 405-412.

Kirschbaum, F. and Schugardt, C. 2002. Reproductive strategies and developmental aspects in mormyrid and gymnotiform fishes. Journal of Physiology Paris 96: 557-566.

Kudo, S. 1980. Sperm penetration and the formation of a fertilization cone in the common carp egg. Development, Growth and Differentiation 22: 403-414.

Kudo, S., Linhart, O. and Billard, R. 1994. Ultrastructural studies of sperm penetration in the egg of the European catfish, Silurus glanis. Aquatic Living Resources 7: 93-98.

Kwon, A. S. and Kim, K. H. 2004. A comparison of ultrastructures of spermatozoa of two catfish, *Silurus asotus* and *S. microdorsalis* (Teleostei: Siluridae). Korean Journal of Limnology 16: 128-134.

Kwon, A. S., Kim K. H. and Lee, Y. H. 1998. Ultrastructure of spermatozoa in the catfish *Silurus asotus*. Development and Reproduction 2: 75-80.

Lahnsteiner, F. and Patzner, R. A. 1991. A new method for electron-microscopical fixation of spermatozoa of freshwater teleosts. Aquaculture 97: 301-304.

Lahnsteiner, F. and Patzner, R. A. 2007. Sperm morphology and ultrastructure in fish. Pp. 1-61. In S. M. H. S. M. H. Alavi, J. Cosson, K. Coward and G. Rafiee (eds), *Fish Spermatology*. Alpha Science International Ltd., Oxford, U. K.

Lahnsteiner, F., Weismann, T. and Patzner, R. A. 1992. Monosaccharides as energy sources during motility of spermatozoa in *Leuciscus cephalus* (Cyprinidae, Teleostei). Fish Physiology and Biochemistry 10: 283-289.

Lauder, G. V. and Liem, K. F. 1983. The evolution and interrelationships of the actinopterygian fishes. Bulletin of the Museum of Comparative Zoology 150: 95-197.

Lavoué, S., Miya, M., Inoue, J. G., Saitoh, K., Ishiguro, N. B. and Nishida, M. 2005. Molecular systematics of the gonorynchiform fishes (Teleostei) based on whole mitogenome sequences: Implications for higher-level relationships within the Otocephala. Molecular Phylogenetics and Evolution 37: 165-177.

Lee Y. H. 1998. The ultrastructure of spermatozoa in the bagrid catfish, *Pseudobagrus fulvidraco* (Teleostei, Siluriformes, Bagridae). Korean Journal of Electron Microscopy 28:39-48.

Lee, Y. H. and Kim, K. H. 1999. Ultrastructure of the south torrent catfish, *Liobagrus mediadiposalis* (Teleostei, Siluriformes, Amblycipitidae) spermatozoon. Korean Journal of Limnology, 32(3): 271-280.

Lee, Y. H. and Kim, K. H. 2001. The ultrastructure of spermatozoa of the slender catfish *Silurus microdorsalis* (Teleostei, Siluriformes, Siluridae) with phylogenetic considerations. Journal of Submicroscopic Cytology and Pathology 33: 329-336.

Lima, F. C. T., Malabarba, L. R., Buckup, P. A., Pezzi da Silva, J. F., Vari, R. P., Harold, A., Benine, R., Oyakawa, O. T., Pavanelli, C. S., Menezes, N. A., Lucena, C. A. S., Malabarba, M. C. S. L., Lucena, Z. M. S., Reis, R. E., Langeani, F., Casatti, L., Bertaco, V. A., Moreira, C., and Lucinda, P. H. F. 2003. Genera *Incertae Sedis* in Characidae. Pp. 106-169. *In*: Reis, R. E., Kullander, S. O., and Ferraris, C. J. (eds). Check list of the freshwater fishes of South and Central America (CLOFFSCA). Edipucrs, Porto Alegre, Brazil. 729 pp.

Lin, D.-J., You, Y.-L. and Su, M. 2003. Studies on testicular histology and spermatogenesis of teleost, *Spinibarbus caldwelli* (Nichols). Acta Hydrobiologica Sinica 27: 563-571.

Lin, G. 1995. Studies on the testis differentiation and ultrastructure of spermatocytes in the catfish, *Clarias lazera*. Journal of Nanchang University (Natural Science) 19: 158-164 (In Chinese).

Loir, M., Cauty, C. Planquette, P. and Le Bail, P. Y. 1989. Comparative study of the male reproductive tract in seven families of South-American catfishes. Aquatic Living Resources 2: 45-56.

Lopes, D. C. J. R., Bazzoli, N., Rito, M. F. G. and Maria, T. A. 2004. Male reproductive system in the South American catfish *Conorhynchus conirostris*. Journal of Fish Biology 64: 1419-1424.

Magalhães, A. L. B., Bazzoli, N., Santos, G. B. and Rizzo, E. 2004. Reproduction of the South American dogfish characid, *Galeocharax knerii*, in two reservoirs from upper Parana a River basin, Brazil. Environmental Biology of Fishes 70: 415-425.

Maggese, M. C., Cukier, M. and Cussac, V. E. 1984. Morphological changes, fertilizing ability and motility of *Rhamdia sapo* (Pisces, Pimelodidae) sperm induced by media of different salinities. Rev Brasil Bio 44: 541-546.

Malabarba, L. R. 1998. Monophyly of the Cheirodontinae, characters and major clades (Ostariophysi: Characidae). Pp. 193-233. In L. R. Malabarba, R. E. Reis, R. P. Vari, Z. M. S. Lucena and C. A. S. Lucena (eds), *Phylogeny and Classification of Neotropical Fishes*. Edipucrs, Porto Alegre, Brazil.

Malabarba, L. R. and Weitzman, S. H. 2003. Description of a new genus with six new species from southern Brazil, Uruguay and Argentina, with a discussion of a putative characid clade (Teleostei: Characiformes: Characidae). Comunicações do Museu de Ciência e Tecnologia da PUCRS, série Zoológica, 16: 67-151.

Mansour, N. and Lahnsteiner, F. 2003. Morphology of the male genitalia and sperm fine structure in siluroid fish. Journal of Submicroscopic Cytology and Pathology 35: 277-285.

Mansour, N., Lahnsteiner, F. and Patzner, R. A. 2002. The spermatozoon of the African catfish: Fine structure, motility, variability and its behaviour in seminal vesicle secretion. Journal of Fish Biology 60: 545-560.

Matos, E., Matos, P., Corral, L. and Azevedo, C. 2000. Estrutura fina do espermatozóide de *Acestrorhynchus falcatus* Bloch (Teleostei,Characidae) da região norte do Brasil. Revista Brasileira de Zoologia 17: 747-752.

Matos, E., Matos, P., Oliveira, E. and Azevedo, C. 1993a. Ultaestrutura do espermatozoide do pacu, *Metynnis maculatus* Kner, 1860 (Pisces, Teleostei) do rio Amazonas. Revista Brasileira de Ciências Morfológicas 10: 7-10.

Matos, E., Matos, P., Oliveira, E. and Azevedo, C. 1993b. Ultrastructure in the spermatogenesis of the fish, *Hoplosternum littorale* (Hancock) (Teleostei, Callichthyidae) of the Amazon River. Revista Brasileira de Zoologia 10: 219-227 (In Portuguese).

Matos, E., Matos, P., Santos, M. N. S. and Azevedo, C. 1998. Morphological and ultrastructural aspects of the spermatozoon of *Curimata inornata* Vari, 1989 (Pisces, Teleostei) from the Amazon river. Acta Amazonica 28: 449-453.

Matos, E., Matos, P., Corral, L. and Azevedo, C. 1999a. Ultrastructure of the spermatozoon of *Leporinus friderici* Bloch, 1794 (Pisces, Teleostei) from the Amazon River. Revista de Ciências Agrárias, Belém **31**: 93-99.

Matos, E., Santos, M. N. S., Corral, L. and Azevedo, C. 1999b. Ultrastructural aspects of the spermatozoon of *Hoplerythrinus unitaeniatus* Spix, 1829 (Pisces, Teleostei, Erythrinidae) from the Amazon Region. Revista de Ciências Agrárias, Belém **32**: 27-32.

Mattei, C. and Mattei, X. 1984. Spermatozoïdes biflagellés chez un poisson téléostéen de la famille Apogonidae. Journal of Ultrastructure Research 88: 223-228.

Mattei, X. 1970. Spermatogenèse comparée des poissons. Pp. 567-569. In B. Baccetti (ed.), *Comparative Spermatology.* Academic Press, New York.

Mattei, X. 1991. Spermatozoon ultrastructure and its systematic implications in fishes. Canadian Journal of Zoology 69: 3038-3055.

Mattei, X., Siau, Y., Thiaw, O. T. and Thiam, D. 1993. Peculiarities in the organization of testis of *Ophidion* sp. (Pisces Teleostei). Evidence for two types of spermatogenesis in teleost fish. Journal of Fish Biology 43: 931-937.

Mattei, X., Marchand, B. and Thiaw, O. T. 1995. Unusual midpiece in the spermatozoon of a teleost fish, *Citharinus* sp. Journal of Submicroscopic Cytology and Pathology 27: 189-191.

Meisner, A. D., Burns, J. R., Weitzman, S. H. and Malabarba, L. R. 2000. Morphology and histology of the male reproductive system in two species of internally inseminating South American catfishes, *Trachelyopterus lucenai* and *T. galeatus* (Teleostei: Auchenipteridae). Journal of Morphology 246:131-141.

Mekkawy, I. A. A. and Osman, A. G. M. 2006. Ultrastructural studies of the morphological variations of the egg surface and envelopes of the African catfish *Clarias gariepinus* (Burchell, 1822) before and after fertilisation, with a discussion of the fertilisation mechanism. Scientia Marina 70: 23-40.

Munoz-Guerra, S., Azorin, F., Casas, M. T., Maristany, M. A., Roca, J., Subirana, J. A. and Marcet, X. 1982. Structural organization of sperm chromatin from the fish *Carassius auratus.* Experimental Cell Research 137: 47-54.

Nath, A. and Chand, G. B. 1998. Ultrastructure of spermatozoa correlated with phylogenetic relationship between *Heteropneustes fossilis* and *Rana tigrina.* Cytobios 95: 161-165.

Nelson, J. S. 2006. *Fishes of the World*, 4th edition. John Wiley and Sons, Inc., Hoboken, NJ. 601 pp.

Ohta, T. 1991. Initial stages of sperm-egg fusion in the freshwater teleost *Rhodeus ocellatus ocellatus.* Anatomical Record 229: 195-202.

Ohta, T. and Iwamatsu, T. 1983. Electron microscopic observations on sperm entry into eggs of the Rose Bitterling, *Rhodeus ocellatus.* Journal of Experimental Zoology 227: 109-119.

Ohta, T., Kato, K. H., Abe, T. and Takeuchi, T. 1993. Sperm morphology and distribution of intramembranous particles in the sperm heads of selected freshwater teleosts. Tissue and Cell 25: 725-735.

Ohta, T., Mizuno, T., Mizutani, M. and Matsuda, M. 1994. Sperm morphology and IMP distribution in membranes of spermatozoa of cyprinid fishes I. Journal of Submicroscopic Cytology and Pathology 26: 181-189.

Ohta, T., Yoshida, M. and Kato, S. 1998. Electron microscopic observations on sperm penetration in cytochalasin-treated eggs of the rose bitterling. Cell Structure and Function 23: 179-186.

Oliveira, C. L. C. de. 2007. Análise comparada da ultraestrutura dos espermatozóides e morfologia da glândula branquial em espécies de Cheirodontinae (Characiformes: Characidae). Doctoral Thesis. Universidade Federal do Rio Grande do Sul, Porto Alegre, Brazil.

Oliveira, L. C. de, Burns, J. R., Malabarba, L. R. and Weitzman, S. H.2008. Sperm ultrastructure in the inseminating *Macropsobrycon uruguayanae* (Teleostei: Characidae: Cheirodontinae). Journal of Morphology 269: 691-697.

Ortí, G. 1997. Radiation of characiform fishes: evidences from mitochondrial and nuclear DNA sequences. Pp. 219-243. In Kocher, T. D. and Stepien, C. A., (eds), *Molecular Systematics of Fishes*, Academic Press, San Diego.

Parreira, G. G. and Godinho, H. P. 2004. Sperm ultrastructure of the cangati *Trachelyopterus galeatus* (Linnaeus, 1766) (Siluriformes, Auchenipteridae), an internal fertilizer fish. Molecular Bilogy of the Cell 15:91A (supplement).

Patil, J. G. and Khoo, H. W. 1995. Ultrastructural in sity hybridization and autoradiographic detection of foreign DNA in zebrafish spermatozoa. Biochemistry and Molecular Biology International? 35: 965-969.

Pecio, A. 2003. Spermiogenesis and fine structure of the spermatozoon in a headstander, *Chilodus punctatus* (Teleostei, Characiformes, Anostomidae). Folia Biologica (Cracow) 51: 55-62.

Pecio, A. and Rafiñski, J. 1994. Structure of the testis, spermatozoa and spermatozeugmata of *Mimagoniates barberi* Regan, 1907 (Teleostei: Caracidae), an internally fertilizing, oviparous fish. Acta Zoologica (Stockholm) 75:179-185.

Pecio, A. and Rafiñski, J. 1999. Spermiogenesis in *Mimagoniates barberi* (Teleostei: Ostariophysi: Characidae), an oviparous, internally fertilizing fish. Acta Zoologica (Stockholm) 80: 35-45.

Pecio, A., Burns, J. R. and Weitzman, S. H. 2005. Sperm and spermatozeugma ultrastructure in the inseminating species *Tyttocharax cochui, T. tambopatensis,* and *Scopaeocharax rhinodus* (Pisces: Teleostei: Characidae: Glandulocaudinae: Xenurobryconini). Journal of Morphology 263: 216-226.

Pecio, A., Burns, J. R. and Weitzman, S. H. 2007. Comparison of spermiogenesis in the externally fertilizing *Hemigrammus erythrozonus* and the inseminating *Corynopoma riisei* (Teleostei: Characiformes: Characidae). Neotropical Ichthyology 5: 457-470.

Poirier, G. R. and Nicholson, N. 1982. Fine structure of the testicular spermatozoa from the Channel Catfish, *Ictalurus punctatus*. Journal of Ultrastructure Research 80: 104-110.

Psenicka, M., Rodina, M., Nebesarova, J. and Linhart, O. 2006. Ultrastructure of spermatozoa of tench Tinca tinca observed by means of scanning and transmission electron microscopy. Theriogenology 66: 1355-1363.

Quagio-Grassiotto, I. and Carvalho, E. D. 2000. Ultrastructure of *Sorubim lima* (Teleostei, Siluriformes, Pimelodidae) spermiogenesis. Journal of Submicroscopic Cytology and Pathology 32: 629-633.

Quagio-Grassiotto, I. and Oliveira. C.2008. Sperm ultrastructure and a new type of spermiogenesis in two species of Pimelodidae with a comparative review of sperm ultrastructure in Siluriformes. *Zoologischer Anzeiger* (in press).

Quagio-Grassiotto, I., Negrao, J. N. C., Carvalho, E. D. and Foresti, F. 2001a. Ultrastructure of spermatogenic cells and spermatozoa in *Hoplias malabaricus* (Teleostei, Characiformes, Erythrinidae). Journal of Fish Biology 59: 1494-1502.

Quagio-Grassiotto, I., Oliveira, C. and Gosztonyi, A. E. 2001b. The ultrastructure of spermiogenesis and spermatozoa in *Diplomystes mesembrinus*. Journal of Fish Biology 58: 1623-1632.

Quagio-Grassiotto, I., Gameiro, M. C., Schneider, T., Malabarba, L. R. and Oliveira, C. 2003. Spermiogenesis and spermatozoa ultrastructure in five species of the Curimatidae with some considerations on spermatozoal ultrastructure in the Characiformes. Neotropical Ichthyology 1: 35-45.

Quagio-Grassiotto, I., Spadella, M. A., De Carvalho, M. and Oliveira, C. 2005. Comparative description and discussion of spermiogenesis and spermatozoal ultrastructure in some species of Heptapteridae and Pseudopimelodidae (Teleostei: Siluriformes). Neotropical Ichthyology 3: 401-410.

Reis, R. E., Kullander, S. O. and Ferraris, C. J. Jr. (eds) 2003. Check list of the freshwater fishes of South and Central America. EDIPUCRS, Porto Alegre, Brazil. 742pp.

Roberts, T. R. 1973. Interrelationships of ostariophysans. Pp. 373-395. In P. H. Greenwood, R. S. Miles and C. Patterson (eds), *Interrelationships of Fishes*. Journal of the Linnean Society of London Zoology 53 (suppl 1) Academic Press, New York.

Romagosa, E., Narahara, M. Y., Borella, M. I., Parreira, S. F. and Fenerich-Verani, N. 1999. Ultrastructure of the germ cells in the testis of matrinxa, *Brycon cephalus* (Teleostei, Characidae). Tissue and Cell 31: 540-544.

Rutaisire, J., Muwazi, R. T. and Booth, A. J. 2006. Ultrastructural description of spermiogenesis and spermatozoa in *Labeo victorianus*, Boulenger, 1901 (Pisces : Cyprinidae). African Journal of Ecology 44: 102-105.

Santos, J. E., Bazzoli, N., Rizzo, E. and Santos, G. B. 2001. Morphofunctional organization of the male reproductive system of the catfish *Iheringichthys labrosus* (Lutken, 1874) (Siluriformes: Pimelodidae). Tissue and Cell 33: 533-540.

Shahin, A. A. B. 2006a. Semicystic spermatogenesis and biflagellate spermatozoon ultrastructure in the Nile electric catfish *Malapterurus electricus* (Teleostei : Siluriformes : Malapteruridae). Acta Zoologica (Copenhagen) 87: 215-227.

Shahin, A. A. B. 2006b. Spermatogenesis and spermatozoon ultrastructure in the Nile pebblyfish *Alestes dentex* (Teleostei: Characiformes: Alestidae) in Egypt. World Journal of Zoology 1: 1-16.

Shahin, A. A. B. 2007a. A novel type of spermiogenesis in the nile catfish *Chrysichthys auratus* (Siluriformes : Bagridae) in Egypt, with description of spermatozoon ultrastructure. Zoological Research 28: 193-206.

Shahin, A. A. B. 2007b. Spermatogenesis and spermatozoon ultrastructure in the Nile catfish *Synodontis schall* (Pisces: Actinopterygii: Synodontidae) in Egypt. Acta Zoologica Sinica 53: 689-699.

Spadella, M. A. 2007. Estudos filogenéticos na superfamília Loricarioidea (Teleostei, Siluriformes) com base na ultraestrutura dos espermatozóides. Doctoral Thesis. Instituto de Biologia. Universidade Estadual de Campinas. Campinas.

Spadella, M. A., Oliveira, C. and Quagio-Grassiotto, I. 2006a. Occurrence of biflagellate spermatozoa in Cetopsidae, Aspredinidae, and Nematogenyidae (Teleostei: Ostariophysi: Siluriformes). Zoomorphology 125: 135-145.

Spadella, M. A., Oliveira, C. and Quagio-Grassiotto, I. 2006b. Spermiogenesis and introsperm ultrastructure of *Scoloplax distolothrix* (Ostariophysi: Siluriformes: Scoloplacidae). Acta Zoologica (Copenhagen) 87: 341-348.

Spadella, M.A., Oliveira, C. and Quagio-Grassiotto, I. 2007. Comparative analysis of spermiogenesis and sperm ultrastructure in Callichthyidae (Teleostei: Ostariophysi: Siluriformes). Neotropical Ichthyology 5: 337-350.

Stanley, H. P. 1965. Electron microscopic observations on the biflagellate spermatids of the teleost fish *Porichthys notatus*. Anatomical Record 151: 477.

Stein, H. 1981. Licht- und elektronenoptische Untersuchungen an den Spermatozoen verschiedener Süsswasserknochenfische (Teleostei). Zeitschrift für Angewandte Zoologie 68: 183-198.

Stoumboudi, M. and Abraham, M. 1996. The spermatogenetic process in *Barbus longiceps*, *Capoeta damascina* and their natural sterile hybrid (Teleostei, Cyprinidae). Journal of Fish Biology 49: 458-468.

Sullivan, J. P., Lundberg, J. G. and Hardman, M. 2006. A phylogenetic analysis of the major groups of cat fshes (Teleostei: Siluriformes)using rag1 and rag2 nuclear gene sequences. Molecular Phylogenetics and Evolution 41: 636-662.

Toledo-Piza, M. 2000. The Neotropical fish subfamily Cynodontinae (Teleostei: Ostariophysi: Characiformes): a phylogenetic study and revision of *Cynodon* and *Rhaphiodon*. American Museum Novitates 3286: 1-88.

Veríssimo-Silveira, R. 2007. Estudo filogenético da sub-ordem Characoidei (Teleostei, Ostariophysi, Characiformes) com base na ultraestrutura dos espermatozóides. Doctoral thesis, Universidade Estadual Paulista, Botucatu.

Verissimo-Silveira, R., Gusmao-Pompiani, P., Vicentini, C. A. and Quagio-Grassiotto, I. 2006. Spermiogenesis and spermatozoa ultrastructure in *Salminus* and *Brycon*, two primitive genera in Characidae (Teleostei : Ostariophysi : Characiformes). Acta Zoologica 87: 305-313.

Vicentini, C. A. 2002. Estrutura comparativa da espermatogênese em peixes neotropicais, *Prochilodus scrofa, Astyanax scabripinnis* e *Phalloceros caudimaculatus* (Pisces: Teleostei). Livre Docência Thesis, Universidade Estadual Paulista, Bauru.

Vicentini, C. A., Franceschini-Vicentini, I. B., Benetti, E. J. and Orsi, A. M. 2001. Testicular ultrastructure and morphology of the seminal pathway in *Prochilodus scrofa*. Journal of Submicroscopic Cytology and Pathology 33: 357-362.

Wei, G., Chen, Y. and Dai, D. 1997. Studies on ultrastructure of spermatocyte and spermoblast in *Leiocassis longirostris*. Xinan Nongye Daxue Xuebao 19: 378-380 (In Chinese).

Weitzman, S. H. 1954. The osteology and relationships of the South American characid fishes of the subfamily Gasteropelecidae. Stanford Ichthyological Bulletin 4: 213-263.

Weitzman, S. H. and Menezes, N. A. 1998. Relationships of the tribes and genera of the Glandulocaudinae (Ostariophysi: Characiformes: Characidae) with a description of a new genus, *Chrysobrycon*. Pp. 171-192. In L. R. Malabarba, R. E. Reis, R. P. Vari, Z. M. S. Lucena and C. A. S. Lucena (eds), *Phylogeny and Classification of Neotropical Fishes*. Edipucrs, Porto Alegre, Brazil.

Weitzman, S. H., Menezes, N. A., Evers, H.-G. and Burns, J. R. 2005. Putative relationships among inseminating and externally fertilizing characids, with a description of a new genus and species of Brazilian inseminating fish bearing an anal-fin gland in males (Characiformes: Characidae). Neotropical Ichthyology 3: 329-360.

Wolenski, J. S. and Hart, N. H. 1987. Scanning electron microscope studies of sperm incorporation into the Zebrafish *Brachydanio* egg. Journal of Experimental Zoology 243: 259-274.

Yin, H. B., Sun, Z. W., Liu, Y. T. and Pan, W. Z. 2000. Ultrastructure of the spermatozoon of *Silurus soldatovi*. Journal of Fisheries China? 24: 302-305 (In Chinese).

Zaiden, S. F. 2000. Morfologia gonadal e metabolismo energético da piraputanga Brycon hilarii (Cuvier e Valenciennes, 1849) (Pisces, Characidae), em cativeiro, durante o ciclo reprodutivo anual. Doctoral thesis. Centro de Aqüicultura, Universidade Estadual Paulista, Jaboticabal.

Zhang, X., Zhang, L. and Shen, X. 1991. Study on the ultrastructure of cryopreserved sperm of fishes. Acta Scientiarum Naturalium Universitatis Normalis Hunanensis 14: 160-164 (In Chinese).

Zhang, Y., Luo, Q. and Zhong, M. 1992. Studies on the developmental stages of testis, spermatogenesis and spermatoleosis in *Leiocassa longirostris*. Zoological Research 13: 281-287 (In Chinese).

Zhang, Y. G., Luo, Q. S., . and Zhong, M. C. 1993. Stuides on the structure of testis and spermatozoon in *Leiocassis longirostris*. .Acta Hydrobiologica Sinica 17: 246-251 (In Chinese).

CHAPTER 12

Ultrastructure of Spermatozoa: Euteleostei: Argentiformes, Esociformes, Salmoniformes and Osmeriformes

Barrie G. M. Jamieson

12.1 EUTELEOST RELATIONSHIPS

The Euteleostei (Argentiformes through Tetraodontiformes) are regarded as the apomorph sister-group of the Clupeomorpha (Clupeiformes and Ostariophysi), as confirmed by Ortí and Li (**Chapter 1**) who also confirm monophyly of the Euteleostei, findings previously supported from mitogenomic analysis (Inoue *et al.* 2003). However, Nelson (2006) found "less than desirable convincing evidence" that the Euteleostei is monophyletic. Spermatologically they are correspondingly diverse and offer no single spermatozoal character indicative of monophyly.

12.2 ARGENTIFORMES, ESOCIFORMES, SALMONIFORMES AND OSMERIFORMES

The grouping Protacanthopterygii is recognized by Nelson (2006) to include the Argentiformes, Osmeriformes, Salmoniformes and Esociformes. However, mitogenomic analyses appear to have shown that protacanthopterygians as currently defined merely represent a collective, polyphyletic group of basal euteleosts, located between the basal teleosts (elopomorphs and below) and neoteleosts (stomiiforms and above) (Ishiguro *et al.* 2003). Lopez *et al.* (2004) give molecular evidence for a close relationship of Osmeriformes and Stomiiformes. These are also found to be sister-groups by Ortí and Li who

School of Integrative Biology, University of Queensland, Brisbane, Queensland 4072, Australia

recognize a lower clade for Argentiformes and their sister clade Esociformes + Salmoniformes (**Chapter 1**) (Fig. 12.1). They confirm the non-monophyly of the Protacanthopterygii, a group which is not here recognized.

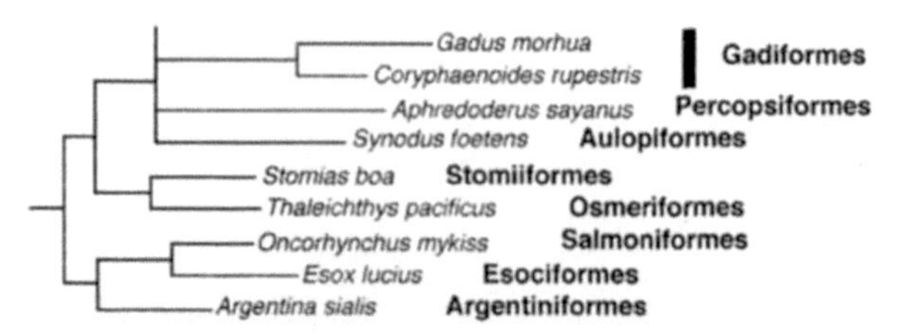

Fig. 12.1 Phylogeny of Salmoniformes and related orders based on partitioned analyses of ten nuclear genes. Note that Argentiformes are the sister group of the Esociformes + Salmoniformes and that the Osmeriformes, often associated with salmoniforms, show instead a sister-group relationship with the Stomiiformes. Adapted from Ortí and Li, **Chapter 1**, Fig. 1.5.

12.3 ORDER ARGENTIFORMES

The Argentiformes, marine smelts, have a complex posterior branchial structure ("epibranchial organ") termed the crumenal organ (Nelson 2006). The group contains the argentinoid and alepocephaloid fishes. From analysis of nuclear sequences, the Argentiformes appear to be the sister-taxon of Esociformes + Salmoniformes (Ortí and Li) (Fig. 12.1).

Sperm literature: No argentinoids have been examined for sperm ultrastructure. The five examined argentiform species are all in the Alepocephaloidei: Alepocephalidae, *Xenodermichthys* sp. (Mattei *et al.* 1981); *Xenodermichthys copei* and *Bajacalifomia megalops* (Mattei 1991) and this account (Fig. 12.2); and Platytroctidae = Searsiidae, *Searsia* sp. (Mattei *et al.* 1981).

Sperm ultrastructure: Mattei *et al.* (1981) show that *Searsia* sp. and *Xenodermichthys* sp. there is a single ring-shaped mitochondrion (as in salmonoids) and that the A subtubule is septate in axonemal doublets 1, 2, 5, 6, and 7 (contrast 1, 2, 3, 5, 6, and 7 in *Salmo*). This axoneme configuration appeared to be a synapomorphy of the Alepocephalidae and Platytroctidae (= Searsiidae) (see (Jamieson 1991), and therefore of the Argentiformes., so far as this small sample showed. However, there is some uncertainty as to the situation in *Xenodermichthys* as Mattei (1991) states that this genus, *Searsia* and *Salmo* have in common ITDs in doublets 1, 3, 5, 6, and 7, in addition to sharing a pseudoacrosome.

The 1, 2, 5, 6 and 7 arrangement is also seen in the Galaxiidae (Marshall 1989) and possibly indicates relationship of galaxioids with the Argentiformes rather than the Salmoniformes or the Osmeriformes in which latter they are currently placed, though homoplasy cannot be ruled out.

A further point of uncertainty is the arrangement of the mitochondrion relative to the axoneme. In Mattei *et al.* (1981) the mitochondrion of

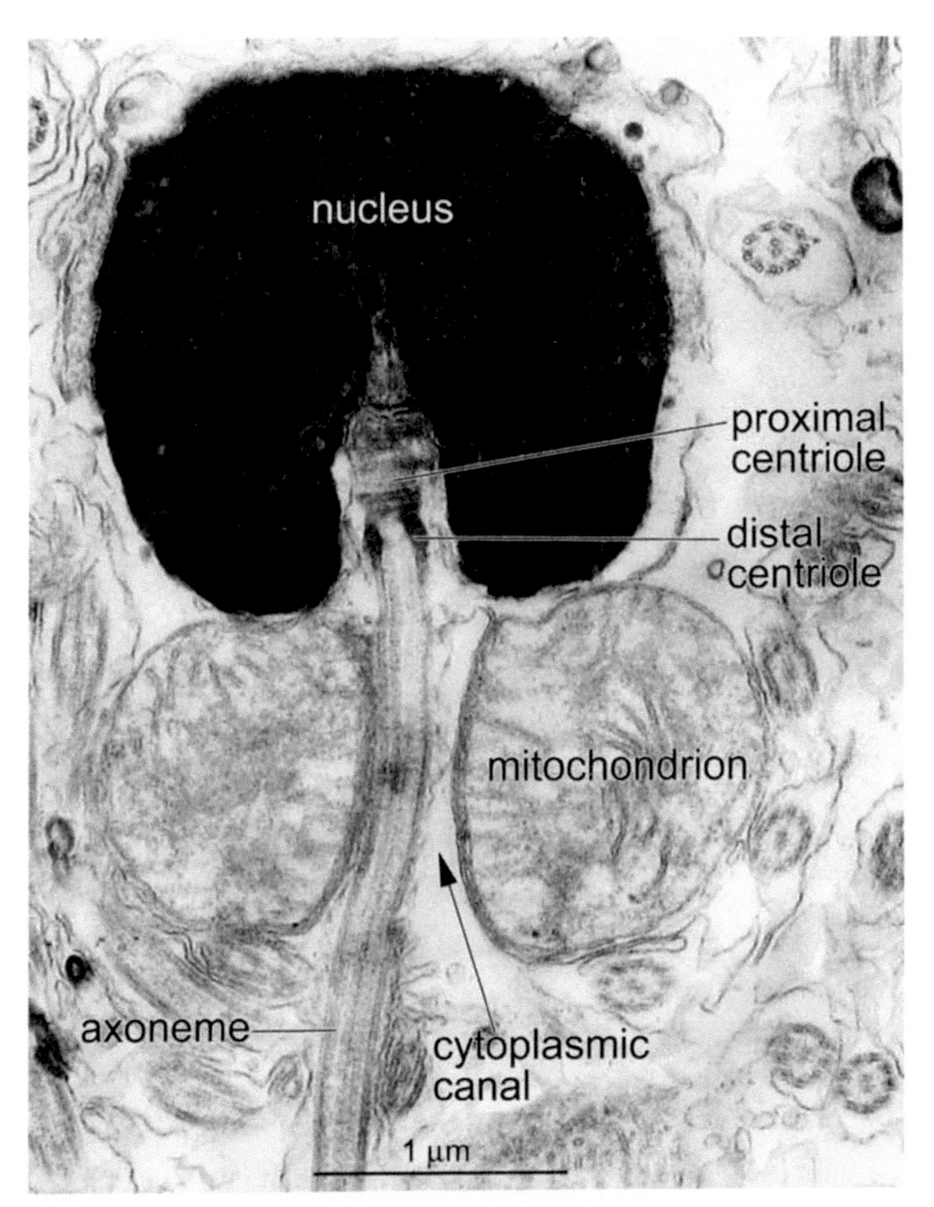

Fig. 12.2 *Bajacalifornia megalops* (Alepocephalidae). Longitudinal section of a spermatozoon. Original, from an unpublished TEM negative courtesy of Professor Xavier Mattei.

Xenodermichthys sp. is clearly illustrated in a micrograph as having a closed C-shape encircling the axoneme whereas in the later study (Mattei 1991) *Xenodermichthys* (there including *X. copei* in addition to *X.* sp.) the mitochondrion is said to be placed laterally in relation to the flagellum as in

Searsia and *Hucho* and in contrast to the encircling condition in *Salmo* and in the alepocephalid *Bajacalifornia*.

Remarks: The ultrastructure of argentinoid sperm, particularly the ring-shaped mitochondrion, suggested a closer linkage with salmonoids in a phylogram than was usual (Jamieson 1991), a linkage now supported by molecular analysis, though with intervention of the Esociformes (Ortí and Li, **Chapter 1**) (see also Fig. 12.1).

12.4 ORDER ESOCIFORMES

From spermatozoal characters Jamieson (1991) recognized the independent phylogenetic position of the esocids and confirmed their removal from the Salmoniformes which has been advocated by Fink and Weitzman (1982) and by Lauder and Liem (1983). These authors consider the "Esocae" to be the plesiomorphic sister-group of an ostariophysan-salmoniform - neoteleostean assemblage (Fig. 12.1). However, mitogenomic analysis (Lavoué *et al.* 2005) has found the Esociformes and Salmoniformes to be sister-groups in a clade which is the sister of the Neoteleostei; this contraindicates a gonorynchiform-clupeiform sister grouping found in a mitogenomic study by Ishiguro *et al.* (2003), though the issue cannot be considered settled. The esociform-salmonoid sister relationship was also supported from nuclear sequences (Lopez *et al.* 2004; Ortí and Li, **Chapter 1**).

All esociforms occur in temperate and arctic freshwaters of the Northern Hemisphere and are predatory (Lauder and Liem 1983). The group contains the pikes and pickerels (Esocidae) and the mudminnows (Umbridae).

For relationships of the Western Australian *Lepidogalaxias salamandroides* and its possible relationships with the Esocidae see **Section 12.6.2.2** below.

***Esox lucius*:** The adult is shown in Figure 12.3. The spermatozoon of the Pike, *Esox lucius*, has been briefly described (Billard 1970; Stein 1981) (Fig. 12.4). Billard (1970) considers it to show a strong resemblance to the aquasperm of Cyprinidae (see Fig. 11.9), being placed in his group I spermatozoa which

Fig. 12.3 *Esox lucius* (Esocidae). After Jordan, D.S. 1907. *Fishes*. Henry Holt, New York. Fig. 96.

include the Carp but which also embraces the sperm of the labroid perciform *Oreochromis (=Tilapia)*. Stein (1981) finds it similar to the sperm of *Perca* and *Cottus*, both of which it resembles in the form of the head.

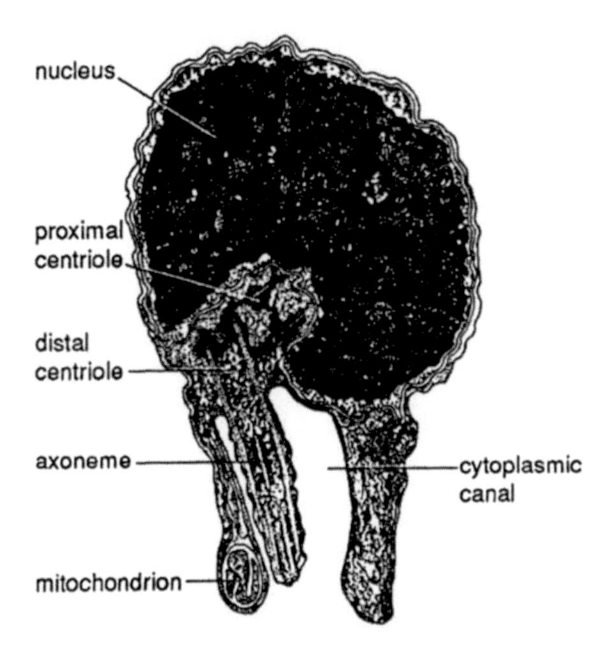

Fig. 12.4 *Esox lucius* (Esocidae). Longitudinal section of spermatozoon. Drawn from a micrograph of Stein 1981. Zeitschrift fuer Angewandte Zoologie 68: 183-198, Fig. 1. From Jamieson, B. G. M. 1991. *Fish Evolution and Systematics: Evidence from Spermatozoa.* Cambridge University Press, Cambridge. Fig. 12.3.

The nucleus (Fig. 12.4) is rounded in one plane but flattened in the other, imparting bilateral symmetry (Billard 1970; Stein 1981). It is 2 μm long, 1.8 μm in maximum diameter and 1.5 μm in minimum diameter. The chromatin is in the form of dense masses (Billard 1970), a perciform rather than cyprinid condition. The nuclear membrane is strongly undulated. Stein states that the pike shows no relationship to the other salmoniforms which belong to his group 1 sperm in which the head is somewhat elongated and in which the midpiece is central behind the head whereas it somewhat overlaps the head in the pike as it does in cyprinids.

Dilatation of intermembrane spaces and fusion of mitochondria is rudimentary compared with the trout. The mitochondria are disposed asymmetrically in the posterior region of the head, as in the Carp.

The flagellum is inserted laterally where one extremity of the proximal centriole is engaged in a nuclear fossa situated on the ventral face of the head while the other is applied to the distal centriole. Inner and outer dynein arms are present in the classical 9+2 axoneme. Anteriorly, as in the trout, the doublets are connected by an internal dense ring and a further peculiarity is the presence of an additional outer, approximately radial arm. A cytoplasmic expansion [flagellar fin] is present on one side only (Billard 1970).

***Esox masquinongy*:** The sperm of the Muskellunge, *Esox masquinongy*, analyzed by scanning and transmission electron microscopy, has an acrosomeless spherical head, 1.5 μm in diameter. The chromatin consists of many clumps of dense granules surrounded by a nuclear envelope bordering the cytoplasmic membrane (Fig. 12.5B). The sperm differs markedly from that of *Esox lucius*, and of most other teleost fishes, in having an elongated midpiece with abundant mitochondria. It forms a sleeve separated from the flagellum by an invagination of the cell membrane, forming a cytoplasmic canal. The proximal centriole is perpendicular to the distal centriole. Both centrioles are located within the nuclear fossa and have a conventional 9+0 microtubular triplet construction. The flagellum has the classical 9+2 axoneme structure. As in *E. lucius*, there is a single flagellar fin. There is possibly a relationship between abundant mitochondria and the long duration of sperm movement in muskellunge (Lin *et al.* 1996).

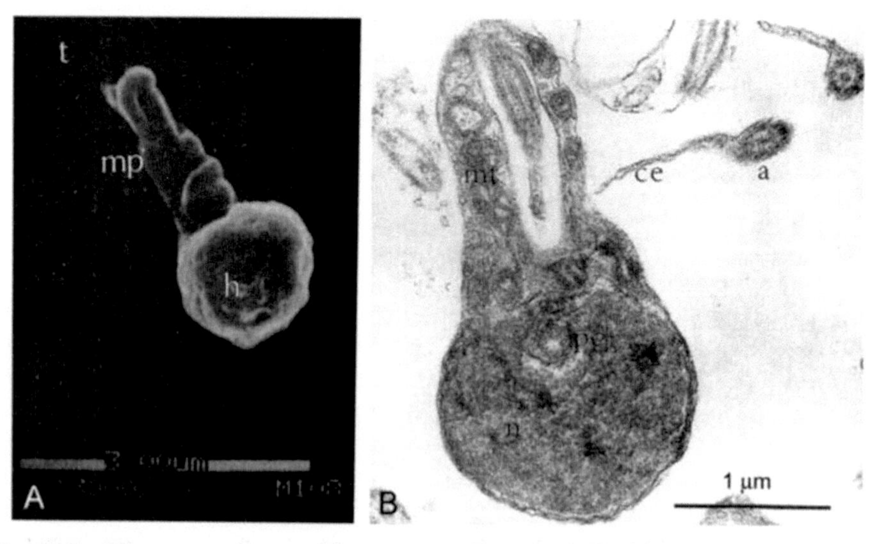

Fig. 12.5 *Esox masquinongy*. Muskellunge (Esocidae). **A**. SEM of sperm. **B**. TEM longitudinal section. a, axoneme; ce, single axonemal fin; mp, midpiece; mt, mitochondria; n, nucleus; pc, proximal centriole; t, tail. Adapted from Lin, F., Liu, L. and Dabrowski, K. 1996. Transactions of the American Fisheries Society 125: 187-194, Fig. 1. With permission of the American Fisheries Society.

Remarks: *Esox* differs from cyprinids, and agrees with salmonids, in having flagellar fins (Billard 1970; Stein 1981) (albeit unpaired in this genus) but presence of fins is a plesiomorphy for Euteleostomi (=Osteichthyes) and does not in itself indicate esocid-salmonid affinity. It was considered (Jamieson

1991) that similarities of the sperm of *Esox lucius* to those of both cyprinids and perciforms gave support to the argument that esocoids are not salmoniforms. Failure of the mitochondria in *Esox* sperm to form a ring was considered plesiomorphic relative to salmoniforms. However, molecular analysis has suggested that the Esociformes is the sister-group of the Salmoniformes (see **Chapter 1**) (Fig. 12.1).

12.5 ORDER SALMONIFORMES

It is not proposed here to deal at length with the vexed history of the classification of the Salmoniformes. In Nelson (1984) the Salmoniformes contained the suborders Esocoidei, Argentinoidei, Lepidogalaxioidei and Salmonoidei. The Salmonoidei contained the superfamilies Osmeroidea, Galaxioidea and Salmonoidea. Later, the osmeroids, as Osmeriformes (including the galaxioids and *Lepidogalaxias*), and argentinoids, as Argentiformes, were excluded Nelson (2006). DNA analysis supports separation of these orders, the osmeriforms being closer to the Stomiiformes than to an argentiform + (esociform+salmoniform) clade (see **Chapter 1**) (Fig. 12.1).

The Salmoniformes (trouts and salmons) are now restricted to the family Salmonidae (Nelson 2006). This contains three subfamilies: the Coregoninae, Salmoninae and Thymallinae. The Coregoninae contain the genus *Coregonus* (whitefish), Salmoninae the genera *Oncorhynchus*, *Salmo* (trout) and *Salvelinus* (charr) and Thymallinae the genus *Thymallus* (grayling). Salmonids occur mainly in freshwater. They have external fertilization, spawning in a redd (Salmoninae and Thymallinae) or pelagically (Coregoninae).

Sperm literature: Salmonid species, in all three subfamilies, investigated for spermatozoal ultrastructure are as follows.

Salmoninae: *Brachymystax lenok* (Zhang and Wang 1992; Yoon *et al.* 1996; Drokin *et al.* 1998, Hara and Okiyama 1998); *Hucho hucho* (Stein 1981; Radziun and Tomasik 1985); *Oncorhynchus gorbuscha* (Drozdov *et al.* 1981), *O. keta* (Kobayashi and Yamamoto 1987; Hara and Okiyama 1998); *O. kisutch* (Lowman 1953); *Oncorhynchus masou masou* (Hara and Okiyama 1998); *O. masou formosanus* (Gwo *et al.* 1996, 1999); *O. mykiss (=Salmo gairdneri)* (Billard 1970; 1983a; Fribourgh and Soloff 1976; Mattei *et al.* 1981, Stein 1981; Malejac *et al.* 1990; Gwo *et al.*, 1993; Lahnsteiner *et al.* 1996), *O. tshawytscha* (Zirkin 1975); *Salmo salar* (Nicander 1968); *Salmo trutta* (Furieri 1962); *S. trutta fario* (Billard 1983a; Stein 1981, Drokin *et al.* 1998); *Salvelinus alpinus* (Stein 1981; Lahnsteiner and Patzner 2007); *Salvelinus fontinalis* (Fribourgh 1978, Stein 1981, Lahnsteiner and Patzner 1991). For adults of some Salmoninae and Osmeridae in which sperm ultrastructure has been studied see Figures 12.6 and 12.7.

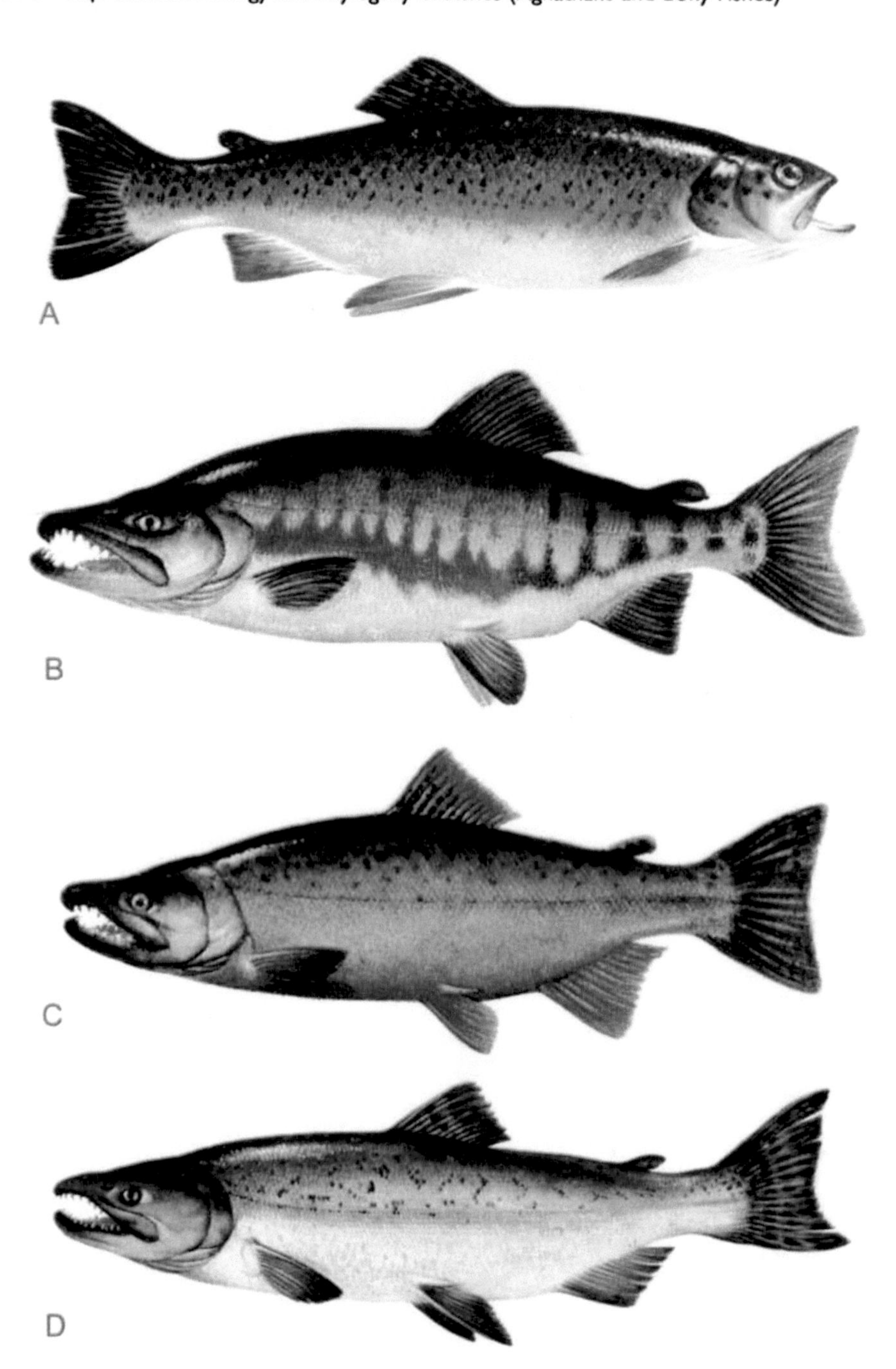

Fig. 12.6 Some species of *Oncorhynchus* (Salmoninae). **A**. Rainbow trout (*O. mykiss*). **B**. Chum salmon (*O. keta*). **C**. Coho salmon (*O. kisutch*). **D**. Chinook or Quinat salmon (O. *tschawytscha*). By Norman Weaver. With kind permission of Sarah Starsmore.

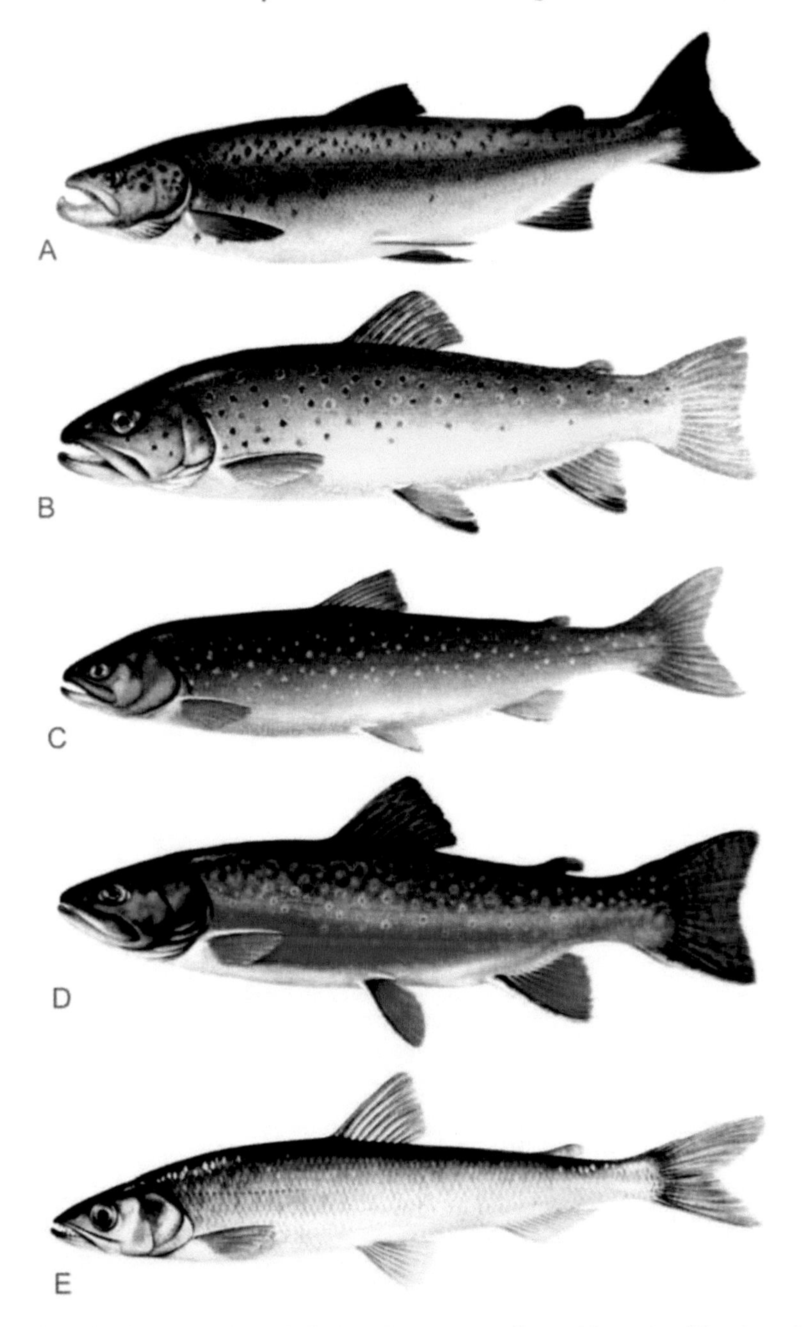

Fig. 12.7 Further species of Salmoninae, and Osmeridae. **A**. Atlantic salmon (*Salmo salar*). **B**. Brown trout (*Salmo trutta*). **C**. Arctic char (*Salvelinus alpinus*). **D**. Brook trout (*S. fontinalis*). (All Salmoninae). **E**. European smelt (*Osmerus eperlanus*). (Osmeridae). By Norman Weaver. With kind permission of Sarah Starsmore.

Coregoninae: *Coregonus lavaretus* (Lahnsteiner *et al.* 1992b; Lahnsteiner and Patzner 2007); *C. wartmanni* (Stein 1981).

Thymallinae: *Thymallus thymallus* (Stein 1981; Lahnsteiner and Patzner 1991, 2007; Lahnsteiner *et al.* 1991; 1992a).

Sperm ultrastructure: The spermatozoa of the three subfamilies are compared in Fig. 12.8.

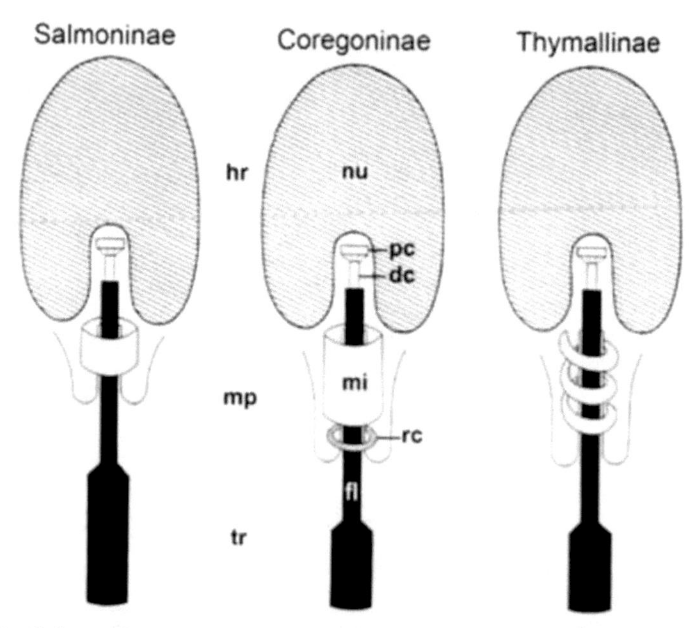

Fig. 12.8 Schematic reconstruction of types of spermatozoa in salmonid fish. Dc, distal centriole; fl, flagellum; hr, head region; mi, mitochondrion; mp, midpiece; nu, nucleus; pc, proximal centriole; rc, ring of cytoskeletal filaments; tr, tail region. Slightly modified after Lahnsteiner, F. and Patzner, R. A. 2007. Pp. 1-61. In S. M. H. Alavi, J. Cosson, K. Coward and G. Rafiee (eds), Fig. 16.

Putative acrosome: A putative acrosome (pseudoacrosome *sensu* Mattei 1991) has been reported in spermiogenesis of *Oncorhynchus mykiss (=Salmo gairdneri)*, the Rainbow Trout (Billard 1983a). This was described merely as a dark layer produced by adhesion of nuclear material to its envelope by Billard (1983b) but an acrosome-like vesicle was ascribed only to the mid-stage of spermiogenesis in this species by Hara and Okiyama (1998) and is not seen in the mature sperm illustrated (Fig. 12.9A-D). A pseudoacrosome is also reported for the argentiform *Xenodermichthys*. A similar structure in *Lepidogalaxias salamandroides* is here considered to be a true acrosome in the absence in that species of an egg micropyle (see **Section 12.5.2.2** below).

Nucleus: Their nucleus is elongate ovoid, less than twice as long as wide, and is surrounded by a thin layer of cytoplasm. The chromatin is heterogeneous

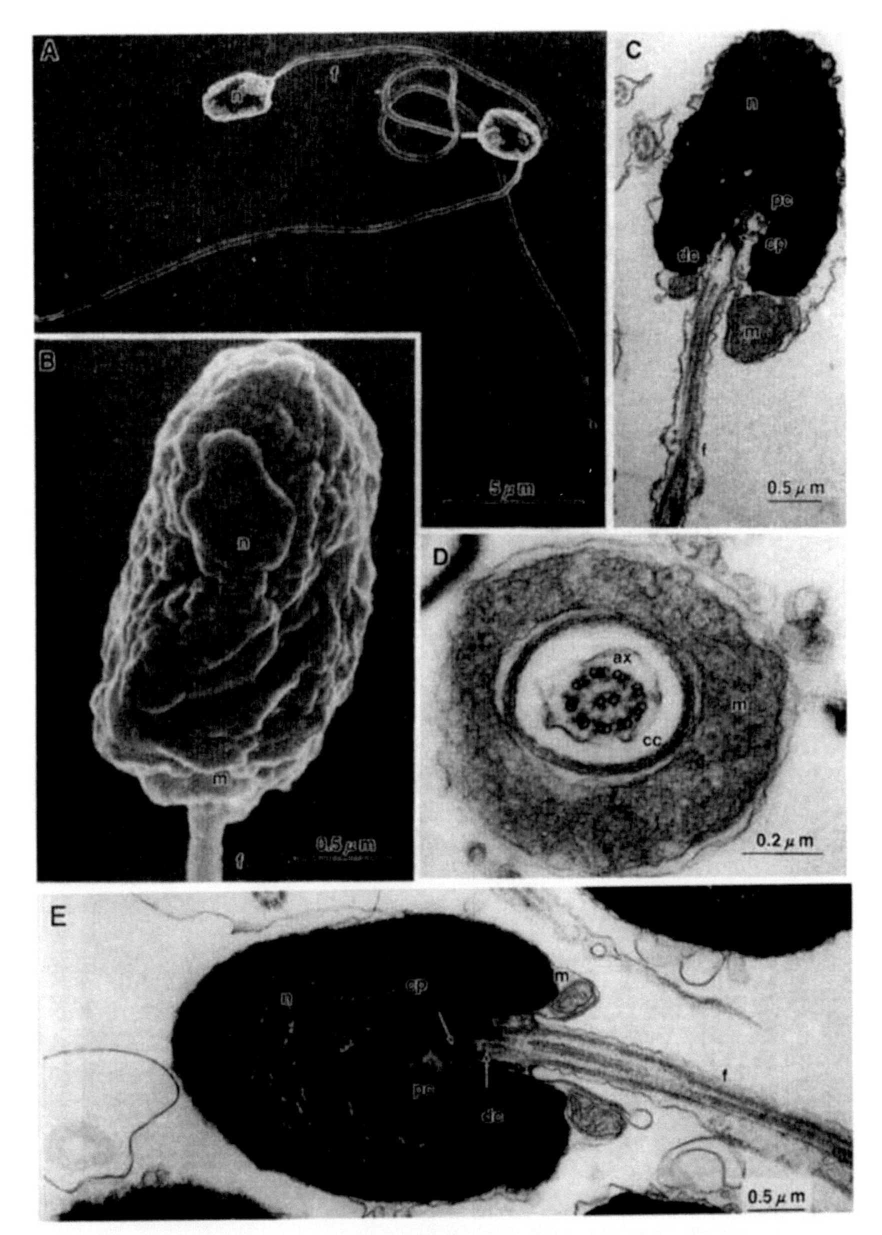

Fig. 12.9 Spermatozoa of Salmoniformes, Salmonidae, Salmoninae. **A-D**. *Oncorhynchus mykiss*. A. SEM of spermatozoa. **B**. SEM of sperm head. **C**. TEM longitudinal section of a spermatozoon. **D**. TEM transverse section through the single, annular mitochondrion. **E**. *Oncorhynchus keta*. TEM longitudinal section of spermatozoon. cp, centriolar plug; dc, distal centriole; f, flagellum; m, mitochondrion; n, nucleus; pc, proximal centriole. Adapted from Hara, M. and Okiyama, M. 1998. Bulletin of The Ocean Research Institute, University of Tokyo 33: 1-138, Figs. 22A-D and 23B. With permission.

and may have electron-lucent patches (e.g. *Oncorhynchus tshawytscha*, Fig. 12.10). A moderately deep basal nuclear fossa houses the two centrioles.

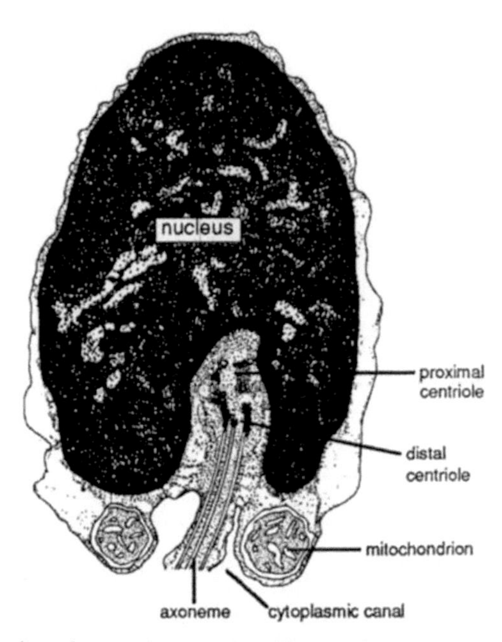

Fig. 12.10 *Oncorhynchus tshawytscha* (Salmoninae) Longitudinal section of spermatozoon. The nucleus is extremely electron-dense but clumps of material can often be seen. From Jamieson 1991, after a micrograph by Zirkin, B.R. 1975. Journal of Ultrastructure Research 50: 174-184, Fig. 13.

An osmiophilic body at the base of the nucleus in *Coregonus wartmanni* and *Salvelinus alpinus* (Stein 1981) is possibly homologous with the round bodies seen at the anterior border of the nucleus in some Percichthyidae (see **Chapter 15**).

Midpiece: The midpiece is small in the Salmoninae, medium-sized in the Thymallinae and is largest in the Coregoninae (Fig. 12.8). In the Salmoninae, the mitochondrion surrounds the cytoplasmic canal in the form of an incomplete ring, in the form of a closed ring or helically in one and a half gyres. In the Thymallinae, the mitochondrion is helical, surrounding the midpiece in 2 or 3 gyres as detected in scanning electron micrographs (Lahnsteiner and Patzner 2007). In the Coregoninae, the midpiece is cylindrical and forms a complete ring around the cytoplasmic canal.

For *Salmo trutta* Furieri (1962) reported that sometimes several mitochondria are present but this may have been an immature condition as it is known for spermatids. Fribourgh (1978) also suggests the presence of separate mitochondria in the sperm of the Brook Trout, *Salvelinus fontinalis*, but Stein (1981) states that there is a single mitochondrial ring in this species and in *Salmo*, *Coregonus* and *Thymallus*. In *Brachymystax lenok*, mitochondria around the centriolar complex are finally organized into a single annular structure (Zhang and Wang 1992). An exception for the Salmoninae may be *Hucho*

hucho in which a single unilateral mitochondrion is depicted (Radziun and Tomasik 1985) *fide* (Mattei 1991); however, Stein (1981) reports a single ring-like mitochondrion for this species.

In *Oncorhynchus mykiss* and *Salmo trutta fario*, as in other salmonids, the base of the flagellum is surrounded by a short mitochondrial collar which surrounds the cytoplasmic canal. The cytoplasmic canal, bounded by the usual fold of the plasma membrane, separates the midpiece from the flagellum. A second membrane is present throughout the length of the midpiece between the mitochondrion and the plasma membrane which externally borders the cytoplasmic canal (Billard 1983b). This additional membrane is seen in other orders, for instance, the characiform *Paracheirodon* (see **Chapter 11**).

Centriolar complex: The two centrioles are mutually perpendicular in the three subfamilies and each consists of the normal nine triplets of microtubules. The distal may be slightly displaced relative to the proximal centriole, the extent of the displacement varying from sperm to sperm (Billard 1983b). Various connecting fibrils and densities are summarized by Lahnsteiner and Patzner (2007). The distal centriole is closed anteriorly by an osmiophilic disc, the centriolar plug.

Flagellum: Usual metazoan flagellar structure is described by Billard (1983b). The triplets of the distal centriole give way posteriorly to doublets. They are conjoined by a ring, and each is linked to the plasma membrane at the level of the base of the implantation fossa by an arm [Y-link equivalent] shortly in front of the osmiophilic mass. The arms disappear at the commencement of the two central singlets. The plane of the two singlets passes through the plane of the longitudinal axis of the proximal centriole. Distally the B tubules disappear first while the A tubules lose their [dynein] arms. At the tail tip only one or two tubules remain (Billard 1983b).

Shortly behind the midpiece (Billard 1983b), extensions of the flagellar plasma membrane [flagellar fins], appear. ITDs are present: the A tubule of the 1, 2, 3, 5, 6, and 7 doublets appears darker and septate in the region of the fins. Two lateral fins were also observed on the 9+2 flagellum in *Salmo, Coregonus, Salvelinus* and *Thymallus* by Stein (1981) and are normal for salmonids (Lahnsteiner and Patzner 2007) though Billard (1983b) observed about 12% of flagellar cross sections to exhibit three fins in *Oncorhynchus mykiss*and *Salmo trutta fario* sperm.

The three subfamilies exhibit clear fine structural differences mainly in the dimensions of the midpiece and the organization of the mitochondrion. Furthermore osmiophilic bodies, located laterally to each centriole, were observed only in the Salmoninae. These are regarded as additional stabilization structures of the centriolar complex. Coregoninae are distinguished by a filamentous ring, thought to be as a cytoskeletal stabilization structure, at the caudal end of the midpiece (Lahnsteiner and Patzner 2007).

Sperm motility: Sperm motility studies (Lahnsteiner *et al.* 1999, 2007) reveal no differences between the three subfamilies (other than individual variation) and thus indicate that the observed structural differences between the three subfamilies are not coupled with functionality (Lahnsteiner *et al.* 1999; Lahnsteiner and Patzner 2007), at least in terms of motility.

Salmonoid sperm and phylogeny: Relationship of the Salmonoidei and Argentinoidei is endorsed by the synapomorphic ring-shaped single mitochondrion and some elongation of the nucleus. The presence of flagellar fins is a symplesiomorphy but contrasts with the apomorphic loss of these in most of the related Ostariophysi.

12.6 ORDER OSMERIFORMES

These are the freshwater smelts. They spawn in fresh water, except for *Osmerus eperlanus* and perhaps some salangins (Nelson 2006).

12.6.1 Superfamily Osmeroidea

The osmeroids usually have an adipose fin (Nelson 2006). They are the true smelts, marine anadromous or landlocked and freshwater forms in the Pacific, Arctic and Atlantic oceans and their drainages (Hearne 1984).

Fig. 12.11 *Plecoglossus altivelis (Osmeridae). Courtesy of The Fish Database of Taiwan. WWW electronic publication, version 2005/5.* http://fishdb.sinica.edu.tw, *(4/1/2008).*

Osmeroid sperm: The following taxa have been examined for sperm ultrastructure: **Osmeridae**, Hypomesinae, *Hypomesus pretiosus japonicus, H. transpacificus nipponensis*; Osmerinae, Osmerini, *Mallotus villosus; Osmerus eperlanus mordax* (Fig. 12.12); *Spirinchus lanceolatus* (all four by Hara and Okiyama 1998); Osmerinae, Salangini, *Salangichthys ishikawae; S. microdon* (Fig. 12.14) (both (Hara and Okiyama 1998); Plecoglossinae, *Plecoglossus altivelis* (Kudo 1983; Ohta *et al.* 1993; Utsugi 1993; Gwo *et al.* 1994) (Fig. 12.13, adult 12.11).

The sperm of the hypomesines *Hypomesus pretiosus japonicus* and *H. transpacificus nipponensis*, the osmerines *Mallotus villosus, Osmerus eperlanus mordax* (Fig. 12.12) and *Spirinchus lanceolatus*, and the plecoglossine *Plecoglossus altivelis* (Fig. 12.13) described by Hara and Okiyama (1998) differ notably from those of salmonids in having a single, unilateral mitochondrion in contrast to the annular, cylindrical or spiral mitochondrion of the latter family. They also differ in the very deep nuclear fossa which gives the nucleus an inverted U-shape in longitudinal section.

In contrast with the hypomesines, osmerins, and plecoglossines, with their unilateral mitochondrion, in the sperm of the two examined salangins the

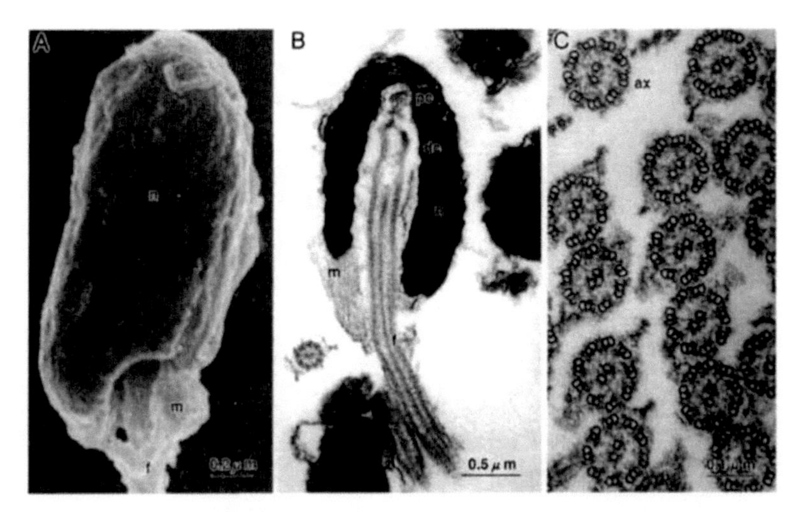

Fig. 12.12 *Osmerus eperlanus mordax* (Osmeridae, Osmerinae, Osmerini) spermatozoa. **A**. SEM; note the single, discrete, asymmetrically located mitochondrion. **B**. TEM longitudinal section. **C**. TEM transverse sections of flagella with lateral fins. ax, axoneme; dc, distal centriole; f, flagellum; m, mitochondrion; n, nucleus; pc, proximal centriole. From Hara, M. and Okiyama, M. 1998. Bulletin of The Ocean Research Institute, University of Tokyo 33: 1-138, Fig. 13B-D. With permission.

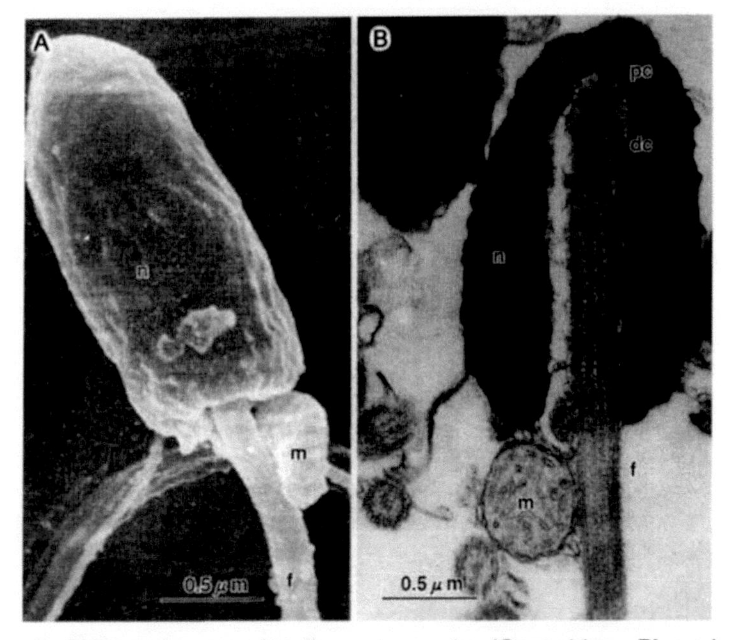

Fig. 12.13 M*Plecoglossus altivelis ryukyuensis*. (Osmeridae, Plecoglossinae). Spermatozoa. **A**. SEM of the nuclear region. **B**. TEM longitudinal section. dc, distal centriole; f, flagellum; m, single unilateral mitochondrion; n, nucleus; pc, proximal centriole. Adapted from Hara, M. and Okiyama, M. 1998. Bulletin of The Ocean Research Institute, University of Tokyo 33: 1-138, Fig. 19B and C. With permission.

axoneme is encircled by a single mitochondrion or several mitochondria in *Salangichthys microdon* (Fig. 12.14), and *S. ishikawae*, respectively. A further distinguishing characteristic of selangin sperm is the indistinct boundary between the single or multiple mitochondrion and the axoneme. They also differ from salmonids in that although the nuclear fossa is deep, as in other osmerids, its outline is irregular and the nucleus is subspherical. These distinctive features support monophyly of the tribe Salangini (Hara and Okiyama 1998).

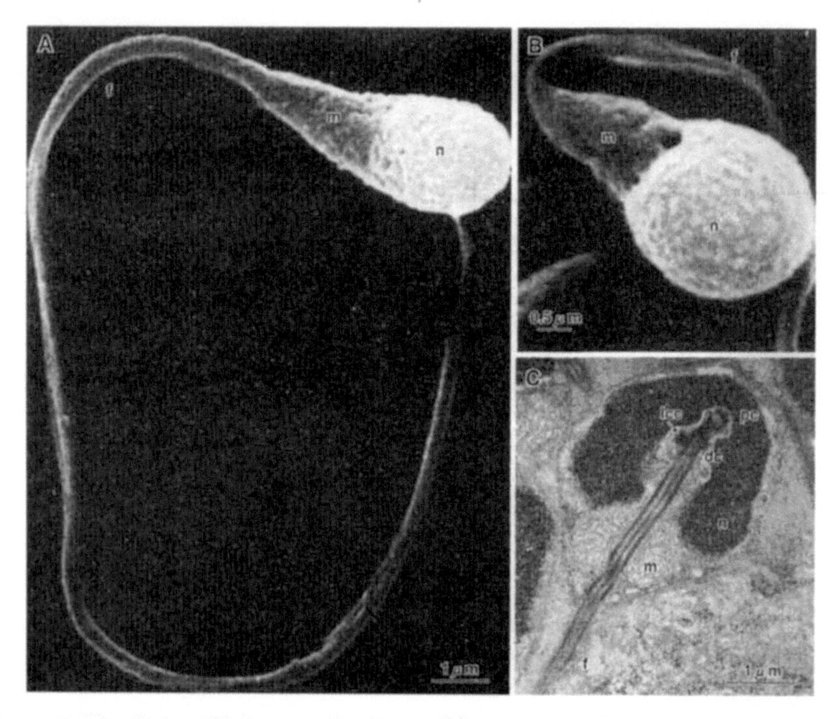

Fig. 12.14 *Salangichthys microdon.* (Osmeridae, Osmerinae, Salangini). Spermatozoa. **A**. SEM of whole sperm. **B**. SEM of sperm head. **C**. TEM longitudinal section. Adapted from Hara, M. and Okiyama, M. 1998. Bulletin of The Ocean Research Institute, University of Tokyo 33: 1-138, Fig. 20A-C. With permission.

In all eight of these hypomesines and osmerines, the two centrioles are mutually perpendicular and are located close to the apex of the nucleus but the proximal centriole is slightly displaced relative to the basal body (Hara and Okiyama 1998).

12.6.2 Superfamily Galaxioidea

12.6.2.1 Family Galaxiidae, Subfamily Galaxiinae

Galaxiids are mainly freshwater and some are anadromous. An example of the genus *Galaxias* is shown in Figure 12.15.

Fig. 12.15 A galaxiid, *Galaxias brevipinnis*. From Norman, J. R. 1937. *Illustrated Guide to the Fish Gallery*. Trustees of the British Museum, London, Fig. 20.

***Galaxias olidus* sperm:** A single species of the superfamily Galaxioidea, *Galaxias olidus*, has been examined for sperm ultrastructure (Marshall 1989, unpublished, in Jamieson 1991).

The nucleus is moderately elongate, with a length of 2.0-2.3 μm (mean 2.2 μm) and a width of 0.7 μm (n=4). The basal fossa is deep to the extent that the longitudinal sagittal section appears inverted U-shaped and most cross sections doughnut-shaped. The chromatin consists of an electron-dense matrix containing scattered darker particles, and closely conforming to the double nuclear membrane.

There is a single, annular mitochondrion subdivided by few large cristae. This surrounds a short, narrow cytoplasmic canal. There appears to be a short extension of the mitochondrion along one side of the basal region of the nucleus.

Probably because of the moribund condition of examined sperm, no centrioles have been observed. The axoneme, inserted in the longitudinal axis of the head, has a 9+2 pattern of microtubules and, in different regions, has 1, 2 or no fins. The fins are orientated in the plane of the central singlets. The lumina of the subtubules of the doublets are complete in many sections and in all of those which are cut through the nucleus. In some sections the A subtubule of doublets 1, 2, 5, 6 and 7 appears dark and occluded because it is partitioned by a septum (Fig. 12.16). Each doublet has two dynein arms (Marshall, 1989).

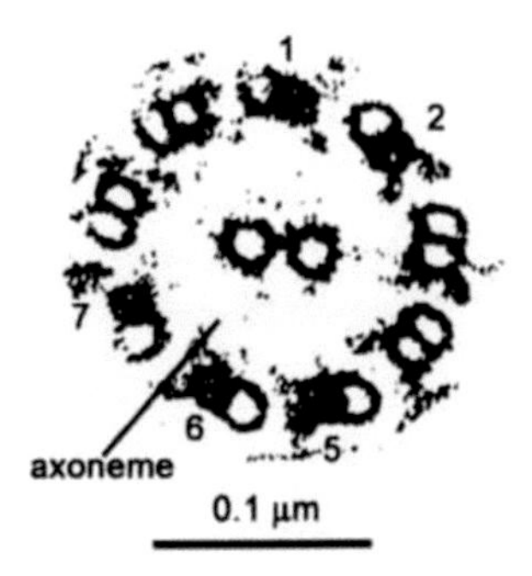

Fig. 12.16 *Galaxias olidus* (Galaxiidae) TEM transverse section of a sperm axoneme, showing occluded (septate) A microtubules in doublets 1,2,5,6 and 7. Lateral fins are present but not shown. Courtesy of C.J. Marshall.

Sperm relationships: The sperm of *Galaxias olidus* thus closely resembles a salmonid sperm but the nucleus, though no longer, is more slender and therefore appears more elongate and is more deeply penetrated by the basal fossa (as in osmerids), and the axoneme has a different pattern of occlusion of the doublets. As noted above, the galaxiid pattern of occlusion agrees with that in argentinoids. Salmonid occlusion differs only in septation of doublet 3 in addition to those occluded in *Galaxias* and it is uncertain what phylogenetic significance should be attached to this difference. The encircling mitochondrion is also seen in the salangin osmerine *Salangichthys microdon*.

12.6.2.2 Family Galaxiidae, Subfamily Lepidogalaxiinae

The single species of the suborder Lepidogalaxioidei and family Lepidogalaxiidae, *Lepidogalaxias salamandroides*, the Salamander Fish (Fig. 12.17), is known only from a small creek in South-western Australia. It was first described as a galaxiid (see Pusey 1983) but was considered by Rosen (1974) to be more closely related to the esocoids than to galaxioids and osmeroids despite similarities with these. Fink and Weitzman (1982) questioned inclusion of *Lepidogalaxias* in the Esocae. Fink (1984) found *Lepidogalaxias* to be "a potpourri of contradictory and reductive characters". He paired it with salmonids but saw it as the sister-group of the Neoteleostei (Stomiiformes, Aulopiformes, Myctophiformes and Acanthomorpha). Jamieson (1991), with reservations, relegated it to the Salmoniformes. Nelson (2006) tentatively places the monotypic subfamily Lepidogalaxiinae in the Galaxiidae within the Osmeriformes. Nevertheless, from analysis of mitochondrial DNA, Waters *et al.* (2000) consider that *Lepidogalaxias* is not a galaxioid and that an esocid relationship cannot be rejected.

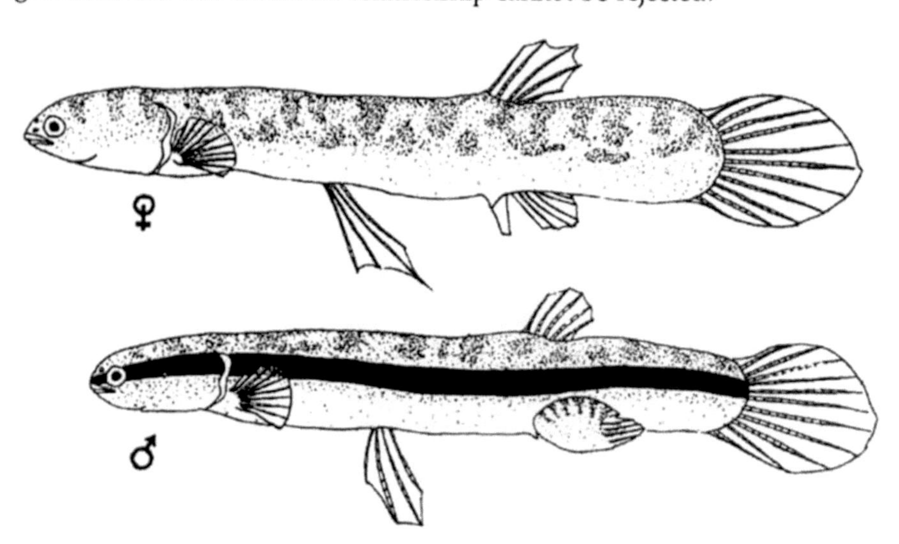

Fig. 12.17 *Lepidogalaxias salamandroides* Galaxiidae, Lepidogalaxiinae From Jamieson 1991, after Pusey, B.J. 1983. Journal of the Australia New Guinea Fishes Association. 1, 9-11, Fig. 1.

Sperm literature: The sperm of *Lepidogalaxias salamandroides* has been investigated ultrastructurally by Leung (1988), see also Jamieson (1991).

***Lepidogalaxias*:** The spermatozoon (Fig. 12.18) differs from all other neopterygian sperm in having an acrosome (excepting putative homologues noted for a few species) with one or two perforatoria, and in lacking triplet centrioles (Leung 1988). It is unusual but not unique in attachment of the flagellum anterior to the nucleus, a condition seen (with no phylogenetic construction), for instance, in the biflagellate sperm of *Polypterus* and in some characiform sperm.

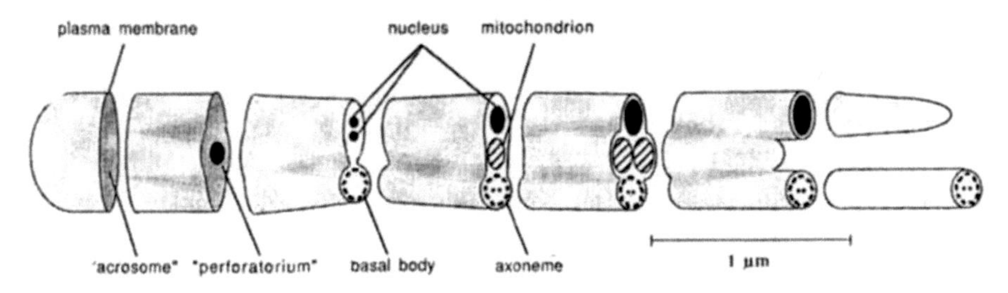

Fig. 12.18 *Lepidogalaxias salamandroides* (Galaxiidae, Lepidogalaxiinae). Schematic model of the spermatozoon. Courtesy of L. K. P. Leung.

A flattened structure filled with a homogeneous matrix on the anterior end of the spermatozoon resembles an acrosome vesicle. It has a dense cylindrical core about 0.6 µm long (range 0.49-0.71 µm, n=6) identified as a perforatorium. In some spermatozoa two perforatoria are present.

The nucleus is exceptionally long for the Pisces, at about 20 µm, a length exceeded only by the sperm nuclei of sharks and *Neoceratodus*. It is 0.3 µm in maximum diameter. Its chromatin is dense and homogeneous with occasionally a single axial lacuna. In some sperm the anterior end of the nucleus bifurcates. Its tapered posterior end, bounded by the nuclear envelope and plasma membrane, projects freely parallel to the anterior region of the axoneme.

Two elongate mitochondria, 10-13 µm long and 0.28 µm wide, are situated between the nucleus and the axoneme, parallel to these. The anterior ends of the mitochondria are usually at some distance (0-1.3 µm) behind the anterior end of the nucleus. The cristae are predominantly longitudinal. Two to four microtubules are occasionally associated with the mitochondria.

No typical triplet centrioles have been observed. A basal body, 0.46-0.54 µm long, located at the anterior end of the flagellum consists of 9 doublets. It is amorphous, lacking the circular configuration, in cross section, of the flagellum with which it is continuous. Electron-dense material is associated with the tubules and radial links of the basal body. Some tubules narrower than these tubules are present on the inner side of the doublets.

The 9+2 axoneme is approximately 53 µm long; dynein arms, radial links, central sheath and central cross bridges are present. Towards its posterior end, as is usual in flagella, A and B subtubules separate from each other and the arrangement of the tubules becomes disorientated (Leung 1988).

Spermatozeugmata: In the tubules of the tissue surrounding the testis spermatozeugmata occur in which the sperm are parallel and approximately in register; the anterior ends of the sperm are less compactly arrayed than the posterior ends.

The acrosome and phylogeny: The presence of an acrosome in this teleost is intriguing. An acrosome is known from the Ceratodontiformes, Lepidosireniformes, Coelacanthiformes, Polypteriformes and Acipenseriformes but not in taxa of presumed more recent origin, the Lepisosteiformes, Amiiformes or teleostean groups with the exception that perinuclear intermembranous scales in *Lepadogaster lepadogaster* (Gobiesocidae) have tentatively been identified as a vestigial acrosome (Mattei and Mattei 1978) and "pseudoacrosomes' are reported in *Oncorhynchus mykiss* and *Xenodermichthys*, as noted above. Acrosome-like vesicles are present in *Paracheirodon* and *Melanotaenia* sperm and immature *Gambusia* sperm (see **Chapter 14**).

Presence of an acrosome in *Lepidogalaxias* correlates with absence of a micropyle from the egg, an exceptional condition for a teleost. This in turn is presumably an adaptation to some (specialized?) feature of its internal fertilization. It is possible that the acrosome also functions in the known embedment of the sperm in the wall of the crypt region of the oviduct (Pusey and Stewart 1989).

An acrosome is here considered a basic (plesiomorphic) feature of fish sperm irrespective of whether fertilization was primitively internal or external in fish. The presence of the acrosome in *Lepidogalaxias*, a condition shared, in salmoniforms, with the spermatid of *Salmo* (Billard 1983a) contributes little to the hypothesis of relationship of their two suborders. In view of the phylogenetic position of salmoniforms suggested from molecular analysis (see **Chapter 1**) presence of an acrosome-like structure would not appear to indicate a basal position for the Salmoniformes.

Leung (1988) suggested that internal fertilization in *Lepidogalaxias*, whether primitive or not, is adaptive to the highly acidic waters in which it lives. The adverse effects of acidity on teleost sperm had been demonstrated by Urho *et al.* (1984). Leung considered the alternatives that internal fertilization and the complex form of the sperm (including presence of the acrosome) were retentions from an ancestral condition or were secondarily acquired in response to the acid environment and concluded that the former hypothesis was more parsimonious. Pusey and Stewart (1989) perhaps rightly consider it more parsimonious to regard internal fertilization in *Lepidogalaxias* as secondary. This view is here considered to be endorsed by the clearly secondary absence of the micropyle, a structure already present in the Chondrostei. Pusey and Stewart (1989) suggest that internal fertilization has arisen in relation to sperm competition while also suggesting that it may allow the female to confine the laying of fertilized eggs to optimal conditions for their development.

12.7 CHAPTER SUMMARY

Molecular evidence indicates that Protacanthopterygii, as currently defined, constitute a polyphyletic group, that Osmeriformes have a close relationship with Stomiiformes and that a lower clade contains Argentiformes and their sister clade Esociformes+Salmoniformes.

In the Argentiformes the mitochondrion is reported to be C-shaped as in Salmoniformes; the A subtubule is septate in axonemal doublets 1, 2, 5, 6, and 7, a synapomorphy of the Alepocephalidae and Platytroctidae (= Searsiidae), contrasting with 1, 2, 3, 5, 6, and 7 in *Salmo*, though some uncertainty exists as to the presence or absence of septation in A subtubule 3 in argentiforms. *Searsia* and *Salmo* share presence of a pseudoacrosome, at least before maturity. The 1, 2, 5, 6 and 7 arrangement is also seen in the Galaxiidae.

Esocid sperm differ from those of argentiforms and salmoniforms in having a number of separate (in E. *masquinongy* abundant) mitochondria and approach cyprinid or cichlid sperm in morphology. They are unusual in having a single flagellar fin.

The Salmoniformes contain the single family Salmonidae with three subfamilies. In the Salmoninae, the mitochondrion surrounds the cytoplasmic canal as an incomplete ring, a closed ring or helically in one and a half gyres. In the Thymallinae, the mitochondrion is helical, surrounding the midpiece in 2 or 3 gyres. In the Coregoninae, the midpiece is cylindrical and forms a complete ring. The flagellum has two, or in some sperm of *Oncorhynchus mykiss* and *Salmo trutta fario*, three fins.

In the Osmeriformes, the sperm of osmeroids differ notably from those of salmonids in having a single, unilateral mitochondrion. They also differ in the very deep nuclear fossa which gives the nucleus an inverted U-shape in longitudinal section. In the Galaxioidea, *Galaxias olidus* has a single, annular mitochondrion (as in salangin osmerine *Salangichthys microdon*). The flagellum has 1, 2 or no fins. ITDs occur in A subtubules 1, 2, 5, 6 and 7, as in osmerids. In the Galaxiidae, *Lepidogalaxias salamandroids*, of uncertain placement, differs from all other neopterygian sperm in having an acrosome (excepting putative homologues noted for a few species) with one or two perforatoria, and in lacking triplet centrioles. It differs from all other members of the Argentiformes through Osmeriformes, in attachment of the flagellum anterior to the nucleus. Presence of the acrosome and perforatoria correlates with the absence, exceptional for Neopterygii, of an egg micropyle.

12.8 LITERATURE CITED

Billard, R. 1970. Ultrastructure comparée de spermatozoïdes de quelques poissons Téléostéens. Pp. 71-79. In B. Baccetti (ed.), *Comparative Spermatology*. Academic Press, New York.

Billard, R. 1983a. Spermiogenesis in rainbow trout (*Salmo gairdneri*). Cell and Tissue Research 233: 265-284.

Billard, R. 1983b. Ultrastructure of trout spermatozoa: changes after dilution and deep-freezing. Cell and Tissue Research 228: 205-218.

Drokin, S. I., Stein, H. and Bartscherer, H. 1998. Effect of cryopreservation on the fine structure of spermatozoa of rainbow trout (*Oncorhynchus mykiss*) and brown trout (*Salmo trutta* F. *fario*). Cryobiology 37: 263-270.

Drozdov, A. L., Kolotukhina, N. K. and Maksimovich, A. A. 1981. Histological Structure of Testes and Ultrastructure of Spermatozoa in the Salmon *Oncorhynchus gorbuscha*. Biologiya Morya (Vladivostok): 49-53.

Fink, W. L. 1984. Basal euteleosts; relationships. Pp. 202-206. In H. G. Moser, W. J. Richards, D. M. Cohen, M. P. Fahay, J. Kendall, A.W. and S. L. Richardson (eds), *Ontogeny and Systematics of Fishes*. Special Publication Number 1 American Society of Ichthyologists and Herpetologists.

Fink, W. L. and Weitzman, S. H. 1982. Relationships of the stomiiform fishes (Teleostei):with a description of *Diplophos*. Bulletin of the Museum of Comparative Zoology Harvard 150: 31-93.

Fribourgh, J. H. 1978. Morphology of the brook trout spermatozoon as determined by scanning and transmission electron microscopy. The Progressive Fish Culturist 40: 26-29.

Fribourgh, J. H. and Soloff, B. L. 1976. Scanning electron microscopy of the Rainbow Trout (*Salmo gairdneri* Richardson) spermatozoon. Arkansas Academy of Science Proceedings 30: 41-43.

Furieri, P. 1962. Prime osservazioni al microscopio elettronico sullo spermatozoo di *Salmo trutta* L. Bollettino della Societa Italiana Biologia Sperimentale 38: 1030-1032.

Gwo, J. C., Kurokura, H. and Hirano, R. 1993. Cryopreservation of spermatozoa from rainbow trout, common carp, and marine puffer. Nippon Suisan Gakkaishi 59: 777-782.

Gwo, J. C., Lin, X. W., Gwo, H. H., Wu, H. C. and Lin, P. W. 1996. The ultrastructure of Formosan landlocked salmon, *Oncorhynchus masou formosanus*, spermatozoon (Teleostei, Salmoniformes, Salmonidae). Journal of Submicroscopic Cytology and Pathology 28: 33-40.

Gwo, J. C., Lin, X. W., Kao, Y. S. and Chang, H. H. 1994. The ultrastructure of ayu, *Plecoglossus altivelis*, spermatozoon (Teleostei, Salmoniformes, Plecoglossidae). Journal of Submicroscopic Cytology and Pathology 26: 467-472.

Gwo, J. C., Ohta, H., K., O., Wu, H. C. and Lin, P. W. 1999. Cryopreservation of fish sperm from the endangered Formosan landlocked salmon (*Oncorhynchus masou formosanus*). Theriogenology 51: 569-582.

Hara, M. and Okiyama, M. 1998. An ultrastructural review on the spermatozoa of Japanese fishes. Bulletin of The Ocean Research Institute, University of Tokyo 33: 1-138.

Hearne, M. E. 1984. In Ontogeny and Systematics of Fishes. Special Publication Number 1. American Society of Ichthyologists and Herpetologists, pp 153-155.

Inoue, J. G., Miya, M., Tsukamoto, K. and Nishida, M. 2003. Basal actinoperygian relationships: a mitogenomic perspective on the phylogeny of the "ancient fish". Molecular Phylogenetics and Evolution 26: 110-120.

Ishiguro, N. B., Miya, M. and Nishida, M. 2003. Basal euteleostean relationships: A mitogenomic perspective on the phylogenetic reality of the "Protacanthopterygii". Molecular Phylogenetics and Evolution 27: 476-488.

Jamieson, B. G. M. 1991. *Fish Evolution and Systematics: Evidence from Spermatozoa*. Cambridge University Press, Cambridge. 319 pp.

Jordan, D. S. 1907. *Fishes*. Henry Holt, New York. 789 pp.

Kobayashi, W. and Yamamoto, T. S. 1987. Light and Electron Microscopic Observations of Sperm Entry in the Chum Salmon Egg. Journal of Experimental Zoology 243: 311-322.

Kudo, S. 1983. Response to sperm penetration of the cortex of eggs of the fish, *Plecoglossus altivelis*. Development, Growth and Differentiation 25: 163-170.

Lahnsteiner, F., Berger, B. and Weismann, T. 1999. Sperm metabolism of the teleost fishes *Oncorhynchus mykiss* and *Chalcalburnus chalcoides* and its relation to motility and viability. Journal of Experimental Zoology 284: 454-465.

Lahnsteiner, F., Berger, B., Weismann, T. and Patzner, R. A. 1996. Changes in morphology, physiology, metabolism, and fertilization capacity of rainbow trout semen following cryopreservation. Progress in Fish Culture 58: 149-159.

Lahnsteiner, F. and Patzner, R. A. 1991. A new method for electron-microscopical fixation of spermatozoa of freshwater teleosts. Aquaculture 97: 301-304.

Lahnsteiner, F. and Patzner, R. A. 2007. Sperm morphology and ultrastructure in fish. Pp. 1-61. In S. M. H. Alavi, J. Cosson, K. Coward and G. Rafiee (eds.). *Fish Spermatology*. Alpha Science International Ltd./Morgan and Claypool, USA.

Lahnsteiner, F., Patzner, R. A. and Weismann, T. 1991. The fine structure of spermatozoa of the grayling, *Thymallus thymallus* (Pisces, Teleostei). Journal of Submicroscopic Cytology and Pathology 23: 373-377.

Lahnsteiner, F., Patzner, R. A. and Weismann, T. 1992a. The fine structural changes in spermatozoa of the grayling, *Thymallus thymallus* (Pisces, Teleostei), during routine cryopreservation. Aquaculture 103: 73-84.

Lahnsteiner, F., Weismann, T., Tinzl, M. and Patzner, R. A. 1992b. Fine structure of spermatozoa of the whitefish *Coregonus* sp. (Salmonidae, Teleostei). A third organization type of spermatozoa in salmonid fishes. Zeitschrift fuer Fishkunde 1: 105-115.

Lauder, G. V. and Liem, K. F. 1983. The evolution and interrelationships of the actinopterygian fishes. Bulletin of the Museum of Comparative Zoology 150: 95-197.

Lavoue, S., Miya, M., Inoue, J. G., Saitoh, K., Ishiguro, N. B. and Nishida, M. 2005. Molecular systematics of the gonorynchiform fishes (Teleostei) based on whole mitogenome sequences: Implications for higher-level relationships within the Otocephala. Molecular Phylogenetics and Evolution 37: 165-177.

Leung, L. K. P. 1988. The ultrastructure of the spermatozoon of *Lepidogalaxias salamandroides* and Its phylogenetic significance. Gamete Research 19: 41-50.

Lin, F., Liu, L. and Dabrowski, K. 1996. Characteristics of muskellunge spermatozoa: I. Ultrastructure of spermatozoa and biochemical composition of semen. Transactions of the American Fisheries Society 125: 187-194.

Lopez, J. A., Chen, W.-J. and Ortí, G. 2004. Esociform phylogeny. Copeia: 449-464.

Lowman, F. G. 1953. Electron microscope studies of silver salmon spermatozoa (*Oncorhynchus kisutch* [Walbaum]). Experimental Cell Research 5: 335-360.

Malejac, M. I., Loir, M. and Maisse, G. 1990. Membrane properties of spermatozoa from rainbow trout (*Oncorhynchus mykiss*) and their relationship to the freezing ability of sperm. Aquatic Living Resources 3: 43-54.

Marshall, C. J. 1989. *Cryopreservation and ultrastructural studies on teleost fish gametes*. Honours thesis, University of Queensland.

Mattei, C. and Mattei, X. 1978. Spermiogenesis in a Teleost Fish Lepadogaster-Lepadogaster Part 1 the Spermatid. Biologie Cellulaire (Ivry Sur Seine) 32: 257-266.

Mattei, C., Mattei, X., Marchand, B. and Billard, R. 1981. Réinvestigation de la structure des flagelles spermatiques: cas particulier des spermatozoïdes à mitochondrie annulaire. Journal of Ultrastructure Research 74: 307-312.

Mattei, X. 1991. Spermatozoon ultrastructure and its systematic implications in fishes. Canadian Journal of Zoology 69: 3038-3055.

Nelson, J. S. 1984. *Fishes of the World*, 2nd edition. John Wiley and Sons, New York. 523 pp.

Nelson, J. S. 2006. *Fishes of the World*, 4th edition. John Wiley and Sons, Inc., Hoboken, NJ. 601 pp.

Nicander, L. 1968. Gametogenesis and the ultrastructure of germ cells in vertebrates. Proceedings of the 6th International Congress on Artificial Reproduction in Animals, Paris 1: 89-107.

Ohta, T., Kato, K. H., Abe, T. and Takeuchi, T. 1993. Sperm morphology and distribution of intramembranous particles in the sperm heads of selected freshwater teleosts. Tissue and Cell 25: 725-735.

Pusey, B. J. 1983. The Shannon Mud Minnow. Journal of the Australia New Guinea Fishes Association 1: 9-11.

Pusey, B. J. and Stewart, T. 1989. Internal fertilization in *Lepidogalaxias salamandroides* Mees (Pisces: Lepidogalaxiidae). Zoological Journal of the Linnean Society 97: 69-79.

Radziun, K. and Tomasik, L. 1985. Ultrastructure of *Hucho hucho* (L.) spermatozoa. Acta Ichthyologica et Piscatoria 15: 130-140.

Rosen, D. E. 1974. Phylogeny and zoogeography of salmoniform fishes and relationships of *Lepidogalaxias salamandroides*. Bulletin of the American Museum of Natural History 153: 265-326.

Stein, H. 1981. Licht- und elektronenoptische Untersuchungen an den Spermatozoen verschiedener Süsswasserknochenfische (Teleostei). Zeitschrift für Angewandte Zoologie 68: 183-198.

Urho, L., Hudd, R. and Hildén, M. 1984. Kalojen sittiöden liikkumisaika pH:n funktiona. Memoranda Societatis pro Fauna et Flora Fennica 60: 41-42.

Utsugi, K. 1993. Motility and morphology of the sperm of the ayu, Plecoglossus altivelis, at different salinities. Japanese Journal of Ichthyology 40: 273-278.

Waters, J. M., Lopez, J. A. and Wallis, G. P. 2000. Molecular phylogenetics and biogeography of galaxiid fishes (Osteichthyes: Galaxiidae): Dispersal, vicariance, and the position of *Lepidogalaxias salamandroides*. Systematic Biology 49: 777-795.

Yoon, J. M., Chung, K. Y., Reu, D. S., Lew, I. D., Roh, S. C. and Kim, G. W. 1996. Electron microscopic observations on micropyle after sperm penetration in rainbow trout, *Oncorhynchus mykiss*. Korean Journal of Zoology 39: 173-181.

Zhang, X.-C. and Wang, S.-A. 1992. Ultrastructure of the testis and spermatogenesis in *Brachymystax lenok* (Pallas). Acta Zoologica Sinica 38: 355-358 (In Chinese).

Zirkin, B. R. 1975. The ultrastructure of nuclear differentiation during spermiogenesis in the salmon. Journal of Ultrastructure Research 50: 174-184.

CHAPTER 13

Ultrastructure of Spermatozoa: Neoteleostei: Stenopterygii, Cyclosquamata, Scopelomorpha and Paracanthopterygii

Barrie G. M. Jamieson

13.1 NEOTELEOSTEI

The Neoteleostei, not formally recognized by Nelson (2006) contains eight superorders: Stenopterygii, Ateleopodomorpha, Cyclosquamata, Scopelomorpha, Lampriomorpha, Polymixiomorpha, Paracanthopterygii and Acanthopterygii. This chapter deals with the Stenopterygii, Cyclosquamata, Scopelomorpha and Paracanthopterygii.

13.2 SUPERORDER STENOPTERYGII

As for the single order, Stomiiformes.

13.3 ORDER STOMIIFORMES

These include the dragon fishes; mostly tropical to temperate with many deep-sea (Nelson 2006). Partioned analysis of ten nuclear genes supports the unusual stance that the Stomiiformes are the sister group of the Osmeriformes (Ortí and Li, **Chapter 1**, Fig. 1.5).

13.3.1 Families Sternoptychidae, Phosichthyidae and Stomiidae

Examined sternoptychids are *Argyropelecus gigas* (Thiaw *et al.* 1990) and *Maurolicus japonicus* (Hara and Okiyama 1998).

School of Integrative Biology, University of Queensland, Brisbane, Queensland 4072, Australia

Sternoptychidae: In the sperm of *Maurolicus japonicus* (Fig. 13.1A,B), the nucleus is spherical and has a deep nuclear fossa. The chromatin forms dense masses. The two, mutually perpendicular centrioles lie in the nuclear fossa. The distal centriole has a centriole plug. A bilateral rootlet-like structure (striated rootlet) associated with the distal centriole observed during the middle stage of spermiogenesis disappears when the spermatid is fully formed. The number of mitochondria has not been confirmed: there is either a single large annular mitochondrion or multiple mitochondria, which collectively exceed the size of the nucleus, located at its base, and are exceptionally large relative to those of other teleost sperm. The flagellum has a tube-like fin [raised plasma membrane] and a classical 9+2 axoneme. The A-tubules of doublets 1, 2, 3, 5, 6, 7 and 8 contain septa due to ITDs (Hara and Okiyama 1998).

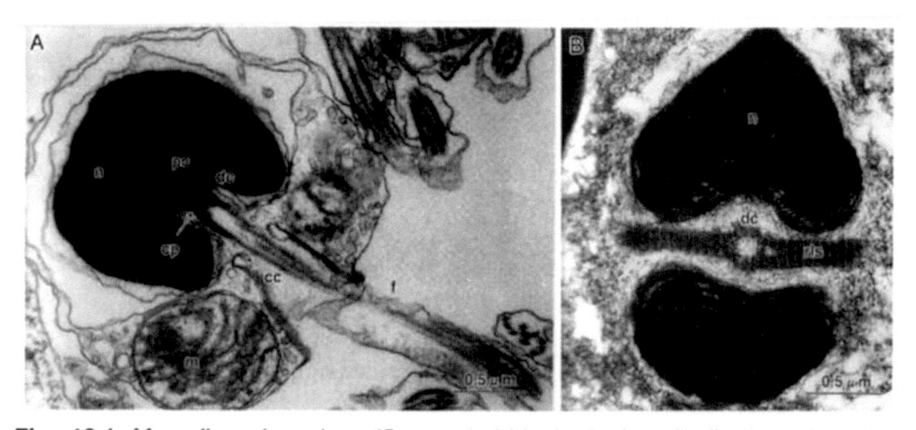

Fig. 13.1 *Maurolicus japonicus* (Sternoptychidae). **A**. Longitudinal section of a spermatozoon. **B**. Transverse section of a late spermatid through the nuclear fossa, showing the transient striated rootlet at the distal centriole. cc, cytoplasmic canal; cp, centriolar plug; dc, distal centriole; m, mitochondrion; n, nucleus; pc, proximal centriole; rls, striated rootlet. After Hara, M. and Okiyama, M. 1998. Bulletin of the Ocean Research Institute, University of Tokyo 33: 1-138, Figs. 25A and E. With permission.

Argyropelecus gigas sperm differ in modifications of the proximal centriole, such as the microtubular prolongation at one end ("fibrous body" *sensu* Afzelius 1978) (Fig. 13.2A,B) coupled with dislocation of mitochondria as well as occurrence of a large, finely granular mass in the midpiece, and development of a dense perinuclear deposit at the base of the nucleus (Thiaw *et al.* 1990; Hara and Okiyama 1998). The prolongation of the proximal centriole is an extension of the A and B microtubules. Very close to the centriole these doublets are organized like a 9+0 axoneme but more distally this axoneme becomes disorganized (Thiaw *et al.* 1990).

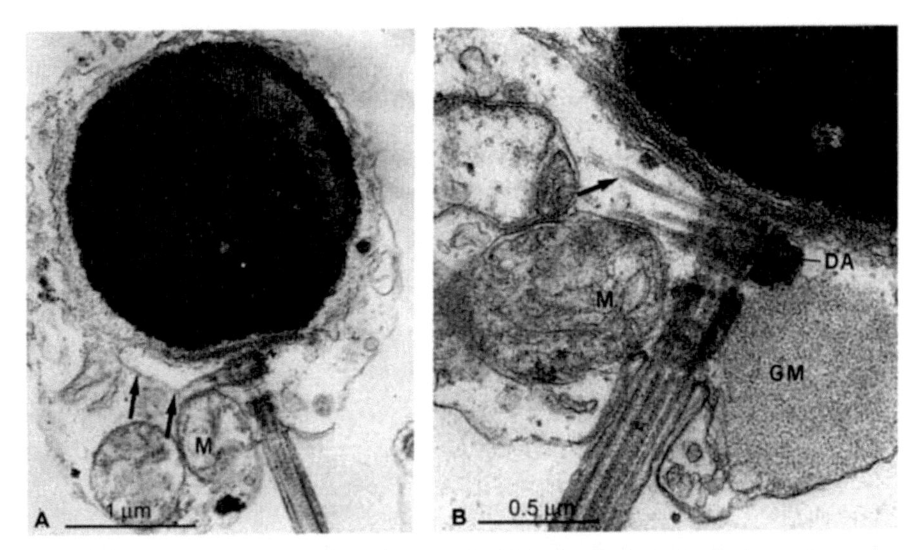

Fig. 13.2 *Argyropelecus gigas* (Sternoptychidae). **A**. Longitudinal section of a spermatozoon, showing the microtubular prolongations (arrows) of the proximal centriole. **B**. The proximal centriole shows a dense area (DA) at one end. At the level of the midpiece it faces a fine granular mass (GM). At the other end the centriole is prolonged by microtubules (arrow) turned towards the mitochondria (M), at the level of the midpiece. Adapted from Thiaw, O. T., Thiam, D. and Mattei, X. 1990. Journal of Submicroscopic Cytology and Pathology 22: 357-360, Figs. 3 and 4.

Phosichthyidae and Stomiidae: Like *Maurolicus japonicus*, the related *Polymetme corythaeola* in the Phosichthyidae (=Photichthyidae) referred to in a discussion by Thiaw *et al.* (1990) and *Melanostomias* sp., in the Stomiidae, Melanosomiinae, both briefly described by Mattei (1991), do not show these peculiarities (see also Fig. 13.3). *P. corythaeola* is unusual in sporadic occurrence of a pseudoacrosome. Mattei (1991) considers *P. corythaeola* and *Melanostomias* sp. to have the same spermatic features as the Argentinoidei and Salmonidae: a single mitochondrion with a spherical or annular shape [as in *Maurolicus japonicus* ?] and ITDs though the ITDs are not always evident.

Remarks: Thiaw *et al.* (1990) consider the extension of the proximal centriole in *Argyropelecus gigas* to be an aborted second axoneme and they list those families of fish which have a functional biflagellate condition (see also **Chapter 7, Section 7.3**). They propose that the aborted ciliary outgrowth described by Afzelius (1978) for the sperm of *Lepisosteus* is the same structure as that of *A. gigas*. They compare the persistence of a vestigial bundle of microtubules from the proximal centriole in the poeciliids *Poecilia latipinna*, *Gambusia affinis* and *Xiphophorus helleri*; the split bundle in Elopomorpha; and a single microtubule arising from the proximal centriole in the adrianychthyiid *Oryzias latipes*. In contrast to *A. gigas*, the closely related phosichthyid *Polymetme corythaeola* lacks the centriolar extension. It is only in

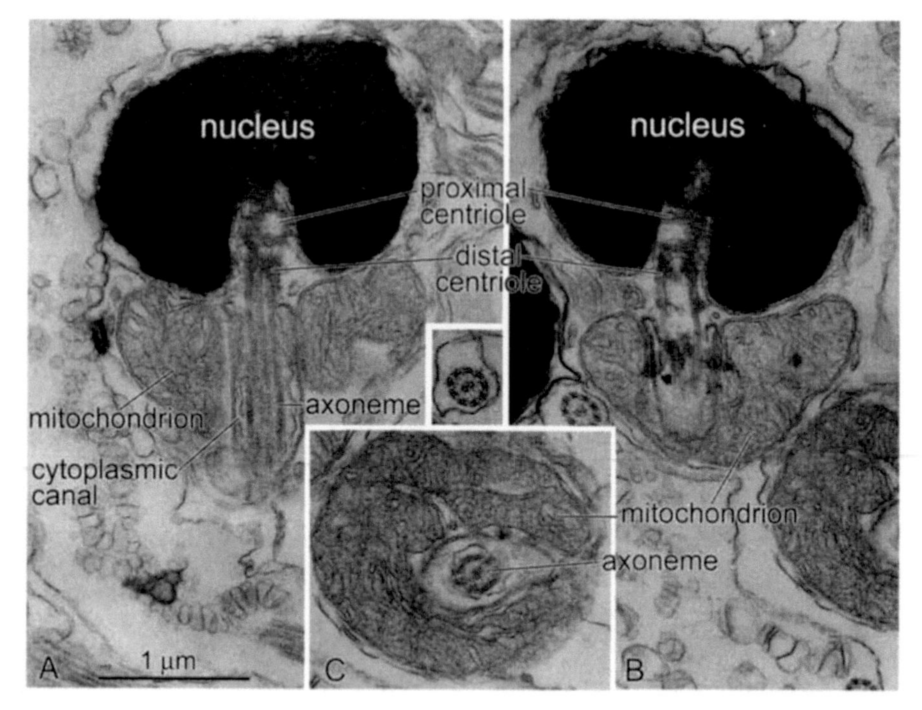

Fig. 13.3 *Melanostomias* sp. (Stomiidae). **A** and **B**. Longitudinal sections of spermatozoa. **A**. Transversely cutting the proximal centriole. **B**. At right angles through the proximal centriole. **C**. Transverse section of the midpiece, showing the single C-shaped mitochondrion almost completely encircling the axoneme. **Small inset**: Transverse section of an axoneme; ITDs are not evident. Original, from unpublished TEM negatives courtesy of Professor Xavier Mattei.

cases where there is no distinction between proximal centriole and distal centriole, as in true biflagellarity, that the two centrioles of the spermatid each produce an intact 9+2 axoneme (Thiaw *et al.* 1990).

13.4 SUPERORDER CYCLOSQUAMATA

As for the single order Aulopiformes

13.5 ORDER AULOPIFORMES

Aulopiformes are marine, benthic to pelagic fish. Members of the order have a specialization of the gill arches which is apparently unknown in other teleosts. Many are synchronous hermaphrodites (Roberts 1973; Lauder and Liem 1983; Nelson 2006). They are paraphyletically close to the Stomiiformes in the molecular analyses of Miya *et al.* (2003) and Ortí and Li (**Chapter 1**, Fig. 1.5).

13.5.1 Family Aulopidae

Aulopids are the flagfins, in tropical and subtropical seas, the Atlantic (including the Mediterranean) and the Pacific (Nelson 2006).

Aulopus japonicus: The sperm of *Aulopus japonicus* (Fig. 13.4A,B) has a spherical nucleus with granular chromatin with electron-lucent patches and a deep basal fossa. Two mutually perpendicular centrioles are located in the fossa approximately at the center of the nucleus. A single ring-shaped mitochondrion encircles the base of the axoneme, though much more developed on one side of the axoneme than the other. Lateral fins are said to be present on the single flagellum (Hara and Okiyama 1998) but do not appear to be well defined.

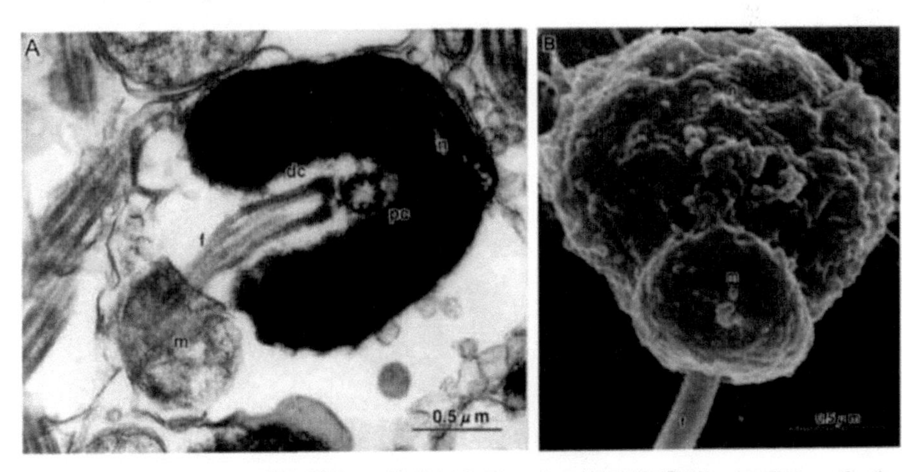

Fig. 13.4 *Aulopus japonicus* (Aulopidae). **A**. Longitudinal section of the spermatozoon. **B**. SEM of a spermatozoon. From Hara, M. and Okiyama, M. 1998. Bulletin of The Ocean Research Institute, University of Tokyo 33: 1-138, Fig. 29C and B. With permission.

13.5.2 Family Synodontidae

Synodontids, lizardfish (adult, Fig. 13.5), are marine (rarely brackish) in the Atlantic, Indian and Pacific Oceans (Roberts 1973; Lauder and Liem 1983; Nelson 2006).

Synodon saurus: The head region of the anacrosomal spermatozoon of *Synodon saurus* is elongated and approximately sickle shaped (Fig. 13.6A,B). The mitochondrion is integrated into the head region as it lies caudally in a ca 0.40 to 0.50 μm deep curvature of the nucleus. The head/midpiece complex has a length of 3.20 ± 0.t0 μm and a maximal diameter of 1.40 ± 0.65 μm. The nucleus has a length of 2.95 ± 0.1 μm and maximal diameter of 1.30 ± 0.05 μm. The chromatin is granular and the envelope lacks pores. Apart from the caudal fold in which the mitochondrion is located the nucleus has a second deep invagination, the basal fossa.

Fig. 13.5 *Trachinocephalus myops* (Synodontidae). Photo by John E. Randall.

The centriolar complex is located at the cranial end of the nuclear notch. Its proximal and distal centrioles (=basal body) are mutually perpendicular (Fig. 13.6A,B) and are composed of 9 pairs [not triplets] of peripheral microtubules,

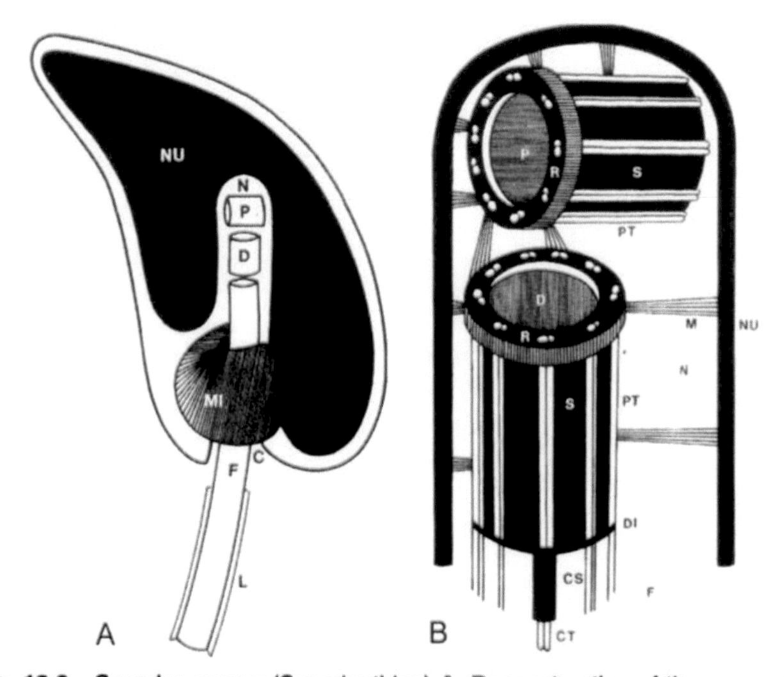

Fig. 13.6 *Synodon saurus* (Synodontidae) **A**. Reconstruction of the spermatozoon. **B**. Centriolar complex. C, cytoplasmic canal; CS, central sleeve of the flagellum; CT, central microtubules; D, distal centriole; DI, disk of electron-dense material; F, flagellum; L, lateral fins of the flagellum; M, microfibrils; MI, mitochondrion; N, nuclear notch (basal fossa); NU, nucleus; P, proximal centriole; PT, peripheral microtubules; R, ring of electron-dense material; S, sleeve of electron-dense material interconnecting the peripheral tubules. After Lahnsteiner, F. and Patzner, R. A. 1996. Journal of Submicroscopic Cytology and Pathology 28: 297-303, Figs. 15 and 16.

interconnected (as in the perciforms *Trachinus draco* and *Uranoscopus scaber*) by a thin sleeve of osmiophilic material. A ring of osmiophilic material adheres to the proximal end of the distal centriole and to one end of the proximal centriole. Proximal and distal centrioles are interconnected with each other by microfibrils via the two rings of electron-dense material. Both centrioles are attached to the nuclear fossa by microfibrils which insert at the centriolar microtubules as well as at the electron-dense rings (Fig. 13.6B). A single U-shaped mitochondrion is located in the basal fold of the nucleus (Fig. 13.6A). The cytoplasmic canal separates the mitochondrion and the flagellum. The 9+2 flagellum has a length of about 40 µm and a diameter of 0.29 ± 0.05 µm. Very small lateral fins are present. For further details see Lahnsteiner and Patzner (1996).

***Trachinocephalus* and *Saurida*:** The spermatozoa of the lizardfishes *Trachinocephalus myops* and *Saurida parri* are anacrosomal aquasperm known only from line diagrams (Mattei 1970, 1991) (Fig. 13.7). In both the sperm morphology is closely similar to that of *Synodon*; the nucleus is an asymmetrical cone with a deep basal fossa which contains the proximal centriole at right angles to the basal body. The large mitochondrion is depicted on one side only of the base of the flagellum.

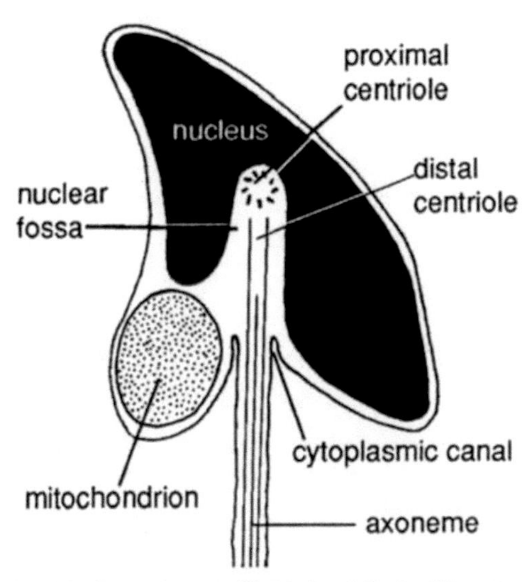

Fig. 13.7 *Trachinocephalus myops* (Synodontidae). Diagrammatic longitudinal section of the spermatozoon. After Mattei, X. 1970. Pp. 59-69. In B. Baccetti (ed.). *Comparative Spermatolog*, Academic Press, New York, Fig. 4:12.

Remarks: Little can be determined with regard to aulopiform relationships from sperm morphology beyond noting that the mitochondrion is single as in the argentinoid-salmonoid assemblage but differs in *Synodon*, *Trachinocephalus* and *Saurida* in being unilateral. Furthermore, the relative uniformity and considerable modification of the sperm of these three genera supports monophyly of the Synodontidae and is divergent from the condition in the Aulopidae, exemplified by *Aulopus* with its ring-shaped, albeit asymmetrical, mitochondrion which approaches the salmoniform condition. All aulopiform sperm examined to date have a deep nuclear fossa.

13.6 SUPERORDER SCOPELOMORPHA

The grouping Scopelomorpha, now contains a single order Myctophiformes, all of which are deep-sea pelagic or benthopelagic (Nelson 2006).

13.7 ORDER MYCTOPHIFORMES

Myctophiforms are the lanternfishes, with two families, the Neoscopelidae and Myctophidae, of which only myctophids have been examined for sperm ultrastructure.

13.7.1 Family Myctophidae

The type-family, the Myctophidae (adult, Fig. 13.8) is characterized by luminescent photophores arranged along the sides of the body, hence the name lantern fish; they are amongst the most common deep-sea fish.

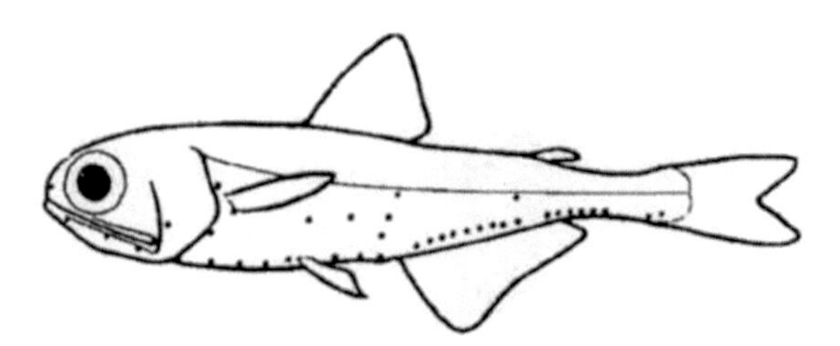

Fig. 13.8 A lantern fish, *Symbolophorus barnardi* (Myctophidae). After Tony Ayling in Ayling T. and Cox, G. 1982. *Guide to the Sea Fishes of New Zealand.* William Collins Publishers Ltd., Auckland, New Zealand. With permission.

***Lampanyctus* sperm:** The African, Senegalese, *Lampanyctus* sp., (Myctophidae, Lampanictinae) has a biflagellate spermatozoon (Fig. 13.9A) with a 9+0 axonemal structure (Fig.13.9B).

The nucleus is S-shaped and about 6 µm long; the mitochondria are not organized into a true midpiece; and the two flagella, inserted laterally and considerably proximal to the base of the nucleus, are 55 µm long. The two centrioles (described only for the spermatid) are parallel (Mattei and Mattei 1976).

In marked contrast, *Lampanyctus jordani* has uniflagellate sperm (Fig. 13.10). The nucleus is spherical. The chromatin consists of dense masses with scattered pale lacunae. Only a distal centriole has been observed (Hara and Okiyama 1998). Lateral flagellar fins appear to be present. This variation intragenerically suggests that monophyly of the genus may be suspect.

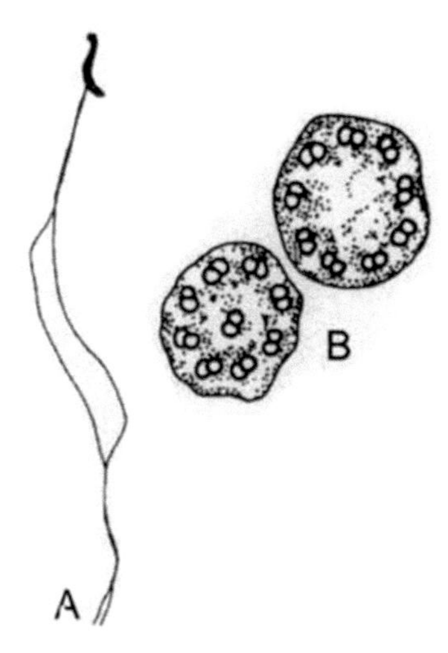

Fig. 13.9 *Lampanyctus* sp. (Myctophidae). **A**. Spermatozoon *in toto*. From Mattei, X. and Mattei, C. 1976a. Journal de Microscopie et de Biologie Cellulaire 25: 187-188, Fig. 1e. **B**. Transverse sections of axonemes, showing 9 doublets but no central singlets. One of the sections is abnormal in having one of the doublets displaced to the center. After a micrograph of Mattei, X. and Mattei, C. 1976a. Journal de Microscopie et de Biologie Cellulaire 25:187-188, Fig. 1d.

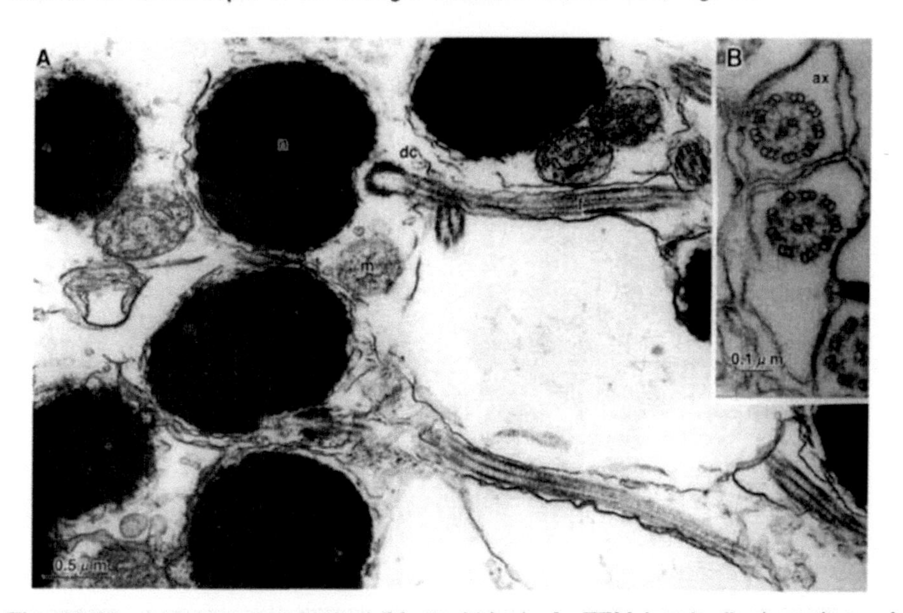

Fig. 13.10 *Lampanyctus jordani* (Myctophidae). **A.** TEM longitudinal sections of spermatozoa. **B.** Transverse sections of flagella. ax, axoneme; dc, distal centriole; f, flagellum; m, mitochondrion; n, nucleus. Adapted from Hara, M. and Okiyama, M. 1998. Bulletin of The Ocean Research Institute, University of Tokyo 33: 1-138, Fig. 28A and C. With permission.

***Symbolophorus californiensis* and *Notoscopelus* sp.** The spermatozoa of *S. californiensis* and *Notoscopelus* sp. have been examined by SEM and TEM (Hara 2007).

The sperm of both species, illustrated here for *Symbolophorus californiensis* (Fig. 13.11), have (1) numerous small spherical mitochondria in close contact with the nuclear membrane which covers an S-shaped nucleus. The nucleus

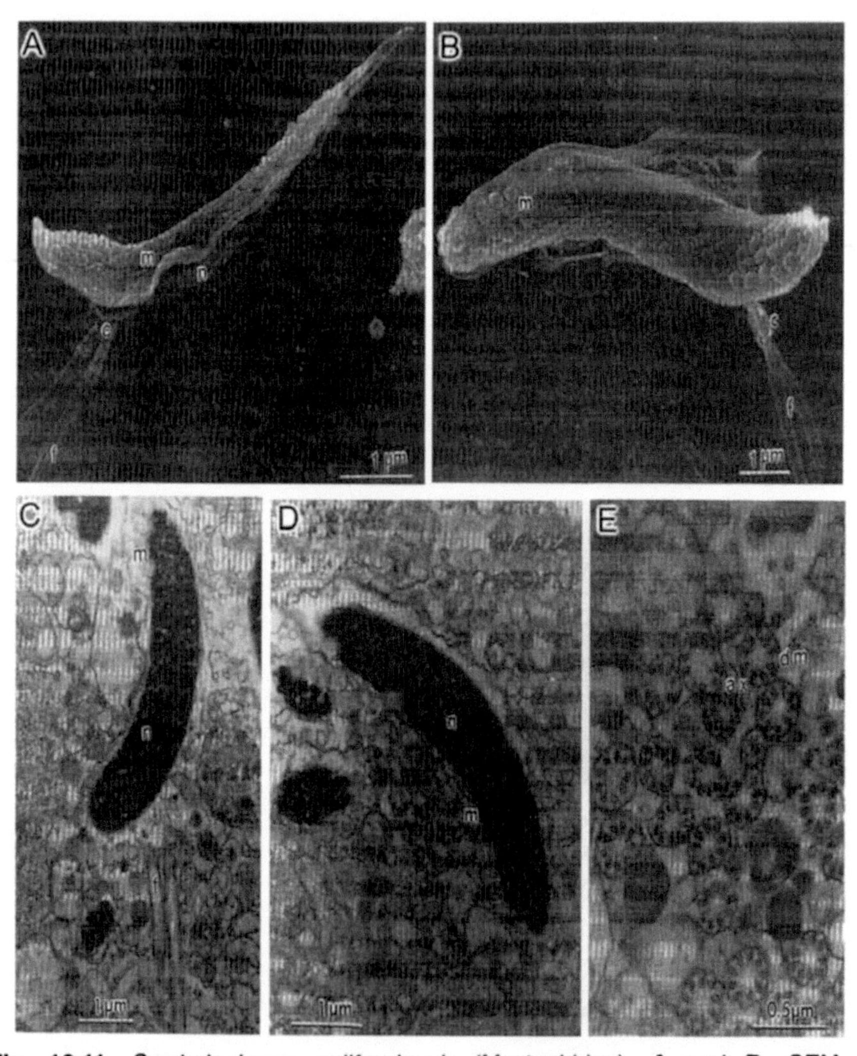

Fig. 13.11 *Symbolophorus californiensis* (Myctophidae). **A** and **B**. SEM of spermatozoa, showing the crescentic nucleus fringed on the concave face by mitochondria, and the biflagellate condition. **C** and **D**. TEM, approximate longitudinal sections. **C**. Showing the two flagella. **E**. Transverse sections of flagella, showing the 9+0 axonemes. ax, axoneme; dm, doublet microtubule; f, flagellum; m, mitochondria; n, nucleus; Adapted from Hara, M. 2007. Japanese Journal of Ichthyology 54: 41-46, Fig. 1, A,B,E-G. With permission.

is 5 µm long and ca 1 µm wide. (2) As in *Lampanyctus,* two centrioles are located about one third along the length of the nucleus and each gives rise to a flagellum. The flagella are ca 47 µm long. (3) Each has a 9+0 axonemal structure. Hara (2007) suggested that these three apomorphic characters [in combination] may be peculiar to the Myctophidae. Although they appear to define a section of the family they do not define the uniflagellate *Lampanyctus jordani* nor the supposedly aflagellate *Lampanyctodes hectoris* and *Diaphus danae* (see below).

***Lampanyctodes* and *Diaphus* sperm:** Scanning electron microscopy of the sperm of the lampanyctines *Lampanyctodes hectoris* and unspecified observations on *Diaphus danae* are said by Young *et al.* (1987) to have revealed aflagellate spermatozoa but this is not ascertainable from the micrograph which they provide.

Remarks: The reported diversity in the few lampanyctid sperm examined, from the uniflagellate to biflagellate to, supposedly, aflagellate condition is unparalleled in other fish families. Biflagellarity is known elsewhere in 32 species in 17 families of the Euteleostomi (Table 13.1) but only the uniflagellate elopomorph sperm share with myctophids the 9+0 axoneme (and have nuclear mitochondria). No phylogenetic significance can be attached to the latter resemblance. Sperm ultrastructure (biflagellarity and/or spiral nucleus) may, nevertheless, support relationship of batrachoidiforms + gobiesociforms. It remains to be ascertained whether the diversity of sperm structure in the Myctophidae indicates polyphyly.

Table 13.1 Occurrence of biflagellate spermatozoa in Euteleostomi

Order *Family* *Species*	*Common Name*	*Reference*
Polypteriformes		
Polypteridae		
Polypterus senegalus	Gray bichir	Mattei 1970
Myctophiformes		
Myctophidae		
Lampanyctus sp.	lantern fish	Mattei and Mattei 1976
Batrachoidiformes		
Batrachoididae		
Opsanus tau	Oyster toadfish	Hoffman 1963
Porichthys notatus	Plainfin midshipman	Stanley 1965
Perciformes		
Apogonidae		
Apogon affinis	Bigtooth cardinal fish	Mattei and Mattei 1984, 1991
Apogon annularis	Ringtail cardinalfish	Fishelson *et al.* 2006

Table 13.1 Contd. ...

Table 13.1 Contd. ...

Order *Family* *Species*	*Common Name*	*Reference*
Apogon crassiceps	Transparent cardinalfish	Fishelson *et al.* 2006
Apogon evermanni	Evermann's cardinalfish	Fishelson *et al.* 2006
Apogon hungi	Cardnal fish	Fishelson *et al.* 2006
Apogon imberbis	Cardinal fish	Lahnsteiner 2003
Apogon lineolata	Shimmering cardinal	Fishelson *et al.* 2006
Apogon semilineatus	Half-lined cardinal	Hara and Okiyama 1998
Apogon semiornatus	Oblique-banded cardinalfish	Fishelson *et al.* 2006
Apogon talboti	Flame cardinalfish	Fishelson *et al.* 2006
Cheilodipterus lineatus	Tiger cardinal	Fishelson *et al.* 2006
Siphamia mossambica	Sea urchin cardinal	Fishelson *et al.* 2006
Siphamia permutata	Cardnal fish	*Fishelson et al.* 2006
Cichlidae		
Satanoperca jurupari	Demon eartheater	Matos *et al.* 2002
Gobiesocidae		
Lepadogaster lepadogaster	Shore clingfish	Mattei and Mattei 1978
Gobiidae		
Leucopsarion petersii	Shiro-uo	Hara 2004
Zoarcidae		
Macrozoarces americanus	Ocean Pout	Yao *et al.* 1995
Zoarces elongatus	Nagagaji	Koya *et al.* 1993
Siluriformes		
Ariidae		
Genidens genidens	Guri sea catfish	Burns *et al.*, Chapter 11
Aspredinidae		
Bunocephalus amazonicus	Banjo catfish	Spadella *et al.* 2006
Cetopsidae		
Cetopsis coecutiens	Piracatinga catfish	Spadella *et al.* 2006
Doradidae		
Anadoras wedellii	Carataí, Thorny catfish	Burns *et al.*, Chapter 11
Heptapteridae		
Rhamdia sapo	South American catfish	Maggese *et al.* 1984
Ictaluridae		
Ictalurus punctatus	Channel catfish	Poirier and Nicholson 1982
Malapteruridae		
Malapterurus electricus	Nile electric catfish	Shahin 2006
Liobagrus mediadiposalis	South torrent catfish	Lee and Kim 1999
Nematogenyidae		
Nematogenys inermis	Mountain catfish	Spadella *et al.* 2006
Lepidosireniformes		
Protopteridae		
Protopterus annectens	West African lungfish	Boisson *et al.* 1967
Protopterus aethiopicus	Marbled lungfish	Purkerson *et al.* 1974

13.8 SUPERORDER PARACANTHOPTERYGII

Paracanthopterygii are predominantly marine; only five genera of the percopsiforms, the gadiform *Lota*, a few brotulids, some batrachoidiforms and the fluviatile gobiesocids are freshwater fish (Lauder and Liem 1983). They may represent a side branch in teleost evolution, the remaining fishes (Acanthopterygii) either sharing a common ancestry within or being derivatives of a related lineage (Nelson 1984). Lauder and Liem (1983) list a large set of synapomorphies for the Paracanthopterygii and support the view that this group is the plesiomorphic sister-group of the Acanthopterygii. However, monophyly of the Paracanthopterygii remains highly questionable (Miya *et al.* 2003; Nelson 2006, Ortí and Li, Fig. 13.12). The high diversity of sperm morphologies found among its different orders (Jamieson 1991; Mattei 1991), with the occurrence of many autapomorphies, supports the suspicion that the currently accepted classifications of the Paracanthopterygii may include unrelated groups (Medina *et al.* 2003).

Spermatozoal ultrastructure has not been studied in the Percopsiformes but is known for the Ophidiiformes, Gadiformes, Batrachoidiformes, Gobiesociformes and Lophiiformes. As noted by Medina *et al.* (2003), if monophyly of the Paracanthopterygii be assumed (but this is questioned above), the ancestral (presumably simple) spermatozoal pattern must have undergone a great radiation in sperm morphologies. These include, *inter alia*, apart from the gadiform sperm patterns described below, simple Type I spermatozoa (Antennariidae and Chaunacidae), biflagellate sperm (Batrachoidiformes) and, if included, Gobiesociformes and spermatozoa with well developed, long, midpiece and nucleus (*Neoceratias*) (Jamieson 1991; Mattei 1991).

13.9 ORDER OPHIDIIFORMES

These are the cusk-eels. The Bythitidae, in the Ophidiiformes, are live-bearing. In molecular analyses, ophidioforms are seen as the sister-group of a partial percomorph assemblage (Miya *et al.* 2003) or of a batrachoidiform+ atherinomorph+percomorph assemblage (Ortí and Li, **Chapter 1**).

In a valuable light microscopical paper on ophidioid spermatophores, Nielsen *et al.* (1968) suggest that the development of these structures may be a device ensuring that the spermatozoa are kept alive in the female until the eggs mature.

***Ophidion* sperm:** The spermatozoon of *Ophidion* sp.(suborder Ophidioidei, Ophidiidae, Ophidiinae) has been examined ultrastructurally by Mattei *et al.* (1989) and that of *O. barbatum* by Hernández *et al.* (2005). That of *Lamprogrammus exutus* was illustrated diagrammatically by Mattei (1991) (Fig. 13.22).

The nucleus in *Ophidion* sp. is 8 μm long and circular in cross section. The base is 0.8 μm wide and its anterior end is drawn out into a very slender process, 2.5 μm long and 0.05 μm wide (Mattei *et al.* 1989). The nucleus is slenderly conical in *Ophidion barbatum* (Hernández *et al.* 2005) despite the fact that this species is externally fertilizing (Breder and Rosen 1966). The nucleus in *Lamprogrammus exutus* is drawn as rounded conical but not attenuated by Mattei (1991). The "intermediate piece" is short, measuring about 0.6 µm long and 1 µm wide. It contains two centrioles, 7 to 8 mitochondria and a vesicular cytoplasm. In places the mitochondria appear from a micrograph to be in two tiers of which the more posterior encloses a short cytoplasmic canal. The flagellum is about 100 µm long. It has the 9+2 structure and two lateral fins which are generally inclined at 10° and 30° relative to the plane of the central singlets. As many as eight 10 nm wide longitudinal microtubules are present under the plasma membrane near the lateral extremity of each fin and appear to act as a cytoskeleton (Mattei *et al.* 1989). The organization of the testis was shown by Mattei *et al.* (1993) to be of the semi-cystic type corresponding to the unrestricted type of Grier (Grier *et al.* 1980; Grier 1981) and synonymous with the lobular type of Billard (1986, 1990).

Sperm phylogeny: *Ophidion* shows elongation of the nucleus which is common on adoption of internal fertilization (denied, however, for *O. barbatum*) but is otherwise unmodified apart from the filamentous form of the apex of the nucleus. These modifications must be interpreted as independent developments which do not cast light on the phylogenetic affinities of ophidioids. According to Mattei (1991) *Ophidion*, *Neoceratias*, and *Lepadogaster*, all in different orders, are the only known teleosteans in which the sperm nucleus shows a "twined" front end. However, this similarity does not appear to reflect relationship.

13.10 ORDER GADIFORMES

Gadiformes, cods, were seen as the sister-group of the percopsiforms by Lauder and Liem (1983) and they have a close relationship with these in the independent molecular analysis of Miya *et al.* (2003), where the relationship is paraphyletic, and Ortí and Li (**Chapter 1**, Fig. 1.5, and 13.12 below), where it is unresolved.

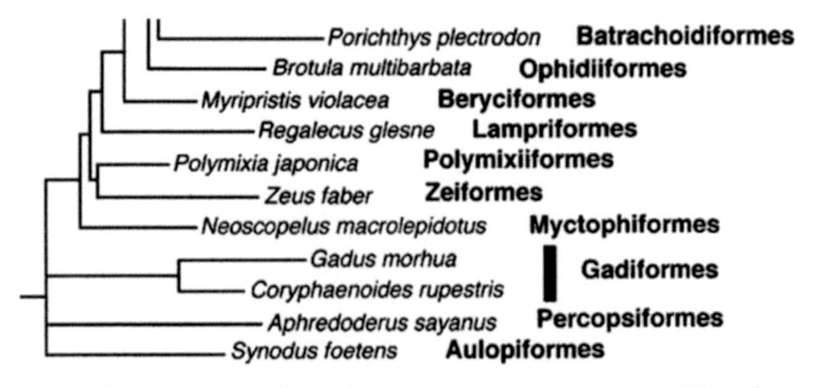

Fig. 13.12 Phylogenetic position of the Gadiformes based on partitioned analyses of ten nuclear genes. Adapted from Ortí and Li (**Chapter 1**, Fig. 1.5).

Sperm literature: In Jamieson (1991) the only gadiform species for which sperm had been studied was the gadid *Lota vulgaris*, examined optically by Retzius (1905), with a uniflagellate anacrosomal aquasperm. Sperm ultrastructure is now known in the Moridae - *Laemonema laureysi* Mattei (1991); Merlucciidae - *Laemonema laureysi* and *Merluccius polli* (Mattei 1991) and *M. merluccius* (Medina *et al.* 2003); and Gadidae - *Trisopterus minutus capelanus* (Malatesta *et al.* 1991), *Lota lota* (Lahnsteiner *et al.* 1994).

13.10.1 Family Moridae

Morids are the deep-sea cods.

Sperm ultrastructure: Mattei (1991) examined the sperm of *Laemonema laureysi*. He observed in it a synapomorphy of the Moridae and Merlucciidae that supported their being placed together in the suborder Gadoidei: the flagellar axis is situated laterally in relation to the nucleus (resembling the Type II perciform spermatozoon).

13.10.2 Family Merlucciidae

These are the merlucciid hakes, in the Atlantic and Pacific (Nelson 2006), exemplified by *Merluccius merluccius*, the European hake (Fig. 13.13E).

Sperm ultrastructure: In the acrosomeless spermatozoon of *Merluccius merluccius*, described by Medina *et al.* (2003), the rounded head is followed by an elongate midpiece and the tail (Fig. 13.13A). The head and the midpiece together measure about 3.5 μm in length and 1.75 μm in width. The nucleus, which shows a homogeneous electron-dense chromatin, lacks nuclear pores and is ovoid in shape, somewhat flattened at its apical surface (Fig. 13.13A-C). The midpiece is conical, tapering at the distal end (Fig. 13.13A). By SEM the flagellum, which measures about 30 μm in length, displays a smooth cylindrical shape throughout its length; no flagellar fins are observed.

The two centrioles (250 nm long and ca 180 nm wide) are parallel and located in one plane perpendicular to the longitudinal axis of the sperm cell. They are ca 450 nm apart (center-to-center distance) and show the conventional 9 triplets + 0 configuration. Centrally, at the base of the nucleus, the nuclear envelope forms two shallow invaginations in coincidence with the location of the centrioles, but these remain outside the nuclear depressions (Fig. 13.13B). The flagellar apparatus is attached to the nucleus by means of a lateral plate made of an electron-dense material that keeps the lateral surface of the axonemal basal body connected to the nuclear envelope (Fig. 13.13A). Therefore, the axoneme is orientated parallel to the basal surface of the nucleus (as also reported by Mattei 1991 for *Merluccius polli*), and the insertion of the flagellar apparatus in the spermatozoon is asymmetrical. The proximal centriole is also linked to the nuclear surface by granular material. The two centrioles are mutually parallel. In the midpiece, an array of microtubules is occasionally seen projecting from the proximal centriole. Nine radial fibers project from the basal body triplets and contact the plasma membrane at the caudal part of the centriole (Figs. 13.13A). A cytoplasmic canal surrounds the proximal segment of the flagellum for ca 2 μm. Throughout the cytoplasmic canal the plasma membrane is thickened with a dense layer on its inner side.

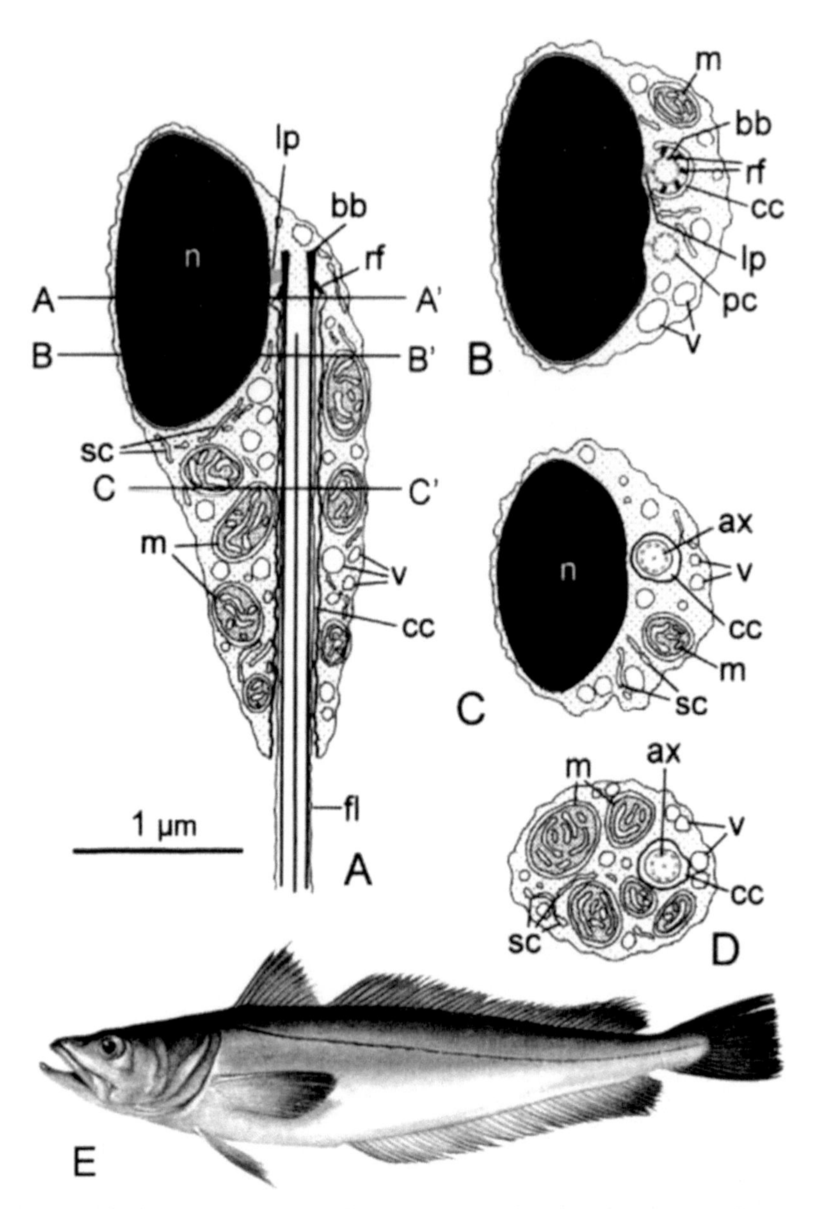

Fig. 13.13 *Merluccius merluccius*, European hake (Merlucciidae). **A-D**. Schematic drawings of the spermatozoon. **A**. Sagittal section at the level of the distal centriole (basal body). **B-D**. Consecutive transverse sections through the head [at the level of the centrioles (approximately A–A') and axoneme (B- B')], and through the midpiece (C- C') (all frontal views). ax, axoneme; bb, basal body (distal centriole); cc, cytoplasmic canal; fl, flagellum; lp, lateral plate; m, mitochondria; n, nucleus; pc, proximal centriole; rf, radial fibers; sc, smooth membrane cisternae; v, midpiece vesicles. From Medina, A., Megina, C., Abascal, F. J. and Calzada, A. 2003. Acta Zoologica (Copenhagen) 84: 131-137, Fig. 2. With permission of Blackwell Publishing. **E**. Adult. By Norman Weaver. With kind permission of Sarah Starsmore.

A considerable amount of cytoplasm is retained in the elongate midpiece, which contains numerous electron-lucent vesicles, several round or ovoid mitochondria and smooth tubular membrane cisternae. The mitochondria measure ca 0.5 μm in diameter, and show irregular cristae and a moderately electron-dense matrix. Up to five mitochondrial profiles can be seen in cross-sections of the midpiece close to the nucleus, and more than 10 mitochondrial units appear to be present throughout the midpiece.

The axoneme has the usual 9+2 arrangement except adjacent to the basal body, where the pair of central microtubules is absent. At this level the peripheral doublets are attached to the flagellar membrane by Y-links. The flagellum has neither intratubular differentiations nor lateral fins (Medina *et al.* 2003).

Remarks: As noted by Medina *et al.* (2003) the spermatozoon of *Merluccius merluccius* have Type II sperm [or resemble these], thus confirming previous observations on three gadiform species (Mattei 1991; Lahnsteiner *et al.* 1994). As in the other gadiforms studied, intratubular differentiations, inner septa located in the A microtubules of doublets 1, 2, 5 and 6, originally described as a specific perciform-type sperm feature (Mattei, C. *et al.* 1979), are not found in *M. merluccius*. However, they are also absent in some other perciform-type spermatozoon (Gwo *et al.* 1994; Hara and Okiyama 1998; Abascal *et al.* 2002).

Other features shared by all the known gadiform spermatozoa are the elongate midpiece and deep cytoplasmic canal, and the absence of flagellar fins. These characteristics suggest a relatively high homogeneity in the sperm ground plan of the Gadiformes studied. This is the only known paracanthopterygian order possessing a typical perciform-type sperm but interspecific differences are seen (Medina *et al.* 2003).

As also noted by Medina *et al.* (2003), *Merluccius merluccius* has mutually parallel proximal and distal centrioles, an apomorphic condition that seems unique among the few investigated gadiforms. A parallel orientation of the sperm centrioles is rare in fish sperm but has been observed in the agnathan *Lampetra*, the osteoglossomorphs *Pantodon* and *Gymnarchus*, and the pomacentrid perciform *Chromis chromis* whereas the most frequent perpendicular arrangement (the one normally found in somatic cells) is considered plesiomorphic in animal sperm cells (Jamieson 1991).

Satellite rays have not been reported for other gadiforms, but are present in *Merluccius merluccius*, radiating from the distal centriole and attaching it to the plasma membrane at the proximal end of the cytoplasmic canal as in many other fish groups, apparently as a plesiomorphic feature.

While several mitochondria are found in the spermatozoa of *Laemonema laureysi*, *Merluccius polli* (Mattei 1991) and *M. merluccius* (Medina *et al.* 2003), the sperm of the gadid *Lota lota* differs from these in that the midpiece bears a single horseshoe-shaped mitochondrion (Lahnsteiner *et al.* 1994). As Medina *et al.* (2003) observe, the size of the midpiece depends to some extent on the number of mitochondria stored by the sperm cell, which also determines the depth of the post-nuclear canal (Baccetti *et al.* 1984). In agreement with this observation, the sperm cytoplasmic canal in *Lota lota* is shallower than it is in *M. merluccius*.

If the Type I sperm, which is present in a large number of acanthopterygian families including several families of perciforms, is considered the plesiomorphic condition for the Paracanthopterygii–Acanthopterygii assemblage, a perciform-type (Type II) spermatozoon would have evolved independently in both taxa (Medina *et al.* 2003) (see also **Chapter 15**).

13.10.3 Family Gadidae

The spermatozoa of the gadids *Trisopterus minutus capelanus* and *Lota lota* and have been described ultrastructurally by Malatesta *et al.* (1991) and Lahnsteiner *et al.* (1994) respectively.

Lota lota: The spermatozoon of *Lota lota*, the Burbot (Fig. 13.14A, adult; B, sperm), described by Lahnsteiner *et al.* (1994), has a spherical head with a diameter of 1.54 ± 0.11 μm containing the nucleus which lacks acrosome and

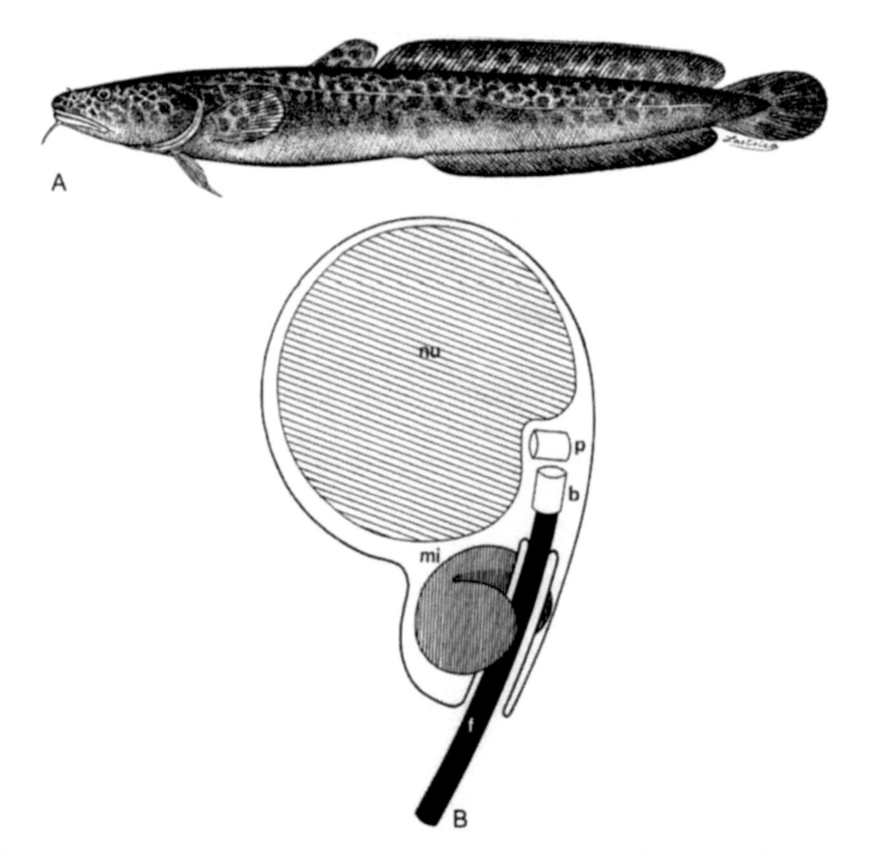

Fig. 13.14 *Lota lota* (Gadidae). **A**. Adult. Courtesy of FAO© **B**. Schematic reconstruction of the spermatozoon. b, distal centriole (basal body); cy, cytoplasmic canal; f, flagellum; mi, mitochondrion; nu, nucleus; p, proximal centriole. From Lahnsteiner, F., Weismann, T. and Patzner, R. A. 1994. Journal of Submicroscopic Cytology and Pathology 26: 445-447, Fig. 1.

nuclear pores and consists of granular heterogeneous condensed chromatin. The nuclear fossa, situated mediolaterally, houses the centriolar complex. The complex is formed by the proximal and the distal centriole (basal body) located at a right angle to each other and consisting of nine triplets (length 0.26 ± 0.07 µm; diameter 0.25 ± 0.04 µm). The anterior portion of the distal centriole and the whole proximal centriole are embedded in electron-dense material.

The midpiece forms an asymmetrical truncated cone (length 0.55 ± 0.10 µm; diameter - measured in the middle of the cone - 0.49 ± 0.09 µm) (Fig. 13.14) and is penetrated sagittally by the cytoplasmic canal (diameter 0.43 ± 0.08 µm). Unlike many other teleosts, no membrane underlies the invaginated portion of the plasmalemma which forms the cytoplasmic canal One mitochondrion is located in the midpiece region surrounding the canal in the form of an incomplete ring (Fig. 13.14); the cristae are irregularly arranged or sometimes in parallel and have a diameter of 0.020 ± 0.004 µm.

The typical 9+2 flagellum is 23.9 ± 4.08 µm long, and 0.25 ± 0.02 µm wide. The two central microtubules are surrounded by the central sheath which is interconnected with the peripheral microtubules by the centrifugal extending radiations. For the first 0.39 ± 0.02 µm from its origin from the basal body the flagellum lacks central microtubules. The flagellum has no lateral fins and ends in an endpiece in which the microtubules are reduced in number and irregularly arranged (Lahnsteiner *et al.* 1994).

Remarks: The C-shaped mitochondrion approaches the condition in salmoniforms but the orientation of the flagellum relative to the nucleus is that of a Type II perciform sperm.

***Trisopterus minutus capelanus*:** Malatesta *et al.* (1991) in a study of spermatogenesis in *T. minutus*, confirm the absence of an acrosome.

13.11 ORDER BATRACHOIDIFORMES

In batrachoids, an order with a single family, the Batrachoididae, the body is usually scaleless (or in some has small cycloid scales) and the head is large with eyes tending to dorsal (for additional characters, see Nelson 2006). They are mostly marine (Lauder and Liem 1983; Nelson 2006). The Batrachoidiformes (Fig. 13.15A,B) were envisaged by Lauder and Liem (1983) as the sister-group of the Lophiiformes (but see Fig. 13.12), a view which receives some support from the mutual possession of helical sperm nuclei, a rare condition. The two orders were seen by them as the sister-group of the Gobiesociformes. However, in molecular analysis the three orders are far separated (Ortí and Li, **Chapter 1**), though Lophiiformes and Batrachiformes are close in the phylogeny presented by Nelson (2006).

Biflagellarity of batrachoidiform and gobiesociform sperm, as far as they are known, is a remarkable similarity and if the two orders are not closely related must be regarded as a homoplasy by virtue of relationship (symparamorphy *sensu* Jamieson 1984).

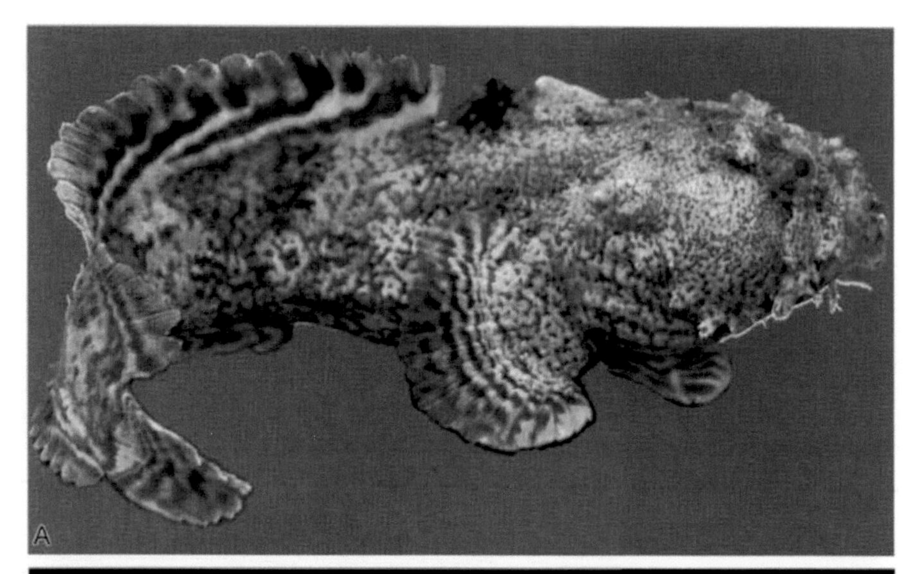

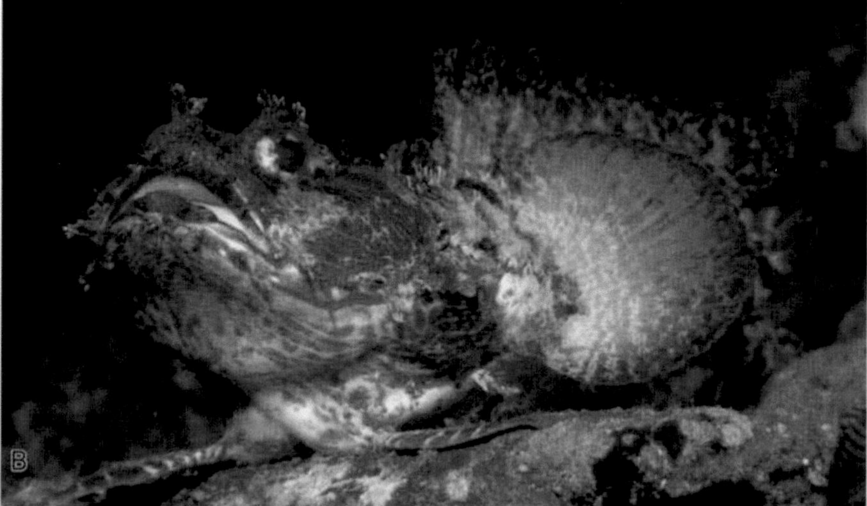

Fig. 13.15 Toad fishes (subfamily Batrachoidinae). **A**. *Opsanus tau*. Adapted from NASA. **B**. *Halophryne diemensis*, Triton Bay, West Papua (Indonesia). Courtesy of Dr. Gerry Allen.

Batrachoid sperm: The three investigated species of the Batrachoididae are *Opsanus tau*, the Oyster Toadfish (Hoffman 1963; Casas *et al.* 1981) (adult, Fig. 13.15A) and *Halobatrachus (=Batrachus) didactylus*, the Lusitanian Toadfish (Mattei 1991 and this account), both of the Batrachoidinae, and *Porichthys notatus*, the Plainfin (Stanley 1965), in the Porychthyinae. At least *Opsanus* and *Porichthys* are spawners with parental care (Breder and Rosen 1966) and all three have biflagellate aquasperm, simple in *Opsanus* and *Halobatrachus*, more complex in *Porichthys*.

***Opsanus*:** The light microscope study of Hoffman (1963) provides little information beyond demonstrating the biflagellarity of the sperm of *Opsanus tau*. The ultrastructural account of Casas *et al.* (1981) is directed to an investigation of the chromatin but yields additional data. The sperm (Fig. 13.16) is a typical teleostean aquasperm except for its biflagellarity.

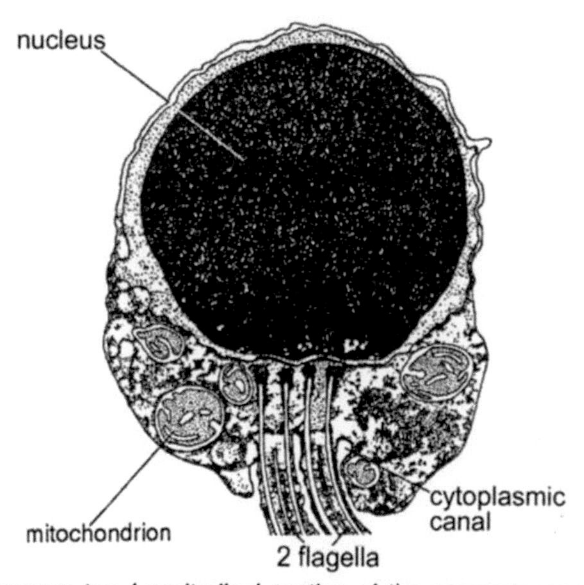

Fig. 13.16 *Opsanus tau*. Longitudinal section of the spermatozoon. Drawn from micrographs after Casas, M. T., Munoz-Guerra, S. and Subirana, J. A. 1981. Biology of the Cell 40: 87-92, Fig. 2e and f.

The nucleus is spheroidal. Upon maturation of the spermatozoon the chromatin fibers become less apparent than during spermiogenesis. The chromatin is uniformly distributed through the nucleus. It appears as granules and filaments of various sizes (50-100 Å). Thus this histone-containing nucleus does not show the high degree of compaction achieved in other species which contain protamines in their sperm nuclei. The individuality of the 200 Å fibers, seen in spermatids, is lost. *O. tau* may represent an intermediate case between similarly histone-containing sperm in which the 200 Å diameter fibers are preserved at maturity (*Holothuria* and *Limulus*) and those in which complete condensation is achieved through the effect of protamines (Casas *et al.* 1981). The midpiece contains many irregularly arranged round mitochondria and there is a short cytoplasmic canal around the base of each flagellum. The two centrioles are mutually parallel and each is associated with a separate slight indentation of the nucleus (Casas *et al.* 1981). It is not stated whether the 9+2 flagella possess fins. These are not present in other biflagellate batrachoid sperm.

Halobatrachus didactylus: In a diagram of the sperm of *H.* (=*Batrachus*) *didactylus*, Mattei shows a somewhat elongate nucleus, bilateral mitochondria, and two flagella. This morphology is confirmed here (Fig. 13.17).

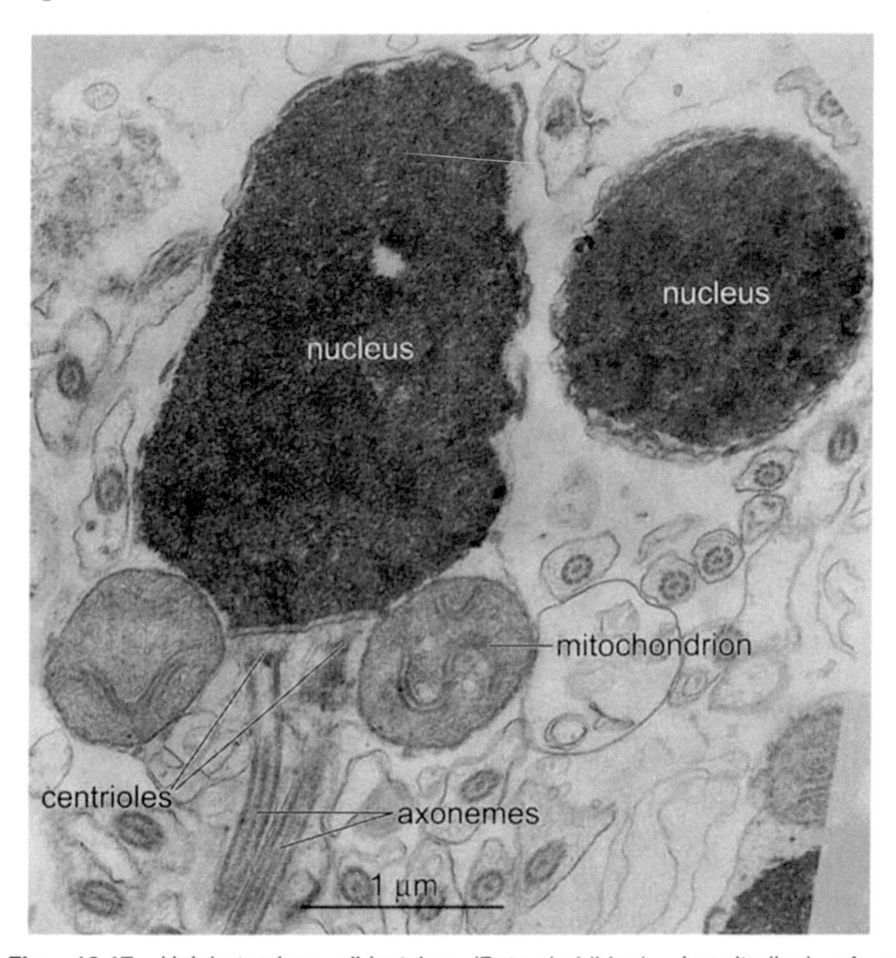

Fig. 13.17 *Halobatrachus didactylus* (Batrachoididae). Longitudinal of a spermatozoon showing elongate nucleus and biflagellate condition. A transverse section of an adjacent nucleus reveals its circular profile. Original, from an unpublished negative courtesy of Professor Xavier Mattei.

Porichthys: In the late spermatid of *Porichthys notatus* (not illustrated), in contrast with *Opsanus*, the nucleus is spiral, although again with no evident acrosome, and the mitochondria are elongate, lying close around the basal portions of the flagella to form a distinct midpiece. The parallel centrioles are each connected by a fan-like array of filaments to the nuclear membrane which is slightly indented (Stanley 1965). There is a suggestion of a similar connection in micrographs of the *Opsanus* sperm.

Batrachoid sperm and phylogeny: The high cellular DNA content led Hinegardner (1968) to regard *Opsanus tau* as primitive, a view which Casas *et al*. (1981) consider might be supported by the presence of histones and the low degree of condensation of the sperm nucleus. Somatically batrachoids are in fact highly modified if basal euteleosts. The histone-rich sperm nuclei in *O. tau* perhaps indicate their primitive origins while their biflagellarity, with that of *Porichthys notatus* and *Halobatrachus didactylus*, is a further apomorphy shared (homplasically?) with gobiesociforms. Whether this biflagellarity is a true synapomorphy or a homoplasy, the fact that biflagellarity is, seen, independently, in basal pre-teleostean lineages (lungfishes and *Polypterus*) indicates that this condition cannot be regarded *per se* as indicating an advanced position for batrachoids.

13.12 ORDER LOPHIIFORMES

The anglerfishes. Sixteen families; all marine; most species in deep water (Lauder and Liem 1983; Nelson 1984, 2006); typically with an illicium (line) and esca (bait) that attract prey to the mouth. They have been considered to be closely related to the Batrachoidiformes (Nelson 2006), and the two were considered sister-groups by Lauder and Liem (1983). However, molecular analyses (Miya *et al*. 2003; Ortí and Li, **Chapter 1**, Fig. 1.5), suggest that the Lophiiformes and Tetraodontiformes are sister-groups and that lophiiforms should therefore be placed in the Percomorpha. Credence is added to the molecular analyses as they utilized different genetic systems, the former (Fig. 13.18) used mitogenomic while the latter used nuclear sequences. The great

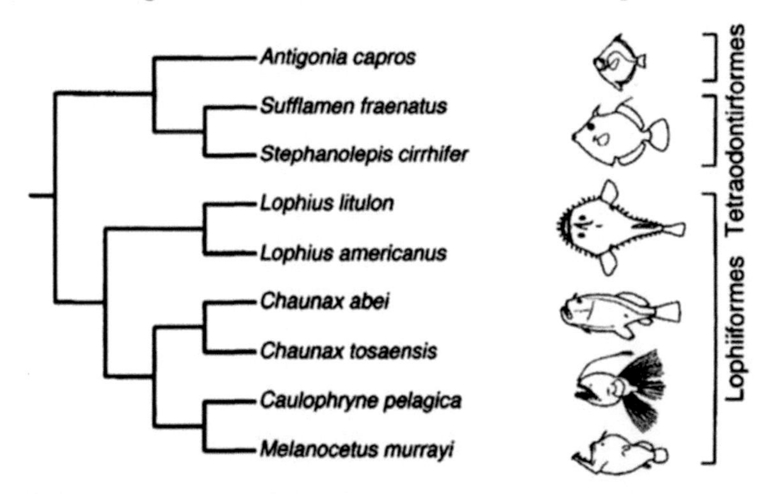

Fig. 13.18 Sister-group relationship of Lophiiformes and Tetraodontiformes indicated by analysis of complete mitochondrial DNA sequences. Adapted from Miya, M., Takeshima, H., Endo, H., Ishiguro, N. B., Inoue, J. G., Mukai, T., Satoh, T. P., Yamaguchi, M., Kawaguchi, A., Mabuchi, K., Shirai, S. M. and Nishida, M. 2003. Molecular Phylogenetics and Evolution 26: 121-138, Fig. 5 (part). With permission.

morphological differences between lophiiforms and tetraodontiforms are echoed by major differences in spermatozoal ultrastructure between Neoceratiidae and tetraodontiforms but the sperm of the chaunacid *Dibranchus atlanticus* (Fig. 13.21) resembles those of the tetraodontids *Lagocephalus laevigatus* (Fig. 15.128) and *Sphoeroides spengleri* (Fig. 15.129), particularly in the two tiers of mitochondria. No spermatozoal synapomorphies were found by Jamieson (1991) between lophiiforms and batrachoidiforms.

13.12.1 Suborder Antennerioidea

13.12.1.1 Family Antennariidae

Antennariids are the frogfishes (adult, Fig. 13.19), in all tropical and subtropical seas except the Mediterranean, occasionally temperate (Nelson 2006).

Fig. 13.19 A frog fish, *Antennarius striatus* (Antennariidae, Subfamily Antennariinae). Photo Jeff Wright. Courtesy of the Queensland Museum.

In contrast with *Neoceratias* (below), the Frogfish *Antennarius senegalensis* (Antennariidae) has a basic teleostean ect-aquasperm classifiable as Type I, known only from an illustration by Mattei (1970) (Fig. 13.20).

13.12.2 Suborder Ogcocephalioidei

13.12.2.1 Families Chaunacidae and Ogcocephalidae

Chaunacids are the coffin fishes or sea toads, in the Atlantic, Indian and Pacific Oceans (Nelson 2006). Ogcocephalids are the batfish in all tropical and many subtropical seas but absent from the Mediterranean (Nelson 2006).

Chaunax pictus (Chaunacidae) and *Dibranchus atlanticus* (Ogcocephalidae) are briefly referred to and illustrated in line diagrams by Mattei (1991) and *D. atlanticus* in the present account (Fig. 13.21). Both appear to be Type I sperm.

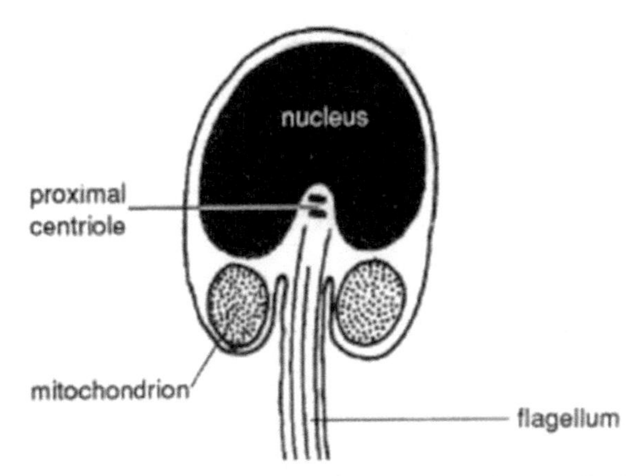

Fig. 13.20 *Antennarius senegalensis.* Diagrammatic longitudinal section of the spermatozoon. After Mattei, X. 1970. Pp. 567-569. In B. Baccetti (ed.). *Comparative Spermatology,* Academic Press, New York, Fig. 4: 15.

The sperm of *C. pictus* has a "simple structure" with a moderate basal fossa containing the proximal centriole, one tier of mitochondria on each side of the midpiece, and is similar to that of *Antennarius*. That of *D. atlanticus* is more complex, having two tiers of mitochondria (Fig. 13.21A) arranged in a circlet one tier of which has eight, radially adpressed mitochondria (Fig. 13.21B). An indistinct profile of the proximal centriole is observable in the nuclear fossa; only the anterior region of the distal centriole lies in the fossa (Fig. 13.21A). It has 9 triplets and, at the transition to the single axoneme, two central singlets (Fig. 13.21B, and inset).

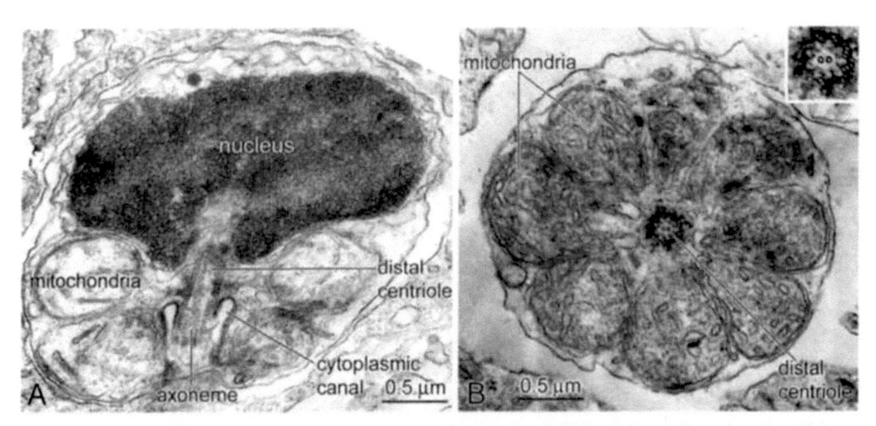

Fig. 13.21 *Dibranchus atlanticus* (Ogcocephalidae), Atlantic batfish. **A**. Longitudinal section of a spermatozoon, showing the depressed nucleus and two tiers of mitochondria. **B**. Transverse section of the midpiece through the distal centriole at its transition to the axoneme and the circlet of eight mitochondria. Inset: detail of the centriole. Original, from TEM negatives courtesy of Professor Xavier Mattei.

13.12.2.2 Family Neoceratiidae

Neoceratiids are the toothed seadevils, in the Atlantic, Indian and Pacific Oceans (Nelson 2006).

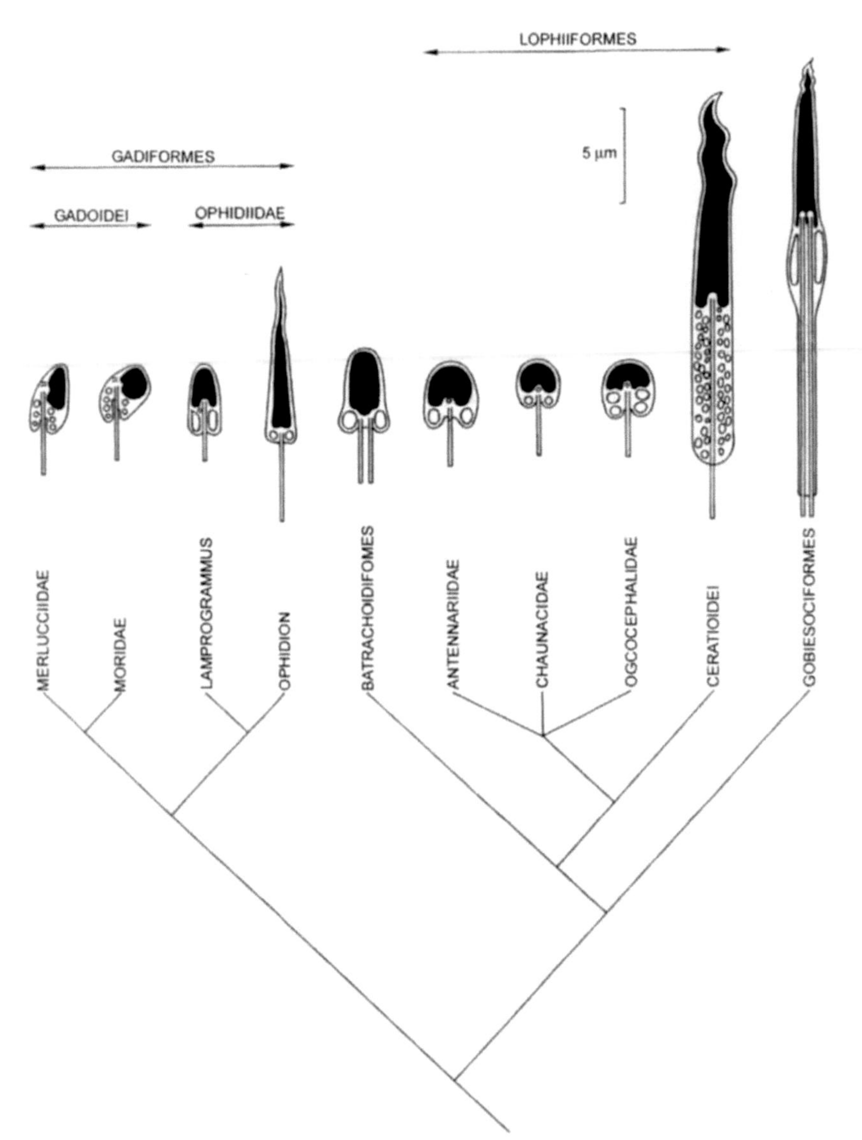

Fig. 13.22 Phylogenetic relationships and spermatozoal structure in the Paracanthopterygii envisaged by Mattei, X. 1991. Canadian Journal of Zoology 69: 3038-3055, Fig. 10. With permission.

Neoceratias spinifer, the Toothed seadevil, and only species of the Neoceratiidae (adult, Fig. 13.23), is a deep-sea, North Atlantic species with a dwarf parasitic male. It lacks the illicium. The complex sperm, briefly described from a recently dead fish by Jespersen (1984) differs from that of batrachoids and gobiesociforms in being uniflagellate but shows similarities to those of the batrachoid *Porichthys* and the gobiesociform *Lepadogaster lepadogaster*. It differs notably from these, apart from its single flagellum, in having numerous small mitochondria. This and the elongate nucleus are notable differences from other described lophiiforms. The spermatozoon is needle-shaped (see Ceratioidei, now Ceratioidea, as redrawn by Mattei 1991, in Fig. 13.22). The head is about 11 μm long with an helical anteriorly tapering nucleus. A small acrosome and acrosomal filament present in the spermatid are lost in spermiogenesis. The midpiece, containing uniformly scattered small spherical unmodified mitochondria, forms a cylinder, 8 μm long and 2 μm wide, around the flagellum from which, throughout its length, it is separated by a cytoplasmic canal. The two mutually perpendicular centrioles lie in a deep basal nuclear fossa, the posterior giving rise to the 9+2 flagellum (Jespersen 1984).

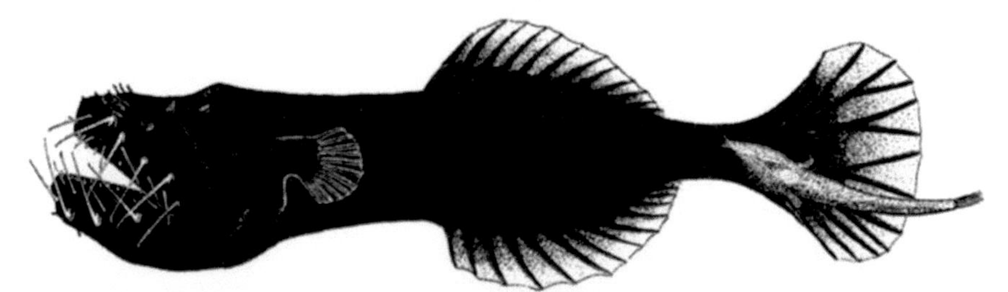

Fig. 13.23 *Neoceratias spinifer* (Neoceratiidae). Adult female with attached dwarf male. From Bertelsen, E. 1951. The ceratioid fishes. Ontogeny, taxonomy, distribution and biology. Dana Report, 39, 276 pp.

Remarks: The co-occurrence of complex spermatozoa with absence of copulatory organs in *Porichthys, Lepadogaster* and *Neoceratias* (Briggs 1955; Jespersen 1984) is rare in teleosts. Nothing is known about the mode of fertilization in *Lepadogaster* and *Neoceratias* though *Porichthys* is supposedly free spawning. However, the spawning of the deep-sea angler fish *Linophryne arborifera*, a species closely related to *N. spinifer*, has been described (Bertelsen 1980). The eggs are suspended from the genital pore of the female in long sheets of a mucoid substance in which they are embedded. These sheets are presumed to be brought into close contact with the attached male. Fertilization of *L. lepadogaster* and *N. spinifer* is probably similar to that in *L. arborifera*. The spermatozoa of these fishes are probably adapted for penetrating the mucoid covering of the eggs.

13.13 CHAPTER SUMMARY

In the superorder Stenopterygii, stomiiform sperm are round-headed anacrosomal aquasperm but that of *Argyropelecus* is highly unusual for teleosts in having a short flagellum-like extension of the proximal centriole. A much more highly developed extension occurs elsewhere, in the Elopomorpha, with no suggestion of close relationship. In the Cyclosquamata, aulopiforms have aquasperm with a lateral ring-or U-shaped mitochondrion and a deep basal nuclear fossa, this mitochondrial structure being seen in the phylogenetically close but distinct salmoniforms. In the Scopelomorpha, the myctophids *Lampanyctus* sp., *Symbolophorus* and *Notoscopelus* have a highly distinctive biflagellate sperm type, with a 9+0 axonemal structure known elsewhere in fishes only in the Elopomorpha. However, polyphyly of the Myctophidae, or at least diversity in fertilization biology, is suggested by occurrence of uniflagellate sperm with a 9+2 axoneme in *Lampanyctus jordani* and, though requiring confirmation, of aflagellate sperm in the sperm of *Lampanyctodes* and *Diaphus*. No spermatozoal support for the questionable grouping Paracanthopterygii has been found. Within this group the Ophidiiformes show modest to great elongation of the nucleus despite external fertilization. Gadiform spermatozoa are unified in elongation of the midpiece (moderate, however, in *Lota*), with a deep cytoplasmic canal; intratubular differentiations; flagellar fins seen in many perciform sperm are absent. The known sperm of Batrachoidiformes are biflagellate, thus resembling those of gobiesociforms; the nucleus may be rounded or somewhat elongate. The Lophiiformes, considered from molecular evidence to be related to the percomorph Tetraodontiformes, have sperm varying from the simple anacrosomal aquasperm of *Antennarius*, through sperm with two tiers of mitochondria resembling those of the tetraodontids *Dibranchus atlanticus* and *Lagocephalus laevigatus*, to the highly modified elongate sperm of *Neoceratias spinifer*, with its helically tipped nucleus and elongate midpiece with multiple small mitochondria, very different from known tetraodontiform sperm.

13.14 LITERATURE CITED

Abascal, F. J., Medina, A., Megina, C. and Calzada, A. 2002. Ultrastructure of *Thunnus thynnus* and *Euthynnus alletteratus* spermatozoa. Journal of Fish Biology 60: 147-153.

Afzelius, B. A. 1978. Fine structure of the garfish spermatozoon. Journal of Ultrastructure Research 64: 309-314.

Baccetti, B., Burrini, A. G., Callaini, G., Gibertini, G., Mazzini, M. and Zerunian, S. 1984. Fish Germinal Cells 1. Comparative Spermatology of 7 Cyprinid Species. Gamete Research 10: 373-396.

Bertelsen, E. 1980. Notes on Linphrynidae. V: A revision of the deepsea anglefishes of the *Linophryne arborifera* - group (Pisces, Ceratioidei). Steenstrupia 6: 29-70.

Billard, R. 1986. Spermatogenesis and spermatology of some teleost fish species. Reproduction, Nutrition and Développement 26: 877-1024.

Billard, R. 1990. Spermatogenesis in teleost fish. Pp. 183-212. In G. E. Lamming (ed.), *Marshall's Physiology of Reproduction, Vol. 2, Reproduction in Males*. Churchill Livingston, Edinburgh.

Boisson, C., Mattei, C. and Mattei, X. 1967. Troisième note sur la spermiogenèse de *Protopterus annectens* (Dipneuste) du Sénégal. Institut Fondamental d'Afrique Noire Bulletin Série A (Sciences Naturelles) 29: 1097-1121.

Breder, C. M. and Rosen, D. E. 1966. *Modes of reproduction in fishes*. The American Museum of Natural History, The Natural History Press, New York. 941 pp.

Briggs, J. C. 1955. A monograph of the clingfishes (Order Xenopterygii). Stanford Ichthyological Bulletin 6: 1-224.

Casas, M. T., Munoz-Guerra, S. and Subirana, J. A. 1981. Preliminary Report on the Ultrastructure of Chromatin in the Histone Containing Spermatozoa of a Teleost Fish. Biology of the Cell (Paris) 40: 87-92.

Cerisola, H. 1990. Structural and ultrastructural aspects of the testis of the Clingfish *Sicyases sanguineus*. Revista de Biologia Marina 25: 81-98.

Grier, H. J. 1981. Cellular organization of the testis and spermatogenesis in fishes. American Zoologist 21: 345-357.

Grier, H. J., Linton, J. R., Leatherland, J. F. and De Vlaming, V. L. 1980. Structural evidence for two different testicular types in teleost fishes. The American Journal of Anatomy 159: 331-345.

Gwo, J.-C., Gwo, H. H., Ko, Y. S., Lin, B. H. and Shih, H. 1994. Spermatozoan ultrastructure of two species of grouper *Epinephelus malabaricus* and *Plectropomus leopardus* (Teleostei, Perciformes, Serranidae) from Taiwan. Journal of Submicroscopic Cytology and Pathology 26: 131-136.

Hara, M. 2007. Ultrastructure of spermatozoa of two species of Myctophidae; *Symbolophorus californiensis* and *Notoscopelus* sp. Japanese Journal of Ichthyology 54: 41-46.

Hara, M. and Okiyama, M. 1998. An ultrastructural review on the spermatozoa of Japanese fishes. Bulletin of The Ocean Research Institute, University of Tokyo 33: 1-138.

Hernández, M. R., Sàbat, M., Muñoz, M. and Casadevall, M. 2005. Semicystic spermatogenesis and reproductive strategy in *Ophidion barbatum* (Pisces, Ophidiidae) l. Acta Zoologica (Stockholm) 86: 295–300.

Hinegardner, R. 1968. Evolution of cellular DNA content in teleost fishes. American Naturalist 102: 517-523.

Hoffman, R. A. 1963. Gonads, spermatic ducts, and spermatogenesis in the reproductive system of male toadfish, *Opsanus tau*. Chesapeake Science 4: 21-29.

Jamieson, B. G. M. 1984. Spermatozoal ultrastructure in *Branchiostoma moretonensis* Kelly, a comparison with *B. lanceolatum* (Cephalochordata) and with other deuterostomes. Zoologica Scripta 13: 223-229.

Jamieson, B. G. M. 1991. *Fish Evolution and Systematics: Evidence from Spermatozoa*. Cambridge University Press, Cambridge. 319 pp.

Jespersen, Å. 1984. Spermatozoans from a parasitic dwarf male of *Neoceratias spinifer* Pappenheim, 1914. Videnskabelige Meddelelser Dansk Naturhistorisk Forening 145: 37-42.

Lahnsteiner, F. 2003. The spermatozoa and eggs of the cardinal fish. Journal of Fish Biology 62: 115-128.

Lahnsteiner, F. and Patzner, R. A. 1996. Fine structure of spermatozoa of three teleost fishes of the Mediterranean Sea: *Trachinus draco* (Trachinidae, Perciformes), *Uranuscopus scaber* (Uranuscopidae, Perciformes) and *Synodon saurus* (Synodontidae, Aulopiformes). Journal of Submicroscopic Cytology and Pathology 28: 297-303.

Lahnsteiner, F., Weismann, T. and Patzner, R. A. 1994. Fine structure of spermatozoa of the burbot, *Lota lota* (Gadidae, Pisces). Journal of Submicroscopic Cytology and Pathology 26: 445-447.

Lauder, G. V. and Liem, K. F. 1983. The evolution and interrelationships of the actinopterygian fishes. Bulletin of the Museum of Comparative Zoology 150: 95-197.

Lee, Y. H. and Kim, K. H. 1999. Ultrastructure of the south torrent catfish, *Liobagrus mediadiposalis* (Teleostei, Siluriformes, Amblycipitidae) spermatozoon. Korean Journal of Limnology, 32(3): 271-280. Korean Journal of Limnology 32: 271-280.

Maggese, M. C., Cukier, M. and Cussac, V. E. 1984. Morphological changes, fertilizing ability and motility of *Rhamdia sapo* (Pisces, Pimelodidae) sperm induced by media of different salinities. Rev Brasil Bio 44: 541-546.

Malatesta, M., Petrini, E., Froglia, C. and Gazzanelli, G. 1991. Prime osservazioni sulla organizzazione del testicolo di *Trisopterus minutus capelanus*. Bolletino dei Societa Italiano di Biologia Spermientale 67: 853-860.

Matos, E., Santos, M. N. S. and Azevedo, C. 2002. Biflagellate spermatozoon structure of the hermaphrodite fish *Satanoperca jurupari* (HECKEL, 1840) (Teleostei, Cichlidae) from the Amazon River. Brazilian Journal of Biology 62: 847-852.

Mattei, C., Mattei, X. and Marchand, B. 1979. Réinvestigation de la structure des flagelles spermatiques: les doublets 1:2:5 et 6. Journal of Ultrastructure Research 69: 371-377.

Mattei, X. 1969. *Contribution à l'étude de la spermiogenèse et des spermatozoïdes de poissons par les méthodes de la microscopie électronique*. D.Sc. thesis, Université de Montpellier.

Mattei, X. 1970. Spermatogenèse comparée des poissons. Pp. 567-569. In B. Baccetti (ed.) *Comparative Spermatology*. Academic Press, New York.

Mattei, X. 1991. Spermatozoon ultrastructure and its systematic implications in fishes. Canadian Journal of Zoology 69: 3038-3055.

Mattei, X. and Mattei, C. 1976. Spermatozoïdes à deux flagelles de type 9+0 chez *Lampanyctus* sp. (Poisson Myctophidae). Journal de Microscopie et de Biologie Cellulaire 25: 187-188.

Mattei, X., Siau, Y., Thiaw, O. T. and Thiam, D. 1993. Peculiarities in the organization of testis of *Ophidion* sp. (Pisces Teleostei). Evidence for two types of spermatogenesis in teleost fish. Journal of Fish Biology 43: 931-937.

Medina, A., Megina, C., Abascal, F. J. and Calzada, A. 2003. The sperm ultrastructure of *Merluccius merluccius* (Teleostei, Gadiformes): Phylogenetic considerations. Acta Zoologica (Copenhagen) 84: 131-137.

Miya, M., Takeshima, H., Endo, H., Ishiguro, N. B., Inoue, J. G., Mukai, T., Satoh, T. P., Yamaguchi, M., Kawaguchi, A., Mabuchi, K., Shirai, S. M. and Nishida, M. 2003. Major patterns of higher teleostean phylogenies:a new perspective based on 100 complete mitochondrial DNA sequences. Molecular Phylogenetics and Evolution 26: 121-138.

Nelson, J. S. 1984. *Fishes of the World*, 2nd edition. John Wiley and Sons, New York. 523 pp.

Nelson, J. S. 2006. *Fishes of the World*, 4th edition. John Wiley and Sons, Inc., Hoboken, NJ. 601 pp.

Poirier, G. R. and Nicholson, N. 1982. Fine structure of the testicular spermatozoa from the Channel Catfish, *Ictalurus punctatus*. Journal of Ultrastructure Research 80: 104-110.

Purkerson, M. L., Jarvis, J. U. M., Luse, S. A. and Dempsey, E. W. 1974. X-ray analysis coupled with scanning and transmission electron microscopic observations of spermatozoa of the African lungfish, *Protopterus aethiopicus*. Journal of Zoology (London) 172: 1-12.

Retzius, G. 1905. Die Spermien der Leptokardier, Teleostier und Ganoiden. Biologische Untersuchungen von G Retzius Nf 12: 103-115.

Roberts, T. R. 1973. Interrelationships of ostariophysans. Pp. 373-395. In P. H. Greenwood, R. S. Miles and C. Patterson (eds), *Interrelationships of Fishes*. Journal of the Linnean Society of London Zoology 53 (suppl 1) Academic Press, New York.

Shahin, A. A. B. 2006. Semicystic spermatogenesis and biflagellate spermatozoon ultrastructure in the Nile electric catfish *Malapterurus electricus* (Teleostei : Siluriformes : Malapteruridae). Acta Zoologica (Copenhagen) 87: 215-227.

Spadella, M. A., Oliveira, C. and Quagio-Grassiotto, I. 2006. Occurrence of biflagellate spermatozoa in Cetopsidae, Aspredinidae, and Nematogenyidae (Teleostei : Ostariophysi : Siluriformes). Zoomorphology 125: 135-145.

Stanley, H. P. 1965. Electron microscopic observations on the biflagellate spermatids of the teleost fish *Porichthys notatus*. Anatomical Record 151: 477.

Thiaw, O. T., Thiam, D. and Mattei, X. 1990. Extension of proximal centriole in a teleost fish spermatozoon. Journal of Submicroscopic Cytology and Pathology 22.

Yao, Z., Emerson, C. J. and Crim, L. W. 1995. Ultrastructure of the spermatozoa and eggs of the ocean pout (*Macrozoarces americanus* L.), an internally fertilizing marine fish. Molecular Reproduction and Development 42: 58-64.

Young, J. W., Blaber, S. J. M. and Rose, R. 1987. Reproductive biology of three species of midwater fishes associated with the continental slope of eastern Tasmania, Australia. Marine Biology 95: 323-332.

CHAPTER 14

Ultrastructure of Spermatozoa: Acanthopterygii: Mugilomorpha and Atherinomorpha

Barrie G. M. Jamieson

14.1 ACANTHOPTERYGII

The Acanthopterygii comprise 13 orders, in three series (Mugilomorpha, Atherinomorpha and Percomorpha) and as many as 267 families, 2,422 genera and 14,797 species (Nelson 2006). The basic sperm type is the anacrosomal aquasperm but several groups have secondarily developed similarly anacrosomal introsperm.

14.2 SERIES MUGILOMORPHA

14.3 ORDER MUGILIFORMES

Mugiliforms are the mullets (Fig. 14.6A). They occur in oceans, bays, estuaries and freshwater in all except polar regions. Relationship with atherinids has previously been refuted (see de Sylva 1984) but the Mugiliformes are seen to be the sister-group of the Atherinomorpha ((Beloniformes + Atherinformes) + Cyprinidontiformes) in the present work (Ortí and Li, **Chapter 1**, Fig. 1.5, and Fig. 14.1 below).

14.3.1 Family Mugilidae

Mugiliforms which have been examined for ultrastructure of the spermatozoon or at least spermatogenesis are the mugilids *Liza aurata (=Mugil auratus)* (Brusle and Brusle 1978; Brusle 1981; Eiras-Stofella and Gremski

School of Integrative Biology, University of Queensland, Brisbane, Queensland 4072, Australia

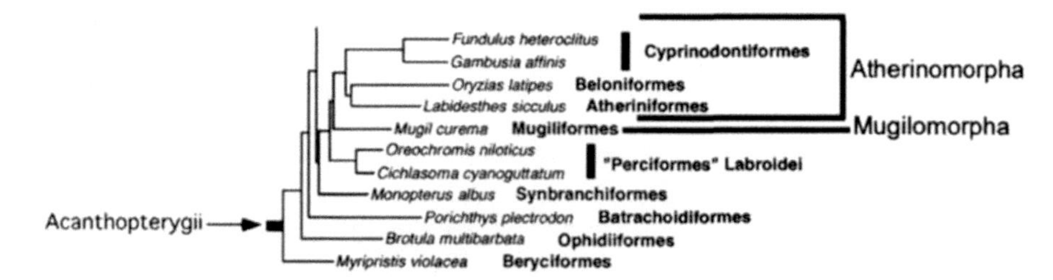

Fig. 14.1 Phylogeny of the Mugiliformes and Atherinomorpha (Atheriniformes, Beloniformes and Cyprinodontiformes) from analysis of 10 nuclear genes. Adapted from Ortí and Li (**Chapter 1**, Fig. 1.5).

1991); *Liza ramada* (El-Gammal *et al.* 2003); *M. liza* and *M. platanus* (Eiras-Stofella and Gremski 1991); *M. curema* (Eiras-Stofella *et al.* 1993) and, as *M. metzelaari* (Mattei 1991, see also this account); and *M. cephalus* (Mousa *et al.* 1998) (adult, Fig. 14.5A). The Polynemidae, formerly associated with the Mugilidae, have been placed in the Perciformes by Nelson (2006).

Liza: The sperm of *Liza aurata* (Brusle 1981) (Fig. 14.2) and *L. dumerilii* (van der Horst 1976; van der Horst and Cross, 1978), both Mugilidae; are similar. That

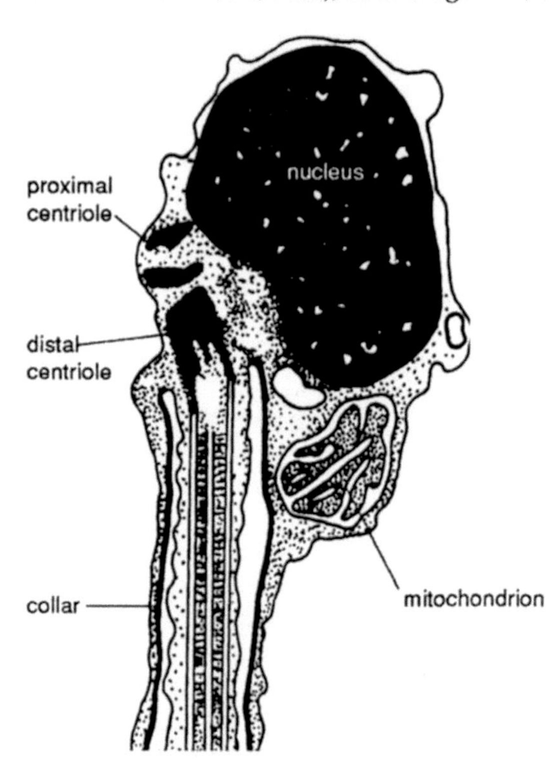

Fig. 14.2 *Liza aurata* (Mugilidae). Longitudinal section of spermatozoon. After a micrograph by Brusle, S. 1981. *Cell and Tissue Research* 217: 415-424, Fig. 13.

of *L. ramada*, described by El-Gammal *et al.* (2003) appears to show some small differences.

The nucleus, bilobed or kidney-shaped in longitudinal section, is tilted relative to the axoneme. The chromatin is very coarsely granular. In *L. dumerilii* a dense body (round body) has been observed lying in contact with the nucleus in some sperm; the nuclear and overlying plasma membrane covering the "ventral" or "dorsal" tip of the nucleus forms bulbous evaginations containing chromatin. The nucleus is said to be ovoidal with densely granular chromatin in *L. ramada*.

Four subspherical mitochondria, in *Liza aurata* and *L. dumerilii*, are arranged in a ring but eccentrically around the base of the axoneme from which they are separated by a long cytoplasmic canal. In *L. dumerilii* mitochondria cristae are sparse and are plate-like and tubular, the former sometimes arranged in a circular pattern as in brown adipose tissue. The midpiece in *L. ramada* contains a few spherical mitochondria, separated from the axoneme by the cytoplasmic canal (El-Gammal *et al.* 2003).

Two fully developed centrioles are present. Despite the eccentric emergence of the flagellum, the proximal centriole, and in one plane the distal centriole, lies in the nuclear fossa in *Liza dumerilii*; during spermiogenesis nuclear rotation occurs; a cross striated structure above the proximal centriole is reminiscent of the axial body of *Poecilia reticulata* (Van der Horst and Cross 1978). Both centrioles are said to lie outside the shallow nuclear fossa, at right angles to each other, in the spermatozoon of *L. ramada*.

In *Liza dumerilii* two classes of flagella were observed: 80 percent varied between 39 and 43 µm in length (mean 41 µm) while the remainder, considered to indicate abnormal sperm, were much shorter and varied more in length (21- 30 µm, mean 26 µm) (van der Horst and Cross, 1978). No fins are described or illustrated for the 9+2 axoneme. In *L. ramada* it was noted that (as is normal in flagella) the axoneme had a 9+0 configuration at the transition from the basal body.

Mugil curema, M. liza and M. platanus: The spermatozoa of *M. curema* (Eiras-Stofella *et al.* 1993) (This account, Fig. 14.3), *M. liza* and *M. plantanus* (Eiras-Stofella and Gremski 1991) (Fig. 14.4) have compact, strongly electron-dense granular chromatin. The head is rounded but slightly flattened; the nucleus is covered by an undulating membrane. The head and midpiece have a mean diameter of ca 1.56 µm in *M. curema* and 1.50 µm in *M. liza* and *M. platanus*, the nucleus in *M. curema* being ca 1.08 µm wide. The midpiece, not divided from the head, is characterized by the presence of two centrioles, a cytoplasmic canal, and few vesicles. For *M. curema*, an unusually large number of mitochondria for mugilids (9-10), approximately 0.59 µm in diameter, is reported (Eiras-Stofella *et al.* 1993), contrasting with *M. liza* and *M. platanus* said to have at most four mitochondria (Eiras-Stofella and Gremski 1991). The diameter of the mitochondria is 0.55-0.66 µm in *M. curema*, but is smaller, 0.35 µm, in *M. liza* and *M. platanus*.

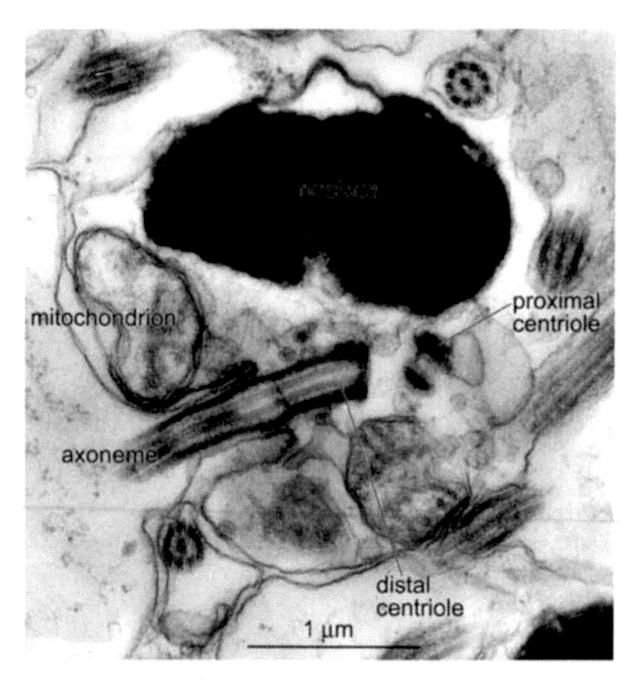

Fig. 14.3 *Mugil curema*. Longitudinal section of a spermatozoon. Note the Type II characteristics: centrioles outside the nuclear fossa; axoneme oblique relative to the nucleus. Original, from an unpublished TEM negative courtesy of Professor Xavier Mattei.

A thin transverse lamella divides the distal centriole from the axoneme. A basal foot (not shown in Fig. 14.3) extends obliquely from this centriole towards the nucleus in *Mugil curema*. The proximal centriole is at about 60° to the distal centriole. The centrioles in some planes appear to lie loosely within the wide, shallow nuclear hilus [fossa] but in other profiles are seen to be external to it. The cytoplasmic canal is strengthened by a small amount of

Fig. 14.4 A-D *Mugil* spermatozoa (Mugilidae). **A**. SEM of *M. platanus* semen. **B**. TEM Longitudinal section (LS) of *M. platanus* sperm head. **C**. LS *M. liza* spermatozoon. **D**. Transverse section of the flagellum. Arrowheads indicate densities (ITDs) in A subtubules 1, 2, 5 and 6. Note the asymmetry of the cell, the direction of chromatin maturation (open arrow). **E**. LS spermatozoon of *M. liza*. **F**. LS *M. liza* sperm head. **G**. *M. platanus* spermatozoon. **H**. Section of the sperm head in the flagellum transitional area of *M. liza*. **I**. Mitochondria of *M. platanus*. bf, basal foot; Chc, compact chromatin; CH_g, chromatin granules; dm, dense material; f, flagellum; h, nuclear hilus; H, head; L, lamella; * microtubules of the 9 + 0 type; ** microtubules of the 9 + 2 type; M, mitochondria;.N, nucleus; pC_a, cytoplasmic canal; v, vesicles; y, Y-links. After Eiras-Stofella, D. R. and Gremski, W. 1991. Microscopia Electronica y Biologia Celular 15: 173-177, Figs. 1-7a.

Fig. 14.4A-D Contd. ...

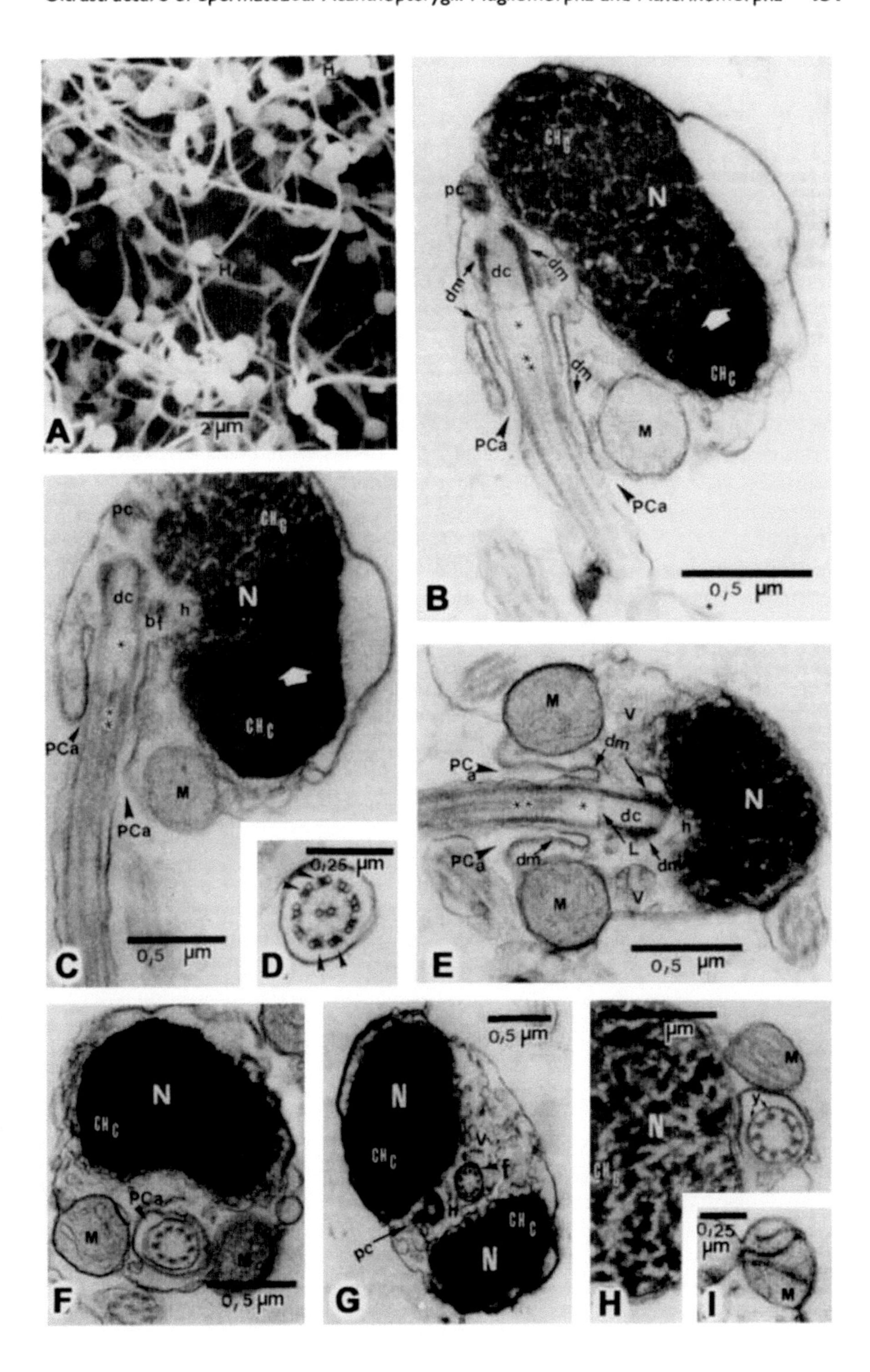
A
H
2 µm
B
pc
N
dc
dm
PCa
M
0,5 µm
C
pc
dc
bf
h
N
PCa
M
0,5 µm
D
0,25 µm
E
M
V
dm
PCa
dc
h
L
N
0,5 µm
F
N
PCa
M
0,5 µm
G
N
V
f
pc
H
N
0,5 µm
H
µm
M
N
I
0,25 µm
M

dense material. The flagellum conforms to the 9+0 flagellar pattern near the transition region in its midpiece and 9+2 from there to its distal region. Its apical region in *M. curema* was shown to have Y-links. In all three species, subtubules 1, 2, 5 and 6 are electron-dense in more than 50% of the spermatozoa (Eiras-Stofella and Gremski 1991; Eiras-Stofella *et al.* 1993). The sperm of this species, under the junior synonym, *M. metzelaari*, was attributed to Type II by Mattei (1991) (This account, Fig. 14.3).

Remarks: The spermatozoon of *Liza dumerilii* appears to be intermediate between the Type I and II categories of Mattei although it is classified as type I by van der Horst and Cross (1978). *L. ramada, Mugil curema, M. liza* and *M. platanus* closely resemble Type II peciform spermatozoa in the exclusion of the centrioles from the nuclear fossa, the approximately tangential flagellum and the presence of intratubular differentiations (ITDs) in A subtubules 1, 2, 5 and 6. It is difficult to envisage evolutionary pathways which would connect these mugilids with perciforms with Type II spermatozoa. Labroids are the percomorphs which are most closely related to mugilids (see Ortí and Li, **Chapter 1**, Fig. 1.5) but they possess Type I and not Type II spermatozoa. The close similarity of the sperm of mugilids and Type II sperm of perciforms may be due to convergence (homoplasy), though the detailed similarity including presence of ITDs is remarkable. Possibly there is a genetic propensity for modified, Type II spermiogenesis in Mugilomorpha and Percomorpha (see also Ostariophysi) which may be an example of parallelism by virtue of relationship which the author (Jamieson 1984, 1991) has previously designated paramorphy as distinct from homoplasy.

The presence of a round body in *L. dumerilii* is an interesting resemblance to the percichthyid *Maccullochella* (see **Section 14.4**).

14.4 SERIES ATHERINOMORPHA

Atherinomorphs differ from acanthopterygians, *inter alia*, in lacking a ball and socket joint between the palatine and maxilla. Their monophyly has convincingly been demonstrated (Rosen and Parenti 1981; Lauder and Liem 1983; Parenti 2005). A sister-group relationship with the Mugilidae is widely supported (see Nelson 2006) and is indicated in the phylogenetic analysis of ten nuclear genes by Ortí and Li (**Chapter 1**) (Figs. 1.5, and Fig. 14.1). However, contrary to the classification employed by Nelson (2006) Ortí and Li find the Atheriniformes to be the sister-group of the Beloniformes in a clade which is sister to the Cyprinodontiformes; the Mugiliformes, represented by *Mugil curema*, forms the plesiomorph sister-group of this combined antherinomorph clade. Nelson (2006) accepts the view that the Atheriniformes are sister to the remaining Atherinomorpha and therefore placement of the Atheriniformes in a superorder Atherinea, whereas the Beloniformes and Cyprinidontiformes are placed in the superorder Cyprinodontea. In view of the conflict between these two systems, the superordinal ranks will not be used in the present account.

Fig. 14.5 Mugiliformes and Atheriniformes. **A**. *Mugil cephalus* (Mugilidae). **B**. *Craterocephalus stercusmuscarum* (Atherinidae). **C**. *Melanotaenia duboulayi* (Melanotaeniidae), female **D**. *Poecilia reticulata* (Poeciliidae), male left, female right. **E**. *Xiphophus helleri* (Poeciliidae), female above, male below. In these poeciliids, the anal fin is modified as an intromittent organ. Photos: **A** and **D**, Bruce Cowell; **B**,**C**,**E**, Gary Cranitch. All courtesy of the Queensland Museum.

Most atherinomorph species are surface-feeding and about 75% live in fresh or brackish water (Nelson 2006). Atherinomorphs include the killifishes (Cyprinodontidae), live-bearing top minnows (Poeciliidae), silversides (Atherinidae), four-eyed fishes (Anablepidae), rice fishes (Adrianichthyidae), half beaks (Hemiramphidae), needle fishes (Belonidae) and marine flying fishes (Exocoetidae). The Beloniformes contain the suborders Adrianichthyoidei and Exocoetoidei, both previously subsumed in the Cyprinodontiformes by Nelson (1984).

In all three atherinomorph orders spermatogonia are restricted to the distal end of the testicular tubules. This is a notable contrast with the Salmoniformes, Perciformes, and Cypriniformes in which spermatogonia are distributed along the entire length of the tubules (Grier *et al.* 1980; Grier 1981; Grier and Collette 1987; Parenti 2005). This dichotomy underlines the phylogenetic unity and discreteness of the Atherinomorpha indicated, from other characters, above. The telogonic atherinomorph condition is presumably apomorphic relative to the hologonic condition general in teleosts, as the conditions are termed here after Jamieson (1991). Billard (1986), in a review of spermatogenesis and spermatology of some teleost species, referred to the two types of testis as lobular [hologonic] and tubular [telogonic].

Reproductive modes in the Atherinomorpha range from egg scattering to, in some Cyprinodontiformes and Beloniformes, varying degrees of viviparity (Grier 1976; Uribe and Grier 2006). Internal fertilization is considered to have evolved from free-spawning independently among the cyprinodontiforms in several families and is attended by production of live young or rarely (*Zenarchopterus*) laying of fertilized eggs. It has been directly observed or has been inferred from the possession of spermatozeugmata, of spermatophores, of ovarian sperm, or of internal embryos, in the aplocheilids *Rivulus marmoratus* and *Cynolebias*; in poeciliids; in goodeids; in *Jenynsia lineata*; in *Anableps anableps* and *A. dowi*; in *Horaichthys setnai* (all Cyprinodontiformes) and in some exocoetoid hemiramphids (Beloniformes) *viz.* the viviparous *Dermogenys pusillus*, *Hemirhamphodon pogonognathus*, and *Nomorhamphus hageni* and at least 11 species of the egg-laying genus *Zenarchopterus*; and in 19 species of the atherinoid family Phallostethidae (Grier and Parenti 1994; Downing and Burns 1995; Grier 1981; Parenti 1981; Grier and Collette 1987; Parenti and Grier 2004; Parenti 2005; Uribe and Grier 2006; Evans and Meisner, **Volume 8B, Chapter 4**). Most exocoetoids are, however, externally fertilizing.

14.5 ORDER ATHERINIFORMES

The Order Atheriniformes was first named by Rosen (1964) for exocoetoids and cyprinodontoids and these included all taxa here placed in the Atherinomorpha. Members of the Atheriniformes usually have two dorsal fins, the first, if present, with flexible dorsal spines; the anal fin is usually preceded by a spine. Additional characteristics are listed by Nelson (2006). Despite the lack of firm apomorphies in adult somatic morphology for the Atheriniformes, the telogonic testicular condition is a unique synapomorphy (autapomorphy) of the Atheriniformes - Cyprinodontiformes - Beloniformes assemblage, setting them apart from other teleosts.

Families Atherinopsidae, Atherinidae and Melanotaeniidae: Thirteen species in three families of the Atheriniformes, Atherinopsidae, Atherinidae and Melanotaeniidae) have been examined for sperm ultrastructure: *Labidesthes sicculus vanhyningi* (Grier *et al.* 1990) (Fig. 14.6), *Chirostoma jordani* (Reygadas and Escorcia 1998) (Atherinopsidae); *Craterocephalus; stercusmuscarum* (adult, Fig. 14.5B; sperm, Fig. 14.7D-F), *Craterocephalus helenae*, *Craterocephalus marjoriae* (Fig. 14.7A-C), *Quirichthys stramineus* (sometimes placed in *Craterocephalus*) (Atherinidae), *Cairnsichthys rhombosmoides*, *Iriatherina werneri* (Fig. 14.7G, H), *Melanotaenia duboulayi* (adult, Fig. 14.5C; sperm, Fig. 14.7I, J), *M. maccullochi* (adult, Fig. 14.8N; sperm, Fig. 14.8A-C), *Pseudomugil mellis* (Fig. 14.8D-F), *P. signifer* (Fig. 14.8G-I) and *P. tenellus* (Fig. 14.8J-M) (Melanotaeniidae) (Marshall 1989). Of these only *Labidesthes sicculus vanhyningi*, the Florida subspecies, is internally fertilizing, though it is possible that more northerly subspecies do not have internal fertilization (Grier *et al.* 1990). Otherwise in the Atheriniformes only Phallostethidae are known to be inseminating (Grier and Parenti 1994) but they have not been examined for sperm ultrastructure.

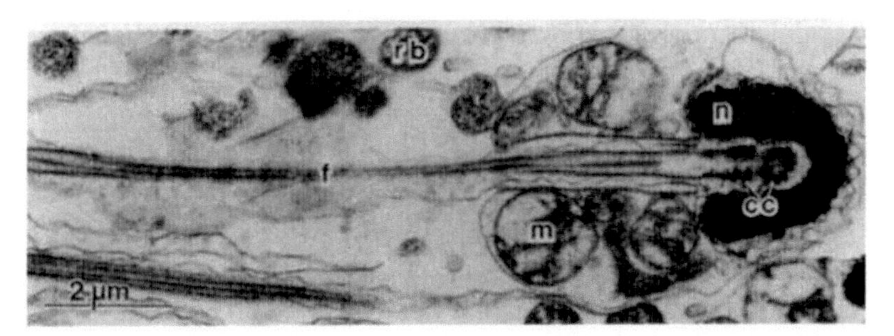

Fig. 14.6 Atheriniform spermatozoa. *Labidesthes sicculus*. Longitudinal section of a spermatozoon. Note the little modified, Type I morphology despite internal fertilization. cc, centriolar complex in a deep nuclear fossa; f, flagellum; m, mitochondrion; n, nucleus. From Grier, H. J., Moody, H. J. and Cowell, B. C. 1990. Copeia 1: 221-226, Fig. 2D. With permission of the American Society of Ichthyologists and Herpetologists.

Length: The spermatozoon of *Chirostoma jordani* is 15.4 ± 0.3 µm long (Reygadas and Escorcia 1998).

Acrosomoid: Present only in the two *Melanotaenia* species, anterior to the nucleus in longitudinal sections of many sperm, there is a single large, empty vesicle (Fig. 14.8A), here termed the acrosomoid after Jamieson (1991) (see Remarks). This is visible as a distinct protrusion in live, actively motile sperm, under the light microscope.

Nucleus: The nucleus is approximately isodiametric, with a diameter of 1.1 µm in *Craterocephalus marjoriae* and *Quirichthys stramineus*, 1.2 µm in

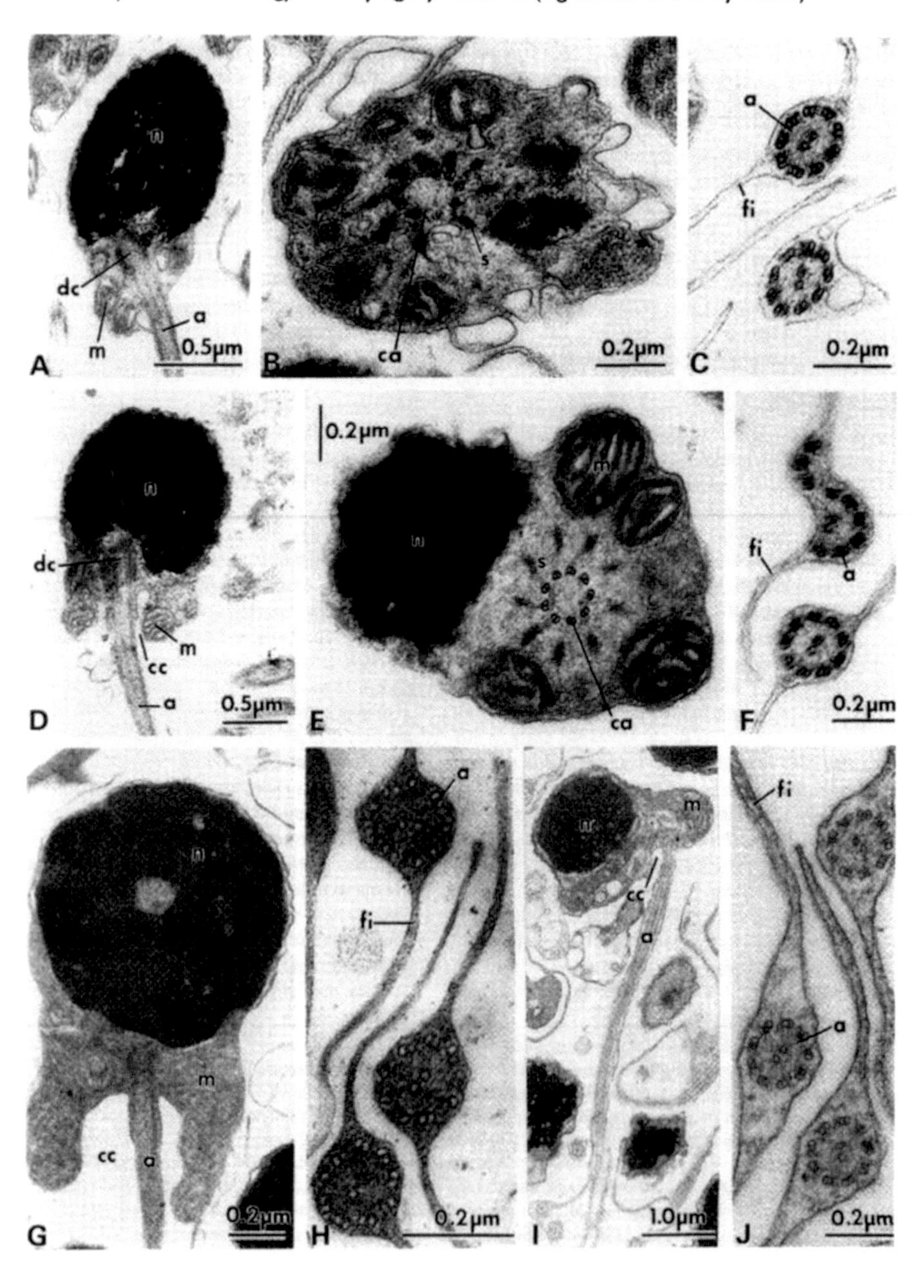

Fig. 14.7 Atheriniform spermatozoa continued. **A-C**. *Craterocephalus marjoriae*. **A**. Longitudinal section (LS) of spermatozoon showing transverse proximal centriole with satellite rays. **B**. Transverse section *(TS)* of the proximal, 9+0 region of the axoneme with satellite rays and ring of mitochondria. **C**. TS of 9+2 region of flagellum with lateral fins. **D-F**. *Craterocephalus stercusmuscarum*. **D**. LS showing distal centriole partly enclosed in nuclear fossa. **E**. TS of proximal, 9+0, region of

Fig. 14.7 Contd. ...

Melanotaenia maccullochi and 1.5 µm in *Iriatherina werneri* and *Melanotaenia duboulayi*. It is longer (1.3 µm) than wide (1.1 µm) in *Craterocephalus marjoriae* but is wider than long in *Craterocephalus helenae* (0.8 µm long and 1.0 µm wide) and *Cairnsichthys* rhombosmoides (0.7 µm long and 0.9 µm wide). In *Pseudomugil* it is crescentic, being 1.8 µm wide in *P. mellis* and 1.4 µm wide in *P. signifer*, while in *P. tenellus* it is hemispheroidal with a diameter of 1.1 µm. It is isodiametric at slightly over 2 µm in *Labidesthes sicculus*.

The chromatin is usually flocculent or at least has numerous lacunae, and is sufficiently coarse grained to conform only irregularly with the double nuclear envelope; in *Iriatherina werneri*, alone, it is finely granular and conforms to the envelope. Only in this species, in *Craterocephalus helenae* and in the two *Melanotaenia* species is a nuclear fossa absent. The fossa includes the distal centriole but only in *Craterocephalus marjoriae*, the three *Pseudomugil* species and *Labidesthes sicculus* is type I morphology attained by inclusion of the proximal centriole.

Midpiece: The mitochondria form a ring around the axoneme. There is a single layer of several mitochondria in *Melanotaenia duboulayi* and *M. maccullochi* (Fig. 14.8A) but in the 9 other examined atheriniforms there is more than one tier. In cross section there are 4 in *Quirichthys stramineus;* 4 or 5 in *Craterocephalus stercusmuscarum* and *Iriatherina werneri;* 5 or 6 in *Craterocephalus helenae* and *C. marjoriae;* 5-7 in *Pseudomugil tenellus;* 6 or 7 in *Pseudomugil mellis;* 7 in *Cairnsichthys rhombosmoides;* and 6-10 in *P. signifer*. In longitudinal section there are 1 or 2 in *Craterocephalus helenae, C. marjoriae* and *Cairnsichthys rhombosmoides;* 2 or 3 in *Iriatherina werneri;* 3 in *Pseudomugil signifer* and *P. tenellus;* and 4 or 5 in *Pseudomugil mellis*.

There is no collar or cytoplasmic canal in *Melanotaenia duboulayi* and *M. maccullochi* but this is present in the other species. It is only 0.2 µm long in *Craterocephalus stercusmuscarum* and *Cairnsichthys rhombosmoides;* 0.3 µm in *Craterocephalus marjoriae;* 0.7 µm in *Iriatherina werneri,* and *Pseudomugil tenellus;* 0.9 µm in *Pseudomugil signifer;* 1.0 µm in *Craterocephalus helenae;* and 1.5 µm long in *Pseudomugil mellis*.

Centrioles: The proximal and distal centrioles are mutually perpendicular but not in the same longitudinal axis in *Craterocephalus marjoriae,* and in the same

---

Fig. 14.7 Contd. ...

axoneme with satellite rays. **F**. TS of 9+2 region of flagellum with lateral fins. **G, H**. *Iriatherina werneri.* **G**. LS spermatozoon. **H**. TS of 9+2 region of flagellum with lateral fins. **I, J**. *Melanotaenia duboulayi.* **I**. LS sagittal plane of spermatozoon. **J**. TS flagella with lateral fins. a, axoneme; ca, transition from centriole to axoneme; cc, cytoplasmic canal; de, distal centriole; fi, fin; m, mitochondrion; n, nucleus; s, satellite ray. From Jamieson, B. G. M. 1991. *Fish Evolution and Systematics: Evidence from Spermatozoa*. Cambridge University Press, Cambridge, Fig. 15.2. Courtesy of C. J. Marshall.

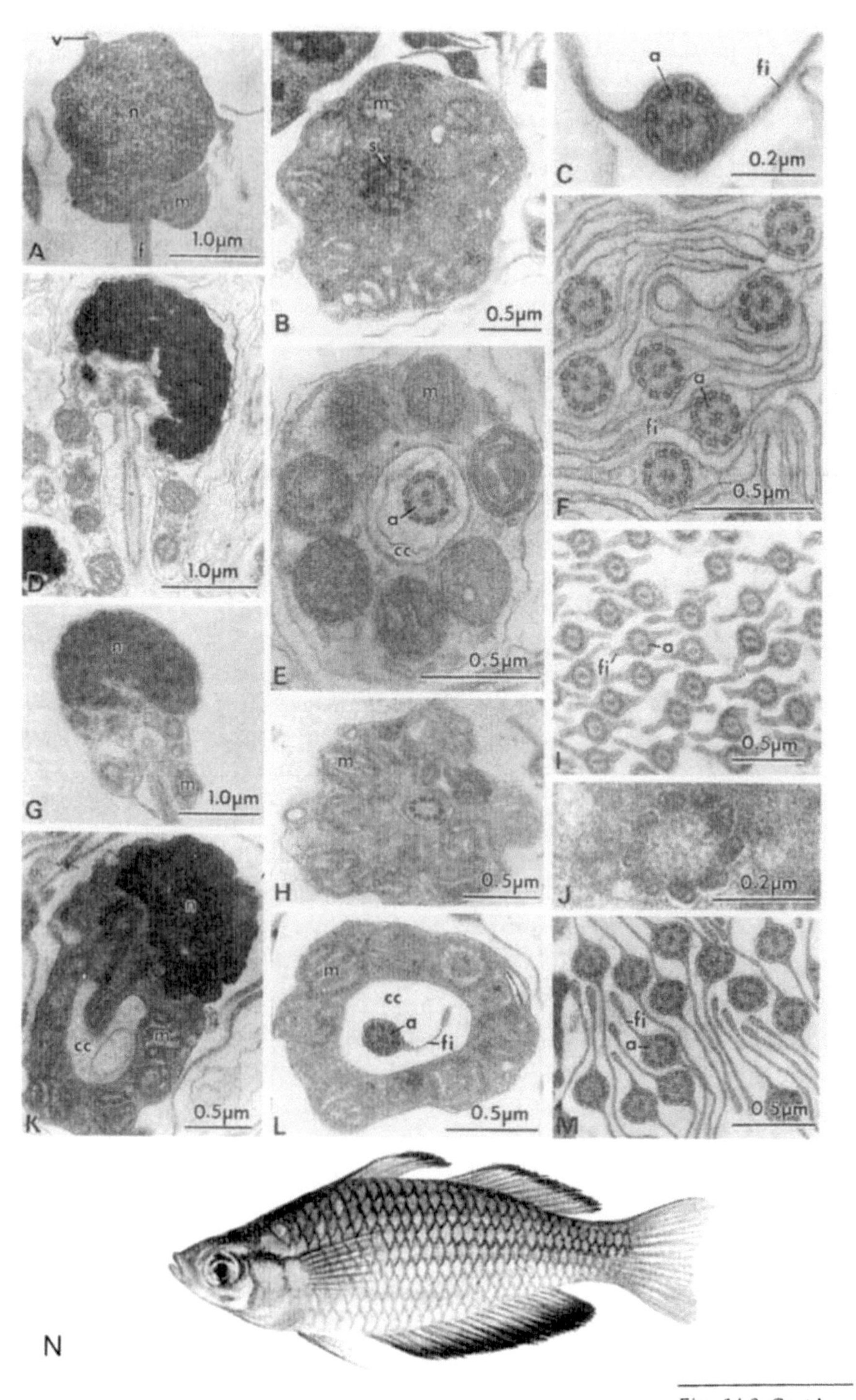

Fig. 14.8 Contd. ...

axis in *Pseudomugil mellis, P. signifer* and *P. tenellus*. Their long axes are parallel to each other in *Craterocephalus stercusmuscarum, C. helenae, Quirichthys stramineus, Iriatherina werneri* and *Melanotaenia maccullochi* but only in *M. duboulayi* are they also serial, in the same longitudinal axis. The proximal is at about 30° to the distal centriole in *Cairnsichthys rhombosmoides*.

Satellite apparatus: Nine well developed satellite rays surround the distal centriole in *Craterocephalus helenae, C. marjoriae* (Fig. 14.7B) and *C. stercusmuscarum* (Fig. 14.7E); *Quirichthys stramineus, Cairnsichthys rhombosmoides*, and *Melanotaenia maccullochi* but they appear to be weakly if at all developed in all three *Pseudomugil species*, in *Iriatherina werneri* and in *Melanotaenia duboulayi*.

Flagellum: In all species the flagellum bears two fins and is of the 9+2 pattern, apparently always with inner and outer dynein arms present. In *Iriatherina werneri* the flagellum is in the longitudinal axis of the nucleus but in all other species it subtends a considerable angle to the head, varying from about 40° in *Craterocephalus stercusmuscarum* to 90° in the two *Melanotaenia* species.

Remarks: The structure seen in both *Melanotaenia* species and here termed an acrosomoid is in the position occupied by the acrosome in most animal sperm but, with the exception of the internally fertilizing salmoniform *Lepidogalaxias* (Leung 1988), an acrosome is absent from mature teleost sperm. An acrosome is present in the sperm of agnathans through chondrosteans. Evidence that the telostean aquasperm has undergone simplification is seen in the transient existence of an acrosome vestige in the spermatid of *Oncorhynchus mykiss* (=*Salmo gairdneri*) (Billard 1983) and *Neoceratias spinifer* (Jespersen 1984), suggesting the occurrence of acrosome-bearing (but not necessarily internally fertilizing) sperm in the immediate ancestry of the Neopterygii. It would appear that a plesiomorphic genetic propensity to produce an acrosome has

Fig. 14.8 Contd. ...

Fig. 14.8 A-M. Atheriniform spermatozoa continued. **A-C.** *Melanotaenia macullochi.* **A.** Longitudinal section (LS) of spermatozoon showing anterior vesicle (acrosomoid). **B.** Transverse section of distal centriole with satellite rays and ring of mitochondria. **C.** TS of 9+2 region of flagellum with lateral fins. **D-F.** *Pseudomugil mellis.* **D.** LS spermatozoon. **E.** TS of 9+2 region of axoneme through mitochondria and cytoplasmic canal. F. TS of 9+2 region of flagellum with lateral fins. **G-I.** *Pseudomugil signifer.* **G.** LS spermatozoon. **H.** TS of 9+2 region of axoneme through mitochondria and cytoplasmic canal. **I.** TS of 9+2 region of flagellum with lateral fins. **I-M.** *Pseudomugil tenellus.* **J.** TS of distal centriole. **K.** LS spermatozoon showing transverse proximal centriole and longitudinal distal centriole. L. TS of 9+2 region of axoneme through mitochondria and cytoplasmic canal. M. TS flagella with lateral fins. A, axoneme; cc, cytoplasmic canal; f, fin; m, mitochondrion; n, nucleus; s, satellite ray; v, acrosome-like vesicle. From Jamieson, B. G. M. 1991. *Fish Evolution and Systematics: Evidence from Spermatozoa*. Cambridge University Press, Cambridge, Fig. 17.3. Courtesy of C. I. Marshall. **N.** *Melanotaenia macullochi* adult. By Norman Weaver. With kind permission of Sarah Sarsmore.

been retained in these species although its expression, which is fullest in the mature lepidogalaxiid sperm, is presumably secondary and an adaptation to a secondary internal fertilization. That the acrosomoid in *Melanotaenia* is a true homologue of an acrosome is, nevertheless, open to question and requires further investigation.

Sperm relationships: Several features seen in some or all of the investigated atheriniforms are deduced to be plesiomorphic for the order and, indeed, for the Acanthopterygii. All of these, except the flocculent chromatin and presence of flagellar fins, appear to be plesiomorphies for the Actinopterygii as a whole. These are the absence of an acrosome; the flocculent externally irregular nature of the chromatin; probably the presence of a basal nuclear fossa enclosing both centrioles (Type I condition); a perpendicular arrangement of the proximal centriole relative to the distal centriole (basal body), this being the plesiomorphic condition for the Metazoa; presence of satellite rays around the basal body, as in many lower Metazoa; a moderately long cytoplasmic canal; perhaps four or five mitochondria; two dynein arms to each axonemal doublet; presence of two flagellar fins (at the same time an autapomorphy for the Actinopterygii); and insertion of the flagellum in or near the longitudinal axis of the head.

***Craterocephalus*:** On this basis, perhaps the most plesiomorphic atheriniform sperm are seen in the atherinid *Craterocephalus marjoriae*, with perpendicular centrioles, and perhaps in *Quirichthys stramineus*, though material of the latter species was too poorly preserved for certain evaluation. *Quirichthys* has been placed in the Melanotaeniidae or in the Atherinidae (see Allen 1980) but is placed in the Atherinidae, subfamily Craterocephalinae, by Nelson (2006), being placed in *Craterocephalus* in Fishbase.. The apomorphic location of the proximal centriole parallel to and lateral to the distal centriole in *Quirichthys stramineus, Craterocephalus helenae* and *C. stercusmuscarum* may be a synapomorphy of the two genera (or of *Quirichthys* with the more evolved members of *Craterocephalus*) indicating that *Quirichthys* is correctly placed in the Melanotaeniidae.

Marshall (1989) points out that taxonomic studies have indicated that there are two groups of hardyheads within *Craterocephalus*, one containing *C. marjoriae* and the other *C. stercusmuscarum*, and that sperm structure supports this and indicates that *C. helenae* should be placed in the *stercusmuscarum* group.

***Cairnsichthys*:** Several resemblances between the sperm of *Cairnsichthys*, conventionally placed in the Melanotaeniidae (Nelson 2006), and sperm of *Craterocephalus*, in the Atherinidae, appear to be symplesiomorphy (presence of satellites and of a cytoplasmic canal; location of the distal centriole in the nuclear fossa) and do not therefore justify a change in attribution.

***Melanotaenia*:** The most modified spermatozoa in the Atheriniformes are seen in *Melanotaenia* where the cytoplasmic canal has been lost; the mitochondria have apparently secondarily been reduced to a single layer; the centrioles are mutually parallel; the satellite apparatus is reduced in *M. duboulayi*, though not in *M. maccullochi*; and extreme displacement of the nucleus relative to the centrioles has occurred so that the axoneme is tangential to the nucleus, the complete type II condition; the anterior vesicle (acrosomoid) is a unique apomorphic possibly representing "endogenous" retention of an acrosome. The relatively modified condition of *Melanotaenia* sperm is consistent with taxonomic placement of the genus in the most advanced section of the Melanotaeniidae (Allen 1980).

Pseudomugil: In some ways as modified are the sperm of the three *Pseudomugil* (Blue-eye) species. Here the nucleus has become crescentic but plesiomorphic location of both centrioles in the fossa is retained; satellites are apomorphically absent; the large number of mitochondria is probably apomorphic, as is elongation of the cytoplasmic canal, with 3 to 5 tiers of mitochondria along the axoneme. The blue eyes have been shuffled between the Atherinidae and the Melanotaeniidae but are regarded as the most primitive melanotaeniids by Allen (1980). The perpendicular arrangement of centrioles in *Pseudomugil* is inconsistent with an origin from other examined melanotaeniids all of which have parallel or (*Cairnsichthys*) angled centrioles but is consistent with a basal position in the order occupied also by the atherinid *Craterocephalus marjoriae*. In other respects, as indicated, *Pseudomugil* sperm are highly modified.

Iriatherina: This has some advanced features (parallel centrioles, weak or no satellite apparatus) but is distinctive in the finely granular, compact chromatin and absence of a nuclear fossa. Absence of the fossa may be a synapomorphy with *Melanotaenia*, thus supporting current placement in the Melanotaeniidae. Generally, however, sperm structure in atheriniforms suggests that familial classification of the atherinid-melanotaeniid section of the order may require revision.

14.6 ORDER CYPRINODONTIFORMES

The Exocoetoidei and Adrianichthyoidei included with the Cyprinodontoidei in the Cyprinodontiformes by Nelson (1984) were placed by Jamieson (1991) in the order Beloniformes, as advocated by Rosen and Parenti (1981) and later endorsed by Nelson (2006), the Exocoetoidei there being termed the Belonoidei. The Cyprinodontiformes have been regarded as the sister-group of the Beloniformes, sharing with them several apomorphies (Parenti 2005). However, we have seen above that phylogenetic analysis of ten nuclear genes indicates that Cyprinodontiformes are the sister group of the Beloniformes+Atheriniformes (Ortí and Li, **Chapter 1**, Fig. 1.5, and Fig. 14.1). Internal synapomorphies of the Cyprinodontiformes indicate that the order is monophyletic (Lauder and Liem 1983) (see also Fig. 14.1).

The restricted Cypriniformes consists of the killifishes and their relatives, small to medium-sized fishes which live in shallow fresh and brackish water. Two suborders, the Aplocheiloidei and Cyprinodontoidei are recognized (Parenti 1981; Nelson 2006).

Reproductive modes: Reproduction in the group is exceptionally varied with oviparity, ovoviparity, viviparity and, in *Rivulus marmoratus*, functional hermaphroditism (see also Aulopiformes). In *R. marmoratus*, the individual has simultaneously functional ovary and testis; internal self fertilization occurs and then eggs are laid (Harrington 1961). Viviparity is deduced to have evolved three times in the order (Blackburn 2006). Among the viviparous forms there is a vast array of schedules and morphological modifications for internal development such as the trophotaeniae of the goodeids and the intra- and extra-follicular gestation and superfetation of some poeciliids (Parenti 1981; Able 1984; Grier *et al.*, **Chapter 3**).

Sperm literature: Sperm ultrastructure has been examined in many species of the Cyprinodontiformes (Asai 1971; Billard 1970; Brummett and Dumont 1979; Dadone and Narbaitz 1967; Grier 1973a; b; 1975; Grier *et al.* 1978; Jamieson 1991; Jonas-Davies *et al.* 1983; Mattei 1970; Mattei and Boissin 1966; Mattei *et al.* 1967; Mizue 1969; Nicander 1968; Porte and Follenius 1960; Russo and Pisanó 1973; Selman and Wallace 1986; Thiaw *et al.* 1986; Yasuzumi 1971).

14.6.1 Suborder Cyprinodontoidei

Cyprinodontoidei are pantropical and in temperate Laurasia from N. America as far east as Iran occurring in fresh, brackish and saltwater (Parenti, 1981).

The Cyprinodontoidei contain the families Profundulidae, Goodeidae, Fundulidae, Valenciidae, Cyprinodontidae, Anablepidae (including *Jenynsia* previously placed in the Jenynsiidae) and Poeciliidae (Parenti 2005). Individual species are mentioned only in the context of a review of comparative sperm morphology of the investigated families.

Families Cyprinodontidae and Fundulidae: The cyprinodontid *Cyprinodon variegatus*, the Sheepshead Minnow, and *Fundulus heteroclitus*, the Mummichog have been briefly examined for spermiogenesis by Yasuzumi (1971).

***Fundulus heteroclitus* sperm:** Yasuzumi (1971) describes some features of the mature sperm of *F. heteroclitus*. Brief SEM and TEM data on the sperm of this species are provided by Brummett and Dumont (1979). Gametogenesis and the mature spermatozoon have been described by Selman and Wallace (1986) (Fig. 14.9).

The nucleus is uniformly dense. It is horseshoe-shaped in longitudinal section in one plane, the deep fossa so formed being open along one surface and containing the proximal and distal centrioles and the base of the 9+2 flagellum (Yasuzumi 1971; Selman and Wallace 1986). From SEM and TEM observations the head is 1.3 μm wide, 1.0 μm thick and 2.25 μm long; the median groove extends from the posterior end almost to the anterior end; the flagellum inserts into the groove approximately 0.5 μm from the posterior end and at the point of insertion is surrounded by a low collar of a few very large

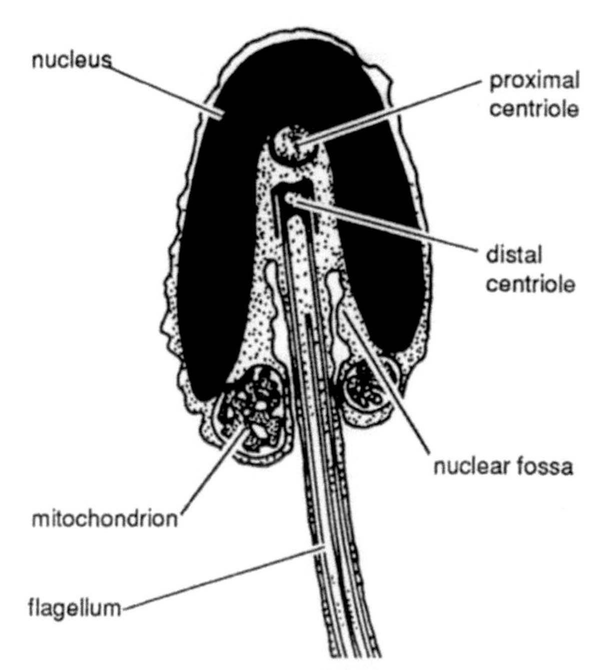

Fig. 14.9 *Fundulus heteroclitus*, the Mummichog (Fundulidae). Diagrammatic longitudinal section of the spermatozoon. Drawn from a micrograph of Selman, K. and Wallace, R. B. 1986. American Zoologist 26: 173-192, Fig. 12d.

mitochondria; the 9+2 pattern flagellum is 30-50 µm long and has a pair of lateral fins (Brummett and Dumont 1979). Fins are not described or illustrated by other workers. The "central doublet" [2 singlets] terminates shortly behind the distal centriole (Selman and Wallace 1986). This externally fertilizing cyprinodont sperm is thus similar to the internally fertilizing goodeid sperm (below).

Flagellum in other cyprinodontoids: Flagellar structure in the cyprinodontoids, the nothobranchiids *Nothobranchus (=Fundulosoma) thierryi, Epiplatys ansorgei, E. bifasciatus, E. chaperi* and *E. fasciolatus* has been described by Thiaw *et al.* (1986; see also aplocheilids, below). The testis is typical of the Atheriniformes, with spermatogonia restricted to the most distal, upper ends of the tubules (Grier 1981; Selman and Wallace 1986).

Family Goodeidae: Goodeidae for which sperm ultrastructure is briefly summarized are *Ameca splendens, Ataenobius toweri, Characoden lateralis* (Grier *et al.* 1978) and *Xenotoca eiseni* (Grier *et al.* 1978; Grier *et al.* 2006) (Fig. 14.10).

Although live-bearing, all have sperm with the basic anacrosomal aquasperm structure. As in the perciform *Zoarces viviparus*, this is here considered to suggest relatively recent acquisition of internal fertilization. Testis-structure, ovarian structure, embryonic adaptations for internal gestation, and the form of the spermatozeugmata and of the spermatozoa all indicate that goodeids have evolved internal fertilization independently of the poeciliids a view supported by the failure to develop a true intromittent organs (see also Evans and Meisner, **Volume 8B, Chapter 4**). Goodeid sperm, exemplified by *Xenotoca eiseni* (Fig. 14.10), have a slightly elongated nucleus, hollowed "ventrally" to form a fossa in which the centriolar complex, basal body and associated satellites and the base of the flagellum reside. The midpiece consists of a single ring of discrete mitochondria. The central flagellar doublets are said to terminate a short distance behind the basal body. The 9+2 flagellum has two lateral fins (Grier *et al*. 1978). This sperm structure is remarkably like that of internally fertilizing cottid scorpaenids (**Chapter 15**).

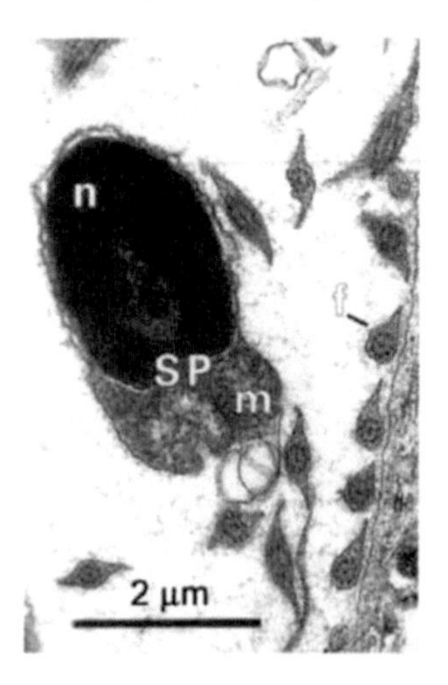

Fig. 14.10 *Xenotoca eiseni* (Goodeidae). A spermatozoon in a spermatocyst; f, sperm flagella, with lateral fins, associated with the Sertoli cell plasmalemma; m, mitochondrion; n, nucleus; SP, spermatozoon. Adapted from Grier, H. J., Uribe, M. C., Parenti, L. R. and De La Rosa-Cruz, G. 2006. Fecundity, the germinal epithelium, and folliculogenesis in viviparous fishes. Pp. 191-216. In M. C. Uribe and H. Grier (ed.), *Viviparous Fishes*, vol. New Life Publications. Homestead, Florida, Fig. 7 (part).

Families Anablepidae and Poeciliidae: Viviparous poeciliids (*Gambusia affinis, Poecilia latipinna, Xiphophorus helleri, X. maculatus, P.* (=*Lebistes*) *reticulata*) and examined anablepid species (*Anableps anableps* and *Jenynsia lineata*) differ from goodeids and cyprinodontids in having complex sperm with elongated nucleus and midpiece.

Anableps anableps: The adult of the Four-eyed fish is shown in Figure 14.11. The sperm nucleus has compact chromatin and forms an elongate cone with a deep implantation fossa containing both centrioles in perpendicular arrangement. The distal centriole is very long and has an amorphous electron-dense structure on one side. The midpiece consists of a long mitochondrial sheath containing many separate mitochondria with longitudinal cristae. They lie in a cytoplasmic sleeve in longitudinal rows. In a transverse section of this sleeve, two to four C-shaped mitochondria, surrounding the flagellum are located along the cytoplasmic canal. The 9+2 flagellum has two lateral fins present only at the posterior part in the cytoplasmic canal and just beyond (Pecio *et al.* 2006).

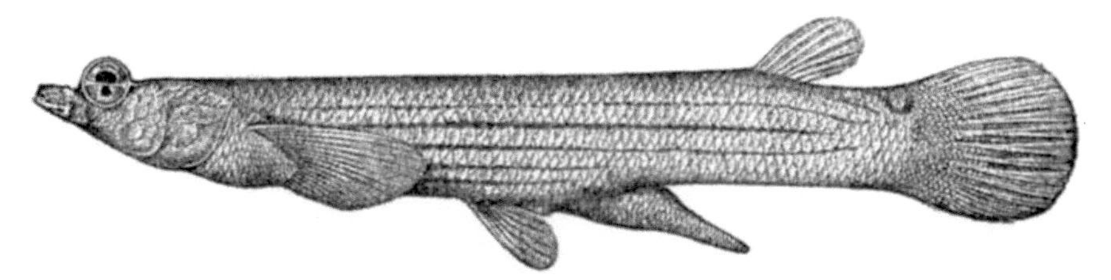

Fig. 14.11 *Anableps anableps* (Anablepidae). From Norman, J. R. 1937. *Illustrated Guide to the Fish Gallery*. Trustees of the British Museum, London, Fig. 28.

Jenynsia lineata: In *Jenynsia lineata*, the One-sided Live bearer, formerly in the Jenynsiidae, a viviparous fish according to Breder and Rosen (1966), the nucleus has a deep fossa apparently (as in poeciliids) open along one surface and containing the two centrioles (Fig. 14.12A).

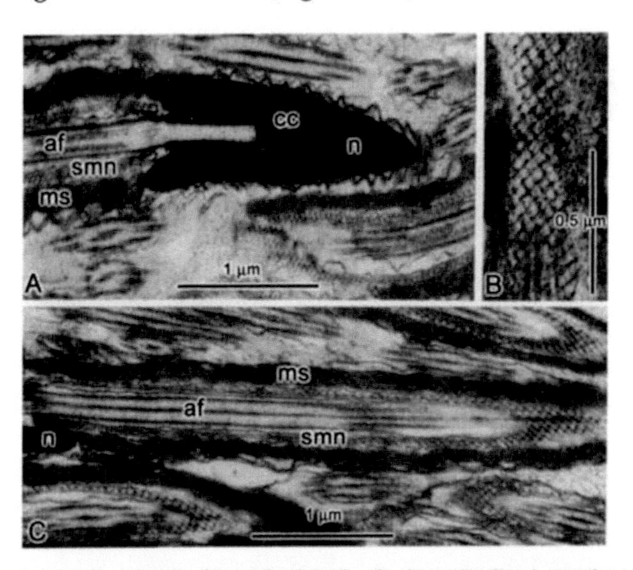

Fig. 14.12 *Jenynsia lineata* (Anablepidae). **A**. Longitudinal section (LS) of the anterior end of a mature spermatozoon. **B**. LS showing geometrical pattern of the submitochondrial net. **C**. LS of the midpiece of a mature spermatozoon. Adapted from Dadone, L. and Narbaitz, R. 1967. Zeitschrift für Zellforschung 80: 214-219, Figs. 1-3. With kind permission of Springer Science and Business Media.

The mitochondria of the spermatid differ from those of poeciliids and *Anableps* in fusing to form a mitochondrial sheath containing at maturity a single, cylindrical mitochondrial derivative. This is narrowly open along one side and has longitudinal cristae. Between the mitochondrial derivative and the cytoplasmic canal around the axoneme there is a "submitochondrial net" (Fig. 14.12B,C) which consists of filaments enclosed between two membranes. The filaments are separated by a constant distance of about 3 Å and form a regular three dimensional pattern (Dadone and Narbaitz 1967). In micrographs the 9+2 flagellum shows short lateral fins in only a few sections.

The presence of a tubular intromittent organ in Anablepidae may account for the failure to develop spermatozeugmata as sperm packaging is not advantageous (see also Evan and Meisner, **Volume 8B, Chapter 4**).

Poeciliidae: Five species of the live-bearing family Poeciliidae have been examined for sperm ultrastructure: *Gambusia affinis*, the Mosquito Fish (Jamieson 1991) (Fig. 14.13A-E); *Poecilia latipinna*, the Sailfin Mollie (Mizue 1969) (Black Mollie variety, Fig. 14.13J-L); (Grier 1973a,b) (Wild form); Jamieson 1991 (Aquarium fish); *Poecilia reticulata*, the Guppy (Asai 1971; Billard 1970b; Mattei, 1970; Mattei and Boissin 1966; Porte and Follenius 1960; Jamieson 1991) (adult, Fig. 14.13D; sperm, Fig. 14.13F,G); *Xiphophorus helleri*, the Swordtail (Jonas-Davies *et al.*, 1983; Azevedo and Corral 1983; Jamieson 1991; Grier *et al.* 2006) (adult, Fig. 14.13E; sperm, Fig. 14.13H, I); *Xiphophorus* (=*Poecilia*) *maculatus*, the Southern Platy (Russo and Pisano 1973. The account of Azevedo and Corral (1983) is chiefly a valuable investigation of spermatogenesis and the development of the spermatozeugmata in *X. helleri*.

The evolution of viviparity in poeciliids has involved testicular modification in which spermatozeugmata are transferred to the female reproductive tract. The sperm heads are characteristically external with the tails forming the center of the spermatozeugma (Grier 1976), as shown for *Xiphophorus helleri* (Fig. 14.13H). *Poecilia formosa* consists almost entirely of females which use males of other species to stimulate development of the egg without any genetic contribution (Lauder and Liem 1983). Only the subfamily Poeciliinae possesses gonopodia, exhibits internal fertilization and (with the

Fig. 14.13 Spermatozoa of Poeciliidae. **A-E**. *Gambusia affinis*. **A**, **D** and **E**. Longitudinal sections of spermatozoa. **B**. Detail of the acrosome-like vesicles on the apices of the nuclei embedded in Sertoli cells. **C**. Transverse sections (LS) of the nuclear fossa and of the midpiece. **F**, **G**. *Poecilia reticulata*. **F**. TS of midpiece and axonemes, showing alteration of doublets and loss of central singlets at endpiece. **G**. LS of two spermatozoa. That on the left shows the greater width of the nucleus, cut in the "lateral" plane, with rounded tip. That on the right shows the nucleus cut in the "dorso-ventral" plane (i.e. viewed "laterally") and appearing narrower and with pointed tip. **H**, **I**. *Xiphophorus helleri*. **H**. Section of part of

Fig. 14.13 Contd. ...

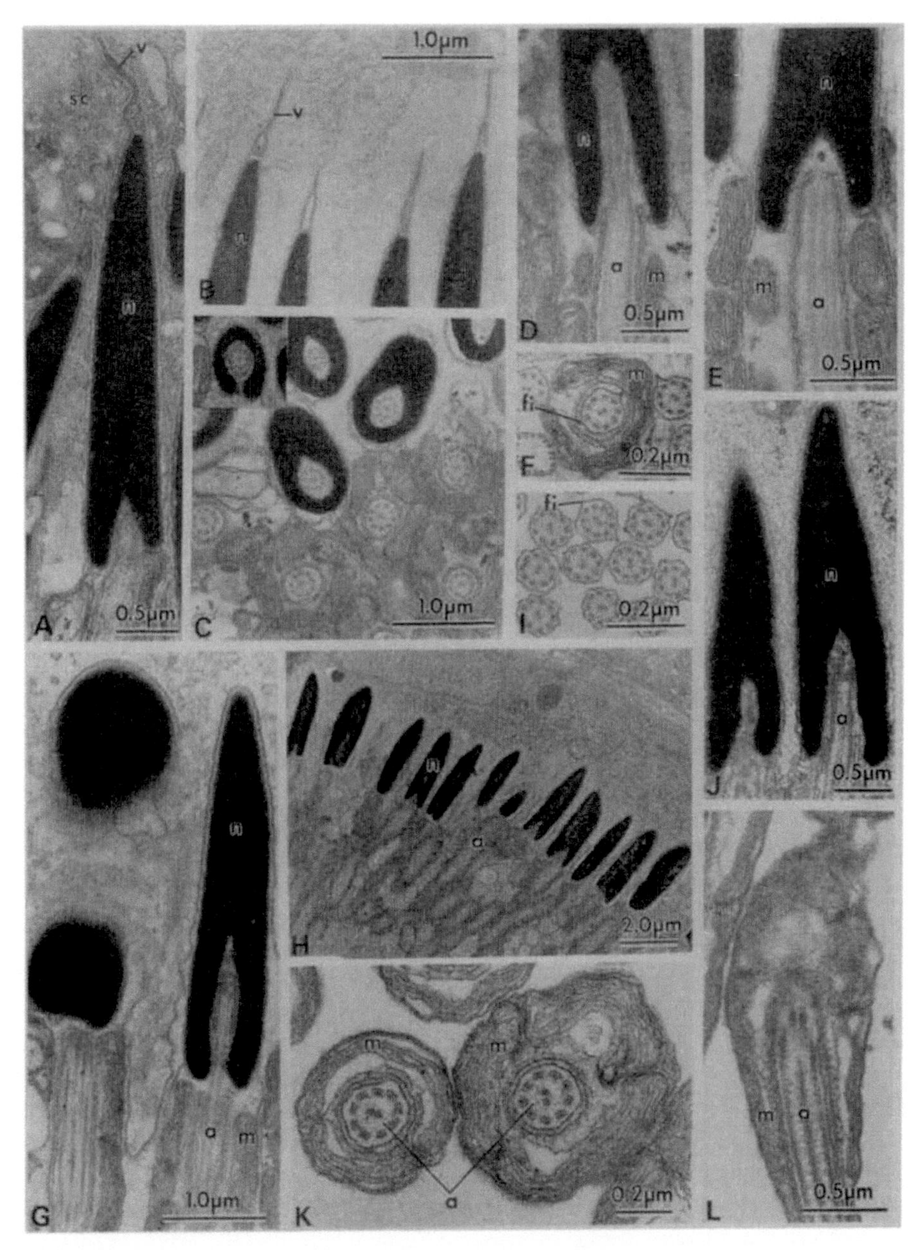

spermatozeugma of late spermatids, showing centrifugal orientation of the nuclei and central mass of axonemes. **I.** TS axonemes showing the two short lateral fins. J-L. *Poecilia latipinna*. J. Head and anterior midpiece. K. TS midpiece. Posterior end of midpiece. a. axoneme; de. distal centriole; m. mitochondrion; n. nucleus; sc. Sertoli cell; v. acrosome-like vesicle. From Jamieson, B. G. M. 1991. *Fish Evolution and Systematics: Evidence from Spermatozoa*. Cambridge University Press, Cambridge, Fig. 17.6.

exception of the oviparous *Tomeurus*, sperm ultrastructure of which is unknown) give birth to live young (see Evans and Meisner, **Chapter 22**).

All of the poeciliids listed above have been studied principally for spermiogenesis. Jamieson (1991) attempted to augment knowledge of spermatozoal ultrastructure by observations on the sperm of *Poecilia reticulata, Gambusia affinis, Xiphophorus helleri* and *P. latipinna* (the Black Mollie variety).

Acrosome: An acrosome is absent in poeciliids, as in other atherinomorphs. However, in the spermatozoa or very late spermatids of *Gambusia affinis*, when their heads are still embedded in the Sertoli cells, there is an elongate vesicle on the apex of the nucleus (Fig. 14.13A,B), separated from the latter by the nuclear envelope and covered by the general plasma membrane of the sperm cell. Although this vesicle does not persist in the free spermatozoon it is possibly homologous with an acrosome and/or with the anterior vesicle of sperm of the atheriniform *Melanotaenia*. It is not known whether it occurs in other poeciliids.

The nucleus in all species is an elongate cone, apex anteriorly, strongly compressed in one plane which conventionally has been termed the lateral plane. In *P. reticulata*, at its greatest width, at the commencement of basal narrowing, its width is 0.74 μm "laterally" and 1.2 μm "dorsoventrally". Although the nucleus appears pointed when viewed dorsally, it is blunt and rounded when viewed laterally, as shown for *P. reticulata* in Fig. 14.13G. The chromatin is so strongly condensed as to reveal no substructure and is electron opaque. Basally the nucleus is rounded and slightly narrower. The length of the nucleus is relatively uniform among the different species: 3.5 μm in *G. affinis* and 3.7 μm in *P. reticulata* and *X. helleri*. The nucleus is thus in fact shaped like a parallel-sided spatula or thick, apically rounded blade. Nuclear pores are absent at maturity (Grier 1975).

The base of the nucleus is penetrated by a deep implantation fossa which occupies approximately its posterior third and houses the basal body and anterior region of the axoneme (*Gambusia affinis*, Fig. 14.13A-C; *Poecilia reticulata*, Fig. 14.13G; *P. latipinna*, Fig. 14.13J). Anteriorly the fossa is circular in cross section and is enclosed on all sides by nuclear material but in its posterior region it is open "ventrally" through a posteriorly widening slit (Fig. 14.13C, inset).

There is some doubt as to whether the ventral opening of the fossa persists at maturity, though the advanced and apparently mature state of the sperm examined by Jamieson (1991) suggested that it did. A similar, Sertoli embedded stage is considered mature for *Poecilia latipinna* by Grier (1973b). A deep fossa open "ventrally" has also been described for *Gambusia affinis* by Grier (1975) and for *P. latipinna* by Grier (1973) and Mizue (1969). In *Xiphophorus helleri*, it is described as "a deep notch in the base of the nucleus" (Jonas-Davies *et al.*, 1983). However, in *Poecilia reticulata*, the groove is shown as closing off by maturity (Asai 1971; Billard 1970b) or, in contrast, is drawn as opening dorsally and ventrally, a doubtful observation (Porte and Follenius

1960). It is agreed that in all species the fossa is much deeper and the surrounding nuclear material much thinner in the spermatid than in the spermatozoon. There is also a deep posteroventral ventral nuclear fossa in the sperm of *Jenynsia*, and, indeed, in goodeids.

Mitochondria: In *Poecilia latipinna*, the long mitochondrial sheath appears to contain many separate mitochondria in longitudinal rows (Mizue 1969). This longitudinal array of mitochondria, with longitudinal cristae, has been clearly demonstrated for *Xiphophorus helleri* (Jonas-Davies *et al.* 1983; Azevedo and Corral 1983), who report ca 21-25 mitochondria, and was shown by Jamieson (1991) for all examined species, including *Gambusia affinis* (Fig. 14.13A,D,E); *Poecilia latipinna* (Fig. 14.13J) and *P. reticulata* (Fig. 14.13G). This region of the sperm constitutes the midpiece. The mitochondria in the species here examined lie in a very long cytoplasmic sleeve which invests the axoneme but is separated from it by the so-called cytoplasmic canal ubiquitous in osteichthyan sperm. The length of the sleeve is about 8 μm in *P. reticulata* and of the same order, though not precisely determined, in the other species. The inner and outer (plasma) membranes of the canal are continuous at its anterior limit. In the mature sperm this limit is shortly behind the nucleus (*G. affinis*, Fig. 14.13D; *P. reticulata*, Fig. 14.13G) or just within the fossa (*P. latipinna*, Fig. 14.13J; *X. helleri*, see also Jonas-Davies *et al.* 1983). A local dilatation of the cytoplasmic canal occurs at its anterior commencement, as noted by Jonas-Davies *et al*. (1983).

In cross section of the mitochondrial sleeve, two to four mitochondria embrace the flagellum, and are predominantly C-shaped, or at least shortly crescentic, in section (Fig. 14.13C, F, K). Each has several regularly disposed cristae, all parallel with the inner and outer mitochondrial membranes, joining the wall of the mitochondrion at the tips of the C-shape, and seen in longitudinal section of the sperm to run longitudinally (Fig. 14.13A, D, E, G, J).

In longitudinal section of the axoneme, the mitochondria appear as a single row on each side, arranged end to end, there being about 7 or 8, with some variation, in a longitudinal row. Because of variations in the length and circumferential extent of individual mitochondria, the lines of apposition of adjacent mitchondria both longitudinally and around the axoneme are somewhat irregular as, therefore are the longitudinal mitochondrial columns. Although in longitudinal section most mitochondria appear several times longer than wide, rarely being isodiametric, each entire mitochondrion, in three-dimensions, is typically a C-shaped cylinder of greater diameter than length. Variation in numbers of mitochondrial columns reported in the literature may partly be ascribed to variation in the number of circumaxonemal mitochondria along the length of the midpiece. In *Poecilia reticulata*, as noted by Billard (1970b), there are usually four longitudinal columns of mitochondria and at the distal extremity of the "intermediate piece" only two, contiguous crescentic columns. Five columns of mitochondria are shown for this species by Porte and Follenius (1960). The sperm of *Xiphophorus maculatus* is similar but some cross sections, at least, show only three mitochondria, representing the longitudinal columns of separate mitochondria (Russo and Pisanó 1973). Azevedo and Corral (1983) cite a total of 21-25 mitochondria in *X. helleri*. Jamieson (1991) confirmed that dense granules present in the mitochondria early in spermiogenesis do not persist [or are infrequent] in the mature spermatozoon (Billard 1970b).

Submitochondrial net: Densification of the plasma membrane lining the inner surface of the mitochondria, closest to the axoneme, has been recognized by some workers as a submitochondrial net in poeciliid sperm. It is considered the equivalent of the submitochondrial net of *Jenynsia*. The mesh of the net appears as a regular diamond lattice. In sagittal sections the vertices appear as electron-dense profiles, about 7.5 nm in diameter, which are spaced approximately 42 nm apart. The electron-dense material at the vertices appears to attach both to the outer surface of the mitochondria and to the cytoplasmic surface of the overlying plasma membrane (Jonas-Davies *et al.* 1983).

Centrioles: Of the two spermatid centrioles, only the distal centriole, forming the basal body for the flagellum persists intact in the mature spermatozoon in the four species here examined and this appears general for the other examined poeciliids. The complex structure of the centriolar apparatus in *Poecilia reticulata* has been described by Mattei and Boissin (1966) and is shown in Fig. 14.14.

In late spermiogenesis the triplets of the proximal centriole become occluded and it is reduced to a remnant by maturity. In *Poecilia reticulata*, as in other poeciliids, an intercentriolar lamellated body disappears after having contributed to an electron-dense cap on the proximal end of the basal body. In *Gambusia affinis*, as in other poeciliids, rodlike structures appear on opposite sides of the basal body during spermiogenesis. In *G. affinis* and *P. latipinna* (Grier 1973a,b) and probably *P. reticulata* (Billard 1970b) these appear to have a dual origin: there is an apparent fusion of two striated satellites, on one side of the basal body, to produce a long striated satellite, and in addition an amorphous electron-dense structure is present on the opposite side of the basal body, in the spermatid and spermatozoon (Grier 1975).

Flagellum: Within the nuclear fossa, the flagellum has what Grier (1973b), for *Poecilia latipinna*, has termed a specific flagellar-nuclear geometric orientation. When viewed in cross section, a line drawn perpendicular to the plane of the two central singlets, and passing between them, will always approximately bisect the nucleus in the wide, dorsoventral plane. A considerable inclination of this plane nevertheless occurs, as shown for *Gambusia affinis* in Fig. 14.13C.

The 9+2 flagellum has two short lateral fins in the four species examined here: *Poecilia reticulata* (Fig. 14.13F), as also shown by Mattei *et al.* (1967b), *Gambusia affinis*, *Xiphophorus helleri* (Fig. 14.13I), and shown by Azevedo and Corral (1983), and the Black Mollie, and in *Xiphophorus* (=*Poecilia*) *maculatus*

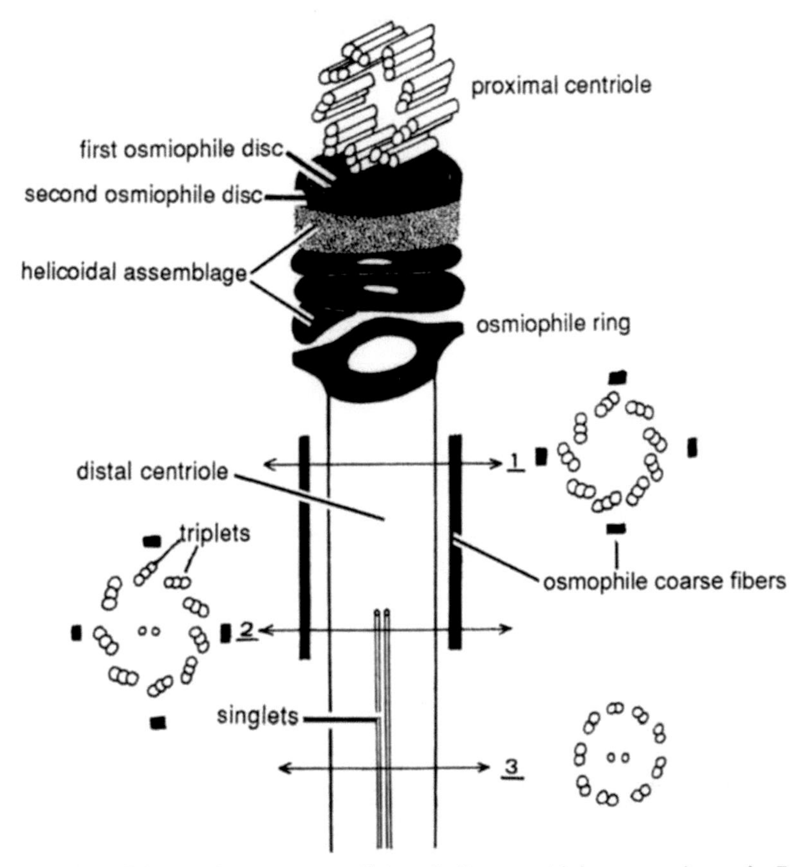

Fig. 14.14 Schematic reconstruction of the centriolar complex of *Poecilia reticulata*. 1. Transverse section (TS) of the distal centriole in the midregion. 2. TS of the base of the distal centriole. 3. TS of the flagellum. From Mattei, X. and Boissin, C. 1966. Comptes Rendus Hebdomadaires des Séances de l'Académie des Sciences D 262: 2620-2622, Fig. 10.

as demonstrated by Russo and Pisanó (1973). They are probably present in all poeciliids. They are very short, rarely (as in some profiles of *X. helleri* sperm) each equaling the width of the axoneme, but their presence is taken here to attest the ect-aquasperm ancestry of the poeciliid sperm, long fins being typical of externally fertilizing species. Their orientation varies in neighboring testicular sperm from the plane of the central singlets to about 45° to this (Fig. 14.13I), probably reflecting a dynamic situation.

At the posterior end of the flagellum, as shown in *Poecilia reticulata* and confirmed here (Fig. 14.13F), the lateral fins are lost, the central singlets disappear, and the doublets approach the plasma membrane, reducing their distance from this to 100 Å. More posteriorly the doublets disrupt into 18 singlets; ultimately the flagellum loses its circular cross section and the number of microtubules is progressively reduced (Mattei *et al.* 1967b).

Remarks: The elongate nuclei and, as Nicander (1968) has noted, the long mitochondrial sheaths of these poeciliids are characteristic of [though not constant in] internally fertilizing sperm.

14.6.2 Suborder Aplocheiloidei

Pelvic fins inserted close together; metapterygoid present; three basibranchials; a dorsal ray on each of the first two dorsal radials. This definition, for the family Aplocheilidae, including all rivulines, of Nelson (1984) is here applied to the suborder *sensu* Parenti (1981). Rivulines are freshwater fishes of Africa, southern Asia, southern N. America, and S. America. Asian rivulines are the family Aplocheilidae; African rivulines are the family Nothobranchiidae.

Flagellar ultrastructure: A large number of cyprinodontiforms, mostly aplocheiloids, has been investigated by Thiaw *et al.* (1986) for flagellar structure in the spermatozoon. They include *Notobranchius steinforti, Scriptaphyosemion (=Aphyosemion) guignardi, Scriptaphyosemion(=Aphyosemion) nigrifluvi, Aphyosemion herzogi; Aphyosemion riggenbachi* (both known to be externally fertilizing, non-guarding egg scatterers) and *Aphyosemion splendopleure* (all Nothobranchiidae); *Procatopus (=Aplocheilichthys) lamberti* and *Aplocheilichthys normani* (both now Poeciliidae); *Aplocheilus lineatus* (Aplocheilidae); and *Nematolebias whitei*? (lapsus *Cynolebias wittei?)* (Rivulidae). A similar spermatozoal morphology was reported, although not described, but a wide diversity of flagellar structure was encountered. The flagellar membrane has one, two or three lateral expansions [fins] depending on the species. Peripheral doublets of the axonemes show only the external arm, excepting two species (the nothobranchiid *Scriptaphyosemion guignardi* and the aplocheilid *Aplocheilus lineatus*) that completely lack the arms though, like all the other species, retaining motility. Intratubular differentiations (ITDs) are present in A or B tubules of the doublets, as in central tubules of some species, whereas others are totally devoid of such differentiations. The ITDs can affect all doublets or preferentially doublets 1, 5 and 6, rarely with intraspecific variation. It is suggested that these variations may be due to neutral mutations (Thiaw *et al.* 1986).

***Fundulopanchax (=Aphyosemion) gardneri*:** (Fig. 14.15). The nucleus of the sperm of *Fundulopanchax gardneri* is spherical to slightly ellipsoid with long axis anteroposterior, 2.2-2.5, mean 2.4, μm long (n=4). It has a short basal fossa, narrower than the flagellum. The chromatin is electron-dense and moderately condensed, there being many lacunae; its outline is smooth with only slight flocculence.

The midpiece is irregular in form, differing between individual spermatozoa. A broad anterior portion contains the centrioles and several irregularly arranged mitochondria. The cytoplasmic canal, with a maximum recorded length of 2.3 µm, commences behind the distal centriole. The cytoplasm enclosing it contains subspheroidal or elongate bodies, with concentric cristae, which appear to be modified mitochondria but consists mostly of a long trailing sleeve.

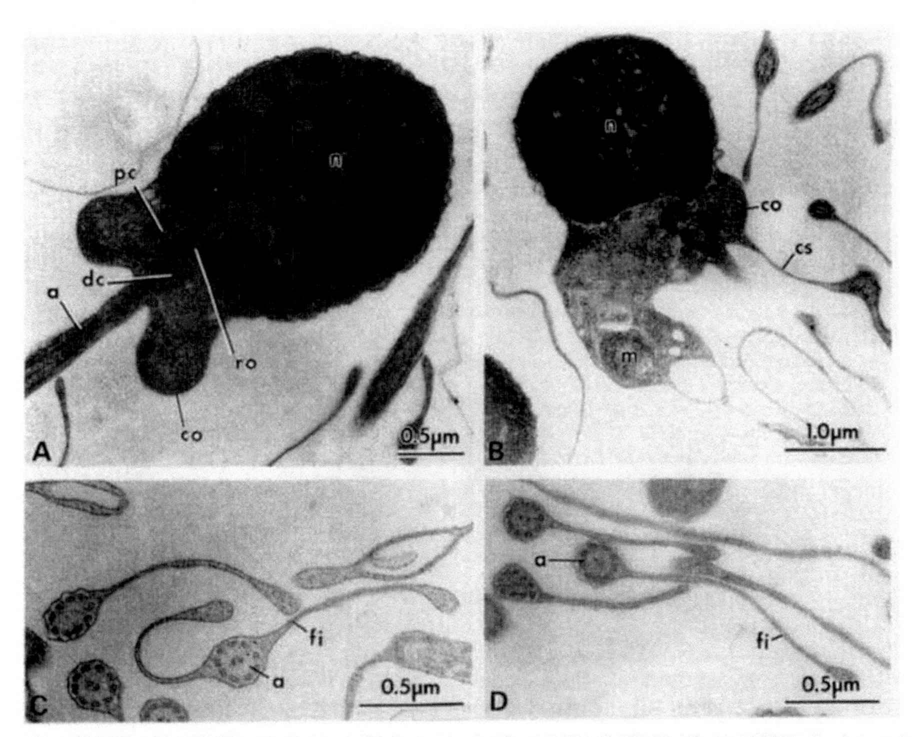

Fig. 14.15 *Fundulopanchax (=Aphyosemion) gardneri* (Nothobranchiidae). **A** and **B**. Longitudinal sections of spermatozoa. **A**. Showing proximal centriole parallel to distal centriole (basal body) and rootlet-like structure projecting from the basal body into the nuclear fossa. **B**. LS showing apparent shedding of the posterior part of the collar, seen in many sperm. **C** and **D**. Transverse sections of the flagellum. There are two lateral fins throughout most of its length but only one fin near its posterior extremity where the 9+2 pattern is disrupted. a, axoneme; co, cytoplasmic collar; cs, cytoplasmic sleeve; de, distal centriole; fi, fm; m, mitochondrion; n, nucleus; pc, proximal centriole; ro, rootlet-like structure. From Jamieson, B. G. M. 1991. *Fish Evolution and Systematics: Evidence from Spermatozoa*. Cambridge University Press, Cambridge, Fig. 17.8.

The longitudinal axes of the two centrioles are parallel, with the proximal located immediately anterior to the distal centriole. A rootlet-like structure, with longitudinal not transverse striae, extends obliquely from the anterior end of the distal centriole to the nuclear fossa (Fig. 14.15A).

The 9+2 flagellum has, as in other aplocheiloids, unusually long lateral fins approximately in the plane of the central singlets, giving the entire flagellum a width up to 6 μm. Each fin is slightly expanded terminally. Posteriorly there is only as single, unilateral fin (Fig. 14.15C,D) (Jamieson 1991).

14.6.2.1 Family Rivulidae

The monogeneric family Rivulidae contains about 100 nominal species of which all except *Rivulus marmoratus* are externally fertilizing gonochores (Huber 1992). Only this internally fertilizing species has been examined for general sperm ultrastructure (Kweon *et al.* 1998), though *Nematolebias whitei?* (*lapsus Cynolebias wittei?*) has been examined for flagellar ultrastructure (see above). Facultative internal fertilization has been reported in *Cynolebias melanotaenia* and *C. bruccei* (see Evans and Meisner, **Volume 8B, Chapter 4**).

***Rivulus marmoratus*:** The mature spermatozoon is ca 28 μm long and has an acrosomeless spherical head (Fig. 14.16A,B), ca 1.8 mm wide, with an electron-dense nucleus and a bell-shaped axial nuclear fossa. The chromatin consists of numerous, electron-dense, flocculent masses (Fig. 14.16A).

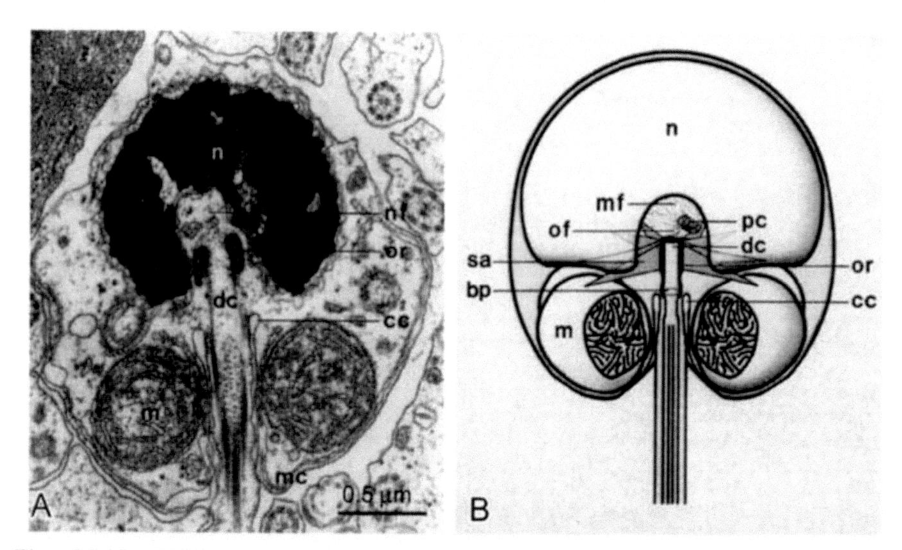

Fig. 14.16 *Rivulus marmoratus* (Rivulidae). **A**. TEM longitudinal section of a spermatozoon, showing electron-dense nucleus (n) with a bell-shaped fossa (nf), distal centriole (dc) surrounded by an osmiophilic ring (or), spherical mitochondria (m) with well developed cristae and the proximal region of the tail separated from the mitochondrial collar (mc) by the narrow cytoplasmic canal (cc). **B**. Diagram of the spermatozoon. Additional abbreviations: bp, basal plate; mf, microtubules, of, osmiophilic filaments; sa, satellite apparatus. Adapted from Kweon, H.-S., Park, E.-H. and Peters, N. 1998. Copeia 1998: 1101-1106, Fig. 2A,B. With permission of the American Society of Ichthyologists and Herpetologists.

The midpiece contains seven equisized mitochondria. The mitochondrial collar surrounds the base of the flagellum from which it is separated by a cytoplasmic canal.

The proximal centriole fills most of the nuclear fossa, whereas the distal centriole is located at the base of head. Numerous netlike microfibrils interconnect the proximal centriole with the nucleus. The proximal centriole is orientated at 45° to the distal centriole and both centrioles appear to have a 9 doublets+0 configuration. The anterior end of the distal centriole is surrounded by an osmiophilic ring and radial satellite rays are present.

The 9+2 flagellum is ca 26 μm, the A subtubules of doublets 1, 5 and 6 are darker and occluded. A pair of lateral fins is aligned with the central singlets but is absent proximally and distally. The proximal part of the flagellum is devoid of central singlets (Kweon *et al.* 1998).

Remarks: Internally fertilizing sperm in fishes usually show considerable modification, including elongation of the nucleus. *Rivulus marmoratus* is unusual in having a spermatozoon of the simple teleostean aquasperm type (*sensu* Jamieson 1991), showing no evident modifications of the morphology seen in externally fertilizing fish sperm, despite internal fertilization in this species. As the other species of the family are external fertilizers it is clear that internal fertilization has been achieved independently of other internally fertilizing atherinomorphs and it likely that the unmodified morphology of the sperm of *R. marmoratus* is due to relatively recent acquisition of internal fertilization.

Such anacrosomal aquasperm are known from other internally fertilizing fishes: the perciform *Zoarces viviparus* (though there biflagellate), the viviparous goodeids *Ameca splendens*, *Ataenobius toweri* and *Characoden lateralis* (references in Jamieson 1991); the oviparous atherinid *Labidesthes sicculus* (Grier *et al.* 1990), and several species of characids with internal fertilization (see Burns *et al.*, **Chapter 11** and Evans and Meisner, **Volume 8B, Chapters 4**).

14.7 ORDER BELONIFORMES

The Beloniformes (Synentognathi), include the needlefishes, rice fishes, half beaks, and flying fishes. They dominate the epipelagic region of the tropics and subtropics (Nelson 2006). For a discussion of their controversial placements in the Cyprinodontiformes or Beloniformes, and a diagnosis, see that work.

Species of the suborder Adrianichthyoidei, family Adrianichthyidae, inhabit fresh and/or brackish waters. Most species of the other four families, comprising the suborder Belonoidei (=Exocoetoidei), are epipelagic marine fishes but, in these, several genera of the Belonidae and Hemiramphidae are restricted to freshwaters and a few other genera contain estuarine and freshwater as well as marine species (Collette *et al.* 1984). It is noteworthy that internal fertilization is limited to non-marine beloniforms (see Evans and Meisner, **Volume 8B, Chapter 4**).

Beloniform sperm literature: Sperm ultrastructure has been described for the Adrianichthyidae-*Oryzias* latipes (see Grier 1976; Sakai 1976; and, SEM,

Iwamatsu and Ohta 1981); the Exocoetidae - *Fodiator acutus* (Mattei 1970); the Hemiramphidae - the live-bearing *Hemirhamphodon pogonognathus* (Jamieson 1989; Jamieson and Grier 1993), an externally fertilizing hemiramphid, *Arrhamphus sclerolepis* (Jamieson 1991; Jamieson and Grier 1993) and *Zenarchopterus dispar, Nomorhamphus celebensis and Dermogenys pusillus* (Jamieson and Grier 1993). In addition, Downing and Burns (1995) have described the spermatozeugmata of three genera of viviparous halfbeaks (Hemiramphidae): *Nomorhamphus, Dermogenys* and *Hemirhamphodon* by light microscopy. Strangely, no sperm of the Belonidae have been described ultrastructurally. Relationships of beloniforms on the basis of sperm ultrastructure are shown in Fig. 14.17.

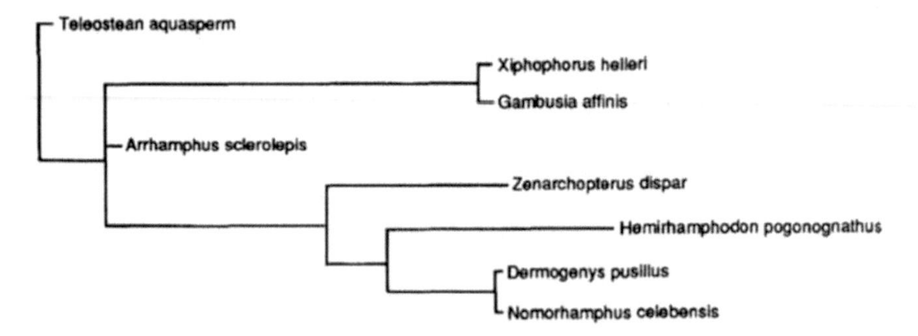

Fig. 14.17 Tree derived by branch and bound parsimony analysis of 8 spermatozoal characters (some multistate) for 7 studied atherinomorphs plus a basic teleostean aquasperm. Branch lengths reflect numbers of apomorphic changes in spermatozoal ultrastructure. The novel sister-group relationship of *Dermogenys* and *Nomorhamphus* and unity of the Poeciliidae are well supported by the spermatozoal data. Jamieson, B. G. M. and Grier, H. J. 1993. Hydrobiologia 271: 11-25, Fig. 5. With kind permission of Springer Science and Business Media.

14.7.1 Suborder Adrianichthyoidei

Rosen and Parenti (1981) defined the adrianichthyoids on five characters, including great expansion of the articular surface of the fourth epibranchial and possession of a reduced autopalatine with posterior articular cartilage. They enlarged the Adrianichthyidae to include the Horaichthyidae and Oryziidae. Collette *et al.* (1984) added a further, larval, character distinguishing them from exocoetoids. The four genera, *Adrianichthys, Horaichthys, Oryzias* and *Xenopoecilus,* totaling only 11 species, inhabit fresh and/or brackish waters from India and Japan to the Indo-Australian Archipelago (Collette *et al.* 1984).

14.7.1.1 Family Adrianichthyidae

Oryzias latipes: The Japanese Medaka, *Oryzias latipes* (adults, Fig. 14.18, see also Matsuda and Sakaizumi, **Chapter 26**), is the only adrianichthyoid

Fig. 14.18 *Oryzias latipes*, the Japanese Medaka (Hd-rR inbred strain). Above, a male. Below, at a smaller scale, females. The female at bottom right is carrying fertilized eggs in the pelvic region. Courtesy of Dr. Masaru Matsuda.

species examined for sperm ultrastructure. Spermiogenesis was investigated by TEM by Sakai (1976); Iwamatsu and Ohta (1981) made a valuable SEM study of fertilization and the micropyle; and Grier (1976) elucidated the ultrastructure of the spermatozoon by TEM (Fig. 14.19). The spermatozoon is nearer to the basic teleostean aquasperm than that of the poeciliids as befits its external fertilization. Thus the nucleus is rounded, the basal nuclear fossa is less deep and contains a persistent proximal centriole of triplet construction, and although the midpiece continues as a long "sleeve" around the base of the flagellum, the mitochondria are scattered and unmodified; the intercentriolar body differs significantly from that of poeciliids, being both smaller and disappearing earlier. Two flagellar fins are again present (Grier 1976). Although the rounded nucleus and basal location of the fossa give some support to exclusion of *O. latipes* from the Cyprinodontidae in which it is sometimes placed, sperm structure does not yet allow a decision between placement of *Oryzias* and other adrianichthyids in the Beloniformes, as indicated in a phylogram by Collette *et al.* (1984) and advocated by Rosen and

Parenti (1981) and accepted by Nelson (2006). This species has a form of parental care in which fertilized eggs adhere to the pelvic region of the female (Fig. 14.18) (see Kolm, **Volume 8B, Chapter 9**).

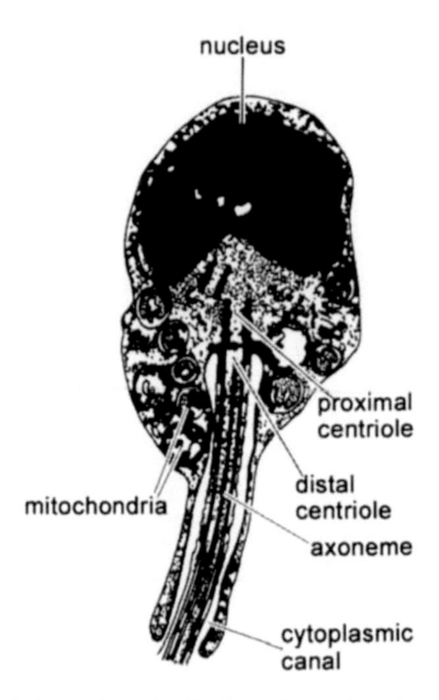

Fig. 14.19 *Oryzias latipes*. Longitudinal section of mature spermatozoon. From Jamieson, B. G. M. 1991. *Fish Evolution and Systematics: Evidence from Spermatozoa*. Cambridge University Press, Cambridge, Fig. 17.11. After a micrograph of Grier, R.I. 1976. *Cell and Tissue Research* 168: 419-431, Fig. 14.

14.7.2 Suborder Belonoidei (= Exocoetoidei)

Among other features, scales large, usually 38-60 in the lateral line; mouth opening small; no isolated finlets; dorsal and anal fins usually with 8-16 rays each; teeth small. This is the superfamily Exocoetoidea, within the Exocoetoidei (equivalent to the Beloniformes here) placed in the Cyprinodontiformes by Nelson (1984).

14.7.2.1 Family Exocoetidae

These are the flying fishes; marine, tropical to warm temperate in the Atlantic, Indian and Pacific Oceans.

Fodiator acutus: The spermatozoon of the flying fishes *Fodiator acutus* (Exocoetidae) is a simple teleostean anacrosomal aquasperm (Fig. 14.20) (Mattei 1970).

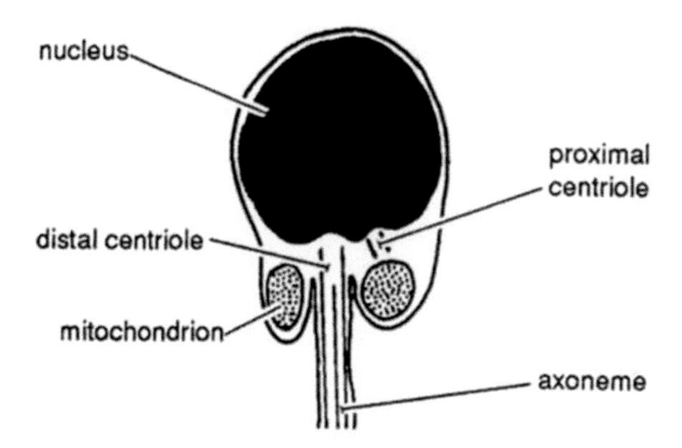

Fig. 14.20 *Fodiator acutus*. Diagrammatic longitudinal section of the spermatozoon. After Mattei, X. 1970. In B. Baccetti (ed.), *Comparative Spermatology*. Academic Press, New York, Fig. 4:19.

14.7.2.2 Family Hemiramphidae

These are the halfbeaks, marine and freshwater; tropical to warm temperate in the Atlantic, Indian and Pacific oceans (Nelson 2006), with two subfamilies, the Hemiramphinae and Zenarchopterinae.

Trends in the evolution of beloniform and cyprinodontiforms spermatozoa, so far as they are known, are tentatively indicated in Fig. 14.21.

14.7.2.3 Externally fertilizing belonoids (=exocoetoids)

Hemiramphidae, Hemiramphinae: *Arrhamphus sclerolepis*. In the *Arrhamphus* sperm (Jamieson 1991; Jamieson and Grier 1993), the nucleus (Fig. 14.22A) is subspheroidal and 1.6 μm long. Basally it is indented as a poorly defined fossa, sufficient to house only part of the proximal centriole. The chromatin consists of numerous large, separate, electron-dense, flocculent masses in a pale matrix. The two nuclear membranes remain separated by a considerable perinuclear cisterna. The mitochondria are spherical and cristate and are arranged in two tiers longitudinally (Fig. 14.22A). Ten mitochondria are seen in transverse section of a tier, symmetrically arranged around the central axis (Fig. 14.22B). Those of the posterior tier lie in the short (0.8 μm long) mitochondrial collar, around the periaxonemal invagination (cytoplasmic canal) but those of the anterior tier surround the distal centriole (basal body) anterior to the collar (Fig. 14.22A). There is some thickening of the wall of the collar surrounding the canal (subplasmalemmal densification) which may be the equivalent of the submitochondrial dense layer of the internally fertilizing, zenarchopterid species.

The proximal centriole (Fig. 14.22A) is located anteriorly to and slightly to one side of the distal centriole and is tilted at an angle of approximately 45° to its long axis and, therefore, to that of the axoneme. The greater diameter of the nucleus is at right angles to the longitudinal axis of the proximal centriole; the

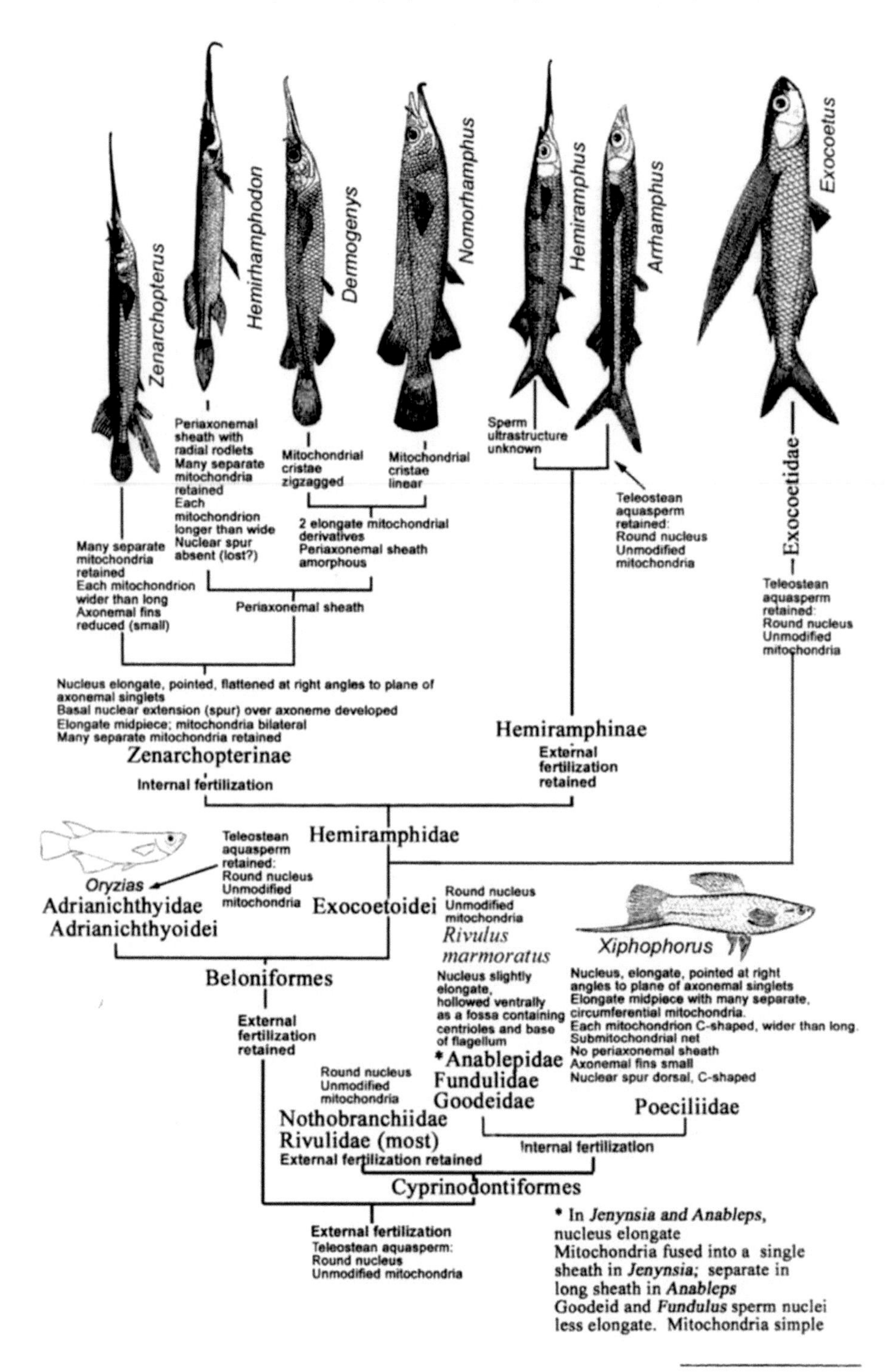

Fig. 14.21 Contd. ...

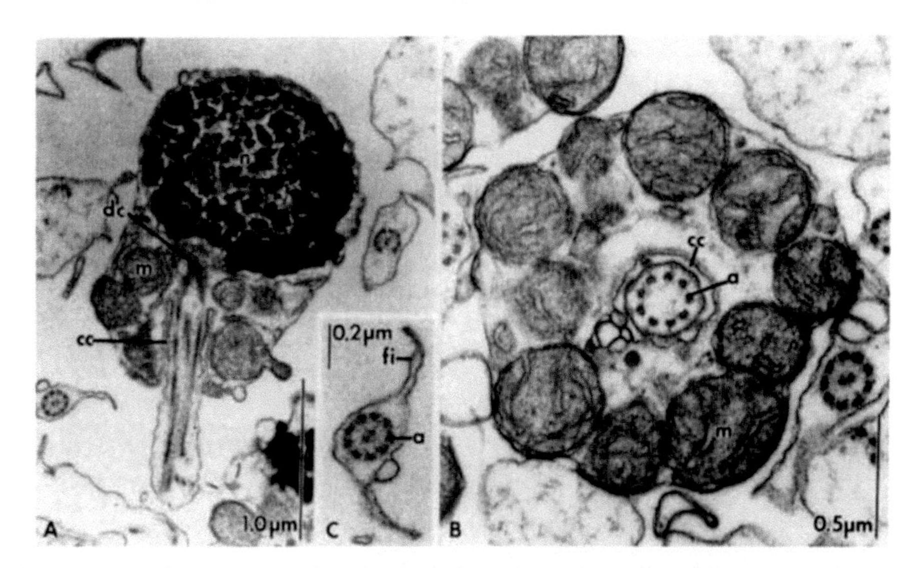

Fig. 14.22 *Arrhamphus sclerolepis*. **A**. Longitudinal section of the spermatozoon. **B**. Transverse section through the midpiece and the initial 9+0 region of axoneme. **C**. Transverse section of the axoneme, showing the two lateral fins. From Jamieson, B. G. M. and Grier, H. J. 1993. Hydrobiologia 271: 11-25, Fig. 1.

nucleus therefore appears tilted relative to the longitudinal axis of basal body and axoneme.

The plasma membrane investing the 9+2 flagellum is extended as a pair of axonemal fins approximately in the plane of the two central singlets and doublet radii 3-4 (Fig. 14.22C). The fins run longitudinally along a large portion of the flagellum, as indicated by their frequency in transverse sections. The fins show no particular apical differentiation. A region with 9 doublets but no central singlets intervenes, in the collar region, between the basal body and the axoneme proper (Jamieson 1991; Jamieson and Grier 1993).

Hemiramphidae, Hemiramphinae: *Hemiramphus balao*. The sperm of *H. balao* (Fig. 14.23) closely resembles that of *Arrhamphus* but the number of mitochondria has not been ascertained.

Fig. 14.21 Contd. ...

Fig. 14.21 Trends in the evolution of beloniform and cyprinodontiform spermatozoa. Illustrations of fishes, representative of their genera, are *Xiphophorus helleri* from Jordan (1907), *Arrhamphus brevis, Zenarchopterus dispar, Hemirhamphodon pogonognathus, Dermogenys orientalis, Nomorhamphus celebensis, Hemiramphus far* and *Exocoetus volitans* from Weber & DeBeaufort (1922) and *Oryzias latipes* from Nelson (1984). Greatly modified after Jamieson, B. G. M. and Grier, H. J. 1993. Hydrobiologia 271: 11-25, Fig. 6.

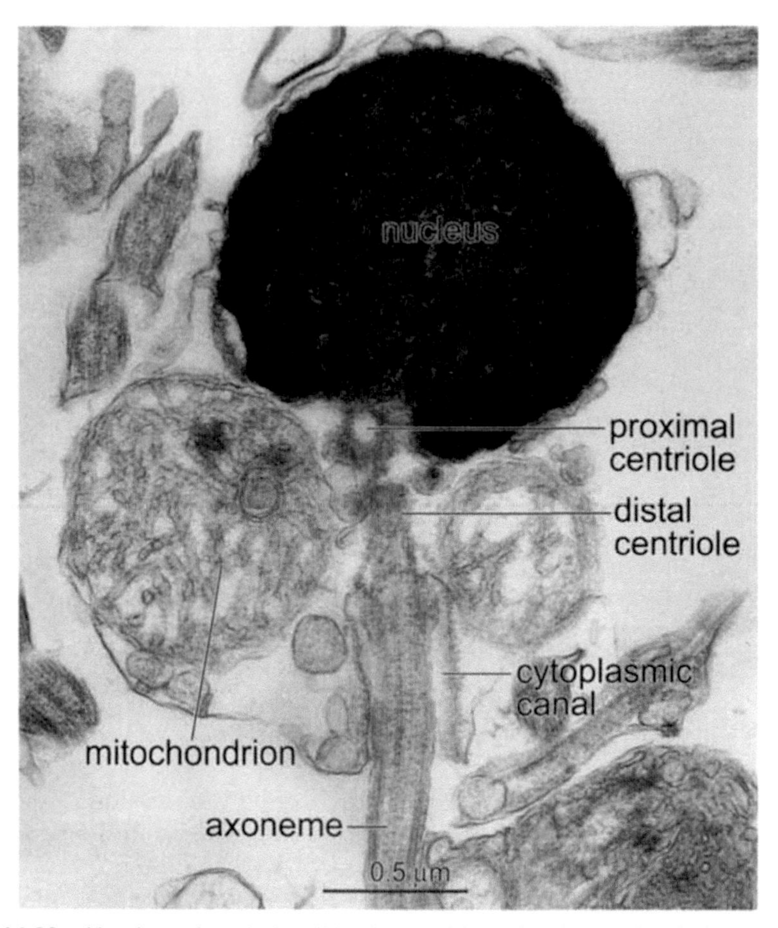

Fig. 14.23 *Hemiramphus balao* (Hemiramphidae, Hemiramphinae). Longitudinal section of a spermatozoon. Original, from an unpublished TEM negative courtesy of Professor Xavier Mattei.

14.7.2.4 Internally fertilizing belonoids (=exocoetoids)

Zenarchopterinae: Mature sperm of *Dermogenys pusillus* (Fig. 14.27), *Hemirhamphodon pogonognathus* (adult, Fig. 14.24; sperm Figs. 14.25, 14.26) and *Nomorhamphus celebensis* (Fig. 14.28) and late spermatids of *Zenarchopterus dispar* (Fig. 14.29) all internally fertilizing species, have been examined (Jamieson 1989; 1991; Jamieson and Grier 1993).

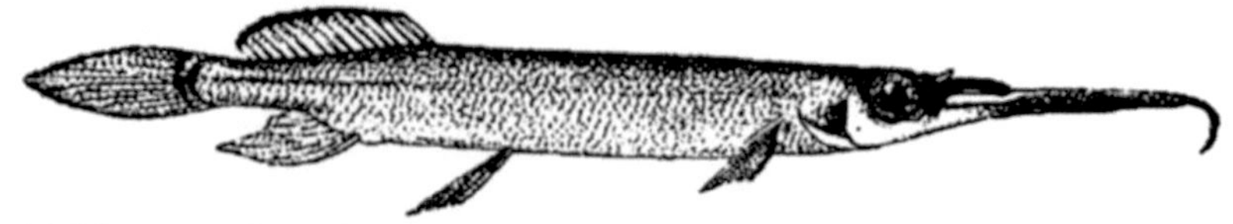

Fig. 14.24 *Hemirhamphodon pogonognathus* (Hemiramphidae, Hemiramphinae). After Weber, M. and De Beaufort, L. F. 1922. *The Fishes of the Indo-Australian Archipelago.* Brill, Leiden, Fig. 54.

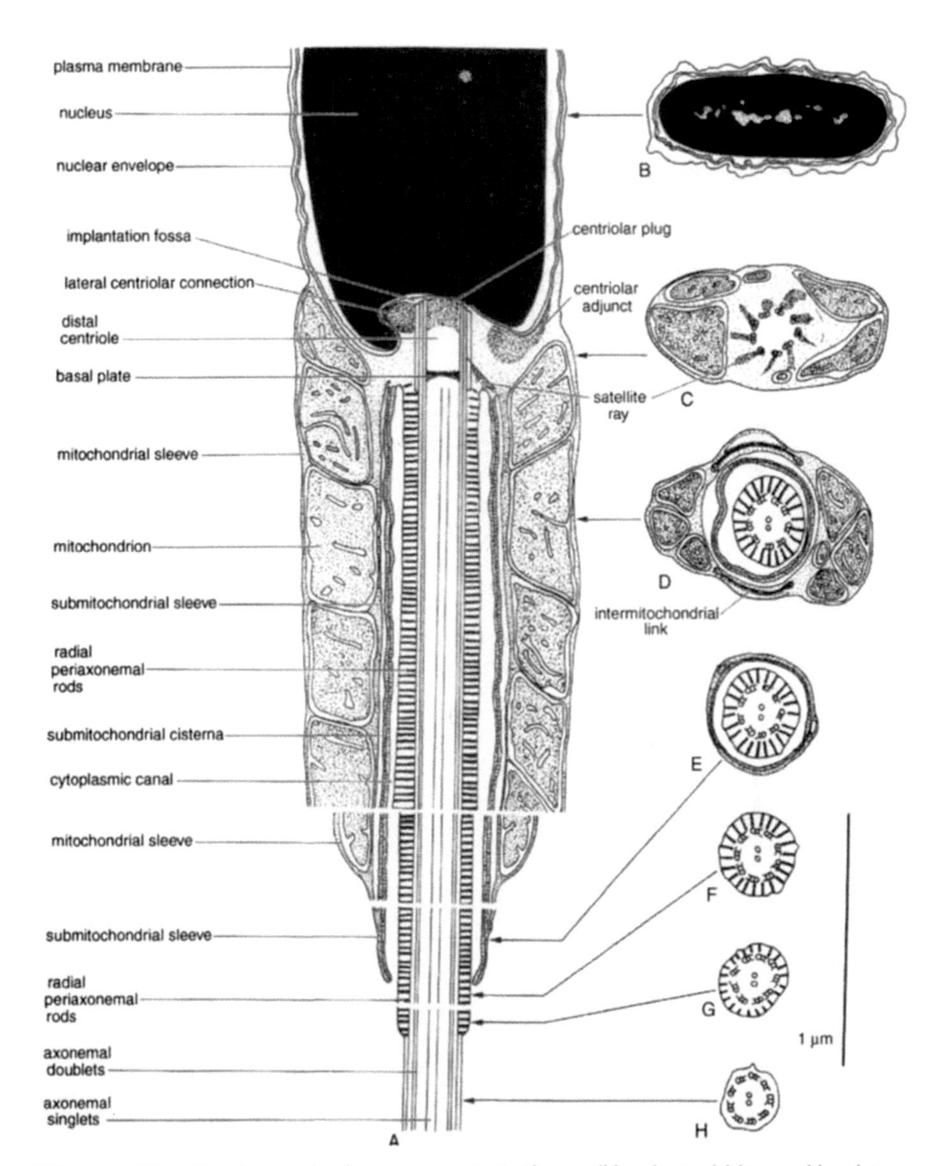

Fig. 14.25 *Hemirhamphodon pogonognathus* (Hemiramphidae, Hemiramphinae). Diagrammatic longitudinal and transverse sections of the spermatozoon. **A**. Frontal longitudinal section, showing the bilateral mitochondria. The posterior two portions are deduced from transverse sections. **B-H**. Transverse sections through the regions indicated in **A**. **B**. Nucleus. **C**. Centriole. **D**. Midpiece. **E**. Region of united mitochondrial and submitochondrial sleeves. **F**, **G**. Successive posterior regions, behind the sleeve, with periaxonemal rods. **H**. The simple axoneme. From Jamieson, B. G. M. 1989. Gamete Research 24: 247-260, Fig.1.

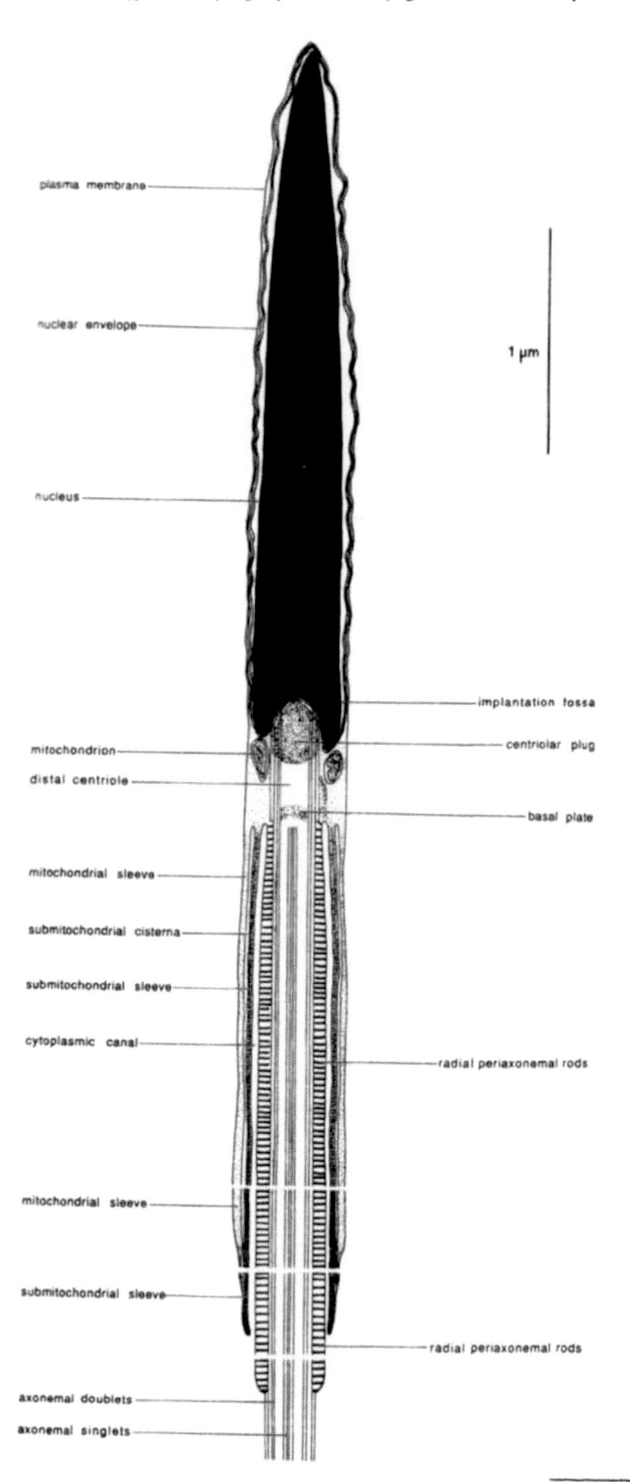

Fig. 14.26 Contd. ...

Nucleus: Sectional views show that in the four zenarchopterine species the sperm nucleus has the form of an approximately parallel-sided round-tipped blade as indicated by longitudinal sections (Figs. 14.27A,B,D; 14.28A,C; 14.29A,B,C). Its strongly depressed form is clearly seen in transverse sections (Figs. 14.27f-J; 14.28D-I; 14.29D). Nuclear flattening is always at right angles to the plane of the two central singlets of the axoneme or within a few degrees of this. The base of the nucleus is indented by an implantation fossa. This is least developed in *Hemirhamphodon* in which it is less than one tenth of the length of the nucleus. The fossa is strongly eccentric but its disposition is different in *Hemirhamphodon*, on the one hand, compared with *Dermogenys* (Fig. 14.27B-D), *Nomorhamphus* (Fig. 14.28A-C) and *Zenarchopterus* (Figs 14.29A-C), on the other. In *Hemirhamphodon* it appears eccentric only in section through the broader, frontal plane, whereas in the other three species its eccentricity is seen in sagittal section of the nucleus where it incises the 'posteroventral' face of the nucleus (Figs. 14.27A-C; 14.28B,C; 14.29A,B). Only the anterior half of the distal centriole (basal body of the flagellum) is contained within the fossa in *Hemirhamphodon* and *Zenarchopterus* (Fig. 14.29A-C) but in *Dermogenys* (Fig. 14.27B-D) and *Nomorhamphus* (Fig. 14.28A-C) the initial part of the flagellum is also included. In *Dermogenys and Nomorhamphus,* therefore, the nucleus has a long, spur-like continuation behind, and overarching, the anterior region of the flagellum and its basal body. The spur is about 0.23 of the total length of the nucleus in *Dermogenys* and about 0.3 of its length in *Nomorhamphus.* In *Zenarchopterus* this posterior extension is bulkier and less elongate while in *Hemirhamphodon* the nucleus does not extend posteriorly over the axoneme. Although the spur is 'dorsal' in terms of the nomenclature adopted for the nucleus, it is approximately in the radius of axonemal doublet number 3 (in Afzelius notation) and therefore lateral in terms of the usual convention of regarding axonemal doublet number 1 as dorsal (Figs. 14.27IJ; 14.28H, I; 14.29D). In this account, as in Jamieson and Grier (1993) from which it is derived, I term flattening of the nucleus dorsoventral and the spur as dorsal, however, and impose this orientation on the sperm as a whole.

The length of the nucleus is 3.13-3.35, mean 3.22 μm (number of sperm = 3) in *Hemirhamphodon;* 3.3-4.1, mean 3 .65 μm *(n* = 4) in *Dermogenys;* 3.0 μm *(n* = 2) in *Nomorhamphus;* and 2.4- 2.7 μm, mean 2.5 μm *(n* = 4) in *Zenarchopterus.* In *Hemirhamphodon* and *Zenarchopterus* (Fig. 14.29A-E), at maturity the chromatin is strongly electron-dense with the exception, in *Hemirhamphodon,* of occasional

---

Fig. 14.26 Contd. ...

Fig. 14.26 *Hemirhamphodon pogonognathus* (Hemiramphidae, Hemiramphinae). Diagrammatic longitudinal of the spermatozoon, at right angles to that shown in the previous figure. Note the absence of mitochondria, excepting rudiments in the vicinity of the centriole, in this plane. The variably delineated intermitochondrial link is not shown. From Jamieson, B. G. M. 1989. Gamete Research 24: 247-259, Fig. 2.

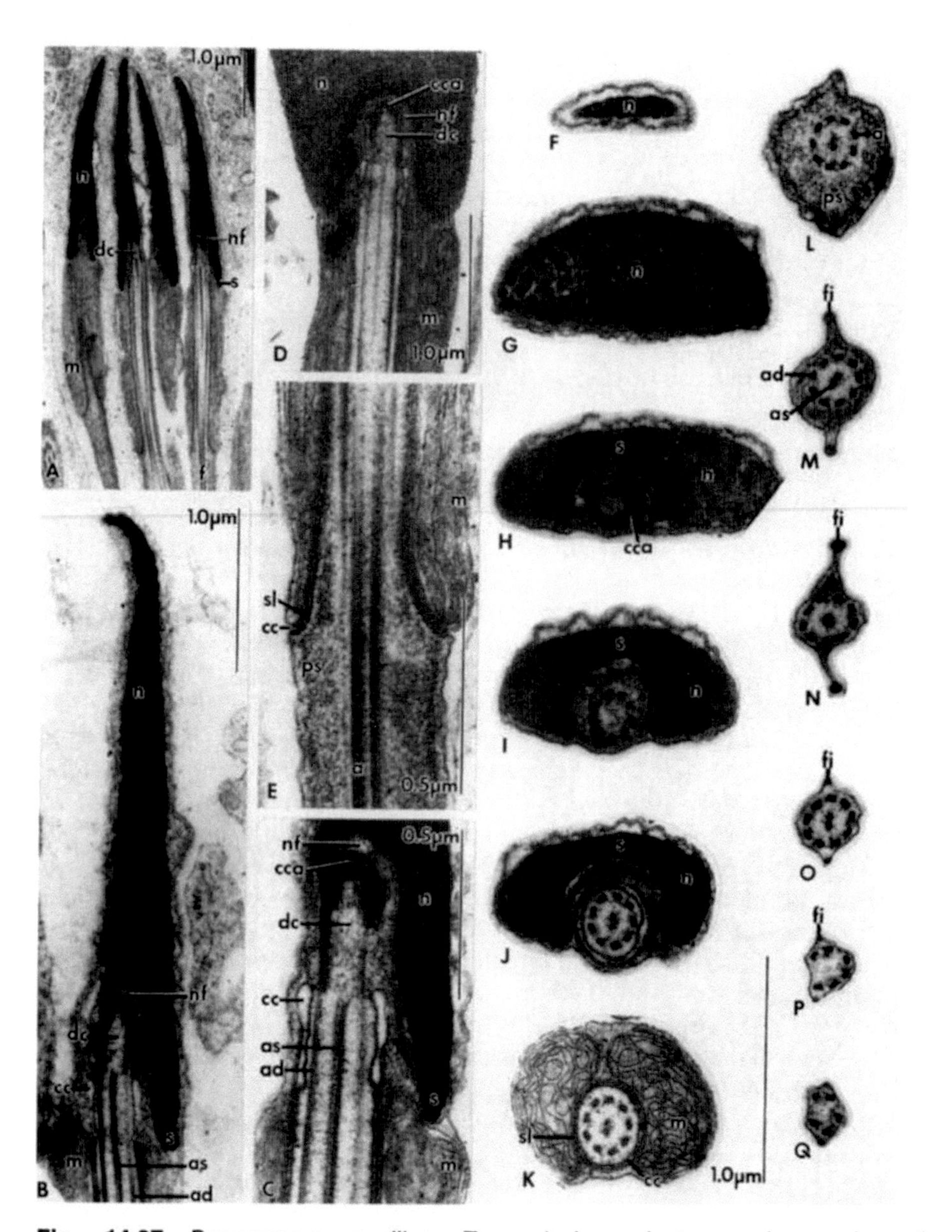

Fig. 14.27 *Dermogenys pusillus.* Transmission electron micrographs of spermatozoa. **A-E**. Longitudinal sections. **A**. Four spermatozoa showing entire nuclear and midpiece regions and anterior region of the axoneme. Note the 'dagger-like' profile of the nucleus when sectioned in or near the plane of the central axonemal singlets. **B**. Nucleus and adjacent region of midpiece. **C**. Detail of the nuclear-centriolar junction. **D**. Base of the nucleus showing flattening apparent when sectioned in the plane at right angles to that of the central axonemal

Fig. 14.27 Contd. ...

small clear lacunae. In *Dermogenys* (Fig. 14.27A-J) and *Nomorhamphus* (Fig. 14.28A-I) the chromatin consists of closely packed coarse granules with pale matrix material between.

Midpiece: As frequently occurs during teleostean spermiogenesis, the midpiece, grows posteriorwards as a mitochondrial sleeve around the axoneme (Figs. 14.27A-E, K; 14.28C, J; 14.29A-C, F). In all four zenarchopterid species the sleeve attains an unusually great length compared with externally fertilizing teleost sperm here exemplified by the hemiramphine *Arrhamphus*. The mitochondrial sleeve is separated from the axoneme by the cytoplasmic canal (periaxonemal space). The inner wall of the mitochondrial sleeve, lining the canal, is modified in all species, having the appearance of two parallel membranes, with some densification, and may be termed the submitochondrial dense layer. In *Hemirhamphodon*, but not recognizably in the other three species, this thickening is separated from the mitochondria by a narrow cisterna and has been referred to as the submitochondrial sleeve. The cisterna does not, however, open posteriorly, unlike the cytoplasmic canal, and therefore a single sleeve, the mitochondrial sleeve, including the submitochondrial dense layer, is recognized (Figs. 14.25, 14.26). In the four species the plasma membrane of the head continues posteriorly over the outer surface of the mitochondrial sleeve and turns anteriorly to line the inner surface of the sleeve, as far forward as the basal plate of the distal centriole, before turning posteriorly to cover the axoneme.

In *Hemirhamphodon*, *Dermogenys*, *Nomorhamphus*, and, although the definitive condition is unknown, *Zenarchopterus*, the arrangement of the mitochondria is unique for investigated atherinomorphs (Jamieson 1989). The mitochondria are grouped bilaterally, on opposing sides of the axoneme, on each side of a plane which passes through doublet 3 and between doublets 7 and 8. In *Hemirhamphodon* mitochondrial segregation is extreme and mitochondria, are absent 'dorsally and 'ventrally', in this plane (Fig. 14.25). As a result, in *Hemirhamphodon*, mitochondria are fully visible in frontal

---

Fig. 14.27 Contd. ...

singlets. **E**. Posterior end of the midpiece and of the two mitochondrial derivatives, showing virtual occlusion of the cytoplasmic canal by widening of the cytoplasmic zone around the anterior region of the axoneme (peri-axonemal sheath). **F-Q**. Transverse sections in 'anteroposterior' sequence. **F-J** show the depression of the nucleus in the plane at right angles to that of the central axonemal singlets and (**H-J**) the basal nuclear fossa and accompanying 'dorsal spur'. **K**. Shows the bilateral distribution of mitochondria relative to the plane of the axonemal singlets. **L**. Section through peri-axonemal sheath. **M-Q**. Sections through the axoneme, showing posteriorwards reduction of the axonemal fins and progressive disruption of the 9+2 pattern of microtubules. From Jamieson, B. G. M. 1991. *Fish Evolution and Systematics: Evidence from Spermatozoa*. Cambridge University Press, Cambridge. Jamieson, B. G. M. and Grier, H. J. 1993. Hydrobiologia 271: 11-25, Fig. 2. With kind permission of Springer Science and Business Media.

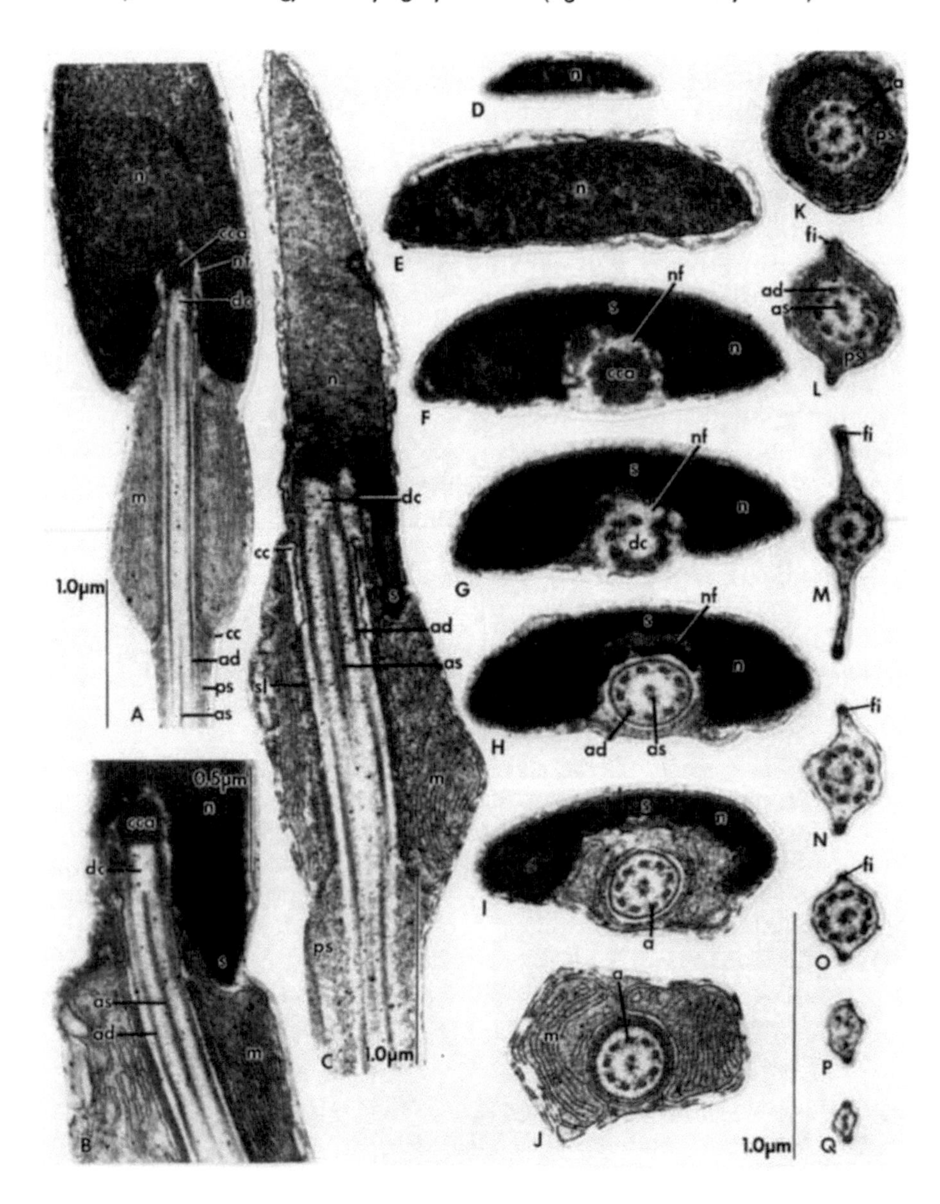

Fig. 14.28 *Nomorhamphus celebensis* (Hemiramphidae, Zenarchopterinae). Transmission electron micrographs of spermatozoa. **A-C**. Longitudinal sections. **A**. Base of the nucleus showing flattening apparent when sectioned in the plane at right angles to that of the central axonemal singlets. **B**. Detail of the nuclear-centriolar junction. **C**. Entire nuclear and midpiece region and anterior region of the axoneme. Note the 'dagger-like' profile of the nucleus when sectioned in or near the plane of the central axonemal singlets and virtual occlusion of the cytoplasmic canal by widening of the cytoplasmic zone around the anterior region of the

Fig. 14.28 Contd. ...

sections, where the mitochondrial sleeve is wide, but are not seen in sagittal sections, where the sleeve is narrow. In this species, in cross section two to four mitochondria are seen on each side, usually of very unequal sizes. Each is angular in section, approximately triangular to trapezoidal and has well developed irregularly arranged cristae and a moderately dense intercristal matrix. The individuality of each mitochondrion is maintained in longitudinal section of the sperm in which each is usually two or more times longer than wide; there are several discrete mitochondria in longitudinal succession on each side. In *Dermogenys*, although the mitochondria are bilateral they are contiguous or nearly so dorsally. *Dermogenys* and *Nomorhamphus* differ from *Hemirhamphodon* in having only a single very elongate mitochondrion, which may be termed a mitochondrial derivative, on each side (Figs. 14.27A; 14.28A, C). The later spermatids of *Zenarchopterus* resemble *Hemirhamphodon* sperm in having discrete mitochondria in longitudinal succession (Fig. 14.29A-C) but in cross section as many as nine are seen spaced around the axoneme (Fig. 14.29F), much as occurs in the spermatozoon of *Arrhamphus*. However, in transverse sections in which the nuclear spur is sectioned (Fig. 14.29E), this is seen to interrupt the circle of mitochondria in the radius passing between doublet 8 and 9, confirming a bilateral arrangement. In some, tangential longitudinal sections (Fig. 14.29G) individual mitochondria can be seen to have lost their initially spherical form and to have elongate transversely around part of the circumference of the axoneme. It is unlikely from the advanced state of the spermatids that the mitochondria lose their discrete condition.

The arrangement of the cristae in the mitochondria differs remarkably in the four internally fertilizing species. In the numerous mitochondria of *Hemirhamphodon*, the few cristae are chiefly transverse and in transverse section of the midpiece are vertical and parallel; in *Zenarchopterus* three to five or more cristae are chiefly parallel to the long axis of the mitochondrion and therefore transverse relative to the whole spermatozoon. In *Dermogenys and Nomorhamphus*, although the sperm are very similar, there are major differences in the cristae. There is no reason at present to believe that the differences are artifacts of fixation. In *Dermogenys*, each mitochondrial

---

Fig. 14.28 Contd. ...

axoneme (peri-axonemal sheath). **D-Q**. Transverse sections in 'anteroposterior' sequence. **D-I** show the depression of the nucleus in the plane at right angles to that of the central axonemal singlets and (**F-I**) the basal nuclear fossa and accompanying 'dorsal spur'. **J**. Shows the bilateral distribution of mitochondria relative to the plane of the axonemal singlets. Cristae are linear in contrast with their zigzagged appearance in *Dermogenys*. **L**. Through peri-axonemal sheath. **M-Q**. Sections though the axoneme, showing posteriorwards reduction of the axonemal fins and progressive disruption of the 9+2 pattern of microtubules. From Jamieson, B. G. M. and Grier, H. J. 1993. Hydrobiologia 271: 11-25, Fig. 3. With kind permission of Springer Science and Business Media.

derivative shows several (commonly about five) slightly wavy longitudinal cristae, each running most of the length of the derivative (Fig. 14.27A-E), but in transverse section of the midpiece and of the derivative, the mitochondria have the appearance of a continuous zigzag, with rounded angles, meandering through the matrix (Fig. 14.27K). In *Nomorhamphus*, in contrast, the cristae are almost straight; several run the length of the longitudinal profile of the mitochondrial derivative (Figs 14.28A-C) and, in transverse section of the organelle, appear as parallel parenthesis-shaped structures, orientated vertically (Fig. 14.28J). An 'intermitochondrial link' extending between the mitochondrial masses of opposing sides, near the adaxial border of the mitochondrial sleeve on each side of the axoneme in *Hemirhamphodon* (Fig. 14.25) has not been identified in the other species, though a dense horizontal bracket shaped structure seen in transverse section of the nuclear spur in *Nomorhamphus* (Fig. 14.28H), but anterior to the mitochondrial derivatives, is conceivably equivalent. In the single longitudinal section of *Hemirhamphodon* sperm in which a complete profile of the midpiece has been obtained, the approximate length of the mitochondrial sleeve (measured along the mitochondria only) is 3.5 μm. It is 1.8 μm in *Nomorhamphus* (n = 1), 2.7-2.9 μm in *Dermogenys* (n = 2), and 2.9-3.0 μm in *Zenarchopterus* (n = 2).

Centriolar apparatus: In all four zenarchopterid species the distal centriole extends into the implantation fossa and is capped by a dense mass (centriolar cap) which possibly contains remnants of the otherwise unrecognizable proximal centriole (Figs. 14.27A-D; 14.28A-C; 14.29A-C). The distal centriole consists of nine elements, which only in *Hemirhamphodon* have been resolved into triplets of microtubules; it is probable that more favorable fixation would reveal triplets in the other species. The distal centriole is continuous with, and in the same longitudinal axis as, the doublets of the axoneme. A conspicuous basal plate at the base of the centriole in *Hemirhamphodon* (Figs. 14.25, 14.26) has not been identified in the other species. In *Hemirhamphodon* each of the triplets gives rise, near its base, to a stout posterolaterally directed satellite ray tilted, in transverse section of the centriole, at about 30° to the radius, and in the same direction as the dynein arms of the axoneme. Satellites have also been identified in *Zenarchopterus* (Fig. 14.29D)

Fig. 14.29 *Zenarchopterus dispar* (Hemiramphidae, Zenarchopterinae). Transmission electron micrographs of late spermatid. **A** and **B**. Longitudinal section (LS) of the nuclear and midpiece regions and anterior region of the axoneme. Note the 'dagger-like' profile of the nucleus when sectioned in or near the plane of the central axonemal singlets. **C**. LS nucleus and much of midpiece, showing flattening apparent when nucleus is sectioned in the plane at right angles to that of the central axonemal singlets. **D**. TS through the nuclear spur which interrupts the mitochondrial circlet. **E**. Transverse section (TS) of the nucleus. **F**. TS

Fig. 14.29 Contd. ...

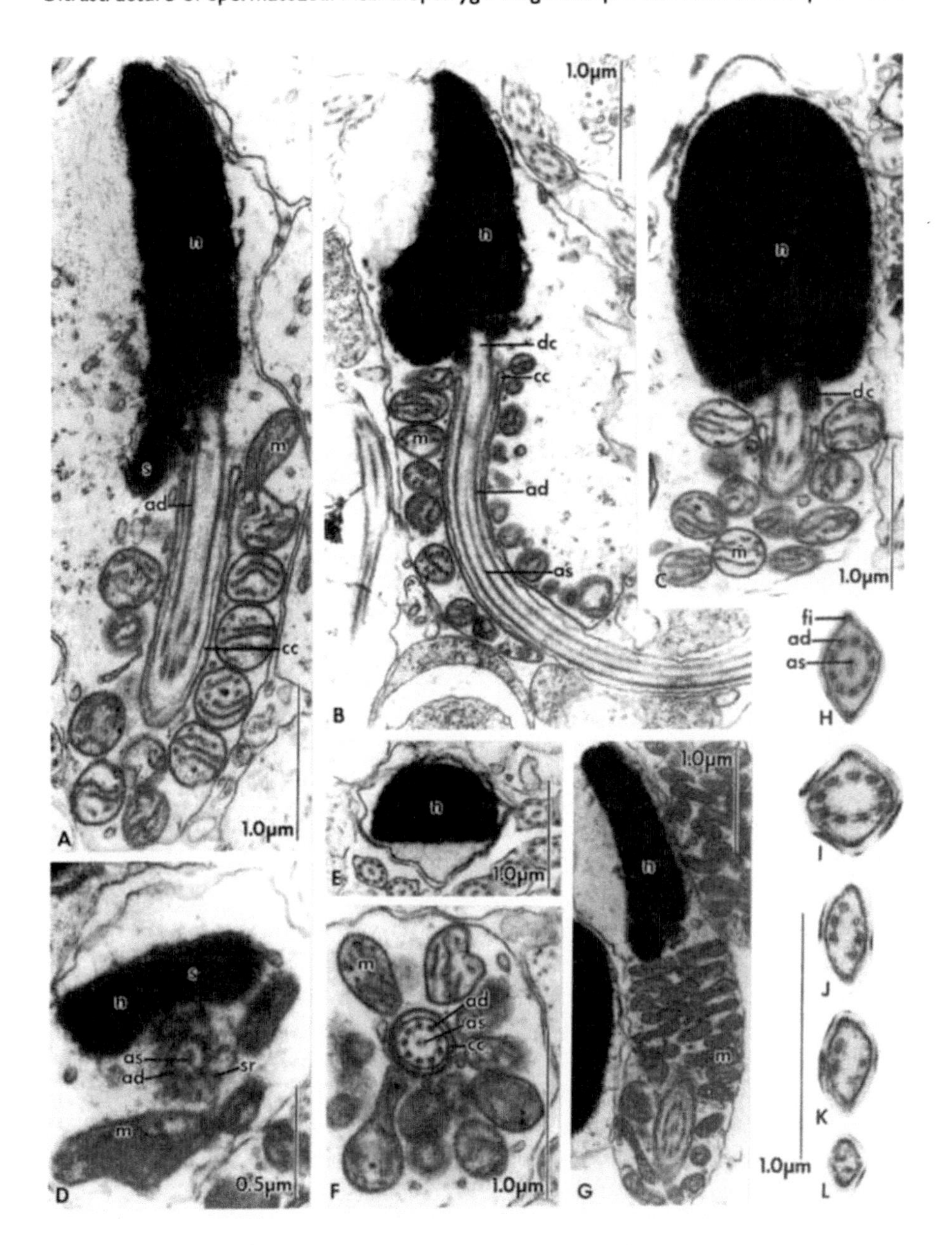

midpiece. **G**. LS midpiece showing mitochondria wider than long. **H-L**. TS through the axoneme, showing axonemal fins, reduced relative to *Dermogenys* and *Nomorhamphus*, and progressive disruption of the 9+2 pattern of microtubules. From Jamieson, B. G. M. 1991. *Fish Evolution and Systematics: Evidence from Spermatozoa*. Cambridge University Press, Cambridge. Jamieson, B. G. M. and Grier, H. J. 1993. Hydrobiologia 271: 11-25, Fig. 4. With kind permission of Springer Science and Business Media.

Axoneme: The axoneme has the microtubular pattern usual for teleostean sperm of nine doublets, each with two dynein arms, and two inner singlets (Figs. 14.27J-0; 14.28H-O; 14.29F, H). In *Hemirhamphodon,* from the anterior limit of the cytoplasmic canal, near the anterior end of the midpiece, to well behind the midpiece, the axonemal doublets are separated from the flagellar plasma membrane by a broad zone of cytoplasm. In cross section of the axoneme, this cytoplasmic zone is seen to contain 23 radial rodlets (decreasing slightly in number posteriad) (Fig. 14.25D-G). These rodlets, which are negative for glycogen in the Thiery test, are not seen in the other species. In *Dermogenys* and *Nomorhamphus,* a wide zone of cytoplasm is present around the doublets in the portion of the flagellum immediately behind the midpiece (Figs. 14.27L; 14.28K). In both species, the submitochondrial dense layer curves out laterally in contact with the shoulder-shaped commencement of the cytoplasmic zone (Figs. 14.27E; 14.28A-C). The opening of the cytoplasmic canal to the exterior is almost occluded by this approximation of the two structures. In both species the submitochondrial layer appears to consist of two parallel membranes. The plasma membrane of the axoneme, from the anterior limit of the cytoplasmic canal to the posterior limit of the canal is thickened and dense (Figs. 14.27E; 14.28A, C), its electron density being more pronounced that that of the submitochondrial layer. No particular widening of the cytoplasm around the axonemal doublets is apparent in the late spermatids of *Zenarchopterus* and it seems unlikely that this develops by maturity.

In *Zenarchopterus, Dermogenys* and *Nomorhamphus,* as in *Arrhamphus,* there is a pair of flagellar fins in the plane of the central singlets. In *Dermogenys* and *Nomorhamphus,* the tips of the fins, as seen in transverse section of the flagellum, are filled with dense material and the overlying plasma membrane is also dense, giving an appearance in *Nomorhamphus* resembling a matchhead while in *Dermogenys* there is a circular swelling. Each density represents a continuous longitudinal rod-like thickening of the free edge of the fin. The length (radial extent) of a fin is variable along its longitudinal course but is maximally about 0.25 μm, a little wider than the 0.2 μm diameter of the doublet circlet in the axoneme, in both genera. In *Zenarchopterus* sperm, the fins are not so well defined, being wide-based and short, and the apical density is less developed, forming merely a dense cap of thickened membrane. It seems unlikely that at full maturity they reach the dimensions seen in *Dermogenys* and *Nomorhamphus.* Axonemal fins are absent in *Hemirhamphodon.* Further posteriorly along the *Hemirhamphodon* flagellum the radial rods are absent from the axoneme and the plasma membrane is closely approximated to the doublets. In all species, at the posterior end of the axoneme the arrangement of the microtubules becomes progressively disrupted (Figs. 14.27P, Q; 14.28P, Q; 14.29I-L). The B microtubules become open, C-shapes before ending.

Conclusion for beloniforms and cyprinodontiforms: Trends in the evolution sperm of the Hemiramphidae are indicated in Fig. 14.21, adapted from a parsimony analysis by Jamieson and Grier (1993) to conform with current taxonomy. A trend occurring homplasically in internally fertilizing cyprinodontiforms (poeciliids and anablepids) and zenarchopterines (*Zenarchopterus*, *Dermogenys* and *Nomorhamphus*) is elongation but chiefly flattening of, and development of a 'ventral' groove, containing the centrioles and base of the flagellum, on the nucleus. This also occurs in *Jenynsia* and *Anableps*, both now in the Anablepidae. (Although elongation and flattening of the nucleus is highly developed in *Hemirhamphodon*, it has a short nuclear fossa.) However, some internally fertilizing sperm are little modified. Thus *Rivulus marmoratus* (Rivulidae) has a round-headed sperm with unmodified mitochondria resembling a simple teleostean aquasperm, suggesting that internal fertilization in this species has been acquired relatively recently. Intermediate between the simple teleostean aquasperm and the elongate modified sperm of poeciliids are the sperm of *Fundulus* (Fundulidae) and *Xenotoca* (Goodeidae) in which there is only moderate elongation of the nucleus, with development of the ventral centriolar groove, and the mitochondria remain few and simple.

For computational purposes (Jamieson and Grier 1993) arbitrarily attributed a single tier of mitochondria to the basic teleostean aquasperm, though it is possible that more than one tier is basic. It is noteworthy that the aquasperm of the externally fertilizing hemiramphine studied, *Arrhamphus sclerolepis*, already possesses more than one tier of mitochondria, as is characteristic of the internally fertilizing exocetoids, the Zenarchopterinae, though modified to a pair of longitudinal mitochondrial derivatives in *Dermogenys* and *Nomorhamphus*. The possession and form of the mitochondrial derivatives in the latter two genera appears to be an important synapomorphy between them. Although poeciliids and zenarchopterines have independently elongated the midpiece and multiplied the number of mitochondria in it, a noteworthy difference between the two groups is the circumferential distribution of mitochondria in poeciliids whereas in zenarchopterines the mitochondria or mitochondrial derivatives are bilateral. Even where most nearly circumferential, in *Zenarchopterus dispar*, they are interrupted 'dorsally' by the posterior extension of the nucleus. In the plesiomorphic condition for teleosts, as in *Arrhamphus*, the mitochondria are subspheroidal. They have departed radically from the subspheroidal form in the internally fertilizing fishes and have multiplied along the lengthened midpiece. Although in longitudinal section of poeciliid sperm the individual mitochondria appear to be elongate, they are seen in cross section of the midpiece to be C-shaped and considerably more extensive circumferentially than longitudinally. In *Zenarchopterus* they have independently also become more extensive circumferentially than longitudinally. In *Hemirhamphodon* they are moderately elongate, being two or three times as long as wide. In *Dermogenys* and *Nomorhamphus*, as noted above, they have been modified as a pair of mitochondrial derivatives, an apparently monophyletic event. A wide cytoplasmic zone around the anterior region of the axoneme, the periaxonemal sheath, appears to have developed monophyletically in the ancestry of *Hemirhamphodon*, *Dermogenys* and *Nomorhamphus*. *Hemirhamphodon* has taken this development further by the acquisition of radial rodlets, in a circlet of 23 with a longitudinal repeat. The periaxonemal sheath is a distinctive zenarchopterine development, though apparently plesiomorphically absent in *Zenarchopterus*, and is not seen in poeciliids. A distinctive development in poeciliids, and apparently in *Jenynsia*, is the submitochondrial net though the submitochondrial dense layer of zenarchopterine sperm is possibly its homologue.

A pair of fin-like extensions of the flagellum, approximately in the plane of the central singlets of the axoneme, is a synapomorphy of, and simultaneously plesiomorphic for, the Actinopteri, i.e. Actinopterygii excepting the Cladistia (see Jamieson 1991, and **Chapter 7**). Poeciliids have greatly reduced the fins. Hemiramphines have retained well-developed fins and fertilization is external, in the only taxon studied, *Arrhamphus*. In *Dermogenys* and *Nomorhamphus* there is some modification of the tips of the fins. Reduction of the fins has occurred, apparently independently, in *Zenarchopterus*, in which they are small, and in *Hemirhamphodon* in which they are absent.

In basic teleostean aquasperm, as also in *Arrhamphus*, the nucleus does not extend posteriorly over the base of the axoneme. Such an extension, loosely termed a 'spur', has developed in poeciliids and independently in zenarchopterines. Whereas in poeciliids the posterior extension of the nucleus is a C-shaped structure embracing almost the entire circumference of the axoneme, in zenarchopterines it is a 'dorsal' plate. Its absence in *Hemirhamphodon* may represent a loss but it is possible that this genus basically lacks this extension.

Bearing in mind that modifications from the basic teleostean aquasperm have occurred independently in poeciliids and zenarchopterines, there are similar and different responses to internal fertilization in the two families. Similar changes have been elongation and flattening of the nucleus in the same plane, elongation of the midpiece and multiplication of the number of mitochondria, though with further modification of these two a pair of elongate mitochondrial derivatives in two zenarchopterines; a tendency to reduction of the axonemal fins; and development of a posterior extension of the nucleus over the base of the axoneme. Different responses have been a circumferential distribution of mitochondria in poeciliids against a bilateral arrangement in zenarchopterines; development of a periaxonemal sheath of cytoplasm exclusively in zenarchopterines, above *Zenarchopterus*; and of a submitochondrial net only in poeciliids (also seen independently in *Jenynsia*).

All of these modifications relative to the aquasperm condition are regarded (Jamieson 1989, 1991; Jamieson and Grier 1993; see also Javonillo and Burns, **Chapter 17**) as having been occasioned by the adoption of internal fertilization. They are thus directly attributable to a change in fertilization biology but to what extent their details are constrained by features of the genome peculiar to poeciliids, zenarchopterines or atherinomorphs or are demanded by minute differences in fertilization biology, or by a combination of the two, seems indeterminable. The functional significance of modifications of the sperm is partly treated in Jamieson (1991). Some of the modifications, particularly those pertaining to elongation of the nucleus, elongation of the midpiece, and development of lateral elements to the axoneme, are seen in the phylogenetically very distinct Chondrichthyes (see Jamieson 1991, 2005) but there have different expressions. Differences in sperm morphology between atherinomorph groups and between these and chondrichthyans may be due to phylogenetic constraints and/or subtle differences in fertilization biology (Jamieson 1989, 1991; Jamieson and Grier 1993; see also Javonillo and Burns, **Chapter 17**).

14.7.2.5 Family Scomberesocidae

These are the sauries, in tropical and temperate seas (Nelson 2006). From a morphocladistic analysis Aschliman *et al.* (2005) concluded that that sauries are most closely related to needlefishes (Belonidae), supporting the historical concept of a superfamily Scomberesocoidea as a monophyletic assemblage. The analysis strongly supported Zenarchopterinae as a valid group sister to the Scomberesocoidea.

***Cololabis saira*:** The sperm of *C. saira*, the Pacific Saury, has been described (Hara *et al.* 1997; Hara. and Okiyama 1998). It is an oviparous, externally fertilizing, non-guarding species but has a specialized form of fertilization in

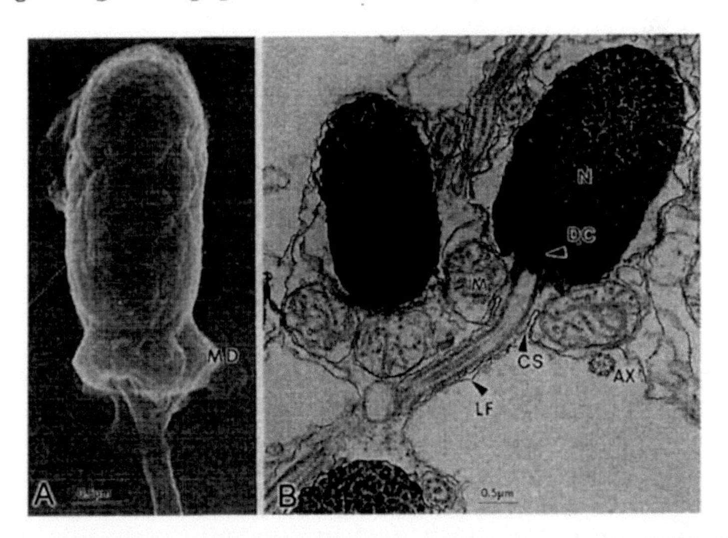

Fig. 14.30 *Cololabis saira* (Scomberesocidae) spermatozoa. **A**. SEM. **B**. TEM longitudinal sections. AX, axoneme; CS, cytoplasmic sleeve; DC, distal centriole; LF, lateral flagellar fin; M, mitochondrion; MD, midpiece; N, nucleus. Adapted from Hara, M., Kurita, Y., Watanabe, S., Watanabe, Y. and Okiyama, M. 1997. Bulletin of Tohoku National Fisheries Research Institute 59: 139-147, Figs. 12 and 13. With permission.

which it appears that filaments which attach the egg to the substrate may also aid entry of the sperm through the micropyle.

The spermatozoon (Fig. 14.30A,B) is an anacrosomal aquasperm but, in apparent correlation with the modified fertilization mode, has an elongate, oblong ovoid ("bell-shaped"), nucleus (about twice as long as wide). The chromatin is electron-dense, consisting of evenly distributed dense masses. There is only a shallow basal fossa. The short midpiece contains about six mitochondria encircling the basal body of the flagellum as a sleeve. No proximal centriole has been observed. The flagellum bears lateral fins.

Remarks: Hara and Okiyama (1998) note that elongation of the nucleus in *Cololabis saira* is an apomorphic departure from the basic *Arramphus*-like condition though less conspicuous than that seen in the introsperm of *Hemirhamphodon*.

14.8 CHAPTER SUMMARY

The Mugilomorpha, represented by the Mugilidae, are seen from molecular analysis to be the plesiomorphic sister-group of the Atherinomorpha. Mugilids are externally fertilizing and have sperm resembling Type II sperm of the Perciformes (including the presence of ITDs) or intermediate between Types I and II. It is suggested that the Type II sperm represent parallelism with perciforms by virtue of genetic relationship (paramorphy).

In the Atherinomorpha, the Cyprinodontiformes are the apomorphic sister-group of a clade contain the Atheriniformes+Beloniformes in molecular analysis. All atherinomorphs are united by the unique telogonic development of the ovary. Atheriniform families which have been examined for sperm ultrastructure are the Atherinopsidae, Atherinidae and Melanotaeniidae. Of these only *Labidesthes sicculus vanhyningi*, in the Atherinopsidae, and the (uninvestigated) Phallostethidae are known to be internally fertilizing. *L. s. vanhyningi* has a Type I spermatozoon modified only in the deep nuclear fossa. The sperm of genera of the Atherinidae and Melanotaeniidae are discussed. The most plesiomorphic atheriniform sperm studied appears be that of the atherinid *Craterocephalus marjoriae*, with perpendicular centrioles. A more advanced Type I sperm is that of *Iriatherina*, with parallel centrioles, in which absence of a nuclear fossa may be a synapomorphy with *Melanotaenia*, thus supporting current placement in the Melanotaeniidae.

Reproductive modes in the Cyprinodontiformes are exceptionally varied, with oviparity, ovoviparity (in *Rivulus marmoratus* with functional hermaphroditism) and viviparity. Within the order, insemination or internal fertilization are known to occur in four families: Rivulidae, Goodeidae, Anablepidae and Poeciliidae (Poeciliinae). Although internally fertilizing, Rivulidae and Goodeidae, lacking true intromittent organs, have simple round-headed sperm but those of the Anablepidae (*Anableps anableps* and *Jenynsia lineata*), with a tubular intromittent organ, have more modified sperm with somewhat elongate nuclei and midpieces. The internally fertilizing Poeciliidae (Poeciliinae) have taken elongation of the nucleus, development of a ventral nuclear fossa containing the centrioles, and elongation of the midpiece further. Although poeciliines possess a gonopodium the sperm are packaged in spermatozeugmata.

In the order Beloniformes fertilization is external or internal. Where it is external the spermatozoa are correspondingly round-headed aquasperm, as in the Adrianichthyidae (*Oryzias latipes*), the Exocoetidae (*Fodiator acutus*) and some Hemiramphidae (*Arrhamphus sclerolepis, Hemiramphus balao*). The sperm of the internally fertilizing belonoids studied, all in the Zenarchopterinae, are profoundly modified (*Dermogenys pusillus, Hemirhamphodon pogonognathus, Nomorhamphus celebensis* and (late spermatids) *Zenarchopterus dispar*). Nuclei are strongly elongate and mitochondria in different arrays form elongate midpieces. In *Zenarchopterus, Dermogenys* and *Nomorhamphus* (as in *Arrhamphus*) there is a pair of flagellar fins in the plane of the central singlets. The flagellum in *Hemirhamphodon* lacks fins but the axoneme is unique in being surrounded by 23 radial rodlets. *Cololabis saira*, in the belonoid family Scomberesocidae, has an aquasperm but the specialized form of external fertilization involving egg filaments which has been inferred correlates with approximate doubling of the length of the nucleus. The fertilization biology and evolution of spermatozoa in the Atheriniformes, Beloniformes and Cyprinodontiformes spermatozoa is discussed.

14.9 LITERATURE CITED

Able, K. W. 1984. Cyprinodontiformes: development. Pp. 362-368. In H. G. Moser, W. J. Richards, D. M. Cohen, M. P. Fahay, A. W. Kendall Jr., and S. L. Richardson (eds), *Ontogeny* and *Systematics of Fishes*. Special Publication Number 1 American Society of Ichthyologists and Herpetologists.

Allen, G. R. 1980. A generic classification of the rainbowfishes (Family Melanotaeiniidae). Records of the West Australian Museum 8: 449-490.

Asai, T. 1971. Fine structure of centriolar complex in spermiogenesis of the viviparous teleost fish *Lebistes reticulatus*. Journal of Nara Medical Association 22: 371-382.

Aschliman, N. C., Tibbetts, I. R. and Collette, B. B. 2005. Relationships of sauries and needlefishes (Teleostei: Scomberesocoidea) to the internally fertilizing halfbeaks (Zenarchopteridae) based on the pharyngeal jaw apparatus. Proceeding of the Biological Society of Washington 118: 416-427.

Azevedo, C. and Corral, L. 1983. Ultrastructural study of spermatogenesis and of spermatozeugmogenesis in *Xiphophorus helleri* (viviparous teleost). Folia Anatomica Universitatis Conimbrigensis 48: 45-70 (In Portuguese).

Billard, R. 1970. La spermatogenèse de *Poecilia reticulata*. IV. La spermiogenèse. Étude ultrastructurale. Annales de Biologie Animale Biochimie Biophysique 10: 493-510.

Billard, R. 1983. Spermiogenesis in rainbow trout (*Salmo gairdneri*). Cell and Tissue Research 233: 265-284.

Billard, R. 1986. Spermatogenesis and spermatology of some teleost fish species. Reproduction, Nutrition and Développement 26: 877-1024.

Blackburn, D. G. 2006. Evolutionary origin of viviparity in fishes. Pp. 283-297. In M. C. Uribe and H. Grier (eds), *Viviparous Fishes*, New Life Publications. Homestead, Florida.

Brummett, A. R. and Dumont, J. N. 1979. Initial stages of sperm penetration into the egg of *Fundulus heteroclitus*. Journal of Experimental Zoology 210: 417-434.

Brusle, S. 1981. Ultrastructure of Spermiogenesis in *Liza aurata* Teleostei Mugilidae. Cell and Tissue Research 217: 415-424.

Brusle, S. and Brusle, J. 1978. An Ultrastructural Study of Early Germ Cells in Mugil-Auratus Teleostei Mugilidae. Annales de Biologie Animale Biochimie Biophysique 18: 1141-1154.

Collette, B. B., Mcgowen, G. E., Parin, N. V. and Mito, S. 1984. Beloniformes: development and relationsips. Pp. 335-354. In H. G. Moser, W. J. Richards, D. M. Cohen, M. P. Fahay, A. W. Kendall Jr. and S. L. Richardson (eds), *Ontogeny and Systematics of Fishes*. Special Publication Number 1 American Society of Ichthyologists and Herpetologists.

Dadone, L. and Narbaitz, R. 1967. Submicroscopic structure of spermatozoa of a cyprinodontiform teleost, *Jenynsia lineata*. Zeitschrift für Zellforschung 80: 214-219.

De Sylva, D. P. 1984. Mugiloidei: development and relationships. Pp. 530-533. In H. G. Moser, W. J. Richards, D. M. Cohen, M. P. Fahay, A. W. Kendall and S. L. Richardson (eds), *Ontogeny and Systematics of Fishes*. Special Publication Number 1 American Society of lchthyologists and Herpetologists. vol. 1. Allen Press, Lawrence, Kansas.

Downing, A. L. and Burns, J. R. 1995. Testis morphology and spermatozeugma formation in three genera of viviparous halfbeaks: *Nomorhamphus*, *Dermogenys*, and *Hemirhamphodon* (Teleostei: Hemiramphidae). Journal of Morphology 225: 329-343.

Eiras-Stofella, D. R. and Gremski, W. 1991. Ultrastructural analysis of the mullet *Mugil liza* and *Mugil platanus* (Teleostei, Mugilidae) spermatozoa. Microscopia Electronica y Biologia Celular 15: 173-177.

Eiras-Stofella, D. R., Gremski, W. and Kuligowski, S. M. 1993. The ultrastructure of the mullet *Mugil curema* Valenciennes (Teleostei, Mugilidae) spermatozoa. Revista Brasileira de Zoologia 10: 619-628.

El-Gammal, H. L., Bahnasawy, M. H., El-Sayyad, H. I. and El-Gammal, I. M. 2003. The ultrastructure of spermatozoa and spermiogenesis in *Sarotherodon galilaeus* and *Liza ramada* (Teleostei, Perciformes, Cichlidae and Mugilidae). Journal of the Egyptian German Society of Zoology 40: 69-80.

Grier, H. J. 1973a. Aspects of germinal cyst and sperm development in *Poecilia latipinna* (Teleostei: Poeciliidae). Journal of Morphology 146: 229-250.

Grier, H. J. 1973b. Ultrastructure of the testis in the teleost *Poecilia latipinna*. Spermiogenesis with reference to the intercentriolar lamellated body. Journal of Ultrastructure Research 45: 82-92.

Grier, H. J. 1975. Aspects of germinal cyst and sperm development in *Poecilia latipinna* (Teloestei: Poecilidae). Journal of Morphology 146: 229-250.

Grier, H. J. 1976. Sperm development in the teleost *Oryzias latipes*. Cell and Tissue Research 168: 419-431.

Grier, H. J. 1981. Cellular organization of the testis and spermatogenesis in fishes. American Zoologist 21: 345-357.

Grier, H. J. and Collette, B. B. 1987. Unique spermatozeugmata in testes of halfbeaks of the genus *Zenarchopterus* (Teleostei: Hemiramphidae). Copeia 1987: 300-311.

Grier, H. J., Fitzsimons, J. M. and Linton, J. R. 1978. Structure and ultrastructure of the testis and sperm formation in goodeid teleosts. Journal of Morphology 156: 419-438.

Grier, H. J., Linton, J. R., Leatherland, J. F. and De Vlaming, V. L. 1980. Structural evidence for two different testicular types in teleost fishes. The American Journal of Anatomy 159: 331-345.

Grier, H. J., Moody, H. J. and Cowell, B. C. 1990. Internal fertilization and sperm morphology in the brook silverside, *Labidesthes sicculus*. Copeia 1: 221-226.

Grier, H. J. and Parenti, L. R. 1994. Reproductive biology and systematics of phallostethid fishes as revealed by gonad structure. Environmental Biology of Fishes 41: 287-299.

Grier, H. J., Uribe, M. C., Parenti, L. R. and De La Rosa-Cruz, G. 2006. Fecundity, the germinal epithelium, and folliculogenesis in viviparous fishes. Pp. 191-216. In M. C. Uribe and H. Grier (eds), *Viviparous Fishes*, New Life Publications. Homestead, Florida.

Hara, M., Kurita, Y., Watanabe, S., Watanabe, Y. and Okiyama, M. 1997. Reproductive biology of the Pacific saury, *Cololabis saira*: Fine structures of the egg and spermatozoon. Bulletin of Tohoku National Fisheries Research Institute 59: 139-147 (In Japanese with English abstract).

Hara, M. and Okiyama, M. 1998. An ultrastructural review on the spermatozoa of Japanese fishes. Bulletin of The Ocean Research Institute, University of Tokyo 33: 1-138.

Harrington, R. W. 1961. Oviparous hermaphroditic fish with internal self-fertilization. Science 134: 1749-1750.

Huber, J., H. 1992. Review of *Rivulus*, ecobiogeography relationships. Société Francaise d'Ichthyologie, Paris.

Iwamatsu, T. and Ohta, T. 1981. Scanning electron microscopic observations on sperm penetration in teleostean fish. Journal of Experimental Zoology 218: 261-277.

Jamieson, B. G. M. 1984. Spermatozoal ultrastructure in *Branchiostoma moretonensis* Kelly, a comparison with *B. lanceolatum* (Cephalochordata) and with other deuterostomes. Zoologica Scripta 13: 223-229.

Jamieson, B. G. M. 1989. Complex spermatozoon of the live-bearing half-beak *Hemirhamphodon pogonognathus* Bleeker Ultrastructural Description Euteleostei Atherinomorpha Beloniformes. Gamete Research 24: 247-260.

Jamieson, B. G. M. 1991. *Fish Evolution and Systematics: Evidence from Spermatozoa*. Cambridge University Press, Cambridge. 319 pp.

Jamieson, B. G. M. 2005. Chondrichthyan spermatozoa and phylogeny. Pp. 201-236. In (W. C. Hamlett (ed.), Reproductive biology and phylogeny of Chondrichthyes: sharks, batoids and chimaeras. [Reproductive Biology and Phylogeny. Volume 3.]. Science Publishers Inc.

Jamieson, B. G. M. and Grier, H. J. 1993. Influences of phylogenetic position and fertilization biology on spermatozoal ultrastructure exemplified by exocoetoid and poeciliid fish. Hydrobiologia 271: 11-25.

Jespersen, Å. 1984. Spermatozoans from a parasitic dwarf male of *Neoceratias spinifer* Pappenheim, 1914. Videnskabelige Meddelelser Dansk Naturhistorisk Forening 145: 37-42.

Jonas-Davies, J. A. C., Winfrey, V. and Olson, G. E. 1983. Plasma membrane structure in spermatogenic cells of the Swordtail Teleostei *Xiphophorus helleri*. Gamete Research 7: 309-324.

Kweon, H.-S., Park, E.-H. and Peters, N. 1998. Spermatozoon ultrastructure in the internally self-fertilizing hermaphroditic teleost, *Rivulus marmoratus* (Cyprinodontiformes, Rivulidae). Copeia 1998: 1101-1106.

Lauder, G. V. and Liem, K. F. 1983. The evolution and interrelationships of the actinopterygian fishes. Bulletin of the Museum of Comparative Zoology 150: 95-197.

Leung, L. K. P. 1988. The ultrastructure of the spermatozoon of *Lepidogalaxias salamandroides* and Its phylogenetic significance. Gamete Research 19: 41-50.

Marshall, C. J. 1989. *Cryopreservation and ultrastructural studies on teleost fish gametes*. Honours thesis, University of Queensland.

Mattei, X. 1970. Spermiogenèse comparée des poissons. Pp. 57-69. In B. Baccetti (ed.), *Comparative Spermatology*. Academic Press, New York.

Mattei, X. 1991. Spermatozoon ultrastructure and its systematic implications in fishes. Canadian Journal of Zoology 69: 3038-3055.

Mattei, X. and Boissin, C. 1966. Le complexe centriolaire du spermatozoïde de *Lebistes reticulatus*. Comptes Rendus Hebdomadaires des Séances de l'Académie des Sciences D 262: 2620-2622.

Mattei, X., Mattei, C. and Boissin, C. 1967. L'extrémité flagellaire du spermatozoïde de *Lebistes reticulatus* (Poeciliidae). Comptes Rendus des Séances de la Scociété de Biologie de l'ouest Africain 161: 884-887.

Mizue, K. 1969. Electron-microscopic study on spermiogenesis of black sailfin molley, *Mollienesia latipinna* Le Sverp. Bulletin Faculty of Fisheries, Nagasaki University 28: 1-17.

Mousa, M. A., El-Gamal, A. E. H. S. and Khalil, M. B. A. 1998. Comparative study on the spermatogenesis in the testis of *Mugil cephalus* (Teleostei, Mugilidae) in different habitats. Journal of the Egyptian German Society of Zoology 27: 213-224.

Nelson, J. S. 1984. *Fishes of the World*, 2nd edition. John Wiley and Sons, New York. 523 pp.

Nelson, J. S. 2006. *Fishes of the World*, 4th edition. John Wiley and Sons, Inc., Hoboken, NJ. 601 pp.

Nicander, L. 1968. Gametogenesis and the ultrastructure of germ cells in vertebrates. Proceedings of the 6th International Congress on Artificial Reproduction in Animals, Paris 1: 89-107.

Parenti, L. R. 1981. A phylogenetic and biogeographic analysis of cyprinodontiform fishes (Teleostei, Atherinomorpha). Bulletin of the American Museum of Natural History 168: 335-557.

Parenti, L. R. 2005. The phylogeny of atherinomorphs: evolution of a novel fish reproductive system. Pp. 13-30. In M. C. Uribe and H. J. Grier (eds), *Viviparous Fishes*. New Life Publications, Homestead.

Parenti, L. R. and Grier, H. J. 2004. Evolution and phylogeny of gonad morphology in bony fishes. Integrative and Comparative Biology 44: 333-348.

Pecio, A., Burns, J. R. and Grier, H. J. 2006. Testis structure, spermatogenesis and spermatozoon ultrastructure in a four-eyed fish *Anableps anableps* L. 1758 (Teleostei : Atherinomorpha : Cyprinodontiformes : Anablepidae). Acta Biologica Cracoviensia Series Botanica 48: 24-24.

Porte, A. and Follenius, E. 1960. La spermiogénèse chez *Lebistes reticulatus* Étude au microscope électronique. Bulletin de la Société Zoologique de France 85: 82-88.

Reygadas, R. C. and Escorcia, H. B. 1998. Testis histology and ultrastructure of *Chirostoma jordani* (Osteichthyes: Atherinidae) (SPA). Revista de Biologia Tropical 46: 943-949.

Rosen, D. E. 1964. The relationships and taxonomic position of the halfbeaks, killifishes, silversides, and their relatives. Bulletin of the American Museum of Natural History 127: 219-267.

Rosen, D. E. and Parenti, L. R. 1981. Relationships of *Oryzias*, and the groups of atherinomorph fishes. American Museum Novitates 2719: 1-25.

Russo, J. and Pisanó, A. 1973. Some ultrastructural characteristics of *Platypoecilus maculatus* spermatogenesis. Bollettino di Zoologia 40: 201-207.

Sakai, Y. T. 1976. Spermiogenesis of the teleost, *Oryzias latipes*, with special reference to the formation of the flagellar membrane. Development, Growth and Differentiation 18: 1-13.

Selman, K. and Wallace, R. B. 1986. Gametogenesis in *Fundulus heteroclitus*. American Zoologist 26: 173-192.

Thiaw, O. T., Mattei, X., Romand, R. and Marchand, B. 1986. Reinvestigation of spermatic flagella structure: the teleostean Cyprinodontidae. Journal of Ultrastructure and Molecular Structure Research 97: 109-118.

Uribe, M. C. and Grier, H. J. 2006 (eds), *Viviparous Fishes.*. New Life PublicationsHomestead, Florida. 609 pp.

Van Der Horst, G. 1976. *Aspects of the reproductive biology of Liza dumerili (Steindachner, 1869) (Teleostei: Mugilidae) with special reference to sperm*. (*Fide* Van der Horst, G. and Cross, R.H.M. 1978). Ph. D. thesis, University of Port Elizabeth.

Van Der Horst, G. and Cross, R. H. M. 1978. The structure of the spermatozoon of *Liza dumerili* (Teleostei) with notes on spermiogenesis. Zoologia Africana 13: 245-258.

Yasuzumi, F. 1971. Electron microscope study of the fish spermiogenesis. Journal of Nara Medical Association 22: 343-355.

CHAPTER 15

Ultrastructure of Spermatozoa: Acanthopterygii Continued: Percomorpha

Barrie G. M. Jamieson

15.1 SERIES PERCOMORPHA

No definition is available for the Percomorpha as they do not appear to have any unique autapomorphies not already present in their presumed common ancestry with the atherinomorphs and related groups. This difficulty in definition of the Percomorpha is not surprising in view of their evident polyphyly, as seen also in the order Perciformes, from molecular analysis (Fig. 15.1; see also Ortí and Li, **Chapter 1**, Fig. 1.5). Constituent orders of the Percomorpha are the Lampriformes*, Beryciformes, Zeiformes, Gasterosteiformes, Indostomiformes*, Pegasiformes*, Syngnathiformes, Dactylopteriformes, Synbranchiformes, Scorpaeiniformes, Perciformes, Pleuronectiformes, and Tetraodontiformes (*sperm ultrastructure unknown). The various groups will be treated here chiefly in the sequence adopted by Nelson (2006) in which definitions are given.

15.1.1 Spermatozoal diversity and relationships.

There is great uncertainty as to internal percomorph relationships and the considerable spermatozoal diversity known does not notably contribute to assessment of affinities. Spermatozoal ultrastructure of only a small fraction of species has been studied, however, with whole orders awaiting investigation, and there are indications that spermatology will contribute substantially to classification and phylogeny on fuller enquiry. Thus, peculiarities of sperm ultrastructure are at least compatible with a relationship between some orders, for instance the Dactylopteriformes and Perciformes and the Zeiformes and

School of Integrative Biology, University of Queensland, Brisbane, Queensland 4072, Australia

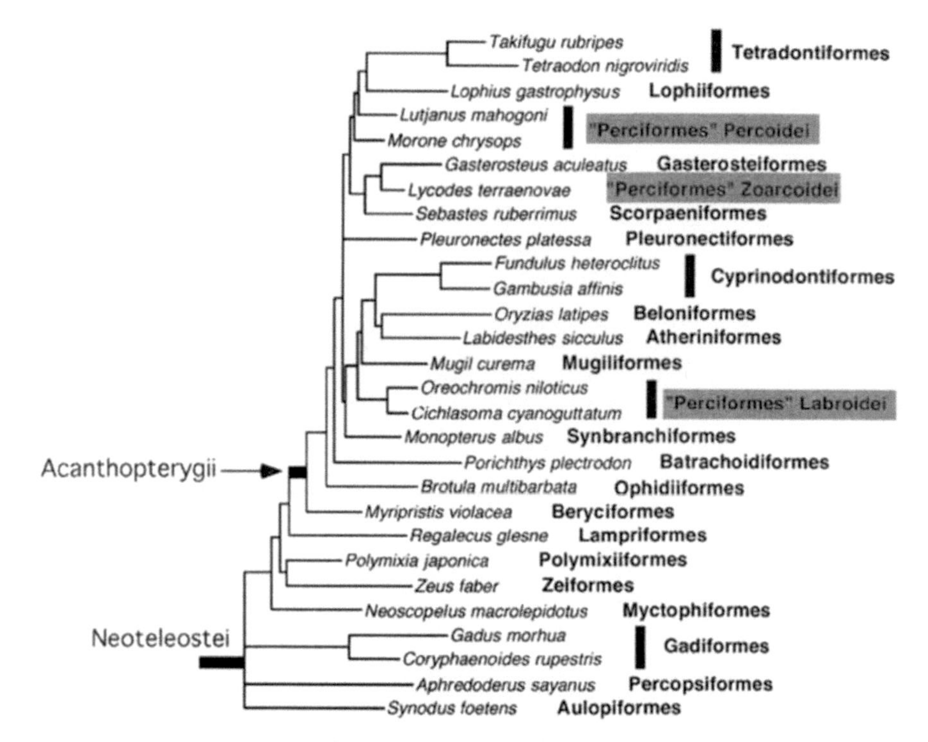

Fig. 15.1 Phylogeny of selected Neoteleostei from an analysis of 10 nuclear genes, to show the relationships of the Percomorpha. Note the polyphyly of the Perciformes (shaded) which form three distinct clades: 1. Labroidei, 2. Zoarcoidei and 3. Pleuronectiformes through Tetraodontiformes. Adapted from Ortí and Li (Chapter 1, Fig. 1.5).

Tetraodontiformes and there is a widespread perciform type, with flocculent chromatin and a thin-walled sleeve trailing behind the midpiece, and with it investing the cytoplasmic canal, seen in several families exemplified by the Centropomidae, Percichthyidae, Mugiloididae, Cichlidae and others (see below). Most percomorphs have simple anacrosomal aquasperm, with or without some asymmetry, and have external fertilization.

Internal fertilization has evolved independently in a few percomorph groups (see Breder and Rosen 1966; Thresher 1984; Nelson 2006). In these groups, the sperm though again derivable from the anacrosomal aquasperm, usually have elongated nuclei and differ between groups. Elongation of the nucleus is seen in the scorpaenids *Helocolenus dactylopterus*, *Oligocottus maculosus* and *Sebastiscus marmoratus*, the embiotocids *Cymatogaster aggregata* and *Ditrema temminckii*, the anarhichadid *Anarhichas lupus* and the cottids *Hemilepidotus gilberti* and *Alcichthys alicornis*. The nucleus is disclike in the hemitripterid *Blepsias cirrhosis* which, though internally inseminating, has external sperm transfer. There is little or no elongation of the nucleus in the internally fertilizing, male mouth brooding apogonids.

15.2 ORDER BERYCIFORMES

All that is known of the ultrastructure of beryciform sperm is a diagram of a longitudinal section of the sperm of *Holocentrus* sp. (Holocentridae) by Mattei (1991). This shows a rounded sperm head, nucleus indented to about midlength by the nuclear fossa, which contains the two mutually perpendicular centrioles, and a mitochondrion on each side behind the nucleus.

15.3 ORDER ZEIFORMES

The Zeiformes (dories and boarfish) are a heterogeneous and probably paraphyletic assemblage of marine, mainly deep-sea fish. Patterson (1964) considered them to be the sister-group of the Beryciformes but Rosen (1984) relegated them to the Tetraodontiformes. However, in the present molecular analyses (Orti and Li, **Chapter 1**, Fig. 1.5 and Fig. 15.1) Zeiformes appear to be the sister-group of the Polymixiiformes.

15.3.1 Family Zeidae

Zeus faber: In the Zeiformes, *Zeus faber* (adult, Fig. 15.2) has a broadly conical asymmetrical sperm (but nevertheless Type I) with a very deep basal nuclear

Fig. 15.2 *Zeus faber* (Zeidae), the John Dory. By Norman Weaver. With kind permission of Sarah Starsmore.

invagination in which the proximal centriole is perpendicular to the basal body. There appear to be several mitochondria disposed asymmetrically relative to the axoneme in an illustration by Mattei (1970) (Fig. 15.3).

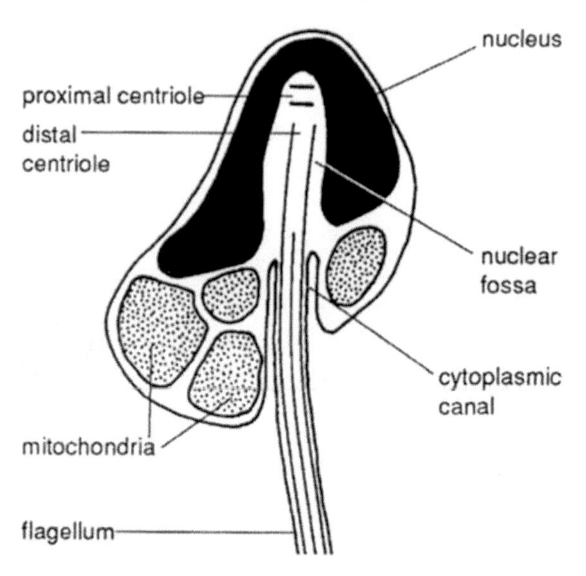

Fig. 15.3 *Zeus faber* (Zeidae). Diagrammatic longitudinal section of the spermatozoon. After Mattei, X. 1970. Pp. 59-69. In B. Baccetti (ed.) *Comparative Spermatology*, Academic Press, New York, Fig. 4: 18.

Renarks: *Zeus faber* is a serial spawner with pelagic eggs. The deep invagination of the nucleus in *Zeus* and in Tetraodontiformes was considered to be a synapomorphy (Jamieson 1991) which, although not restricted to these forms, added some support to placement by Rosen (1984) of Zeiformes in the Tetraodontiformes. However, the asymmetrical distribution of the mitochondria differs from the tetraodontiform condition and the molecular analysis suggests that the two orders are not closely related.

15.4 ORDER SYNBRANCHIFORMES

Synbranchiforms are the swamp eels. There are three families of which only the Synbranchidae have been examined for sperm ultrastructure. Synbranchids occur in tropical and subtropical fresh or occasionally brackish water and are rarely marine (Nelson 2006).

15.4.1 Family Synbranchidae

Synbranchus marmoratus: This is a protogynous, diandric species. Primary males develop directly whereas secondary males arise by the sex reversal of females. They have unrestricted, lobular testes (references in Lo Nostro *et al.* 2003). The eggs are very large, ca 3 mm in diameter.

The spermatozoa have a nuclear diameter of 2.5 ± 0.49 μm; the chromatin is highly condensed (Fig. 15.4C,D). They are biflagellate cells, each of the two centrioles giving rise to a flagellum (Fig. 15.4C,D). The midpiece is short, forming a small collar (Fig. 15.4B), and containing two basal bodies (Fig. 15.4C,D,E), each composed of nine triplets. The basal bodies are jointly surrounded by a ring of four mitochondria of various sizes (Fig. 15.4E). The flagella are approximately 50 μm long and have the typical 9+2 configuration (Fig. 15.4E, inset). Each axoneme is continuous with its basal body and the two central singlets commence shortly behind this. There is no cytoplasmic canal. The flagella are orientated are right angles to the nucleus. The spermatozoa are not inserted into Sertoli cells (Lo Nostro *et al.* 2003).

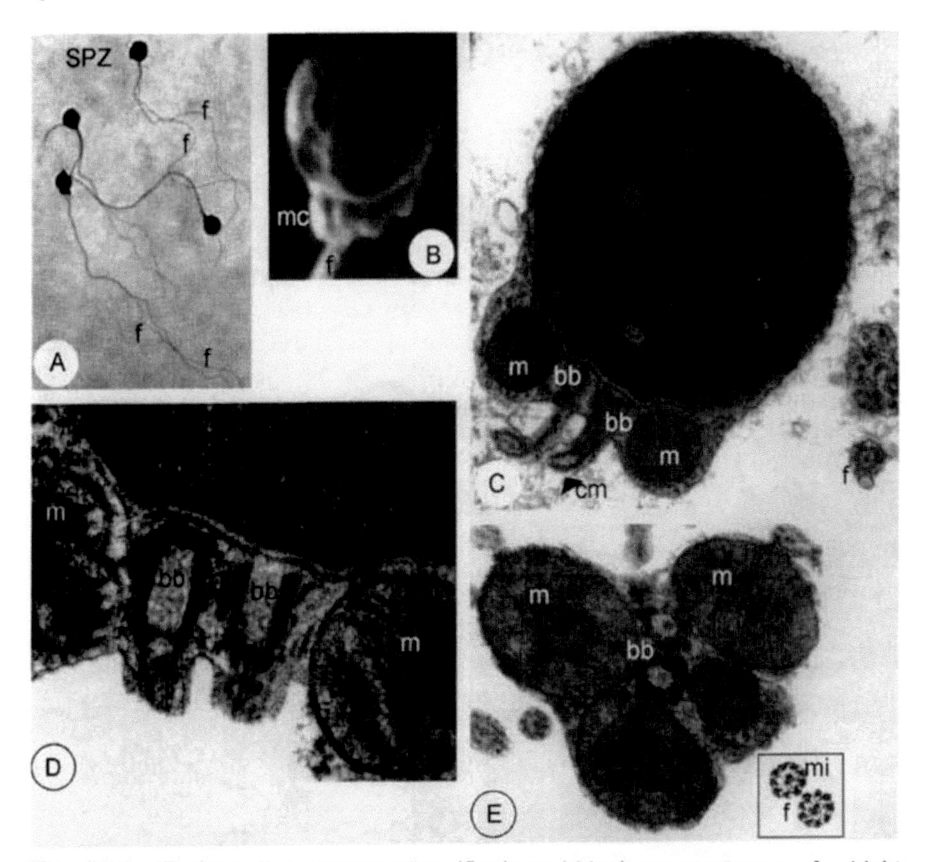

Fig. 15.4 *Synbranchus marmoratus* (Synbranchidae) spermatozoa. **A**. Light micrograph. x 140. **B**. SEM; note the mitochondrial collar. x 6,900. **C**. TEM longitudinal section, showing the rounded nucleus and two flagella. x 9,950. **E**. TEM transverse section. Inset: flagella in transverse section. x 14,000. bb, basal body; cm, cytoplasmic membrane; f, flagellum; m, mitochondrion; mc, mitochondrial collar; mi, microtubules. From Lo Nostro, F. L., Grier, H., Meijide, F. J. and Guerrero, G. A. 2003. Tissue and Cell 35: 121-132, Fig. 6. Copyright Elsevier. With permission.

Remarks: The orientation of the flagella indicates that the spermatozoon is of a modified Type I. The occurrence of biflagellate sperm in euteleostomes is listed in **Chapter 13**, Table 13.1.

15.5 ORDER GASTEROSTEIFORMES

Gasterosteiforms are the sticklebacks and relatives. Nelson (2006) accepts them as the sister-group of the Syngnathiformes. They contain two suborders with 11 families (Nelson 2006).

15.5.1 Suborder Gasterosteoidei

Gasterosteoids produce a glue-like substance from the kidneys which is used by males to construct a nest of plant materials. There are four families, with nine genera (Nelson 2006).

15.5.1.1 Family Hypoptychidae

These are the sand eels, marine in Japan and Korea to the Sea of Okhotsk, with a single species (Nelson 2006).

***Hypoptychus dybowskii*:** The spermatozoon of *H. dybowskii* (Fig. 15.5) has an elongate elliptic nucleus with a shallow nuclear fossa and chromatin in dense masses. Only the distal centriole (basal body) has been observed. A single annular row of spherical mitochondria surrounds the centriolar region. Only a small cytoplasmic canal is illustrated. The 9+2 axoneme is loosely surrounded by the cytoplasmic membrane but no lateral fins are present (Hara and Okiyama 1998).

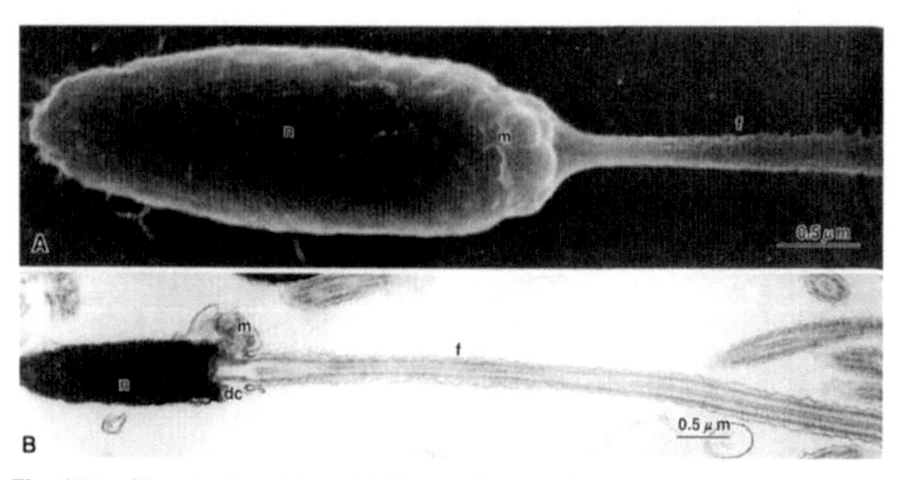

Fig. 15.5 *Hypoptychus dybowskii*, Korean Sandeel (Hypoptychidae) spermatozoa. **A**. SEM. **B**. TEM longitudinal section. dc, distal centriole; f, flagellum; m, mitochondria; n, nucleus. Adapted from Hara, M. and Okiyama, M. 1998. Bulletin of The Ocean Research Institute, University of Tokyo 33: 1-138, Fig. 31B,C. With permission.

Remarks: As noted by Hara and Okiyama (1998), the elongate nucleus is unusual for an anacrosomal aquasperm.

15.5.1.2 Family Aulorhynchidae

In the spermatozoon of *Aulichthys japonicus* (Fig. 15.6), the nucleus is elongate elliptic with a shallow nuclear fossa and chromatin in dense masses. Only the distal centriole (basal body) has been observed. A complex mitochondrial sheath, with elongate tortuous mitochondria, and a very deep cytoplasmic canal which forms a long sleeve surround the proximal region of the flagellum. ITDs, in the form of septa, are present in the B subtubules of all doublets (Hara and Okiyama 1998).

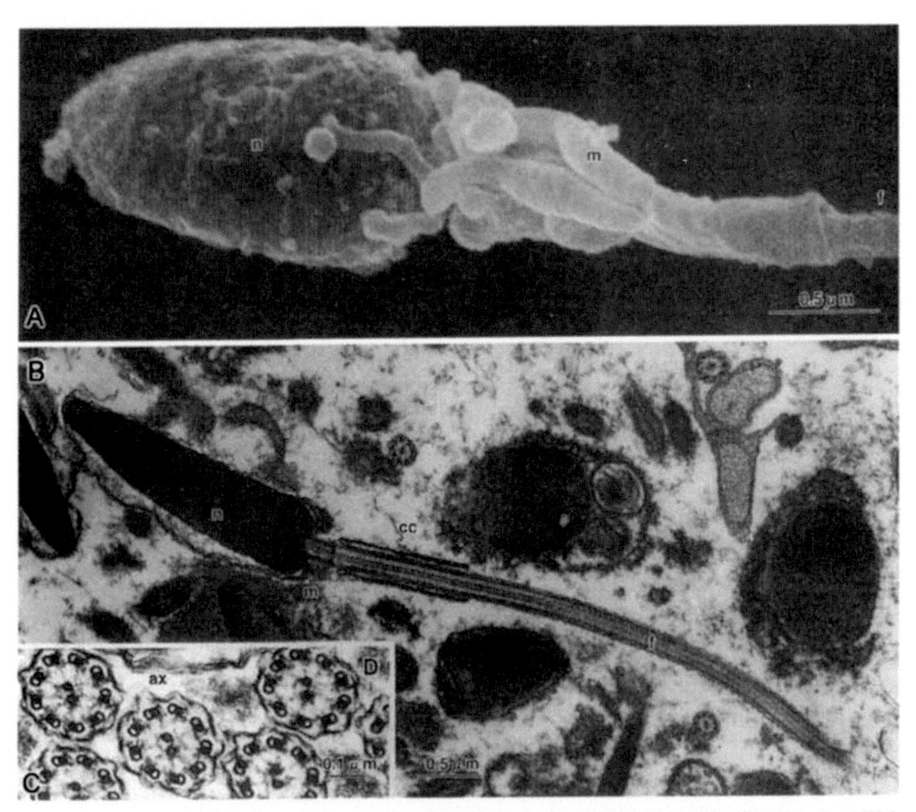

Fig. 15.6 *Aulichthys japonicus*, Tubesnout (Aulorhynchidae) spermatozoa. **A**. SEM. **B**. TEM longitudinal section. **C**. TEM transverse sections of axonemes. ax, axoneme; cc, cytoplasmic canal; f, flagellum; m, mitochondrion; n, nucleus. Adapted from Hara, M. and Okiyama, M. 1998. Bulletin of The Ocean Research Institute, University of Tokyo 33: 1-138, Fig. 32B,C. With permission.

Remarks: *Aulichthys japonicus* is a "hider in the live invertebrate" and has internal gametic association (in which sperm contact the eggs internally but fertilization is delayed until the eggs are spawned), in contrast with *A.*

flavidus which has normal external fertilization. The eggs of *A. japonicus* were shown to be deposited in the peribranchial cavity of ascidians by Balon (1975) prior to fertilization. The elongate nucleus and midpiece and occurrence of ITDs are considered to be typical of complex anacrosomal introsperm (Hara and Okiyama 1998).

15.5.1.3 Family Gasterosteidae

Gasterosteids are the true sticklebacks, marine, brackish and freshwater in the Northern Hemisphere (Nelson 2006). We owe to Hara and Okiyama (1998) our knowledge of the sperm of *Gasterosteus aculeatus, Pungitius sinensis* and *P. tymensis.*

***Gasterosteus aculeatus, Pungitius pungitius, P. sinensis* and *P. tymensis*.** The nucleus in *Gasterosteus aculeatus* is approximately spherical (Fig. 15.7A); in *Pungitius pungitius, P. sinensis* (Fig. 15.7D,F) and *P. tymensis* it is pyriform; in all four species it has a shallow nuclear fossa and the chromatin is in the form of dense masses with scattered pale lacunae (Fig. 15.7B,D). A proximal centriole has not been observed. The distal centriole (basal body) has a centriolar plug. A centriolar adjunct and lateral centriolar connection is reported for *P. pungitius* and *P. tymensis*. In all three species many small mitochondria are contained in a mitochondrial sheath which forms a deep cytoplasmic canal. The flagellum, with a 9+2 axoneme, has two lateral fins (Fig. 15.7C,E,G).

Remarks: Aspects of reproductive biology in *Pungitius* which might correlate with differences from *Gasterosteus aculeatus* require clarification. In both taxa eggs are fertilized externally, in the nest.

15.5.2 Suborder Synganthoidei

15.5.2.1 Family Syngnathidae

Syngnathoids, previously placed in a separate order Syngnathiformes have a small mouth at the end of a tube-shaped snout, excepting *Enchelyocampus* which lacks even a short tubiform snout (Nelson 1984, 2006). They include

Fig. 15.7 Spermatozoa of three species of Gasterosteidae. **A-C**. *Gasterosteus aculeatus.* **A**. SEM of the spermatozoon. **B**. TEM longitudinal section. C. TEM of flagella each with a 9+2 axoneme and two lateral fins. **D** and **E**. *Pungitius sinensis.* **D**. TEM longitudinal section of a spermatozoon. **E**. Transverse section of a flagellum with two lateral fins. **F** and **G**. *Pungitius tymensis.* **F**. TEM longitudinal sections of spermatozoa. **G**. Transverse sections of flagella showing lateral fins. ax, axoneme; ca, centriolar adjunct; cc, cytoplasmic canal; cp, centriolar globe; dc, distal centriole; f, flagellum; fi, lateral fin; lcc, lateral centriolar connection; m, mitochondrion; n, nucleus; Adapted from Hara, M. and Okiyama, M. 1998. Bulletin of The Ocean Research Institute, University of Tokyo 33: 1-138, Figs. 33B-D, 35B,C, 34B,C. With permission.

Fig. 15.7 Contd. ...

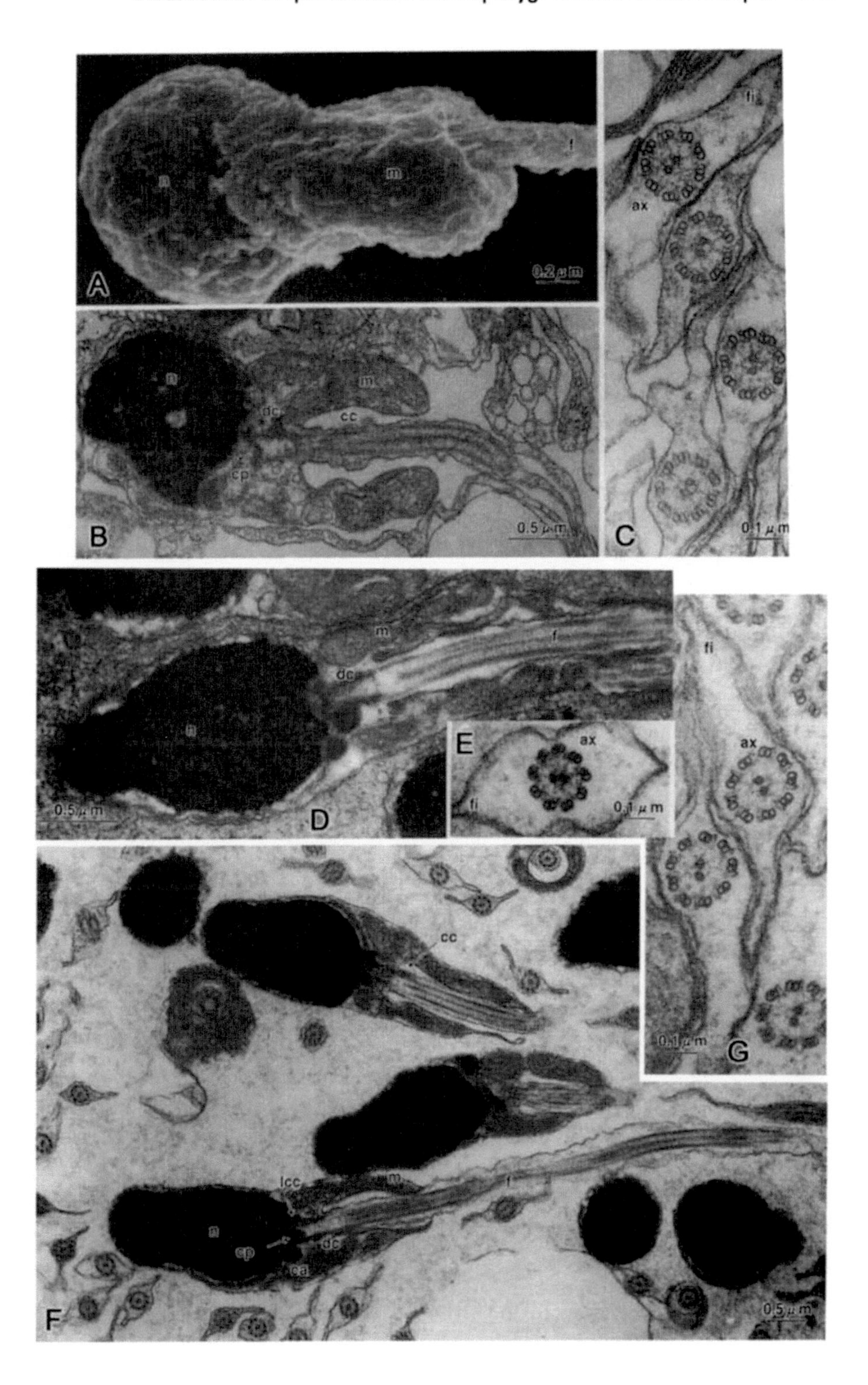
A
n
m
f
0.2 μm
B
n
m
dc
cc
cp
0.5 μm
C
fi
ax
0.1 μm
D
m
f
dc
n
0.5 μm
E
ax
fi
0.1 μm
G
fi
ax
0.1 μm
F
cc
lcc
m
f
n
cp
dc
ca
0.5 μm

the sea horses, pipe fishes and comet fishes, among others. Parental care in representatives of the family Syngnathidae is carried out exclusively by the males. The reversal of the roles of the sexes is specific to polygamous species of the family; it has not been recorded in monogamous species nor in the polygamous *Syngnathus acusimilis* (Kornienko 2001).

Sperm literature. Accounts exist for the sperm of *Syngnathus schlegeli* (Hara and Okiyama 1998; Watanabe *et al*. 2000) (Fig. 15.9A-C; adult 15.9D), *S. acus* and *S. abaster* (Carcupino *et al*. 1999) (Fig. 15.8) and *S. acusimilis* (Kornienko and Drozdov 2001).

***Syngnathus acus, S. abaster* and *S. schlegeli*:** The sperm of these species are elongated cells (>15 µm; 85 µm in *S. schlegeli* according to Watnabe *et al*. 2000) with a cylindrical head entirely occupied by a nucleus ca 2·5 µm long and 0.5 µm wide (Fig. 15.8H) (3 µm and 0.6 µm, respectively, in *S. schlegeli*). No acrosome was observed. As shown in the first two species, at its basal end, the nucleus has a deep nuclear fossa containing two centrioles. A classic 9+2 axoneme originates from the distal centriole (15.8D inset and G). The midpiece of the sperm tail is demarcated clearly behind the nucleus by small cristate mitochondrial rings, which are situated in a cytoplasmic collar around the base of the flagellum and separated from it by a periaxonemal space (Figs 15.8B–D and H). The number of the mitochondrial rings is different in the two species: two in *S. abaster* (Fig. 15.8B,D,H) and four in *S. acus* (Fig. 15.8C) (Carcupino *et al*. 1999). It appears from Figure 15.8D, inset, that there is a continuous mitochondrial ring around at least part of the flagellum and not a ring of separate mitochondria. The sperm of *S. schlegeli* (Hara and Okiyama 1998; Watnabe *et al*. 2000) conforms with *S. acus* and *S. abaster* in external features; two tiers of annular mitochondria are reported (Hara and Okiyama 1998) but Watanabe *et al*. (2000) observed two or three spiral rows. The two centrioles, in the deep basal fossa, are mutually at 45°.The flagellum has two ill-defined lateral fins (Fig. 15.9). Fins were not reported for the other two species but lateral swelling of the axoneme in *S. abaster* is consistent with their presence (Fig. 15.8G).

Watanabe *et al*. (2000) report a second type of testicular spermatozoon, termed atypical sperm, in which the head is about three times the size of that of typical sperm. These seem to the present author to have the appearance of spermatids.

Remarks: Elongation of the nucleus is a frequent feature of internally fertilizing sperm. In pipefish the eggs are inseminated when deposited in the male brood pouch where mucous material keeps the eggs pasted together on the pouch walls (Carcupino *et al*. 1999). These authors nevertheless regarded fertilization as external and considered that the elongate sperm head may promote motility in a viscous medium. Watanabe *et al*. (2000) also observe that the sperm have to swim in viscous ovarian fluid during external fertilization and that the environment for fertilization is therefore equivalent to that for internal fertilization.

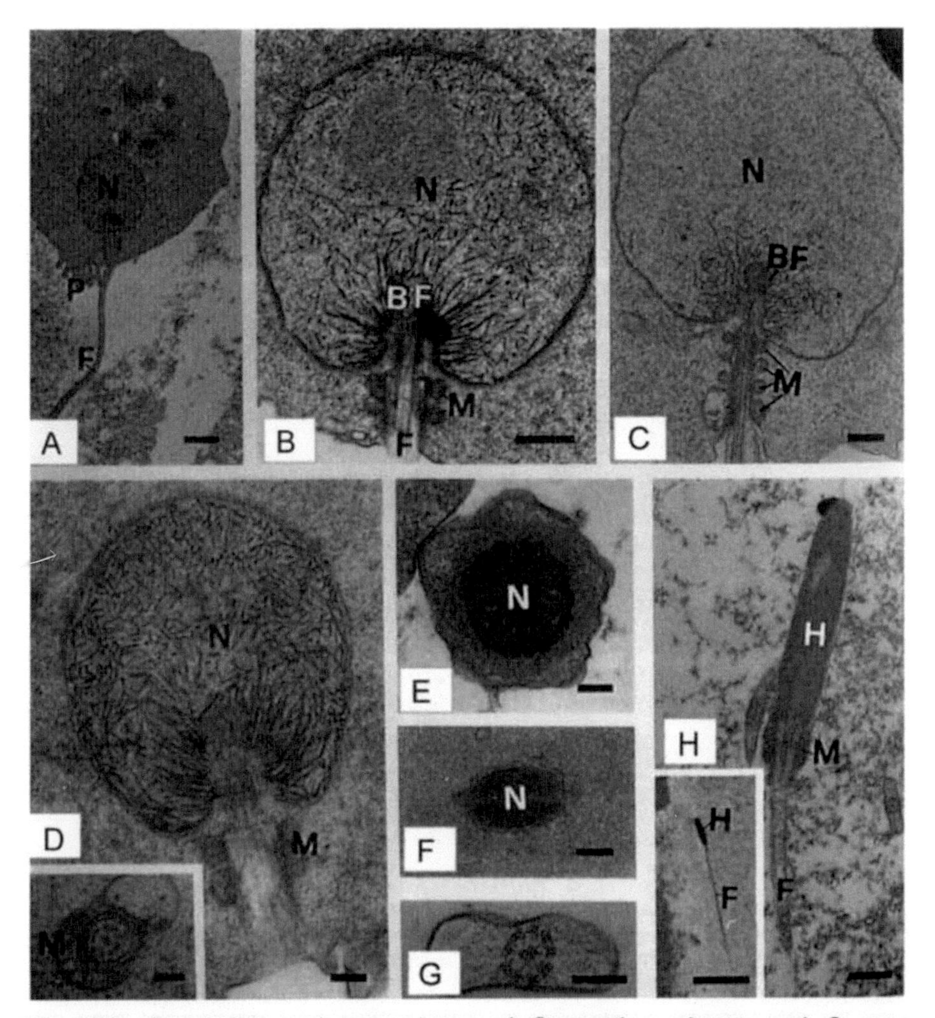

Fig. 15.8 Spermatids and spermatozoa of *Syngnathus abaster* and *S. acus* (Syngnathidae). **A** and **B**. Spermatid of *S. abaster* recognizable by two mitochondrial rings around the first portion of axoneme. **C**. Spermatid of *S. acus* with four mitochondrial rings. **A**. Scale bar 1 µm; **B** and **C**. Scale bar 500 nm. BF, Basal fossa; F, flagellum; M, mitochondrial rings; N, nucleus; P, pseudopodia-like cytoplasmic processes. **D**. Spermatid of *S. abaster* in a more advanced stage of development, characterized by more condensed chromatin. Inset shows the ring-shaped mitochondria and 9+2 axoneme. Scale bar 200 nm. M, Mitochondrial rings; N, nucleus. **E** and **F**. Transverse sections of more condensed spermatids nucleus. Scale bar 200 nm. N, Nucleus. **G**. Transverse section of sperm tail with 9+2 axoneme. Scale bar 200 nm. **H**. Longitudinal section of mature sperm of *S. abaster*. Inset shows whole mature sperm at low magnification. Scale bar 500 nm; inset Scale bar 5 µm. F, Flagellum; H, sperm head; M, mitochondrial rings. Adapted from Carcupino, M., Baldacci, A., Corso, G., Franzoi, P., Pala, M. and Mazzini, M. 1999. Journal of Fish Biology 55: 344-353, Figs. 14-21. With permission of Blackwell Publishing.

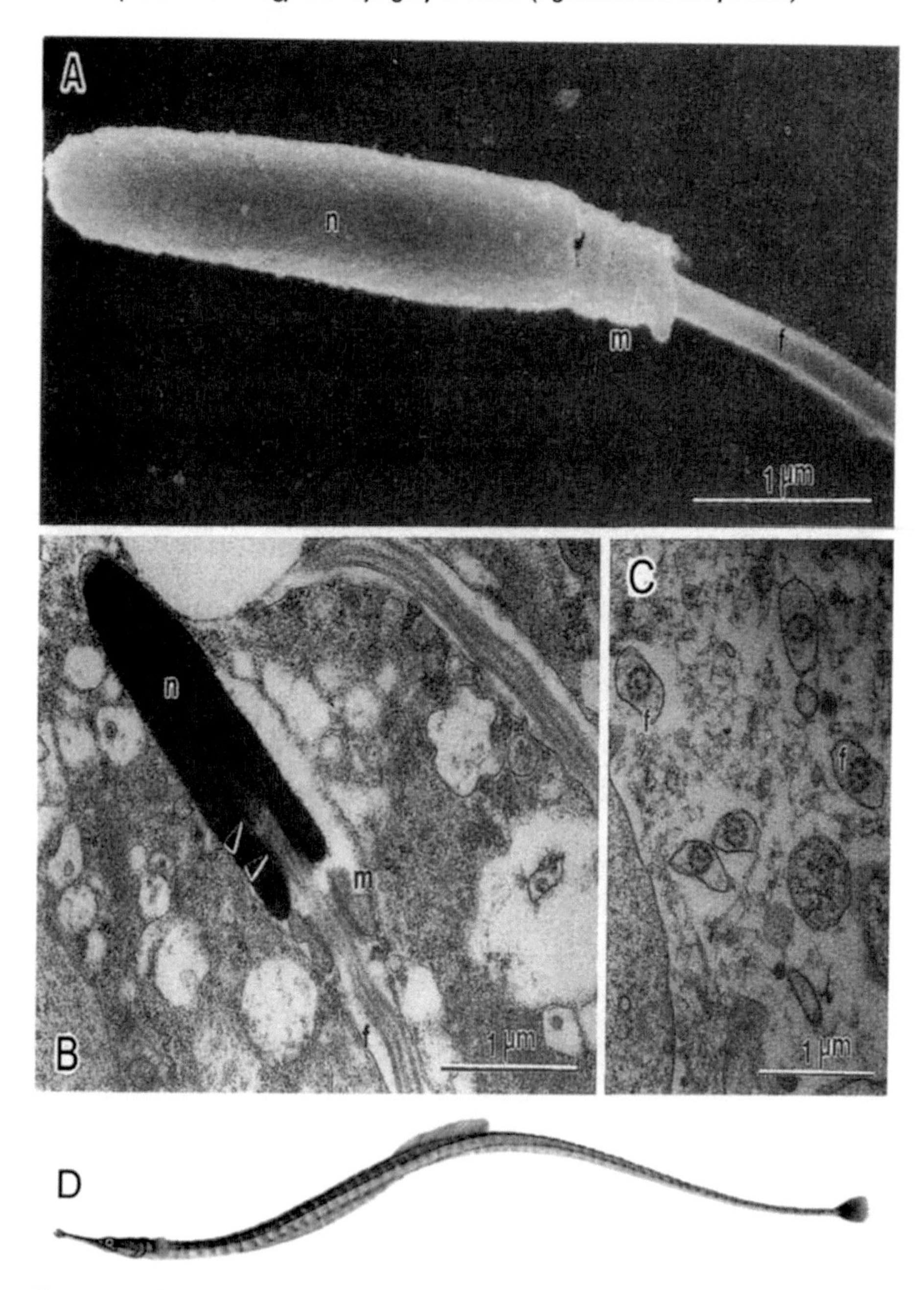

Fig. 15.9 *Syngnathus schlegeli* (Syngnathidae). **A**. SEM of a spermatozoon. **B**. TEM longitudinal section of a spermatozoon. **C**. Transverse sections of flagella. f, flagellum; m, mitochondrion; n, nucleus; arrowheads indicate centrioles. Adapted from Watanabe, S., Hara, M. and Watanabe, Y. 2000. Zoological Science (Tokyo) 17: 759-767, Fig. 5B,C,D. Copyright 2000. The Zoological Society of Japan. With permission. **D**. Adult. By Norman Weaver. With kind permission of Sarah Starsmore.

15.5.2.2 Family Fistulariidae

Fistularia tabacaria: In the syngnathoid family Fistulariidae (Fig. 15.10A), *Fistularia tabacaria*, the Blue-spotted Comet Fish, has a basic anacrosomal aquasperm but the proximal centriole is oblique and to the side of the basal body, each centriole in its own small embayment of a wide, deep basal nuclear fossa. The mitochondria, symmetrically disposed, appear to be in two tiers (illustration by Mattei 1970) (Fig. 15.10B) but whether annular is not known.

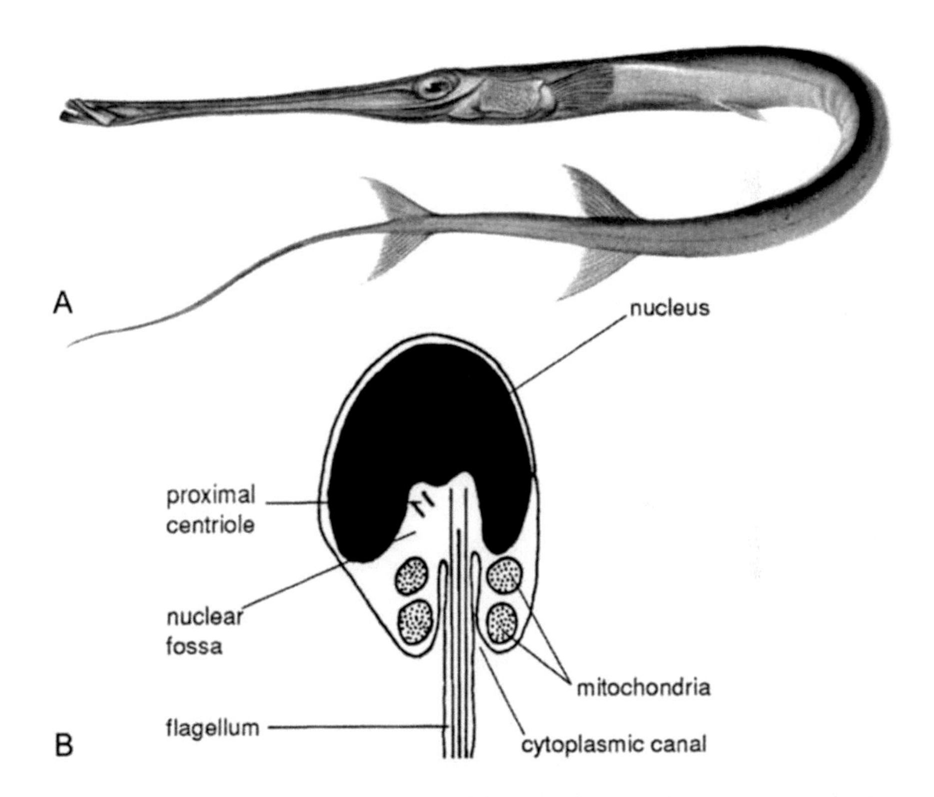

Fig. 15.10 **A**. *Fistularia petimba* (Fistulariidae). By Norman Weaver. With kind permssion of Sarah Starsmore. **B**. *F. tabacaria.* Diagrammatic longitudinal section of the spermatozoon. After Mattei, X. 1970. Pp. 567-569. In B. Baccetti (ed.). *Comparative Spermatology,* Academic Press, New York.

15.6 ORDER SCORPAENIFORMES

Scorpaeniforms, the "mail-cheeked fish" constitute the fourth largest order of fish, with over 1000 species. They are benthic or epibenthic in tropical to polar waters with some freshwater representatives. The suborbital stay is the sole defining character (autapomorphy) but is possibly not indicative of monophyly (see Washington *et al*. 1984a, 1984b); in addition, the hypurals are fused into two distinct, large plates (Lauder and Liem 1983).

Modes of reproduction: These vary widely. Many families spawn individual pelagic eggs while others spawn demersal clusters of adhesive eggs but most produce pelagic egg masses enclosed in a gelatinous matrix. There are strong trends towards internal fertilization as is *Helicolenus* in which internal fertilization occurs but zygotes are spawned (Munoz *et al.* 2000) while *Sebastes* and the comephorids of Lake Baikal are live-bearing (Washington *et al.* 1984a,b).

Sperm ultrastructure: The Scorpaeniformes exhibit two types of gamete: a form that deviates slightly from the simple gamete by a slight elongation of the nucleus and the presence of two mitochondrial layers, as in *Sebastiscus*, and an evolved form with a well-developed, elongate nucleus and midpiece, as in *Oligocottus* These two forms are found in species that employ internal fertilization. Hann (1927) studied the spermatozoa of 32 species of Cottidae and observed the enormous variability in size and shape of the gamete in this family (Mattei 1991). Mattei lists as examined but does not describe the sperm of the scorpaenid *Helicolenus dactylopterus* (Sebastinae) and *Pontinus kuhlii* (Scorpaeninae) and the peristediid (=triglid) *Peristedion cataphractum*.

15.6.1 Suborder Dactylopteroidei

These are the flying gurnards, benthic fish which produce sounds by stridulation, using the hyomandibular bone, and which "walk" on the sea bed by alternately moving the pelvic fins. Affinity with the pegasids and syngnathiforms has been suspected (Nelson 1984). Their inclusion in the Scorpaeniformes has been questioned but spermatozoal ultrastructure supports the decision of Nelson (2006), chiefly on the basis of molecular evidence, to retain them in this order.

15.6.1.1 Family Dactylopteridae

***Dactylopterus volitans*:** The only species of the Dactylopteroidei examined, *Dactylopterus (=Cephalacanthus) volitans*, the Flying Gurnard (adult, Fig. 15.11) Dactylopteridae), has a uniflagellate anacrosomal aquasperm, described by Boisson *et al.* (1968) and Mattei (1970) (Fig. 15.12). It is strongly modified but is clearly derivable from the usual teleostean type. The mode of reproduction is unknown but is presumably by external fertilization.

The moderately elongate nucleus is penetrated so deeply that only a thin layer of chromatin covers the anterior end of the expanded fossa. The nucleus is grossly asymmetrical about the fossa.

Three or four large mitochondria are situated at the base of the large posterior prolongation of the nucleus on the same side of the axoneme. A cytoplasmic collar, surrounding the base of the flagellum, is thin on one side but houses the mitochondria on the other.

The nuclear fossa houses two mutually perpendicular triplet centrioles (Type I arrangement) of which one, the distal centriole, forms the basal body. From an osmiophilic mass associated with the tip of the distal centriole, arise

Fig. 15.11 *Dactylopterus (=Cephalacanthus) volitans*, the Flying Gurnard. Photo Robert A. Patzner, Austria.

bundles of fibrils which reach the proximal centriole where each is capped by a dense lamina.

The 9+2 flagellum is exceptional in having nine very thin accessory fibers peripheral to the doublets and two circlets, each of nine, internal to these. Presence of rudimentary fins is indicated in the illustration (for further details see Boissin *et al.* 1968a).

Sperm relationships: Sperm ultrastructure is at least compatible with a relationship between Dactylopteriformes, e.g. *Dactylopterus (=Cephalacanthus) volitans* (Fig. 15.12), and the Eleotridae in the Perciformes, as exemplified by *Hypseleotris galii* (Jamieson 1991) (Fig. 15.93) though it is possible that resemblances are homoplasies by virtue of relationship (symparamorphies of Jamieson 1984). In both of these species the nucleus is so deeply invaded by the basal fossa as to retain only a very thin layer of chromatin (in *Hypseleotris* subapically no chromatin) between the outer and inner nuclear envelope over the head of the flagellum; two centrioles are present; and the mitochondria are restricted to one side of the axoneme. Asymmetry of the head is more pronounced in *Dactylopterus volitans* of the two species. The mode of fertilization of both species, in which sperm morphology is considerably modified, is unknown but there is no reason to doubt that it is external. Axonemal fins, apparently small in *D. volitans*, are, in contrast, very well developed in *Hypseleotris galii*.

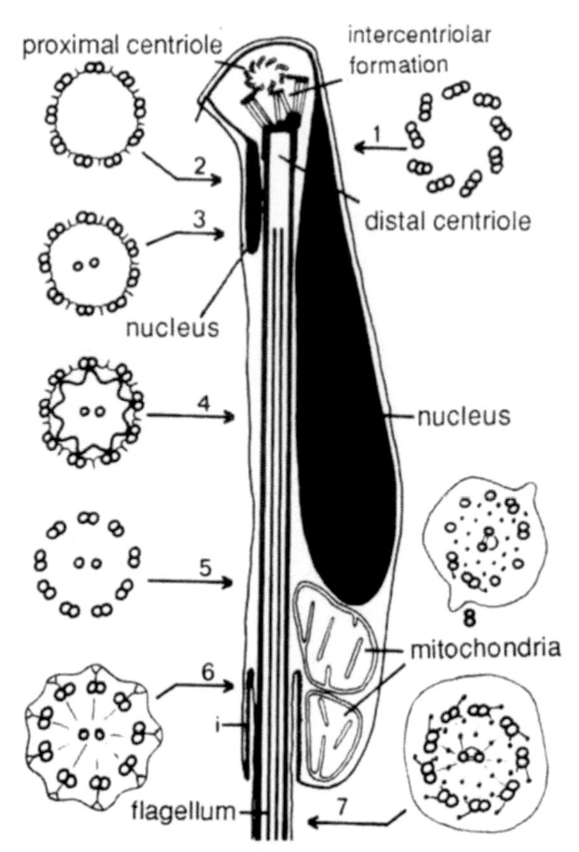

Fig. 15.12 *Dactylopterus (=Cephalacanthus) volitans*. Diagrammatic sections of the spermatozoon. From Jamieson (1991), after Boisson, C., Mattei, X. and Mattei, C. 1968. Comptes Rendus des Séances de la Société de Biologie de l'ouest Africain 162: 820-823, Fig. 1.

15.6.2 Suborder Scorpaenoidei

15.6.2.1 Family Scorpaenidae

Subfamily Scorpaeninae: *Scorpaena angolensis*. A scorpaenid is depicted in Figure 15.13A. The sperm of *S. angolensis* Mattei (1970, 1991) (Fig. 15.13B) appears somewhat similar to that of *Cottus* (below) but it is not known whether the nucleus is depressed in one plane nor is the mode of fertilization known.

***Scorpaena notata*:** All that is stated for the sperm of the ovuliparous, externally fertilizing *S. notata* is that they have a round head about 1.39 µm (± 0.04 SE, n = 25) in diameter; the midpiece is relatively long (1.35 µm) and is made up of two periflagellar layers, each of which contains three or four spherical mitochondria (Munoz *et al.* 2002b).

Mattei (1991) examined but gives no data on the sperm of *Pontinus kuhli*.

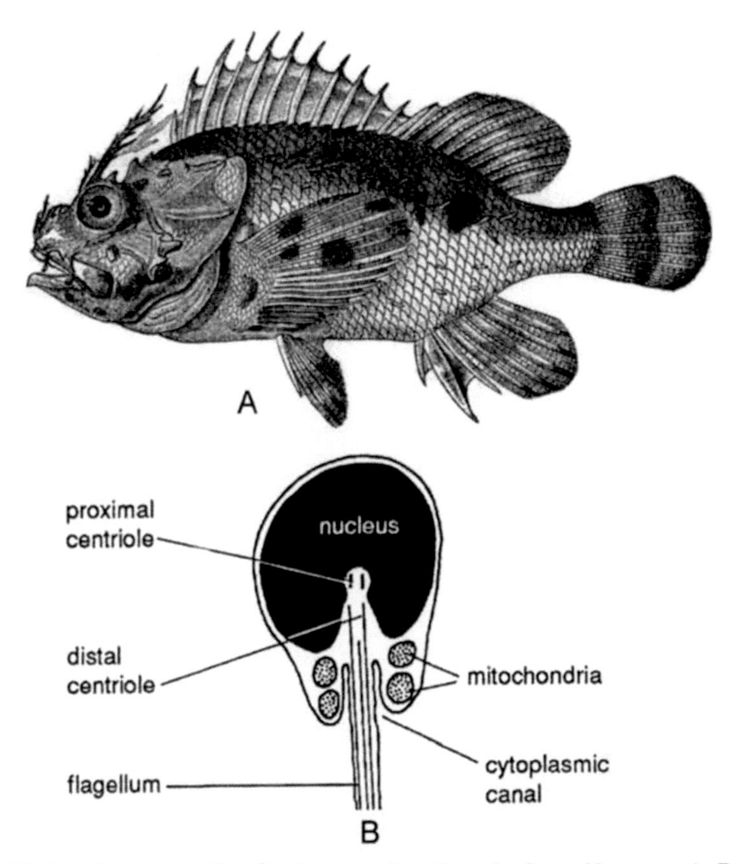

Fig. 15.13 A. A scorpaenid, *Scorpaena grandicornis*, from Norman, J. R. 1937. *Illustrated Guide to the Fish Gallery*. Trustees of the British Museum, London, Fig. 42. **B.** *Scorpaena angolensis*. Diagrammatic longitudinal section of the spermatozoon. From Jamieson 1991, after Mattei, X. 1970. Pp. 567-569. In B. Baccetti (ed.), *Comparative Spermatology*, Academic Press, New York, Fig. 4: 3.

Remarks: It appears that the sperm of *Scorpaena* are little modified from the Type I condition, the two tiers of mitochondria and approximately longitudinally aligned centrioles being notable apomorphies. The simple form is in keeping with external fertilization known for *Scorpaena notata*.

Subfamily Sebastinae: *Helicolenus dactylopterus*. The spermatozoa of *Helicolenus dactylopterus dactylopterus* have been briefly described in a study of gametogenesis (Munoz *et al.* 2002a). They have an elongated and slightly concave head, ca 2.02 ± 0.4 µm long. The acrosomeless nucleus is uniformly electron-dense. The midpiece is long, ca 2 µm, and consists of four periflagellar mitochondrial layers, each of which contains between five and nine mitochondria, a variation possibly relating to the maturation stage. The flagellum as the usual 9+2 configuration of microtubules and has two lateral fins.

Mattei (1970; 1991) notes that the Scorpaeniformes include species with internal fertilization which exhibit two types of gamete: a form that deviates slightly from the simple gamete by a slight elongation of the nucleus and the presence of two mitochondrial layers in *Sebasticus* (referring to Mizue 1968; contrast four layers in *Sebasticus marmoratus*) and an evolved form with a well-developed nucleus and midpiece in *Oligocottus*.

Helicolenus d. dactylopterus is a zygoparous species. The male urogenital papilla acts as the copulating organ. The fertilized ova are retained in the female reproductive tract for a up to 10 months prior to their release (Munoz *et al.* 2002a). After insemination of the female, the spermatozoa are stored in cryptal structures that are situated in the interlamellar gaps, very near the muscular-connective tissue rachis of the ovary, and are connected to the ovarian lumen by a duct. Within the crypt the sperm heads usually point towards the cryptal epithelium and the tails towards the cavity (Munoz *et al.* 2000, 2002a). At fertilization, the sperm are expelled by means of spasmodic contraction from the crypts (Munoz *et al.* 2000). *H. dactylopterus* is considered to occupy an intermediate position, within the family Scorpaenidae, between strictly ovuliparous species, such as some *Scorpaena* and *Sebastolobus*, and viviparous species with several levels of matrotrophy, such as those of the genus *Sebastes*. The elongation of the sperm nucleus and of the midpiece is characteristic of the transport and storage of sperm in internally fertilizing species. No intraovarian embryos have been detected, and it is deduced that spawning [of fertilized eggs] probably takes place shortly after fertilization (Munoz *et al.* 2002a). Mattei (1991) examined but did not describe the sperm of this species.

Sebasticus marmoratus. *S. marmoratus* also has internal fertilization and is ovoviparous. The late spermatid, indicative of the mature morphology, has been described (Mizue 1968; Lin and You 1998; You *et al.* 2002). The nucleus is about twice as long as wide and is depressed on one ('dorsal') face. The mitochondria, in a sleeve surrounding the cytoplasmic canal, are illustrated as lying in four tiers and are estimated to number 30-40 (Lin and You 1998). The figure of 8 cited by Mizue (1968) possibly refers to a single tier. It is stated that the proximal centriole is lost by maturity (Fig. 15.14). The 9+2 flagellum has two lateral fins.

Only the normal axonemal structure is described for *Sebastes schlegeli* (Chung and Chang 1995).

Subfamily Tetraroginae: Tetrarogins are the extremely venomous sailback scorpion fishes or wasp fishes.

***Paracentropogon (=Hypodytes) rubripinnis*:** The spermatozoon of *P. (=Hypodytes) rubripinnis* (Fig. 15.15A,B) has a spherical nucleus with a shallow apical "dip" owing to an hiatus in the chromatin. The chromatin consists of dense masses with scattered pale lacunae. The distal centriole (basal body) but not the proximal centriole has been observed. A single tier of small spherical mitochondria surrounds the base of the flagellum. The flagellum has a pair of lateral fins (Hara and Okiyama 1998).

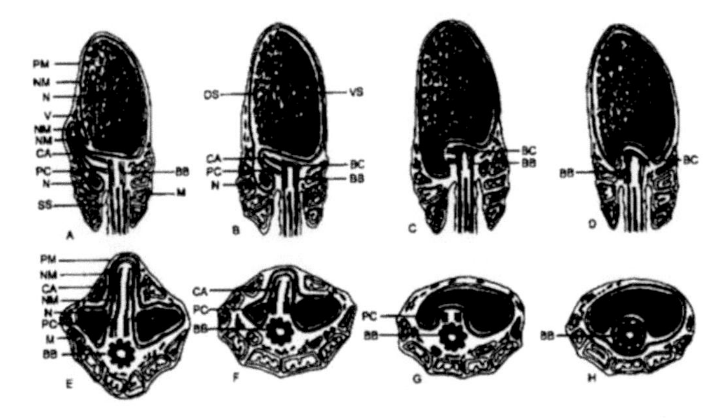

Fig. 15.14 *Sebasticus marmoratus* (Scorpaenidae). Diagrams of centriolar changes in the late spermatid. **A-D**. Median sagittal sections. **E-H**. Transverse sections through the distal centriole (basal body). BB, basal body; BC, basal body cap; CA, 'centriolar adjunct'; DS, dorsal face of nucleus; M, mitochondrion; N, nucleus; NM, nuclear membrane; PC, proximal centriole; PM, plasma membrane; SS, cytoplasmic canal; V, vesicle; VS, ventral face of nucleus. Lin, D. J. and You, Y. I. 1998. Zoological Research 19: 359-366.

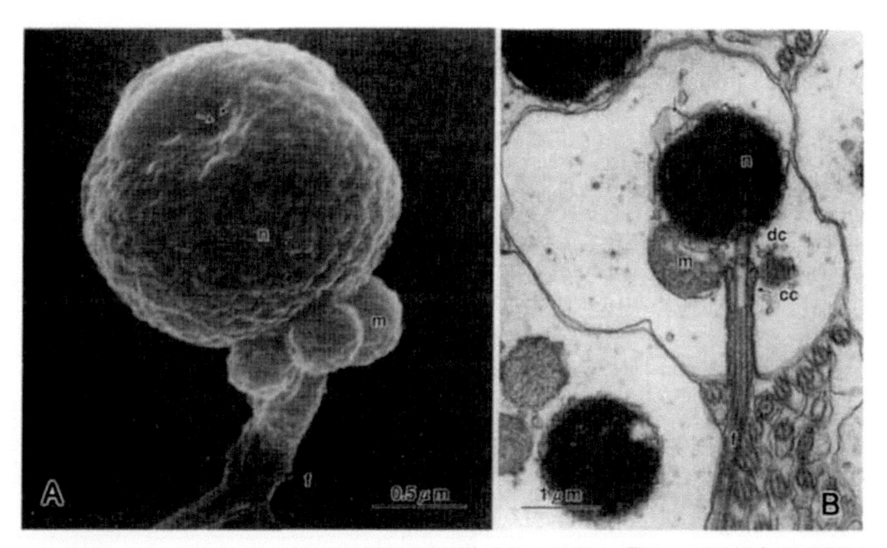

Fig. 15.15 *Paracentropogon rubripinnis* (Scorpaenidae, Tetraroginae) spermatozoa. **A**. SEM of the nuclear region. **B**. TEM longitudinal section. Adapted from Hara, M. and Okiyama, M. 1998. Bulletin of the Ocean Research Institute, University of Tokyo 33: 1-138, Fig. 37B and C. With permission.

Remarks: The round nucleus with an apical "dip" is unknown in other scorpaeniforms but is seen, presumably with no phylogenetic implications beyond a common genetic propensity, in the cichlid *Cichla intermedia* and in the pleuronectiforms *Paralichthys olivaceus* and *Scophthalmus maximus*.

15.6.2.2 Family Triglidae

Triglids are the sea robins or gurnards, in tropical and temperate seas. Mattei (1991) listed but did not describe the sperm of the triglid *Peristedion cataphractum*. That of *Trigla lyra* has been briefly described, in an account of gametogenesis, by Munoz *et al.* (2002c). It has a round acrosomeless head of uniform density. The midpiece is 0.9 μm long and contains a ring of three or four spherical mitochondria. The flagellar axoneme has the 9+2 pattern of microtubules (Munoz *et al.* 2002c). Micrographs reveal a moderate, compact basal nuclear fossa from which the flagellum projects, indicating a Type I sperm *sensu* Mattei.

15.6.3 Suborder Hexagrammoidei

15.6.3.1 Family Hexagrammidae

These are the greenlings; marine, North Pacific. The sperm of *Hexagrammos agrammus* (Fig. 15.16), *H. octogrammus* and *H. otakii* have been described by Hara and Okiyama (1998). The nucleus is elliptic or elongate ovoid and incompletely surrounds the axoneme. The proximal centriole is at 45° to the distal centriole and both are exposed near the tip of the nucleus. In H. *agrammus* they are slightly displaced relative to each other. A centriolar plug has been demonstrated on the distal centriole in *H. octogrammus*, a single (*H. otakii, H. agrammus*) or four (*H. octogrammus*) spherical mitochondria are situated as the posterior end of the nucleus. Tube-like fins (not lateral laminae) are present on the single flagellum. The axoneme has the usual 9+2 pattern of microtubules. ITDs are said to be present in all A subtubules but this perhaps needs confirmation.

Remarks: A similar arrangement, with the nucleus enwrapping the flagellum and exposure of centrioles anteriorly is seen in the trichodontid *Arctoscopus japonicus* (adult, Fig. 15.16F) and is considered a synapomorphy of the Hexagrammidae and the Trichodontidae (Hara and Okiyama 1998).

15.6.4 Suborder Cottoidei

As noted by Hayakawa and Munehara (2004) the reproductive modes in cottoids can be divided into non-copulatory and copulatory, although they are oviparous (Breder and Rosen 1966; Munehara *et al.* 1989). Copulation of cottoids is further divisible into two types: 1) true copulation whereby the male transfers spermatozoa to the female reproductive tract by inserting an interomittent organ and 2) internal insemination following external sperm

Fig. 15.16 *Hexagrammos agrammus* (Hexagrammidae). **A**. SEM of whole spermatozoon. **B**. SEM of spermatozoon. **C**. TEM longitudinal section through nucleus near surface of sperm. **D**. TEM longitudinal section of spermatozoon. **E**. TEM transverse section through nucleus. c, centriole; cc, cytoplasmic canal; cp,

Fig. 15.16 Contd. ...

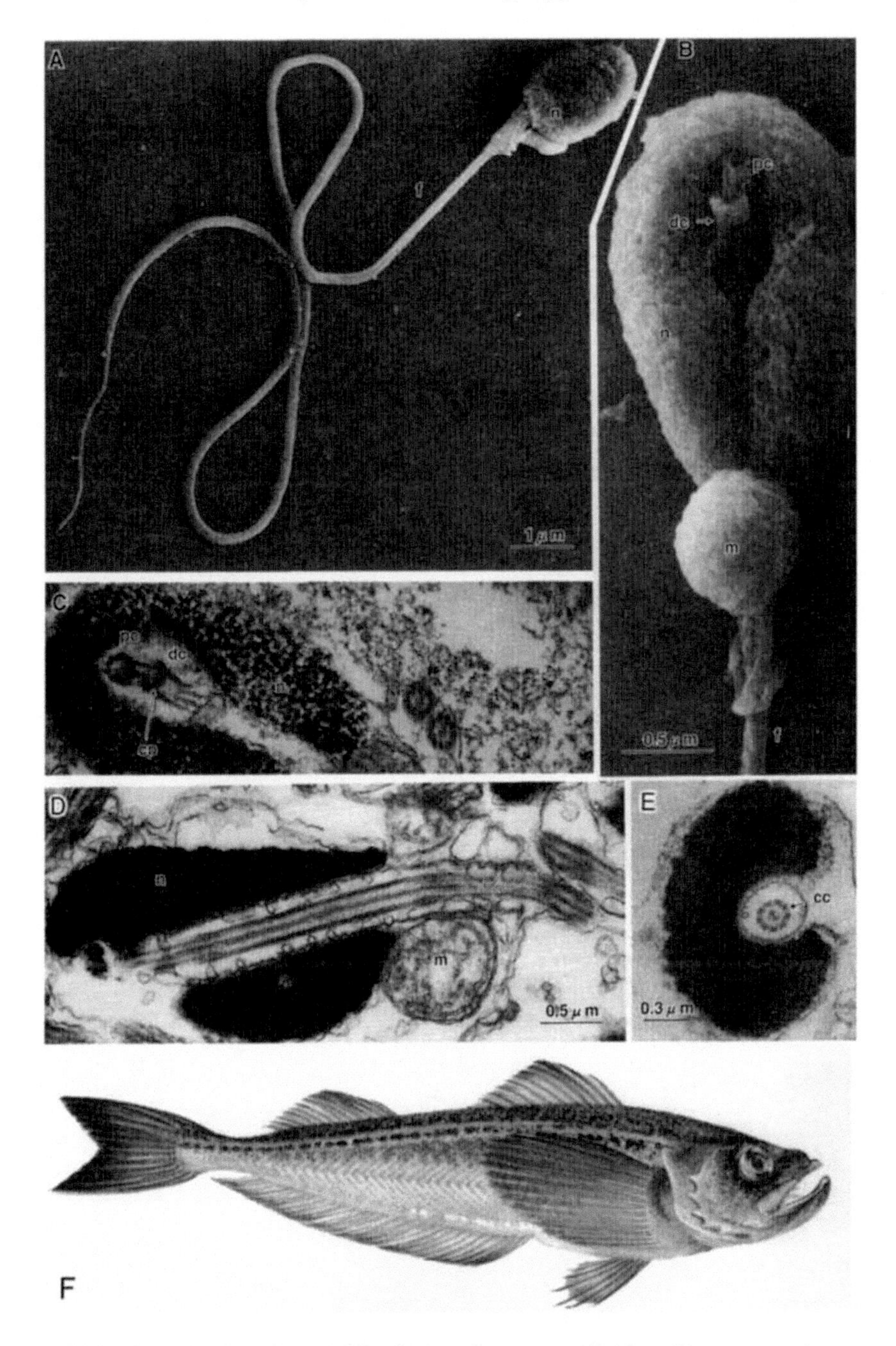

centriolar plug; dc, distal centriole; f, flagellum; m, mitochondrion; n, nucleus; pc, proximal centriole. Adapted from Hara, M. and Okiyama, M. 1998. Bulletin of The Ocean Research Institute, University of Tokyo 33: 1-138, Fig. 40A-A-C, E, F. With permission. **F**. *Arctoscopus japonicus*, Sailfin sandfish. By Norman Weaver. With kind permission of Sarah Starsmore.

transfer, as in *Blepsias cirrhosus*. In the latter mode, the male emits semen towards the ovarian fluid emerging from the tip of the female reproductive organ (Munehara *et al.* 1989; Munehara 1996).

15.6.4.1 Family Cottidae

Sperm structure: Hann (1930) examined, by light microscopy, the sperm of thirty-two species of Cottidae, from twenty genera. Within the family he found two general types of sperm, the oval and the slender, with intergrading forms, in various species, between. Those having the oval type include the genus *Cottus* while the slender-sperm forms are chiefly from the subfamily, Oligocottinae. Some of the species with oval sperm had additional 'spermatid masses' which were discharged in the milt. The spermatid masses were found in seven species and subspecies of *Cottus* and in three other genera of the family, but have not been found outside the Cottidae.

Ultrastructural descriptions of sperm exist for *Alcichthys alcicornis* (Koya *et al.* 1997, 2002); *Cottus gobio* (Stein 1981; Lahnsteiner *et al.* 1997); *Hemilepidotus gilberti* (Hayakawa *et al.* 2002a,c; Hayakawa, Youichi *et al.* 2002b; Hayakawa 2003; Hayakawa *et al.* 2004; Hayakawa and Munehara 2004); and *Oligocottus maculosus* (Stanley 1966; 1969).

***Cottus gobio*:** The sperm of *Cottus gobio*, the Miller's Thumb or River Bullhead (adult, Fig. 15.17), has been described by Stein (1981) and Lahnsteiner *et al.* (1997).

Fig. 15.17 *Cottus gobio*, the Miller's Thumb or River Bullhead. After Buckland, F. 1891. *Natural History of British Fishes*. SPCK, London, p. 24.

Cottus gobio: The spermatozoon (Fig. 15.18) of *Cottus gobio* is asymmetrical as the flagellum inserts laterally (here termed ventrally) to the nucleus. It has no acrosome and the head region is elongated (length 2.24 ± 0.12 μm, diameter 0.79 ± 0.09 μm, n = 10), cf 2.8 μm (Stein 1981). The long midpiece, which the posterior region of the nucleus overlaps, is approximately cylindrical in shape (length 1.98 ± 0.12 μm, diameter 0.79 ± 0.09 μm, n=10) (Lahnsteiner *et al.* 1997), cf. 1.7 μm long (Stein 1981) and contains five to six large, ovoid mitochondria (length 0.71 ± 0.11 μm, diameter 0.41 ±.0.09 μm, n=10) (Lahnsteiner *et al.* 1997). As many as three mitochondria are seen in a longitudinal row (Stein 1981). It is penetrated by the cytoplasmic canal (diameter 0.48 ± 0.02 μm, n=10). The cytoplasm contains numerous electron-dense particles, consisting of rosette-shaped subparticles with diameters of 0.003-0.005 μm (n = 10), indicative of glycogen α-particles. The flagellum has the 9+2 pattern of microtubules. A single lateral axonemal fin is present according to Lahnsteiner *et al.* (1997) but Stein (1981) reports two very long fins. These authors do not report the presence of parasperm in this species.

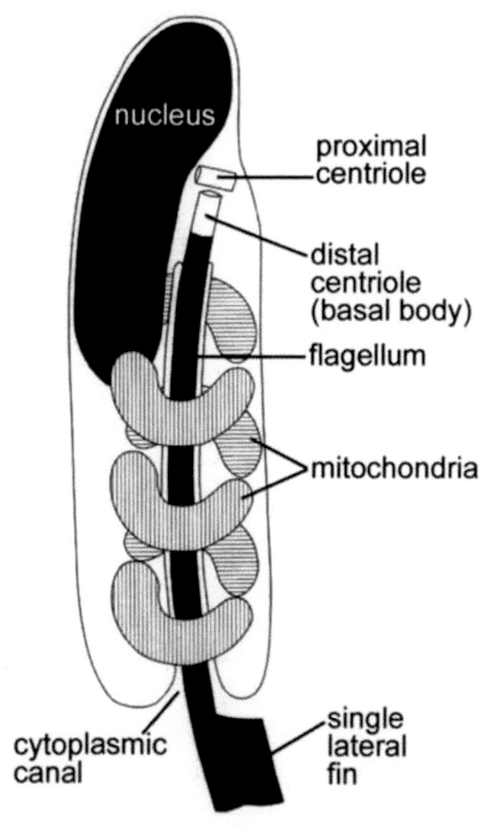

Fig. 15.18 *Cottus gobio* (Cottidae). Schematic reconstruction of the spermatozoon. After Lahnsteiner, F., Berger, B., Weismann, T. and Patzner, R. A. 1997. Journal of Fish Biology 50: 564-574, Fig. 5. With permission of Blackwell Publishing.

Remarks: The structure of the spermatozoon of *Cottus gobio* (Fig. 15.18) which is externally fertilizing, is similar to that of the internally fertilizing *Oligocottus maculosus* (Fig. 15.20) though less elongate Its structure, physiology and metabolism appear adapted for internal fertilization but, although the partners appose their ventral surfaces, they spawn in sheltered caves with low water flow (Lahnsteiner *et al.* 1997). As the sperm have the structure associated with internal fertilization, one is tempted to speculate that there has been a reversal to external from internal fertilization.

***Oligocottus maculosus*:** The relationship with *Cottus* of *Oligocottus maculosus*, the Tide Pool Sculpin, of which the sperm is described by Stanley (1966, 1969) (adult Fig. 15.19, spermatozoon Fig. 15.20), is seen in the spade-like form of the nucleus with the flagellar basal body and proximal centriole inserted into a groove along of the flattened sides of the nucleus, parallel with its long axis.

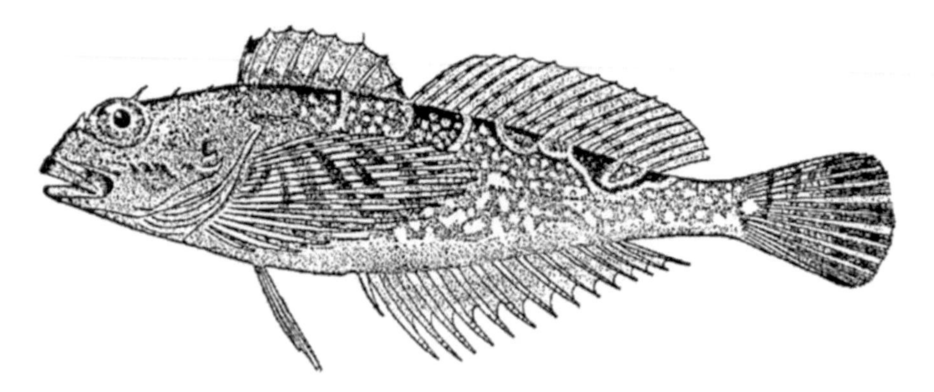

Fig. 15.19 *Oligocottus maculosus.* After Jordan, D. S. 1907. *Fishes.* Henry Holt, New York, Fig. 566.

The nucleus is more elongate in this species, however. A further difference is transformation of the mitochondria of the spermatid into two elongate bodies surrounding the base of the tail in the cytoplasmic sleeve common to both species.

The elongation of the nucleus and development of elongate mitochondrial derivatives are features frequently seen in internally fertilizing animal sperm. It is therefore noteworthy that this species has a penis and that the existence of internal fertilization has been established (Breder and Rosen 1966).

***Hemilepidotus gilberti*:** There is a series of papers, using electron microscopy, on the non-copulatory marine sculpin *Hemilepidotus gilberti* (Cottidae) (Hayakawa *et al.* 2002a; Hayakawa, Youichi *et al.* 2002b; Hayakawa *et al.* 2002c; Hayakawa 2003; Hayakawa *et al.* 2004; Hayakawa and Munehara 2004), which has external fertilization, and *Blepsias cirrhosis* (Hemitripteridae), a species with external sperm transfer but with internal insemination (Hayakawa and Munehara 2004). These authors have recognized the aflagellate cells in the spermatid masses as parasperm, coexisting with the flagellate eusperm, terms coined for such dimorphism by Healy and Jamieson (1981) Hayakawa *et al.* (2002c, 2004) attributed to the parasperm a role in preventing dispersal of eusperm and impeding fertilization by the eusperm of other males.

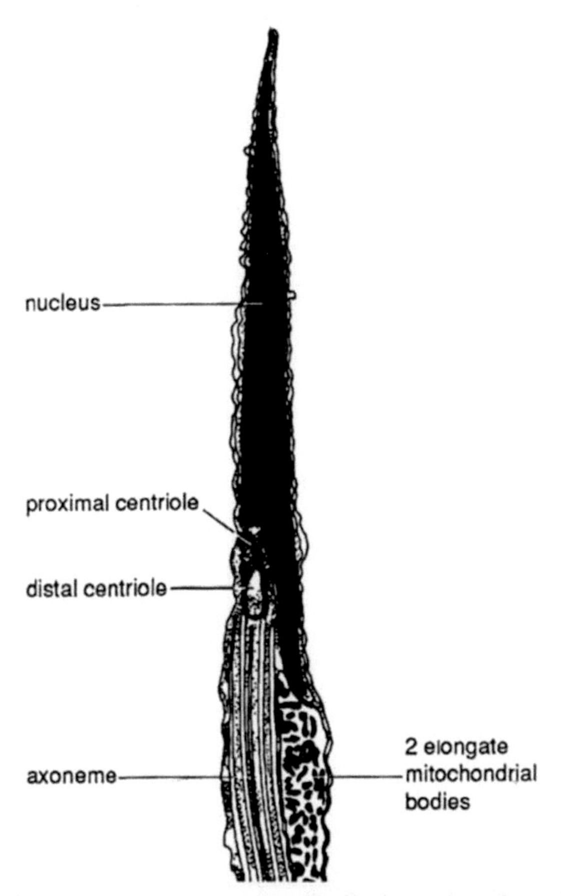

Fig. 15.20 *Oligocottus maculosus.* Longitudinal section through the "sagittal" plane of a late spermatid. From Jamieson (1991) after a micrograph by Stanley, H. P. 1969. Journal of Ultrastructure Research 27: 230-243, Fig. 12.

The euspermatozoon of *Hemilepidotus gilberti*, described by Hayakawa, Youichi *et al.* (2002b), consist of a head, lacking an acrosome, a short midpiece and a long flagellum. The nucleus is almost round in dorsal or ventral (frontal) view (Fig. 15.21C) but is seen to be disc-like when viewed in sagittal longitudinal section, measuring 2.7-3.0 μm in length and 1.6-1.8 μm in width. An elongated groove in the nucleus (seen in transverse section in the upper sperm in Fig. 15.21A) extends from the "caudal" portion of the midpiece. The nucleus appears (Fig. 15.21A) to end caudally anterior to the midlength of the midpiece, however. The axoneme is inserted, within this groove, near the tip of the nucleus, where it originates from the distal centriole which in turn is surmounted by the proximal centriole (Fig. 15.21A). Two or three relatively large mitochondria, elongate on one side of the axoneme, constitute the midpiece. The portion of the midpiece posterior to the nucleus measures about 1.4 μm and surrounds the proximal, but not the most anterior part, of the flagellum. The flagellum is ca 20 μm long and has the usual 9+2 configuration of microtubules. The lateral flagellar fin is said to be unpaired but is shown in Fig. 15.21B and in the corresponding legend to be paired.

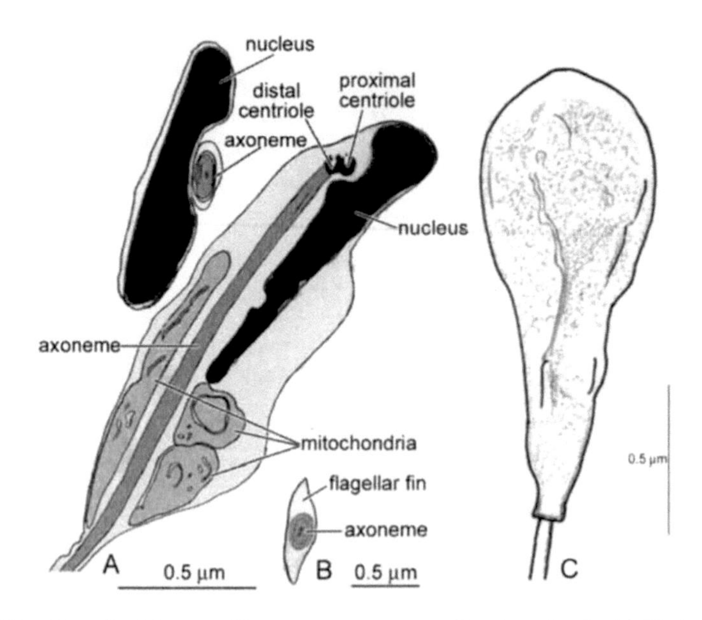

Fig. 15.21 *Hemilepidotus gilberti* spermatozoa (Cottidae). **A**. TEM longitudinal section of spermatozoon, showing lateral view (below) and approximate transverse section (above). **B**. TEM transverse section of flagellum showing lateral fins. **C**. SEM of spermatozoon in frontal view. dc, distal centriole; h, sperm head; m, mitochondria; n, nucleus; pc, proximal centriole. Semidiagrammatic representation based on micrographs of Hayakawa, Y., Komaru, A. and Munehara, H. 2002. Journal of Morphology 253: 243-254, Fig. 2B,C,D.

Alcichthys alcicornis: In the copulating sculpin, *Alcichthys alcicornis*, the sperm are stored in the ovary but float freely in the ovarian fluid, there being no specialized storage. This species is ovuliparous and shows internal gametic association (as, also, in *Oligocottus maculatus*), a term signifying intimate association of male and female gametes in the female reproductive tract followed by release of the eggs and subsequent fertilization in the external environment (Munehara 1988; Koya *et al.* 1997; Koya *et al.* 2002; see also Evans and Downing, **Volume 8B, Chapter 4, Section 4.3.7**) (Fig. 15.22C). In two most remarkable micrographs Koya *et al.* (2002) show spermatozoa entering the micropyle (Fig. 15.22A) and traversing it to the egg surface (Fig. 15.22B), before extrusion of the egg and its associated sperm into the external seawater.

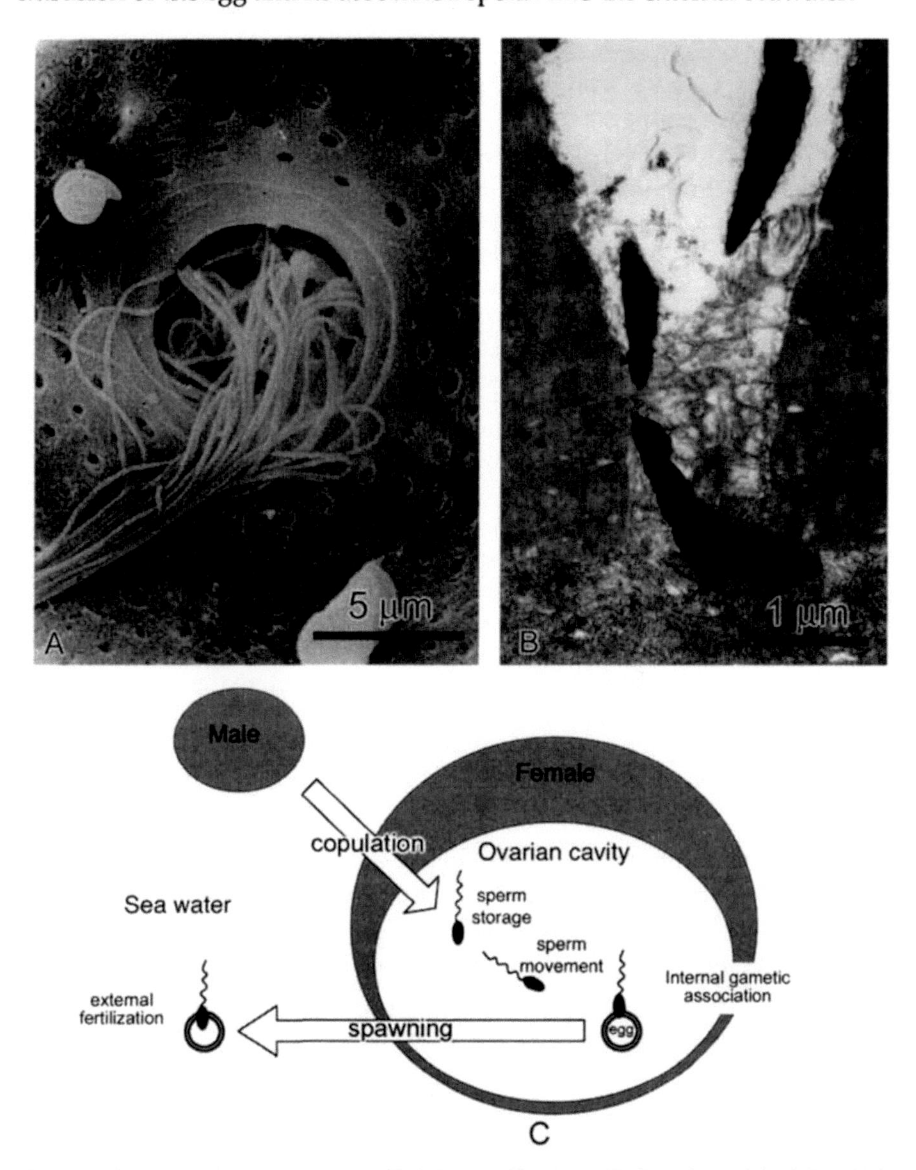

Fig. 15.22 *Alcichthys alcicornis* (Cottidae). Sperm entering the micropyle of the egg in the ovarian cavity. **A**. SEM surface view. **B**. TEM of sperm in the micropyle. Electron-dense sperm nuclei are seen. **C**. Demonstrating internal gametic association. Sperm are introduced into the ovary by copulation and later enter the micropyle of ovulated eggs but membrane fusion, breakdown of cortical alveoli and resumption of the second meiotic division does not occr until eggs are relased into the sea water. Courtesy of Dr. Yasunori Koya.

The spermatozoon of *Alcichthys alcicornis* is ca 37 μm long. The nucleus and midpiece are elongated and this region is thick and spatula-shaped, ca 5.1 μm long, 1.6 μm wide and 0.9 μm thick Two centrioles are present slightly posterior to the midlength of the flattened, acrosomeless nucleus. From the distal centriole, a groove with an axoneme runs posteriorly along the long axis of the head, as in *Cottus gobio*, *Oligocottus* and Hexagrammidae all of which are Scorpaeniformes. In *A. alcicornis,* a midpiece joining the posterior end of the head, with no appreciable delimitation, has many small mitochondria surrounding the axoneme (as many as 10 mitochondrial sections being seen in one section). On the centriolar side of the nucleus, several mitochondria are distributed anteriorly along the axoneme in the head portion. A short, thin cytoplasmic sleeve surrounds the proximal portion of the sperm tail, and a 9+2 arrangement of microtubules in the flagellum are seen, as characteristic of the spermatozoa of most teleost species (Koya *et al.* 2002).

Remarks: Internal gametic association occurs in the three families Cottidae, Hemitripteridae and Agonidae (Koya *et al.* 2002) and is also reported for the aulorhynchid *Aulichthys japonicus* (Hara and Okiyama 1998) (see **Section 15.5.1.2**). As they note, elongation of the nucleus correlates with internal gametic association, in which the eggs are inseminated internally but fertilization is delayed until the eggs have been shed to the exterior, as also seen in *Oligocottus maculosus*. Such elongation has also been observed in *O. snyderi* and *O. rubellio* (Fink and Haydon 1960). In contrast, in the externally fertilizing *Cottus gobio* the sperm nuclei are rounded though the midpiece, with as many as three mitochondria, is elongated (Stein 1981). The large number of mitochondria found in *Alcichthys alcicornis* is thought to contribute to the extraordinarily long period of motility of the sperm of this fish (Koya *et al.* 2002).

15.6.4.2 Family Hemitripteridae

Hemitripterids are the sea ravens, in the northwestern Atlantic and North Pacific (Nelson 2006).

***Blepsias cirrhosus*. The euspermatozoon** of *Blepsias cirrhosus* is described by Hayakawa and Munehara (2004) (Fig. 15.23). The head, lacking an acrosome and occupied by the nucleus, is disk-like in shape, 1.6 –2.0 μm long, 1.3–1.6 μm wide and 0.6–1.08 μm thick (Fig. 15.23B-D). The flattened form of the head is apparent in lateral views (Fig. 15.23C) and in transverse section (Fig. 15.23E–G). An elongated groove extends laterally from the base to the apical end of the nucleus and in it is lodged the anterior portion of the axoneme. Thus, the proximal centriole is located at the anterior part of the nucleus (Fig. 15.23E). The long midpiece (ca 3 μm long) contains several mitochondria and surrounds the proximal portion of the flagellum from which it is separated by a deep cytoplasmic canal (Fig. 15.23E,H–J). The flagellum is ca 30 μm long and the axonemal complex shows the usual 9+2 arrangement within the plasma membrane and a small amount of cytoplasm (Fig. 15.23K).

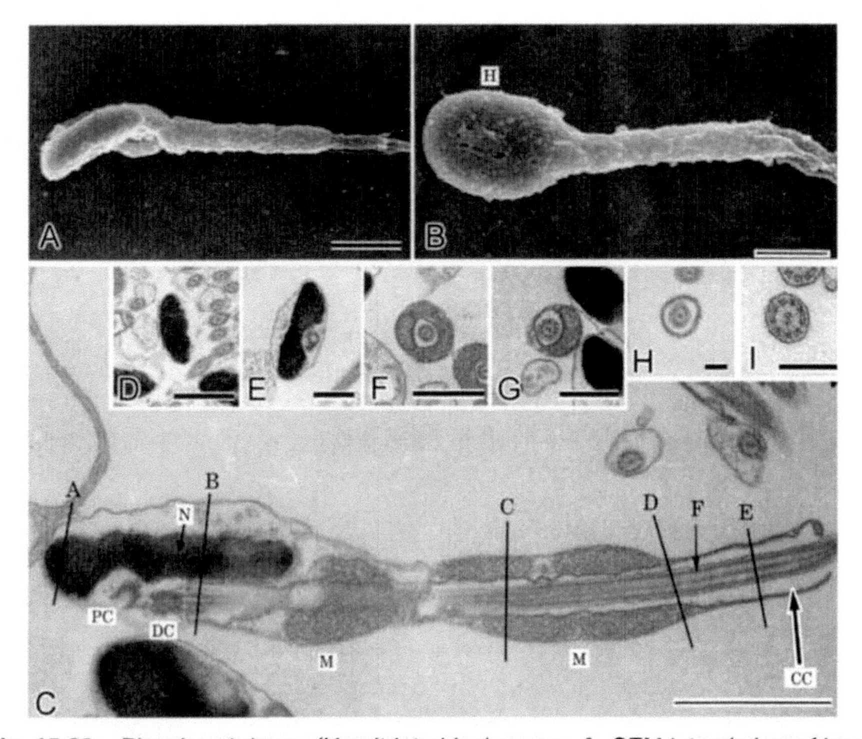

Fig. 15.23 *Blepsias cirrhosus* (Hemitripteridae) sperm. **A**. SEM lateral view of head and midpiece. **B**. Dorsal (or ventral) view of head and midpiece. **C**. TEM sagittal section of a euspermatozoon. **D-I**. Transverse sections at levels A-E in **C**. Scale bars in **A-C** 1 μm; in **D-F** 0.5 μm; in **H** and **I** 0.2 μm. CC, cytoplasmic canal; DC, distal centriole; F, flagellum; M, mitochondria; N, nucleus; PC, proximal centriole. Adapted from Hayakawa, Y. and Munehara, H. 2004. Journal of Fish Biology 64: 1530-1539, Fig1. With permission of Blackwell Publishing.

Paraspermatozoa: Parasperm confirmed by ultrastructural investigation have been reported for *Blepsias cirrhosus* (Hayakawa and Munehara 2004), *Cottus hangiongensis* (Quinitio and Takahashi 1992a,b); *C. nozawae* (Quinitio 1989, *fide* Hayakawa *et al*. 2007) and *Hemilepidotus gilberti* (Hayakawa *et al*. 2002a,b, 2007). At least nine other species have been suggested, from light microscopy, to have parasperm (Hayakawa *et al*. 2007). Hayakawa and Munehara (2004) recognized certain aberrant spermatids in *Blepsias cirrhosus* as developing paraspermatozoa. In the testis undergoing spermiogenesis, aberrant spermatids can be observed with normal euspermatids in the same cyst. The aberrant spermatids differ from normal euspermatids in the abundance of cytoplasm. Aberrant spermatids found in the same cyst along with late euspermatids are binuclear and this is also true for aberrant spermatids seen in the sperm duct. Nuclei of the cells are observed as masses of highly electron-dense chromatin globules, ranging from 0 1to 0 6 μm in diameter. Aberrant spermatids are connected to euspermatids by intercellular bridges.

At the completion of development, the cells are spherical in shape, ranging from 5.0 to 5.3 µm in diameter. All aberrant spermatids have two flagella that are shorter than those of euspermatozoa. They are considered to be hyperpyrenic. These cells do not degenerate after spermiation and are observed in the common sperm duct as a component of semen that would be released at ejaculation. Aberrant spermatids were observed in the testis of all the males examined.

Remarks: As noted by Hayakawa and Munehara (2004), the male *Blepsias cirrhosus* has no intromittent organ for copulation, because sperm transfer is external although insemination occurs internally. The female everts the genital duct and emits gelatinous ovarian fluid from its tip. The male emits semen towards the gelatinous ovarian fluid, and the female retracts it into the ovary. Semen clinging to the ovarian fluid is stored in the ovary (Munehara, 1996).

15.7 ORDER PERCIFORMES

The Perciformes (perch-like fishes) is the most diverse of all fish orders and, indeed, is the largest vertebrate order. Perciforms dominate in vertebrate ocean life and among fish of many tropical and subtropical fresh waters. These and other statements about the group have little meaning in view of the doubtful monophyly of the group for which uniquely diagnostic features are not known (Lauder and Liem, 1983; Nelson 2006). Polyphyly and paraphyly of the perciforms is confirmed by molecular analysis (Ortí and Li, **Chapter 1**; see also Fig. 15.1).

Table 15.1 Occurrence of Type I and Type II spermiogenesis *sensu* Mattei 1970 in the Percomorpha. bp = brood pouch; ef = external fertilization; ef[1] putative external fertilization; If = internal fertilization. Iga = internal gametic association.

Taxon	*Type I*	*Type II*	*Other*	*Mode*
Beryciformes	I			ef
Holocentridae				
Zeiformes	I			ef
Zeidae				
Synbranchiformes			I biflagellate	ef
Synbranchidae				
Gasterosteiformes				
Gasterosteoidei				
Hypoptychidae			I Modified	ef[1]
Aulorhynchidae			I Modified	iga
Gasterosteidae			I Modified	ef
Synganthoidei				

Table 15.1 Contd. ...

Table 15.1 Contd. ...

Taxon	*Type I*	*Type II*	*Other*	*Mode*
Fistulariidae	I			ef[1]
Syngnathidae			I Modified	bp
Scorpaeniformes				
Dactylopteroidei				
Dactylopteridae			I Modified	ef[1]
Scorpaenoidei				
Scorpaenidae				
Scorpaeninae	I			ef
Sebastinae			I Modified	if
Tetraroginae	I			ef[1]
Triglidae	I		I Modified	ef[1]
Hexagrammoidei				
Hexagrammidae			I	ef
Cottoidei				
Cottidae			I? Modified, some dispermic	ef iga if
Hemitripteridae			Eu- and para-sperm	if
Perciformes				
Percoidei				
Acropomatidae		II		ef[1]
Apogonidae			Biflagellate	iga
Carangidae		II		ef[1]
Centracanthidae	I			ef
Centropomidae		II		ef
Cepolidae		II		ef[1]
Chaetodontidae		II		ef
Echeneidae		II		ef[1]
Emmelichthyidae		II		ef[1]
Gerreidae		II		ef[1]
Haemulidae		II		ef[1]
Kuhlidae			Intermediate I-II	ef
Latidae		II		ef
Lutjanidae		II		ef[1]
Malacanthidae		II		ef[1]
Monodactylidae		II		ef[1]
Moronidae		II		ef
Mullidae	I		Deep fossa	ef[1]
Opistognathidae		II		ef
Percichthyidae		II		ef
Percidae		II		ef

Table 15.1 Contd. ...

Table 15.1 Contd. ...

Taxon	Type I	Type II	Other	Mode
Pomatomidae		II		ef
Polynemidae		II		ef
Priacanthidae		II		ef[1]
Scianidae		II		ef[1]
Serranidae		II		ef
Sparidae	I			ef
Labroidei				
Cichlidae, most				
Satanoperca	I		I Modified	ef
			Biflagellate	if
Embiotocidae			Modified	if
Pomacentridae	I		Slightly modified	ef
Labridae	I	II		ef
Zoarcoidei				
Zoarcidae			Modified, biflagellate	if
Anarhichadidae			I Modified, facultatively biflagellate	if
Trachinoidei				
Ammodytidae			I Modified	ef[1]
Pinguipedidae	I			
Percophidae	I		Slightly modified	
Trachinidae		II		ef
Trichodontidae			I Modified	ef
Uranoscopidae		II		ef[1]
Blennioidei				
Blenniidae	I	II	See modifications	ef
Clinidae		II	Or strongly modified	ef
				if
Gobiesocoidei				
Gobiesocidae			Strongly modified, biflagellate	ef
Gobioidei			See Table 15.5	
Eleotridae			I modified	ef
Gobiidae	I		And modifications of I	ef
Odontotubidae			I Variously modified	ef
Pterelotridae			Greatly modified	ef
Rhyacichthyidae	I			ef
Schindleridae			Highly modified	ef
Acanthuroidei				
Acanthuridae		II		ef[1]
Ephippidae		II		ef[1]

Table 15.1 Contd. ...

Table 15.1 Contd. ...

Taxon	Type I	Type II	Other	Mode
Siganidae	I			ef
Scombroidei				
Istiophoridae	I			ef
Scombridae		II		ef
Sphyraenidae		II		ef[1]
Trichiuridae		II		ef
Stromateoidei				
Ariommatidae		II		ef[1]
Anabantoidei				
Anabantidae		II		ef
Osphronemidae	I			ef
Pleuronectiformes				
Pleuronectoidei				
Citharidae	I			ef[1]
Scophthalmidae	I			ef
Paralichthyidae	I			ef
Pleuronectidae	I		sl. elongate nucleus	ef
Bothidae	I?			ef[1]
Soleidae	I			ef[1]
Tetraodontiformes				
Balistoidei				
Balistidae	I		Deep fossa	ef
Diodontidae	I			ef[1]
Monacanthidae	I		sl. elongate nucleus	ef
Tetraodontoidei				
Tetraondontidae	I		sl. or greatly elongate nucleus. Fossa deep or not	ef

The perciform sperm type: Mattei (1991) defines a 'perciform type' of spermatozoon, characterized by a rounded nucleus with a depression at its base; the flagellar axis parallel to the base of the nucleus; the centrioles situated outside the nuclear depression; the mitochondria situated at the base of the flagellum; and the axoneme possessing ITDs in the A tubules of doublets 1, 2, 5 and 6, as already defined by Mattei *et al.* (1979). Mattei (1991) attributes the perciform sperm to the families Acropomatidae, Carangidae, Centropomidae, Cepolidae, Chaetodontidae, Echeneidae, Emmelichthyidae, Ephippidae, Gerreidae, Haemulidae, Kuhlidae, Lutjanidae, Malacanthidae, Percichthyidae, Pomatomidae, Priacanthidae, Scianidae and Serranidae. He also attributes this sperm type to the Acanthuroidei, Anabantoidei, Mugiloidei, Polynemoidei, Sphyraenoidei and Stromateoidei. In all, he states that the perciform type was

Fig. 15.24 Perciform (Type II) spermatozoon, according to Mattei, X. 1991. Canadian Journal of Zoology 69: 3038-3055, Fig. 13 (part). With permission.

encountered, in the Perciformes, in 29 of the 41 families studied, in 10 suborders, mainly the Percoidei, in which 19 out of 23 families exhibited it. It was regarded as an evolved type. This is the Type II sperm recognized by Mattei (1969, 1970), though the Type II desgination was abandoned by Mattei (1991). However, it may be noted that Mattei (1969) regarded Type I as the evolved type and Type II as the primitive type. I concur with the later view that suppression of nuclear rotation in Type II spermiogenesis is evolved relative to the full rotation, with movement of the centrioles into the nuclear fossa, which occurs in Type I spermiogenesis (Mattei 1970). The two types of spermiogenesis are further defined in **Section 15.7.1**.

The perciform (Type II) sperm type shows interspecific variation in the size and shape of the nucleus and the number of mitochondria (Mattei 1991).

15.7.1 Suborder Percoidei

Many decades of research in systematic ichthyology have failed to find a meaningful definition of, or boundaries for, the Percoidei (Johnson 1984). Percoids are best represented in the near shore marine environment and form a significant component of the reef-associated fish fauna of tropical and subtropical seas. Association with brackish water occurs in many such families and some have one or more exclusively freshwater members, but some families are restricted to freshwaters: the north temperate Percidae and Centrarchidae, the south temperate Percichthyidae and the tropical Nandidae (Johnson 1984). I follow Kaufman and Liem (1982), Lauder and Liem (1983), Johnson (1984) and Nelson (2006) in excluding the Pomacentridae, Cichlidae and Embiotocidae, placing these, with the Labridae, in the Labroidei.

Chief sperm types: Within the percoids but supposedly generalizable to all teleosts, Mattei (1970) recognizes two types of spermiogenesis and therefore two types of sperm. These are exemplified by the Mullidae and Haemulidae, respectively.

Spermiogenesis in the mullid *Pseudupeneus (=Upeneus) prayensis* is illustrated as an example of what Mattei (1970) designates "Type I" spermiogenesis (Fig. 15.25). In this four stages are recognized: (1) the young spermatid, with central nucleus, basal centrioles, and mitochondria scattered in the cytoplasm; (2) migration of the centrioles in the direction of the nucleus, drawing with them the flagellum and the cell membrane to give a cytoplasmic canal; (3) formation of the basal nuclear fossa and a 90° rotation [of the nucleus into the same axis as the flagellum] bringing the centrioles into the nuclear fossa; and (4) migration of the mitochondria to the base of the nucleus. The resultant anacrosomal aquasperm is approximately symmetrical (Fig. 15.25G). For further details of the Mullidae, see that family (**Section 15.7.1.14**).

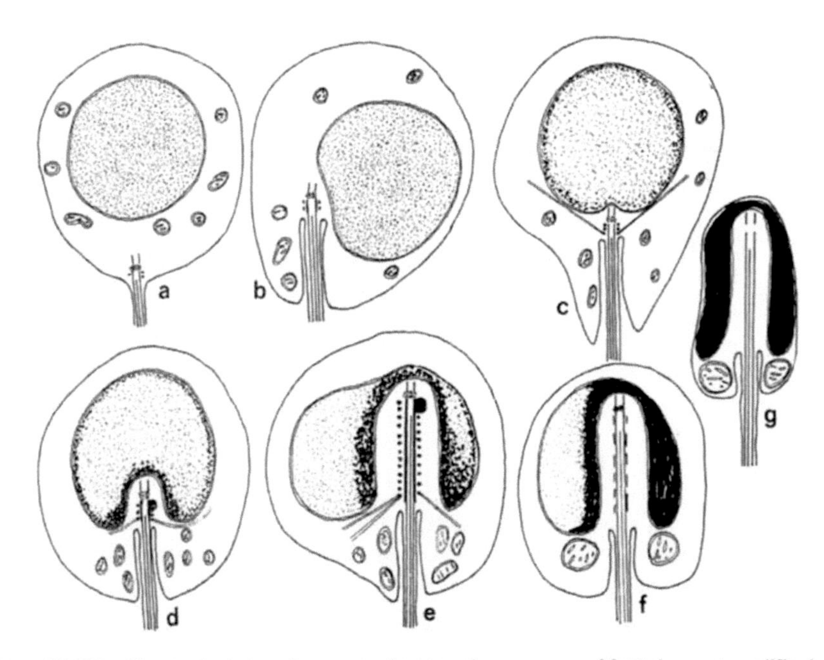

Fig. 15.25 Type I teleost spermiogenesis *sensu* Mattei, exemplified by *Pseudupeneus (=Upeneus) prayensis* (Mullidae). Note that the deep nuclear fossa is not a required condition for recognition of this sperm type. From Mattei, X. 1970. Pp. 567-569. In B. Baccetti (ed.), *Comparative Spermatology*, Academic Press, New York, Fig. 5.

The type I spermiogenesis of *Pseudupeneus (=Upeneus)* (Fig. 15.25) contrasts with "type II" spermiogenesis, exemplified by the haemulid (=Pomadasyid) *Parapristipoma octolineatum* (Fig. 15.26) (see also Mattei 1970).

In Type II spermiogenesis, at stage (2) the centrioles behave similarly but the axis of the flagellum becomes tangential relative to the nucleus. At (3) a nuclear fossa again forms but no rotation of the nucleus occurs and the mitochondria surround the centrioles which do not enter the fossa (Mattei 1970). In the axoneme of this type the A tubules of the doublets 1,2,3 and 6 have intratubular differentiations (ITDs). This type is restricted to highly evolved teleosts and questionably only the perciforms, where 25 families out of the 39 studied exhibited a spermatozoon of this model (Mattei 1988). For further details of the Haemulidae, see that family below (**Section 15.7.1.1**).

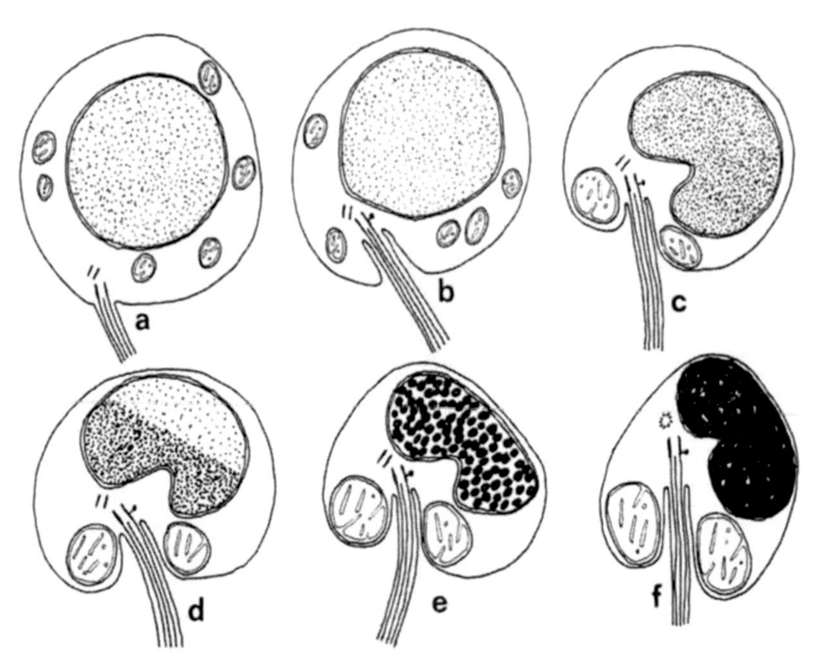

Fig. 15.26 Type II teleost spermiogenesis *sensu* Mattei, exemplified by *Parapristipoma octolineatum* (Haemulidae), African Striped grunt. From Mattei, X. 1970. Pp. 567-569. In B. Baccetti (ed.), *Comparative Spermatology*, Academic Press, New York, Fig. 6.

Operationally this classification has significant utility for mature spermatozoa, even if the development stages are not known, as insertion of the centrioles into the nucleus (Type I) or their isolation from this (Type II) are observable. It is does not appear, however, that all variants of even simple anacrosomal aquasperm (particularly with regard to the orientation of the axoneme) can be described by only two categories. Mattei (1970) recognizes exceptions, citing the existence of an intermediate condition in *Galeoides decadactylus* (Fig. 15.50) and the inapplicability of this classification (outside percomorphs) to the aflagellate sperm of *Gymnarchus niloticus*. Furthermore, in erecting the classification into two sperm types, he does not attempt to place more than 70 teleost sperm investigated into either category.

15.7.1.1 Family Haemulidae

In addition to *Parapristipoma octolineatum* (above), Mattei (1969) examined the sperm of *Pomadasys jubelini* (Fig. 15.27) and Mattei (1991) the sperm of *Brachydeuterus auritus* and *Plectorhinchus mediterraneus*, without description. All three are assignable to the perciform sperm (Type II).

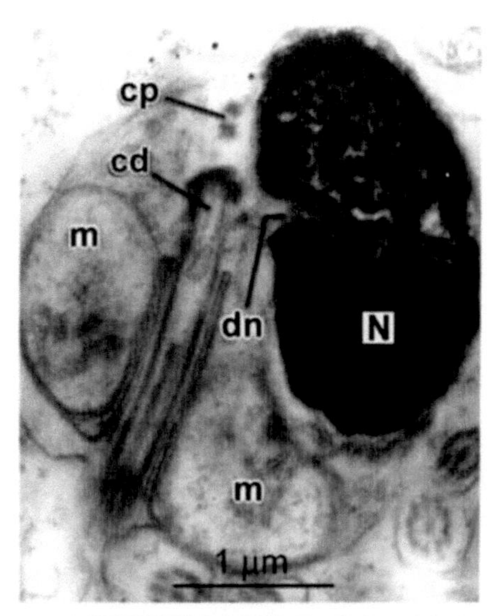

Fig. 15.27 *Pomadasys jubelini* (Haemulidae) spermatozoon. cd, distal centriole; cp, proximal centriole; dn, nuclear fossa; m, mitochondrion, N, nucleus. Adapted from Mattei, X. 1969. *Contribution à l'étude de la spermiogenèse et des spermatozoïdes de poissons par les méthodes de la microscopie électronique.* D.Sc. thesis, Université de Montpellier, Fig. XVIIIb. With permission.

15.7.1.2 Families Emmelichthyidae, Lutjanidae, Gerreidae, Monodactylidae, Chaetodontidae, Priacanthidae, Malacanthidae, Nandidae (Badinae), Pomatomidae, Echeneidae, Acropomatidae

Mattei (1991) has examined the sperm of *Erythrocles monodi* (Emmelichthyidae), *Lutjanus goreensis* (Lutjanidae), *Eucinostomus (=Gerres) melanopterus* (Gerreidae), *Monodactylus (=Psettus) sebae* (Monodactylidae), *Chaetodon hoefleri, C. luciae, C. mancellae* (Chaetodontidae), *Priacanthus arenatus* (Priacanthidae), *Branchiostegus (=Latilus) semifasciatus* (Malacanthidae), *Pomatomus saltatrix* (Pomatomidae), *Echeneis naucrates* (Echeneidae), *Synagrops microlepis* (Acropomatidae, attributed to the Percichthyidae), *Cepola paucinaradiata* (*Cepolidae*) and ascribes them to the perciform type (Type II).

Chaetodon hoefleri. The rootlet extending from the proximal centriole and the element of putative endoplasmic reticulum extending from the vicinity of the axoneme in the sperm cell of *C. hoefleri* are immature features (Fig. 15.28). The Four-banded butterflyfish is a dioecious spawner with external fertilization (Breder and Rosen 1966).

***Pomatomus saltatrix*:** The diffuse nature of the chromatin and the wide zone of perinuclear cytoplasm in the illustrated sperm cell of *P. saltatrix* (Fig. 15.29) suggest that it, too, is incompletely mature. The proximal centriole is at right angles to the distal centriole and there is a single tier of mitochondria and a moderately long cytoplasmic canal (Mattei 1969 and present study). The Bluefish is a dioecious batch spawner with external fertilization (van der Elst 1976).

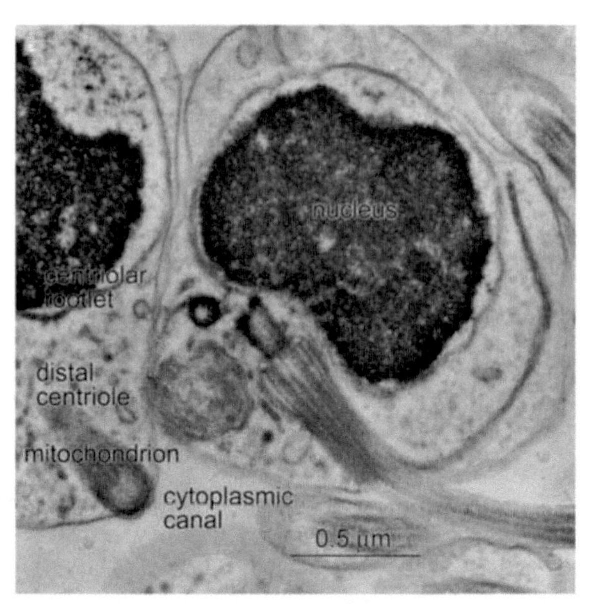

Fig. 15.28 *Chaetodon hoefleri* (Chaetodontidae), Four-banded butterfly fish. Longitudinal section of a late Type II spermatid. Arrow, endoplasmic reticulum. Original, from an unpublished negative courtesy of Professor Xavier Mattei.

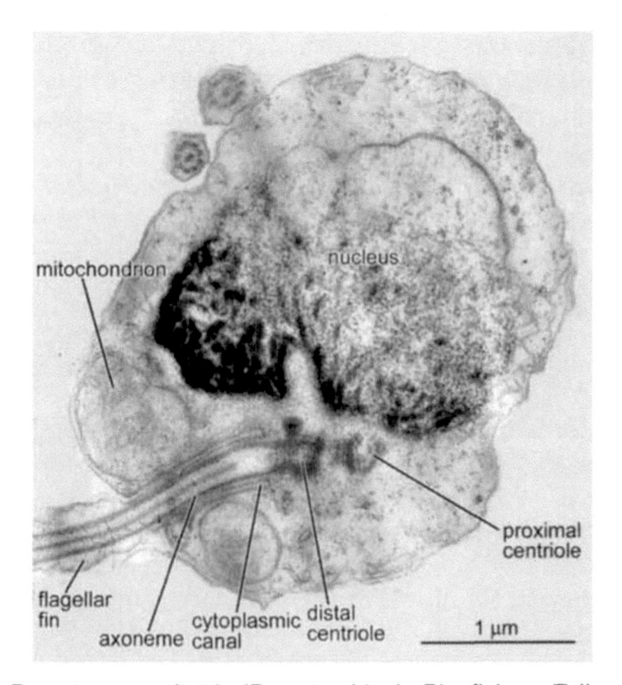

Fig. 15.29 *Pomatomus saltatrix* (Pomatomidae), Bluefish or Tailor. Longitudinal section of the Type II spermatozoon. Original, from an unpublished TEM negative courtesy of Professor Xavier Mattei.

Echeneis naucrates: *E. naucrates* (Fig. 15.30A) has a Type II spermatozoon, illustrated by Mattei (1969) (Fig. 15.30B). It has two or three tiers of mitochondria and a correspondingly long cytoplasmic canal. The proximal centriole is at about 45° to the distal centriole.

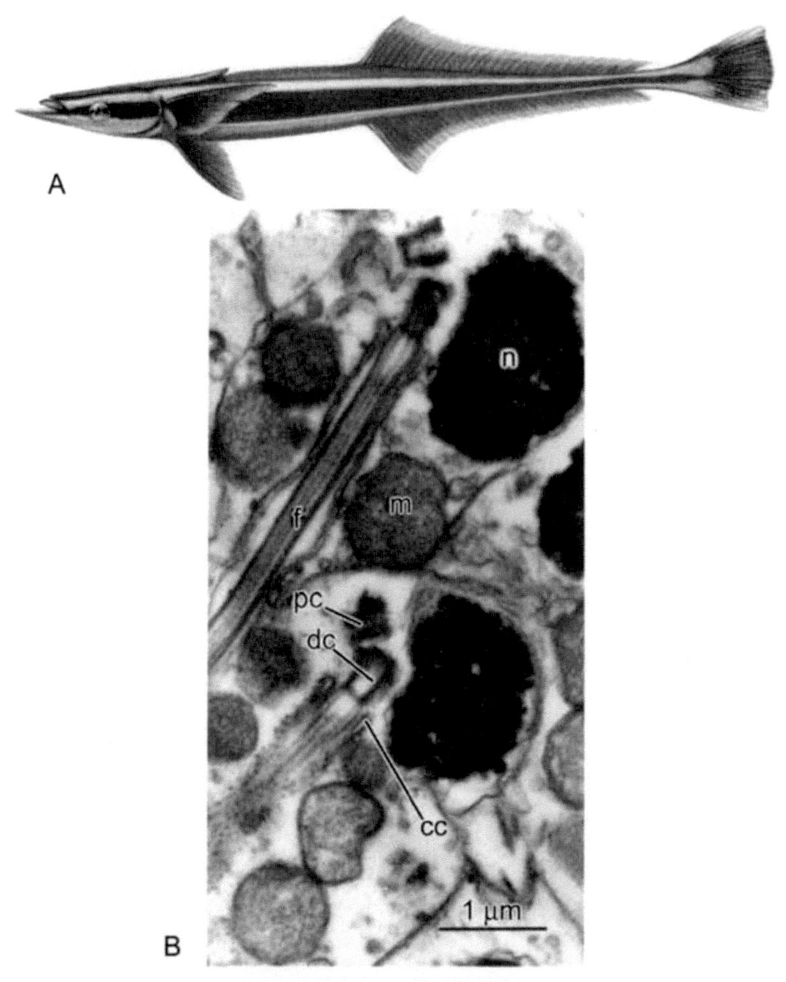

Fig. 15.30 *Echeneis naucrates* (Echeneidae), Live sharksucker. **A**. Adult. By Norman Weaver. With kind permission of Sarah Starsmore. **B**. Longitudinal section of two spermatozoa. cc, cytoplasmic canal; dc, distal centriole; f, flagellum; pc, proximal centriole; m, mitochondrion. Adapted from Mattei, X. 1969. *Contribution à l'étude de la spermiogenèse et des spermatozoïdes de poissons par les méthodes de la microscopie électronique*. D.Sc. thesis, Université de Montpellier, Fig. XVIIIc. With permission.

Badis badis: In an ultrastructural examination of Type II spermatogenesis in *B. badis*, placed in the subfamily Badinae of the Nandidae by Nelson (2006),

Fishelson and Delarea (1995) showed that the flagellum has a long unilateral fin, as in the Percichtyidae (Jamieson 1991).

15.7.1.3 Family Latidae

Lates has been removed from the Centropomidae to the Latidae (references in Nelson 2006), though the resemblance to the centropomids is exceedingly close. The Australian *Lates calcarifer*, the Barramundi (adult, Fig. 15.31), has an anacrosomal aquasperm (Fig. 15.32) possessing an elliptical nucleus with a narrow, moderately deep basal fossa.

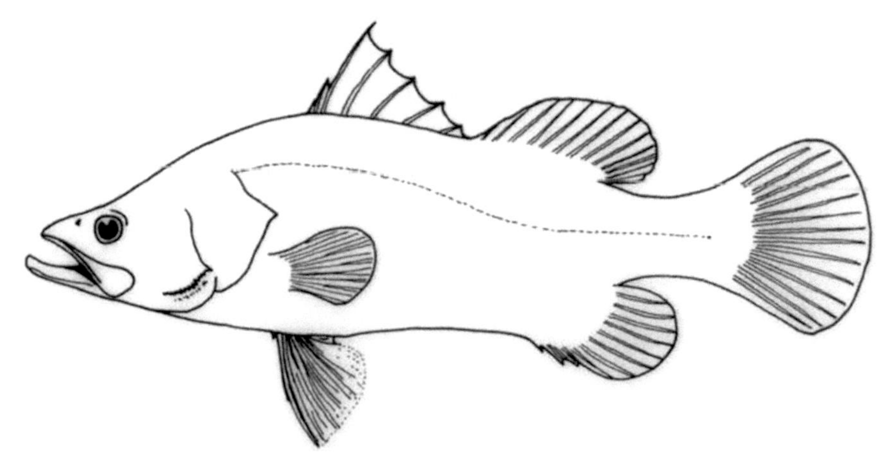

Fig. 15.31 *Lates calcarifer*, the Barramundi. From Jamieson, B. G. M. 1991. Fish Evolution and Systematics: Evidence from Spermatozoa. Cambridge University Press, Cambridge, Fig. 15.13. Drawn by Chistopher Tudge.

The chromatin consists of coarse masses narrowly separated by clear material, giving it a coarsely flocculent appearance. Round bodies (see Percichthyidae) are absent.

There are 4 to 6 mitochondria distributed asymmetrically around a short cytoplasmic canal. The long axis of the proximal centriole is at 110° to that of the axoneme and about 45° to the sagittal plane; in some sections this centriole lies in a moderately deep nuclear fossa (not visible in Fig. 11.32) but the distal centriole is wholly outside it. Each centriole is of the triplet type; the proximal has nine dense peripheral accessory densities; an undetermined number of fibers accompanies the distal centriole. There is a long cytoplasmic canal, surrounded for most of its length by a narrow cytoplasmic sleeve. The axoneme is of the 9+2 type, with a pair of lateral fins, but as usual lacks central singlets for a short region below the basal body (Leung, in Jamieson 1991).

This spermatozoon of *Lates calcarifer* appears (Jamieson 1991) to be intermediate between types I and II of Mattei (1970). However, Mattei (1991) examined the sperm of *Lates niloticus* and ascribed it to the perciforrr sperm (Type II).

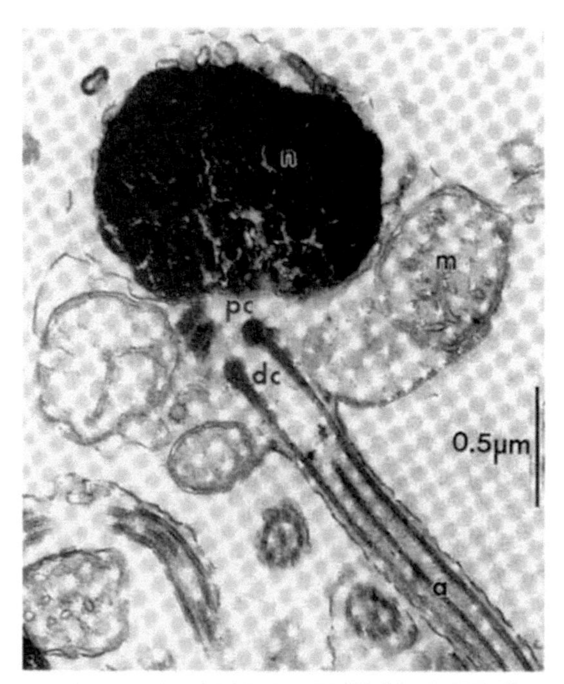

Fig. 15.32 *Lates calcarifer* (Latidae), the Barramundi. Longitudinal section of a spermatozoon (thawed after cryopreservation with 20% glycerol). ax, axoneme; dc, distal centriole; m, mitochondrion; n, nucleus; pc, proximal centriole. Courtesy of Luke K.-P. Leung in Jamieson, B. G. M. 1991. *Fish Evolution and Systematics: Evidence from Spermatozoa*. Cambridge University Press, Cambridge, Fig. 15.14.

15.7.1.4 Family Moronidae

These are the temperate basses, exemplified by *Morone chrysops* (Fig. 15.33A).

***Dicentrarchus* and *Lateolabrax*:** *Dicentrarchus punctatus (=Morone punctata)* (Mattei 1991 and this account), and *Lateolabrax japonicus* (Hara and Okiyama 1998) placed by the latter authors in the Percichthyidae, have the perciform Type II spermatozoon. *L. japonicus* is an externally fertilizing egg scatterer. That of *D. punctatus* (Fig. 15.33B) demonstrates the classical Type II features, with both centrioles located outside the shallow nuclear fossa and the flagellum tangential to the nucleus. In both species, small mitochondria surround the base of the flagellum and there is a moderately long cytoplasmic canal. In *L. japonicus* the flagellum is said to bear a tube-like fin and although the plasma membrane is also widely separated from the axoneme in *D. punctatus*, no transverse sections demonstrate laminar fins. The A subtubules in *L. japonicus* of doublets 1, 2, 5 and 6 have been shown to be septate, as typical of the perciform sperm (Hara and Okiyama 1998). A scanning electron micrograph reveals an ovoid-ellipsoidal head (He and Woods 2004).

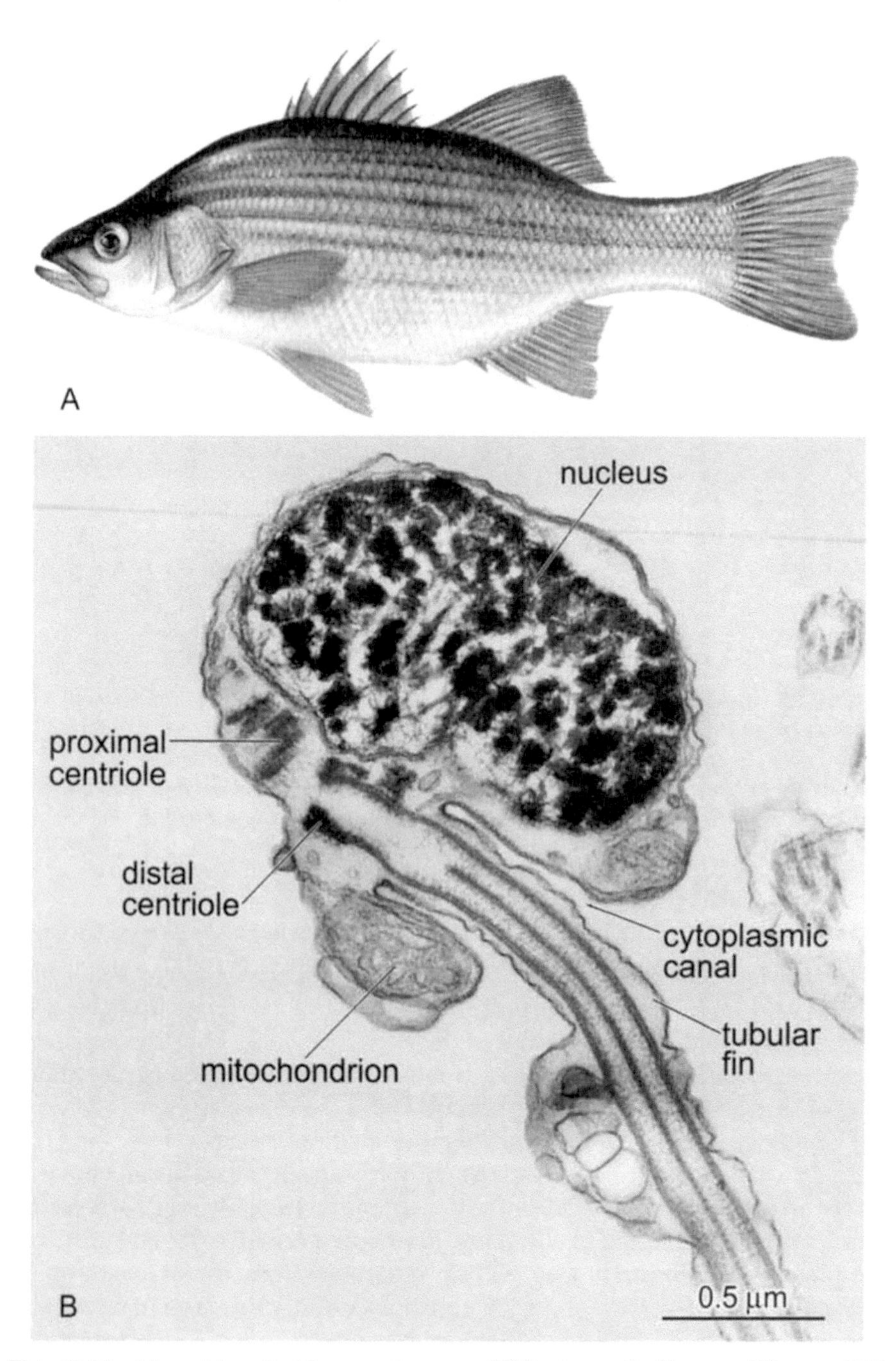

Fig. 15.33 Moronidae. **A**. *Morone chrysops*, White bass. By Norman Weaver. With kind permission of Sarah Starsmore. **B**. *Dicentrarchus punctatus*, Spotted seabass. Demonstrating the Type II spermatozoon. Original, from an unpublished TEM micrograph courtesy of Professor Xavier Mattei.

15.7.1.5 Family Percichthyidae

Percichthyids are the so-called temperate basses, in freshwater, rarely brackish habitats in Australia and South America (Nelson 2006).

Sperm of the Australian taxa *Maccullochella* (*Macc. macquariensis*, Fig. 15.34I-L; *Macc. peeli*, Fig. 15.34M,N) and *Macquaria* (*Macq. ambigua*, Figs. 15.34D-H, 15.35; *Macq. australasica*, Fig. 15.34A-C; *Macq. novemaculeata*) so far investigated correspond generally to the above description for *Lates* but show some notable exceptions and demonstrate what appear to be genus-specific characters with relationship to numbers of mitochondria, and of axonemal fins, and presence or absence of nuclear "round bodies" (Fig. 15.34J) and of a cytoplasmic canal. *Maccullochella* differs in the presence of only a single, large mitochondrion (Fig. 15.34K) (occasionally two mitochondria in *M acc. peeli*) and the presence of "round bodies" within the nuclear envelope but discrete from the general chromatin. There is a single round body in *Macc. macquariensis* (Fig. 15.34J), and there are three or four in *Macc. peeli*. The nuclear fossa is deep in *Macc. peeli*, is a moderate concavity in *Macc. macquariensis*, and is weakly developed in *Macquaria*. *Macquaria ambigua* has several mitochondria, some of which are sometimes fused as an incomplete ring (Fig. 15.34E); in *Macq. australasica* there are 3 to 6 mitochondria as some may fuse; in *Macq. novemaculata* there are approximately 5. In both genera the basal body and 9+2 axoneme are at 90° or more to the proximal centriole and tangential or nearly so to the nucleus (Fig. 15.34C,D,F,J,M). Various densities occur around the centrioles. *Maccullochella* sperm have only one lateral fin (Fig. 15.34L,N) and have no cytoplasmic canal while those of *Macquaria*, like *Lates*, have two fins (Fig. 15.34G) and a long cytoplasmic canal most of which is bounded by a thin, trailing cytoplasmic sleeve which only anteriorly expands as the mitochondrion-containing midpiece (Fig. 15.34D). Terminally, the axoneme lacks fins (Fig. 15.34H) (Leung, unpublished, Jamieson 1991).

15.7.1.6 Family Serranidae

These are the sea basses; marine, in tropical and temperate seas; a few in freshwater (Nelson 2006).

Serranus atricauda, S. cabrilla and S. scriba. *Serranus* species are simultaneous hermaphrodites with external fertilization. Spermatozoa of *Serranus atricauda*, *S. cabrilla* (adult, Fig. 15.36E; sperm, Fig. 15.36C,D) and *S. scriba* (Fig. 15.36A,B) described by Garcia-Diaz *et al.* (1999) are anacrosomal aquasperm in present terminology. In all three species the spermatozoon has an ovoid head, a short midpiece consisting of several mitochondria, and a long flagellum. The head and midpiece are nor clearly separated and contain only very small amounts of cytoplasm with a granular appearance. The nucleus contains highly electron-dense granular chromatin. The nuclear envelope lacks pores. The nuclear fossa is not deep and its dimensions are similar to the contained centriolar complex.

The midpiece includes the centriolar complex and is separated from the flagellum by a cytoplasmic canal. The size and number of mitochondria differs among the three examined species. The proximal and distal centrioles are mutually perpendicular. In *S. atricauda* and *S. cabrilla* the spermatozoon is

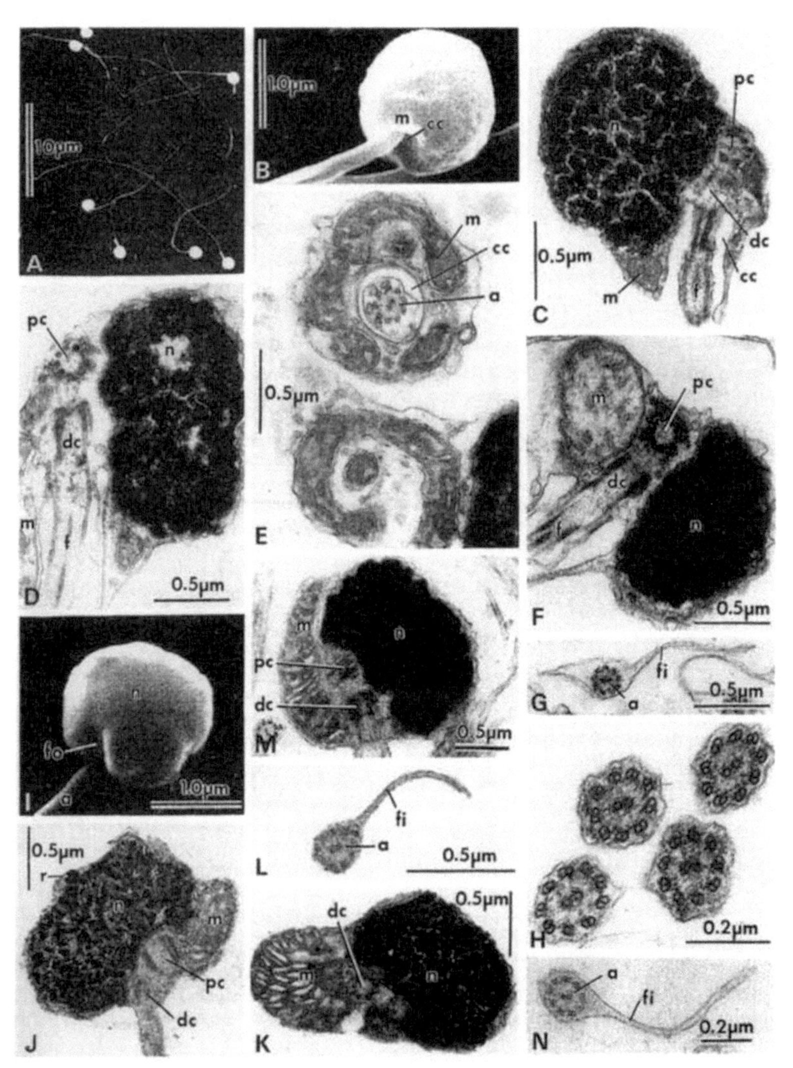

Fig. 15.34 Sperm of Australian Percichthyidae. **A-C**. *Macquaria australasica*. **A**. Scanning electron micrograph (SEM) of spermatozoa. **B**. SEM of head, showing cytoplasmic canal. **C**. Longitudinal section (LS) of spermatozoon close to the sagittal plane. **D-H**. *Macquaria ambigua*. **D**. LS sperm. **E**. Transverse section (TS) of two midpieces, showing variable shape of mitochondria. **F**. LS through centrioles. **G**. TS flagellum through fins. **H**. TS flagellum behind fins. **I-L**. *Maccullochella macquariensis*. **I**. SEM, showing nuclear fossa. **J**. LS sperm, showing apical round body. **K**. TS head through initial, 9+0 region of axoneme and single mitochondrion. **L**. TS flagellum, showing single fin. **M-N**. *Maccullochella peeli*. **M**. LS sperm. **N**. TS tail and lateral fin. a, axoneme; cc, cytoplasmic canal; dc, distal centriole; fi, fin; fo, nuclear fossa; m, mitochondrion; n, nucleus; pc, proximal centriole; r, round body. Courtesy of L. K.-P. Leung. From Jamieson, B. G. M. 1991. *Fish Evolution and Systematics: Evidence from Spermatozoa*. Cambridge University Press, Cambridge, Fig. 15.15.

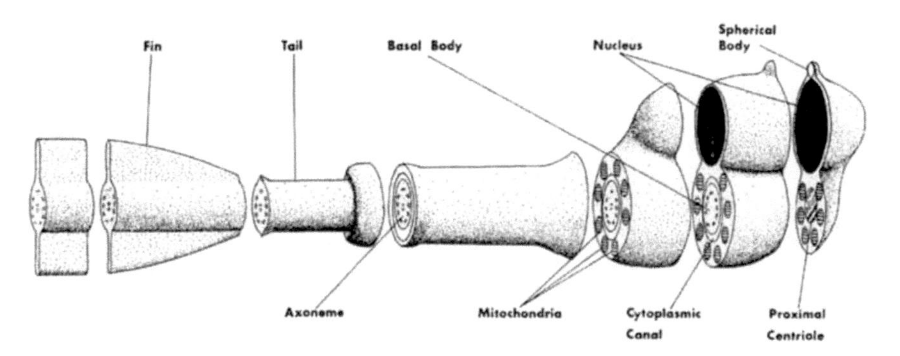

Fig. 15.35 *Macquaria ambigua*. Schematic diagram of spermatozoon. Courtesy of A. Marshall, in Jamieson, B. G. M. 1991. *Fish Evolution and Systematics: Evidence from Spermatozoa*. Cambridge University Press, Cambridge, Fig. 15.16.

symmetrical and the centrioles and flagellum lie at a right angle to the base of the head [the Type I condition] but in *S. scriba* the sperm is asymmetrical and the flagellum has a lateroposterior insertion, approaching the Type II condition, as stated by Mattei (1991) for *S. cabrilla*. The proximal centriole has nine peripheral pairs of microtubules [not triplets] and, as usual, no central microtubules. The distal centriole (basal body) consists of nine pairs of peripheral microtubules and, it is stated, two central microtubules. The flagellum shows a typical structure of 9+2 structure.

The head and midpiece together measure 2.33 ± 0.25 and, 1.58 ± 0.10 µm in length, with a maximal diameter of 1.92 ± 0.15 and 1.52 ± 0.15 µm in *Serranus atricauda* and *Serranus cabrilla*, respectively. The ovoid nucleus is 1.19 ± 0.12 µm in longitudinal axis and 1.39 ± 0.13 um in lateral axis for *S. atricauda*; it is 0.98 ± 0.09 um in length and 7.34 ± 0.11 µm in width for *S. atricauda*. The nuclear fossa is 0.28 ± 0.04 µm long and has a diameter of 0.39 ± 0.05 µm in *S. atricauda*; it reaches values of 0.27 ± 0.06 µm in length and 0.37 ± 0.04 µm in width for *S. cabrilla*. In *S. scriba*, the head and midpiece together have a length of 2.83 ± 0.20 µm and a width of 1.77 ± 0.15 um. The ovoid nucleus is asymmetrical and measures 1.39 ± 0.15 µm maximum in length and1.22 ± 0.13 µm in width; it consists of densely packed granular chromatin material. Despite the asymmetry of the sperm of *S. scriba* the nuclear fossa is located in the longitudinal axis of the spermatozoon; it has a length of 0.34 ± 0.07 µm and a diameter of 0.2 1 ± 0.05 µm.

Although the acrosome is missing in all three species, the spermatozoon in *Serranus scriba* exhibits a stack of membranes apposed to the nuclear envelope in its anterior region (Fig. 15.36A). It consists of two membranes separated by a distinct perinuclear cisterna, which contains electron-lucent material.

The midpiece contains five equal-size spherical mitochondria with a diameter of 0.54 ± 0.08 µm in *Serranus atricauda*, and four mitochondria with a diameter of 0.57± 0.06 µm in *S. cabrilla* (Fig. 15.36D). In *S. scriba* the midpiece contains four spherical mitochondria with a diameter of 0.71 ± 0.18 µm: three appear to form a right angle triangle, and one is closer to the centriolar complex (Fig. 15.36A).

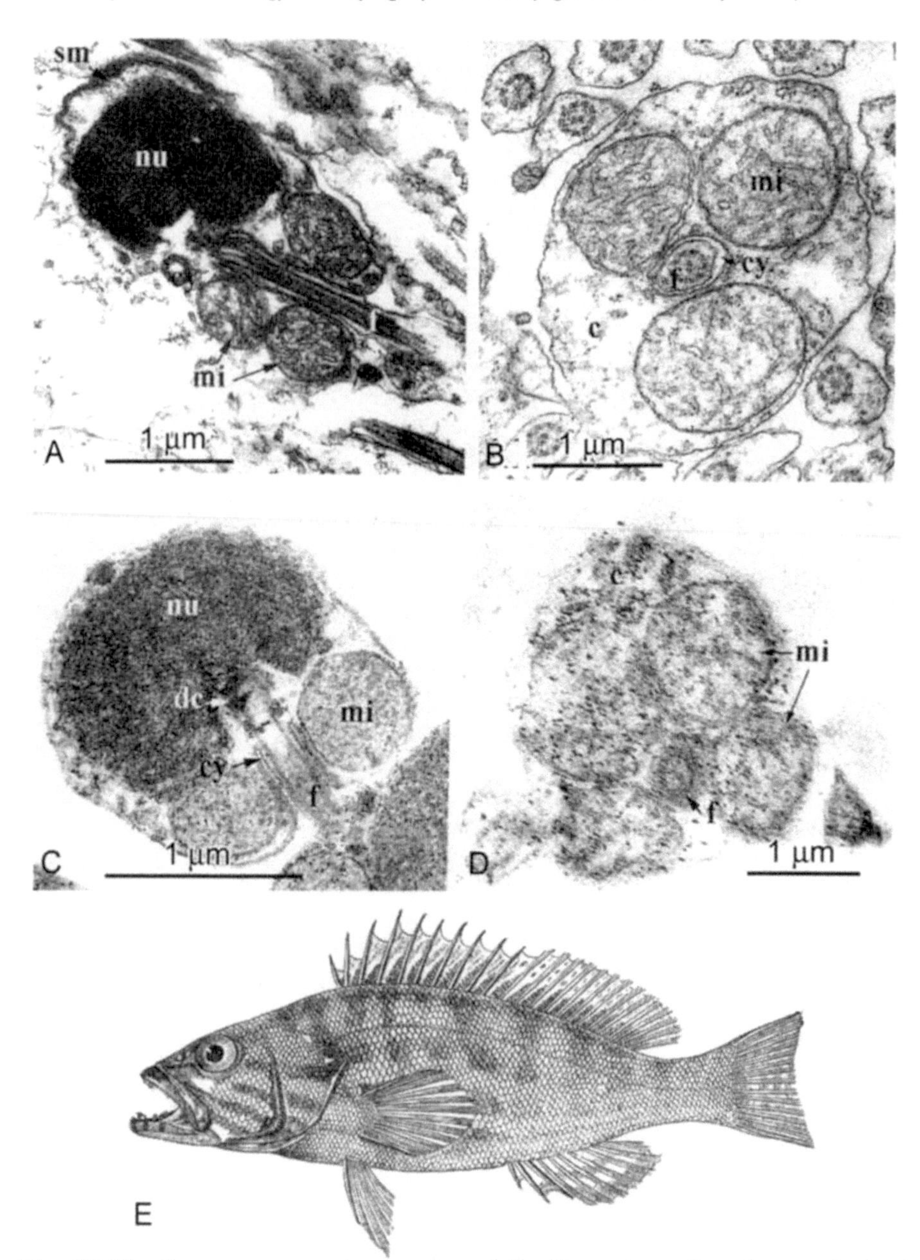

Fig. 15.36 *Serranus* spermatozoa. **A** and **B**. *Serranus scriba*. **A**. Longitudinal section. **B**. Transverse section of the midpiece. **B** and **C**. *S. cabrilla*. **C**. Longitudinal section. **D**. Transverse section of midpiece. c, cytoplasm; cy, cytoplasmic canal; dc, distal centriole; f, flagellum; mi, mitochondrion; nu, nucleus; sm. stacked membrane. Adapted from Garcia-Diaz, M. M., Lorente, M. J., Gonzalez, J. A. and Tuset, V. M. 1999. Journal of Submicroscopic Cytology and Pathology 31: 503-508, Figs. 9, 11, 2 and 3. **E**. *Serranus cabrilla*. From Norman, J. R. 1937. *Illustrated Guide to the Fish Gallery*. Trustees of the British Museum, London, Fig. 34.

The cytoplasmic canal is 0.32 ± 0.02 µm in width and 0.41 ± 0.04 µm in length in *Serranus atricauda*; and 0.37 ± 0.05 µm long and 0.21 ± 0.03 µm wide in *S. cabrilla*. The proximal centriole presents a diameter of 0.21 ± 0.02 µm and a length of 0.18 ± 0.03 µm in *S. atricauda*, while it is 0.21 ± 0.03 µm in length and 0.19 ± 0.03 µm in width for *S. cabrilla*. The distal centriole measures 0.20 ± 0.02 µm and 0.17 ± 0.02 µm in diameter for *S. atricauda* and *S. cabrilla*, respectively. In *S. scriba* the cytoplasm of the midpiece shows several lipid droplets; the cytoplasmic canal has a diameter of 0.33 ± 0.03 µm and its length is 1.13 ± 0.08 µm; the proximal centriole is 0.25 ± 0.02 µm long and 0.30 ± 0.02 µm wide. The distal centriole has a diameter of 0.30 ± 0.02 µm.

The flagellum has a length of about 43 µm and a diameter of 0.19 ± 0.03 µm in *Serranus atricauda*; ca 40 µm in length and 0.18 ± 0.05 µm in width in *S. cabrilla*; and ca 47 µm and a diameter of 0.30 ± 0.04 µm in *S. scriba*.

***Plectropomus leopardus*.** The account of the sperm of the serranid *Plectropomus leopardus*, the Coral Trout, given by Jamieson (1991) is here slightly augmented from Gwo *et al.* (1994). The nucleus is approximately 1.4 µm long (average diameter 1.4-1.7 µm, Gwo *et al.* 1994) and is almost flat posterolaterally so that there is no appreciable fossa (Fig. 15.37A). The distal centriole (basal body of the flagellum) is juxtaposed to the middle of the flattened region so that the flagellum is at about 45° to the flattened surface. The sperm thus approximates to a Type II sperm *sensu* Mattei (1970). In this respect it resembles the sperm of *Serranus scriba*, described above. The chromatin is poorly condensed, consisting of many irregular flocculent dense masses separated by pale matrix.

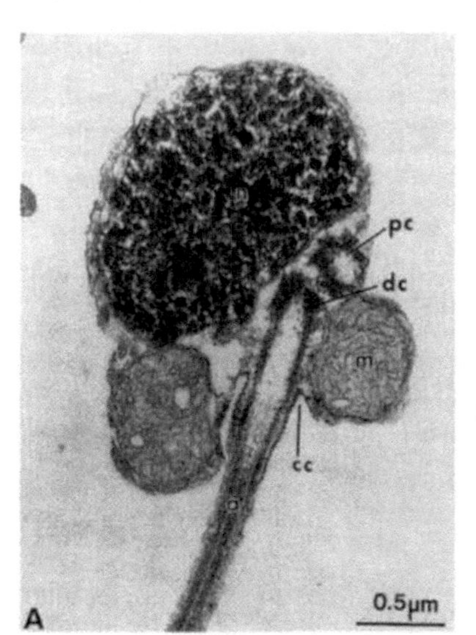

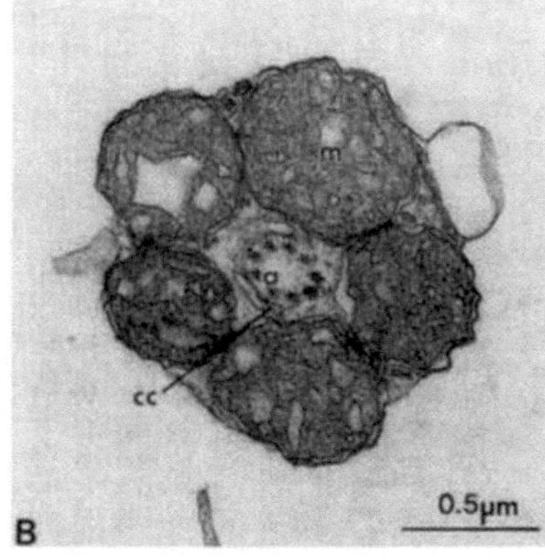

Fig. 15.37 *Plectropomus leopardus,* Coral Trout (Serranidae). **A**. Longitudinal section of spermatozoon. **B**. Transverse section of midpiece. a. axoneme; cc. cytoplasmic canal; dc. distal centriole; fi, fin; fo. nuclear fossa; m. mitochondrion; n. nucleus; pc. proximal centriole. From Jamieson, B. G. M. 1991. *Fish Evolution and Systematics: Evidence from Spermatozoa*. Cambridge University Press, Cambridge, Fig. 15.17.

The midpiece is of simple construction. Five (or rarely six, Gwo) subspherical cristate mitochondria are regularly arranged around the distal centriole and the base of the axoneme occupying the entirety of the short collar-like midpiece (Fig. 15.37A,B). They are seen surrounding the cytoplasmic canal in Figure 15.37B and, as judged from the longitudinal section (Fig. 15.37A), the central structure with 9 doublets and no central singlets is therefore the proximal end of the axoneme. The condition of the centriolar microtubules, whether triplets or doublets in both centrioles is unknown. There is no trailing sleeve to the mitochondrial collar. Densification occurs between contiguous outer membranes of adjacent mitochondria. Small beadlike masses or vesicles (considered by Gwo to possibly be mitochondrial remnants) are present in the vicinity of some mitochondria. The two centrioles are mutually at right angles, with the proximal displaced to one side of the longitudinal axis of the distal centriole (Fig. 15.37A). The 9+2 flagellum, develops central microtubules in the posterior region of the midpiece. It has two lateral fins relative to which the two central singlets are tilted.

***Epinephelus*:** An adult of this genus is illustrated (Fig. 15.38). The spermatozoon of *Epinephelus malabaricus*, a giant protogynous hermaphrodite fish, described and illustrated by Gwo *et al.* (1994), has a roundish head. The ovoid nucleus, measuring about 2.0 μm on its longest axis and 1.8 μm in diameter, is laterally offset relative to the flagellar axis. The chromatin is coarsely granular and several irregularly shaped lacunae occur within it. The nuclear fossa is poorly developed. Both the proximal and distal centrioles are wholly outside the nuclear fossa. The proximal centriole is inclined at 45° to the distal centriole (basal body). Each centriole is composed of triplet microtubules. The proximal centriole has nine dense peripheral accessory densities.

Fig. 15.38 A serranid. *Epinephelus maculatus*. Courtesy of Jean-Lou Justine.

The midpiece contains six equal-sized spherical mitochondria, encircles the basal body of the flagellum and is completely separated from it by a deep cytoplasmic canal. A second membrane is in close vicinity to the plasma membrane in the cytoplasmic canal region. Both basal body and 9+2 axoneme are nearly tangential to the nucleus. A striated basal root attaches laterally in the midregion of the basal body. Its broad base, attaching to the triplet rays of distal centriole, tapers peripherally to end in a relatively densely stained knob. The alar sheets, radiating from the basal body triplets, attach the basal body to the plasma membrane. At the neck of the flagellum a necklace composed of seven electron-dense particles is present. Y-shaped bridges link the doublets to the flagellar membrane. Neither ITDs (intratubular differentiations) nor lateral flagellar fins were observed.

Zhao *et al.* (2003) imply in a brief account of the sperm of *Epinephelus coioides*, the Orange-spotted Grouper, that the proximal centriole is at 60° to the distal centriole.

Remarks: As noted by Gwo *et al.* (1994), the spermatozoa of *Plectropomus* and *Epinephelus* approximate to the Type II spermatozoon *sensu* Mattei (1970) which is characterized by the flagellum being parallel to the base of the nucleus, both centrioles situated outside or eccentric to the nuclear fossa, and an axoneme in which the A tubules of doublets 1, 2, 5 and 6 exhibit an ITD (Mattei 1988, 1991). However, as Gwo notes, spermatozoa of both serranids, and it may be added, all examined serranids, do not show ITDs in their flagella.

In *Plectropomus leopardus* the two centrioles are mutually at right angles with the proximal centriole displaced to one side of the longitudinal axis of the distal centriole, contrasting with the arrangement in *Epinephelus malabaricus*. The midpiece in *P. leopardus* contains five spherical mitochondria (rarely the six seen in *E. malabaricus*) and is separated from the flagellum by a shallow cytoplasmic canal unlike the deep canal of the latter species. In *P. leopardus* the average nuclear diameter, at 1.4 -1 7 μm, is smaller than that in *E. malabaricus*. More distinct cytoplasmic vesicles with various sizes and clear spaces are located under the plasma membrane of the flagellum and surround the axoneme at the proximal part of the flagellum of *P. leopardus* compared with *E. malabaricus*.

As also noted by Gwo *et al.* (1994), of all their spermatozoal characteristics, nuclear size, depth of the cytoplasmic canal and centriolar geometry seem to be the only characters that are species-specific in serranids; the similarity of the sperm of *Plectropomus* and *Epinephelus* supports or at least does not conflict with, a close relationship.

15.7.1.7 Family Opistognathidae

Spermatozoa of the jawfish *Opistognathus whitehurstii*, an externally fertilizing mouth-brooder, are round in shape with a diameter of about 1.8 μm. They show no differentiation into a head and a neck region, but mitochondria are located at the caudal cell pole. There are four spherical mitochondria, approximately of the same size, but no acrosome is present. The centriolar complex, which consists of the proximal centriole and the basal body of the flagellum, is located in the nuclear fossa (Manni and Rasotto 1997).

15.7.1.8 Family Percidae

Percidae are a widely represented freshwater fish group, in lakes and rivers. Fertilization is external by spawning and eggs are deposited on plants or in nests. Curiously, sperm ultrastructure has been investigated in only two genera, the Perch, *Perca fluviatilis* (Lahnsteiner and Patzner 1995) and the Pikeperch, *Sander lucioperca* (Lahnsteiner and Mansour 2004). The following account is based largely on the review of Lahnsteiner and Patzner (2007).

***Sander lucioperca* and *Perca fluviatilis*:** The spermatozoa of *Sander lucioperca* (Fig. 15.39D-I) and *Perca fluviatilis* (Fig. 15.39A-C) are similar to each other in the occurrence of an ovoid nucleus and the lateral insertion of the flagellum. The flagellum inserts mediolaterally on the nucleus, and therefore the spermatozoa are asymmetrical. The spermatozoa of *P. fluviatilis* also exhibit a differentiated midpiece region (Fig. 15.39B,C), whereas in *S. lucioperca* the midpiece is fused with the head region (Fig. 15.39E,F). The ovoid nucleus consists of condensed, granular chromatin material and lacks nuclear pores. Laterally, the nucleus has a shallow depression, where the mutually perpendicular proximal and distal centrioles are located (Fig. 15.39C,G,H). These consist of nine triplets of microtubules. The anterior portion of the distal centriole (basal body) is embedded in a ring of osmiophilic material (Fig. 15.39C,G,H). Microfibrils interconnect the centrioles with the nucleus and with each other. The midpiece is penetrated by a cytoplasmic canal, which is bordered by a double membrane. The 9+2 flagellum is 30 to 35 μm long and has paired lateral fins, giving it a width of 0.7–0.9 μm (Fig. 15.39D,I).

The sperm of *Sander lucioperca* has been shown by Lahnsteiner and Patzner (2007) to differ from that of *Perca fluviatilis* in several parameters: (1) In *P. fluviatilis*, the head is bigger than in *S. lucioperca*. (2) in *P. fluviatilis* the midpiece is clearly differentiated from the head (Fig. 15.39C) but is incorporated in the head region in *S. lucioperca* (Fig. 15.39E). (3) in *P. fluviatilis* the midpiece contains one mitochondrion surrounding the cytoplasmic canal in the form of an incomplete ring (Fig. 15.39A) whereas in *S. lucioperca*, the mitochondria are arranged irregularly (Fig. 15.39F). It is uncertain whether there is only one extremely folded mitochondrion or several mitochondria located closely together. (4) The shallow nuclear fossa of *S. lucioperca* contains two additional invaginations (Fig. I4.39F),"satellite nuclear notches" *sensu* Gwo (6) Both species exhibit special structures which interconnect the basal body with the nucleus. This structure is cylindrically shaped in *P. fluviatilis* (Fig. L4.39C) but is an electron-dense plate in *S. lucioperca* (Fig. 15.39G).

As reviewed by Lahnsteiner and Patzner (2007), in *Perca fluviatilis* spawning occurs in shallow waters. The females twist the eggs around plants, roots or logs in the form of an egg-strand. Several males follow the females and fertilize the released eggs. In *Sander lucioperca* brood care is observed as the males build nests and guard the eggs, deposited by an individual female, until hatching of larvae.

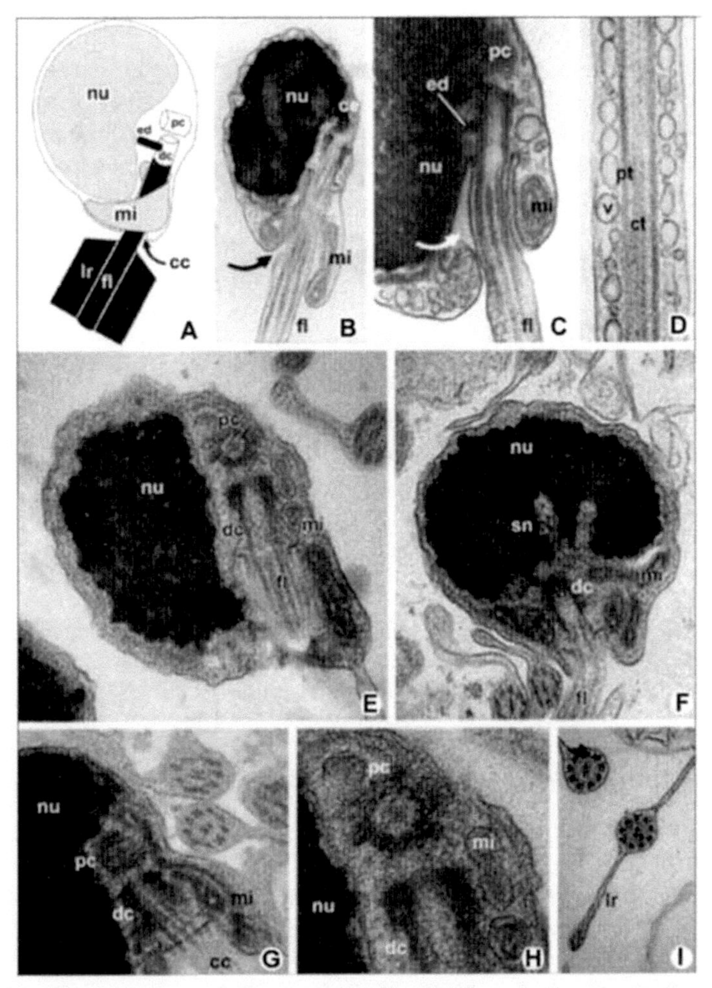

Fig. 15.39 Spermatozoa of *Perca fluviatilis* (**A-D**) and *Sander lucioperca* (**E-I**) (Percidae). **A**. Schematic reconstruction, *Perca*. **B**. Longitudinal section in the plane of a mitochondrion, *Perca*. x 12,000. **C**. Longitudinal section of midpiece, *Perca*. x 35,000. **D**. Longitudinal section of flagellum with vesicles in the lateral fins, *Perca*. x 90,000. **E**. Longitudinal section of head region, *Sander*. x 36,600. **F**. Sagittal section with the invaginations of the nucleus ("satellite nuclear notches"), *Sander*. x 35,500. **G**. Centriolar complex with proximal centriole in longitudinal section. Note fingerlike protrusion of mitochondria, *Sander*. x 43,500. **H**. Centriolar complex with proximal centriole in cross section, *Sander*. x 78,300. **I**. Cross section of the flagellum with lateral fin. x 43,500. arrow = cc, cytoplasmic canal; ce, centriolar complex; ct, central microtubules of the flagellum; dc, distal centriole; ed, electron dense material adjacent to the distal centriole; fl , flagellum, lr, lateral flagellar fin; mi, mitochondrion; nu, nucleus; pc, proximal centriole; sn, "satellite nuclear notch"; v, vacuole; pt, peripheral microtubules. After Lahnsteiner, F. and Patzner, R. A. 2007. Pp. 1-61. In S. M. H. Alavi, J. Cosson, K. Coward and G. Rafiee (eds), *Fish Spermatology*. Alpha Science International Ltd., Oxford, U.K., Fig. 11.

15.7.1.9 Family Carangidae

Selene setapinnis: In the sperm of the carangid *Selene (=Vomer) setapinnis* (Mattei *et al.* 1979) (Fig. 15.40), the nucleus is lateral relative to the flagellar axis (type II *sperm sensu* Mattei). A moderately long cytoplasmic canal is present and separate mitochondria lie on each side of the axoneme in a line at right angles to the nuclear axis.

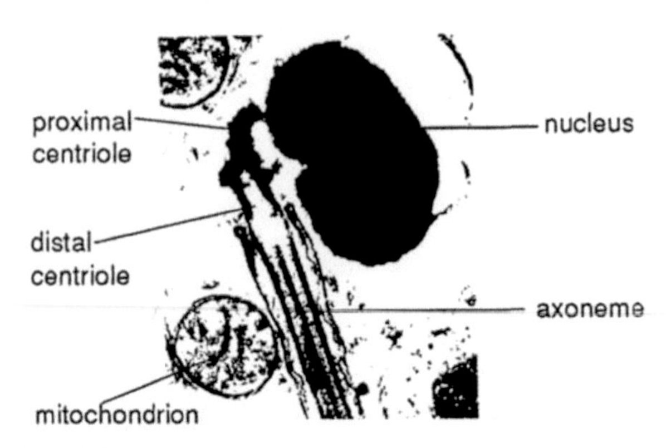

Fig. 15.40 *Selene (=Vomer) setapinnis* (Carangidae), Atlantic Moonfish. Longitudinal section of the spermatozoon. From Jamieson 1991 after Mattei, C., Mattei, X. and Marchand, B. 1979. Journal of Ultrastructure Research 69: 371-377, Fig. 1.

A transverse section of the flagellum is illustrated for *Selar crumenophthalmus*, showing septate A subtubules 1, 2, 5 and 6 (Mattei *et al.* 1979).

Trachurus mediterraneus: The sperm head in *Trachurus mediterraneus* differs from that in *Diplodus sargus* and *Boops boops* in being wider than long (Lahnsteiner and Patzner 1998) (Fig. 15.41A). Unlike the latter two species, which have deep nuclear fossae, the centriolar complex is located outside the small fossa and the flagellum, with the usual 9+2 arrangement of microtubules, is inserted craniolaterally in the classical Type II configuration. The centrioles are orientated at an acute angle to each other. The midpiece is partly integrated into the head region and contains four mitochondria. Flagellar fins are absent. For dimensions, see Lahnsteiner and Patzner (1998).

Seriola dumerelii: The spermatozoon of *Seriola dumerelii*, the Amberjack (adult Fig. 15.41B), have been described and illustrated by Maricchiolo *et al.* (2002). The ovoid, acrosomeless head measured 1.96 ± 0.09 μm long and 1.36 ± 0. 11 μm wide (n = 50) The chromatin is condensed and finely granular. It has a shallow nuclear fossa. The proximal and distal centrioles are approximately at right angles to each other at the base of the nucleus. The distal centriole (basal body) is external to the fossa and is surrounded by electron-dense material. Both centrioles are formed by nine peripheral triplets of microtubules; with no central microtubules.

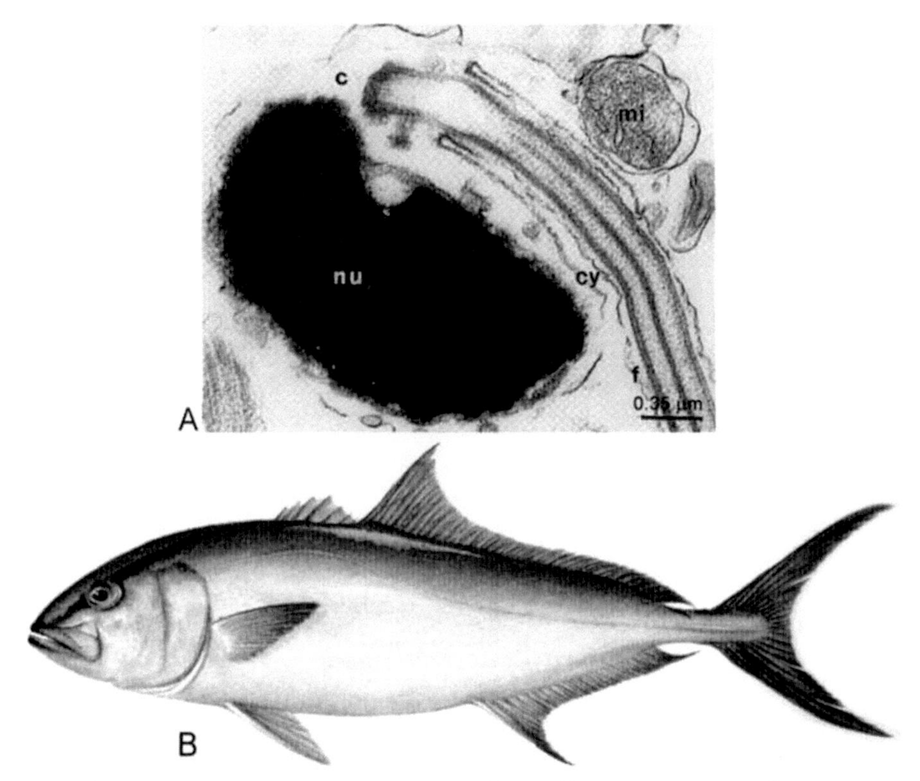

Fig. 15.41 *Trachurus mediterraneus* (Carangidae), Horse mackerel. TEM longitudinal section of spermatozoon. c, centriolar complex; cy, cytoplasmic canal; f, flagellum; mi, mitochondrion; nu, nucleus. After Lahnsteiner, F. and Patzner, R. A. 1998. Journal of Fish Biology 52: 726-742, Fig. 1c. With permission of Blackwell Publishing. **B**. *Seriola dumerelii*, the Amberjack. By Norman Weaver. With kind permission of Sarah Starsmore.

The midpiece is short, irregularly-shaped and characterized by a collar at its distal end. Although up to eight mitochondria were present in the midpiece as seen in scanning electron micrographs, a smaller number were visible in sections, as they did not all lie in the same plane. As many as three are seen in a stack on one side of the flagellum. A narrow cytoplasmic canal separates the mitochondria from the flagellum.

In the 9+2 axoneme, lumina of doublets 1,2,5 and 6 in A microtubules are electron-dense, probably because of the presence of ITDs. The flagellum, which measures 33.43 ± 3.28 µm (n = 50) in length, is inserted eccentrically into the head, tangential to the base of the nucleus and has a pair of lateral fins

Remarks: The *Seriola* sperm is said to differ from a Type II sperm in having the proximal centriole within, rather than outside, the nuclear fossa but in a micrograph it appears to be external to the very shallow fossa.

***Chloroscombrus chrysurus* and *Psettus sebae*:** Mattei (1991) examined the sperm of *Chloroscombrus chrysurus* and *Monodactylus (=Psettus) sebae* and ascribes them to the perciform type (Type II).

15.7.1.10 Family Sparidae

Sparids are the porgies; predominantly marine, in the Atlantic, Indian and Pacific oceans (Nelson 2006). They have external fertilization with pelagic mass spawning. The family has about 112 species in 35 genera (Froese and Pauly 2004) in six subfamilies: the Boopsinae, Denticinae, Diploinae, Pagellinae, Pagrinae and Sparinae (Smith and Smith 1986) of which all but the Denticinae have been examined for sperm ultrastructure. Placement in the various subfamilies has little meaning as much paraphyly and polyphyly has been demonstrated (Orrell *et al.* 2002).

Sperm literature: Investigated sparids are *Acanthopagrus australis* (Fig. 15.42A,C,E,G, I,K,M,O) and *A. berda* (Fig. 15.42B,D,F,H,J,L,N,P) (Gwo *et al.* 2005), *A. schlegeli* (Hong *et al.* 1991; Gwo *et al.* 1993a; Gwo and Gwo 1993; Gwo 1995b), *A. (=Sparus) latus* (Gwo 1995b; Dong *et al.* 1998), *Lithognathus mormyrus* (Mattei 1991); and *Rhabdosargus (=Sparus) sarba* (Fig. 15.42D,E) (Gwo *et al.* 2004a) (Sparinae); *Diplodus cervinus*, (Mattei 1991 and this account, Fig. 15.46), *D. puntazzo* (Taddei *et al.* 1999, 2001), *D. sargus* (Fig. 15.45) (Lahnsteiner

---

Fig. 15.42 Sparid spermatozoa. **A**, **C**, **E**, **G**, **I**, **K**, **M**, **O**. *Acanthopagrus australis.* **B**, **D**, **F**, **H**, **J**, **L**, **N**, **P**. *Acanthopagrus berda*. Scales 0.2 µm. **A** and **B**. Sagittal longitudinal section of a spermatozoon showing the round nucleus with the axial nuclear fossa containing the centriolar complex. Scanning electron micrographs of spermatozoa showing the head (h), midpiece (mp) and flagellum (f); m, mitochondrion; n, nucleus; pc, proximal centriole. **C** and **D**. The proximal centriole (pc) is connected by electron-dense filament (of; arrow) to the electron-dense ring of the distal centriole (dc). The mitochondrial collar encircles the proximal region of the flagellum (f) and is separated from it completely by the cytoplasmic canal (cc); as, alar sheet; bp: basal plate; n, nucleus. **E** and **F**. A nuclear notch (nn) extends distally from the posterior nuclear invagination (fossa). The distal centriole (dc) forms a basal body (bb) of the flagellum (f). The alar sheets (as) link the basal body to plasma membrane; bb; basal body; bf, basal foot. **G** and **H**. The region of proximal centriole (pc). Note the walls appear barrel-shaped, i.e., the outside diameter appears greater in the middle than at the ends of the structure. **I** and **J**. The distal centriole (dc) showing the 9+0 microtubular construction; bf, basal foot. **K** and **L**. Electron-dense filament and alar sheets (as) are visible around distal centriole (dc); bf, basal foot. **M**. The transition region of the distal centriole. Note the alar sheets (as) and Y-shaped bridge (y). Each doublet is connected to the plasma membrane by a Y-shaped bridge. **N**. A flagellum with the 9+2 axonemal doublet configuration; cc, cytoplasmic canal. **O** and **P**. The mitochondria ring with mitochondrion (m); cc, cytoplasmic canal. From Gwo, J. C., Chiu, J. Y., Lin, C. Y., Su, Y. and Yu, S. L. 2005. Tissue and Cell 37: 109-115, Fig. 1. Copyright Elsevier. With permission.

Fig. 15.42 Contd. ...

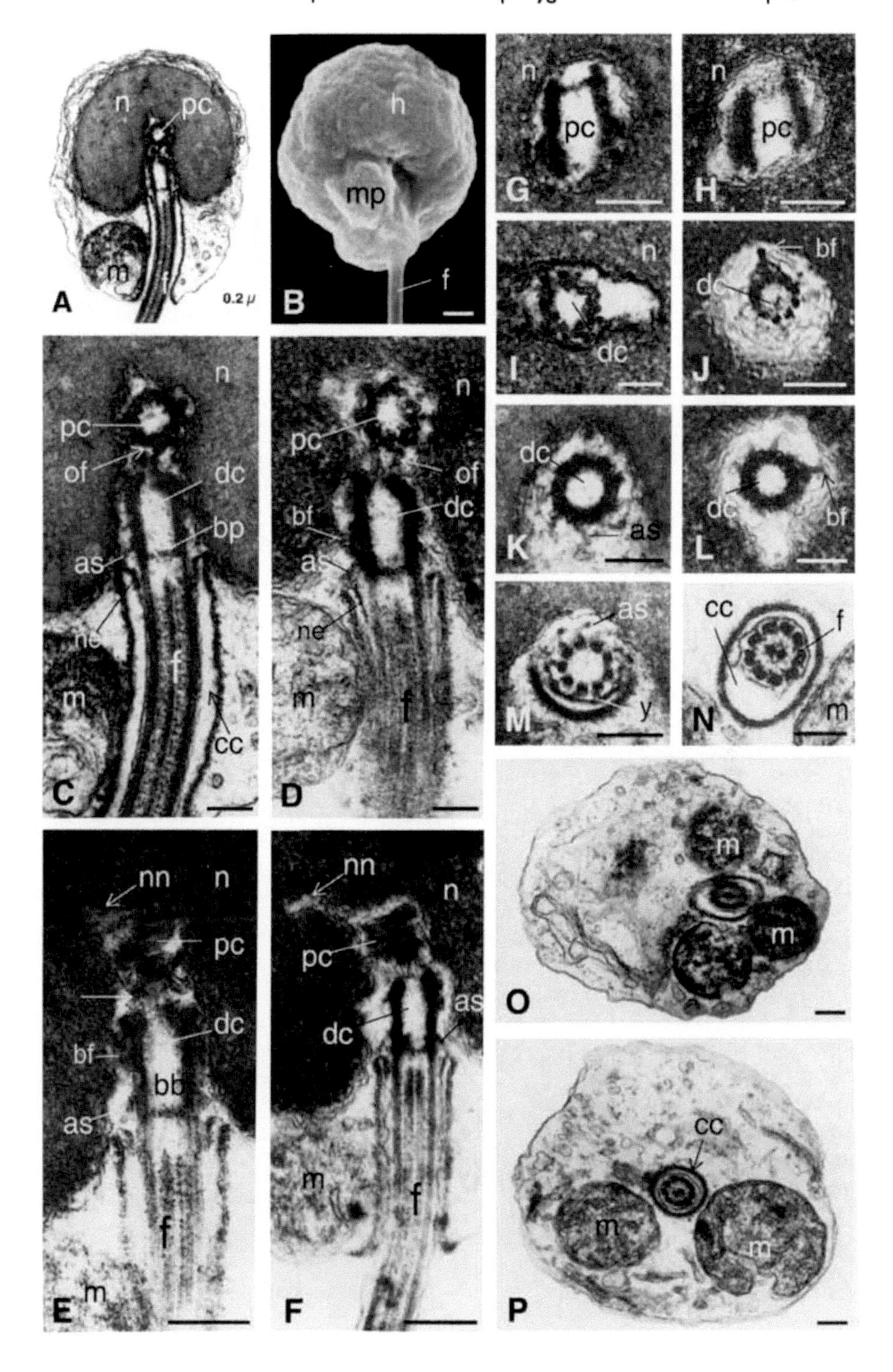
n
pc
m
A
0.2 μ
h
mp
f
B
G
H
I
J
bf
dc
K
as
L
C
of
bp
ne
cc
D
E
nn
bb
F
M
y
N
O
P

and Patzner 1998, 2007), *Lagodon rhomboides* and *Archosargus probatocephus* (Gwo *et al.* 2005) (Diplodinae); *Boops boops* (Fig. 15.47) (Mattei 1970; Zaki *et al.* 2005; Lahnsteiner and Patzner 1998) (Boopsinae), *Pagellus erythrinus* (Assem 2003; Maricchiolo *et al.* 2004) (Pagellinae) (Fig. 15.45B,C), *Pagrus major* (Fig. 15.44A-C) (Hara and Okiyama 1998; Gwo *et al.* 2004a) (Pagrinae) and *Sparus aurata* (Maricchiolo *et al.* 2007) (Sparinae).

Sperm ultrastructure: The following account of sparid sperm is partly drawn from the review of Lahnsteiner and Patzner (2007).

Sparidae have simple anacrosomal aquasperm with a spherical or ovoid head. Table 6 in Lahnsteiner and Patzner (2007) should be consulted for details, particularly of dimensions. In the species belonging to the Diplodinae and Boopsinae the head is ovoid with the lateral axis greater than the longitudinal axis; in the other subfamilies the head is spherical. The nucleus consists of electron-dense granular chromatin and the nuclear envelope lacks pores. The nuclear fossa (termed the nuclear notch by Lahnsteiner and Pastner 2007, a term perhaps better restricted to the extension of the fossa, the satellite nuclear notch, described by Gwo *et al.* 2005) conforms closely with the shape of the centriolar complex and anterior region of the axoneme. Gwo (2005) recognizes two morphotypes for 10 species within the Sparidae on the basis of the extent of the nuclear fossa: the fossa in sperm of the diplodines *Lagodon rhomboides* and *Archosargus probatocephus* is shallow (less than half of the nuclear diameter); in contrast, the fossa in the remaining species, *viz. Acanthopagrus australis, A. berda, A. latus, A. schlegeli, Rhabdosargus sarba, Pagrus major, Diplodus cervinus, D. sargus, D. puntazzo* and *Sparus aurata*, is deep (greater than one half of the nuclear diameter. Thus both morphotypes are seen in the Diplodinae, a subfamily which appears polyphletic from weighted nucleotide analysis (Orrell *et al.* 2002). As examples of variation in form of the nuclear fossa, in *Acanthopagrus australis*, it is small and shaped like a triangle, but in *A. berda* is a long cylinder.

Attention has been called (Gwo and Chang 1993; Gwo and Gwo 1993; Gwo 1995a; Gwo *et al.* 2005) to the presence of an extension, which Gwo terms the satellite nuclear notch, from the nuclear fossa in the six Pacific sparids, *Acanthopagrus australis* (nn in Fig. 15.42E), *A. berda, A. latus, A. schlegeli, Pagrus major* and *Rhabdosargus sarba*, also seen in *Sparus aurata* (Fig. 15.43A,B). The distribution of this feature in other sparids is uncertain but it appears to be unique to the family. In *Sparus aurata* the satellite nuclear notch contains a cross-striated cylindrical body considered by Maricchiolo *et al.* (2007) to resemble the striated rootlet of elopmorphs and the cross-striated fibrous body of *Lepisosteus osseus*.

The midpiece is conical in all species investigated and is penetrated by a cytoplasmic canal (Fig. 15.44A-E, 15.45A). There is one mitochondrion in the midpiece of *Boops boops, Diplodus puntazzo, D. sargus* (Fig. 15.45A), and *Pagrus major*, two in *Rhabdosargus sarba, Acanthopagrus australis*, and *Lagodon rhomboides*, three in *A. berda* and four in *A. latus, A. schlegeli* and *Archosargus probatocephus*. Variations occur in the size of the mitochondria, for instance

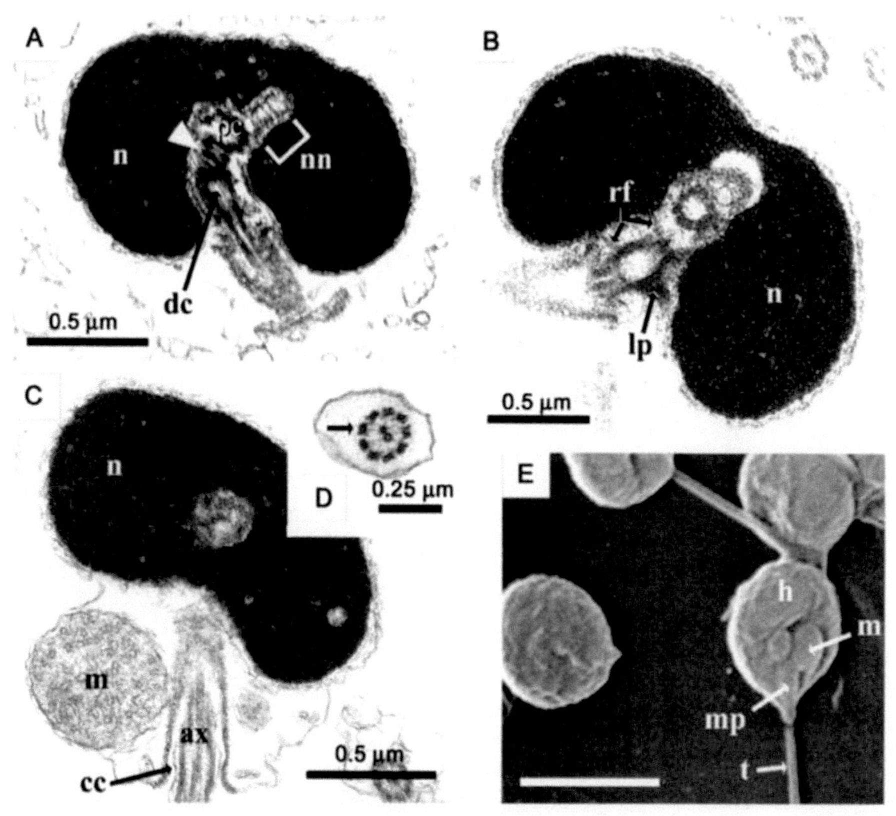

Fig. 15.43 *Sparus aurata* spermatozoa (Sparidae). **A**. Sagittal section through the centriolar complex showing the nuclear fossa containing the proximal and distal centrioles linked to each other by a disk of electron-dense material (arrowhead). The proximal centriole is associated with a cross-striated cylindrical body lying inside a satellite nuclear notch. **B**. Oblique section through the centriolar complex. **C**. Sagittal section through head and midpiece showing the mitochondrion. **D**. Transverse section through the flagellum showing the 9+2 axonemal pattern (arrow). **E**. Scanning electron micrograph of spermatozoa. Adapted from Maricchiolo, G., Genovese, L., Laurà, R., Micale, V. and Muglia, U. 2007. Histology and Histopathology 22: 79-83, Fig. 1.

those of the *A. australis* are much smaller than those of *A. berda*. Reported diameters of mitochondria are as follows: *Acanthopagrus latus* and *A. schlegeli*, 0.3-0.4 µm (Lahnsteiner and Patzner 2007b) or *A. latus* 0.4-0.5 µm (Gwo 1995a) and *A. schlegeli* 0.62 µm (Gwo and Chang 1993); *A. australis*, 0.67 µm and *A. berda*, 0.82 µm (Gwo *et al.* 2005); *Rhabdosargus sarba*, 0.71 ± 0.02 µm; *Diplodus puntazzo* and *D. sargus*, 1.05 ± 0.10 µm; *Boops boops*, 0.91 ± 0.05 µm; *Pagrus major*, ca 1.03 ± 0.11 µm (Lahnsteiner and Patzner 2007b; Gwo *et al.* 2005).

The proximal and distal centrioles lie within the nuclear fossa, as characteristic of Type I sperm, and are mutually at right angles (Figs. 15.44A,

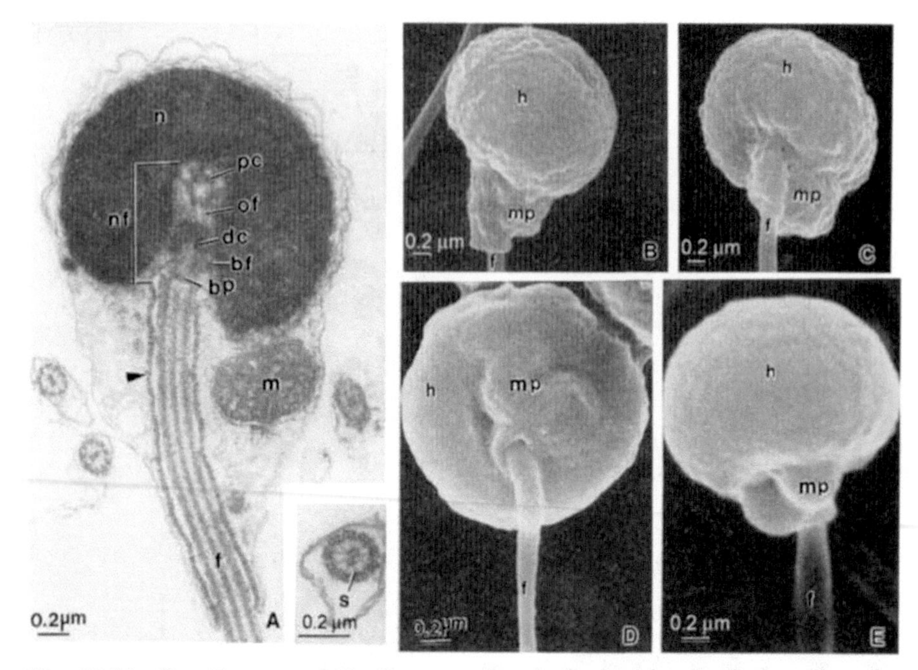

Fig. 15.44 Sparid sperm. **A-C**. *Pagrus major*. **A**. Sagittal longitudinal section of a spermatozoon showing the round nucleus with the axial nuclear fossa (nf) which contains the centriolar complex. The proximal centriole (pc) is connected by electron-dense filament (of) to the electron-dense ring of the distal centriole (dc). Arrowhead indicates two closely apposed membranes run along between the cytoplasmic canal membrane and the outer layer of the mitochondrion (m). A mitochondrion lies near the proximal region of the flagellum (f); bf, basal foot; bp, basal plate; n, nucleus; nf, nuclear fossa; of, electron-dense filament. Inset shows a transverse section of the flagellum, with radial spokes (s). **B** and **C**. Scanning electron micrographs of spermatozoa showing the head (h), midpiece (mp), and flagellum (f). **D** and **E**. *Rhabdosargus sarba*. Scanning electron micrographs of spermatozoa showing the same features. No lateral fins on the flagellum are observed. Adapted from Gwo, J. C., Kuo, M. C., Chiu, J. Y. and Cheng, H. Y. 2004. Tissue and Cell 36: 141-147, Fig 1A, N, P, O; 2A, B. Copyright Elsevier. With permission.

15.45A,C) except, at least, in *Lagodon rhomboides* in which the proximal centriole is inclined at 30° to the distal centriole. Considerable variation exists in the fine structure of the centriolar complex. The proximal and distal centrioles lie in the same axis in the Boopsinae, Diplodinae and, Pagrinae, but are not in the same axis in the Sparinae nor in the pagellid *Pagellus erythrinus* (Fig. 15.45C). In *Acanthopagrus australis* (Gwo *et al.* 2005), both centrioles have a classic 9+0 microtubular triplet construction, as also demonstrated for *Diplodus sargus* (Lahnsteiner *et al.* 1995; Lahnsteiner and Patzner 1998, 2007) and is probably general in the family. In *A. australis*, at the neck of the

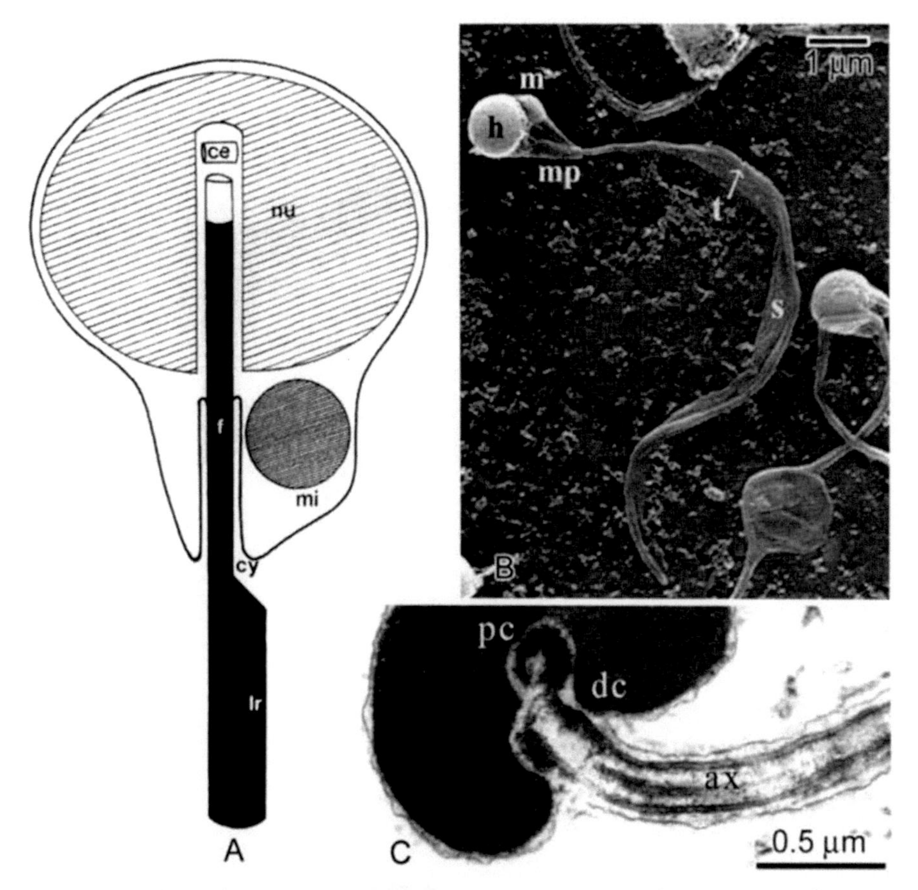

Fig. 15.45 Sparid sperm. **A**. *Diplodus sargus*. Schematic representation. ce, centriolar complex; cy, cytoplasmic canal; f, flagellum; lr, lateral flagellar fin; mi, mitochondrion; nu, nucleus. From Lahnsteiner, F., Wieismann, T. and Patzner, R. A. 1995. Journal of Submicroscopic Cytology and Pathology 27: 259-266, Fig. 14. **B** and **C**. *Pagellus erythrinus* spermatozoa. **B**. SEM of whole sperm; note the single flagellar fin. **C**. Longitudinal section through the centriolar complex. h, sperm head; m, mitochondrion; mp, midpiece; s, flagellar fin; t, tail. Adapted from Maricchiolo, G., Genovese, L., Laura, R., Micale, V. and Muglia, U. 2004. Histology and Histopathology 19: 1237-1240, Fig1a and d.

flagellum there is a necklace of three electron-dense particles. A necklace is also known (together with alar sheets and basal foot) in *A. berda* and other Pacific Ocean sparids but has not been observed in sparids from the Western hemisphere (*L. rhomboides* and *Archosargus probatocephus*) (references in Gwo 2005). The transitional region of the axoneme in *A. australis* lacks central singlets, as also shown for the diplodine *Lagodon rhomboides;* the doublets are connected to the plasma membrane by Y-links. Inner and outer dynein arms have been shown to be present in *L. rhomboides* (Gwo *et al.* 2005).

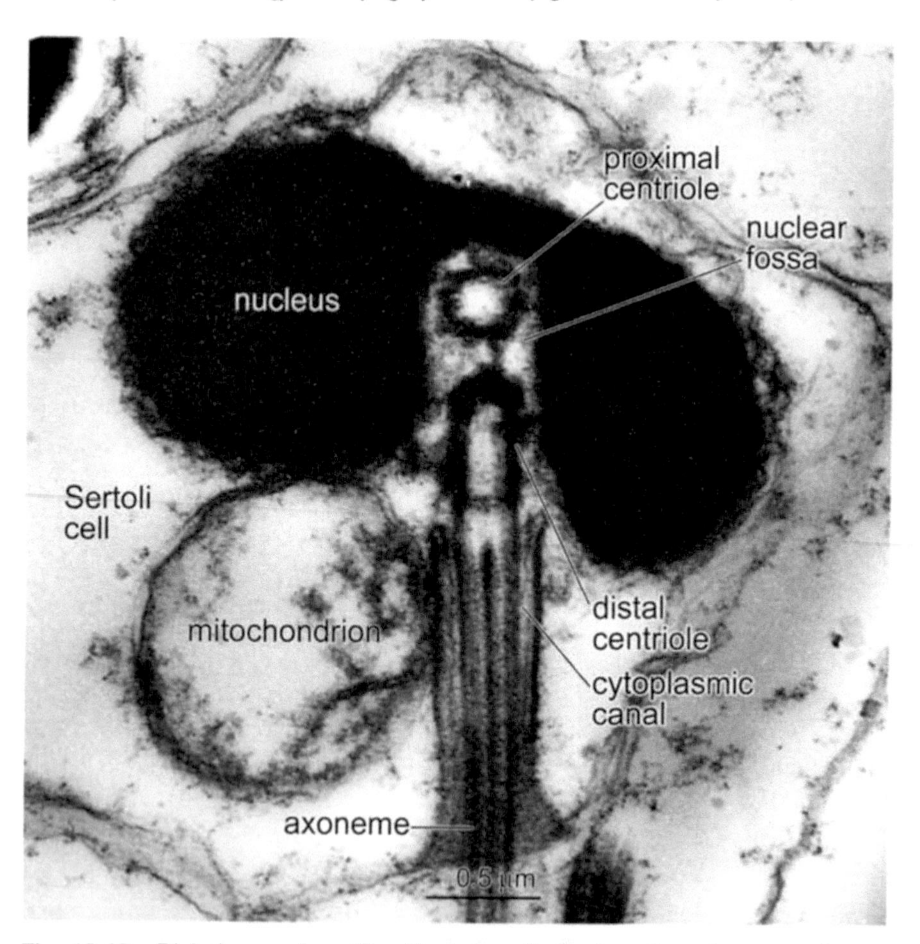

Fig. 15.46 *Diplodus cervinus* (Sparidae). Longitudinal section of a spermatozoon. Original, from an unpublished TEM micrograph courtesy of Professor Xavier Mattei.

Although serranid sperm are predominantly symmetrical, with a central nuclear fossa which houses the centrioles and with the flagellum in the longitudinal sperm axis, the sperm of *Boops boops* illustrated (Fig. 15.47) although referable to Type I, shows an asymmetry and a 'skewing' of the flagellum which approaches the Type II condition. It resembles the sperm of the gobiid *Periophthalmus* in its chief features (Mattei 1970). This is true also of the sperm of *Diplodus sargus* (Fig. 15.45) and *D. puntazzo* (Taddei *et al.* 2001). The asymmetry is correlated with presence of only a single mitochondrion. The sperm of *Pagrus major*, described and illustrated by Hara and Okiyama (1998), also with a single mitochondrion, departs less from the Type I condition, however. The fossa in the pagellid *Pagellus erythrinus* has a suggestion of the nuclear notch *sensu* Gwo but there housing the proximal centriole (Fig. 15.45C).

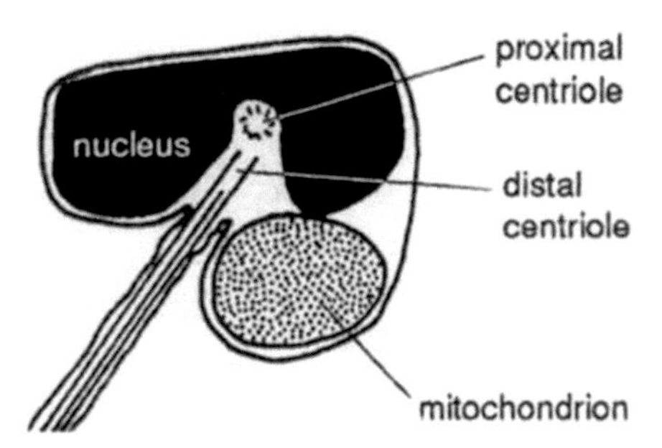

Fig. 15.47 *Boops boops.* Diagrammatic longitudinal section of the spermatozoon. After Mattei, X. 1970. Pp. 567-569. In B. Baccetti (ed.), *Comparative Spermatology*, Academic Press, New York, Fig. 4: 14.

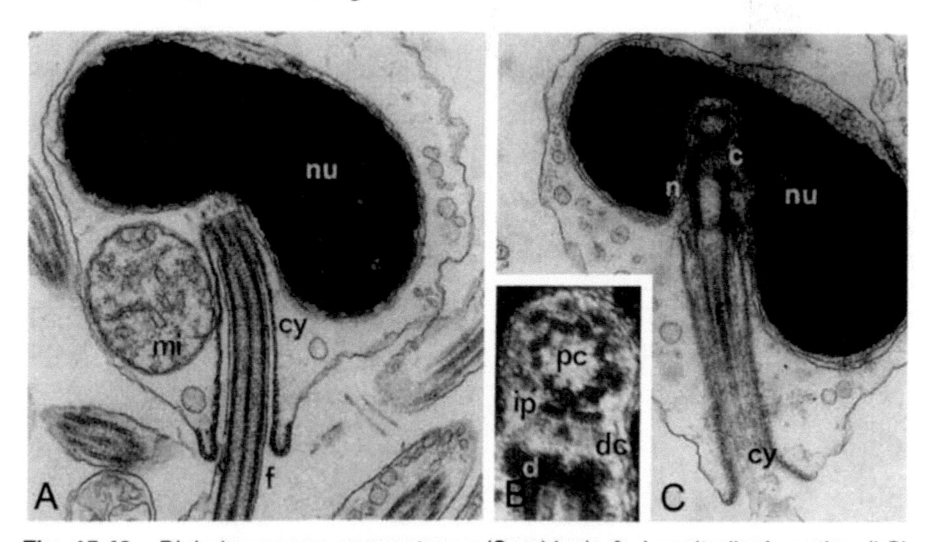

Fig. 15.48 *Diplodus sargus* spermatozoa (Sparidae). **A**. Longitudinal section (LS) in the plane of the mitochondrion. **B**. Transverse section of the flagellum. **C**. LS in the plane of the centriolar complex. cy, cytoplasmic canal; d, disk of electron-dense material; dc, distal centriole; f, flagellum; ip, insertion plate of centriolar complex; mi, mitochondrion; n, nuclear fossa; nu, nucleus; pc, proximal centriole. After Lahnsteiner, F., Wieismann, T. and Patzner, R. A. 1995. Journal of Submicroscopic Cytology and Pathology 27: 259-266, Figs. 10 and 11.

In the Sparinae and Pagrinae electron-dense structures, the basal foot and the rootlet are located laterally and opposite to each other at the distal centriole whereas in the Boopsinae and Diplodinae this structure is absent. In the Boopsinae and Diplodinae an approximately quadratic plate of electron dense material is found adjacent to the side of the proximal centriole facing the distal centriole. These electron-dense structures are considered to be stabilization structures, as microfilaments insert in this area and connect the centrioles to each other or to the nucleus. In all species investigated, the two outermost microtubules of each triplet of the distal centriole elongate to form the peripheral tubules of the flagellum. The flagellum consists of the usual 9+2 microtubules. It has one lateral fin (as in the Percichthyidae and Nandidae) in *Diplodus sargus* (Lahnsteiner *et al.* 1995; Lahnsteiner and Patzner 1998, 2007), *Boops* (Lahnsteiner and Patzner 1998) and *Pagellus erythrinus* (Maricchiolo *et al.* 2004, not two as in the abstract); and apparently no fins in *Acanthopagrus* (Gwo and Gwo 1993; Gwo 1995a), *Rhabdosargus sarba* and *Pagrus major* (Gwo *et al.* 2004a). Contrary to some accounts, no sparid is known to have two flagellar fins. For an account of motility, see Lahnsteiner and Patzner (2007).

15.7.1.11 Family Centracanthidae

Centracanthids are the picarel porgies, marine perciforms of the eastern Atlantic and South Africa (Nelson 2006).

Smaris melanurus: Mattei (1991) the sperm of *Smaris melanurus* (Mattei 1969, 1991) appears to be an asymmetrical Type I spermatozoon, as the centrioles are lodged in the nuclear fossa (Fig. 15.49).

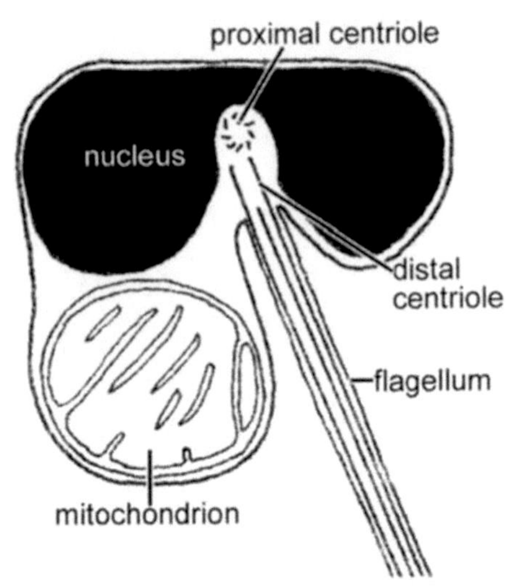

Fig. 15.49 *Smaris melanurus* (Centracanthidae). Diagrammatic longitudinal section of the spermatozoon. Adapted from Mattei, X. 1969. *Contribution à l'étude de la spermiogenèse et des spermatozoïdes de poissons par les méthodes de la microscopie électronique*. D.Sc. thesis, Université de Montpellier, Fig. 16e. With permission.

The condition of a single mitochondrion seen in *Smaris melanurus* was also reported for *Boops boops, Clinus nuchipinnis, Ophioblennius atlanticus, Saurida parri, Trachinocephalus myops* and *Periophthalmus papillio* by Mattei (1969).

Spicara chrysalis: appears to have a type I sperm *sensu* Mattei. In a micrograph of late spermatids the nucleus is ovoid but is penetrated to above its equator by the basal fossa which contains the two centrioles; three or four mitochondria in two tiers are seen (Carrillo and Zanuy 1977). The genus *Spicara* is referred to the Sparidae on the basis of molecular analysis (Orrell *et al.* 2002).

15.7.1.12 Family Polynemidae

Relationships of this family to the mugilids and Sphyraenids is no longer considered probable. It may be the sister group of the Sciaenidae (Nelson 2006). Mattei (1991) examined the sperm of *Pentanemus quincarinus* and ascribed it to the perciform type (Type II). He had earlier described the sperm of *Galeoides dedactylus* and shown it to be intermediate between types I and II (Mattei 1969, 1970).

***Galeoides dedactylus* and *Polydactylus virgnianus*:** In the spermatozoon (Fig. 15.50B,C) the nucleus is asymmetrical relative to the flagellar axis and incompletely envelops the centriolar complex; it seems only to have effected a partial rotation of about 45°. Four to five mitochondria constitute a periflagellar ring.

The spermatozoon of the polynemid *Polydactylus virginicus*, described by Gusmao-Pompiani *et al.* (2005), resembles that of *Galeoides dedactylus* and in again being of an intermediate type, between Mattei's type I and type II. The head is spherical with a hemi-arched nucleus containing highly condensed filamentous clusters of chromatin interspersed by electron lucent areas, and no acrosome (Fig. 15.51A). The centriolar complex is close to the upper end of the nucleus, with the proximal centriole lateral and oblique to the distal centriole. The distal centriole (basal body of the axoneme) is very long and extends from the nuclear fossa to the initial portion of the cytoplasmic canal (Fig. 15.51C). The flagellum axis is eccentric to the nucleus (Fig. 15.51A).

The centrioles are coated by a sleeve of homogeneous electron-dense material. At the proximal end the distal centriole, the sleeve forms an incomplete cap. The proximal and distal centrioles are laterally attached to each other by microfibrils. No additional stabilization structures are observable between the nuclear envelope and the centrioles or the plasma membrane and the distal end of the distal centriole. The short midpiece has no tubular cisternae and has a short cytoplasmic canal (Figs. 15.51A,D). In the midpiece, close to the nucleus, one single, large, ring-shaped mitochondrion (differing from the separate mitochondria of *Galeoides dedactylus*) surrounds the initial portion of the axoneme, separated from it by the cytoplasmic canal (Fig. 15.51A,C,F). In the transition zone between the basal body and the axoneme, the basal plate has nine peripheral doublets and no central microtubules (Fig. 15.51B). The flagellum (Fig. 15.51E) has the classic 9+2 axonemal structure. Intratubular differentiations (*sensu* Mattei *et al.* 1979) are absent. The flagellar membrane has two irregular short lateral fins (Fig. 15.51E) (Gusmao-Pompiani *et al.* 2005).

15.7.1.13 Family Sciaenidae

Mattei (1991) examined the sperm of *Sciaena umbra*, *Umbrina canariensis* and "*Cybium tritor*" and ascribed them to the perciform type (Type II). Further

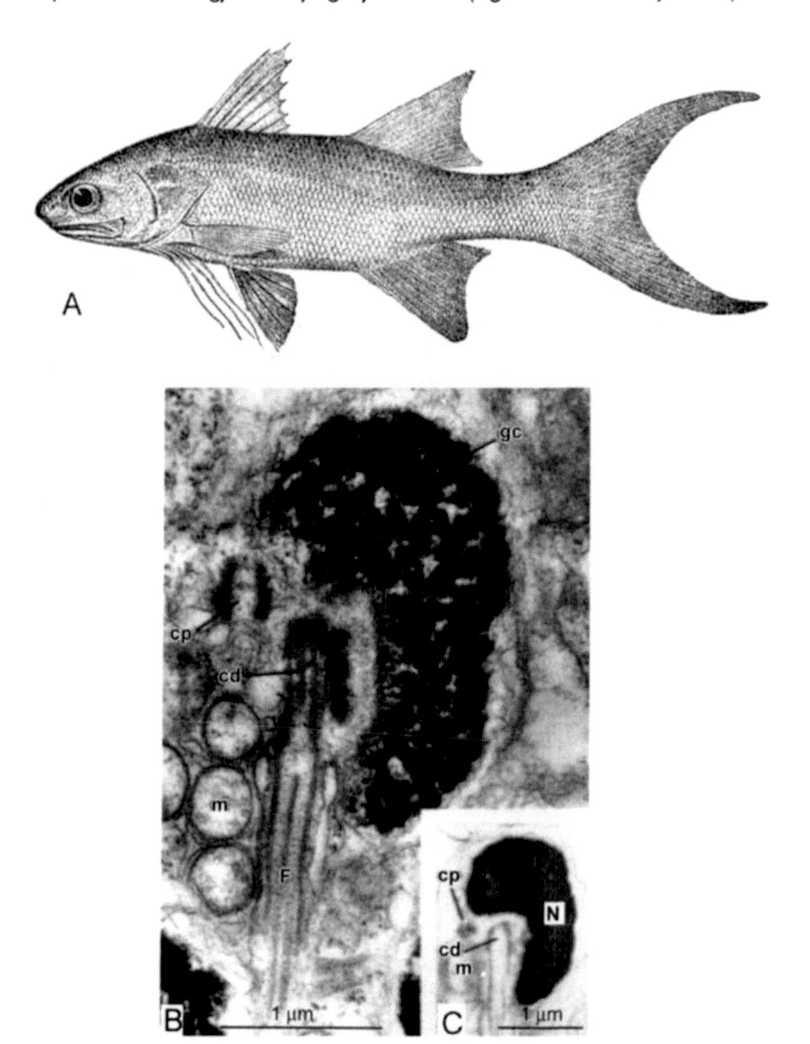

Fig. 15.50 A A polynemid, *Polynemus quadrifilis*. From Norman, J. R. 1937. *Illustrated Guide to the Fish Gallery*. Trustees of the British Museum, London, Fig. 41. **B** and **C**. *Galeoides dedactylus* (Polynemidae), Lesser African threadfin. Showing incomplete rotation. **A**. Late spermatid. **B**. Mature spermatozoon. cd, distal centriole; F, flagellum; m, mitochondrion; N, nucleus; cp, proximal centriole. Adapted from Mattei, X. 1969. *Contribution à l'étude de la spermiogenèse et des spermatozoïdes de poissons par les méthodes de la microscopie électronique*. D.Sc. thesis, Université de Montpellier, Fig. XVIIIg,h. With permission.

sperm described were: *Cynoscion striatus; Larimus breviceps; Menticirrhus americanus; Micropogonias furnieri*, all by Gusmao-Pompiani *et al.* (2005); *Micropogonias undulatus* (Gwo and Arnold 1992; Gwo 1995b); *Paralonchurus brasiliensis* (Gusmão-Pompiani *et al.* 2005); *Plagioscion squamosissimus* (Gusmão *et al.*, 1999, 2002; Gusmão-Pompiani *et al.* 2005); *Pseudosciaena crocea* (You and

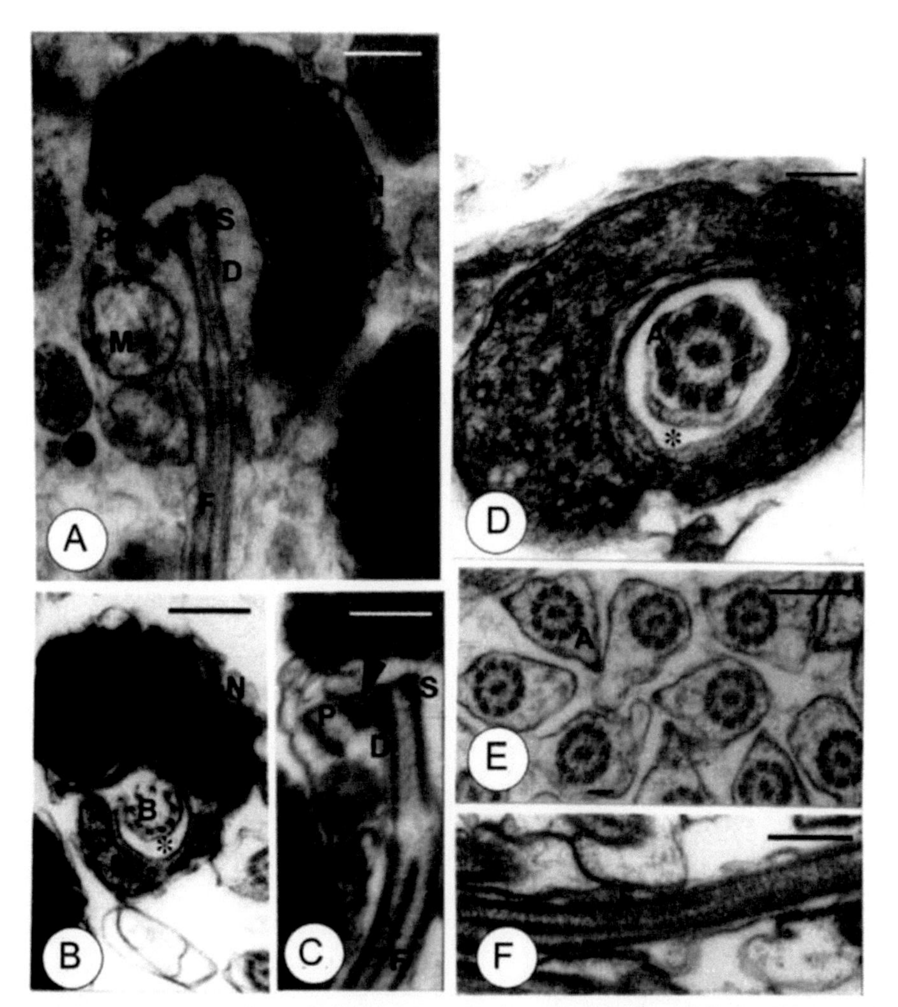

Fig. 15.51 *Polydactylus virginicus* spermatozoon (Polynemidae), Barbu. **A**. Spermatozoon longitudinal section (bar 0.33 µm). **B**. Spermatozoon cross section (bar 0.33 µm). **C**. Centriolar complex (bar 0.27 µm). **D**. Midpiece cross section (bar 0.15 µm). **E**. Flagellum cross section (bar 0.27 µm). **F**. Flagellum longitudinal section (bar 0.33 µm). A, axoneme; B, basal plate; C, centriolar cap; D, distal centriole; F, flagellum; L, electron lucent nuclear area. M, mitochondria; N, nucleus; P, proximal centriole; S, centriolar sleeve; asterisk, cytoplasmic canal; black arrowhead, centriolar stabilization structures; white double arrow, nuclear fossa. After Gusmao-Pompiani, P., Oliveira, C. and Quagio-Grassiotto, I. 2005. Tissue and Cell 37: 177-191, Figs. 42-47. Copyright Elsevier. With permission.

Lin 1997); *Stellifer rastrifer* (Gusmão-Pompiani *et al.* 2005); *Umbrina cirrosa* (Taddei *et al.* 1999) and *Umbrina coroides* (Gusmão-Pompiani *et al.* 2005).

The following account is taken largely from Gusmao-Pompiani *et al.* (2005).

Table 15.2 Spermatozoa characteristics observed in Sciaenidae and Polynemidae. From Gusmao-Pompiani, P., Oliveira, C. and Quagio-Grassiotto, I. 2005. Tissue and Cell 37: 177-191, Table 1.

Characteristics	*Sciaenidae*	*Exceptions in Sciaenidae*	*Polynemidae*
Spermatozoa type	Type I		Intermediate
Head form	Spherical		Spherical
Nucleus form	Ovoid		Hemi-arc
Presence of large electron–lucent area in the nucleus	Absent	Present in *P. brasiliensis* and *S rastrifer*	Absent
Chromatin	Highly condensed filamentous clusters		Highly condensed filamentous clusters Absent
Nuclear fossa	Double arched. Lateral and shallow	Double arched, lateral, deep and narrow in *P brasiliensis*	
Centriolar complex	Outside of the nuclear fossa		Close to the upper end of the nucleus
Relative position of the centrioles	Proximal centriole anterior and perpendicular to the distal		Proximal centriole clusters and oblique to the distal
Centriolar stabilization structures	Conspicuous		Not so conspicuous
Position of flagellum axis	Parallel to the nucleus		Eccentric to the nucleus
Midpiece	Short with a short cytoplasmic canal and some tubular vesicles		Short with a short cytoplasmic canal and no vesicles
Mitochondria number, size and shape	No more than 10, large and spherical	No more than 10, elongate and fused in *P. squamosissimus*	One, large and ring-shaped
Intratubular differentiations	Present	Absent in *P. squamosissimus* and *C. striatus*	Absent
Fins	Short and irregular		Short and irregular

Sperm ultrastructure is very similar in the different sciaenid species which have been investigated. All are of Type II and only small differences have been observed among the samples (Table 15.2). They have a spherical head with an ovoid nucleus containing highly condensed filamentous clusters of chromatin interspersed by electron-lucent areas, and, as usual, no acrosome. Sometimes, as in *Stellifer rastrifer* (Fig. 15.52) and in *Plagioscion brasiliensis,* the nucleus has a large electron-lucent area opposite to the nuclear fossa. The nuclear (implantation) fossa is usually double-arched, lateral and shallow, but in *P.*

brasiliensis the double arc is deep and narrow and in *Pseudosciaena crocea,* according to You and Lin (1997) it is situated on the dorsal surface of the nucleus and is groove-like, running from the anterior end to the posterior end of the nucleus. The centriolar complex lies outside the nuclear fossa except in *Pseudosciaena crocea* (You and Lin 1997). The proximal centriole is perpendicular to the distal centriole (Fig. 15.52A).

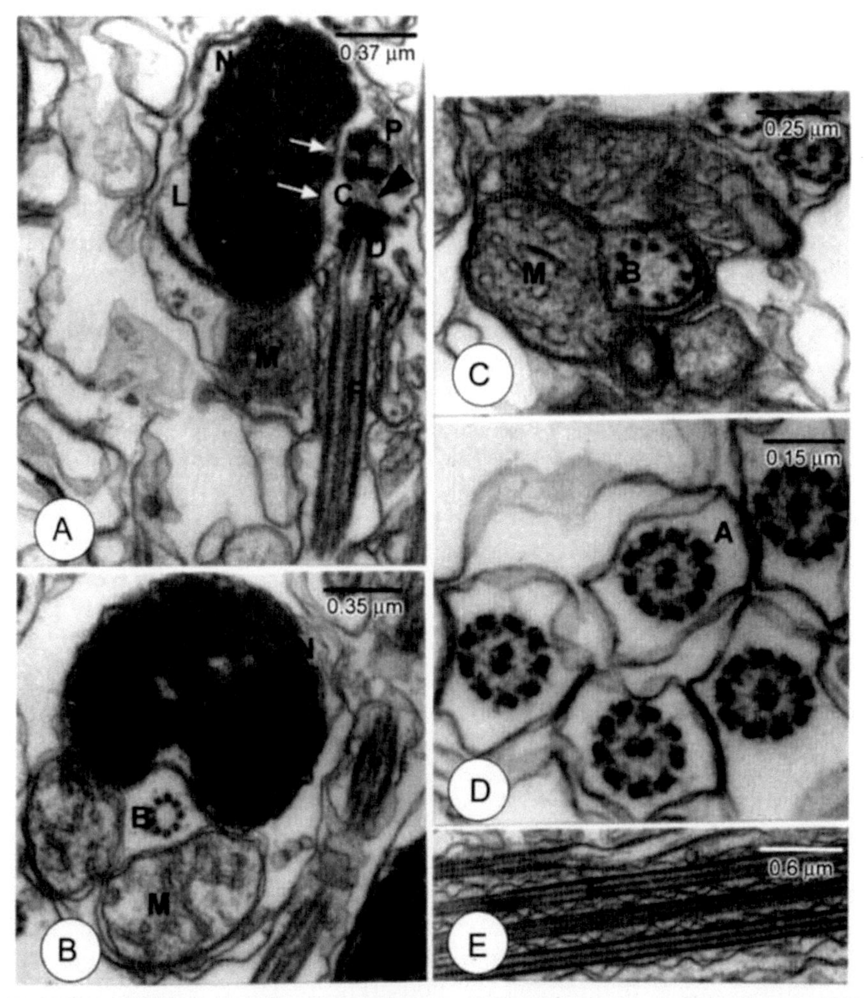

Fig. 15.52 *Stellifer rastrifer* spermatozoon (Sciaenidae). **A**. Longitudinal section (LS). **B**. Transverse section (TS). **C**. TS midpiece. **D**. TS flagella. **E**. LS flagellum. A, axoneme; B, basal body; D, distal centriole; F, flagellum L, electron-lucent nuclear area; M, mitochondrion; N, nucleus; P, proximal centriole; arrow, centriolar stabilization structures; asterisk, cytoplasmic canal. After Gusmao-Pompiani, P., Oliveira, C. and Quagio-Grassiotto, I. 2005. Tissue and Cell 37: 177-191, Figs 1-5. Copyright Elsevier. With permission.

In general in the Sciaenidae species the centrioles are coated by a sleeve of homogeneous electron-dense material and have many stabilization structures. The proximal end of the distal centriole is covered by a complete or incomplete cap of homogeneous electron-dense material (Fig. 15.52A). Proximal and distal centrioles are attached to each other by microfibrils. Various microfibrillar stabilization structures are present (see Gusmao-Pompiani *et al.* 2005). The distal centriole, forming the basal body of the axoneme, extends from the nuclear fossa to the initial segment of the cytoplasmic canal. The flagellum is parallel to the nucleus (Fig. 15.52A).

The midpiece is short, with some tubular cisternae and a short cytoplasmic canal. No more than ten large spherical (marine species) or elongate (freshwater species, *Plagioscion squamosissimus*) mitochondria surrounded the initial portion of the axoneme and are separated from it by the cytoplasmic canal (Fig. 15.52A). An additional, thin membrane lines the inner membrane of the cytoplasmic canal in, at least, *Pseudosciaena crocea* (You and Lin 1997).

In the transition zone between the basal body and the axoneme, the basal plate has nine peripheral doublets and no central microtubules (Fig. 15.52B,C) and, at least, in *Pseudosciaena crocea* (You and Lin 1997) has no dynein arms on the doublets. The flagellum exhibited the classic 9+2 axonemal structure. ITDs, i.e. septa in microtubule A of the peripheral doublets 1, 2, 5 and 6 (Fig. 15.52D), were observed, except in *Cynoscion striatus* and *Plagioscion squamosissimus*. The flagellar membrane is distant from the axoneme and has poorly defined, irregular short fins (Fig. 15.52D,E) (Gusmao-Pompiani *et al.* 2005). At the end of the principal piece, the microtubules, which constitute the axoneme, spiral to become a basket-like structure; several microtubules extend into the endpiece in *Pseudosciaena crocea* (You and Lin 1997)

15.7.1.14 Family Mullidae

Mullids are the goatfish, marine (rarely brackish) in the Atlantic, Indian and Pacific Oceans (Nelson 2006)

Sperm ultrastructure: The spermatozoon of the mullid *Pseudupeneus (=Upeneus) prayensis* was used in **Section 15.7.1** (Fig. 15.25) to exemplify the Type I spermatozoon *sensu* Mattei (1970). However, a specific feature of *Peudupeneus* sperm (Boisson *et al.* 1969) and also of *Paraupeneus spilurus* (Gwo *et al.* 2004b) which is not general for type I sperm is the very deep penetration of the nucleus by its basal fossa so that in longitudinal section the nucleus has the form of an inverted U (Boisson *et al.* 1969; Mattei 1970). Unusually, also, in these two species, the proximal centriole, which is connected to the nuclear membrane at the summit of the fossa, has its central axis in the same longitudinal axis as that of the basal body (Fig. 15.53).

Fig. 15.53 Mullidae (goatfishes). **A.** *Pseudupeneus (=Upeneus) prayensis*, West African goatfish. Transverse section of the base of the distal centriole; **B.** Longitudinal section of a spermatozoon. Nucleoplasm is no longer apparent but has left its imprints on the nucleus (arrows). cc, cytoplasmic canal; dc, distal centriole; f, flagellum; pc, proximal centriole; m, mitochondrion. Adapted from Mattei,

Fig. 15.53 Contd. ...

X. 1969. *Contribution à l'étude de la spermiogenèse et des spermatozoïdes de poissons par les méthodes de la microscopie électronique.* D.Sc. thesis, Université de Montpellier, Fig. XVII, f,g. With permission. **C**. *Pseudupeneus maculatus*, Spotted goatfish. By Norman Weaver. With kind permssion of Sarah Starsmore.

Both of these features, deep penetration of the nucleus and serial coaxial centrioles, are also reported by Boissin *et al.* (1969) in *Pegusa triophthalmus* (Soleidae) (see also Mattei 1970) (Fig. 15.119), *Balistes forcipatus* (Balistidae), *Aluterus punctatus* (Monacanthidae), *Chilomycterus reticulatus* (Diodontidae), and *Scorpaena angolensis* (Scorpaenidae) (Mattei 1970; and this chapter) for *Pseudobalistes fuscus* (Balistidae) (Fig. 15.125A,B). In the cyprinodontid *Fundulus heteroclitus* the nucleus is also U-shaped but the centrioles appear to be mutually perpendicular (Yasuzumi 1971). In longitudinal section of the *Pseudupeneus* sperm a separate mitochondrion is seen on each side in contact with the corresponding rim of the nucleus peripheral to an anteriorly widening but fairly short cytoplasmic canal.

The mature spermatozoon of *Paraupeneus spilurus*, described by Gwo *et al.* (2004b) has a computer-mouse-shaped head, a short midpiece consisting of six mitochondria, and a flagellum (Fig. 15.54A,M-P). The nucleus is long, flattened in on one side, tapering anteriorly (Fig. 15.54A,M-P). The apical margin of the head is elevated (Fig. 15.54O). Neither an acrosomal complex nor vesicle is present (Fig. 15.54A-C). The nucleus, measuring about 1.61 μm in length and 1.11 μm in width, has the form of an inverted U in longitudinal section (Fig. 15.54A-C). Its anterior end is covered by a nuclear envelope and plasma membrane (Fig. 15.54A-C). The posterior region of the nucleus is very deeply penetrated by a nuclear fossa, circular in transverse section (Fig. 15.54E-J), which contains the centriolar complex and proximal portion of the axoneme (Fig. 15.54A). The chromatin, within a double nuclear membrane, is electron-dense and homogenous (Fig. 15.54A-J,M). The two centrioles (proximal and distal centrioles) lie within the anterior region of the nuclear fossa and are arranged coaxially in the longitudinal axis of the spermatozoon (Fig. 15.54A-C). They are not connected to each other by osmiophilic filaments (Fig. 15.54A-C). Each centriole consists of the classical arrangement 9+0

Fig. 15.54 A–P Goatfish, *Paraupeneus spilurus* (Mullidae) spermatozoon. **A–D.** Longitudinal sections of a spermatozoon showing the inverted U-shaped nucleus (n) with the deep axial nuclear fossa which contains the centriolar complex and the proximal portion of the flagellum (f). The proximal centriole is located at the summit of the nuclear fossa and is connected to the nuclear membrane. No central tubules are present in the transitional region of the flagellum. The mitochondrion (m), bounded by an outer and inner membrane, is located laterally to the flagellum. The cytoplasmic canal separates the flagellum from the mitochondrion in the midpiece region. Transverse sections of the centrioles do not show the supposed triplet microtubular structure (Fig. A1, A2). **E–L**. Successive, anteroposterior, transverse sections of the spermatozoon. **E** and **F**. Transverse sections through the region of the proximal centriole (pc). Arrows indicate the flattened depression of nucleus. **G–I**. The distal centriole at the level of the osmiophilic ring showing the classic 9 peripheral triplets of microtubules construction. Triplets are often obscured by the osmiophilic ring. Arrows indicate the flattened depression of nucleus. **J.** The basal

Fig. 15.54 Contd. ...

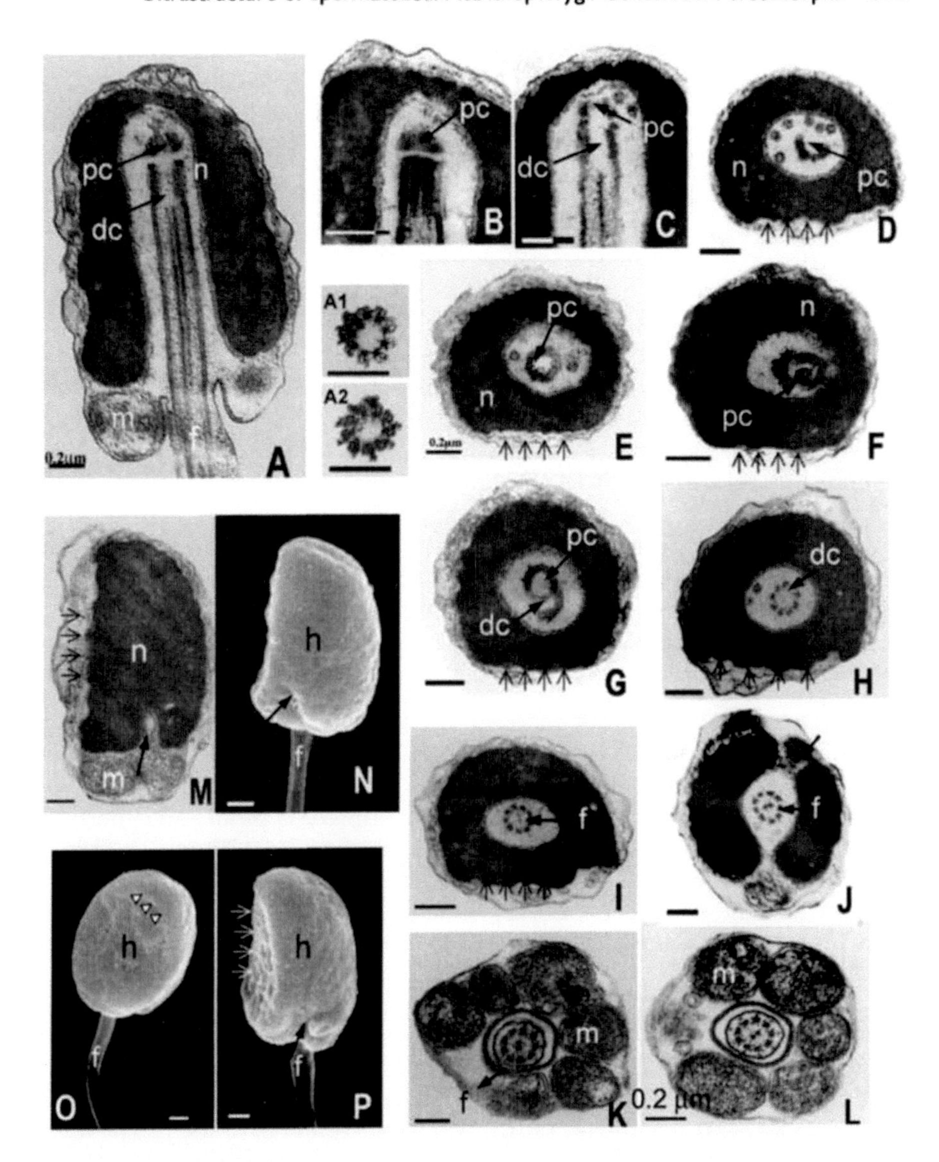

region of the distal centriole showing the typical 9+2 microtubular doublet construction. Note the indention of nucleus (arrowhead). **K** and **L**. The mitochondrion (m) has cristae and a medium electron-dense matrix. The flagellum (f) is separated from the mitochondrion by the invagination of plasmalemma. **M–P**. Transmission and scanning electron micrographs of three different rotations of the spermatozoa. The nucleus (n) is long, flattened in one side (arrows), and tapering anteriorly. The apical margin of the head is elevated (blank triangles). Note the indentation of the nucleus (arrow head). dc, distal centriole; f: flagellum; h: head; m: mitochondrion; n, nucleus; pc, proximal centriole. Scale bar 0.2 μm in all figures. From Gwo, J. C., Yang, W. T., Kuo, M. C., Takemura, A. and Cheng, H. Y. 2004. Tissue and Cell 36: 63-69, Fig. 2. Copyright Elsevier. With permission.

triplets (Fig. 15.54A-H). The distal centriole forms the basal body of the flagellum (Fig. 15.54A). An osmiophilic ring surrounds the anterior end of the distal centriole and the posterior end of the proximal centriole, and covers the outer surface of each triplet in transverse section (Fig. 15.54D-H). The A and B innermost tubules of each triplet are connected to each other by an inner ring, and continue posteriorly as the nine peripheral axonemal doublets (Fig. 15.54E–L).

Six spherical mitochondria, that appear round (about 0.44 µm in diameter) in longitudinal and cross sections, are tightly packed together, resulting in flattened surfaces where these organelles touch (Fig. 15.54K,L). Often the nuclear envelope makes contact with the exterior mitochondrial membrane (Fig. 15.54A,M). The mitochondrial matrix is electron-dense with irregularly arranged cristae. The narrow cytoplasmic canal separates the flagellum from the midpiece (Fig. 15.54A,K,L). The flagellum (axoneme) has a classical 9+2 structure, with inner and outer dynein arms (Fig. 15.54I-L). In the transition region, no central tubules are present (Fig. 15.54A,C). The central tubules are connected by a single strand (Fig. 15.54L). Five electron-dense triangular particles, termed the necklace, are present in the plasma membrane surrounding the neck of the flagellum (Fig. 15.54A). Neither ITDs nor lateral flaagellar fins are present (Gwo *et al.* 2004a).

***Mullus barbatus*:** The spermatozoon of *Mullus barbatus* has been described by and its motility investigated by Lahnsteiner *et al.* (1995) and Lahnsteiner and Patzner (1998). The head is ovoid and has a length of 3.4 ± 0.3 µm and a diameter of 1.90 ± 0.2 µm (Fig. 15.55A). The midpiece is small and cylindrical with a length of 0.52 ± 0.08 µm and a diameter of 1.50 ± 0.20 µm. The length of the flagellum is 38 to 45 µm. The nucleus consists of electron-dense granular chromatin and has a deep nuclear fossa (length 2.51 ± 0.21 µm, diameter 0.59 ± 0.14 µm) running caudocranially in the longitudinal axis of the spermatozoon (Fig. 15.55A) and contains the centriolar complex.

The proximal centriole is located at the cranial end of the nuclear fossa and is formed by nine peripheral pairs of microtubules, with no central singlets, and has a length of 0.25 ± 0.05 µm. Cranially its microtubules contact the nuclear fossa; caudally a disk of electron-dense material with a thickness of 0.12 ± 0.07 µm and a diameter of 0.32 ± 0.10 µm adheres to them. The distal centriole (basal body of the flagellum) measures 0.33 ± 0.06 µm in length and 0.26 ± 0.07 µm in diameter and consists of 9 pairs of peripheral microtubules and of two central singlet microtubules. The peripheral tubules are surrounded by a sleeve of electron-dense material; nine coarse fibers adhere to the exterior side of this sleeve and are interconnected with the nuclear fossa by microtubular structures. Cranially the distal centriole ends in a disk of electron-dense material with a thickness of about 0.1 µm. This disk is interconnected with the electron-dense disk of the proximal centriole by microfibrils. The flagellum has a diameter of 0.31 ± 0.05 µm and consists of two central and nine pairs of peripheral microtubules. The two central microtubules are surrounded by the central sheath which is interconnected with the peripheral microtubules by the centrifugal extending rays. The flagellum has a pair of lateral fins with a width of 039 ± 0.08 μm (Lahnsteiner *et al.* 1995).

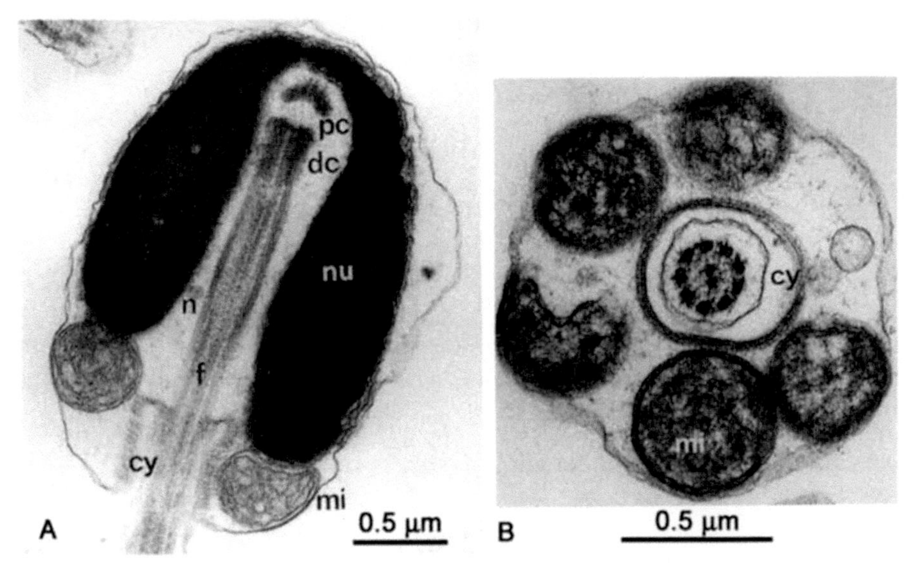

Fig. 15.55 *Mullus barbatus* spermatozoa (Mullidae). **A**. Longitudinal section of spermatozoon. **B**. Transverse section of midpiece. cy, cytoplasmic canal; dc, distal centriole; f, flagellum; mi, mitochondrion; n, nuclear fossa; nu, nucleus; pc, proximal centriole. Adapted from Lahnsteiner, F., Wieismann, T. and Patzner, R. A. 1995. Journal of Submicroscopic Cytology and Pathology 27: 259-266, Figs. 1 and 2.

The midpiece contains five spherical mitochondria with a diameter of 0.59 ± 0.14 μm and with irregularly arranged tubular cristae (Fig. 15.55B). The nuclear envelope often contacts the exterior mitochondrial membrane (Fig. 15.55A). The midpiece is traversed by the cytoplasmic canal (diameter of about 0.4 μm) in which the flagellum is located (Fig. 15.55B) and which runs longitudinally from the caudal to the cranial end of the midpiece. A second membrane is located underneath the invaginated portion of the plasmalemma (Fig. 15.55B) and is cranially fused with the plasmalemma by electron-dense material (Lahnsteiner *et al.* 1995).

15.7.1.15 Family Kuhlidae

Kuhlids, the flagtails, are marine, brackish and freshwater fishes of the Indo-Pacific region (Nelson 2006).

Nannoperca oxleyana: This southeastern Australian freshwater species is alternatively placed in the family Nannopercidae, between the Tetraponidae and Centrarchidae while the Kuhlidae are placed after the Percichthyidae (Johnson 1975). It is an externally fertilizing non-guarder.

The sperm nucleus (Fig. 15.56) is 1.3 μm long and 0.6 μm wide and is mostly located laterally to the centrioles and anterior portion of the axoneme. The

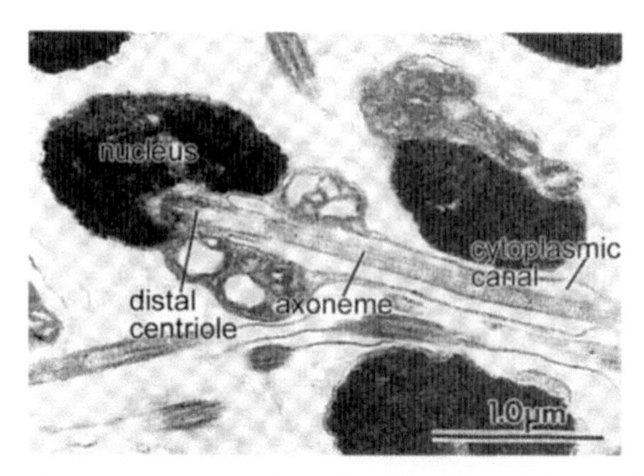

Fig. 15.56 *Nannoperca oxleyana* (Kuhlidae), Oxleyan pygmy perch. Longitudinal section of the spermatozoon. From Marshall, C. J. 1989 in Jamieson, B. G. M. 1991. *Fish Evolution and Systematics: Evidence from Spermatozoa.* Cambridge University Press, Cambridge, Fig. 15.18.

centrioles nevertheless are deeply embedded in it giving a form intermediate between the Type I and II spermatozoon *sensu* Mattei. The chromatin appears flocculent, consisting of large dark particles and scattered pale lacunae, and irregularly conforms to the double nuclear membrane. Three to four irregularly shaped distinct, cristate mitochondria form a ring around the long (2.4 µm) cytoplasmic canal. The two triplet centrioles are mutually perpendicular. There is, as in *Macquaria* and *Lates*, a long trailing cytoplasmic sleeve behind the midpiece. The 9+2 axoneme initially has a 9+0 configuration. The A subtubules of doublets 1 and 6 are septate. Both dynein arms are present. Two long fins, at least twice the width of the flagellum, lie approximately in the plane of the central singlets (modified from Marshall 1989 in Jamieson 1991).

15.7.1.16 Family Apogonidae

Apogonids are the cardinal fish; predominantly marine, in the great oceans. They copulate and have internal gametic association but the males are mouth-brooders.

Sperm ultrastructure: The fine structure of the spermatozoon has been studied in *Apogon* (=*Paroncheilus*) *affinis* (Mattei and Mattei 1984; referred, Mattei, 1991, to the junior synonym *A. stauchi*), *A. semilineatus* (Hara and Okiyama 1998), *A. imberbis* (Lahnsteiner 2003) and, in a profound study of testis structure and spermatogenesis with brief reference to spermatozoa, *A. annularis, A. apogonides, A. cookii, A. crassiceps, A. cyanosoma, A. evermanni, A. fleurieu, A. fraenatus, A. hungi, A. lineolata, A. multitaeniatus, A. pseudotaeniatus, A. semiornatus, Cheilodipterus lineatus, Siphamia mossambica* and *S. permutata* (Fishelson *et al.* 2006).

Apogonidae have anacrosomal aquasperm which have a small midpiece and are biflagellate but, in several species, have varying percentages of uniflagellate sperm (see below).

In several species, e.g. *A. semilineatus* (Fig. 15.57) the sperm heads are round (isodiametric), 1.8-2.0 µm in diameter. However, in *A. affinis* (Fig. 15.58), *A. imberbis, A. fleurieu, A. apogonoides* and *Apogon talboti*, the heads are oval-elongated (anisodiametric), 1.6 × 2.2 µm (Fishelson *et al.* 2006), contrast 1 × 1.5 µm for *A. affinis* (Mattei and Mattei 1984). The nuclear chromatin is described as not compact but containing enclaves of clear nucleoplasm (Mattei and Mattei 1984) but Fishelson *et al.* (2006) found it to be homogenous and opaque, with a dense basal line at the attachment point of the sperm tail. In at least *Apogon affinis, A. imberbis* and *A. semilineatus*, the nuclear envelope is detached from the nuclear mass and retains two distinct membranes. In the spermatozoa of all three species, the midpiece is penetrated by a cytoplasmic canal (Mattei and Mattei 1984; Lahnsteiner and Patzner 2007).

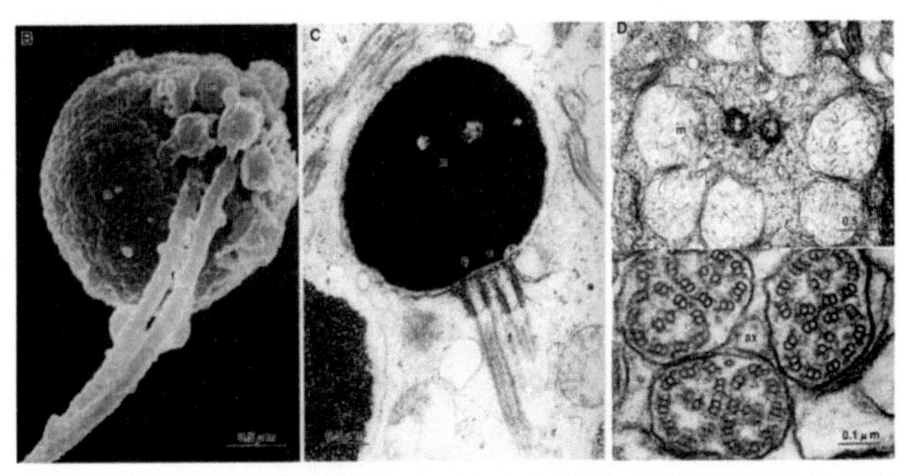

Fig. 15.57 *Apogon semilineatus* (Apogonidae). **A**. SEM of spermatozoon. **B**. TEM longitudinal section of spermatozoon, showing the two flagella. **C**. Transverse section of the midpiece, showing mitochondria. **D**. Transverse sections of flagella. ax, axonemes; c, centrioles; f, flagellum, m, mitochondria; n, nucleus. Adapted from Hara, M. and Okiyama, M. 1998. Bulletin of The Ocean Research Institute, University of Tokyo 33: 1-138, Fig. 50B-E. With permission.

Uniflagellate sperm, with a single fossa, correspond with the Type I perciform spermatozoon, having a 9+2 flagellum 200 nm wide, 12–18 µm long, and attached to a basal body lacking central singlets. This body connects apically to the distal centriole bound by microtubules to the nucleus. The midpiece has a short collar and 8–12 mitochondria (Fishelson *et al.* 2006).

In biflagellate sperm, each centriole, constituting a basal body, is lodged in a separate shallow basal nucler fossa. The basal bodies of both flagella end anteriorly on an electron-dense plate at the sperm head; the two 9+2 flagella, 10–12 µm long, are enclosed in a joint cytoplasmic membrane, forming together a ca 400 nm wide sperm tail (Fishelson *et al.* 2006). A canal is formed between the flagella inside this cytoplasm shaft, often bearing a few elongated vacuoles (Fishelson *et al.* 2006). However, in *A. affinis* and, the two flagella are separate (Lahnsteiner and Patzner (2007). They are covered by a single membrane in *A. semilineatus* (Fig. 15.57B) in which no appreciable nuclear fossae are apparent (Fig. 15.57D) (Hara and Okiyama 1998).

The relationship between the numbers of bi- or monoflagellate sperm in a male cardinal fish differs in the various species and also changes with age (length) of the fish. For example, in *Cheilodipterus lineatus* of 110 mm SL, 70% of sperm are biflagellate, whereas in males of 90 mm SL, 50% are bi- and 50% are monoflagellate; in *Siphamia permutata, S. mossambica, Apogon annularis,* and *A. evermanni* almost 70% of sperm are biflagellate; and in *A. crassiceps* and *A. talboti* approximately 80% are biflagellate (Fishelson *et al.* 2006).

In *A. imberbis* and *A. affinis*, there are two tiers of mitochondria and although Hara and Okiyama (1998) and Lahnsteiner (2003) state that there is only one row of about seven mitochondria in *A. semilineatus*, SEM (Hara and Okiyama 1998) reveals that one or more mitochondria may form a second tier (Fig. 15.57A, and Fig 50A of the latter authors). Mattei and Mattei (1984) state that the midpiece in *A.* affinis (=*Paroncheilus* sp.) contains about 15 mitochondria, in two tiers, located around the two axonemes. It has a posterior region where the mitochondria are separated by a common cytoplasmic canal from the two flagella. This contrasts with the biflagellate

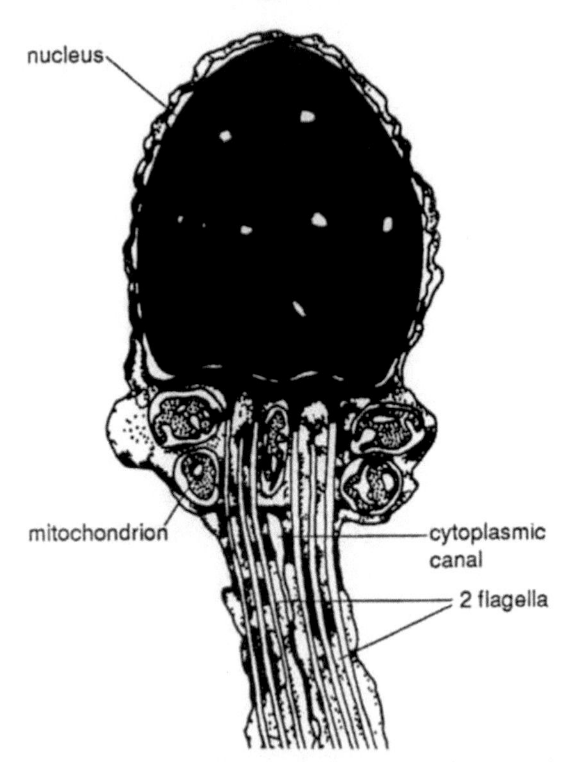

Fig. 15.58 *Apogon affinis* (=*Paronocheilus* sp.) (Apogonidae). Biflagellate spermatozoon. From Jamieson 1991, after a micrograph of Mattei, C. and Mattei, X. 1984. Journal of Ultrastructure Research 88: 223-228, Fig. 7.

sperm of gobiesociforms and siluriforms in which each flagellum has its own cytoplasmic canal. Mitochondrial cristae are sparse and the matrix is dense.

In *A. affinis*, the two parallel centrioles are each composed of nine triplets. Each centriole is continued as a transition region, consisting of armless doublets lacking bridges and unaccompanied by central singlets; in each axoneme a satellite is associated with one of the doublets; the two affected doublets are diametrically opposite within the intermediate piece. This region is followed by a normal 9+2 axoneme in which each doublet has two dynein arms. A saccule formed from the external membrane of the nuclear envelope passes through the transition zone of each axoneme, sometimes accompanied by a mitochondrion. Usually the saccule penetrates at the level of the doublet provided with a satellite so that each axoneme appears in cross section to be divided into two groups of 3 and 6 elements. Somewhat similar division of the axoneme by the outer nuclear membrane is seen in Acanthocephala (Mattei and Mattei 1984). This saccule does not occur in, for instance, *A. imberbis* and *A. semilineatus* (Lahnsteiner and Patzner 2007).

Lahnsteiner (2003) confirmed that the basal body in *A. affinis* consists of 9 triplets but Lahnsteiner and Patzner (2007) reported that that of *A. imberbis* consists of nine doublets; this difference perhaps requires confirmation from serial sections. A proximal centriole is unknown in apogonid sperm, whether uni- or biflagellate. In *A. imberbis* sperm; the cranial end of the basal body is embedded in electron dense material which fastens this structure to the nucleus and this appears to be the case in *A. semilineatus* (Hara and Okiyama 1998) (Fig. 15.57B) and is clearly the basal line described by Fishelson *et al*. (2006).

Remarks: In the Apogonidae, insemination is internal in the form of "internal gametic association" (Munehara 1997) in which sperm are transferred to the oviduct of the female via the ventral fins of the male. In *Apogon imberbis*, spawning is characterized by three main behavioral traits. First the male and female swim side by side, then the male wraps his anal fin around the abdomen of the female covering the genital papilla. Fertilized eggs are released by the female and mouth-bred by the males (references in Lahnsteiner 2003). It is noteworthy that in these apogonids, despite internal insemination, there is little or only moderate elongation of the sperm nucleus.

15.7.2 Suborder Labroidei

The Labroidei consist of the Cichlidae (tilapias and relatives), Pomacentridae (Damsel fishes) and Embiotocidae (surf perches) together with the Labridae (wrasses), Odacidae (cales) and Scaridae (parrot fishes) (Greenwood and Liem 1981; Kaufman and Liem 1982; Nelson 2006). Kaufman and Liem (1982) and Lauder and Liem (1983) considered the Pomacentridae to be the plesiomorph sister-group of the other labroids, (as earlier suggested by Stiassny 1980); the cichlids to be the plesiomorph sister group of embiotocids and labrids; and embiotocids that of labrids. This precise sequence of families is not supported by the shared occurrence of demersal eggs and parental care of hatched young in cichlids and some pomacentrids (Richards and Leis 1984) unless, as seems unlikely, these features are plesiomorphic for labroids. Furthermore, molecular data have not corroborated the monophyly of the Labroidei (Streelman and Karl 1997). From the molecular analyses of Ortí and Li labroids, though represented only by cichlids, are separate from other perciforms and lie in a clade in which they form the sister-group of a clade containing the Mugiliformes, Beloniformes, Atheriniformes and Cyprinidontiformes (Ortí and Li, **Chapter 1**, Figs. 1.5, and 15.1).

Sperm of the families Pomacentridae, Cichlidae and Embiotocidae and Labridae have been examined ultrastructurally (see these families below). As Quagio-Grassiotto *et al.* (2003) observes, from the data available it appears that there are no spermiogenic or spermatozoal characteristics exclusively found in the members of the suborder Labroidei; thus, either these characteristics may have evolved independently in the families of the Labroidei or, as suggested by Streelman and Karl (1997), it is not a natural [monophyletic] group.

Mattei (1991) recognizes the perciform type sperm for the Labridae but different sperm types for the Pomacentridae, Cichlidae and Embiotocidae (see below).

15.7.2.1 Family Cichlidae

Cichlids are freshwater and brackish water fish of Central and South America (one species north to Texas), West Indies, Africa, Madagascar, Israel, Syria, Iran, coastal India and Sri Lanka (Nelson 2006). Species flocks characteristic of the African lakes would be an interesting subject for the study of species-specificity of spermatozoal ultrastructure and steps in this direction have been taken by Fishelson (2003).

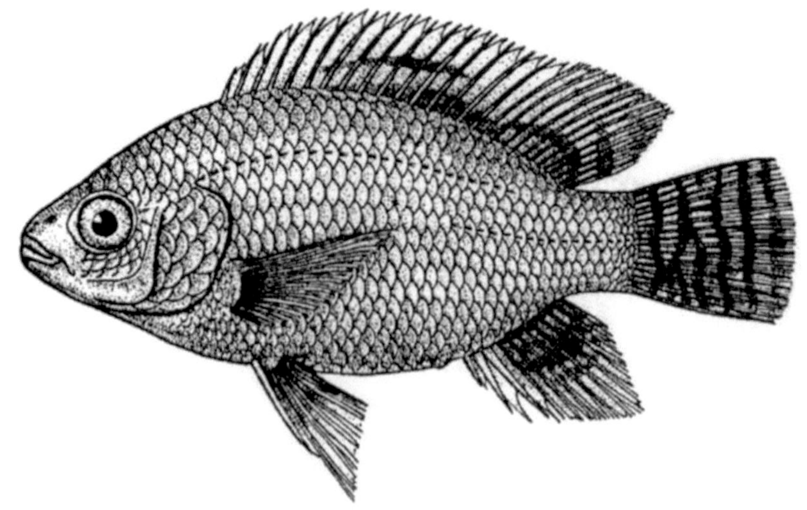

Fig. 15.59 *Oreochromis (=Tilapia) niloticus.* From Grassé, P.-P. 1958. *Traité de Zoologie.* XIII Agnathes et poissons. Masson et Cie, Fig. 1732.

Cichlid species studied for sperm ultrastructure are *Hemichromis fasciatus* and *Oreochromis (=Tilapia) niloticus* (Mattei 1970); *Oreochromis niloticus* (Guha *et al.* 1988; Lou and Takahashi 1989; Silva and Godhino 1991; Don and Avtalion 1993; You and Lin 1998); *O. mossambicus* (Pieterse 2006); *Crenicichla saxatilis* (Matos *et al.* 1995); *Satanoperca jurupari* (Matos *et al.* 2002b), *Crenicara punctulatum* (Matos *et al.* 2002a), *Cichla intermedia* (Quagio-Grassiotto *et al.* 2003) and *Sarotherodon galilaeus* (El-Gammal *et al.* 2003). In addition, a profound study of spermatogenesis and fertilization biology, with considerable detail of spermatozoal ultrastructure, was made by Fishelson (2003) on *Tilapia zillii* (Fig. 15.60AI), *Oreochromis aureus* (Fig. 15.61C), *O. niloticus, Sarotherodon galilaeus* (Tilapinae) and *Archocentrus (=Cichlasoma) nigrofasciatum* (Fig. 15.61A,B), *Neolamprologus brichardi, Aulonocara nyassae, Maylandia (=Pseudotropheus) zebra, Pseudotropheus lombardoi, Labeotropheus trewavasae, L. fuellborni, Melanochromis auratus* (Fig. 15.60AII) and *Haplochromis (=Astatotilapia) flaviijosephi* (Fig. 15.60AIII), (Haplochrominae); for sperm dimensions see Table 15.3. Spermatogenesis in *Cichla ocellaris* has been investigated by Cruz-Landim and Cruz-Hoflung (1986/1987).

Table 15.3 Dimensions of sperm components in cichlids (in μm and SD) and number of mitochondria in the midpiece (max., min. and average). n = 15 per species. Slightly modified after Fishelson, L. 2003. Journal of Morphology 256: 285-300.

Species	*Nucleus (μm)*	*Collar (μm)*	*Tail (μm)*	*Number of mitochondria*
A. nigrofasciatum	1.7 (±0.06)	0.8 (±0.15)	12 (±2.6)	8–10 (9.4)
T. zillii	2.1 (±0.1)	1.2 (±0.3)	18 (±2.0)	8–10 (9)
N. brichardi	2.0 (±0.2)	1.4 (±0.2)	20 (±1.18)	8–10 (8.6)
Mel. auratus	1.4 (±0.2)	1.3 (±0.2)	20 (±2.0)	8–10 (9.6)
Aul. nyassae	1.5 (±0.25)	1.4 (±0.15)	20 (±1.2)	12–14 (12.4)
May. zebra	1.4 (±0.18)	1.5 (±0.2)	20 (±1.2)	10–12 (11.6)
P. lombardoi	1.6 (±0.2)	1.5 (±0.15)	22 (±2.0)	8–10 (9.2)
L. trewavasae	1.5 (±0.16)	1.2 (±0.05)	24 (±1.2)	12–14 (12.6)
L. fuellbornii	1.8 (±0.2)	1.5 (±0.8)	22 (±2.6)	12–14 (13.2)
H. flaviijosephi	1.6 (±0.18)	1.6 (±0.2)	24 (±2.0)	12–14 (13.2)
O.niloticus	2.0 (±0.2)	1.4 (±0.8)	26 (±2.4)	8–10 (8.5)
O. auratus	1.8 (±0.16)	1.2 (±0.4)	24 (±2.2)	8–10 (8.4)
S.galilaeus	1.8 (±0.12)	1.4 (±0.4)	24 (±1.2)	10–12 (11.2)

The sperm of *Cichla intermedia* (Quagio-Grassiotto *et al.* 2003) (Fig. 15.62) and *Oreochromis niloticus* (adult, Fig. 15.59; sperm, Figs. 15.64, 15.65, 15.66B), *Hemichromis fasciatus* (Mattei 1970) (Fig. 15.66A), *Crenicichla saxatilis* (Matos et al. 1995); *Crenicara punctulatum* (Matos *et al.* 2002a), *Haplochromis flaviijosephi (=Astatotilapia flaviijosefi)* (Fig. 15.60AIII) *Archocentrus (=Cichlasoma) nigrofasciatum* (Fig. 15.61A,B), and others listed in Table 15.3 (Fishelson 2003) have a Type I spermatozoon *sensu* Mattei (1970) (Fig. 15.25), though more asymmetrical in *O. niloticus* (Fig. 15.64) than is characteristic of this type. Fishelson (2003) states that asymmetry of the collar as reported for

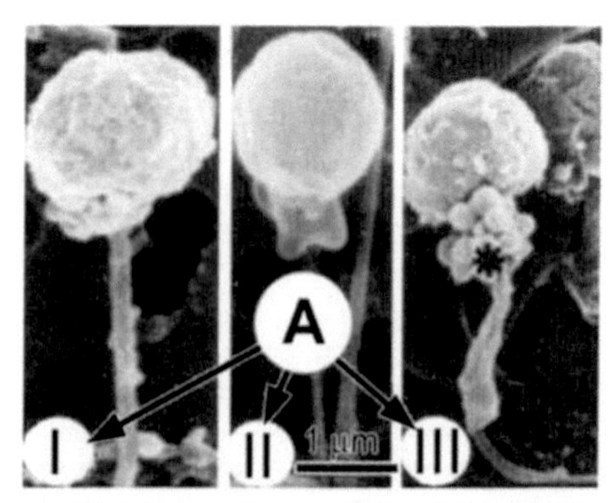

Fig. 15.60 SEM of cichlid spermatozoa. **A I**. *Tilapia zillii*. **A II**. *Melanochromis auratus*. **A III**. *Haplochromis (=Astatotilapia) flaviijosephi*. Asterisk, mitochondria. Courtesy of Professor Lev Fishelson.

Oreochromis niloticus, with more mitochondria on one side of the axoneme, was not found in the sperm of any cichlids which he examined, including *O. niloticus*, and seemed to be an artifact. However, such asymmetry is illustrated for *O. niloticus* by Guha *et al.* (1988) and independently by Lou and Takahashi (1989) (see Figs. 15.63A-C, 15.64). The apparent discrepancy is possibly due to different planes of section.

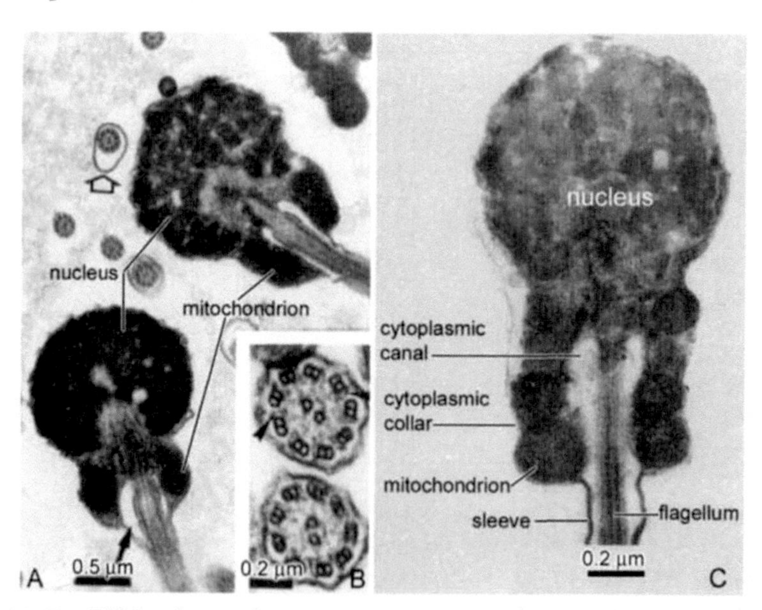

Fig. 15.61 TEM of cichlid spermatozoa. **A**. *Archocentrus (=Cichlasoma) nigrofasciatum*. Spermatid above and mature spermatozoon below. Broad arrow shows a transverse section of the flagellum in the cytoplasmic sleeve. Narrow arrow, sleeve. **B**. Transverse sections of flagella of same species. Arrows indicate dynein arms. **C**. *Oreochromis aureus*. Spermatozoon. Relabeled, courtesy of Professor Lev Fishelson.

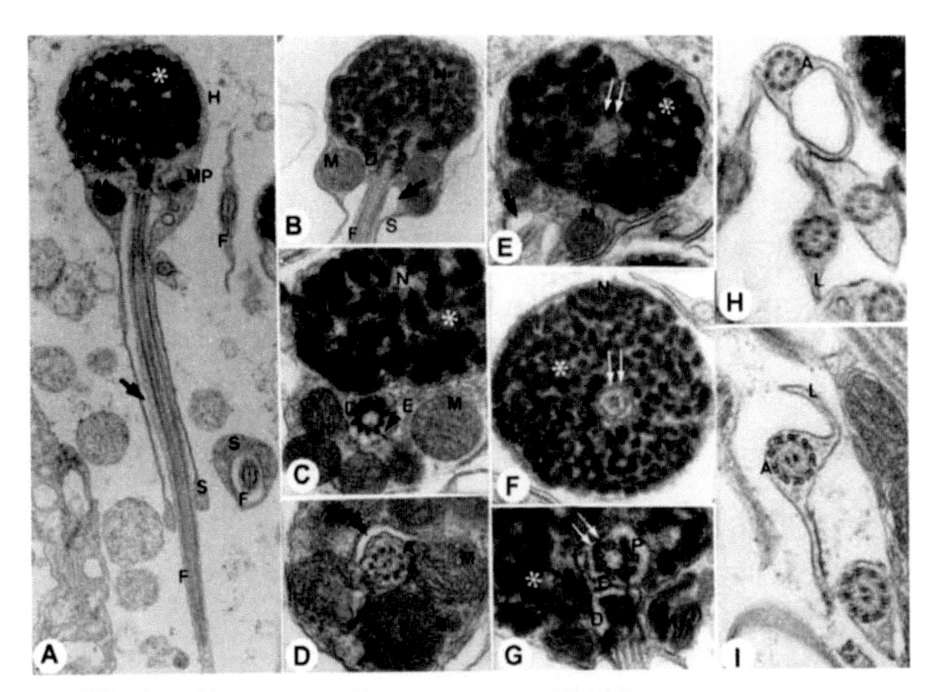

Fig. 15.62 A. *Cichla intermedia* spermatozoon (Cichlidae). x 14,200. **B–D**. Head and midpiece initial region of spermatozoa. **B**. x 12,900; **C**. x 19,200; **D**. x 28,000. **E–G**. Head and midpiece region. **E**. x 15,500. **F**. x 18,700. **G**. x 22,600. **H** and **I**. Flagellum of spermatozoa (cross-sections). H. x 29,500. I x 39,500. Abbreviations: A, axoneme; D, distal centriole; E, cisternae of the endomembrane system. F, flagellum; H, head; L, lateral fins; M, mitochondria; MP, midpiece; N, nucleus; P, proximal centriole; S, cytoplasmic sheath; arrow, cytoplasmic canal; asterisk, clusters of compacted chromatin; double arrow, nuclear fossa; arrow, cytoplasmic canal; arrowhead, radial fibrils. From Quagio-Grassiotto, I., Antoneli, F. N. and Oliveira, C. 2003. Tissue and Cell 35: 441-446, Fig. 3. Copyright Elsevier. With permission.

As usual in teleosts, there is no acrosome. The head is formed by a round or slightly elongated nucleus. It is 2.0-2.6 µm (*Oreochromis niloticus*) (Guha *et al.* 1988; Silva and Godhino 1991) or 1.8 µm (*Cichla intermedia*) in diameter and has a length ranging from 1.4–2.1 µm (Fishelson 2003) (see additional data in Table 15.3). It is encircled by a narrow strip of cytoplasm with no organelles. In *O. niloticus*, the small, eccentric midpiece is ca 1.5 µm long, the flagellum ca 20.3 µm long (Lou and Takahashi 1989) (17 µm long according to Guha *et al.* 1988) and the head + sleeve to tail length ratio is ca 0.15 (Mattei 1970; Don and Avtalion 1993), comparing well with data in Table 15.3.

The nucleus is occupied by many electron-dense chromatin globules, which form a mass by close adhesion leaving irregular interspaces among them (*Oreochromis niloticus*) or highly condensed filamentous clusters of chromatin (*Cichla intermedia*) (15.62A–C). The chromatin in *O. niloticus* was considered to be in its most condensed form (Mattei 1970; Don and Avtalion 1993). In *C. intermedia* the chromatin is scalloped anteriorly in a way that appears to the author to be somewhat reminiscent of the 'dip' or hiatus in the nucleus of the pleuronectiforms *Paralichthys olivaceus* and *Scophthalmus maximus*. In *O. niloticus* the anterior surface of the nucleus is somewhat lobulated. The resulting uneven contour of the head portion is evident in the scanning electron microscopic aspect of the spermatozoon (Fig. 15.63B) (Lou and Takahashi 1989). Lobulation of the anterior surface of the nucleus was also noted by Guha *et al.* (1988) (Fig. 15.64) in *O. niloticus* and Matos *et al.* (1995) observed irregular projections of the nucleus in *Crenicichla saxatilis*. All species have a basal nuclear fossa which contain the proximal and often the distal centriole (e.g. Fishelson 2003).

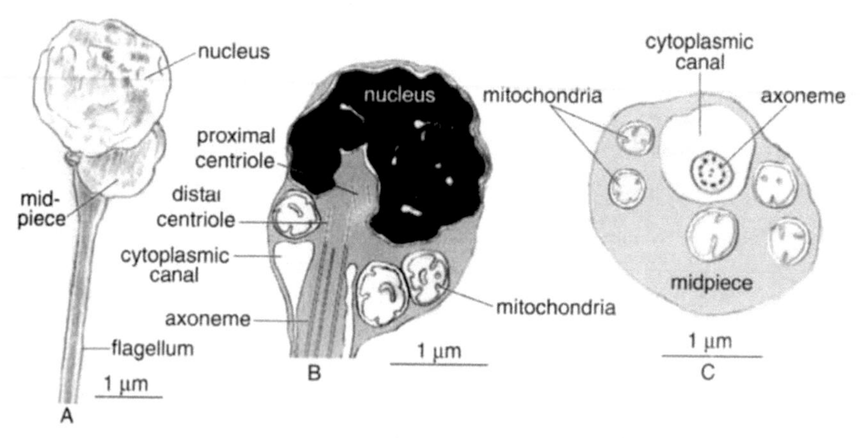

Fig. 15.63 *Oreochromis niloticus* (Cichlidae). Mature spermatozoa. **A**. Scanning electron micrograph of a mature spermatozoon from milt, showing uneven contour of the head portion. **B**. Longitudinal section, showing aggregated globular chromatin in the head and separated mitochondria in the midpiece. **C**. Transverse section of the midpiece, showing an eccentric configuration; mitochondria are more numeous than is indicated. Semidiagrammatic after micrographs of Lou, Y. H. and Takahashi, H. 1989. Journal of Morphology 200: 321-330, Figs. 12-14.

There is widespread agreement that, in teleost sperm, the term collar should be applied to the posterior mitochondrion-containing region of the midpiece where this region surrounds a cytoplasmic canal (e.g. Jaspers *et al.* 1976; An *et al.* 1999; Matos *et al.* 2002b; Assem 2003; Hu *et al.* 2005; Pecio *et al.* 2005), whether or not the midpiece is well defined from the nuclear part of the sperm head. The term sleeve, in the strict sense, has been applied to an extension, from the collar, of a sheath in which the outer plasma membrane and the membrane surrounding the cytoplasmic canal are close together and do not enclose mitochondria (e.g. Jamieson 1991; Hu *et al.* 2005). Thus Hu (2005) recognized, in the sperm of the cyprinid *Puntius conchonius*, "a collar, containing several irregular mitochondria embedded in the cytoplasmic mass,

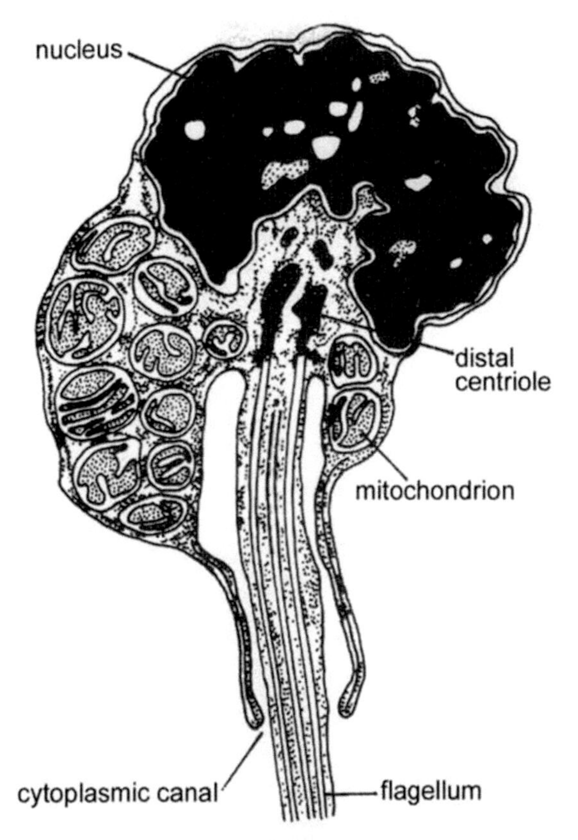

Fig. 15.64 *Oreochromis niloticus* (Cichlidae). Longitudinal section of spermatozoon. Showing asymmetry of the midpiece around the tail axis. This asymmetry is not general for cichlids. After a micrograph by Guha *et al.* 1988. Proceedings of the 46th Annual Meeting of the Electron Microscopy Society of America. San Francisco Press, San Francisco. pp. 278-279, Fig. 1.

and a slender caudal segment of variable length, the sleeve, behind the collar." However, as the caudal segment, or "trailing sleeve" (*sensu* Jamieson 1991), may be present or absent within a family, as in the Cichlidae (for instance, in *Hemichromis fasciatus* (Fig. 15.66A) (Mattei 1970) although a cytoplasmic collar defines the cytoplasmic canal, there is no extension as a sleeve) it is understandable that the two components are grouped by some authors as the collar (e.g. Fishelson 2003) (Table 15.3) or, less appropriately, the sleeve (Don and Avtalion 1993). A well-developed sleeve [or collar] is exemplified by *Crenicara punctulatum* (Matos *et al.* 2002a) and *Cichla intermedia* (Quagio-Grassiotto *et al.* 2003) (Fig. 15.62A). The sleeve may invest the anterior two thirds of the flagellum (Guha *et al.* 1988; Silva and Godhino 1991) and is sometimes longer than the sperm head (Fishelson 2003).

Fishelson (2003) relates the length of the collar to the number of mitochondria, and shows for 13 species, that in tilapins and in some haplochromins there are 8–10 mitochondria and the collar is 0.7–1.2 μm in length, whereas in *Haplochromis (=Astatotilapia) flaviijosephi* and some related species, there are 14–20 mitochondria and the collar is 1.2–1.6 μm long, longer than the sperm head in certain species (Table 15.3). Scattered data for numbers of mitochondria may be added. In *Oreochromis niloticus,* the base of the midpiece, generally eight to ten mitochondria are closely aggregated (Fig. 15.63) (Lou and Takahashi 1989). In *Sarotherodon galilaeus* the midpiece is said to contain few spherical mitochondria, which are separated from the

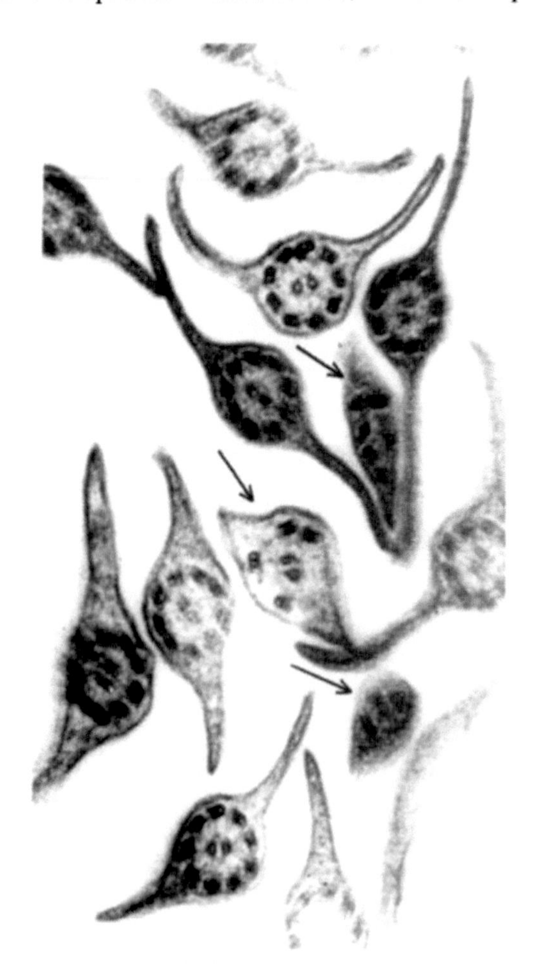

Fig. 15.65 *Oreochromis (=Tilapia) niloticus* (Cichlidae). Transverse sections of flagella, showing the long lateral fins. Arrows indicate three endpieces with disrupted microtubular configurations. From Mattei, X. 1969. *Contribution à l'étude de la spermiogenèse et des spermatozoïdes de poissons par les méthodes de la microscopie électronique.* D.Sc. thesis, Université de Montpellier, Fig. XXIXh. With permission.

axoneme by the cytoplasmic canal (El-Gammal *et al.* 2003); however, Fishelson (2003) lists 10-12 mitochondria for this species. There are 7 to 8 spherical mitochondria in *Crenicichla saxatilis* (Matos *et al.* 1995) and 6 or 7 in *Crenicara punctulatum* (Matos *et al.* 2002a).

In *Cichla intermedia* the mitochondria, concentrated near the nucleus, are round or slightly elongated and contain numerous cristae and an electron-dense matrix. They are disposed in two layers around the initial segment of the flagellum and are separated from it by the cytoplasmic canal. Each layer has about five mitochondria (Fig. 15.62B,C). In it and *Oreochromis niloticus* a cytoplasmic sleeve, flanking a cytoplasmic canal, and slightly dilated at its posterior end, extends caudal to the mitochondrial mass and investing the anterior region of the flagellum (Fig. 15.64), being about 1.2 μm long in *O. niloticus* (Lou and Takahashi 1989).

In *Cichla intermedia*, the centriolar complex is surrounded by electron-dense material and cisternae of the endomembrane system. The proximal centriole penetrates into the initial (basal) portion of the nuclear fossa (implantation fossa) and is oblique to the flagellum (Fig. 15.62B). Nine radial fibrils, one for each microtubule doublet, anchor the distal centriole (Quagio-Grassiotto *et al.* 2003) (Fig. 15.62C). In *Oreochromis niloticus* the proximal and the anterior region of the distal centriole are housed in the fossa; they are depicted a mutually perpendicular (Mattei, 1970; Guha *et al.* 1988) (Figs. 15.64, 15.66B), though in some micrographs they appear to be mutually at an oblique angle (Lou and Takahashi 1989) (Fig. 15.63B). In *S. galilaeus* they are perpendicular but only the proximal centriole lies in the fossa (El-Gammal *et al.* 2003). In *Crenicichla saxatilis* they are illustrated as almost perpendicular (Matos *et al.* 1995).

The flagellum shows a classical 9+2 axoneme and has two lateral fins (Mattei 1969) (Fig. 15.63) (Guha *et al.* 1988; Lou and Takahashi 1989; Silva and Godhino 1991; Matos *et al.* 1995; Quagio-Grassiotto *et al.* 2003) (Fig. 15.62H,I), as also in *Crenicara punctulatum* (Matos *et al.* 2002a). It is therefore remarkable that flagellar fins were not observed in the 13 species, including some of these, studied by Fishelson (2003). The fins are of moderate length and are in line, as usual, with the central singlets; the axoneme has outer dynein arms but the inner arms are not conspicuous. No intratubular differentiations are present (Guha *et al.* 1988).

The dimensions of the sperm tail and the consequent total dimensions of mature sperm differ in the various cichlids (Table 15.3). The general trend is toward longer sperm in the substrate brooders than in the mouth brooders, a phenomenon that Fishelson (2003) considered might be related to the more active swimming required during fertilization in moving waters. There was an interesting variability in intraspecific sperm length: e.g., in *Tilapia zillii* (18–24 μm) and in *Haplochromis flaviijosephi* (15–18 μm). As noted by Fishelson, the longest sperm in a cichlid (42.21 μm) was observed by Stockley *et al.* (1997) in *Julidochromis ornatus*, a substrate brooder. This variability was not correlated with body length. Study of the 13 cichlids also confirmed the view of Stockley *et al.* (1997) that polygynous species have shorter sperm than monogamous ones and that variability is related to sperm competition (Fishelson 2003).

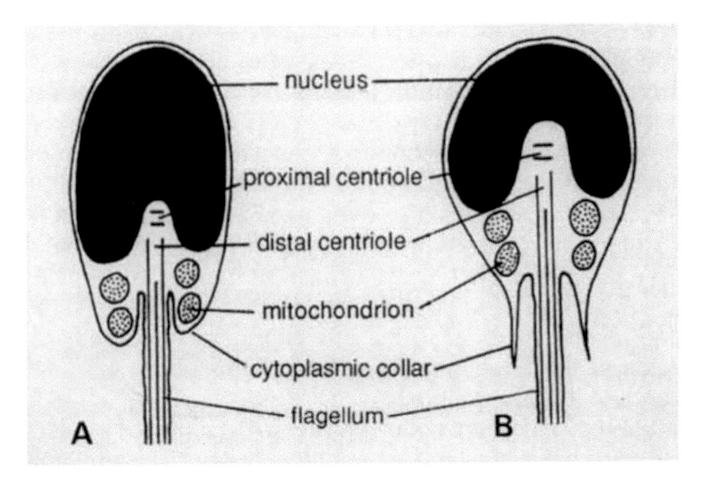

Fig. 15.66 Cichlid spermatozoa. **A**. *Hemichromis fasciatus*. **B**. *Oreochromis (=Tilapia) nilotica* (Cichlidae). Diagrammatic longitudinal section of the spermatozoa. Adapted from Mattei, X. 1970. After Mattei, X. 1970. Pp. 59-69. In B. Baccetti (ed.), *Comparative Spermatology*. Academic Press, New York. Fig. 4: 11 and 4.29.

Satanoperca jurupari: In the Amazonian self-fertilizing hermaphrodite cichlid *S. jurupari*, a mouth brooder, the spermatozoon is of the introsperm type. It is further exceptional for investigated cichlids in having two flagella.

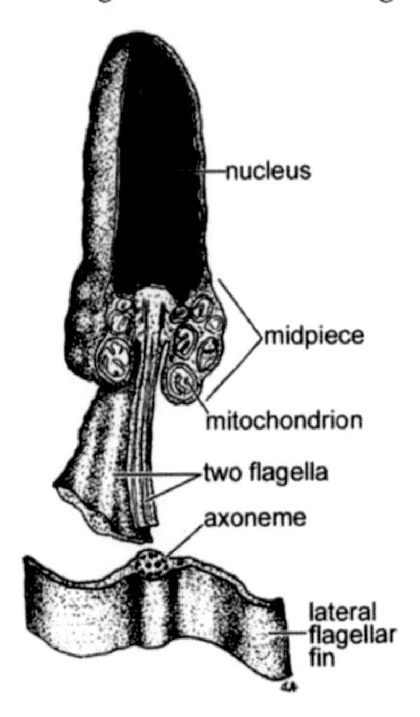

Fig. 15.67 *Satanoperca jurupari* (Cichlidae, Geophaginae). Schematic drawing of the spermatozoon. Relabeled after Matos, E., Santos, M. N. S. and Azevedo, C. 2002. Brazilian Journal of Biology 62: 847-852, Fig. 11. With permission.

The mature spermatozoa consist of a short head containing a nucleus but without acrosome, a midpiece and two tails (Figs. 15.67, 15.68).

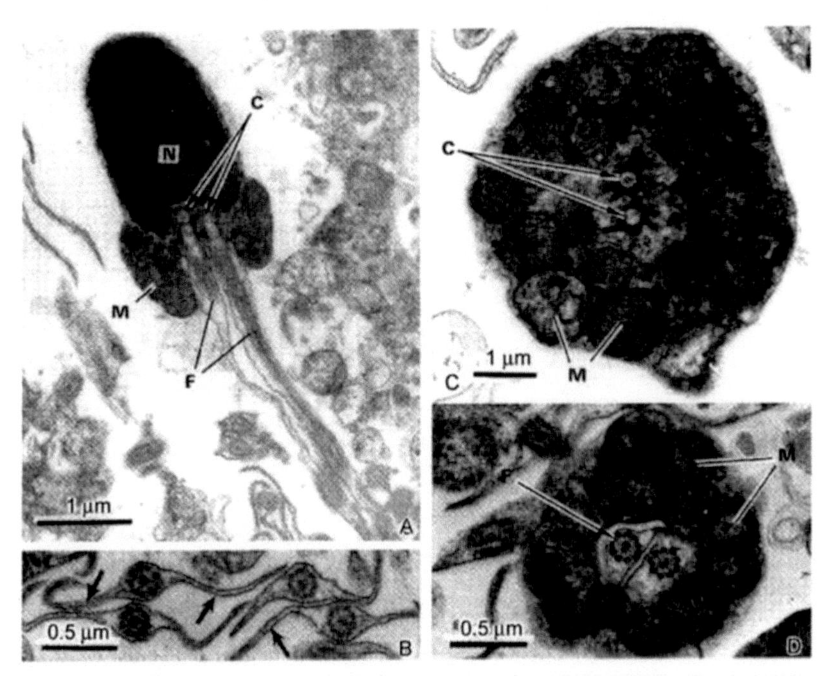

Fig. 15.68 *Satanoperca jurupari* (Cichlidae, Geophaginae). **A**. Longitudinal section of a spermatozoon showing a dense nucleus, lacking an acrosome, two centrioles in continuity with the two flagella, the mitochondrial collar of the midpiece, and the anterior region of the flagellum. **B**. Four flagella with lateral fins (arrows) and 9+2 axoneme. **C**. Transverse section (TS) of midpiece, showing the two centrioles. **D**. TS of midpiece through two axonemes. C, centrioles; F, flagellum; M, mitochondria; N, nucleus. From Matos, E., Santos, M. N. S. and Azevedo, C. 2002. Brazilian Journal of Biology 62: 847-852, Figs. 6-9. With permission.

The electron-dense nucleus presents two forms. Most of the nuclei were cylindrical 3 µm (3 mm in original account) long and 1.3 µm wide and with a rounded anterior end (Fig. 15.68A). Some other spermatozoa contain a circular nucleus 3.0-3.4 µm in diameter. The midpiece consists of a collar or ring-shaped mitochondrion containing several mitochondria situated at different levels, and two centrioles (Fig. 15.68A,C,D). At its posterior end the midpiece forms a curved flange delimiting a cytoplasmic collar formed by the mitochondria (Fig. 15.68A). This collar is 0.5 µm high.

In the collar region of the midpiece the plasmalemma of each flagellum develops two lateral cytoplasmic extensions (fins) on opposite sides of the flagellum (Fig. 15.68B) which extend along most of the length of the tail. The largest extension, including the axoneme, is 3 µm in diameter. The axoneme has the usual 9+2 microtubular pattern (Fig. 15.68B,D) (Matos *et al.* 2002b).

Remarks: Biflagellate spermatozoa, seen in *Satanoperca jurupari*, have been described in a disparate minority of teleost species (e.g. Mattei and Mattei 1984; Jamieson 1991; see Table 13.1). Lateral fins on the flagellum are somewhat more common. The occurrence of two nuclear sizes in the sperm of *S. jurupari* possibly indicates presence of fertilizing eusperm and genetically non-fertilizing parasperm, though this requires confirmation. Elongation of the nucleus, though here not greatly pronounced, is characteristic of internally fertilizing species and therefore of introsperm, see, for instance, Poeciliidae (**Chapter 14**).

At least three characteristics seem to be characteristic of Cichlidae: (1) compact chromatin [though often with lacunae]; (2) slightly eccentric nuclear fossa; and, (3) a high number (about 10) of mitochondria around the initial segment of the flagellum (Quagio-Grassiotto *et al.* 2003).

It is seen that cichlids have type I spermatozoa, *sensu* Mattei (1970) as defined by orientation of the flagellum perpendicular to the nucleus following nuclear rotation in spermiogenesis.

15.7.2.2 Family Embiotocidae

The Embiotocidae is a family of coastal marine, rarely freshwater, North Pacific sea perch which are fully viviparous, delivering large, well-developed young. The male has a small intromittent organ which represents a modified anterior end of the anal fin (see also Evans and Downing, **Volume 8B, Chapter 4, Section 4.3.8.3**). Monophyly of the embiotocids is indicated by a suite of specialized mechanisms for viviparity, including modified and vascularized median fins used in prenatal young for exchange with the convoluted and vascularized ovarian lining (Webb and Brett 1972; Lauder and Liem 1983).

Cymatogaster aggregata, the White Surf-fish or Shiner Surfperch (adult, Fig. 15.69), has been shown (Gardiner 1978) to produce bundles of sperm

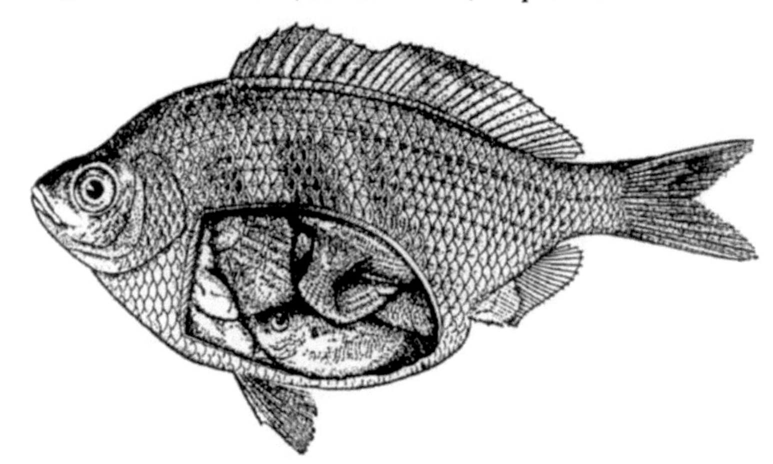

Fig. 15.69 *Cymatogaster aggegata*, the White Surf-fish or Shiner Surfperch (Embiotocidae). Viviparous female dissected to show enclosed young. After Jordan, D.S. 1907. *Fishes*. Henry Holt, New York. Fig. 22.

which have an extracellular capsule and are therefore true spermatophores. Each spermatophore contains some 600 parallel spermatozoa and thus differs notably from the spermatozeugmata of poeciliids, in which internal fertilization is clearly an independent development, which have the sperm heads located peripherally to a core of flagella. In the female they release the spermatozoa within an hour of insemination.

The spermatozoon of *C. aggregata* is an anacrosomal introsperm (Fig. 15.70), approximately 50 μm long. The head (chiefly nucleus) is elongate (4 μm long) with condensed chromatin and is strongly depressed in one longitudinal plane (1 μm wide and 0.4 μm deep). Both centrioles occupy depressions in the nucleus, anterior to its midlength, which are interconnected by a thin isthmus of cytoplasm. The distal centriole is cappe ' by an electron dense cone-shaped body. The elongate midpiece (3.5 μm long) contains 6 mitochondria, arranged three on each side; an organelle-free cytoplasmic sleeve continues 1 mm behind them. From its origin in the head and continuing through the midpiece, the anterior 4.5 μm of the 9+2 flagellum runs in a long cytoplasmic canal. Lateral protuberances of the flagella [fins] are irregularly arranged but mostly paired, in the plane of the two central singlets (Gardiner 1978).

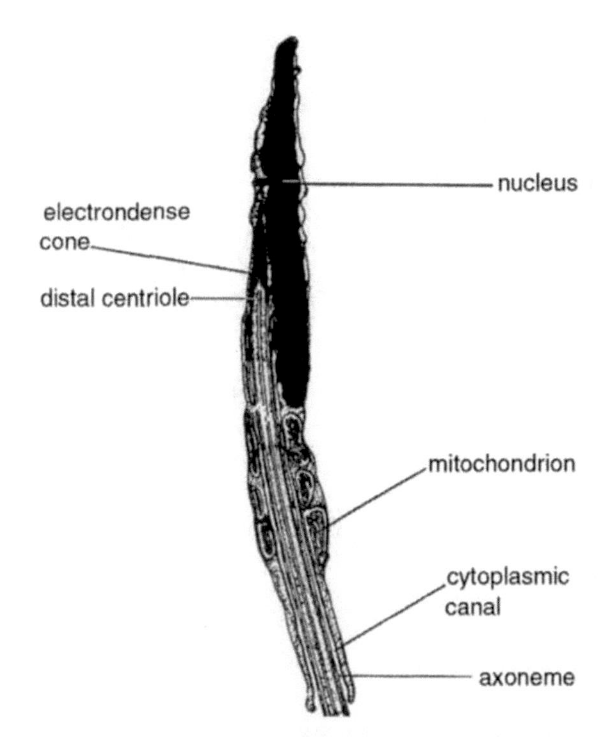

Fig. 15.70 *Cymatogaster aggregata* (Embiotocidae). This longitudinal section near the midline passes through the proximal centriole within the depression of the upper surface of the head and through the electron dense cap on the anterior end of the flagellum. After a micrograph by Gardiner, D.M. 1978. Journal of Fish Biology 13: 435-438, Fig. IB.

Ditrema temminckii: The spermatozoon of *D. temminckii* described by Hara and Okiyama (1998) (Fig. 15.71) closely resembles that of *Cymatogaster aggregata*. The elongate, pointed nucleus has dense chromatin. The long mitochondrial sleeve contains 5 to 7 mitochondria, arranged 2 to 4 on each side. The proximal centriole has not been observed. The distal centriole (basal body), and therefore the flagellum, is inserted at about midlength of the nucleus and is again capped by an electron-dense cone.

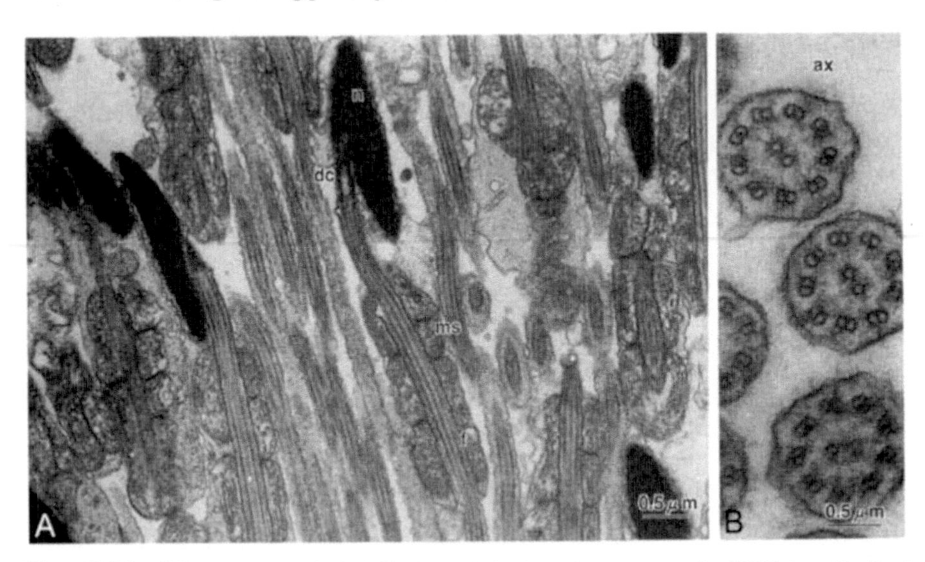

Fig. 15.71 *Ditrema temminckii* (Embiotocidae) spermatozoa. **A**. TEM longitudinal sections of spermatozoa. **B**. Transverse sections of flagella. ax, axoneme; dc, distal centriole; f, flagellum; ms, mitochondrial sleeve (midpiece); n, nucleus. Adapted from Hara, M. and Okiyama, M. 1998. Bulletin of The Ocean Research Institute, University of Tokyo 33: 1-138, Fig. 52A,D. With permission.

Remarks: The elongate, modified form of the spermatozoa of these two embiotocids correlates with their internal fertilization. The electron-dense cone on the distal centriole may be a synapomorphy of the family (Hara and Okiyama 1998).

15.7.2.3 Family Pomacentridae

These are the damsel fishes; marine, rarely brackish, in all tropical seas but primarily Indo-Pacific (Nelson 2006) (Figs. 15.72, 15.73).

Stegastes leucostictus: In the sperm of the pomacentrid *Stegastes (=Pomacentrus) leucostictus* a stack of membranes (not equivalent to the Golgi-derived structure in *Protopterus)* applied to the nuclear envelope of the spermatid and originating at the end of meiosis is said to persist into the spermatozoon (Mattei and Mattei 1976a) but no mature spermatozoon is illustrated.

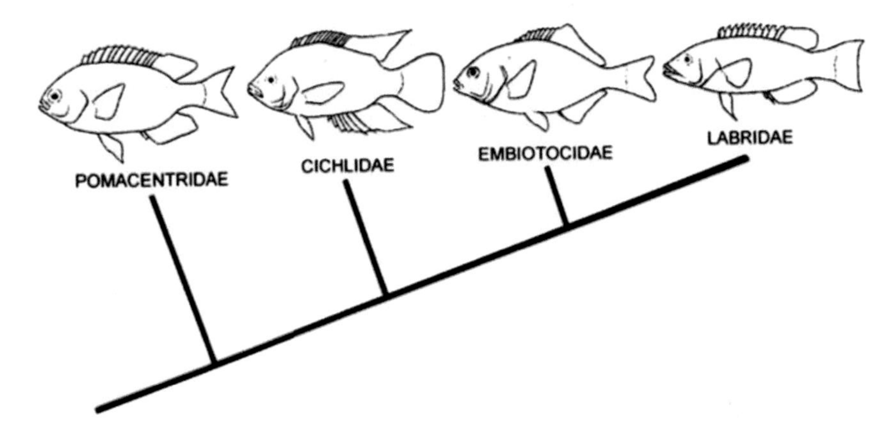

Fig. 15.72 Interrelationships of the major labroid lineages. From Jamieson 1991, modified from Lauder and Liem 1983. Bulletin of the Museum of Comparative Zoology 150: 95-197, Fig. 52.

Fig. 15.73 An example of the Pomacentridae, *Dascyllus aruanus* (Perciformes). Photo Barrie Jamieson.

Mattie (1991) illustrated diagrammatically a sperm of *Abudefduf analogus* and *Chromis* sp. and showed them to be a slight modification of the "simple type" [Type I], also found in the Istiophoridae, Mullidae, Cichlidae and Percophidae, distinct from the perciform type (Type II).

Chromis chromis: The following account is drawn from and slightly modifies that of Lahnsteiner and Patzner (1997). The acrosomeless spermatozoon of the pomacentrid *Chromis chromis* is symmetrical (Fig. 15.74) and is also here considered to represent a slight modification of the Type I sperm. The head region is approximately kidney-shaped. The nucleus consists of granular chromatin and has a wide moderately deep central invagination, the nuclear fossa. The proximal and distal centriole (basal body of the flagellum), lie within the fossa with their long axes in the long axis of the sperm though the proximal is slightly displaced relative to the basal body. The basal body is bordered by an electron-dense disk.

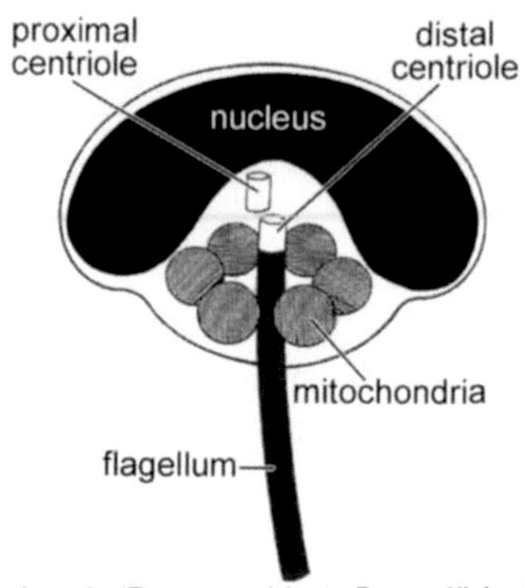

Fig. 15.74 *Chromis chromis* (Pomacentridae), Damselfish. Schematic reconstruction of spermatozoon. After Lahnsteiner, F. and Patzner, R. A. 1997. Journal of Submicroscopic Cytology and Pathology 29: 477-485, Fig. 19 (part).

The centrioles are composed of nine pairs [not triplets] of peripheral microtubules which adhere to an electron-dense sleeve. Nine electron-dense coarse fibers which run parallel to the microtubules adhere to the exterior side of the entire proximal centriole and to the cranial half of the distal centriole. Microfibrils arise from these coarse fibers and fasten the centrioles to the nuclear envelope.

In the caudal half of the distal centriole centrifugally extending fibers (satellite lamellae) arise from the coarse fibers and anchor the basal body in the cytoplasm. Microfibrils fasten the centrioles to each other and to the nuclear envelope (see illustrations in Lahnsteiner and Patzner 1997). The midpiece is approximately ellipsoidal. As a special feature the midpiece has no cytoplasmic canal and the proximal portion of the flagellum, normally within the canal, is located within the cytoplasm. Six mitochondria are arranged around the basal body. They are spherical, have a moderately dark matrix and cristae which are irregular in cross sections and are also irregularly arranged.

The axoneme has the usual 9+2 configuration of microtubules but in the very proximal portion of the flagellum (about 0.2 μm from its origin) central microtubules are absent. The flagellum has no lateral fins (Lahnsteiner and Patzner 1997).

15.7.2.4 Family Labridae

Labrids are the wrasses, in the Atlantic, Pacific and Indian Oceans.They are broadcast spawners.

***Coris julis, Thalassoma pavo, T. bifasciatum, Symphodus ocellatus* and *Lachnolaimus maximus*:** The spermatozoa of *Coris julis, Thalassoma pavo* and *Symphodus ocellatus* are described and extensively illustrated by Lahnsteiner and Patzner (1997) (Fig. 15.75) and those of *Thalassoma bifasciatum* and *Lachnolaimus maximus* by Robinson and Prince (2003) (Fig. 15.76). Mattei (1991) examined but did not describe *C. julis* and *T. pavo*, beyond ascribing labrid sperm to the perciform type. They have no acrosome. In *C. julis, T. pavo, T. bifasciatum* and *L. maximus*, the flagellum inserts craniolaterally at the nucleus, as characteristic of Type II sperm. In the Type I sperm of *Symphodus ocellatus* the flagellum inserts at the centriolar complex in a deep, central invagination of the nucleus that is about 1.8 μm and 0.3 μm diameter (for spermatozoal dimensions see Table 1 in Lahnsteiner and Patzner (1997). In *C. julis, T. bifasciatum* and *L. maximus*, the head is spherical, in *T. pavo* and *S. ocellatus*

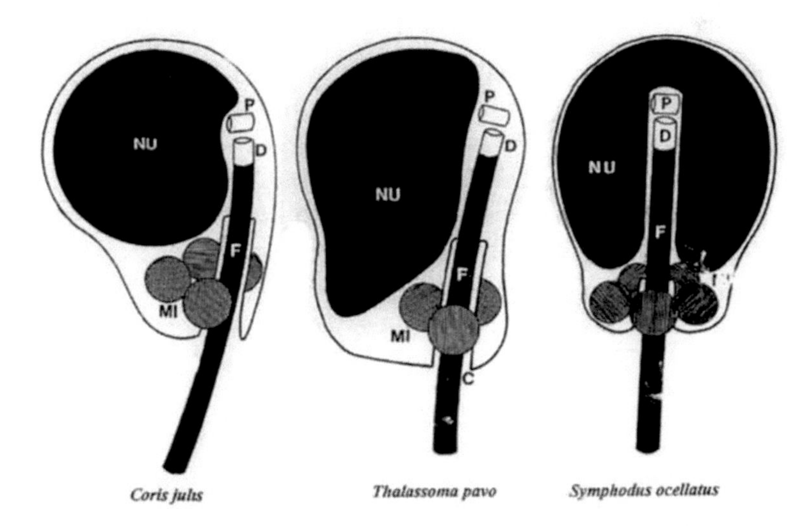

Fig. 15.75 Schematic reconstructions of the sperm of *Coris julis, Thalassoma pavo* and *Symphodus ocellatus* (Labridae). C, cytoplasmic canal; D, distal centriole; F, flagellum; Mi, mitochondria; Nu, nucleus; P, proximal centriole. After Lahnsteiner, F. and Patzner, R. A. 1997. Journal of Submicroscopic Cytology and Pathology 29: 477-485, Fig. 19 (part).

ovoid. The chromatin of the nucleus is electron-dense and (Lahnsteiner and Patzner 1997) granular and densely packed; with no nuclear pores. The centriolar complex is located craniolaterally in a slight invagination of the nucleus except in *S. ocellatus* in which it is located in a narrow but very deep, symmetrically located fossa. In all five species both centrioles consist of nine pairs [not triplets] of peripheral microtubules; they are mutually perpendicular or, in *L. maximus*, the proximal is orientated at approximately 135° to the distal centriole. In *T. pavo* the entire proximal centriole is embedded in a ring of osmiophilic material, in *C. julis* only that part of the proximal centriole facing the nucleus. The cranial portion of the distal centriole, and in *S. ocellatus* that part of the proximal centriole facing the nucleus, is embedded in a ring of osmiophilic material. In *C. julis* the caudal end of the distal centriole is bordered by a ca 0.05 µm thick electron-dense disk. In the three species the centrioles are involved in a system of microfibrils (for further details see Lahnsteiner and Patzner 1997).

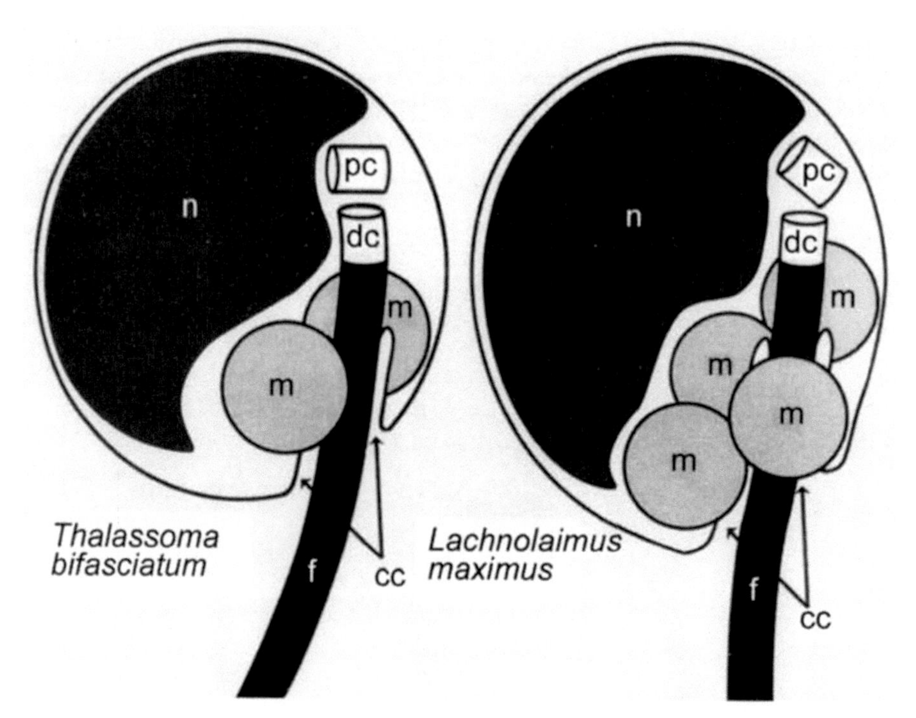

Fig. 15.76 Schematic representations of sperm of *Thalassoma bifasciatum* and *Lachnolaimus maximus* (Labridae). cc, cytoplasmic canal; dc, distal centriole; f, flagellum; m, mitochondrion; n, nucleus; pc, proximal centriole. From Robinson, M. P. and Prince, J. S. 2003. Bulletin of Marine Science 72: 247-252, Fig. 3. With permission.

The midpiece region of spermatozoa of *Coris julis, Thalassoma pavo* and *S. ocellatus* is approximately cylindrical, and is penetrated by the cytoplasmic canal. A second membrane lies underneath the invaginated portion of the plasmalemma (Lahnsteiner and Patzner 1997). The midpiece of contains two spherical mitochondria in *Thalassoma bifasciatum*, three in *T. pavo*, four in *C. julis* and *Lachnolaimus maximus*, and five in *S. ocellatus*. The mitochondrial cristae are approximately tubular in cross sections and irregularly arranged, and in *S. ocellatus* may also be parallel. The 9+2 flagellum lacks lateral fins. In the very proximal portion of the flagella (up to about 0.1-0.2 µm from its origin) central microtubules are absent (Lahnsteiner and Patzner 1997; Robinson and Prince 2003).

The sperm of *Thalassoma bifasciatum* and *Lachnolaimus maximus* are similar to those of *T. pavo, Symphodus ocellatus* and *Coris julis*. All five wrasses are unusual, though not unique, in having only doublets, rather than triplets, in their centrioles (Lahnsteiner and Patzner 1997, Robinson and Prince 2003). A difference is that the head of *T. pavo* is ovoid versus the spherical head of *T. bifasciatum* and that the centrioles of *L. maximus* are mutually at an angle of 135° rather than the 90° seen in the four other labrids. Neither *T. bifasciatum* nor *L. maximus* has the deep basal nuclear fossa seen in the sperm of *S. ocellatus*. The greater number of mitochondria in the sperm of *L. maximus* relative to that of *T. bifasciatum* is tentatively ascribed to the greater distance apart of individuals of the former species during spawning (Robinson and Prince 2003).

Remarks: As noted by Lahnsteiner and Patzner (1997) the sperm of *Coris julis* and *Thalassoma pavo* are most similar to spermatozoa of the perciform families Trachinoidae, Uranoscopidae (Lahnsteiner and Patzner 1996), Serranidae (Gwo *et al.* 1994) and Sciaenidae (Gwo and Arnold 1992) in the lateral insertion of the flagellum, the organization of the centriolar complex and the occurrence of several single and spherical mitochondria. The central, deep nuclear fossa of spermatozoa of *Symphodus ocellatus*, in contrast, is a feature similar to spermatozoa of Mullidae (Lahnsteiner *et al.* 1995), Balistidae, Soleidae, Scorpaenidae, Diodontidae and Monacanthidae (Boisson *et al.* 1969; Jamieson 1991).

***Halichoeres tenuispinnis*:** The sperm of *H. tenuispinnis* described by Hara and Okiyama (1998) is of the perciform type (Type II). It departs, however, from Type II in having a moderately deep, though small, nuclear fossa. The chromatin forms dense masses with scattered pale lacunae. A proximal centriole has not been reported but appears to be visible in some of the micrographs; the distal centriole is situated outside the nucleus. A ring of spherical mitochondria surrounds the base of the flagellum. A cytoplasmic canal bounded by the usual fold of the plasma membrane separates the mitochondria from the typical 9+2 flagellum which is loosely surrounded by the plasma membrane, apparently in the absence of lateral fins.

15.7.3 Suborder Zoarcoidei

There are nine zoarcoid families, occurring mainly in the North Pacific. All are marine. Zoarcoids have only a single nostril (Nelson 2006).

15.7.3.1 Family Zoarcidae

Zoarces elongatus: (15.77E) is a viviparous species. After insemination the sperm are held in the ovarian cavity. The sperm (Koya *et al.* 1993) (Figs. 15.77A-D, 15.78) have a discoidal head (appearing blade-like in one plane),

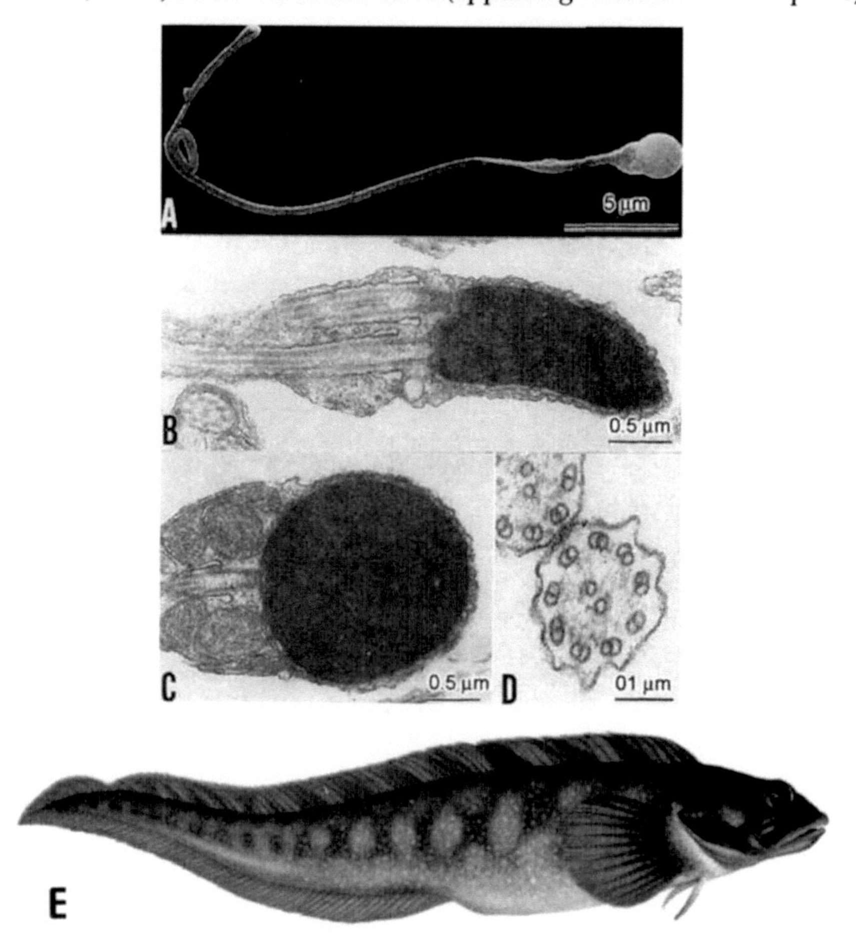

Fig. 15.77 *Zoarces elongatus* sperm (Zoarcidae). **A**. SEM of sperm, with discoidal head and biflagellate tails. **B**. Sagittal section of sperm head and the two flagella. **C**. Frontal section of sperm head and midpiece, showing one flagellum between the right and left mitochondria. **D**. Transverse section of a flagellum. Showing 9+2 arrangement of microtubules. After Koya, Y., Ohara, S., Ikeuchi, T., Adachi, S., Matsubara, T. and Yamauchi, K. 1993. Bulletin of the Hokkaido National Fisheries Research Institute 57: 21-31, Fig. 4. **E**. Adult. By Norman Weaver. With kind permission of Sarah Starsmore.

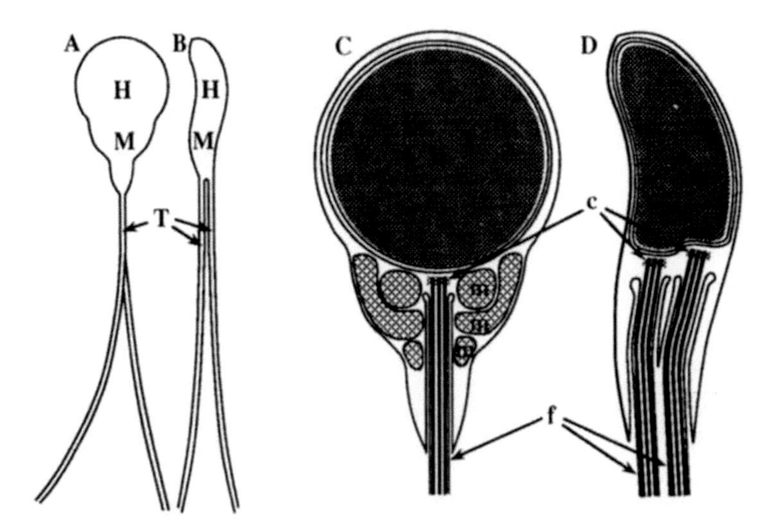

Fig. 15.78 *Zoarces elongatus* spermatozoa, diagrams (Zoarcidae). **A**. 'Dorsal' view. **B**. Lateral view. **C**. Internal structure in the horizontal plane. **D**. Internal structure in the sagittal plane. H, head; M, midpiece; T, tail; c, centriole; f, flagella; m, mitochondrion; n, nucleus. After Koya, Y., Ohara, S., Ikeuchi, T., Adachi, S., Matsubara, T. and Yamauchi, K. 1993. Bulletin of the Hokkaido National Fisheries Research Institute 57: 21-31, Fig. 5

six mitochondria (not enumerated but described as being of various sizes and shapes by Hara and Okiyama (1998) and two flagella. The two centrioles are parallel but, as an autapomorphy, anchor at slightly different levels to the base of the nucleus which each slightly indents (Hara and Okiyama 1998). No flagellar fins are apparent. The sperm were said not to differ in morphology, apart from their biflagellarity, from those of oviparous [zoarcid] species (Koya *et al.* 1993). The round appearance of the nucleus in fontal section may explain the apparently simple form of the sperm described for *Zoarces viviparus* by Retzius (1905).

Macrozoarces americanus: Ocean Pout sperm (Fig. 15.79A-H) have an oval head and a long midpiece measuring 2 × 1.5 µm and 2.4 × 0.7 µm (length × width), respectively (Yao *et al.* 1995). In some sperm, the head had a swollen anterior portion (Yao *et al.* 2000) (Fig. 15.79G,H).

The nucleus is electron-dense and an acrosome is absent. The long midpiece has many well-developed, elongate mitochondria containing a dense matrix (Fig. 15.79A,B). Up to nine mitochondria form a sheath around the proximal region of the tails in the midpiece (Fig. 15.79D,E). Two parallel flagella extend posteriorly from the centrioles at the base of the head and are associated with each other by an enclosing cytoplasmic membrane at the end of the midpiece but are separated into two individual tails further posteriorly (Fig. 15.79F). Each tail is composed of nine peripheral and one central pair

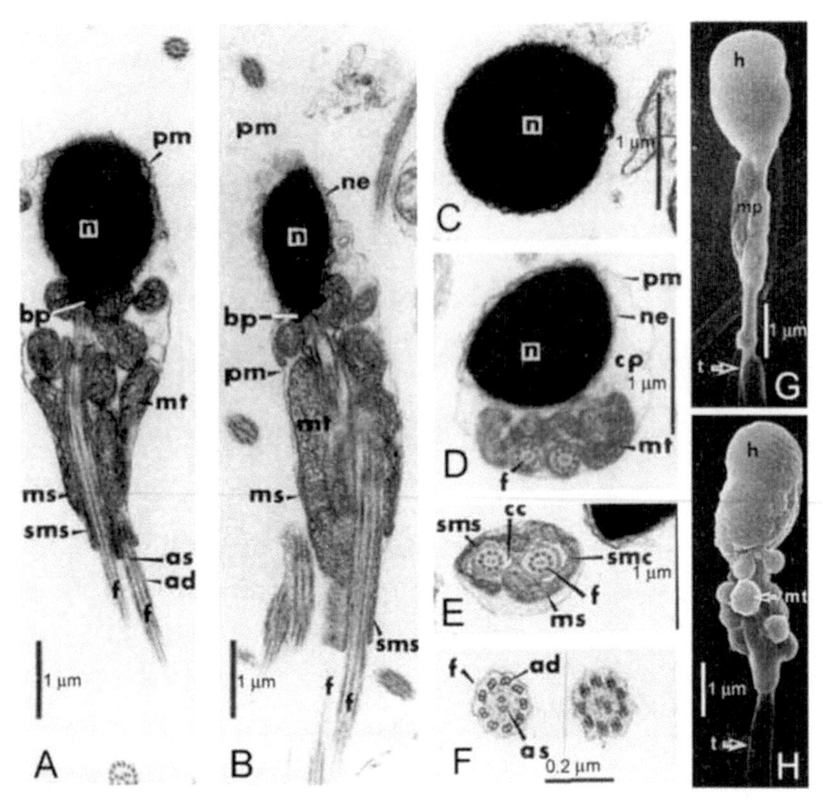

Fig. 15.79 *Macrozoarces americanus* (Zoarcidae). TEM of mature spermatozoa. **A**. Longitudinal section (LS) and **B**. Sagittal LS of spermatozoa showing the sperm head without acrosome, midpiece with nine round and/or elongate mitochondria, and two axonemes (tails). **C**. Transverse section (TS) of the sperm head. **D**. Postnuclear TS of sperm showing sperm head and two axonemes in mitochondrial sleeve. **E.** TS of midpiece showing two axonemes in mitochondrial sleeve. **F.** TS of the axoneme consisting of 9+2 microtubules. **G**. EM of a fresh spermatozoon. **H**. SEM of a post-thawed spermatozoonl although the mitochondria are severely swollen their distribution is well shown. ad, axonemal doublet; as, axonemal singlet; bp, basal plate of centriole; cc, cytoplasmic canal; cp, cytoplasm; df, double flagellum (biflagellate condition); f, flagellum; f', abnormal flagellum with more than 9+2 pattern of microtubules; h, sperm head; mp, midpiece; ms, mitochondrial sleeve; mt, mitochondria; n, nucleus; ne, nuclear envelope; pm, plasma membrane; smc, submitochondrial cisterna; sms, submitochondrial sleeve; t, sperm tail. **A**-**F**. Adapted from Yao, Z., Emerson, C. J. and Crim, L. W. 1995. Molecular Reproduction and Development 42: 58-64, Fig. 2A-F. **G**, **H**. Adapted from Yao, Z., Crim, L. W., Richardson, G. F. and Emerson, C. J. 2000. Aquaculture 181: 361-375, Fig. 6A,B.

(9+2) of microtubules. However, a varied microtubule arrangement (more than 9+2 microtubules) was noted in some sections. The length of the sperm tail is 80 µm (Yao *et al.* 1995).

Remarks: The long midpiece and the biflagellate nature of the sperm are considered by Yao *et al.* (1995) to be associated with the long life-span of the sperm and with sperm dispersal in the ovary to fertilize the eggs internally. The ocean pout eggs are enveloped by a porous chorionic membrane similar to that found in other teleosts but have two micropyles, a condition likely related to a mechanism of egg fertilization which increases the egg fertlity in the presence of low sperm numbers. Following insemination, the egg cohere and form a tightly associated egg mass in sea water.

15.7.3.2 Family Anarhichadidae

Anarhichas lupus: In *A. lupus* (Common Wolffish) (Fig. 15.80A), the spermatozoon (Pavlov *et al.* 1997; Pavlov 2005) (Fig. 15.80B) has an elongate oval head, 3.3 µm long (SD 1.4, n = 465) in two males and with a width about 1.47-fold less. Sperm size varied with time. The nucleus has densely packed chromatin containing fibrillar structures and granules and lacks nuclear pores. A basal nuclear fossa extends deeply into the nucleus for about two thirds of its length. The centriolar complex is situated in the upper part of the fossa. An electron-dense homogeneous spindle-shaped structure, with its long axis perpendicular to the axoneme, is situated above the basal plate of the basal body (distal centriole) which consists of nine triplets of microtubules with particles of electron dense material between them. Location of centrioles in the centriolar complex is uncertain. Sagittal sections usually show one basal plate and the spindle-shaped structure but in several cases, an additional centriole is situated above the first one, with a small angle between two basal plates. Several spermatozoa with microtubules originating from the second centriole were observed and two axonemes were then distinguishable (Fig. 15.80). One axoneme with a plasmalemma forms a flagellum. The other axoneme probably never reaches the external part of the midpiece (see Remarks). The fate of the additional axoneme is not clear.

The midpiece has the form of a half-oval by SEM. It is situated eccentrically relative to the nuclear invagination and flagellum (Fig. 15.80). Many oval mitochondria (usually 10, up to 15) and vesicles of different size are distributed in the midpiece. The 9+2 axoneme apparently lacks ITDs. A cytoplasmic sheath, sometimes with vesicles, surrounds the axoneme. The width of the sheath varies along the axoneme and transverse sections of the flagellum show 1-2 lateral ridges (fins), or these may be absent at some levels. The length of the flagellum ranges from 11.2-31.7 (mean 21.2 µm) (SD 5.5, n = 40).

Spermatids are usually found in the ejaculate together with mature spermatozoa, the ratio varying individually.

Remarks: As noted by Pavlov *et al.* (1997) the spermatozoon head of *Anarhichas lupus* (Common Wolffish) is similar to that of *Zoarces elongatus* (Eel Pout) and *Macrozoarces americanus* (Ocean Pout) but differs in having a nuclear fossa. Both of the latter two species have biflagellate sperm (Koya *et al.* 1993; Yao *et al.* 1995). Thus, the variation in number of tails within the suborder Zoarcoidei, from one flagellum basically in *Anarhichas lupus*, to two flagella, is

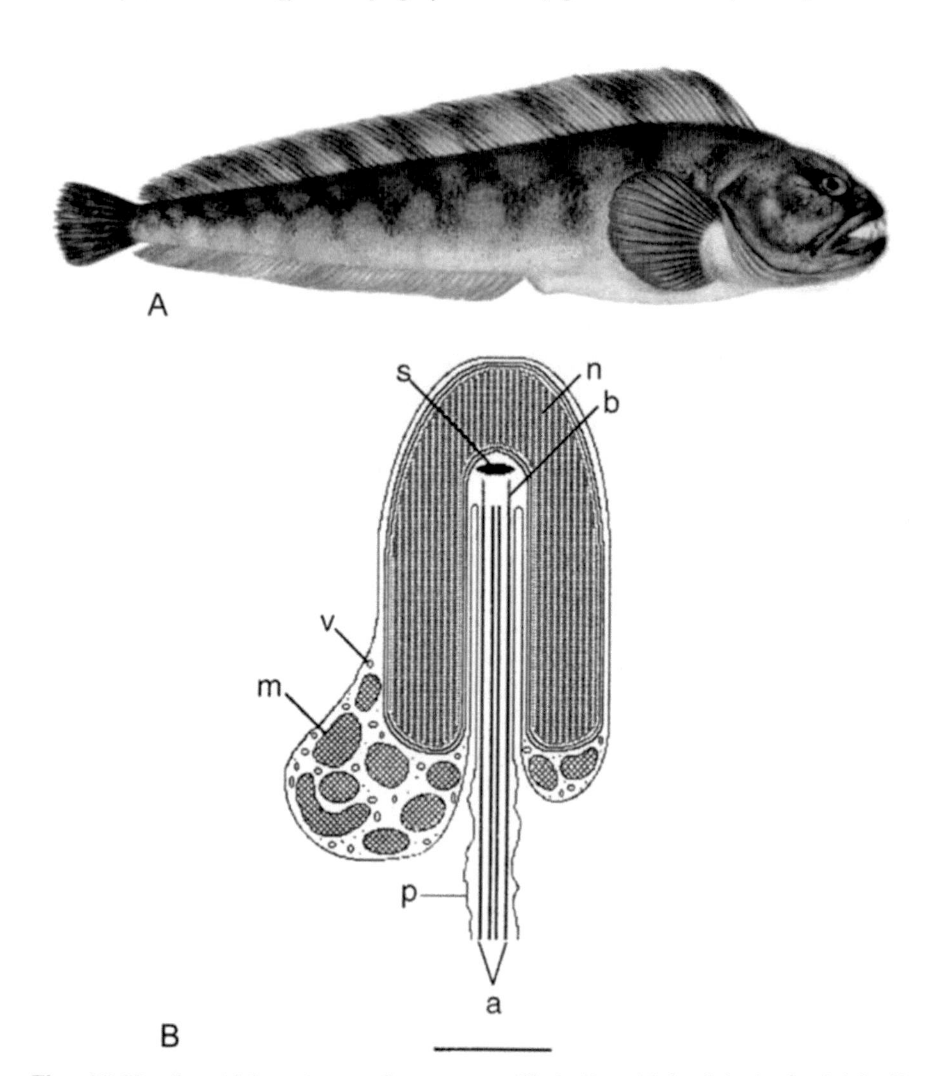

Fig. 15.80 *Anarhichas lupus*, Common wolffish (Anarhichadidae). **A**. Adult. By Norman Weaver. With kind permission of Sarah Starsmore. **B**. Diagram of internal structure of a spermatozoon. a, axoneme; b, basal plate; m, mitochondria; n, nucleus; p, plasmalemma; s. electron dense structure; v, vesicle. (Bar = 1 μm, based on the measurements of spermatozoa with optical microscope). Note that two free flagella were never observed. From Pavlov, D. A., Knudsen, P., Emel'yanova, N. Y. G. and Moksness, E. 1997. Aquatic Living Resources 10: 187-194, Fig. 2. With permssion.

similar to that observed in several families of the order Siluriformes (see that order). Despite the biflagellate condition shown in Figure 15.80), Pavlov *et al.* (1997) state that spermatozoa with biflagellate tails were not observed but could not be totally excluded in Wolffish. They note, nevertheless, that a polymorphism in spermatozoal ultrastructure and size apparently occurs in the Wolffish and that polymorphism was also found in Wolffish eggs: small proportions of eggs from several females had from 2 to 5 micropyles. Similar variability in the number of micropyles was described in ocean pout (Yao *et al.* 1995). Pavlov *et al.* (1997) discuss the significance of presence of spermatids in the ejaculate of the Wolffish. They also note that in spite of internal fertilization, the general morphology of the Wolffish spermatozoon is close to that in oviparous fishes; the ultrastructure of spermatozoa in the species from the family Zoarcidae is more specialized having a better developed midpiece and biflagellate tail, corresponding with a more derived type of reproduction in some species of the family that are viviparous. The mode of reproduction in Wolffish with internal insemination but release of eggs into the external medium and the corresponding spermatozoon morphology could be regarded as a first step towards viviparity within the suborder Zoarcoidei (Pavlov *et al.* 1997).

15.7.4 Suborder Trachinoidei

Trachinoids are generally small, primarily shallow-living temperate and tropical marine demersal or burrowing fish, including the jaw fishes, eel blennies, sand perches, weaver fish and others, some of which have affinities with the Blennioidei or Percoidei. The suborder is heterogeneous and probably polyphyletic. Of families for which sperm are described here, Nelson (2006) questions inclusion of the Pinguipedidae in the suborder but by definition the Trichodontidae are included.

Sperm types: Hara and Okiyama (1998) note that there are two sperm types possibly supporting polyphyly of the suborder: the perciform type (Type II) in Trachinidae, Uranoscopidae and Ammodytidae and the "primitive' non-perciform type (Type I) in the Percophidae [and Pinguipedidae]. The Trichodontidae have highly modified sperm.

15.7.4.1 Family Trichodontidae

The sperm of the sandfish *Arctoscopus japonicus* is highly unusual as the nucleus incompletely wraps around a long basal part of the flagellum, leaving the centrioles exposed at the anterior end (Fig. 15.81). The chromatin forms dense masses. The two centrioles are at 45° to each other. The distal centriole is parallel to the nucleus and gives rise to the flagellum. The A-tubules are dense. Two similar sized mitochondria are located posterior to the nucleus (Hara and Okiyama 1998).

Remarks: The sperm are aquasperm produced in mass spawning in the sea. They have an exceptionally long nucleus for an aquasperm and the long nuclear fissure appears to be unique for the Perciformes. However, Hara and Okiyama (1998) note that a nuclear fissure, and densities of the A subtubules of the axoneme, also occur in the Hexagrammidae in the Scorpaeniformes (see

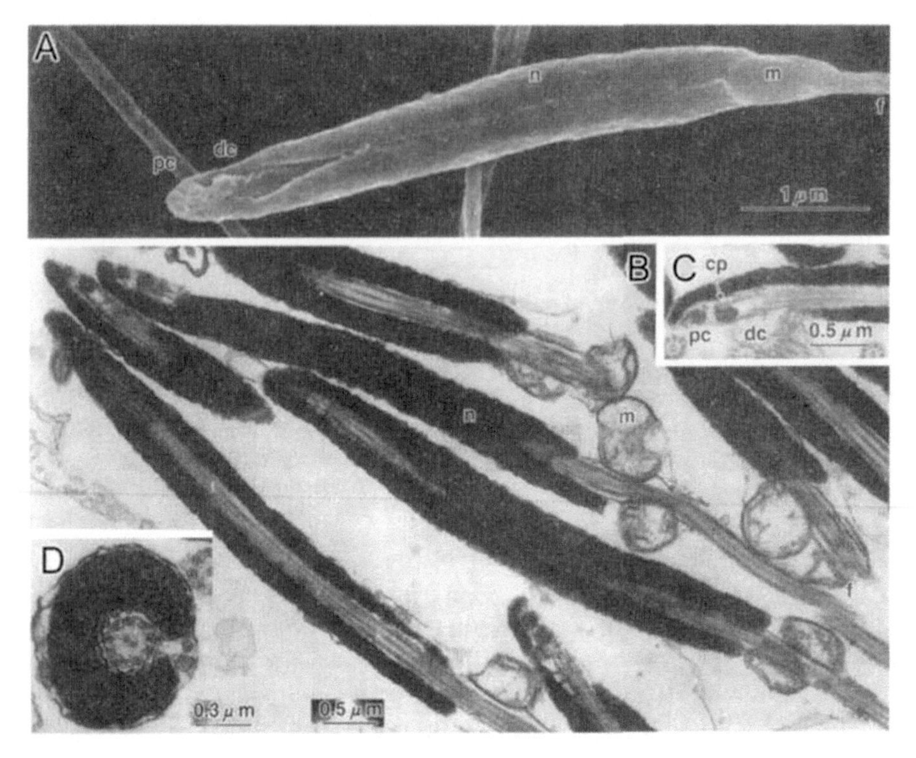

Fig. 15.81 *Arctoscopus japonicus* (Trichodontidae). **A**. SEM of the nuclear region of the spermatozoon. **B**. TEM of longitudinal sections of sperm. **C**. LS tip of nucleus at the centrioles. **D**. TEM of a transverse section of the nucleus and flagellum with axoneme. dc, distal centriole; f, flagellum; m, mitochondria; n, nucleus; pc, proximal centriole. Adapted from Hara, M. and Okiyama, M. 1998. Bulletin of The Ocean Research Institute, University of Tokyo 33: 1-138, Fig. 58B-E. With permission.

Section 15.6.3.1). It is pertinent that Nelson (2006) notes that some trichodontoids may belong to the latter order.

15.7.4.2 Family Pinguipedidae

The sandperches, marine; Atlantic coast of South America and Africa, Indo-Pacific and off Chile (Nelson 2006); formerly placed in the Mugiloididae (Nelson 1984).

***Parapercis* sp.** The sperm of the sandperch *Parapercis* sp. (adult, Fig. 15.82; sperm, Fig. 15.83) is of the type II *(Parapristipoma)* kind of Mattei (1970): the proximal centriole is slightly displaced relative to the basal nuclear fossa, and the distal centriole is lateral to and at an angle, here about 45°, to the other; and there is a moderately long cytoplasmic canal. The small, ovoid mitochondria form two or more tiers in the broad cytoplasmic collar. An unusual feature for perciforms, but shared with Labridae and other trachinotids, is that there are no fins on the 9+2 flagellum.

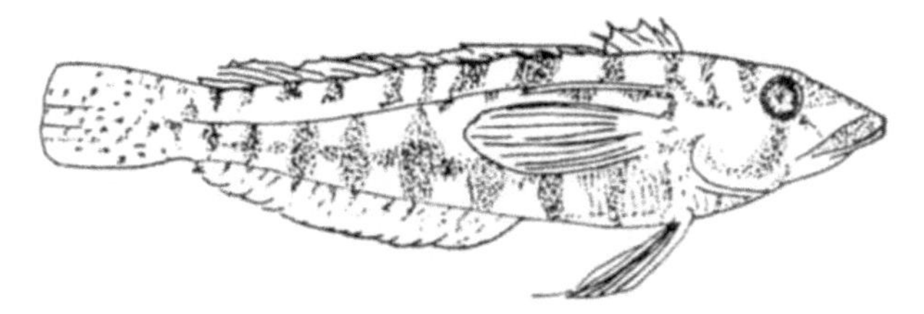

Fig. 15.82 *Parapercis* sp. (Pinguipedidae). Specimen from which sperm are described. From Heron Island, Great Barrier Reef, Australia. From Jamieson, B. G. M. 1991. *Fish Evolution and Systematics: Evidence from Spermatozoa*. Cambridge University Press, Cambridge, Fig. 16.8.

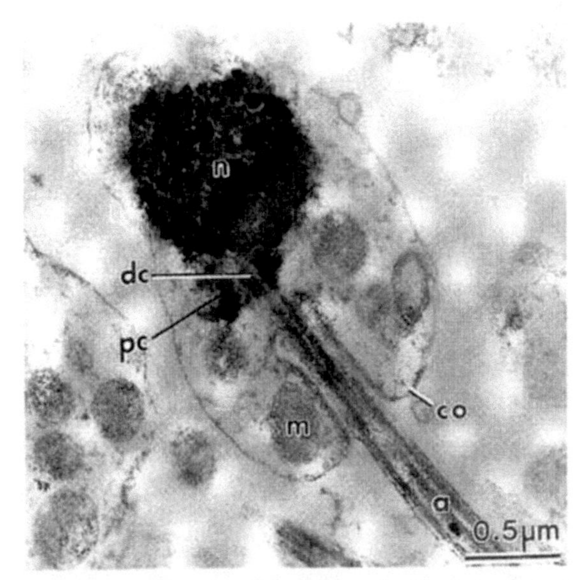

Fig. 15.83 *Parapercis* sp. Pinguipedidae). Longitudinal section of spermatozoon. a, axoneme; co, cytoplasmic collar; dc, distal centriole; m, mitochondrion; n, nucleus; pc, proximal centriole; pc, proximal centriole. From Jamieson, B. G. M. 1991. *Fish Evolution and Systematics: Evidence from Spermatozoa*. Cambridge University Press, Cambridge, Fig. 16.9.

15.7.4.3 Family Percophidae

***Bembrops caudimaculata*:** Mattei (1991) examined the sperm of the duckbill, *B. caudimaculata*. He regarded it as a slight modification of the simple [Type I] sperm and gave a diagrammatic illustration showing a rounded nucleus with a wide, shallow basal fossa, proximal centriole perpendicular to the distal centriole (basal body), and two tiers of mitochondria. He ascribed the modified simple sperm to only five families: Istiophoridae, Mullidae, Pomacentridae, Cichlidae, and Percophidae.

15.7.4.4 Family Ammodytidae

These are the Sand lances, cold to tropical in the Arctic, Atlantic, Indian and Pacific Oceans.

Ammodytes personatus: The spermatozoon of *A. personatus* (Hara and Okiyama 1998; Hara 2003) (Fig. 15.84) is highly unusual in the posterior tilting of the elongate nucleus which has a spatulate, posteriorly narrowing outline. The flagellum is inserted in a deep fossa near the posterior end of the nucleus. The two mutually perpendicular centrioles are situated in the fossa; the distal being slightly displaced relative to the proximal centriole so that the two are not in the same plane. A single, small mitochondrion is located on the posterior face of the nucleus abutting the flagellum. The flagellum is of the usual 9+2 type but is remarkably swollen as the plasma membrane is separated from the axoneme by a considerable zone of cytoplasm. The A subtubules of doublets 1, [2], 5 and 6 are septate.

Hara and Okiyama (1998) consider that the asymmetry of the sperm and the presence of ITDs at 1, 2, 5 and 6 indicate derivation of this sperm from the perciform type.

Attention is here called to the striking similarity of the sperm to that of elopomorphs, including the elongate, posteriorly tilted nucleus and the presence on this of a singe mitochondrion. It does not, however appear that these similarities indicate phylogenetic affinity and the other apomorphies of

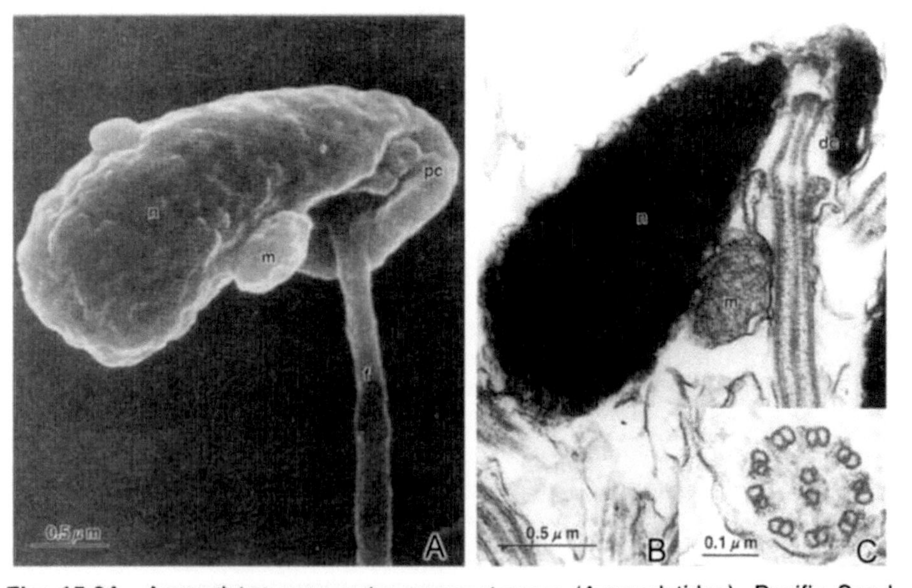

Fig. 15.84 *Ammodytes personatus* spermatozoon (Ammodytidae), Pacific Sand eel. A. SEM of spermatozoon. B. TEM longitudinal section. C. Transverse section of an axoneme. Adapted from Hara, M. and Okiyama, M. 1998. An ultrastructural review on the spermatozoa of Japanese fishes. Bulletin of The Ocean Research Institute, University of Tokyo 33: 1-138, Fig. 54B,C,D. With permission.

elopomorphs (split extended proximal centriole; centriolar rootlet and the 9+0 configuration of the axoneme) are absent.

15.7.4.5 Families Trachinidae and Uranoscopidae

Trachinids are the weever fish and uranoscopids are the star gazers.

Mattei (1991) examined the sperm of *Trachinus armatus*, the Guinean weever, and *T. radiatus*, in the Trachinidae, and *Uranoscopus polli*, in the Uranoscopidae, and ascribed them to the perciform spermatozoon. That of *U. polli* is here considered intermediate between Types I and II and those of *Trachinus* as Type II. The sperm of these closely related families have been compared by Lahnsteiner and Patzner (1996) who confirm that the general organization of spermatozoa of *Trachinus draco* and *Uranoscopus scaber* is similar (Fig. 15.85A).

The sperm head is ovoid, the midpiece region approximately cylindrical and not clearly separated from the head region (Fig. 15.85A). Head and midpiece together are 2.45 ± 0.22 μm long in *Trachinus draco* and 2.32 ± 0.23 μm long in *Uranoscopus scaber*. The maximal diameter of the head region is 2.04 ± 0.20 μm in *T. draco* and 1.72 ± 022 μm in *U. scaber*. The flagellum inserts laterally on the nucleus and therefore the spermatozoa of both species are asymmetrical. The shape of the nucleus is ovoid, the chromatin granular and densely packed and the nuclear envelope has no nuclear pores. In *U. scaber* the nucleus is 1.53 ± 0.13 μm long and 1.19 ± 0.11 μm wide, in *T. draco* it is 1.70 ± 0.09 μm long and 1.00 ± 0.03 μm wide. The centriolar complex is located on the lateral side of the nucleus in a slight 0.4 to 0.5 μm deep invagination. It consists of a proximal and a distal centriole which are mutually perpendicular (Fig. 15.85A). Both centrioles are composed of nine pairs of peripheral microtubules. The centrioles have a length of 0.35 ± 0.06 μm and a diameter of 0.27 ± 0.04 μm. A ring of osmiophilic material is attached to the proximal end of the distal centriole which has a diameter of 0.30 ± 0.05 μm and is 0.12 ± 0.01 μm thick. At the distal end the distal centriole is bordered by an about 0.05 μm thin electron-dense disk. The proximal centriole reveals no specialized structures. Proximal and distal centrioles are fastened to each other by microfibrils: these interconnect the electron-dense ring of the distal centriole with the lateral side of the proximal centriole (Fig. 15.85B). The proximal centriole itself is fastened to the nuclear envelope by microfibrils. The interconnection between distal centriole and nucleus differs between *U. scaber* and *T. draco* (Fig. 15.85B). In *U. scaber* the side of the distal centriole as well as the electron-dense ring are interconnected with the nuclear envelope by microfibrils. In *T. draco* there is an additional stabilization structure, an approximately quadratic plate of electron-dense material. It is located between the lateral side of the basal body and the nuclear envelope and has a side length of 0.60 ± 0.05 μm and a thickness of 0.06 ± 0.02 μm. It is attached to the distal centriole as well as to the nuclear envelope by numerous microfibrils. In *U. scaber* as well as in *T. draco* the cytoplasmic canal is 0.62 ±.04 μm long. The midpiece contains six spherical mitochondria in *T.*

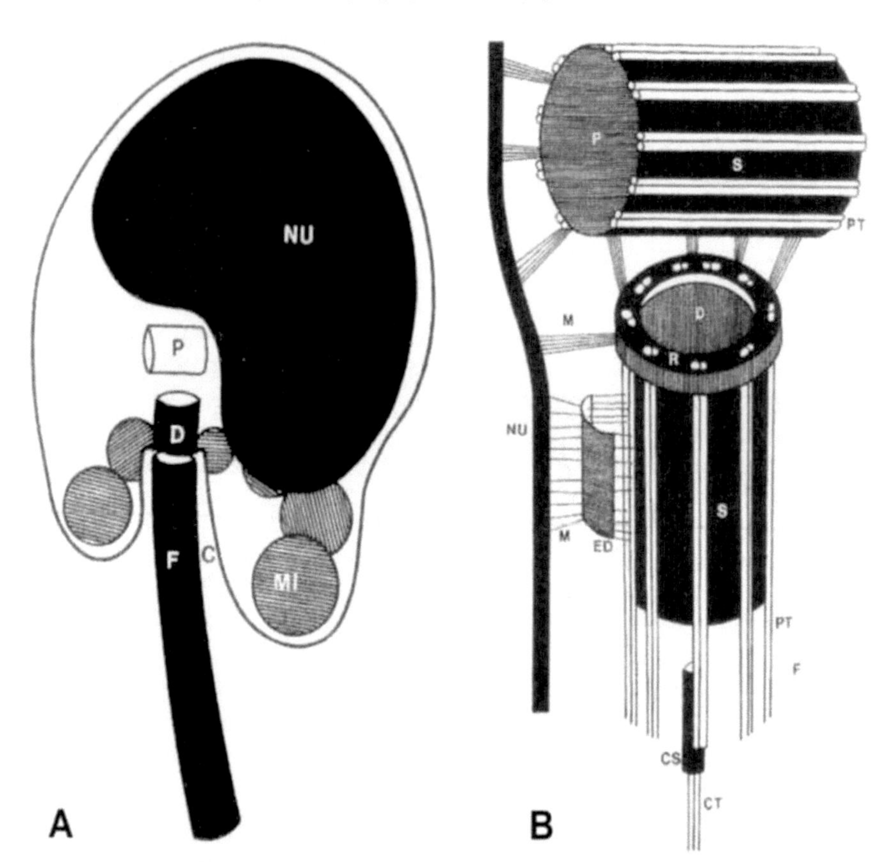

Fig. 15.85 A. Reconstruction of the sperm organization of Trachinoidei. The midpiece of *Uranoscopus scaber* (Uranoscopidae) contains six mitochondria (shown in the figure); that of *Trachinus draco* (Trachinidae) has seven to eight. **B**. Centriolar complex of spermatozoa of Trachinoidei. The electron-dense stabilization structure (ED), located between the distal centriole and the nucleus is found only in *Trachinus draco*; in *Uranoscopus scaber* the distal centriole is directly interconnected with the nuclear envelope by microfibrils. C, cytoplasmic canal; CS, central sleeve of the flagellum; CT, central microtubules; D, distal centriole; ED, electron-dense stabilization structure of the distal centriole; F, flagellum; M, microfibrils; MI, mitochondrion; NU, nucleus; P, proximal centriole; PT, peripheral microtubules; R, ring of electron-dense material; S, sleeve of electron-dense material interconnecting the peripheral tubules. After Lahnsteiner, F. and Patzner, R. A. 1996. Journal of Submicroscopic Cytology and Pathology 28: 297-303, Figs. 5 and 6.

draco and seven to eight in *U. scaber*, arranged around the cytoplasmic canal. The 9+2 flagellum has a length of about 40 μm and a diameter said to be 0.26 ± 0.03 μm. No flagellar fins are reported (Lahnsteiner and Patzner 1996).

Micrographs of the sperm of *Uranoscopus polli*, the Whitespotted stargazer (Mattei unpublished), also reveal a tilted nucleus with eccentric fossa, as in *U.*

scaber, but the proximal centriole is displaced out of the axis of the distal centriole, both centrioles lying well outside the nuclear fossa.

Trachinus armatus: *T. armatus*, like *T. draco*, has a Type II spermatozoon (Fig. 15.86). Centrioles are again outside the shallow nuclear fossa (observable in some sections) and mutually perpendicular. The number of mitochondria has not been ascertained but two are visible in some sections. One or two fins, not observed in *T. draco*, are present in different regions of the flagellum.

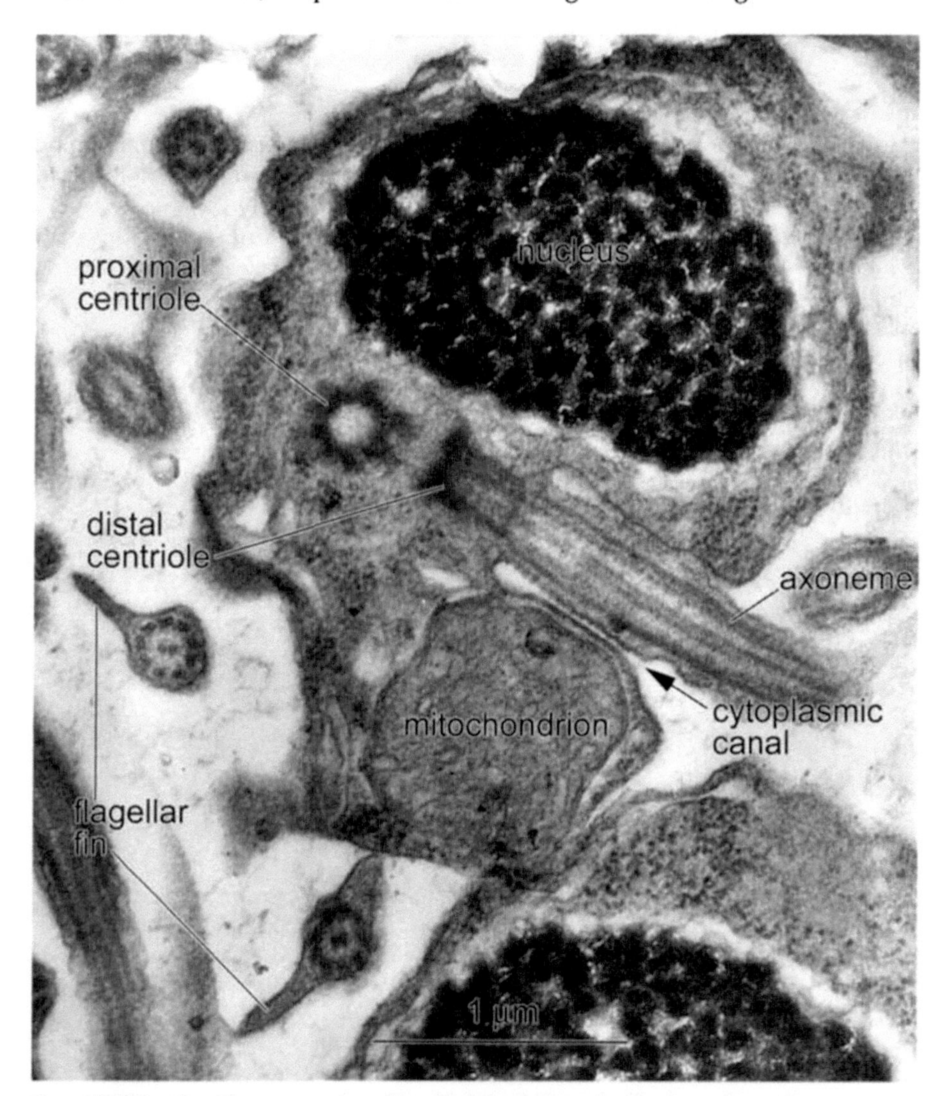

Fig. 15.86 *Trachinus armatus* (Trachinidae). Longitudinal section of spermatozoon. Original, from an unpublished TEM micrograph courtesy of Professor Xavier Mattei.

15.7.5 Suborder Blenniodei

Blennioids, worldwide in marine habitats, have less deep bodies than percoids, with a short trunk and a relatively attenuated caudal region. The dorsal and anal fins are long and low, terminating near the caudal fin, and the pectoral and usually the caudal fins are rounded. There is an exact correspondence in number between dorsal and posterior anal soft rays and the vertebrae supporting them (Rosenblatt 1984). They are divisible into two, northern and tropical, predominantly marine groups. It is doubtful that the suborder is monophyletic. Many are characterized by demersal eggs, parental care and an advanced state of the newly hatched larvae (Matarese *et al.* 1984).

15.7.5.1 Family Blenniidae

Bleniids are common marine littoral, benthic fish with external fertilization by spawning. Semen is adhered to the substrate and eggs are deposited on it. There are some 345 species in 53 genera (Froese and Pauly 2004).

In the Blenniidae, the fine structure of sperm has been studied in 15 species: *Aidablennius sphynx* (Lahnsteiner and Patzner 1990a), *Lipophrys adriaticus, L. dalmatinus, L. canevae, Parablennius tentacularis, P. incognitus, Salaria pavo*, and *Scartella cristata (=Blennius cristata)* by Lahnsteiner and Patzner (1990b,c; 2007), in *Hypleurochilus (=Blennius) bananensis, Parablennius pilicornis (=Blennius vandervekeni), Ophioblennius atlanticus* by Mattei (1991) (Fig. 15.87D), in *Entomacrodus striatus, Blenniella* (=*Istiblennius*) *bilitonensis*, and *Rhabdoblennius ellipes* by Hara and Okiyama (1998), and in *Lipophrys (=Blennius) pholis* by Silveira *et al.* (1990). Spermatozoal ultrastructure in the family has been reviewed by Lahnsteiner and Patzner (2007).

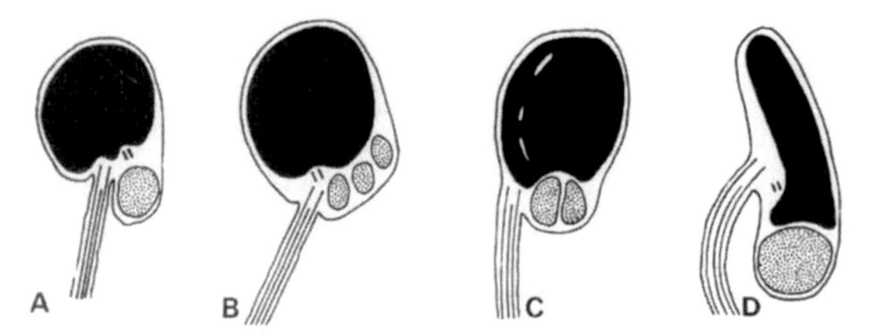

Fig. 15.87 Blennioidei. Diagrammatic longitudinal sections of the spermatozoa. **A**. *Clinus nuchipinnis* (Clinidae). **B**. *Scartella (=Blennius) cristatus.* **C**. *Parablennius pilicornis (=Blennius vanderwekeni).* **D**. *Ophioblennius atlanticus* (Blenniidae). After Mattei, X. 1970. Pp. 59-69. In B. Baccetti (ed.), *Comparative Spermatology*. Academic Press, New York. Fig. 4: 2. 20, 26, and 16.

Sperm ultrastructure: Spermatozoa of the Blenniidae are uniflagellate, anacrosomal aquasperm. Mattei (1970) recognized common features of the anacrosomal aquasperm of the clinid, *Clinus nuchipinnis* and bleniids, *Scartella (=Blennius) cristata, Parablennius pilicornis (=Blennius vanderveken*i) and

Ophioblennius atlanticus then investigated (Fig. 15.87). These features, representing a bleniid sperm type, shared with clinids, were tilting of the axonemal relative to the nuclear axis, location of the mitochondrial material on only one side of the base of the axoneme (at least as seen in one plane), and presence of a proximal centriole in addition to the basal body. In *Ophioblennius atlanticus* (Fig. 15.87D) the curved nucleus is elongated in a direction virtually at right angles to the base of the flagellum.

Lahnsteiner and Patzner (1990c) recognized Type I spermiogenesis *sensu* Mattei (1970) in the bleniids *Parablennius incognitus, P. tentacularis* and *Scartella pavo*. Type II spermiogenesis was identified in *Aidablennius sphinx, Lipophrys adriaticus, L. canevae, L. dalmatinus* and *Scartella cristata*.

Lahnsteiner and Patzner (1990c, 2007) have taken the analysis of sperm types in the Blenniidae further and recognize four sperm types in the family (Fig. 15.88: Types A,B,C,D). The reasons for this great diversity in sperm structure in externally fertilizing fishes are not clear.

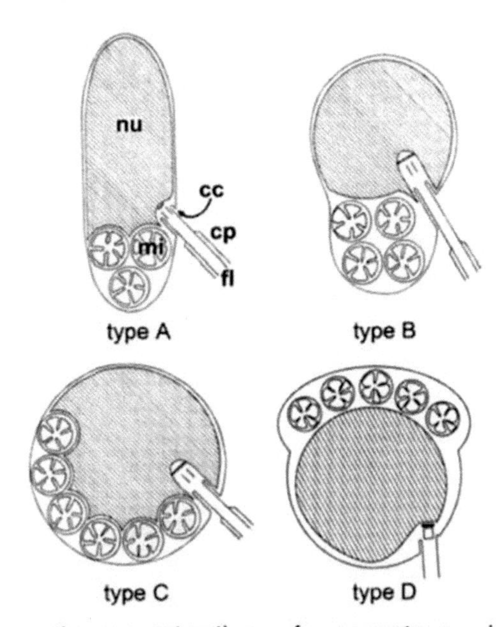

Fig. 15.88 Schematic reconstruction of spermatozoa in blenniid fishes (Blenniidae). cc, cytoplasmic canal; cp, cytoplasmic protrusions; fl, flagellum, mi, mitochondrion, nu, nucleus. From Lahnsteiner, F. and Patzner, R. A. 2007. Pp. 1-61. In S. M. H. S. M. H. Alavi, J. Cosson, K. Coward and G. Rafiee (eds.), *Fish Spermatology*. Alpha Science International Ltd., Oxford, U.K., Fig. 4.

Type A is found in *Ophioblennius atlanticus, Parablennius incognitus, P tentacularis*, and *Salaria pavo*. These spermatozoa have an elongated, ellipsoid form (see Lahnsteiner and Patzner 2007) (Fig. 15.88A). The head + midpiece complex is 3.10 ± 0.10 µm long. The midpiece is incorporated in the head region. A very specific feature is that the mitochondria are located in [caudal] invaginations of the nucleus (Fig. 15.88A). Spermatozoa of *S. pavo* have three spherical mitochondria while the spermatozoa of *P incognitus* and *P. tentacularis* have four. *O. atlanticus* has only one mitochondrion. Almost no cytoplasm is visible in the head and midpiece region. The proximal and distal centrioles are located caudolaterally in a shallow invagination of the nucleus (nuclear notch, here termed nuclear fossa) at right angles to each other and are interconnected by electron-dense fibers.

Lipophrys (=Blennius) pholis: Lahnsteiner and Patzner (2007) include *Lipophrys (=Blennius) pholis* in type A but this placement cannot be sustained as in this species the mitochondria rest in indentations at the pole of the nucleus opposite to that of the centrioles and flagellum (see Figs. 11 and 12 of Silveira *et al.* (1990). Its sperm in fact has some characteristics of Types A, C and D: the elongate approximately rectangular nucleus of Type A, the embedding of the mitochondria in indentations of the nucleus seen in Type C, and location of the mitochondria at the opposite pole to the centrioles seen in Type D. Nevertheless, the four types of Lahnsteiner and Patzner remain useful reference points, though these authors considered Types D and B to be similar and that differences described depend only on subjective interpretation of the fine structure.

The morphology of the sperm of *Lipophrys (=Blennius) pholis* may be summarized as follows. Five or six mitochondria are arranged in a circle, almost contiguous centrally, in indentations at the tip of the nucleus. The head is 3.7-4.8 μm long and the nucleus is 0.6 μm wide anteriorly and 0.9 μm wide posteriorly and is therefore considerably elongated. It has electron-transparent areas (vacuoles). The proximal centriole lies in a small, shallow nuclear fossa. The 9+2 flagellum is 29 μm long. Although spermiogenic stages are illustrated by Silveira *et al.* (1990) it is not clear whether it is the mitochondria or the basal body (proximal centriole) which has migrated to the "anterior" pole of the nucleus, i.e. whether the free, mitochondrial pole of the mature sperm is the original apical or basal pole of the nucleus. It may also be noted that the testis in *L. pholis* is supposedly both of the "unrestricted spermatogonia testis type" of Grier (1981) and the lobular type of Billard (1986) (Silveira *et al.* 1990).

Type B is found in *Aidablennius sphynx* and *Hypleurochilus bananensis*. These spermatozoa are differentiated into a separate head and midpiece region; both have a spherical shape (Fig. 15.88B) and are 1.10 ± 0.05 μm wide. Four spherical mitochondria are found in the midpiece region of *A. sphynx* and three in *H. bananensis* which, in contrast to sperm type A, are free and not located in invaginations of the nucleus. The centriolar complex is located caudolaterally in a bell-shaped nuclear fossa and consists of a proximal and distal (basal body) centriole orientated at right angles to each other. The two centrioles are interconnected by electron-dense fibers (Lahnsteiner and Patzner 2007).

Type C. occurs in *Lipophrys adriaticus, L. canevae, L. dalmatinus, Parablennius pilicornis* and *Scartella cristata* (Fig. 15.88C). The midpiece region is integrated into the head region which are jointly 1.60 ± 0.01 μm long. The spermatozoa have a spherical head/midpiece complex. Six spherical or slightly ovoid mitochondria are located in caudal invaginations of the nucleus in *L. adriaticus, L. canevae, L. dalmatinus*, and *S. cristata*, and two mitochondria are present in *P. pilicornis*. The centriolar complex is located caudolaterally in a bell-shaped nuclear fossa exhibiting similar features to *Aidablennius sphynx*; the centrioles are mutually perpendicular (Lahnsteiner and Patzner 2007).

Type D spermatozoa (Fig. 15.88D) are found in *Entomacrodus striatus*, *Blenniella bilitonensis*, and *Rhabdoblennius ellipes* (Fig. 15.89). The head and nucleus are spherical and jointly 2.30 ± 0.1 µm long. Type D sperm have no midpiece, but the mitochondria are located apically (cranial) at the nucleus and free in the cytoplasm (Lahnsteiner and Patzner 2007) or, in *E. striatus*, only very slightly indent the nucleus (Hara and Okiyama 1998). The centriolar complex consists of proximal and distal centrioles, although the proximal centriole was not observed in *B. bilitonensis*. The centriolar complex is located in a shallow nuclear fossa and the centrioles lie approximately in the same longitudinal axis (Hara and Okiyama 1998) (not as Lahnsteiner and Patzner, 2007, state, parallel to each other).

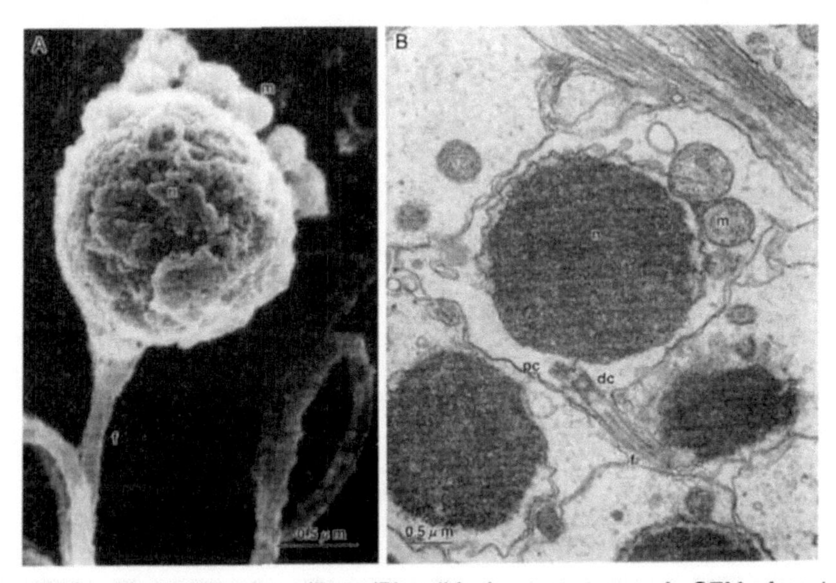

Fig. 15.89 *Rhabdoblennius ellipes* (Blenniidae) spermatozoa. A. SEM of nucleus and mitochondria. B. TEM longitudinal section. dc, distal centriole; f, flagellum; m, mitochondrion; n, nucleus; pc, proximal centriole. Adapted from Hara, M. and Okiyama, M. 1998. Bulletin of The Ocean Research Institute, University of Tokyo 33: 1-138, Fig. 48, B, C. With permission.

In *Aidablennius sphynx*, *Lipophrys canevae*, *L. dalmatinus*, *Parablennius incognitus*, *P tentacularis*, *Salaria pavo*, and *S. cristata*, the flagellum has a length of 16.5 ± 0.4 µm and has paired lateral flagellar fins. In the spermatozoa of *Entomacrodus striatus*, *Blenniella bilitonensis*, *Rhabdoblennius ellipes* and *Lipophrys pholis* the flagellum is ca 30 µm long. No information is available concerning flagellar fins.

Blenniidae exhibit a special mode of spermiogenesis as the final maturation processes of spermatozoa occur in the testicular glands and spermatic ducts (Lahnsteiner and Patzner 1990c; Lahnsteiner *et al.* 1990; see also Grier 1981; Billard 1986; Silveira *et al.* 1990).

15.7.5.2 Family Clinidae

Clinids are the kelp blennies. A large number of clinids have internal fertilization in the ovaries (see Evans and Downing, **Volume 8B, Chapter 4, Section 4.3.8.1**).

***Clinus nuchipinnis*:** The sperm of *C. nuchipinnis* was examined by Mattei (1970) (Fig. 15.87A). It resembles the Type B blenniid spermatozoon as defined by Lahnsteiner and Patzner (2007) (Fig. 15.88) though there appears to be only a single, unilateral mitochondrion and the proximal centriole is at an acute angle to the basal body. In other clinids (below) the nucleus is greatly elongated and there are numerous mitochondria.

Other species: Fishelson *et al.* (2007) investigated spermatogenesis and spermatocytogenesis in 16 species of viviparous clinid fishes (Clinidae, Blennioidei) by means of light and electron microscopy: *Clinus superciliosus; C. arborescens; C. cottoides; C. acuminatus; C. taurus; Muraenoclinus dorsalis; Cirrhibarbis capensis; Pavoclinus pavo; P. graminis; Blennophis striatus; Cristiceps aurantiacus; C. australis; Heteroclinus roseus; H. adelaidae; H. wilsoni*; and *Springeratus xanthosoma*. Maturation of the spermatids takes place in spermatocysts formed by Sertoli cells. Anisodiametric and slightly flattened sperm heads are eventually formed, 0.4–0.5 μm in diameter and 7.5 ± 1 μm long, bearing 80 ± 15 μm long flagella. The sperm are packed into spermatozeugmata (*C. acuminatus*, Fig. 15.90F) within the spermatocysts, enveloped and penetrated by the mucoid material of the Sertoli cells. On maturation of the spermatids, the spermatocyst dimensions increase, attaining 40 ± 8 μm in diameter in the smaller species of *Heteroclinus*, and up to 90 ± 10 μm in the larger males of *Clinus superciliosus* and *C. cottoides*. As sperm head volume is ca. 2.24 μm^3, the number of sperm in the smallest mature spermatocysts reaches ca 440 and in the largest over 2,900. Upon release from the cysts, the spermatozeugmata are transported along the sperm ducts to the posterior ampullae where they are stored in the epididymis. During copulation, the spermatozoa are transported from there to the female via the intromittent organ. Fertilization in the examined clinids occurs in the egg follicles as in some Poeciliidae.

During maturation of the late spermatids, the cell plasma membrane becomes closely attached to the nucleus. The sperm head is flattened (*Clinus taurus*, Fig. 15.90D) and bent apically (late spermatids, *C. cottoides*, Fig. 15.90A,B). The flagellum greatly elongates, terminating at the basal body deep within the nucleus and with the usual 9+2 microtubules (Fig. 15.90C and

Fig. 15.90 Micrographs of spermatids and sperm in clinids. **A**. Nuclei of stage 5 spermatids. SEM. Magnification x 5,000. Inset: ibid. TEM. x 7,000. **B**. Heads of maturing spermatids of *Clinus cottoides*. x 20,000. **C**. Sections of sperm heads and flagella of the same. TEM. x 20,000. Inset: cross-sections of flagellum with collar. TEM. x 35,000. **D**. Cross-section of sperm heads of *C. taurus*. TEM. x 20,000. **E**. Mitochondria along the sperm flagella (the collar membrane not visible) of *C.*

Fig. 15.90 Contd. ...

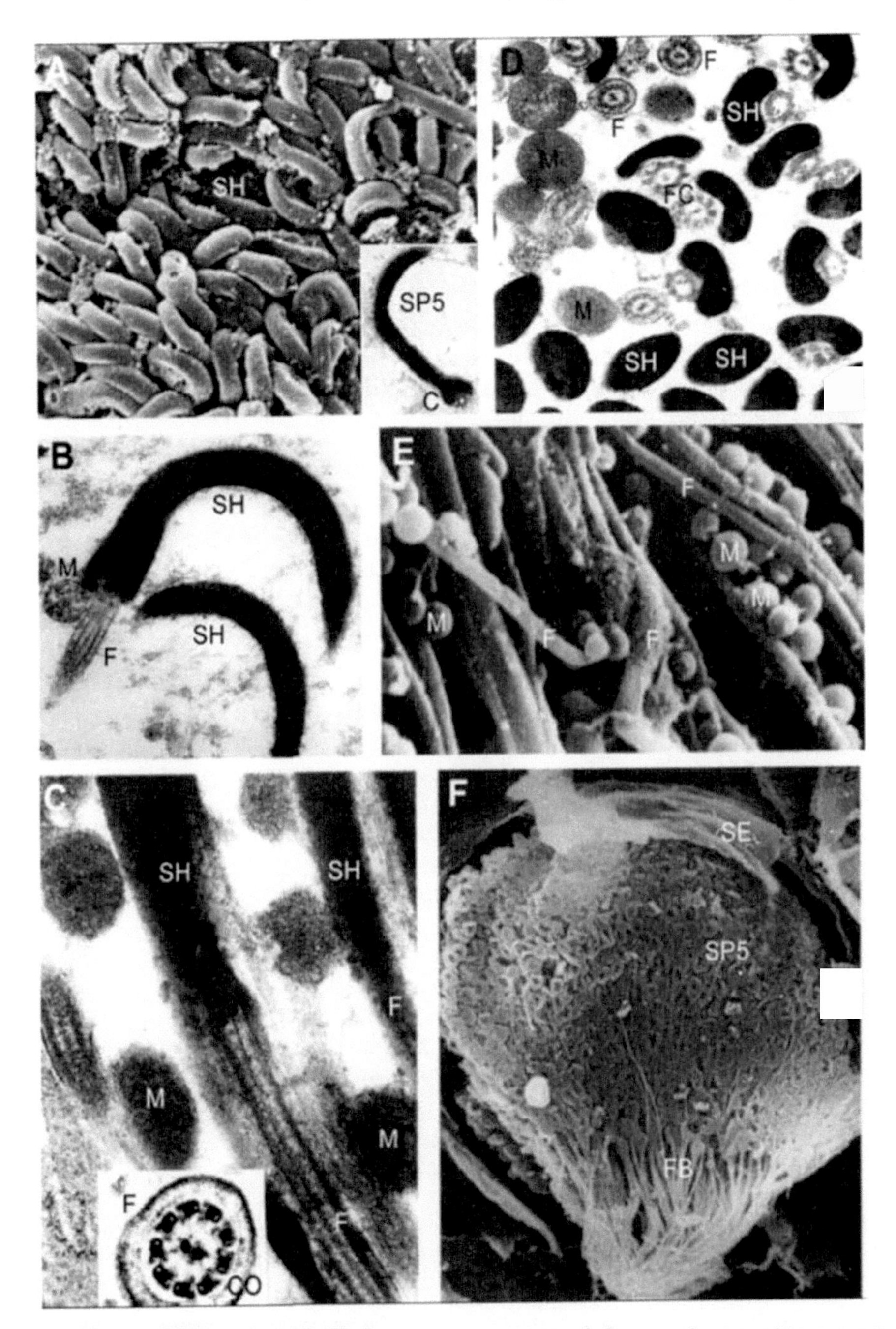

superciliosus. SEM. x 15,000. **F**. Spermatozeugmata of *C. acuminatus* with stage 5 spermatids. SEM. x 900. C, centriole; CO, collar (sleeve); F, flagellum; FB, bundle of flagella; FC, flagellum at the centriole; M, mitochondria; SE, envelope of spermatocyst; SH, sperm head; SP 5, stage 5 spermatids. Courtesy of Professor Lev Fishelson.

inset). Formation of the elongated collar around the sperm base and flagellum occurs (*C. cottoides*, Fig. 15.90C and inset; *C. taurus*, Fig. 15.90D). Round mitochondria, 0.2–0.3 μm in diameter, extend within the collar along the flagellum (*C. superciliosus*, Fig. 15.90E). The spermatozoa in the spermatocyst cohere to form spermatozeugmata, with sperm heads directed to the apical pole of the cyst, and the flagella directed posteriorly (*C. acuminatus*, Fig. 15.90F) (Fishelson *et al.* 2007).

15.7.6 Suborder Gobiesocoidei

Clingfish and their relatives were previously (Nelson 1984) placed in a separate order Gobiesociformes. They are primarily marine bottom dwellers in shallow waters and occur worldwide in tropical and temperate seas. They have a scaleless head and body and no swim bladder (for discussion of other characters, (see Lauder and Liem 1983, Jamieson 1991; and Nelson 1984, 2006).

The former gobiesociforms were tentatively regarded by Jamieson (1991) as the sister-group of the Batrachoidiformes on the basis of shared biflagellarity of sperm in conjunction with batrachoidiform + gobiesociform + lophiiform similarities. Sperm ultrastructure (biflagellarity and/or spiral nucleus) appeared to confirm relationship of these three groups.

15.7.6.1 Family Gobiesocidae

Lepadogaster: Like the batrachoids, the gobiesocids have biflagellate sperm, as seen in the clingfish *Lepadogaster lepadogaster* (Mattei and Mattei, 1978), resembling the complex sperm of the batrachoid *Porichthys*. More species need to be studied to confirm if biflagellarity is general for these sister-groups.

The spermatozoon of *Lepadogaster lepadogaster* is about 90 μm long of which the nucleus constitutes 8 μm, the intermediate piece 13 μm and the free portion of the flagella 70 μm with an endpiece of 1 μm (Mattei and Mattei 1978) (Fig. 15.91). An acrosome remnant is recognized in the spermatid (Mattei and Mattei 1978). The elongate nucleus is straight except for a pointed anterior region which forms a helix with three turns. Nuclear pores are present basally. The "intermediate piece" consists of two regions: an anterior portion in which 6-10 elongate mitochondria are parallel to the flagella from which they are separated by two cytoplasmic canals; and a longer, 10 μm long, region in which the two flagella are enclosed in a common sheath, consisting of the wall of the cytoplasmic canals, which contains fine granules.

The two centrioles are mutually parallel. The outer dynein arms are absent from the doublets of the 9+2 axonemes. No flagellar fins are present in numerous micrographs of cross sections of the tails. In the distal portion of each flagellum the doublets lose their secondary fibers [B subtubules] and the 9 peripheral and two central singlets become disordered. The terminal piece is enlarged (Mattei and Mattei 1978).

The sperm of *Lepadogaster l. lepadogaster* use the spiral nucleus to penetrate other cells, including spermatids, while in the testes. This was not observed in the supposed subspecies *L. lepadogaster purpurea*. This difference was considered by Mattei and Mattei (1978) to vindicate the former separation of these entities as distinct species.

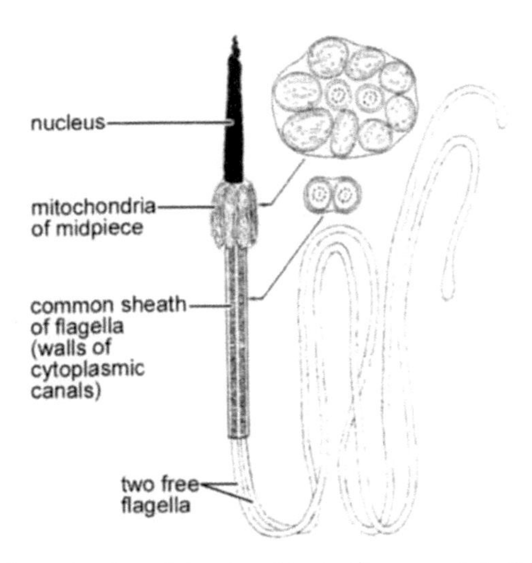

Fig. 15.91 *Lepadogaster lepadogaster.* Spermatozoon. Adapted from Mattei, C. and Mattei, X. 1978. Biologie Cellulaire 32: 267-274, Fig. 2.

Sicyases sanguineus: Structure and ultrastructure features of the testis of *S. sanguineus* were examined by Cerisola (1990) in order to characterize spermatogenic cells, Sertoli cells, interstitial cells and epithelial cells of the efferent duct system.

15.7.7 Suborder Gobioidei

Gobioids are one of the most speciose group of fishes, with approximately 270 genera, 2211 species or 10% of all teleosts (Rupple 1984; Nelson 2006). They appear to form a monophyletic group with the Rhyacichthyidae as their most primitive representatives. They have unique sperm glands.

Sperm literature: Publications on gobioid sperm are listed under the various families below. However, a highly significant paper by Hara (2004) which has not been cited by other reviewers and which deals with no less than 38 gobioid species in six families deserves special mention. Considerations of space allow only a summary of the paper.

Gobioid spermatozoa were found to be exceptionally diversified comprising most of the known teleostean sperm morphs. They were classified into six types (see Tables 15.4, 15.5) based on a combination of morphological characters of the sperm. Type I and Type II of Hara (2004) should not be confused with Types I and II in the terminology of Mattei (1970). The types are briefly defined below, with examples of species thus characterized. A full list of species examined by Hara (2004) is given in Table 15.5.

Type 1 (20 species, including *Rhyacichthys aspro*) - nucleus ovoidal to spherical, with distinct or indistinct hiatus in chromatin on lateral surface; single mitochondrion, located opposite chromatin in all but one species; single flagellum (see Rhyacichthyidae, below).

Table 15.4 Six types of gobioid spermatozoa and their diagnostic features. From Hara, M. 2004. Japanese Journal of Ichthyology 51: 1-32, Table 2.

Type	Nucleus (Head)	Mitochondria	Flagellum	Spermatozoon
I	round-ovoidal; hiatus in chromatin absent, moderate or deep	single, round, unilateral	single	
II	round	numerous, single layer	single	
III	round-cone shaped	numerous, multi-layer	single	
IV	round with special anterior accessory	numerous, multi-layer	double	
V	Bell-shaped gradually tapering off	numerous, multi-layer	single	
VI	elongate, cylindrical, entirely penetrated by flagellum	single, enclosing flagellum at base	single	

Type II (8 species, including *Sicyopterus japonicus*) - spherical nucleus without hiatus in chromatin; mitochondria numerous, in several longitudinal columns around base of flagellum; single flagellum (see Gobiidae, Gobiinae, below).

Type III (3 species of Odontobutidae and 4 species of Gobionellinae) - nucleus spherical to cone-shaped; numerous, multilayered mitochondria; single flagellum.

Type IV (*Leucopsarion petersi*)-spherical nucleus with special anterior "accessory"; numerous irregularly-shaped mitochondria, around base of flagellum; two flagella. The accessory structure is unique in the examined Pisces and this is one of the most bizarre sperm in the class. (See Gobiidae, Gobionellidae, below).

Type V (*Ptereleotris hanae*) - nucleus bell-shaped, anteriorly attenuated; mitochondria numerous, spherical, around base of flagellum; single flagellum (see Ptereleotridae below).

Type VI (*Schindleria* sp.) - nucleus elongate, cylindrical, penetrated by flagellum to well beyond anterior tip; single mitochondrion enclosing flagellum at base; single flagellum. (See Schindleriidae, below).

Comparison between these types and current gobioid systematics by Hara (2004) has revealed the following points:

(1) Rhyacichthyidae and the two subfamilies of Eleotridae share uniform sperm morphology within Type I.
(2) Types IV, V and VI are distinct in having exceptional features.
(3) Of six families of Gobioidei, Gobiidae has the most diversified sperm morphology.

Table 15.5 Spermatozoal structure in gobioids (classification of Nelson 1994). After Hara, M. 2004. Japanese Journal of Ichthyology 51: 1-32, Table 3.

Family	*Subfamily*	*Species*	*Spermatozoon structure*	*Type*
Rhyacichthyidae		*Rhyacichthys aspro*		I
Odontobutidae		*Odontobutis obscura*	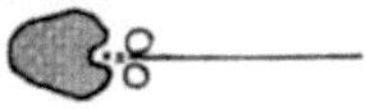	III
		Odontobutis interrupta		III
		Odontobutis platycephala		III
Eleotridae	Eleotrinae	*Electris fusca*		I
		Eleotris acanthopoma		I
		Electris melanosoma		I
	Butinae	*Butis butis*		I
		Butis amboinensis	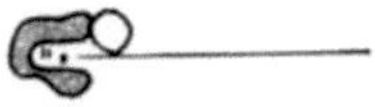	I
		Bostrchus sinensis	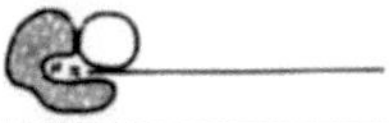	I
Gobiidae	Oxiidercinae	*Periophthalmus modestus*		I
		Scartelaos histophorus		I
		Boleophthalmus pectinirosris		I
	Sicydiinae	*Sicyopterus japonicus*		II
	Amblyopinae	*Odntamblyopus lacepedii*	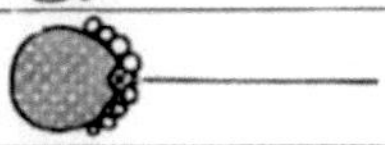	II
	Gobionellinae	*Acanthogobius flavimanus*		II
		Rhingobius sp. CB	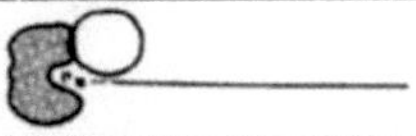	I
	Gobionellinae	*Rhinogobius* sp. DA	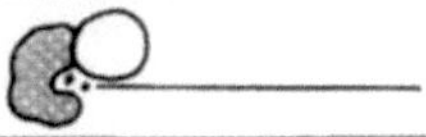	I
		Tridentiger kuroiwae		I

Table 15.5 Contd. ...

Table 15.5 Contd. ...

		Tridentiger brevispinis		I
		Gymnogobius breunigii		III
		Gymnogobius urotaenia		III
		Gymnogobius petschiliensis	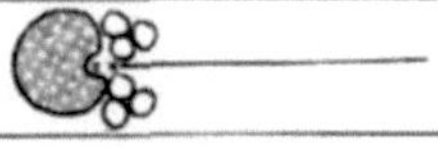	III
		Luciogobius guttatus		III
		Leucopsarion petersii		IV
	Gobiinae	*Eviota prasina*		I
		Amblygobius phalaena		I
		Acentrogobius viridipunctatus		I
		Acentrogobius pflaumii		I
		Pleurosicya muscarum		I
		Valenciennea longipinnis	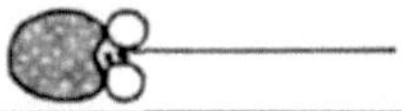	II
		Paragobiodon echinocephalus		II
		Cryptocentrus caeruleomaculatus		II
		Bryaninops yongei		II
	Gobiinae	*Fusigobius humeralis*		II
		Favonigobius gymnauchen		I
Microdesmidae	Ptereleotrinae	*Ptereleotris hanae*	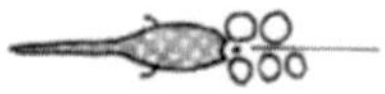	V
Schindleriidae		*Schindleria* sp.		VI

In addition, present groupings of sperm morphs have been compared by Hara with gobioid phylogenies derived from molecular evidence (Akihito *et al.* 2000) (Fig. 15.99). The results suggest that spermatozoa are potentially useful in evaluating the relationships within the Gobioidei (Hara 2004).

Gobioid families will now be discussed though consideration of space will limit the number of individual species that can be detailed.

15.7.7.1 Family Rhyacichthyidae

These are the loach gobies, in freshwater. *Rhyacichthys aspro* has a Type I sperm *sensu* Hara (2004) (see Table 15.4).

15.7.7.2 Family Odontobutidae

These are the freshwater sleepers. *Odontobutis obscura, O. interrupta* and *O. platycephala* have Type III sperm *sensu* Hara (2004) (see Tables 15.4, 15.5).

15.7.7.3 Family Eleotridae

Eleotrids are the sleepers and sand gobies; those examined for sperm ultrastructure are *Hypseliotris galii* (Jamieson 1991), *Eleotris prasina* (Hara and Okiyama 1998), *E. fusca, E. acanthopoma* and *E. melanosoma* (Hara 2004) and *Oxeleotris marmoratus* (Suwanjarat *et al.* 2005). They conform to Type I of Hara (2004) with the exception of *O. marmoratus* which conforms with Type II of Hara reported to have a few mitochondria; micrographs of this species show a single lateral mitochondrion or bilateral mitochondria. Some details for *H. galii*, from Jamieson (1991), are given below.

***Hypseleotris galii*:** The nucleus in the eleotrid *Hypseleotris galii*, the Fire-tailed Gudgeon (adult, Fig. 15.92; sperm, Fig. 15.93), is elongate pyriform and is totally impaled by the axoneme and centriolar apparatus, with persistence of only a very thin layer of chromatin and investing nuclear envelope over the apical portion of the implantation fossa (Fig. 15.93E). The centriolar fossa is slightly eccentric (Fig. 15.93B), and correspondingly the nuclear layer is thicker on one side than elsewhere. A striking feature, now known for other eleotrids (Hara 2004), is interruption of the chromatin in one to three loci apically and subapically. As a result only apposed nuclear envelopes

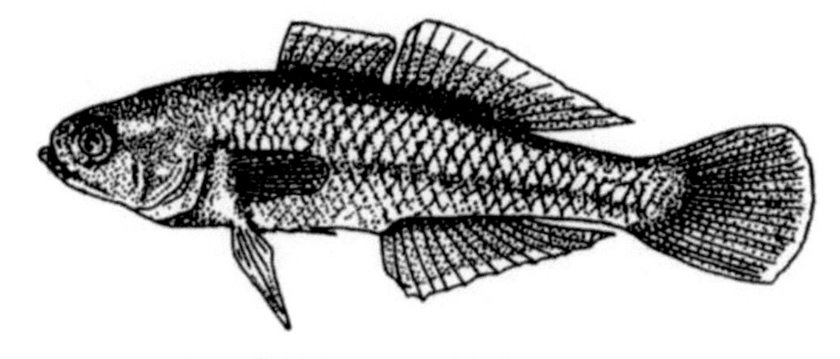

Fig. 15.92 *Hypseleotris galii*, the Fire-tailed Gudgeon. From Jamieson, B. G. M. 1991. *Fish Evolution and Systematics: Evidence from Spermatozoa*. Cambridge University Press, Cambridge, Fig. 16.12. Drawn by Christopher Tudge.

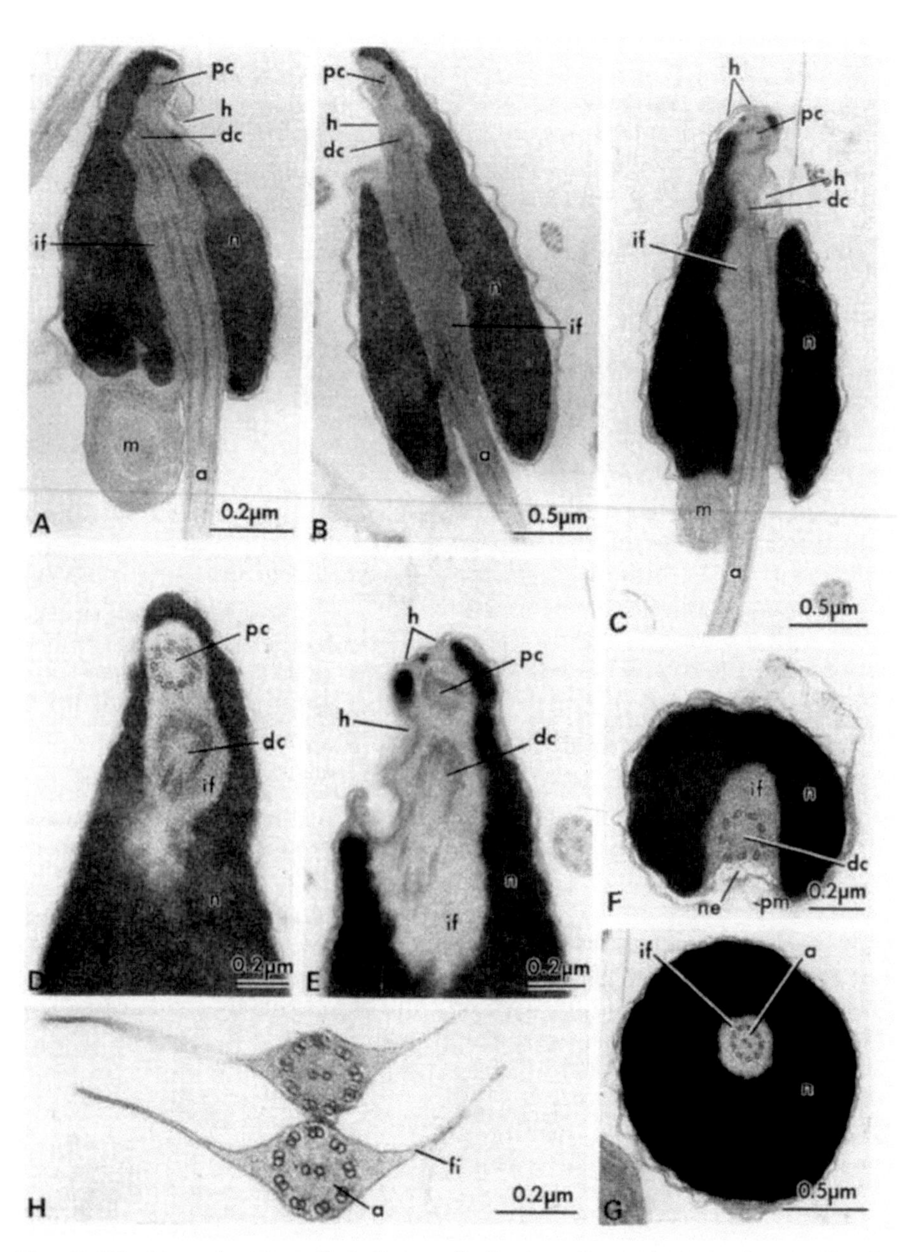

Fig. 15.93 *Hypseleotris galii.* **A-C**. Longitudinal sections of spermatozoa, showing subapical interruption (hiatus) of the chromatin of the nucleus. **A, C**. "Lateral" view showing single, unilateral mitochondrion. **B**. Approximately "frontal" view. **D, E**. Frontal and lateral views, respectively, of centrioles (only doublets have been seen) in nuclear fossa. **F**. Transverse section (TS) through distal centriole, showing its proximity to the collapsed nuclear envelope and plasma membrane. **G**. TS of

Fig. 15.93 Contd. ...

intervene between the deep implantation fossa and the plasma membrane (Fig. 15.93C, see label h). The largest hiatus (0.4 μm long) in the chromatin is subapical and on the face of the nucleus to which the longitudinal axis of the proximal centriole is directed (Jamieson 1991).

The midpiece is highly asymmetrical. A single large, transversely cristate mitochondrion is situated in a unilateral collar, abutting the wide side of the nucleus, on one side of the axoneme (Fig. 15.93C). As in *Periophthalmus* (Fig, 15.94), the nucleus is straight-edged where it abuts the mitochondrion.

The basal body and 9+2 flagellum are in the long axis of the nuclear fossa. The proximal centriole is perpendicular though slightly upturned relative to the basal body. A pair of lateral flagellar fins is present (Jamieson 1991).

Remarks: Molecular analysis would be instructive in determining whether similarities in sperm ultrastructure noted by Jamieson (1991) between Eleotridae and Dactylopteriformes, e.g. *Dactylopterus (=Cephalacanthus) volitans* (Mattei 1970) are homoplasic or indicative of relationship. In both entities the nucleus is so deeply invaded by the basal fossa as to retain only a very thin layer of chromatin (in *Hypseleotris* subapically no chromatin) between the outer and inner nuclear envelope over the head of the flagellum; two centrioles are present; and the mitochondria are restricted to one side of the axoneme. Asymmetry of the head is more pronounced in *D. volitans* of the two genera. A functional explanation of the deep penetration of the fossa and attenuation of chromatin is not at present apparent.

A single mitochondrion, as in all eleotrids, is seen, *inter alia,* in the rhyacicthyid *Rhyacicthys,* and the gobiids *Scartelaos, Boleophthalmus, Rhinogobius, Tridentiger, Eviota, Amblygobius, Pleurosicya* and *Favonigobius* (Hara 2004) as in the percomorphs *Maccullochella* (Percichthyidae) (Jamieson 1991) and *Periophthalmus papilio* (Mattei 1970). At the risk of using the supposedly discredited "common is primitive" criterion, it might be deduced that the single mitochondrion is plesiomorphic for gobioids. If not, the condition has arisen independently several times in gobioids (see also Phylogeny, Fig. 15.99).

The nuclear penetration and prenuclear extension of the flagellum in *Schindleria* sp. appears to be an extreme development from the condition seen in Hypseleotris; Hara and Okiyama (1998) saw close relationship of *Schindleria* with the Eleotridae and/or the Gobiidae.

Fig. 15.93 Contd. ...

nucleus basally through the fossa and 9+2 axoneme. **H**. TS two axonemes, showing fins. a, axoneme; dc, distal centriole; fi, fin; h, hiatus in chromatin; if, implantation fossa; m, mitochondrion; n, nucleus; ne, nuclear envelope; pc, proximal centriole; pc, proximal centriole; pm, plasma membrane. From Jamieson, B. G. M. 1991. *Fish Evolution and Systematics: Evidence from Spermatozoa*. Cambridge University Press, Cambridge, Fig. 16.13.

15.7.7.4 Family Gobiidae

The gobies are almost ubiquitous in marine, brackish and occasionally freshwater tropical and subtropical areas (Nelson 2006).

Species of Gobiidae for which spermatozoal ultrastructure has been examined are: *Acanthogobius flavimanus*, Morisawa (1982) (Fig. 15.95); *Glossogobius giuris* (Zutshi and Murthy 2001); *Gobius bucchichi* and *Lesuerigobius fiesii* (Fishelson 1990) (restricted to an investigation of endoplasmic reticulum in spermiogenesis); *Periophthalmus papilio* (Mattei 1970; Jamieson 1991, Hara 2004) (Fig. 15.94); *Rhinogobius* sp. and *Tridentiger kuroiwae* (Hara and Okiyama 1998; Hara 2004). Many further species are described by Hara (2004) (see Table 15.5). Gobiids have Type I and Type II spermatozoa *sensu* Hara.

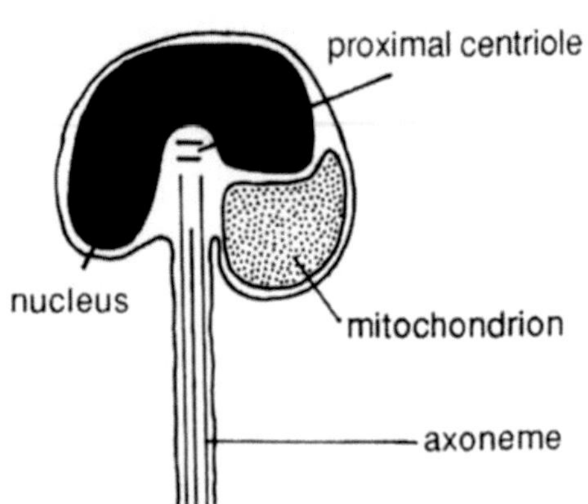

Fig. 15.94 *Periophthalmus papilio* (Gobiidae). Diagrammatic longitudinal section of the spermatozoon. After Mattei, X. 1970. Pp. 59-69. In B. Baccetti *(ed.)*, *Comparative Spermatology*, Academic Press, New York. Fig. 4: 23.

Periophthalmus papilio: In the sperm of the gobiid *Periophthalmus papilio*, the Mudskipper, (Fig. 15.94) (family Gobiidae), the mitochondrial material, forming a single mass, again lies to one side only of the axoneme.

The nucleus is asymmetrical with a considerable basal fossa containing the proximal centriole (Mattei 1970), It can be considered a modified Type I sperm *sensu* Mattei (1970) and accords with Type I of Hara (2004).

The number of gobiid sperm described by Hara and Okiyama (1998) and Hara (2004) is too great to be described individually here but they are illustrated diagrammatically in Table 15.5 and some of the variation will be outlined.

The type I sperm *sensu* Hara (2004) is known for some 21 species being seen in *Periophthalmus* (as noted above), *Scartelaos, Boleophthalmus* (Oxudercinae), *Rhinogobius, Tridentiger* (Gobionellinae), *Eviota* (placed in the Eleotridae by Hara and Okiyama 1998), *Amblygobius, Pleurosicya* and *Favonigobius* (Gobiinae).

The Type II sperm *sensu* Hara (2004) is known for eight gobiid species: *Sicyopterus japonicus* (Sicydiinae), *Odontamblyopus lacepedii* (Amblyopinae),

Acanthogobius flavimanus (Gobionellinae) (Fig. 15.95), *Valenciennea longipinnis, Parogobiodon echinocephalus, Cryptocentrus caeruleomaculatus, Bryaninops yongei* and *Fusigobius humeralis* (Gobiinae).

The Type III sperm *sensu* Hara (2004) is known for four gobiid species: *Gymnogobius breunigii, G. urotaenia, G. petschiliensis* (Fig. 15.96) and *Luciogobius guttatus* (all Gobionellinae) (Hara 2004). The sperm of *G. breunigi* and *Luciogobius guttatus* are very similar yet distinctive, with long, conical truncated conical nucleus and multiple mitochondria (see Table 15.5) and

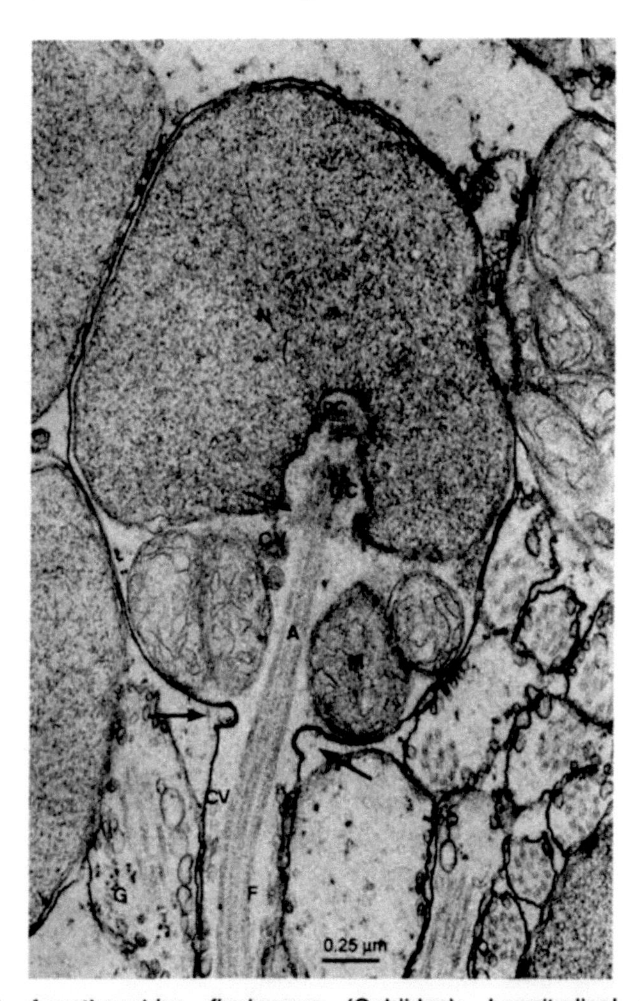

Fig. 15.95 *Acanthogobius flavimanus* (Gobiidae). Longitudinal section of spermatozoon. A, axoneme; CV, cytoplasmic vesicles; DC, distal centriole; F, flagellum; G, cytoplasmic granules; M, mitochondrion; N, nucleus; PC, proximal centriole. Small arrow, electron-dense material on side of distal centriole; large arrows, highly electron-dense plasma membrane forming characteristic constriction at the border between midpiece and tail. After Morisawa, M. 1982. Bulletin of St. Marianna University 11: 31-37, Fig.3.

suggest that the two species should not be generically separate. The other two species of *Gymnogobius*, *G. urotaenia*, *G. petschiliensis*, have round-headed sperm with multiple mitochondria and support their congeneric status.

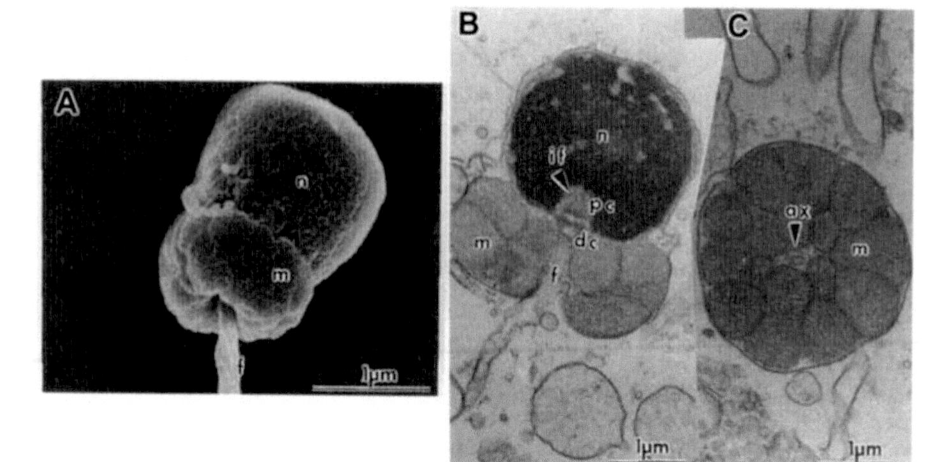

Fig. 15.96 *Gymnogobius petschiliensis* (Gobiidae) spermatozoa. **A**. SEM of sperm head. **B**. Longitudinal section of spermatozoon. **C**. Transverse section of midpiece. Hara, M. 2004. Japanese Journal of Ichthyology 51: 1-32, Fig. 8B, C, D. With permission.

The Type IV spermatozoon *sensu* Hara (2004) (Tables 15.4, 15.5) is known only for *Leucopsarion petersii* (Gobionellinae) (Fig. 15.97). The apical accessory structure is unknown in any other fish spermatozoon and the biflagellate condition is not known in other gobioids. There are two well-developed lateral flagellar fins, an unusual feature in biflagellate sperm. One has to go as far afield as the sperm of the polychaete *Idanthyrsus pennatus* to see something resembling its accessory structure, albeit superficially.

15.7.7.5 Family Microdesmidae

The Type V spermatozoon *sensu* Hara (2004) is known only for the dartfish, *Ptereleotris hanae* (Ptereleotrinae). The nucleus is fusiform with a long pointed tip and is described by Hara as bell-shaped. There are numerous multilayered mitochondria and a single flagellum.

15.7.7.6 Family Schindleriidae

The Schindleriidae are the infantfishes, small neotenic oceanic fish limited to a single genus with three species. The genus *Schindleria* is highly paedomorphic, one species being the lightest, possibly shortest lived and most paedomorphic vertebrate (Nelson 2006). *S. praematura* is dioecious with external fertilization but distinct pairing (Breder and Rosen 1966).

One may speculate that the presumed genetic modification attendant on this paedomorphism may account for one of the most unusual spermatozoa in

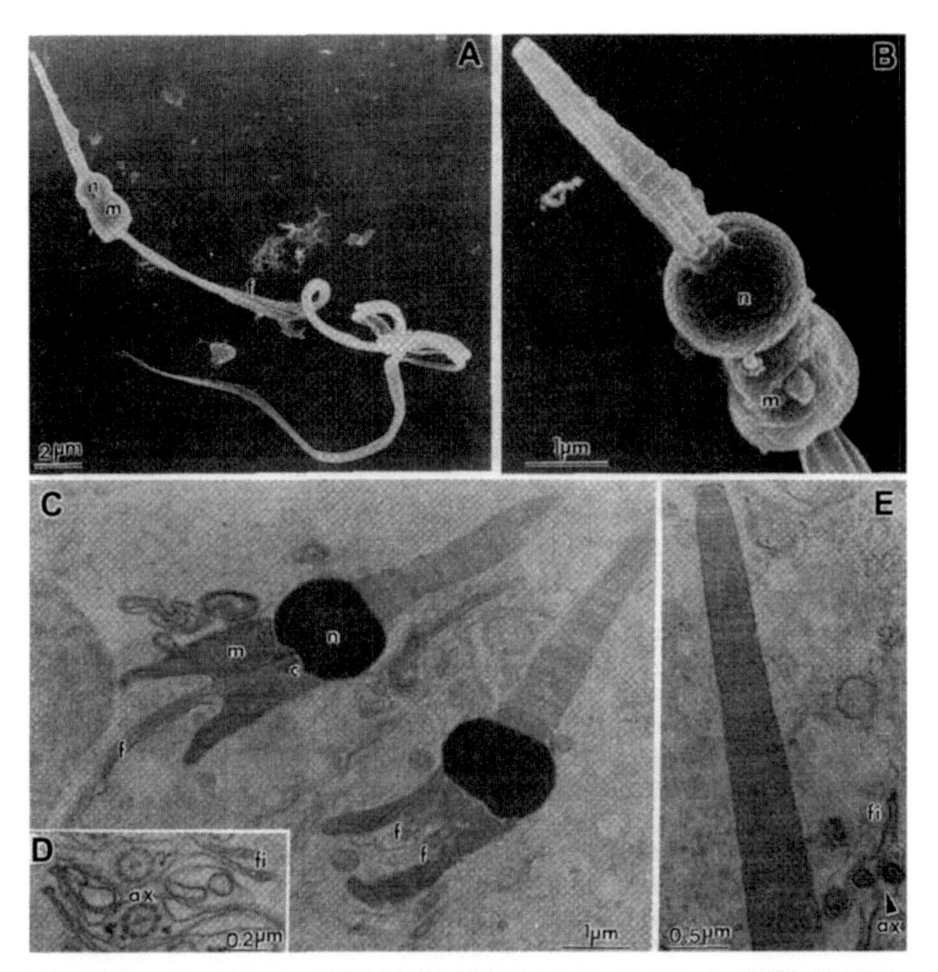

Fig. 15.97 *Leucopsarion petersii* (Gobiidae) spermatozoa. **A**. SEM of whole spermatozoon. **B**. SEM of sperm head, with accessory apical accessory structure, and midpiece. **C**. TEM longitudinal section of two spermatozoa showing the accessory structure and the biflagellate condition. **D**. TEM transverse sections through flagella, showing well developed lateral fins. **E**. TEM longitudinal section of the accessory structure. Adapted from Hara, M. 2004. Japanese Journal of Ichthyology 51: 1-32, Figs. 8G, H and 9A-C. With permission.

fish and vertebrates in general. Thus in sperm described by Hara and Okiyama (1998) and Hara (2004) of a species resembling *S. praematura* the flagellum penetrates the nucleus, extending far anteriorly and posteriorly of it; centrioles have not been observed but a cap-like structure is situated at the tip of the flagellum (Fig. 15.98). The single mitochondrion is located posterior to the nucleus and surrounds the flagellum (see also Table 15.5). This is the unique Type VI spermatozoon of Hara (2004).

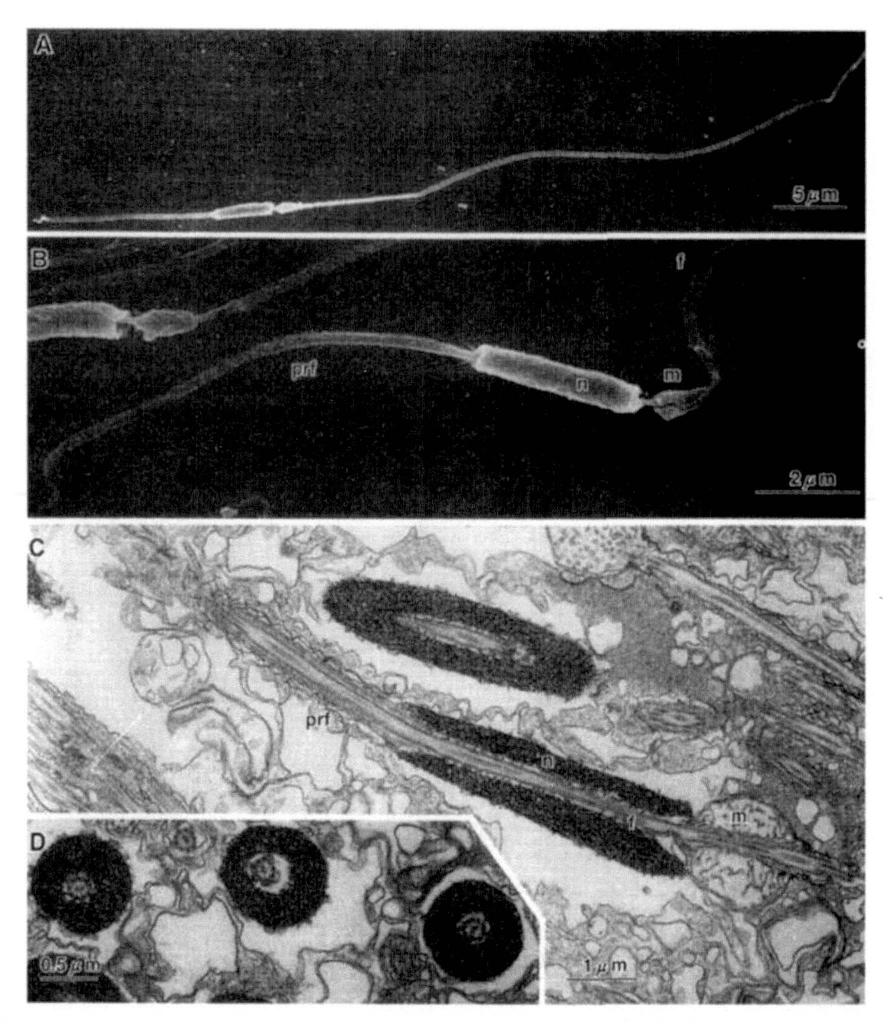

Fig. 15.98 *Schindleria* sp. **A,B**. SEM of spermatozoa. **C**. TEM of a longitudinal section of a spermatozoon. **D**. TEM of transverse sections of the nucleus showing penetration by the flagellum. f, flagellum; m, mitochondrion; n, nucleus; prf, prenuclear flagellum. After Hara, M. and Okiyama, M. 1998. Bulletin of The Ocean Research Institute, University of Tokyo 33: 1-138, Fig. 45A-D. With permission.

Gobioid sperm and phylogeny: Akihito *et al.* (2000) computed a phylogeny based on the nucleotide sequences and the deduced 380-residue-long amino acid sequences of the mitochondrial cytochrome b genes from 28 representative gobioid species. These species included 18 of the 38 species examined for sperm ultrastructure by Hara (2004) who have superimposed diagrams of the sperm on the molecular phylogram (Fig. 15.99). The sperm fall into six clusters, corresponding with clades in the phylogeny as shown in Fig. 15.99 and Table 15.6.

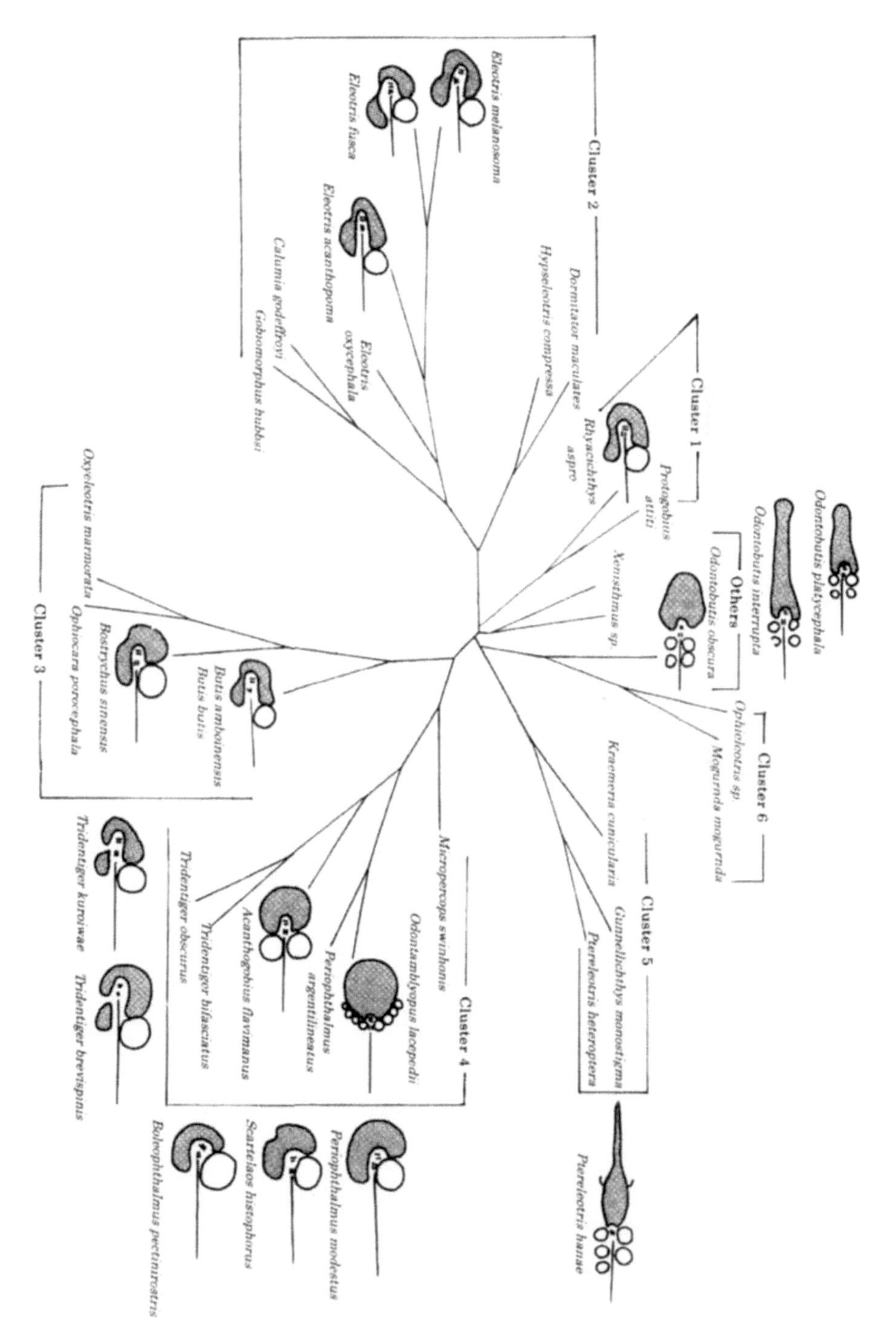

Fig. 15.99 Phylogenetic relationships of gobioid fish (Akihito *et al.* 2000) and spermatozoal ultrastructure of selected species. After Hara, M. 2004. Japanese Journal of Ichthyology 51: 1-32, Fig. 14. With permission.

It can be seen in Fig. 15.99 and Table 15.5, that the Type III sperm occurs in three distinct clades.

It is here considered that the single mitochondrion is plesiomorphic for gobioids. It is usual to consider a multimitochondrial condition to be plesiomorphic for fish sperm (Jamieson 1991) and while that view is maintained here it appears that the multimitochondrial condition in gobioids is a reacquisition of that condition, i.e. a reversal. This is evidenced by the highly unusual crowded condition of multiple mitochondria in *Odontamblyopus lacepedii, Acanthogobius flavimanus, Gymnogobius petschiensis* (Fig. 15.96B), and *Luciogobius guttatus* and by the simple, more typically teleostean or perciform-like sperm of Type I.

Table 15.6 Comparison between the molecular phylogeny of Akihito et al. (2000) and types of spermatozoa recognized by Hara (2004) in gobioids. Slightly adapted from Hara, M. 2004. Japanese Journal of Ichthyology 51: 1-32, Table 4.

Molecular *Phylogeny*	*Sperm type*	*Species common to both studies (Akihito et al. 200 and Hara 2004)*	*Related species in Hara (2004)*
Cluster 1	I	*Rhyacichthys aspro*	
2	I	*Eleotris melanosoma* *Eleotris fusca* *Eleotris acathopoma*	
3	I	*Butis amboinensis* *Bostrychus sinensis*	*Butis butis*
4	I & II	*Odontamblyopus lacepedii* *Acanthogobius flavimanus*	*Periophthalmus modestus* *Scartelaos histophorus* *Boleophthalmus pectinirostris* *Tridentiger kuroiwae* *Tridentiger brevispinis*
5	V		*Ptereleotris hanae*
6			
Others	III	*Odontobutis obscura*	*Odontobutis interrupta* O. platycephala

The various morphologies outlined above for Types II-VI and diagrammatically illustrated in Table 15.4 are modifications of a basic teleostean type, with only minor modification, including more than one circlet of mitochondria in the Type II sperm of *Acanthogobius flavimanus* but greatly modified in Types III-VI, with elongation of the nucleus as a truncated cone in *Gymnogobius breuniggi* and *Luciogobius guttatus* (Type III) and culminating with the bizarre sperm of *Leucopsarion petersii* (Type IV), *Ptereleotris hanae* (Type V) and *Schindleria* (Type VI).

15.7.8 Suborder Acanthuroidei

15.7.8.1 Family Ephippidae

These are the spadefishes, in the Atlantic, Pacific and Indian Oceans. Mattei (1991) examined but did not describe the sperm of *Drepane africana* beyond ascribing it to the perciform [Type II] spermatozoon. Examination of an unpublished micrograph of four spermatozoa (courtesy of X. Mattei) reveals a broad nucleus with flattened anterior end and a deep, narrow, symmetrically placed fossa which leaves only a narrow zone of chromatin anteriorly; two centrioles orientated at ca 45° to each other and lying well outside the nuclear fossa; at least three asymmetrically disposed mitochondria; and a 9+2 flagellum almost transverse to the base of the nucleus.

15.7.8.2 Family Siganidae

Siganids are the rabbitfishes, marine in the tropical Indo-West Pacific and eastern Mediterranean (Nelson 2006).

Siganus: *S. fuscescens* is a dioecious and externally fertilizing, non-guarding egg scatterer (Woodland 1990). The sperm nucleus is spherical and measures 1.51 ± 0.15 μm in diameter in *S. fuscescens* (Gwo *et al.* 2004b) or is slightly ovoid and 1.37 ± 0.15 μm long by 1.21 ± 0.13 μm wide (n=20) in *S. rivulatus* (Lahnsteiner and Patzner 1999). The basal nuclear fossa is shallow and not well developed (Fig. 15.100D–F). The chromatin is electron-dense, granular and heterogeneous (Fig. 15.100D–F). Nuclear pores are absent (Lahnsteiner and Patzner 1999).

The two centrioles (Fig. 15.100F-I) are orientated at an acute angle to each other (Fig. 15.100F,I), more strongly in *S. rivulatus* than in *S. fuscescens*, each with nine triplets (the proximal centriole has a length of 0.29 ± 0.05 μm and a diameter of 0.22 ± 0.06 μm in *S. fuscescens* and 0.27 ± 0.04 μm in *S. rivulatus*). The distal centriole (basal body of the axoneme), extends from the level of the anterior end of the cytoplasmic canal to the shallow basal nuclear fossa (Fig. 15.100D). The fossa is scarcely apparent in *S. rivulatus* in which the distal centriole is about twice as long as the proximal centriole. In transverse sections, an osmiophilic ring, embedded in an electron-dense material, surrounds the anterior end of both centrioles (Figs. 15.100D–F, 15.101). Osmiophilic filaments connect the pair of centrioles (Fig. 15.100F,I).

The midpiece contains seven (*S. fuscescens*) or six (*S. rivulatus*) (Fig. 15.101) distinct spherical mitochondria, with a diameter of 0.51 ± 0.04 μm (Fig. 15.100A–C) or 0.39 ± 0.04 μm respectively, an electron-lucent matrix, and irregularly arranged tubular cristae (Fig. 15.100E,H). The mitochondria are arranged in two rows and surround the flagellum at the midpiece region in S. *fuscescens* (Fig. 15.100A–D) or lie in one plane with occasional displacement of two mitochondria towards the cranial end in *S. rivulatus* (Fig. 15.101). The

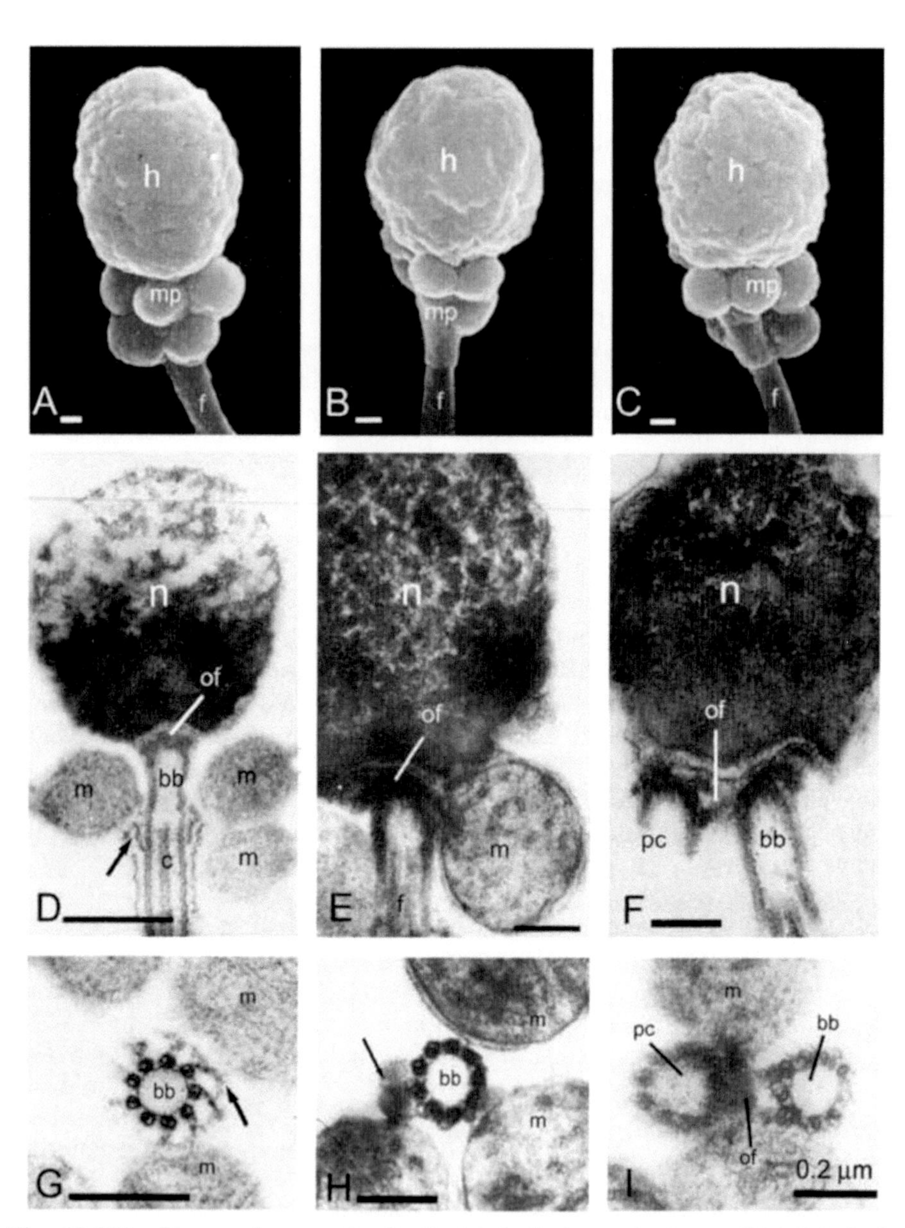

Fig. 15.100 *Siganus fuscescens*, Mottled spinefoot, spermatozoa (Siganidae). **A-C**. SEM of three different rotations of the spermatozoon. **D-I**. TEM longitudinal sections (three different rotations of the spermatozoon) of the basal diplosome (centriolar) ultrastructure. The paired, parallel centrioles nestle in a shallow nuclear fossa. Arrowhead indicates the cytoplasmic canal. **G**. Transverse section of the midpiece showing the classic 9+2 arrangement of microtubules. Arrows indicate centrifugally extending alar sheets anchoring the basal body. At their cranial end

Fig. 15.100 Contd. ...

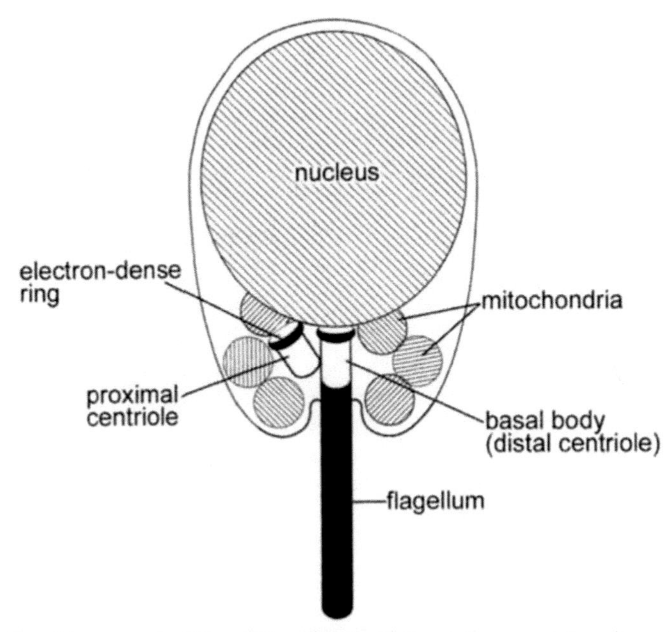

Fig. 15.101 *Siganus rivulatus* (Siganidae). Schematic reconstruction of spermatozoon. Adapted from Lahnsteiner, F. and Patzner, R. A. 1999. Journal of Fish Biology 55: 820-835, Fig. 7.

midpiece is 0.55 ± 0.08 μm by 0.74 ± 0.20 μm wide in *S. rivulatus*. A second membrane lies under the inner plasmalemma of the cytoplasmic canal (Lahnsteiner and Patzner 1999).

The flagellum has a classical 9+2 axoneme. Four electron-dense triangular particles, termed the necklace, were observed in the plasma membrane surrounding the neck of the flagellum (Fig. 15.100D). In both species, neither lateral fins nor vesicles are present in the flagellum. It is ca 40 μm long in *S. rivulatus* in which has been shown to have an endpiece with microtubules reduced in number and arranged irregularly (Lahnsteiner and Patzner 1999).

15.7.8.3 Family Acanthuridae

Mattei (1991) ascribed the sperm of the Surgeon fish, *Acanthurus monroviae*, to the perciform [Type II] sperm (This account, Fig. 15.102).

Fig. 15.100 Contd. ...

the centriolar microtubules are embedded in rings of electron-dense material. bb, basal body (distal centriole); c, central microtubules; f, flagellum; h, head; m, mitochondrion; mp, midpiece; of, osmiophilic fibers; pc, proximal centriole. Scale bar 0.2 μm in all figures. From Gwo, J. C., Yang, W. T., Kuo, M. C., Takemura, A. and Cheng, H. Y. 2004. Tissue and Cell 36: 63-69, Fig. 1. Copyright Elsevier. With permission.

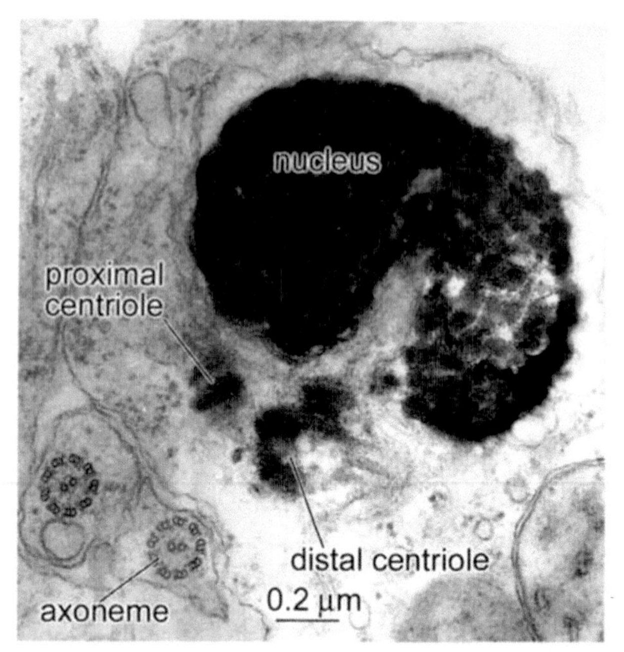

Fig. 15.102 *Acanthurus monroviae*. Type II spermatozoon. Original, from an unpublished TEM negative courtesy of Professor Xavier Mattei.

15.7.9 Suborder Scombroidei

Scombroids are perciform fish with a fixed premaxilla; the premaxillae are not only united with each other but also with the maxillae, forming a rigid non-protrusible upper jaw which can become elongate to form a rostrum. They include the fastest swimming fish, sailfish, swordfish and bluefin tuna (Lauder and Liem, 1983; Nelson 2006).

15.7.9.1 Family Sphyraenidae

Sphyraenids are the barracudas. Mattei (1991) ascribed the sperm of *Sphyraena sphyraena* to the perciform [Type II] sperm (This account, Fig. 15.103).

15.7.9.2 Family Trichiuridae

Trichiurids are the cutlassfishes; marine in the Atlantic, Indian and Pacific Oceans and are exemplified by *Trichiurus lepturus* (Fig. 15.104A).

Trichiurus: Like those of the clinids, bleniids, and eleotrids, the sperm of *Trichiurus lepturus* (Mattei 1969; Figs. 15.104B, 15.105), also described as the junior synonym *T. japonicus* (Hara and Okiyama 1998) (Fig. 15.105A-C), are asymmetrical, Type II perciform sperm. This species is a dioecious, externally fertilizing, non-guarding egg scatterer.

The nucleus in both species is on one side of the basal portion of the flagellum which lies in a very long cytoplasmic canal (Fig. 15.105A,B). The chromatin consists of dense masses with scattered lacunae (Hara and

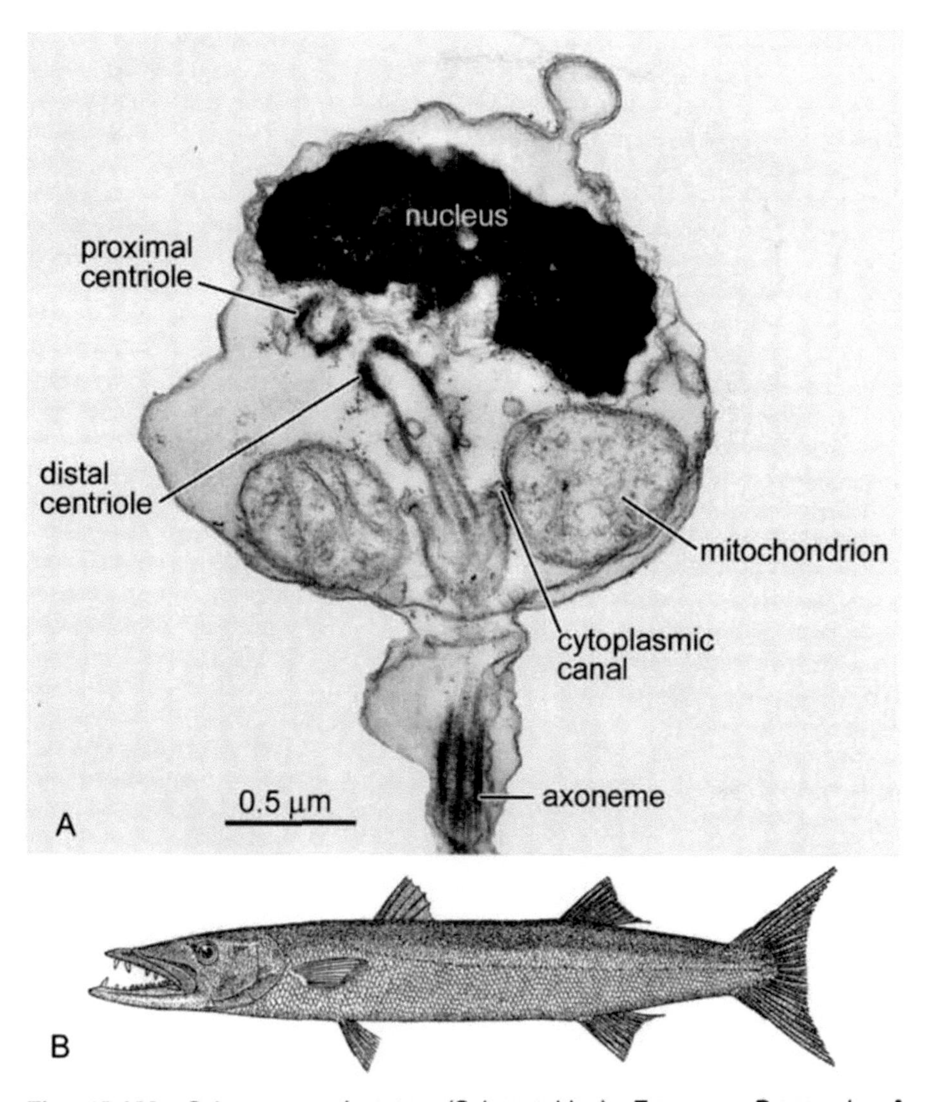

Fig. 15.103 *Sphyraena sphyraena* (Sphyraenidae), European Barracuda. **A**. Longitudinal section of the spermatozoon. Original, from an unpublished TEM micrograph courtesy of Professor Xavier Mattei. **B**. A barracuda, *Sphyraena barracuda (= S. commersoni)*. From Norman, J. R. 1937. *Illustrated Guide to the Fish Gallery*. Trustees of the British Museum, London, Fig. 67.

Okiyama 1998). Small rounded mitochondria occur on both sides of the base of the axoneme, but in *T. lepturus*, at least, though described as multiple annular tiers, are reduced in number as seen in longitudinal section on the nuclear side. The persistence of the cytoplasmic canal in the spermatozoon is shown (Mattei and Mattei 1976b) to be due to the presence of an anchoring

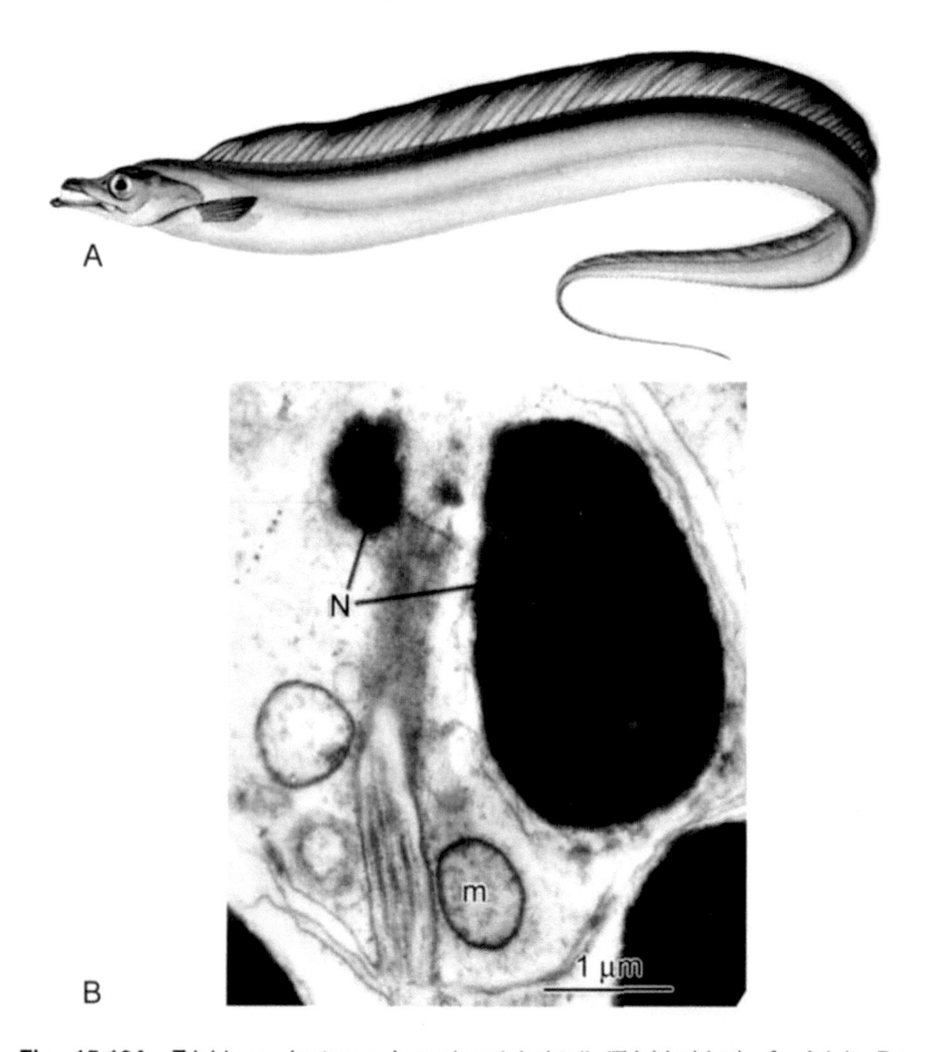

Fig. 15.104 *Trichiurus lepturus*, Largehead hairtail (Trichiuridae). **A**. Adult. By Norman Weaver. With kind permission of Sarah Starsmore. **B**. Longitudinal section of spermatozoon. m, mitochondrion; N, nucleus. Adapted from Mattei, X. 1969. *Contribution à l'étude de la spermiogenèse et des spermatozoïdes de poissons par les méthodes de la microscopie électronique*. D.Sc. thesis, Université de Montpellier, Fig. XVIIId. With permission.

system between the distal centriole and the bottom of the membrane invagination constituting the canal. A helix of three granulations occurs at the base of the axoneme and is considered equivalent to the collar described in oligochaete and prosobranch sperm (Mattei and Mattei 1976b). The centrioles are mutually perpendicular. The 9+2 axoneme (Fig. 15.105C) is surrounded by a loose plasma membrane (Hara and Okiyama 1998) but no lateral fins are present.

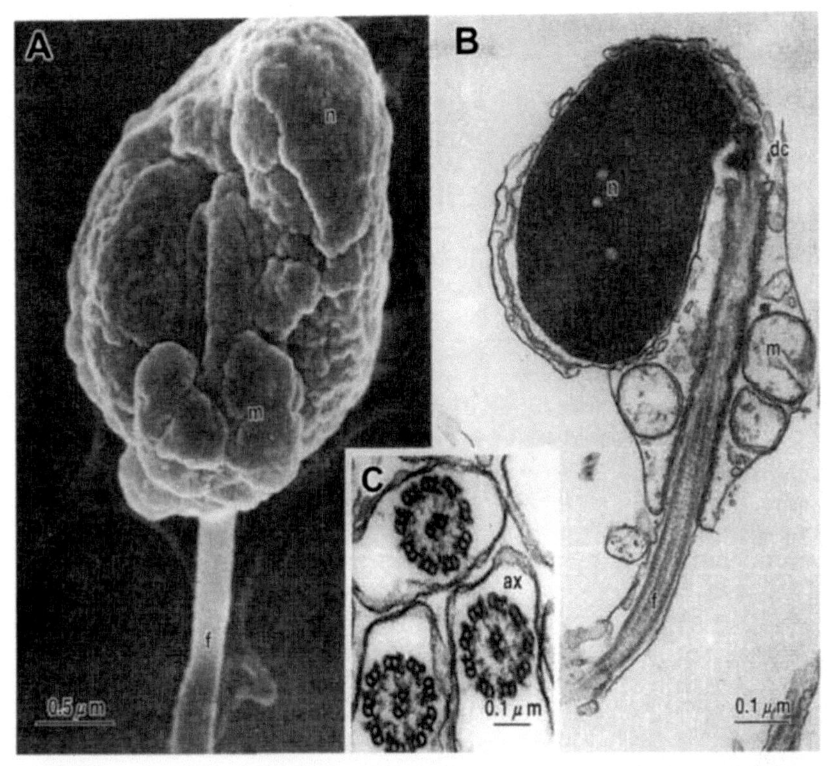

Fig. 15.105 *Trichiurus lepturus (=T. japonicus)* (Trichiuridae) spermatozoon. **A**. SEM of head. **B**. TEM longitudinal section. **C**. Transverse sections of flagella. ax, axoneme; f, flagellum; m, mitochondrion; n, nucleus; dc, distal centriole [possibly the proximal centriole]. Adapted from Hara, M. and Okiyama, M. 1998. Bulletin of The Ocean Research Institute, University of Tokyo 33: 1-138, Fig. 55B-D. With permission.

Remarks: Hara and Okiyama (1998) point out that the *Trichiurus* spermatozoon differs from the perciform type *sensu stricto* in not having ITDs in the axoneme and that four scombrids described by Mattei (1991) and Hara and Okiyama (1998) have the same type of sperm while the istiophorid *Istiophorus americaus* differs in having, the "primitive" [Type I] spermatozoon. They suggest that an incomplete rotation of the nucleus (i.e. Type II spermiogenesis) may be synapomorphic for scombroids.

15.7.9.3 Family Scombridae

Scombrids are the mackerels and tunas; marine, in tropical and subtropical seas; rarely in freshwater. Scombrid sperm species for which sperm ultrastructure has been examined are: *Scomberomorus (=Cybium) tritor* (Mattei 1991); *Euthynnus alletteratus* (Abascal *et al.* 2002); *Scomber australasicus* (Hara and Okiyama 1998); *S. japonicus* (Mattei 1991; Hara and Okiyama 1998) and *Thunnus thynnus* (Abascal *et al.* 2002, 2004).

***Scomber australicus* and *S. japonicus*.** The sperm of *S. australicus* and *S. japonicus* (Fig. 15.106) are described and extensively illustrated by Hara and Okiyama (1998). They are dioecious, externally fertilizing non-guarding egg scatterers (Collette and Nauen 1983). The sperm are of the perciform type, Type II, but again lacking axonemal ITDs. The chromatin consists of dense masses with scattered pale lacunae. The two mutually perpendicular centrioles are exposed from the nucleus and the distal is slightly displaced. Four (*S. japonicus*) (Fig. 15.106) or five (*S. australicus*) spherical mitochondria surround the base of the flagellum (Hara and Okiyama 1998).

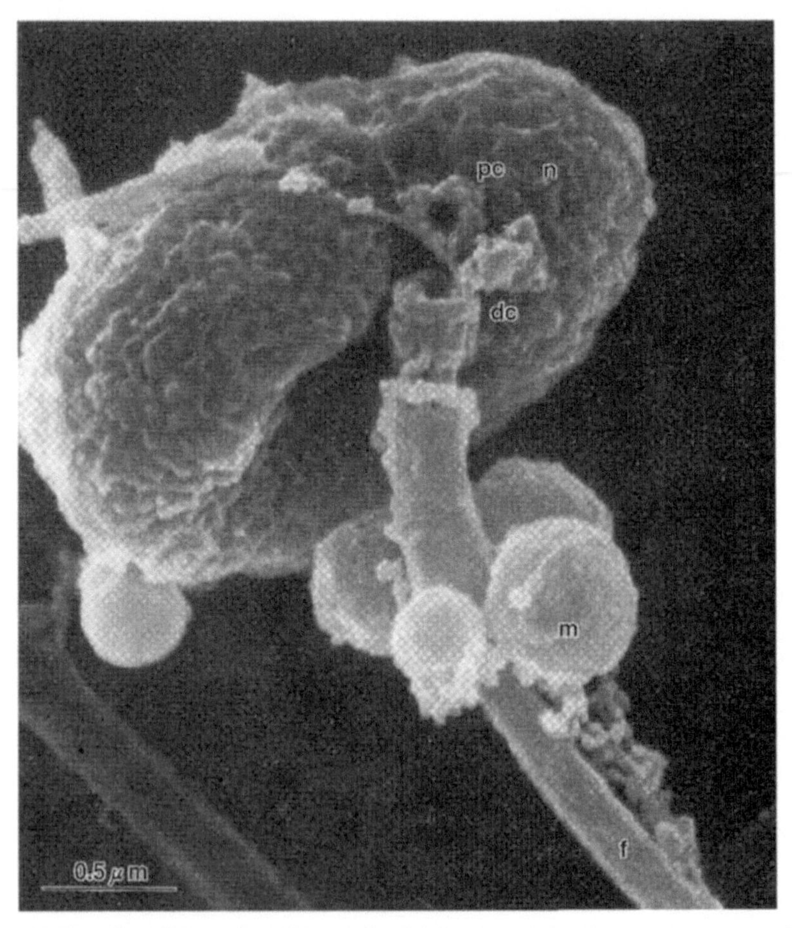

Fig. 15.106 *Scomber japonicus* (ScomJbridae). SEM of spermatozoon. From Hara, M. and Okiyama, M. 1998. Bulletin of The Ocean Research Institute, University of Tokyo 33: 1-138, Fig. 56B. With permission.

***Scomberomorus tritor*:** The sperm of *Scomberomorus (=Cybium) tritor*, the West African Spanish Mackerel, is assigned to the perciform type by Mattei (1991). Examination of an unpublished micrograph of two spermatozoa (courtesy of X. Mattei) reveals a depressed, oval nucleus with narrow fossa extending to midlength; two centrioles outside the fossa, of which the distal has a distinct 'adjunct'; mitochondria of uncertain number, and a flagellum transverse to the base of the nucleus.

***Thunnus thynnus* and *Euthynnus alletteratus*:** The spermatozoon of *T. thynnus*, the Bluefin Tuna (adult, 15.107A; sperm, Fig. 15.108), and *Euthynnus alletteratus*, the Little tunny (adult, Fig. 107B; sperm, Fig. 15.109) have an acrosomeless head and a long flagellar tail. Both are externally fertilizing non-guarding egg scatters.

Fig. 15.107 *Thunnus thynnus*, the Common Tunny or Bluefin Tuna. **B.** *Euthynnus alletteratus*, Little Tunny (Scombridae). By Norman Weaver. With kind permission of Sarah Starsmore.

The sperm head of *Thunnus thynnus* is ca 1·9 µm long and 1·8 µm wide, and comprises the apical nucleus and a few spherical mitochondria located at its base. The ovoid nucleus contains electron-dense, granular chromatin and has a very shallow basal fossa (Fig. 15.108A,B). The two triplet centrioles are approximately perpendicular to each other and lie outside the basal fossa (Fig. 15.108B-D). In *T. thynnus* the distal centriole (basal body) is much longer (ca 300 nm long) than the proximal centriole (ca 150 nm long) (Fig. 15.108B,D). Anteriorly, the distal centriole is surrounded by dense material (Fig. 15.108C,D), and laterally an electron-dense plate projects from one side of the basal body to the nuclear envelope (Fig. 15.108B,D). The distal centriole and the axoneme are

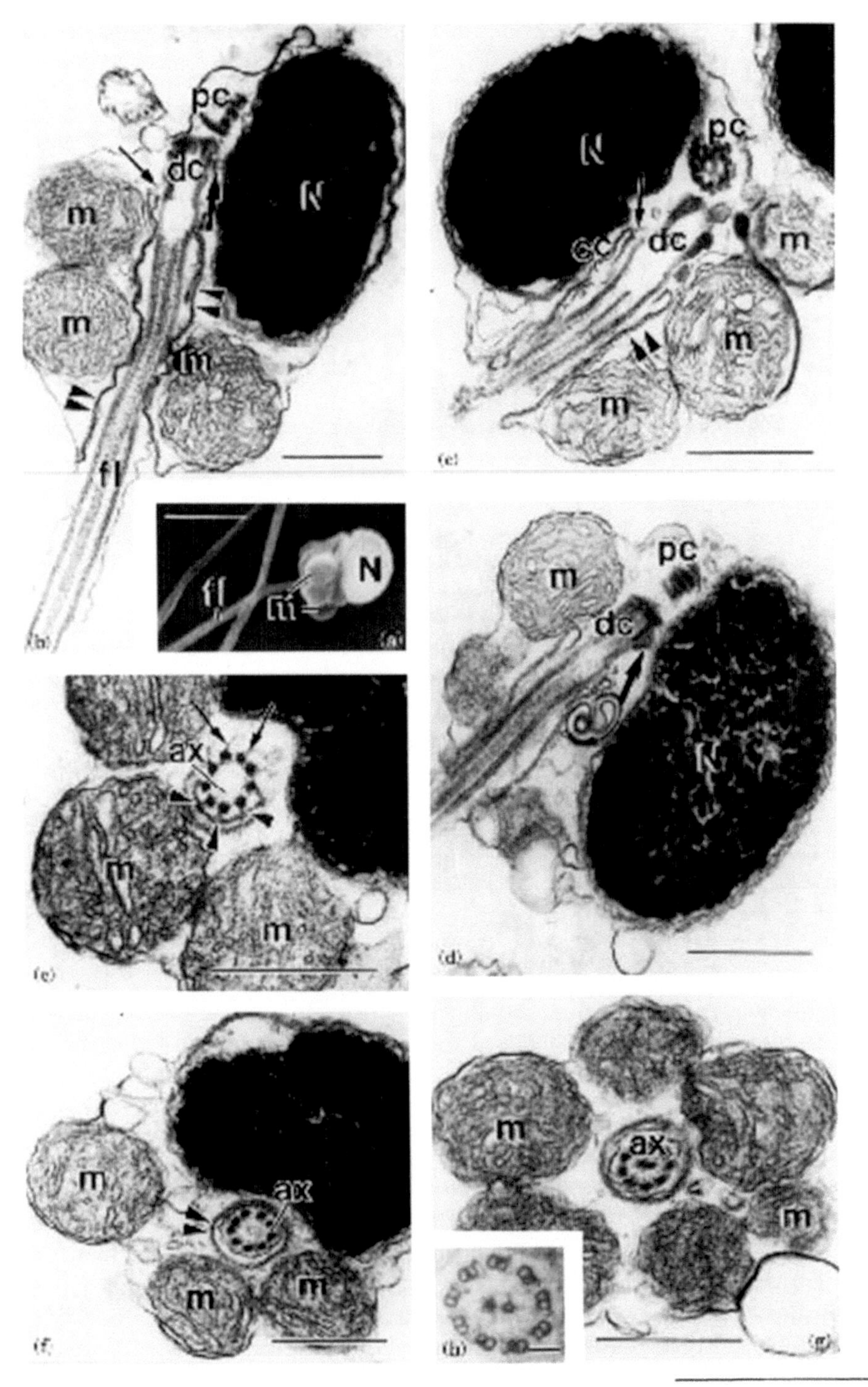

Fig. 15.108 Contd. ...

orientated parallel to the basal surface of the nucleus, as characteristic of a Type II sperm (Fig. 15.108B-D). Nine radial fibers project from the basal body to the plasma membrane at the bottom of the cytoplasmic canal (Fig. 15.108B-E). The structural pattern of the centrioles and axial filament in *Euthynnus alletteratus* is identical (Fig. 15.109B-D,F,H). Furthermore, accessory structures similar to those described for *T. thynnus* are present, such as the lateral plate that connects the distal centriole to the nucleus (Fig. 15.109B,C), the radial fibers projecting from the basal body to the plasma membrane at the anterior end of the cytoplasmic canal (Fig. 15.109B,C), and the bridges connecting the axonemal doublets to the flagellar membrane at the commencement of the flagellum (Fig. 15.109E) (Abascal *et al.* 2002).

The 9+2 flagellum lacks lateral fins (Fig. 15.108H); the central singlet microtubules are absent at the transition between the basal body and the axoneme (Fig. 15.108E,F). In this short region, the lumen of the peripheral microtubules appear filled with osmiophilic material and Y-shaped bridges attach each doublet to the flagellar membrane (Fig. 15.108E). Transverse sections at different levels of the flagellum show that microtubules A and B of the axonemal doublets are both hollow (Fig. 15.108H; 15.108H).

In *Thunnus thynnus*, the maximum of six spherical mitochondria, each ca 550 nm in diameter, are observable around the flagellum at the midpiece region (Fig. 15.108B,G), separated from the axoneme by the deep, narrow cytoplasmic canal (Fig. 15.108E,B,F). A dense layer is present beneath the plasma membrane in the region of the cytoplasmic canal (Fig. 15.107B-F) The only obvious interspecific difference between *Euthynnus alletteratus* and *T. thynnus* is the dissimilar shape of the mitochondria, which in *E. alletteratus* are somewhat more elongate and irregular in shape (Fig. 15.108D,F,G). (Abascal *et al.* 2002; Abascal *et al.* 2004).

Remarks: The spermatozoa of the investigated species of scombrids are closely similar. Between *Thunnus thynnus* and *Euthynnus alletteratus* the only apparent dissimilarity lies in the shape and arrangement of the mitochondria. The sperm mitochondrial number also seems to be slightly variable within the family: four spherical mitochondria are present in *Scomber japonicus*, five in *S.*

Fig. 15.108 Contd. ...

Fig. 15.108 *Thunnus thynnus* sperm (Scombridae). **A**. Scanning electron micrographs (SEM) of spermatozoa showing the head, comprising the nuclear and mitochondrial regions, and the proximal part of the tail. **B-H**. Transmission electron micrographs (TEM) of sagittal (**B**,**D**), frontal (**C**), and transverse (**E**-**H**) sections at various levels of the sperm cell, including the nucleus (**E**,**F**), mitochondrial ring (**G**), and flagellum (**H**). ax, axoneme; cc, cytoplasmic canal; dc, distal centriole (basal body); fl, flagellum; m, mitochondria; N, nucleus; pc, proximal centriole; arrow-head, Y-shaped bridges at the proximal region of the axoneme; double arrow-heads, dense layer underlying the cytoplasmic canal membrane; large arrow, lateral plate; small arrow, radial fibers projecting from the base of the distal centriole. Scale bars: 0·5 µm, except in (**A**) 2 µm, and (**H**) 50 nm. From Abascal, F. J., Medina, A., Megina, C. and Calzada, A. 2002. Journal of Fish Biology 60: 147-153, Fig. 1. With permission of Blackwell Publishing.

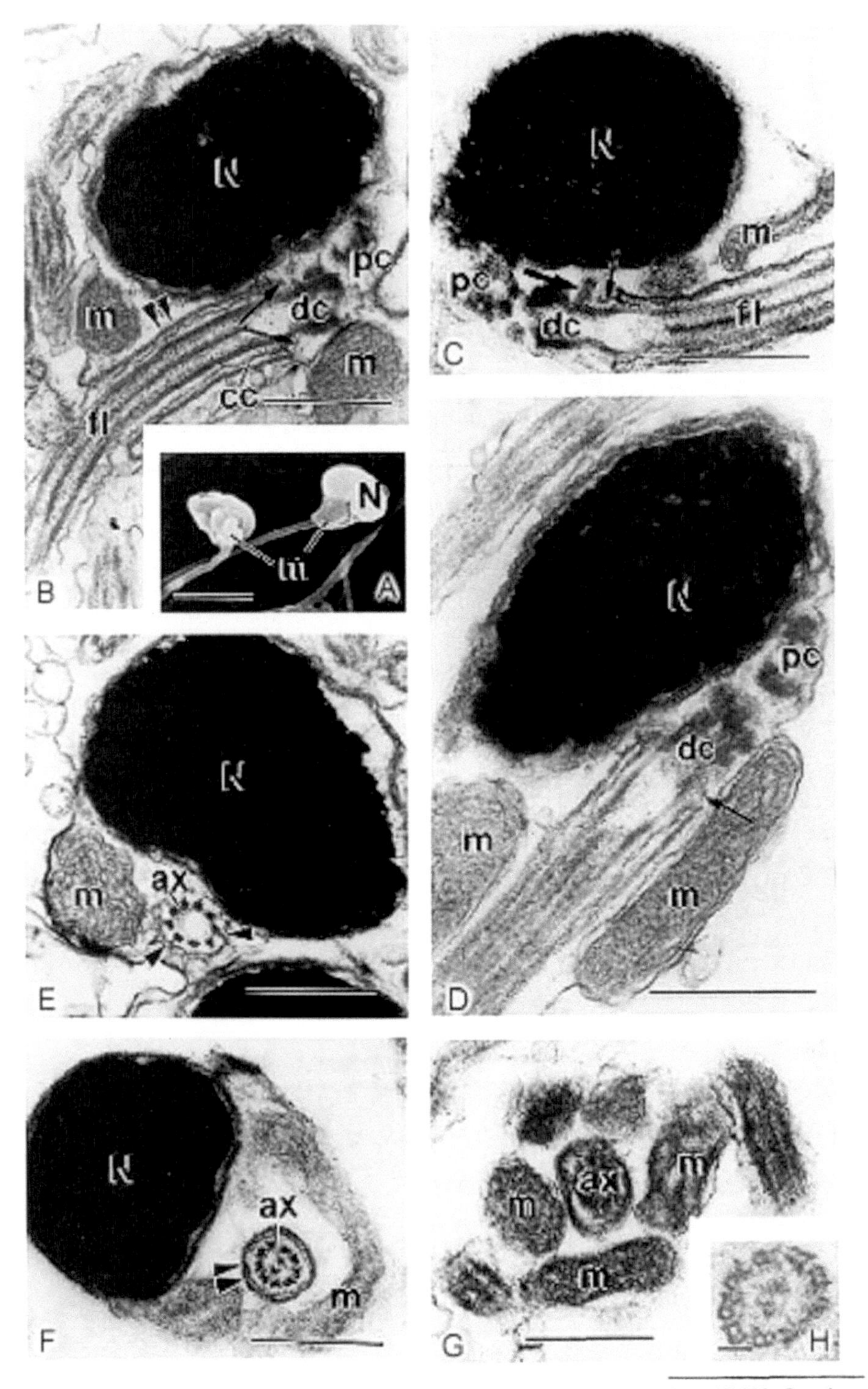

Fig. 15.109 Contd. ...

australasicus (Hara and Okiyama 1998) and six in *T. thynnus*. This character could be useful in scombrid taxonomy (Abascal *et al.* 2002).

15.7.9.4 Family Istiophoridae

These are the tropical and subtropical billifishes. The spermatozoon of *Istiophorus albicans (=I. americanus)*, an externally fertilizing pelagic spawner, is diagrammatically illustrated but not otherwise described by Mattei (1991). It is depicted as having a round nucleus, mutually perpendicular centrioles, two tiers of mitochondria and the flagellum in the long axis of the sperm and is thus excluded from the perciform Type II.

15.7.10 Suborder Stromateoidei

15.7.10.1 Family Ariommatidae

These are the tropical and subtropical ariommatids. Mattei (1991) ascribed the sperm of *Ariomma melanum* to the perciform [Type II] sperm.

15.7.11 Suborder Anabantoidei

Family Anabantidae

Anabantids, externally fertilizing non-guarding egg scatterers, are the climbing gouramies. Mattei (1991) ascribed the sperm of *Ctenopoma* sp. to the perciform [Type II] sperm.

15.7.11.1 Family Osphronemidae

These are the freshwater gouramies, in the Indian subcontinent and Southeast Asia.

Macropodus opercularis: The Paradise fish; is externally fertilizing and the male guards the fertilized eggs in a bubble-nest. In *M. opercularis* the sperm head consists of an oval nucleus, with condensed chromatin and a moderate fossa containing the centrosome. A cytoplasmic collar containing several small mitochondria surrounds the base of the 9+2 flagellum (Lee *et al.* 2006). From illustrations this appears to be a Type I spermatozoon although only the proximal centriole lies within the basal nuclear fossa.

Fig. 15.109 Contd. ...

Fig. 15.109 *Euthynnus alletteratus* sperm (Scombridae). **A**. SEM of two spermatozoa. **B**-**H**. TEM of sagittal (**B**,**D**), frontal (**C**), and transverse (**E**-**H**) sections at various levels of the spermatozoon, including the nuclear (**E**,**F**), mitochondrial (**G**), and flagellar (**H**) regions. ax, axoneme; cc, cytoplasmic canal; dc, distal centriole (basal body); fl, flagellum; m, mitochondria; N, nucleus; pc, proximal centriole; arrow-head, Y-shaped bridges at the proximal region of the axoneme; double arrow-head, dense layer underlying the cytoplasmic canal membrane; large arrow, lateral plate; small arrow, radial fibers projecting from the base of the distal centriole. Scale bars: 0·5 µm, except in (**A**) 2 µm, and (**H**) 50 nm. From Abascal, F. J., Medina, A., Megina, C. and Calzada, A. 2002. Journal of Fish Biology 60: 147-153, Fig. 2. With permission of Blackwell Publishing.

15.8 ORDER PLEURONECTIFORMES

Pleuronectiforms include over 500 species of flatfish, flounders and soles, diagnosed as monophyletic by absence of bilateral symmetry in the adults owing to location of both eyes on one side. The body is highly compressed and they lie on the eyeless side (Lauder and Liem, 1983; Nelson 2006).

15.8.1 Suborder Pleuronectoidei

15.8.1.1 Family Citharidae

These are the large scale flounders. The spermatozoon of *Citharus liguatula (=C. macrolepidotus)* was illustrated in a TEM micrograph by Mattei (1969) (Fig. 15.110) and diagrammatically by Mattei (1991). It has an inverted U-shaped

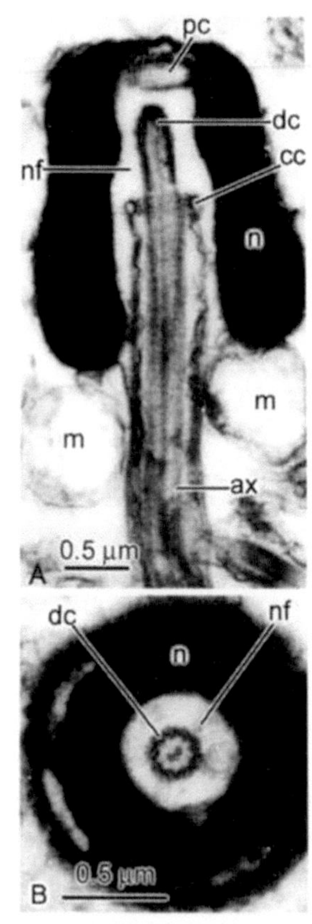

Fig. 15.110 *Citharus linguatula* (Citharidae), Atlantic spotted flounder **A**. Sagittal section of spermatozoon. **B**. Transverse section through the nucleus at the level of the distal centriole with its nine triplets. ax, axoneme; cc, cytoplasmic canal; dc, distal centriole; m, mitochondrion; n, nucleus; nf, nuclear fossa; pc, proximal centriole. Adapted from Mattei, X. 1969. *Contribution à l'étude de la spermiogenèse et des spermatozoïdes de poissons par les méthodes de la microscopie électronique.* D.Sc. thesis, Université de Montpellier, Fig. XXIIc,d. With permission.

nucleus with a very deep basal fossa, in which lie the two centrioles, mutually at right angles near the apex of the fossa, and the base of the flagellum. There is a single tier of mitochondria flanking a long cytoplasmic canal which commences at approximately half of the length of the fossa.

15.8.1.2 Family Scophthalmidae

Psetta maximus: The Turbot, *Psetta maxima* (=*Scophthalmus maximus*), is an externally fertilizing sequential spawner. Its spermatozoon has been described by Suquet *et al.* (1993, 1994) whose accounts are summarized here (Fig. 15.111A-C). It is about 43-45 µm long. The head, lacking an acrosome, is approximately spherical, measuring about 1.7 µm at maximum diameter and 1.3 µm in length. A deep depression in the anterior end of the nucleus does not appear to be an artifact as it is present in SEM (Fig. 15.111B) and TEM (Fig. 15.111C) micrographs. There is also a deep basal nuclear fossa so that the nucleus appears bilobed. Both centrioles lie in the basal (implantation) fossa. The proximal centriole, said to be approximately perpendicular to the distal, appears to be tandem in the long axis of the sperm (Fig. 15.111C). The distal centriole has dense projections connecting it with the plasma membrane. Eight to, frequently, ten mitochondria form a ring around the cytoplasmic canal (Fig. 15.111A). The 9+2 flagellum is about 43 µm long; proximally (for about 1 µm) circular in transverse section, at ca 0.2 µm wide. It is said to project from a "muff' which is presumably the border of the cytoplasmic canal. In its midregion the flagellum has "widespread lateral ridges", giving it a width of ca 0.6 µm. Distally it decreases abruptly in diameter.

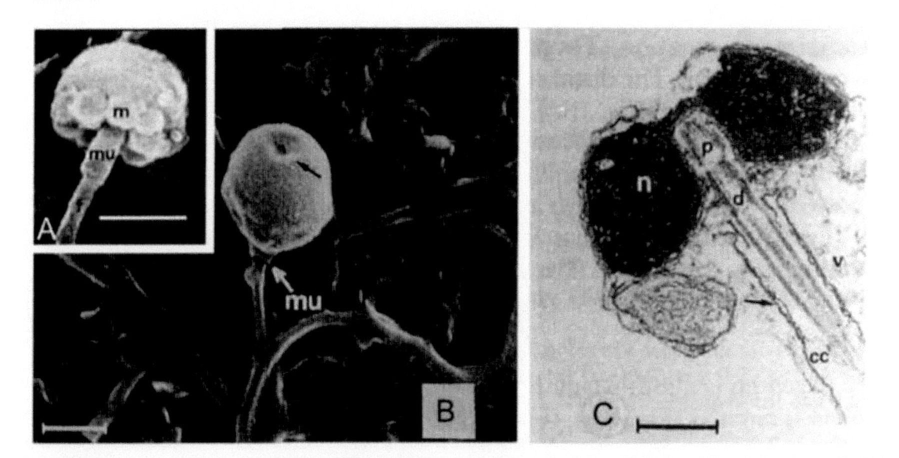

Fig. 15.111 *Psetta maxima* (Scophthalmidae) spermatozoa, Turbot. **A**. SEM showing ring of mitochondria. Bar 1 µm. **B**. SEM showing depression (arrow) at apex of head. Bar 1 µm. **C**. TEM longitudinal section of the head and midpiece, showing the anterior depression and deep basal fossa, giving the nucleus a bilobed appearance. Bar 0.5 µm. cc, cytoplasmic canal; d, distal centriole; m, mitochondrion; n, nucleus; v, vesicles in midpiece. Adapted from Suquet, M., Dorange, G., Omnes, M. H., Normant, Y., Le Roux, A. and Fauvel, C. 1993. Journal of Fish Biology 42: 509-516, Fig. 1b,c,d. With permission of Blackwell Publishing.

15.8.1.3 Family Paralichthyidae

Paralichthyids are the sand flounders.

***Paralichthys olivaceus*:** The spermatozoon of *P. olivaceus* (termed the Lefteye Flounder) described and illustrated by Hara and Okiyama (1991) (Fig. 15.112) (see also Wang *et al.* 1999, 2002) has a spherical nucleus with an apical 'dip' or hiatus in the chromatin. The chromatin forms dense masses. The two centrioles are mutually perpendicular, with the proximal displaced slightly relative to the long axis of the distal centriole. The chromatin tightly envelops the proximal centriole as a narrow nuclear fossa. A single ring of small spherical mitochondria surround the basal region of the flagellum (Hara and Okiyama 1991). "Wing-like lateral fins" are present (Wang *et al.* 2002).

The apical hiatus in the chromatin suggests close relationship with *Scophthalmus*).

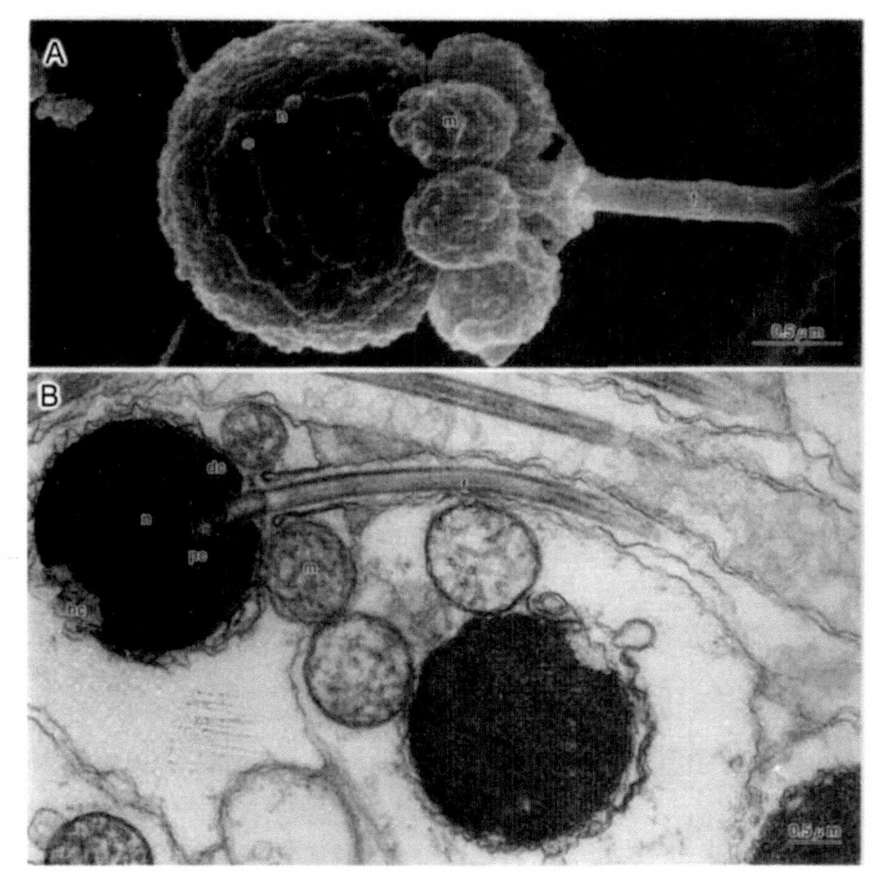

Fig. 15.112 *Paralichthys olivaceus* (Paralichthyidae) spermatozoa, Bastard Halibut. **A**. SEM of the nucleus and midpiece region **B**. Above, TEM longitudinal section of a spermatozoon. Adapted from Hara, M. and Okiyama, M. 1998. Bulletin of the Ocean Research Institute, University of Tokyo 33: 1-138, Fig. 62 B and C. With permission.

15.8.1.4 Family Pleuronectidae

Pleuronectids are the righteye flounders. They are externally fertilizing non-guarding egg scatteres. Sperm ultrastructure has been described for *Platichthys flesus* (Jones and Butler 1988), *Eopsetta grigorjewi* (An *et al.* 1999), *Pseudopleuronectes (=Limanda) yokohamae*, Marbled Sole, *Pseudopleuronectes (=Limanda) herzensteini*, Brown Sole and *Platichthys stellatus*, Starry Flounder (Chang and Chang 2002).

Platichthys flesus, the flounder, like all pleuronectiforms, has an anacrosomal aquasperm (Retzius 1905, light microscopy; Jones and Butler 1988, TEM) (Figs. 15.113, 113).

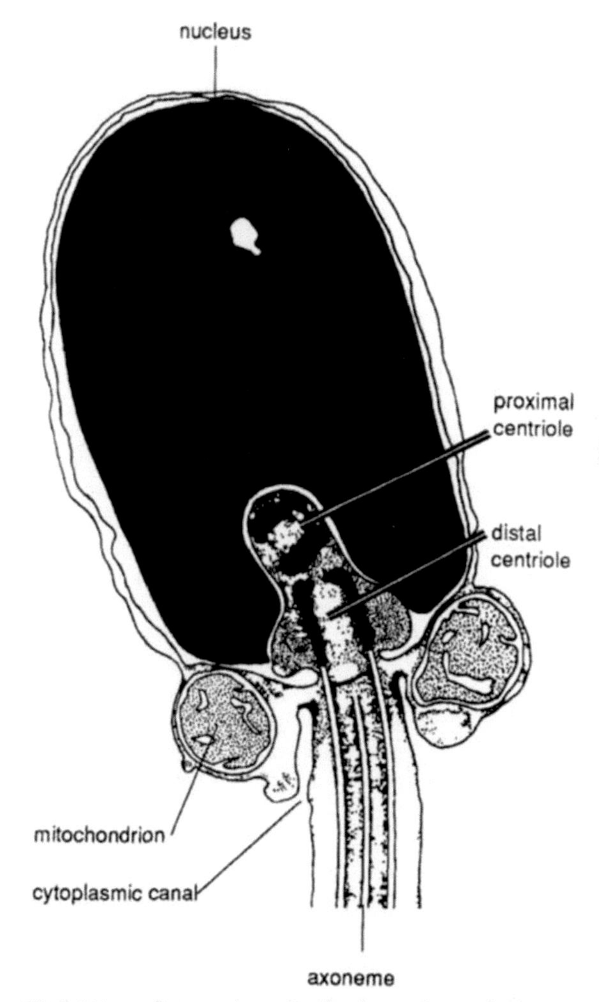

Fig. 15.113 *Platichthys flesus.* Longitudinal section of the spermatozoon. a, axoneme; co, cytoplasmic collar; dc, distal centriole; m, mitochondrion; n, nucleus; pc, proximal centriole. From Jamieson 1991, after micrographs by Jones, P. R. and Butler, R. D. 1988. Journal of Ultrastructure and Molecular Structure Research 98: 71-82, Figs. 8 and 12.

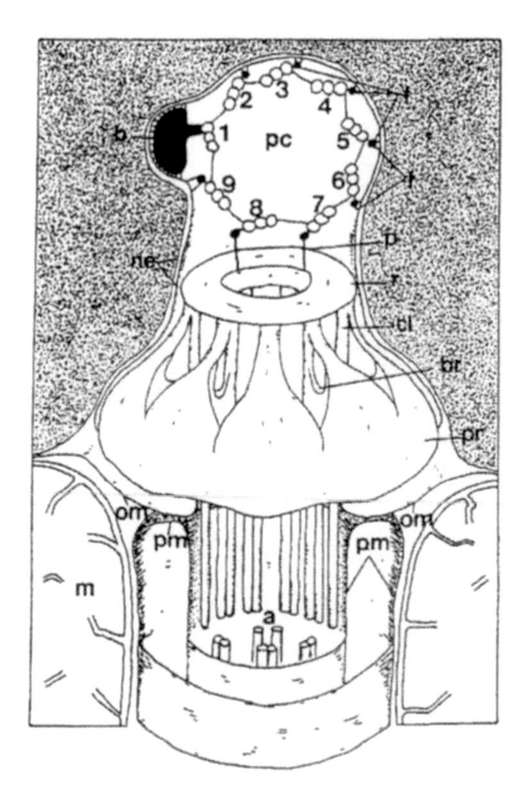

Fig. 15.114 *Platichthys flesus.* Schematic reconstruction of midpiece. The proximal centriole (pc), in the anterior region of the implantation fossa, has an associated mushroom-shaped body (b) which is the initial reference point for numbering the triplets (1-9) of the centriole. Triplets 2-9 have associated fibers (f). Fibers 2-6 and 9 are linked by short projections (l) to the nuclear envelope (ne). Fibers 7 and 8 are linked by longer projections (p) to the anterior ring (r) of the collar. This gives rise to 9 branched columns (cl) associated with the triplets of the distal centriole. Posteriorly a bridge (br) links the two branches, leaving an electron-lucent window, and the outer branches [satellite rays] thicken to form a continuous posterior ring (pr). The posterior surface of the collar is bounded by a loop of the outer membrane (om) of the nuclear envelope. This separates the collar from the apex of a similar loop of the plasma membrane (pm) which itself separates the axoneme (a) and the mitochondria (m). From Jones, P. R. and Butler, R. D. 1988. Journal of Ultrastructure and Molecular Structure Research 98: 71-82, Chart 1.

The bullet-shaped, homogeneously electron-dense nucleus has a deep basal fossa. The anterior surfaces of 8 spherical cristate mitochondria form a ring within shallow depressions of the caudal surface of the nucleus. The two centrioles (Fig. 15.113) are located within the basal fossa (Type I arrangement). Both are embedded in and linked by pericentriolar material which is intimately associated with the nuclear envelope. Around the proximal centriole the material is organized into a dense ring that bears nine fibers associated with the triplets, and a large fibrous body, all connected to the nuclear envelope. The material around the distal centriole is organized into a complex collar. The 9+2 flagellum is separated from the midpiece by a short cytoplasmic canal (Jones and Butler, 1988), and (e.g. Fig. 15.113) appears to be expanded but whether as true fins is uncertain.

Eopsetta grigorjewi: The sperm of the Roundnose Flounder has homogeneous chromatin, a cytoplasmic collar containing seven mitochondria; a 9+2 axoneme; and well-developed lateral flagellar fins (An *et al.* 1999).

Limanda, Platichthys* and *Paralichthys: The fresh sperm of *Limanda yokohamae*, Marbled Sole (Fig. 15.115A,B), *L. herzensteini*, Brown Sole (Fig. 15.116A-C) and *Platichthys stellatus*, Starry Flounder (Fig. 15.116D,E) are ellipsoidal and that of *Paralichthys olivaceus*, Olive Flounder (Fig. 15.116F,G) have been described and illustrated in a study of the effects of cryopreservation by Chang and Chang (2002). The heads of the sperm are

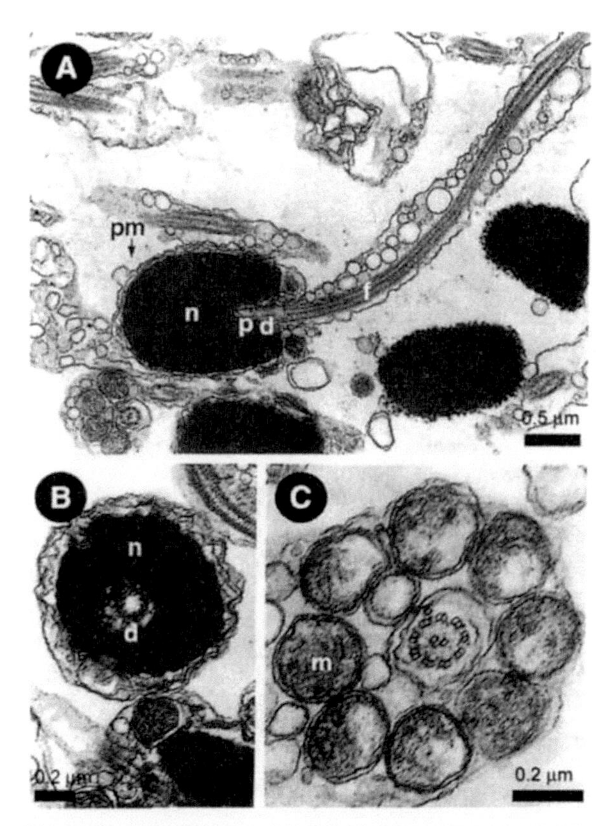

Fig. 15.115 *Limanda yokohamae*, Marbled Sole (Pleuronectidae). **A**. Sagittal section of a spermatozoon. **B**. Transverse section of the nucleus and distal centriole. **C**. Transverse section of the midpiece, showing the eight mitochondria. d, distal centriole; f, flagellum; m, mitochondrion; n, nucleus; p, proximal centriole; pm, plasma membrane. After Chang, Y. J. and Chang, Y. J. 2002. Journal of Fisheries Science and Technology 5: 87-96, Fig. 1.

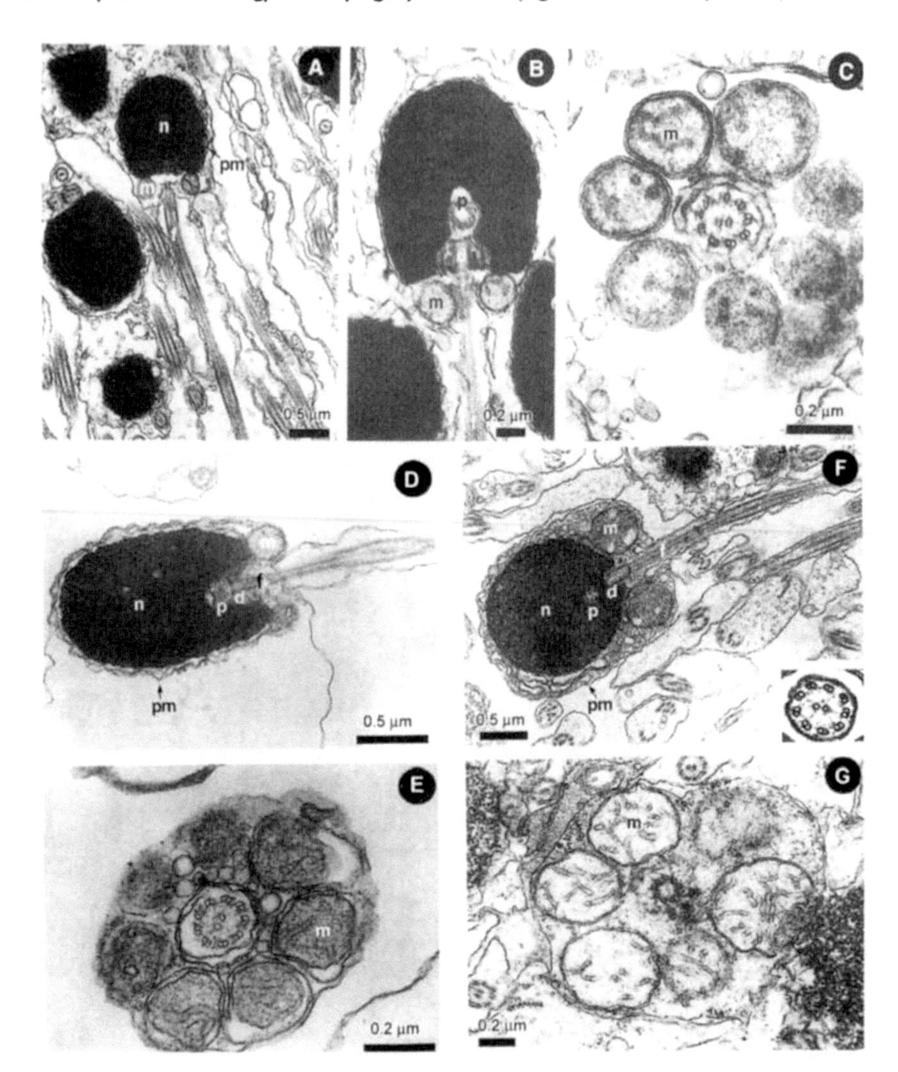

Fig. 15.116 Spermatozoa of Pleuronectidae, continued. **A-C**. *Limanda herzensteini*. **A**. Sagittal section of spermatozoon. **B**. Sagittal section of spermatozoon, showing proximal and distal centrioles. **C**. Transverse section (TS) of midpiece, showing the seven mitochondria. **D, E**. *Platichthys stellatus*. **D**. Sagittal section of spermatozoon. **E**. TS of midpiece, showing the seven mitochondria. **F, G**. *Paralichthys olivaceus*. **F**. Sagittal section of spermatozoon, with (insert) TS axoneme. **G**. TS of midpiece, showing the six mitochondria. Adapted from Chang, Y. J. and Chang, Y. J. 2002. Journal of Fisheries Science and Technology 5: 87-96, Figs. 3 and 4.

ellipsoidal excepting that of the Olive Flounder, which is round. The lengths and widths of heads in the four species are 1.30~1.73 µm (1.53 ± 0.18 µm) and 0.85~1.22 µm (1.05 ± 0.16 µm) in Marbled Sole, 1.23~1.42 µm (1.32 ± 0.08 µm) and 1.15~1.20 µm (0.17 ± 0.02 µm) in Brown Sole, 1.22~1.44 µm (1.35 ± 0.09 µm) and 0.89~0.97 µm (0.92 ± 0.03 µm) in Starry Flounder, 1.25~1.67 µm (1.52

± 0.16 μm) and 1.15~1.63 μm (1.40 ± 0.17μm) in Olive Flounder, respectively. The chromatin is compact and homogeneous, except in the Olive Flounder in which it is granular and forms non-homogeneous clumps. The nucleus in each species has a basal fossa housing the mutually perpendicular proximal centriole and distal centriole (basal body). The numbers of independent mitochondria encircling the base of the flagellum in the midpiece is eight in Marbled Sole, seven in Brown Sole and Starry Flounder and six in Olive Flounder. The flagellum has the typical 9+2 axoneme structure in the four species (Chang and Chang 2002). From micrographs it appears that the axoneme is surrounded by a wide layer of vacuolated cytoplasm but that lateral flagellar fins may be absent.

15.8.1.5 Family Bothidae

Bothids are the lefteye flounders, in the Atlantic, Indian and Pacific Oceans. Mattei (1991) examined but did not describe the sperm of *Chascanopsetta lugubris*. Because of the rarity of the material a spermatid is illustrated here (Fig. 15.117). Condensation of the chromatin is commencing around the basal nuclear fossa and the mitochondria are still scattered posterior to the nucleus.

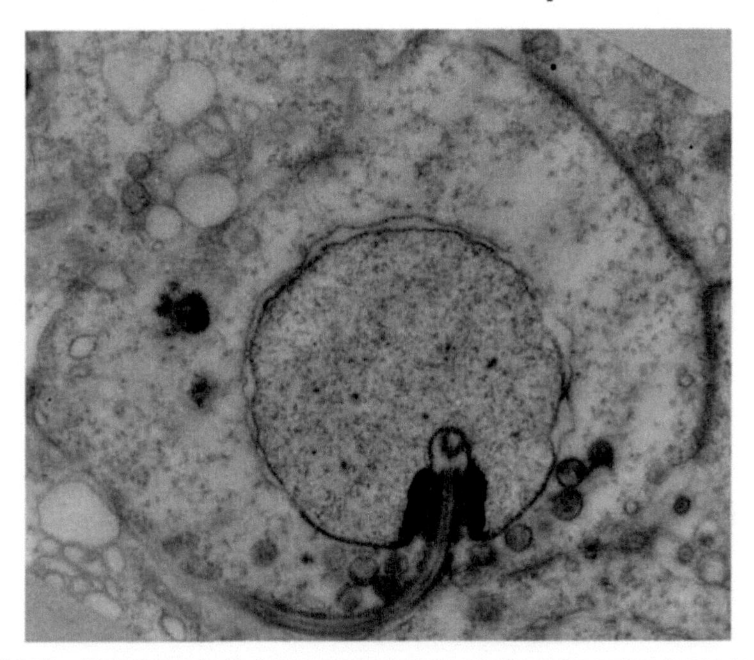

Fig. 15.117 *Chascanopsetta lugubris* (Bothidae), Pelican flounder. Sagittal section of a late spermatid. Original, from an unpublished TEM negative courtesy of Professor Xavier Mattei.

15.8.1.6 Family Soleidae

The soles. Mattei (1991) listed the sperm of *Synaptura* sp. The sperm of *Pegusa triophthalmus* was briefly described by Mattei (1970, 1991). That of *Solea senegalensis* was illustrated diagrammatically by Mattei (1991) and described in detail by Medina *et al.* (2004). Vallisneri *et al.* (2001) present a light microscope plate of the sperm of *Solea solea* (adult, Fig. 15.118C); their ultrastructural micrographs of the eggs show numerous pores which have the appearance of multiple micropyles.

***Solea senegalensis*:** The diagrammatic illustration of the spermatozoon of *Solea senegalensis* (Mattei 1991) (Fig. 15.118A,B) accords with the later description by Medina *et al.* (2004) (Fig. 15.118A, B). It has a total length of approximately 44 μm. The ovoid head measures 2.25 ± 0.12 μm in length (n = 7) and 1.58 ± 0.12 μm in width (n=6) and the tail 41.30 ± 0.82 μm (n = 4).

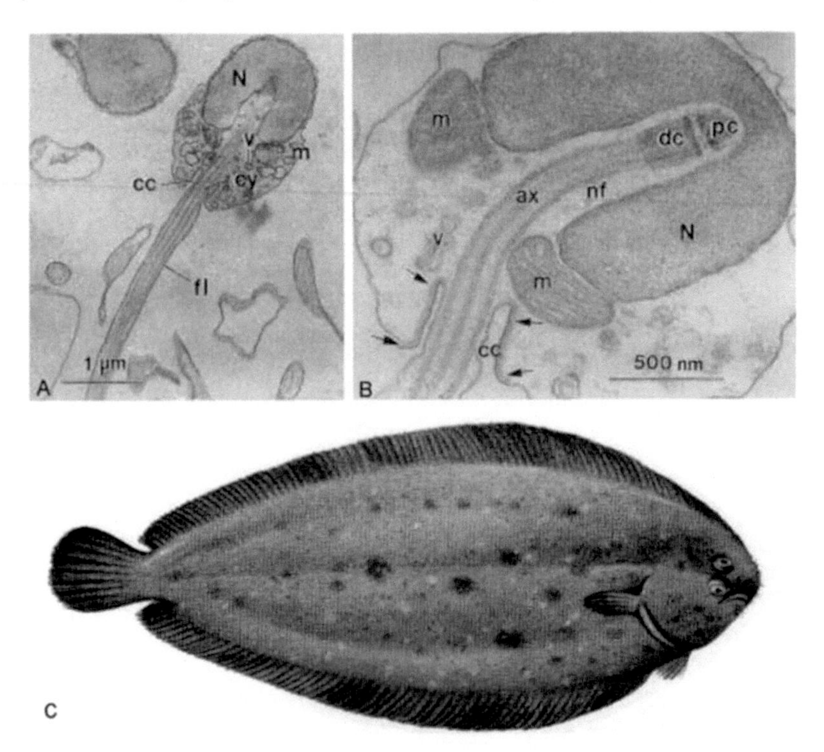

Fig. 15.118 *Solea* (Soleidae). **A**. *Solea senegalensis* Senegalese sole. **A** and **B**. Longitudinal sections of the spermatozoon show a homogeneous nucleus (N) with a deep nuclear fossa (nf) where the coaxially arranged proximal and distal centrioles (pc and dc) are housed. The distal centriole (basal body) gives rise to the axoneme (ax) of the flagellum (fl). A ring of mitochondria (m) with parallel cristae and dense matrix lies at the base of the nucleus and is enclosed in a conspicuous mass of cytoplasm (cy) that contains numerous vesicles (v). The cytoplasmic mass is separated from the flagellum by a shallow cytoplasmic canal (cc) where a three-layered, membrane-like structure is apposed to the plasma membrane (arrows in B). Beyond the cytoplasmic canal, the flagellar membrane separates from the axoneme to form the paired lateral fins. Adapted from Medina, A., Megina, C., Abascal, F. J. and Calzada, A. 2000. Journal of Submicroscopic Cytology and Pathology 32: 645-650, Figs. 3 and 4. **C**. *Solea solea*, European sole. By Norman Weaver. With kind permission of Sarah Starsmore.

The nucleus has homogeneous chromatin, lacks nuclear pores, and has a pronounced caudal fossa (1,032. ± 68.0 nm long by 309.7 ± 43.4 nm wide, n = 5) through which the axoneme penetrates deeply to the anterior end of the nucleus. The caudal region of the head is occupied by the bulk of the cytoplasm, which contains numerous electron-clear vesicles and several mitochondria distributed in a non-protuberant ring around the opening of the nuclear fossa and comprising the midpiece. Some of the smaller vesicles penetrate into the nuclear fossa around the axoneme and centrioles. The mitochondria are rather irregular in shape and more than eight mitochondrial profiles often occur in transverse sections of the midpiece. The two centrioles are arranged coaxially, in tandem (confirming Mattei 1991). The proximal centriole is shorter (~135 nm long) than the distal centriole (~287 nm long), both showing a dense plate on the adjacent surfaces. Transverse sections of the centrioles do not reveal the usual triplet microtubular structure, since the microtubules are embedded in homogeneous material, thus forming a continuous hollow cylinder or short tube. There is no evidence of pericentriolar anchoring structures. The inner membrane of the cytoplasmic canal is lined internally by a layer of dense material. A pair of lateral fins gives the tail a maximum width of ~1.1 μm. Sometimes only one lateral fin is visible in cross-sectioned tails; in the region of the cytoplasmic canals and at the extremity of the tail no fins are seen (Medina *et al.* 2000).

Pegusa triophthalmus, a soleid, also has a very deep nuclear fossa (Fig. 15.119), resembling that of *Solea senegalensis* and also the mullid *Pseuupneus (=Upeneus) prayensis* (Fig. 15.25) in this respect and in location of the proximal centriole in the same longitudinal axis and the basal body. However, in a later diagram (Mattei 1991) depicted the proximal centriole as oblique to the basal body. It is a classical Type I sperm *sensu* Mattei (1970).

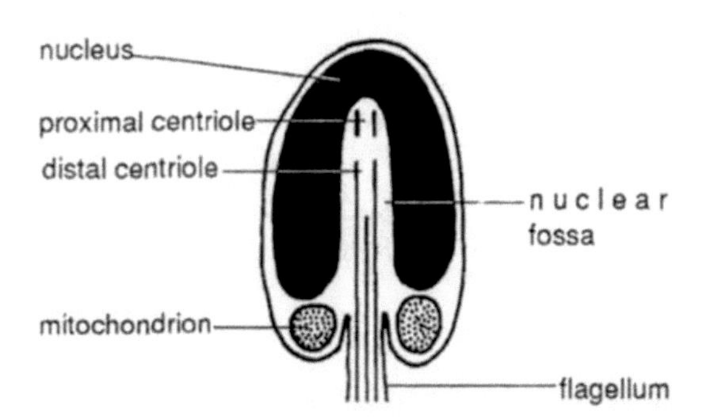

Fig. 15.119 *Pegusa triophthalmus* (Soleidae) spermatozoon. After Mattei, X. 1970. Pp. 567-569. In B. Baccetti (ed.), *Comparative Spermatology*, Academic Press, New York, Fig. 4: 13.

Remarks: As shown above, spermatozoal ultrastructure has been examined in representatives of the families Citharidae (*Citharus linguatula*); Scophthalmidae (*Scophthalmus maximus*); Paralichthyidae (*Paralichthys olivaceus*); Pleuronectidae (*Platichthys flesus, Eopsetta grigorjewi, Limanda yokohamae, L. herzensteini* and

Platichthys stellatus); Bothidae (*Chascanopsetta lugubris*) and Soleidae (*Synaptura* sp., *Solea solea, S. senegalensis* and *Pegusa triophthalmus*). The considerable diversity of sperm ultrastructure shown in the examined pleuronectiforms is summarized in Figure 15.120, below.

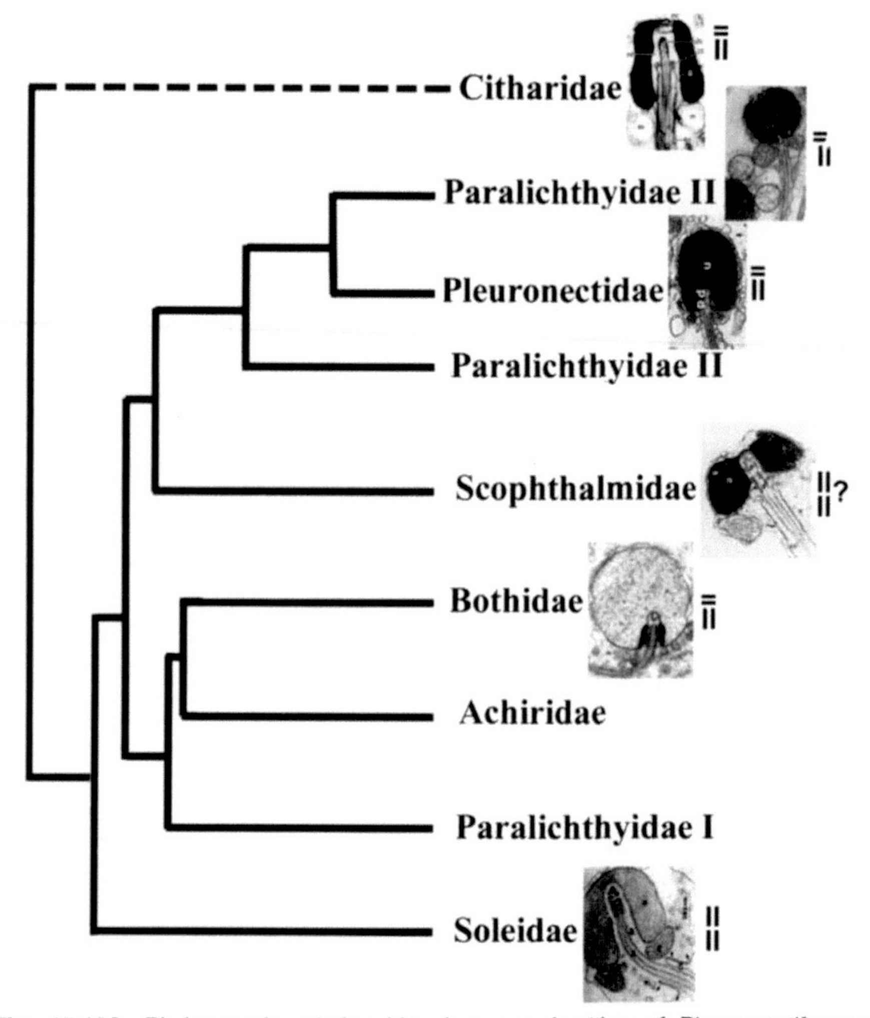

Fig. 15.120 Phylogenetic relationships between families of Pleuronectiformes, based on the analysis of mitochondrial 16s rDNA sequences by Pardo, B., Machordom, A., Foresti, F., Porto-Foresti, F., Azevedo, F. C., Banón, R., Sánchez, L. and Martínez, P. 2005. Scientia Marina 69: 531-543, Fig.3. Thumb-nail micrographs of sperm have been added from the present volume. Orientation of the proximal and distal centriole is indicated to the right of each micrograph. The Citharidae, widely regarded as the sister group of the families analyzed, have here been added at the base of the dendrogram. Bothidae are represented by a spermatid and their mature sperm morphology is unknown. Original.

A striking feature is the very deep nuclear fossa, leaving only a thin layer of chromatin at the tip of the nucleus, in Citharidae and Soleidae. The two centrioles and the base of the axoneme are contained in this fossa, with the proximal centriole close to the tip of the nucleus. As these two families appear to be successive basal groups in the dendrogram (Fig. 15.120) it is here deduced that this sperm morphology is plesiomorphic for the Pleuronectiformes. It is noteworthy, as also observed by Medina *et al.* (2000), that this morphology is also seen in the Mullidae, (exemplified by *Pseudupeneus (=Upeneus) prayensis* and *Mullus barbatus*) (Mattei 1970; Lahnsteiner *et al.* 1995; Lahnsteiner and Patzner 1998) and some tetraodontiforms, e.g. the Balistidae, in both of which the two centrioles are in the unusual coaxial, tandem, clearly apomorphic, configuration that occurs in the Soleidae (the centrioles are mutually perpendicular in citharids).

If it be accepted that the bell-shaped nucleus with its deep fossa in the Citharidae and Soleidae, which we may term the mulliform condition, is plesiomorphic for pleuronectiforms, though in itself apomorphic relative to basic teleostean anacrosomal aquasperm, the following trends may be discerned in other families (see Fig, 15.120). In a scophthalmid through paralichthyid clade, the basal *Scophthalmus* has retained the deep nuclear fossa but the nucleus has become greatly widened and depressed longitudinally and its chromatin is eroded apically. The centrioles appear from an illustration to retain the mutually perpendicular condition seen in all investigated pleuronectiforms except soleids. The Scophthalmidae is the sister of the Pleuronectidae and the *P. olivaceus* section of the triphyletic Paralichthyidae in the dendrogram (Fig. 15.120). Pleuronectidae have retained the plesiomorphic, mutually perpendicular arrangement of the centrioles; the nuclear fossa is less deep and the surrounding chromatin much wider. Their sister group, Paralichthyidae (*sensu* II of Pardo *et al.* 2005), in the clade represented by *Paralichthys olivaceus*, have reverted to a rounded nucleus, with small fossa, typical of the teleostean anacrosomal aquasperm. However, in *P. olivaceus*, the apical region of the chromatin is indented (Hara and Okiyama 1998) though not in the example of this species illustrated by Chang and Chang (2002). This rare condition of erosion of the apical chromatin seen, more highly developed, in *Scophthalmus maximus*, is presumably a homoplasy, albeit labile. Such parallelism might nevertheless be due to relationship, a paramorphy *sensu* Jamieson (1984, 1991.

In the sister of the above clade (Fig. 15.120), only a bothid has been studied and the two other members have not been investigated for sperm ultrastructure. Little can be said of bothid sperm as we know only a spermatid of *Chascanopsetta lugubris* in which nuclear condensation is only commencing. It seems likely that the nuclear fossa does not extend more deeply than half way into the nucleus at maturity.

The above interpretation of spermatozoal evolution is dependent on the cladogram used. Hoshino (2001) has provided a cladogram, based on somatic characters, in which *Citharus* and the Soleidae are advanced groups, though in separate clades. If this were accepted, the mulliform sperm morphology in pleuronectiforms would appear to be derived, a view perhaps countered by its appearance in the Mullidae. Hoshino, again, groups the Scophthalmidae in the same clade as the Pleuronectidae but differs in placing the Bothidae in this clade. In this interpretation the simple sperm of the Paralichthyidae would be a plesiomorphic retention. There appears to be widespread agreement that the Psettodidae, which have unfortunately not been examined for sperm ultrastructure, are the sister group of all other pleuronectiforms. It should be added that the sperm of the cyprinodontoid *Fundulus heteroclitus* is also of the "mulliform" type, with perpendicular centrioles as in *Citharus*, and the possibility has to be considered that this sperm type is a homoplasic adaptation to peculiar fertilization biology, though the nature of this is not known beyond its being external fertilization.

As Medina *et al.* (2000) have stated, the alternative arrangements of centrioles justify the phylogenetic separation of 'soles' (suborder Soleoidei) from 'flounders' (suborder Pleuronectoidei).

The presence of lateral fins is considered by Medina *et al.* (2000) to be common to all pleuronectiforms and was clearly demonstrated for *Solea senegalensis* by those authors. However, the "widespread lateral ridges" in *Scophthalmus* (Suquet *et al.* 1993) are questionably sufficiently well defined to be termed flagellar fins and the expansion of the flagellum in *Platichthys* (Jones and Butler 1988) is also not definitely fin-like. Nevertheless the expansions of the flagellum illustrated in longitudinal section for *Paralichthys olivaceus* by Hara and Okiyama (1998) are confirmed as wing-like lateral fins by Wang *et al.* (2002). There are few examples of cross sections of pleuronectiform flagella in the literature and further examples are needed in order to establish the distribution of flagellar fins.

A further issue is the suggestion from sperm ultrastructure that the Tetraodontiformes may be derived from a common ancestor with the Pleuronectiformes as balistoides (see below) have the mulliform sperm morphology, as also noted by Medina *et al.* (2000).

15.9 ORDER TETRAODONTIFORMES

The Tetraodontiformes contain about 320 species of mostly shallow water, circumtropical and subtropical marine puffers, porcupine fish and their relatives.

Tetraodontiformes (plectognaths) (Fig. 15.121) show striking examples of extreme reductive evolution and represent one of the main end lines of teleost radiation. Monophyly is suggested by a suite of somatic synapomorphies (Lauder and Liem 1983). Their sister-group is possibly the Acanthuroidei among the Perciformes from which they may have been derived by paedomorphosis (Patterson 1964; Tyler 1968; Winterbottom 1974; Lauder and Liem 1983). The phylogeny of Lauder and Liem (1983) differs considerably from a molecular phylogeny by Holcroft (2005) (Fig. 15.131) which, nevertheless, confirms their monophyly.

Fig. 15.121 Examples of Tetraodontiformes. Above: A puffer, *Tetractenos hamiltoni* (Tetraodontidae). Below: A porcupine fish, *Diodon holocanthus* (Diodontidae). Photos Bruce Cowell. Courtesy of the Queensland Museum.

Living Tetraodontiformes include three suborders (Nelson 2006): the Triacanthodoidei (sperm ultrastructure unknown); Balistoidei and Tetraodontoidei. The Balistoidei include three superfamilies Triacanthoidea, Ostracioidea (sperm ultrastructure in both unknown) and Balistoidea with two families, the Balistidae (leatherjackets) and Monacanthinae (filefishes), for both of which sperm ultrastructure has been described. The Tetraodontoidei contains four families, of which the Tetraodontidae (puffers) and Diodontidae (porcupine fishes) (Fig. 15.121) have been examined for sperm ultrastructure.

15.9.1 Suborder Balistoidei

15.9.1.1 Families Balistidae and Monacanthidae

The sperm of the investigated balistids, *Balistes capriscus (=B. forcipatus)* (Mattei 1970), *Pseudobalistes fuscus* (Jamieson 1991) (Fig. 15.125A,B) and the investigated monocanthids, *Aluterus schoepfi (=A. punctatus)* (Mattei 1991 and this account) (Fig. 15.124) and the two investigated monocanthids *Rudarius ercodes* and *Brachaluteres ulvarum* (Hara and Okiyama 1998), externally fertilizing egg guarders, differ notably from the spermatozoon of the tetraodontid *Tetractenos (=Gastrophysus) hamiltoni* (adult, Fig. 15.121; sperm, Fig.15.125C-G), and other species of *Takifugu* (see below), in the deep penetration of the nuclear fossa, to give an inverted U-shaped nucleus. They accord with *Takifugu* in symmetrical arrangement of simple mitochondria in a ring around the cytoplasmic canal but differ in orientation of the two centrioles in the same longitudinal axis whereas they are mutually perpendicular in at least *Tetractenos (=Gastrophysus) hamiltoni*. However, the supposed tetraodontid *Ephippion*, the Prickly puffer (Fig. 15.129) has the deep nuclear fossa of balistids and monacanthids but the mutually perpendicular centrioles of *Takifugu*. As shown below (**Section 15.9.2.2**) the diodontid *Chilomycterus antennatus* (Fig. 15.122B) is intermediate in having the longitudinal centrioles of balistids but a shallower nuclear fossa as in *Takifugu*.

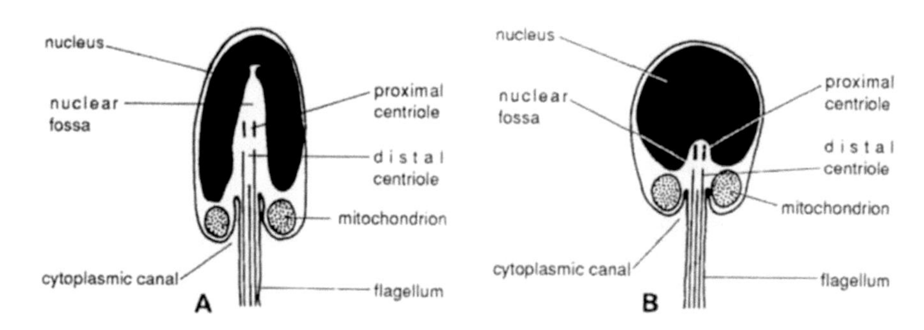

Fig. 15.122 Tetraodontiformes. **A.** *Balistes capriscus* (Balistidae), Grey triggerfish. **B.** *Chilomycterus antennatus* (Diodontidae), Bridled burrfish. Diagrammatic longitudinal sections of the spermatozoa After Mattei, X. 1970. Pp. 567-569. In B. Baccetti (ed.), *Comparative Spermatology*, Academic Press, New York, Fig. 4: (6 and 4).

Pseudobalistes fuscus: In the sperm of *P. fuscus* the nucleus, deeply penetrated by the implantation fossa, is approximately horseshoe-shaped in longitudinal section (Fig. 15.125A,B). It is 1.3-1.6, mean 1.5 µm long and 1.0-1.3, mean 1.1 µm, wide. The chromatin is electron-dense and compact though with some indication of approximation of large masses between which several lacunae are usually present. The profile of the nucleus at its envelope has corresponding indentations but is otherwise smooth. A small number of large, sparsely cristate, irregular, mutually adpressed mitochondria is grouped in a single layer around the cytoplasmic canal. In longitudinal section, the cytoplasmic collar continues as a short prolongation behind the mitochondria on each side (Fig. 15.125B). The cytoplasmic canal is approximately 0.6-0.8 μm long. The two centrioles lie within the anterior half of the deep implantation fossa (Fig. 15.125B). The proximal centriole is unusual not only in being in the same axis as the basal body but also having its longitudinal axis similarly orientated (the so-called coaxial condition); it is near but not at the anterior limit of the nuclear fossa. At least one, apparently the distal, centriole consists of 9 triplets (Fig. 15.125B, inset). The 9+2 flagellum has long lateral fins (Fig. 15.125A) the plane of which is slightly tilted relative to that of the two central singlets.

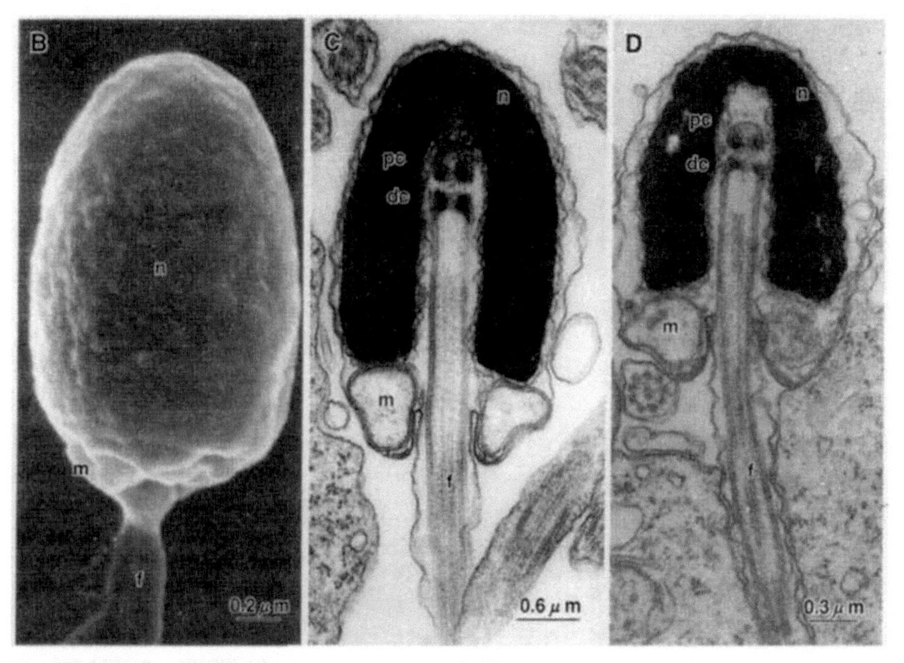

Fig. 15.123 A. SEM of a spermatozoon of *Rudarius ercodes* (Monacanthidae). **B.** TEM longitudinal section of same. **C.** TEM longitudinal section of a spermatozoon of *Brachaluteres ulvarum* (Monacanthidae). dc, distal centriole; f, flagellum; m, mitochondrion; n, nucleus; pc, proximal centriole; After Hara, M. and M. Okiyama 1998. Bulletin of The Ocean Research Institute, University of Tokyo 33: 1-138, Fig. 61B-D. With permission.

***Rudarius, Brachaluteres* and *Aluterus*:** The sperm of the monacanthids *Rudarius ercodes, Brachaluteres ulvarum* (Hara and Okiyama 1998) and *Aluterus* *schoepfi (=A. punctatus)* (Mattei 1991, and this account) have a bell-shaped nucleus with a very deep basal nuclear fossa.

The chromatin is in the form of dense masses with scattered pale lacunae in *Rudarius ercodes* (Fig. 15.123B) and, though, less frequent, in *Aluterus schoepfi* (Fig. 15.124A). The two centrioles lie in tandem, with axes longitudinal, in the fossa. The space anterior to the proximal centriole appears empty or, in *B. ulvarum* contains dense material. There is a single annular tier of separate small mitochondria, with indistinct cristae, at the base of the nucleus, numbering five in *A. schoepfi*. A moderately deep cytoplasmic canal is present. The sperm resemble those of the Balistidae in the deep nuclear fossa and the concentric, tandem (coaxial) orientation of the centrioles.

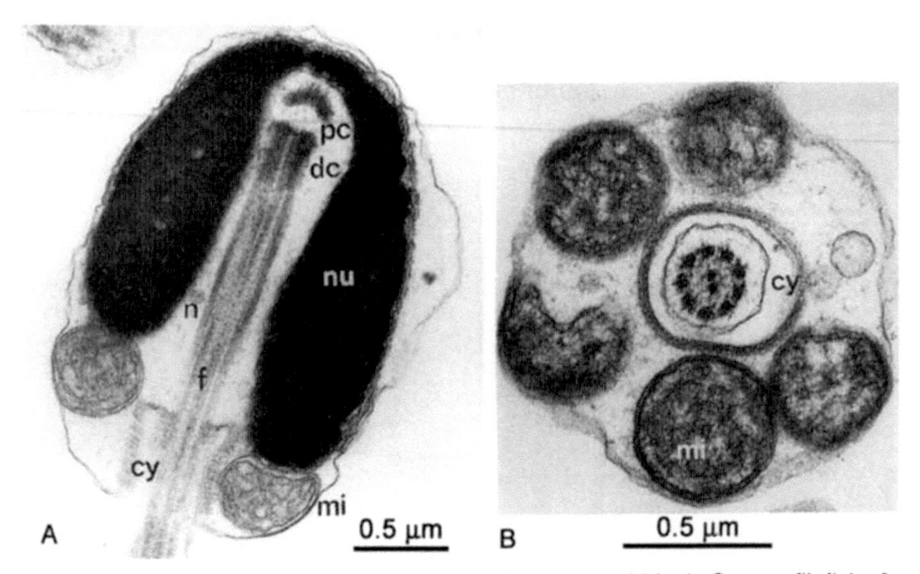

Fig. 15.124 *Aluterus schoepfi (=A. punctatus)* (Monacanthidae), Orange filefish. **A**. Longitudinal section of spermatozoon. **B**. Transverse sections of two midpieces and an adjacent nucleus. Original, from TEM micrographs courtesy of Professor Xavier Mattei.

15.9.2 Suborder Tetraodontoidei

15.9.2.1 Family Tetraodontidae

Tetradontids are the puffers. An adult is exemplified by *Tetractenos hamiltoni* (Fig. 15.121).

Sperm literature: The following tetraodontid species have been examined for sperm ultrastructure: *Ephippion guttifer* (Mattei 1991); *Lagocephalus laevigatus* (Mattei 1991); *Sphoeroides (=Sphaeroides) spengleri* (Mattei 1991 and this account); *Takifugu niphobles** (Gwo *et al.* 1993b; Miyaki and Yoshikoshi 1993; Hara and Okiyama 1998; Morisawa 2001); *Takifugu (=Fugu) obscurus* (Lu *et al.* 1999); *T. pardalis**; *T. poecilonotus**; *T. porphyreus*; *T. radiatus** (all Miyaki *et al.*

1996); *T. rubripes** (Miyaki *et al.* 1996; Chang *et al.* 1998; Zhang *et al.* 1999); and *T. xanthopterus** (Miyaki *et al.* 1996). Asterisked species were illustrated by scanning electron micrographs by Miyaki and Yoshikoshi (1993) (Fig. 15.126) who give a table of head and tail width dimensions, augmented in Table 15.6. According to these authors there are no significant differences (P>0.01) in any of the measured parts of the spermatozoa of the first six species listed. They are (all?) externally fertilizing egg scatterers.

Table 15.6 Spermatozoal dimensions (mostly mean ± SD) for seven species of Takifugu and Tetractenos.

Species	*Head length (μm)*	*Head width (μm)*	*Tail width (μm)*	*Reference*
T. niphobles	1 .92 ± 0.07	0.83 ± 0.02	0.18 ± 0.04	Miyaki and Yoshikoshi 1993
	2.5-2.7	1.0-1.4		Gwo et al. 1993b
	1.7	0.84		Hara and Okiyama 1998
	1.5	0.75		Morisawa, S. 2001
T. pardalis	2.01 ± 0.10	0.86 ± 0.04	0.21 ± 0.03	Miyaki and Yoshikoshi 1993
T. poecilonotus	1.90 ± 0.11	0.82 ± 0.04	0.18 ± 0.05	Miyaki and Yoshikoshi 1993
T. radiatus	2.05 ± 0.11	0.84 ± 0.04	0.21 ± 0.05	Miyaki and Yoshikoshi 1993
T. rubripes	1.97 ± 0.08	0.76 ± 0.03	0. 18 ± 0.02	Miyaki and Yoshikoshi 1993
T. rubripes	1.35 ± 0.03	0.65 ± 0.10		Chang, Y.J. et al. 1998
T. xanthopterus	1.94 ± 0.10	0.83 ± 0.03	0.16 ± 0.04	Miyaki and Yoshikoshi 1993
Te. hamiltoni	2.4	1.0		Jamieson 1991

***Tetractenos hamiltoni*:** The nucleus of the sperm of *Tetractenos (=Gastrophysus) hamiltoni* (adult, Fig. 15.121; sperm Fig. 15.125C-G) is unusually elongate for an externally fertilizing spermatozoon. It is approximately cylindrical, 2.4 μm long and 1.0 μm wide at its greatest (midlength) diameter, with slightly convex sides and a rounded tip. Basally it is indented by a club-shaped implantation fossa, 0.6 μm long and maximally 0.3 μm wide. The chromatin is coarsely granular with some pale lacunae.

The eight, spherical cristate mitochondria are symmetrically arranged in a ring around the proximal region of the flagellum, shortly behind the distal centriole (Fig. 15.125D). They are housed in a cytoplasmic collar maximally 0.6 μm long, which is separated from the flagellum by a periaxonemal space bounded by the invaginated plasma membrane, the so-called cytoplasmic canal. Immediately within the plasma membrane lining the canal, in

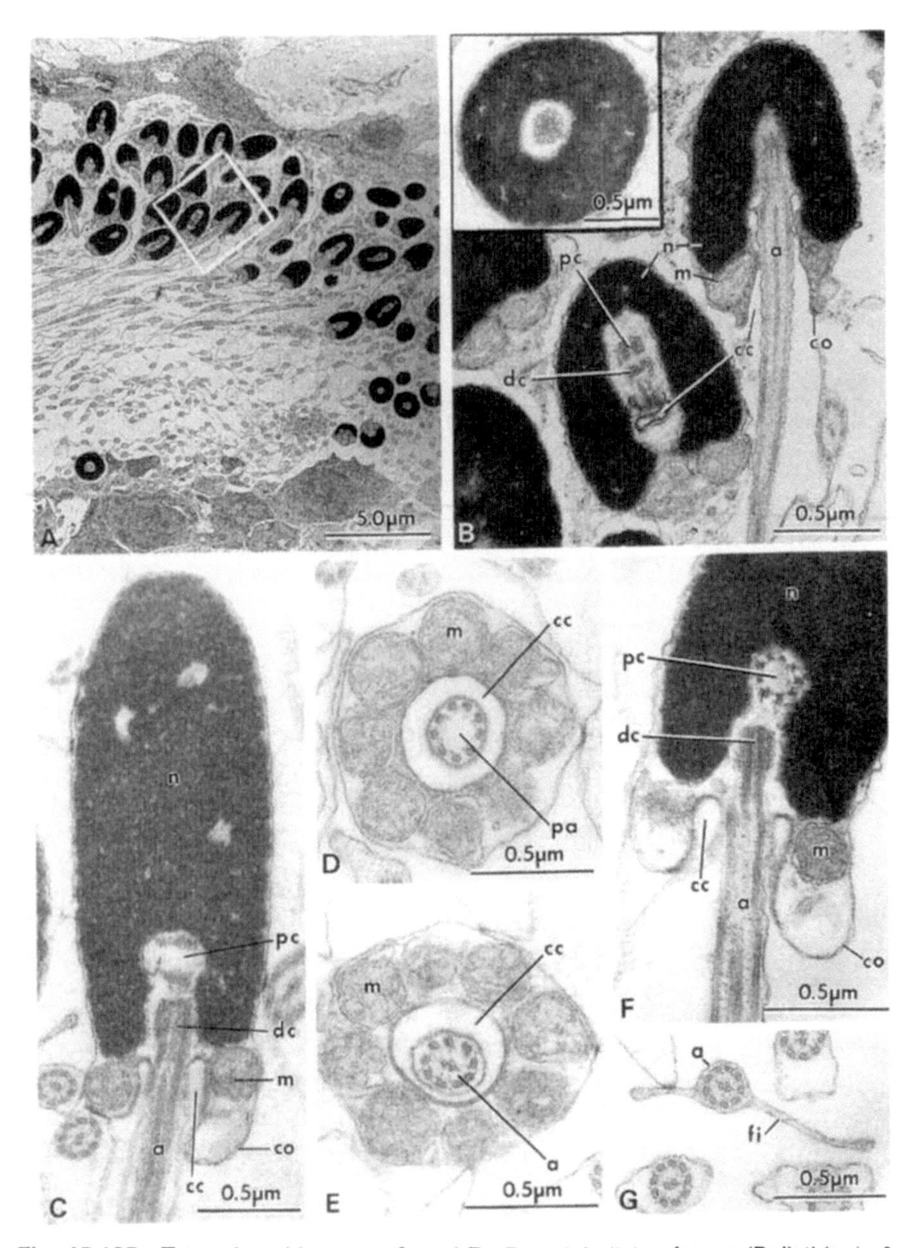

Fig. 15.125 Tetraodontoid sperm. **A** and **B**. *Pseudobalistes fuscus* (Balistidae). **A**. Spermatozoa in the testis, showing heads orientated towards the testicular epithelium. The flagella show paired lateral fins. **B**. Detail showing deep nuclear (implantation) fossa and serially coaxial centrioles. Inset, transverse section (TS) of a centriole, showing triplets. **C-G**. *Tetractenos (= Gastrophysus) hamiltoni*. **C**. Longitudinal section of spermatozoon, showing mutually perpendicular centrioles

Fig. 15.125 Contd. ...

transverse section, there is a narrow, dense layer with some indication of transverse striation resembling that of a septate junction. This layer may be the equivalent of the submitochondrial net of poeciliid sperm. The two mutually perpendicular centrioles lie within the implantation fossa in the same longitudinal axis (Fig. 15.125C,F). Only doublets surrounding and contiguous with a delicate ring have been demonstrated for the proximal centriole (Fig. 15.125F) which is housed in the apical expansion of the fossa. The distal centriole (basal body) consists of 9 triplets, which distally become doublets. The proximal part of the axoneme within the midpiece also consists of 9 doublets (Fig. 15.125D) behind which two central singlets are added (Fig. 15.125E). The axoneme has two lateral fins relative to which the plane of the two central singlets is slightly tilted (Fig. 15.125G). The lumina of the A and B subtubules are patent; weakly developed inner and outer dynein arms are visible in some sections (Jamieson 1991).

***Takifugu niphobles, T. pardalis, T. poecilonotus, T. porphyreus, T. radiatus, T. rubripes, and T. xanthopterus*:** Spermatozoa of these seven species resemble one another in their head dimensions and structure, and midpiece structure (Figs. 125-127). They closely resemble that of *Tetractenos (=Gastrophysus) hamiltoni* (Fig. 15.125C-G).

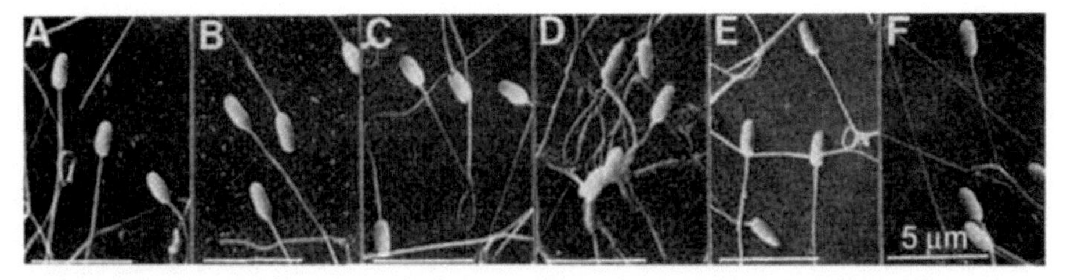

Fig. 15.126 *Takifugu* spermatozoa. Scanning electron micrographs. **A**. *T. niphobles*. **B**. *T. pardalis*. **C**. *T. poecilonotus*. **D**. *T. radiatus*. **E**. *T. rubripes*. **F**. *T. xanthopterus*. After Miyaki, K. and Yoshikoshi, K. 1993. Nippon Suisan Gakkaishi 59: 891, Fig. 1.

The nucleus of the spermatozoa is elongate, being approximately bullet-shaped or elongate ovoid. Dimensions are given in Table 15.6. The chromatin

---

Fig. 15.125 Contd. ...

in moderately deep nuclear fossa. **D**. TS of the midpiece showing 9+0 pattern of microtubules at transition from centriole to axoneme. **E**. TS of the midpiece through 9+2 region of axoneme. **F**. Longitudinal section of spermatozoon at right angles to C., showing mutually perpendicular centrioles in moderately deep nuclear fossa. **G**. TS of flagellum showing pair of lateral fins. a, axoneme; cc, cytoplasmic canal; co, collar; dc, distal centriole; m, mitochondrion; n, nucleus; pa, proximal, 9+0 region of axoneme. From Jamieson, B. G. M. 1991. *Fish Evolution and Systematics: Evidence from Spermatozoa*. Cambridge, Cambridge University Press, Fig. 16.20.

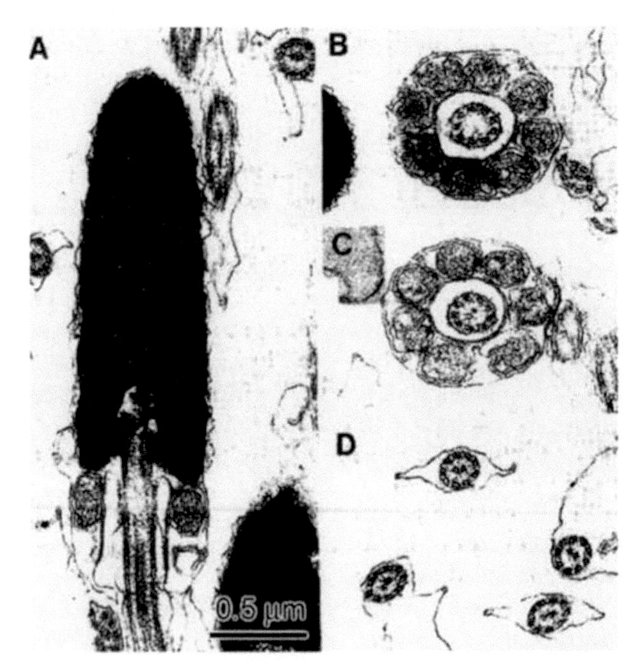

Fig. 15.127 *Takifugu porphyreus* (Tetraodontidae). TEM of spermatozoa. **A**. Longitudinal section. **B**,**C**. Transverse section (TS) of the midpiece. **D**. TS of the flagella. From Miyaki, K., Yoshikoshi, K. and Tabeta, O. 1996. Fisheries Science (Tokyo) 62: 543-546, Fig. 4.

is electron-dense. Basally the nucleus is indented by a clubshaped implantation fossa, 0.3-0.5 µm long and maximally 0.3 µm wide. In the midpiece, 7-9 independent mitochondria, 0.21-0.33 µm in diameter, encircle the flagellum. Thus the number of mitochondria in *Takifugu* spermatozoa is variable. It is six or seven in *T. vermicularis* sperm (Kurokura 1992, *fide* Miyaki 1996) as in *T. niphobles* (confirmed by Gwo *et al.* (1993b) (but eight or nine according to (Morisawa 2001); eight in *T. pardalis*, and *T. poecilonotus* (as in *Tetractenos hamiltoni*, Jamieson, 1991, and *T. (=Fugu) obscurus*, Lu 1999); seven in *T. rubripes* (Fig. 15.128C); seven or eight in *T. porphyreus* (Fig. 15.127) and *T. xanthopterus*; and nine in *T. radiatus*. They are housed in a cytoplasmic collar maximally 0.8 µm long. The cytoplasmic canal is 0.4- 0.7 µm long. The centrioles lie within the implantation fossa in the same longitudinal axis (coaxial). The proximal part of the axoneme within the midpiece consists of nine doublets behind which two central singlets are added. The flagellum has two lateral fins (Miyaki *et al.* 1996) as also noted for *T. niphobles* in a careful investigation by Morisawa (2001) in the region behind the collar in a study of the effects of various osmolalities.

The orientation of the centrioles is not usually described but at least in the plate for *T. pecilonotus* the proximal centriole appears to lie at a large angle to the distal centriole and not to be coaxial with it. The orientation of the two

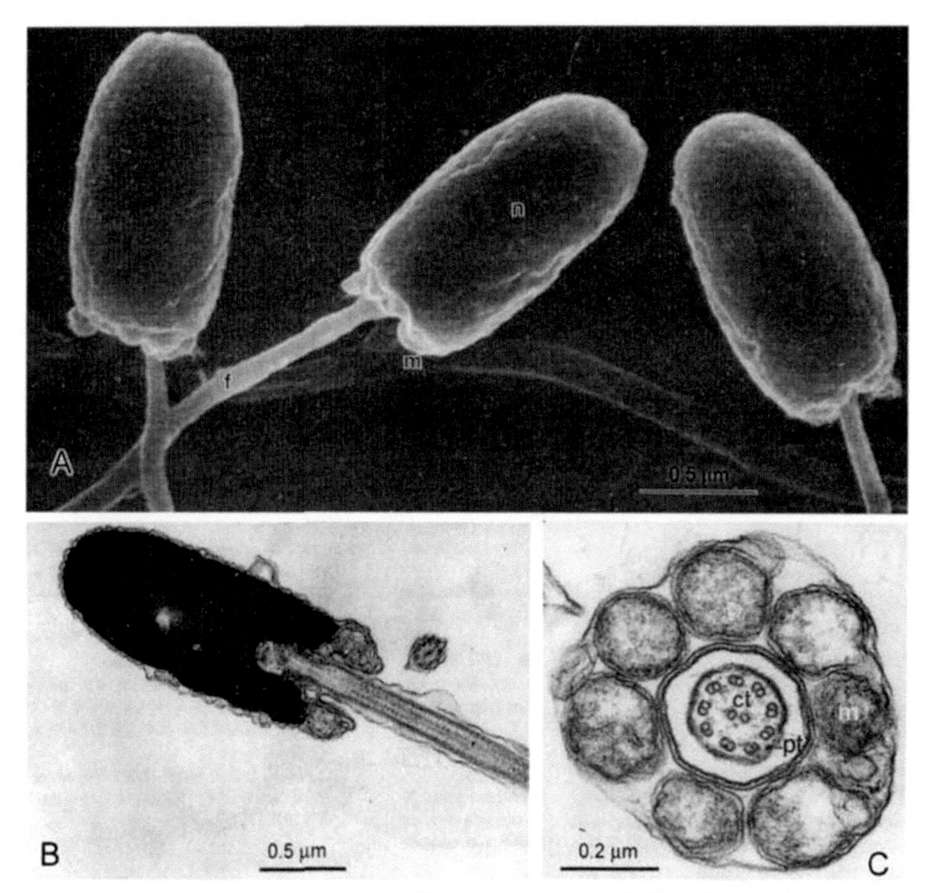

Fig. 15.128 *Takifugu* spermatozoa (Tetraodontidae). **A**. SEM of three spermatozoa of *Takifugu niphobles*. **B**. TEM longitudinal section of cryopreserved spermatozoon of *T. rubripes*. **C**. Transverse section of midpiece of *T. rubripes*. A. After Hara, M. and Okiyama, M. 1998. Bulletin of The Ocean Research Institute, University of Tokyo 33: 1-138, Fig. 60B. With permission. B and C. Adapted from Chang, Y. J., Chang, Y. J. and Lim, H. K. 1998. Journal of the Korean Fisheries Society 31: 353-358, Figs. 2A and 1C.

centrioles in *T. niphobles* is also given by Hara and Okiyama (1998) as mutually perpendicular, as appears to be confirmed in a micrograph (their Fig. 60C) but this observation is at odds with their comment that *Tetractenos (=Gastrophysus) hamiltoni*, as described by Jamieson (1991), is exceptional in that the centrioles are mutually perpendicular. The proximal centriole is clearly perpendicular to the distal centrioles in micrographs of *T. niphobles* by Morisawa (2001). It seems probable that the latter condition is normal for the genus *Takifugu*.

In regarding *Gastrophysus* as a genus separate from *Takifugu*, Miyaki (1996) concluded that the similar ultrastructure of the sperm in the two taxa indicated that it was a characteristic of the family Tetraodontidae rather than of the genus *Takifugu* but in view of the synonymy of the two taxa it may now be considered characteristic and even diagnostic of *Takifugu*.

Miyaki (1996) ascribes the ready hybridization of *Takifugu* species to the similar external and internal morphology of the spermatozoa.

***Ephippion* and *Lagocephalus*.** The sperm of *Lagocephalus laevigatus* illustrated diagrammatically (Fig. 15.129) has the relatively shallow basal fossa and mutually perpendicular centrioles typical of tetraodontids but differs from others of the family in having two tiers of mitochondria (Mattei 1991). In contrast, the sperm of *Ephippion guttifer* (Fig. 15.129) has the deep nuclear fossa of a balistid, as noted above but differs from the latter family in having mutually perpendicular centrioles.

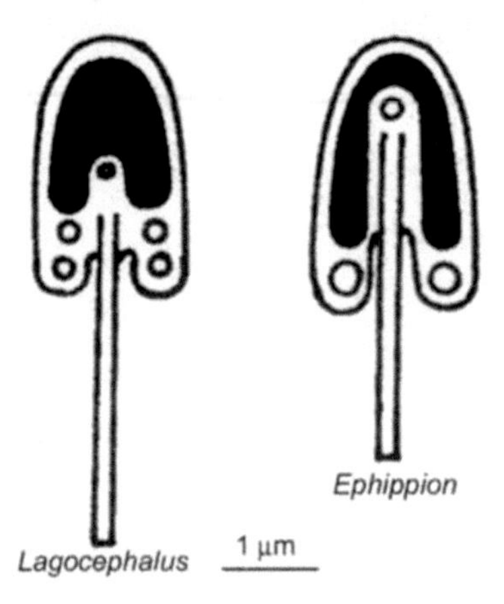

Fig. 15.129 Comparison the sperm of the tetraodontids *Lagocephalus laevigatus* and *Ephippion guttifer*. Adapted from Mattei, X. (1991). Canadian Journal of Zoology 69(12): 3038-3055, Fig. 12 (part). With permission.

Sphoeroides spengleri: The sperm of *S. spengleri* (Mattei 1991 and this account) (Fig. 15.130) closely resembles the sperm of *Takifugu*. It is noteworthy that the proximal centriole has nine doublets rather than the nine triplets usual in centrioles. It is not known whether this is a general feature also of *Takifugu* sperm but it is seen in *Tetractenos hamiltoni*.

15.9.2.2 Family Diodontidae

Diodontids are the porcupine fishes (burrfishes) (adult, Fig. 15.121).

***Chilomycterus*:** The spermatozoa of the diodontid *Chilomycterus antennatus* (see Mattei 1970) (Fig. 15.122B) are intermediate between balistids and tetraodontids in having the longitudinal centrioles of balistids but a shallower nuclear fossa as in *Takifugu*. The condition in *C. reticulatus*, listed by Mattei (1991), is unknown.

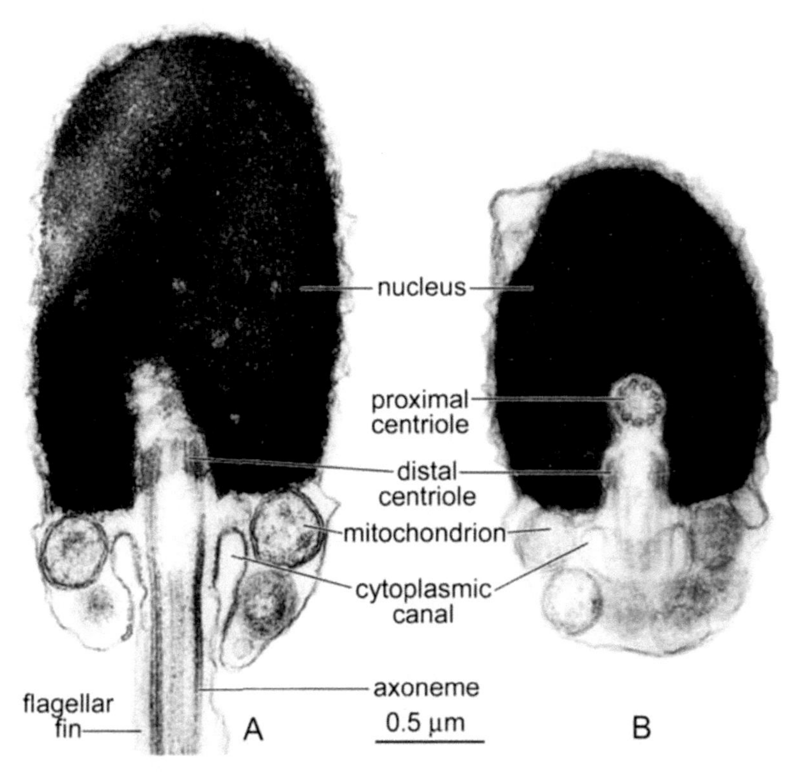

Fig. 15.130 *Sphoeroides spengleri* (Tetraodontidae) spermatozoa in longitudinal section. **A**. Showing the elongate nucleus. **B**. Slightly oblique, section at right angles to **A**, showing the mutually perpendicular centrioles, the proximal centriole with nine doublets. Original, from unpublished TEM micrographs courtesy of Professor Xavier Mattei.

Tetraodcntiform spermatozoal phylogeny: In the tetraodontiform taxa for which spermatozoal ultrastructure is known, two major groups are discernible and these agree well with two sister clades identified by Holcroft (2005) in an analysis of the nuclear gene RAG1 (Fig. 15.131): (1) a monocanthid + balistid group which has mulliform sperm with very deep nuclear fossa and longitudinal tandem (coaxial) centrioles and (2) a tetraodontid + diodontid clade. In the latter clade the diodontid has a round headed Type I spermatozoon, differing from the norm for this type in having tandem, coaxial centrioles; the tetraodontid sperm have perpendicular centrioles; nuclear fossae are less than half the length of the nucleus, with the exception of *Ephippion*. The sperm of *Ephippion* resembles that of the monacanthids and balistids but differs in the perpendicular centrioles. Its structure differs sufficiently from the other examined tetraodontids to prompt one to query whether this genus is a true tetraodontid. It is noteworthy, therefore, that Tyler and Bannikov (1992) note that *Ephippion* is exceptional

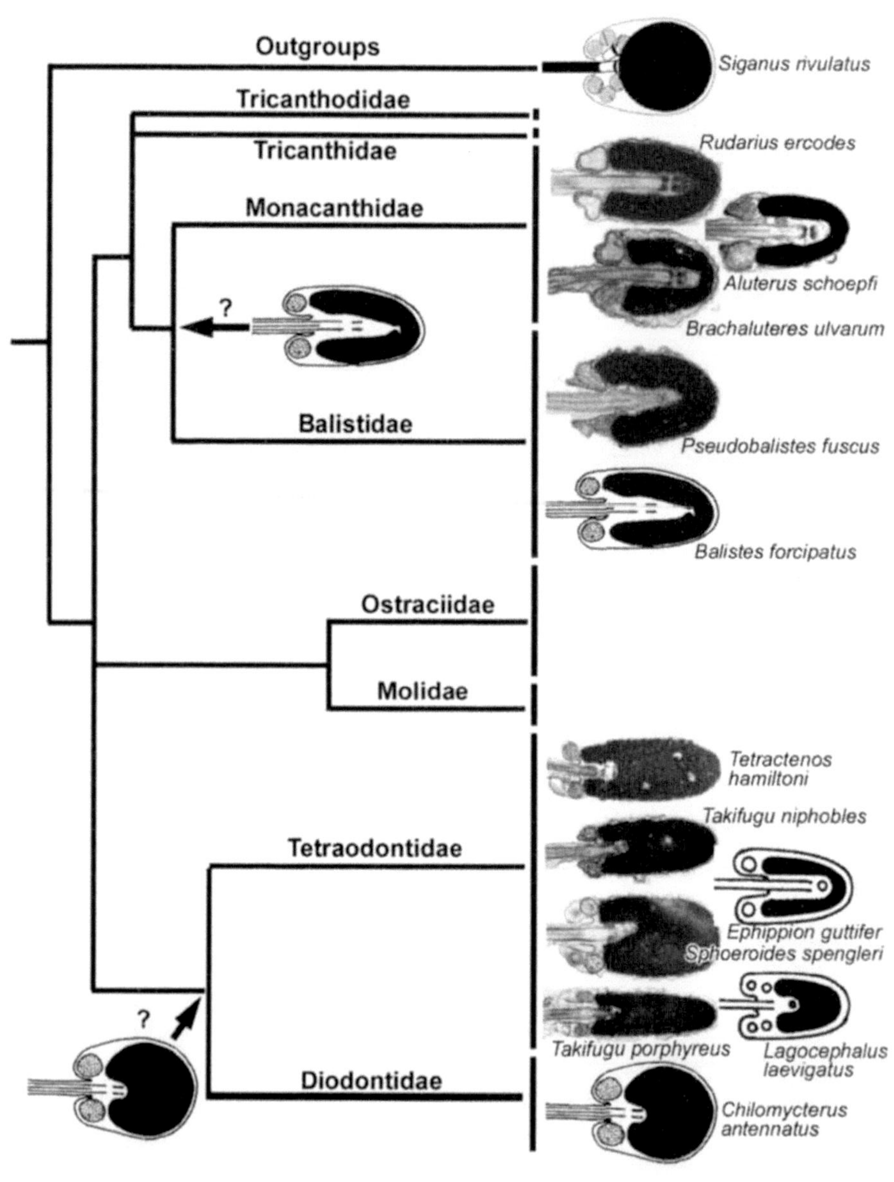

Fig. 15.131 Tetraodontiform spermatozoa from sources listed in this chapter superimposed on a majority-rule consensus dendrogram resulting from Bayesian analyses of the nuclear gene RAG1 by Holcroft, N. I. 2005. Molecular Phylogenetics and Evolution 34: 525-544, Fig. 5. With permission of Elsevier. Of the several outgroup genera employed in the analysis only a siganid is represented here. Note that, of the taxa for which sperm are known, there are two major clades: (1) a monocanthid + balistid clade which has mulliform sperm with very deep nuclear fossa and longitudinal tandem centrioles and (2) a tetraodontid + diodontid clade.

Fig. 15.131 Contd. ...

among tetraodontids in having plate-like scales similar to those of the Ostraciidae. Spermatozoal ultrastructure in ostraciids and molids has yet to be studied.

If, as Holcroft (2005) indicates, the diodontids are the sister-group of the tetraodontids, the fact that the examined diodontid species, *Chilomycterus antennatus*, has simple aquasperm, albeit with longitudinal, coaxial centrioles, suggests that tetraodontids have independently elongated the sperm nucleus, while, except in *Ephippion*, retaining a nuclear fossa of only moderate extent.

The constancy, as far as is known from the small sample of species, of the very deep nuclear fossa with longitudinal, tandem, coaxial centrioles in the investigated monacanthids and balistids suggests a common ancestor for the two groups with these spermatozoal characteristics. This morphology may in turn have originated from a simple aquasperm as exemplified in the outgroup by *Siganus*. Species used as the outgroup by Holcroft (2005) were *Lophius americanus*, *Siganus doliatus*, *Antigonia capros*, *Chaetodipterus faber*, *Drepane punctata*, *Morone chrysops*, *Zebrasoma scopas* and *Chaetodon striatus*. It would have been instructive if a mullid had been included. However, the neat division between the monacanthid + balistid clade on the one hand and the tetraodontid + diodontid clade on the other remains confounded by the enigmatic *Ephippion*. Much additional sampling of tetraodontiform species for sperm ultrastructure is needed.

If origin of tetraodontiform sperm from a simple anacrosomal aquasperm be accepted, similarity of citharid and soleid spermatozoa to those of *Ephipphion* and, excepting the centriolar orientation, to those of the balistid + monacanthid assemblage would appear to be due to convergence unless evidence for inclusion of all of these entities in a single clade were forthcoming. Molecular evidence to date is equivocal as to relationships between the Pleuronectiformes and Tetraodontiformes (Ortí and Li, **Chapter 1**, Fig. 1.5 and Fig. 15.1; Chen *et al.* 2007).

15.10 CHAPTER SUMMARY

The distribution of sperm types in the Percomorpha reflects the demonstrated polyphyly of the group (see Ortí and Li, **Chapter 1**) and its diversity of fertilization modes (see Table 15.1). The following summary of sperm types and fertilization biology in relation to taxa is confined to the species for which spermatozoal ultrastructure is described in this chapter.

Fig. 15.131 Contd. ...

In the latter clade the diodontid has a round headed Type I spermatozoon, differing from the norm for this type in having tandem centrioles; the tetraodontid sperm have perpendicular centrioles; nuclear fossae are less than half the length of the nucleus, with the exception of *Ephippion*. The sperm of *Ephippion* resembles that of the monacanthids and balistids but differs in the perpendicular centrioles. Original.

The vast majority of percomorphs are externally fertilizing but internal fertilization has arisen, in most cases independently, in the Scorpaeniformes (Sebastinae, some Cottidae and the Hemipteridae), Labroidei (some Cichlidae and the Embiotocidae), Zoarcoidei (Zoarcidae and Anarhichadidae) and Blennioidei (some Clinidae). Internal gametic association occurs in some Gasterosteiformes (Aulorhynchidae) and some Perciformes (Apogonidae in the Percoidei).

External fertilization occurs in almost all orders and suborders of the Percomorpha and involves one or other of two sperm types (Type I and Type II, *sensu* Mattei 1970) or presumed modifications of these. In Type I, during spermiogenesis a basal nuclear fossa forms, a 90° rotation of the nucleus into the same axis as the flagellum occurs which brings the centrioles into the nuclear fossa; and the mitochondria migrate to the base of the nucleus. Type I is considered the plesiomorphic sperm type for the Percomorpha, though not for the Euteleostomi (=Osteichthyes) nor for agnathans. It is seen in other telostean superorders, such as the Mugilomorpha and Atherinomorpha. It is the simple anacrosomal aquasperm *sensu* Jamieson (1991). The unmodified Type I spermatozoon is the only one known, in the albeit limited ultrastructural studies, for the Beryciformes, and Zeiformes, and some families or subfamiliesof other orders (Fistulariidae in the Gasterosteiformes, Scorpaeninae and Tetraroginae in the Scorpaenoidei, Centracanthidae, Sparidae in the Percoidei, Pinguipedidae in the Trachinoidei, Rhyacichthyidae in the Gobioidei (though there with a very deep basal fossa), Siganidae in the Acanthuroidei, Istiophoridae in the Scombroidei, Osphronemidae in the Anabantoidei, Citharidae, Scophthalmidae, Paralichthyidae, and Soleidae in the Pleuronectiformes, and Diodontidae and some Tetraodontidae in the Tetraodontiformes. These are always externally fertilizing. Slightly or greatly modified sperm also ascribable to Type I occur in the Gasterosteiformes (Hypoptychidae, Aulorhynchidae, Gasterosteidae and Syngnathidae); Scorpaeniformes (Dactylopteridae, Sebastinae, Triglidae, Hexagrammidae and some Cottidae) Labroidei (all Cichlidae), some Trachinoidei (Percophidae and Trichodontidae) where they are usually externally fertilizing.

The Type II spermatozoon is an evolved type which it is the commonest sperm type in the Perciformes. In Type II spermiogenesis, the final stage of nuclear rotation is omitted, or is limited, so that the flagellar axis is approximately parallel to the base of the nucleus and the centrioles are situated outside the nuclear depression. Sperm resembling Type II occur in some Mugilidae and a minority of Ostariophysi. These appear to be due to convergence (homoplasy) relative to the perciform type, as is supported by their absence from those percomorphs which, from analysis of molecular sequences, appear to be phylogenetically more basal, the Beryciformes, Zeiformes, Synbranchiformes, Gasterosteiformes and Scorpaeniformes. They may, alternatively, represent parallelism by virtue of genetic relationship (paramorphy, see **Chapter 14, Section 14.3.1**).

There seem to be no cases where internally fertilizing spermatozoa are derivable from Type II but derivation from Type I is based on inclusion of the centrioles in the nuclear fossa and this could conceivably be a concomitant modification from the Type II condition in which rotation into the fossa was achieved.

The various modifications from the basic euteloestean aquasperm are detailed in the text and cannot be comprehensively reiterated here but some modifications may be noted. One of the most obvious is the development of two flagella. Such biflagellarity (with fertilization mode in parentheses) is seen, in the Percomorpha, in the Synbranchiformes (ef) and, within the Perciformes, in the Apogonidae (iga), the cichlid *Satanoperca* (if), the Zoarcidae (if), facultatively in the Anarhichadidae (if) and in the Gobiesocidae (ef). There is thus a considerable correlation with internal fertilization, considering the relative rarity of this in percomorphs. We have seen that biflagellarity also occurs in Synbranchiformes and some members of the Siluriformes (Nematogenyidae, Cetopsidae, Ictaluridae, Malapteruridae, Doradidae, Plotosidae, Amblycipitidae and Asperdinidae), Batrachoidiformes (Batrachoididae) and Myctophiformes (some Myctophidae) in which fertilization is external. (Some siluriforms in the families Auchenipteridae and Scoloplacidae, with uniflagellate sperm, are, however, inseminating (see Evans and Downing, **Volume 8B, Chapter 4**).

A frequent modification is elongation of the nucleus and/or the midpiece. Although in percomorphs elongation of the nucleus appears to be a requirement of internal fertilization (but contrast Goodeidae, for instance, in the Atherinomorpha), elongation also commonly occurs in externally fertilizing perciform sperm. In the Gasterosteiformes, slight elongation of the nucleus is seen in Gasterosteidae whereas elongation is pronounced in the Hypoptychidae, Aulorhynchidae, and Syngnathidae. In the Scorpaeniformes, nuclear elongation is slight in the Sebastinae, which are internally fertilizing, but is strong and emphasized by asymmetry in the Dactylopteridae, Hexagrammidae, Hemitripteridae, with external fertilization, and the Cottidae, with internal gametic association or with true internal fertilization. The Hemitripteridae, and some Cottidae, are highly unusual in having two sperm types, euspermatozoa and paraspermatozoa. Elongation of the nucleus does not occur in the examined Percoidei (in which 27 families have been examined for spermatozoal ultrastructure). As noted, they have Type II spermatozoa, excepting the biflagellate Apogonidae, the Kuhlidae, in which the sperm appears intermediate between Types I and II, and the Mullidae which have a very deep basal nuclear fossa (see also some Tetraodontiformes). Nuclear elongation is also absent from the Acanthuroidei (Acanthuridae, Ephippidae and Siganidae), Scombroidei (Istiophoridae, Scombridae, Sphyraenidae and Trichiuridae), and Stromateoidei (Ariommatidae), all examined species of which, like the Percoidei, have demonstrated or putative external fertilization.

The Blennioidei are remarkable for variation from Type I or Type II sperm in some Blenniidae, or Type II in some Clinidae, to major modifications detailed in section **15.7.5**. In modified sperm, nuclear elongation is absent or moderate in the Blenniidae but may be great in the Clinidae but only in the latter is it accompanied by internal fertilization.

In the Gobiesocoidei, of which only *Lepadogaster*, in the Gobiesocidae, has been examined, this species has a highly modified spermatozoon, with elongation of the nucleus and midpiece and biflagellarity, though externally fertilizing.

In terms of spermatozoal ultrastructure, and its variability, the Gobioidei, despite being external fertilizing, are the most remarkable of the Perciformes and Percomorpha. The Rhyacichthyidae endorse their basic phylogenetic position in having a relatively simple Type I spermatozoon though the basal fossa is very deep and there is a single, lateral mitochondrion. This type is also seen in Eleotridae and some Gobiidae but there are bizarre modifications in the gobiids and in the examined pterotrinid, culminating, in the Schindleriidae, in which *Schindleria* has a spermatozoon in which the flagellum completely penetrates and extends well beyond the sheath-like nucleus.

Pleuronectiformes have Type I sperm in which the chief modification is deep penetration of the nucleus by the basal fossa, thus resembling the mullid spermatozoon (though penetration is moderate in pleuronectids). A similar type of spermatozoon occurs in the Tetraodontiformes in the Balistidae, Monacanthidae and some Tetraodontidae. However, *Diodon*, in the Diodontidae, has a simple Type I sperm. Other tetraodontids have moderate fossae and nuclei elongated to about twice their width.

15.11 ACKNOWLEDGMENTS

I am most grateful to Professor Xavier Mattei for providing unpublished TEM micrographs and negatives which have been reproduced in this chapter. Special thanks are also due to Drs. Hara and Okiyama for freely making their excellent review of the spermatozoa of Japanese fishes available to me and to Dr. Yasunori Koya for his admirable micrographs. Other authors have also been most generous in providing reprints. I am also deeply grateful to publishers who allowed free reproduction of illustrations.

15.12 LITERATURE CITED

Abascal, F. J., Medina, A., Megina, C. and Calzada, A. 2002. Ultrastructure of *Thunnus thynnus* and *Euthynnus alletteratus* spermatozoa. Journal of Fish Biology 60: 147-153.

Abascal, F. J., Megina, C. and Medina, A. 2004. Testicular development in migrant and spawning bluefin tuna (*Thunnus thynnus* (L.) from the eastern Atlantic and Mediterranean. Fishery Bulletin (Seattle) 102: 407-417.

Afzelius, B. A. 1978. Fine structure of the garfish spermatozoon. Journal of Ultrastructure Research 64: 309-314.

Akihito, Akihisa, I., Iwata, A., Kobayashi, T., Ikeo, K., Imanishi, T., Ono, H., Umehara, Y., Hamamatsu, C., Sugiyama, K., Ikeda, Y., Sakamoto, K., Fumihito, A., Susumu, O. S. and Gojobori, T. 2000. Evolutionary aspects of gobioid fishes based upon a phylogenetic analysis of mitochondrial cytochrome b genes. Gene 259: 5-15.

An, C. M., Lee, J. S. and Huh, S.-H. 1999. Spermiogenesis and spermatozoal ultrastructure of the roundnose flounder, *Eopsetta grigorjewi* (Teleostei: Pleuronectidae). Journal of the Korean Fisheries Society 32: 730-736.

Assem, S. S. 2003. Reproductive biology, spermatogenesis and ultrastructure of the testes of the sparid fish, *Pagellus erythrinus*. Journal of the Egyptian German Society of Zoology 42.

Balon, E. K. 1975. Reproductive guilds of fishes: a proposal and definition. Journal of the Fisheries Research. Board of Canada 32: 821-864.

Billard, R. 1986. Spermatogenesis and spermatology of some teleost fish species. Reproduction, Nutrition and Développement 26: 877-1024.

Boisson, C., Mattei, X. and Mattei, C. 1968. Le spermatozoïde de *Dactylopterus volitans*, Linné (Poisson Cephalacanthidae):étudié au microscope électronique. Comptes Rendus des Séances de la Société de Biologie de l'ouest Africain 162: 820-823.

Boisson, C., Mattei, X. and Mattei, C. 1969. Mise en place et évolution du complexe centriolaire au cours de la spermiogenèse *d'Upeneus prayensis* C.V. (Poisson Mullidae). Journal de Microscopie (Paris) 8: 103-112.

Breder, C. M. and Rosen, D. E. 1966. *Modes of reproduction in fishes*. The American Museum of Natural History, The Natural History Press, New York. 941 pp.

Carcupino, M., Baldacci, A., Corso, G., Franzoi, P., Pala, M. and Mazzini, M. 1999. Testis structure and symplastic spermatid formation during spermatogenesis of pipefishes. Journal of Fish Biology 55: 344-353.

Carrillo, M. and Zanuy, S. 1977. Quelques observations sur le testicule chez *Spicara chryselis* C. V. Investigación Pesquera 41: 121-146.

Cerisola, H. 1990. Structural and ultrastructural aspects of the testis of the Clingfish *Sicyases sanguineus*. Revista de Biologia Marina 25: 81-98.

Chang, Y. J. and Chang, Y. J. 2002. Milt properties of four flatfish species and fine structure of their cryopreserved spermatozoa. Journal of Fisheries Science and Technology 5: 87-96.

Chang, Y. J., Chang, Y. J. and Lim, H. K. 1998. Physico-chemical properties of milt and fine structure of cryopreserved spermatozoa in tiger puffer (*Takifugu rubripes*). Journal of the Korean Fisheries Society 31: 353-358 (In Korean).

Chen, W.-J., Ruiz-Carus, R. and Orti, G. 2007. Relationships among four genera of mojarras (Teleostei: Perciformes: Gerreidae) from the western Atlantic and their tentative placement among percomorph fishes. Journal of Fish Biology 70 (Supplement B): 202-218.

Chung, E.-Y. and Chang, Y. J. 1995. Ultrastructural changes of germ cell during the gametogenesis in Korean rockfish, Sebastes schlegeli. Journal of the Korean Fisheries Society 28: 736-752.

Collette, B. B. and Nauen, C. H. 1983. *FAO species catalogue. Vol. 2. Scombrids of the world. An annotated and illustrated catalogue of tunas, mackerels, bonitos and related species known to date. FAO Fisheries Synopses.* 137pp.

Cruz-Landim, C. and Cruz-Hoflung, M. A. 1986/1987. Aspectos de espermatogénese de tecunaré (*Cichla ocellaris* Schneider, 1801) (Teleostei, Cichlidae). Acta Amazónica 16/17: 65-72.

Don, J. and Avtalion, R. R. 1993. Ultraviolet irradiation of tilapia spermatozoa and the Hertwig effect: electron microscope analysis. Journal of Fish Biology 42: 1-14.

Dong, X., Ye, Y. and Wu, C. 1998. Spermogenesis in the yellowfin porgy (*Sparus latus* Houttuyn), with emphasis on the associated mitochondrion. Chinese Journal of Oceanology and Limnology 16: 144-148.

El-Gammal, H. L., Bahnasawy, M. H., El-Sayyad, H. I. and El-Gammal, I. M. 2003. The ultrastructure of spermatozoa and spermiogenesis in *Sarotherodon galilaeus* and *Liza ramada* (Teleostei, Perciformes, Cichlidae and Mugilidae). Journal of the Egyptian German Society of Zoology 40: 69-80.

Fink, B. D. and Haydon, G. B. 1960. Sperm morphology of two cottid fishes in electron micrographic silhouettes. Copeia 1960: 319-322.

Fishelson, L. 1990. Unusual cell organelles during spermiogenesis in two species of gobies (Gobiidae, Teleostei). Cell and Tissue Research 262: 397-400.

Fishelson, L. 2003. Comparison of testes structure,spermatogenesis, and spermatocytogenesis in young aging,and hybrid cichlid fish (Cichlidae, Teleostei). Journal of Morphology 256: 285-300.

Fishelson, L. and Delarea, Y. 1995. Unilateral winged flagellum of sperm of *Badis badis* (Pisces: Teleostei). Copeia 1995: 241-243.

Fishelson, L., Gon, O., Holdengreber, V. and Delarea, Y. 2007. Comparative spermatogenesis, spermatocytogenesis, and spermatozeugmata formation in males of viviparous species of clinid fishes (Teleostei: Clinidae, Blennioidei). The Anatomical Record 290: 311-323.

Froese, P. and Pauly, R. 2004. Fish Base. World Wide Web electronic publication www.fishbase.org.

Garcia-Diaz, M. M., Lorente, M. J., Gonzalez, J. A. and Tuset, V. M. 1999. Comparative ultrastructure of spermatozoa of three marine teleosts of the genus *Serranus*: *Serranus atricauda, Serranus cabrilla* and *Serranus scriba*. Journal of Submicroscopic Cytology and Pathology 31: 503-508.

Gardiner, D. M. 1978. The origin and fate of spermatophores in the viviparous teleost *Cymatogaster aggregata* (Perciformes: Embiotocidae). Journal of Morphology 155: 157-172.

Grau, A., Sarasquete, M. C. and Crespo, S. 1991. Spermiogenesis in *Serioila dumerelii*, Riss 1810: an ultrastructural study. Pp. 187. In *Research for aquaculture: fundamentals and applied aspect. Proceedings of the European Society of Comparative Physiology and Biochemistry*, Antibes.

Grier, H. J. 1981. Cellular organization of the testis and spermatogenesis in fishes. American Zoologist 21: 345-357.

Guha, T., Siddiqui, A. Q. and Prentis, P. F. 1988. Ultrastructure of testicular spermatozoon of the fish Oreochromis niloticus. Pp. 278-279. In G. W. Bailey (ed.), *Proceedings of the 46th Annual Meeting of the Electron Microscopy Society of America*. San Francisco Press, San Francisco.

Gusmao, P., Foresti, F. and Quagio-Grassiotto, I. 1999. Ultrastructure of spermiogenesis in *Plagioscion squamosissimus* (Teleostei, Perciformes, Sciaenidae). Tissue and Cell 31: 627-633.

Gusmao, P., Foresti, F. and Quagio-Grassiotto, I. 2002. The ultrastructure of the pre-meiotic and meiotic stages of spermatogenesis in *Plagioscion squamosissimus* (Teleostei, Perciformes, Scianidae). Journal of Submicroscopic Cytology and Pathology 34: 159-165.

Gusmao-Pompiani, P., Oliveira, C. and Quagio-Grassiotto, I. 2005. Spermatozoa ultrastructure in Sciaenidae and Polynemidae (Teleostei : Perciformes) with some consideration on Percoidei spermatozoa ultrastructure. Tissue and Cell 37: 177-191.

Gwo, J., C., Gwo, H. H. and Chang, S. L. 1993a. Ultrastructure of the spermatozoon of the teleost fish *Acanthopagrus schlegeli* (Perciformes: Sparidae). Journal of Morphology 216: 29-33.

Gwo, J. C. 1995a. Spermatozoan ultrastructure of the teleost fish *Acanthopagrus latus* (Perciformes: Sparidae) with special reference to the basal body. Journal of Submicroscopic Cytology and Pathology 27: 391-396.

Gwo, J. C. 1995b. Ultrastructural study of osmolality effect on spermatozoa of three marine teleosts. Tissue and Cell 27: 491-497.

Gwo, J. C. and Arnold, C. R. 1992. Cryopreservation of Atlantic Croaker spermatozoa: evaluation of morphological changes. Journal of Experimental Zoology 264: 444-453.

Gwo, J. C. and Chang, S. L. 1993. The ultrastructure of the spermatozoon of the teleost fish *Acanthopagrus schlegeli* (Perciformes, Sparidae). Journal of Morphology 216: 29-33.

Gwo, J. C., Chiu, J. Y., Lin, C. Y., Su, Y. and Yu, S. L. 2005. Spermatozoal ultrastructure of four Sparidae fishes: *Acanthopagrus berda, Acanthopagrus australis, Lagodon rhomboids* and *Archosargus probatocephus*. Tissue and Cell 37: 109-115.

Gwo, J. C. and Gwo, H. H. 1993. Spermatogenesis in the black porgy, *Acanthopagrus schlegeli* (Teleostei: Perciformes: Sparidae). Molecular Reproduction and Development 36: 75-83.

Gwo, J. C., Gwo, H. H., Ko, Y. S., Lin, B. H. and Shih, H. 1994. Spermatozoan ultrastructure of two species of grouper *Epinephelus malabaricus* and *Plectropomus leopardus* (Teleostei, Perciformes, Serranidae) from Taiwan. Journal of Submicroscopic Cytology and Pathology 26: 131-136.

Gwo, J. C., Kuo, M. C., Chiu, J. Y. and Cheng, H. Y. 2004a. Ultrastructure of *Pagrus major* and *Rhabdosargus sarba* spermatozoa (Perciformes: Sparidae: Sparinae). Tissue and Cell 36: 141-147.

Gwo, J. C., Kurokura, H. and Hirano, R. 1993b. Cryopreservation of spermatozoa from rainbow trout, common carp, and marine puffer. Nippon Suisan Gakkaishi 59: 777-782.

Gwo, J. C., Yang, W. T., Kuo, M. C., Takemura, A. and Cheng, H. Y. 2004b. Spermatozoal ultrastructures of two marine perciform teleost fishes, the goatfish, *Paraupeneus spilurus* (Mullidae) and the rabbitfish, *Siganus fuscescens* (Siganidae) from Taiwan. Tissue and Cell 36: 63-69.

Hahn, H. M. 1927. The history of the germ cell of *Cottus baridii* Girard. Journal of Morphology and Physiology 43: 427-498.

Hann, H. M. 1930. Variation in spermiogenesis in the teleost family Cottide. Journal of Morphology and Physiology 50: 393-411.

Hara, M. 2003. Fine structure and functional characteristics of the gamete of the Japanese sand lance, *Ammodytes personatus*. Japanese Journal of Ichthyology 50: 35-45 (In Japanese).

Hara, M. 2004. Diversity of the sperm morphology in gobioid fishes. Japanese Journal of Ichthyology 51: 1-32.

Hara, M. and Okiyama, M. 1998. An ultrastructural review on the spermatozoa of Japanese fishes. Bulletin of The Ocean Research Institute, University of Tokyo 33: 1-138.

Hayakawa, Y. 2003. Sperm polymorphism. Pp. 66-88. In A. Nakazono (ed.), *Sex and Behavioral Ecology of Aquatic Bioresources*. Koseisha-Koseikaku, Tokyo.

Hayakawa, Y., Akiyama, R. and Munehara, H. 2004. Antidispersive effect induced by parasperm contained in semen of a cottid fish *Hemilepidotus gilberti*: estimation by models and experiments. Japanese Journal of Ichthyology 51: 31-42.

Hayakawa, Y., Akiyama, R., Munehara, H. and Komaru, A. 2002a. Dimorphic sperm influence semen distribution in a non-copulatory sculpin *Hemilepidotus gilberti*. Environmental Biology of Fishes 65: 311-317.

Hayakawa, Y., Komaru, A. and Munehara, H. 2002b. Ultrastructural observations of eu- and paraspermiogenesis in the cottid fish *Hemilepidotus gilberti* (Teleostei: Scorpaeniformes: Cottidae). Journal of Morphology 253: 243-254.

Hayakawa, Y. and Munehara, H. 2004. Ultrastructural observations of euspermatozoa and paraspermatozoa in a copulatory cottoid fish *Blepsias cirrhosus*. Journal of Fish Biology 64: 1530-1539.

Hayakawa, Y., Munehara, H. and Komaru, A. 2002c. Obstructive role of the dimorphic sperm in a non-copulatory marine sculpin Hemipedotus gilberti to prevent other males' eusperm from fertilization. Environmental Biology of Fishes 64: 419-427.

He, S. and Woods, L. C., III 2004. Effects of dimethyl sulfoxide and glycine on cryopreservation induced damage of plasma membranes and mitochondria to striped bass (*Morone saxatilis*) sperm. Cryobiology 48: 254-262.

Healy, J. M. and Jamieson, B. G. M. 1981. An ultrastructural examination of developing and mature paraspermatozoa in *Pyrazus ebeninus* (Mollusca, Gastropoda, Potamididae). Zoomorphology 98: 101-119.

Holcroft, N. I. 2005. A molecular analysis of the interrelationships of tetraodontiform fshes (Acanthomorpha: Tetraodontiformes). Molecular Phylogenetics and Evolution 34: 525-544.

Hong, W., Zhang, Q. and Ni, Z. 1991. Spermatogenesis in spermiogenesis of yellowfish seabream in Xipu Bay. Journal of Fisheries, China 15: 302-307. (In Chinese).

Hoshino, K. 2001. Monophyly of the Citharidae (Pleuronectoidei: Pleuronectiformes: Teleostei) with considerations of pleuronectoid phylogenyy. Ichthyological Research 48: 391-404.

Hu, J.-H., Zhang, Y.-Z., Fu, C.-L. and Liu, X.-Z. 2005. Ultrastructure of rosy barb Puntius conchonius spermatozoon. Acta Zoologica Sinica 51: 892-897.

Jamieson, B. G. M. 1984. Spermatozoal ultrastructure in *Branchiostoma moretonensis* Kelly, a comparison with *B. lanceolatum* (Cephalochordata) and with other deuterostomes. Zoologica Scripta 13: 223-229.

Jamieson, B. G. M. 1991. *Fish Evolution and Systematics: Evidence from Spermatozoa*. Cambridge University Press, Cambridge. 319 pp.

Jaspers, E. J., Avault, J. W. J. and Roussel, J. D. 1976. Spermatozoal morphology and ultrastructure of Channel catfish *Ictalurus punctatus*. Transactions of the American Fisheries Society 105: 475-480.

Johnson, G. D. 1975. The procurrent spur: an undescribed perciform caudal character and its phylogenetic implications. Occasional Papers of the Californian Academy of Sciences 121: 1-23.

Johnson, G. D. 1984. Percoidei: development and relationships. Pp. 464-498. In H. G. Moser, W. J. Richards, D. M. Cohen, M. P. Fahay, A. W. Kendall and S. L. Richardson (eds), *Ontogeny and Systematics of Fishes*. Special Publication Number 1 American Society of Ichthyologists and Herpetologists. vol. 1. Allen Press, Lawrence, Kansas.

Jones, P. R. and Butler, R. D. 1988. Spermatozoon ultrastructure of *Platichthys flesus*. Journal of Ultrastructure and Molecular Structure Research 98: 71-82.

Jordan, D. S. 1907. *Fishes*. Henry Holt, New York.

Kornienko, E. S. 2001. The Spawning Behavior of the Pipefish *Syngnathus acusimilis*. Russian Journal of Marine Biology 27: 54-57.

Kornienko, E. S. and Drozdov, A. L. 2001. Fine structure of the spermatozoa of the pipefish *Syngnathus acusimilis*. Voprosy Ikhtiologii 41: 851-854 (In Russian).

Koya, Y., Munehara, H. and Takano, K. 1997. Sperm storage and degradation in the ovary of a marine copulating sculpin, *Alcichthys alcicornis* (Teleostei: Scorpaeniformes): role of intercelluar junctions between ovarian epithelial cells. Journal of Morphology 233: 153-163.

Koya, Y., Munehara, H. and Takano, K. 2002. Sperm storage and motility in the ovary of the marine sculpin *Alcichthys alcicornis* (Teleostei: Scorpaeniformes), with internal gametic association. Journal of Experimental Zoology 292: 145-155.

Koya, Y., Ohara, S., Ikeuchi, T., Adachi, S., Matsubara, T. and Yamauchi, K. 1993. Testicular development and sperm morphology in the viviparous teleost, *Zoarces elongatus*. Bulletin of the Hokkaido National Fisheries Research Institute 57: 21-31.

Lahnsteiner, F. 2003. The spermatozoa and eggs of the cardinal fish. Journal of Fish Biology 62: 115-128.

Lahnsteiner, F., Berger, B., Weismann, T. and Patzner, R. A. 1997. Sperm structure and motility in the freshwater teleost *Cottus gobio*. Journal of Fish Biology 50: 564-574.

Lahnsteiner, F. and Mansour, M. 2004. Sperm fine structure of the pikeperch, *Sander lucioperca* and its differences from *Perca fluviatilis*. Journal of Submicroscopic Cytology and Pathology 36: 309-312.

Lahnsteiner, F. and Patzner, R. A. 1990a. Fuctions of the testicular gland of blenniid fish: Structural and histochemical investigations. Experientia 46: 1005-1007.

Lahnsteiner, F. and Patzner, R. A. 1990b. The mode of male germ cell renewal and ultrastructure of early spermatogenesis in *Salaria (= Blennius) pavo* (Teleostei: Blenniidae). Zoologischer Anzeiger 224: 129-139.

Lahnsteiner, F. and Patzner, R. A. 1990c. Spermiogenesis and structure of mature spermatozoa in bleniid fishes (Pisces, Bleniidae). Journal of Submicroscopic Cytology and Pathology 22: 565-576.

Lahnsteiner, F. and Patzner, R. A. 1995. Fine structure of spermatozoa of two marine teleost fishes, the red mullet, *Mullus barbatus* (Mullidae) and the white sea bream, *Diplodus sargus* (Sparidae). Journal of Submicroscopic Cytology and Pathology 27.

Lahnsteiner, F. and Patzner, R. A. 1996. Fine structure of spermatozoa of three teleost fishes of the Mediterranean Sea: *Trachinus draco* (Trachinidae, Perciformes), *Uranuscopus scaber* (Uranuscopidae, Perciformes) and *Synodon saurus* (Synodontidae, Aulopiformes). Journal of Submicroscopic Cytology and Pathology 28: 297-303.

Lahnsteiner, F. and Patzner, R. A. 1997. Fine structure of spermatozoa of four littoral teleosts, *Symphodus ocellatus, Coris julis, Thalassoma pavo* and *Chromis chromis*. Journal of Submicroscopic Cytology and Pathology 29: 477-485.

Lahnsteiner, F. and Patzner, R. A. 1998. Sperm motility of the marine teleosts *Boops boops, Diplodus sargus, Mullus barbatus* and *Trachurus mediterraneus*. Journal of Fish Biology 52: 726-742.

Lahnsteiner, F. and Patzner, R. A. 1999. Characterization of spermatozoa and eggs of the rabbitfish. Journal of Fish Biology 55: 820-835.

Lahnsteiner, F. and Patzner, R. A. 2007. Sperm morphology and ultrastructure in fish. Pp. 1-61. In S. M. H. Alavi, J. Cosson, K. Coward and G. Rafiee (eds), *Fish Spermatology*. Alpha Science International Ltd., Oxford, UK.

Lahnsteiner, F., Richtarski, U. and Patzner, R. A. 1990. The functions of the testicular gland of *Salaria pavo* and *Lipophrys dalmatinus* (Teleostei, Bleniidae) as revealed by electron microscopy and enzyme histochemistry. Journal of Fish Biology 37: 85-97.

Lahnsteiner, F., Wieismann, T. and Patzner, R. A. 1995. Fine structure of spermatozoa of two marine teleost fishes, the red mullet, *Mullus barbatus* (Mullidae) and the white sea bream, *Diplodus sargus* (Sparidae). Journal of Submicroscopic Cytology and Pathology 27: 259-266.

Lauder, G. V. and Liem, K. F. 1983. The evolution and interrelationships of the actinopterygian fishes. Bulletin of the Museum of Comparative Zoology 150: 95-197.

Lee, T.-H., Chiang, T.-H., Huang, B.-M., Wang, T.-C. and Yang, H.-Y. 2006. Ultrastructure of spermatogenesis of the paradise fish, *Macropodus opercularis*. Taiwania 51: 170-180.

Lin, D. J. and You, Y. I. 1998. A study on the changes during maturation of the late spermatid and the structure of the spermatozoon of *Sebasticus marmoratus*. Zoological Research 19: 359-306 (In Chinese).

Lo Nostro, F. L., Grier, H., Meijide, F. J. and Guerrero, G. A. 2003. Ultrastructure of the testis in *Synbranchus marmoratus* (Teleostei, Synbranchidae): The germinal compartment. Tissue and Cell 35: 121-132.

Lou, Y. H. and Takahashi, H. 1989. Spermiogenesis in the Nile tilapia *Oreochromis niloticus* with notes on a unique pattern of nuclear chromatin condensation. Journal of Morphology 200: 321-330.

Lu, M., Ge, Z., Ni, J., Gao, W. and Ye, J. 1999. Ultrastructura [sic] observation on sperm egg and early sperm-penetration in *Fugu obscurus*. Journal of Fisheries Science of China 6: 5-8.

Manni, L. and Rasotto, M. B. 1997. Ultrastructure and histochemistry of the testicular efferent duct system and spermiogenesis in *Opistognathus whitehurstii* (Teleostei, Trachinoidei). Zoomorphology (Berlin) 117: 93-102.

Maricchiolo, G., Genovese, L., Laura, R., Micale, V. and Muglia, U. 2002. The ultrastructure of amberjack (*Seriola dumerilii*) sperm. European Journal of Morphology 40: 289-292.

Maricchiolo, G., Genovese, L., Laura, R., Micale, V. and Muglia, U. 2004. Fine structure of spermatozoa in the common pandora (*Pagellus erythrinus* Linnaeus, 1758) (Perciformes, Sparidae). Histology and Histopathology 19: 1237-1240.

Maricchiolo, G., Genovese, L., Laurà, R., Micale, V. and Muglia, U. 2007. Fine structure of spermatozoa in the gilthead sea bream (*Sparus aurata* Linnaeus, 1758) (Perciformes, Sparidae). Histology and Histopathology 22: 79-83.

Marshall, C. J. 1989. *Cryopreservation and ultrastructural studies on teleost fish gametes.* Honours thesis, University of Queensland.

Matos, E., Batista, C., P., M. and Azevedo, C. 2002a. Ultrastructure of the spermatozoon of *Crenicara punctulatum* (Teleostei, Cichlidae) from the Amazon Basin of Brazil. Revista de Ciências Agrárias, Belém 38: 151-160 (In Portuguese).

Matos, E., Matos, P., Corral, L. and Azevedo, C. 1995. Ultrastructural study of the spermatozoon of *Crenicichla saxatilis* Linnaeus, 1758 (Pisces, Teleostei) from the Amazon region. Brazilian Journal of Morphological Science 12: 109-114.

Matos, E., Santos, M. N. S. and Azevedo, C. 2002b. Biflagellate spermatozoon structure of the hermaphrodite fish *Satanoperca jurupari* (Heckel, 1840) (Teleostei, Cichlidae) from the Amazon River. Brazilian Journal of Biology 62: 847-852.

Matsubara, K. 1943. Studies on the scorpaenid fishes of Japan. In Transactions of the Sigenkagaku Kenyusho 1 and 2 (*Fide* Washington *et al.*, 1984).

Mattei, C. and Mattei, X. 1976a. Présence d'un système membranaire associé au noyau dans le gamète de Pomacentrus leucostictus (Poisson Téléostéen). Comptes Rendus des Seances de la Société de Biologie 170: 234-240.

Mattei, C. and Mattei, X. 1984. Spermatozoïdes biflagellés chez un poisson téléostéen de la famille Apogonidae. Journal of Ultrastructure Research 88: 223-228.

Mattei, C. and Mattei, X. 1978. La spermiogenèse d'un poisson téléostéen (*Lepadogaster lepadogaster*). II. Le spermatozoide. Biologie Cellulaire 32: 267-274.

Mattei, C., Mattei, X. and Marchand, B. 1979. Réinvestigation de la structure des flagelles spermatiques: les doublets 1:2:5 et 6. Journal of Ultrastructure Research 69: 371-377.

Mattei, X. 1969. *Contribution à l'étude de la spermiogenèse et des spermatozoïdes de poissons par les méthodes de la microscopie électronique.* D.Sc. thesis, Université de Montpellier.

Mattei, X. 1970. Spermiogenèse comparée des poissons. Pp. 57-69. In B. Baccetti (ed.), *Comparative Spermatology*. Academic Press, New York.

Mattei, X. 1988. The flagellar apparatus of spermatozoa in fish Ultrastructure and evolution. Biology of the Cell 63: 151-158.

Mattei, X. 1991. Spermatozoon ultrastructure and its systematic implications in fishes. Canadian Journal of Zoology 69: 3038-3055.

Mattei, X. and Mattei, C. 1976b. Ultrastructure du canal cytoplasmique des spermatozoïdes de téléostéens illustrée par l'étude de la spermiogenèse de *Trichiurus lepturus*. Journal de Microscopie et de Biologie Cellulaire 25: 249-258.

Medina, A., Megina, C., Abascal, F. J. and Calzada, A. 2000. The spermatozoon morphology of *Solea senegalensis* (Kaup, 1858) (Teleosti, Pleuronectiformes). Journal of Submicroscopic Cytology and Pathology 32: 645-650.

Miyaki, K. and Yoshikoshi, K. 1993. Scanning electron microscope observation on cryopreserved spermatozoa in six species of pufferfishes. Nippon Suisan Gakkaishi 59: 891.

Miyaki, K., Yoshikoshi, K. and Tabeta, O. 1996. Transmission electron microscopic observations on spermatozoa in seven species of puffers genus *Takifugu* (Tetraodontidae, Tetraodontiformes). Fisheries Science (Tokyo) 62: 543-546.

Mizue, K. 1968. Studies on a scorpaenous fish *Sebastiscus marmoratus* Cuvier et Valenciennes -VI Electron-microscopic study of spermatogenesis. Bulletin of the Faculty of Fisheries, Nagasaki University 25: 9-24.

Morisawa, M. 1982. The fine structure of the spermatozoon of the goby, *Acanthogobius flavimanus*. Bulletin of St. Marianna University 11: 31-37.

Morisawa, S. 2001. Ultrastructural studies of late-stage spermatids and mature spermatozoa of the puffer fish, *Takifugu niphobles* (Tetraodontiformes) and the effects of osmolality on spermatozoan structure. Tissue and Cell 33: 78-85.

Munehara, H. 1988. Spawning and subsequent copulating behavior of the elkhorn sculpin *Alcichthys alicornis* in an aquarium. Japanese Journal of Ichthyology 35: 358-364.

Munehara, H. 1996. Sperm transfer during copulation in the marine sculpin *Hemitripterus villosus* (Pisces: Scorpaeniformes) by means of a retractable genital duct and ovarian secretion in females. Copeia 1996: 452–454.

Munehara, H. 1997. The reproductive biology and early life stages of *Podothecus sachi* (Apogonidae). Fishery Bulletin 95: 612-619.

Munehara, H., Takano, K. and Koya, Y. 1989. Internal gametic association and external fertilization in the elkhorn sculpin, *Alcichthys alcicornis*. Copeia 1989: 673–678.

Munoz, M., Casadevall, M. and Bonet, S. 2002a. Gametogenesis of *Helicolenus dactylopterus* dactylopterus (Teleostei, Scorpaenidae). Sarsia 87: 119-127.

Munoz, M., Casadevall, M. and Bonet, S. 2002b. Testicular structure and semicystic spermatogenesis in a specialized ovuliparous species: *Scorpaena notata* (Pisces, Scorpaenidae). Acta Zoologica (Stockholm) 83: 213-219.

Munoz, M., Casadevall, M., Bonet, S. and Quagio-Grassiotto, I. 2000. Sperm storage structures in the ovary of *Helicolenus dactylopterus dactylopterus* (Teleostei: Scorpaenidae): An ultrastructural study. Environmental Biology of Fishes 58: 53-59.

Munoz, M., Koya, Y. and Casadevall, M. 2002c. Histochemical analysis of sperm storage in *Helicolenus dactyolpterus dacylopterus* (Teleostei: Scorpaenidae). Journal of Experimental Zoology 292: 156-164.

Munoz, M., Sabat, M., Mallol, S. and Casadevall, M. 2002d. Gonadal structure and gametogenesis of *Trigla lyra* (Pisces: Triglidae). Zoological Studies 41: 412-420.

Nelson, J. S. 1984. *Fishes of the World*, 2nd edition. John Wiley and Sons, New York. 523 pp.

Nelson, J. S. 2006. *Fishes of the World*, 4th edition. John Wiley and Sons, Inc., Hoboken, New Jersey. 601 pp.

Orrell, T. M., Carpenter, K. E., Musick, J. A. and Graves, J. E. 2002. Phylogenetic and biogeographic analysis of the Sparidae (Perciformes: Percoidei) from cytochrome b sequences. Copeia 2002: 618-631.

Pardo, B., Machordom, A., Foresti, F., Porto-Foresti, F., Azevedo, F. C., Banón, R., Sánchez, L. and Martínez, P. 2005. Phylogenetic analysis of flatfish (Order Pleuronectiformes) based on mitochondrial 16s rDNA sequences. Scientia Marina 69: 531-543.

Patterson, C. 1964. A review of Mesozoic Acanthopterygian fishes, with special reference to those of the English Chalk. Philosophical Transactions of the Royal Society London Ser B 247: 213-482.

Pavlov, D. A. 2005. Reproductive biology of wolffishes (Anarhichadidae) and the transition from oviparity to viviparity in the Suborder Zoarcoidei. In M. C. Uribe and H. J. Grier (eds), *Vivparous Fishes*. New Life Publications, Homestead, Fl.

Pavlov, D. A., Knudsen, P., Emel'yanova, N. G. and Moksness, E. 1997. Spermatozoon ultrastructure and sperm production in wolffish (*Anarhichas lupus*), a species with internal fertilization. Aquatic Living Resources 10: 187-194.

Pecio, A., Burns, J. R. and Weitzman, S. H. 2005. Sperm and spermatozeugma ultrastructure in the inseminating species *Tyttocharax cochui*, *T. tambopatensis*, and *Scopaeocharax rhinodus* (Pisces: Teleostei: Characidae: Glandulocaudinae: Xenurobryconini). Journal of Morphology 263: 216-226.

Pieterse, G. M. 2006. *Histopathological changes in the testis of Oreochromis mossambicus (Cichlidae) as a biomarker of heavy metal pollution*. Philosophiae Doctor in Aquatic Health thesis, Rand Afrikaans University.

Quagio-Grassiotto, I., Antoneli, F. N. and Oliveira, C. 2003. Spermiogenesis and sperm ultrastructure in *Cichla intermedia* with some considerations about Labroidei spermatozoa (Teleostei, Perciformes, Cichlidae). Tissue and Cell 35: 441-446.

Quinitio, G. F. and Takahashi, H. 1992. Occurrence of aberrant spermatids in the freshwater sculpin, *Cottus hangionensis*. Pp. 802-805. In L. M. Chou, A. D. Munro *et al.* (eds), *The third Asian fisheries forum: Proceedings of the Third Asian Fisheries Forum Singapore 26-30*. Asian Fisheries Society, Manila, Philippines.

Quinitio, G. F. and Takahashi, H. 1992. An ultrastructural study on the occurrence of aberrant spermatids in the testis of the river sculpin, *Cottus hangiongensis*. Japanese Journal of Ichthyology 39: 235-241.

Retzius, G. 1905. Die Spermien der Leptokardier, Teleostier und Ganoiden. Biologische Untersuchungen von G Retzius Nf 12: 103-115.

Richards, W. J. and Leis, J. M. 1984. Labroidei: development and relationships. Pp. 542-547. In H. G. Moser, W. J. Richards, D. M. Cohen, M. P. Fahay, A. W. Kendall and S. L. Richardson (eds), *Ontogeny and Systematics of Fishes. Special Publication Number 1 American Society of Ichthyologists and Herpetologists*, vol. 1. Allen Press, Lawrence, Kansas.

Robinson, M. P. and Prince, J. S. 2003. Morphology of the sperm of two wrasses, *Thalassoma bifasciatum* and *Lachnolaimus maximus* (Labridae, Perciformes). Bulletin of Marine Science 72: 247-252.

Rosen, D. E. 1984. Zeiforms as primitive plectognath fishes. American Museum Novitates 2782: 1-45.

Rupple, D. 1984. Gobioidei: development. Pp. 582-587. In H. G. Moser, W. J. Richards, D. M. Cohen, M. P. Fahay, A. W. Kendall and S. L. Richardson (eds), *Ontogeny and Systematics of Fishes*. Special Publication Number 1 American Society of Ichthyologists and Herpetologists. Vol. 1. Allen Press, Lawrence, Kansas.

Samira, S. A. 1999. Reproductive biology, spermatogenesis and ultrastructure of testes *Caranx crysos* (Mitchill, 1815). Bulletin of the National Institute of Oceanography and Fisheries, Alexandria 25: 311-329.

Silva, M. and Godhino, H. P. 1991. O spermatozóide de *Oreochromis niloticus* (Peixe, Teléosteo). Revista Brasileira de Biologia 51: 39-43.

Silveira, H., Rodrigues, P. and Azevedo, C. 1990. Fine structure of the spermatogenesis of *Blennius pholis* (Pisces, Blenniidae). Journal of Submicroscopic Cytology and Pathology 22: 103-108.

Smith, J. L. B. and Smith, M. M. 1986. Family No. 183: Sparidae. Pp. 1050. In M. M. Smith and P. C. Heemstra (eds), *Smith's Sea Fishes*. Macmillan, Johannesburg, South Africa.

Stanley, H. P. 1966. A fine structural study of spermiogenesis in the teleost fish *Oligocottus maculosus*. Anatomical Record 154: 426-427.

Stanley, H. P. 1969. An electron microscope study of spermiogenesis in the teleost fish *Oligocottus maculosus*. Journal of Ultrastructure Research 27: 230-243.

Stein, H. 1981. Licht- und elektronenoptische Untersuchungen an den Spermatozoen verschiedener Süsswasserknochenfische (Teleostei). Zeitschrift für Angewandte Zoologie 68: 183-198.

Stiassny, M. L. J. 1980. *The anatomy and phylogeny of two genera of African cichlid fishes*. PhD thesis, University of London (*Fide* Lauder and Liem 1983).

Stockley, M. T., Gage, M. J. G., Parker, G. A. and Moller, A. P. 1997. Sperm competition in fishes: the evolution of testis size and ejaculation characteristics. American Naturalist 149: 933-954.

Streelman, J. T. and Karl, S. A. 1997. Reconstructing labroid evolution with single-copy nuclear DNA. Proceedings of the Royal Society B 264: 1011-1020.

Suquet, M., Billard, R., Cosson, J., Dorange, G., Chauvaud, L., Mugnier, C. and Fauvel, C. 1994. Sperm features in turbot (*Scophthalmus maximus*): a comparison with other freshwater and marine species. Aquatic Living Resources 7: 283-294.

Suquet, M., Dorange, G., Omnes, M. H., Normant, Y., Le Roux, A. and Fauvel, C. 1993. Composition of the seminal fluid and ultrastructure of the spermatozoon of turbot (*Scophthalmus maximus*). Journal of Fish Biology 42: 509-516.

Suwanjarat, J., Amornsakun, T., Thongboon, L. and Boonyoung, P. 2005. Seasonal changes of spermatogenesis in the male sand goby *Oxyeleotris marmoratus* Bleeker, 1852 (Teleostei, Gobiidae). Songklanakarin Journal of Science and Technology 27.

Taddei, A. R., Abelli, L., Baldacci, A., Fausto, A. M. and Mazzini, M. 1999. Ultrastructure of spermatozoa of marine teleosts *Umbrina cirrosa* L. and *Diplodus puntazzo* (Cetti) and preliminart observations on effects of cryopreservation. Pp. 407-413. In G. Enne and G. F. Greppi (ed.), *New Species for Mediterranean Aquaculture*. Elsevier, Paris.

Taddei, A. R., Barbato, F., Abelli, L., Canese, S., Moretti, F., Rana, K. J., Fausto, A. M. and Mazzini, M. 2001. Is cryopreservation a homogeneous process? Ultrastructure and motility of untreated, prefreezing, and postthawed spermatozoa of *Diplodus puntazzo* (Cetti). Cryobiology 42: 244-255.

Thresher, R. E. 1984. *Reproduction in Reef Fishes*. TFH Publications, Brookvale-NSW.

Tyler, J. C. 1968. A monograph on plectognath fishes of the superfamily Triacanthoidea. Monographs of the Academy of Natural Sciences Philadelphia 16: 1-364.

Tyler, J. C. and Bannikov, A. F. 1992. A remarkable new genus of tetraodontiform fish with features of both balistids and ostraciids from the Eocene of Turkmenistan. Smithsonian Contributions to Paleobiology 72: 1-17.

Vallisneri, M., Tinti, F., Stagni, A. and Tommasini, S. 2001. Assessment of gonadal cycle in *Solea solea* L. by a histo-cytological study. Rivista di Idrobiologia 40: 27-36.

Van Der Elst, R. 1976. Game fish of the east coast of South Africa. I. The biology of the elf, *Pomatomus saltatrix* (Linnaeus), in the coastal waters of Natal. Ocean Research Institute. Invest. Reports 44: 1-59.

Wang, H. T., Xu, Y. L. and Zhang, P. J. 1999. Ultrastructure of flounder (*Paralichthys olivaceus*) sperm. Marine Science (Beijing) 6: 5-7 (In Chinese).

Wang, H. T., Zhang, P. J., Xie, J. L. and Jiang, M. 2002. Further observation of the spermatozoa of left-eye flounder *Paralichthys olivaceus* by electronic microscopy. Chinese Journal of Oceanology and Limnology 20.

Washington, B. B., Moser, H. G., Laroche, W. A. and Richards, W. J. 1984a. Scorpaeniformes: development. Pp. 405-428. In H. G. Moser, W. J. Richards, D. M. Cohen, M. P. Fahay, A. W. Kendall and S. L. Richardson (eds), *Ontogeny and Systematics of Fishes. Special Publication Number 1 American Society of Ichthyologists and Herpetologists*, vol. 1. Allen Press, Lawrence, Kansas.

Washington, B. B., Eschmeyer, W. N. and Howe, K. M. 1984b. Scorpaeniformes: relationships. Pp. 438-447. In H. G. Moser, W. J. Richards, D. M. Cohen, M. P. Fahay, A. W. Kendall and S. L. Richardson (eds), *Ontogeny and Systematics of Fishes. Special Publication Number 1. American Society of Ichthyologists and Herpetologists*, vol. 1. Allen Press, Lawrence, Kansas.

Watanabe, S., Hara, M. and Watanabe, Y. 2000. Male internal fertilization and introsperm-like sperm of the Seaweed pipefish (*Syngnathus schlegeli*). Zoological Science (Tokyo) 17: 759-767.

Webb, P. W. and Brett, J. R. 1972. Respiratory adaptations of prenatal young in the ovary of two species of viviparous seaperch *Rhacochilus vacca* and *Embiotoca lateralis*. Journal of the Fish Research Board of Canada 29: 1525-1542.

Winterbottom, R. 1974. The familial phylogeny of the Tetraodontiformes (Acanthopterygii: Pisces) evidenced by their comparative myology. Smithsonian Contributions to Zoology 155: 1-201.

Woodland, D. J. 1990. *Revision of the fish family Siganidae with descriptions of two new species and comments on distribution and biology*. Bernice P. Bishop Museum, Honolulu, Hawaii.

Yao, Z. and Crim, L. W. 1995. Copulation, spawning and parental care in captive ocean pout. Journal of Fish Biology 47: 171-173.

Yao, Z., Emerson, C. J. and Crim, L. W. 1995. Ultrastructure of the spermatozoa and eggs of the ocean pout (*Macrozoarces americanus* L.), an internally fertilizing marine fish. Molecular Reproduction and Development 42: 58-64.

Yao, Z., Crim, L. W., Richardson, G. F. and Emerson, C. J. 2000. Motility, fertility and ultrastructural changes of ocean pout (*Macrozoarces americanus* L.) sperm after cryopreservation. Aquaculture 181: 361-375.

Yasuzumi, F. 1971. Electron microscope study of the fish spermiogenesis. Journal of Nara Medical Association 22: 343-355.

You, Y. I. and Lin, D. J. 1998. The release of vesicles from the nucleus of *Tilapia nilotica* spermatid during spermiogenesis. Acta Zoologica Sinica 44: 257-263.

You, Y.-L. and Lin, D.-J. 1997. The ultrastructure of the spermatozoon of the teleost, *Pseudosciaena crocea* (Richardson). Acta Zoologica Sinica 43: 119-126.

You, Y.-L., Lin, D.-J. and Zhong, X.-R. 2002. Ultrastructure of spermatogenesis of ovoviviparous teleost, Japanese stingfish *Sebastiscus marmoratus* Cuvier et Valenciennes. Journal of Tropical Oceanography 21: 70-75.

Zaki, M. I., Negm, R. K., El-Agamy, A. and Awad, G. S. 2005. Ultrastructure of male germ cells and character of spermatozoa in *Boops boops* (family Sparidae) in Alexandria Coast, Egypt. Egyptian Journal of Aquatic Research 31.

Zhang, X., Jiang, M., Yao, F., Cong, J., Guo, E., Gao, L., Fan, R. and Yao, S. 1999. On the ultrastructure of the spermatozoa of the Oriental red-globefish (*Takifugu rubripres*). Journal of Oceanography of the University of QIngdao 29: 255-258.

Zhao, H., Liu, X., Lin, H. and Liufu, Y. 2003. Ultrastructure of spermatozoa and the effects of salinity, temperature and pH on spermatozoa motility of *Epinephelus coioides* orange-spotted grouper. Journal of Fishery Sciences of China 10: 286-290.

Zutshi, B. and Murthy, P. S. 2001. Ultrastructural changes in testis of gobiid fish *Glossogobius giuris* (Ham) induced by fenthion. Indian Journal of Experimental Biology 39: 170-173.

CHAPTER 16

Ultrastructure of Spermatozoa: Sarcopterygii

Barrie G. M. Jamieson

16.1 THE SARCOPTERYGII

The classification of the Sarcopterygii (lobe-finned fishes and derived relatives) employed here is chiefly that of Cloutier and Ahlberg (1996) which was adopted by Nelson (2006). Despite some equivocal results, molecular studies appear to confirm the finding (e.g. Meyer and Wilson 1990; Meyer and Dolven 1992) that lungfishes are the sister-group of the Tetrapoda though these two groups are also the closest relatives of *Latimeria* (Yokobori *et al.* 1994; Zardoya and Meyer 1996, 1997a; b). From analysis of sequences of 44 nuclear genes Takezaki *et al.* (2004) concluded that the coelacanth, lungfish, and tetrapod lineages diverged within such a short time interval that their relationships appear to be an irresolvable trichotomy. All of these studies endorse the sarcopterygian status of *Latimeria* demonstrated by Hillis *et al.* (1991). The sister relationship between the Sarcopterygii and Acinopterygii is endorsed by Ortí and Li (**Chapter 1**, Fig. 1.3) in a consideration of phylogenetic analyses of morphology and molecular data depicting the Coelacanthimorpha as the sister group of the Ceratodontiformes +Tetrapoda. In a molecular analysis of 8 genes, including 11,766 base pairs (Fig. 1.4) they find the Sarcopterygii to be the sister of the Actinopterygii, though sarcopterygian fishes are not represented.

16.1.1 Sperm Literature

The spermatozoa of the sarcopterygian fishes have been described ultrastructurally for the coelacanthimorphan *Latimeria chalumnae* (Mattei *et al.* 1988a); and for Ceratodontimorpha represented by the Australian Lungfish,

School of Integrative Biology, University of Queensland, Brisbane, Queensland 4072, Australia

Fig. 16.1 The coelacanth, *Latimeria chalumnae* (Latimeriidae). Natural History Museum London specimen. Photo B. G. M. Jamieson.

Neoceratodus forsteri (Jespersen 1971; Jamieson 1995, 1999); *Protopterus* (Boisson 1963; Boisson *et al.* 1967; Purkerson *et al.* 1974); and *Lepidosiren* (Matos *et al.* 1988; Matos and Azevedo 1989). Tetrapod sperm literature relevant to this study is briefly summarized in Jamieson (1995, 1999).

16.2 SUBCLASS COELACANTHIMORPHA (ACTINISTIA)

The relationships of the Coelacanthimorpha, and specifically *Latimeria,* were analyzed by Jamieson (1991) in the context of gnathostome evolution. Two, albeit preliminary, parsimony analyses on different though partly overlapping data indicated that coelacanths are the plesiomorph sister-group of the Ceratodontiformes (= extant Dipnoi) + tetrapods, the triplet comprising the Sarcopterygii.

16.3 ORDER COELACANTHIFORMES

Coelacanthiformes have the caudal fin diphycercal, consisting of three lobes; external nostrils and no choana (for additional characters, see Nelson 2006). The body length in *Latimeria* reaches 1.8 m. The order contains several Devonian to Cretaceous families but one family (Latimeriidae) has two living representatives in the genus *Latimeria* (Fig. 16.1) (Nelson 2006).

Latimeria chalumnae is known from its South African (East London) type locality and a major population off the Comoros. It has since been captured off the coasts of Mozambique, Madagascar, Kenya and Tanzania, all in the western Indian Ocean, and the Jesser Canyon at Sodwana Bay off the northeast coast of South Africa (References in Sasaki *et al.* 2007). The second known species, *Latimeria menadoensis* occurs in Indonesian waters (References in Erdmann *et al.* 1999).

16.3.1 Phylogenetic Relationships of Coelacanths

There are two chief theories of coelacanth relationships with other groups. The first may be termed the sarcopyterygian theory and the second the chondrichthyan theory.

Sarcopyterygian theory: In the sarcopyterygian theory, lungfishes, coelacanths and tetrapods form a monophyletic Sarcopterygii. As stated by Brinkmann, Denk *et al.* (2004), these three groups all originated within a short time window of about 20 million years, in the early Devonian (about 380 to 400 million years ago) and this short divergence time makes determination of the phylogenetic relationships among the three lineages difficult.

Three alternative evolutionary pathways are usually envisaged (Fig. 16.2A-C). All three envisage Sarcopterygii as the sister-group to the Actinopterygii but differ as to whether lungfishes are the sister of the tetrapods (Fig. 16.2A), coelacanths are the sister of the tretrapods (Fig. 16.2B) or lungfishes and coelacanth are sisters in a clade which is sister to the tetrapods (Fig. 16.2C).

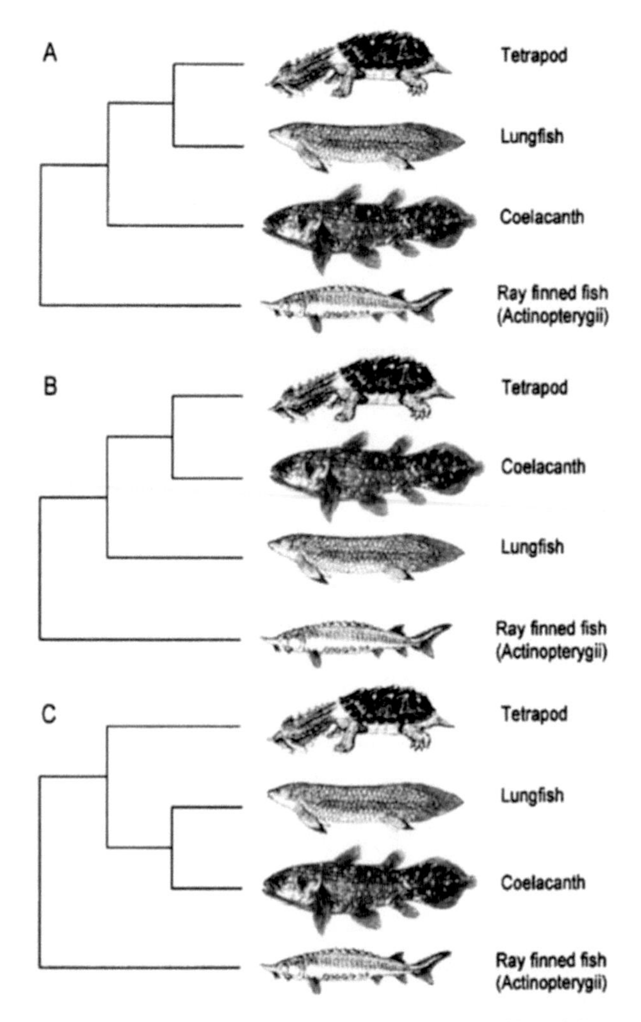

Fig. 16.2 The sarcopterygian theory of coelacanth relationships, showing the three alternative phylogenies. After Brinkmann, H., Denk, A., Zitzler, J., Joss, J. J. and Meyer, A. 2004. Journal of Molecular Evolution 59: 834-848, Fig. 1. With kind permission of Springer Science and Business Media.

Morphological similarities between lungfishes and tetrapods are briefly referred to in **Section 16.5**, below.

Forey (1988) notes similarities associated with the brain between *Latimeria* and lungfishes, giving some support to **C**. On the other hand it has been suggested that the unique structure of the basilar papillae of the inner ear is an homology [synapomorphy] between *Latimeria* and tetrapods (Fritsch 1987), supporting **B**. Bernstein (2004) refutes the view that the organ in *Latimeria* functions as a tympanic ear but it is not clear that this excludes homology.

The possession of ectodermal tooth enamel is a further resemblance (Miller 1969). The presence of a notochord in the absence of vertebral centra is a resemblance between *Latimeria* and lungfishes (references in Locket 1980), which might be regarded as supporting **C** but is probably a symplesiomorphy. The relatively small size of its nucleated erythrocytes and the striking resemblance of the ultrastructural features of its leukocytes to those of higher vertebrate leukocytes is considered to support the view that *Latimeria* is a close living relative of tetrapods (Jarial 2005), giving some support to **B**.

Some molecular mitogenomic and/or nuclear analyses relevant to the three alternative phylograms **A-C**, will be briefly outlined here.

Zardoya and Meyer (1996) supported **C**, monophyly of three lungfish genera + *Latimeria chalumnae* as the joint sister of the tetrapods but no chondrichthyans were included. Brinkmann, Denk *et al.* 2004, rejected **B** but results were ambiguous between **A** and **C**. Brinkmann, Venkatesh *et al.* (2004) unequivocally supported **A,** and both studies showed chondrichthyes as the plesiopmorph sister to all other gnathostomes. Kikugawa *et al.* (2004) supported **A**, with basal Chondrichthyes, but with no coelacanths. Takezaki *et al.* (2004) despite analysing sequences of 44 nuclear genes encoding amino acid residues at 10,404 positions found that the results did not support conclusively any one of the three hypotheses (Fig. 16.2A,B,C); Chondrichthyes were basal. Alternative **A** was supported in a maximum parsimony and a maximum likelihood tree, while **C** was supported [less convincingly] in a neighbor-joining tree. They concluded as noted above that the coelacanth, lungfish, and tetrapod lineages formed an irresolvable trichotomy. Hedges *et al.* (1993) supported **A** (with no cartilaginous fish represented) but Blair and Hedges (2005) supported **C**, with cartilaginous fish as the most basal gnathostomes (though neither was unambiguously placed). Noonan *et al.* (2008) placed *Latimeria menadoensis* near the tetrapods but did not include lungfishes. Hillis and Dixon (1989) found support for regarding *Latimeria chalumnae* as the sister of the tetrapods but lungfishes were not represented in the analysis.

It is thus seen that, where the sarcopterygian theory is upheld, most molecular analyses support **A**: lungfishes are the sister-taxon of tetrapods and *Latimeria* is the sister of lungfishes + tetrapods. Where chondrichthyans are included in these analyses they are seen as basal relative to an actinopterygian + osteichthyan (including tetrapod) assemblage. These views are provisionally accepted here.

Inoue *et al.* (2005), showed the two *Latimeria* species, representing the Sacopterygii, to form a clade above the shark *Scyliorhinus* and as the sister of a large range of Actinopterygii. However, no lungfishes or tetrapods were included in the analysis.

Chondrichthyan theory: Locket (1980) has given a review of coelacanth biology. Morphological evidence proposed in support of a sister relationship between *Latimeria* and the Chondrichthyes was summarized by Forey (1988) and cannot be given in detail here. Løvtrup (1977) recognized ten soft anatomical and physiological characters shared by *Latimeria* and Chondrichthyes. Some of these were rejected by Forey in terms of relationship of the two taxa on the grounds that they might be primitive vertebrate features: large eggs, fatty liver, simple structure of the thymus and thyroid, and salt-secretion through a cloacal gland and we may add the spiral valve in the intestine. However, it may be argued that in the early evolution of gnathostomes these were gnathostome synapomorphies which would have linked the coelacanths and cartilaginous fishes. Whether retention of high levels of urea and trimethylamine oxids in both groups is synapomorphic is debatable. Lagios (1979) added apparently apomorphic characters of the pituitary gland shared by the two groups. Characteristics of the haemoglobin of *Latimeria* appear to be intermediate between those of elasmobranchs and chondrosteans on the one hand and higher bony fish on the other Wood *et al.* (1972; see also references in Locket 1980).

A valuable discussion of features of *Latimeria* relative to bony fishes, particularly lungfishes, and tetrapods is given by Forey (1988) who concludes that that the seemingly primitive structure of the brain, heart (lacking S-curve) and intestine are not expected in a tetrapod ancestor and that some characters, such as urea retention and large eggs are more like those of cartilaginous fishes. He poses the question, if *Latimeria* is the closest relative of tetrapods, why are there so few tetrapod-like features? To attempt some answer to this query we will here turn to molecular evidence supposedly supporting relationship with cartilaginous fishes.

Molecular evidence for the theory that coelacanths are the closest relatives of chondrichthyans is sparse. Arnason *et al.* (2004), in an analysis of amino acid sequences of 12 mitochondrial protein-coding genes rejected the hypothesis that gnathostome fish, including coelacanth and dipnoans, were precursory to tetrapods, but the relationships of coelacanths, represented by *Latimeria chalumnae*, were poorly resolved. They claimed, however, in a re-analysis of nuclear 18S and 28S rRNA genes, support for the sister-group relationship of tetrapods to gnathstome fish and portrayed Chondrichthyes and coelacanth (Actinistia) as mutual sister-taxa. Mallatt and Winchell (2007), in an analysis of 28S and 18S rRNA genes, criticized the analyses of Arnason *et al.* (2004) on the grounds that the lamprey, the only outgroup used by the latter authors, could be too distant an outgroup to allow resolution among the basal gnathostomes. It is, however, difficult to envisage what outgoup could have been closer to the base of the gnathostomes. Signal loss in the fast evolving mitochondrial genomes might, indeed, as they suggested, have obscured relationships but the alternative Arnason *et al.* analysis based on the more slowly evolving nuclear genome, with Chondrichthyes and Actinistia as sisters, would not be subject to this caveat. Nevertheless, Mallatt and Winchell (2007) (their Fig. 3) showed *Latimeria chalumnae* as the plesiomorph sister of the Chondrichthyes (elasmobranchs and Holocephali) though with lungfishes basal to all other ganthostomes including *Latimeria* and chondricthyes. They

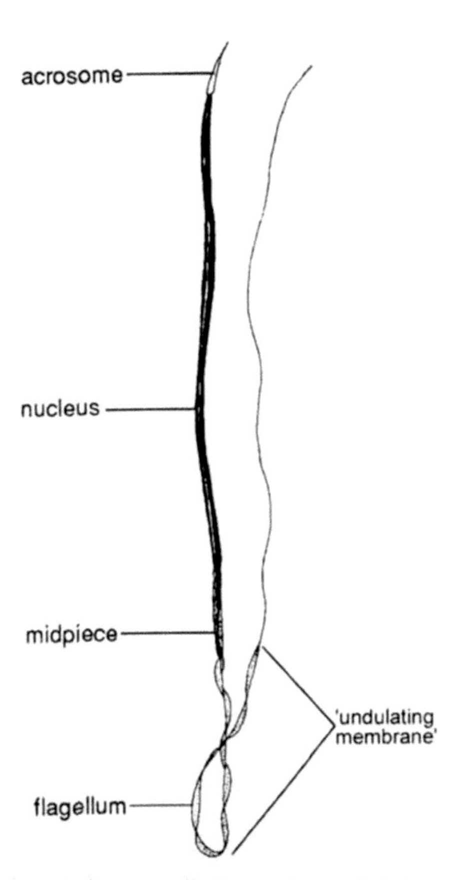

Fig. 16.3 *Latimeria chalumnae* (Latimeriidae). Light microscope drawing of spermatozoon labeled by the present author. After Tuzet, O. and Millot, J. (1959). Annales des Sciences Naturelles Zoologie 12 Ser 1: 61-69, Fig.3C.

thus implied paraphyly of Euteleostomi (=Osteichthyes) owing to the isolated position of *Latimeria*. Arguments from molecular data for a sister relationship of *Latimeria* and chondrichthyans have not been widely accepted.

Conclusion: To date molecular analysis has not been more successful than morphocladistics in elucidating the relationships of *Latimeria* and it is to be regretted that the former appears to have largely displaced the latter. It is to be hoped, however, that analysis of larger gene sequences will provide a definitive phylogenetic classification. At present a sarcopterygian status, and alternative A, seem the most acceptable. A tentative attempt will be made in **Section 16.3.3** to resolve chondrichthyan, bony fish and tetrapod affinities of *Latimeria* in terms of features of the spermatozoon and particularly to examine aspects of the axonemal ultrastructure which it shares with cartilaginous fishes and sarcopterygians.

16.3.2 Ultrastructure of *Latimeria* Spermatozoon

The spermatozoa of coelacanths are known only from formalin-fixed material of *Latimeria chalumnae*, examined by light microscopy (Tuzet and Millot (1959) (Fig. 16.3), and by TEM (Mattei 1988, a brief mention of the flagellum; and Mattei *et al.* 1988a) (Fig. 16.4).

The spermatozoon of *Latimeria* is filiform (Tuzet and Millot 1959). The ultrastructure of sperm in the testis is the same as that in the renal canal (Mattei *et al.* 1988b) (Fig. 16.4A-E).

The gently curved acrosome is 2.8 μm (Tuzet and Millot 1959), confirmed as about 3 μm long (Mattei *et al.* 1988b). Presence, almost constantly, of a "siderophilic" granule, thought to correspond with a centrosome, at is anterior extremity (Tuzet and Millot 1959) has not been confirmed, The anterior end of the acrosome is prolonged into a slender point, 0.2 μm in diameter, and its base is situated laterally on the tip of the nucleus (Fig. 16.4C). The acrosome contains three longitudinal rodlets (putative perforatoria) which extend from deep within the nucleus (see below) as far as its anterior extremity (Mattei *et al.* 1988b) (Fig. 16.4A,C,D).

The head [acrosome + nucleus] is 26 μm long (Tuzet and Millot 1959). A length of approximately 25 μm is reported for the nucleus by Mattei *et al.* (1988b). The contents of the nucleus are dense and compact, though in the aged spermatid concentric laminae are visible. The anterior extremity of the nucleus is drawn out into a bevel and penetrated by a central cavity, like a

---

Fig. 16.4 A-D *Latimeria chalumnae* (Latimeriidae). Formalin-fixed material. **A**. Longitudinal section of base of nucleus, showing accompanying mitochondria. **B**. Transverse section (TS) near base of nucleus, showing C-shaped mitochondrial formation. **C** and **D**. Longitudinal sections of a spermatozoon with transverse sections at levels indicated. **E**. TS of axonemes, enlarged in the inset, showing accessory axonemal columns which are present only proximally. a, acrosome vesicle; aa, anterior end of acrosome; cs, cytoplasmic sleeve; dc, distal centriole; dg, dense granules; ec, endonuclear canal; p, putative perforatorium; pc, proximal centriole; rb?, putative retronuclear body; sr, satellite ray. Courtesy of Professor Xavier Mattei. **F-J**. Sections of spermatozoa of Chondrichthyes. **F-I**. *Chiloscyllium punctatum* (Hemiscyllidae). **F**. Longitudinal sections of the acrosome and tip of nucleus. **G**. Longitudinal section of flagellum showing accessory axonemal columns. **H** and **I**. TS axonemes, each with two accessory axonemal columns. **J**. *Dasyatis fluviorum* (Dasyatidae). TS of a testicular spermatozoon through the transient cytoplasmic sleeve which surrounds the axoneme and paired accessory axonemal columns. The sleeve contains small coated vesicles, large vesicle with double membranes (putative modified mitochondria) and separate membranes. **C-H** adapted from Jamieson, B. G. M. 2005. Chondrichthyan spermatozoa and phylogeny. Pp. 201-236. In W. M. Hamlett (ed.), *Reproductive biology and phylogeny of Chondrichthyes: sharks, batoids and chimaeras.*, vol. 3. Science Publishers, Enfield, Figs. 7.7A-E.

Fig. 16.4A-D Contd. ...

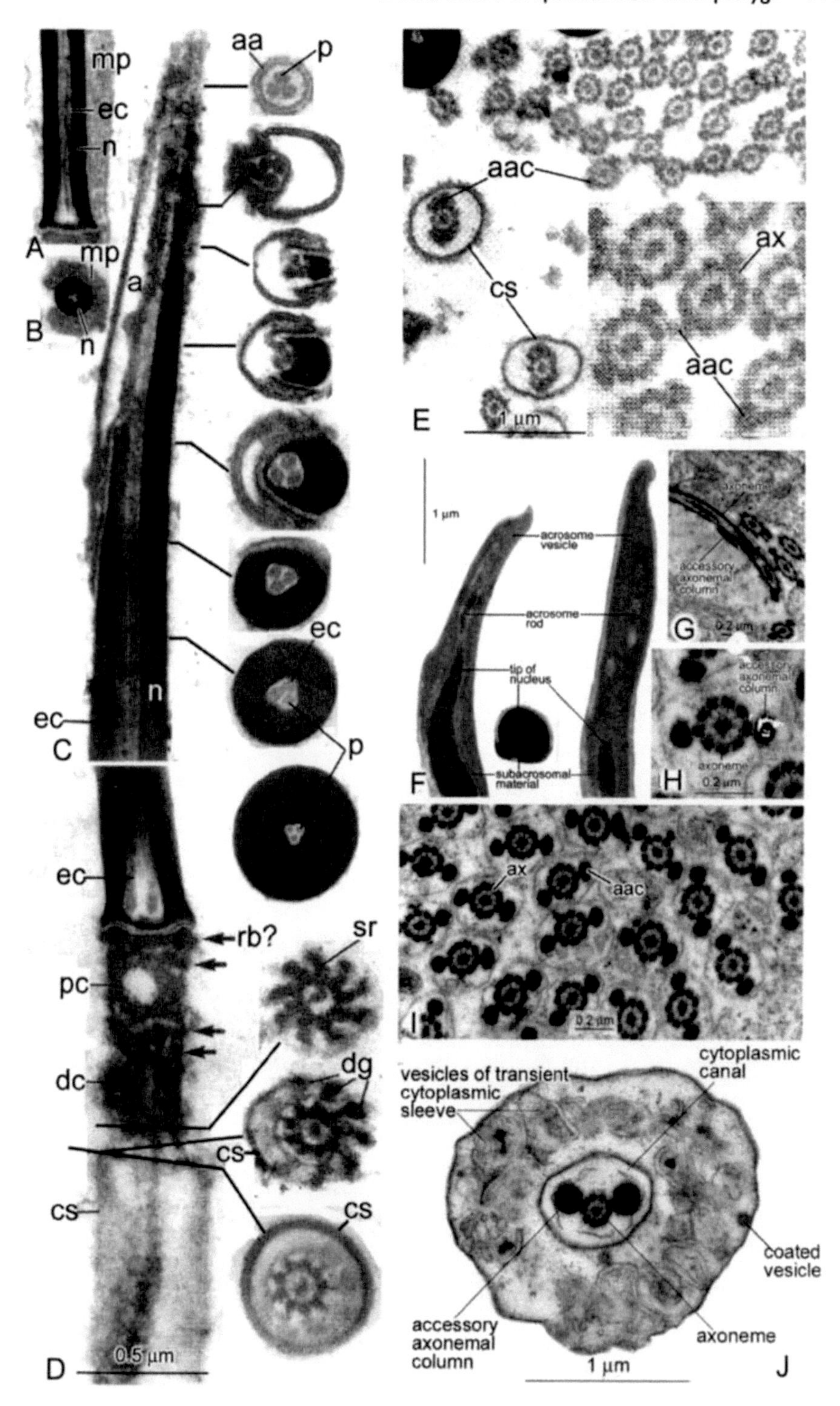

mp
ec
n
A
mp
B
n
a
aa
p
ec
n
p
ec
C
ec
rb?
pc
dc
sr
dg
cs
cs
cs
0.5 µm
D
aac
cs
ax
aac
1 µm
E
1 µm
acrosome vesicle
acrosome rod
tip of nucleus
subacrosomal material
F
axoneme
accessory axonemal column
0.2 µm
G
accessory axonemal column
axoneme
0.2 µm
H
ax
aac
0.2 µm
I
vesicles of transient cytoplasmic sleeve
cytoplasmic canal
coated vesicle
accessory axonemal column
axoneme
1 µm
J

hypodermic needle (Fig. 16.4C). This intranuclear [endonuclear] canal penetrates the entire length of the nucleus but does not perforate its base Fig. 16.4A,D).

The diameter of the nucleus is 0.4 μm at the base of the acrosome, 0.6 μm in its midregion, and 0.3 μm for a posterior region of about 5 μm in length. The posterior extremity of the nucleus is enlarged, as is the endonuclear canal, in this region (Fig. 16.4A,D).

The transverse section of the intranuclear canal is trilobed as it contains the three rodlets (here regarded as perforatoria) which are prolonged into the acrosome, and comprise the "intranuclear filament" of Tuzet and Millot (1959). The rodlets are each 50 nm in diameter and are very regularly arranged. Two (abnormal) giant nuclei were observed, among the very many spermatozoa observed, in one of which the intranuclear canal had an ovoid cross section and contained only two rods and the other of which had three canals and four rods.

The posterior region of the nucleus is surrounded by a cylindrical sleeve (manchon perinucleaire) which is usually not complete (Fig. 16.4A,B). This sleeve appears to be homogeneous and to be limited by a double membrane. It is absent from some sperm, probably because of its fragility. It is implied that the sleeve is mitochondrial as spermatozoa observed by light microscopy after staining for mitochondria show a nuclear length of only 20 μm and an intermediate piece 5 μm long. In a TEM micrograph, what appears to be an indubitable mitochondrion extends from below the centriolar complex, reaching almost to the beginning of the cytoplasmic canal, and extends anteriorly up the side of the basal region of the nucleus (Mattei *et al.* 1988b). This corresponds with the report by Tuzet and Millot (1959) of an elongated midpiece which, with the two posterior "centrosomes", constituted an intermediate segment 5 μm long.

The 'neck', about 0.8 μm long, contains two centrioles and paracentriolar structures which are disposed between the proximal centriole and the nucleus and around the two centrioles (Fig. 16.4D). The proximal centriole, consisting of 9 triplets, is at right angles to the distal centriole which latter forms the basal body of the flagellum. The distal centriole is provided with 9 satellite rays. Satellite rays, arising from the distal centriole, are common in the aquasperm of invertebrates but are rare in fishes, being seen in *Latimeria*, dipnoans (Boisson *et al.* 1967; Jespersen 1971; Mattei *et al.* 1988) and in some perciforms and atheriniforms.

"Paracentriolar" masses (arrows in Fig. 16.4D) flank the centriolar region. One of these, anterior to the proximal centriole, is here considered to closely resemble the retronuclear body of *Protopterus* sperm.

The region of the neck is prolonged as a membranous collar ('manchon membranaire') or cytoplasmic sleeve, 0.6 to 1 μm long. It is connected [distally only?] to the distal centriole by the satellite rays. Nine dense granules are present at this level (Mattei *et al.* 1988). The collar appears in micrographs (at least anteriorly, Fig. 16.4E) to be separated from the axoneme by a space equivalent to the so-called cytoplasmic canal typical of the sperm of gnathostome fishes (Fig. 16.4D,E).

The flagellum begins within the membranous collar (cytoplasmic sleeve). It is cylindrical for a length of 0.1 μm after which two lateral elements (here termed accessory axonemal columns) are associated with the axoneme, still within the cytoplasmic sleeve (Mattei *et al.* 1988). On the free flagellum these elements are not as wide as at the level of the collar. From micrographs, it appears that within the collar or sleeve each element is applied to almost half of the periphery of the axoneme whereas on the free flagellum it is not much greater in diameter than an axonemal doublet (Fig. 16.4E). The two elements are associated with doublets 3 and 8; most sections of axonemes lack the elements (Mattei *et al.* 1988) which presumably, therefore, extend through only a small portion of the flagellum. These elements clearly correspond with the short undulating membrane, in the anterior part of the flagellum, recognized by Tuzet and Millot (1959). Apart from these lateral elements, neither these authors nor Mattei (1988) give any evidence of lateral flagellar fins. The axoneme has the 9+2 pattern. The doublets have two arms and rays extend to the two central singlets (Mattei *et al.* 1988).

16.3.3 Latimeria Sperm and Phylogeny

Comparison with chondrichthyan sperm: Some features of shark and ray sperm are shown in Fig. 16.4F-I. The spermatozoon of *Latimeria* shows a number of similarities to the sperm of holocephalans and elasmobranchs. The sperm are elongate introsperm. The acrosome vesicle is elongate and pointed, its base on the nucleus is slanted, and it encloses a subacrosomal putative perforatorium (usually three in *Latimeria*) (see Fig. 16.4F for shark sperm). In chondrichthyans (Figs. 16.4G-I) and *Latimeria*, the basal portion of the axoneme is accompanied by two accessory axonemal columns, one near doublets 3 the other near 8. I have previously (Jamieson 1999) tentatively regarded the common presence of these columns as homoplasic but from the **A** type phylogeny (Fig. 16.2), it appears more probable that the two accessory axonemal columns have been inherited in *Latimeria* from an osteichthyan ancestor shared with the Chondrichthyes. The columns thus appear to be symplesiomorphic for Sarcopterygii, they occur in modified form in Ceratodontimorpha (*Neoceratodus*) and in tetrapods (Urodela and some Anura) (Fig. 16.5).

Regarding differences, the endonuclear canal, housing the greater length of the perforatoria in *Latimeria* is absent from chondrichthyans sperm. In chondrichthyans the midpiece, containing many mitochondria, is highly unusual for animal sperm in being interpolated between the base of the nucleus and the centrioles, of which the distal centriole forms the basal body of the axoneme. *Latimeria* differs, but is also unusual, in having the mitochondria wrapped around the base of the nucleus.

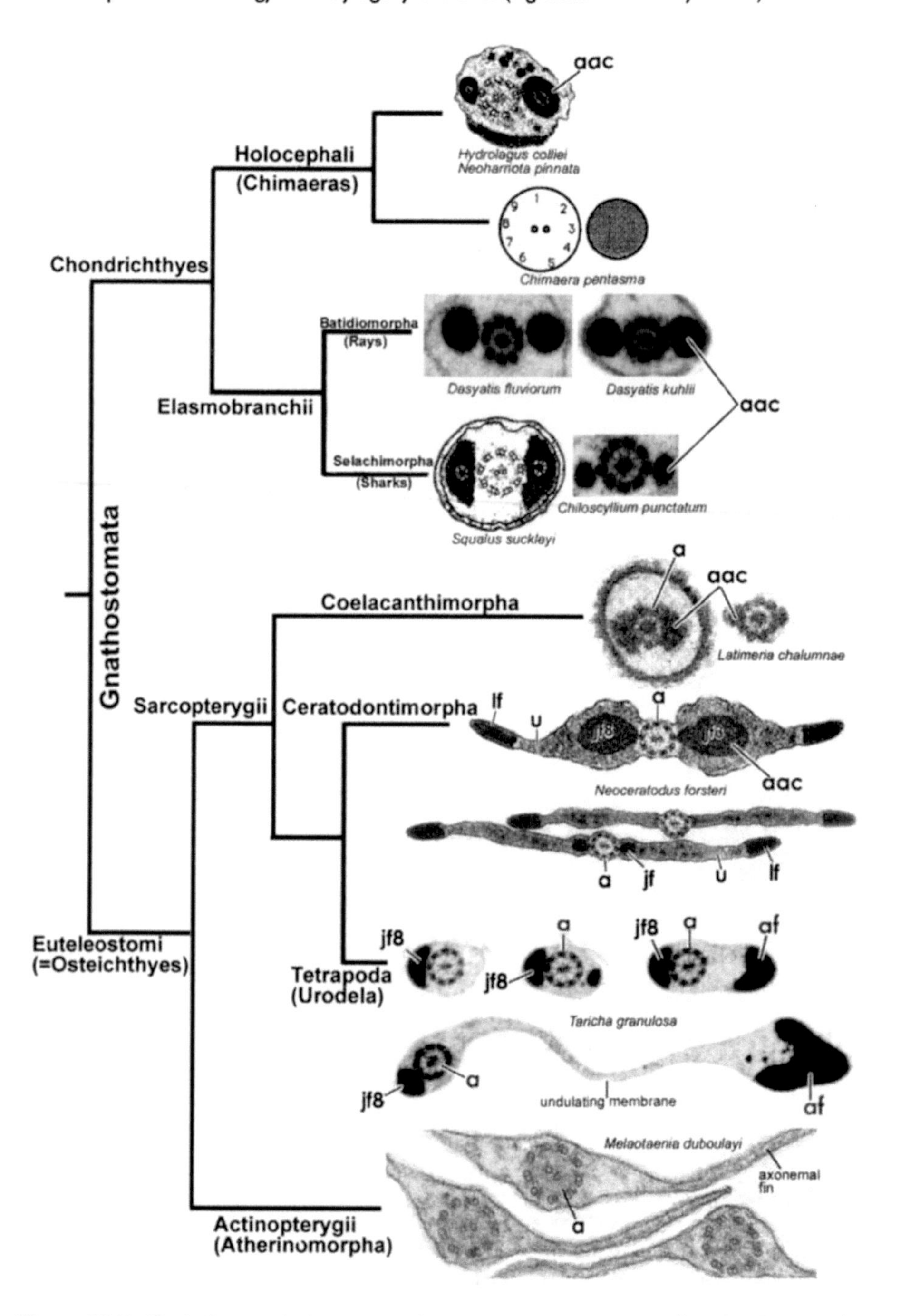

Fig. 16.5 Tentative phylogeny of accessory axonemal structures in Gnathostomata, showing the apparent homology of these structures of *Latimeria*

Fig. 16.5 Contd. ...

Comparison with Dipnotetrapodomorpha (Ceratodontimorpha + Tetrapoda) sperm: The spermatozoon of *Latimeria* resembles that of *Neoceratodus* in the Ceratodontiformes in important respects: the elongate nucleus with one or more endonuclear canals housing putative perforatoria which extend anterior to the nucleus; presence of a retronuclear body (also seen in *Lepidosiren* and *Protopterus*); and presence of a pair of accessory axonemal columns opposite doublets 3 and 8 (not known in *Lepidosiren* and *Protopterus*) (see also **Section 16.9**). It is no longer correct to say that *Latimeria* and lungfishes share the presence of lateral flagellar fins as though these are well developed in *Neoceratodus* (Figs. 16.5, 16.11N,O), *Lepidosiren* and *Protopterus*, reassessment of the published accounts indicates that they are absent from *Latimeria* sperm.

A notable difference of *Latimeria* is location of mitochondria laterally ensheathing the base of the nucleus and not, as in *Neoceratodus*, behind the nucleus. Mattei *et al.* (1988a) state that location of the mitochondria around the posterior region of the nucleus has been observed in a nemertean. It is also seen in the Australian lamprey *Mordacia praecox* (Hughes and Potter 1969) and is an independent apomorphy of non-appendicularian urochordates. No phylogenetic significance is attached to this mitochondrial location relative to other phyla.

16.4 REPRODUCTIVE BIOLOGY IN LATIMERIA

Latimeria is a live-bearer of a type termed a matrotrophous oophage or is possibly an adelphophage. This energetically efficient type of viviparity is also known in most mackerel sharks (Lamniformes); the large and dense yolk facilitates the early development of a definitive phenotype within the oviduct. This allows the large young to begin oral feeding on other less advanced siblings and ova in the same oviduct. Thus one or a few large, fully

---

Fig. 16.5 Contd. ...

with those of Chondrichthyes, lungfishes and tetrapods (urodeles). a, axoneme; aac, accessory axonemal column; af, axial fiber; jf, juxta-axonemal fiber; jf8, juxta-axonemal fiber at 8; l, lf, lateral fiber; u, undulating membrane. Chondrichthyes after Jamieson, B. G. M. 2005. Chondrichthyan spermatozoa and phylogeny. Pp. 201-236. In W. M. Hamlett (ed.), *Reproductive biology and phylogeny of Chondrichthyes: sharks, batoids and chimaeras.*, vol. 3. Science Publishers, Enfield, Fig. 7.15. Coelacanthimorpha after Mattei, X., Siau, Y. and Seret, B. 1988. Journal of Ultrastructure and Molecular Structure Research 101: 243-251, Fig. 18. Ceratodontimorpha after Jamieson, B. G. M. 1999. Spermatozoal phylogeny of the Vertebrata. Pp. 303-331. In C. Gagnon (ed.), *The Male Gamete. From Basic Science to Clinical Applications.* Cache River Press, Vienna, USA, Fig. 1M,N. Urodela after Scheltinga, D. M. and Jamieson, B. G. M. 2003. The Mature Spermatozoon. Pp. 203-274. In D. M. Sever (ed.), *Reproductive Biology and Phylogeny of Urodela*, vol. 1. Science Publishers, Enfield, New Hampshire, U.S.A, Fig. 7.17G-J. Actinopterygii after Jamieson, B. G. M. 1991. *Fish Evolution and Systematics: Evidence from Spermatozoa.* Cambridge University Press, Cambridge, Fig. 17.2J. Original.

developed, urea-retaining predatory young are born (see review by Balon *et al.* 1988). The eggs are among the largest known in the non-amniote vertebrates (eggs of some sharks reach 10 cm or more in diameter) exceeding 9 cm in diameter (Anthony 1980) (Fig. 16.6). Ovoviviparity was established by Smith *et al.* (1975) (Fig. 16.7).

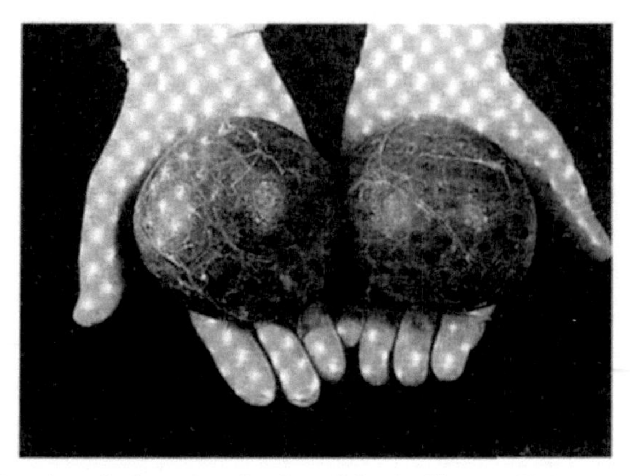

Fig. 16.6 *Latimeria chalumnae* (Latimeriidae). Two eggs, among the largest known in the non-amniote vertebrates. From Anthony, I. 1980. Proceedings of the Royal Society of London B 208: 349-367, Plate 2.

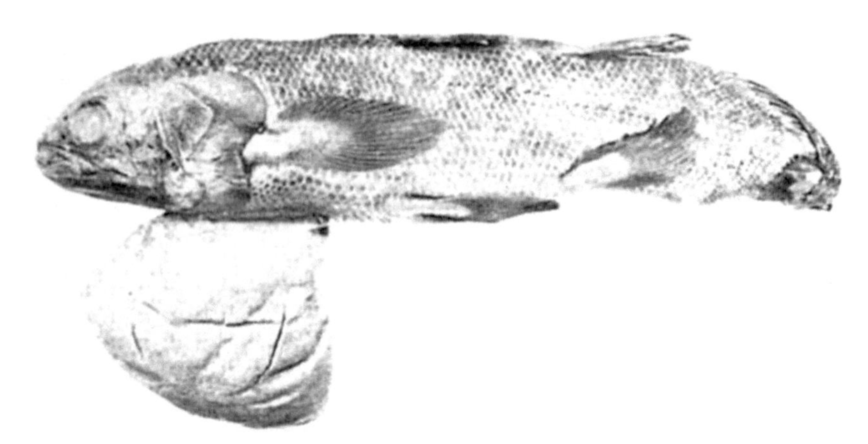

Fig. 16.7 *Latimeria chalumnae* (Latimeriidae). Yolk-sac young. Note fins and tail compressed by oviducal wall and morphological similarity to the adult. From Smith, C. L., Rand, C. S., Schaeffer, B. and Atz, I. W. 1975. Science 190: 1105-1106, Fig. 2. Copyright 1975 by the American Association for the Advancement of Science. With permission.

An acrosome reaction has not been the subject of study in *Latimeria.* It is accompanied by extrusion of perforatorial rod(s) in *Lampetra* (Afzelius and Murray 1957; Jaana and Yamamoto 1981); the hagfish *Eptatretus* (Morisawa

1999; Morisawa and Cherr 2002); *Branchiostoma* (Morisawa *et al.* 2004); and *Acipenser* (Detlaf and Ginzburg 1963; Ginsburg 1977; Cherr and Clark 1984). The fate of the rods during fertilization is also unknown in the cladistian *Polypterus* nor in lungfishes.

16.5 SUPERORDER CERATODONTIMORPHA

The three extant genera of the Ceratodontiformes (= extant Dipnoi, the lungfishes) are illustrated in Figs. 16.8, 16.9. The old view (e.g. Bischoff 1840), that dipnoans are more closely related to tetrapods, specifically the Amphibia, than are any other fish has been revived by several workers, notably Patterson (1980); Gardiner (1980); Rosen *et al.* (1981); Lauder and Liem (1983) (see also **Chapter 6**, Fig. 6.11); Northcutt (1986) and Forey (1980, 1986, 1988; Clack (2002). It hinged on the conclusion that the internal excurrent nostril of Recent lungfishes is a true choana such as exists in tetrapods.

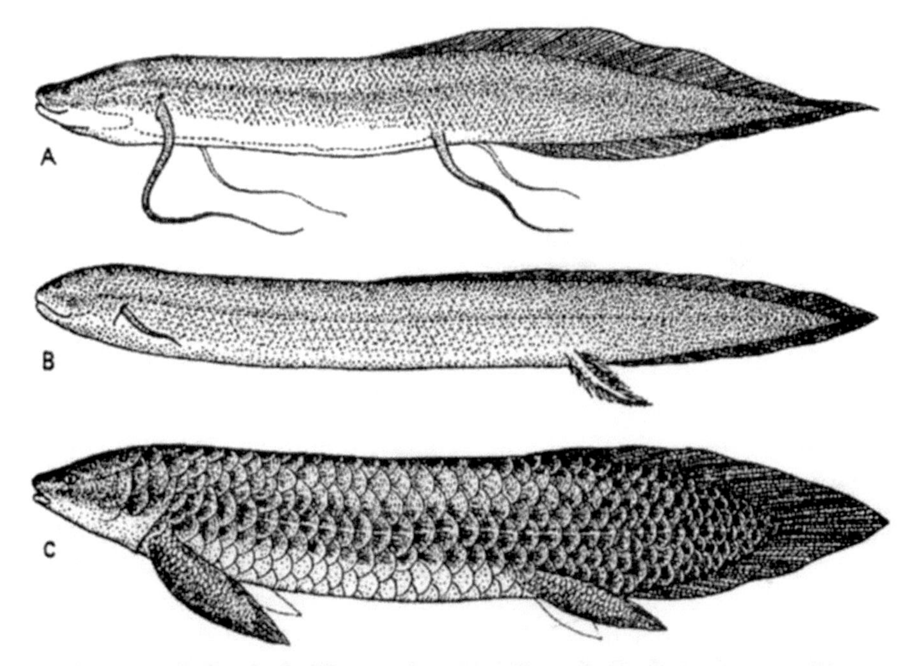

Fig. 16.8 Adult Ceratodontiformes (= extant Dipnoi). **A**. *Protopterus aethiopicus* (Africa). **B**. *Lepidosiren paradoxa* (S. America). **C**. *Neoceratodus forsteri* (Australia). From Norman, I. R. 1937. Illustrated Guide to the Fish Gallery. British Museum (Natural History). Trustees of the British Museum, London, Fig. 50.

As Patterson (1980) states, the synapomorphies between lungfishes and tetrapods listed by Bischoff in 1840 appear uncontested and to them may be added (*inter alia*) about 34 derived characters of the 57 listed by Kesteven

(1951), the gill-arch muscles discussed by Wiley (1979)) and ciliation of the larvae described by Whiting and Bone (1980). The synapomorphies between Ceratodontiformes and Tetrapoda also include the presence of the choana; the structure of the pelvic girdle (Rosen *et al.* 1981); and numerous further features of the soft anatomy such as the partially divided conus arteriosus, an atrial septum; the dermal bone pattern covering the braincase; loss of the interhyal; and the structure of the pelvic and pectoral appendage Rosen *et al.* (1981).

Embryologically Ceratodontiformes resemble amphibians and differ from other fishes in that cleavage is total and gastrulation produces a yolk plug. The presence in the larvae of *Lepidosiren* and *Protopterus* of a sucker and external gills is a further resemblance to the Amphibia (Young 1981). Molecular evidence has been discussed above.

The sperm of three species of Ceratodontiformes are now well studied ultrastructurally. Sperm ultrastructure reinforces separation of *Neoceratodus* (Ceratodontidae) from *Protopterus* and *Lepidosiren* while offering an autapomorphy (retronuclear body or its derivative) linking the three genera. Separation of *Protopterus* and *Lepidosiren* in distinct families (Protopteridae and Lepidosirenidae) is not supported and both should be referred to the Protopteridae. The retronuclear body and the two outer axonemal columns are possible synapomorphies of Ceratodontimorpha (including the Ceratodontiformes) and Coelacanthimorpha (=Actinistia).

16.6 ORDER CERATODONTIFORMES

16.6.1 Suborder Ceratodontoidei

Family Ceratodontidae

This family contains the Australian Lungfish, *Neoceratodus forsteri* (adult, Figs. 16.8C, 16.9A). Unlike *Lepidosiren* and *Protopterus*, the pectoral and pelvic fins are flipper-like and it lacks external gills (Nelson 2006). For a valuable account of the biology of *N. forsteri*, see Kemp (1986).

***Neoceratodus* spermatozoon** The spermatozoon of the Australian Lungfish has been described ultrastructurally (Jespersen 1971, Fig. 16.10; Jamieson 1995, 1999, Fig. 16.11A-R).

As noted by Jespersen (1971) the sperm of *Neoceratodus* and *Protopterus* show notable similarities: the heads are almost equally large; the acrosome and nucleus are of similar proportions; and the acrosome develops from a single Golgi saccule. Features distinguishing the spermatozoon from that of *Protopterus* are the uniflagellate condition of *Neoceratodus;* the absence of perforatoria in *Protopterus;* and the postmitochondrial rather than premitochondrial location of the retronuclear body in *Neoceratodus.* Homology of this body in the two genera seems assured by its initial premitochondrial location in the spermatid of *Neoceratodus.*

The head is about 70 μm long, by far the longest known in fishes. Its proportions appear similar to those of *Protopterus* for which lengths are not available and it similarly tapers very gradually to the pointed tip, being 2 μm wide at its greatest width, near its posterior end.

Fig. 16.9 Adult Ceratodontiformes (= extant Dipnoi). **A**. *Neoceratodus forsteri*, the Australian lungfish (Ceratodontidae). Photo Gary Cranitch. Courtesy of the Queensland Museum. **B**. *Lepidosiren paradoxa*, the South American lungfish (Lepidosirenidae). From *Wikipedia, The Free Encyclopedia*. **C** and **D**. *Protopterus aethiopicus*. Photo Beth Vosoba. Courtesy of Guillermo Orti. **E**. *P. annectens*. By Norman Weaver. Courtesy of Sarah Starsmore.

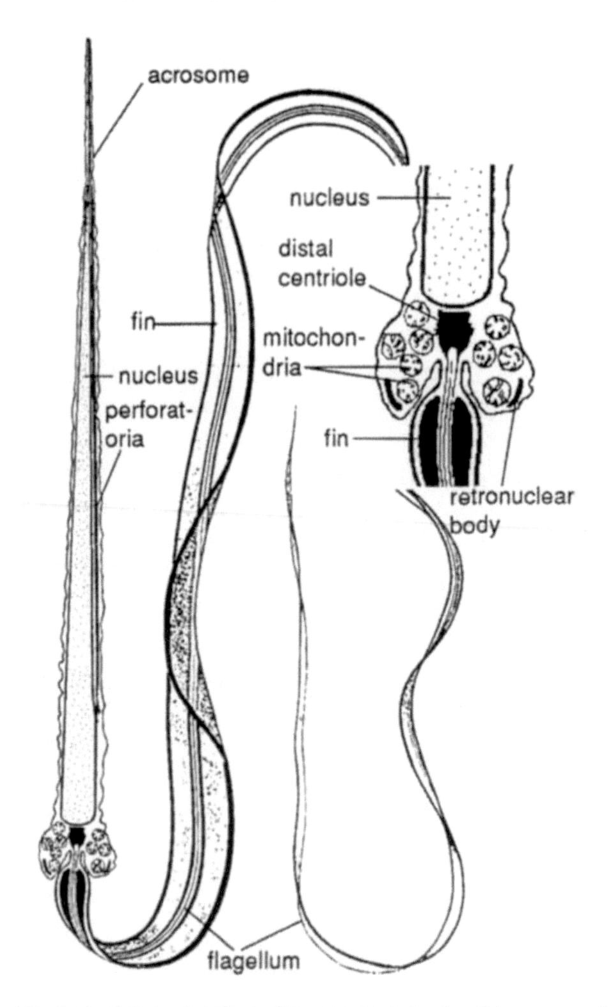

Fig, 16.10 *Neoceratodus forsteri* (Ceratodontidae). Diagram of the mature spermatozoon. Inset shows detail of midpiece region. After Jespersen, A. 1971. Journal of Ultrastructure Research 37: 178-185, Fig. 18.

A slender elongate acrosome is present on, but unlike that of *Protopterus,* does not appear to ensheath, the tip of the nucleus. The acrosome again forms, as in *Protopterus,* from a single Golgi vesicle in close contact with the nuclear membrane (Jespersen 1971).

Unlike *Protopterus,* in which perforatoria are absent, in *Neoceratodus* there are two, three (Fig. 16.11C-F,H,I) or sometimes four perforatoria initially in one canal but more posteriorly (Fig. 16.9H) in as many canals as there are perforatoria. *Neoceratodus* is exceptional in sarcopterygians in that the perforatoria re-emerge from the nucleus at their posterior ends (Fig. 16.11I) (Jespersen 1971; Jamieson 1995, 1999).

The 70 μm long nucleus is lanceolate with condensed chromatin.

The short, almost globular midpiece is about 2 μm long. It contains numerous irregularly arranged spherical mitochondria and, axially, two triplet centrioles differing from those of *Protopterus* in being mutually perpendicular. A "button" extends from the proximal centriole into the basal nuclear fossa. Behind the longitudinal distal centriole (basal body) the midpiece surrounds an invagination, deeper than that in *Protopterus*, containing the base of the 9+2 axoneme. This invagination is, clearly, the equivalent of the cytoplasmic canal of teleost sperm.

At the posterior limit of the midpiece there is a dense, granular, ring-shaped body. In the spermatid this surrounds the centrioles and Jespersen appears correct in equating it with the retronuclear body of the *Protopterus* sperm.

The tail is a remarkable 200 μm long. Where the flagellum is still within the cytoplasmic canal it is bordered by two oblong, longitudinal dense bodies, one at each of doublets 3 and 8 (Jespersen 1971; Jamieson 1995, 1999), as in *Latimeria*. As in *Protopterus*, after an initial strictly cylindrical region where the flagellum emerges from the midpiece, the cell membrane is expanded as two equal sized "wings", the flagellar fins.

Remarks: In *Neoceratodus* the anterior region of the sperm axoneme (not merely within the cytoplasmic canal as reported by Jespersen 1971) has a large dense rod on each side [accessory axonemal column], at doublets 3 and 8, and each rod is continuous with a lateral flagellar fin which has a further, smaller density, termed the lateral fiber, near its lateral extremity (Figs. 16.5, 16.11L-O). Jamieson (1995) tentatively recognized homology between each fin and an amphibian sperm undulating membrane, between the large dense rods at doublets 3 and 8 and the amphibian juxta-axonemal fibers and between the terminal densities and the axial fiber. This, if valid, would suggest that lissamphibians have retained, from an ancestor shared with dipnoans only one of two former, bilateral, undulating membranes, and only one of a former pair of axial rods (accessory axonemal columns), but that the two juxta-axonemal fibers of urodeles are a persistence of the paired ancestral condition (Fig. 16.5). The fiber at doublet 8 is usually but not always lost in the Anura (Jamieson 1995, 1999).

It seems likely that the flagellar fins in many Actinopterygii, in which they may be a basal synapomorphy, have developed independently in the ray-finned fishes relative to those of the sperm of lungfishes. Such homoplasic origin is acceptable as fins are known spasmodically in polychaete and echinoderm sperm. Nevertheless, the alternative that fins are retained in lungfishes from the ancestor of the Euteleostomi (=Osteichthyes) and were lost in the Coelacanthimorpha deserves consideration. If, however (as is likely) internal fertilization in coelacanths is primitive (as in Chondrichthyes), lateral flagellar fins may never have developed in coelacanth sperm. Their presence in lungfishes and (as the undulating membrane) in amphibians would then be a derived condition.

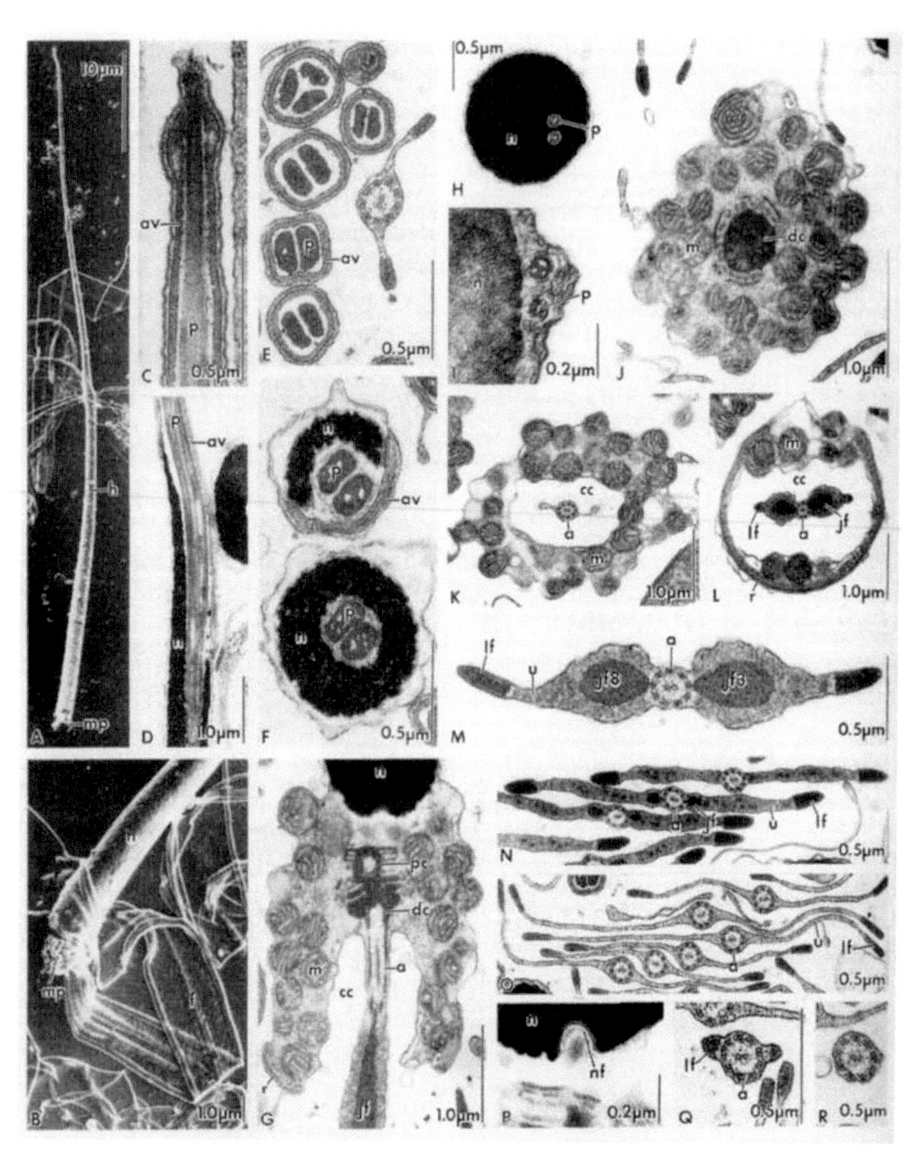

Fig. 16.11 *Neoceratodus forsteri*, the Australian lungfish (Ceratodontidae). Ultrastructure of the spermatozoon. **A**. Scanning electron micrograph (SEM) of the head and midpiece. **B**. SEM of the base of the nucleus and midpiece and the flagellum with its lateral fins or undulating membranes. **C**. Longitudinal sections (LS) of the perforatoria, showing their extension anterior to the nucleus and, in **D**, in an endonuclear canal. **E**. Transverse sections (TS) of the perforatoria within the acrosome. **F**. TS of perforatoria entering and within the nucleus. **G**. LS of midpiece, showing mitochondrial collar. **H**. TS of two perforatoria, each in a separate endonuclear canal. **I**. TS of two perforatoria posteriorly emergent from the nucleus.

Fig. 16.11 Contd. ...

16.6.2 Suborder Lepidosirenoidei

These are the S. American and African lungfishes. The pectoral and pelvic fins are filamentous, without rays and the larvae with external gills. Cloutier and Ahlberg (1996) have placed *Lepidosiren* and *Protopterus* in the same family, Lepidosirenidae, but Nelson (2006) retains them in separate families. Spermatozoal ultrastructure indicates that both should be placed in the same family.

16.6.2.1 Family Lepidosirenidae

This family Lepidosirenidae consists only of the South American lungfish, *Lepidosiren paradoxa* (adult, Figs. 16.8B, 16.9B) unless *Protopterus* is included.

Sperm ultrastructure: The sperm of *Lepidosiren paradoxa* (Fig. 16.12B-E) has an elongate nucleus with condensed chromatin, and is capped by an acrosome, giving a sharply pointed sperm head (Fig. 16.12B,C). Endonuclear canals and perforatoria are not reported. Immediately behind the nucleus is a very large, granular body, the width of the base of the nucleus, which is here termed the retronuclear body (Fig. 16.12D). Posterior to this body lies a compact, globose group of mitochondria (number unspecified) followed by two centrioles and two 9+2 flagella; each flagellum has two lateral fins with expanded lateral ends which have dense contents (Fig. 16.12E). No accessory axonemal

---

Fig. 16.11 Contd. ...

J. TS of midpiece through the distal centriole. **K**. TS of far anterior region of axoneme, within the cytoplasmic canal. **L**. Same further distally, showing beginning of lateral and juxta-axonemal fibers on each side. **M**. Same shortly behind the midpiece. **N**. Further distally, showing more slender undulating membranes still with lateral and greatly reduced juxta-axonemal fibers. **O**. Still further distally the slender undulating membranes now lacking the juxta-axonemal fibers. **P**. LS of basal nuclear fossa. **Q**. Axoneme far distally, shortly before the endpiece, with greatly reduced undulating membranes. **R**. TS of endpiece. a, axoneme; af, axial fiber; an, annulus; av, acrosome vesicle; b, barb; cc, cytoplasmic canal; cd, cytoplasmic droplet; cy, cytoplasm; d3, density (juxta-axonemal fiber?) at 3; db, dense, intermitochondrial body; dc, distal centriole; el, electron-lucent space; ec, endonuclear canal; f, flagellum; fs, fibrous sheath; h, head; hmt, helical microtubules; jf, juxta-axonemal fiber; jf3, juxta-axonemal fiber at 3; jf8, juxta-axonemal fiber at 8; lc, longitudinal column; lf, lateral fiber; m, mitochondrion; mp, midpiece; mts, sheath of microtubules; n, nucleus: nf, basal nuclear fossa; nk, neck; ni, infolding of nucleus into neckpiece (retronuclear body); nr, nuclear rostrum; nri, nuclear ridge; p, perforatorium; pa, paraxonemal rod; pc, proximal centriole; pf, peripheral fiber; r, retronuclear body; sc. subacrosomal cone; sdb, small dense body; stc, striated (segmented) column; su, subacrosomal material; u, undulating membrane. From Jamieson, B. G. M. 1999. Pp. 303-331. In C. Gagnon (ed.), *The Male Gamete. From Basic Science to Clinical Applications*. Cache River Press, Vienna, USA, Fig. 1.

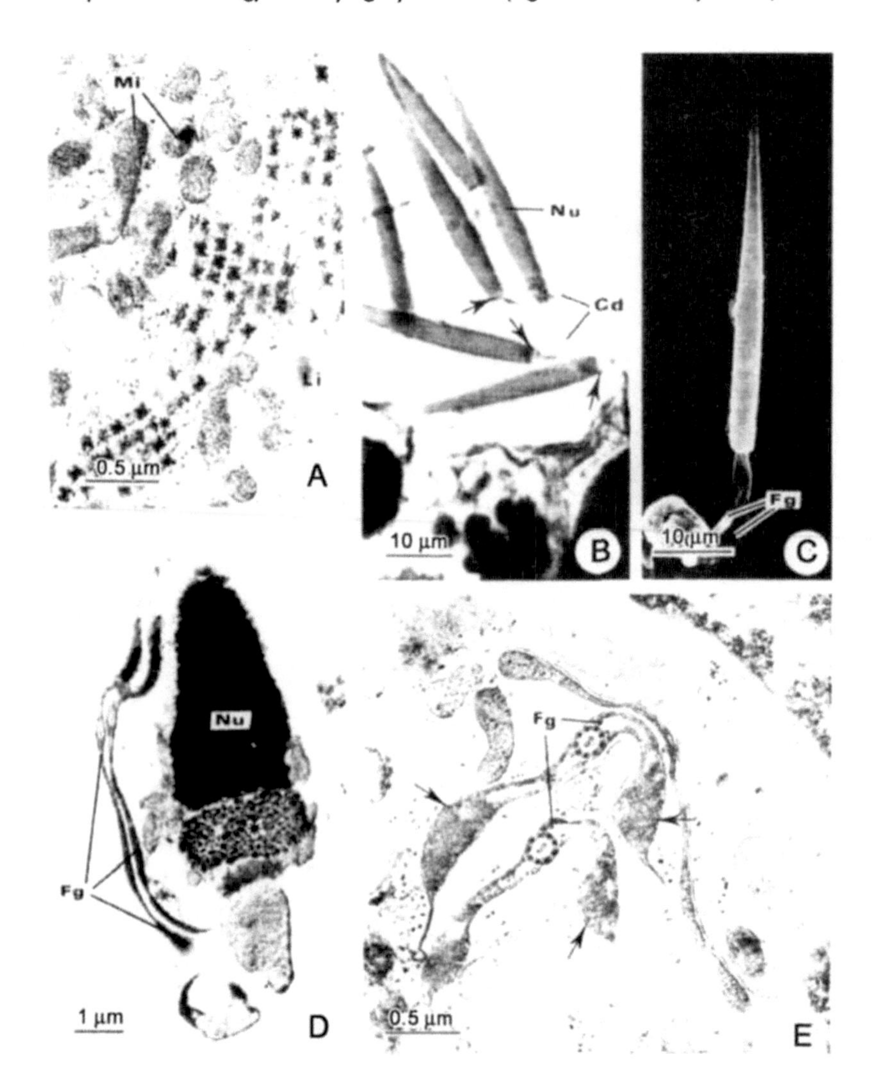

Fig. 16.12 *Lepidosiren paradoxa* (Lepidosirenidae). **A**. TEM of late spermatids, showing annulate lamellae. **B**. Light micrographs of spermatozoa, showing nucleus, midpiece (arrows) and tail. **C**. SEM of anterior portion of spermatozoon, showing the biflagellate condition. **D**. TEM of oblique section through the zone of transition between the nucleus, *granular body (retronuclear body) and tail with two flagella with lateral membranous expansions (fins). **E**. TEM of section through the two flagella, with 9+2 configuration and lateral expansions (fins) (asterisk). cd, sperm tail; Fg, flagella; Li, lipids; mi, mitochondria; Nu, nucleus; After Matos, E. and Azevedo, C. 1989. Revista Brasileira de Ciências Morfológicas 6: 67-71, Figs. 6-10.

columns are reported. During spermiogenesis, annulate lamellae are seen (Fig. 16.12A) (Matos *et al.* 1988; Matos and Azevedo 1989).

16.6.2.2 Family Protopteridae

Protopterids consist of a single genus, *Protopterus*, African lungfishes, with at least four species. Although listed here, the family is not considered distinct from the Lepidosirenidae.

***Protopterus* sperm ultrastructure:** Both examined species of Protopteridae, *Protopterus annectens* (Boissin *et al.* 1967; Mattei 1969, Mattei 1970) and *P. aethiopicus* (Purkerson *et al.* 1974), have biflagellate sperm, like *Lepidosiren*, whereas that of *Neoceratodus* is uniflagellate. The spermatozoa of the two *Protopterus* species are uniform. Adults of *Protopterus* are illustrated in Figures 16.8A and 16.9C,D.

The very long tapering nucleus is capped, and deeply penetrates, a pointed acrosome (Figs. 16.13, 16.15A,B) which (Boisson *et al.* 1967; Mattei 1969, 1970) arises not from Golgi secretions, as in Chondrichthyes, but apparently from a single Golgi saccule. According to Purkerson *et al.* (1974) the nucleus remains filamentous but the other workers indicate final condensation. Endonuclear canals and perforatoria are not reported.

The midpiece consists of the retronuclear body, the mitochondria and the two centrioles. The midpiece is approximately a twentieth of the length of the nucleus from which it is delimited by a circumferential groove. There are numerous small spherical mitochondria (Figs. 16.13, 16.14, 16.15D-G).

Between the base of the nucleus and the two parallel centrioles there is a large, structure, the retronuclear body (Figs. 16.13, 16.14, 16.15E,F). In *Protopterus aethiopicus* this is said to display excessively dense regions some of which are in striated array just proximal to what is unaccountably termed the "distal" centriole. Purkerson *et al.* (1974) consider the retronuclear body to be strikingly comparable with the striated columns of mammalian sperm (see also homology of this body in **Section 16.7.2**). In *P. annectens*, although there are concentric rings during development, the mature body later consists of spherules disseminated in a less osmiophilic mass and finally becomes a compact formation (Boisson *et al.* 1967; Mattei 1969) (Fig. 16.15E).

The remainder of the arbitrarily defined midpiece or intermediate segment is short, containing numerous small cristate mitochondria associated with the two centrioles. In one longitudinal section of the sperm of *P. aethiopicus* in the plane of the flagella, nine circular mitochondrial profiles can be counted posterior to the very large retronuclear body.

The two triplet centrioles are parallel and longitudinal, separated by 1 μm. Each continues as a flagellum. The proximal extremities of the centrioles are lodged in the center of an osmiophilic cylinder. Extending from each cylinder are satellites (Figs. 16.l4A,B, 16.15D,F,G) consisting of 9 thin lamellae, 0.2 μm long, situated in the perpendicular plane relative to the axis of the centriole and inserted between the triplets (Boisson 1963; Mattei 1969).

In both species each 9+2 flagellum (stated by Mattei, 1988, to be initially of 9+0 construction) is circular in cross section on leaving the midpiece but further distally has two lateral expansions (flagellar fins) giving a total flagellar width of 50 μm (Boissin 1963) (Figs. 16.14, 16.15A). As is frequent in

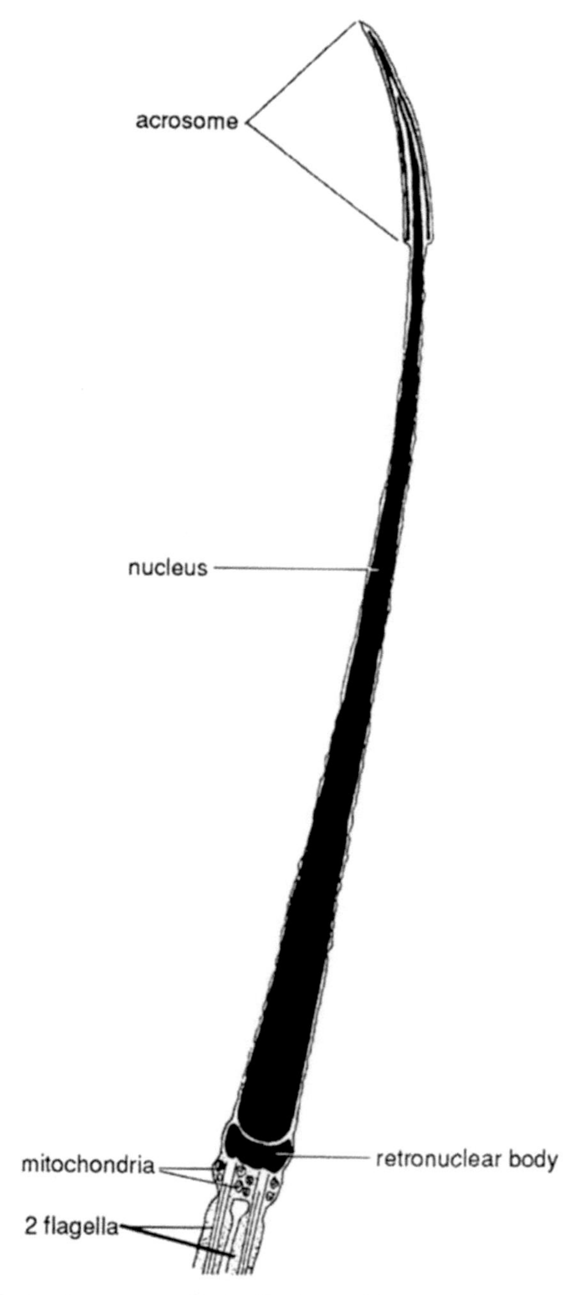

Fig. 16.13 *Protopterus annectens* (Protopteridae). Mature spermatozoon. Fro Jamieson 1991, fter Boisson, C., Mattei, C. and Mattei, X. 1967. Institut Fondamental d'Afrique Noire Bulletin Série A (Sciences Naturelles) 29: 1097-1121, Fig. If.

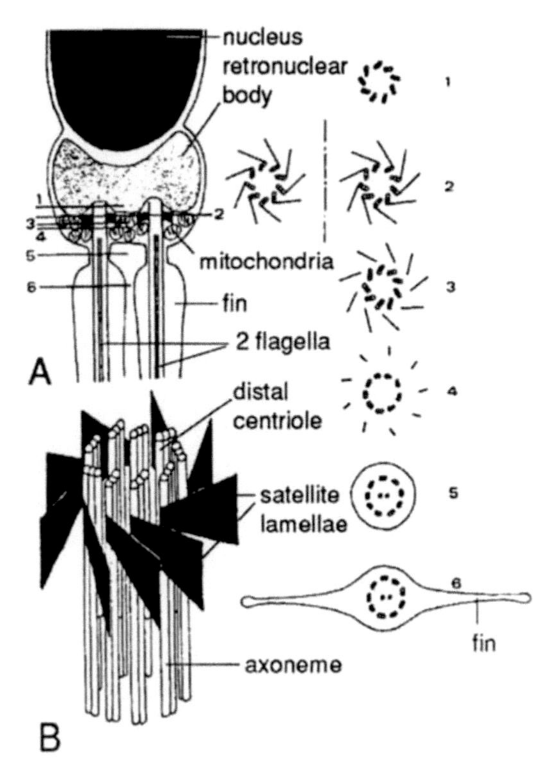

Fig. 16.14 *Protopterus annectens* (Protopteridae). Mature spermatozoon. **A**. Intermediate piece. **B**. Diagrammatic reconstruction of a centriole. Right column: transverse sections through levels indicated by numerals in **A**. From Jamieson 1991, after Boisson *et al.* 1967. Institut Fondamental d'Afrique Noire. Bulletin Serie A. (Sciences Naturelles) 29: 1097-1121, Fig. 2.

fishes, the fins lie in the plane of the two central singlets. In *Protopterus annectens,* at least, the fins have pointed ends in the proximal region of the flagellum, where they are about 1 µm wide, but in the distal region, where they are about 2 µm wide, they are terminated by a vesiculate expansion; the vesicles correspond with longitudinal tubules. Fine trabeculae unite the two surfaces of each fin (Boissin 1963). Scanning electron microscopy reveals that the flagella are about twice the length of the head (Purkerson *et al.* 1974). No accessory axonemal columns are reported.

In vivo the flagella are seen to beat slowly relative to flagella of most animal sperm and this is ascribed by Boissin (1963) to the presence of the fins, an interesting contrast with the questionable view that from the standpoint of fluid-dynamic efficiency it is virtually immaterial whether the cross sectional shape of the sperm tail is circular or of another shape (Flower 1967).

Features of spermiogenesis: According to Boissin *et al.* (1963) in *Protopterus* the germinal cells are immense although the nuclei, mitochondria and Golgi

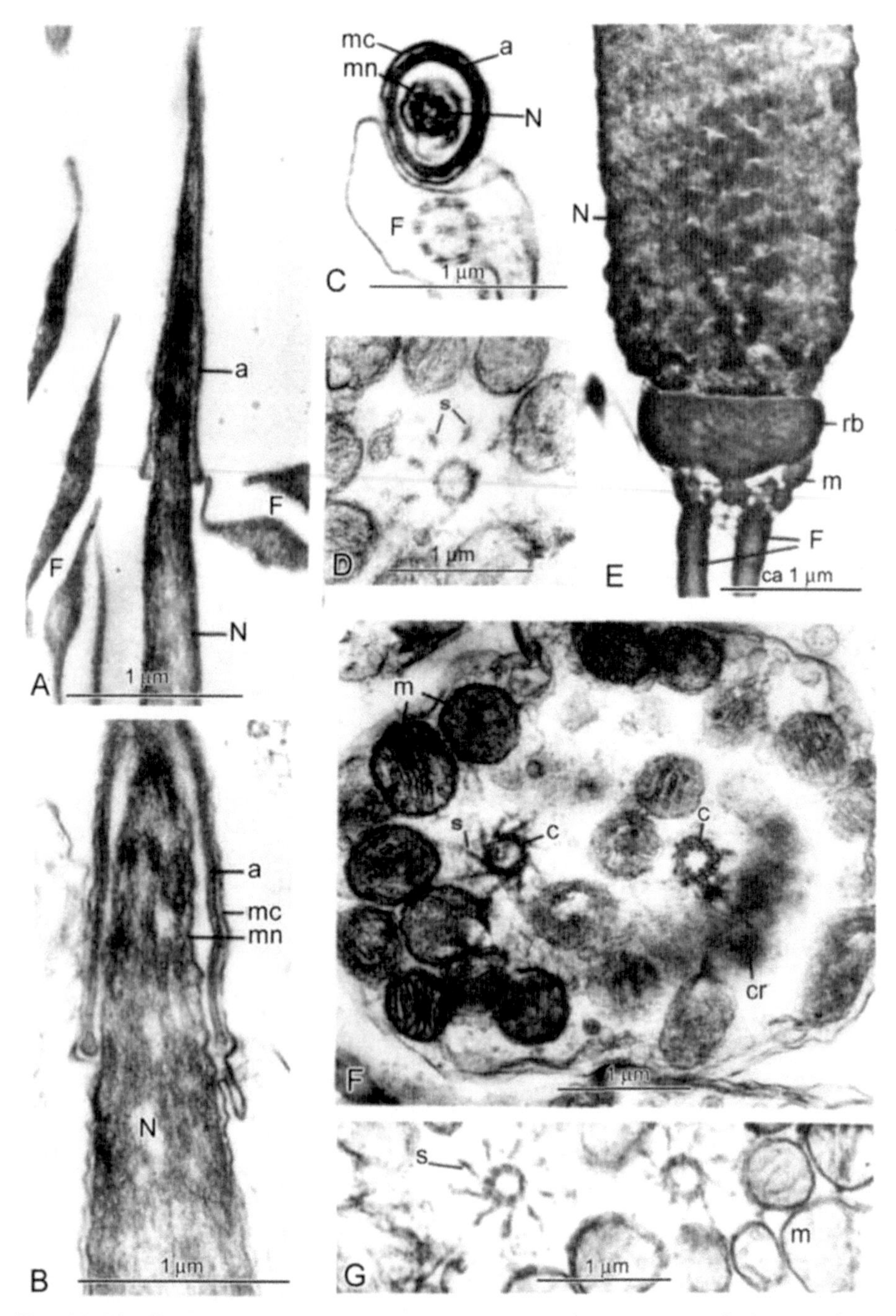

Fig. 16.15 *Protopterus annectens* (Protopteridae). Spermatozoa. **A**. Longitudinal section (LS) of the acrosome on the tip of the nucleus. **B**. The same. **C**. Transverse

Fig. 16.15 Contd. ...

apparatus are of normal size; the cytoplasm of the spermatid is frothy, being invaded by innumerable vacuoles; the Golgi apparatus, well developed in spermatocytes I and II, resolves in the spermatid into secretory vesicles of which only one survives as a saccule in contact with the nuclear membrane and becomes the somewhat reduced acrosome; and the retronuclear body forms from osmiophilic granules said to originate from the cell membrane.

16.7 SARCOPTERYGIAN SPERM AND PHYLOGENY

A discussion of symplesiomorphies and synapomorphies of sarcopterygian sperm follows, with some emendment of a previous account (Jamieson 1999).

16.7.1 Symplesiomorphies of Sarcopterygian Sperm

As deduced from the sperm of sarcopterygian fishes (*Latimeria* and the Ceratodontiformes) and a survey of tetrapod sperm, the following features appear plesiomorphic (carried over from a common ancestor shared with the Actinopterygii) for the spermatozoa of Sarcopterygii.

Absence of subacrosomal cone: Sarcopterygian fish sperm have a very long, slenderly conical acrosome vesicle (*Latimeria*, Figs. 16.3, 16.4; *Neoceratodus*, Figs. 16.10, 16.11A-E; *Protopterus*, Figs. 16.13, 16.15A,B; extent uncertain in *Lepidosiren*, Fig. 11.12B-E) but lack the subacrosomal cone that is basic to tetrapods (see Jamieson 1999).

More than one perforatorium: External to the Sarcopterygii a perforatorium which is extruded on reaction as an acrosomal filament is present in lampreys (Afzelius and Murray 1957; Nicander 1968; Nicander and Sjoden 1968) and the hagfish *Eptatretus* (Morisawa 1999; Morisawa and Cherr 2002). Within the Actinopterygii, the cladistian *Polypterus senegalus*, has an axial endonuclear canal but a perforatorium remains to be identified (Mattei 1970) and the chondrostean *Acipenser* has one to three endonuclear canals (DiLauro *et al.* 1998, 1999, 2000, 2001; Zarnescu 2005). In *Acipenser*, an acrosome reaction

Fig. 16.15 Contd. ...

section (TS) of the acrosome on the nucleus above a section of a flagellum. **D**. TS of a flagellum in the transition region; nine doublets are present but the central singlets are not sectioned; the satellites are reduced to nine elongate punctuations. **E**. LS through the nucleus, retronuclear body, mitochondria and the two flagella. **F**. TS through two centrioles, the centriole on the right lodged in a depression of the retronuclear body; that on the left sectioned at the base of retronuclear body, has nine satellites associated with the triplets. **G**. TS of the two centrioles at the base of the retronuclear body; each has nine satellites orientated in the same sense. Adapted from Mattei, X. 1969. *Contribution à l'étude de la spermiogenèse et des spermatozoïdes de poissons par les méthodes de la microscopie électronique*. D.Sc. thesis, Université de Montpellier, Figs. VIIb,c,d,e,f;h VIII,f. With permission.

involving subacrosomal material has been demonstrated though the role of the material in the endonuclear canals is uncertain (Detlaf and Ginzburg 1963; Cherr and Clark 1984).

Within the Sarcopterygii, there is only one endonuclear canal in the coelacanth, *Latimeria chalumnae*, but this contains three (less commonly two, or rarely, four) perforatoria. Endonuclear canals and recognizable perforatoria are absent (presumed lost) in *Lepidosiren* and *Protopterus*. The number of endonuclear canals and of enclosed perforatoria is one in the lissamphibians Urodela, though absent from cryptobranchids and proteids (Scheltinga and Jamieson 2003a), Gymnophiona (Scheltinga *et al.* 2003) and basal Anura (*Ascaphus* and *Leiopelma*) (Scheltinga and Jamieson 2003b), in the caiman (though poorly substantiated by micrographs), tinamou, rhea and non-passerines (e.g. galliforms) (Jamieson 2006). In the Chelonia and *Crocodylus johnstoni* there are two or three canals and there are two in *Sphenodon* (Jamieson 1999 and references therein). Presence of multiple perforatoria in both of the euteleostome (=osteichthyan) lineages, the Actinopterygii and the Sarcopterygii represented by *Latimeria*, suggests that the presumed common sarcopterygian ancestor of Lissamphibia and amniotes possessed more than one perforatorium and possibly more than one endonuclear canal. A single canal occurs in the Lissamphibia, excepting where lost in most Anura (Jamieson *et al.* 1993), and appears basic to all amniotes above turtles and *Sphenodon* (Jamieson and Healy 1992) being seen in phylogenetically less derived birds (Jamieson 2006). In squamates the single perforatorium is wholly prenuclear (Jamieson 1995, 1999).

In *Acipenser* the canals are spiralled around each other as they are in turtles, *Sphenodon* and *Crocodylus johnstoni* (Jamieson *et al.* 1997). The spiral arrangement or at least the presence of one or more endonuclear canals may well be a synapomorphy for the Euteleostomi, (Actinopterygii, Sarcopterygii and, within the latter, the Tetrapoda). The canals are absent (presumed lost) in the highly simplified sperm of the Holostei and Neopterygii (Jamieson 1991, 1999).

Elongation of the nucleus: The nucleus is long in the sarcopterygian fishes (*Latimeria*, Figs. 16.3, 16.4; *Neoceratodus*, Figs. 16.10, 16.11A,B; *Lepidosiren*, Fig. 16.12B,C; *Protopterus*, Fig. 16.13). This may be a plesiomorphic retention from basal euteleostome (osteichthyan) fishes as it is also long in *Acipenser*, a basal actinopterygian, and in Chondrichthyes (see Jamieson 1995, 2005). Further elongation in sarcopterygian fishes appears apomorphic (see below).

Midpiece: A simple midpiece, as in *Neoceratodus* (Fig. 16.10, 16.11G,J-L), and in *Acipenser*, with some of the mitochondria in a cytoplasmic collar, is presumably plesiomorphic for the Sarcopterygii. Location of a putatively mitochondrial sleeve, usually incomplete, lateral to the nucleus in *Latimeria* is clearly apomorphic.

Axoneme: A 9+2 axoneme is plesiomorphic for the Sarcopterygii. It has been argued above that presence of a pair of accessory axonemal columns in some Sarcopterygii (*Latimeria*, lungfishes and urodeles and some anurans in the Lissamphibia is symplesiomorphic as it also characterizes chondrichthyans sperm. Though columns are absent from Actinopterygii, they may, however, be a synapomorphy for the Gnathostomata.

16.7.2 Spermatozoal Synapomorphies of the Sarcopterygii

At least four synapomorphies can be proposed for the Sarcopterygii on the basis of sperm ultrastructure (Jamieson 1999). These pertain to the perforatoria, the nucleus, the retronuclear body and the structure of the flagellum.

Further elongation of the nucleus: The great length of the nucleus in Coelacanthimorpha (acrosome and nucleus 25-26 μm long) and Ceratodontiformes (nucleus 70 μm long in *Neoceratodus*; similarly elongate in *Protopterus*; head very approximately 35 μm long in *Lepidosiren paradoxa*) may be an initial synapomorphy of the Sarcopterygii.

Prenuclear perforatoria: The portions of the perforatoria (where perforatoria are present) within the sarcopterygian nucleus lie in one or more endonuclear canals (plethodontid urodeles are exceptional in having acrosomal rods which do not extend into the nucleus). Acipensiforms differ from Sarcopterygii in having the unreacted perforatoria restricted to the endonuclear canals. Endonclear canals may be a sarcopterygian synapomorphy but, again, as they occur in the Acipenseriformes they may be, if not homoplasies, an osteichthyan synapomorphy lost in most actinopteryagians. It is probable, in view of the presence of rods and endonuclear canals in *Latimeria* and *Neoceratodus*, that their absence in the lepidosirenoids *Protopterus* (Boisson 1963; (Mattei 1970) and apparently, though not stated, *Lepidosiren* (Matos *et al.* 1988; Matos and Azevedo 1989) is secondary. Alternatively, as an endonuclear canal occurs in lamprey sperm, these canals may be primitive in gnathostomes.

The extension, anterior to the nucleus, of rod-like structures, the perforatoria, has been regarded as a synapomorphy, and autapomorphy, for the Sarcopterygii (Jamieson 1991, 1999) but chondrichthyans have a solely prenuclear perforatorium. If it were not a homoplasy between the two groups it would have to be regarded as sarcopterygian symplesiomorphy but polarity of this character appears indeterminable.

Retronuclear body: A large dense body, between the nucleus and centrioles and termed the retronuclear body has been described for *Protopterus annectens* (Boisson 1963; Mattei 1969) (Figs. 16.13, 16.14, 16.15E) and *P. aethiopicus* (Purkerson *et al.* 1974). It has been homologized above with the granular body of the sperm of *Lepidosiren paradoxa* (Fig. 16.10B), with a smaller structure which, though postmitochondrial, originates behind the nucleus, in *Neoceratodus* (Figs. 16.10, inset, 16.11G) (Jespersen 1971; Jamieson 1991, 1995,

1999) and with a postnuclear structure (termed by Mattei *et al.* 1988, the 'paracentriolar body') in *Latimeria chalumnae* (Jamieson 1991) (Fig. 16.4D). Its cross striation in *P. aethiopicus* has led to its being compared with the striated columns of mammalian sperm (Purkerson *et al.* 1974). It is tentatively considered homologous with the neck region of urodele and anuran sperm (Jamieson 1991, 1995, 1999), being very well developed in salamandroids and some other urodeles (see Scheltinga and Jamieson 2003a). The retronuclear body is therefore considered to be a synapomorphy, and autapomorphy, of the Sarcopterygii (Jamieson 1999).

Undulating membranes: In *Neoceratodus*, as in all other lungfishes, lateral flagellar fins are present at doublets 3 and 8 (Figs. 16.5, 16.11L-O,Q), excepting at the endpiece (Fig. 16.11R). Shortly behind the distal centriole, within the mitochondrial collar (Fig. 16.11L), and behind this (Fig. 16.11M), for a short distance, each fin is supported by a large dense juxta-axonemal rod and by a smaller lateral rod within its free extremity (Jamieson 1995, 1999). The fin becomes more extensive behind this short anterior region but the rods are reduced in size (Fig. 16.11N,O) and soon only the lateral fiber persists (Fig. 16.11O,Q). Lateral fins are absent in *Latimeria* sperm. Fins (also at doublets 3 and 8) occur in many actinopterygian fishes (Mattei 1988; Jamieson 1991) and could conceivably have been precursory to coelacanthimorphan and dipnoan undulating membranes but homoplasy cannot be ruled out as lateral axonemal fins occur also in some echinoderms and protostomes (Jamieson 1995).

Ceratodontiform axonemal fins are considered homologous with the undulating membrane of lissamphibian sperm (Jamieson 1995, 1999). It is thus proposed that presence of two undulating membranes (as distinct from accessory axonemal columns) is a sarcopterygian synapomorphy. Even if lateral axonemal fins are sarcopterygian symplesiomorphies carried over from a euteleostome (osteichthyan) ancestor, their elaboration in lungfishes and amphibians is clearly synapomorphic.

16.8 CHAPTER SUMMARY

Mitogenomic and nuclear gene analyses and sperm ultrastructure suggest that lungfishes (Ceratodontiformes) are the sister-group of the Tetrapoda and that these two groups are also the closest relatives of *Latimeria*. However, an alternative analysis of nuclear genes has suggested that the coelacanth, lungfish, and tetrapod lineages diverged within such a short time interval that their relationships are an irresolvable trichotomy.

Latimeria is a live-bearer of a type termed a matrotrophous oophage or is possibly an adelphophage whereas lungfishes are oviparous. In the Ceratodontiformes, sperm ultrastructure reinforces separation of *Neoceratodus* (Ceratodontidae) from *Protopterus* and *Lepidosiren* while offering an autapomorphy (retronuclear body or its derivative) linking the three genera.

Separation of *Protopterus* and *Lepidosiren* in distinct families (Protopteridae and Lepidosirenidae) is not supported. The retronuclear body and the two outer axonemal columns are possible synapomorphies of Ceratodontimorpha (including the Ceratodontiformes) and Coelacanthimorpha (=Actinistia). *Protopterus annectens* and *P. aethiopicus* have biflagellate sperm, like *Lepidosiren*, whereas the sperm of *Neoceratodus* and *Latimeria* are uniflagellate.

The following features of the sperm sarcopterygian fish appear plesiomorphic (carried over from a common ancestor with the Actinopterygii). They have a very long, slenderly conical acrosome vesicle (*Latimeria, Neoceratodus, Lepidosiren; Protopterus,* as also in the Chondrostei) but lack the subacrosomal cone that is basic to tetrapods. There is only one endonuclear canal in the coelacanth, *Latimeria chalumnae,* but this contains two, three or rarely four perforatoria. Presence of multiple perforatoria in both of the euteleostome (=osteichthyan) lineages, the Actinopterygii and the Sarcopterygii represented by *Latimeria,* suggests that the presumed common sarcopterygian ancestor of Lissamphibia and amniotes possessed more than one perforatorium and possibly more than one endonuclear canal. The spiral arrangement of, or at least the presence of one or more, endonuclear canals may be a synapomorphy for the Euteleostomi (Actinopterygii, Sarcopterygii and, within the latter, the Tetrapoda). Elongation of the sperm nucleus in sarcopterygian fishes may be a plesiomorphic retention from euteleostome (osteichthyan) fishes (see *Acipenser* and Chondrichthyes) but further elongation in sarcopterygian fishes appears apomorphic. A simple midpiece, as in *Neoceratodus,* with some of the mitochondria in a cytoplasmic collar, is deduced to be plesiomorphic for the Sarcopterygii. Location of a putatively mitochondrial sleeve, usually incomplete, lateral to the nucleus in *Latimeria* is clearly apomorphic. A 9+2 axoneme is plesiomorphic for the Sarcopterygii. The lateral flagellar fins are probably an independent, homoplasic development in sarcopterygians relative to actinopterygians.

The great length of the nucleus in Coelacanthimorpha and Ceratodontimorpha (70 μm long in *Neoceratodus*) is a synapomorphy of the Sarcopterygii. The extension, anterior to the nucleus, of the perforatoria constitutes a synapomorphy, and autapomorphy, of the group. The absence of perforatoria and endonuclear canals in the lepidosirenoids *Protopterus* and apparently *Lepidosiren* is deduced to be secondary. The retronuclear body is considered to be a synapomorphy, and autapomorphy, of the Sarcopterygii. It is tentatively considered homologous with the neck region of urodele and anuran sperm.

The paired ceratodontiform flagellar fins are considered to be homologous with the single undulating membrane of lissamphibian sperm. It is thus proposed that presence of two undulating membranes is a sarcopterygian synapomorphy and of one a lissamphibian synapomorphy. Accessory axonemal columns, constant for chondrichthyans and occurring in *Latimeria, Neoceratodus* and lissamphibians are interpreted as a gnathostome synapomorphy and their apparent absence in *Protopterus* or *Lepidosiren* is presumably a loss. Even if lateral axonemal fins (as distinct from columns) are sarcopterygian symplesiomorphies carried over from a euteleostome (osteichthyan) ancestor, their elaboration in lungfishes and amphibians is considered synapomorphic.

16.9 ACKNOWLEDGMENTS

I am grateful to the Australian Research Council for former grants which supported my investigations into ultrastructure and phylogeny of the spermatozoa of invertebrates and vertebrates.

16.10 LITERATURE CITED

Afzelius, B. A. and Murray, A. 1957. The acrosomal reaction of spermatozoa during fertilization or treatment with egg water. Experimental Cell Research 12: 325-337.

Ahlberg, P. E. 1989. Paired fin skeletons and relationships of the fossil group Porolepiformes (Osteichthyes: Sarcopterygii). Zoological Journal of the Linnean Society 96: 119-166.

Anthony, J. 1980. Évocation des travaux français sur *Latimeria* notamment depuis 1972. Proceedings of the Royal Society of London B 208: 349-367.

Arnason, U., Gullberg, A., Janke, A., Joss, J. and Elmerot, C. 2004. Mitogenomic analyses of deep gnathostome divergence: a fish is a fish. Gene 333: 61-70.

Balon, E. K., Bruton, M. N. and Fricke, H. 1988. A fiftieth anniversary reflection on the living coelacanth, *Latimeria chalumnae*: some new interpretations of its natural history and conservation status. Environmental Biology of Fishes 23: 241-280.

Bernstein, P. 2004. The inner ear anatomy of the coelacanth revealed by high-resolution X-ray CT. Integrative and Comparative Biology 44.

Bischoff, T. L. W. V. 1840. Description anatomique du *Lepidosiren paradoxa*. Annales de Science Naturelle 14: 116-159.

Blair, J. E. and Hedges, S. B. 2005. Molecular phylogeny and divergence times of deuterostome animals. Molecular Biology and Evolution 22: 2275-2284.

Boisson, C. 1963. La spermiogenèse de *Protopterus annectens* (Dipneuste) du Sénégal étudiée au microscope optique et quelques détails au microscope électronique. Annales de la Faculté des Sciences Université de Dakar 10: 43-72.

Boisson, C., Mattei, C. and Mattei, X. 1967. Troisième note sur la spermiogenèse de *Protopterus annectens* (Dipneuste) du Sénégal. Institut Fondamental d'Afrique Noire Bulletin Série A (Sciences Naturelles) 29: 1097-1121.

Brinkmann, H., Denk, A., Zitzler, J., Joss, J. J. and Meyer, A. 2004. Complete mitochondrial genome sequences of the South American and the Australian lungfish: Testing of the phylogenetic performance of mitochondrial data sets for phylogenetic problems in tetrapod relationships. Journal of Molecular Evolution 59: 834-848.

Brinkmann, H., Venkatesh, B., Brenner, S. and Meyer, A. 2004. Nuclear protein-coding genes support lungfish and not the coelacanth as the closest living relatives of land vertebrates. Proceedings of the National Academy of Sciences of the United States of America 101: 4900-4905.

Cherr, G. N. and Clark, W. H. 1984. An acrosome reaction in sperm from the White Sturgeon, *Acipenser transmontanus*. The Journal of Experimental Zoology 232: 129-139.

Clack, J. A. 2002. *Gaining ground: the origin and evolution of tetrapods.* Indiana University Press, Bloomington. 369 pp.
Cloutier, R. and Ahlberg, P. E. 1996. Morphology, characters and interrelationships of basal sarcopterygians. Pp. 445-479. In M. L. Stiassny, L. R. Parenti and G. D. Johson (ed.), *Interrelationships of fishes.* Academic Press, San Diego.
Detlaf, T. A. and Ginzburg, A. S. 1963. Acrosome reaction in sturgeons and the role of calcium ions in the union of gametes. Doklady Akademii Nauk Sssr 153: 1461-1464.
Dilauro, M. N., Kaboord, W., Walsh, R. A., Krise, W. F. and Hendrix, M. A. 1998. Sperm-cell ultrastructure of North American sturgeons. I. The Atlantic sturgeon (Acipenser oxyrhynchus). Canadian Journal of Zoology 76: 1822-1836.
Dilauro, M. N., Kaboord, W. S. and Walsh, R. A. 1999. Sperm-cell ultrastructure of North American sturgeons. II. The shortnose sturgeon (Acipenser brevirostrum, Lesueur, 1818). Canadian Journal of Zoology 77: 321-330.
Dilauro, M. N., Kaboord, W. S. and Walsh, R. A. 2000. Sperm-cell ultrastructure of North American sturgeons. III. The lake sturgeon (*Acipenser fulvescens* Rafinesque, 1817). Canadian Journal of Zoology 78: 438-447.
Dilauro, M. N., Walsh, R. A., Peiffer, M. and Bennett, R. M. 2001. Sperm-cell ultrastructure of North American sturgeons. IV. The pallid sturgeon (*Scaphirhynchus albus* Forbes and Richardson, 1905). Canadian Journal of Zoology 79: 802-808.
Erdmann, M. V., Caldwell, R. L., Jewett, S. L. and Tjakrawidjaja, A. 1999. The second recorded living coelacanth from north Sulawesi. Environmental Biology of Fishes 54: 445-451.
Flower, J. W. 1967. The effect of tail shape on the propulsive efficiency of spermatozoa. Annals of the Entomological Society of America 60: 639-640.
Forey, P. L. 1980. *Latimeria*: a paradoxical fish. Proceedings of the Royal Society of London B 208: 369-384.
Forey, P. L. 1986. Relationships of lungfishes. Journal of Morphology Supplement 1: 75-91.
Forey, P. L. 1988. Golden jubilee for the coelacanth *Latimeria chalumnae.* Nature 336: 727-732.
Fritzsch, B. 1987. Inner ear of the coelacanth fish *Latimeria* has tetrapod affinities. Nature 327: 153-154.
Gardiner, B. G. 1980. Tetrapod ancestry: a reappraisal. Pp. 177-185. In A. L. Panchen (ed.), *The terrestrial environment and the origin of land vertebrates. Systematics Association Special Volume 15.* Academic Press, London and New York.
Ginsburg, A. S. 1977. Fine structure of the spermatozoan and acrosome reaction in *Acipenser stellatus.* Pp. 246-256. In D. K. Beljaev (ed.), *Problemy eksperimental noj biiologii.* Nauk, Moscow.
Hedges, S. B., Hass, C. A. and Maxson, L. R. 1993. Relations of fish and tetrapods. Nature 363: 501-502.
Hillis, D. M., Dixon, M. T. and Ammerman, L. K. 1991. The Relationships of the Coelacanth Latimeria-Chalumnae Evidence from Sequences of Vertebrate 28s Ribosomal Rna Genes. Environmental Biology of Fishes 32: 119-130.
Hughes, R. L. and Potter, I. C. 1969. Studies on gametogenesis and fecundity in the lampreys *Mordacia praecox* and *M. mordax* (Petromyzonidae). Australian Journal of Zoology 17: 447-464.

Inoue, J. G., Miya, M., Tsukamoto, K. and Nishida, M. 2004. Mitogenomic evidence for the monophyly of elopomorph fishes (Teleostei) and the evolutionary origin of the leptocephalus larva. Molecular Phylogenetics and Evolution 32: 274-286.

Jaana, H. and Yamamoto, T. S. 1981. The ultrastructure of spermatozoa with a note on the formation of the acrosomal filament in the Lamprey *Lampetra japonica.* Japanese Journal of Ichthyology 28: 135-147.

Jamieson, B. G. M. 1991. *Fish Evolution and Systematics: Evidence from Spermatozoa.* Cambridge University Press, Cambridge.

Jamieson, B. G. M. 1995. Evolution of tetrapod spermatozoa with particular reference to amniotes. Memoires du Museum National d'Histoire Naturelle 166: 343-358.

Jamieson, B. G. M. 1999. Spermatozoal phylogeny of the Vertebrata. Pp. 303-331. In C. Gagnon (ed.), *The Male Gamete. From Basic Science to Clinical Applications.* Cache River Press, Vienna, USA.

Jamieson, B. G. M. 2005. Chondrichthyan spermatozoa and phylogeny. Pp. 201-236. In W. C. Hamlett (ed.), *Reproductive biology and phylogeny of Chondrichthyes: sharks, batoids and chimaeras.* Science Publishers, Enfield, NH, USA.

Jamieson, B. G. M. 2006. Avian Spermatozoa: Structure and Phylogeny. Pp. 349-511. In B. G. M. Jamieson (ed.), *Reproductive Biology and Phylogeny of Birds,* vol. 6A. Science Publishers, Enfield, New Hampshire.

Jamieson, B. G. M. and Healy, J. M. 1992. The phylogenetic position of the tuatara *Sphenodon* (Sphenodontida, Amniota), as indicated by cladistic analysis of the ultrastructure of spermatozoa. Philosophical Transactions of the Royal Society of London B 335: 207-219.

Jamieson, B. G. M., Lee, M. S. Y. and Long, K. 1993. Ultrastructure of the spermatozoon of the internally fertilizing frog *Ascaphus truei* (Ascaphidae: Anura: Amphibia) with phylogenetic considerations. Herpetologica 49: 52-65.

Jamieson, B. G. M., Scheltinga, D. M. and Tucker, A. D. 1997. The ultrastructure of spermatozoa of the Australian freshwater crocodile, *Crocodylus johnstoni* Krefft, 1873 (Crocodylidae, Reptilia). Journal of Submicroscopic Cytology and Pathology 29.: 265-274.

Jarial, M. S. 2005. Ultrastructural study of the blood cells of the coelacanth Latimeria chalumnae (Rhipidistia: Coelacanthini). Journal of Submicroscopic Cytology and Pathology 37: 83-92.

Jespersen, Å. 1971. Fine structure of the spermatozoon of the Australian Lungfish *Neoceratodus forsteri* (Krefft). Journal of Ultrastructure Research 37: 178-185.

Kemp, A. 1986. The biology of the Australian lungfish, *Neoceratodus forsteri* (Krefft 1870). Journal of Morphology Supplement 1: 181-198.

Kesteven, H. L. 1951. The origin of the tetrapods. Proceedings of the Royal Society of Victoria 59: 93-138.

Kikugawa, K., Katoh, K., Kuraku, S., Sakurai, H., Ishida, O., Iwabe, N. and Miyata, T. 2004. Basal jawed vertebrate phylogeny inferred from multiple nuclear DNA-codes genes. BMC Biology 2: 1-11.

Lagios, M. D. 1979. The coelacanth and the Chondrichthyes as sister groups: a review of shared apomorph characters and a cladistic analysis and reinterpretation. Occasional Papers of the California Academy of Sciences 134: 25-44.

Locket, N. A. 1980. Some advances in coelacanth biology. Proceedings of the Royal Society B 208: 265-307.

Løvtrup, S. 1977. *The Phylogeny of Vertebrata*. Wiley, London. 330 pp.

Mallatt, J. and Winchell, C. J. 2007. Ribosomal RNA genes and deuterostome phylogeny revisited: More cyclostomes, elasmobranchs, reptiles, and a brittle star. Molecular Phylogenetics and Evolution 43: 1005-1022.

Matos, E. and Azevedo, C. 1989. Ultrastructural study of the spermatogenesis of Lepidosiren paradoxa (Pisces, Dipnoi) in the Amazon region. Revista Brasileira de Ciências Morfológicas 6: 67-71 (In Portuguese).

Matos, E., Gusmao, S., Azevedo, C. and Corral, L. 1988. Ultraestrutura da espematogenese de *Lepidosiren paradoxa* (Pisces, Dipnoi) da Amazonia. Pp. 711. In Editor *40 Reunniao Anual da Societa Brasileira para o Progresso da Ciencia, Sao Paulo.*

Mattei, X. 1969. *Contribution à l'étude de la spermiogenèse et des spermatozoïdes de poissons par les méthodes de la microscopie électronique*. Thesis. Faculté des Sciences Montpellier (*Fide* Cassas *et al.* 1981) thesis.

Mattei, X. 1970. Spermiogenèse comparée des poissons. Pp. 57-69. In B. Baccetti (ed.), *Comparative Spermatology*. Academic Press, New York.

Mattei, X. 1988. The flagellar apparatus of spermatozoa in fish Ultrastructure and evolution. Biology of the Cell 63: 151-158.

Mattei, X., Siau, Y. and Seret, B. 1988. Étude ultrastructurale du spermatozoïde du coelacanthe: *Latimeria chalumnae*. Journal of Ultrastructure and Molecular Structure Research 101: 243-251.

Meyer, A. and Dolven, S. I. 1992. Molecules, fossils and the origin of tetrapods. Journal of Molecular Evolution 35: 102-113.

Meyer, A. and Wilson, A. C. 1990. Origin of tetrapods inferred from their mitochondrial DNA affiliation in lungfish. Journal of Molecular Evolution 31: 359-364.

Morisawa, S. 1999. Acrosome reaction in spermatozoa of the hagfish *Eptatretus burgeri* (Agnatha). Development Growth and Differentiation 41: 109-112.

Morisawa, S. and Cherr, G. N. 2002. Acrosome reaction in spermatozoa from hagfish (Agnatha) *Eptatretus burgeri* and *Eptatretus stouti*: Acrosomal exocytosis and identification of filamentous actin. Development Growth and Differentiation 44: 337-344.

Morisawa, S., Mizuta, T., Kubokawa, K., Tanaka, H. and Morisawa, M. 2004. Acrosome reaction in spermatozoa from the amphioxus *Branchiostoma belcheri* (Cephalochordata, Chordata). Zoological Science (Tokyo) 21: 1079-1084.

Nelson, J. S. 2006. *Fishes of the World*, 4th edition. John Wiley and Sons, Inc., Hoboken, NJ. 601 pp.

Noonan, J. P., Grimwood, J., Danke, J., Schmutz, J., Dickson, M., Amemiya, C. T. and Myers, R. M. 2004. Coelacanth genome sequence reveals the evolutionary history of vertebrate genes. Genome Research 14: 2397-2405.

Northcutt, R. G. 1986. Lungfish neural characters and their bearing on sarcopterygian phylogeny. Journal of Morphology Supplement 1: 277-297.

Patterson, C. 1980. Origin of tetrapods: historical introduction to the problem. Pp. 159-175. In A. L. Panchen (ed.), *The Terrestrial Environment and the Origin of Land Vertebrates*. Systematics Association Special Publication 15. Academic Press, London and New York.

Picheral, B. 1967. Structure et organization du spermatozoïde de *Pleurodeles waltlii* Michah (Amphibien Urodèle). Archives de Biologie, Paris 78: 193-221.

Purkerson, M. L., Jarvis, J. U. M., Luse, S. A. and Dempsey, E. W. 1974. X-ray analysis coupled with scanning and transmission electron microscopic observations of spermatozoa of the African lungfish, *Protopterus aethiopicus*. Journal of Zoology (London) 172: 1-12.

Rosen, D. E., Forey, P. L., Gardiner, B. G. and Patterson, C. 1981. Lung fishes, tetrapods, paleontology, and plesiomorphy. Bulletin of the American Museum of Natural History 167: 159-276.

Sasaki, T., Sato, T., Miura, S., Bwathondi, P. O. J., Ngatunga, B. P. and Okada, N. 2007. Mitogenomic analysis for coelacanths (*Latimeria chalumnae*) caught in Tanzania. Gene (Amsterdam) 389: 73-79.

Scheltinga, D. M. and Jamieson, B. G. M. 2003a. The Mature Spermatozoon. Pp. 203-274. In D. M. Sever (ed.), *Reproductive Biology and Phylogeny of Urodela*, vol. 1. Science Publishers, Enfield, NH, USA.

Scheltinga, D. M. and Jamieson, B. G. M. 2003b. Spermatogenesis and the Mature Spermatozoon: Form, Function and Phylogenetic Implications. Pp. 119-251. In B. G. M. Jamieson (ed.), *Reproductive Biology and Phylogeny of Anura*, vol. 2. Science Publishers, Enfield, NH, USA.

Scheltinga, D. M., Wilkinson, M., Jamieson, B. G. M. and Oommen, O. V. 2003. Ultrastructure of the mature spermatozoa of caecilians (Amphibia: Gymnophiona). Journal of Morphology 258: 179-192.

Smith, C. L., Rand, C. S., Schaeffer, B. and Atz, J. W. 1975. *Latimeria*, the living coelacanth, is ovoviviparous. Science 190: 1105-1106.

Takezaki, N., Figueroa, F., Zaleska-Rutczynska, Z., Takahata, N. and Klein, J. 2004. The phylogenetic relationship of tetrapod, coelacanth, and lungfish revealed by the sequences of forty-four nuclear genes. Molecular Biology and Evolution 21: 1512-1524.

Tuzet, O. and Millot, J. 1959. La spermatogenèse de *Latimeria chalumnae* Smith (Crossoptérygien coelacanthidé). Annales des Sciences Naturelles Zoologie 1: 61-69.

Whiting, H. P. and Bone, Q. 1980. Ciliary cells in the epidermis of the larval Australian dipnoan *Neoceratodus*. Zoological Journal of the Linnean Society 68: 125-137.

Wiley, E. O. 1979. Ventral gill arch muscles and the interrelationships of gnathostomes, with a new classification of the Vertebrata. Zoological Journal of the Linnean Society 67: 149-179.

Wood, S. C., Johansen, K. and Weber, R. E. 1972. Haemoglobin of a coelacanth. Nature 239: 283-285.

Yokobori, S.-I., Hasegawa, M., Ueda, T., Okada, N., Nishikawa, K. and Watanabe, K. 1994. Relationship among coelacanths, lungfishes and tetrapods: A phylogenetic analysis based on mitochondrial cytochrome oxidase I gene sequences. Journal of Molecular Evolution 38: 602-609.

Young, J. Z. 1981. *The Life of Vertebrates*, 3rd edition. Clarendon Press, Oxford. 820 pp.

Zardoya, R. and Meyer, A. 1996. Evolutionary relationships of the coelacanth, lungfishes, and tetrapods based on the 28S ribosomal RNA gene. Proceedings of the National Academy of Sciences of the United States of America 93: 5449-5454.

Zardoya, R. and Meyer, A. 1997a. The complete DNA sequence of the mitochondrial genome of a "living fossil", the coelacanth (*Latimeria chalumnae*). Genetics 146: 995-1010.

Zardoya, R. and Meyer, A. 1997b. Molecular phylogenetic information on the identity of the closest living relative(s) of land vertebrates. Naturwissen 84: 389-397.

Zarnescu, O. 2005. Ultrastructural study of spermatozoa of the paddlefish, *Polyodon spathula*. Zygote 13: 241-247.

CHAPTER 17

Sperm Modifications Related to Insemination, with Examples from the Ostariophysi

Robert Javonillo[1], *John R. Burns*[1], *Stanley H. Weitzman*[2]

17.1 INSEMINATING TELEOSTS

The vast majority of teleosts, estimated at 97%, are externally fertilizing whereby eggs and sperm are shed into the surrounding aqueous medium and fertilization takes place outside the body of the female (Pecio *et al.* 2007). This contrasts with insemination, whereby ripe males are able to transfer sperm to the reproductive tract of the female (Burns and Weitzman 2005). Insemination must take place in order for true internal fertilization to occur, and internal fertilization, in turn, is required for viviparity, where at least some stage of embryonic development occurs inside the female reproductive tract (Wourms 1981; see also Uribe Aranzábal *et al.*, **Chapter 3**). Prior to the work of Munehara *et al.* (1989), it had been assumed that all inseminating fishes were also internally fertilizing: for example, see tables in Breder and Rosen (1966). However, a novel reproductive mode termed "internal gametic association" was first shown to occur in species of the scorpaeniform family Cottidae (Munehara *et al.* 1989, 1991; Koya *et al.* 1993; see also Abe and Munehara, **Volume 8B, Chapter 6**). In the case of internal gametic association, males are able to transfer sperm to the female reproductive tract (insemination) where the spermatozoa enter (i.e., associate with) the micropyles of the "eggs"

[1]Department of Biological Sciences, George Washington University, Washington, DC, 20052, USA
[2]Division of Fishes, Department of Vertebrate Zoology, National Museum of Natural History, Smithsonian Institution, Washington, DC, 20560-0109, USA

(oocytes). The eggs, with their associated spermatozoa, are subsequently shed into the surrounding seawater which apparently stimulates the fertilization event (Munehara *et al.* 1989, 1991; Koya *et al.* 1993). Thus, in spite of there being internal delivery of the sperm (insemination), fertilization itself (i.e., union of sperm and egg) actually occurs externally in the seawater. Given this novel reproductive method, for those species whose female reproductive tracts have simply been shown to contain spermatozoa (Figs. 17.1 and 17.2), the more accurate term to describe this would be "insemination" until the actual time of fertilization can be determined. For many inseminating species, fertilization may indeed occur internally as ovulated eggs enter the lumen of the ovary where spermatozoa are stored. However, this entire process may take place very rapidly just prior to oviposition. Therefore confirmation of fertilization, as with histological methods, would require sacrifice of females

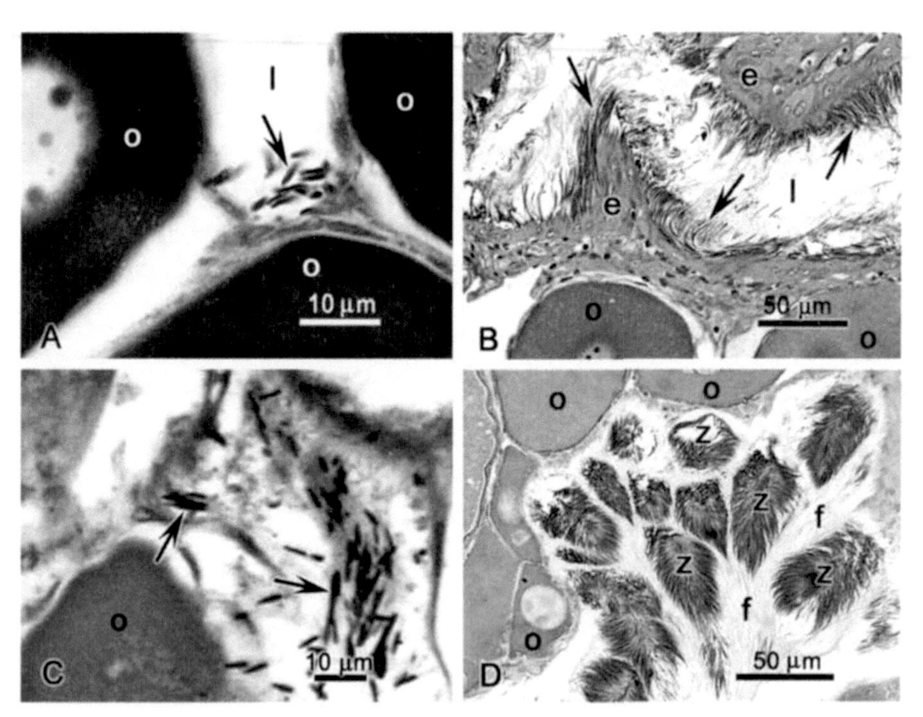

Fig. 17.1 Light micrographs of inseminated ostariophysan ovaries. **A**. *Gephyrocharax valenciae* (Characiformes, Characidae, Stevardiinae, Stevardiini). Standard length (SL) 32.6 mm, ANSP 112230, Venezuela. This ovary contained only previtellogenic oocytes. **B**. *Trachelyopterus lucenai* (Siluriformes, Auchenipteridae). SL 150 mm, MCP 18469, Brazil. **C**. *Rachoviscus crassiceps* (Characiformes, Characidae, *incertae sedis*). SL 24.1 mm, USNM 220756, Brazil. **D**. *Xenurobrycon macropus* (Characiformes, Characidae, Stevardiinae, Xenurobryconini). SL 14.9 mm, USNM 317053, Paraguay. Note the presence of intact spermatozeugmata. e, ovarian epithelium; f, flagellar portions of spermatozeugmata; l, ovarian lumen; o, oocyte cytoplasm; z, spermatozeugma; arrow, spermatozoa. Original.

at precisely the time of ovulation, something which has yet to be successfully accomplished for any species.

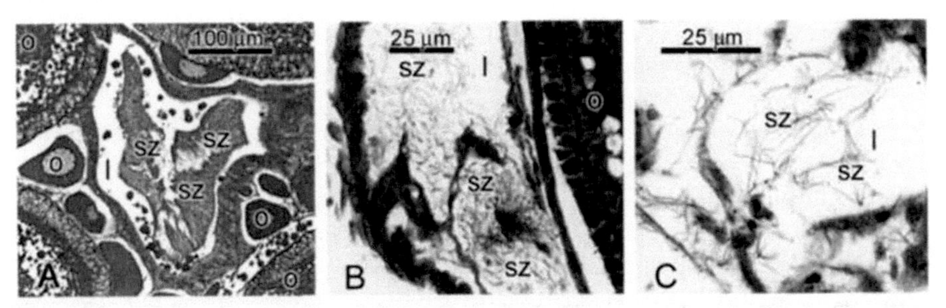

Fig. 17.2 Light micrographs of inseminated siluriform ovaries. **A**. *Scoloplax dicra* (Scoloplacidae). SL 13.8 mm, MUSM 5458, Peru. First published micrograph of an inseminated scoloplacid. Note the large clumps of spermatozoa. **B**. *Astroblepus sabalo* (Astroblepidae). SL 90.9 mm, USNM 167877, Peru. First documented case of insemination in this species. **C**. *Astroblepus choatae* (Astroblepidae). SL 63.2 mm, USNM 121129, Venezuela. First documented case of insemination in this species. I, ovarium lumen; o, oocyte; sz, spermatozoa.

Jamieson (1991) discussed various hypotheses regarding whether internal or external fertilization is the more primitive mode in chordates. Although internal fertilization is clearly the primitive mode in some fish lineages (e.g. Chondrichthyes and Actinistia), most evidence points to external fertilization as the ancestral mode in the Teleostei (Jamieson 1991). When considering whether insemination is either a derived or primitive condition for a taxon, a phylogeny should be brought to bear. Of ostariophysans, inseminating species are known for members of the characiform family Characidae and the siluriform families Astroblepidae, Auchenipteridae and Scoloplacidae (Table 17.1). As progress is made in resolving the relationships among major lineages of siluriforms (e.g., Sullivan *et al.* 2006) and characiforms (Calcagnotto *et al.* 2005), hypotheses about the evolution of insemination can be scaled down to the level of relationships among closely related species and genera. According to a recent phylogenetic hypothesis for major characiform lineages (Calcagnotto *et al.* 2005), those genera that were both included in the analysis and contain inseminating species (*Bryconamericus, Creagrutus, Gephyrocharax, Knodus, Mimagoniates*) are not basal within the order, thus insemination is not the plesiomorphic reproductive mode for Characiformes. Similarly, Astroblepidae, Auchenipteridae and Scoloplacidae are not basal lineages and are not all closely related families within Siluriformes (Sullivan *et al.* 2006). Thus insemination is most parsimoniously interpreted as having multiple origins during the evolution of catfishes.

Many externally fertilizing animals, including invertebrates, produce fairly simple spermatozoa often referred to as "aquasperm" (Jamieson 1987). Among teleosts, aquasperm are often characterized as having spherical to ovoid nuclei, short midpieces containing several cristate mitochondria, and one flagellum per cell (Jamieson 1991). Although some externally fertilizing groups present modified spermatozoa, these cells appear to be adapted to reproductive conditions unique to these species. The great majority of inseminating teleosts produce cells that appear to be modifications of aquasperm adapted to this unique mode of reproduction (Jamieson 1991; Pecio *et al.* 2007).

The main purpose of this chapter is to discuss the possible selective advantages of morphological specializations of reproductive organs seen in inseminating teleosts. Although many of these modifications are present in a number of inseminating taxa, the examples presented herein will concentrate on ostariophysan species. Prior to Jamieson's (1991) work on sperm ultrastructure in fishes, no sperm ultrastructural data were available for any inseminating ostariophysan species. Since that time sperm ultrastructure in over 30 inseminating ostariophysan species has been analyzed (Burns *et al.* **Chapter 11**). In addition, basic gonadal anatomy and secondary sex character descriptions have become available for many inseminating ostariophysans (e.g., Pecio and Rafiński 1994; Burns *et al.* 1995; Meisner *et al.* 2000b; Javonillo *et al.* 2007). This chapter expands on the work presented at the II International Symposium on Livebearing Fishes in Querétaro, Mexico in 2002 (Burns and Weitzman 2005).

Museum numbers, when available, are given in figure legends and Table 17.1. Institutional abbreviations follow Leviton *et al.* (1985) except as follows: ICNMHN, Instituto de Ciencias Naturales, Museo de Historia Natural, Universidad Nacional de Colombia, Bogotá, Colombia; LBP, Laboratório de Biologia e Genética de Peixes, Departamento de Morfologia, Instituto de Biociências, UNESP, Botucatu, São Paulo, Brazil; LIRP, Laboratório de Ictiologia, Faculdade de Filosofia, Ciências e Letra de Ribeirão Preto, Universidade de São Paulo, São Paulo, Brazil; MBUCV-V, Museo de Biología, Instituto de Zoología Tropical, Universidad Central de Venezuela, Caracas, Venezuela; MUSM, Museo de Historia Natural de la Universidad Nacional Mayor de San Marcos, Lima, Peru; NAM, Museu Anchieta Porto Alegra, Porto Alegre, Brazil; SU, specimens formerly part of Stanford University collections, now deposited at California Academy of Sciences, San Francisco, California, USA. Listed below are teleost taxa for which insemination is known or reported. This compilation is not intended to serve as an exhaustive bibliography but does highlight that insemination is not a diagnostic feature for just one lineage within Teleostei. In addition, Table 17.1 presents further details on inseminating ostariophysan species.

A. Species in the characiform family Characidae, subfamily Glandulocaudinae (Table 17.1; Nelson 1964a,b,c; Burns *et al.* 1995; Weitzman and Menezes 1998; Castro *et al.* 2003). Note that the subfamily Glandulocaudinae was recently split into two subfamilies (Weitzman *et al.* 2005), with Glandulocaudinae now containing only *Glandulocauda*, *Mimagoniates*, and *Lophiobrycon*, and Stevardiinae containing the remainder of genera from the Glandulocaudinae *sensu* Weitzman and Menezes (1998).

B. Species in the characiform family Characidae, subfamily Stevardiinae (Table 17.1; Kutaygil 1959; Nelson 1964a,b,c; Burns *et al.* 1995; Weitzman and Menezes 1998; Weitzman *et al.* 2005).

C. Species in the tribe Compsurini (Table 17.1; Malabarba 1998) of the characiform family Characidae, subfamily Cheirodontinae (Burns *et al.* 1997).

D. Several species of the characiform family Characidae having uncertain subfamilial designations (i.e., *incertae sedis*; Table 17.1; Burns *et al.* 2000; Weitzman *et al.* 2005).

E. Species in the siluriform family Auchenipteridae (Table 17.1; von Ihering 1937; Loir *et al.* 1989; Burns *et al.* 2000; Meisner *et al.* 2000b).

F. Species in the siluriform family Scoloplacidae (Table 17.1; Schaefer *et al.* 1989; Burns *et al.* 2000; Spadella *et al.* 2006). Fig. 17.2A is the first published histological demonstration of insemination in a scoloplacid, *Scoloplax dicra*. Insemination is implied in other species based on the production of highly modified sperm (Spadella *et al.* 2006).

G. Species in the siluriform family Astroblepidae (Table 17.1). Insemination is herein confirmed in this family for the first time, in *Astroblepus sabalo* (Fig. 17.2B) and *A. chotae* (Fig. 17.2C). Insemination is implied in other species based on the production of highly modified sperm (Spadella 2007).

H. *Lepidogalaxias salamandroides* of the osmeriform family Galaxiidae (Pusey and Stewart 1989; Jamieson **Chapter 12**). The familial designation of this species is the subject of controversy (Nelson 2006). Some authors contend that it is the sole member of Lepidogalaxiidae, but we follow Nelson's (2006) placement of *L. salamandroides* in Galaxiidae.

I. Species in the ophidiiform family Bythitidae (Wourms and Bayne 1973; Wourms and Cohen 1975; Cohen and Wourms 1976; Jamieson 1991 and **Chapter 13**).

J. Species in the genus *Barathronus* of the ophidiiform family Aphyonidae (Nielsen 1984).

K. Species in the ophidiiform family Parabrotulidae (Turner 1936).

L. *Labidesthes sicculus* of the atheriniform family Atherinopsidae (Grier *et al.* 1990). In older classifications, this monotypic genus was placed in a more inclusive Atherinidae (Nelson 2006).

M. Species in the atheriniform family Phallostethidae (Grier and Parenti 1994).

N. *Horaichthys setnai* in the beloniform family Adrianichthyidae (Kulkarni 1940; Grier 1984). *Oryzias latipes* and species of *Xenopoecilus* are reported to be inseminating (Amemiya and Murayama 1931). (For beloniform sperm ultrastructure, see Jamieson **Chapter 14**).

Table 17.1 Inseminating ostariophysan species. Insemination was confirmed histologically by presence of sperm in ovary (C), or inferred by sperm with elongate nuclei (I). References where specimens have been reported previously are as follows: 1 = Burns et al. 1997, 2 = Burns et al. 1995, 3 = Castro et al. 2003, 4 = Burns and Weitzman 2005, 5 = Menezes et al. 2003, 6 = Burns et al. 2000, 7 = Spadella 2007, 8 = Meisner et al. 2000b, 9 = Spadella et al. 2006. Specimens not previously reported are listed as "new."

Taxon	*Country*	*Insemination*	*Museum Number*	*SL (mm), male*	*SL (mm), female*	*Ref.*
Order Characiformes						
Family Characidae						
Subfamily Cheirodontinae						
Tribe Compsurini						
Acinocheirodon melanogramma	Brazil		MCP 18596	27.6		1
	Brazil	C	MCP 18598		26.3	1
	Brazil	C	MCP 18597		37.7	1
Compsura gorgonae	Panama	C	USNM 293228	22.3	25.4	1
				23.4		1
Compsura heterura	Brazil	C	MZUSP 39117	26.5	26.7	1
Kolpotocheirodon theloura	Brazil	C	MZUSP 38839	29.9	24.8	1
Macropsobrycon uruguayanae	Brazil	C	MCP 11997	33.2	34.3	1
		C			33.3	1
Odontostilbe dialeptura	Panama	C	USNM 208525	27.8	28.6	1
Odontostilbe mitoptera	Panama	C	USNM 208541	32.0	32.6	1
Saccoderma melanostigma	Venezuela	C	SU 18145	23.8	23.4	1
Subfamily Glandulocaudinae						
Tribe Glandulocaudini						
Glandulocauda melanogenys	Brazil	C	USNM 236415	42.1	39.5	2
				41.4		2
Glandulocauda melanopleura	Brazil	C	MZUSP 48511	33.8	34.3	new
Lophiobrycon weitzmani	Brazil	C	LIRP 4338		25.5	3
Mimagoniates barberi	Paraguay	C	USNM 327586	29.3	26.0	2

		C			26.8	2
		C			25.9	2
Mimagoniates inequalis	Brazil	C	MZUSP 75515	32.2	24.7	4
Mimagoniates lateralis	Brazil	C	USNM 326250	31.4	37.3	2
Mimagoniates microlepis	Brazil	C	MZUSP 26899	60.8	41.6	2
Mimagoniates rheocharis	Brazil	C	USNM 306339	48.6	34.3	2
Mimagoniates sylvicola	Brazil	C	USNM 276557	27.2	26.6	2
Subfamily Stevardiinae						
Tribe Corynopomini						
Corynopoma riisei	Venezuela		USNM 216889	38.6		2
				44.4		2
	Venezuela	C	USNM 219619	43.0	46.8	2
Gephyrocharax atracaudatus	Panama	C	MCZ 109976	43.3	41.4	2
				45.0		2
	Panama		USNM 293217	41.3		2
	Columbia	C	USNM 329755	45.8	44.2	2
Gephyrocharax chocoensis	Columbia	C	FMNH 56017	50.6	51.3	2
Gephyrocharax intermedius	Panama	I	FMNH 12515	33.8		2
Gephyrocharax melanocheir	Colombia		FMNH 56049	34.6		4
	Colombia	C	USNM 79209		36.0	4
Gephyrocharax valencia	Venezuela	C	USNM 326251		29.2	2
	Venezuela	C	ANSP 112230		32.6	2
Gephyrocharax venezuelae	Venezuela	C	USNM 121363	32.2	36.5	2
	Venezuela	C	USNM 121366		35.3	2
					37.2	2
Gephyrocharax sp.	Colombia	C	USNM 309218	39.7	42.3	4
Pterobrycon landoni	Colombia	C	ICNMHN 1691	17.8	16.3	4
Pterobrycon myrnae	Costa Rica		USNM 211459	30.8		2
	Costa Rica	C	USNM 211462		35.7	2

Tribe Diapomini						
Acrobrycon ipanquianus	Peru	C	CAS 45837	77.5	82.2	4
Acrobrycon sp.	Bolivia	C	AMNH 16242	47.6	46.8	2
	Bolivia	C	USNM 319277	51.9	63.3	2
Diapoma speculiferum	Brazil	C	USNM 221154	31.5	33.5	2
	Brazil	C	NAM 9115	42.5	39.0	2
Diapoma terofali	Brazil	C	USNM 270284	42.7	38.4	2
Planaltina britskii	Brazil	C	MZUSP 26911	29.8	32.0	4
Planaltina glandipedis	Brazil	C	USNM 362136	26.2	27.0	5
Planaltina myersi	Brazil	C	MNRJ 10634	31.6	30.9	2
	Brazil	C	USNM 278989	37.4	35.0	2
Tribe Hysteronotini						
Hysteronotus megalostomus	Brazil	C	USNM 236399	28.5	28.5	2
	Brazil	C	MCP 16460	33.6	31.2	2
Pseudocorynopoma doriae	Brazil	C	USNM 287144	50.7	48.5	2
Pseudocorynopoma heterandria	Brazil	C	USNM 270647	55.4	69.2	2
Tribe Landonini						
Landonia latidens	Ecuador	I	MCZ 48663	38.4		2
Tribe Phenacobryconini						
Phenacobrycon henni	Ecuador	I	MCZ 48660	27.3		2
	Ecuador	I	CAS 39035	33.8		2
Tribe Xenurobryconini						
Argopleura chocoensis	Columbia		USNM 76943	41.2		2
	Columbia	C	CAS 39030	46.2	47.9	2
Argopleura magdalensis	Columbia	C	ANSP 127516	43.9	41.5	2
	Columbia	C	USNM 235922		43.9	2
Chrysobrycon hesperus	Columbia	C	USNM 326247	67.5	58.3	2
Chrysobrycon myersi	Peru	I	USNM 235928	44.0		2
Iotabrycon praecox	Ecuador	C	USNM 216802	16.1	18.8	2
	Ecuador	C	USNM 236064	16.2	16.5	2

Ptychocharax rhyacophila	Venezuela	C	MBUCV-V 18564	55.0	55.0	2
				59.3		2
Scopaeocharax rhinodus	Peru	C	USNM 329756	24.7	24.9	2
Scopaeocharax sp.	Peru	C	USNM 368089	17.5	19.0	4
Tyttocharax tambopatensis	Peru	C	USNM 328936	15.0	13.8	4
Tyttocharax sp.	Ecuador	C	USNM 309251	12.9	14.7	2
		C		17.5	16.4	2
Xenurobrycon macropus	Paraguay	C	USNM 317053	14.8	14.9	new
Xenurobrycon polyancistrus	Bolivia	C	USNM 278191	13.1	13.9	2
Characids of uncertain affinity:						
New genus and species	Brazil	C	MCP 12638	34.7	28.4	4
Attonitus bounites	Peru	C	USNM 349701	43.0	43.0	4
Attonitus ephimeros	Peru	C	USNM 349696	45.7	33.1	6
Attonitus irisae	Peru		USNM 349698	46.1		4
	Peru	C	USNM 349704		36.6	4
Brittanichthys axelrodi	Brazil	C	USNM 221991	30.9	20.5	6
Bryconadenos tanaothoros	Brazil	C	USNM 352061	37.5	36.5	6
Bryconamericus pectinatus	Peru	C	MUSM 3708	30.4	40.2	6
"*Cheirodon*" *ortegai*	Peru	C	MZUSP 25993	34.0	25.5	6
Creagrutus lepidus	Venezuela	C	USNM 325045	41.8	35.4	4
Creagrutus melasma	Venezuela	C	USNM 349411	26.0	26.5	4
Hollandichthys multifasciatus	Brazil	C	USNM 320271	70.3	61.2	4
Knodus sp.		C	USNM 362836	34.0	32.7	2
Note: incorrectly identified as "*Planaltina* sp." in Burns *et al.* (1995)						
Knodus sp.	Brazil	C	MZUSP 38004	34.0	35.5	4
Monotocheirodon pearsoni	Bolivia	C	CAS 59792	30.4	32.1	new
Monotocheirodon sp.	Peru		ANSP 143791	33.9		4
	Peru	C	ANSP 143792		37.8	4
Monotocheirodon sp.	Peru	C	MUSM 6756	39.0	38.0	6

Monotocheirodon sp.	Peru	C	MUSM 11082	35.0	36.7	4
Rachoviscus crassiceps	Brazil	C	USNM 220756		24.1	new
Order Siluriformes						
Family Astroblepidae						
Astroblepus chotae	Venezuela	C	USNM 121129	93.1	63.2	new
Astroblepus sabalo	Peru	C	USNM 167877	94.9	90.9	new
Astroblepus cf. *mancoi*	Peru	I	LBP 3284			7
Family Auchenipteridae						
Trachelyopterus galeatus	Brazil	C	MCP 11779	149	130	6,8
Trachelyopterus lucenai	Brazil	C	MCP 18469	147	150	6,8
Family Scoloplacidae						
Scoloplax dicra	Peru	C	MUSM 5458	19.3	13.8	6
Scoloplax distolothrix	Brazil	I	LBP 1424			9
	Brazil	I	LBP 1938			9

O. Species in the beloniform subfamily Zenarchopterinae (included in the family Hemiramphidae by Nelson 2006; Grier and Collette 1987; Downing and Burns 1995; Meisner and Burns 1997a,b).

P. *Rivulus marmoratus* (self-fertilizing), *Campellolebias brucei, Cynopoecilus melanotaenia* in the cyprinodontiform family Rivulidae (Harrington 1961; Parenti 1981; Jamieson 1991; Kweon *et al.* 1998; Lazara 2000; Nelson 2006; Burns and Weitzman 2005). (For sperm ultrastructure in the Antherinomorpha, see Jamieson **Chapter 14**)

Q. Species in the genera *Anableps* and *Jenynsia* in the cyprinodontiform family Anablepidae (Turner 1938a, 1940c,d, 1950; Burns and Flores 1981; Grier *et al.* 1981; Knight *et al.* 1985; Burns 1991; Meisner *et al.* 2000a).

R. Species in the cyprinodontiform family Poeciliidae (Turner 1940a; Scrimshaw 1944, 1945; Grier *et al.* 1980; Grier 1981; Jamieson 1991; Grove and Wourms 1991, 1994).

S. Species in the cyprinodontiform family Goodeidae (Turner 1933, 1937, 1940b; Grier *et al.* 1978).

T. *Aulichthys japonicus* in the gasterosteiform family Aulorhynchidae is reported to exhibit internal gametic association (Jamieson **Chapter 15**).

U. *Sebasticus marmoratus, Helicolenus dactylopterus* and species in the genus *Sebastes* of the scorpaeniform family Scorpaenidae (Jamieson 1991; Shimizu *et al.* 1991; Wourms 1991; Muñoz *et al.* 2000, 2002; Collins *et al.* 2001). The monophyly of Scorpaeniformes was rejected by Smith and Wheeler (2004), but we arbitrarily utilize the classification scheme of Nelson (2006), wherein Scorpaenidae includes *Sebasticus, Helicolenus,* and *Sebastes*. (For sperm ultrastructure in the Percomorpha, see Jamieson **Chapter 15**).

V. Certain genera in the scorpaeniform families Cottidae (Bolin 1944; Koya *et al.* 1993), Hemitripteridae (Munehara *et al.* 1991), and Agonidae (Munehara 1997; Koya *et al.* 2002; See also Abe and Munehara, **Volume 8B, Chapter 6**).

W. Species of the genus *Comephorus* in the scorpaeniform family Comephoridae (Breder and Rosen 1966; Nelson 2006).

X. Species in the perciform family Embiotocidae (Turner 1938b, 1952; Tarp 1952; Gardiner 1978a,b,c,d).

Y. Species in the perciform family Apogonidae (Thresher 1984; Balon 1985).

Z. Species in the genus *Zoarces* of the perciform family Zoarcidae (Kristofferson *et al.* 1973; Jamieson 1991; Korsgaard *et al.* 2002).

AA. *Anarhichas lupus* of the perciform family Anarhichadidae is reported to be internally fertilizing (Pavlov 1994; Moksness and Pavlov 1996).

BB. Species in the genera *Starksia* and *Xenomedea* of the perciform family Labrisomidae (Hubbs 1952; Böhlke and Springer 1961; Rosenblatt and Taylor 1971).

CC. Some species of the perciform family Clinidae (Veith 1979; Nelson 2006).

In addition, Harms (1935) reported viviparity (which requires insemination) in some Indonesian species of *Boleophthalmus* and *Periophthalmus* (perciform family Gobiidae). However, Clayton (1993) noted that these observations needed verification. The osteoglossiform, *Pantodon buchholzi*, should also be checked for insemination, based on the presence of highly modified sperm (Jamieson 1991). Finally, many species in the siluriform family Loricariidae exhibit marked sexual dimorphism (Breder and Rosen 1966) and should be examined as well. As discussed below, evolutionary modification of sperm and sexual dimorphism often accompany the reproductive mode of insemination.

17.2 POSSIBLE SELECTIVE ADVANTAGES OF INSEMINATION

17.2.1 Increases the Probability of Fertilization

Eggs and sperm of non-inseminating teleosts are shed into the aqueous environment, resulting in dilution of the gametes. These cells are often numerous and some eggs, which are metabolically expensive to produce, may go unfertilized. Insemination may reduce this effect of dilution and therefore increase the probability of fertilization, all else being equal. If such is the case, selective pressure to maintain the production of high numbers of eggs can be relaxed concomitant with the evolution of insemination, and so females would become evolutionarily "free" to produce fewer eggs. Fecundity reflects the number of eggs per female and is an important demographic trait connected to life-history strategy. Based on the above scenario, we might expect to find lower fecundities in lineages of inseminating ostariophysans compared to more basal externally fertilizing relatives (Azevedo *et al.* 2000). This rationale is akin to explaining strategies along the *r*-*K* selection continuum.

Our prediction receives tentative support when relative fecundities (oocytes per mg total body mass) of small characids are taken from the literature and compared. When relative fecundities for the externally fertilizing species *Aphyocharax anisitsi* (0.68, Gonçalves *et al.* 2005), *Bryconamericus iheringii* (0.36, Lampert *et al.* 2004), *B. stramineus* (0.35, Lampert 2003), *Cheirodon ibicuhiensis* (0.50, Oliveira *et al.* 2002), *Odontostilbe pequira* (0.71, Gonçalves *et al.* 2005), *Serrapinnus calliurus* (0.63, Azevedo *et al.* 2000), and *S. piaba* (0.74, Silvano *et al.* 2003) are pooled, the mean value is 0.57. For the inseminating species *Compsura heterura* (0.55, Oliveira 2003), *Diapoma speculiferum* (0.41, Azevedo *et al.* 2000), *D. terofali* (0.57, Gonçalves *et al.* 2005), *Mimagoniates microlepis* (0.27, Azevedo 2000), *M. rheocharis* (0.36, Azevedo 2000), and *Pseudocorynopoma doriae* (0.34, Azevedo *et al.* 2000; 0.51, Ferriz *et al.* 2007), mean relative fecundity is 0.43. Although this result (0.57 > 0.43) lacks statistical rigor and does not account for phylogeny, it nevertheless invites further scrutiny into the relationship between insemination and fecundity.

17.2.2 Protects the Gametes from a Potentially Harmful Environment

Upon exiting the male body, spermatozoa encounter an environment that is osmotically challenging. Spermatozoa can be activated by contact with the external environment, but sperm cells do not survive indefinitely in the water column. Within several minutes after transfer to freshwater, a spermatozoon may become hydrated to such an extent that its plasma membrane ruptures, while the egg is protected from hypo-osmotic shock by the cortical reaction (Billard 1986). Differences are anticipated when comparing freshwater to saltwater species, given the two dissimilar osmotic regimes. Data indicate that sperm longevity tends to be shorter in freshwater than in seawater for externally fertilizing species. Sperm motility times for some saltwater spawners vary from 50 sec to more than 24 hrs (Hogan and Nicholson 1987; Lahnsteiner and Patzner 1998, 1999; Molony and Sheaves 2001). On the other hand, sperm motility or survival times for freshwater spawners have generally been calculated to be less than 2-3 min (Billard 1986; Hogan and Nicholson 1987; Billard and Cosson 1992; Lahnsteiner *et al.* 1995), although longer motility periods (3.8-5.6 min) have been reported by some authors (Vladić and Järvi 1997; Molony and Sheaves 2001). During an expedition to the Peruvian Amazon in 1996 by one of us (JRB), measurements of total dissolved solids in the water at many of the collecting sites were at the lower limit of the sensitivity of the meter, indicating extremely low solute concentrations. Duration of sperm motility can also correlate with duration of time eggs may be fertilized (Ginzburg 1968). Recent studies have demonstrated that the majority (>75%) of spermatozoa from *Boops boops* (Sparidae) and *Thalassoma pavo* (Labridae), marine external fertilizers, stop progressive movement within 15 minutes, but a small percentage exhibit local motility lasting greater than 120 minutes (Lahnsteiner and Patzner 1998, 2007). The sperm from *Perca fluviatilis* (Percidae), a freshwater or brackish externally fertilizing species, are immobile after 60 seconds (Lahnsteiner *et al.* 1995). On the other hand, the spermatozoa of *Alcichthys alcicornis* (Cottidae), a marine inseminating species with internal gametic association, were motile for 7-14 days when maintained in ovarian fluid (Lahnsteiner and Patzner 2007). Also noteworthy is the finding that ovarian fluid prolongs the motility of sperm in *Gasterosteus aculeatus* (Gasterosteidae), an external fertilizer (Elofsson *et al.* 2006). Thus, secretions of the female reproductive tract appear to be a temporary safe haven for spermatozoa. In support of this, Kutaygil (1959) found that an inseminated female *Corynopoma riisei* (Characidae) could lay fertile eggs up to ten months after being separated from a male, indicating that sperm can survive in the ovary for an extended period.

Water pH may also affect sperm motility. Billard (1986) found that spermatozoa were immotile at a pH of 7 or less with maximal motility at pH 9 for a freshwater salmonid. Low pH, especially below 5, decreased duration of sperm motility in five freshwater species (Urho *et al.* 1984). Leung (1988) posited that internal fertilization in an osmeriform, *Lepidogalaxias salamandroides*, may have been secondarily acquired in response to the highly acidic water in which this species occurs. Some species listed in Table 17.1 can be found in acidic waters where the pH may be as low as 4.0. Thus, insemination has the potential to reduce contact time between spermatozoa and detrimental effects of acidic water.

Because every lost spermatozoon is a lost chance at reproductive success, natural selection might alter the average number of spermatozoa released during each spawning event. On one hand, insemination could provide a means to minimize the number of wasted male gametes. Thus small ejaculate size might be expected for inseminating species as compared to external fertilizers. On the other hand, a comparatively large ejaculate size might be expected for inseminating species, if a large ejaculate ensures insemination. Future comparative studies, analogous to Hosken's (1998) study on bats (Megachiroptera), may provide data to reject one of these hypotheses.

17.2.3 Permits the Temporal and Spatial Separation of Mating and Oviposition

During courtship and mating, male and female may be more vulnerable than usual to predators. Insemination followed by some degree of sperm storage, whether it be fractions of a second or months, permits temporal and spatial separation of mating and fertilization. Among inseminating ostariophysans, the timing of insemination and oviposition may have evolved to follow patterns of resource seasonality. To that end, it would be informative to 1) characterize the reproductive cycles of inseminating species as following a seasonal pattern or not, 2) determine the range of time that females of a given species can store spermatozoa, 3) determine if mating and oviposition occur in different microhabitats, and 4) determine whether eggs tend to hatch during times that maximize resources (food, shelter from predators) for hatchlings. Given that the males of some highly seasonal inseminating characids (*Diapoma* spp., *Mimagoniates microlepis*, *Pseudocorynopoma doriae*) mature earlier than females (Azevedo *et al.* 2000; Azevedo 2004; Ferriz *et al.* 2007), insemination may actually occur before eggs mature and are able to be fertilized. For example, we have found ovaries containing spermatozoa and only previtellogenic oocytes (Fig. 17.1A; Burns and Weitzman 2005; Javonillo *et al.* 2007). Even if these sperm were remnants from a prior reproductive season, their presence would indicate that insemination allowed long term survival of sperm which might have the ability to fertilize the new generation of eggs.

In the viviparous perciform surfperch, *Cymatogaster aggregata* (family Embiotocidae), the male and female reproductive cycles are actually six months out of phase (Gardiner 1978a), requiring periods of sperm storage by the female. Sperm storage has been documented in a number of inseminating teleost taxa, generally by the observation of the laying of fertilized eggs by isolated females, or the production of multiple broods by isolated females in viviparous species. Examples include the osmeriform, *Lepidogalaxias salamandroides* (Pusey and Stewart 1989); the beloniform, *Horaichthys setnai* (Kulkarni 1940); the Poeciliidae (Ginzburg 1968; Potter and Kramer 2000); the scorpaeniforms, *Helicolenus dactylopterus* (Muñoz *et al.* 2002) and *Alcichthys alcicornis* (Koya *et al.* 2002); and the perciform, *Cymatogaster aggregata* (Gardiner 1978a). Evidence for sperm storage is also known for a number of inseminating ostariophysan fishes, including *Trachelyopterus striatulus* in the catfish family Auchenipteridae (von Ihering 1937) and several species in the characid subfamilies Stevardiinae and Glandulocaudinae (Weitzman *et al.* in Weitzman and Fink 1985). Our observations of live species have also confirmed sperm storage in the following stevardiine and glandulocaudine species: *Corynopoma riisei, Mimagoniates barberi, M. lateralis, M. microlepis, Tyttocharax tambopatensis* and *Pseudocorynopoma doriae*. Therefore, females potentially exert some control over when to mate and when to release eggs.

17.2.4 Allows for Sperm Competition

Sperm competition and cryptic female choice are also feasible consequences of sperm storage by the female but it remains to be seen whether any ostariophysan female may be inseminated by multiple males. Interestingly, the opening to the reproductive tract of a female of an inseminating species is not always open, but rather may be found covered by a layer of tissue. Closed female gonopores have been observed not only for ostariophysan fishes but also for livebearing poeciliids and anablepids. Open gonopores have been observed in female specimens of the characids *Pseudocorynopoma doriae* and *Diapoma* sp. (RJ and JRB unpublished), but in a mature female *Brittanichthys axelrodi* (Characidae) the reproductive tract appeared to be separated from the exterior by a thin membrane (Javonillo *et al.* 2007). Serial sections of a female *Scoloplax distolothrix* (Scoloplacidae) also failed to reveal an opening in the gonopore region (JRB unpublished). Peters and Mäder (1964) found that the oviducts of virgin female *Xiphophorus helleri* (Poeciliidae) remained closed until rupture by the male gonopodium during mating. The opening then healed over, opened for parturition, and then again closed. Weishaupt (1925) reported similar findings for the cofamilial *Poecilia reticulata*. As early as 1875, Weyenbergh wrote that the genital opening also apparently closes after each birth in *Jenynsia* (Anablepidae). Throughout most of the gestation period of female *Anableps dowi* (Anablepidae), the gonopore area is completely covered by skin and is resistant to puncture by a blunt probe, opening only after the birth of a brood (Burns and Flores 1981). Other inseminating fishes may exhibit such closures of the female reproductive tract. The consequences of closing of the female gonopore are unknown with respect to sperm competition in inseminating species.

17.3 SPERMATOZOAL MODIFICATIONS ASSOCIATED WITH INSEMINATION

17.3.1 Nucleus

17.3.1.1 Elongation of the Nucleus

By far, the most frequently observed sperm modification in inseminating fishes is elongation of the nucleus. Most externally fertilizing teleosts produce "aquasperm," generally characterized by a spherical to ovoid nucleus and short midpiece (Fig. 17.3A,F; Jamieson 1987, 1991; Mattei 1991). Deviations

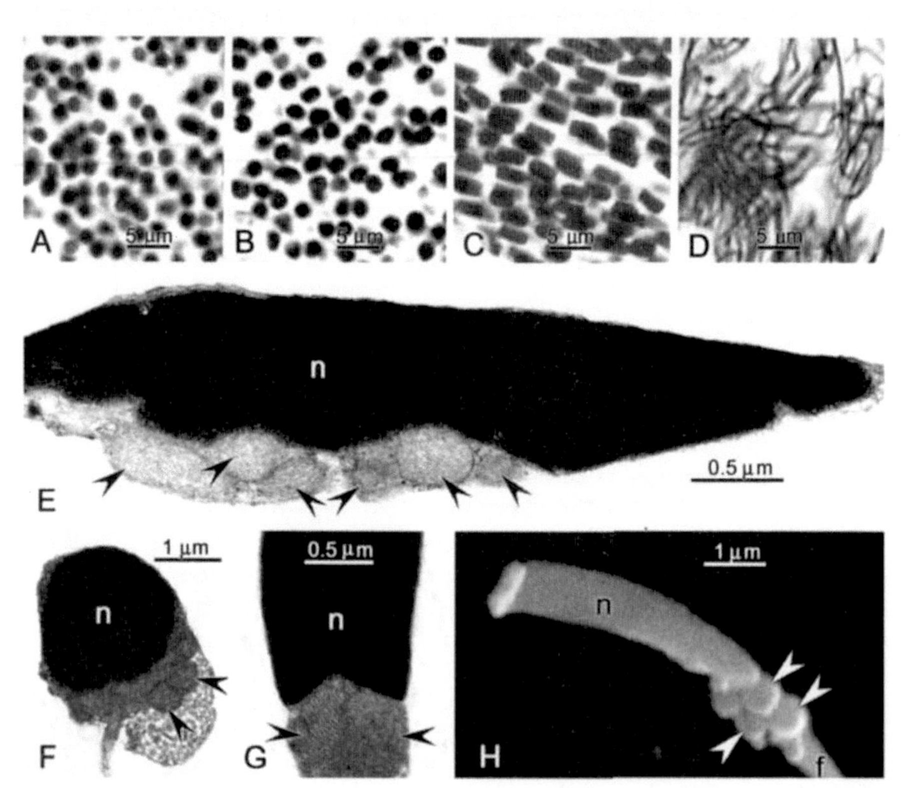

Fig. 17.3 Spermatozoa of externally fertilizing (**A**,**F**) and inseminating (**B**-**E**,**G**,**H**) species (Characiformes, Characidae). **A**-**D**, Light micrographs; **E**-**G**, Longitudinal sections, TEMs; **H**, SEM. **A**. *Knodus meridae* (*incertae sedis*). SL 31.5 mm, USNM 121473, Venezuela. **B**. *Planaltina myersi* (Stevardiinae, Diapomini). SL 37.4 mm, USNM 278989, Brazil. **C**. *Compsura heterura* (Cheirodontinae, Compsurini). SL 26.5 mm, MZUSP 39117, Brazil. **D**. *Pseudocorynopoma doriae* (Stevardiinae, Hysteronotini). SL 60.5 mm, MCP 18470, Brazil. **E**. *Bryconadenos tanaothoros* (*incertae sedis*). SL 50.0 mm, USNM 352061, Brazil. **F**. *Serrapinnus kriegi* (Cheirodontinae, Cheirodontini). SL 24.2 mm, pet shop specimen. **G**. *Odontostilbe dialeptura* (Cheirodontinae, Compsurini). SL 35.5 mm, USNM 348763, Panama. **H**. *Brittanichthys axelrodi* (*incertae sedis*). SL 27.8 mm, courtesy of L. Chao, Brazil. f, flagellum; n, nucleus; arrowhead, mitochondrion. Original.

from this pattern among externally fertilizing species are thought to be adaptations related to factors such as penetration of thick egg coats, etc. (Jamieson 1991).

Within the Ostariophysi, inseminating species are known from three families in Siluriformes (Astroblepidae, Auchenipteridae, Scoloplacidae) and a single family in Characiformes (Characidae). No species in Gymnotiformes, Cypriniformes, or Gonorynchiformes is known to be inseminating.

Thus far all species analyzed microscopically in the siluriform families Auchenipteridae (Fig. 17.1B; Meisner *et al.* 2000b; Burns *et al.* 2002; Parreira and Godinho 2004), Scoloplacidae (Fig. 17.2A) Burns *et al.* 2000; Spadella *et al.* 2006), and Astroblepidae (Fig. 17.2B,C; Spadella 2007) produce spermatozoa with markedly elongate nuclei. In the characiform family Characidae, on the other hand, a number of inseminating species produce sperm cells indistinguishable from classic aquasperm. These species include *Planaltina myersi* (Fig. 17.3B), *P. glandipedis* and *P. britskii* (Menezes *et al.* 2003), *Kolpotocheirodon theloura* ("species A" in Burns *et al.* 1997), *Knodus* sp. (Burns and Weitzman 2005), and *Attonitus irisae* and *A. ephimeros* (Weitzman *et al.* 2005). All other inseminating characids examined to date produce spermatozoa with nuclei ranging from slightly to extremely elongate (Fig. 17.3C-E,G,H; Fig. 17.4A-D; Fig. 17.5A,F).

A number of possible selective advantages have been proposed for nuclear elongation in inseminating fishes.

Movement through the female gonopore: Cells with elongate nuclei often have nuclear widths that are less than the nuclear diameters of aquasperm (Burns et al. 1995, 1997). Hence, at any one time a greater number of elongate cells may be able to fit in the opening of the female gonopore, thus increasing the number of spermatozoa transferred to the female reproductive tract (Burns and Weitzman 2005).

Movement within the female reproductive tract: Cells with elongate nuclei have more streamlined sperm heads. This may facilitate movement through viscous secretions and narrow, tortuous pathways within the female reproductive tract, particularly the ovary (Gardiner 1978a). This potential adaptation would pertain once insemination had been effected.

Directional movement: Sperm head elongation may facilitate the forward movement of the sperm cell, thus increasing its directional movement toward the female gonopore (Burns and Weitzman 2005). Since the drag force, i.e., the force resisting motion through a fluid, is proportional to the frontal area of the cell, a cell with a spherical head would encounter more resistance to moving forward than a more elongate cell. An elongate cell has less resisting force in the forward direction, but greater resisting force in the side direction. The work of Cosson *et al.* (2000) on the movement of paddlefish and sturgeon spermatozoa support this view. Both sturgeon (Jamieson 1991) and paddlefish

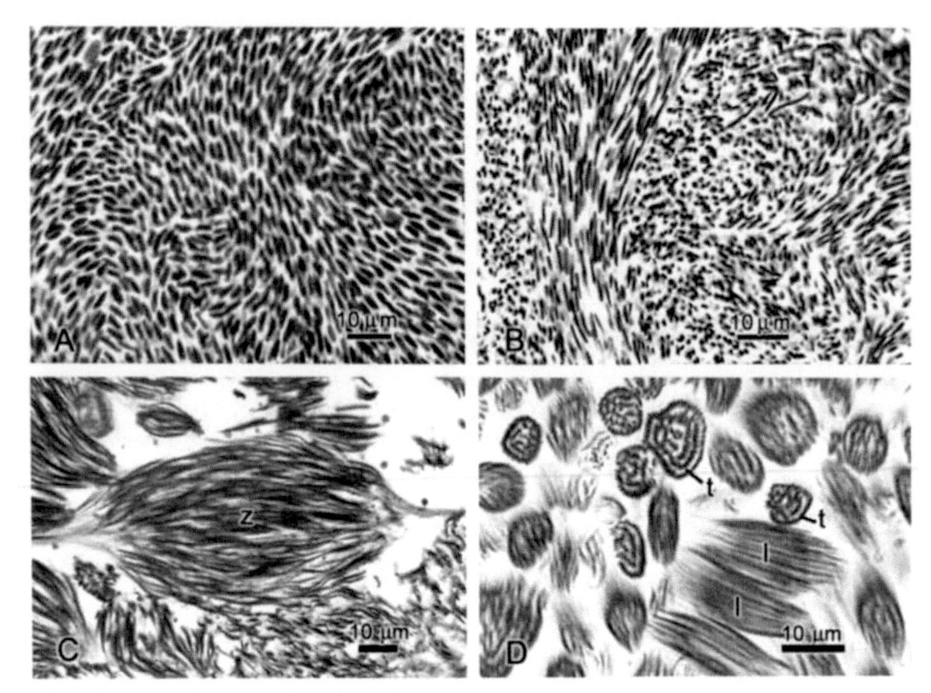

Fig. 17.4 Light micrographs though testis ducts of inseminating species (Characiformes, Characidae). **A**. *Acinocheirodon melanogramma* (Cheirodontinae, Compsurini). SL 27.6 mm, MCP 18596, Brazil. Note that many spermatozoa tend to align parallel to one another. **B**. *Gephyrocharax venezuelae* (Stevardiinae, Stevardiini). SL 32.2 mm, USNM 121363, Venezuela. Note the spermatozoa forming a flowing pattern at the left side of the micrograph. **C**. *Lophiobrycon weitzmani* (Glandulocaudinae). SL 25.7 mm, MZUSP 83353, Brazil. Longitudinal section through a spermatozeugma (z). **D**. *Scopaeocharax rhinodus* (Stevardiinae, Xenurobryconini). SL 24.7 mm, USNM 329756, Peru. Transverse (t) and longitudinal (l) sections through spermatozeugmata. Original.

(Zarnescu 2005) produce sperm cells with elongate nuclei and their sperm trajectories remained close to linear (Cosson *et al.* 2000) (For spermatozoa of Acipenseriformes, see Jamieson **Chapter 8**). On the other hand, the work of Lahnsteiner and Patzner (2007) suggests that sperm head shape has no effect on the swimming patterns of selected species from the teleost families Salmonidae, Cottidae, Percidae and Sparidae, where motility was predominantly linear in all specimens. The sperm-nucleus shape in these species varied from spherical to ovoid, and all species appear to be externally fertilizing, although the reproductive mode of the cottid, *Cottus gobio*, is debated (Lahnsteiner and Patzner 2007). Unfortunately, no species with significantly elongate sperm nuclei was analyzed in this study. Therefore, whether or not nuclear elongation significantly increases directionality of sperm cell movement remains to be demonstrated.

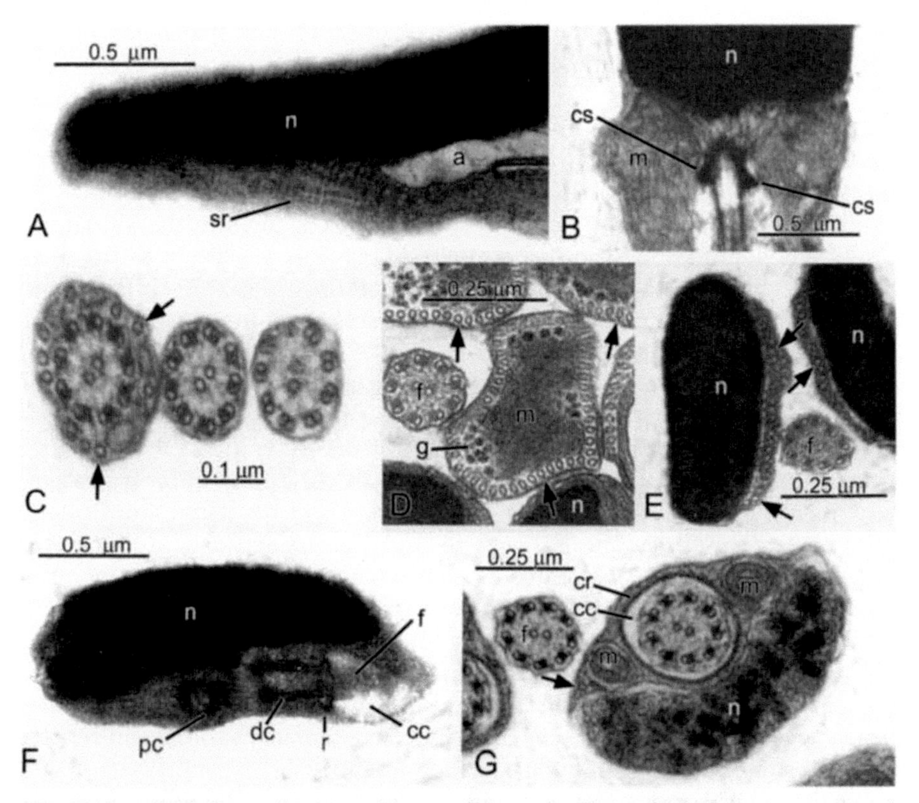

Fig. 17.5 TEMs through spermatozoa of inseminating ostariophysan species. **A**. *Brittanichthys axelrodi* (Characiformes, Characidae, *incertae sedis*). SL 27.8 mm, courtesy of L. Chao, Brazil. Longitudinal section through anterior tip of spermatozoon. **B,C**. *Macropsobrycon uruguayanae* (Characiformes, Characidae, Cheirodontinae, Compsurini). SL 39.0 mm, MCP 18588, Brazil. **B**. Longitudinal section through posterior nucleus and midpiece region. **C**. Transverse sections through progressively more posterior (left to right) regions of flagellum. **D**. *Trachelyopterus lucenai* (Siluriformes, Auchenipteridae). SL 147 mm, MCP 18469, Brazil. Transverse section through midpiece region and flagellum. **E**. *Mimagoniates barberi* (Characiformes, Characidae, Glandulocaudinae). Transverse sections through nuclei and flagellum. **F**. *Diapoma* sp. (Characiformes, Characidae, Stevardiinae, Diapomini). SL 41.0 mm, MCP 18478, Brazil. Oblique section through nucleus and centrioles. **G**. *Pseudocorynopoma doriae* (Characiformes, Characidae, Stevardiinae, Hysteronotini). SL 60.5 mm, MCP 18470, Brazil. Transverse section through nuclear region and flagellum. a, artifact space; cc, cytoplasmic canal; cr, dense ring around cytoplasmic collar; cs, centriolar spur; dc, distal centriole; f, flagellum; g, glycogen; m, mitochondrion; n, nucleus; pc, proximal centriole; r, dense ring around entrance to cytoplasmic canal; sr, striated rootlet; arrow, accessory microtubule. Original.

Clumping: Many inseminating teleosts have been shown to produce distinct sperm packets within the testes that in turn are transferred to the female reproductive tract through the gonopore. Unencapsulated sperm bundles are referred to as spermatozeugmata, whereas those encapsulated by either cells or acellular material are called spermatophores (Grier 1981; Grier and Parenti 1994). The transfer of distinct packets of spermatozoa to the female appears to increase the probability and effectiveness of insemination by maintaining high sperm densities during transport (Ginzburg 1968). Examples of inseminating teleosts that produce sperm packets include the siluriform family Auchenipteridae (Loir *et al.* 1989; Meisner *et al.* 2000b; Burns *et al.* 2002); the characiform subfamily Glandulocaudinae (Fig. 17.4C; Pecio and Rafiński 1994, 1999; Burns *et al.* 1995); the characiform subfamily Stevardiinae (Fig. 17.4D; Burns *et al.* 1995; Pecio *et al.* 2005); the *incertae sedis* characid species *Brittanichthys axelrodi* (Javonillo *et al.* 2007) and *Bryconadenos tanaothoros* (Weitzman *et al.* 2005); the osmeriform species *Lepidogalaxias salamandroides* (Leung 1988); the atheriniform family Phallostethidae (Grier and Parenti 1994); *Horaichthys setnai* in the beloniform family Adrianichthyidae (Grier 1984); the genera *Zenarchopterus* (Grier and Collette 1987) and *Hemirhamphodon, Dermogenys* and *Nomorhamphus* (Downing and Burns 1995) of the beloniform family Hemiramphidae, subfamily Zenarchopterinae; the cyprinodontiform families Anablepidae (*Anableps dowi* only) (Grier *et al.* 1981), Poeciliidae (Vaupel 1929; De Felice and Rasch 1969; Hurk, van den *et al.* 1974; Grier 1981) and Goodeidae (Grier *et al.* 1978); and the perciform family Embiotocidae (Gardiner 1978c). Fig. 17.1D demonstrates that intact sperm packets can be transferred to a female characid and remain intact for some period within the ovarian lumen.

On the other hand, not all inseminating teleosts produce histologically demonstrable sperm bundles. However, even in these species nuclear elongation may still play a role in facilitating some degree of clumping (Atwood and Chia 1974). For example, spermatozoa of *Corynopoma riisei* (Characidae, Stevardiinae) appear to be released as mucilaginous "droplets" (Kutaygil 1959). In the study of Burns *et al.* (1995) it was noticed that on histological sections of testes from inseminating characids that did not produce sperm packets, spermatozoa appeared to be arranged into "flowing patterns" within the sperm ducts (Fig. 17.4A,B). These patterns were evident even on testis sections from species that produce spermatozoa with only slightly elongate nuclei (Fig. 17.4A). Thus, it was suggested that even a slight degree of nuclear elongation may increase the likelihood of sperm clumping. As a consequence, passage of clumps of sperm cells to the female would presumably result in loss of fewer cells to the surrounding aqueous environment, and a higher sperm density would increase the probability of more cells entering the female gonopore.

In hydrodynamics, the ratio of a fluid's inertial forces to its viscous forces is represented by the Reynolds number (Purcell 1977). For large objects or organisms such as ships or fishes moving through water, the Reynolds number is high and inertial forces dominate. For example, if a fish suddenly stops swimming, it will continue to move forward a certain distance due to inertia before coming to a stop. This contrasts sharply with small objects or organisms, such as spermatozoa or protozoa, moving through an aqueous medium. In this case "inertia is totally irrelevant" (Purcell 1977) and the viscosity of the water is the dominant factor. So, if a spermatozoon ceases flagellar movement, the cell will come to a complete stop due to the effects of viscosity. Therefore, the principles that govern movement in aqueous media at high and low Reynolds number may be quite different.

At high Reynolds numbers, even non-motile, elongate bodies tend to clump together in an aqueous environment. An example of this would be the familiar side-to-side clumping of untethered canoes or boats in a marina. However, given that the hydrodynamic principles that govern such movement at high Reynolds numbers may not apply at low Reynolds numbers, a different explanation for the observed "flowing patterns" of the spermatozoa may be necessary. The recent work of Narayan *et al.* (2007) may shed some light on this. These workers designed an experiment to test the movement of self-driven objects with anisotropic shape. Rods of copper wire, 0.8 mm in diameter, were cut into lengths of 4.6 mm with each end having been etched to produce a final form resembling a rolling pin. For uncertain reasons, similar results were not obtained with cylindrical rods (Hecke, van 2007). Rods at different densities were placed into a circular chamber 13 cm in diameter and 1 mm high and periodically agitated (to simulate self-propulsion). The results showed that at a range of densities, the rods exhibited "macroscopic swirls" remarkably similar to those seen among the spermatozoa in Fig. 17.4A,B. A rod was found to be approximately 2.3 times more likely to move along its length than transverse to its length. The main factors at play appeared to be complex connections among alignment, density, and flow (Hecke, van 2007). Such "swarming" in driven collections of elongate (non-spherical) particles has been observed at critical densities in systems ranging from the macroscopic (schools of fish) to the microscopic (bacterial colonies). Even cultures of keratinocytes obtained from fish scales moved in coherent groups above a certain density of cells (Szabó *et al.* 2006). Therefore, as the commentary by van Hecke (2007) emphasized, "shape matters."

Sperm cells with elongate nuclei also have an "anisotropic" shape (i.e., different dimensions for length and width). Thus, their behavior at high densities may reflect that observed in the study of Narayan *et al.* (2007). Riedel *et al.* (2005) found that above a certain density, sea urchin (genus *Strongylocentrotus*) spermatozoa self-organized into arrays of vortices. The spermatozoa of these sea urchins also have elongate sperm heads. These authors conclude that "large-scale coordination of cells can be regulated hydrodynamically" (Riedel *et al.* 2005). Finally, in a study of hypothetical swimming performance of elastic swimmers at low Reynolds numbers, Lauga (2007) found that swimming performance was always better for long slender bodies than for those with spherical bodies. This was the case for both swimming speed and efficiency.

In conclusion, despite the dearth of studies on the hydrodynamics of movement at low Reynolds numbers, a more elongate shape does appear to have a profound effect on the performance of a cell. Clumping, especially at higher cell densities, appears to be enhanced. Additionally, both swimming speed and swimming efficiency may be increased. Thus, even a slight degree of elongation of the nucleus of a spermatozoon may result in a significant selective advantage to the individual population.

17.3.1.2 Flattening of the nucleus

Aside from evolutionary elongation of the sperm nucleus, some inseminating species produce sperm nuclei that appear relatively flattened in some planes. For example, the anterior portions of sperm nuclei from *Brittanichthys axelrodi* (*incertae sedis* in Characidae; Fig. 17.3.H) and *Mimagoniates barberi* (Glandulocaudinae; Fig. 17.5E) are flattened, as are those of *Scopaeocharax rhinodus* and *Tyttocharax tambopatensis* (Stevardiinae; Pecio *et al.* 2005). All four of these species produce spermatozeugmata. There are other inseminating ostariophysans that do not produce spermatozeugmata but still produce sperm with elongate nuclei that are somewhat flattened (rather than rounded) along one side. Evolutionary flattening of the nucleus, like elongation, may therefore also aid in packing of sperm into high density aggregations.

17.3.2 Enlargement of the Midpiece (mitochondrial region)

Data from externally fertilizing teleosts indicate that spermatozoa rely heavily on ATP produced from glycolysis, oxidative phosphorylation and the tricarboxylic cycle for their energy needs, both during the motile and immotile stages (Lahnsteiner *et al.* 1999; Dreanno *et al.* 2000). Indeed, the short sperm motility time measured in some externally fertilizing freshwater teleosts appears to correlate with a decrease in intracellular ATP and an inadequate energy supply (Billard 1986; Christen *et al.* 1987; Billard and Cosson 1992). Fawcett (1970) suggested that an enlargement of the midpiece increases the capacity of the sperm's energy-generating mechanism. A study of midpiece length in the salmonid, *Salmo salar*, showed that longer midpieces were associated with higher ATP concentrations, indicating that the longer midpieces have a greater capacity for ATP formation via mitochondrial oxidative phosphorylation (Vladić *et al.* 2002). Lahnsteiner and Patzner (1998) also suggest that the number of mitochondria may determine the efficiency of energy supply for spermatozoa.

The midpiece, the region containing the mitochondria, tends to be limited in size in the sperm of externally fertilizing species (Fig. 17.3F), whereas in inseminating species it is often enlarged (Jamieson 1991; Mattei 1991). The elongate midpieces often present in the sperm of inseminating teleosts are thought to help prolong the lifespan of the spermatozoon during storage in the ovary, as well as provide energy for sperm dispersal throughout the ovary

(Pecio and Rafiński 1994: Yao *et al.* 1995). Data on the inseminating scorpaeniform, *Alcichthys alcicornis*, indicate that spermatozoa may indeed move within the ovary during a major part of the spawning season from March to May (Koya *et al.* 2002). These authors suggest that the sperm cells in this species may be able to continue to move for about one month, even without extracellular energy substrates, due to the abundant energy generated by the numerous mitochondria in the elongate midpiece.

Enlarged midpieces of inseminating ostariophysans tend to be due either to an increase in the number of mitochondria (Fig. 17.3E) or to enlargement of individual mitochondria (Fig. 17.3G,H; Fig. 17.5G). Within Characiformes, spermatozoa of glandulocaudines (Pecio and Rafiński 1994, 1999), a number of stevardiines (Burns *et al.* **Chapter 11**; Burns *et al.* 1998; Pecio *et al.* 2007), and *Bryconadenos tanaothros* (Fig. 17.3E, *incertae sedis* in Characidae; Weitzman *et al.* 2005) have multiple mitochondria located along one side of the elongate nucleus, and in some cases mitochondria continue posterior to the nucleus. In Siluriformes, multiple mitochondria are located to one side of the cytoplasmic canal in auchenipterids (Burns *et al.* 2002) and surround the cytoplasmic canal in scoloplacids (Spadella *et al.* 2006) and astroblepids (Spadella 2007). Extremely elongate mitochondria are characteristic of *Pseudocorynopoma doriae* (Characidae, Stevardiinae, Fig. 17.5G; Burns *et al.* **Chapter 11**). Fig. 17.3 allows a comparison of mitochondrial size between two species in the same subfamily (Characiformes, Cheirodontinae), the externally fertilizing *Serrappinis kriegi* (Fig. 17.3F) and the inseminating *Macropsobrycon uruguayanae* (Fig. 17.3G), which has enlarged mitochondria located posterior to the nucleus (Oliveira *et al.* 2008; Burns *et al.* **Chapter 11**). The spermatozoon of another inseminating characid, *Brittanichthys axelrodi*, bears several large spherical mitochondria (each ~0.6 μm in diameter) in an area ventral to and continuing beyond the posterior portion of an elongate nucleus (Fig. 17.3H; Javonillo *et al.* 2007). By comparison, four externally fertilizing characids (*Serrapinnus kriegi*, *S. calliurus*, *Aphyocharax anisitsi*, and *Knodus* sp.) have mitochondria of diameter 0.2-0.6 μm (Fig. 17.3F; Javonillo *et al.* 2007). In summary, many inseminating ostariophysan species increase the mitochondrial capacity by increasing the size or number of these organelles.

17.3.3 Basal Body Reinforcement

Other modifications observed in, but not necessarily restricted to (Lahnsteiner and Patzner 2007), inseminating species relate to the centrioles, particularly the distal centriole that serves as the basal body for the flagellum. Given that it is likely that most inseminated spermatozoa remain immotile until after transfer to the female, especially those that are bundled into some type of packet, possible selective advantages of these modifications probably relate more to spermatozoal movement within the female reproductive tract. In all inseminated ostariophysan females analyzed to date, spermatozoa are found throughout the lumina of the ovarian ducts and ovaries, and often squeezed into the folds of these structures (Fig. 17.1A-D). In addition to having to negotiate these narrow, winding passageways, it is also likely that the ovarian lumen contains secretions that make the liquid environment more viscous. Therefore, passage through this internal environment would apparently put more demands on the flagellum than swimming in a strictly aqueous medium. Most of the centriolar modifications relate to the distal centriole that serves as the basal body or "anchor" of the flagellum, although the proximal centriole may also be involved. Modifications that may serve to reinforce, strengthen or stabilize the flagellar base include electron-dense material among or within the microtubules of the centrioles (Fig. 17.5B,F) and electron-dense "spurs" (Fig. 17.5B).

Finally, more elaborate stabilizers include striated rootlets. Not exclusive to inseminating ostariophysans, a striated rootlet is found in the anterior portion of the *Brittanichthys axelrodi* (Characidae, *incertae sedis*) spermatozoon (Fig. 17.5A; Javonillo *et al.* 2007), as in the sperm of *Macropsobrycon uruguayanae* (Characidae, Cheirodontinae; Burns *et al.* 1998). In ciliated cells, striated rootlets are composed of the protein rootletin (Yang *et al.* 2002, 2005). Rootlets do not appear necessary for movement of cilia but their absence in mutant cells results in decreased survival and decreased ability to withstand mechanical stress compared to normal ciliated cells (Yang *et al.* 2005). The striated rootlets of spermatozoa may therefore stabilize cells during flagellar swimming. Survival and movement of spermatozoa through the ovarian cavity would be assisted by robust anchorage of the flagellum through a striated rootlet.

17.3.4 Flagellum

To date, no inseminating ostariophysan is known to have biflagellate sperm. Lateral projections or fins, on either side of the sperm flagellum, are present in the siluriform families Scoloplacidae (Spadella *at al.* 2006) and Astroblepidae (Spadella 2007), these being the only inseminating ostariophysans thus far analyzed to have such structures (Burns *et al.* **Chapter 11**). All else equal, if the force for flagellar propulsion is proportional to the surface area of the flagellum, perhaps axonemal fins evolved to provide increased surface area for propulsion. Increased area could also be obtained by increasing the length of the flagellum or number of flagella per spermatozoon. One recent account of sperm morphology for an externally fertilizing teleost (*Pelvicachromis taeniatus*, Cichlidae) highlighted the unusual overall length of the cell: mean total sperm length was 69.7 μm, with flagella accounting for 65.8 μm (94.4%) of mean spermatozoon length (Thünken *et al.* 2007). Those authors pointed to sperm competition (both between-male and among sperm of the same male) as a possible factor in the evolution of long sperm flagella. The lengths of sperm flagella have been measured for some inseminating ostariophysans (e.g, Pecio *et al.* 2005; Javonillo *et al.* 2007) but the impact of flagellum length on probability of fertilization has not.

The location of the single flagellum relative to the nucleus, like those of mitochondria, varies for inseminating ostariophysans. In the siluriform families Scoloplacidae (Spadella *e al.* 2006), Astroblepidae (Spadella 2007) and Auchenipteridae (Burns *et al.* 2002), as well as in *Macropsobrycon uruguayanae* (Characidae; Fig. 17.5B), the flagellum originates posterior to the nucleus. This configuration would appear to result in the nucleus being pushed forward by the flagellum. Sperm of other inseminating species such as the stevardiine characids *Scopaeocharax rhinodus* and *Tyttocharax tambopatensis* have a flagellum that originates anterior to the tip of an elongate nucleus (Pecio *et al.* 2005). Similarly, in the inseminating characids *Brittanichthys axelrodi* (Fig. 17.5A; Javonillo *et al.* 2007), *Bryconadenos tanaothoros* (Weitzman *et al.* 2005), *Diapoma speculiferum* (Burns *et al.* 1998) and *Chrysobrycon* sp. (Burns *et al.* **Chapter 11**) the sperm flagellum originates ventral to the anterior half of an elongate nucleus. In these configurations, the nucleus appears to be more pulled than pushed by the flagellum.

The extent to which the anterior portion of the flagellum is contained within a cytoplasmic collar has also been noted for most inseminating ostariophysans for which there are ultrastructural data (Burns *et al.* **Chapter 11**). In one arrangement, a long cytoplasmic collar develops during spermiogenesis. This long collar persists in the spermatozoon, as in *Diapoma*, *Corynopoma*, and *Pseudocorynopoma* (Burns *et al.* 1998), but its posterior portion degenerates before spermiation in *Tyttocharax cochui*, *T. tambopatensis* and *Scopaeocharax rhinodus* (Stevardiinae, Xenurobryconini; Pecio *et al.* 2005), and *Mimagoniates barberi* (Glandulocaudinae; Pecio and Rafiński 1994, 1999). In cases where an elongate collar is "bound" to a nucleus that is also elongate, motion of the flagellum might cause the whole nucleus to undulate as well. This kind of motion is again an alternative to cases in which the nucleus is either pushed by a flagellum that is posterior to the nucleus or pulled forward by the flagellum that originates anterior to the nucleus, as in *Tyttocharax* and *Scopaeocharax* (Pecio *et al.* 2005). The functional implications for the location of the flagellum and the relative length of the cytoplasmic collar have yet to be examined.

In one specimen of an undescribed species of *Diapoma* (Stevardiinae), an electron-dense ring that is positioned at the distal end of the distal centriole (Fig. 17.5F) may serve to reinforce the opening of the cytoplasmic collar. In the spermatozoon of *Pseudocorynopoma doriae* (Stevardiinae), the electron-dense ring in the proximal portion of the cytoplasmic collar may be another mechanism to gird the cell's structural integrity (Fig. 17.5G). Finally, in many inseminating ostariophysans the A-tubules of the axonemal doublets are electron-dense, at least at some point along the flagellum (Fig. 17.5C-E,G; Burns *et al.* **Chapter 11**). Although the functional significance of this is unknown, this adaptation may serve to strengthen the axoneme in some manner.

17.3.5 Accessory Microtubules

In spermatozoa of some ostariophysan species, microtubules are not limited to the axoneme. For example the proximal portion of the axoneme of the *Macropsobrycon uruguayanae* (Characidae, Cheirodontinae) spermatozoon is surrounded by a ring of unpaired microtubules that is concentric with the axoneme (Fig. 17.5C). In *Trachelyopterus lucenai* (Auchenipteridae) sperm, the midpiece region of the cell is completely encircled by a peripheral row of accessory microtubules (Fig. 17.5D, Burns *et al.* 2002). Unpaired microtubules are also situated along one side of the nucleus of *Mimagoniates barberi* (Glandulocaudinae) sperm (Fig. 17.5E). In cross sections of sperm from *Pseudocorynopoma doriae* (Stevardiinae), we also observe occasional unpaired microtubules that are not part of the axoneme (Fig. 17.5G). In most of these cases, the accessory microtubules lie just within the cell membrane rather than more medially. A likely advantage of accessory microtubules is that they help stiffen the part of the cell in which they are located, making the cell more resistant to bending. Additional stiffness imparted to the cell by accessory microtubules could reduce drag as the cell moves through viscous ovarian fluids (Fauci 1996), thereby increasing efficiency of movement.

17.3.6 Glycogen

Spermatozoa of some inseminating ostariophysans have been found to hold substantial glycogen deposits (Fig. 17.5D). These inclusions could represent energy stores, especially for sperm that endure for a relatively long time, as in cases where spermatozoa survive in inseminated females for months (Kutaygil 1959). Most inseminating characiform species studied do not seem to show abundant glycogen in the cytoplasm, however transmission electron microscopy reveals glycogen in the cytoplasm of spermatozoa of the stevardiines *Diapoma speculiferum* and *Diapoma* sp., and the glandulocaudine *Mimagoniates barberi*. In an inseminating auchenipterid catfish, *Trachelyopterus lucenai*, abundant glycogen stores are present throughout the elongate midpiece (Fig. 17.5D, Burns *et al.* 2002).

17.4 OTHER MORPHOLOGICAL MODIFICATIONS ASSOCIATED WITH INSEMINATION

17.4.1 Testis Modifications

Another modification sometimes observed in inseminating species is the development of a posterior region of the testis for use in sperm storage, although such areas may also be present in some externally fertilizing teleosts (Ginzburg 1968; Lahnsteiner *et al.* 1994; Rasotto and Sadovy 1995). Such areas have been reported for inseminating species in the following teleost families: Galaxiidae (Pusey and Stewart 1989), Phallostethidae (Grier and Parenti 1994), Zenarchopteridae (Hurk, van den, 1973; Grier and Collette 1987; Downing and Burns 1995), Anablepidae (Burns 1991), Poeciliidae (Pandey

1969; Hurk, van den *et al.* 1974), Goodeidae (Grier *et al.* 1978), and Embiotocidae (Gardiner 1978c). Of ostariophysans, distinct sperm storage areas have been described in the inseminating auchenipterid catfishes, *Trachelyopterus lucenai* and *T. galeatus* (Meisner *et al.* 2000b). In mature males of subfamilies Glandulocaudinae and Stevardiinae (Characidae) and certain other *incertae sedis* characid species, the posterior testis is aspermatogenic and stores sperm or spermatozeugmata (Pecio and Rafiński 1994; Burns *et al.* 1995; Pecio *et al.* 2001; Weitzman *et al.* 2005; Burns and Weitzman 2006; Javonillo *et al.* 2007). Successful fertilization or multiple inseminations may thus require large reservoirs of sperm (Burns *et al.* 1995). This point is especially true for glandulocaudines and stevardiines because they lack evident intromittent organs. Richter (1986) suggested that in *Corynopoma riisei* (Stevardiinae), the anal fin is used to direct spermatozoa towards the female gonopore and observed a "cloud of sperm" in the water just after separation of a mating pair. Thus, in these species, substantial sperm loss may accompany mating and large reserves of sperm may therefore be required.

17.4.2 Secondary Sex Characters

The behavioral and chemical cues elicited by courting individuals are also reproductively significant traits because mates must be in close proximity to effect insemination. Visual cues depend on both the visible and ultraviolet (UV) wavelengths in some inseminating species (White *et al.* 2003). Mate choice is affected by whether female *Girardinichthys multiradiatus* (Goodeidae) detect the UV-reflective markings of males (Macías Garcia and Burt de Perera 2002). Similar results obtain for the poeciliid *Xiphophorus nigrensis* (Cummings *et al.* 2003). Other features are used in visual signaling: scales, fins, and opercles for example. From males of the stevardiine characid genera *Corynopoma* and *Pterobrycon* such structures (elongate "gill paddles" and humeral scales, respectively) are abducted from the long axis of the body and displayed to females during courtship (Bussing and Roberts 1971; Richter 1986).

Known only from characid fishes thus far, the glands found on the anteriormost gill arches and caudal fins of mature males are thought to produce chemical signals (pheromones) that influence reproductive encounters (Weitzman and Fink 1985; Bushmann *et al.* 2002). These "gill glands" develop on the ventral portion of each first gill arch in males of some characid species (Burns and Weitzman 1996; Bushmann *et al.* 2002). Although there is some correlation between the presence of insemination and presence of gill glands, not all inseminating characid species exhibit gill glands and not all species with gill glands inseminate. Caudal glands, or caudal organs, are of two varieties (Weitzman *et al.* 2005). A large scale forms a pocket, lined with hypertrophied mucous cells, on either side of the peduncle on males of *Corynopoma riisei* and other stevardiines (Atkins and Fink 1979). The caudal organ of male glandulocaudines, on the other hand, is formed by modification of caudal-fin rays and scales on the dorsolateral portion of the caudal peduncle, and the secretory cells appear to be club cells similar to the alarm substance cells characteristic of the Ostariophysi (Pfeiffer 1977; Weitzman *et al.* 2005).

17.5 CONCLUSIONS

The spermatozoon is a highly specialized cell and above, in separate sections, we have addressed morphological and functional aspects of spermatozoa. Although we discuss portions of ostariophysan reproductive systems separately above, such traits could be genetically linked and selection pressure could act on multiple traits simultaneously. It is necessary to study form, function, and phylogeny as pertinent to an integrative knowledge of insemination. In the same way that others investigate feeding behavior as encompassing such matters as biomechanics, anatomy, ontogeny, and cladistic relationships (e.g., Westneat 2004), so should we strive to synthesize such kinds of information about insemination in fishes. Opportunities and challenges lie ahead.

Because cladograms can be used as predictive tools for inferring the attributes of organisms yet to be sampled (Smith and Wheeler 2006), reproductive mode can be predicted for species. For example, because insemination is known for some members of Cheirodontinae (Characidae; Table 17.1; Burns *et al.* 1997) and *Trachelyopterus* (Auchenipteridae; Table 17.1; Meisner *et al.* 2000b), it is not unreasonable to hypothesize that other members of that subfamily and genus, respectively, are also inseminating. Furthermore, given 1) that the production of elongate sperm nuclei seems to correlate with insemination, 2) that multiple members of Scoloplacidae inseminate with elongate sperm nuclei (Spadella *et al.* 2006), 3) that Scoloplacidae is the sister group to Astroblepidae + Loricariidae (Sullivan *et al.* 2006), and 4) that *Astroblepus chotae* and *A. sabalo* have been found to inseminate, we suspect that insemination will be found for other members of Astroblepidae.

The family Characidae is attractive as a model system in which to investigate reproductive evolution, especially insemination. This group includes a number of subfamilies, such as Cheirodontinae, Stevardiinae, and Glandulocaudinae, and each of these three taxa includes inseminating species. The latter two subfamilies plus some *incertae sedis* genera constitute a group referred to as "Clade A" (Weitzman *et al.* 2005). Cheirodontinae is not part of Clade A, but once the relationships among cheirodontines and Clade A genera (including stevardiines and glandulocaudines) are better resolved, we can infer the number of times insemination evolved in Characidae. At this time we cannot correlate insemination with other putatively derived traits. If insemination originated once in Characidae, then a cladogram can be used to determine which character-state changes either preceded or followed the appearance of insemination. If insemination is inferred to have multiple origins, statistical methods, such as the concentrated-changes test of Maddison (1990), could be employed to detect whether insemination was more likely than not to have originated in the presence of another given character state.

The microscopic scale of sperm anatomy and insemination presents challenges for investigating certain aspects of insemination among ostariophysans. For example, it could prove difficult to quantify the effect of variation in sperm morphometrics on reproductive success. Do individual males produce highly variable sperm? (Evidence for much within-individual variation in sperm nucleus length or diameter has been lacking in species surveyed [Burns *et al.* 1995].) Is there variation among individuals of a population? If so, does the variation affect reproductive success? Answering such questions might require measurement of spermatozoa and an ability to track them individually before and after insemination, as well as experimentally controlling for other aspects of mates' phenotype and environment.

Above we describe reproductive morphology and propose evolutionary hypotheses regarding insemination. For other vertebrate groups, similar hypotheses (about sperm competition, sperm biomechanics, etc.) have been tested but the conclusions of those studies do not necessarily extend to the biology of ostariophysan fishes. The sheer diversity of this lineage surely encourages further study of reproduction in its members.

17.6 CHAPTER SUMMARY

Most teleost fishes produce eggs that are fertilized externally, but insemination has been demonstrated for a growing number of teleosts. We maintain a distinction between insemination and internal fertilization because internal gametic association (insemination followed by external fertilization) is known for some species. A list is provided of groups within Teleostei for which insemination is known. This enumeration also includes internally fertilizing and viviparous species. A table of inseminating ostariophysans is also provided, which includes species for which insemination had not been histologically demonstrated previously: *Glandulocauda melanopleura, Xenurobrycon macropus, Monotocheirodon pearsoni, Rachoviscus crassiceps, Astroblepus chotae* and *A. sabalo.* The precise location of fertilization (external or internal) is not known for any inseminating ostariophysan.

We review hypotheses regarding the selective advantage(s) of insemination. Briefly, the reproductive mode of insemination may have arisen to increase the probability of fertilization, protect gametes from a potentially harmful environment, permit temporal and spatial separation of mating and oviposition, or to allow for sperm competition. Evolutionary modifications of spermatozoa are associated with insemination. Most inseminating teleosts produce spermatozoa with elongate nuclei, in contrast to the spermatozoa of externally fertilizing fishes, which usually have spherical nuclei. Evolutionary elongation of the nucleus may aid one or more of the following:

movement of sperm through the female gonopore, movement within the female reproductive tract, directional movement, or clumping. Flattening of the sperm nucleus might also help to achieve side-by-side alignment of those cells. The longevity of the spermatozoon outside of the male body may be ensured by increasing the size of the sperm midpiece, glycogen storage within the spermatozoon, as well as secretions of both the male and female reproductive tracts. Natural selection may also act on additional structures of the spermatozoon, such as the flagellum and its basal body and supporting cytoskeletal elements such as accessory microtubules. Morphological modifications often associated with insemination also include the development of an aspermatogenic storage region in the testis, sexual dimorphism, intromittent organs, communication by chemical signals and certain mating behaviors that bring mates very close together.

Comparisons of sperm biology may be undertaken at the populational and interspecific levels of diversity, especially for Characidae and, more generally, Ostariophysi. Ongoing phylogenetic work will test homologies of reproductive characters and will create a more rigorous historical framework for biological comparisons. We predict that histology will reveal that more teleost species are inseminating.

17.7 ACKNOWLEDGMENTS

Thanks are also extended to all of our collaborators, past and present, and to others who have helped us with this research. Most are acknowledged in our prior publications. Funding in the past has been provided in part by George Washington University, Tropical Fish Hobbyist Magazine, Dr. Herbert R. Axelrod, Smithsonian Institution, and CNPq (Brazil).

17.8 LITERATURE CITED

Amemiya, I. and Murayama, S. 1931. Some remarks on the existence of developing embryos in the body of an oviparous cyprinodont, *Oryzias* (*Aplocheilus*) *latipes* (Temminck and Schlegel). Proceedings of the Imperial Academy of Japan 7: 176-178.

Atkins, D. L. and Fink, W. L. 1979. Morphology and histochemistry of the caudal gland of *Corynopoma riisei* Gill (Teleostei, Characidae). Journal of Fish Biology 14: 465-469.

Atwood, D. L. and Chia, F.-S. 1974. Fine structure of an unusual spermatozoon of a brooding sea cucumber, *Cucumaria lubrica*. Canadian Journal of Zoology 52: 519-523.

Azevedo, M. A. 2000. Biologia reprodutiva de dois glandulocaudíneos com inseminação, *Mimagoniates microlepis* e *Mimagoniates rheocharis* (Teleostei: Characidae), e características de seus ambientes. Unpublished M.S. thesis. Universidade Federal do Rio Grande do Sul, Porto Alegre, Brazil.

Azevedo, M. A. 2004. *Análise comparada de caracteres reprodutivos em três linhagens de characidae (Teleostei: Ostariophysi) com inseminação*. Unpublished Ph.D. dissertation. Universidade Federal do Rio Grande do Sul, Porto Alegre, Brazil.

Azevedo, M. A., Malabarba, L. R. and Fialho, C. B. 2000. Reproductive biology of the inseminating glandulocaudine *Diapoma speculiferum* Cope (Teleostei: Characidae). Copeia 2000: 983-989.

Balon, E.K. (ed.) 1985. Early life histories of fishes: New developmental, ecological and evolutionary perspectives. *Developments in Environmental Biology of Fishes* 5. Dr. W. Junk Publishers, Dordrecht, The Netherlands, pp. 280.

Billard, R. 1986. Spermatogenesis and spermatology of some teleost fish species. Reproduction Nutrition Développement 26: 877-920.

Billard, R. and Cosson, M. P. 1992. Some problems related to the assessment of sperm motility in freshwater fishes. Journal of Experimental Zoology 261: 122-131.

Böhlke, J. E., and Springer, V. G. 1961. A review of the Atlantic species of the clinid genus *Starksia*. Proceedings of the Academy of Natural Sciences of Philadelphia 113: 29-60.

Bolin, R. L. 1944. A review of the marine cottid fishes of California. Stanford Ichthyological Bulletin 3: 1-135.

Breder, C. M., Jr., and Rosen, D. E. 1966. *Modes of Reproduction in Fishes: How Fishes Breed*. T.F.H. Publications, Jersey City, New Jersey, 941 pp.

Burns, J. R. 1991. Testis and gonopodium development in *Anableps dowi* (Pisces: Anablepidae) correlated with pituitary gonadotropic area development. Journal of Morphology 210: 45-53.

Burns, J. R. and Flores, J. A. 1981. Reproductive biology of the cuatro ojos, *Anableps dowi* (Pisces: Anablepidae), from El Salvador and its seasonal variations. Copeia 1981: 25-32.

Burns, J. R. and Weitzman, S. H. 2005. Insemination in ostariophysian fishes. Pp.107-134. In H. J Grier and M. C. Uribe (eds), *Viviparous Fishes*. New Life Publications, Homestead, Fl.

Burns, J. R. and Weitzman, S. H. 1996. Novel gill-derived gland in the male swordtail characin, *Corynopoma riisei* (Teleostei: Characidae: Glandulocaudinae). Copeia 1996: 627-633.

Burns, J. R. and Weitzman, S. H. 2006. Intromittent organ in the genus *Monotocheirodon* (Characiformes: Characidae). Copeia 2006: 529-534.

Burns, J. R., Weitzman, S. H., Grier, H. J. and Menezes, N. A. 1995. Internal fertilization, testis and sperm morphology in glandulocauline fishes (Teleostei: Characidae: Glandulocaulinae). Journal of Morphology 224: 131-143.

Burns, J. R., Weitzman, S. H. and Malabarba, L. R. 1997. Insemination in eight species of cheirodontine fishes (Teleostei: Characidae: Cheirodontiae). Copeia 1997: 433-438.

Burns, J. R., Weitzman, S. H., Lange, K. R. and Malabarba, L. R. 1998. Sperm ultrastructure in characid fishes (Teleostei, Ostariophysi). Pp. 235-244. In L. R. Malabarba, R. E. Reis, R. P. Vari, Z. M. S. Lucena and C. A. S. Lucena (eds), *Phylogeny and Classification of Neotropical Fishes*. EDIPUCRS, Porto Alegre, Brazil.

Burns, J. R., Weitzman, S. H., Malabarba, L. R. and Meisner, A. D. 2000. Sperm modifications in inseminating ostariophysan fishes, with new documentation of inseminating species. P. 255. In B. Norberg, O. S. Kjesbu, G. L. Taranger, E. Andersson and S. O. Stefansson (eds), *Proceedings of the 6th International Symposium on the Reproductive Physiology of Fish*. Institute of Marine Resources and University of Bergen, Bergen, Norway.

Burns, J. R., Meisner, A. D., Weitzman, S. H. and Malabarba, L. R. 2002. Sperm and spermatozeugma ultrastructure in the inseminating catfish, *Trachelyopterus lucenai* (Ostariophysi: Siluriformes: Auchenipteridae). Copeia 2002: 173-179.

Bushmann, P. J., Burns, J. R. and Weitzman, S. H. 2002. Gill-derived glands in glandulocaudine fishes (Teleostei: Characidae: Glandulocaudinae). Journal of Morphology 253: 187-195.

Bussing, W. A. and Roberts, T. R. 1971. Rediscovery of the glandulocaudine fish *Pterobrycon* and hypothetical significance of its spectacular humeral scales (Pisces: Characidae). Copeia 1971: 179-181.

Calcagnotto, D., Schaefer, S. A. and DeSalle, R. 2005. Relationships among characiform fishes inferred from analysis of nuclear and mitochondrial gene sequences. Molecular Phylogenetics and Evolution 36: 135-153.

Castro, R. M. C., Ribeiro, A. C., Benine, R. C. and Melo, A. L. A. 2003. *Lophiobrycon weitzmani*, a new genus and species of glandulocaudine fish (Characiformes: Characidae) from the rio Grande drainage, upper rio Paraná system, southeastern Brazil. Neotropical Ichthyology 1: 11-19.

Christen, R., Gatti, J. L. and Billard, R. 1987. Trout sperm motility. The transient movement of trout sperm is related to changes in the concentration of ATP following the activation of the flagellar movement. European Journal of Biochemistry 166: 667-671.

Clayton, D. A. 1993. Mudskippers. Oceanography and Marine Biology: An Annual Review 31: 507-577.

Cohen, D. M. and Wourms, J. P. 1976. *Microbrotula randalli*, a new viviparous ophidioid fish from Samoa and New Hebrides whose embryos bear trophotaeniae. Proceedings of the Biological Society of Washington 89: 81-98.

Collins, P. M., O'Neill, D. F., Barron, B. R., Moore, R. K. and Sherwood, N. M. 2001. Gonadotropin-releasing hormone content in the brain and pituitary of male and female grass rockfish (*Sebastes rastrelliger*) in relation to seasonal changes in reproductive status. Biology of Reproduction 65: 173-179.

Cosson, J., Linhart, O., Mims, S. D., Shelton, W.L. and Rodina, M. 2000. Analysis of motility parameters from paddlefish and shovelnose sturgeon spermatozoa. Journal of Fish Biology 56: 1348-1367.

Cummings, M. E., Rosenthal, G. G. and Ryan, M. J. 2003. A private ultraviolet channel in visual communication. Proceedings of the Royal Society of London. Series B: Biological Sciences 270:8 97-904.

De Felice, P. A. and Rasch, E. M. 1969. Chronology of spermatogenesis and spermiogenesis in poeciliid fishes. Journal of Experimental Zoology 171: 191-208.

Downing, A. L. and Burns, J. R. 1995. Testis morphology and spermatozeugma formation in three genera of viviparous halfbeaks: *Nomorhamphus, Dermogenys* and *Hemirhamphodon* (Teleostei: Hemiramphidae). Journal of Morphology 225: 329-343.

Dreanno, C., Seguin, F., Cosson, J., Suquet, M. and Billard, R. 2000. ^{1}H-NMR and ^{31}P-NMR analysis of energy metabolism of quiescent and motile turbot (*Psetta maxima*) spermatozoa. Journal of Experimental Zoology 286: 513-522.

Elofsson, H., Van Look, K. J. W., Sundell, K., Sundh, H. and Borg, B. 2006. Stickleback sperm saved by salt in ovarian fluid. Journal of Experimental Biology 209: 4230-4237.

Fauci, L. J. 1996. A computational model of the fluid dynamics of undulatory and flagellar swimming. American Zoologist 36: 599-607.

Fawcett, D. W. 1970. A comparative view of sperm ultrastructure. Biology of Reproduction Supplement 2: 90-127.

Ferriz, R. A., Fernández, E. M., Bentos, C. A. and López, G. R. 2007. Reproductive biology of *Pseudocorynopoma doriai* [sic] (Pisces: Characidae) in the High Basin of the Samborombón River, province of Buenos Aires, Argentina. Journal of Applied Ichthyology 23: 226-230.

Gardiner, D. M. 1978a. Cyclic changes in fine structure of the epithelium lining the ovary of the viviparous teleost, *Cymatogaster aggregata* (Perciformes: Embiotocidae). Journal of Morphology 156: 367-379.

Gardiner, D. M. 1978b. Fine structure of the spermatozoon of the viviparous teleost, *Cymatogaster aggregata*. Journal of Fish Biology 13: 435-438.

Gardiner, D. M. 1978c. The origin and fate of spermatophores in the viviparous teleost *Cymatogaster aggregata* (Perciformes: Embiotocidae). Journal of Morphology 155: 157-172.

Gardiner, D. M. 1978d. Utilization of extracellular glucose by spermatozoa of two viviparous fishes. Comparative Biochemistry and Physiology 59A: 165-168.

Ginzburg, A. S. 1968. Fertilization in Fishes and the Problem of Polyspermy. Academy of Sciences (USSR), Moscow.

Gonçalves, T. K., Azevedo, M. A., Malabarba, L. R. and Fialho, C. B. 2005. Reproductive biology and development of sexually dimorphic structures in Aphyocharax anisitsi (Ostariophysi: Characidae). Neotropical Ichthyology 3: 433-438.

Grier, H. J. 1981. Cellular organization of the testis and spermatogenesis in fishes. American Zoologist 21: 345-357.

Grier, H. J. 1984. Testis structure and formation of spermatophores in the atherinomorph teleost *Horaichthys setnai*. Copeia 1984: 833-839.

Grier, H. J. and Collette, B. B. 1987. Unique spermatozeugmata in testes of halfbeaks of the genus *Zenarchopterus* (Teleostei: Hemiramphidae). Copeia 1987: 300-311.

Grier, H. J. and Parenti, L. R. 1994. Reproductive biology and systematics of phallostethid fishes as revealed by gonad structure. Environmental Biology of Fishes 41: 287-299.

Grier, H. J., Fitzsimmons, J. M. and Linton, J. R. 1978. Structure and ultrastructure of the testis and sperm formation in goodeid teleosts. Journal of Morphology 156: 419-438.

Grier, H. J., Horner, J. and Mahesh, V. V. 1980. The morphology of enclosed testicular tubules in a teleost fish, *Poecilia latipinna*. Transactions of the American Microscopical Society 99: 268-276.

Grier, H. J., Burns, J. R. and Flores, J. A. 1981. Testis structure in three species of teleosts with tubular gonopodia. Copeia 1981: 797-801.

Grier, H. J., Moody, D. P. and Cowell, B. C. 1990. Internal fertilization and sperm morphology in the brook silverside, *Labidesthes sicculus* (Cope). Copeia 1990: 221-226.

Grove, B. D. and Wourms, J. P. 1991. The follicular placenta of the viviparous fish *Heterandria formosa*. I. Ultrastructure and development of the embryonic absorptive surface. Journal of Morphology 209: 265-284.

Grove, B. D. and Wourms, J. P. 1994. Follicular placenta of the viviparous fish *Heterandria formosa*. II. Ultrastructure and development of the follicular epithelium. Journal of Morphology 220: 167-184.

Harms, J. W. 1935. Die Realisation von Genen und die consecutive Adaptation. 4. Mitteilung. Experimentell hervorgerufener Medienwechsel: Wasser zu Feuchtluft bsw. zu Trockenluft bei Gobiiformes. (*Gobius, Boleophthalmus* und *Periophthalmus*). Zeitschrift für wissenschaftliche Zoologie 146: 417-463.

Harrington, R. W., Jr. 1961. Oviparous hermaphroditic fish with internal self-fertilization. Science 135: 1749-1750.

Hecke, M. van. 2007. Materials science: Shape matters. Science 317: 49.

Hogan, A. E. and Nicholson, J. C. 1987. Sperm motility of sooty grunter, *Hephaestus fuliginosus* (Macleay), and jungle perch, *Kuhlia rupestris* (Lacépède), in different salinities. Australian Journal of Marine and Freshwater Research 38: 523-528.

Hosken, D. J. 1998. Testes mass in megachiropteran bats varies in accordance with sperm competition theory. Behavioral Ecology and Sociobiology 44: 169-177.

Hubbs, C. 1952. A contribution to the classification of the blennioid fishes of the family Clinidae, with a partial revision of the eastern Pacific forms. Stanford Ichthyological Bulletin 4: 41-165.

Hurk, van den, R. 1973. The localization of steroidogenesis in the testes of oviparous and viviparous teleosts. Proceedings of the Koninklijke Nederlandse Academie van Wetenschappen, Series C 76: 270-279.

Hurk, van den, R., Meek, J. and Peute, J. 1974. Ultrastructural study of the testis of the black molly (*Mollienisia latipinna*). II. Sertoli cells and Leydig cells. Proceedings of the Koninklijke Nederlandse Academie van Wetenschappen, Series C 77: 470-475.

von Ihering, R. 1937. Oviductal fertilization in the South American catfish *Trachycorystes*. Copeia 1937: 201-205.

Jamieson, B. G. M. 1987. A biological classification of sperm types, with special reference to annelids and mollusks, and an example of spermiocladistics. Pp. 311-332. In H. Mohri (ed.), *New Horizons in Sperm Cell Research*. Gordon and Breach, New York.

Jamieson, B. G. M. 1991. *Fish Evolution and Systematics: Evidence from Spermatozoa*. Cambridge University Press, Cambridge. 319 pp.

Javonillo, R., Burns, J. R. and Weitzman, S. H. 2007. Reproductive morphology of *Brittanichthys axelrodi* (Teleostei: Characidae), a miniature inseminating fish from South America. Journal of Morphology 268: 23-32.

Knight, F. M., Lombardi, J., Wourms, J. P. and Burns, J. R. 1985. Follicular placenta and embryonic growth of the viviparous four-eyed fish (*Anableps*). Journal of Morphology 185: 131-142.

Korsgaard, B., Andreassen, T. K. and Rasmussen, T. H. 2002. Effects of an environmental estrogen, 17α-ethinyl-estradiol, on the maternal-fetal trophic relationship in the eelpout *Zoarces viviparus* (L). Marine Environmental Research 54: 735-739.

Koya, Y., Munehara, H., Takano, K. and Takahashi, H. 1993. Effects of extracellular environments on the motility of spermatozoa in several marine sculpins with internal gametic association. Comparative Biochemistry and Physiology, A 106A:25-29.

Koya, Y., Munehara, H. and Takano, K. 2002. Sperm storage and motility in the ovary of the marine sculpin *Alcichthys alcicornis* (Teleostei: Scorpaeniformes), with internal gametic association. Journal of Experimental Zoology 292: 145-155.

Kristofferson, R., Broberg, S. and Pekkarinen, M. 1973. Histology and physiology of embryotrophe formation, embryonic nutrition, and growth in the eelpout, *Zoarces viviparus*. Annales Zoologici Fennici 10: 457-477.

Kulkarni, C. V. 1940. On the systematic position, structural modifications, bionomics and development of a remarkable new family of cyprinodont fishes from the province of Bombay. Records of the Indian Museum 42: 379-423.

Kutaygil, N. 1959. Insemination, sexual differentiation and secondary sex characters in *Stevardia albipinnis* Gill. Istanbul Üniversitesi Fen Fakültesi Mecmuasi, Série B. 24: 93-128.

Kweon, H.-S., Park, E.-H. and Peters, N. 1998. Spermatozoon ultrastructure in the internally self-fertilizing hermaphroditic teleost, *Rivulus marmoratus* (Cyprinodontiformes, Rivulidae). Copeia 1998: 1101-1106.

Lahnsteiner, F. and Patzner, R. A. 1998. Sperm motility of the marine teleosts *Boops boops, Diplodus sargus, Mullus barbatus* and *Trachurus mediterraneus*. Journal of Fish Biology 52: 726–742.

Lahnsteiner, F. and Patzner, R. A. 1999. Characterization of spermatozoa and eggs of the rabbitfish. Journal of Fish Biology 55: 820-835.

Lahnsteiner, F. and Patzner, R. A. 2007. Sperm morphology and ultrastructure in fish. Pp. 1-61. In S. M. H. Alavi, J. Cosson, K. Coward and G. Rafiee (eds), *Fish Spermatology*. Alpha Science International Ltd., Oxford, U. K.

Lahnsteiner, F., Patzner, R. A. and Weismann, T. 1994. The testicular main ducts and the spermatic ducts in some cypinid fishes – II. Composition of the seminal fluid. Journal of Fish Biology 44: 459-467.

Lahnsteiner, F., Berger, B., Weismann, T. and Patzner, R. 1995. Fine structure and motility of spermatozoa and composition of the seminal plasma in the perch. Journal of Fish Biology 47: 492–508.

Lahnsteiner, F., Berger, B. and Weismann, T. 1999. Sperm metabolism of the teleost fishes *Chalcalburnus chalcoides* and *Oncorhynchus mykiss* and its relation to motility and viability. Journal of Experimental Zoology 284: 454-465.

Lampert, V. R. 2003. Biologia reprodutiva de duas espécies do gênero *Bryconamericus* (Characidae: Tetragonopterinae) dos sistemas dos rios Jacuí e Uruguai, RS. Unpublished M.S. thesis. Universidade Federal do Rio Grande do Sul, Porto Alegre, Brazil.

Lampert, V. R., Azevedo, M. A. and Fialho, C. B. 2004. Reproductive biology of *Bryconamericus iheringii* (Ostariophysi: Characidae) from rio Vacacaí, RS, Brazil. Neotropical Ichthyology 2: 209-215.

Lauga, E. 2007. Floppy swimming: viscous locomotion of actuated elastica. Physical Review E 75: 041916-1- 041916-16

Lazara, K. J. 2000. The killifishes: an annotated checklist, synonymy and bibliography of recent oviparous cyprinodontiform fishes. The Killifish Master Index, 4th edition. American Killifish Association, Cincinnati, Ohio.

Leung, L. K.-P. 1988. Ultrastructure of the spermatozoon of *Lepidogalaxias salamandroides* and its phylogenetic significance. Gamete Research 19: 41-49.

Leviton, A. E., Gibbs, R. H., Jr., Heal, E. and Dawson, C. E. 1985. Standards in herpetology and ichthyology: Part I. Standard symbolic codes for institutional resource collections in herpetology and ichthyology. Copeia 1985: 802-832.

Loir, M., Cauty, C., Planquette, P. and Le Bail, P. Y. 1989. Comparative study of the male reproductive tracts in seven families of South-American catfishes. Aquatic Living Resources 2: 45-56.

Macías Garcia, C. and Burt de Perera, T. 2002. Ultraviolet based female preferences in a viviparous fish. Behavioral Ecology and Sociobiology 52: 1-6.

Maddison, W. P. 1990. A Method for Testing the Correlated Evolution of Two Binary Characters: Are Gains or Losses Concentrated on Certain Branches of a Phylogenetic Tree? Evolution 44: 539-557.

Malabarba, L. R. 1998. Monophyly of the Cheirodontinae, characters and major clades (Ostariophysi: Characidae). Pp. 193-233. In L. R. Malabarba, R. E. Reis, R. P. Vari, Z. M. S. Lucena and C. A. S. Lucena (eds), *Phylogeny and Classification of Neotropical Fishes*. EDIPUCRS, Porto Alegre, Brazil.

Mattei, X. 1991. Spermatozoon ultrastructure and its systematic implications in fishes. Canadian Journal of Zoology 69: 3038-3055.

Meisner, A. D. and Burns, J. R. 1997a. Testis and andropodial development in a viviparous halfbeak, *Dermogenys* sp. (Teleostei: Hemiramphidae). Copeia 1997: 44-52.

Meisner, A. D. and Burns, J. R. 1997b. Viviparity in the halfbeak genera *Dermogenys* and *Nomorhamphus* (Teleostei: Hemiramphidae). Journal of Morphology 234: 295-317.

Meisner, A. D., Burns, J. R. and Ghedotti, M. J. 2000a. Mode of embryonic nutrition in four species of *Jenynsia* (Teleostei: Atherinomorpha: Anablepidae). P. 178. In B. Norberg, O. S. Kjesbu, G. L. Taranger, E. Andersson and S. O. Stefansson (eds), *Proceedings of the 6th International Symposium on the Reproductive Physiology of Fish*. Institute of Marine Resources and University of Bergen, Bergen, Norway.

Meisner, A. D., Burns, J. R., Weitzman, S. H. and Malabarba, L. R. 2000b. Morphology and histology of the male reproductive system in two species of internally inseminating South American catfishes, *Trachelyopterus lucenai* and *T. galeatus* (Teleostei: Auchenipteridae). Journal of Morphology 246:131-141.

Menezes, N. A., Weitzman, S. H. and Burns, J. R. 2003. A systematic review of *Planaltina* (Teleostei: Characiformes: Characidae: Glandulocaudinae: Diapomini) with a description of two new species from the upper Rio Paraná, Brazil. Proceedings of the Biological Society of Washington 116: 557-600.

Moksness, E. and Pavlov, D. A. 1996. Management by life cycle of wolffish, *Anarhichas lupus* L., a new species for cold-water aquaculture: a technical paper. Aquaculture Research 27: 865-883.

Molony, B. W. and Sheaves, M. J. 2001. Challenges of external insemination in a tropical sparid fish, *Acanthopagrus berda*. Environmental Biology of Fishes 61: 65-71.

Munehara, H. 1997. The reproductive biology and early life stages of *Podothecus sachi* (Pisces: Agonidae). Fishery Bulletin 95: 612-619.

Munehara, H., Takano, K. and Koya, Y. 1989. Internal gametic association and external fertilization in the elkhorn sculpin, *Alcichthys alcicornis*. Copeia 1989: 673-678.

Munehara, H., Takano, K. and Koya, Y. 1991. The little dragon sculpin *Blepsias cirrhosus*, another case of internal gametic association and external fertilization. Japanese Journal of Ichthyology 37: 391-393.

Muñoz, M., Casadevall, M., Bonet, S. and Quagio-Grassiotto, I. 2000. Sperm storage structures in the ovary of *Helicolenus dactylopterus dactylopterus* (Teleostei: Scorpaenidae): an ultrastructural study. Environmental Biology of Fishes 58: 53-59.

Muñoz, M., Koya, Y. and Casadevall, M. 2002. Histochemical analysis of sperm storage in *Helicolenus dactylopteus dactylopterus* (Teleostei: Scorpaenidae). Journal of Experimental Zoology 292: 156-164.

Narayan, V., Ramaswamy, S. and Menon, N. 2007. Long-lived giant number fluctuations in a swarming granular nematic. Science 317: 105-108.

Nelson, J. S. 2006. *Fishes of the World*, 4th edition. John Wiley and Sons, Inc., Hoboken, NJ. 601 pp.

Nelson, K. 1964a. Behavior and morphology in the glandulocaudine fishes (Ostariophysi, Characidae). University of California Publications in Zoology 75: 59-152.

Nelson, K. 1964b. The temporal patterning of courtship behaviour in the glandulocaudine fishes (Ostariophysi, Characidae). Behaviour 14: 90-146.

Nelson, K. 1964c. The evolution of a pattern of sound production associated with courtship in the characid fish *Glandulocauda inequalis*. Evolution 18: 526-540.

Nielsen, J. G. 1984. Two new, abyssal *Barathronus* spp. from the North Atlantic (Pisces: Aphyonidae). Copeia 1984: 579-584.

Oliveira, C. L. C. 2003. *Análise comparada de caracteres reprodutivos e da glândula branquial de duas espécies de Cheirodontinae (Teleostei: Characidae)*. Unpublished M.S. thesis. Universidade Federal do Rio Grande do Sul, Porto Alegre, Brazil.

Oliveira, C. L. C., Fialho, C. B. and Malabarba, L. R. 2002. Período reprodutivo, desova e fecundidade de *Cheirodon ibicuhiensis* Eigenmann, 1915 (Ostariophysi: Characidae) do arroio Ribeiro, Rio Grande do Sul, Brasil. Comunicações do Museu de Ciências e Tecnologia PUCRS, Série Zoologia 15: 3-14.

Oliveira, C. L. C. de, Burns, J. R., Malabarba, L. R. and Weitzman, S. H. 2008. Sperm ultrastructure in the inseminating *Macropsobrycon uruguayanae* (Teleostei: Characidae: Cheirodontinae). Journal of Morphology 269: 691-697.

Pandey, S. 1969. Effects of hypophysectomy on the testis and secondary sex characters of the adult guppy *Poecilia reticulata* Peters. Canadian Journal of Zoology 47: 775-781.

Parenti, L. R. 1981. A phylogenetic and biogeographic analysis of cyprinodontiform fishes (Teleostei, Atherinomorpha). Bulletin of the American Museum of Natural History 168: 335-557.

Parreira, G. G. and Godinho, H. P. 2004. Sperm ultrastructure of the cangati *Trachelyopterus galeatus* (Linnaeus, 1766) (Siluriformes, Auchenipteridae), an internal fertilizer fish. Molecular Biology of the Cell 15:91A (supplement). (CD ROM)

Pavlov, D. A. 1994. Fertilization in the wolffish, *Anarhichus lupus*: external or internal? Journal of Ichthyology 34: 140-151.

Pecio, A. and Rafiński, J. 1994. Structure of the testis, spermatozoa and spermatozeugmata of *Mimagoniates barberi* Regan, 1907 (Teleostei: Caracidae), an internally fertilizing, oviparous fish. Acta Zoologica (Stockholm) 75: 179-185.

Pecio, A. and Rafiński, J. 1999. Spermiogenesis in *Mimagoniates barberi* (Teleostei: Ostariophysi: Characidae), an oviparous, internally fertilizing fish. Acta Zoologica (Stockholm) 80: 35-45.

Pecio, A., Lahnsteiner, F. and Rafiński, J. 2001. Ultrastructure of the epithelial cells in the aspermatogenic part of the testis in *Mimagoniates barberi* (Teleostei: Characidae: Glandulocaudinae) and the role of the their secretions in spermatozeugmata formation. Annals of Anatomy 183: 427-435.

Pecio, A., Burns, J. R. and Weitzman, S. H. 2005. Sperm and spermatozeugma ultrastructure in the inseminating species *Tyttocharax cochui, T. tambopatensis*, and *Scopaeocharax rhinodus* (Pisces: Teleostei: Characidae: Glandulocaudinae: Xenurobryconini). Journal of Morphology 263: 216-226.

Pecio, A., Burns, J. R. and Weitzman, S. H. 2007. Comparison of spermiogenesis in the externally fertilizing *Hemigrammus erythrozonus* and the inseminating *Corynopoma riisei* (Teleostei: Characiformes: Characidae). Neotropical Ichthyology 5: 457-470.

Peters, G. and Mäder, B. 1964. Morphologische Veränderungen der Gonadensführgänge sich fortpflanzenden Schwerttragerweibchen (*Xiphophorus helleri* Heckel). Zoologischer Anzeiger 173: 243-257.

Pfeiffer, W. 1977. The distribution of fright reaction and alarm substance cells in fishes. Copeia 1977: 653-665.

Potter, H. and Kramer, C. R. 2000. Ultrastructural observations on sperm storage in the ovary of the platyfish, *Xiphophorus maculatus* (Teleostei: Poeciliidae): the role of the duct epithelium. Journal of Morphology 245: 110-129.

Purcell, E. 1977. Life at low Reynolds number. American Journal of Physics 45: 3–10.

Pusey, B. J. and Stewart, T. 1989. Internal fertilization in *Lepidogalaxias salamandroides* Mees (Pisces: Lepidogalaxiidae). Zoological Journal of the Linnean Society 97: 69-79.

Rasotto, M. B. and Sadovy, Y. 1995. Peculiarities of the male urogenital apparatus of two grunt species (Teleostei, Haemulidae). Journal of Fish Biology 46: 936-948.

Richter, H-J. 1986. The swordtail characin—*Corynopoma riisei.* Tropical Fish Hobbyist 35: 44-49.

Riedel, I. H., Kruse, K. and Howard, J. 2005. A self-organized vortex array of hydrodynamically entrained sperm cells. Science 309: 300-303.

Rosenblatt, R. H. and Taylor, L. R., Jr. 1971. The Pacific species of the clinid tribe Starksiini. Pacific Science 25: 436-463.

Schaefer, S. A., Weitzman, S. H. and Britski, H. A. 1989. Review of the Neotropical catfish genus *Scoloplax* (Pisces: Loricariodea: Scoloplacidae) with comments on reductive characters in phylogenetic analysis. Proceedings of the Academy of Natural Sciences of Philadelphia 141: 181-211.

Scrimshaw, N. S. 1944. Embryonic growth in the viviparous poeciliid, *Heterandria formosa*. Biological Bulletin 87: 37-51.

Scrimshaw, N. S. 1945. Embryonic development in poeciliid fishes. Biological Bulletin 88: 233-246.

Shimizu, M., Kusakari, M., Yoklavich, M. M., Boehlert, G. W. and Tamada, Y. 1991. Ultrastructure of the epidermis and digestive tract in *Sebastes* embryos, with special reference to the uptake of exogenous nutrients. Environmental Biology of Fishes 30: 155-163.

Silvano, J., Oliveira, C. L. C., Fialho, C. B., Gurgel, H. C. B. 2003. Reproductive period and fecundity of *Serrapinnus piaba* (Characidae: Cheirodontinae) from the rio Ceará Mirim, Rio Grande do Norte, Brazil. Neotropical Ichthyology 1: 61-66.

Smith, W. L. and Wheeler, W. C. 2004. Polyphyly of the mail-cheeked fishes (Teleostei: Scorpaeniformes): evidence from mitochondrial and nuclear sequence data. Molecular Phylogenetics and Evolution 32: 627-646.

Smith, W. L. and Wheeler, W. C. 2006. Venom evolution widespread in fishes: a phylogenetic road map for the bioprospecting of piscine venoms. Journal of Heredity 97: 206-217.

Spadella, M. A. 2007. Estudos filogenéticos na superfamília Loricarioidea (Teleostei, Siluriformes) com base na ultraestrutura dos espermatozóides. Unpublished Ph.D. dissertation. Instituto de Biologia, Universidade Estadual de Campinas, Campinas, Brazil.

Spadella, M. A., Oliveira, C. and Quagio-Grassiotto, I. 2006. Spermiogenesis and introsperm ultrastructure of *Scoloplax distolothrix* (Ostariophysi: Siluriformes: Scoloplacidae). Acta Zoologica (Copenhagen) 87: 341-348.

Sullivan, J. P., Lundberg, J. G. and Hardman, M. 2006. A phylogenetic analysis of the major groups of catfishes (Teleostei: Siluriformes) using rag1 and rag2 nuclear gene sequences. Molecular Phylogenetics and Evolution 41: 636-662.

Szabó, B., Szöllösi, G. J., Gönci, B., Jurányi, Zs., Selmeczi, D. and Vicsek, T. 2006. Phase transition in the collective migration of tissue cells: experimental and model. Physical Review E 74: 061908-1-061908-5

Tarp, F. H. 1952. A revision of the family Embiotocidae (the surfperches). State of California Department of Fish and Game, Bureau of Marine Fisheries, Fish Bulletin 88: 1-99.

Thresher, R.E. 1984. *Reproduction in Reef Fishes*. TFH Publications, Neptune City, New Jersey.

Thünken, T., Bakker, T. C. M. and Kullman, H. 2007. Extraordinarily long sperm in the socially monogamous cichlid fish *Pelvicachromis taeniatus*. Naturwissenschaften 94: 489-491.

Turner, C. L. 1933. Viviparity superimposed upon ovo-viviparity in the Goodeidae, a family of cyprinodont teleost fishes of the Mexican Plateau. Journal of Morphology 55: 207-251.

Turner, C. L. 1936. The absorptive processes in the embryos of *Parabrotula dentiens*, a viviparous, deep-sea brotulid fish. Journal of Morphology 59: 313-325.

Turner, C. L. 1937. The trophotaeniae of the Goodeidae, a family of viviparous cyprinodont fishes. Journal of Morphology 61: 495-523.

Turner, C. L. 1938a. Adaptations for viviparity in embryos and ovary of *Anableps anableps*. Journal of Morphology 62: 323-349.

Turner, C. L. 1938b. Histological and cytological changes in the ovary of *Cymatogaster aggregata* during gestation. Journal of Morphology 62: 351-373.

Turner, C. L. 1940a. Pseudoamnion, pseudochorion and follicular pseudoplacenta in poeciliid fishes. Journal of Morphology 67: 59-89.

Turner, C. L. 1940b. Follicular pseudoplacenta and gut modifications in anablepid fishes. Journal of Morphology 67: 91-105.

Turner, C. L. 1940c. Adaptations for viviparity in jenynsiid fishes. Journal of Morphology 67: 291-297.

Turner, C. L. 1940d. Pericardial sac, trophotaeniae, and alimentary tract in embryos of goodeid fishes. Journal of Morphology 67: 274-289.

Turner, C. L. 1950. The skeletal structure of the gonopodium and gonopodial supensorium of *Anableps anableps*. Journal of Morphology 86: 329-366.

Turner, C. L. 1952. An accessory respiratory device in embryos of the embiotocid fish *Cymatogaster aggregata* during gestation. Copeia 1952: 146-147.

Urho, L., Hudd, R. and Hildén, M. 1984. The motility of fish sperms as a function of pH. Memoranda Societatis pro Fauna et Flora Fennica 60: 41-42.

Vaupel, J. 1929. The spermatogenesis of *Lebistes reticulatus*. Journal of Morphology 47: 555-587.

Veith, W. J. 1979. Reproduction in the live bearing teleost *Clinus superciliosus*. South African Journal of Zoology 14: 208-214.

Vladić, T. and Järvi, T. 1997. Sperm motility and fertilization time span in Atlantic salmon and brown trout—the effect of water temperature. Journal of Fish Biology 50: 1088-1093.

Vladić, T , Afzelius, B. A. and Bronnikov, G. E. 2002. Sperm quality as reflected through morphology in salmon alternative life histories. Biology of Reproduction 66: 98-105.

Weishaupt, E. 1925. Die Ontogenie der Genitalorgane von *Girardinus reticulatus*. Zeitschrift für wissenschaftliche Zoologie 126: 571-611.

Weitzman, S. H. and Fink, S. V. 1985. Xenurobryconin phylogeny and putative pheromone pumps in glandulocaudine fishes (Teleostei: Characidae). Smithsonian Contributions to Zoology 421: i-iii,1-121.

Weitzman, S. H. and Menezes, N. A. 1998. Relationships of the tribes and genera of the Glandulocaudinae (Ostariophysi: Characiformes: Characidae) with a description of a new genus *Chrysobrycon*. Pp. 159-180. In L. R. Malabarba, R. E. Reis, R. P. Vari, Z. M. S. Lucena and C. A. S. Lucena (eds), *Phylogeny and Classification of Neotropical Fishes*. EDIPUCRS, Porto Alegre, Brazil.

Weitzman, S. H., Menezes, N. A., Evers, H.-G. and Burns, J. R. 2005. Putative relationships among inseminating and externally fertilizing characids, with a description of a new genus and species of Brazilian inseminating fish bearing an anal-fin gland in males (Characiformes: Characidae). Neotropical Ichthyology 3: 329-360.

Westneat, M. W. 2004. Evolution of levers and linkages in the feeding mechanisms of fishes. Integrative and Comparative Biology 44: 378-389.

Weyenbergh, H. 1875. Contribución al conocimiento del género *Xiphophorus* Heck. Un género de pescados vivíparos. Periódico Zoológico (Argentina) 2: 9-28.

White, E. M., Partridge, U. C. and Church, S. C. 2003. Ultraviolet dermal reflection and mate choice in the guppy, *Poecilia reticulata*. Animal Behaviour 65: 693–700.

Wourms, J. P. 1981. Viviparity: the maternal-fetal relationship in fishes. American Zoologist 21: 473-515.

Wourms, J. P. 1991. Reproduction and development of *Sebastes* in the context of evolution of piscine viviparity. Environmental Biology of Fishes 30: 111-126.

Wourms, J. P. and Bayne, O. 1973. Development of the viviparous brotulid fish, *Dinematichthys ilucoeteoides*. Copeia 1973: 32-40.

Wourms, J. P. and Cohen, D. M. 1975. Trophotaeniae, embryonic adaptations in the viviparous ophidioid fish *Oligopus longhursti*: a study of museum specimens. Journal of Morphology 147: 385-401.

Yang, J., Liu, X., Yue, G., Adamian, M., Bulgakov, O. and Li, T. 2002. Rootletin, a novel coiled-coil protein, is a structural component of the ciliary rootlet. Journal of Cell Biology 159: 431-440.

Yang, J., Gao, J., Adamian, M., Wen, X-H., Pawlyk, B., Zhang, L., Sanderson, M. J., Zuo, J., Makino, C. L., Li, T. 2005. The ciliary rootlet maintains long-term stability of sensory cilia. Molecular and Cellular Biology 25: 4129-4137.

Yao, Z., Emerson, C. J. and Crim, L. W. 1995. Ultrastructure of the spermatozoa and eggs of the ocean pout (*Macrozoarces americanus* L.), an internally fertilizing marine fish. Molecular Reproduction and Development 42: 58-64.

Zarnescu, O. 2005. Ultrastructural study of spermatozoa of the paddlefish, *Polyodon spathula*. Zygote 13: 241-247.

Index

B

C

F

G

H

M

N

O

T

U

V

W

X

Y

Z